ADAPTATION AND LEARNING IN CONTROL AND SIGNAL PROCESSING 2001

(ALCOSP 2001)

A Proceedings volume from the IFAC Workshop, Cernobbio-Como, Italy, 29 - 31 August 2001

Edited by

S. BITTANTI

Dipartimento di Elettronica e Informazione,
Politecnico di Milano, Milano, Italy

Published for the

INTERNATIONAL FEDERATION OF AUTOMATIC CONTROL

by

PERGAMON

An Imprint of Elsevier Science

ELSEVIER SCIENCE Ltd
The Boulevard, Langford Lane
Kidlington, Oxford OX5 1GB, UK

Elsevier Science Internet Homepage
http://www.elsevier.com

Consult the Elsevier Homepage for full catalogue information on all books, journals and electronic products and services.

IFAC Publications Internet Homepage
http://www.elsevier.com/locate/ifac

Consult the IFAC Publications Homepage for full details on the preparation of IFAC meeting papers, published/forthcoming IFAC books, and information about the IFAC Journals and affiliated journals.

First edition 2002

Library of Congress Cataloging in Publication Data

A catalogue record for this book is available from the Library of Congress

British Library Cataloguing in Publication Data

A catalogue record for this book is available from the British Library

ISBN 0-08-043683 8
ISSN 1474-6670

Transferred to digital printing 2005

To Contact the Publisher

Elsevier Science welcomes enquiries concerning publishing proposals: books, journal special issues, conference proceedings, etc. All formats and media can be considered. Should you have a publishing proposal you wish to discuss, please contact, without obligation, the publisher responsible for Elsevier's industrial and control engineering publishing programme:

Dr Martin Ruck
Publishing Editor
Elsevier Science Ltd
The Boulevard, Langford Lane
Kidlington, Oxford
OX5 1GB, UK

Phone: +44 1865 843230
Fax: +44 1865 843920
E.mail: m.ruck@elsevier.co.uk

General enquiries, including placing orders, should be directed to Elsevier's Regional Sales Offices – please access the Elsevier homepage for full contact details (homepage details at the top of this page).

IFAC WORKSHOP ON ADAPTATION AND LEARNING IN CONTROL AND SIGNAL PROCESSING 2001

Sponsored by
International Federation of Automatic Control (IFAC)
Technical Committees on:
- Adaptive Control and Learning
- Stochastic Systems

Organized by
Italian National Member Organization of IFAC: CNR (Consiglio Nazionale delle Ricerche)

Supported by
Dipartimento di Elettronica e Informazione, Politecnico di Milano, Italy
CNR - Centro Studio Tecnologie Informazione e Automazione, Milano, Italy
European TMR Research Network *Non-Linear and Adaptive Control*

FOREWORD

In control and signal processing, adaptation is a natural tool to cope with real-time changes in the dynamical behaviour of signals and systems. In this area, strongly connected with prediction and identification, there has been an increasing interest in switching and supervising methods. Moreover, in recent years, special attention has been paid to the ideas evolving around the theory of statistical learning as a potential tool of improved adaptation.

The IFAC workshop on *Adaptation and Learning in Control and Signal Processing* held in Cernobbio-Como, Italy, August 29-3 2001 aimed at presenting the full picture of the area by gathering the experts in the field and all interested researchers from universities and industries.

The program consisted of technical sessions, organized in two parallel streams and three plenary lectures, with two of the plenary lectures appearing in this Proceedings volume. The technical sessions included 78 papers covering the following subjects:

- Model Reference and Predictive Control
- Multiple Model Control
- Adaptive Control I/II
- Adaptive Control and Learning
- Learning
- Adaptive Control of Nonlinear Systems I/II
- Supervisory Control
- Neural Networks for Control
- PID Design Methods
- Sliding Mode
- Adaptive Filtering and Estimation
- Identification Methods I/II

Many of the papers referred to specific applications. In particular, three sessions were fully devoted to specific areas of engineering, namely Applications to Mechanical and Bio-Mechanical Systems, Application to Mechanical Systems, Applications to Power Plants and Software Tools.

There were 107 participants from 26 Countries: Australia (4), Brazil (2), Canada (3), Czech Republic (4), Denmark (1), France (1), Finland (1), Germany (2), Hungary (1), India (1), Italy (23), Japan (28), Korea (2), Malta (2), Mexico (1), Poland (2), Portugal (3), Romania (1), Russia (1), Spain (2), Sweden (2), Switzerland (2), The Netherlands (1), UK (11), USA (5).

The participants had the opportunity to be exposed to new ideas, meet friends and enjoy the lake atmosphere.

This workshop has a long-standing history within IFAC. The first meeting (with the title Adaptive Control and Signal Processing) took place back in 1983, San Francisco with subsequent meetings taking place in Lund (1986), Glasgow (1989), Grenoble (1992), Budapest (1995) and again, Glasgow (1998). The next meeting is planned for Yokohama, Japan, between August 30 and September 1, 2004.

Sergio Bittanti (Editor)

CONTENTS

PLENARY PAPER I

MODEL REFERENCE AND PREDICTIVE CONTROL

APPLICATIONS TO MECHANICAL AND BIO-MECHANICAL SYSTEMS

ADAPTIVE CONTROL AND LEARNING

NEURAL NETWORKS FOR CONTROL

MULTIPLE MODEL CONTROL

ADAPTIVE CONTROL I

LEARNING

PID DESIGN METHODS

ADAPTIVE CONTROL OF NONLINEAR SYSTEMS I

APPLICATIONS TO POWER PLANTS AND SOFTWARE TOOLS

SUPERVISORY CONTROL

SLIDING MODE

PLENARY PAPER II

ADAPTIVE FILTERING AND ESTIMATION

APPLICATION TO MECHANICAL SYSTEMS

IDENTIFICATION METHODS I

ADAPTIVE CONTROL OF NONLINEAR SYSTEMS II

IDENTIFICATION METHODS II

ADAPTIVE CONTROL II

NONLINEAR IDENTIFICATION AND ADAPTIVE CONTROL OF COMBUSTION ENGINES

Rolf Isermann and Norbert Müller

Darmstadt University of Technology
Institute of Automatic Control
Laboratory of Control Systems and Process Automation
Landgraf-Georg-Str. 4, D-64283 Darmstadt, Germany
Tel: +49/6151-162114, Fax: +49/6151-293445
{RIsermann,NMueller}@iat.tu-darmstadt.de

Abstract: Advanced engine control systems require accurate models of the thermodynamic-mechanical process, which are substantially nonlinear and often time-variant. After briefly introducing the identification of nonlinear processes with grid-based look-up tables and a special local linear Radial Basis Function network (LOLIMOT), a comparison is made with regard to computation effort, storage requirements and convergence speed. A new training algorithm for online adaptation of look-up tables is introduced which reduces the convergence time considerably. Application examples and experimental results are shown for a multi-dimensional nonlinear model of NO_X emissions of a Diesel engine, and for the adaptive feedforward control of the ignition angle of a SI engine.*Copyright* © *2001 IFAC*

Keywords: Engine control, Learning systems, Identification, Neural networks, Feedforward control

1. INTRODUCTION

The last two decades in the automotive industry have seen an ever increasing usage of electronics. In the middle 1970s car manufacturers introduced microprocessor-based engine control systems to meet the conflicting state regulations and customer demands of high fuel economy, low emissions, best possible engine performance and ride comfort. Especially the American regulations (clean air act 1963, on-board diagnosis (OBD I 1988, OBD II 1994)) and the corresponding European regulations EURO 1 (1992), EURO 2 (1996), EURO 3 (2000) had a considerable effect on the development of new engine control strategies. New sensors with electrical output and new actuators had to be developed. Also auxiliary units like electro-mechanical controlled carburetors and low pressure injection systems for SI engines and high pressure injection systems for Diesel engines have shown a development from pure mechanical to electro-mechanical devices with electronic control. New actuators were added like for exhaust gas recirculation, camshaft positioning and variable geometry turbochargers. Today's combustion engines are completely microcomputer controlled with many *actuators* (e.g. electrical, electro-mechanical, electro-hydraulic or electro-pneumatic actuators, influencing spark timing, fuel-injector pulse widths, exhaust gas recirculation valves), several *measured output variables* (e.g. pressures, temperatures, engine rotational speed, air mass flow, camshaft position, oxygen-concentration of the exhaust gas), taking into account different *operating phases* (e.g. start-up, warming-up, idling, normal operation, overrun, shut down). The microprocessor-based control has grown up to a rather complicated control unit with 50-120 look-up tables, relating about 15 measured inputs and about 30 manipulated variables as outputs. Because many output variables like torque and emission concentrations are mostly not available as measurements (too costly or short life time) a majority of control functions is feedforward.

In the future new electronically controlled actuators and new sensors entail additional control functions for new engine technologies (variable turbine geometry (VTG) turbo chargers, swirl control, dynamic manifold pressure, variable valve timing (VVT) of inlet valves). Increasing computational capabilities using floating point processors will allow advanced estimation techniques for non-measurable quantities like engine torque or exhaust gas properties and precise feedforward and feedback control over large ranges and with small tolerances.

2. CONTROL STRUCTURE FOR INTERNAL COMBUSTION ENGINES: STATE OF THE ART

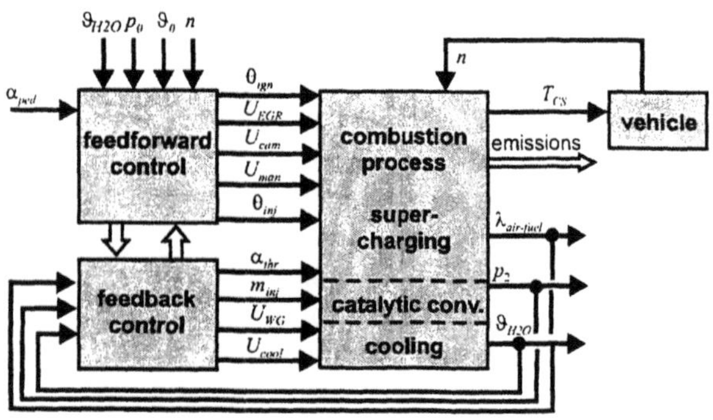

Fig. 1. Simplified control structure of a SI engine

Modern IC engines obtain increasingly more actuating elements. SI engines have - except the classical inputs like amount of injection m_{inj}, ignition angle θ_{ign}, injection angle θ_{inj} - additionally controlled air/fuel ratio, exhaust gas recirculation and variable valve timing, see **Fig. 1**. Diesel engines were until 1987 only manipulated by injection mass and injection angle. Now they have additional manipulated inputs like exhaust gas recirculation, variable geometry turbochargers, common rail pressure and modulated injection, see **Fig. 2**. The engine control system has therefore to be designed for 5-10 main manipulated variables and 5-8 main output variables, leading to a complex nonlinear multiple-input multiple-output system. Because some of the design requirements contradict each other suitable compromises have to be made. Since the majority of control functions are realized by feedforward structures, precise models are required. Feedback control is used in the case of SI engines for example for the λ-control (keeping a stoichiometric air/fuel ratio λ=1 for the catalyst), for

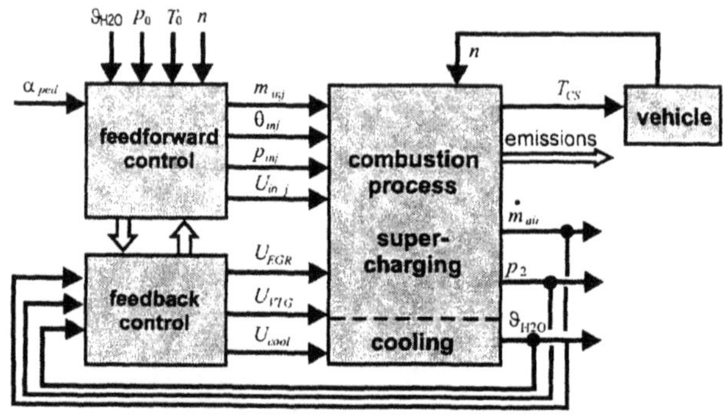

Fig. 2. Simplified control structure of a Diesel engine with turbocharger

the electronic throttle control and for the ignition angle in case of knock. Diesel engines with turbochargers possess a charging pressure control with waste gate or variable vanes and speed overrun break away. Both engines have idling speed control and coolant water temperature control. Additionally, there are several auxiliary closed loop controls like fuel pressure control and oil pressure control. All control functions have to be defined for all possible operating phases. Most feedforward control functions are implemented as grid-based two-dimensional (i.e. two input signals) look-up tables or as one dimensional characteristics. This is because of the strongly nonlinear static and dynamic behavior of the IC engines and the direct programming and fast processing in microprocessors with fixed point arithmetics. Some of the functions are based on physical models with correction factors, but many control functions can only be obtained by systematic experiments on dynamometers and with vehicles.

3. IDENTIFICATION OF NONLINEAR SYSTEMS: LOOK-UP TABLES

Due to the extremely nonlinear dynamics of internal combustion engines, and due to the fact that often measured data is the only available information on the connection between cause and effect, engine control units rely heavily upon the calibration of black-box models. In spite of their limitations to problems with one- and two-dimensional input spaces, grid-based look-up tables are by far the most common type of nonlinear static models used in practice. The reason for this lies in their simplicity and their extremely low computational evaluation demand. Modern engine control units of spark ignition engines contain more than 100 look-up tables (Adler, 2000).

In the following the basics of grid-based look-up tables are given. The off-line estimation of the heights of the interpolation nodes will be explained, followed by the presentation of an on-line adaptation algorithm using the

NLMS approach and a newly developed, modified RLS algorithm.

3.1 *Structure of two-dimensional look-up tables*

Grid-based look-up tables can also be interpreted as a 2nd order B-spline network (Brown and Harris, 1994). They consist of a set of data points or interpolation nodes positioned on a multi-dimensional grid. Each node comprises two components. The scalar data point heights are estimates of the approximated nonlinear function at their corresponding data point position. All nodes are stored e.g. in the ROM of the control unit. For model generation usually all data point positions are fixed a-priori.

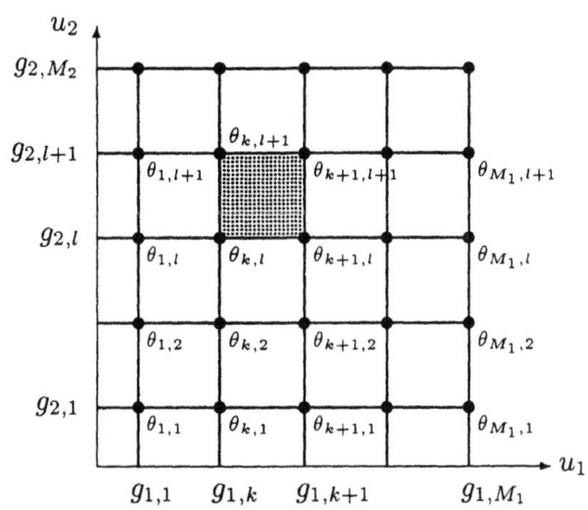

Fig. 3. grid-based placement of the interpolation nodes of a two-dimensional look-up table.

In the following we consider a two-dimensional look-up table of the size $M_1 \times M_2$, see **Fig. 3**. It consists of interpolation nodes, located on the grid lines $g_{1,1}, \ldots, g_{1,M_1}$ and $g_{2,1}, \ldots, g_{2,M_2}$. The model output $\hat{y}(u_1, u_2)$ for a given input (u_1, u_2) is calculated by considering the closest points to the bottom left, bottom right, top left, and top right of the model input, see **Fig. 4**. The heights of the interpolation nodes $\theta_{k,l}$, $\theta_{k+1,l}$, $\theta_{k,l+1}$, and $\theta_{k+1,l+1}$ are weighted with the corresponding opposite area:

$$\begin{aligned}\hat{y} = \; & \theta_{k,l} \cdot \frac{A_{k+1,l+1}}{A} + \theta_{k+1,l} \cdot \frac{A_{k,l+1}}{A} \\ & + \theta_{k,l+1} \cdot \frac{A_{k+1,l}}{A} + \theta_{k+1,l+1} \cdot \frac{A_{k,l}}{A}\end{aligned} \qquad (1)$$

with the areas

$$\begin{aligned}A_{k+1,l+1} &= (g_{1,k+1} - u_1)\,(g_{2,l+1} - u_2)\\ A_{k,l+1} &= (u_1 - g_{1,k})\,(g_{2,l+1} - u_2)\\ A_{k+1,l} &= (g_{1,k+1} - u_1)\,(u_2 - g_{2,l})\\ A_{k,l} &= (u_1 - g_{1,k})\,(u_2 - g_{2,l})\end{aligned}$$

The total area A is calculated as

$$A = (g_{1,k+1} - g_{1,k})\,(g_{2,l+1} - g_{2,l}),$$

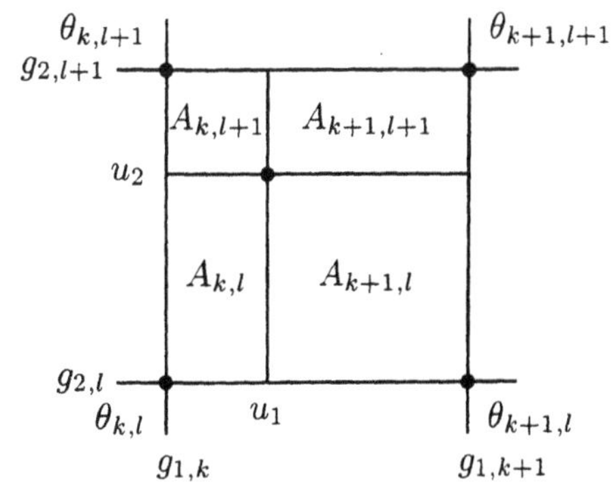

Fig. 4. Areas for interpolation within a two-dimensional look-up table.

with k and l denoting the closest interpolation node which is smaller than u_1 and u_2, respectively. (1) can be described in a basis function formulation

$$\hat{y} = \sum_{i=1}^{M} w_i \cdot \Phi_i(\underline{u}, \mathbf{c}) \qquad (2)$$

with the M-dimensional weighting vector

$$\begin{aligned}\underline{w} &= [\theta_{1,1} \;\ldots\; \theta_{M_1,M_2}]^T\\ &= [w_1 \; w_2 \;\ldots\; w_M]^T\end{aligned} \qquad (3)$$

which represents the heights of the interpolation nodes. With regard to (1) for each input vector $\underline{u} = [u_1 \; u_2]^T$ in general only 4 basis functions Φ_i are non-zero. The basis functions depend on the positions $\mathbf{c}$ of the interpolation nodes and on the input vector $\underline{u}$.

3.2 *Off-line estimation of the look-up table heights*

For generating a look-up table, the most widely applied method to obtain the heights of the interpolation nodes is to position measurement data points directly on the grid points. Then, an additional approximation step can be avoided. However, if available measurement data points do not correspond to the desired positions of the grid, or if the locations of the grid have to be changed due to memory restrictions, estimation techniques have to be applied in order to derive the heights of the interpolation nodes. This allows an arbitrary positioning of the grid lines, irrespective of the location of the measurement data points. In the following an approach for off-line estimation of the heights of the interpolation nodes is explained.

The optimization of the grid location represents a nonlinear optimization problem and is not addressed in this paper, in principle, any nonlinear optimization technique can be applied, (Nelles, 2000).

The off-line estimation of the heights of the interpolation nodes can be performed by least squares method (Isermann, 1992), in order to fit to measurement data that does not lie on the pre-specified grid (Nelles and

Fink, 2000). The loss function $\sum_{i=1}^{N} e^2(i)$ is minimized, where $e(i) = y(i) - \hat{y}(i)$ represent the model errors for data sample $\{\underline{u}(i), y(i)\}$. For a given data set $\{\underline{u}(i), y(i)\}_{i=1}^{N}$, $N > M$, the regression matrix $\boldsymbol{X}$ is

$$\boldsymbol{X} = \begin{bmatrix} \Phi_1(\underline{u}(1), \boldsymbol{c}) & \Phi_2(\underline{u}(1), \boldsymbol{c}) & \cdots & \Phi_M(\underline{u}(1), \boldsymbol{c}) \\ \Phi_1(\underline{u}(2), \boldsymbol{c}) & \Phi_2(\underline{u}(2), \boldsymbol{c}) & \cdots & \Phi_M(\underline{u}(2), \boldsymbol{c}) \\ \vdots & \vdots & & \vdots \\ \Phi_1(\underline{u}(N), \boldsymbol{c}) & \Phi_2(\underline{u}(N), \boldsymbol{c}) & \cdots & \Phi_M(\underline{u}(N), \boldsymbol{c}) \end{bmatrix}$$

where N is the number of data samples of the input data $\{\underline{u}(i)\}_{i=1}^{N}$, and $\boldsymbol{c}$ contains the coordinates of the interpolation nodes.

The parameter vector $\underline{w}$, which contains the heights of the interpolation nodes, see (3), is given as

$$\underline{w} = (\boldsymbol{X}^T \boldsymbol{X})^{-1} \boldsymbol{X}^T \underline{y}, \qquad (4)$$

with $\underline{y} = [y(1)\, y(2) \ldots y(N)]^T$.

The regression matrix $\boldsymbol{X}$ is typically sparse. In the two-dimensional case each row of $\boldsymbol{X}$ contains only 4 non-zero elements. If the measurement data does not cover the whole input space, some basis functions will not be activated at all, $\boldsymbol{X}$ will contain columns of zeros and may become singular. In order to make the least squares problem solvable the corresponding regressors and their associated heights have to be removed vom $\boldsymbol{X}$ and $\underline{w}$ (Nelles, 2000).

3.3 *On-line adaptation of look-up table heights using NLMS algorithm*

On-line -or instantaneous- learning means that information about the desired input/output behavior is given only by isolated samples $(\underline{u}(t), y(t))$. On-line adaptation is of special importance for time-variant systems, e.g. due to changing ambient conditions or aging. The adaptation method with the lowest computational requirements is the *normalized least mean square* algorithm (NLMS) (Brown and Harris, 1994), it can be applied as follows for the adaptation of the heights of the interpolation nodes:

$$w_i(t+1) = w_i(t) + \beta \cdot e(t) \cdot \frac{\Phi_i(\underline{u}(t), \boldsymbol{c})}{\sum_{j=1}^{M} \Phi_j^2(\underline{u}(t), \boldsymbol{c})},$$

$e(t)$ denotes the error between the correct value $y(t)$ and the old network output $\hat{y}(\underline{u}(t))$. The learning rate β must be within the range $0 < \mu < 2$, however, appropriate values vary between 1 for fast learning and $\mu << 1$ for robustness against measurement noise. In order to further improve robustness against noise and to avoid over-fitting, a dead zone with minimal error value for parameter adaptation can be set (Heiss, 1996).

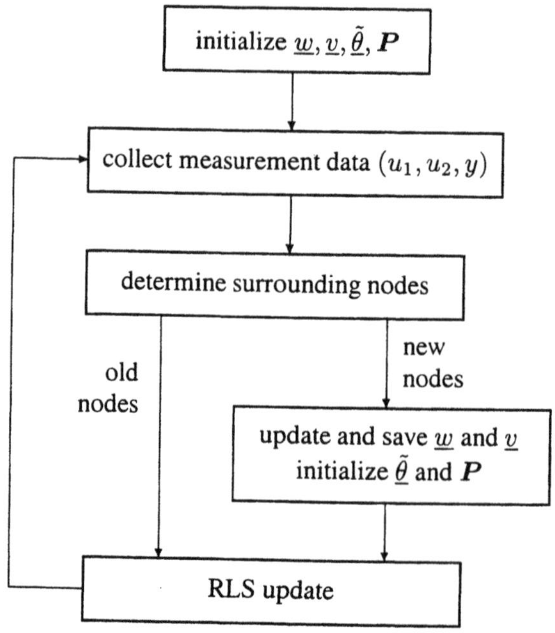

Fig. 5. flow chart of modified RLS algorithm

3.4 *On-line adaptation of look-up table heights using RLS algorithm*

In relation to the usually used NLMS algorithm, the convergence time during on-line adaptation of the heights of the interpolation nodes can be considerably reduced by applying recursive least squares (RLS) algorithms (Isermann, 1992). The look-up table consists of $M_1 \times M_2$ interpolation nodes, however numerical stability can only be obtained, if the problem is reduced to the estimation of fewer parameters. The algorithm proposed in the following estimates only the heights of the surrounding interpolation nodes of a query point, i.e. instead of the M parameters in $\underline{w}$ only 4 parameters $\tilde{\underline{\theta}}$ are estimated at each time instant. During adaptation, the algorithms requires for each height w_i the storage of a variance value σ_i^2. Without prior knowledge the variances $\underline{v} = [\sigma_1^2 \ldots \sigma_M^2]^T$ can be initialized as $\sigma_i^2 = 100 \ldots 1000$.

We consider the estimation of the parameter vector $\tilde{\underline{\theta}} = [\theta_{k,l}, \theta_{k+1,l}, \theta_{k,l+1}, \theta_{k+1,l+1}]^T = [w_j \ldots w_{j+3}]^T$, see **Fig. 4**. The matrix $\boldsymbol{P}$ is initialized as

$$\boldsymbol{P} = \begin{pmatrix} \sigma_j^2 & 0 & 0 & 0 \\ 0 & \sigma_{j+1}^2 & 0 & 0 \\ 0 & 0 & \sigma_{j+2}^2 & 0 \\ 0 & 0 & 0 & \sigma_{j+3}^2 \end{pmatrix}, \qquad (5)$$

standard RLS estimation can then be applied:

$$\underline{\gamma}(t+1) = \frac{1}{\lambda + \underline{\psi}^T(t+1)\, \boldsymbol{P}(t)\, \underline{\psi}(t+1)} \boldsymbol{P}(t) \underline{\psi}(t+1)$$

$$\tilde{\underline{\theta}}(t+1) = \tilde{\underline{\theta}}(t) + \underline{\gamma}(t+1) \left(y(t+1) - \underline{\psi}^T(t+1) \tilde{\underline{\theta}}(t) \right)$$

$$\boldsymbol{P}(t+1) = \frac{1}{\lambda} \left(\boldsymbol{I} - \underline{\gamma}(t+1) \underline{\psi}^T(t+1) \right) \boldsymbol{P}(t)$$

with

$$\underline{\psi}(t) = [\Phi_j(\underline{u}(t), \boldsymbol{c}) \ \dots \ \Phi_{j+3}(\underline{u}(t), \boldsymbol{c})].$$

λ is a forgetting factor that allows exponential forgetting of the old measurements. It allows a trade-off between tracking performance and noise suppression. Φ_j are the weighting functions as in (2).

Before each RLS update the algorithm verifies that the surrounding interpolation nodes did not change as a result of a changing operating condition. However, if the surrounding interpolation nodes changed since the last update, the diagonal elements of $\boldsymbol{P}$ are retrieved and saved in the variance vector $\underline{v}$. $\boldsymbol{P}$ is re-initialized as in (5), the variances of the new interpolation nodes are used as the new diagonal elements. **Fig. 5** shows the flow chart of the proposed RLS algorithm for updating look-up tables.

In order to avoid deterioration of the already estimated model parameters, a supervisory level freezes the identification algorithm if excitation is insufficient, e.g. if deviations between model output and desired output are incremental.

4. FAST LOCAL LINEAR NEURAL NETWORKS

Look-up tables are restricted to one- and two-dimensional applications since memory and computational requirements grow exponentially with the number of input signals. For high dimensional problems, neural networks have been successfully applied. Well known neural networks for the identification of nonlinear processes are multilayer perceptrons (MLP) and radial basis functions (RBF), (Isermann *et al.*, 1997). In the following local linear model networks are considered which have several advantages for engine measurements. Local linear model tree (LOLIMOT) is an extended radial basis function network. It is extended by replacing the output layer weights with a linear function of the network inputs (Nelles, 1999). Thus, each neuron represents a local linear model with its corresponding validity function. Furthermore, the radial basis function network is normalized, that is the sum of all validity functions for a specific input combination sums up to one. The Gaussian validity functions determine the regions of the input space where each neuron is active. The input space of the net is divided into M hyper-rectangles, each represented by a linear function.

The output y of a LOLIMOT network with n inputs $x_1, \dots, x_n$ is calculated by summing up the contributions of all M local linear models

$$\hat{y} = \sum_{i=1}^{M} (w_{i0} + w_{i1}x_1 + \dots + w_{in}x_n) \cdot \Phi_i(\underline{x}) \quad (6)$$

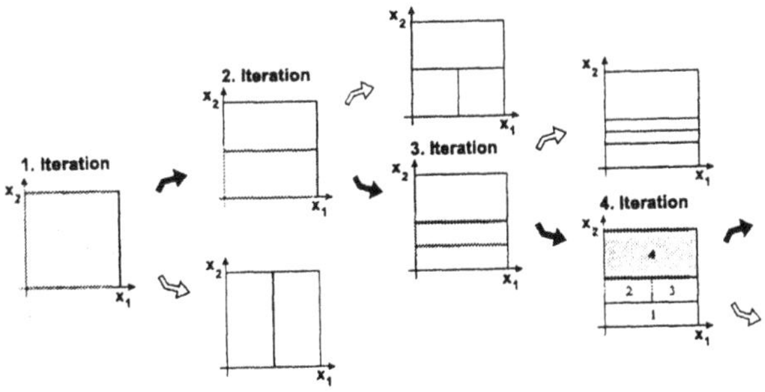

Fig. 6. First four iterations of LOLIMOT training algorithm

where w_{ij} are the parameters of the i-th linear regression model and x_i is the model input. The validity functions Φ_i are typically chosen as normalized Gaussian weighting functions.

$$\Phi_i(\underline{x}) = \frac{\mu_i}{\sum_{j=1}^{M} \mu_j} \quad (7)$$

with

$$\mu_i = \exp\left(-\frac{1}{2}\frac{(x_1 - c_{i1})^2}{\sigma_{i1}^2} - \dots - \frac{1}{2}\frac{(x_n - c_{in})^2}{\sigma_{in}^2}\right)$$

with center coordinates $\underline{c}_{ij}$ and dimension individual standard deviations $\underline{\sigma}_{ij}$. LOLIMOT is trained as follows. In an outer loop the network structure is optimized. It is determined by the number of neurons and the partitioning of the input space, which is defined by the centers and standard deviations of the validity functions. An inner loop estimates the parameters and possibly the structure of the local linear models. The network structure is optimized by a tree construction algorithm that determines the centers and standard deviations of the validity functions (**Fig. 6**).

The LOLIMOT algorithm partitions the input space in hyper-rectangles. In the center of each hyper-rectangle, the validity function of the corresponding linear model is placed. The standard deviations are chosen proportional to the size of the hyper-rectangle. This makes the size of the validity region of a local linear model proportional to its hyper-rectangle extension. Thus, a model may be valid over a wide operating range of one input variable but only in a small area of another one. At each iteration of the outer loop, one additional neuron is placed in a new hyper-rectangle, which is derived from the local linear model with the worst local error measure

$$J_{local} = \sum_{j=1}^{N} \Phi_i(\underline{x}(j)) \cdot (y(j) - \hat{y}(j))^2.$$

In other words, the local error is the sum of the squared errors weighted with the corresponding validity function Φ_i over all data samples N. The new hyper-rectangle is

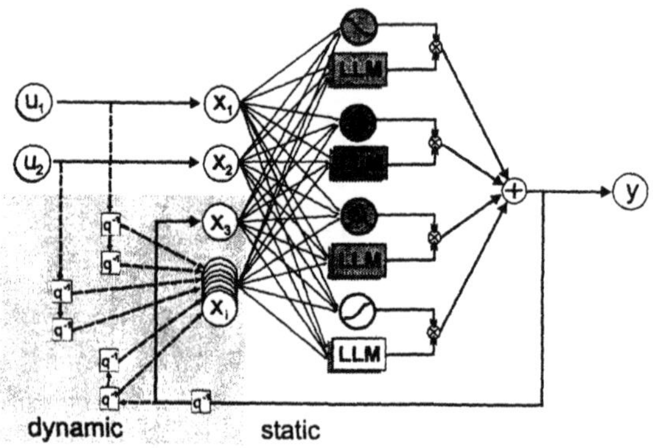

Fig. 7. LOLIMOT net with external dynamics, LLM: Local Linear Models The filters are chosen as simple time delays q^{-1}

then chosen by testing the possible cuts in all dimensions and taking the one with the highest performance improvement.

In an inner loop the parameters of the local linear models are estimated by a local weighted least-squares technique. The prediction errors are weighted with the corresponding validity function. Each local linear model is estimated separately, that is the overlap between neighbored local models is neglected. This approach is very fast and robust. It is especially important to note that due to the local estimation approach the computational demand increases only linearly with the number of local models.

In addition to approximating stationary relations of nonlinear processes, LOLIMOT is also capable of simulating the dynamic behavior of processes. In order to model multivariate, nonlinear, dynamic processes the following time-delay neural network approach can be taken.

$$\hat{y}(t) = f(\underline{x}(t))$$
$$\underline{x}(t) = [u_1(t-1), ..., u_1(t-m), ..., u_p(t-1)..., u_p(t-m), \hat{y}(t-1), ..., \hat{y}(t-m)]^T$$

where m is the dynamic order of the model, $u_1(t), \ldots, u_p(t)$ are the p model inputs and $\hat{y}(t)$ is the model output at time t, see **Fig. 7**.

Some of the main advantages of the LOLIMOT approach is its fast training time of some 10..30 s, compared to many minutes to hours with other neural networks, its applicability to adaptive problems (Fischer *et al.*, 1998) and the interpretability of the net structure and parameters in a physical sense.

4.1 *On-line Adaptation of the LOLIMOT Parameters*

For on-line adaptation of the LOLIMOT network, the validity functions Φ_i are kept fixed and only the parameters of the local linear models are adapted.

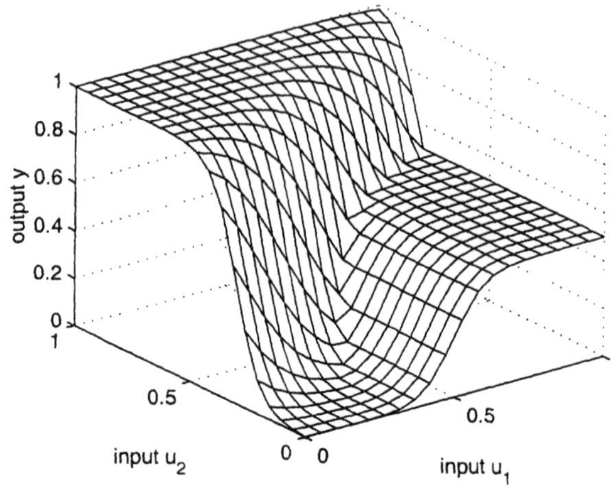

Fig. 8. Nonlinear test function with two input signals.

The following *recursive weighted least-squares* (RWLS) algorithm with exponential forgetting is utilized to estimate the parameters of each local linear model. For the j-th local linear model, it computes a new parameter estimate $w_j(t)$ at the time instant t as follows, see (Isermann *et al.*, 1992):

$$\underline{\gamma}(t+1) = \frac{1}{\frac{\lambda}{\Phi_j(\underline{x}(j))} + \underline{\psi}_j^T(t+1)\, \boldsymbol{P}_j(t)\, \underline{\psi}_j(t+1)} \boldsymbol{P}_j(t) \underline{\psi}_j(t+1)$$

$$\underline{w}_j(t+1) = \underline{w}_j(t) + \underline{\gamma}_j(t+1) \left(y(t+1) - \underline{\psi}_j^T(t+1) \underline{w}_j(t) \right)$$

$$\boldsymbol{P}_j(t+1) = \frac{1}{\lambda} \left(\boldsymbol{I} - \underline{\gamma}_j(t+1) \underline{\psi}_j^T(t+1) \right) \boldsymbol{P}_j(t)$$

λ is a forgetting factor that implements exponential forgetting of the old measurements and Φ_j is the weighting of the actual data with the validity function value. The role of the forgetting factor as well as supervisory tasks for the on-line adaptation are discussed in (Fink *et al.*, 1999).

5. COMPARISON OF LOOK-UP TABLE AND LOCAL LINEAR NEURAL NETWORK FOR THE TWO-DIMENSIONAL CASE

For the comparison of the on-line adaptation algorithms of look-up table and local linear neural network, a nonlinear test function with two input signals was considered, see **Fig. 8**. The testing structure shown in **Fig. 9** was used to train the look-up table using NLMS and RLS adaptation algorithms and the LOLIMOT neural network using RWLS adaptation algorithm. u_1 and u_2 have been random numbers within $[0, 1]$, before training the network parameters of look-up table and LOLIMOT neural network have been initialized to zero.

Fig. 10 shows the comparison of memory requirements and computation time, and of time of convergence for the on-line training. In the first and second diagrams,

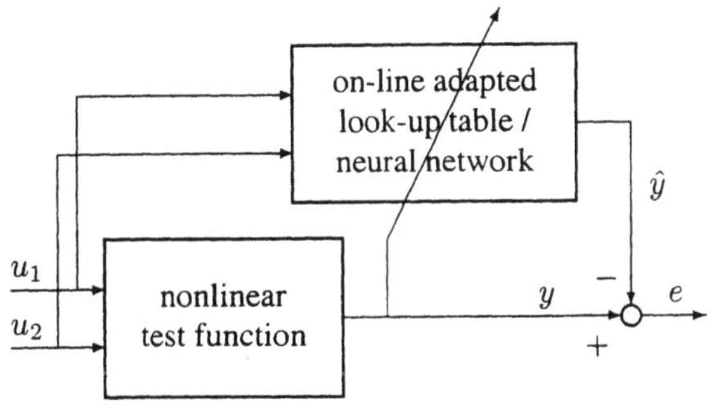

Fig. 9. Testing structure used for on-line adaptation of look-up tables and neural networks.

values in black refer to *generalization*, values in white refer to the case that also *adaptation* is activated. All data is normalized to look-up table values without adaptation. The first column shows values of the look-up table, using the NLMS adaptation algorithm. Memory and computational requirements are small, however time of convergence is large. Additional memory and computational requirements of the NLMS adaptation is fractional and not visible in the diagram. If the look-up table is adapted using the proposed RLS algorithm, memory expense doubles since in addition to the heights also their covariances have to be stored. However, time of convergence can be considerably reduced. Memory requirements are large for the LOLIMOT neural network, also the computational requirements, which is due to the evaluation of the exponential functions. However, due to the fast RWLS algorithm, time of convergence during adaptation is very fast.

For the two dimensional case the conventional look-up table outperforms the LOLIMOT neural network, if fast convergence is required the application of RLS adaptation is advantageous.

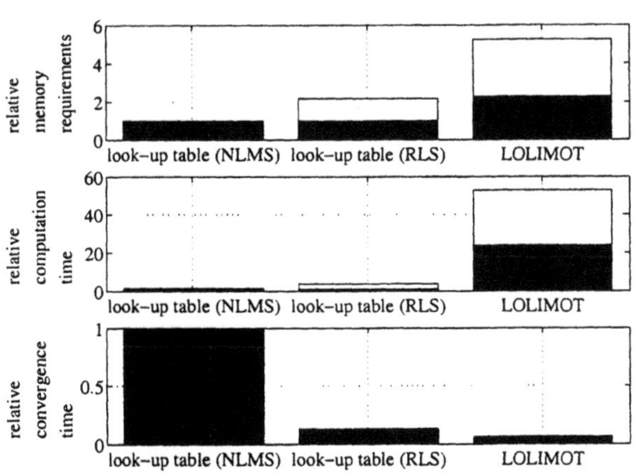

Fig. 10. Comparison of memory and computational requirements, as well as time of convergence for the on-line training of a two-dimensional look-up and a two-dimensional neural network.

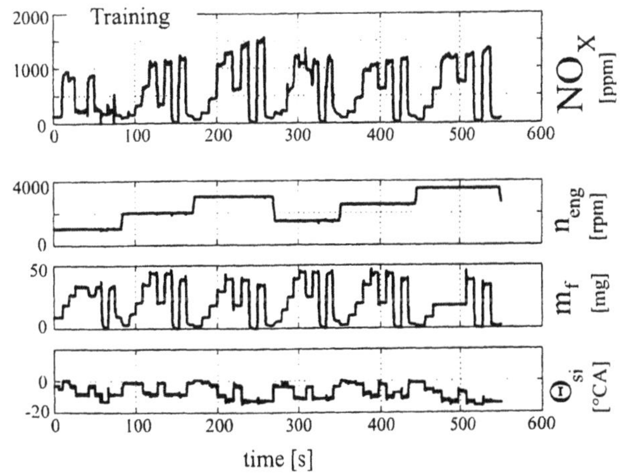

Fig. 11. Training data set for a dynamic NO_X model

Notice however, that for the high dimensional case memory and computational requirements of the look-up table grow exponentially with the number of input signals, which is not the case for the LOLIMOT algorithm (Nelles, 1999). Therefore in the high dimensional case the LOLIMOT algorithm would be the better choice.

6. NONLINEAR IDENTIFICATION OF EXHAUST GAS COMPONENTS

Theoretical models of exhaust gas concentrations are too complex and labor-intensive for control design. The reason is that complex thermodynamic and chemical equations, side-effects like swirl, tumble, quenching, local temperatures etc. have to be taken into account (Heywood, 1988). Therefore, exhaust gas models are mostly experimental.

In contrast to two-dimensional look-up tables, neural nets can easily be applied for high order problems in order to represent all relevant variables influencing the emissions. Depending on the complexity of the engine, more than 5-7 inputs might be necessary to consider the most relevant variables concerning the exhaust gas formation. Among these are e.g. the engine speed, load, injection angle and -pressure, EGR and the position of the guide blades of turbochargers with variable turbine geometry (VTG). Another advantage of dynamic neural nets is the ability to dynamically measure the engine's behavior and calculate its static maps by means of the dynamic neural network. Consequently, there is no need to measure the whole look-up table using measurement points on equidistant grids, which helps to save expensive test stand time. As an example for engine emission models, a dynamic NO_X model of a 1.9l direct injection diesel engine is described. When recording the training data for dynamic models, one should always keep in mind, that the measured data has to contain the whole range of amplitudes and dynamics of the process in

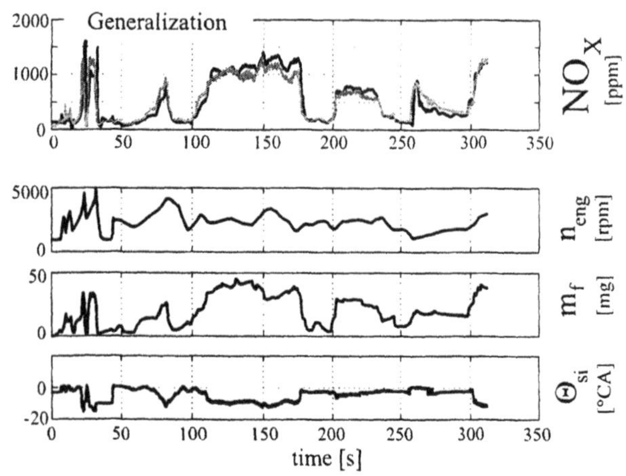

Fig. 12. Generalization of the NO_X model trained with the data in Fig. 11

order to get satisfying results. APRBS signals (amplitude modulated pseudo random binary signal) are often very suitable as process inputs because they excite the process at a wide range of amplitudes and frequencies. In this example, step responses of differing amplitudes have proven to be suitable for the training of the NO_X-models. **Fig. 11** shows the dynamically measured NO_X and the three net-inputs m_f, n_{eng} and Θ_{si} of the training data set (EGR and VTG were disabled during this test).

After building the neural model according to (8), the model can be applied to a new (unknown) data set and simulates the NO_X according to the respective input data (on-line in the vehicle, if necessary).

$$\hat{NO_X}(k) = f_{LOLIMOT}(m_f(k-1), \quad n_{eng}(k-1), \Theta_{si}(k-1), \hat{NO_X}(k-1)) \tag{8}$$

Fig. 12 illustrates the good generalization performance of a first order dynamic net with 15 neurons for the NO_X emissions. Despite the excitation of the net-inputs

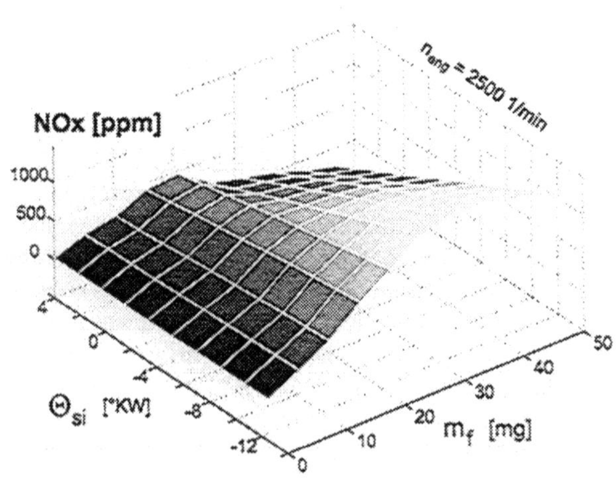

Fig. 13. NO_X-Map, calculated from dynamic neural net

in Figs. 11 and 12 were of quite different quality, the simulated NO_X is able to follow the measured values with an error of just a few percent, which was quite satisfying.

Another important feature of these dynamic neural models is the possibility of deriving stationary maps from dynamic models. The data in **Fig. 11** led to a dynamic model according to (8). This model was then used to calculate the stationary two-dimensional look-up table shown in **Fig. 13**. Thus, it is possible to get a good visual overview on the engine's characteristics by maps derived from dynamic measurements. Even more importantly, this feature allows a fast dynamic measurement of the engine's characteristics and a calculation of the static engine maps by means of the dynamic neural model.

The obtained dynamic model for the NO_X emissions can be used for online optimization tasks for modern engine control units (Hafner *et al.*, 2000). If sensors for the NO_X emissions should be available as they are used in latest direct injection gasoline engines (Glück *et al.*, 2000), the RLWS adaptation method, see Section 4.1 could be used for online adaptation. This could be used for compensation of varying fuel qualities or manufacturing tolerances.

7. ADAPTIVE FEEDFORWARD CONTROL OF IGNITION ANGLE

Cylinder pressure signals contain valuable information for closed loop engine control. For using this information low cost combustion pressure sensors with high long-term stability have been developed (Anastasia and Pestana, 1987; Herden and Küsell, 1994) and are already installed into certain production engines (Inoue *et al.*, 1993).

The objective of *ignition control* is to achieve optimum engine efficiency for each combustion event. Generally factors that influence the optimal ignition angle are engine specifications like configuration of the combustion chamber, operating conditions like engine speed, load, temperature, and exhaust gas recirculation flow rate, as well as ambient conditions such as air temperature, air pressure and humidity in the atmosphere.

Standard ignition control systems are based on feedforward control and therefore rely heavily upon calibration of look-up tables. The database values are initially calibrated from an analysis of a nominal engine under fixed environmental conditions. However, aging effects, manufacturing tolerances or a changing environment usually change the engine characteristics and lead to deteriorating performance.

Based on cylinder pressure sensors, Learning Feed-Forward Control (LFFC) can be used in order to opti-

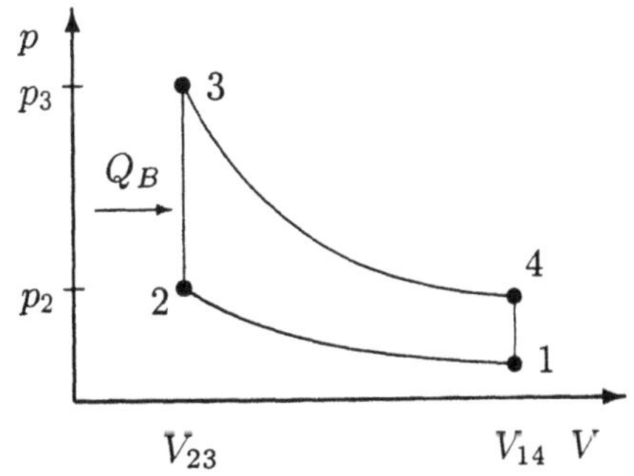

Fig. 14. Constant volume diagram of an ideal constant-volume combustion cycle

mize the point of ignition of each cylinder. The LFFC consists of a function approximator, the latter is represented by a two-dimensional look-up table. The adaptive look-up table is trained on-line using the RLS training algorithm. The learning process is performed during normal operation of the engine, without determining special excitation signals. The learning system is just collecting and processing input-output samples at the individual operating points.

7.1 *Cylinder Pressure Evaluation*

Not only thermodynamic analysis but also extensive experimental results suggest that the optimum efficiency of each combustion event is achieved if 50 % energy conversion occurs at 8° crankshaft angle (CA) after Top Dead Center (TDC), irrespective of the operating point, (Bargende,. 1995; Matekunas, 1986). Therefore the crank angle location of 50% energy conversion is to be calculated and to be controlled by modifying the point of ignition accordingly.

The energy conversion during a combustion cycle can be described by the Mass Fraction Burned (MFB) $x_{MFB}(\theta)$ of the cylinder charge at a specific crank angle θ. The MFB, in turn, can be approximated by

$$x_{MFB}(\theta) = \frac{p(\theta)\,V(\theta)^{\kappa} - p_i\,V_i^{\kappa}}{p_{eoc}\,V_{eoc}^{\kappa} - p_i\,V_i^{\kappa}} \tag{9}$$

where $p(\theta)$ and $V(\theta)$ denote the measured cylinder pressure and the cylinder volume at a certain crank angle, respectively. The indices *eoc* and i denote specific crank angle locations at the *e*nd *o*f *c*ombustion and before *i*gnition, respectively. κ stands for the polytropic index.

This approximation can be derived by considering the pressure-volume diagram of an ideal constant-volume combustion cycle, see **Fig. 14**. For the compression stroke (1 → 2), as well as for the power stroke (3 → 4) an polytropic behavior can be assumed, i.e.

$$p(\theta) \cdot V(\theta)^{\kappa} = C_1 \tag{10}$$

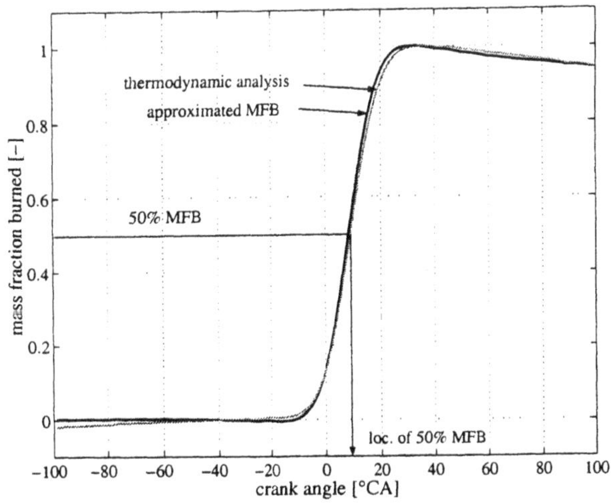

Fig. 15. Approximation of MFB in comparison to thermodynamic analysis. The location of 50% MFB is to be controlled by appropriate settings of the point of ignition.

for the compression stroke and

$$p(\theta) \cdot V(\theta)^{\kappa} = C_2 \tag{11}$$

for the power stroke. Therefore the amount of energy Q_B, released between ignition (point 2) and the end of combustion (point 3) is

$$Q_B \sim C_2 - C_1 = p_{eoc} \cdot V_{eoc}^{\kappa} - p_i \cdot V_i^{\kappa}. \tag{12}$$

During combustion the energy released is proportional to $p(\theta)\,V(\theta)$, normalized to the values between end of combustion and ignition, which leads to equation (9).

In **Fig. 15** for an arbitrary cycle the approximation of MFB according to (9) is compared to MFB calculated by thermodynamic analysis of the cylinder pressure. In the presented control system the crank angle location of $x_{MFB}(50\%)$ is calculated by means of the approximation mentioned above.

7.2 *Control Structure*

Applying closed-loop ignition control with for example standard PI-controllers, this cannot provide acceptable control performance under fast changing operating conditions since the controllers cannot be tuned for high control effort. This is due to the fact, that after ignition of the air-fuel mixture, the combustion cannot be influenced any further. Therefore the ignition angle can only be computed for the next cycle, based on measurements from the present engine cycles. Therefore a dead time of one cycle is inherent. Moreover as there exist significant cycle fluctuations even under steady operating conditions, the results of cylinder pressure evaluation

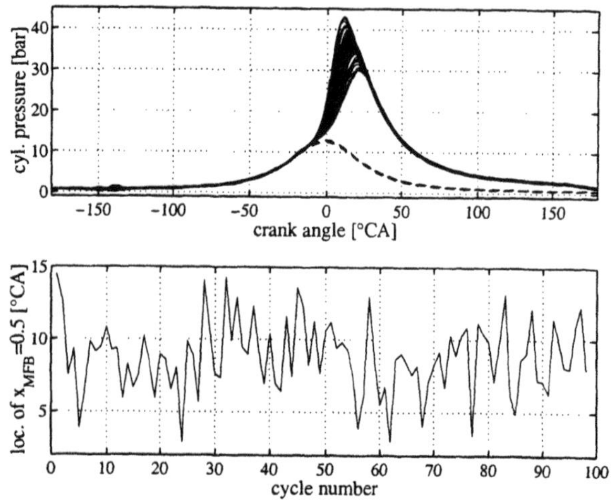

Fig. 16. Upper diagram: measured cylinder pressure signals and reconstructed polytrope; lower diagram: calculated crank angle locations of 50% MFB. $n_{mot} = 3000$ rpm, 50% load

of several engine cycles have to be averaged (e.g. a moving average over 10 cycles is used). Because the system error usually differs from one engine operating condition to another, during engine transients it takes a certain amount of time for the PI controller to „integrate" to a new ignition angle. The stochastic nature of the combustion events can be seen in **Fig. 16**. The upper diagram shows the measured cylinder pressure signals of the compression and the power strokes of 100 consecutive cycles of an arbitrary cylinder. The dotted lines represent the reconstructed polytrope, which is the component of the cylinder pressure which is due to the piston motion (towed or motored pressure) . The lower diagram depicts the corresponding crank angle locations of 50% MFB, which are to be controlled to 8° CA. The stochastic nature of the combustion events is clearly visible.

This motivates the use of Learning Feedforward Control (LFFC), whose general structure is shown in **Fig. 17**. The linear controller C is used to compensate random disturbances and to provide a reference signal for the learning feed-forward controller. It does not need to have a high performance and can be designed in such a way that it provides a robust stability. Since the learning feedforward controller acts instead of a controller's integral term, the linear controller is preferably a simple proportional gain which can be deactivated for noise suppression.

For the ignition control system the function approximator is divided into the conventional, fixed ignition look-up table and into the adaptive offset map, see **Fig. 18**. The operating condition is determined by the engine load (a normalized value calculated from the intake manifold pressure signal) and the engine's rotational speed.

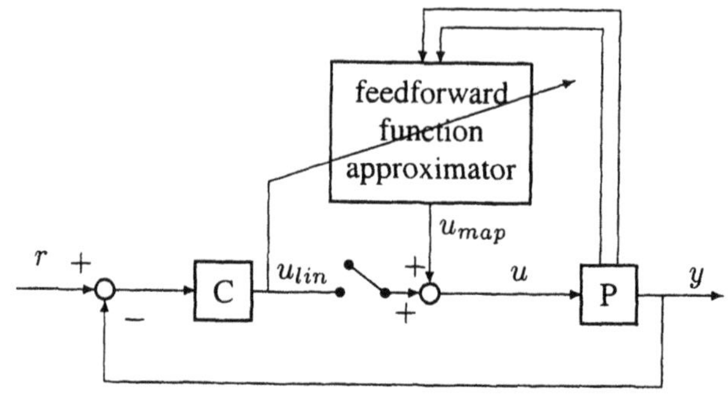

Fig. 17. General structure of a LFFC (Learning Feedforward Controller). The controller C is basically used for the adaptation of the feedforward map

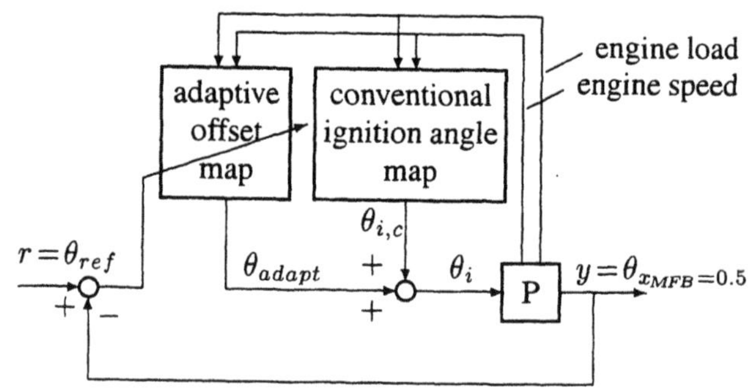

Fig. 18. Structure of the adaptive feedforward ignition control system

The conventional look-up table determines an approximate value of the ignition angle and is valid for all cylinders, it is depicted in **Fig. 20**. For each cylinder the location of 50% MFB is calculated and compared to the reference location of 8° CA. Correction values are calculated and memorized by the adaptive offset look-up table of each cylinder, at the corresponding operating condition. In order to reduce noise in the controller output signal θ_i, no linear proportional controller C is used in parallel to the function approximator, see **Fig. 17**.

7.3 *Measurement Results*

In **Fig. 19** at a constant engine speed of 2500 rpm a certain load change sequence is repeated for three times, see the third diagram. The first diagram shows the crank angle locations of 50% MFB for two cylinders, which are to be controlled to 8° CA after TDC. The second diagram depicts the correction values $\theta_{z,adapt}$ for the ignition points of the corresponding cylinders. For the first 50 sec only the conventional feedforward control is active. The considerable deviations of the crank angle locations of 50% MFB from the optimal value for best engine efficiency at 8° CA after TDC is visible, see the upper diagram. Between 50 and 168 sec the adaptive feedforward controller is active. Since the adaptive input-output map of each cylinder had been initialized to

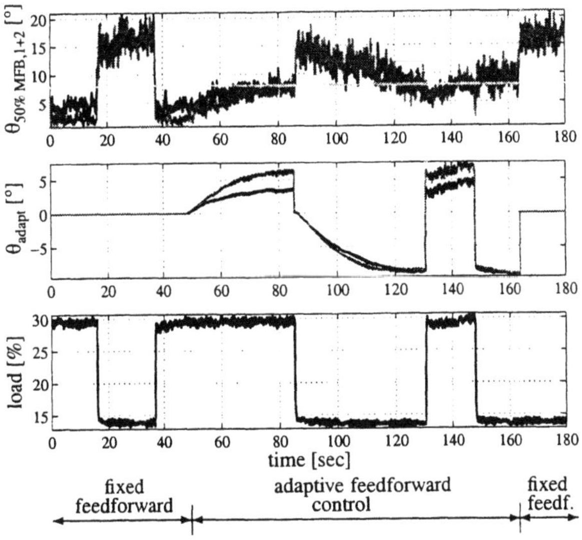

Fig. 19. Control performance of the ignition control system during three equivalent load changes. The adaptive look-up table is initialized with zero, after 50 sec the adaptive feedforward control is activated.

zero, it first has to be adapted for both operating conditions. The control performance for the already adapted feedforward control is shown between 130 and 160 sec. Then it is switched back to the conventional, fixed feedforward control. **Fig. 20** shows the conventional, fixed ignition angle look-up table (a.)) and the adaptive offset map of the second cylinder after a training sequence (b.)).

In spite of the considerable, stochastic nature of the combustion events, the adaptive feedforward control allows to maintain the mean crank angle location of 50% MFB around its optimal value at about 8° CA after TDC.

8. REAL-TIME COMPUTER SYSTEM AND EXPERIMENT CONTROL

As a matter of course the presented cylinder pressure evaluation and the adaptive control algorithms require a considerable calculation effort. State-of-the-art engine

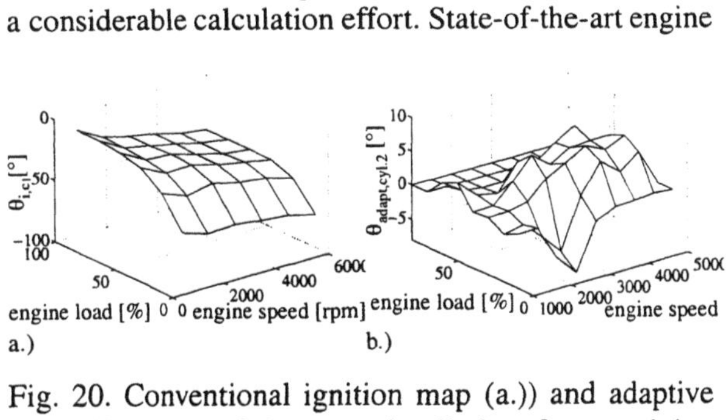

Fig. 20. Conventional ignition map (a.)) and adaptive offset map of the second cylinder after a training sequence (b.)).

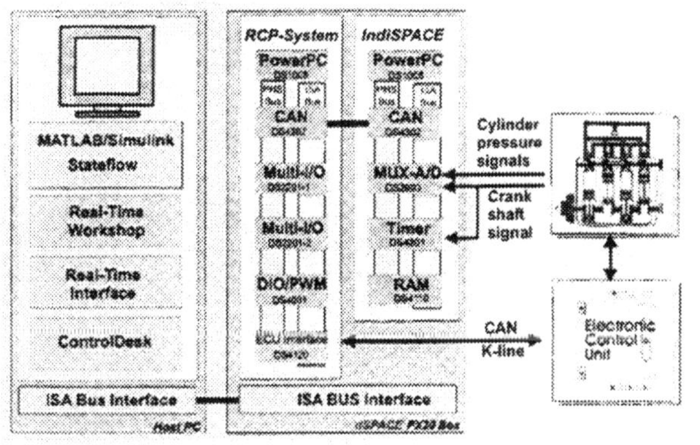

Fig. 21. Rapid control prototyping system configuration

control units (ECU) would not be able to calculate the algorithms in an appropriate time. Assuming that the growth in calculating power proceeds at the high speed of the past, future ECUs should make sufficient calculation time available within only a few years.

However, special real-time computer systems already allow an implementation and testing of the proposed algorithms in vehicle or engine test stands. In order to operate the control algorithms under realistic conditions, an indicating system was implemented on the basis of a PowerPC, type Motorola MPC 750, 480MHz, and coupled with a rapid control prototyping (RCP) system of the company dSPACE, which is also based on a PowerPC. Both systems operate in parallel to the production car's ECU. The goal of the RPC system is to enable a very fast end easy implementation and testing of new control concepts on real-time hardware. The user is enabled to code newly developed algorithms from block diagrams (e.g. MATLAB/Simulink) and download the code by means of an automatic code generation software to the real-time hardware. A complete design iteration can thus be accomplished within a few minutes (Hanselmann, 1996).

The indicating system allows to evaluate the cylinder pressure in real time with a resolution of 1° crankshaft angle for four cylinders. Thermodynamic and signal-based characteristic cylinder pressure features are transmitted to the RCP system. Based on this information, appropriate correction values for the injection duration, the ignition angle and the exhaust gas recirculation rate are calculated and sent to the production car's ECU via CAN bus interface, see **Fig. 21**. The described real-time hardware system enables very fast and easy implementation and testing of complex algorithms including extensive data preprocessing for the cylinder pressure, even under harsh real-time conditions of combustion engines. **Fig. 22** shows the dynamometer, consisting of an 160kW asynchronous motor coupled to the combustion engine.

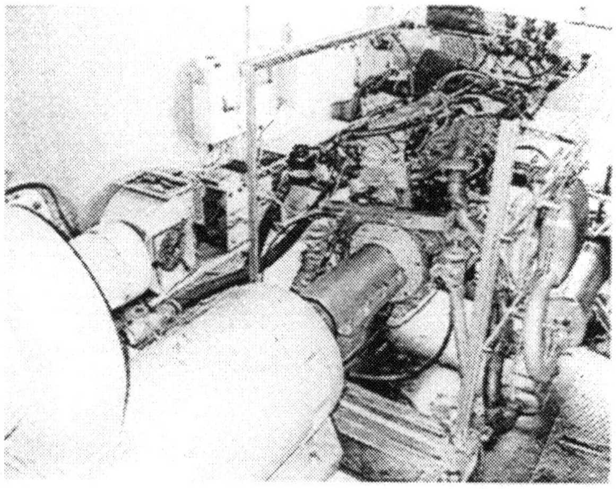

Fig. 22. Asynchronous motor coupled to the combustion engine

9. CONCLUSIONS

The demands for improved engine emissions, performance and efficiency require the application of advanced identification and control strategies. The proposed adaptation algorithms for grid-based look-up tables allow the online training of the heights of the interpolation nodes with improved convergence time. Fast neural networks with local linear models can approximate multi-dimensional problems within small training times of mostly less than a minute. The latter has been shown for the dynamic modelling of the NO_X emissions of a direct injection Diesel engine.

Recent computational capacities and sensor technologies allow the online evaluation of cylinder pressure signals for ignition control. Using cylinder-selective adaptive feedforward controllers, the 50% point of energy conversion can be kept near to the optimal point. The proposed control system is capable to compensate for manufacturing tolerances, fuel quality variations and long-term effects such as aging or wear of the engine. Moreover, the complexity and cost of engine calibration can thus be reduced.

10. REFERENCES

Adler, Ulrich (Ed.) (2000). *Automotive Handbook*. 5 ed.. Robert Bosch GmbH.

Anastasia, Ch.M. and G.W. Pestana (1987). A cylinder pressure sensor for closed loop engine control. In: *SAE Technical Paper Series*. number 870288.

Bargende, M. (1995). Most optimal location of 50% mass fraction burned and automatic knock detection components for automatic optimization of SI-engine calibrations. *MTZ Worldwide*.

Brown, M. and C. Harris (1994). *Neurofuzzy Adaptive Modelling and Control*. Prentice Hall. New York.

Fink, F., M. Fischer and O. Nelles (1999). Supervision of nonlinear adaptive controllers based on fuzzy models. In: *14th IFAC World Congress*. Beijing, China.

Fischer, M., O. Nelles and R. Isermann (1998). Adaptive predictive control of a heat exchanger based on a fuzzy model. *Control Engineering Practice* **6**(2), 259–269.

Glück, K.-H., U. Göbel, H. Hahn, J. Höhne, R. Krebs, T. Kreuzer and E. Pott (2000). Die Abgasreinigung der FSI-Motoren von Volkswagen. *MTZ Motortechnische Zeitschrift* **61**(6), 402–412.

Hafner, M., M. Schüler and R. Isermann (2000). Model-based optimization of ic engines by means of fast neural networks - part 2: Static and dynamic optimization of fuel consumption versus emissions. *MTZ Worldwide*.

Hanselmann, H. (1996). Automotive control: From concept to experiment to product. In: *IEEE International Symposium on Computer-Aided Control System Design*.

Heiss, M. (1996). Error-minimizing dead-zone for basis function networks. *IEEE Transactions on Neural Networks* pp. 1503–1506.

Herden, W. and M. Küsell (1994). A new combustion pressure sensor for advanced engine management. In: *SAE Technical Paper Series*. number 940379.

Heywood, J. B. (1988). *Internal Combustion Engine Fundamentals*. McGraw-Hill.

Inoue, T., S. Matsushita, K. Nakanishi and H. Okano (1993). Toyota lean combustion system - the third generation system. In: *SAE Technical Paper Series*. number 930873.

Isermann, R. (1992). *Identifikation dynamischer Systeme*. Vol. 1. 2 ed.. Springer Verlag. Berlin.

Isermann, R., K.-H. Lachmann and D. Matko (1992). *Adaptive Control Systems*. Prentice Hall. Englewood Cliffs.

Isermann, R., S. Ernst and O. Nelles (1997). Identification with dynamic neural networks - architectures, comparisons, applications -. In: *IFAC Symposium on System Identification*. Fukuoka, Japan.

Matekunas, F. A. (1986). Engine combustion control with ignition timing by pressure ratio management. *U.S. Patent*.

Nelles, O. (1999). Nonlinear System Identification with Local Linear Neuro-Fuzzy Models. PhD thesis. Darmstadt University of Technology. Darmstadt, Germany.

Nelles, O. (2000). *Nonlinear System Identification*. Springer Verlag. Berlin, Heidelberg.

Nelles, O. and A. Fink (2000). Tool zur Optimierung von Rasterkennfeldern. *atp Automatisierungstechnische Praxis* **42**(5), 58–66.

IDENTIFICATION-ORIENTED PREDICTIVE CONTROL

Leonardo C. Kammer [1], **Guy A. Dumont**

*Pulp and Paper Centre, University of British Columbia,
2385 East Mall, Vancouver, BC V6T 1Z4 Canada*

Abstract: This paper introduces a model-based predictive-control strategy that improves identifiability of the process under closed-loop operation. With this strategy in action there is no loss of identifiability when the only signal exciting the loop is a stochastic load disturbance. The main novelty in this work is the usage of a pseudo multi-rate control action. In essence the system alternates between open-loop and closed-loop operation at every sampling time. *Copyright © 2001 IFAC*

Keywords: Closed-loop identification; predictive control; model-based control; iterative methods; robust design.

1. INTRODUCTION

Even though stochastic load disturbances provide persistent excitation to the closed-loop system, identification of the process dynamics under low order, linear time-invariant control is of limited applicability (Åström and Wittenmark, 1995; Ljung, 1999). The suggestions for solving this problem, other than exciting the setpoint, are to increase the complexity of the controller, make the control action time-dependent (or nonlinear) or add a probing signal to the control action, as in dual control strategies (Åström and Wittenmark, 1995). The solution we seek in this work is to use a pseudo multirate control strategy, which can be classified as a time-dependent control action.

Truly multirate systems have different sampling rates for the control action and the process output. This is not the case in our scheme, but we choose not to use all output measurements for control purposes. The mechanism we propose is the following: continuously alternate between a normal control action, $u_t = f(r_t, y_t)$, and a control action that utilizes a prediction of the plant output given the signals up to the previous sampling time, $u_{t+1} = f(r_{t+1}, \hat{y}_{t+1|t})$.

Notice that a standard model-based predictive controller computes a sequence of control actions, at each sampling time, based on the optimization of a criterion that uses the predicted behaviour of the closed-loop signals. Only the first of these control actions is actually applied to the process. Here we propose to implement the first two initial control moves, as explained above, before computing the next set of control actions. As shown below, this will ensure closed-loop identifiability for use e.g. in adaptive or iteratively-tuned predictive control.

The identifiability of processes with an ARX structure is shown on Section 2, while Section 3 draws a few considerations on the control design and analysis of stability. An example and a comparative study are presented on Section 4.

2. SYSTEM IDENTIFICATION

The plants we consider in this paper are described by a generic scalar ARX structure:

[1] Current address: Paprican, 3800 Wesbrook Mall, Vancouver, BC V6S 2L9 Canada. Fax: +1 (604) 222-3207. Email: lckammer@ieee.org

$$y_t = \frac{b_1 q^{-1} + \cdots + b_{n_b} q^{-n_b}}{1 + a_1 q^{-1} + \cdots + a_{n_a} q^{-n_a}} q^{-n_d} u_t + \frac{1}{1 + a_1 q^{-1} + \cdots + a_{n_a} q^{-n_a}} e_t, \quad (1)$$

where $\{e_t\}$ is a zero-mean white-noise process with covariance σ_e^2.

In order to analyze the identifiability of (1), let us consider the worst case scenario: The loop set-point is constant (taken as zero without loss of generality) and the controller is linear and of minimal order, that is,

$$u_t = -k\, y_t. \quad (2)$$

Under this control law not even a simple model like

$$\hat{y}_t = \frac{\hat{b}_1 q^{-1}}{1 + \hat{a}_1 q^{-1}} u_t \quad (3)$$

can be uniquely identified (Åström and Wittenmark, 1995). Nevertheless, if the control action follows the sequence

$$\begin{aligned} &\vdots \\ u_0 &= -k\, y_0 \\ u_1 &= -k\, \hat{y}_1 = k(\hat{b}_1 k + \hat{a}_1) y_0 = -k^* y_0 \\ u_2 &= -k\, y_2 \\ u_3 &= -k^* y_2 \\ &\vdots \end{aligned} \quad (4)$$

then the plant output presents two alternating characteristics (suppose n_b is even):

$$y_{\tau+n_d} = -\sum_{l=1}^{n_a} a_l\, y_{\tau+n_d-l} - b_1\, k\, y_{\tau-1} - b_2\, k^* y_{\tau-3} - b_3\, k\, y_{\tau-3} - \cdots - b_{n_b}\, k^* y_{\tau-n_b-1} + e_{\tau+n_d},$$

for τ odd, and

$$y_{\tau+n_d} = -\sum_{l=1}^{n_a} a_l\, y_{\tau+n_d-l} - b_1\, k^* y_{\tau-2} - b_2\, k\, y_{\tau-2} - b_3\, k^* y_{\tau-4} - \cdots - b_{n_b}\, k\, y_{\tau-n_b} + e_{\tau+n_d},$$

for τ even.

These expressions show that a_{m+n_d} and b_m are not always associated with the same lag of $\{y_t\}$, as would be the case under the control action in (2). Therefore a least-squares estimator can provide a unique solution to the identification problem.

Once the identifiability of complex models is established, the actual estimation of the model parameters is performed with a standard least-squares algorithm; the closed-loop signals $\{y_t, u_t\}$ are used as if they were collected in open loop. An analysis of this type of closed-loop identification, called "direct identification", can be found in Ljung (1999). Our approach inherits all the benefits and drawbacks of this type of closed-loop identification, except that we do not have control over the excitation being provided to the loop. Although the amount of data necessary to estimate models within reasonable confidence margins is comparatively high, this is actually a minor issue since all data used in the identification are obtained from normal operating conditions.

Given that new models can be identified without external excitation (except for the stochastic load disturbance), several different models can be continuously created and stored in a database. Such a collection of models can then be used for robustness analysis and design of a new controller, in an iterative fashion.

3. CONTROL DESIGN AND ANALYSIS

Our developments are currently confined to the application of the alternating control sequence in a model-based predictive controller. Nevertheless, the fact that the loop behaves as a multirate system could be explored in the design of the new controller. Some literature on this topic are papers by Al-Rahmani and Franklin (1990), Truman (1996*b*), Sågfors *et al.* (1998) and Truman and Govan (2000), for instance.

The following analysis of the closed-loop characteristic polynomial is in agreement with the statement: "the avoidance of multirate sampling appears to arise from the algebraic complexities involved" (Truman, 1996*b*). Due to the distinct behaviour of the control action in each pair of consecutive samples, an analysis of the closed loop requires that the signals and dynamics be split into two parts. Therefore, from a standard control action, $u_t = r_t + S(q)/R(q)\, y_t$, we derive

$$\begin{aligned} u_t &= R(q)\, r_t + S_o(q_2)\, y_t + S_e(q_2)\, \hat{y}_{t-1} - R_o(q_2)\, u_{t-1} - R_e(q_2)\, u_{t-2} \\ u_{t+1} &= R(q)\, r_{t+1} + S_o(q_2)\, \hat{y}_{t+1} + S_e(q_2)\, y_t - R_o(q_2)\, u_t - R_e(q_2)\, u_{t-1}, \end{aligned}$$

where q_2 is the forward 2-shift operator ($q_2\, u_t = u_{t+2}$). The process and the model dynamics are respectively given by

$$\begin{aligned} y_t &= B_o(q_2)\, u_{t-1} + B_e(q_2)\, u_{t-2} - A_o(q_2)\, y_{t-1} - A_e(q_2)\, y_{t-2} + A(q)\, v_t \\ \hat{y}_t &= \hat{B}_o(q_2)\, u_{t-1} + \hat{B}_e(q_2)\, u_{t-2} - \hat{A}_o(q_2)\, y_{t-1} - \hat{A}_e(q_2)\, y_{t-2}, \end{aligned}$$

where v_t is a filtered white noise and $A(q) = 1 + q^{-1} A_o(q_2) + q^{-2} A_e(q_2)$. As pointed out in (Truman, 1996*a*), explicit transfer functions relating the outputs y_t and u_t to the inputs r_t and v_t cannot [always] be obtained, but an assessment of absolute and relative stability can be made by evaluating the position of the closed-loop poles in the z-plane. The derivation of the closed-loop characteristic polynomial is straightforward, although messy. For example, the characteristic polynomial of the response from v_t to y_t is

$$\begin{aligned} T_1T_5T_{11} - q_2^{-1}T_5T_{12}(B_oT_1 + q_2^{-1}B_eT_3) \\ - q_2^{-1}T_5T_{11}(B_eT_2 - A_eT_1) \\ + q_2^{-1}(T_7T_{11} + q_2^{-1}T_6T_{12})(A_oT_1 - q_2^{-1}B_eT_4), \end{aligned}$$

where

$$\begin{aligned} T_1 &= 1 + q_2^{-1}R_e - q_2^{-1}S_e\hat{B}_o \\ T_2 &= S_o - q_2^{-1}S_e\hat{A}_o \\ T_3 &= -R_o + q_2^{-1}S_e\hat{B}_e \\ T_4 &= -q_2^{-1}S_e\hat{A}_e \\ T_5 &= T_1 + q_2^{-1}A_eT_1 - q_2^{-1}B_oT_4 \\ T_6 &= B_oT_3 + B_eT_1 \\ T_7 &= B_oT_2 - A_oT_1 \\ T_8 &= T_1 + q_2^{-1}(R_e - S_o\hat{B}_e)T_1 \\ &\quad + q_2^{-1}(R_o - S_o\hat{B}_o)T_3 \\ T_9 &= (S_e - S_o\hat{A}_o)T_1 + (S_o\hat{B}_o - R_o)T_2 \\ T_{10} &= (S_o\hat{B}_o - R_o)T_4 - S_o\hat{A}_eT_1 \\ T_{11} &= T_5T_8 - q_2^{-2}T_6T_{10} \\ T_{12} &= T_5T_9 + q_2^{-1}T_7T_{10} \end{aligned}$$

4. EXAMPLE

This section presents a simulation analysis of the following ARX process:

$$\begin{aligned} (1 - q^{-1})(1 - 1.9q^{-1} + 1.08q^{-2} - 0.09q^{-3})\, y_t \\ = (0.1q^{-5} - 0.025q^{-6})\, u_t + e_t, \quad (5) \end{aligned}$$

with $\sigma_e^2 = 0.01$. This process contains an integrator, which is known to the user. It is also known that the static gain of the plant without the integrator is approximately 0.8 and the delay is around 5 samples. Therefore the initial model is simply taken as

$$(1 - q^{-1})\, y_t = 0.8q^{-5}u_t + e_t. \quad (6)$$

Given the crude approximation provided by (6), the control law is conservatively designed as

$$(1 - q^{-1})\, u_t = -(0.1 - 0.09q^{-1})\, \tilde{y}_t, \quad (7)$$

where

$$\tilde{y}_t = \begin{cases} y_t, & \text{if } t \text{ is even} \\ \hat{y}_{t|t-1}, & \text{if } t \text{ is odd.} \end{cases}$$

Notice the presence of an integrator in (7), which is a common practical requirement.

The estimator adopted in this example is the one-step-ahead predictor:

$$\hat{y}_{t|t-1} = \hat{B}(q)\, u_t + [1 - \hat{A}(q)]\, y_t. \quad (8)$$

This choice is made in order to keep things simple, even though a Kalman predictor is likely to provide better results.

An initial simulation is performed and 10000 points are collected under stationary closed-loop behaviour. These data are employed in the identification of a new ARX model, using the least-squares method. Since the process is known to contain an integrator, the output signal is filtered through $1 - q^{-1}$, therefore the final model must be augmented by this same polynomial. Figure 1 shows the sequence of parameters estimated for $\hat{A}(q)$ and $\hat{B}(q)$ as the orders of these polynomials are simultaneously incremented while n_d remains fixed at 0 (see (1)).

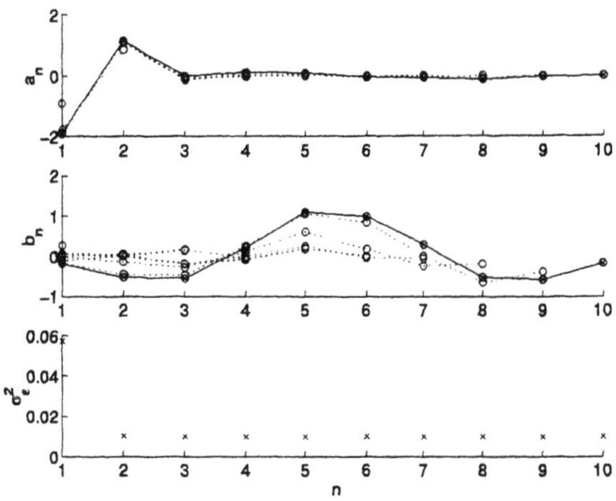

Fig. 1. Estimated parameters of the polynomials $\hat{A}(q)$ and $\hat{B}(q)$, for $n(= n_a = n_b)$ from 1 to 10 (top and medium, respectively). Covariance of the residuals for each value of n (bottom).

The parameters of the polynomial $\hat{A}(q)$ converge as n_a increases. This does not happen to the parameters of $\hat{B}(q)$ due to the low level of excitation provided by the control action. That is, the auto-regression of the output signal is a very strong signal compared to the control action. Under this control law, sufficient accuracy is only obtained with a much larger number of points. From Figure 1 we conclude that $\hat{A}(q)$ contains two significant parameters and for $\hat{B}(q)$ we can only estimate a single parameter (and the time delay) with reasonable accuracy. The newly estimated model is

$$\begin{aligned} (1 - q^{-1})(1 - 1.8246q^{-1} + 0.9215q^{-2})\, y_t \\ = 0.0777q^{-5}u_t + e_t \quad (9) \end{aligned}$$

and another conservative controller design (based on a Linear Quadratic criterion) leads to

$$\begin{aligned} (1 - q^{-1})(1 + 0.383q^{-1} + 0.436q^{-2} \\ + 0.478q^{-3} + 0.509q^{-4})\, u_t \\ = -(1.70 - 4.78q^{-1} + 4.59q^{-2} - 1.51q^{-3})\, \tilde{y}_t. \\ (10) \end{aligned}$$

As shown in Figure 2, the control action in (10) provides enough excitation to the loop and both $\hat{A}(q)$ and $\hat{B}(q)$ converge to particular polynomials as their orders are incremented. The full-order polynomials $A(q)$ and $B(q)$ can be identified with good accuracies under these circumstances.

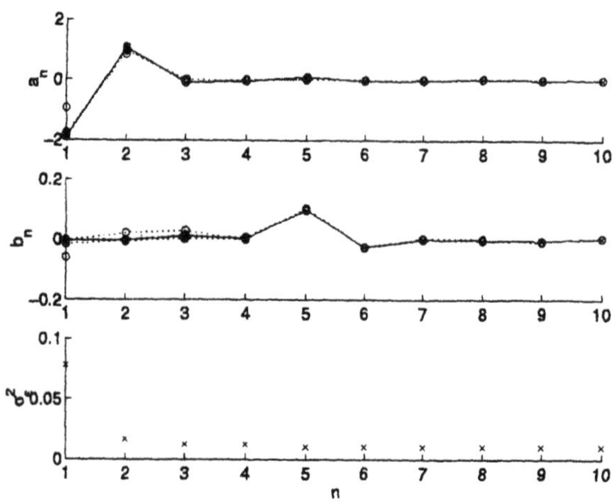

Fig. 2. Estimated parameters of the polynomials $\hat{A}(q)$ and $\hat{B}(q)$ under the control action in (10).

4.1 *Comparative study*

Three distinct control schemes are compared in terms of closed-loop identification of the plant given by (6). The controller parameters are the ones in (10):

$$R(q) = (1 - q^{-1})(1 + 0.383q^{-1} + 0.436q^{-2} + 0.478q^{-3} + 0.509q^{-4})$$
$$S(q) = 1.70 - 4.78q^{-1} + 4.59q^{-2} - 1.51q^{-3},$$

and the control actions are

$$M_0: \quad u_t = -\frac{S(q)}{R(q)} y_t$$
$$M_1: \quad u_t = -\frac{S(q)}{R(q)} \tilde{y}_t$$
$$M_2: \quad u_t = -\frac{S(q)}{R(q)} y_t + z_t,$$

where z_t is an independent source of white noise with covariance σ_z^2.

Given that the covariances of y_t and u_t under M_1 cannot be simultaneously replicated with a single σ_z^2 in M_2, this latter scheme is used twice with distinct values of σ_z^2: M_{2a} produces the same covariance in y_t as M_1 does; M_{2b} produces the same covariance in u_t as M_1 does. The covariances of y_t and u_t under these control actions are shown on Table 1.

Table 1. Covariances of y_t and u_t under various control schemes

	M_0	M_1	M_{2a}	M_{2b}
σ_y^2	26.18	26.24	26.24	28.16
σ_u^2	0.119	0.244	0.121	0.244

One thousand experiments are performed under each of the four control strategies. Each experiment contains a collection of 1000 points of stationary closed-loop data. These data used for estimating an ARX model with $n_a = 4$, $n_b = 3$ and $n_d = 5$. Table 2 contains the mean values and standard deviations of the parameters estimated for the polynomials $A(q)$ and $B(q)$ under each control scheme.

Table 2. Mean values and standard deviations (in parentheses) of the estimated parameters

	actual	M_0	M_1
a_1	-1.9	-1.90 (0.031)	-1.90 (0.031)
a_2	1.08	1.08 (0.068)	1.08 (0.068)
a_3	-0.09	-0.091 (0.069)	-0.092 (0.069)
a_4	0	0.0008 (0.031)	0.0010 (0.031)
b_1	0.1	0.102 (0.0162)	0.101 (0.0070)
b_2	-0.025	-0.0255 (0.0218)	-0.0247 (0.0070)
b_3	0	0.0003 (0.0174)	0.0004 (0.0077)

	actual	M_{2a}	M_{2b}
a_1	-1.9	-1.90 (0.031)	-1.90 (0.031)
a_2	1.08	1.08 (0.068)	1.08 (0.068)
a_3	-0.09	-0.091 (0.069)	-0.091 (0.069)
a_4	0	0.0008 (0.031)	0.0007 (0.031)
b_1	0.1	0.102 (0.0157)	0.101 (0.0076)
b_2	-0.025	-0.0256 (0.0205)	-0.0250 (0.0083)
b_3	0	0.0005 (0.0167)	0.0005 (0.0081)

The results of the estimation of $\hat{A}(q)$ are nearly identical in all schemes, whereas the standard deviation of the parameters of $\hat{B}(q)$ are inversely related to the covariance of u_t. Although σ_u^2 is the same under M_1 and M_{2b}, the former mechanism presents slightly better results in the identification of $\hat{B}(q)$.

The main disadvantage of M_0 is the loss of identifiability as the complexity of the model increases. The same effect is observed under a controller of low complexity.

The mechanism M_1 introduces an excitation that is proportional to the plant-model mismatch, which is a desirable feature. The degree of loop excitation can be manipulated by using the predictor at every n steps, instead of every second step.

The mechanism M_2 is the most flexible of them, although choosing the right level does require some insight about the process. Not only the level of σ_z^2 can be chosen freely, but also z_t can have its frequency modulated in order to emphasize parts of the model. Despite its flexibility, this mechanism has a negative psychological impact on industrial plant personnel. In this respect it may be easier to negotiate a trial with mechanism M_1.

An expansion of the model identification to, say, an ARMAX structure is relatively simple under the mechanism M_2, with the usage of instrumental variables. This is not a simple issue under the mechanism M_1 and an iterative method is very likely to be needed.

5. CONCLUSION

The novel control approach introduced in this article is shown to be capable of maintaining the identifiability of closed-loop systems when the controller has low complexity and the only signal driving the loop is a stochastic noise. A comparative study is made of this approach with one that injects an independent external excitation into the control action. Pros and cons for each mechanism were identified.

There are very many fronts still to be investigated under this proposed strategy. Its extension into multivariable systems and its application to performance monitoring/assessment are probably the most relevant from a practical point of view. Another important issue is its extension to the identification of more generic models, which demands the usage of less robust iterative search algorithms. Once iterative methods are needed, it might be worthwhile integrating the goals of the identification with those of the predictive control strategy, as proposed by Shook *et al.* (1992).

6. REFERENCES

Al-Rahmani, H. M. and G. F. Franklin (1990). A new optimal multirate control of linear periodic and time-invariant systems. *IEEE Transactions on Automatic Control* **AC-35**(4), 406–415.

Åström, K. J. and B. Wittenmark (1995). *Adaptive Control.* 2nd ed.. Addison-Wesley. Reading, U.S.A.

Ljung, L. (1999). *System Identification: Theory for the User.* 2nd ed.. PTR Prentice Hall. Upper Saddle River, U.S.A.

Sågfors, M. F., H. T. Toivonen and B. Lennartson (1998). H_∞ control of multirate sampled-data systems: A state-space approach. *Automatica* **34**(4), 415–428.

Shook, D. S., C. Mohtadi and S. L. Shah (1992). A control-relevant identification strategy for GPC. *IEEE Transactions on Automatic Control* **AC-37**(7), 975–980.

Truman, A. W. (1996*a*). An analysis of subrate control systems. *International Journal of Control* **63**(2), 365–392.

Truman, A. W. (1996*b*). A subrate weighted minimum-variance control law. *International Journal of Control* **65**(1), 93–115.

Truman, A. W. and M. Govan (2000). Multirate-sampled digital feedback system design via a predictive control approach. *IEE Proceedings-Control Theory and Applications* **147**(3), 293–302.

AN ALTERNATE NEW GAP METRIC FOR ROBUSTNESS MEASURE

L. Keviczky and **Cs. Bányász**

Computer and Automation Research Institute, Hungarian Academy of Sciences
H-1111 Budapest, Kende u 13-17, HUNGARY
Phone: +361-466-5435; Fax: +361-466-7503
e-mail: keviczky@sztaki.hu; banyasz@sztaki.hu

Abstract: Vinnicombe's ν-gap metrics is more and more widely used for measuring the distance between two transfer functions. Some alternate new metrics are introduced amplifying the principal frequency domain where the sensitivity equals to one and/or where the closed-loop frequency gain equals to one. Examples are shown, when these metrics are more sensitive to parameter variations. *Copyright© 2001 IFAC*

Keywords: perturbation, robustness, ν-gap metrics

1. INTRODUCTION

Consider first a simple closed-loop system given in Fig. 1, where R denotes the regulator and P the plant transfer function, respectively. Let us summarize some basic formulas from the classical control theory. The complementary sensitivity function (CSF) S_c (or closed-loop output transfer function) can be computed by

$$S_c = \frac{RP}{1+RP} = \frac{Y}{1+Y} \tag{1}$$

and the sensitivity function (SF) S (or closed-loop error transfer function) is given by

$$S = \frac{1}{1+RP} = \frac{1}{1+Y} \tag{2}$$

Here $Y = RP$ is the open-loop transfer function of the closed-loop.

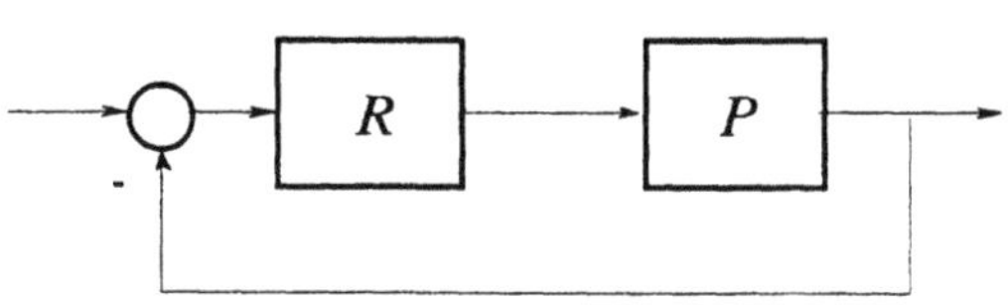

Fig. 1. Block-scheme of a simple closed-loop system

2. $M-\alpha$ AND $E-\beta$ CURVES

Consider the unity feedback system shown in Fig. 1. The overall transfer function S_c is given by (1) and

$$S_c(j\omega) = M(\omega)e^{j\alpha(\omega)} \tag{3}$$

Hence the closed loop magnitude, by convention is given by the symbol M and often referred to as the magnification of the system.

Since every point in the complex plane of a Nyquist diagram has a value of closed-loop magnitude and phase associated with it, then points with the same value of magnitude or phase can be joined to form loci of constant closed-loop magnitude and phase. These are referred to as M contours and α contours respectively, and can be shown to be two families of circles (Horowitz, 1963), with center

$\left[M^2/(1-M^2), j0\right]$ and radius $\left|M/(1-M^2)\right|$.

A family of, so-called $M-\alpha$, circles can thus be drawn on the Nyquist diagram as loci of constant M, as shown in Fig. 2. $M=1$ is a special case giving a locus which is a straight line at $[0,-0.5j]$. As M becomes very large the circles become small and tend towards the critical $[-1, j0]$ point. When an open loop Nyquist plot $Y(j\omega)$ for a unity feedback system is superimposed on this, each intersection with an M circle gives a point on the closed-loop magnitude plot, the value of ω being obtained by interpolation on the $Y(j\omega)$ locus. The M circle which is just tangential to the $Y(j\omega)$ locus gives a value of M^*, and the corresponding ω^* can be obtained from the tangent point of $Y(j\omega)$.

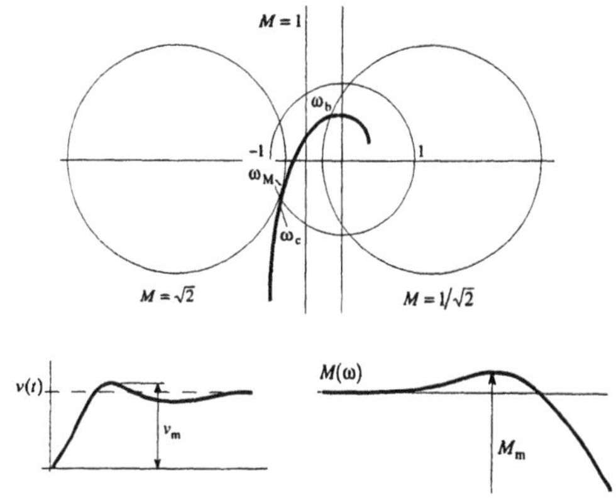

Fig. 2. M contours on the Nyquist diagram

In this way one can also determine the frequency $\omega^* = \omega_M$ where the maximal value M_m is obtained. If the $\mathcal{H}_\infty$ criterion is formulated for the closed-loop output transfer function S_c (or CSF) then

$$\|S_c\|_\infty = M_m \tag{4}$$

A simple approximate relationship is well known between the overshoot (ν_m) of the step response of the closed loop and the $\mathcal{H}_\infty$ limit (M_m) reached, under certain conditions (see also Fig. 3). The ω_b frequency where the Nyquist plot intersects the $M = 1/\sqrt{2}$ circle called bandwidth by definition.

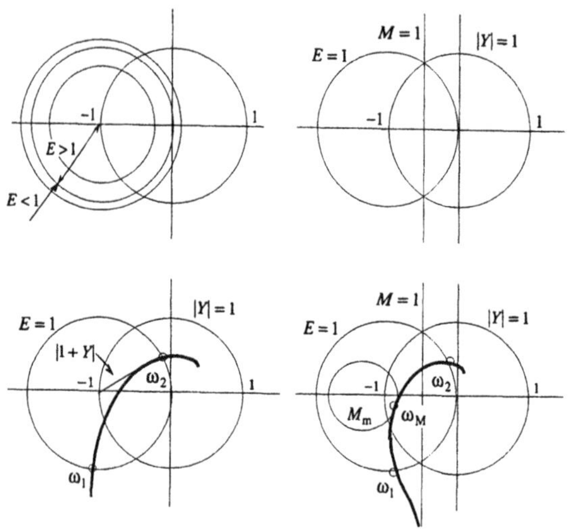

Fig. 3. E contours on the Nyquist diagram

In a similar way we can introduce another family of circles for the error transfer function (or SF)

$$S(j\omega) = E(\omega)e^{j\beta(\omega)} \tag{5}$$

that we call $E-\beta$ curves. The loci of constant E (shown in Fig. 3) are circles centered at $[-1, j0]$ with radius of $1/E$. The $\mathcal{H}_\infty$ norm can similarly be read off and given by

$$\|S\|_\infty = E_m \tag{6}$$

at $\omega = \omega_E \neq \omega_M$. The two intersections (if there are any) ω_1 and ω_2 of the $E=1$ circle indicate the frequency bounds that comprise the frequency interval where the magnitude of the sensitivity is higher than unity.

The above $M-\alpha$ and $E-\beta$ circles give better understanding of the meaning of the minimization of either $\|S\|_\infty$ or $\|S_c\|_\infty$ or even simultaneously both criteria. These curves explain the interplay between these twin design requirements, too.

3. VINNICOMBE'S ν-GAP METRIC

One of the most important contribution of the past decade in control theory has been the explicit incorporation of model uncertainties in the design of high performance control systems. These uncertainties can result from a lack of precision in mathematical modeling of the plant and/or changes in the plant parameters by time. Two main techniques for design of controllers for systems with significant uncertainties are the robust and adaptive control theories. System identification is an important tool in both techniques as well as in many other control design methods. The $\mathcal{H}_\infty$ framework leads to methods for designing robust stability and performance, therefore development of system identification techniques compatible with this technique is essential. For all above areas the interpretation of the $\mathcal{H}_\infty$ criteria and ν-gap metrics are very important.

Robustness analyses and design techniques are based on the comparison and relationships of two distances: the first is the distance of the nominal system from the stability limit; the second is the limiting distance between the nominal system and all possible actual ones. The first distance is not very sophisticated, practically it is always some variation of the minimal distance from the stability limit: the $(-1, j0)$ point. The second distance can be much more diverse because of: the special diverse character of the frequency functions (Nyquist diagrams), the special behaviors of the complex plane and functions (the infinite circle is one point) and the specific frequency and corresponding complex domains of stability, performance and robustness properties.

Besides the autonomous internal development of the robustness issues in control theory there were two major factors in this subject area. The first was the new developments in perturbation theory of linear operators (e.g. Kato, 1980) in the past decade. The second was the excellent idea of Vinnicombe (1993) introducing his special ν-gap metric.

Inspite of that Vinnicombe's ν-gap metric is based on the classical Riemann-sphere (R-sphere) transformation (or representation) of the complex plane it was a great idea, because: it normalizes all above mentioned distances to the interval $[0,1]$, furthermore his ν-gap metric amplifies the specific frequency domain around the cutting frequency (ω_c), which is the principal domain in stability analysis. Using Vinnicombe's ν-gap metric a long list of new achievements has been reached and published in the past few years (Belanger, 1995; Gevers, *et al.*, 1999; Vinnicombe, 1997, 1998).

Denote the coordinates of a point of the open-loop frequency function $Y(j\omega)$ on the complex plane by

$$\xi = \operatorname{Re} Y(j\omega)\ ;\ \eta = \operatorname{Im} Y(j\omega)\ ;\ \zeta = 0\ ;\ |Y|^2 = \xi^2 + \eta^2 \tag{7}$$

The transformation to an $Y'(j\omega)$ represented on the surface of a unity (diameter) R-sphere is given by

$$\xi' = \frac{\xi}{1+\xi^2+\eta^2} = \frac{\xi}{1+|Y|^2} \quad ; \quad \zeta' = |Y|^2 / \left(1+|Y|^2\right)$$

$$\eta' = \frac{\eta}{1+\xi^2+\eta^2} = \frac{\eta}{1+|Y|^2} \tag{8}$$

The perpendicular projection of $Y'(j\omega)$ to the complex plane gives the point, where

$$|Y'|^2 = (\xi')^2 + (\eta')^2 = \frac{|Y|}{1+|Y|^2} \tag{9}$$

The distance between two transfer functions Y_1 and Y_2 on the plane is

$$|\Delta| = |Y_1 - Y_2| \tag{10}$$

The "string" distance between the projections of the two transfer functions Y_1 and Y_1 on the R-sphere is

$$\kappa_V(Y_1, Y_2) = \frac{|Y_1 - Y_2|}{\sqrt{1+\xi_1^2+\eta_1^2}\sqrt{1+\xi_2^2+\eta_2^2}} = \\ = \frac{|Y_1 - Y_2|}{\sqrt{1+|Y_1|^2}\sqrt{1+|Y_2|^2}} = \frac{|\Delta|}{\sqrt{\left(1+|Y_1|^2\right)\left(1+|Y_2|^2\right)}} \tag{11}$$

Finally the Vinnicombe's ν-gap metric is given by

$$\delta_V = \delta_V(Y_1, Y_2) = \begin{cases} \max \kappa_V(Y_1, Y_2) & ; \ \delta_V(Y_1, Y_2) \le 1 \\ 1 \end{cases} \tag{12}$$

4. STABILITY AND MODEL UNCERTAINTY

The principal frequency domain continuously changed in the history of the development of modeling and control theory. At the very beginning the low frequency domain was the most important and these early investigations resulted in the type 0, 1 and 2 control-loops eliminating the static errors for type 0, 1 and 2 external deterministic disturbances and/or reference signals. Later the medium frequency domain, the vicinity of the cutting frequency (ω_c: where the open-loop frequency function $Y(j\omega)$ fulfills the condition $|Y(j\omega_c)| = 1$) became most important, when most of the stability and some of the performance properties of closed-loop systems were explored. The last decade concentrated to another frequency domain around ω_E, where the sensitivity function $S(j\omega)$ reaches its maximum: $|S(j\omega_E)| = E(\omega_E) = \max$. Another important frequency domain would be around ω_M, where the closed-loop transfer function $S_c(j\omega)$ reaches its maximum: $|S_c(j\omega_M)| = M(\omega_M) = \max$. Unfortunately this frequency domain got the least interest, because of the increasing complexity of the necessary analysis techniques in the above order of paradigms. While $M(\omega_M)$ is mainly connected to performance properties ($\omega_M \ne \omega_E \ne \omega_c$ and there is a good relationship between $M(\omega_M)$ and the $v_m = \max\{v(t)\}$, i.e., the overshoot of the step response in the closed-loop control system), then $E(\omega_E)$ is connected to the most natural robustness distance $\rho = \min|1 + Y(j\omega)| = 1/E(\omega_E)$. Here ρ is the nearest distance of $Y(j\omega)$ from the stability limit point $(-1, j0)$. (It is easy to see that δ_V is a certain string analogy of ρ.)

The role of the point $[-1, j0]$ and the geometric interpretation of stability on the Nyquist plot is well known from the classical literature. This stability view is mostly based on the uncertainty in the open-loop transfer function $Y = RP$ and not in the plant P itself. The CSF S_c given by (1) can be simply rearrange as

$$S_c = \frac{P}{1/R + P} \tag{13}$$

which formally shows that the inverse regulator transfer function $1/R$ can take over the role of the point $[-1, j0]$ and in this case the uncertainty in P and stability directly can be investigated geometrically by the distance

$$d(\omega) = d\left(P, -R^{-1}\right) = \left|R^{-1} + P\right| \tag{14}$$

The corresponding robust stability margin is now

$$\mu = \min d\left(P, -R^{-1}\right) = \min\left|R^{-1} + P\right| \tag{15}$$

which takes over the role of ρ (see formerly). Fig. 4 shows $d(\omega_M)$ for a third order plant controlled by a *PID* regulator.

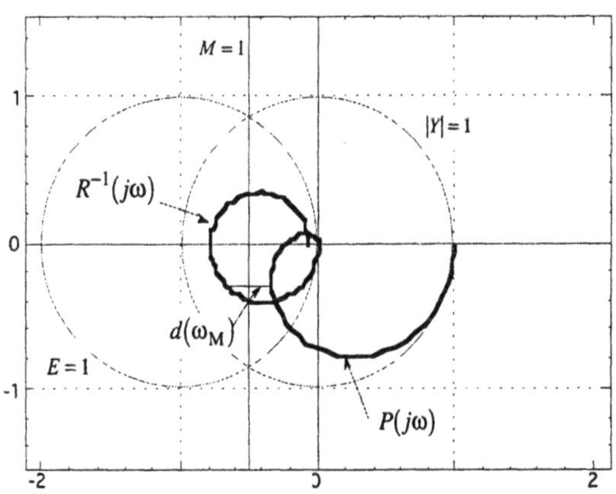

Fig. 4. The robustness distance d at ω_M

This simple approach was generalized by Vinnicombe, (cited by Gevers, *et al.*, 1999) to a "chordal" stability margin:

$$b_{PR} = \begin{cases} \min \kappa_V\left(P, -R^{-1}\right) & \text{if } R \text{ and } P \text{ is stable} \\ 0 & \text{otherwise} \end{cases} \tag{16}$$

Consider a model P_1 and a controller R that stabilizes P_1 with a stability margin b_{P_1R}. Then R stabilizes all P_2 such that

$$\delta_V(P_1, P_2) < b_{P_1R} \tag{17}$$

Because (17) is a min-max (necessary) condition one can use a less cautious (sufficient) condition pointwise (by frequency, see Gevers, *et al.*, 1999)

$$\kappa_V(P_1, P_2) < \kappa_V(P_1, -R^{-1}) \quad \forall\, \omega$$
$$\delta_V(P_1, P_2) < 1 \tag{18}$$

In this application it is quite natural that δ_V is used, because the principal frequency domain is the vicinity of the cutting frequency ω_c. There are other paradigms of robustness analysis, when similar amplification could be useful in other frequency domains.

5. ALTERNATE NEW ν-GAP METRICS

The specific maximum frequencies ω_E and ω_M are mostly greater than ω_c for the "well behaving" type 1 control systems, but it is difficult to give a general relationship for this triple. This is why one can not say that the higher interest in robustness issues increased the interest for the higher frequency domains in controller design. Inspite that this is the case in most of the situation we should rather tell that the behavior of the loop-frequency characteristics became more interesting in special (principal) frequency domains (loop-shaping). These methods based on $|S_c(j\omega)| = M(\omega)$ are more difficult, because of the special Archimedes-circles corresponding to the so-called M - α curves. Loop-shaping is the most widely used technique based on $|S(j\omega)| = E(\omega)$ and the corresponding E - β curves are also easy to use (Bányász and Keviczky, 1996, 1999). It is obvious that from robustness point of view the principal frequency domain is the vicinity of ω_E. In our paper an alternate new gap metric is introduced that naturally amplifies the frequency region around $\omega_{E=1}$, i.e., $|S(j\omega)| = E(\omega) = 1$ and reserves all other good properties of Vinnicombe's ν-gap metric, so all previous results can easily be reformulated, if necessary.

Principal frequency domain $\omega_{E=1}$

Let us shift the R-sphere left by unity thus the vertical ax of the sphere is located now in the stability limit point $(-1, j0)$ and name this sphere K-sphere. It is easy to see that the whole complex plane can be mapped onto the surface of the K-sphere in a similar manner as in case of the R-sphere. The necessary co-ordinate transformation to get $Y'_K(j\omega)$ is very simple

$$\xi_K = \xi + 1 \quad ; \quad \eta'_K = \eta' \quad ; \quad \zeta'_K = \zeta' \tag{19}$$

$$\xi'_K = \frac{\xi}{1+\xi_K^2+\eta^2} - 1 \quad ; \quad \eta'_K = \frac{\eta}{1+\xi_K^2+\eta^2}$$
$$\zeta'_K = \frac{\xi_K^2+\eta^2}{1+\xi_K^2+\eta^2} \tag{20}$$

The "string" distance between two transfer functions Y_1 and Y_2 on the K-sphere is

$$\kappa_K(Y_{K,1}, Y_{K,2}) = \frac{|(1+Y_1)-(1+Y_2)|}{\sqrt{1+|1+Y_1|^2}\sqrt{1+|1+Y_2|^2}} =$$
$$= \frac{|Y_1 - Y_2|}{\sqrt{1+|1+Y_1|^2}\sqrt{1+|1+Y_2|^2}} = \kappa_K(Y_1, Y_2) \tag{21}$$

Formulate the new ν-gap metric now based on the string distance using the K-sphere similarly to (12)

$$\delta_K = \delta_K(Y_1, Y_2) = \begin{cases} \max \kappa_K(Y_1, Y_2) & ; \ \delta_K(Y_1, Y_2) \le 1 \\ 1 \end{cases} \tag{22}$$

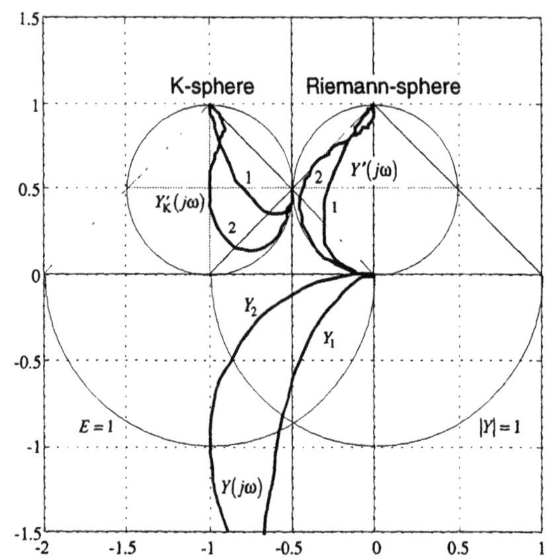

Fig. 5. Complex figure representing the two mappings used in the two gap metrics

The complex Fig. 5 shows the most important relationships. The upper part of the figure is the front view of the R and K-spheres with the mapped frequency functions $Y'(j\omega)$ and $Y'_K(j\omega)$. The limiting lines of the two cones correspond to the two circles $|Y(j\omega)| = 1$ and $E(\omega) = 1$. The bottom part is the lower half of the complex plane with Nyquist curves of Y_1 and Y_2, furthermore with the "lower: Im≤0" parts of the two circles mentioned. The "equator" of the R-sphere corresponds to ω_c and the "equator" of the K-sphere corresponds to $E(\omega) = 1$ (i.e., $\omega_{E=1}$).

Consider a very simple example of type 1 control system with

$$Y_1(j\omega) = \frac{K_1}{s(1+sT)} = \frac{b_o}{a_2 s^2 + a_1 s + a_o} = \frac{b_o}{a_2 s^2 + a_1 s}$$
$$b_o = 0.8 \ ; \ a_2 = 0.8 \ ; \ a_1 = 1 \ ; \ a_o = 0 \tag{23}$$

Assume that $Y_2(j\omega)$ is given by the variation of b_o in the interval

$$Y_2(j\omega) \ : \ 0.1 \le b_o \le 1.0 \tag{24}$$

Assume that $Y_2(j\omega)$ is given now by the variation of a_2 in the interval

$$Y_2(j\omega) \quad : \quad 0.1 \le a_2 \le 1.0 \qquad (25)$$

These two cases show the parameter sensitivity of δ_K, which is considerably greater than δ_V (see Figs. 6 and 7).

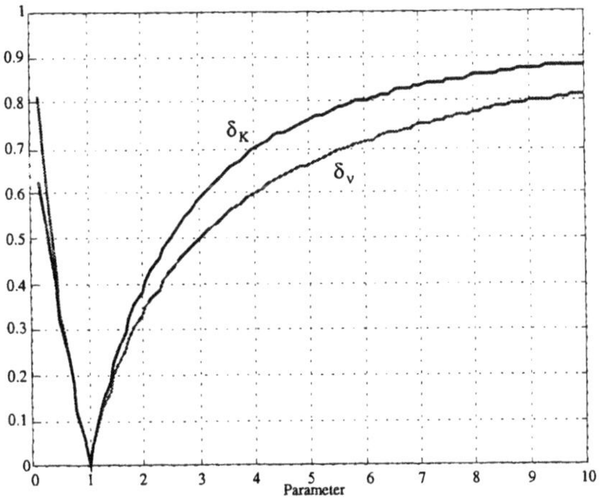

Fig. 6. The parameter sensitivity of δ_K and δ_V to the variation in b_0

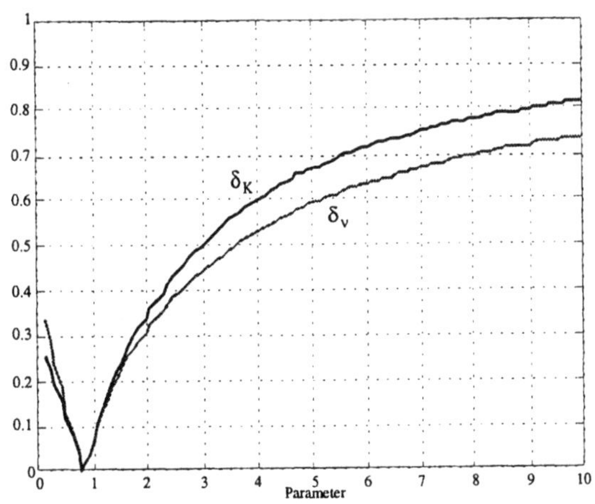

Fig. 7. The parameter sensitivity of δ_K and δ_V to the variation in a_2

It is interesting to display the mapping of the circles $E(\omega)=1$, $|Y(j\omega)|=1$ and the line $M(\omega)=1$ to the corresponding sphere sections (see Fig. 8). The figure shows that the regions corresponding to each other on the Nyquist (complex) plane and on the R- and K-sphere are considerably different. These regions are shown in Fig. 9.

It is probably well known that the most important regions in control system design are: 1, 2 and 3. Regions 4, 5 and 6 have very little influence on the medium and high frequency properties of closed-loop stability, performance and robustness. (The only exception is the low frequency, or static gain determining the steady state errors at different type systems.) The region 1 is considerably, the region 2 is visibly bigger on the K-sphere, however the region 3 is smaller. This is why it can not be stated that δ_K is better than δ_V at all cases, but we could show up considerable improvement in many practical examples.

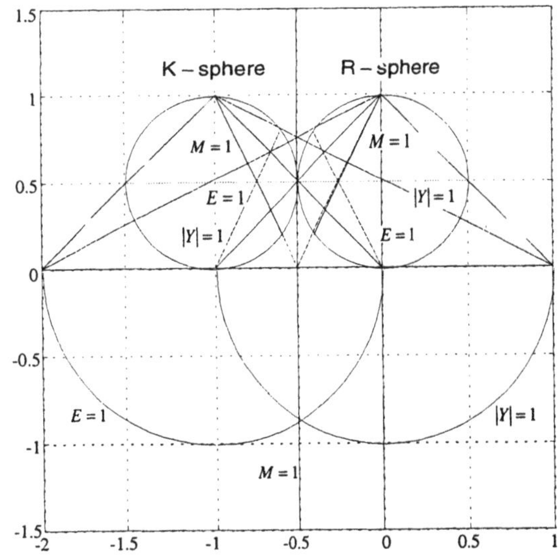

Fig. 8. Mapping of the circles $|E|=1$, $|Y|=1$ and the line $|M|=1$ to R- and K-sphere sections

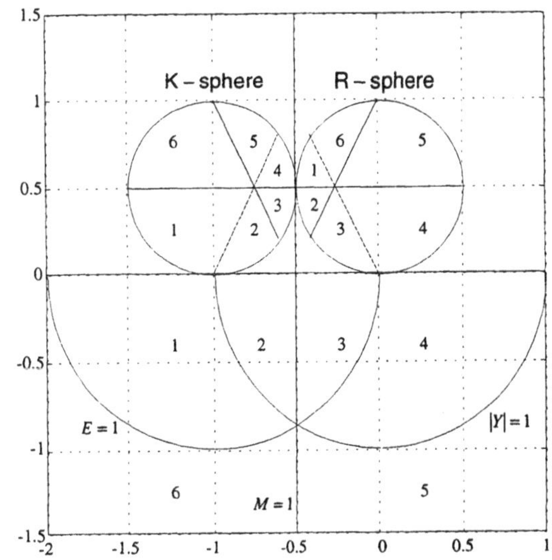

Fig. 9. Corresponding regions on the complex plane and on the R- and K-sphere

Principal frequency domain $\omega_{M=1}$

We would need such gap metric, which produces bigger region 2 and 3. This is the third metric discussed here. Similarly to the previous cases we can introduce such alternate sphere, which amplifies the principal frequency domain around $M(\omega)=1$ (i.e., $\omega_{M=1}$).

Let us shift the R-sphere left by 0.5 thus the vertical ax of the sphere is located now in the point -0.5 and name this sphere KM-sphere. It is easy to see that the whole complex plane can be mapped onto the surface of the KM-sphere in a similar manner as in case of the R-sphere or the K-sphere. The necessary co-ordinate transformation to get $Y'_{KM}(j\omega)$ is very simple

$$\xi_{KM} = \xi + 0.5 \quad ; \quad \eta'_{KM} = \eta' \quad ; \quad \zeta'_{KM} = \zeta' \qquad (26)$$

$$\xi'_{KM} = \frac{\xi}{1+\xi_{KM}^2+\eta^2} - 0.5 \quad ; \quad \eta'_K = \frac{\eta}{1+\xi_{KM}^2+\eta^2} \quad ;$$

$$\zeta'_{KM} = \frac{\xi_{KM}^2+\eta^2}{1+\xi_{KM}^2+\eta^2} \tag{27}$$

The "string" distance between two transfer functions Y_1 and Y_2 on the KM-sphere is

$$\kappa_{KM}\left(Y_{KM,1}, Y_{KM,2}\right) = \frac{\left|(0.5+Y_1)-(0.5+Y_2)\right|}{\sqrt{1+\left|0.5+Y_1\right|^2}\sqrt{1+\left|0.5+Y_2\right|^2}} =$$

$$= \kappa_K(Y_1, Y_2) = \frac{\left|Y_1 - Y_2\right|}{\sqrt{1+\left|0.5+Y_1\right|^2}\sqrt{1+\left|0.5+Y_2\right|^2}} \tag{28}$$

Formulate the new ν-gap metric now based on the string distance using the KM-sphere similarly to (11) and (21)

$$\delta_{KM} = \begin{cases} \max \kappa_{KM}(Y_1, Y_2) & ; \ \delta_{KM}(Y_1, Y_2) \le 1 \\ 1 \end{cases} = \delta_{KM}(Y_1, Y_2) \tag{29}$$

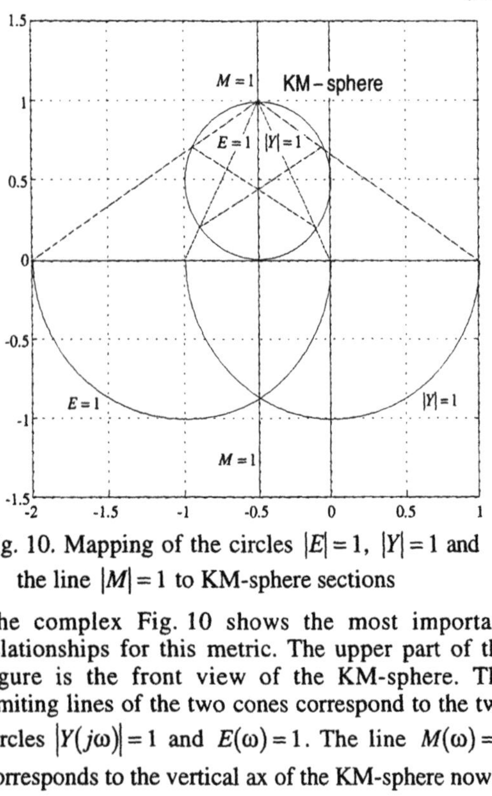

Fig. 10. Mapping of the circles $|E| = 1$, $|Y| = 1$ and the line $|M| = 1$ to KM-sphere sections

The complex Fig. 10 shows the most important relationships for this metric. The upper part of the figure is the front view of the KM-sphere. The limiting lines of the two cones correspond to the two circles $|Y(j\omega)| = 1$ and $E(\omega) = 1$. The line $M(\omega) = 1$ corresponds to the vertical ax of the KM-sphere now.

The corresponding regions on the complex plane and on the KM-sphere are shown in Fig. 11. It is easy to see that both the regions 2 and 3 are bigger now, than on the R-sphere at the original case.

Because the projections to the K- and KM-sphares mean only a frequency domain weighting comparing to the R-sphere the generalized stability margins (16)-(18) can be similarly formulated to δ_K and κ_K, furthermore δ_{KM} and κ_{KM} having defined b_{PR} on these sphares.

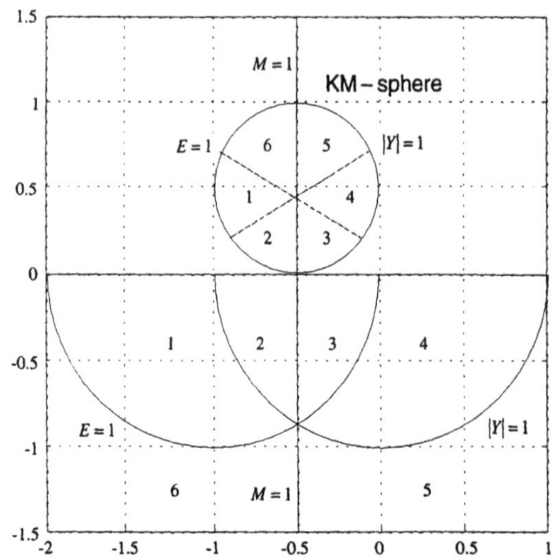

Fig. 11. Corresponding regions on the complex plane and on the KM-sphere

6. CONCLUSIONS

New alternate new ν-gap metrics are introduced amplifying the $\omega_{E=1}$ and $\omega_{M=1}$ principal frequency domains as a modification of Vinnicombe ν-gap metric amplifying the frequency domain around the cutting frequency. Examples are shown, when the new metrics are more sensitive to parameter variations.

REFERENCES

Bányász, Cs. and L. Keviczky (1996). $\mathcal{H}_\infty$ criteria in frequency domain. *Proc. 2nd World Automation Congress, WAC'96*, Montpellier, France, TSI Press Series, Vol. 4, pp. 41-48.

Bányász, Cs. and L. Keviczky (1999). Geometric tools to design robust control systems, *Proc. 14th IFAC World Congress*, Beijing, Vol. G, pp. 277-282.

Bélanger, I. (1995). *Control Engineering*, Saunders College Publishing.

Gevers, M., X. Bombois, B. Codrons, F. De Bruyne and G. Scorletti (1999). *The role of experimental conditions in model validation for control*, Springer Lecture Notes, (Ed. Garulli, A., A. Tesi and A. Vicino), pp. 72-86.

Horowitz, I. M. (1963). *Synthesis of Feedback Systems*, Academic Press, New York.

Kato, T. (1980). *Perturbation Theory for Linear Operators*, Springer-Verlag, Berlin Heidelberg New York, pp. 619.

Vinnicombe, G. (1993). Frequency domain uncertainty and the graph topology, *IEEE Trans. Automatic Control*, 37(6), pp. 331-336.

Vinnicombe, G. (1997). Approximating uncertainty representations using the ν-gap metric.

Vinnicombe, G. (1998). On IQCs and the ν-gap metric, *Proc. 37th IEEE Conf. Decision and Control*, Tampa, FL, USA.

This work was supported by the Hungarian NSF (OTKA).

Model Reference Adaptive Control of Non-linear Multivariable Systems using an Interactor Structure Estimation

Yasuhiko MUTOH

Department of Mechanical Engineering, Sophia University
7-1 Kioicho, Chiyoda-ku, Tokyo, Japan

This paper concerns the model reference adaptive control of non-linear multivariable systems. For this purpous, we make use of the interactor structure estimation and the adaptive control technique for linear systems. *Copyright © 2001 IFAC*

Keywords : model reference adaptive control, nonlinear systems, multivariable systems, interactor matrix

1. INTRODUCTION

The adaptive control for nonlinear systems has attracted a lot of researchers in this decade (Krstic et al., 1995 ; Marino and Tomei, 1995). In this paper, the model reference adaptive control of multi-input multi-output nonlinear systems is concerned. Generally, in the model reference technique for SISO systems, the relative degree of an unknown plant plays a basic role as a prior information. However, for MIMO system, an interactor matrix should be used instead of the relative degree.

An interactor was originally defined for linear systems as a lower left triangular polynomial matrix with monomials at its diagonal positions (Wolovich, et al., 1976). In (Mutoh, et al., 1993), the class of interactors was extended, i.e., an interactor was regarded as any polynomial matrix whose coefficient matrices satisfy a certain algebraic equation.

As for the linear case, the concept of an interactor was proposed for nonlinear systems (Benedetto, et al., 1994). In (Mutoh, 1998), it was shown that an interactor of nonlinear systems can be also regarded as any polynomial matrix whose coefficient matrices satisfy a certain algebraic equation. This implies that it is also possible to obtain the stable interactor parametrization for unknown nonlinear systems. This paper shows how to make use of this interactor structure estimate in the design of the model reference adaptive control for MIMO nonlinear unknown plants. For this purpose, some adaptive control design techniques for linear systems is directly applied to nonlinear systems for simplicity.

2. MRACS FOR SISO SYSTEMS

In this section, the design method of the MRACS for SISO nonlinear systems is briefly summarized. Consider a single-input single-output affine nonlinear plant described as

$$\begin{aligned} \dot{x} &= f(x) + g(x)u \\ y &= h(x) \end{aligned} \qquad (1)$$

where $x \in R^n$, $u \in R$ and $y \in R$ are the state vector, input and output respectively. It is assumed that

1. The zero dynamics of this system is ISS.
2. This system is input-output linearizable by the static state feedback.
3. The relative degree of this system is known as a prior information.
4. Function, $f(x)$, $g(x)$, and $h(x)$ have unknown constant parameters which appear linearly.

5. The state vactor, x, can be used in the adaptive controller.

The purpose of control is to design the plant input u which makes the plant output, y, follow the desired reference output signal, y_d.

Suppose that the relative degree of eq.(1) is r, which implies that

$$\begin{aligned} L_g h(x) &= 0 \\ L_g L_f h(x) &= 0 \\ &\vdots \\ L_g L_f^{r-2} h(x) &= 0 \\ L_g L_f^{r-1} h(x) &\neq 0 \end{aligned} \quad (2)$$

Let p be a differential operator and $D(p)$ be an arbitrary stable polynomial of degree r as follows.

$$D(p) = p^r + a_{r-1}p^{r-1} + \cdots + a_1 p + a_0 \quad (3)$$

From eq.(2), the output y satifies the following equation.

$$D(p)y = q(x) + \lambda(x)u \quad (4)$$

where

$$q(x) = [\ a_0, a_1, \cdots, a_{r-1}, 1\ ] \begin{bmatrix} h(x) \\ L_f h(x) \\ \vdots \\ L_f^{r-1} h(x) \\ L_f^r h(x) \end{bmatrix} \quad (5)$$

and

$$\lambda(x) = L_g L_f^{r-1} h(x) \quad (6)$$

Since unknown constant parameters of $f(x)$, $g(x)$, and $h(x)$ are contained linearly, unknown parameters in $w(x)$ and $\lambda(x)$ also appear linearly, i.e.

$$\begin{aligned} q(x) &= \theta_1^T \phi_1(x) \\ \lambda(x) &= \theta_2^T \phi_2(x) \end{aligned} \quad (7)$$

where θ_1 and θ_2 are newly defined unknown parameter vectors, and $\phi_1(x)$ and $\phi_1(x)$ are associated known regressor vectors with appropriate dimension. Then, eq.(4) can be rewritten as follows.

$$D(p)y = \theta_1^T \phi_1(x) + \theta_2^T \phi_2(x)\ u \quad (8)$$

From this, the output error

$$e = y - y_d \quad (9)$$

satisfies the following error equation.

$$D(p)e = -D(p)y_d + \theta_1^T \phi_1(x) + \theta_2^T \phi_2(x)\ u \quad (10)$$

This is the same type of error equation as for the linear case. Hence, there are some possibilities of applying several established adaptive control design techniques to this problem. The followings are some of those design methods.

[Type 1] (Augmented error method)

Rewrite eq.(10) as

$$\begin{aligned} D(p)e &= F(p)[-\frac{D(p)}{F(p)} y_d + \theta_1^T \frac{1}{F(p)} \phi_1(x) \\ &\quad + \theta_2^T \frac{1}{F(p)} \phi_2(x)\ u] \end{aligned} \quad (11)$$

where $F(s)$ is chosen so that $\frac{F(p)}{D(p)}$ is strictly positive real. Define the error augmenting signal η by

$$D(p)\eta = F(p)[\hat{\theta}_1(t)\frac{1}{F(p)}\phi_1(x) - \frac{1}{F(p)}\hat{\theta}_1(t)\phi_1(x)$$

$$+\hat{\theta}_2(t)\frac{1}{F(p)}\phi_2(x)u - \frac{1}{F(p)}\hat{\theta}_2(t)\phi_2(x)u]$$

From this, the following standard adaptive controller which makes $e \to 0$ is obtained.

$$\begin{aligned} \dot{\hat{\theta}}_1(t) &= \Gamma_1(e+\eta)\frac{1}{F(p)}\phi_1(x) \\ \dot{\hat{\theta}}_2(t) &= \Gamma_2(e+\eta)\frac{1}{F(p)}\phi_2(x)\ u \\ u &= \frac{1}{\hat{\theta}_2^T(t)\phi_2(x)}(\hat{\theta}_1(t)^T\phi_1(x) - D(p)y_d) \end{aligned} \quad (12)$$

where Γ_1 and Γ_2 are symmetric positive definite matrices and $\hat{\theta}_2^T(t)\phi_2(x)$ is the estimate of the nonzero signal $\lambda(x)$.

[Type 2] (STR method)

From the assumption 4, the unknown plant eq.(1) can be described as

$$\dot{x}_i = \alpha_i^T \psi_{1,i} + \beta_i^T \psi_{2,i} u \quad (i = 1, \cdots, n) \quad (13)$$

where α_i and β_i are the original unknown parameters. Usually, the number of original unknown parameters is much less than those in eq.(8). Hence, the identifier should be made for the original plant equation, eq.(13), as follows. identifier

$$\begin{aligned} \dot{\hat{x}}_i &= -c(x_i - \hat{x}_i) + \hat{\alpha}_i^T(t)\psi_{1,i} + \hat{\beta}_i^T(t)\psi_{2,i}u \\ &\qquad (i = 1, \cdots, n) \end{aligned} \quad (14)$$

From this and eq.(10), the following parameter adjusting laws and adaptive input are obtained.

$$\begin{aligned} \dot{\hat{\alpha}}_i(t) &= \Gamma_1(x_i - \hat{x}_i)\psi_{1,i}(x) \\ \dot{\hat{\beta}}_i(t) &= \Gamma_2(x_i - \hat{x}_i)\psi_{2,i}(x)u \\ u &= \frac{1}{\hat{\theta}_2^T(\hat{\alpha},\hat{\beta})\phi_2(x)}(\hat{\theta}_1(\hat{\alpha},\hat{\beta})^T\phi_1(x) - D(p)y_d) \end{aligned} \quad (15)$$

[Type 3] (Backstepping method)

If y, $\dot{y}$, $\cdots$, $y^{(r)}$ are available, the following backstepping method can be used. By introducing new state variables ξ_1, ξ_2, $\cdots$, ξ_r, eq.(10) can be rewritten as

$$\begin{aligned} e &= -y_d + \xi_1 \\ \dot{\xi}_1 &= \xi_2 \\ \dot{\xi}_2 &= \xi_3 \\ &\vdots \\ \dot{\xi}_r &= -a_0\xi_1 - a_1\xi_2 - \cdots - a_{r-2}\xi_r \\ &\quad +\theta_1^T\phi_1(x) + \theta_2^T\phi_2(x)\ u \end{aligned} \tag{16}$$

For example, consider the case where the relative degree of the plant is 2, i.e., $r = 2$. Eq.(16) becomes

$$e = -y_d + \xi_1 \tag{17}$$
$$\dot{\xi}_1 = \xi_2 \tag{18}$$
$$\dot{\xi}_2 = \theta_1^T\phi_1(x) + \theta_2^T\phi_2(x)\ u \tag{19}$$

Note that $\xi_1 = y$ and $\xi_2 = \dot{y}$. The MRACS for this system is designed through the following steps.

[STEP 0]

By introducing a new state z_1, set

$$\xi_1 = y_d + z_1 \tag{20}$$

then, eq.(17) becomes

$$e = z_1 \tag{21}$$

[STEP 1]

Set

$$\xi_2 = \alpha_2 + z_2 \tag{22}$$

where z_2 is a new state and α_2 is a virtual input. Using z_1, eq.(18) can be rewritten as

$$\dot{z}_1 = -\dot{y}_d + \alpha_2 + z_2 \tag{23}$$

The candidate of Lyapunov's function in this step is chosen as follows.

$$V_1 = \frac{1}{2}z_1^2 \tag{24}$$

Derivative of V_1 along eq.(23) becomes

$$\dot{V}_1 = z_1(-\dot{y}_d + \alpha_2 + z_2) \tag{25}$$

In this step, α_2 is chosen as

$$\alpha_2 = -c_1 z_1 + \dot{y}_d \tag{26}$$

where $c_1 > 0$, then

$$\dot{V}_1 = -c_1 z_1^2 + z_1 z_2 \tag{27}$$

[STEP 2]

Using z_2, eq.(19) can be rewritten as

$$\dot{z}_2 = -\dot{\alpha}_2 + \theta_1^T\phi_1(x) + \theta_2^T\phi_2(x)\ u \tag{28}$$

Let $\hat{\theta}_1(t)$ and $\hat{\theta}_2(t)$ be the estimate of unknown parameter vector θ_1 and θ_2 respectively. In this step, the following candidate of Lyapunov's function, V_2, is considered.

$$V_2 = V_1 + \frac{1}{2}z_2^2 + \frac{1}{2}\tilde{\theta}_1^T(t)\Gamma_1^{-1}\tilde{\theta}_1(t) + \frac{1}{2}\tilde{\theta}_2^T(t)\Gamma_2^{-1}\tilde{\theta}_2(t) \tag{29}$$

where

$$\begin{aligned} \tilde{\theta}_1(t) &= \theta_1 - \hat{\theta}_1(t) \\ \tilde{\theta}_2(t) &= \theta_2 - \hat{\theta}_2(t) \end{aligned} \tag{30}$$

Then, the derivative of V_2 along eqs.(23) and (28) becomes

$$\begin{aligned} \dot{V}_2 &= -c_1 z_1^2 + z_1 z_2 \\ &\quad + z_2(-\dot{\alpha}_2 + \hat{\theta}_1^T\phi_1(x) + \hat{\theta}_2^T\phi_2(x)\ u) \\ &\quad + \tilde{\theta}_1^T(t)(\Gamma_1^{-1}\dot{\tilde{\theta}}_1(t) + z_2\phi_1(x)) \\ &\quad + \tilde{\theta}_2^T(t)(\Gamma_2^{-1}\dot{\tilde{\theta}}_2(t) + z_2\phi_2(x)u) \end{aligned} \tag{31}$$

Hence, the parameter adjusting laws and the adaptive control,

$$\begin{aligned} \dot{\hat{\theta}}_1(t) &= \Gamma_1 z_2\phi_1(x) \\ \dot{\hat{\theta}}_2(t) &= \Gamma_2 z_2\phi_2(x)\ u \\ \hat{\theta}_2^T(t)\phi_2(x)u &= -c_2 z_2 + \dot{\alpha}_2 - \hat{\theta}_1(t)^T\phi_1(x) - z_1 \end{aligned} \tag{32}$$

where $c_2 > 0$, makes the derivative of V_2 as follows.

$$\dot{V}_2 = -c_1 z_1^2 - c_2 z_2^2 \tag{33}$$

From this, z_1 and z_2 converge to 0, and, hence, the purpose of MRACS is attained ($e \to 0$). The adaptive input u is generated by

$$u = \frac{1}{\hat{\theta}_2^T(t)\phi_2(x)}(-c_2 z_2 + \dot{\alpha}_2 - \hat{\theta}_1(t)^T\phi_1(x) - z_1) \tag{34}$$

[Type 4] (Backstepping with an adaptive observer)

In the above design method, if y, $\dot{y}$, $\cdots$, $y^{(r)}$ are not available, ξ_1, $\cdots$, ξ_r are estimated as follows.

Eq.(11) can be rewritten as follows.

$$D(p)y = F(p)[\theta_1^T\frac{1}{F(p)}\phi_1(x) + \theta_2^T\frac{1}{F(p)}\phi_2(x)\ u]$$

Then, by introducing the following identifier,

$$D(p)\hat{y} = F(p)[\hat{\Theta}_1^T(t)\frac{1}{F(p)}\phi_1(x) + \hat{\Theta}_2^T(t)\frac{1}{F(p)}\phi_2(x)\ u]$$

and parameter adjusting law,

$$\begin{aligned}\dot{\hat{\Theta}}_1(t) &= \Gamma_1(y-\hat{y})\frac{1}{F(p)}\phi_1(x)\\ \dot{\hat{\Theta}}_2(t) &= \Gamma_2(y-\hat{y})\frac{1}{F(p)}\phi_2(x)u\end{aligned}$$

the estimates of $\xi_1, \cdots, \xi_r$ are obtained by

$$\begin{aligned}\dot{\hat{\xi}}_1 &= \hat{\xi}_2\\ \dot{\hat{\xi}}_2 &= \hat{\xi}_3\\ &\vdots\\ \dot{\hat{\xi}}_r &= -a_0\hat{\xi}_1 - a_1\hat{\xi}_2 - \cdots - a_{r-2}\hat{\xi}_r\\ &\quad +\hat{\Theta}_1^T\phi_1(x) + \hat{\Theta}_2^T\phi_2(x)u\end{aligned} \tag{35}$$

3. INTERACTOR OF NONLINEAR SYSTEMS

Consider a m-input m-output affine nonlinear system described by

$$\begin{aligned}\dot{x} &= f(x) + g(x)u\\ y &= h(x)\end{aligned} \tag{36}$$

where $x \in R^n$, $u \in R^m$ and $y \in R^m$. It is assumed that nonlinear function $f(x) \in R^n$, $g(x) \in R^{n\times m}$ and $h(x) \in R^m$ are C^∞.

Definition 1 *A $m \times m$ polynomial matrix with constant coefficient matrices, $N(p)$, is called an interactor of the nonlinear system eq.(36) if $N(p)$ satisfies the following equation for some $m \times m$ nonsingular matrix $\Lambda(x)$ and $q(x) \in R^m$ in some region Ω .*

$$N(p)y = q(x) + \Lambda(x)u \tag{37}$$

This definition implies that if $N(p)$ is an interactor, the total system from u to $N(p)y$ does not include any differentiator (proper) and the coefficient matrix of the direct term, u, is nonsingular. It should be noted that this interactor is well defined if and only if the nonlinear system eq.(36) is input-output linearizable by static state feedback.

Let $T_i(x) \in R^{m\times m}$ $(i = 1, 2, \cdots)$ be defined by

$$\begin{aligned}T_1(x) &= L_g h(x)\\ T_i(x) &= L_g L_f^{i-1} h(x), \quad (i = 2, 3, \cdots).\end{aligned} \tag{38}$$

Then, the Toeplitz matrix $\Gamma_i(x)$ can be defined as follows.

$$\begin{aligned}\Gamma_1(x) &= T_1(x)\\ \Gamma_i(x) &= \begin{bmatrix} T_1(x) & \cdots & T_i(x)\\ & \ddots & \vdots\\ 0 & \cdots & T_1(x)\end{bmatrix}, \quad (i = 2, 3, \cdots).\end{aligned} \tag{39}$$

Theorem 1 *Let $N(p)$ be described by*

$$N(p) = N_w p^w + N_{w-1}p^{w-1} + \cdots + N_0, \quad (N_i \in R^{m\times m}). \tag{40}$$

Then, the necessary and the sufficient condition for $N(p)$ to be an interactor of the nonlinear system eq.(36) is that N_i's satisfy the following equation for some nonsingular matrix $\Lambda(x)$.

$$\Gamma_w^T(x)\begin{bmatrix} N_w^T\\ N_{w-1}^T\\ \vdots\\ N_1^T\end{bmatrix} = \begin{bmatrix} 0\\ \vdots\\ 0\\ \Lambda^T(x)\end{bmatrix} \tag{41}$$

(Brief proof) By differentiating y repeatedly, we have the following equation.

$$\begin{aligned}N(p)y &= [N_0, N_1, \cdots, N_w]\begin{bmatrix} h(x)\\ L_f h(x)\\ \vdots\\ L_f^w h(x)\end{bmatrix}\\
&+[N_1, N_2, \cdots, N_w] \times \left\{ \begin{bmatrix} T_1(x)\\ T_2(x)\\ T_3(x)\\ \vdots\\ T_w(x)\end{bmatrix} u \right.\\
&+\left(\begin{bmatrix} 0\\ pT_1(x)\\ pT_2(x)\\ \vdots\\ pT_{w-1}(x)\end{bmatrix} u + \begin{bmatrix} 0\\ T_1(x)\\ T_2(x)\\ \vdots\\ T_{w-1}(x)\end{bmatrix} u^{(1)}\right)\\
&+\left(\begin{bmatrix} 0\\ 0\\ p^2T_1(x)\\ \vdots\\ p^2T_{w-2}(x)\end{bmatrix} u + 2\begin{bmatrix} 0\\ 0\\ pT_1(x)\\ \vdots\\ pT_{w-2}(x)\end{bmatrix} u^{(1)}\right.\\
&\left.+\begin{bmatrix} 0\\ 0\\ T_1(x)\\ \vdots\\ T_{w-2}(x)\end{bmatrix} u^{(2)}\right) +\\
&\vdots\\
&+\left(\begin{bmatrix} 0\\ \vdots\\ 0\\ p^{w-1}T_1(x)\end{bmatrix} u\right.\\
&+\binom{w-1}{1}\begin{bmatrix} 0\\ \vdots\\ 0\\ p^{w-2}T_1(x)\end{bmatrix} u^{(1)}\end{aligned}$$

$$+ \binom{w-1}{2} \begin{bmatrix} 0 \\ \vdots \\ 0 \\ p^{w-3}T_1(x) \end{bmatrix} u^{(2)} + \cdots$$

$$+ \binom{w-1}{w-2} \begin{bmatrix} 0 \\ \vdots \\ 0 \\ pT_1(x) \end{bmatrix} u^{(w-2)}$$

$$+ \left. \left. \begin{bmatrix} 0 \\ \vdots \\ 0 \\ T_1(x) \end{bmatrix} u^{(w-1)} \right) \right\} \quad (42)$$

In this equation, $\binom{i}{j}$ denotes the binomial coefficient. Since N_i's are assumed to be constant coefficient matrices, it is straightforward to show that eq.(41) implies that $N(p)$ is an interactor and vice versa. (Q.E.D.)

Let $\bar{\rho}$ denote the rank over the field R. It is assumed that, in some region Ω under consideration, $\bar{\rho}\Gamma_i(x)$ equals the rank of $\Gamma_i(x_0)$ $\forall x_0 \in \Omega$. We then have the following Theorem about the degree of $N(p)$.

Theorem 2 *If $N(p)$ in eq.(40) is an interactor of the nonlinear system eq.(36), then*

$$w = k_{min} \quad (43)$$

where k_{min} is the least integer k such that

$$\bar{\rho}\Gamma_k(x) = \bar{\rho}\Gamma_{k-1}(x) + m. \quad (44)$$

(Proof) It can be proven in the same manner of the proof of Theorem 2 of (Mutoh,et al.,1993).

If there exists k_{min}, then eq.(37) always has a solution N_i's such that $\bar{\rho}\Lambda^T(x) = m$. Furthermore, using the output reordering algorithm (Mutoh,et al.,1993), $N(p)$ can be chosen as a lower left triangular and row proper form with arbitrarily stable polynomials of p for its diagonal elements.

From eq.(32), if $N(p)$ is an interactor, we obtain the following equation.

$$N(p)y = q(x) + \Lambda(x)u \quad (45)$$

where

$$q(x) = [N_0, N_1, \cdots, N_w] \begin{bmatrix} h(x) \\ L_f h(x) \\ \vdots \\ L_f^w h(x) \end{bmatrix} \quad (46)$$

$$\Lambda(x) = [N_1, N_2, \cdots, N_w] \begin{bmatrix} L_g h(x) \\ L_g L_f h(x) \\ \vdots \\ L_g L_f^{w-1} h(x) \end{bmatrix} \quad (47)$$

Then, it is obvious that the state feedback

$$u = -\Lambda^{-1}(x)a(x) + \Lambda^{-1}(x)v \quad (48)$$

makes the closed loop transfer matrix from v to y the inverse of the interactor, $N^{-1}(p)$, in Ω.

4. MRACS FOR MIMO SYSTEMS

The design method of MRACS for MIMO systems is basically the same as for SISO systems. The different point is that, for MIMO systems, the interactor matrix is used instead of the relative degree of the unknown plant.

Consider the m-input m-output nonlinear system, eq.(36). For this system, we also have the assumptions $1 \sim 5$ as for SISO systems. But, as stated in the previous section, the assumption 3 should be modified as

3' The interactor matrix, $N(p)$, of this system is known as a prior information.

However, as for the linear case, coefficients of polynomials at its off-diagonal positions depend on the plant unknown parameters. For this reason, we apply the output reordering algorithm to eq.(41) to obtain the structure estimate of the interactor matrix using a prior knowledge about unknown plant parameters in the generic sense. Estimated interactor is a lower left triangular polynomial matrix of p with arbitrary stable polynomial of degree r_i at i-th diagonal position ($i = 1, \cdots, m$). Where (r_1,r_2,$\cdots$,r_m) is the structure at infinity of eq.(36) and $r_1 \leq r_2 \leq \cdots \leq r_m$,which can be also estimated through the algorithm. Further, estimated interactor is row proper and has the minimum number of unknown parameters in the off-diagonal positions.

Now, consider the following example.

[Example] Let $x = [x_1, x_2, x_3, x_4]^T \in R^4$ and $u = [u_1, u_2]^T, y = [y_1, y_2]^T \in R^2$. Consider the plant, eq.(36), with $f(x)$, $g(x)$ and $h(x)$ described by

$$\begin{aligned} f(x) &= \begin{bmatrix} x_1^2 + x_2 \\ k_1 x_1 x_3 \\ k_2(-x_1 + x_3) \\ 0 \end{bmatrix} \\ g(x) &= \begin{bmatrix} 0 & 0 \\ 0 & k_4 \\ 1 & 0 \\ k_3 & 0 \end{bmatrix} \\ h(x) &= \begin{bmatrix} x_3 \\ x_4 \end{bmatrix} \end{aligned} \quad (49)$$

where $k_1 \sim k_4$ are unknown constant parameters. Note that,

$$T_1(x) = L_g h(x) = \begin{bmatrix} 1 & 0 \\ k_3 & 0 \end{bmatrix}$$

$$T_2(x) = L_g L_f h(x) = \begin{bmatrix} k_2 & 0 \\ 0 & 0 \end{bmatrix}$$
$$T_3(x) = L_g L_f^2 h(x) = \begin{bmatrix} k_2^2 & -k_2k_4 \\ 0 & 0 \end{bmatrix}$$
$$\vdots \qquad (50)$$

and

$$h(x) = \begin{bmatrix} x_3 \\ x_4 \end{bmatrix}$$
$$L_f h(x) = \begin{bmatrix} k_2(x_3 - x_1) \\ 0 \end{bmatrix}$$
$$L_f^2 h(x) = \begin{bmatrix} -k_2(x_1^2 + x_2) + k_2^2(x_3 - x_1) \\ 0 \end{bmatrix}$$
$$L_f^3 h(x) = \begin{bmatrix} -(2k_2x_1 - k_2^2)(x_1^2 + x_2) \\ -k_1k_2x_1x_3 + k_2^3(x_3 - x_1) \\ \\ 0 \end{bmatrix}$$
$$\vdots \qquad (51)$$

Using the output reordering algorithm in the generic sense, the following estimate of the interactor matrix can be obtained.

$$N(p) = \begin{bmatrix} p+1 & 0 \\ \theta_1 p^3 + \theta_2 p^2 & p^3 + 6p^2 + 11p + 6 \end{bmatrix} \qquad (52)$$

θ_1 and θ_2 are uknown constant parameters which should be estimated, and the number of these parameters is the minimum.

Hence, eq.(8) for this example becomes

$$\begin{aligned} D_1(p)y_1 &= \theta_3(x_3 - x_1) + x_3 + u_1 \\ D(p)_2 y_2 &= (\theta_1 p^3 + \theta_2 p^2)y_1 + \theta_4(x_1^2 + x_2) \\ &\quad +\theta_5(x_3 - x_2) + \theta_6 x_1 x_3 + \theta_7(x_1^3 + x_1x_2) \\ &\quad +6x_4 + \theta_8 u_1 + \theta_9 u_2 \end{aligned}$$

where

$$\begin{aligned} D_1(p) &= (p+1) \\ D(p)_2 &= (p^3 + 6p^2 + 11p + 6) \end{aligned} \qquad (53)$$

and $\theta_1 \sim \theta_9$ are newly defined unknown constant parameters. In the second equation of the above, the term, $(\theta_1 p^3 + \theta_2 p^2)y_1$, comes from the fact that the structure estimate of the interactor matrix of this unknown plant is not a diagonal matrix. By applying the SISO adaptive design method to the above equations and from eq.(57), it is straightforward to derive the parameter adjusting law and adaptive control input.

5. CONCLUSIONS

The model reference adaptive control for MIMO systems is presented in this paper. By applying the adaptive design method for linear systems to the nonlinear case, very simple nonlinear MRACS technique is obtained. Since, the plant is not required to have the triangular structure, any change of variables is not necessary. By using generically estimated nonlinear interactor, this design method can be extended to MIMO systems.

REFERENCES

Krstic, M., Kanellakopoulos, I. and Kokotovic, P. (1995). Nonlinear and Adaptive Control Design, Wiley

Marino, R. and Tomei, P. (1995). Nonlinear Control Design, Prentice Hall

Wolovich, W.A. and P.L.Falb (1976). Invariants and Canonical Forms under Dynamic Compensation, SIAM J. Control and Optimization, Vol.14, pp.996-1008

Mutoh, Y. and R.Ortega (1993). Interactor Structure Estimation for Adaptive Control of Discrete-Time Multivariable Nondecouplable Systems. Automatica, Vol.29, pp.635-647

Di Benedetto, M.D., A.Glumineau and C.H.Moog (1994). The Nonlinear Interactor and Its Application to Input-Output Decoupling, IEEE, Trans. Automatic Control, Vol. AC-39, pp.1246-1250

Mutoh, Y. (1995). A note on the invariant properties of interactors, Int. J. Control, Vol.62, pp.1247-1252

Mutoh, Y. (1998). Invariant Properties of Interactor for Nonlinear Multivariable Systems, IFAC Conference on Systems Structure and Control, Nantes University, vol.3, pp.783-788, Nantes, France

ROBUST MRAC OF A LINEAR TIME-VARYING PARABOLIC SYSTEM WITH BOUNDED DISTURBANCE

Kyung-Jinn Yang *, Keum-Shik Hong **, and Seung-Hwan Lee*

* *Department of Mechanical and Intelligent Systems Engineering, Graduate College Pusan National University, Email: jinnky@hyowon.pusan.ac.kr*
** *School of Mechanical Engineering, Pusan National University, 30 Changjeon-dong Kumjeong-ku, Pusan, 609-735, Korea. Tel: +82-51-510-2454, Fax: +82-51-514-0685, Email: kshong@hyowon.pusan.ac.kr*

Abstract: In this paper, a robust MRAC (model reference adaptive control) for a linear parabolic partial differential equation with unknown time-varying parameters and bounded disturbance is investigated. In the adaptive control of time-varying plants, the derivative of a Lyapunov function candidate, which allows the derivation of adaptation laws, is not negative semi-definite in general. Under the assumption that the disturbance is uniformly bounded, the proposed robust MRAC scheme guarantees the boundedness of all signals in the closed loop system and the convergence of the state error near to zero. With an additional persistence of excitation condition, the parameter estimation errors are shown to converge near to zero as well. Simulation results are provided. *Copyright © 2001 IFAC.*

Keywords: Robust model reference adaptive control, parabolic partial differential equation, uniform ultimate boundedness, stability, convergence, persistence of excitation, time-varying parameters.

1. INTRODUCTION

In this paper a robust MRAC of a linear time-varying parabolic system with bounded disturbance is investigated. As in the adaptive control of finite dimensional systems a robust MRAC, under the assumption that the structure of the plant is known and only parameters in the system equation are unknown, is derived. Distributed sensing and actuation are also assumed.

Compared to the adaptive control/identification of finite dimensional systems, that of infinite dimensional systems is not well developed and has been recently studied since early 1980's (Hong, 1997; Hong and Bentsman, 1994; Duncan *et al.*, 1994; Bohm *et al.*, 1998; Renou *et al.*, 2000; Orlov and Bentsman, 2000). In the case of adaptive control for time-varying infinite dimensional systems the MRAC of a linear slowly time-varying parabolic system was presented in (Hong *et al.*, 1999).

The mathematical models of physical plants that control engineers adopt for the purpose of designing control systems normally contain some uncertainty. This is due to imperfect knowledge of system parameters and/or disturbances. Parameter time-variations may be due to unmodeled dynamics, for instance, neglected frictions, neglected high order dynamics, etc., and may also arise from linear approximations along different motions over a wide range of operating conditions. Many aspects of the robust adaptive control of lumped parameter systems have been addressed with respect to unknown time variations of plant parameters and additive disturbances (Kreisselmeier and Narendra, 1982; Middleton and Goodwin, 1988; Freeman, *et al.*, 1998; Polycarpou and Mears, 1998).

The objective of a MRAC scheme is to determine a feedback control law which forces the state of a plant to asymptotically track the state of a given reference model. At the same time, the unknown parameters in the plant model are estimated and used to update the control law. In the infinite dimensional systems the state error and the parameter estimation error take the forms of a partial differential equation and ordinary differential equations, respectively. The resulting closed loop system consisting of the plant, the reference model, and the estimator will be nonlinear. This is true even if the underlying plant and reference model, and the estimator are linear. The nonlinearity arises in the coupling of error dynamics. Consequently, the scheme requires a careful stability analysis to ensure that all signals, both input and output, remain in some sense, bounded. It is also desirable, although not necessarily essential, that some sort of parameter convergence is achieved.

The present paper makes the following contributions: To the authors' best knowledge this paper is the first treatment of an infinite dimensional system with unknown time-varying parameters and additive disturbance in the frame of robust MRAC. The unknown time-varying parameters are not required to be slow, which can be allowed to vary arbitrarily fast, and the disturbance is allowed to vary in both time and spatial variables. The well posedness of the closed loop system is established. Using an appropriate Lyapunov function candidate, the tracking error convergence near to zero is established. With the additional assumption of persistence of excitation the convergence of parameter errors near to zero is established as well.

2. PROBLEM FORMULATION

Consider a DPS described by a linear parabolic PDE with time-varying coefficients as

$$\dot{\xi}(x,t) = a(t)\xi_{xx}(x,t) + b(t)\xi(x,t) + u(x,t) + \varphi(x,t) \tag{1}$$

where $x \in [0,1]$, $t > 0$, $a(t)$ and $b(t)$ are unknown time-varying coefficients that are not necessarily slow-varying, $\xi(x,t)$ is the distributed state of the plant (in heat transfer, for example, it represents the temperature distribution on a rod), $\dot{\xi} = d\xi/dt$, $\xi_{xx} = \partial^2\xi/\partial x^2$, $u(x,t)$ is the control input function, and $\varphi(x,t)$ is the inaccessible external disturbance. Note that $\varphi(x,t)$ is a spatiotemporally function. Boundary conditions are given as

$$\xi(0,t) = h_1(t), \quad \xi(1,t) = h_2(t).$$

Initial condition is given as

$$\xi(x,0) = \xi_0(x).$$

The following assumptions are made.

Assumptions: (i) The structure of (1) (plant) is *a priori* known. (ii) Boundary conditions are *a priori* known and $h_1(\cdot), h_2(\cdot) \in C^\infty[0,\infty)$. (iii) Distributed sensing and actuation are available. (iv) Coefficients $a(t)$ and $b(t)$ are unknown and uniformly bounded with uniformly bounded derivatives. However, $a(t) > 0$ is assumed, due to the parabolicity condition of the plant. (v) Disturbance $\varphi(x,t)$ is uniformly bounded.

Now, the reference model with the same boundary conditions is defined as

$$\dot{\xi}_m(x,t) = a_m\xi_{mxx}(x,t) + b_m\xi_m(x,t) + r(x,t), \tag{2}$$
$$\xi_m(0,t) = h_1(t), \quad \xi_m(1,t) = h_2(t), \quad \xi_m(x,0) = \xi_{m0}(x)$$

where the subscript m indicates variables and parameters related to the reference model, and $r(x,t)$ is the reference input which is analytic on $[0,1]\times[0,\infty)$. It is assumed that $a_m > 0$ and $b_m < a_m\pi^2$. It is known that if $r(\cdot,\cdot)$ is analytic in $[0,1]\times[0,\infty)$, the solution of (2) is analytic in $[0,1]\times\{0 < t < T < \infty\}$.

The control objective for MRAC can now be stated as follows: Find a bounded control signal $u(x,t)$ that drives $\xi(x,t)$ to $\xi_m(x,t)$ asymptotically and keeps all signals in the closed-loop bounded. Now adopting the procedure in (Hong and Bentsman, 1994), consider a control law of the form

$$u(x,t) = (a_m - \hat{a}(t))\xi_{xx}(x,t) + (b_m - \hat{b}(t))\xi(x,t) + r(x,t) \tag{3}$$

where $\hat{a}$ and $\hat{b}$ are adaptive estimates to be specified in the sequel. Substituting (3) into (1) yields the following closed loop plant equation

$$\dot{\xi} = (a_m - \tilde{a})\xi_{xx} + (b_m - \tilde{b})\xi + r + \varphi \tag{4}$$

where $\tilde{a} = \hat{a} - a$ and $\tilde{b} = \hat{b} - b$ are parameter estimation errors. Note that if $\tilde{a} = \tilde{b} = 0$ and $\varphi = 0$, then (4) is exactly the same as (2).

Introducing the state error $e = \xi - \xi_m$, the following state error equation is obtained.

$$\dot{e} = (a_m - \tilde{a})e_{xx} + (b_m - \tilde{b})e - (\tilde{a}\xi_{mxx} + \tilde{b}\xi_m) + \varphi,$$
$$e(0,t) = e(1,t) = 0, \quad e(x,0) = \xi_0(x) - \xi_{m0}(x). \tag{5}$$

Now, consider a functional $V : L_2(0,1)\times R^2 \to R^+$ such that

$$V(e,\tilde{a},\tilde{b}) = \frac{1}{2}\langle e,e\rangle + \frac{1}{2\gamma_a}\tilde{a}^2 + \frac{1}{2\gamma_b}\tilde{b}^2 \tag{6}$$

where γ_a and γ_b are positive constants, which will become the adaptation gains later. $\langle\cdot,\cdot\rangle$ is the inner product in the space $L_2(0,1)$ defined as $\langle h,g\rangle = \int_0^1 h(x)g(x)dx$ and with the induced norm $\|\cdot\|$. Differentiating (6) with respect to t along (5) yields:

$$\dot{V}(t) = \langle e,\dot{e}\rangle + \frac{1}{\gamma_a}\tilde{a}\dot{\tilde{a}} + \frac{1}{\gamma_b}\tilde{b}\dot{\tilde{b}}$$

$$= a_m\langle e_x, e_{xx}\rangle + b_m\langle e, e\rangle + \tilde{a}\left(-\langle e, e_{xx}\rangle - \langle e, \xi_{mxx}\rangle + \dot{\tilde{a}}/\gamma_a\right)$$
$$+\tilde{b}\left(-\langle e, e\rangle - \langle e, \xi_m\rangle + \dot{\tilde{b}}/\gamma_b\right) + \langle e, \varphi\rangle. \quad (7)$$

Let the differential equations of the adaptive estimators in (3) be given as

$$\dot{\hat{a}} = \gamma_a\left(\langle e, e_{xx}\rangle + \langle e, \xi_{mxx}\rangle\right), \quad (8a)$$

$$\dot{\hat{b}} = \gamma_b\left(\langle e, e\rangle + \langle e, \xi_m\rangle\right). \quad (8b)$$

Then, substituting (8a,b) into (7) yields:

$$\dot{V} \le \left(-a_m\pi^2 + b_m\right)\|e\|^2 - \frac{1}{\gamma_a}\tilde{a}\dot{a} - \frac{1}{\gamma_b}\tilde{b}\dot{b} + \langle e, \varphi\rangle. \quad (9)$$

Now, from (5), (6), (8a,b), and (9), the following observations are made: Assume that φ=0, but a and b are time-varying. Then, (9) becomes

$$\dot{V} \le \left(-a_m\pi^2 + b_m\right)\|e\|^2 - \frac{1}{\gamma_a}\tilde{a}\dot{a} - \frac{1}{\gamma_b}\tilde{b}\dot{b}. \quad (10)$$

In this case, $\dot{V}$ may take positive values because of $\tilde{a}\dot{a}/\gamma_a + \tilde{b}\dot{b}/\gamma_b$. Hence, no stability conclusion can be drawn from the Lyapunov function candidate (6). But, further assume that

(a) $\tilde{a}$ and $\tilde{b}$ are bounded and $\dot{a}, \dot{b} \in L_1$. Then, the tracking error $e(t)$ converges to zero as $t \to \infty$.

(b) Now, if only the boundedness of $\tilde{a}\dot{a}$ and $\tilde{b}\dot{b}$ is assumed, then no asymptotic convergence to zero can be asserted. However, if $\tilde{a}\dot{a}/\gamma_a + \tilde{b}\dot{b}/\gamma_b$ is sufficiently small, then it is guaranteed that e is uniformly ultimately bounded within an arbitrarily small neighborhood of zero.

Remark 1: In order to have the boundedness of $\tilde{a}\dot{a}/\gamma_a + \tilde{b}\dot{b}/\gamma_b$, the boundedness of individual $\tilde{a}$, $\tilde{b}$, $\dot{a}$, and $\dot{b}$ is required. The boundedness of $\dot{a}$ and $\dot{b}$ depends on the plant, which can be assumed. Also, in the case that a and b are slowly varying, the boundedness of $\tilde{a}$ and $\tilde{b}$ was shown through averaging analysis (Hong *et al*, 1999). A successful adaptation is possible as long as the plant parameters vary slowly. Now, more general problem would be one in which a, b, $\dot{a}$ and $\dot{b}$ vary in unknown fashion but bounded (no slow varying assumption). The aim of this paper is to show not only the boundedness of all signals in the closed loop including $\hat{a}$ and $\hat{b}$ but also the convergence of $\tilde{a}$ and $\tilde{b}$ to zero.

3. ROBUST MRAC: STABILITY

In this section, the control and adaptation laws are modified so that the boundedness of all signals in the closed loop system and the convergence of the state error e near to zero can be assured. The main idea is to consider the worst case of the uncertainties in the form of possible bounds. Based upon the worst case, the following control algorithm is proposed.

Control Law:

$$u(x,t) = (a_m - \hat{a}(t))\xi_{xx}(x,t) + (b_m - \hat{b}(t))\xi(x,t))$$
$$+r(x,t) + f(x,t) \quad (11)$$

where the additional term $f(x,t)$ is regarded as a new input signal to be determined based on robust control strategy. Let the additional input $f(x,t)$ is given by

$$f(x,t) = \lambda_e e(x,t) + p(x,t) \quad (12)$$

where $\lambda_e < 0$, $e(x,t) = \xi(x,t) - \xi_m(x,t)$,

$$p(x,t) = \begin{cases} -\dfrac{e(x,t)}{\|e(x,t)\|}\mu_e & \text{for } \|\mu_e e(x,t)\| > \varepsilon_e, \\ -\dfrac{e(x,t)}{\varepsilon_e}\mu_e^2 & \text{for } \|\mu_e e(x,t)\| \le \varepsilon_e, \end{cases}$$

$\varepsilon_e > 0$, and $\mu_e \ge \|\varphi(x,t)\|$.

Note that ε_e is a threshold to be designed for the smallness of the state error e.

Adaptation Laws:

$$\dot{\hat{a}} = -\delta_a\hat{a} + \gamma_a\left(\langle e, e_{xx}\rangle + \langle e, \xi_{mxx}\rangle + g_a\right) \quad (13a)$$

$$\dot{\hat{b}} = -\delta_b\hat{b} + \gamma_b\left(\langle e, e\rangle + \langle e, \xi_m\rangle + g_b\right) \quad (13b)$$

where $\delta_a > 0$, $\delta_b > 0$,

$$g_a = \begin{cases} -\mu_a\hat{a}/|\hat{a}| & \text{for } |\mu_a\hat{a}| > \varepsilon_a, \\ -\mu_a^2\hat{a}/\varepsilon_a & \text{for } |\mu_a\hat{a}| \le \varepsilon_a, \end{cases}$$

$$\varepsilon_a > 0, \quad \mu_a \ge |f_a|, \quad f_a \triangleq -\frac{\delta_a}{\gamma_a}\left(a + \frac{\dot{a}}{\delta_a}\right),$$

$$g_b = \begin{cases} -\mu_b\hat{b}/|\hat{b}| & \text{for } |\mu_b\hat{b}| > \varepsilon_b, \\ -\mu_b^2\hat{b}/\varepsilon_b & \text{for } |\mu_b\hat{b}| \le \varepsilon_b, \end{cases}$$

$$\varepsilon_b > 0, \quad \mu_b \ge |f_b|, \text{ and } f_b \triangleq -\frac{\delta_b}{\gamma_b}\left(b + \frac{\dot{b}}{\delta_b}\right).$$

Note that the adaptation laws (13a,b) are implementable. The terms $-\delta_a\hat{a}$ and $-\delta_b\hat{b}$ in (13a,b) are purposely inserted to enhance the convergence of $\hat{a}$ and $\hat{b}$, respectively; g_a and g_b are introduced to cope with the variations of a and b, respectively. Since a, $\dot{a}$, b and $\dot{b}$ are assumed to be bounded, μ_a and μ_b can be selected at reasonable values by making γ_a and γ_b sufficiently large. It is also noted that the control magnitudes μ_a, μ_b, and μ_e are to compensate the maximum possible bounds of f_a, f_b, and φ, respectively, for both positive and negative cases.

Substituting (11) into (1) yields the following closed loop plant equation.

$$\dot{\xi} = (a_m - \tilde{a})\xi_{xx} + (b_m - \tilde{b})\xi + r + f + \varphi. \quad (14)$$

Then, the following state error equation is derived.

$$\dot{e} = (a_m - \widetilde{a})e_{xx} + (b_m - \widetilde{b})e - (\widetilde{a}\xi_{mxx} + \widetilde{b}\xi_m) + f + \varphi,$$

$$e(0,t) = e(1,t) = 0, \quad e(x,0) = \xi_0(x) - \xi_{m0}(x). \quad (15)$$

Differentiating (6) with respect to t along (15) yields:

$$\dot{V}(t) = a_m\langle e_x, e_{xx}\rangle + b_m\langle e,e\rangle + \widetilde{a}\left(-\langle e, e_{xx}\rangle - \langle e, \xi_{mxx}\rangle + \dot{\widetilde{a}}/\gamma_a\right) + \widetilde{b}\left(-\langle e,e\rangle - \langle e,\xi_m\rangle + \dot{\widetilde{b}}/\gamma_b\right) + \langle e,f\rangle + \langle e,\varphi\rangle. \quad (16)$$

Substituting (12) and (13a,b) into (16) yields:

$$\dot{V}(t) = -a_m\langle e_x, e_x\rangle + b_m\langle e,e\rangle + \lambda_e\langle e,e\rangle + \langle p,e\rangle + \langle \varphi,e\rangle - \lambda_a\widetilde{a}^2 + \widetilde{a}f_a + \widetilde{a}g_a - \lambda_b\widetilde{b}^2 + \widetilde{b}f_b + \widetilde{b}g_b \quad (17)$$

where $\lambda_a = \delta_a/\gamma_a$ and $\lambda_b = \delta_b/\gamma_b$. From (17), the following inequality is derived.

$$\dot{V}(t) \le -\left(a_m\pi^2 - b_m - \lambda_e\right)\|e\|^2 + \langle p,e\rangle + \langle\varphi,e\rangle - \lambda_a\widetilde{a}^2 + |\hat{a}|\mu_a + \hat{a}g_a - af_a - ag_a - \lambda_b\widetilde{b}^2 + |\hat{b}|\mu_b + \hat{b}g_b - bf_b - bg_b. \quad (18)$$

The Lyapunov derivative (18) can be examined in two extreme cases as follows.

i) In the case that $|\mu_a\hat{a}| > \varepsilon_a$, $|\mu_b\hat{b}| > \varepsilon_b$, and $\|\mu_e e\| > \varepsilon_e$, (18) yields:

$$\begin{aligned}\dot{V} &\le -\left(a_m\pi^2 - b_m - \lambda_e\right)\|e\|^2 - \lambda_a\widetilde{a}^2 - af_a - ag_a - \lambda_b\widetilde{b}^2 - bf_b - bg_b \\ &= -\left(a_m\pi^2 - b_m - \lambda_e\right)\|e\|^2 - \lambda_a\widetilde{a}^2 - \lambda_b\widetilde{b}^2 \\ &\quad + \left\{\delta_a a^2/\gamma_a + a\dot{a}/\gamma_a + a\mu_a\,\hat{a}/|\hat{a}| + \delta_b b^2/\gamma_b + b\dot{b}/\gamma_b + b\mu_b\,\hat{b}/|\hat{b}|\right\}. \quad (19)\end{aligned}$$

ii) In the case that $|\mu_a\hat{a}| \le \varepsilon_a$, $|\mu_b\hat{b}| \le \varepsilon_b$, and $\|\mu_e e\| \le \varepsilon_e$, (18) yields:

$$\begin{aligned}\dot{V} &\le -\left(a_m\pi^2 - b_m - \lambda_e\right)\|e\|^2 + \mu_e\|e\| - \mu_e^2\|e\|^2/\varepsilon_e \\ &\quad - \lambda_a\widetilde{a}^2 + |\hat{a}|\mu_a - \mu_a^2\,\hat{a}^2/\varepsilon_a - af_a - ag_a \\ &\quad - \lambda_b\widetilde{b}^2 + |\hat{b}|\mu_b - \mu_b^2\,\hat{b}^2/\varepsilon_b - bf_b - bg_b \\ &\le -\left(a_m\pi^2 - b_m - \lambda_e\right)\|e\|^2 - \lambda_a\widetilde{a}^2 - \lambda_b\widetilde{b}^2 \\ &\quad + \left\{\varepsilon_a/4 + a^2\delta_a/\gamma_a + a\dot{a}/\gamma_a + a\mu_a\,\hat{a}/|\hat{a}| + \varepsilon_b/4 + b^2\delta_b/\gamma_b + b\dot{b}/\gamma_b + b\mu_b\,\hat{b}/|\hat{b}| + \varepsilon_e/4\right\}. \quad (20)\end{aligned}$$

Comparing (19) and (20), the terms in the curly bracket of (20) gives the largest value that $\dot{V}(t)$ can have, while the terms in the curly bracket of (19) gives the smallest one. All other cases belong between these two extreme cases. Therefore, considering the worst case, the derivative of the Lyapunov function candidate is bounded as follows:

$$\dot{V} \le -\left(a_m\pi^2 - b_m - \lambda_e\right)\|e\|^2 - \lambda_a\widetilde{a}^2 - \lambda_b\widetilde{b}^2 + \upsilon(t) \quad (21)$$

where $\upsilon(t) = \varepsilon_a/4 + a^2\delta_a/\gamma_a + a\dot{a}/\gamma_a + a\mu_a\,\hat{a}/|\hat{a}| + \varepsilon_b/4 + b^2\delta_b/\gamma_b + b\dot{b}/\gamma_b + b\mu_b\,\hat{b}/|\hat{b}| + \varepsilon_e/4$.

Note that $\upsilon(t)$ is bounded because of the assumption that a, b, $\dot{a}$ and $\dot{b}$ are bounded.

Remark 2: The existence and uniqueness of the solutions for coupled nonautonomous dynamical systems (15) and (13a,b) are addressed in Appendix A. Since $\upsilon(t)$ is bounded, the solutions e, $\widetilde{a}$, and $\widetilde{b}$ are uniformly ultimately bounded.

Remark 3: From $\mu_a \ge |f_a|$ and $\mu_b \ge |f_b|$, μ_a and μ_b can be chosen at reasonable values according to $|f_a|$ and $|f_b|$, respectively. Thus, $\upsilon(t)$ can be pushed in an arbitrarily small boundedness region by making sufficiently small ε_a, ε_b, ε_e, δ_a, δ_b and sufficiently large γ_a, γ_b.

All the above developments are now summarized as follows:

Theorem 1: Consider the coupled nonautonomous dynamical system (15) and (13a,b). Then all signals in the system are bounded. Furthermore, the uniform ultimate boundedness region of the state error e can be made arbitrary small near to zero by a suitable choice of ε_a, ε_b, ε_e, δ_a, δ_b, γ_a, and γ_b.

4. PARAMETER ERROR CONVERGENCE

In addition to the state error convergence near to zero, it is also desirable to have an adaptive control scheme to provide parameter estimation error convergence near to zero as well, i.e., the parameters $\hat{a}$ and $\hat{b}$ converge near to the true parameters $a(t)$ and $b(t)$ as quickly as possible. If the parameter error convergence is established, the robustness of the entire adaptive algorithm can be improved. To assure this, the following additional persistence of excitation condition on the reference model is required.

Let $H = H^2(0,1)$ be a Hilbert space such that $\xi, \xi_m \in H$ which is densely and continuously embedded in L_2, and H^* be the continuous dual space of H. Let $A_m \in L(H, H^*)$ be the reference model dynamic operator, i.e., $A_m = a_m \partial^2/\partial x^2 + b_m$, and $A_1(q): H \to H^*$ be a differential operator such that $A_1(q) = q_1 \partial^2/\partial x^2 + q_2$ for each $q = (q_1, q_2) \in Q$, $Q = R^2$. Also, let $Dom(A_m) = \{\psi \in H : A_m\psi \in L_2\}$ and $Dom(A_1(q)) = \{\psi \in H : A_1(q)\psi \in L_2\}$. The following definition is adopted.

Definition: The reference model (2), or the triple $\{A_m, r, \xi_{m0}\}$, is persistently exciting if there exist positive constants τ_0, δ_0, and ε_0, such that for each $q \in Q$ with $|q|_Q = 1$ and $t \geq 0$ sufficiently large, there exists $\bar{t} \in [t, t+\tau_0]$ for which

$$\left\| \int_{\bar{t}}^{\bar{t}+\delta_0} A_1(q)\xi(\tau)d\tau \right\|_{H^*} \geq \varepsilon_0 + 2\mu_e \frac{\delta_0}{\delta}$$

where $\varepsilon_0 > 0$, $\delta > 0$, and μ_e is defined in (12).

Theorem 2: If $r \in L_\infty(0,\infty; H)$ and $\xi_{m0} \in H$, and if the reference model (2) is persistently exciting, then the uniform ultimate boundedness region of the parameter estimation error vector $\theta = (\tilde{a}, \tilde{b})$ can be made arbitrarily small near to zero by a suitable choice of ε_a, ε_b, δ_a, δ_b, γ_a, and γ_b.

5. SIMULATIONS

Let the heat transfer equation be given with known homogeneous boundary conditions as

$$\dot{\xi} = a(t)\xi_{xx} + u + \varphi, \quad x \in [0,1], \quad t > 0$$

$$\xi(0,t) = \xi(1,t) = 0, \quad \xi(x,0) = \xi_0(x) = 0.1\sin(\pi x) \quad (22)$$

where $a(t)$ is the time-varying conductivity and $\varphi(x,t)$ is the spatiotemporally varying latent heat of transformation which is treated as the disturbance. $a(t)$ and $\varphi(x,t)$ are unknown, but for simulation purpose $a(t) = 2 - 1.5\sin(3t)$ and $\varphi(x,t) = 1.5\sin(3\pi x)\sin(15t)$ are assumed. The reference model is chosen as

$$\dot{\xi}_m = a_m \xi_{mxx} + r, \quad x \in [0,1], \quad t > 0$$

$$\xi_m(0,t) = \xi_m(1,t) = 0, \quad \xi_m(x,0) = \xi_{m0}(x). \quad (23)$$

In this case, the boundary conditions establishes $H = H_0^2(0,1)$, which is endowed with its usual inner product and corresponding induced norm and is densely and continuously embedded in L_2. Let $a_m = 0.5$, $\xi_{m0}(x) = -\sin(\pi x)$, and the constant reference input $r = 5.0$. The control gain in (12) is chosen as $\lambda_e = -2$, and others as $\mu_e = 2.5$ and ε_e=0.1. The adaptive gain in (13a) is chosen as $\gamma_a = 45$, and others as $\delta_a = 0.05$, $\mu_a = 0.1$, and ε_a=0.1. Figure 1 shows the convergence of the state error $e(x,t)$ to zero. Figure 2 shows the convergence of the estimated parameter $\hat{a}(t)$ to the plant parameter $a(t)$.

6. CONCLUSIONS

A robust MRAC algorithm for a linear parabolic system with unknown time-varying parameters and an additive bounded disturbance was developed. The time-varying parameters were assumed to be uniformly bounded with uniformly bounded derivatives, but they were allowed to vary arbitrarily fast. The disturbance was a spatiotemporally varying function. Under the external disturbances and unknown plant parameters, the robust MRAC law developed in this paper assured the closed loop system to track a desired signal which comes from a reference model. Also, with the additional persistence of excitation condition, the algorithm guaranteed the convergence of the adjustable controller parameters near to their nominal values.

The practicality of the control algorithm developed may be limited because the algorithm assumes distributed sensing and actuation. However, the authors firmly believe that this work provides one step toward improving the adaptive control of distributed parameter systems. The feasibility of using a finite number of sensors and actuators is under investigation.

ACKNOWLEDGMENT

This work was supported in part by the Basic Research Program of the Korea Science and Engineering Foundation under grant no. 2001-1-30400-006-2.

REFERENCES

Bohm, M., M.A. Demetriou, S. Reich and I.G. Rosen (1998). Model reference adaptive control of distributed parameter systems. *SIAM J. Control Optim.*, **36**, 33-81.

Duncan, T.E., B. Maslowski and B. Pasik-Duncan (1994). Adaptive boundary and point control of linear stochastic distributed parameter systems. *SIAM J. Control Optim.*, **32**, 648-672.

Freeman, R.A., M. Krstic and P.V. Kokotovic (1998). Robustness of adaptive nonlinear control to bounded uncertainties. *Automatica*, **34**, 1227-1230.

Hong, K.S. (1997) Asymptotic behavior analysis of a coupled time-varying system: Application to adaptive systems. *IEEE Trans. Automat. Contr.*, **42**, 1693-1697.

Hong, K.S. and J. Bentsman (1994). Direct adaptive control of parabolic systems: Algorithm synthesis, convergence and stability analysis. *IEEE Trans. Automat. Contr.*, **39**, 2018-2033.

Hong, K.S., V. Solo and J. Bentsman (1999). Adaptive control of a time-varying parabolic system: Averaging analysis. *Proc. 14th World Congress of IFAC*, Beijing, China, Vol. E, pp. 351-356.

Kreisselmeier, G. and K.S. Narendra (1982). Stable model reference adaptive control in the presence of bounded disturbances. *IEEE Trans. Automat.*

Contr., **27**, 1169-1175.
Middleton, R.H. and G.C. Goodwin (1988). Adaptive control of time varying linear plant. *IEEE Trans. Automat. Contr.*, **33**, 150-155.
Polycarpou, M.M. and M.J. Mears (1998). Stable adaptive tracking of uncertain systems using nonlinearly parameterized on-line approximators. *Int. J. Control*, **70**, 363-38.
Renou, S., M. Perrier, D. Dochain and S. Gendron (2000). Direct adaptive control of a linear parabolic system. *IFAC SYSID 2000*, Vol. ThMD1-1, Santa Barbara, California.

APPENDIX A: EXISTENCE AND UNIQUENESS

Rewrite nonlinear error equations (15) and (13a,b) in the following form

$$\dot{z} = A(t)z + F(t,z)\,, \quad z(0) = z_0$$

where $z = \left(e, \hat{a}, \hat{b}\right)^T$, $\hat{a} = \tilde{a} + a$, $\hat{b} = \tilde{b} + b$,

$$A(t) = \begin{bmatrix} A_0(t) & 0 & 0 \\ 0 & -\delta_a & 0 \\ 0 & 0 & -\delta_b \end{bmatrix},$$

$$\text{and} \quad F(t,z) = \begin{bmatrix} (a-\hat{a})\xi_{mxx} + (b-\hat{b})\xi_m + f + \varphi \\ \gamma_a\langle e, e_{xx}\rangle + \gamma_a\langle e, \xi_{mxx}\rangle + \gamma_a g_a \\ \gamma_b\langle e, e\rangle + \gamma_b\langle e, \xi_m\rangle + \gamma_b g_b \end{bmatrix},$$

where $A_0(t) \overset{\Delta}{=} (a_m + a - \hat{a})\dfrac{\partial^2}{\partial x^2} + \left(b_m + b - \hat{b}\right)$. Define a state space as $H \overset{\Delta}{=} L_2(0,1)\times R^2$, and

$$D(A) = \left\{ \begin{matrix} \left(e, \hat{a}, \hat{b}\right) \in H : e \in H^2(0,1) \cap H_0^1(0,1) \\ \text{with } e(0) = 0 = e(1), \text{and } \hat{a}, \hat{b} \in R \end{matrix} \right\}.$$

Note that the boundary conditions of (15) have been incorporated in the space $H^2(0,1) \cap H_0^1(0,1)$, which is the domain of the differential operator A_0. $D(A)$ is dense, and A is a closed operator (Walker, 1980).

For $z \in D(A)$

$$\langle z, Az\rangle_H \le -C_1\langle z, z\rangle_H\,,$$

where $C_1 = \min\left\{(a_m + a - \hat{a})\pi^2 - \left(b_m + b - \hat{b}\right), \delta_a, \delta_b\right\}$ >0, and $(a_m + a - \hat{a}) \ge \Delta a \ge 0$ and $(b_m + b - \hat{b}) \ge \Delta b \ge 0$ are assumed. Now by the linearity of A, we see that $\omega I - A$ is monotone (accretive) for every $\omega \le C_1$. Hence $A: D(A) \subset H \to H$ is the infinitesimal generator of a linear process $\{S(t)\}_{t\ge 0}$ $= \left\{\left(\Phi(t,0), \hat{A}(t), \hat{B}(t)\right)\right\}_{t\ge 0}$ on H, see (Walker, 1980, Theorem 3.2, p. 92). Note that the first component $E(t,0)$ is generated by A_0. Note also that $\Phi(t,0)e_0$ is the strong solution of the evolution equation $\dot{e}(t) = A_0 e(t)$ for every $e_0 \in D(A_0)$.

Now set $z = \left(e, \hat{a}, \hat{b}\right)$ and $z' = \left(e', \hat{a}', \hat{b}'\right)$. Then

$$\left\|F(t,z) - F(t,z')\right\|_H \le C_2\left\|z - z'\right\|_H,$$

where C_2 is a constant. Therefore $F: H \to H$ is locally Lipschitz continuous in H. Thus a unique solution exists. Finally the solution of (15) can be written in the following variation of constant formula (Pazy, 1983)

$$e(t) = \Phi(t,0)e(0) + \int_0^t \Phi(t,\tau)\left(-\tilde{a}(\tau)\xi_{mxx}(\tau) - \tilde{b}(\tau)\xi_m(\tau) + f(\tau) + \varphi(\tau)\right)d\tau$$

where $\Phi(t,s)$ is the evolution operator associated with A_0 in the space $L_2(0,1)$.

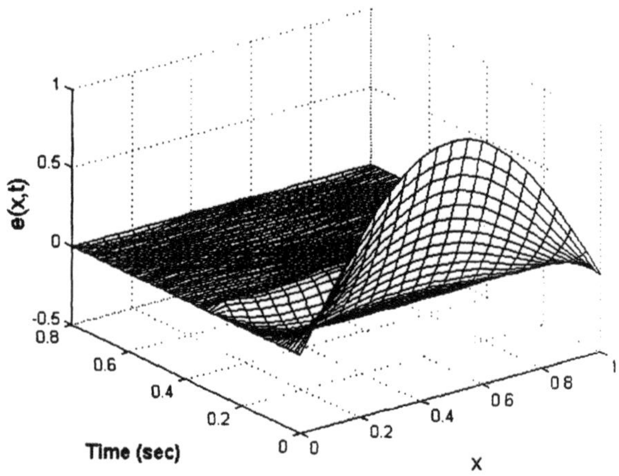

Fig. 1. Convergence of state error $e(x,t)$ to zero.

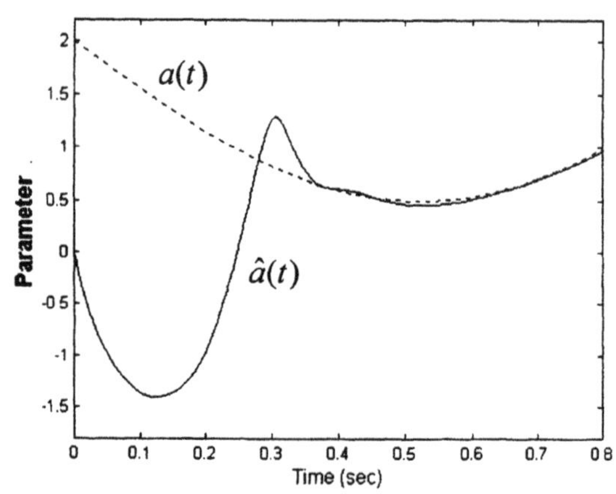

Fig. 2. Convergence of parameter estimation error $\tilde{a}(t)$ to zero.

ROBUST EMM AND ADAPTIVE CONTROL FOR A DISCRETE-TIME PLANT

Wataru Kase

*Department of Electrical Engineering,
Osaka Institute of Technology.
5-16-1 Omiya, Asahi-ku, Osaka 535-8585 JAPAN.*

Yasuhiko Mutoh

*Department of Mechanical Engineering, Sophia University.
7-1 Kioi-cho, Chiyoda-ku, Tokyo 102-0094 JAPAN.*

Abstract: This paper describes a simple method for a design of robust exact model matching controller for discrete-time plant. This method uses the free parameters which arise by employing a nonminimal order observer. Application to MRAC is also discussed. Simulation result are presented which confirm the validity of the design method. *Copyright©2001 IFAC*

Keywords: model following control, discrete-timesystem, robust stability, parametrization, linear programming

1. INTRODUCTION

The concept of exact model matching (EMM) was first proposed by Moore and Silverman (1972) and Wolovich (1974), which has been developed widely for a model reference adaptive control (MRAC) (Elliott and Wolovich, 1979, Goodwin and Sin, 1984, Elliot and Wolovich, 1984). Recently, a simple design method of EMM for a plant with measurement noise was presented by Mutoh and Nikiforuk (1994). This method uses the free parameters which arise by employing a non-minimal order observer. Because of its simplicity, the method is also applicable to MRACS. This paper shows that the same idea can be extended to the design of the robust EMM.

For this purpose we use Rouche's theorem. Ortega (1983) presented a control design method for a plant with possible unstructured uncertainties, in which Rouche's theorem played an essential role. Although, this method is only for stable plants, its basic idea can be used in various types of control problems.

In this paper, by applying the idea of Ortega (1983) to Mutoh and Nikiforuk (1994), a simple design method of robust EMM for a plant with uncertainty will be presented. Although a certain cost function for the robustness is minimized using a parametrization of all stabilizing controllers, it can not be used the same parametrization as in Mutoh and Nikiforuk (1994). So another parametrization which can be obtained by using extended division algorithm by Kase (1999) will be developed. However, since it is difficult to minimize this cost function directly, the problem of minimizing the upper bound of the cost function is considered.

Section 2 provides a statement of problem and derives a minimization problem to increase the robust stability margin. Then, a new parametrization for EMM controller, which gives the constraint condition for the cost in the previous section, is presented Section 3. The robust EMM controller is then given by solving a linear programming problem. Some simulation studies of the EMM will be then presented in Section 4. Concludings remarks will be given in Section 5.

2. PROBLEM STATEMENT

Consider the following transfer function of linear discrete-time plant:

$$G_0(z) := N_0(z)/D_0(z) \tag{1}$$

where $D_0(z)$ and $N_0(z)$ are polynomials in the time-shift operator z. Assume that the mathematical model of the above plant is presented by

$$G(z) := N(z)/D(z) \tag{2}$$

where

$$\begin{aligned} &D(z) = z^n + D_{n-1}z^{n-1} + \cdots + D_0, \\ &N(z) = N_m z^m + N_{m-1}z^{m-1} + \cdots + N_0, \\ &N_m \neq 0, \quad \delta := n - m. \end{aligned} \tag{3}$$

The purpose of the design is to construct an EMM system for the plant model (2) which also stabilizes the plant (1). For this purpose, we make some assumptions as follows:

A1 $N(z)$ and $N_0(z)$ are Hurwitz.
A2 The reference output signals $y_m(t), \ldots, y_m(t+\delta)$ are available at the time instant t.

For the plant model (2), it is well known that the EMM system can be attained by

$$u(t) = \frac{X_d(z)}{z^{d-1}}u(t) + \frac{Y_d(z)}{z^{d-1}}y(t) + \frac{z^\delta}{N_m}y_m(t) \tag{4}$$

where $u(t)$ and $y(t)$ are input and output of the plant respectively. $X_d(z)$ and $Y_d(z)$ are polynomials of the forms

$$\begin{aligned} X_d(z) &= X_{d-2}z^{d-2} + X_{d-3}z^{d-3} + \cdots + X_0 \\ &:= \mathcal{X}_{d-2}S^{d-2}(z), \\ Y_d(z) &= Y_{d-1}z^{d-1} + Y_{d-2}z^{d-2} + \cdots + Y_0 \\ &:= \mathcal{Y}_{d-1}S^{d-1}(z), \\ S^{d-1}(z) &= \left[1 \ z \ \cdots \ z^{d-1}\right]^T, \quad d \geq n \end{aligned} \tag{5}$$

which satisfy the following Diophantine equation

$$\begin{aligned} &X_d(z)D(z) + Y_d(z)N(z) \\ &\quad = z^{d-1}\{D(z) - N_m^{-1}z^\delta N(z)\}. \end{aligned} \tag{6}$$

Applying the control input (4) to the plant (1), the closed-loop transfer function $G_{cl}(z)$ is given by

$$G_{cl}(z) = \frac{N_m^{-1}z^{d+\delta-1}N_0(z)}{z^{d-1}D_0(z) - X_d(z)D_0(z) - Y_d(z)N_0(z)} \tag{7}$$

Therefore, the closed-loop system is stable if the denominator polynomial of the above transfer function is Hurwitz.

Define the following rational function:

$$F(z) = \frac{z^{d-1}D_0(z) - X_d(z)D_0(z) - Y_d(z)N_0(z)}{z^{d-1}N_0(z)} \tag{8}$$

where zeros are the closed-loop poles. This function is analytic outside the unit circle due to the assumption 1. Using eqs.(1), (2) and (6), $F(z)$ can be rewritten as

$$\begin{aligned} F(z) &= \frac{z^{d-1} - X_d(z)}{z^{d-1}}G_0^{-1}(z) - \frac{Y_d(z)}{z^{d-1}} \\ &= \frac{z^{d-1} - X_d(z)}{z^{d-1}}\{G_0^{-1}(z) - G^{-1}(z)\} + G_m^{-1}(z) \end{aligned} \tag{9}$$

where

$$G_m(z) := N_m/z^\delta \tag{10}$$

and thus $G_m^{-1}(z)$ is also analytic. The unstable region $\mathbf{D}$ is given by

$$\mathbf{D} = \{z : \ |z| \geq 1\}. \tag{11}$$

From the assumption A1, $F(z)$ and $G_m^{-1}(z)$ are analytic in $\mathbf{D}$. Therefore, from Rouche's theorem, $F(z)$ and $G_m^{-1}(z)$ have the same number of zeros in $\mathbf{D}$ if the following inequality holds:

$$|G_m^{-1}(z)| > |F(z) - G_m^{-1}(z)|. \tag{12}$$

Since $G_m^{-1}(z)$ does not have any zeros in $\mathbf{D}$, $F(z)$ also has no zeros in $\mathbf{D}$ if eq.(12) holds. Thus, the closed-loop system given by eq.(7) is stable if eq.(12) holds.

Eq.(12) can be written by

$$|1 - z^{1-d}X_d(z)| < \left|\frac{G_m^{-1}(z)}{G_0^{-1} - G^{-1}(z)}\right|. \tag{13}$$

Define the cost function $J(\mathcal{X}_{d-2})$ as follow:

$$J(\mathcal{X}_{d-2}) = \sum_{i=0}^{d-2}|X_i|. \tag{14}$$

Then,

$$|1 - z^{1-d}X_d(z)| \leq J(\mathcal{X}_{d-2}) + 1 \tag{15}$$

holds on the unit circle. Therefore, the purpose of the design is achieved by minimizing $J(\mathcal{X}_{d-2})$ subject to eq.(6). In the following section, the minimizing problem of the above will be discussed using the free parameters included in $\mathcal{X}_{d-2}$.

3. DESIGN OF ROBUST EMM SYSTEM

In this section, we make clear that the degrees of freedom in $\mathcal{X}_{d-2}$, and minimize the cost given in eq.(14) using such freedom.

There are many methods to solve eq.(6). In this paper, a method using state space parameters is presented. At first, the following lemma concerning with an extended division holds.

Lemma 1. Assume that $\alpha(z)$ is monic polynomial of degree n and $\beta(z)/\alpha(z)$ is proper. Then, for a given polynomial $P(z)$ of degree p there exist polynomials $Q(z)$ and $R(z)$ such that

$$\begin{aligned} &P(z)\beta(z) = Q(z)\alpha(z) + R(z), \\ &R(z)/\alpha(z) : \text{strictly proper.} \end{aligned} \tag{16}$$

Let (A, b, c, e) be the controllability canonical realization of $\beta(z)/\alpha(z)$. Then, $Q(z)$ and $R(z)$ are given by the following equation:

$$Q(z) = \mathcal{P}\mathcal{T}_{p-1}S^p(z), R(z) = \mathcal{P}\mathcal{O}_p S^{n-1}(z) \tag{17}$$

where

$$\mathcal{P}S^p(z) = P(z), \quad \mathcal{O}_p(c, A) = \begin{bmatrix} c \\ cA \\ \vdots \\ cA^{p-1} \end{bmatrix}, \tag{18}$$

$$\mathcal{T}_{p-1}(A, b, c, e) = \begin{bmatrix} e & 0 & 0 \cdots & 0 \\ cb & e & 0 & \cdots 0 \\ cAb & cb & e & \cdots 0 \\ \vdots & \vdots & \ddots & \ddots \vdots \\ cA^{p-1}b & cA^{p-2}b & \cdots & cb \;\; e \end{bmatrix}.$$

(Proof). It is well known that there exist $Q(z)$ and $R(z)$ satisfying eq.(16). In the following, it will be shown that $Q(z)$ and $R(z)$ are given by eq.(18).

Since (A, b, c, e) is the realization of $\beta(z)/\alpha(z)$,

$$\begin{aligned} zx(t) &= Ax(t) + bu(t), \\ y(t) &= cx(t) + eu(t) \end{aligned}$$

holds. From the above equations, we have

$$\begin{aligned} S^p(z)y(t) &= \mathcal{O}_p(c, A)x(t) \\ &+\mathcal{T}_{p-1}(A, b, c, e)S^p(z)u(t). \end{aligned} \tag{19}$$

Defining the partial state $\eta(t)$ by

$$x(t) = S^{n-1}(z)\eta(t),$$

we have the following relation

$$D(z)\eta(t) = u(t), \quad N(z)\eta(t) = y(t).$$

Thus, the following equation holds:

$$\begin{aligned} S^p(z)N(z)\eta(t) &= \mathcal{O}_p(c, A)S^{n-1}(z)\eta(t) \\ &+\mathcal{T}_{p-1}(A, b, c, e)S^p(z)D(z)\eta(t). \end{aligned}$$

Premultiplying both side by $\mathcal{P}$ and noticing that the above equation is valid for an arbitrary $\eta(t)$,

$$\begin{aligned} P(z)N(z) &= \mathcal{P}\mathcal{O}_p(c, A)S^{n-1}(z) \\ &+\mathcal{P}\mathcal{T}_{p-1}(A, b, c, e)S^p(z)D(z). \end{aligned}$$

Since $\mathcal{P}\mathcal{O}_p(c, A)S^{n-1}(z)/D(z)$ is strictly proper from the definition of $S^{n-1}(z)$, eq.(18) holds from the uniqueness of the quotient and remainder. □

From eq.(6),

$$\begin{aligned} &X_d(z)D(z)/N(z) + Y_d(z) \\ &= z^{d-1}D(z)/N(z) - N_m^{-1}z^{d+\delta-1}. \end{aligned} \tag{20}$$

From the above equation, it follows that

$$\begin{aligned} &\text{strictly proper part of } z^\delta X_d(z)\frac{D(z)}{z^\delta N(z)} \\ &= \text{strictly proper part of } z^{d+\delta-1}\frac{D(z)}{z^\delta N(z)}. \end{aligned} \tag{21}$$

Let (A, b, c, e) denotes the controllability canonical realization of $D(z)/\{z^\delta N(z)\}$ (Note that $e = N_m^{-1}$). Since the characteristic polynomial of A has z^δ as a factor, A can be given by

$$A = \begin{bmatrix} A_{11} & A_{12} \\ 0 & A_{22} \end{bmatrix}, \quad \begin{aligned} &A_{11} \in \boldsymbol{R}^{\delta\times\delta}, \\ &A_{22} \in \boldsymbol{R}^{m\times m}. \end{aligned} \tag{22}$$

Moreover, since A_{11} is a nilpotent matrix, A^δ can be given by

$$\begin{aligned} &A^\delta = \begin{bmatrix} 0_{n\times\delta} & \boldsymbol{A} \end{bmatrix}, \quad \boldsymbol{A} := \begin{bmatrix} A_{12}(\delta) \\ A_{22}^\delta \end{bmatrix}, \\ &A_{12}(i) = A_{11}A_{12}(i-1) + A_{12}A_{22}^{i-1}, \\ &A_{12}(1) = A_{12}. \end{aligned} \tag{23}$$

Using the above fact, a parametrization of all stabilizing controllers for this case can be given by the following Theorem.

Theorem 1. For $X_d(z)$ and $Y_d(z)$ satisfying the Diophantine equation (6), $z^{1-d}X_d(z)$ is strictly proper and $z^{1-d}Y_d(z)$ is proper if and only if the following equation holds:

$$\begin{aligned} &\boldsymbol{X}_{d1}\mathcal{O}_{d-\delta-1}(c\boldsymbol{A}, A_{22}) \\ &= c\boldsymbol{A}A_{22}^{d-1} - \boldsymbol{X}_{d2}\mathcal{O}_{\delta-2}(c\boldsymbol{A}A_{22}^{d-\delta}, A_{22}), \end{aligned} \tag{24}$$

$$\boldsymbol{X}_{d2}\mathcal{T}_{\delta-3}(A, b, c, e) = \begin{bmatrix} cA^{\delta-2}b \cdots cb \end{bmatrix} \tag{25}$$

where

$$\begin{aligned} &\begin{bmatrix} \boldsymbol{X}_{d1} & \boldsymbol{X}_{d2} \end{bmatrix} := \mathcal{X}_{d-2}, \\ &\boldsymbol{X}_{d1} \in \boldsymbol{R}^{d-\delta}, \quad \boldsymbol{X}_{d2} \in \boldsymbol{R}^{\delta-1}. \end{aligned} \tag{26}$$

(Proof.) If eq.(6) holds, then from Lemma 1 and eq.(21),

$$\left[0_{1\times\delta}\ \mathcal{X}_{d-2} \right] \mathcal{O}_{d+\delta-2}(c,A) = \mathcal{X}_{d-2}\mathcal{O}_{d-2}(cA^\delta, A) = cA^{d+\delta-1}.$$

From eq.(23),

$$\mathcal{O}_{d-2}(cA^\delta, A) = \left[0_{(d-1)\times\delta}\ \mathcal{O}_{d-2}(c\boldsymbol{A}, A_{22}) \right]$$

and eq.(21) can be given by

$$\mathcal{X}_{d-2}\mathcal{O}_{d-2}(c\boldsymbol{A}, A_{22}) = c\boldsymbol{A}A_{22}^{d-1}.$$

Partitioning $\mathcal{X}_{d-2}$ as in eq.(26), eq.(24) holds.

Since the polynomial part of both sides in eq.(20) is equivalent, the following equation holds:

$$Y_d(z) = \left(\left[cA^{d+\delta-2}b \cdots cb \right] - \left[0_{1\times\delta}\ \mathcal{X}_{d-2} \right] \mathcal{T}_{d+\delta-3}(A,b,c,e)\right) S^{d+\delta-2}(z). \tag{27}$$

Since the degree of $Y_d(z)$ is less than d, the elements from $d+1$-th to $d+\delta$-th degree must be zero. Therefore, the eq.(25) holds.

Conversely, if eq.(24) holds, it is clear from Lemma 1 that the following equation holds:

$$\left[0_{1\times\delta}\ \mathcal{X}_{d-2} \right] \mathcal{O}_{d+\delta-2}(c,A) = \left[0_{1\times(d+\delta-1)}\ 1 \right] \mathcal{O}_{d+\delta-1}(c,A).$$

Multiplying the both sides of the above equation by $\sum_{i=1}^{\infty} A^{i-1}bz^{-i}$, it follows that

$$\begin{aligned}
&\left[0_{1\times\delta}\ \mathcal{X}_{d-2} \right] \mathcal{O}_{d+\delta-2}(c,A)\sum_{i=1}^{\infty} A^{i-1}bz^{-i} \\
&= \left[0_{1\times\delta}\ \mathcal{X}_{d-2} \right] \left(\frac{D(z)}{z^\delta N(z)} - \mathcal{T}_{d+\delta-3} \right) S^{d+\delta-2}(z) \\
&= X_d(z)D(z)/N(z) - \left[0_{1\times\delta}\ \mathcal{X}_{d-2} \right] \mathcal{T}_{d+\delta-3} S^{d+\delta-2}(z) \\
&= \left[0_{1\times(d+\delta-1)}\ 1 \right] \mathcal{O}_{d+\delta-1}(c,A)\sum_{i=1}^{\infty} A^{i-1}bz^{-i} \\
&= \left[0_{1\times(d+\delta-1)}\ 1 \right] \left(\frac{D(z)}{z^\delta N(z)} - \mathcal{T}_{d+\delta-2} \right) S^{d+\delta-1}(z) \\
&= z^{d-1}\frac{D(z)}{N(z)} - \left[cA^{d+\delta-2}b \cdots cb\ e \right] S^{d+\delta-1}(z) \\
&= z^{d-1}\frac{D(z)}{N(z)} - \left[cA^{d+\delta-2}b \cdots cb \right] S^{d+\delta-1}(z) \\
&\quad -ez^{d+\delta-1}.
\end{aligned}$$

Defining $Y_d(z)$ as in eq.(27), Diophantine equation (6) can be obtained. It is clear that the degree of $Y_d(z)$ is less than d since eq.(25) holds. □

Since $e = N_m^{-1} \neq 0$, $\mathcal{T}_{d-3}(A,b,c,e)$ is nonsingular. Thus, $\boldsymbol{X}_{d2}$ can be determined uniquely from eq.(25). On the other hand, $\boldsymbol{X}_{d1}$ can be determined if $D(z)$ and $N(z)$ are coprime. The proof of this is given in Appendix.

Corollary 1. Assume that $\delta = 1$. For $X_d(z)$ and $Y_d(z)$ satisfying the Diophantine equation (6), $z^{1-d}X_d(z)$ is strictly proper and $z^{1-d}Y_d(z)$ is proper if and only if the following equation holds:

$$\mathcal{X}_{d-2}\mathcal{O}_{d-2}(c\boldsymbol{A}, A_{22}) = c\boldsymbol{A}A_{22}^{d-1}. \tag{28}$$

It should be noted that eq.(24) and (25) hold not only for the controllability canonical form, but also for any realization of $D(z)/z^\delta N(z)$. So this method is easy to apply for the indirect adaptive control.

From the above discussions, the problem of determining a robust EMM system can be stated as the following simple optimization:

$$\text{minimize } \|\mathcal{X}_{d-2}\|_1 := \sum_{i=0}^{d-2} |X_i| \tag{29}$$

$$\begin{aligned}
\text{subject to } & \boldsymbol{X}_{d1}\mathcal{O}_{d-\delta-1}(c\boldsymbol{A}, A_{22}) \qquad (30)\\
&= c\boldsymbol{A}A_{22}^{d-1} - \boldsymbol{X}_{d2}\mathcal{O}_{\delta-2}(c\boldsymbol{A}A_{22}^{d-\delta}, A_{22}),\\
&\boldsymbol{X}_{d2}\mathcal{T}_{\delta-3} = \left[cA^{\delta-2}b \cdots cb \right]. \qquad (31)
\end{aligned}$$

However, $\boldsymbol{X}_{d2}$ can be determined by eq.(31). Thus, the above optimization problem can be written by the following linear programming problem:

$$\text{minimize } \sum_{i=0}^{d-\delta-1} (X_i^+ + X_i^-) \tag{32}$$

$$\begin{aligned}
\text{subject to } & (\boldsymbol{X}_{d1}^+ - \boldsymbol{X}_{d1}^-)\mathcal{O}_{d-\delta-1}(c\boldsymbol{A}, A_{22}) \\
&= c\boldsymbol{A}A_{22}^{d-1} - \left[cA^{\delta-2}b \cdots cb \right] \\
&\times \mathcal{T}_{\delta-3}^{-1}\mathcal{O}_{\delta-2}(c\boldsymbol{A}A_{22}^{d-\delta}, A_{22}),\\
&\boldsymbol{X}_{d1} = \boldsymbol{X}_i^+ - \boldsymbol{X}_i^-,\\
&\boldsymbol{X}_{d-2}^+ = \left[X_0^+\ X_1^+ \cdots X_{d-2}^+ \right],\\
&\boldsymbol{X}_{d-2}^- = \left[X_0^-\ X_1^- \cdots X_{d-2}^- \right],\\
&X_i^+ \geq 0, \quad X_i^- \geq 0.
\end{aligned}$$

Note that the upper bound of $J_d(\theta)$ can be reduced by increasing the value of d and using the corresponding EMM controller. In fact, define $\mathcal{X}_{d-1}$ by

$$\mathcal{X}_{d-1} = \left[0\ \mathcal{X}_{d-2}^{opt} \right] \tag{33}$$

where $\mathcal{X}_{d-2}^{opt}$ can be obtained by solving the optimization problem eqs.(29) and (30). The above $\mathcal{X}_{d-1}$ satisfies eq.(30) when d is replaced by $d+1$ since

$$\mathcal{X}_{d-1}\mathcal{O}_{d-1}(c\boldsymbol{A}, A_{22}) = \mathcal{X}_{d-2}^{opt}\mathcal{O}_{d-2}(c\boldsymbol{A}, A_{22})A_{22} = c\boldsymbol{A}A_{22}^{d}.$$

Therefore,

$$J_{d+1}(\theta) = \|\mathcal{X}_{d-1}^{opt}\|_1 \leq \|\mathcal{X}_{d-1}\|_1 = \|\mathcal{X}_{d-2}^{opt}\|_1 = J_d(\theta)$$

where $\mathcal{X}_{d-1}^{opt}$ denotes the solution of the optimization problem eqs.(29) and (30) when d is replaced by $d+1$.

4. NUMERICAL EXAMPLE

Consider the following plant $G_0(z)$ and its model $G(z)$:

$$G_0(z) = \frac{z+0.96}{z^2+1.4z+0.5},$$
$$G(z) = \frac{z+0.3}{z^2+1.5z+0.5}.$$

A pulse train function is used as a reference input. Fig.1 shows the output and reference signal of the similation result when $d=2$, which means that there is no degree of freedom in the controller parameters. The output of this case grows unboundedly. Fig.2 and 3 show the output and reference signal of the simulation result when $d=4$ and $d=8$ respectively. The outputs of both case are bounded. The amplitude of the output is slightly small in Fig.3 rather than Fig.2. Fig.4 show the poles location of each cases. In the case where $d=2$, there is a pole outside the unit circle. So the output is unbounded in this case.

5. CONCLUSIONS

In this paper, a design of robust exact model matching systems for a discrete-time plant with possible unstructured uncertainties was presented. Using a parametrization of all stabilizing controllers, the problem of maximizing robust stability region can be reduced to a simple linear programming problem. This method can be easily extended to the indirect adaptive control.

Appendix A.

Lemma 2. If $D(z)$ and $N(z)$ are coprime, then eq.(24) is solvable.

(Proof.) First, assume that $D(z)$ and $z^\delta N(z)$ are coprime. Since $(A,\ b,\ c,\ e)$ is a realization of $D(z)/z^\delta N(z)$,

$$\det \mathcal{O}_{n-1}(c, A) = \det \begin{bmatrix} \mathcal{O}_{\delta-1}(c, A) \\ 0_{m\times\delta}\ \mathcal{O}_{m-1}(c\boldsymbol{A}, A_{22}) \end{bmatrix} \neq 0$$

from eq.(23). The above equation means that $\mathcal{O}_{m-1}(c\boldsymbol{A}, A_{22})$ is nonsingular. Since the number of rows in $\mathcal{O}_{d-\delta-1}(c\boldsymbol{A}, A_{22})$ is $d-\delta = d-n+m$, which is greater than or equals to m, eq.(24) is solvable.

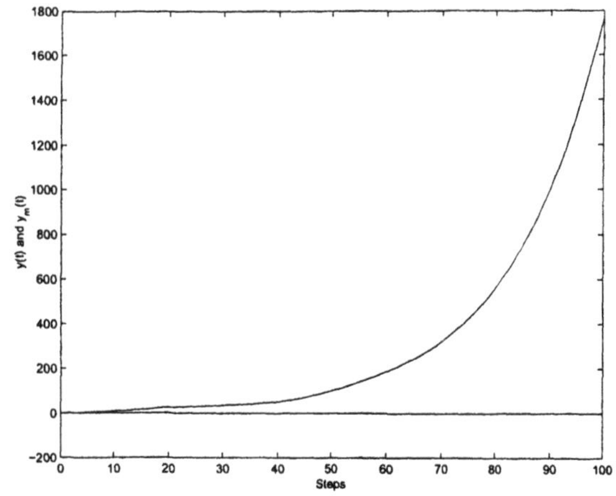

Fig. 1. Simulation of model matching controller with ($d=2$).

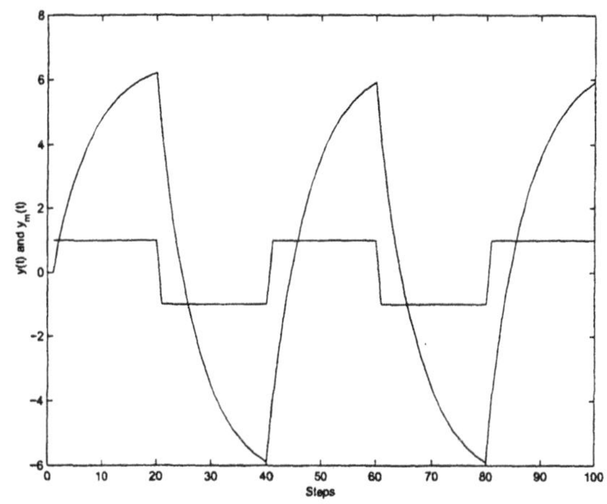

Fig. 2. Simulation result of robust model matching controller ($d=4$).

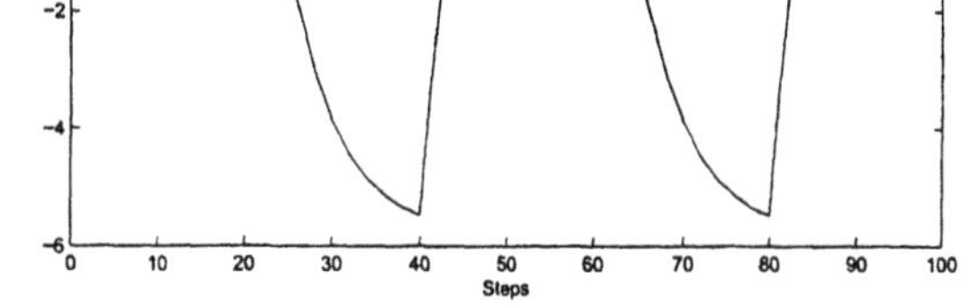

Fig. 3. Simulation result of robust model matching controller ($d=6$).

Next, assume that $D(z)$ and $N(z)$ are coprime, and $D(z)$ and $z^\delta N(z)$ are not coprime (this means $D(z)$ has i-th zeros at the origin). Assume the case where $\delta \le i \le n$. Then, $(c,\ A)$ matrix of the controllability canonical realization of $D(z)/z^{\delta}N(z)$ can be given by

$$A = \begin{bmatrix} A_{11} & A_{12} \\ 0 & A_{22} \end{bmatrix}, \quad \begin{matrix} A_{11} \in \boldsymbol{R}^{\delta\times\delta}, \\ A_{22} \in \boldsymbol{R}^{m\times m}, \end{matrix}$$
$$c = \begin{bmatrix} 0_{1\times\delta} & c_2 \end{bmatrix}.$$

Note that $c\boldsymbol{A} = c_2 A_{22}^{\delta}$ in this case. Since $D(z)$ and $N(z)$ are coprime, $\mathcal{O}_{m-1}(c_2, A_{22})$ is nonsingular. On the other hand,

$$\mathcal{O}_{d-\delta-1}(c\boldsymbol{A}, A_{22}) = \mathcal{O}_{d-\delta-1}(c_2 A_{22}^{\delta}, A_{22}) \qquad \text{(A.1)}$$
$$= \begin{bmatrix} \mathcal{O}_{m-1}(c_2, A_{22}) \\ \mathcal{O}_{d-n-1}(c_2, A_{22})A_{22} \end{bmatrix} A_{22}^{\delta}$$

holds. Since $D(z)$ and $N(z)$ are coprime, and $D(z)$ has i-th zeros at the origin, $N(z)$ has no zeros at the origin, and thus A_{22} is nonsingular. Therefore, eq.(A.1) has column full rank and eq.(24) is solvable.

Finally consider the case where $D(z)$ has i-th zeros at the origin and $0 < i < \delta$. Then, (c, A) matrix of the controllability canonical realization of $D(z)/z^{\delta}N(z)$ can be given by

$$A = \begin{bmatrix} A_{1a} & A_{1b} & 0 \\ 0 & A_{1c} & A_{1d} \\ 0 & 0 & A_{22} \end{bmatrix}, \quad \begin{matrix} A_{1a} \in \boldsymbol{R}^{i\times i}, \\ A_{1c} \in \boldsymbol{R}^{j\times j}, \\ A_{22} \in \boldsymbol{R}^{m\times m}, \end{matrix}$$
$$c = \begin{bmatrix} 0_{1\times i} & c_1 & c_2 \end{bmatrix}, \quad c_1 \in \boldsymbol{R}^{j}, \quad c_2 \in \boldsymbol{R}^{m},$$
$$j := \delta - i,$$

where $\begin{bmatrix} A_{1a} & A_{1b} \\ 0 & A_{1c} \end{bmatrix} = A_{11}$ is a nilpotent matrix. Again, A_{22} is nonsingular since $D(z)$ and $N(z)$ are coprime. Thus, A^j can be written by

$$A^j = \begin{bmatrix} A_{1a}^j & A_{1b}(j) & A_{1e}(j) \\ 0 & 0 & A_{1d}(j) \\ 0 & 0 & A_{22}^j \end{bmatrix},$$

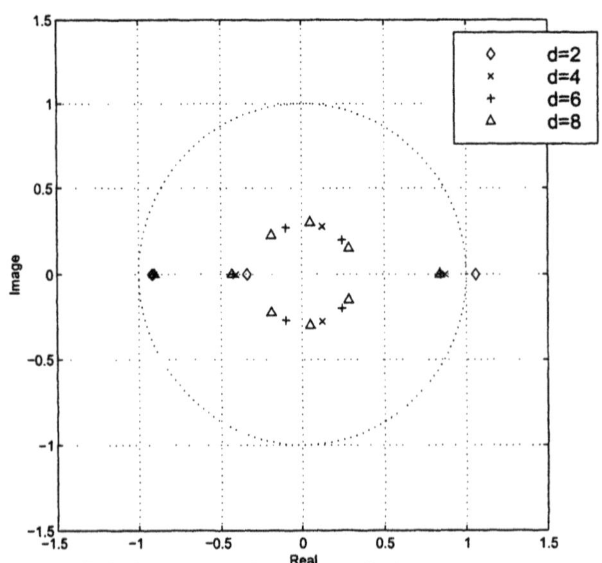

Fig. 4. Pole locations of each simulations.

$$A_{1b}(j) := A_{1a}A_{1b}(j-1) + A_{1b}A_{22}^{j-1},$$
$$\begin{bmatrix} A_{1e}(j) \\ A_{1d}(j) \end{bmatrix} := A_{11}\begin{bmatrix} A_{1e}(j-1) \\ A_{1d}(j-1) \end{bmatrix} + \begin{bmatrix} 0 \\ A_{1d} \end{bmatrix} A_{22}^{j-1},$$
$$A_{1b}(1) = A_{1b}, \quad A_{1e}(1) = 0, \quad A_{1d}(1) = A_{1d}.$$

Since $D(z)$ and $N(z)$ are coprime, $\mathcal{O}_{n-i-1}(c, A)$ has row full rank. Thus,

$$\begin{bmatrix} cA^j \\ cA^{j+1} \\ \vdots \\ cA^{n-i-1} \end{bmatrix} = \begin{bmatrix} 0 & \mathcal{O}_{m-1}(c_1 A_{1d}(j) + c_2 A_{22}^j, A_{22}) \end{bmatrix}$$

is also row full rank. It follows that $\mathcal{O}_{m-1}(c_1 A_{1d}(j) + c_2 A_{22}^j, A_{22})$ is nonsingular. From this fact and the nonsingularity of A_{22}, the solvability of eq.(24) can be obtained as in the case where $\delta \le i \le n$. □

REFERENCES

Moore, B.C., and L.M. Silverman (1972). Model matching by state feedback and dynamic compensation, *IEEE Trans. Automat. Contr.*, **AC-17**, pp.491-497.

Wolovich, W.A. (1974). *Linear multivariable system*, Springer-Verlag, Berlin.

Elliott, H., and W.A. Wolovich (1979). Parameter adaptive identification and control, *IEEE Trans. Automat. Contr.*, **24**, pp.592-599.

Goodwin, G. C., and K. S. Sin (1984). *Adaptive filtering, prediction and control*, Prentice-Hall, NJ.

Elliott, H., and W.A. Wolovich (1984). Parametrazation issues in multivariable adaptive control, *Automatica*, **20**, pp.533-545.

Mutoh, Y., and P.N. Nikiforuk (1994). Suboptimal perfect model matching for a plant with measurement nois and its application to MRACS, *IEEE Trans. Automat. Contr.*, **39**, pp.422-425.

Ortega, R. (1983). Assessment of stability robustness for adaptive controller, *IEEE Trans. Automat. Contr.*, **28**, pp.1106-1109.

Kase, W. (1999). A solution of polynomial matrix equations using extended division algorithmand another description of all-stabilizing controllers, *Int. J. Systems Science*, **30**, pp.95-104.

Identification for the moving objects by using the observed image data

Xinkai Chen* and Guisheng Zhai**
*Department of Intelligent Mechanics, Kinki University, 930 Nishimitani, Naga-gun, Wakayama 649-6493, Japan
E-mail: chen@mec.waka.kindai.ac.jp; Fax: +81-736-77-4754

**Department of Opto-Mechatronics, Wakayama University, 930 Sakaedani, Wakayama 640-8510, Japan
E-mail: zhai@sys.wakayama-u.ac.jp; Fax: +81-73-457-8187

Abstract: In this paper, we consider the problem of estimating the position of an object moving in the space by observing its image with the aid of a CCD camera. The problem can be converted into the observation of a dynamical system with nonlinearities. A new method is proposed to identify the obtained dynamical system. Simulation results show the superiority of the new method to the traditional ones. Further, the proposed method is extended to the estimation of the motion parameters.

Keywords: Nonlinear observer; Identification of Non-linear Systems; Perspective system; *a priori* knowledge; Measurement noise.

1 Introduction

Identification for a moving object in the space by the image data with the aid of a CCD camera has been studied in the past years [1,3-6,8]. A very typical method is the application of the extended Kalman filter (EKF) [4,6,8]. It is well known that the convergence cannot be guaranteed theoretically and it takes a long time to finish the identification. Further, the *a priori* knowledge about the noise is required. A fatal shortcoming of EKF is that the algorithm is very complicated and can hardly be implemented practically with real image data. To overcome these difficulties, Jankovic and Ghosh [3] proposed a new recursive formulation called "identifier based observer" (IBO) to estimate the position and shape of the object, where the idea of the parameter identifier considered in model reference adaptive control is employed (see [7]). The proposed IBO is guaranteed to converge in an arbitrarily large (but bounded) set of initial conditions, and since the convergence is exponential it is believed that the performance of IBO is reliable, robust and would quickly compute the position on real data [3]. For the EKF, no such convergence guarantees are available. It should be noted that the performance of IBO is similar to that of the EKF.

In this paper, we consider identification problem for the perspective system. The formulated problem can be converted into the observation of a dynamical system with nonlinearities. First, a new identification method is proposed to identify the space position of a moving object, where the motion parameters are assumed known. The attraction of the new method lies in that the algorithm is very simple, easy to be implemented practically, and robust to measurement noises. The adaptive algorithm with dead-zone is employed to update the upper bound of the state. Simulation results show that the new method is superior to the traditional ones. Further, the proposed formulation is extended to the parameter identification problems for a class of perspective systems.

The organization of the paper is as follows: Section 2 formulates the problem. In section 3, the robust observer is proposed for the perspective system. In section 4, examples are given to illustrate the new algorithm, and to compare the performance with the traditional methods. In section 5, the proposed method is extended to the parameter identification problems for a class of perspective systems.

2 Problem statement

Consider the movement of the object described by

$$\frac{d}{dt}\begin{bmatrix} x_1(t) \\ x_2(t) \\ x_3(t) \end{bmatrix} = \begin{bmatrix} a_{11} & a_{12} & a_{13} \\ a_{21} & a_{22} & a_{23} \\ a_{31} & a_{32} & a_{33} \end{bmatrix}\begin{bmatrix} x_1(t) \\ x_2(t) \\ x_3(t) \end{bmatrix} + \begin{bmatrix} b_1 \\ b_2 \\ b_3 \end{bmatrix} \quad (1)$$

where $x(t) = [x_1 \quad x_2 \quad x_3]^T$ is the position in the space, $a_{ij}\,(i, j = 1,2,3)$ and $b_i\,(i = 1,2,3)$ are the motion parameters (see [5]).

Also, we suppose the observed position in the image plane is defined by

$$y(t) = [y_1(t), \quad y_2(t)] = \left[\frac{x_1}{x_3}, \quad \frac{x_2}{x_3}\right] \quad (2)$$

Thus, the perspective system is composed of equations (1) and (2).

In this paper, we make the following assumptions.

(A1). The parameters $a_{ij}\,(i, j = 1,2,3)$ and b_i $(i = 1, 2, 3)$ are known.

(A2). $x_3(t)$ meets the condition $x_3(t) > \alpha > 0$, where α is a positive constant.

(A3). $x(t)$ is bounded in finite time.

Remark 1: It is easy to see that assumptions (A2) and (A3) are reasonable by referring to the practical systems.

The purpose of this paper is to estimate the position $x(t)$ by using the observed information $y_1(t)$ and $y_2(t)$ in the image plane.

3 Formulation of the observer

Define

$$y_3(t) = \frac{1}{x_3(t)} \tag{3}$$

Then, equation (1) can be rewritten as

$$\begin{cases} \dot{y}_1(t) = a_{13} + (a_{11} - a_{33})y_1 + a_{12}y_2 - a_{31}y_1^2 \\ \qquad - a_{32}y_1y_2 + (b_1 - b_3y_1)y_3 \\ \dot{y}_2(t) = a_{23} + a_{21}y_1 + (a_{22} - a_{33})y_2 - a_{31}y_1y_2 \\ \qquad - a_{32}y_2^2 + (b_2 - b_3y_2)y_3 \\ \dot{y}_3(t) = -(a_{31}y_1 + a_{32}y_2 + a_{33})y_3 - b_3y_3^2 \end{cases} \tag{4}$$

It is obvious that the position of the object in the space can be calculated as

$$x_1(t) = \frac{y_1(t)}{y_3(t)}, \quad x_2(t) = \frac{y_2(t)}{y_3(t)}, \quad x_3(t) = \frac{1}{y_3(t)} \tag{5}$$

if $y_3(t)$ is available. So, the remaining task is to estimate $y_3(t)$.

In the following, the observer of system (4) is formulated. We consider the system described by

$$\begin{cases} \dot{\hat{y}}_1(t) = a_{13} + (a_{11} - a_{33})y_1 + a_{12}y_2 - a_{31}y_1^2 - a_{32}y_1y_2 \\ \qquad + (b_1 - b_3y_1)\hat{y}_3 + \hat{\lambda}_1(t)\dfrac{e_1}{|e_1| + \delta_1} \\ \dot{\hat{y}}_2(t) = a_{23} + a_{21}y_1 + (a_{22} - a_{33})y_2 - a_{31}y_1y_2 - a_{32}y_2^2 \\ \qquad + (b_2 - b_3y_2)\hat{y}_3 + \hat{\lambda}_2(t)\dfrac{e_2}{|e_2| + \delta_2} \\ \dot{\hat{y}}_3(t) = -(a_{31}y_1 + a_{32}y_2 + a_{33})\hat{y}_3 - b_3\hat{y}_3^2 + \alpha\phi^T(t)w(t), \\ \hat{y}_3(t_i + 0) = M \cdot sign(\hat{y}_3(t_i - 0)) \end{cases} \tag{6}$$

where the initial condition is determined as

$$\hat{y}_1(0) = y_1(0), \ \hat{y}_2(0) = y_2(0), \ \hat{y}_3(0) = \hat{y}_{30}, \tag{7}$$

$\hat{y}_{30}$ is a positive constant; t_i is defined as $t_i = \min\{t : t > t_{i-1} \ and \ |\hat{y}_3| \geq 2M\}$; $M > 0$ is a very large constant; $e_1(t)$ and $e_2(t)$ are respectively defined as

$$e_1 = y_1 - \hat{y}_1, \ e_2 = y_2 - \hat{y}_2; \tag{8}$$

$\delta_i \ (i = 1,2)$ are positive constants; $\phi(t)$ and $w(t)$ are respectively defined as

$$\phi(t) = [(b_1 - b_3y_1), (b_2 - b_3y_2)]^T, \tag{9}$$

$$w(t) = \left[\hat{\lambda}_1(t)\frac{e_1}{|e_1| + \delta_1}, \hat{\lambda}_2(t)\frac{e_2}{|e_2| + \delta_2}\right]^T; \tag{10}$$

$\hat{\lambda}_1(t)$ and $\hat{\lambda}_2(t)$ are respectively defined as

$$\dot{\hat{\lambda}}_1(t) = \begin{cases} 2\alpha_1|e_1| & if \ |e_1| > \delta_1 \\ 0 & otherwise \end{cases}, \tag{11}$$

$$\dot{\hat{\lambda}}_2(t) = \begin{cases} 2\alpha_2|e_2| & if \ |e_2| > \delta_2 \\ 0 & otherwise \end{cases}, \tag{12}$$

α_1 and α_2 are positive constants, $\hat{\lambda}_1(0)$ and $\hat{\lambda}_2(0)$ can be any positive constants.

Remark 2: t_i are the discontinuous points of the observer. $\hat{y}_3(t)$ defined in (6) is bounded.

Remark 3: Since $\hat{y}_3(t)$ is bounded, it can be seen that $|(b_1 - b_3y_1)e_3|$ and $|(b_2 - b_3y_2)e_3|$ are both bounded. These bounds are adaptively updated by (11) and (12).

Now, combining (4) and (6) yields

$$\begin{cases} \dot{e}_1(t) = (b_1 - b_3y_1)e_3 - \hat{\lambda}_1(t)\dfrac{e_1}{|e_1| + \delta_1}, \quad e_1(0) = 0 \\ \dot{e}_2(t) = (b_2 - b_3y_2)e_3 - \hat{\lambda}_2(t)\dfrac{e_2}{|e_2| + \delta_2}, \quad e_2(0) = 0 \\ \dot{e}_3(t) = -(a_{31}y_1 + a_{32}y_2 + a_{33})e_3 - b_3(y_3 + \hat{y}_3)e_3 \\ \qquad - \alpha\phi^T(t)w(t) \end{cases} \tag{13}$$

where $e_3(t)$ is defined as

$$e_3 = y_3 - \hat{y}_3. \tag{14}$$

Let

$$\varepsilon_1(t) = (b_1 - b_3y_1)e_3 - \hat{\lambda}_1(t)\frac{e_1}{|e_1| + \delta_1} \tag{15}$$

and

$$\varepsilon_2(t) = (b_2 - b_3y_2)e_3 - \hat{\lambda}_2(t)\frac{e_2}{|e_2| + \delta_2}. \tag{16}$$

We have the next lemma.

Lemma 1: For $i = 1,2$, $|e_i(t)|$ and $\lambda_i(t)$ are uniformly bounded, and there exist $T_i > 0$ and $\varepsilon_i(\delta_i) > 0$ such that $|\varepsilon_i(t)| \leq \varepsilon_i(\delta_i)$ as $t > T_i$, where $\varepsilon_i(\delta_i) \to 0$ as $\delta_i \to 0$.

Proof: The proof is composed of the following two steps.

Step 1 For $i = 1,2$, $|e_i(t)|$ and $\lambda_i(t)$ are uniformly bounded, and there exist $t_i > 0$ such that $|e_i(t)| \leq \delta_i$ as $t > t_i$.

By remark 3 and the assumptions, it can be seen that $|(b_1 - b_3y_1)e_3(t)|$ is bounded. Suppose its upper bound s λ_1, i.e.

$$|(b_1 - b_3 y_1)e_3(t)| < \lambda_1 . \quad (17)$$

Now, consider the Lyapunov candidate

$$V_1(t) = (e_1(t))^2 + \frac{1}{\alpha_1}\left(\frac{1}{2}\hat{\lambda}_1(t) - \lambda_1\right)^2 \quad (18)$$

If $|e_1(t)| > \delta_1$, then differentiating $V_1(t)$ yields

$$\frac{d}{dt}\left\{(e_1(t))^2 + \frac{1}{\alpha_1}\left(\frac{1}{2}\hat{\lambda}_1(t) - \lambda_1\right)^2\right\}$$

$$= 2e_1(t)(b_1 - b_3 y_1)e_3 - 2\hat{\lambda}_1(t)\frac{e_1^2}{|e_1| + \delta_1} + 2\left(0.5\hat{\lambda}_1(t) - \lambda_1\right)|e_1|$$

$$= 2\left(e_1(t)(b_1 - b_3 y_1)e_3 - \lambda_1|e_1|\right) + 2\hat{\lambda}_1(t)\frac{|e_1|\delta_1}{|e_1| + \delta_1} - \hat{\lambda}_1(t)|e_1|$$

$$\leq -\hat{\lambda}_1(t)|e_1|\left(1 - \frac{2\delta_1}{|e_1| + \delta_1}\right)$$

$$< 0 \quad (19)$$

Thus, $V_1(t)$ decreases monotonically. Further, from (19), it can be seen that the condition $|e_1(t)| \leq \delta_1$ can be satisfied in finite time. Then, there exists $t_1 > 0$ such that $|e_1(t)| \leq \delta_1$ as $t > t_1$, and $V(t)$ is uniformly bounded for $0 \leq t \leq t_1$. Thus, $|e_1(t)|$ and $\hat{\lambda}_1(t)$ are also uniformly bounded for $0 \leq t \leq t_1$. Therefore, $|e_1(t)|$ and $\hat{\lambda}_1(t)$ are also uniformly bounded for all $t \geq 0$ by referring the definition of $\hat{\lambda}_1(t)$.

Similarly, it can be proved that $|e_1(t)|$ and $\hat{\lambda}_2(t)$ are uniformly bounded, and there exists $t_2 > 0$ such that $|e_2(t)| \leq \delta_2$ is valid for all $t > t_2$.

Step 2 $|\dot{e}_1(t)|$ and $|\dot{e}_2(t)|$ are very small after some instant.

For $t > t_1$, differentiating the first equation in (13) yields

$$\ddot{e}_1(t) = \frac{d}{dt}\left((b_1 - b_3 y_1)e_3\right) - \hat{\lambda}_1(t_1)\frac{\dot{e}_1\delta_1}{\left(|e_1| + \delta_1\right)^2} \quad (20)$$

Differentiating $(\dot{e}_1(t))^2$ yields

$$\frac{d}{dt}(\dot{e}_1(t))^2 = 2\dot{e}_1(t)\cdot\frac{d}{dt}\left((b_1 - b_3 y_1)e_3\right) - 2\hat{\lambda}_1(t_1)\frac{\dot{e}_1^2\delta_1}{\left(|e_1| + \delta_1\right)^2}$$

$$\leq 2\dot{e}_1(t)\cdot\frac{d}{dt}\left((b_1 - b_3 y_1)e_3\right) - 2\hat{\lambda}_1(t_1)\frac{\dot{e}_1^2}{4\delta_1} \quad (21)$$

where the fact $|e_1(t)| \leq \delta_1$ is employed. If $|\dot{e}_1(t)| > \frac{4\delta_1}{\hat{\lambda}_1(t_1)}\max_{t \geq t_1}\left(\left|\frac{d}{dt}\left((b_1 - b_3 y_1)e_3\right)\right|\right)$, then $\frac{d}{dt}(\dot{e}_1(t))^2 < 0$, i.e. $|\dot{e}_1(t)|$ decreases until $|\dot{e}_1(t)| \leq \frac{4\delta_1}{\hat{\lambda}_1(t_1)}\max_{t \geq t_1}\left(\left|\frac{d}{dt}\left((b_1 - b_3 y_1)e_3\right)\right|\right)$ is satisfied. Therefore, there exists an instant $T_1 \geq t_1$ such that

$$|\dot{e}_1(t)| \leq \frac{4\delta_1}{\hat{\lambda}_1(t_1)}\max_{t \geq t_1}\left(\left|\frac{d}{dt}\left((b_1 - b_3 y_1)e_3\right)\right|\right) \quad (22)$$

for all $t > T_1$.

Since $\frac{1}{\hat{\lambda}_1(t_1)}\max_{t \geq t_1}\left(\left|\frac{d}{dt}\left((b_1 - b_3 y_1)e_3\right)\right|\right)$ is bounded, it can be concluded that $|\dot{e}_1(t)|$ is very small as $t > T_1$ by choosing very small δ_1. Further, from (21), it can be seen that the decreasing speed of $|\dot{e}_1(t)|$ can be increased by choosing very small δ_1.

Similarly, it can be proved that there exists an instant $T_2 \geq t_2$ such that $|\dot{e}_2(t)|$ is very small as $t > T_2$ by choosing very small δ_2. Thus, the lemma is proved.

We have the next theorem to illustrate our new observer.

Theorem 1. Suppose there exist a positive constant β and a very small positive constant ρ such that

$$\int_t^{t+\rho} \phi^T(\tau)\phi(\tau)d\tau$$
$$= \int_t^{t+\rho}\left((b_1 - b_3 y_1(t))^2 + (b_2 - b_3 y_2(t))^2\right)d\tau \geq \beta \quad (23)$$

for all $t \geq 0$. If the parameter α is chosen large enough, $\hat{y}_3(t)$ can be thought of the approximate estimate of $y_3(t)$.

Proof: Let

$$\gamma = \sup_{t>0}\left\{(a_{31}y_1 + a_{32}y_2 + a_{33}) + b_3\left(y_3 + \hat{y}_3\right)\right\} \quad (24)$$

and

$$\varepsilon(t) = \left|2\alpha e_3(t)\phi^T(t)\begin{bmatrix}\varepsilon_1(t)\\ \varepsilon_2(t)\end{bmatrix}\right|. \quad (25)$$

From Lemma 1, it can be seen that there exists $\varepsilon(\delta_1, \delta_2) > 0$ such that

$$\varepsilon(t) \leq \varepsilon(\delta_1, \delta_2) \quad (26)$$

as $t > \max(T_1, T_2)$, where $\varepsilon(\delta_1, \delta_2) \to 0$ as $\sum_{i=1}^{2}\delta_i \to 0$.

Now, differentiating $(e_3(t))^2$ yields

$$\frac{d}{dt}(e_3(t))^2$$
$$= 2e_3(t)\left(-(a_{31}y_1 + a_{32}y_2 + a_{33})e_3 - b_3\left(y_3 + \hat{y}_3\right)e_3\right)$$
$$- 2\alpha e_3(t)\phi^T(t)w(t)$$

$$\leq 2\gamma(e_3(t))^2 - 2\alpha e_3(t)\phi^T(t)\left(\phi(t)e_3(t) - \begin{bmatrix}\varepsilon_1(t)\\ \varepsilon_2(t)\end{bmatrix}\right)$$

$$\leq -2(\alpha\phi^T(t)\phi(t) - \gamma)(e_3(t))^2 + \varepsilon(t) \tag{27}$$

Let

$$\varsigma(t) = 2(\alpha\phi^T(t)\phi(t) - \gamma). \tag{28}$$

It can be seen that there exist a positive constant μ such that

$$0 < \mu \leq \int_t^{t+\rho} \varsigma(\tau)d\tau \tag{29}$$

for all $t \geq 0$, if α is chosen as $\alpha > \dfrac{0.5\mu + \gamma\rho}{\beta}$. From the assumption (23), it can be easily seen that there exist a constant θ such that

$$\theta \leq \int_t^{t+\vartheta} \varsigma(\tau)d\tau \tag{30}$$

for all $t \geq 0$ and any ϑ satisfying $0 \leq \vartheta \leq \rho$. Let

$$s(t) = \frac{d}{dt}(e_3(t))^2 + \varsigma(t)(e_3(t))^2 - \varepsilon(t) \tag{31}$$

Then, $s(t) \leq 0$.

Solving the differential equation (31) yields

$$(e_3(t))^2 = e^{-\int_T^t \varsigma(\tau)d\tau}(e_3(T))^2 + \int_T^t e^{-\int_\iota^t \varsigma(\tau)d\tau}(s(\iota) + \varepsilon(\iota))d\iota$$

$$\leq e^{-\int_T^t \varsigma(\tau)d\tau}(e_3(T))^2 + \varepsilon(\delta_1, \delta_2)\int_T^t e^{-\int_\iota^t \varsigma(\tau)d\tau}d\iota \tag{32}$$

where $T > \max(T_1, T_2)$.

Now, express $t - T$ as $t - T = k\rho + \sigma$, where $0 \leq \sigma < \rho$. Then, we have

$$\int_T^t \varsigma(\tau)d\tau = \sum_{l=1}^{k}\int_{T+(l-1)\rho}^{T+l\rho}\varsigma(\tau)d\tau + \int_{T+k\rho}^{T+k\rho+\sigma}\varsigma(\tau)d\tau \geq k\mu + \theta. \tag{33}$$

For $T + (l-1)\rho \leq \iota < T + l\rho$, by expressing $\iota = T + (l-1)\rho + \omega$ with $0 \leq \omega < \rho$, we have

$$\int_\iota^t \varsigma(\tau)d\tau = \int_{T+(l-1)\rho+\omega}^{T+k\rho+\sigma}\varsigma(\tau)d\tau$$

$$= \begin{cases}\int_{T+(l-1)\rho+\omega}^{T+k\rho+\omega}\varsigma(\tau)d\tau + \int_{T+k\rho+\omega}^{T+k\rho+\sigma}\varsigma(\tau)d\tau & if\ \sigma \geq \omega\\ \int_{T+(l-1)\rho+\omega}^{T+(k-1)\rho+\omega}\varsigma(\tau)d\tau + \int_{T+(k-1)\rho+\omega}^{T+k\rho+\sigma}\varsigma(\tau)d\tau & if\ \sigma < \omega\end{cases}$$

$$\geq \begin{cases}(k-l+1)\mu + \theta & if\ \sigma \geq \omega\\ (k-l)\mu + \theta & if\ \sigma \geq \omega\end{cases}$$

$$\geq (k-l)\mu + \theta. \tag{34}$$

Thus,

$$\int_T^t e^{-\int_\iota^t \varsigma(\tau)d\tau}d\iota = \sum_{l=1}^{k}\int_{T+(l-1)\rho}^{T+l\rho} e^{-\int_\iota^t \varsigma(\tau)d\tau}d\iota + \int_{T+k\rho}^{t} e^{-\int_\iota^t \varsigma(\tau)d\tau}d\iota$$

$$\leq \sum_{l=1}^{k}\int_{T+(l-1)\rho}^{T+l\rho} e^{-(k-l)\mu-\theta}d\iota + \int_{T+k\rho}^{t} e^{-\theta}d\iota$$

$$\leq \rho\sum_{l=1}^{k} e^{-(k-l)\mu-\theta} + \rho e^{-\theta}$$

$$= \rho e^{-\theta}\frac{e^{-k\mu} - 2e^{\mu} + 1}{1 - e^{\mu}} \tag{35}$$

Then, substituting (33) and (35) into (32) yields

$$(e_3(t))^2 \leq e^{-k\mu}e^{-\theta}(e_3(T))^2 +$$

$$e^{-k\mu}\frac{\rho e^{-\theta}}{1-e^{\mu}}\varepsilon(\delta_1, \delta_2) + \rho e^{-\theta}\frac{1-2e^{\mu}}{1-e^{\mu}}\varepsilon(\delta_1, \delta_2), \tag{36}$$

i.e. $(e_3(t))^2$ decreases exponentially if it is not very small.

By the boundedness of the right hand side of (6), it follows that $t_i - t_{i-1}$ is greater than a positive constant, say χ (As ρ is very small, we can conclude that $\chi >> \rho$), for all i. On every such a time interval, the estimation error $(e_3(t))^2$ decreases exponentially (see (36)) and, at t_i, the estimation error $e_3(t)$ satisfies $|e_3(t_i + 0)| < |e_3(t_i - 0)|$. Therefore, the estimation error $e_3(t)$ decreases exponentially until it becomes very small. The theorem is proved.

Remark 4: The condition (23) can be thought of the observability condition for the perspective system. The condition (23) means that if $(b_1 - b_3y_1(t))^2 + (b_2 - b_3y_2(t))^2$ is very small or zero at some instant, it must increase fast enough thereafter. Further, it can be proved that if $(b_1 - b_3y_1(t))^2 + (b_2 - b_3y_2(t))^2$ is zero in some time interval, it must be identically zero. Moreover, the condition in [3] is much more strict than the condition (23) in this paper. Therefore, it can be seen that the assumption (23) is reasonable for a large class of perspective systems.

Remark 5: The design parameters $\alpha_i > 0$ and $\delta_i > 0$ ($i = 1, 2$) determine the estimating speed and the estimating precision. The parameters $\alpha_i > 0$ ($i = 1, 2$) should be chosen large enough to rapidly adjust $\hat{\lambda}_i(t)$.

Remark 6: The proposed method can be extended to the general n-dimensional perspective systems.

Remark 7: By the analysis in this section, it can be seen that the parameters a_{ij} $(i, j = 1, \cdots, n)$, b_i $(i = 1, \cdots, n)$ and c_i $(i = 1, \cdots, n)$ can be expanded to any known bounded piecewise differentiable functions of time t.

4 Example and simulation results

In this section, we present the simulation results for two examples. The simulation is done by Matlab. The sampling period is chosen as 0.05. Suppose the measured image data is corrupted by 1% with random noise.

Example 1: Consider the movement of the object

described by

$$\frac{d}{dt}\begin{bmatrix} x_1(t) \\ x_2(t) \\ x_3(t) \end{bmatrix} = \begin{bmatrix} -0.2 & 0.4 & -0.6 \\ 0.1 & -0.2 & 0.3 \\ 0.3 & -0.4 & 0.4 \end{bmatrix} \begin{bmatrix} x_1(t) \\ x_2(t) \\ x_3(t) \end{bmatrix} + \begin{bmatrix} 0.5 \\ 0.25 \\ 0.3 \end{bmatrix},$$

$$\begin{bmatrix} x_1(0) \\ x_2(0) \\ x_3(0) \end{bmatrix} = \begin{bmatrix} 1 \\ 1.5 \\ 2.5 \end{bmatrix}. \qquad (37)$$

For this example, $(b_1 - b_3 y_1(t))^2 + (b_2 - b_3 y_2(t))^2$ does not take the value zero identically. Thus, the condition (23) is satisfied.

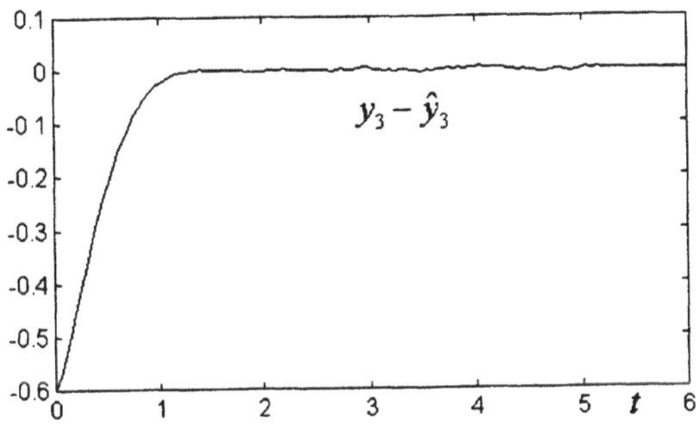

Fig. 1 The difference between y_3 and its estimate $\hat{y}_3$ by using the new observer for Example 1.

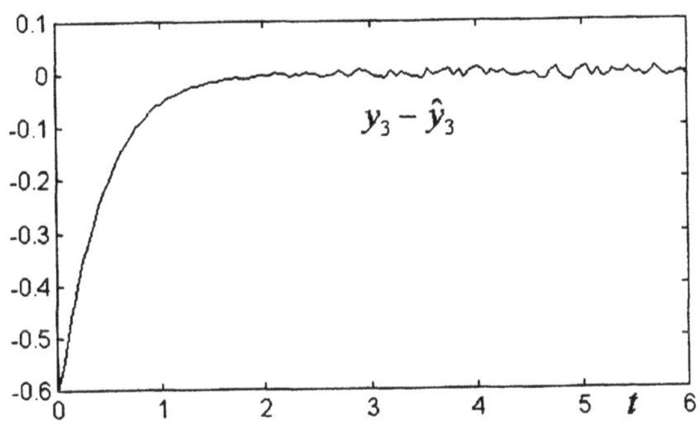

Fig. 2 The difference between y_3 and its estimate $\hat{y}_3$ by using the IBO in [3] for Example 1.

In the observer design, $\hat{y}_3(0)$ is chosen as $\hat{y}_3(0) = 1$. The parameters are chosen as $\alpha = 20$, $\alpha_1 = \alpha_2 = 5$, $\delta_1 = \delta_2 = 0.3$, $M = 100$, $\hat{\lambda}_1(0) = \hat{\lambda}_2(0) = 0.2$.

Comparison between the proposed new observer and the IBO in [3] is performed. Because of the trade-off of the estimating error and the converging speed, the estimating error is compared based on the same converging speed. Figure 1 shows the difference between y_3 and its estimate $\hat{y}_3$ by using the new observer. Figure 2 shows the difference between y_3 and its estimate $\hat{y}_3$ by using the IBO in [3]. It can be seen that the tracking error in Figure 1 is much smaller than that in Figure 2, i.e. the performance of the new observer is better than that of the IBO. Since the performance of the extended Kalman filter (EKF) is similar to that of IBO (see [3]), it can be concluded that the performance of the new observer is also better than that of the EKF. Further, it can be seen that the formulation of the new observer is much simpler than that of the IBO, not to say EKF. And the *a priori* knowledge about the noise is not needed in the new method. Therefore, it can be concluded that the new observer is superior to the traditional IBO and EKF.

Example 2: Consider the period movement of the object described by

$$\frac{d}{dt}\begin{bmatrix} x_1(t) \\ x_2(t) \\ x_3(t) \end{bmatrix} = \begin{bmatrix} 0 & -2\pi & 0 \\ 2\pi & 0 & 0 \\ 0 & 0 & 0 \end{bmatrix} \begin{bmatrix} x_1(t) \\ x_2(t) \\ x_3(t) \end{bmatrix} + \begin{bmatrix} 0 \\ 0 \\ 2\pi\cos(2\pi t) \end{bmatrix},$$

$$\begin{bmatrix} x_1(0) \\ x_2(0) \\ x_3(0) \end{bmatrix} = \begin{bmatrix} 1 \\ 1 \\ 2 \end{bmatrix} \qquad (38)$$

In the observer design, $\hat{y}_3(0)$ is chosen as $\hat{y}_3(0) = 1$. The parameters are chosen as $\alpha = 20$, $\alpha_1 = \alpha_2 = 5$, $\delta_1 = \delta_2 = 2$, $M = 100$, $\hat{\lambda}_1(0) = \hat{\lambda}_2(0) = 0.2$.

Figure 3 shows the difference between y_3 and its estimate $\hat{y}_3$ by using the new observer. It can be easily seen that

$$(b_1 - b_3 y_1(t))^2 + (b_2 - b_3 y_2(t))^2$$
$$= (2\pi\cos(2\pi t))^2\left((y_1(t))^2 + (y_2(t))^2\right)$$

takes the value zero periodically. The method proposed in [3] cannot be applied to this movement.

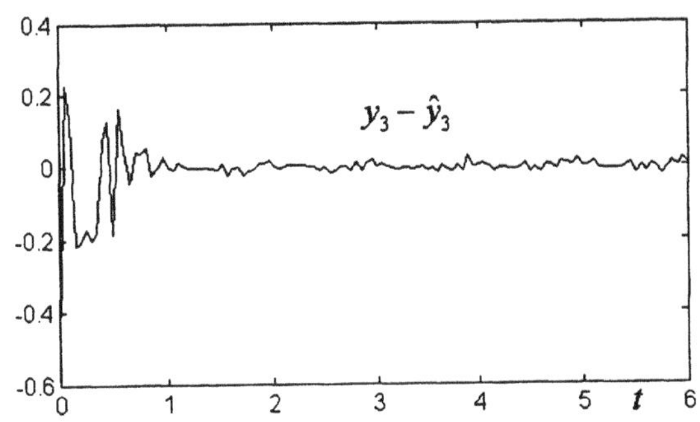

Fig. 3 The difference between y_3 and its estimate $\hat{y}_3$ by using the new observer for Example 2.

5 Motion identification for the perspective systems

In this section, we try to extend the proposed algorithm in Section 3 to the parameter estimation problems for a class of perspective systems. Consider the movement of the object described by

$$\frac{d}{dt}\begin{bmatrix} x_1(t) \\ x_2(t) \\ x_3(t) \end{bmatrix} = \begin{bmatrix} 0 & \omega_1(t) & \omega_2(t) \\ -\omega_1(t) & 0 & \omega_3(t) \\ -\omega_2(t) & -\omega_3(t) & 0 \end{bmatrix}\begin{bmatrix} x_1(t) \\ x_2(t) \\ x_3(t) \end{bmatrix} + \begin{bmatrix} b_1(t) \\ b_2(t) \\ b_3(t) \end{bmatrix} \tag{39}$$

where $x(t) = [x_1 \quad x_2 \quad x_3]^T$ is the position in the space, and $x_3(t) > \eta > 0$; $\omega_i(t)$ and $b_i(t)(i = 1,2,3)$ are the motion parameters. Further, the dynamics of $\omega_i(t)$ and $b_i(t)(i = 1,2,3)$ are known as

$$\dot{\omega}(t) = q(\omega, t), \quad \dot{b}(t) = p(\omega, t) \cdot b \tag{40}$$

The observed image data is

$$y(t) = \begin{bmatrix} y_1(t) \\ y_2(t) \end{bmatrix} = \begin{bmatrix} \frac{x_1}{x_3}, & \frac{x_2}{x_3} \end{bmatrix}^T \tag{41}$$

Purpose: Estimate the position $x(t)$ and the motion parameters by using the image data $y(t)$.

Let $y_3(t) = \frac{1}{x_3(t)}$. Then

$$\begin{bmatrix} \dot{y}_1(t) \\ \dot{y}_2(t) \end{bmatrix} = \begin{bmatrix} 1 & 0 & -y_1 & y_2 & 1+y_1^2 & y_1y_2 \\ 0 & 1 & -y_2 & -y_1 & y_1y_2 & 1+y_1^2 \end{bmatrix}\begin{bmatrix} b_1y_3 \\ b_2y_3 \\ b_3y_3 \\ \omega_1 \\ \omega_2 \\ \omega_3 \end{bmatrix}$$

$$\triangleq \phi(t) \cdot \theta(t) \tag{42}$$

$$\begin{bmatrix} \dot{\theta}_1 \\ \dot{\theta}_2 \\ \dot{\theta}_3 \end{bmatrix} = ((y_1\theta_5 + y_2\theta_6 - \theta_3)I + p(\omega, t))\begin{bmatrix} \theta_1 \\ \theta_2 \\ \theta_3 \end{bmatrix}, \tag{43}$$

$$\begin{bmatrix} \dot{\theta}_4 \\ \dot{\theta}_5 \\ \dot{\theta}_6 \end{bmatrix} = q(\omega, t) \tag{44}$$

Step 1: Estimate $\theta(t)$ based on the equations (42)-(44).

Problem: Only the product $[b_1 \quad b_2 \quad b_3]y_3$ can be estimated.

Step 2: Estimate y_3.

The stereo vision is needed to estimate y_3. Now, we introduce another camera, say Camera 2. Its image data is assumed as

$$y^*(t) = [y_1^*(t), \quad y_2(t)] = \begin{bmatrix} \frac{x_1 - a}{x_3}, & \frac{x_2}{x_3} \end{bmatrix} \tag{45}$$

Then, we have

$$\dot{y}_1^* = [1 \quad 0 \quad -y_1^* \quad y_2 \quad 1+(y_1^*)^2 \quad y_1^*y_2]\theta(t) + ay_1^*\omega_2 y_3 \tag{46}$$

$$\dot{y}_3 = (\omega_2 y_1^* + a\omega_2 + \omega_3 y_2 + b_3 y_3)y_3 \tag{47}$$

Thus, y_3 can be estimated by using the estimate of $\theta(t)$ and the equations (46) and (47).

Step 3: Estimate $[b_1 \quad b_2 \quad b_3]$.

Based on the estimates of y_3 and ω and dynamics

$$\dot{y}_1 = [1 \quad 0 \quad -y_1]y_3 \cdot b + [y_2 \quad 1+y_1^2 \quad y_1y_2]\omega(t) \tag{48}$$

$$\dot{b}(t) = p(\omega, t) \cdot b, \tag{49}$$

b can be estimated.

6 Conclusions

In this paper, firstly, we consider the position identification problem of the perspective system. where the motion parameters are assumed known. The formulated problem can be converted into the observation of a dynamical system with nonlinearties. A new observer is proposed to identify the space position of a moving object. Minor *a priori* knowledge about the system is required. The attraction of the new method lies in that the algorithm is very simple and easy to be implemented. Simulation results show the robustness to measurement noises and the superiority to the traditional ones. Further, the new method is extended to the identification problem for a class of perspective systems with unknown parameters.

References

[1] N. Ayache, *Artificial Vision for Mobile Robots, Stereo Vision and Multisensory Perception* (MIT Press, Cambridge, MA, 1991).

[2] X. Chen, T. Fukuda and K.D. Young, "A new nonlinear disturbance observer," *Systems & Control Letters*, Vol. 41, No. 2, pp. 189-199, 2000.

[3] M. Jankovic and B. K. Ghosh, "Visually guided ranging from observation of points, lines and curves via an identifier based nonlinear observer," *Systems & Control Letters*, Vol. 25, pp. 63-73, 1995.

[4] H. Kano and H. Kanai, "Recursive estimation of motion and shape parameters under perspective projection," *Proceedings of IFAC'99*, Vol. C, pp. 417-422, 1999.

[5] E.P. Loucks, *A perspective System Approach to Motion and Shape Estimation in Machine Vision*, Ph.D Thesis, Washington University, 1994.

[6] L. Matthies, T. Kanade, and R. Szeliski, "Kalman filter-based algorithm for estimating depth from image sequence," *International Journal of Computer Vision*, Vol. 3, pp. 209-236, 1989.

[7] K.S. Narendra and A.M. Annaswamy, *Stable Adaptive Systems* (Prentice-Hall, Englewood Cliffs, NJ, 1989).

[8] B. Sridhar, R. Soursa, and B. Hussein, "Passive range estimation for rotorcraft low-attitude flight," *International Journal of Machine Vision Application*, Vol. 6, pp. 10-24, 1993.

IDENTIFICATION OF A BIOMECHANICAL SYSTEM USING NEURAL NETWORKS

T. Schauer* K. J. Hunt* M. H. Fraser W. Stewart** F. Previdi*****

* *Centre for Systems and Control, University of Glasgow*
Glasgow G12 8QQ, Scotland, UK
Tel. +44-141-3306894, Fax +44-141-3304343
E-mail {t.schauer,k.hunt}@mech.gla.ac.uk
** *National Spinal Injuries Unit, Southern General Hospital*
1345 Govan Road, Glasgow G51 4FT, Scotland, UK
*** *Dipartimento di Elettronica e Informazione,*
Politecnico di Milano
Piazza Leonardo da Vinci, 32 - 20133 Milano, Italy
E-mail previdi@elet.polimi.it

Abstract: This paper describes the modelling of lower-limb dynamics of paraplegics using different neural network structures. A state-space model in form of a Recurrent Neural Network (RNN) and an input-output model involving a Multi-Layer Perceptron (MLP) have been applied to the identification of knee-joint dynamics under electrical stimulation of the quadriceps muscle group. A comparison of these black-box modelling techniques shows that both approaches are suitable for this application in order to achieve an approximation of the nonlinear system. The identification by means of the RNN is described in detail as it represents a new approach for the modelling of this class of system. Advantages of RNNs in comparison to MLPs such as simple structure selection, are highlighted. Additionally, this paper presents a very efficient and novel second-order training technique for RNNs based on the Levenberg-Marquardt optimisation method.

Keywords: Neural Networks, Biocybernetics, Electrical stimulation

1. INTRODUCTION

The development of controlled neuro-prostheses for the restoration of basic motor functions to paraplegics represents a major research area in rehabilitation engineering. These lower-limb prostheses are based on a technique called Functional Electrical Stimulation (FES), in which stimulation of paralysed muscles can generate muscle contraction (Popović and Sinkjær, 2000). FES, together with appropriate sensor and control technology, can provide useful, controlled functional movement. To design controllers for FES systems, it can be useful to have an accurate model which describes the relation between the stimulation parameters and the resulting force, moment or movement.

In this paper we are particularly interested in nonlinear modelling approaches for knee-joint dynamics while the quadriceps muscle group is electrically stimulated. The motivation for modelling this system is founded in the need for sophisticated training methods for the strengthening of the quadriceps muscles. Muscle re-strengthening is indispensable as a preliminary stage before carrying out functional tasks such as FES-cycling or FES-standing. In controlled training an arbitrary

desired cyclic motion of the shank is generated whereas the stimulation patterns are automatically adjusted by a model-based controller subject to fatigue and internal and external disturbances (Schauer *et al.*, 2001).
Black-box modelling techniques have been considered in detail as the human neuro-musculoskeletal system is subject to daily variations. A system identification based on measured open-loop data provides a better model fit to the situation. We believe that the determination of black-box models can be carried out faster than the adaptation of complicated physiologically-based models (Riener, 1999) for a new subject and/or clinical session. However there may be situations where open-loop experiments are not possible, such as standing-up, where full or semi physiologically-based models provide a great advantage.
Existing nonlinear black-box modelling solutions consist mostly of Input-Output (I/O) representations in form of Nonlinear Auto-Regression with eXogenous inputs (NARX) models (Previdi and Bittanti, 2000). This contrasts with the fact that physiologically-based models usually are developed in state-space form.
Fortunately, there always exists a NARX model which possesses an input-output behaviour equivalent to the state-space model. However, the determination of the correct number of present and past inputs as well as the number of delayed outputs in order to predict the present output may represent a complex and time consuming problem. The order of a NARX model may be up to twice that of a state space representation of the same system. Therefore, the input-output mapping of the state-space model is likely to be more nonlinear and to require a larger number of parameters. This is a particular disadvantage when only a limited amount of open-loop data for identification is available.
Recurrent neural networks (RNN) provide a useful framework for the representation and identification of state-space models (Haykin, 1999). In the past, the application of RNNs was limited due to the lack of efficient training algorithms. For this reason mostly NARX models have been investigated where the I/O mapping was approximated by a static Multi-Layer Perceptron (MLP) neural network for which a large pool of efficient training methods exist. In this paper we present a novel training technique for RNNs and identify a "real world" system using a state-space model parameterised by a RNN. Additionally, a comparison to linear and NARX modelling results is performed.

2. METHODS

The system to be modelled is the knee-joint dynamics of a paraplegic while the motion is initiated by electrical stimulation of the knee-extensor muscles. Pulsewidth modulated electrical current pulses are sent to these muscles via surface electrodes and cause contractions and knee-joint motion. The response of the output y (normalised knee-joint angle) to changes in the input u (normalised pulsewidth) is monitored for later identification. This section describes the used experimental setup as well as the investigated neural network structures.

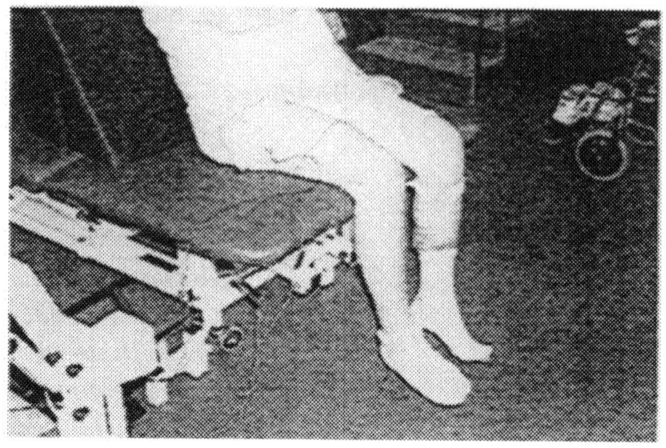

Fig. 1. Position of the subject during the experiments. The three microphones of the ultrasound motion analysis system are visible at the knee, at the ankle and at the hip. The surface electrodes are placed on the thigh. The ultrasound transmitter unit can be seen in the lower left part of the picture.

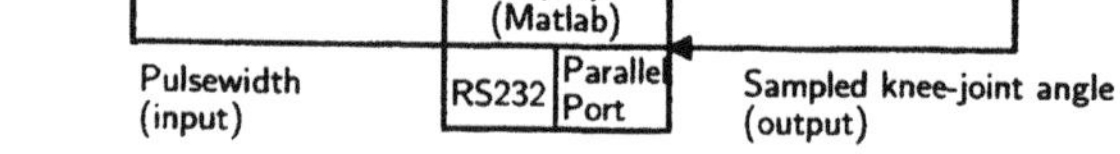

Fig. 2. Experimental set-up and definition of the knee-joint angle

2.1 *Experimental Setup*

The experimental set-up is depicted in Fig. 1 and 2. The subject was seated on a bed with the unloaded shank free to swing. The knee angle y (system output) is measured and sampled every 50 ms by a 3D-motion analysis system (zebris Medizintechnik GmbH, 1999) working with ultrasound. It delivers the angle measurements in realtime via the parallel port to a laptop. The definition of the angle is shown in Fig. 2 ($y = 0 \equiv 90\,\text{deg}$, $y = 1 \equiv 180\,\text{deg}$). The multi-channel stimulator is connected to the laptop via the serial port and delivers constant current rectangular pulses with pulsewidth pw up to 500ms. During the experiments the normalised pulsewidth serves as a variable input signal whereas the current amplitude $I = 80\,\mathrm{mA}$ and the stimulation frequency $f = 20\,\mathrm{Hz}$ are fixed. The pulsewidth at which the knee comes out of low-level saturation (rest position), denoted pw_{sat}^{l}, is taken as $u = 0$ and the pulsewidth at which the knee goes into high-level (extension) saturation, denoted as pw_{sat}^{h}, corresponds to $u = 1$. All implementation was done in MATLAB/SIMULINK using the Realtime Toolbox. A 57 year-old T10 complete paraplegic male subject, 2 years post-injury, training his quadriceps muscles for a FES-cycling project, participated in this study.

2.1.1. *Input Signal* Two successive open-loop tests have been performed using a stimulation signal where the pulsewidth is a multi-level pseudo-random sequence (m-level PRS) (Godfrey, 1993). M-level PRS inputs are periodic, deterministic signals and have an autocorrelation function similar to white noise. These inputs allow the user to highlight nonlinear system behaviour. The m-level PRSs possess 5 and 7 distinct levels in the pulsewidth spanned over the entire input range [0-1]. The first test (5 levels) served for model estimation and the second one (7 levels) for validation. Each test lasted for 40 seconds.

2.2 *Recurrent Neural Network (RNN)*

This section describes the modelling of dynamical systems using RNNs. In section 2.2.1 a general state-space RNN structure will be introduced, which covers several special RNN cases. The training of this type of network will be discussed in sections 2.2.2 and 2.2.3. Finally, the ultimate RNN structure used for identification of the biomechanical system is given in section 2.2.4.

2.2.1. *General RNN Structure* Globally recurrent neural networks have been investigated in this work. Fig. 3 shows a general recurrent network structure proposed in (Werbos, 1990) and analysed below. The network consists of 3 layers: an input layer of $n_{in}+1$ neurons, a hidden layer of n_h neurons and an output layer of n_{out} neurons. Connections between neurons can be either static (solid lines) or dynamic (bold dashed lines). A signal flowing on a static connection from the neuron j to the neuron i is weighted by a scalar w_{ij} whereas on a dynamic connection the weighting is $\tilde{w}_{ij}q^{-1}$. Here, q^{-1} is the backward shift operator, which delays a signal by one time step. The output $o_i(n)$ of the ith neuron at the time instant n is calculated from the corresponding neuron activation $a_i(n)$ by an activation function f_{act_i}:

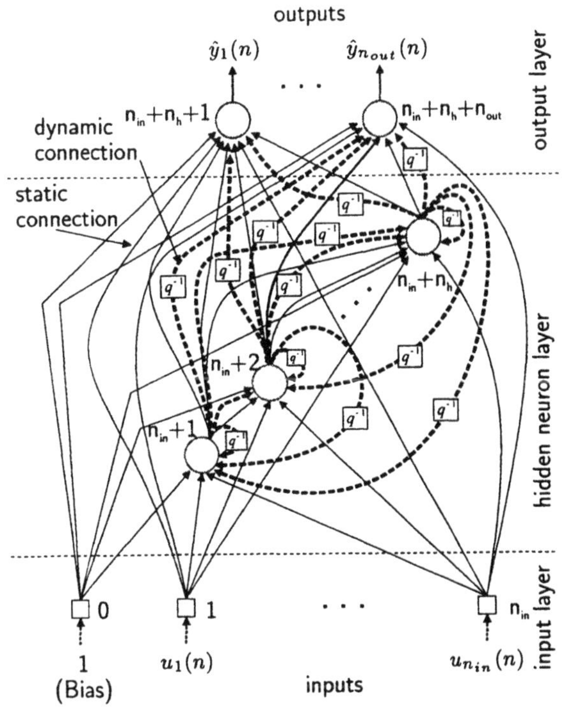

Fig. 3. General recurrent neural network structure

Table 1. Activation functions

Activation function	$o_i(n) = f_{act_i}(a_i(n))$	Layer
linear	$o_i(n) = a_i(n)$	input, output
hyperbolic tangent	$o_i(n) = \frac{e^{a_i(n)}-e^{-a_i(n)}}{e^{a_i(n)}+e^{-a_i(n)}}$	hidden

$$o_i(n) = f_{act_i}(a_i(n)) \tag{1}$$

Activation functions and their assignment to the different layers are listed in Tab. 1. Activation and output of the input neuron i is equal to the external network input u_i:

$$o_i(n) = a_i(n) = u_i(n), \quad i = 1, \cdots, n_{in} \tag{2}$$

Note that the bias of the neurons in the hidden and output layer is implemented as the weight of a static connection from an additional unity input (neuron 0):

$$o_0(n) = a_0(n) = 1 \tag{3}$$

The activation of the ith hidden neuron is calculated by

$$a_i(n) = \sum_{j=0}^{i-1} w_{ij}o_j(n) + \sum_{j=n_{in}+1}^{n_{in}+n_h} \tilde{w}_{ij}o_j(n-1), \tag{4}$$
$$i = n_{in}+1, \cdots, n_{in}+n_h+1.$$

Within the hidden layer, neurons are arranged in a chain. Each hidden neuron can only forward its output to successors in the chain. Dynamic connections within the hidden layer are not limited. The activation of an output neuron i is calculated by

$$a_i(n) = \sum_{j=0}^{n_{in}+n_h} w_{ij} o_j(n) + \sum_{j=n_{in}+1}^{n_{in}+n_h} \tilde{w}_{ij} o_j(n-1),$$
$$i = n_{in} + n_h + 1, \cdots, n_{in} + n_h + n_{out}. \quad (5)$$

There are no static or dynamic connections under the output neurons. The activation of the output neurons is a function of all neuron outputs of the input and hidden layer and the past neuron outputs of the hidden layer. The ith external network output $\hat{y}_i$ is given by the output of the neuron $n_{in} + n_h + i$:

$$\hat{y}_i(n) = o_{n_{in}+n_h+i}(n), \qquad i = 1, \cdots, n_{out}. \quad (6)$$

The dynamical behaviour of the RNN is encoded in the n_h states of the network (outputs of neurons with escaping dynamic connections):

$$x_i(n) = o_{n_{in}+i}(n), \qquad i = 1, \cdots, n_h. \quad (7)$$

The RNN may be formulated as a state-space model

$$\begin{bmatrix} \mathbf{x}(n) \\ \hat{\mathbf{y}}(n) \end{bmatrix} = \psi\left(\begin{bmatrix} \mathbf{x}(n-1) \\ \mathbf{u}(n) \end{bmatrix}, \mathbf{W}, \tilde{\mathbf{W}}, \mathbf{x}(0) \right), \quad (8)$$

where $\hat{\mathbf{y}} = [\hat{y}_1, \cdots, \hat{y}_{n_{out}}]^T$ is the vector of all external outputs, $\mathbf{u} = [u_1, \cdots, u_{n_{in}}]^T$ is the vector of all external inputs, $\mathbf{x} = [x_1, \cdots, x_{n_h}]^T$ is the vector containing the network states, $\mathbf{x(0)}$ is the initial state vector at time instant $n = 0$, $\mathbf{W} = [w_{i,j}]$ is a $(n_{in}+n_h+n_{out}+1 \times n_{in}+n_h+n_{out}+1)$ matrix describing the static connections and $\tilde{\mathbf{W}} = [\tilde{w}_{i,j}]$ is a $(n_{in}+n_h+n_{out}+1 \times n_{in}+n_h+n_{out}+1)$ matrix describing the dynamic connections.

$$\mathbf{W} = \begin{pmatrix} \mathbf{0}_{n_{in}+1 \times n_{in}+1} & \mathbf{0}_{n_{in}+1 \times n_h} & \mathbf{0}_{n_{in}+1 \times n_{out}} \\ \mathbf{W}^{IN \to H} & \mathbf{W}^{H} & \mathbf{0}_{n_h \times n_{out}} \\ \mathbf{W}^{IN \to OUT} & \mathbf{W}^{H \to OUT} & \mathbf{0}_{n_{out} \times n_{out}} \end{pmatrix} \quad (9)$$

$$\tilde{\mathbf{W}} = \begin{pmatrix} \mathbf{0}_{n_{in}+1 \times n_{in}+1} & \mathbf{0}_{n_{in}+1 \times n_h} & \mathbf{0}_{n_{in}+1 \times n_{out}} \\ \mathbf{0}_{n_h \times n_{in}+1} & \tilde{\mathbf{W}}^{H} & \mathbf{0}_{n_h \times n_{out}} \\ \mathbf{0}_{n_{out} \times n_{in}+1} & \tilde{\mathbf{W}}^{H \to OUT} & \mathbf{0}_{n_{out} \times n_{out}} \end{pmatrix} \quad (10)$$

Note that the matrix $\mathbf{W}^H$ is a lower triangular matrix with zero diagonal elements. Let us assume that all free network parameters have been arranged in one vector $\mathbf{\Theta}$. Free parameters are all non-zero elements (weights) of the matrices $\mathbf{W}$ and $\tilde{\mathbf{W}}$ and the initial state vector $\mathbf{x}(0)$. Under this assumption the dimension of the parameter vector becomes

$$\dim(\mathbf{\Theta}) = (n_{in} + 1)n_{out} + (n_{in} + 1)n_h + 2n_h n_{out}$$
$$+ n_h^2 + 1/2(n_h^2 - n_h) + n_h. \quad (11)$$

Note that later not all connections of the described network are employed and consequently the number of parameters is reduced.

2.2.2. *Levenberg-Marquardt Training* This section investigates the question of determining the parameter vector $\mathbf{\Theta}$ in such a way that the RNN is able to predict the output response of another system for a given input sequence. Usually a measured data set $\mathcal{Z}_e = \{\mathbf{u}(n), \mathbf{y}(n)\}_{n=1,\ldots,N}$ is given for the estimation of the parameters where $\mathbf{y}(n)$ is the output vector of the system to be modelled. The network's ability to approximate the system is described by a regularised quadratic performance index:

$$E(\mathbf{\Theta}) = \frac{1}{2}\mathbf{e}(\mathbf{\Theta})^T\mathbf{e}(\mathbf{\Theta}) + \frac{1}{2}\alpha\mathbf{\Theta}^T\mathbf{\Theta} \quad (12)$$
$$= \frac{1}{2}\left[\mathbf{e}(\mathbf{\Theta})^T \ \sqrt{\alpha}\mathbf{\Theta}^T\right]^T \left[\mathbf{e}(\mathbf{\Theta})^T \ \sqrt{\alpha}\mathbf{\Theta}^T\right]$$

Here, $\mathbf{e} \in \mathbb{R}^{n_{out}N}$ is the vector of all instantaneous errors $\varepsilon(n, \mathbf{\Theta})$ over a sequence with length N:

$$\varepsilon(n, \mathbf{\Theta}) = \mathbf{y}(n) - \hat{\mathbf{y}}(n, \mathbf{\Theta})$$
$$\mathbf{e}(\mathbf{\Theta})^T = \left[\varepsilon^T(\mathbf{\Theta}, 1) \ \cdots \ \varepsilon^T(\mathbf{\Theta}, N)\right]. \quad (13)$$

The desired RNN output vector at time instant n is equal to $\mathbf{y}(n)$. The regularisation term $1/2\alpha\mathbf{\Theta}^T\mathbf{\Theta}$ in (12) imposes constraints on the parameters in order to reduce the degrees of freedom in the model, where α is a positive constant. Regularisation improves the generalisation properties of the network (avoids "overfitting"). Training of the RNN means to minimise E with respect to the parameter vector $\mathbf{\Theta}$. Second order optimisation methods such as Gauss-Newton for moderate size problems or conjugate gradients for large scale problems have shown very good results for the training of multi-layer perceptrons. However, only a few attempts have been reported in the literature to apply such training methods to RNN (Li and Yan, 1995; Pedersen, 1997; Puskorius and Feldkamp, 1994). Usually, RNN are trained by gradient descent-like procedures. Such methods are highly ineffective due to long training times and lack of ability to find a good solution. We propose the integration of the Levenberg-Marquard algorithm, which originates from the Gauss-Newton method, for the training of RNNs. It should be noted, that this algorithm appears to be the fastest method for training moderate-sized feedforward neural networks (up to several hundred weights) (The MathWorks, 2000). Second order methods include first and second order derivative information of the performance index. The principle underlying second order methods is to expand the performance index (Taylor series) to second order around the current parameter vector:

$$E(\mathbf{\Theta} + \Delta\mathbf{\Theta}) = E(\mathbf{\Theta}) + \Delta\mathbf{\Theta}\nabla E(\mathbf{\Theta})$$
$$+ \frac{1}{2}\Delta\mathbf{\Theta}\nabla^2 E(\mathbf{\Theta})\Delta\mathbf{\Theta}^T, \quad (14)$$

where $\mathbf{g}(\mathbf{\Theta}) = \nabla E(\mathbf{\Theta})$ is the gradient vector of the performance index and $\mathbf{H}(\mathbf{\Theta}) = \nabla^2 E(\mathbf{\Theta})$ is the Hessian matrix (matrix of second derivatives) of the performance index. To minimise E we calculate the gradient and equate it to zero

$$\nabla E(\Theta + \Delta\Theta) = \mathbf{g}(\Theta) + \Delta\Theta\mathbf{H}(\Theta) = 0. \quad (15)$$

The best weight update is then obtained by

$$\Delta\Theta = -\mathbf{g}(\Theta)\mathbf{H}(\Theta)^{-1}. \quad (16)$$

The gradient is calculated by

$$\mathbf{g} = \left[\mathbf{e}(\Theta)^T \ \sqrt{\alpha}\Theta^T\right]^T \mathbf{J}(\Theta) \quad (17)$$

where $\mathbf{J}(\Theta)$ is the $(n_{out}N \times \dim(\Theta))$ Jacobian matrix of the vector $\left[\mathbf{e}(\Theta)^T \ \sqrt{\alpha}\Theta^T\right]^T$ with respect to the parameter vector Θ.

$$\mathbf{J} = \nabla\left[\mathbf{e}(\Theta)^T \ \alpha\Theta^T\right]^T = \begin{bmatrix} \frac{\partial \mathbf{e}(\Theta)}{\partial \Theta} \\ \sqrt{\alpha}\mathbf{I}_{\dim(\Theta)} \end{bmatrix} \quad (18)$$

Using the Gauss-Newton approximation (omitting the second order terms in $\mathbf{H}$) the Hessian matrix is approximated by

$$\mathbf{H}(\Theta) \approx \mathbf{J}(\Theta)^T\mathbf{J}(\Theta). \quad (19)$$

One problem with the Gauss-Newton method is that the approximated Hessian matrix may not be invertible. To overcome this problem in the Levenberg-Marquardt algorithm a constant μ is added such that

$$\mathbf{H}^*(\Theta) \approx \mathbf{J}(\Theta)^T\mathbf{J}(\Theta) + \mu\mathbf{I}. \quad (20)$$

The parameter update (16) is than calculated by using $\mathbf{H}^*$ (20) (instead of $\mathbf{H}$) and (17) as follows:

$$\Delta\Theta = -\begin{bmatrix} \mathbf{e}(\Theta) \\ \sqrt{\alpha}\Theta \end{bmatrix} \mathbf{J}(\Theta)\left(\mathbf{J}(\Theta)^T\mathbf{J}(\Theta) + \mu\mathbf{I}\right)^{-1} \quad (21)$$

A large constant μ reduces Eq. (21) to the steepest descent method. As μ approaches zero the method becomes Gauss-Newton's method. The parameter μ is initialised with a small value (e.g. 0.01). This value is multiplied by a constant $\mu_{dec} = 0.01$ whenever the performance index is reduced by a step. It is multiplied by a constant $\mu_{inc} = 10$ whenever a step would increase the performance index. If μ gets larger than an upper bound μ_{max}, the algorithm is stopped.

The presented training method is batch learning. The parameters are updated after the presentation of the entire input sequence of $\mathcal{Z}_e$ to the network.

2.2.3. *Dynamic Error Propagation* In order to carry out the Levenberg-Marquardt algorithm, the Jacobian matrix $\mathbf{J}(\Theta)$ has to be calculated. The required partial derivatives of the instantaneous error $\varepsilon_i(n, \Theta)$ for the ith network output with respect to the parameters w_{kl}, $\tilde{w}_{kl}$ and $x_j(0) = o_{n_{in}+j}(0)$ are obtained by dynamic error propagation (Williams and Zipser, 1989). The partial derivatives are continuously calculated forward in time while the input sequence is presented to the network. The calculation of the derivatives at time instant n uses the information about the derivatives available from the previous time step.

The equations given below for the calculation of the partial derivatives of the instantaneous error are valid for all network neurons. The weights w_{ij} and $\tilde{w}_{ij}$ have to be drawn from the matrices $\mathbf{W}$ and $\tilde{\mathbf{W}}$. The partial derivative of the instantaneous error $\varepsilon_i(n, \Theta)$ with respect to the weight w_{kl} is

$$\frac{\partial \varepsilon_i(n)}{\partial w_{kl}} = -\frac{\partial \hat{y}_i(n)}{\partial w_{kl}} = -\frac{\partial o_{n_{in}+n_h+i}(n)}{\partial w_{kl}} \quad (22)$$

with

$$\frac{\partial o_i(n)}{\partial w_{kl}} = \frac{\partial o_i(n)}{\partial a_i(n)}\left(\sum_{p=n_{in}+1}^{i-1} w_{ip}\frac{\partial o_p(n)}{\partial w_{kl}} + \sum_{p=n_{in}+1}^{n_{in}+n_h} \tilde{w}_{ip}\frac{\partial o_p(n-1)}{\partial w_{kl}} + \delta_{i,k}\frac{\partial a_i(n)}{\partial w_{il}}\right). \quad (23)$$

and

$$\frac{\partial a_i(n)}{\partial w_{il}} = \begin{cases} o_l(n) & \text{if } w_{il} \neq 0 \\ 0 & \text{otherwise.} \end{cases}$$

Here, $\delta_{i,k}$ is the Kronecker delta, which is equal to 1 for $i = k$ and zero otherwise. The partial derivative of the instantaneous error $\varepsilon_i(n, \Theta)$ with respect to the weight $\tilde{w}_{kl}$ is

$$\frac{\partial \varepsilon_i(n)}{\partial \tilde{w}_{kl}} = -\frac{\partial o_{n_{in}+n_h+i}(n)}{\partial \tilde{w}_{kl}} \quad (24)$$

with

$$\frac{\partial o_i(n)}{\partial \tilde{w}_{kl}} = \frac{\partial o_i(n)}{\partial a_i(n)}\left(\sum_{p=n_{in}+1}^{i-1} w_{ip}\frac{\partial o_p(n)}{\partial \tilde{w}_{kl}} + \sum_{p=n_{in}+1}^{n_{in}+n_h} \tilde{w}_{ip}\frac{\partial o_p(n-1)}{\partial \tilde{w}_{kl}} + \delta_{i,k}\frac{\partial a_i(n)}{\partial \tilde{w}_{il}}.\right) \quad (25)$$

and

$$\frac{\partial a_i(n)}{\partial \tilde{w}_{il}} = \begin{cases} o_l(n-1) & \text{if } \tilde{w}_{il} \neq 0 \\ 0 & \text{otherwise.} \end{cases}$$

The partial derivative of the instantaneous error $\varepsilon_i(n, \Theta)$ with respect to $x_j(0)$ is

$$\frac{\partial \varepsilon_i(n)}{\partial x_j(0)} = -\frac{\partial o_{n_{in}+n_h+i}(n)}{\partial o_{n_{in}+j}(0)} \quad (26)$$

with

$$\frac{\partial o_i(n)}{\partial o_{n_{in}+j}(0)} = -\frac{\partial o_i(n)}{\partial a_i(n)}\left(\sum_{p=n_{in}+1}^{i-1} w_{ip}\frac{\partial o_p(n)}{\partial o_{n_{in}+j}(0)} + \sum_{p=n_{in}+1}^{n_{in}+n_h} \tilde{w}_{ip}\frac{\partial o_p(n-1)}{\partial o_{n_{in}+j}(0)} + \delta_{n,1}\frac{\partial a_i(n)}{\partial o_{n_{in}+j}(0)}\right) \quad (27)$$

and

$$\frac{\partial a_i(1)}{\partial o_{n_{in}+j}(0)} = \begin{cases} \tilde{w}_{i,n_{in}+j} & \text{if } \tilde{w}_{i,n_{in}+j} \neq 0 \\ 0 & \text{otherwise.} \end{cases}$$

Note that $\partial o_i(n)/\partial a_i(n) = \partial f_{act_i}(a_i(n))/\partial a_i(n)$ is the derivative of the activation function. We assume further that the initial partial derivatives at time instant $n = 0$ are zero. The right order of updating the equations is given in the summary of the training algorithm in Tab. 2.

Table 2. Training algorithm:

Initialization:

- Set the elements of the parameter vector $\boldsymbol{\Theta}$ to small values selected from a uniform distribution.
- Set all partial derivatives $\partial o_i(0)/\partial w_{kl}$, $\partial o_i(0)/\partial \tilde{w}_{kl}$ and $\partial o_i(n)/\partial o_{n_{in}+j}(0)$ to zero.

Training:

Perform the following two steps until a desired performance index, an allowed maximum number of parameter updates, a lower bound for the gradient or an upper bound for the parameter μ is reached:

(1) Presentation of the input sequence to the network and dynamic error propagation:

for time $n = 1 \ldots, N$:
{
  for neuron $i = 0, \ldots, n_{in} + n_h + n_{out}$:
  {
  -calculate $a_i(n)$,$o_i(n)$ using (1)-(6) and $\partial f_{act_i}(a_i(n))/\partial a_i(n)$
  -calculate $\partial o_i(n)/\partial w_{kl}$,$\partial o_i(n)/\partial \tilde{w}_{kl}$ and $\partial o_i(n)/\partial o_{n_{in}+j})(0)$ using (23),(25),(27)
  }
  for output $i = 1, \ldots, n_{out}$:
  {
  -calculate $\partial \varepsilon_i(n)/\partial w_{kl}$,$\partial \varepsilon_i(n)/\partial \tilde{w}_{kl}$ and $\partial \varepsilon_i(n)/\partial x_j(0)$ using (22),(24),(26)
  -calculate $\varepsilon_i(n)$ using (13)
  }
}

(2) Update of the network parameter vector and adaptation of μ

(2.1) Build the Jacobian matrix $\mathbf{J}$ using (18)
(2.2) Calculate $\Delta\boldsymbol{\Theta}$ (21)
(2.3) Update the parameter vector
(2.4) Evaluate the performance index
(2.5) If the new performance index is better than the old one then decrease μ and go to step (1) else decrease μ, undo the parameter changes and go back to step (2.2)

2.2.4. *Ultimate RNN Structure* For the identification of the biomechanical system, a slighty simpler RNN was used. Rivals and Personnaz (1995) proposed a network structure which significantly simplifies the structure selection for the user. The proposed structure is achieved by grouping the neurons of our hidden layer into two groups. The first group consists of the hidden neurons 1 to n_c which give their output only via static connections to the following hidden and output neurons. These neurons are called *computational hidden neurons.* The next n_s hidden neurons build the second group. In contrast, these neurons possess only dynamic output connections to all hidden and output neurons. Additionally, the activation function is chosen to be linear instead of hyperbolic tangent as for the other hidden neurons. The resulting RNN is shown in Fig. 4. We denote the hidden neurons $n_c + 1$ to $n_c + n_s$ as *hidden state neurons.* For simplicity, only one external input

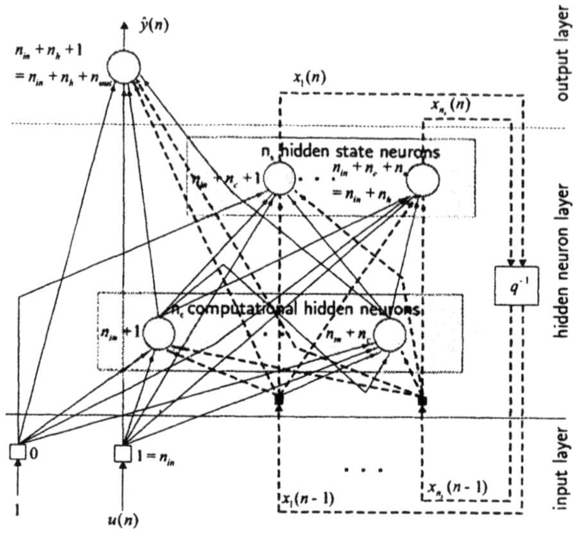

Fig. 4. Ultimate recurrent network structure

and output are shown in Fig. 4. The network is fully described by equations (1)-(6) or (8) as it represents a subclass of the RNN explained before. However several submatrices of the weight matrices $\mathbf{W}$ and $\tilde{\mathbf{W}}$ can be newly partitioned as connections vanish:

$$\tilde{\mathbf{W}}^H = \begin{bmatrix} \mathbf{0}_{n_c \times n_c} & \tilde{\mathbf{W}}^{S \to C} \\ \mathbf{0}_{n_s \times n_c} & \tilde{\mathbf{W}}^S \end{bmatrix}$$

$$\mathbf{W}^H = \begin{bmatrix} \mathbf{W}^C & \mathbf{0}_{n_c \times n_s} \\ \mathbf{W}^{C \to S} & \mathbf{0}_{n_s \times n_s} \end{bmatrix}$$

$$\mathbf{W}^{H \to OUT} = \begin{bmatrix} \mathbf{W}^{C \to OUT} & \mathbf{0}_{n_{out} \times n_s} \end{bmatrix}$$

$$\tilde{\mathbf{W}}^{H \to OUT} = \begin{bmatrix} \mathbf{0}_{n_{out} \times n_c} & \tilde{\mathbf{W}}^{S \to OUT} \end{bmatrix}$$

The dimension of the parameter vector is reduced to

$$\dim(\boldsymbol{\Theta}) = (1 + n_{in})(n_{out} + n_h) + n_s n_c + n_s^2 + 1/2(n_c^2 - n_c) + n_s n_c + n_h n_{out}. \quad (28)$$

The new state vector is given by

$$x_i(n) = o_{n_{in}+n_c+i}(n), \qquad i = 1, \cdots, n_s. \quad (29)$$

Only the number of hidden state neurons is responsible for the state vector dimension. By setting-up the RNN like this, the user can select the structure straightforwardly:

(1) **Number of hidden state neurons** n_s: Often given a-priori, e.g. number of states of a physical and physiological model.

(2) **Number of external input n_{in} and outputs n_{out}**: Predefined by the system to be modelled.
(3) **Number of computational hidden neurons n_c**: *In most cases this is the only structural parameter which has to be determined!* Usually, this is done by trial and error starting from a small value ($n_c = 1$). The nonlinearity of the vector function ψ in Eq. (8) has a strong influence on the required number of computational hidden neurons.

In this paper cross-validation was used to determine the correct numbers n_s and n_c. The trained RNN was tested on a validation data set $\mathcal{Z}_v = \{\mathbf{u}(n), \mathbf{y}(n)\}_{n=N+1,\ldots,N+N_v}$. The numbers $n_c \in \mathbb{Z}^+$ and $n_s \in \mathbb{Z}^+$ were selected in such a way that the performance index E of the RNN for the validation data set becomes minimal. Note that in the case of systems with a time delay of n_k steps, the input of the RNN will be $u(n - n_k)$ instead of $u(n)$.

2.3 *Multi-Layer Perceptron*

In the following, the procedure for the estimation of a nonlinear model based on a Multi-Layer Perceptron (MLP) neural network is presented. In this work, models with only one hidden layer have been considered for the identification of single input-single output systems. As is well known, this model family is described by the following equation

$$\hat{y}(n) = \sum_{i=1}^{n_h^{MLP}} \beta_i \tanh(\mathbf{W}^{\mathbf{MLP}} \boldsymbol{\theta}(n-1) + \mathbf{b}) \quad (30)$$

where $\mathbf{W}^{\mathbf{MLP}}$ is the $(n_h^{MLP} \times n_i^{MLP})$-dimensional matrix of the weights connecting the network input to the hidden layer; n_h^{MLP} and n_i^{MLP} are the dimensions, respectively, of the hidden layer and of the network input; $\boldsymbol{\theta}(n-1)$ is the n_i^{MLP}–dimensional input vector, with $n_i^{MLP} = n_a + n_b$, which is a collection of n_a past measurements of the output and n_b past measurements of the delayed system input $u(n - n_k)$ with n_k steps time delay and $u(n - n_k)$ itself; $\mathbf{b}$ is the n_h^{MLP}–dimensional vector of the biases of the hidden units; β_i with $i = 1, \ldots, n_h$ are the weights connecting the hidden layer to the model output. The two data sets $\mathcal{Z}_e$ and $\mathcal{Z}_v$ previously defined were used for the estimation of the parameters $\mathbf{W}^{\mathbf{MLP}}$, $\mathbf{b}$ and β_i (*parameter estimation issue*) and n_a, n_b and n_h^{MLP} (*structural identification issue*). Concerning the parameter estimation issue, the data set $\mathcal{Z}_e$ was used. A Levenberg-Marquardt algorithm was used for the parameter tuning (The MathWorks, 2000), aiming to minimise the following prediction error performance index

$$E^{MLP}(\mathcal{Z}_e, \Theta) = \frac{1}{2} \sum_{n=1}^{N} (y(n) - \hat{y}(n, \boldsymbol{\theta}, \Theta))^2 \quad (31)$$

Θ is again the set of the non-structural parameters of the neural network. It should be noted that the performance index (13) is based on a simulation of the network instead of a one step ahead prediction used in (31). The regressor vector dimension and n_h were chosen according to the cross-validation procedure proposed in (Previdi and Bittanti, 2000) using the validation data set $\mathcal{Z}_v$ and a simulation performance index.

3. RESULTS

The knee-joint dynamics were modelled by (a) a RNN (b) a MLP and (c) a linear transfer function model (ARX structure). Parameter estimation was carried out using a validation data set of $N = 801$ samples with a sampling time of $T_s = 0.05$ s. The structural identification was done by cross-validation using a validation data set of $N_v = 801$ samples. To compare the results the following simulation Mean Square Error (MSE) was used:

$$MSE(\mathcal{Z}_v, \Theta) = \frac{1}{2N} \sum_{n=1}^{N} (y(n) - \hat{y}(n))^2. \quad (32)$$

The final results of the MSE for all models is given in Tab. 3. The structural identification produced

Table 3. Model Comparison

Structure	dim(Θ)	MSE (est.)	MSE (val.)
RNN	29	$2.7 \cdot 10^{-4}$	$7.6 \cdot 10^{-4}$
MLP	28	$8.2 \cdot 10^{-4}$	$8.0 \cdot 10^{-4}$
ARX	7	$1.1 \cdot 10^{-1}$	$1.0 \cdot 10^{-1}$

the following results:

(a) RNN: $n_s = 3$,n_c=1 (for $\alpha = 0.01$)
(b) MLP: $n_a = 3$,n_b=2,$n_h^{MLP} = 3$
(c) Linear model: 3rd order model with 3 zeros

A time delay $n_k = 5$ was implemented in all models. This time delay is composed of a time delay of 2 steps by the motion analysis system and 3 steps by the biomechanical system itself. An estimation of the initial state of the RNN was omitted. By this step, the dimension of the parameter vector was reduced by the dimension of the state vector so that the number of parameters of MLP and RNN are more similar for model comparison. Fig. 5 shows the estimation data set $\mathcal{Z}_e$ and the simulated nonlinear model responses. The upper graph shows the measured normalised angle (solid line), the RNN response (bold solid line) and the MLP responses (dashed line). The lower part of the figure shows the normalised pulsewidth (m-level PRS input). In Fig. 6, the validation data set $\mathcal{Z}_v$ and the corresponding simulated nonlinear model responses are plotted.

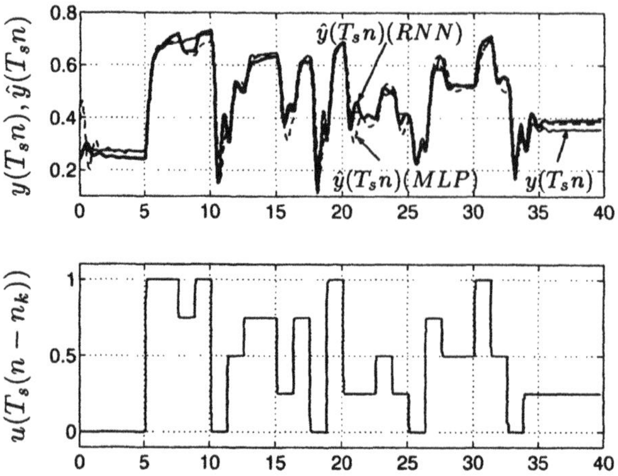

Fig. 5. Estimation data set and simulated model responses

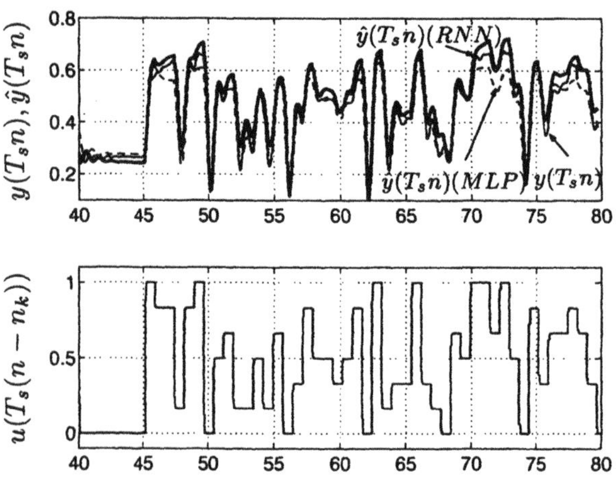

Fig. 6. Validation data set and simulated model responses

4. DISCUSSION AND CONCLUSIONS

The knee-joint dynamics under electrical stimulation of the quadriceps muscle group can successfully be modelled by a state-space model in form of RNNs and by an input-output model in form of MLPs. Both approaches show a significantly better approximation error in comparison to a linear transfer function model. The structural identification for RNNs turns out to be simpler than for MLPs. For this application, the cross-validation strategy for the MLP takes about 10 minutes whereas for the RNN only a fraction of this time is required. The proposed second order training algorithm for RNNs gives a fast and efficient way to tune the network parameters. RNNs together with sophisticated training methods represent an interesting alternative to NARX structures in form of MLPs. The advantages of RNNs will be even more important for multi input-multi output biomechanical systems.

5. REFERENCES

Godfrey, K. (1993). *Perturbation Signals for System Identification.* Prentice Hall.

Haykin, S. (1999). *Neural Networks: A Comprehensive Foundation.* 2 ed.. Prentice Hall.

Li, C.J. and L. Yan (1995). Mechanical system modeling using Recurrent Neural Networks via quasi-Newton learning methods. *Appl. Math. Model.* **19**, 421–428.

Pedersen, M. W. (1997). Optimization of Recurrent Neural Networks for Time Series Modeling. PhD thesis. Technical University of Denmark.

Popović, D. and T. Sinkjær (2000). *Control of Movement for the Physically Disabled.* Springer.

Previdi, F. and S. Bittanti (2000). Identification of Nonlinear Models of Artificial Stimulation of the Quadriceps Muscle. In: *Proc. 4th IFAC Symposium on Modelling and Control in Biomedical Systems.* Greifswald, Germany. pp. 83–89.

Puskorius, G. V. and L. A. Feldkamp (1994). Neuroncontrol of Nonlinear Dynamical Systems with Kalman Filter Trained Recurrent Networks. *IEEE Trans. on Neural Networks* **5**(2), 279–296.

Riener, R. (1999). Model-based development of neuroprostheses for paraplegic patients. *Philosophical Transactions of the Royal Society of London Series B-Biological Scienes* **354**, 877–894.

Rivals, I. and L. Personnaz (1995). Black-box Modeling with State-Space Neural Networks. In: *Neural Adaptive Control Technology I* (R. Zbikowski and K. J. Hunt, Eds.). World Scientific.

Schauer, T., K. J. Hunt, A. Ronchi, M. H. Fraser and W. Stewart (2001). Robust Control of Knee-Joint Motion. In: *Proc. 6th IFESS Conference.* Cleveland, OH.

The MathWorks, Inc. (2000). *Neural Network Toolbox User's Guide, Version 4.*

Werbos, P. J. (1990). Backpropagation through time: what it does and how to do it?. *Proc. of IEEE* **78**, 1550–1560.

Williams, R. J. and D. Zipser (1989). A learning algorithm for continually running fully recurrent neural networks. *Neural Computation* **1**, 270–280.

zebris Medizintechnik GmbH (1999). *Measuring System for 3D-Motion Analysis.* Isny im Allgäu, Germany.

DECENTRALIZED ADAPTIVE CONTINUOUS-TIME CONTROL OF COUPLED DRIVES APPARATUS

Vladimír Bobál, Karel Perůtka, Petr Dostál

*Department of Control Theory, Institute of Information Technologies,
Tomas Bata University, nám. T.G.M. 275, 762 72 Zlín, Czech Republic
Phone: +420 67 754 3217, Fax: +420 67 721 1521, E - mail: bobal@ft.utb.cz*

Abstract: A new, simple solution of decentralized adaptive continuous-time control based on a modified tuning method is presented in this paper. The process is identified by the regression (ARX) continuous–time model using the recursive least squares method (RLSM) with applied restricted directional forgetting. The method proposed has been verified by simulation in Matlab and on the CE108 Coupled Drives Apparatus using the Real Time toolbox in real time conditions. Control of the nonlinear system CE108 was performed using the suboptimal LQ tracking method applied on the external linear model (ELM). *Copyright © 2001 IFAC*

Keywords: Adaptive control; continuous time systems; decentralized control; MIMO; recursive least squares; suboptimal control

1. INTRODUCTION

The decentralized method is frequently used to control MIMO (multi-input, multi output) systems, especially when computers drive the control process. Decentralization means that the problem is broken down into several subtasks which are to a certain extent independent of the solution to the central problem. There are two main advantages to decentralization: a difficult task is divided into easier subtasks; and the resulting controller is more flexible. Vázquez, *et al.* (1999), for example, catalogued several decentralized control techniques and proposed an iterative technique for tuning decentralized PID controllers. Palmor, *et al.* (1995) proposed a sequential method of automatic tuning of decentralized PID controllers for TITO (two input, two output) processes from relay feedback. Macháček and Kotyk (1997) described a method based on sequential tuning with one no-tuning controller at the beginning of the tuning process generated from relay feedback. Wang, *et al.* (1997) introduced a different tuning technique where the tuning process is, again, generated from relay feedback, but by contrast, the local controllers are tuned simultaneously. This effect accelerates control but can only be applied to certain systems.

Adaptation is an alternative when standard controllers with fixed, though adjustable, parameters are unsatisfactory. The fundamental principles of adaptive control theory are presented in Åström and Wittenmark (1989) where different approaches are proposed and used. Successful amongst these are the self-tuning controllers (STC). The main idea on which an STC is founded is a combination of a recursive identification procedure and a choice of controller synthesis. The difference between types of STC lies mainly in the method used to design the controller, the standard being the recursive least squares method of identification (RLSM) applied to a regression model (ARX). Self-tuning systems are described by Welstead and Zarrop (1991). Bobál, *et al.* (1999) presents digital self-tuning controllers based on through computation of the PID parameters, the polynomial solution of control problems and the use of the minimization of the linear quadratic criterion. The controllers in this paper use the recursive continuous-time identification method with restricted directional forgetting (Kulhavý, 1987).

Continuous-time control uses a continuous-time identification algorithm to estimate the parameters of a continuous-time model via derivative filters or linear integral functions. Decentralized adaptive continuous-time control includes all the advantages described above. Control of this type can easily be performed using a personal computer. It has optimal behaviour and uses a continuous-time identification algorithm.

In this paper a new solution to the control of TITO systems is presented. It is based on a method modified to give decentralized adaptive continuous time control. This algorithm will first be described. The results of simulation will be shown then. This is followed by a description of the CE108 Coupled Drives Apparatus. Finally the verification results of the proposed method with order reduction obtained using this apparatus will be given.

2. THE DECENTRALIZED ADAPTIVE CONTINUOUS-TIME CONTROL ALGORITHM

Let us consider the system shown in Fig. 1, where s is the Laplace transformation operator.

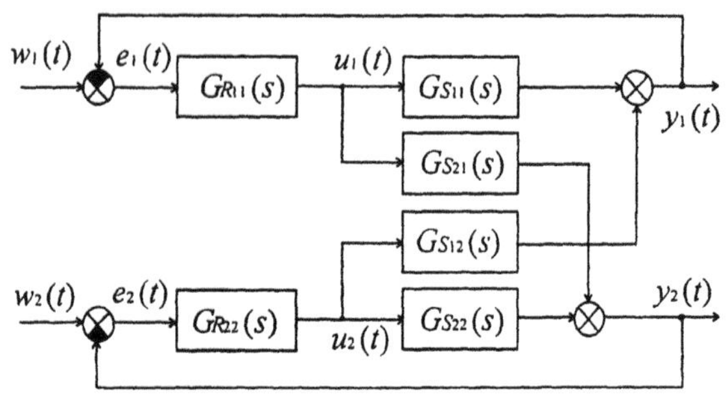

Fig. 1. Diagram of decentralized control of TITO system

The system shown in Fig. 1 has a "P" structure (Palmor, *et al.*, 1995). The approximated transfer functions

$$\bar{G}_{S11}(s) = G_{S11}(s) - \frac{G_{S21}(s)G_{S12}(s)G_{R22}(s)}{1 + G_{S22}(s)G_{R22}(s)} \quad (1)$$

$$\bar{G}_{S22}(s) = G_{S22}(s) - \frac{G_{S21}(s)G_{S12}(s)G_{R11}(s)}{1 + G_{S11}(s)G_{R11}(s)} \quad (2)$$

are derived from Fig. 1. Let us assume that a second order transfer function denominator suffices for most practical applications. This transfer function is described by

$$G_{S_{ij}}(s) = \frac{b_0^{ij}}{s^2 + a_1^{ij}s + a_0^{ij}}, i = 1;2; j = 1;2 \quad (3)$$

The approximated transfer functions have reduced their orders

$$\bar{G}_{S11}(s) = \frac{\bar{b}_0^{11}}{s^2 + \bar{a}_1^{11}s + \bar{a}_0^{11}}$$
$$\bar{G}_{S22}(s) = \frac{\bar{b}_0^{22}}{s^2 + \bar{a}_1^{22}s + \bar{a}_0^{22}} \quad (4)$$

Blocks G_{R11} and G_{R22} in Fig. 1 are controllers transfer functions. Decentralized control used in this paper is based on the method introduced by Palmor, *et al.* (1995). The controllers tune sequentially. The tuning period equals two identification periods. The self-tuning method is based on minimizing the equation error of an ARX model, and the controller parameters are observed using the suboptimal LQ tracking method. In step one the first controller operates as a self-tuning controller over two identification periods whilst the second controller is not self-tuning. In step two the second controller becomes self-tuning for two identification periods whilst the first is not self-tuning. Steps one and two are repeated during the entire control time. The parameters of G_{R11} are calculated from $\bar{G}_{S11}$ and the parameters of G_{R22} are calculated from $\bar{G}_{S22}$. Input and output are filtered before identification, as shown in Fig. 2.

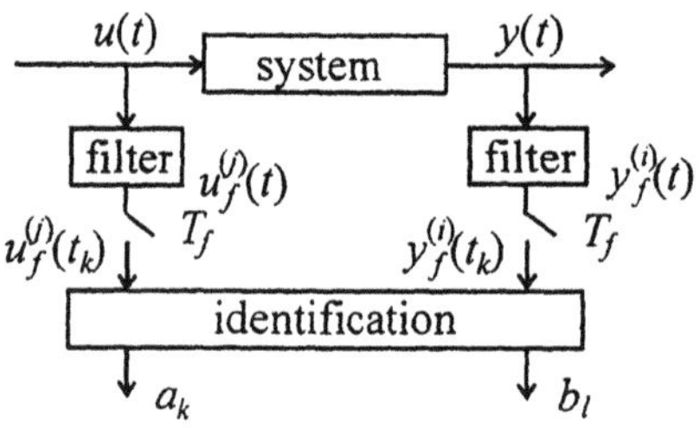

Fig. 2. Block diagram of identification in continuous-time system

In Fig. 2, T_f is the filter period, $y_f^{(i)}$ is the i-th derivation of the filtered output signal, $u_f^{(j)}$ is the j-th derivation of the filtered input signal, and a_k and b_l are the model's estimated parameters. Time domain equation

$$A(\sigma)y_f(t) = B(\sigma)u_f(t) + \varepsilon(t) \quad (5)$$

can be obtained from the filters. Expression (5) is the basic equation to estimate parameters a_k, b_l. Variable $\varepsilon(t)$ expresses the difference between filtered and unfiltered variables. Equation (6) can be expressed in the form

$$\sum_{k=0}^{n} a_k y_f^{(i)}(t) = \sum_{l=0}^{m} b_l u_f^{(j)}(t) + \varepsilon(t) \quad (6)$$

where $n = \deg a$, $m = \deg b$.

The recursive least squares method with restricted directional forgetting used in this paper is described in Bobál, *et al.* (2000). In an ARX process the regressor and vector of the parameters are

$$\begin{aligned}\phi^T(t_k) = [&-y_f(t_k), -y'_f(t_k), \ldots, -y_f^{(n-1)}(t_k),\\ &u_f(t_k), u'_f(t_k), \ldots, u_f^{(m)}(t_k), 1]\end{aligned} \tag{7}$$

$$\Theta^T(t_k) = [a_0, a_1, \ldots, a_{n-1}, b_0, b_1, \ldots, b_m, d] \tag{8}$$

The algorithm of the RLSM can usefully be extended to include adaptive directional forgetting (see Kulhavý, 1987). The value of the directional forgetting factor $\varphi(k)$ basically depends on the level of conformity achieved between the model and the real behaviour of the system. In this case let us minimize criterion

$$J_k(\Theta) = \sum_{i=k_0}^{k} \varphi^{2(k-i)} n^2(i); \quad 0 < \varphi^2 \le 1 \tag{9}$$

where k_0 is the identification start and

$$n(i) = y(i) - \Theta^T \phi(i) = \begin{bmatrix} 1 & -\Theta^T \end{bmatrix} \begin{bmatrix} y(i) \\ \phi(i) \end{bmatrix} \tag{10}$$

When the control law requires identification to provide the values of the time constants and the gain of the system these can be simply calculated from the parameter estimates of the process model using estimates obtained from the actual identified system.

Suboptimal LQ tracking was introduced by Dostál and Bobál (1999).

Let us consider Fig. 3

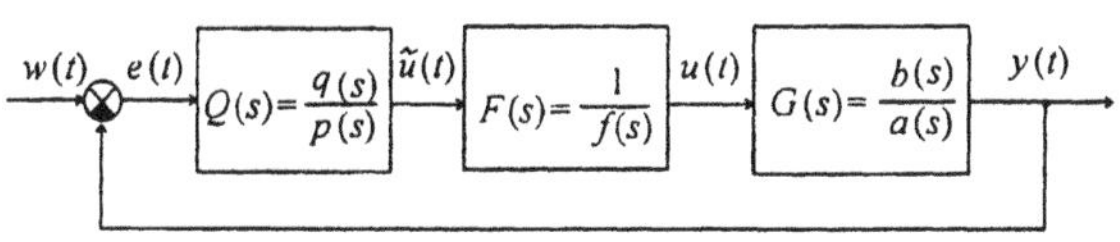

Fig. 3. System with feedback controller

Let us minimize the quadratic functional

$$J = \int_0^{\infty} \left\{ \mu e^2(t) + \varphi \tilde{u}^2(t) \right\} dt \tag{11}$$

where $\mu \ge 0, \varphi > 0$ are the penalization constants. This functional can be written in agreement with the Parseval theorem as

$$J = \frac{1}{2\pi j} \int_{-j\infty}^{j\infty} \left\{ e^*(s) \mu e(s) + \tilde{u}^*(s) \varphi \tilde{u}(s) \right\} ds \tag{12}$$

Let us suppose the Laplace form of the reference signal as

$$w(s) = \frac{h_w(s)}{s\tilde{f}_w(s)} \tag{13}$$

where

$$\deg(h_w) \le \deg(f_w), \quad f_w(0) \ne 0 \tag{14}$$

The controller synthesis with the optimal LQ tracking method gives a controller, which is practically nonproductive because of the dependence of the parameters on the initial values. Therefore, the controller synthesis method giving the suboptimal tracking of the reference signal for a group of reference signals is needed. The spectral factorization of the equations

$$(as)^* \varphi as + b^* \mu b = g^* g, \quad n^* n = a^* a \tag{15}$$

gives the coefficients of g, n which are used to solve diophantine equation

$$asp + bq = gn \tag{16}$$

from which the parameters of the controller are obtained. This equation is a reduction of the following equations

$$g^* q - v^* as = b^* \mu n, \; g^* p - v^* b = (as)^* \varphi n \tag{17}$$

where deg (v) < deg (g), supposing that $\tilde{f}_w$ does not divide a.

3. SIMULATION EXAMPLES

Let us consider the matrix

$$G_S(s) = \begin{pmatrix} \dfrac{1.4}{(21.5s+1)(25.8s+1)} & \dfrac{0.8}{(38.9s+1)(26.3s+1)} \\ \dfrac{0.1}{(17.6s+1)(15.2s+1)} & \dfrac{1.8}{(23.7s+1)(19.9s+1)} \end{pmatrix} \tag{18}$$

The initial values of the approximated transfer functions are

$$\begin{aligned} \bar{G}_{S11}(s) &= \frac{1.7}{(23.1s+1)(24.6s+1)} \\ \bar{G}_{S22}(s) &= \frac{1.9}{(24.6s+1)(20.5s+1)} \end{aligned} \tag{19}$$

Identification period T_0 is 1 s.

The constants in the quadratic functional are

$$\mu_1 = 1,\ \mu_2 = 1,\ \varphi_1 = 2,\ \varphi_2 = 2.$$

The identified values of the approximated transfer functions are

$$\overline{G}_{S11}(s) = \frac{1.5}{(34.4s+1)(15.5s+1)}$$
$$\overline{G}_{S22}(s) = \frac{2.7}{(47.8s+1)(14.8s+1)} \qquad (20)$$

Fig. 4 shows the simulated control process of the method proposed.

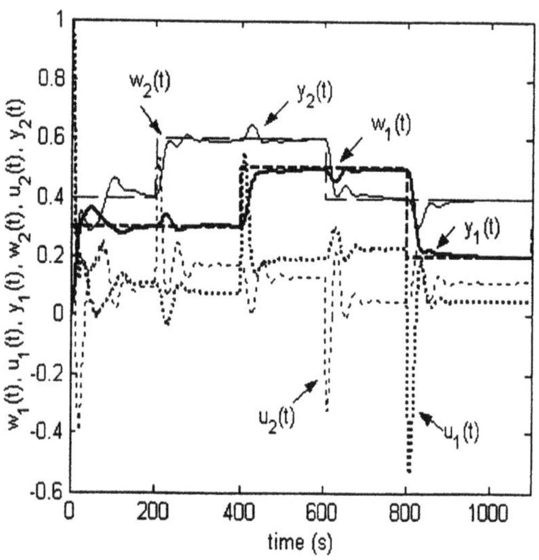

Fig. 4. Simulated control process for first example

As a second example, let us consider the matrix

$$G_S(s) = \begin{pmatrix} \dfrac{2.7}{(0.1s+1)(0.5s+1)} & \dfrac{1.2}{(0.7s+1)(0.2s+1)} \\ \dfrac{1.3}{(0.6s+1)(0.2s+1)} & \dfrac{2.9}{(0.2s+1)(0.4s+1)} \end{pmatrix} \qquad (21)$$

The initial values of the approximated transfer functions are

$$\overline{G}_{S11}(s) = \frac{1}{(2s+1)(3s+1)}$$
$$\overline{G}_{S22}(s) = \frac{1}{(2s+1)(3s+1)} \qquad (22)$$

Identification period T_0 is 1 s. The constants in the quadratic functional are

$$\mu_1 = 1,\ \mu_2 = 1,\ \varphi_1 = 2,\ \varphi_2 = 2.$$

The identified values of the approximated transfer functions are

$$\overline{G}_{S11}(s) = \frac{0.7}{(2.6s+1)(1.6s+1)}$$
$$\overline{G}_{S22}(s) = \frac{1.0}{(3.1s+1)(1.9s+1)} \qquad (23)$$

Fig. 5 shows the second example of the simulated control process of the method proposed.

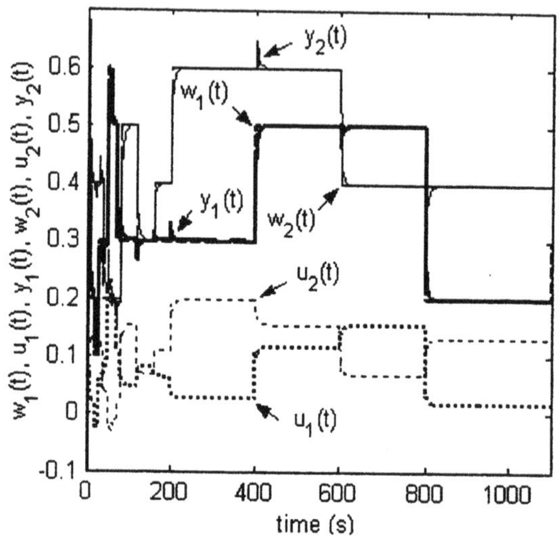

Fig. 5. Simulated control process for second example

4. A DESCRIPTION OF THE CE108 COUPLED DRIVES APPARATUS

Fig. 6 shows the front panel layout of the CE108.

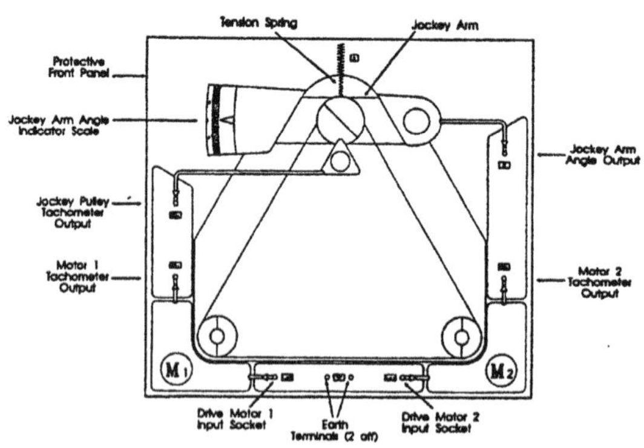

Fig. 6. Front panel layout of CE108

The apparatus consists of three pulleys. Two pulleys are mounted on the shafts of two nominally identical servomotors and the apparatus is controlled by manipulating the drive torques to these servomotors. The third pulley is free to rotate and is mounted on a pivoted arm. The jockey pulley assembly is equipped with a special sensor and tension measuring equipment. Jockey pulley speed and tension form the principle system output. The jockey pulley speed velocity measurement is derived from a pulse counting tachogenerator. The belt tension is measured indirectly by monitoring the angular deflection of the pivoted tension arm to which the jockey pulley is attached. The deflection is detected by servo potentiometer. The inputs to the apparatus are the drive voltages to the servomotor power amplifiers. Symbol u_1 in Fig. 9 represents input value to the amplifier of the first servomotor and u_2 in Fig. 7 and 9 represents input value to the amplifier of the second servomotor. Symbol y_1 in Fig. 8 and 9 is output voltage of the pulse counting tachogenerator and y_2 in Fig. 8 and 9 and θ in Fig. 7 is output voltage which represents the angular deflection of the pivoted tension arm.

Fig. 7 shows the static characteristic of the belt tension, two curves represent the minimum and maximum values during repeated measurements. Fig. 8 shows the step response of the system. The lower curve represents the belt tension and the upper one the jockey pulley speed. Input signals were $u_1 = 0$ V and $u_2 = 9.5$ V.

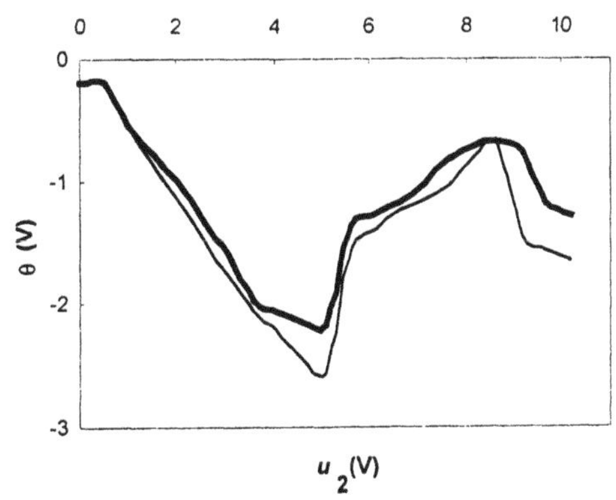

Fig. 7. Static characteristic of belt tension

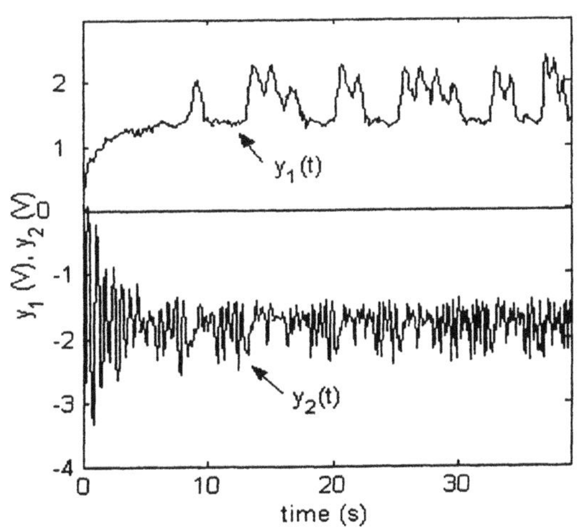

Fig. 8. Step response

5. AN EXAMPLE ON THE CE108 COUPLED DRIVES APPARATUS

The initial values of the approximated transfer functions are

$$\bar{G}_{S11}(s) = \frac{0.29}{(0.61s+1)(0.04s+1)}$$
$$\bar{G}_{S22}(s) = \frac{-0.4}{(0.7s+1)(0.16s+1)} \qquad (24)$$

Identification period T_0 is 0.1 s. The constants in the quadratic functional are

$$\mu_1 = 1,\ \mu_2 = 1,\ \varphi_1 = 2,\ \varphi_2 = 2.$$

The identified values of the approximated transfer functions are

$$\bar{G}_{S11}(s) = \frac{0.44}{(0.81s+1)(0.03s+1)}$$
$$\bar{G}_{S22}(s) = \frac{-0.4}{(0.96s+1)(0.16s+1)} \qquad (25)$$

The input signals are limited to the range 0 to 1, which represents the voltage interval from 0 to 10 Volts. Fig. 9 shows the control process of the proposed method on the CE108 Coupled Drives Apparatus.

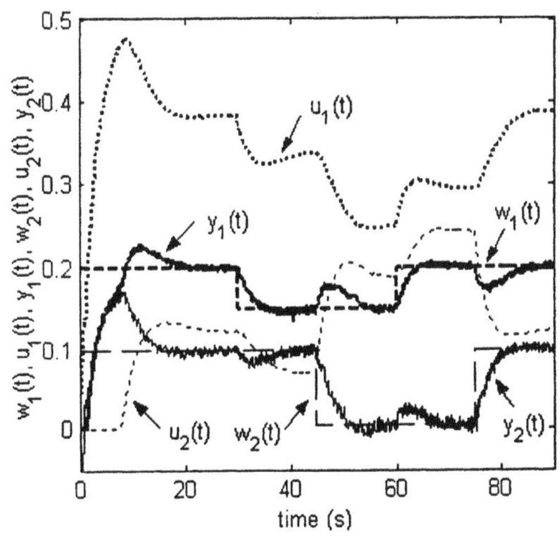

Fig. 9. Control process demonstrated on CE108 Coupled Drives Apparatus

6. CONCLUSIONS

The proposed method which is based on decentralized adaptive continuous-time control with reduced order shows very good dynamic behaviour. The advantages of this method are its simplicity and accuracy which were verified by simulation and on the CE108 Coupled Drives Apparatus. These characteristics demonstrate the usefulness of the suboptimal LQ tracking method in control, and show that a self-tuning algorithm can be used for decentralized control instead of relay feedback .

ACKNOWLEDGEMENTS

This work was supported in part by the Grant Agency of the Czech Republic under grants No.102/99/1292, 102/00/526 and in part by the Ministry of Education of the Czech Republic under grant MSM 281100001.

REFERENCES

Åström, K. J. and B. Wittenmark (1989). *Adaptive Control.* Addison-Wesley, Reading, MA.

Bobál, V., J. Böhm, R. Prokop and J. Fessl (1999). *Practical Aspects of Self-Tuning Controllers: Algorithms and Implementation.* VUTIUM Press, Brno University of Technology (in Czech).

Bobál, V., P. Dostál, J. Macháček and M. Vítečková (2000). Self-tuning PID Controllers Based on Dynamics Inversion Method. In: *Proc. IFAC Workshop on Digital Control*, University of Catalonia, Spain, 183-188.

Dostál, P. and V. Bobál (1999). The suboptimal tracking problem in linear systems. In: *7th Conference on Control and Automation*, Haifa, Israel, 667-673.

Kulhavý, R. (1987). Restricted exponential forgetting in real time identification, *Automatica*, **23**, 586-600.

Macháček, J. and J. Kotyk (1997). Decentralized control of MIMO systems. In: *Proc. 11th Conference PROCESS CONTROL '97*, **2**, pp. 134-138, Slovak University of Technology Bratislava, Tatranske Matliare (in Czech).

Palmor, Z.J., Y. Halevi and N. Krasney (1995). Automatic tuning of decentralized PID controllers for TITO processes. *Automatica*, **35**, 1011-1017.

Shen, S. H. and Ch. Ch. Yu (1994). Use of Relay-Feedback Test for Automatic Tuning of Multivariable Systems. *AIChE Journal*, **40**, 627-646.

Vázquez, F., F. Morilla and S. Dormido (1999). An iterative method for tuning decentralized PID controllers. In: *IFAC World Congress 1999 Proceedings*, Beijing, P. R. China. CD-ROM, C-2a-14-5, Elsevier Science.

Wang Q. G., B. Zou and T. H. Lee (1997). Auto-tuning of Multivariable Controllers from Decentralized Relay Feedback. *Automatica*, **33**, 319-330.

Welstead, P. E. and M. B. Zarrop. (1991). *Self-Tuning Systems: Control and Signal Processing*, Wiley, Chichester.

A SIMPLIFIED MODEL FOR HEAD-NECK SEGMENT FAST MOVEMENTS IN THE FRONTAL PLANE

A. Pedrocchi, A. Lentini, M. Locati, A. Pedotti, G. Ferrigno

Dept. Bioengineering Politecnico di Milano P.zza L. da Vinci, 33- 20133 Milan (Italy)
Bioengineering Center, Fnd Don Gnocchi &Politecnico di Milano V. Capecelatro,64 20148Milan (Italy)

Abstract

The objective of the current work is to develop a simplified model representing the physiological systems acting during fast head movements in frontal plane. All the contributions mechanically concurring to the head movement have been considered: voluntary force generation, viscosity, stiffness, physiological limit to movement range, gravitational field contribution and gamma reflex, depending on spindle activity. Model parameters have been partly derived form literature, when possible, whereas undetermined blocks parameters have been determined by optimising model output correspondence to real data acquired by ELITE™ system in specific designed experimental set-ups. The optimisation have been performed by genetic algorithms. Results show that the model studied is very well representing fast head movements in the whole range of inclination in the frontal plane. Such model is, thus, proposed as a tool for transforming kinematics data on head movements in "neural equivalent data", especially for the assessing head control deasease and planning properly the rehabilitation process. *Copyright © 2001 IFAC*

Keywords: head movements, genetic algorithms, motor control, bio-control

1. INTRODUCTION

The objective of the current work is to develop a model which, taking into account anatomical and physiological considerations, is representative of fast head movement in the frontal plane. The infra-orbital line and the head position are largely proposed in literature as primarily controlled variables by the Central Nervous System, evidences of the stability of their positions during complex movement are known (Berthoz and Pozzo, 1988; Pozzo T., *et al.*, 1990; Pozzo, *et al.*, 1991;Amblard, *et al.*, 1997; Barberini and Macpherson, 1998).

On the basis of the current knowledge in neurophysiology, it is possible to describe the control system of head movements in response to lower segments movements in the frontal plane as reported in Figure 1. The main hypothesis underlining the discussed model relates to the linear superposition of the neural signals concurring at the correction of the infraorbital line alignment with the horizontal. The considered sensorial systems concurring at the control of the neck movement by monitoring in a direct or indirect way the β angle and whose outputs are summed in s_n are (Guitton, *et al.*, 1986):

- proprioception, which is split in two different blocks H_{rp}, representing relative proprioception, depending on joints sensors, calculating the head position as the result of the kinematic chain from the lower segments, and Hap the absolute proprioception which relates with the gravity direction, which depends on graviceptors action;
- vestibular system (Bronstein 1988; Pozzo, *et al.*, 1991), which is represented by the two blocks: H_{sv} static vestibulo, representing the internal ear sensors, and H_{dv} dynamic vestibulo, representing the semicircular channels action;
- vision H_v.

In addition, there are two more contributions. First is the feedforward · control (H_{ff}), representing the anticipatory postural adjustments accompanying

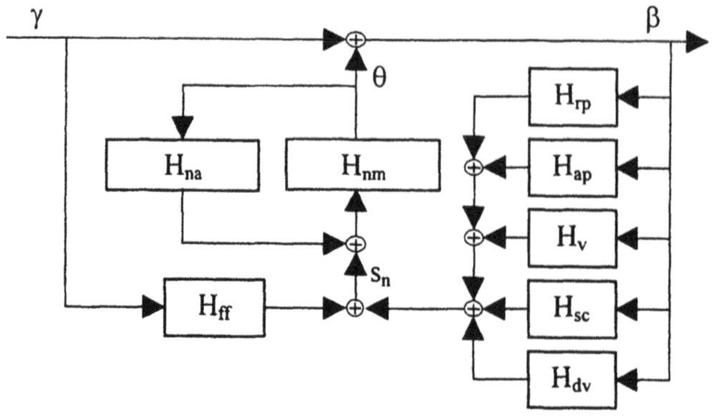

Figure 1. Systems involved in head stabilisation. Where γ is the angle between shoulders and horizontal, θ the angle between head and vertical axis, β the angle between head and shoulders. H_{na} represents the neck articular sensors H_{nm} the neuro–mechanical neck movement, H_{ff} the feed–forward control systems contribution, H_{rp} the relative proprioceptive systems, Hap the absolute proprioceptive systems H_v the visual system, H_{sc} the static vestibular system and H_{dv} the dynamic vestibular system contributions

voluntary movements involving the shoulder axis. The last block represents the neck sensors (H_{ns}), which is still part of proprioception but directly acts on the neck movement by sensing the γ angle.

In this complex scheme of control the first aim of modelling is represented by the transfer function between the global neural input s_n and the correction angle γ. The present paper describes the identification and the validation of this part of the system (H_{nm} and H_{na} in Figure 1). The present work offers a tool for any further investigation aiming at defining quantitatively the "neural contribution" of each control system by the inverse transformation of the movement performed in the neural equivalent input.

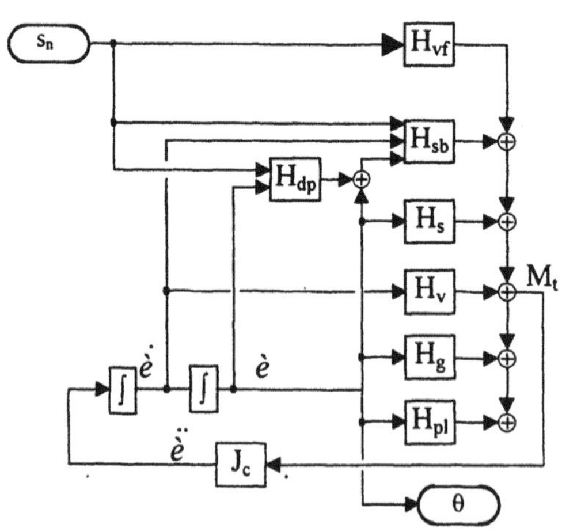

Figure 2 The proposed model. s_n is the input neural input signal, H_{vf} represents the voluntary muscles force, H_{sb} is the stabilisation block, H_s is the stiffness block, H_v is the viscosity block, H_g represents the gravity contribution, H_{pl} represents the physiological limit. H_{dp} determine the desired position, J_c is the moment of inertia of the head.

2. MODEL DESCRIPTION

The starting point in the definition of the presented model has been the six–order system introduced by (Zangemeister, *et al.*, 1981), which represents head movements in the coronal plane.

Analysing the neuro–physiological systems implicated, it has been developed a functional model, presented in Figure 2. The output variable is θ, defined as the angle between the vertical and the median axis of the head. The initial conditions are that the head is perfectly vertical and still, and it's supposed that the shoulders are fixed in the horizontal position, under these hypotheses θ is equal to γ.

The muscular force has been commanded by an equivalent neural input (Hannaford and Stark, 1985) given by algebraic sum of the activation of one right and one left equivalent muscle (Hatze, 1981). This equivalent input s_n ranges between (-1; +1). Its signed value represents the difference of the activation of the two equivalent muscles.

This signal acts directly on the H_{vf} block representing the voluntary force generation, which structure can be approximated by a 1–pole transfer function. The time constant has been imposed to be 50 msec as resulted from Stark's works (Hannaford and Stark, 1985), since the muscles involved in the investigated movements are the same (Kandel, *et al.*, 1994). The gain of this transfer function has been determined by traction experiments accomplished on two normal voluntary subjects (average age 26), which provided a maximum moment generation of 3 N·m.

In parallel to this block other mechanical contributions have been considered. First of all the stiffness (Zangemeister, *et al.*, 1981), which depends mainly on the muscles elongation and on the biological characteristics of the other tissues involved in the movement. The stiffness characteristics have been represented in the model by two separate blocks: the first one implements the physiological limit (H_{pl} block), representing the ligaments and articulations substantial effects as the inclination increases. The H_{pl} block has been modelled by an exponential function, whose two specific parameters (gain G_p and exponential E_p) were not available in the literature and thus have been identified. The second block (Hs) takes in account the linear part of the stiffness characteristic. K_s has been assumed equal to 0.5 N·m/rad as for the head rotation (Zangemeister, *et al.*, 1981).

Another considered element has been the viscosity resistance (H_ϖ) offered by the systems involved in the movement: muscles, connective tissues, ligaments and articulations (Zangemeister, *et al.*, 1981). This contribution has been modeled with a linear block depending on the head angular velocity.

The head stabilisation, once the subject reaches the desired position, has been represented with an other H_{sb} block. This function is physiologically based on the stretch reflex (Kandel, *et al.*, 1994), which originates from the muscular spindles localised within the belly of postural muscles and run in parallel with the main muscle fibres. The value the CNS assigns to the muscle length at the end of the movement depends on the sum of the facilitation and inhibition signals. Then, the spindles detect the muscle length variations: if the length increases they excite the motoneurons, if the length decreases they inhibit the motoneurons. Thus the spindles effect opposes to length variations, when a position is to be stabilized.

It has been supposed, supported by physiology (Kandel, *et al.*, 1994), that the H_{sb} block, thus the stretch reflex, is active when s_n is turn off; it has a dependence on the velocity and on the difference $\Delta\theta$ between the desired position and the actual position θ, and that it has a delay due to the line transmission for the γ–reflex.

The block H_{sb} is thus composed by a function of and $\Delta\theta$, a delay block and at the end a block analogue to the voluntary muscle force generation (H_{vf}), considering that the final effect of the γ–reflex is motoneuron activation.

To define, a memory block H_{dp} has been introduced. It works like a θ-follower when s_n is present and as a maintainer when s_n is inactive. This way it is possible to memorise the position before s_n stops and this position is assumed to be the desired position to be stabilised.

The delay Tr can be estimated through a theoretical approximate calculation, it is possible to estimate a value for Tr between 13–25 ms. Since it is a wide range and considering the numerous variable concurring at its definition, Tr will be identified together with static gain γs, for $\Delta\theta$, and dynamic gain γd.

Lastly, since the head barycentre and the rotation centre are non coincident, but move each other, differently from Zangemeister model in the horizontal plane, it must be considered the influence of the gravitational field (H_g transfer function). To determine the barycentre, the volume of a 3D-head model has been projected on the frontal plane. The head rotation centre has been localised studying the real movements geometry, it resulted to be at 99,4 mm down the line connecting the zigomatic bones in median position.

3. METHODS

3.1 Experimental activity

To validate the model, head movements on 3 voluntary subjects (average age 26) have been acquired using the ELITE system and electromyographic tracings. For the kinematic data we used four markers placed on right and left zygomatic bones and on right and left acromions, while for the electromyographic data we used four couples of surface electrodes placed on sternocleidomastoids (right and left) and on trapezius (right and left). Three protocols have been recorded: fast full-scale movements, fast movements with target angles, and stabilisation after small unpredictable perturbations. For the test related to the target angles, a white inclined bar with a black background has been projected in a dark room in front of the subject, asking him to align his eyes to the bar as fast as he could. The projected image has been chose randomly between a set of 24, 12 in clockwise direction and 12 in anticlockwise, obtained rotating the bar of five degree each time to cover the range from 0° to 60°. Small unpredictable perturbations were produced by pulling unexpected, from an external operator, a nylon wire attached on one side to a helmet on head subject and free on the other.

3.2 Identification algorithms

Genetic algorithms (GAs) have been used for the identification of the viscosity coefficient (B), the parameters of the H_{pl} (G_p and E_p, where $H_{pl}=G_p \cdot \theta^{E_p}$), the gains ($\gamma_s$ and γ_d for the different solutions studied) and delay (T_r) of H_{sb}.

GAs fit particularly well on our optimisation problem because they do not require any particular regularity on the function, they use a stochastic search method avoiding the local minima problem, they converge faster than other optimisation algorithms like Montecarlo and they can be simply adapted to the analysed problem (Davis, 1991; Holland, 1975).

The basic concept of GAs is the survival of a population of individuals. An individual is characterised by his genetic patrimony, represented by the full set of parameters to be determined. Like in nature the individual himself is the expression of his genes, in GA the genotype is represented from the unknown parameters and the phenotype is the output related to a particular genotype.

GAs provide to maintain a set of individuals for some generations. For each generation, every individual phenotype is evaluated, assigning him a fitness value, which represents his adaptation to the environment. Once the GAs have evaluated the actual generation, a new generation is created from the old one using a set of stochastic rules, applied on the fitness values of the previous generation, aimed at generating a set of individual candidates to survive. This process is executed until there's no fitness increment for a fixed amount of generations or until a maximum number of generations. The creation of the new generation follows two steps:

1. the selection of the individuals from the old generation for the creation of the new one
2. the implementation of the genetic operators to modify some of the selected individuals to create new individuals for the next generation

For the selection process, a probability to survive is assigned to each individual, basing on its fitness value. All the individuals are listed by decreasing fitness. The first individual, with the highest fitness value, has a probability q to survive for the next generation while to the i–th individual has a probability Pi to survive according to the follow rule:

$$P_i = q*(1-q)^{n_i-1} \Big/ 1-(1-q)^N \quad (1)$$

Where N is the number of individuals in the generation, n_i is the position of the i–th individual in the list, q is the survival probability assigned to the best individual (n_i=1). The probability q is chose basing the specific problem.

To each individual is associated an interval, whose amplitude is relative to its survival probability P_i such that all the individuals survey the interval between 0 and 1. Generating a random variable uniformly distributed from 0 to 1 a particular interval is identified and its relative individual is chose to be part of the new generation. Genetic operators are applied to some of the selected individuals. There are two classes of genetic operators: recombinations, that modify reciprocally the genotypes of two individuals, and mutations, which change the genotype of the singular individual.

We used all the genetic operators and for each one the same number of individuals is involved.

4. MODEL IDENTIFICATION

The identification process has been divided in two steps: in the first it has been identified the H_v and H_{pl} blocks, in the second the H_{sb} block. The fast full-scale movements have been used to identify the H_v and H_{pl}, hence it has been supposed that, during the raising phase sn is always present, consequently the H_{sb} gives null contribution and can be neglected. This hypothesis has been verified by the analysis of the electromyographic data. EMG recorded for the fast full range movements showed during all the raising front of the movement only the omolateral muscles with significant electric activities.

The movements after small unpredictable perturbations have been used for the identification of the structure and optimal parameter values of H_{sb} block. It has been supposed that being the movements small and unpredictable there isn't any voluntary force generation, i.e. s_n is null, thus the return in the equilibrium position is due only to the stretch-reflex process.

4.1 Determination of the H_v and H_pl blocks

For the identification process we used 43 of the 105 fast full scale movements acquired on all the three subjects, limiting the analysis only to the raising phase, as previously discussed. Movement onset was defined on the leading edge of angular velocity at 0.05 rad/sec; raising end on the trailing edge when velocity returns equal to 0.05 rad/sec.

To synchronise and normalise the curves, they have been resized in time and amplitude. From such processed data, the mean curve and its standard deviation has been calculated.

The input s_n was been drawn from considerations on EMG tracings: we processed the EMGs recorded for fast full–scale movements and we calculated a reference EMG. This reference EMG has been assumed equal to the contribution of the muscular force in the model, i.e. the output of H_{vf}. Hence the corresponding s_n has been estimated with a three normalised rectangle pattern (rect1:amplitude 1 duration 0.14 sec; rect2: amplitude 0.5 duration till 0.30 sec (Δt=0.16 sec); rect3: amplitude 0.22 duration till 0.52 sec (Δt=0.22 sec)).

The fitness η_j for the j-th individual (i.e. the j-th triplet of possible solutions) has been defined as:

$$\eta_i = \sum_i^N d(cj(i);m(i)) \Big/ std(i) \quad (2)$$

Where N is the number of samples; cj(i) is the i-th sample of the j-th output curve corresponding to the j-th individual; m(i) is the i-th sample of the mean curve calculated from the experimental data and std(i) is the i-th sample of the standard deviation determined by the experimental data. d(·,·) is the distance function.

The genotype of the j–th individual is defined by the triplet (Bj, G_pj, E_pj) which corresponds to a phenotype represented by the cj curve.

To fix the number of individual for teach generation and the number of generations needed before the GA ended a preliminary study has been done. We fixed the number of individual equal to 50 and we studied the results of 20 independent GAs let evolve for 200 generations. This way, the fitness improved very fast during the first 30 generations and it increased slower after them. This suggested that the GA was stabilised only after a large number of generations and it didn't guarantee to reach the absolute minimum. To solve this problem, we chose an hybrid implementation: we calculated more independent GAs each one composed of 50 individuals, let them evolve for 100 generations, then we plotted the distribution of the best individuals determined from each GA (one individual for each GA) visualising the trade of the fitness in the space of research. Very broad parameter ranges were initially defined; successive restrictions focused the analysis on the basis of previous cycles results, till the determination of the optimum value for the unknown parameters.

In this way we determined the minimum and maximum edges for the researched parameters. With the last cycle, we determined the optimal value for the unknown parameters: G_p= 79.432 N·m/rad, E_p= 5.394, B= 1.707 N·m/rad, which corresponds a fitness equal to 3.152, corresponding to a mean error for each sample equal to 2.9% with respect to the mean real data curve.

4.2 Determination of the H_{sb} block

Once the viscous and physiological limit blocks have been determined, we proceeded with the second step

of the identification process. In this phase, the hypothesis made is that there isn't any voluntary signal, i.e. sn is null. This is assumed true since we considered small and unpredictable perturbations to head position.
To build the mean curve needed for the identification process we considered the data recorded for stabilisation movement after a small external unpredictable perturbation. The curves have been translated to have all final angles equal to zero, representing the occurred stabilisation. We considered the data only from the first minimum after the first fast return phase to end of the movement.
We implemented GAs similarly to the first identification phase, i.e. a hybrid solution, using the same fitness η. The space of research for the unknown parameters has been fixed by a preliminary consideration: for γ_s and γ_d we tested only the value for which the output moment was of the same order of the voluntary moment, while Tr range has been imposed 13–25 ms as expected by physiological considerations.
The optimal identified value are: γ_s=25 N·m/rad, γ_d=1.704 N·m·sec/rad and Tr= 0.014 sec which corresponds a fitness of 0.069 equal to an mean error for each sample of 0.15% with respect to the mean curve of real data.

5. MODEL VALIDATION

5.1 Model validation for fast full scale movements

The identification process let us determine the optimal value for all the unknown parameters of the model for the two phases of the movement considered. To validate the goodness of the found values we compared the complete model output with a control group of fast full scale movements acquired by the ELITE™ system. The comparison is performed using each of the 103 fast full-scale curves for their total duration.
To have a consistent term of comparison, all the curves were resized so that they had initial angle equal to zero, and we cut them so that they have 1.5 s duration, when all the curves are stabilised. For this validation phase the same input signal as the first phase of the identification, derived by EMG, was used. A very good fitness of the model over the fast full-scale movements is quantitatively confirmed by the square correlation coefficient (r2=0.955±0.053, N=103).
Model validation for fast full range movements is confirmed also by the analysis of two more parameters: the time to reach the maximum angle, and the absolute value of the reached maximum angle. The model presents a mean time to reach the maximum angle equal to 0.24 sec while the median value calculated from the experimental data–set is 0.28 sec and mean value is 0.31 sec. About the maximum reached angle the model shows a mean output equal to 34° while the mean value from experimental data is 33.8°.

5.2 Model generalisation

To generalise the model validity it has been studied its capability to represent not full-scale range fast movements too. For this purpose, the data for fast movements with target angle were considered and the optimal input signal for the model by GAs has been determined. The curves have been divided in four groups putting together the 10 curves which have the final angle closest to the 15°, 20°, 25° and 30° and for each group the mean curve has been calculated.
To determine the optimal input signal sn for each group, it has been used a genetic approach. It has been supposed that the input signal sn is constituted by three contiguous rectangles because of the study made on the EMG and of the results in literature (Kandel, *et al.*, 1994). Since six parameters are to be determined, GAs were adopted in a 6–dimension space. In this s_n identification process, we chose to use the same η fitness and a hybrid implementation of GAs as we done previously. Because of a 6–dimension space of research, it's worthy to use a large number of individuals. We chose to have population of 500 individuals but this choice implied an increase of computational times. To face this problem, we imposed that the GA had to finish if there wasn't an improvement of the fitness for much than 10 consecutive generations. To maintain an high dynamic and to limit the effects of a possible premature termination, a large number of individuals has been involved in the process of genetic modification: in particular 266 individuals for each generation have been modified equally distributed between the genetic operators. For each group it has been calculated 80 GAs, sufficient to have a stabilised fitness as Figure 12 shows. The GAs determined a triphasic input signal for each group with time-amplitude characteristics in accordance with the results in literature (see for example Hannaford and Stark, 1995). Moreover, the morphological comparison between the target angle data and the output curves generated using the identified s_n shows great correlation (Table 1).

Table 1. Mean values of the squared correlation coefficients between real data and model output for each target angle acquired

Target angle	15°	20°	25°	30°
Mean square correlation coeff.	0.975	0.968	0.960	0.983

5.3 Complete model performances

To test the behaviour of the model in the whole range of inclinations in the frontal plane, the four identified sn have been interpolated to build the input signals for angles in the whole range (15°; 30°). For every 2° angles in this range, we used the correspondent interpolated s_n as model input and compared the theoretical angle to the one reached by the model. It has been found that the model generates an output curve whose final angle is very close to the theoretical angle (Table 2).

Table 2. Percent error over the whole range of inclinations of the model output using interpolated inputs

Desired angle	17°	19°	21°	23°	24°	26°	28°
Perc. error	-0.64	-0.93	-0.44	1.34	-0.30	0.41	0.67

6. CONCLUSIONS

The aim of this work is the determination of a model for fast head movements on the frontal plane which represents the main anatomic and physiologic aspects of the head–neck segment control, because of its intrinsic relation with the infraorbital line alignment with the absolute horizontal. The first part of the work was the identification of the principal contributions to head movement and of the relations between them. Thus in parallel with the voluntary force generation we considered the contribution due to gravity field, the viscous resistance and the tissues stiffness, the physiological limit imposed to the movement from the anatomical constraints, and the γ-reflex. Some of the relevant parameters have been derived from literature: namely, we could extend the results of Stark and his workgroup's model about head rotation on the coronal plane. In particular we accepted the dynamic of the voluntary moment generation (for the gain we used experimental traction tests) and the stiffness coefficient proposed while we discarded the viscosity because the slides of the tissues in the two movement are strongly different. Furthermore we introduced new blocks representing the physiological limits, the gravity field and, most of all, the stabilisation due to the effect of the γ–reflex.

Finally the structure of the stabilisation block has been thought taking in account some physiological evidences like the dependency of the γ–reflex from the relative position ($\Delta\theta$) and angular velocity, the delay due to the line transmission and some experimental tests.

The model determined in this way presented six unknown parameters: the viscous coefficient B, the gain G_p and exponent E_p for physiologic limit block, the static γ_s and dynamic γ_d gain for the stabilisation block and the delay Tr for the same.

For the identification of the optimal values for these parameters, three different protocols have been studied: fast full scale movements, fast movements with target angle and movement after a small and unpredictable perturbation. The data have been acquired using ELITE™ and the numerical identification has been performed by GAs.

The successive verifications of model validity and capability to generalisation show the goodness of the whole model. The comparison between the output model curve with the fast full-scale data confirmed how the identified model is representative of the fast full-scale real voluntary movements. In addition, it has been demonstrated that the model is a useful instrument for forward studies: in fact the results of the identifications for the optimal input signal for target angles are completely in accordance with other literature studies and experimental evidences. This characteristic is proved also by the behaviour of the model to an input signal obtained by interpolation of the identified ones. Using an interpolated signal relative to a desired angle the model generated curves with final angles very close to the desired one over the whole range of possible inclinations in the frontal plane.

Three strong hypotheses have been assumed for the definition of the current model, first the summation of the neural inputs, second the absence of a control on the modulation of muscles stiffness, third the absence of co–contraction. The first hypothesis will be discussed on the basis of the future investigations adopting the model in specific protocol set–ups and it's not strongly exploited for the current neuro–muscular model. The second assumption is strongly related to the choice of modelling only fast movements where the modulation of stiffness can be with some extent assumed constant. Active stiffness modulation is currently discussed in literature from the physiological point of view and thus it's still difficult to propose a good modelling of such function, however future investigations will consider this contribution. The assumption of no co–contraction is again resolved in the representation of fast movements and will be discussed in the future possibility to considered slow movements.

In conclusion, through this work we have build a flexible and powerful model for head–neck segment fast movements in the frontal plane, which, respecting the principal anatomical and physiological aspects, represents adequately the neuro–mechanical response to an equivalent neuronal signal, providing thus a useful tool for the investigations about how the various sensorial systems contribute to the modulation of the input signal.

The present model is proposed as a tool available for a future simulation aiming at distinguishing the contributions of the different sensory inputs to the control of head fast movements.

REFERENCES

Amblard B., Assaiante C., Fabre J.–C., Mouchnino, Massion J., Voluntary head stabilization in space during oscillatory trunk movements in the frontal plane performed in weightlessness, Exp Brain Res, vol. 114, pp 214–225, 1997.

Barberini C. L. , Macpherson J. M., Effect of head position on postural orientation and equilibrium, Exp Brain Res, vol. 122, pp 175–184, 1998.

Berthoz B, Pozzo T, Intermittent head stabilization during postural and locomotory tasks in humans. In: Amblard B. Berthoz A, Clarac F (eds), Posture and gait: development, adaptation and modulation, Elsevier, Amsterdam, pp 189–198, 1988.

Bronstein A. M., Evidence for a vestibular input contributing to dynamic head stabilization in man, Acta Otolaryngol (Stockh), vol. 105, pp 1–6, 1988

Davis L., Handbook of genetic algorithms, Van Nostrand Reinhold, New York, 1991

Guitton D., Kearney R. E., Wereley N., Peterson B. W., Visual, vestibular and voluntary contributions to human head stabilization, Exp Brain Res, vol. 64, pp 59–69, 1986

Hannaford B. e Stark L., Roles of the Elements of the triphasic control signal, Experimental Neurology, vol. 90, pp 619–634, 1985

Hatze H., Myocybernetic control models of skeletal muscle: Characteristic and applications, Univ. South Africa, Pretoria, 1981

Holland J. H., Adaptation in natural and artificial systems, MIT press, 1975

Kandel E. R., Scwartz J. H. e Jessell T. M., Principi di Neuroscienze 2° edizione, Ed. Ambrosiana, 1994

Pedotti A., Ferrigno G., ELITE: a digital dedicated hardware system for movement analysis via real–time TV signal processing, IEEE Trans. Biom. Eng., BME–32, pp 943–950, 1985

Pozzo T., Berthoz A., Lefort L., Head stabilization during various locomotor tasks in humans. I. Normal subjects, Exp Brain Res, vol. 82, pp 97–106, 1990.

Pozzo T., Berthoz A., Lefort L., Vitte E. Head stabilization during various locomotor tasks in humans. II. Patients with bilateral peripheral vestibular deficits, Exp Brain Res, vol. 85, pp 208–217, 1991.

Zangemeister W. H. , Lehman S., Stark L., Simulation of head movement trajectories: model anf fit to main sequence, Biol. Cybern, vol. 41, 19–32, 1981.

COMPARISON OF TWO ITERATIVE LEARNING CONCEPTS APPLIED ON A MANIPULATOR WITH COGGING FORCE DISTURBANCES

J. van Dijk, E. Schrijver

Universtity of Twente, Cornelis J. Drebbel Institute for Systems Engineering
Faculty of Mechanical Engineering department of Mechanical Automation
P.O. BOX 217 7500 AE Enschede NETHERLANDS
e-mail: j.vandijk@wb.utwente.nl

Abstract: In laser welding applications, either the laser beam or the workpiece has to be manipulated such that the seam to weld can be tracked. Gantry type of robots driven by linear direct drive synchronous motors can be applied as manipulator(s). In direct drive permanent magnet motor systems, disturbances due to cogging forces are directly noticeable at the end-effector and influence the performance. In the usual P(I)D controlled set-up of these systems, cogging forces induce disturbances which cause errors in seam-tracking accuracy in such a way that laser welding at high speeds (up to 0.5 m/sec.) do not result in qualitatively good weldings. In here it is shown how Iterative Learning feedforward Controller techniques can be designed and applied to achieve a high precision tracking accuracy. Two different ILC techniques, one using a zero phase error approach and one using H_∞ synthesis, are compared. The conclusion is that the ILC using the a-causal zero-phase error approach gives the best results and the ILC using H_∞ synthesis suffers from phase errors. *Copyright © 2001 IFAC*

Keywords: learning control, permanent magnet linear motors, H-infinity, tracking.

1. INTRODUCTION

For laser welding applications a manipulator is needed to move the laser beam relative to the workpiece. It is assumed that the laser is moved by means of a gantry-type of robot. Conventional (general-purpose) drivesystems of these types of manipulators are often implemented with transmissions to link rotational actuators with the desired translational motions. Direct drive systems can eliminate undesirable effects like flexibility and backlash in these conventional systems. However, the influence of disturbances at the actuator side, which are reduced when conventional transmissions are applied, are directly noticeable at the end-effector and influence the performance. Especially in permanent magnet motors the disturbances due to cogging forces can deteriorate the tracking behavior. The tracking accuracy demands for laser welding is ± 60 μm in order to guarantee qualitatively good welding. Figure 6 shows that in the usual PD controlled set up, with permanent magnet direct drive actuators, cogging and friction forces induce disturbances which cause rather big errors in seam-tracking accuracy. In here it is shown how the usual P(I)D controllers can be extended with an Iterative Learning feedforward Control (ILC) concept in order to obtain a much better performance. In section 2 the specifications, test set-up and cogging disturbance are further outlined. In section 3 it is shown how an ILC with current error feedback is designed using the a-causal zero phase error tracking controller algorithm (Tomizuka, 1987) and using the approach of the H_∞ synthesis of iterative learning controller of the Roover (1996) The first method for ILC design is an heuristic method which can be found in different publications (Roover, 1996; Moore, 1999) but we have supplied it with a zero phase error inverse of the transfer-function from feedforward u to the tracking error e (Fig. 2). The second method is robust for uncertainty. For the second design procedure the plant is identified and a worst-case model and a nominal model is derived from experimental data. For short the ILC designs will be called zero-phase ILC and robust-ILC. In order to obtain ILC-implementations for the set-up, both design procedures are applied to the problem of the set-up in section 4. In section 5 we present experimental results and compare the results of both iterative learning strategies. It is shown that both types of ILC's can achieve a tracking accuracy much better then the specifications (see Fig. 6 and Fig. 7) The conclusion is that the zero-phase ILC gives the best results. The robust ILC experiences phase errors, which lead to divergence in the case of continuous learning

2. BACKGROUND INFORMATION

2.1 Specifications

The specifications on a gantry-robot moving the laser beam for welding are as follows. The focused laser beam has a diameter of $300 \cdot 10^{-6}$ m. Experiments have shown that position errors of $60 \cdot 10^{-6}$ m can cause the laser to heat up only one side of the seam resulting in bad metallurgical properties of the weld. Furthermore, for thin plate welding applications, one can apply high velocities, up to 0.5 m/s along the seam. Experiments have shown that tracking errors along the seam less then $\pm\ 60 \cdot 10^{-6}$m result in high quality laser welds when a constant laser power is applied. So, perpendicular to the seam and in the direction of the seam we need the tracking errors to be smaller then $60 \cdot 10^{-6}$m. The test setup we are considering here is one dimensional, only the x-axis of the gantry (see Fig. 1). The moving mass in this setup is 18 kg.

2.2 Cogging

Cogging forces originate from the particular structure of permanent magnet motors. In the usual linear permanent magnet motors the coils are located in the translator. In order to increase flux linkage, the coils are wound around metal teeth. The forces generated by currents through the coils will move the translator in the desired direction. But even without the generated currents through the coils, flux linkage is present. The system tries to minimize the flux path length and as a consequence the resulting forces will try to move the translator to certain preferential positions. This cogging force is a function of the position of the translator relative to the stator.

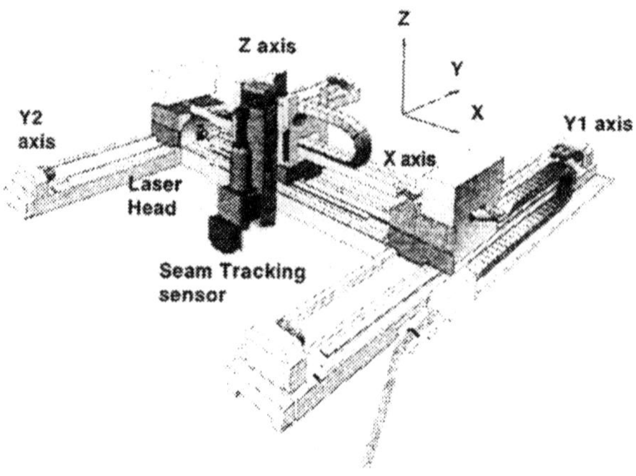

Fig. 1: Gantry robot

3. ITERATIVE LEARNING CONTROL

For controlled systems that have to perform tasks repeatedly, ILC has turned out to be an effective tool for improving the transient performance. The reader is referred to Moore (1999) for an extensive list of references. The approach is motivated by the observation that if the controller C (Fig. 2) is fixed and the process operating conditions are the same each time it executes, then any errors in the output response will be repeated during each operation. These errors can be recorded and can then be used to compute modifications to the input signal that will be applied to the process during the next operation, or trial, of the system (Moore, 1999). The ILC scheme shown in Fig. 2 operates as follows. During the k-th trial at each sample-instance i an input is applied to the process P, which is a sum of a feedforward u_k and a control-action and produces an output $y_k(i)$. At each sample instance, $u_k(i)$ and the error $e_k(i)=y_d(i)-y_k(i)$ are recorded. When the trial is over they are processed off-line by the ILC-algorithm. Based on the $e_k(i)$ the ILC algorithm computes a modified feedforward $u_{k+1}(i)$ for all sample instances that is stored in memory until the next time the system operates, at which time this new feedforward is applied to the process.

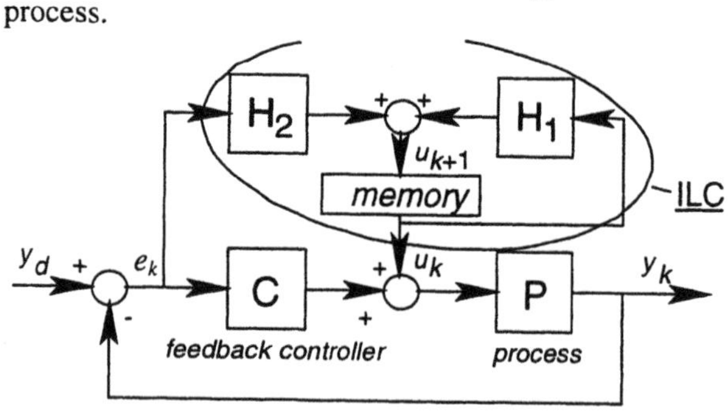

Fig. 2: Block-diagram of feedback and iterative learning controlled process

In this section we will address two different methods of designing model-based ILC-controllers. Both are so called current cycle feedback (CCF) ILC-controllers. This means that the control concepts are a combination of conventional feedback control and ILC feedforward control as shown in Fig. 2. In the following it is assumed that the closed-loop with feedback-controller C is stable and that the openloop process P is linear time invariant. In here we will only focus on the design of the ILC part. Fist of all convergence criteria for CCF ILC are discussed.

3.1 Convergence Criterion

Convergence of an ILC scheme depends on the choice of the algorithm which updates u_{k+1} in Fig. 2 from one trial to the other. The most general update algorithm is presented in Moore, (1999) en represented in Fig. 2:

$$u_{k+1} = H_1 u_k + H_2 e_k \tag{1}$$

where H_1 and H_2 are transfer-functions of filters. Considering Fig. 2 it can easily be found that the transfer function of the process sensitivity U between the error e_k and the input u_k can be written as:

$$U = -\frac{e_k}{u_k} = \frac{P}{1+CP} \tag{2}$$

When using the general update law (1) the error in the next trial (run) can be written as:

$$\begin{aligned} e_{k+1} &= \frac{-P}{1+CP} u_{k+1} = \frac{-P}{1+CP} H_1 u_k + \frac{-P}{1+CP} H_2 e_k \\ &= \left(H_1 - H_2 \frac{P}{1+CP} \right) e_k \end{aligned} \tag{3}$$

Eq. (3) results in the following condition for convergence:

$$\left| \left(H_1 - H_2 \frac{P}{1+CP} \right) \right| < 1 \tag{4}$$

Not only convergence is important, but also the resulting asymptotic error, $e_{k\to\infty} \equiv e^{\infty}$. Moore (1999) has shown that e^{∞} in the SISO case can be written as:

$$e^{\infty} = \frac{(1-H_1)}{\left(1 - H_1 + H_2 \frac{P}{1+CP} \right)} e^0 \tag{5}$$

Eq. (4) and (5) show that the obvious choice is:

$$\begin{aligned} H_1 &= 1 \\ H_2 &= \left(\frac{P}{1+CP} \right)^{-1} = U^{-1} \end{aligned} \tag{6}$$

The following can be observed concerning H_2.

1. H_2 has to be stable. When U has zero's in the right-half plane, the inverse will have unstable poles.
2. H_2 has high gain for high frequencies. Since the plant P is usually strictly proper and the feedback controller C is proper or strictly proper, so the inverse of U is non-proper.
3. The process sensitivity U is uncertain. Especially in the high frequency region. Therefore H_2 will deviate from the real inverse of U at higher frequencies.

3.2 Zero Phase Error method

In order to cope with problems 2 and 3 an additional low-pass filter Q in the update algorithm is introduced:

$$u_{k+1} = Q(u_k + H_2 e_k) \tag{7}$$

which is established by choosing $H_1 = Q$ and $H_2 = Q \cdot U^{-1}$. The price paid for gaining robustness by the introduction of Q is that $e^{\infty} \neq 0$ but will have a fixed value. The so called Zero Phase Error Tracking Control (ZPETC) algorithm will be used in order to cope with problem 1.

Remark, that because all computations of u_k are done off-line, Q and H_2 are allowed to be non-causal and therefore the phase-shift introduced by a usual low-pass Q can be avoided. The ZPETC-algorithm is designed to produce a digital feedforward controller for tracking desired time varying signals. In this design described by Tomizuka (1987) the algorithm in fact produces an inverse of the process-model. In here this algorithm is used to produce a proper inverse of the transfer-function U.

The ZPETC cancels all the poles and cancelable zero's of the modeled transfer-function. For uncancelable zero's, which include zero's outside the unit circle, the ZPETC cancels the phase shift introduced by them. A more extensive explanation of ZPETC can be found in Tomizuka (1987).

Design procedure 1

Based on the results above, the following design procedure for a current cycle feedback (CCF) setup with update equation (7) can be derived:

1. Design the feedback controller C to obtain a robustly stable system
2. Use the ZPETC-agorithm to determine H_2, the best possible inverse of U. This inverse will be valid up to some bandwidth ω_v because of limitations in the a priori plant model P.
3. Design the discrete low-pass-filter Q to cut off near ω_v. With
$$|Q(z)| = 1, \forall \omega \in [0, \omega_v] \text{ and } |Q(z)| \approx 0, \omega > \omega_v$$

In here z is the discrete time shift operator.

In section 4 this procedure will be applied to the setup.

3.3 H_∞ Synthesis Method

Using the update law (7) and the matrix ∞-norm, the convergence criterion (4) changes to:

$$\| Q(I - H_2 \cdot U) \|_{\infty} < 1 \tag{8}$$

Roover (1996) suggested the following. For given U and Q, H_2 can be solved from the (sub)optimal H_∞-synthesis problem:

$$H_2 = \arg\min \|Q(I - H_2 U)\|_\infty \tag{9}$$

Equation (9) describes the well-known 'model-matching problem', i.e. for given Q, H_2 is matched to the inverse of U. An advantage of this robust-ILC design procedure with respect to the zero phase design procedure is the fact that this framework is suitable for designing a multivariable ILC. Another important advantage of formulating the ILC problem in the H_∞ framework, is that uncertainty with respect to U can explicitly be taken into account. Before formulating the algorithm which results in robustness against uncertainty, we will discuss some prerequisites.

Standard process and uncertainty

The standard process format (Zou et al., 1996) as shown in Fig. 3a is used.

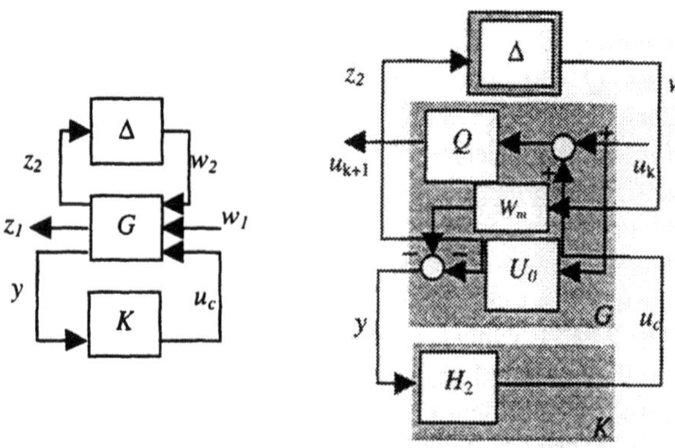

(a) (b)

Fig. 3: Standard process (a), ILC transformed into standard process format (b)

Assume a nominal process sensitivity $U_0(s)$ and a multiplicative output uncertainty represented by a stable and inversely stable weighting function $W_m(s)$. Then the real process sensitivity $U(s)$ can be described by:

$$U(s) = \{(I + W_m(s)\Delta(s)) \cdot U_0(s) \mid \|\Delta(s)\|_\infty < 1\} \tag{10}$$

The ILC-algorithm can now be represented in the standard process format as shown in Fig. 3b by choosing K and G according to (11) and solving K by minimizing the infinity-norm of the closed loop transfer-function between w and z. This closed loop transfer-function can be written as in (12). Combining the ILC convergence criterion with the robustness against uncertainty criterion results in the requirement $\|T_{zw}\|_\infty < 1$.

$$K = H_2, G = \begin{bmatrix} Q & 0 & Q \\ U_0 & 0 & 0 \\ -U_0 & -W_m & 0 \end{bmatrix}$$
$$\begin{bmatrix} z \\ y \end{bmatrix} = G \cdot \begin{bmatrix} w \\ u_c \end{bmatrix}, z = \begin{bmatrix} u_{k+1} \\ z_2 \end{bmatrix}, w = \begin{bmatrix} u_k \\ w_2 \end{bmatrix} \tag{11}$$

$$\begin{bmatrix} u_{k+1} \\ z_2 \end{bmatrix} = T_{zw} \begin{bmatrix} u_k \\ w_2 \end{bmatrix},$$
$$T_{zw} = \begin{bmatrix} Q(I - H_2 U_0) & -H_2 Q W_m \\ U_0 & 0 \end{bmatrix} \tag{12}$$

If the structure of the uncertainty is taken into account then $\|T_{zw}\|_\infty < 1$ is in fact $\mu(T_{zw}) < 1$, with μ the structured singular value of T_{zw} (Zou et al., 1996)

The robust ILC algorithm

A successful iteration can be performed by scaling the off-diagonal elements of T_{zw} and solving the scaled problem by H_∞-synthesis (the so called D-K iteration). Therefore, the following design procedure is presented by the Roover (1996):

Design procedure 2

1. Model the transfer-function U as a nominal model together with an upper bound on the model uncertainty, according to (10)
2. Choose $Q(s)$ to be a low-pass weighting filter.
3. For given Q and U, find an H_2 that minimizes $\|T_{zw}\|_\infty$ using D-K iteration.
4. If an H_2 can be found such that $\|T_{zw}\|_\infty < 1$ increase the bandwidth of Q and perform again step 3; else decrease ω_c of Q. Perform steps 2 and 3 iteratively until the maximum obtainable ω_c is reached.

4. DESIGN OF ILC'S FOR THE APPLICATION

Design of both learning controllers requires knowledge about the transfer-function U from u_k to e_k. For this purpose we have identified the open-loop process P. Figure 4 shows the identified bode magnitude plots of the process P from different data-sets. It also shows a bode magnitude plot of the low-order worst-case model P_p and the nominal model P_0. According to (2) the worst-case transfer-function U_p as well as the nominal model U_0 can now be computed.

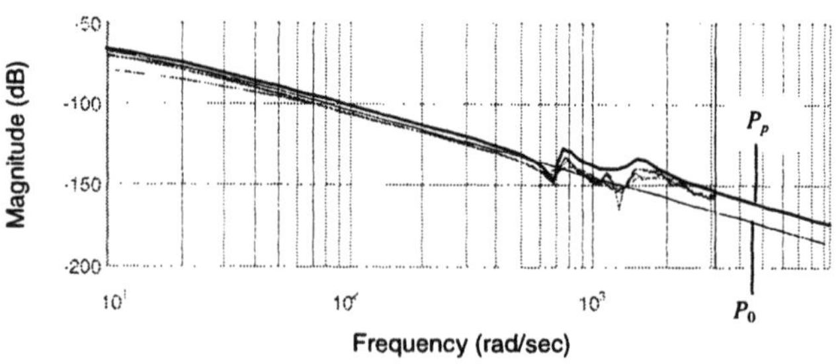

Fig. 4: Models from different data-sets and worst-case model P_p and nominal model P_0

A multiplicative uncertainty weighting function can be derived from U_p and U_0 according to:

$$W_m = \frac{U_p - U_0}{U_0} \tag{13}$$

The worst-case low-order model P_p reads:

$$P_p(s)=\frac{0.184(s^2+35s+4.624\cdot 10^5)(s^2+526s+1.53\cdot 10^6)}{s(s+15)(s^2+63s+5.748\cdot 10^5)(s^2+277s+2.368\cdot 10^6)} \quad \textbf{(14)}$$

The nominal model is described by:

$$P_0(s)=\frac{1/18}{s(s+15)} \quad \textbf{(15)}$$

4.1 Zero phase ILC design

According to the design procedure 1 the controller C is designed as a PD-controller with bandwidth 25 Hz. The filter H_2 is computed by application of the ZPETC-algorithm to the nominal process sensitivity U_0. For this purpose the continuous transfer function U_0 is first transformed into a discrete transfer function by a zero order hold transformation. The result of the ZPETC-algorithm is:

$$H_2(z)=\frac{144478559(z-0.83)(z^2-1.943z+0.9448)}{z(z+0.9974)(z-0.7901)} \quad \textbf{(16)}$$

sampling time: 0.0005sec.

Note the one timestep delay in (16) in order to let H_2 be proper. Since the computation is done off-line it is easy to compensate for this delay and preserve zero phase shift. The filter Q is chosen as a low-pass 4th order Butterworth filter with cutt-off frequency of 75 Hz.

4.2 Robust ILC design

The results of the application of the robust ILC design procedure 2 are as follows. Again the filter Q is chosen as a 4th order low-pass filter. Initially the cut-off frequency is set at 50 Hz. The resulting value of μ of the closed loop interconnection matrix T_{zw} according to (12) is shown in Fig. 5. The result was obtained after 23 D-K iteration steps. Iteration over the steps 2,3 and 4 of design procedure 2 finally resulted in a Q-filter with cut-off frequency of 80 Hz. The value of μ obtained after 56 D-K iterations is also shown in Fig. 5. For higher values of the cut-off frequency, the μ-synthesis did not convergence. The plot of μ against frequency (Fig. 5) indicates that the convergence robustness is worst in the frequency range round 200 rad/sec.

5. TEST RESULTS AND DISCUSSION

In the control of the gantry-robot we are using third order position reference profiles for acceleration and deacceleration parts combined with first order (ramp) position reference profiles for the constant velocity part. The reference profile for the test set-up is constructed as follows. The maximum velocity is 0.5 m/sec., the maximum acceleration allowed is 10 m/sec^2, the test seam length is 0.4 m.

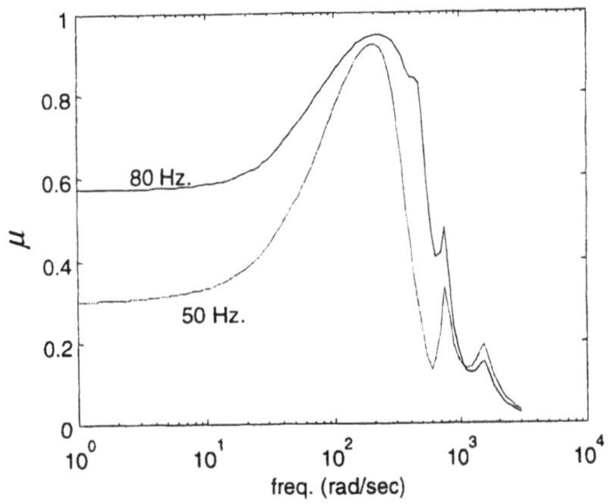

Fig. 5: Plot of $\mu(T_{zw})$ against frequency

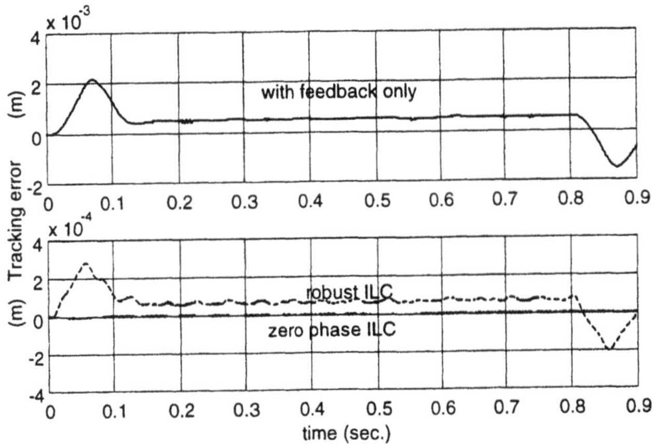

Fig. 6: Tracking error, initial error and error after 5 trials

Figure 6 shows the initial tracking-error with only feedback applied and the error after 5 trials of both the ILC designs. Note that the applied sensor has a resolution of 1.10^{-6} m. The tracking error of the zero-phase ILC is well within the specifications after 5 trials. The tracking error obtained with the robust ILC does not meet the specifications after 5 trials.

A 20 trials learning sequence was performed with both implementations of the ILC's. Figure 7 shows the average tracking error against the number of trials of both ILC algorithms. The average error is defined as $\frac{\sum_i^{n\_samples} abs(e_i)}{n\_samples}$. In here e_i is the tracking error at sample i. It is clear from Fig. 7 that the convergence rate of the zero phase ILC design is higher than the convergence rate of the robust ILC design. This was expected since the gain of H_2 obtained via design procedure 2 is smaller at all frequencies then the gain of H_2 obtained via design procedure 1. The zero phase ILC design has converged after 4 steps to an average error <2 10^{-6} m. The robust ILC obtains the minimal average tracking error after 10 iterations (< 8 10^{-6} m). But, the error is increased again if learning is continued, see Fig. 7.

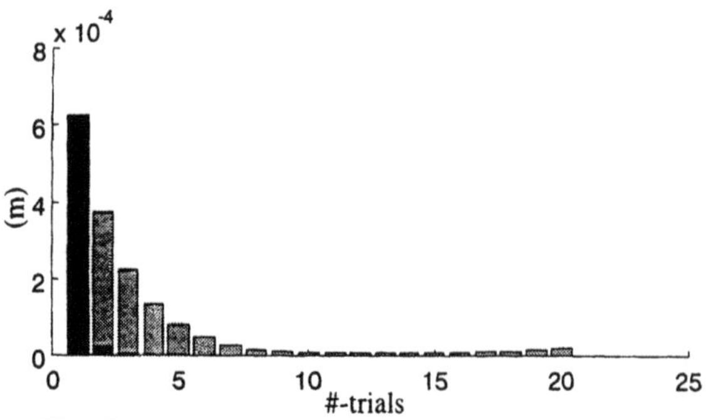

Fig. 7: Average tracking error of zero-phase ILC (dark) and robust-ILC (light)

Figure 8 should be studied to understand the increase in error after 10 trials. Figure 8 shows a bode plot of the worst-case model of the transfer-function , U_p, between the feedforward u and tracking error e. Also is shown the inverse of the filter-transfer H_2, obtained with Q as weighting filter. Note that the inverse of H_2 should be equivalent with U_p.

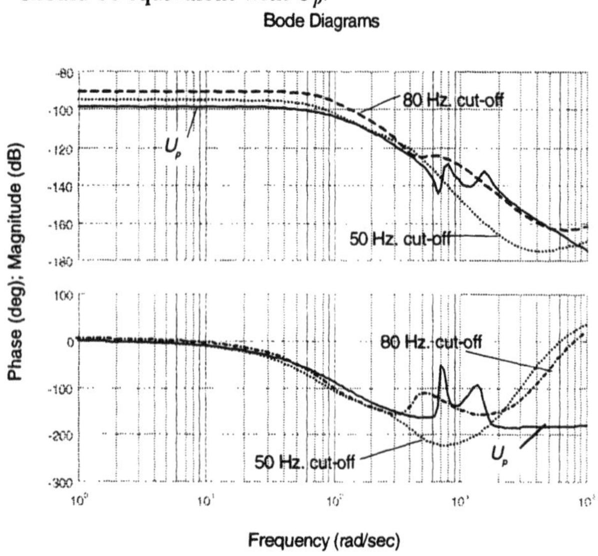

Fig. 8: Bode plot of U_p and inverses of H_2

Figure 8 shows an inverse H_2 with cut-off frequency 50 Hz. applied and an H_2^{-1} with cut-off frequency 80 Hz. applied. It is clear that using Q with 80 Hz. cut-off frequency a better high frequent fit between H_2^{-1} and U_p is obtained. But, this H_2^{-1} shows a phase error, compared to the phase of U_p, for frequencies > 40 rad/sec.

From experiments we have noticed that this phase error causes the increase in error after continued learning. This phenomenon is also registered by Chen and Wen (1999; p114) for non-model based ILC's: "Signal components may be accumulated due to different phase relationships from iteration to iteration. This is the key reason of the divergence for some ILC schemes which may be convergent at the initial few iterations, become divergent as ILC continuously runs". Although, this might not be a problem in practice when iteration is stopped after a fixed number of iterations.

Fig. 9 shows the feedforward u after 5 trials. It clearly shows the force needed for acceleration, friction compensation and cogging compensation.

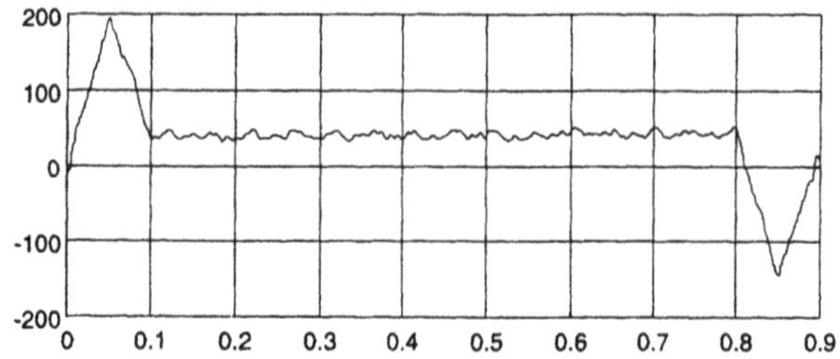

Fig. 9: Learned feedforward u

6. CONCLUSIONS

The paper describes two design procedures for ILC schemes. Measurement results obtained from a laser-welding robot show that as well as the zero phase ILC and the robust ILC, both combined with a PD feedback controller, can track reference paths with tracking errors much less then with feedback alone (Fig. 6) and well within the specifications. The zero-phase ILC converges after 4 trials. The average tracking error of the set-up after 4 trials is $< 2\ 10^{-6}\ m$. The robust ILC converges initially after 10 steps to an average tracking error $< 8\ 10^{-6}\ m$. The robust-ILC is slower in convergence then the zero-phase ILC as expected since it is more cautious at all frequencies to accomplish robustness. The robust-ILC is divergent (Fig. 7) as ILC continuously runs. This is due to the phase error between the computed inverse of the process sensitivity transfer-function H_2, resulting from the D-K iteration, and the inverse of the "real" process sensitivity (Fig. 8).

REFERENCES

Chen Y., and C. Wen, (1999), *Iterative learning control convergence, robustness and applications*, Springer Berlin, ISBN-1-85233-190-9.

Moore, K.L., (1999) "Iterative learning control: An exposi-tory Overview", in *Applied and computational control, signals and circuits*, **1**, ed. Biswa Nath Datta , Birkhauser Boston, ISBN 0-8176-3954-3.

Roover D. de, (1996), Synthesis of a robust iterative learning controller using an H_∞ approach, in *Proc. 35th IEEE Conf. Decision and Control*, Kobe Japan, 3044-3049.

Tomizuka, M., (1987), "Zero phase error tracking algorithm for digital control", *Journal Dyn. Syst. Meas. and Control*, **109**, 65-68.

Zou K., J.C. Doyle, K. Glover, (1996), *Robust optimal Control*, Prentice Hall, Upper Saddle River, ISBN 0-13-456567-3.

A COMPARISON OF OPTIMAL ITERATIVE LEARNING CONTROL SCHEMES

M. Rzewuski * E. Rogers * D. H. Owens **

* *Dept. of Electronics and Computer Science, University of Southampton, Southampton SO17 1BJ UK*
** *Department of Automatic Control and Systems Engineering, University of Sheffield, Sheffield S1 3JD, UK*

Abstract: Recent years has seen much progress in the theory and application of iterative learning control schemes for both linear and (classes of) nonlinear dynamics. In the case of the former, many algorithms based on minimizing a suitable cost function have been reported. Here the interest is in the so-called norm optimal approach where the basic philosophy is to compute the control input on the current trial such that the tracking error is reduced in an optimal way without too much deviation from the control input used on the previous trial. This paper compares the performance of a range of controllers arising from use of the norm optimal approach - both stand alone and against alternatives.

Keywords: iterative learning control, optimal algorithms.

1. INTRODUCTION

Iterative learning control (ILC) is a technique to control systems operating in a repetitive mode with the additional requirement that a specified output trajectory $r(t)$ defined over a finite interval $[0,T]$ is followed to high precision. There are numerous examples of such systems including robot manipulators that are required to repeat a given task to high precision, chemical batch processes or, more generally, the class of tracking systems.

Motivated by human learning, the basic of idea of ILC is to use information from previous executions of the task in order to improve performance from trial to trial in the sense that the tracking error is sequentially reduced ((Arimoto *et al.*, 1984; Moore, 1993)). Typical ILC algorithms construct the input to the plant on a given trial from the input used on the previous trial plus an additive increment that is typically a function of past values of the observed output error, i.e. the difference between achieved and desired plant output. The objective of constructing a sequence of input functions $\{u_k(t)\}_k$, $t \in [0,T]$, such that the performance as the task is repeated is gradually improving, can be refined to a convergence condition on the input and error:

$$\lim_{k\to\infty} ||e_k|| = 0, \quad \lim_{k\to\infty} ||u_k - u_\infty|| = 0 \quad (1)$$

where $e_k(t)$ is the error on trial k, i.e. difference between $r(t)$ and the system output, $y_k(t)$, and $u_k(t)$ is the input to the system on this trial.

The above definition of convergent learning is a stability problem on an infinite-dimensional two-dimensional (2D)-product space. As such it places the analysis of ILC schemes firmly outside the scope of traditional control theory. In particular, ILC must be studied in the context of fixed-point problems or, more precisely, linear repetitive processes ((Rogers and Owens, 1992)).

Since the ILC concept was proposed, a very large number of approaches have been considered but here we consider the very important case when the ILC scheme for linear plant dynamics is designed based on optimal control techniques. One option here is to use a quadratic cost function where the only control input term in this cost function is that to be used on the current trial. Recent years have also seen the emergence of so-called norm optimal ILC which can be realized in terms of current trial mechanisms combined with feedforward of previous trial data. Here the control input term in the cost function is based on the difference between the input signals used on successive trials.

The algorithms which emerge from the norm optimal approach are based on splitting the two-dimensional dynamics into two separate one-dimensional problems. This is done by introducing a performance criterion as the basis of specifying the control input to be used on each trial. In this context, the problem can be interpreted as the determination of the control input on trial $k+1$ that reduces the tracking error in an optimal way but which does not deviate too much from the control input used on trial k. Use of this approach leads to a number of key properties, of which the most important are as follows.

1. Automatic choice of step size (in the iterative procedures needed to solve the optimization problem).

2. Potential for improved robustness through the use of causal feedback of current trial data and feedforward trial data from previous trials.

It is also possible to enhance norm optimal control to include predictive action (Amann *et al.*, 1998), i.e. future predicted error signals are explicitly included in the cost function. In this paper, we use a computer aided analysis tool constructed for optimal control linear model based ILC schemes to undertake a detailed comparative study of norm optimal (including predictive action) ILC controllers for linear plants against those designed using alternative optimal control based approaches. The scope of this study includes the experimental setup reported in (Frueh and Pham, 2000) (thereby giving the option for future experimental verification). We being in the next section by summarizing the necessary background results (for complete details see (Amann *et al.*, 1996*b*; Amann *et al.*, 1998) and the relevant references).

2. BACKGROUND

In this paper we restrict attention to the norm optimal approach in a Hilbert space setting. The 'translation of this setting to the special case of a linear time invariant system is straightforward and will form a starting point for the work reported here. The alternative 'conventional' algorithm for these systems (i.e. a quadratic cost function where the contribution from the control input term is only (explicitly) based on the current trial values) is not detailed here and can be found in, for example, (Frueh and Pham, 2000) and the relevant cited references.

The following is the formal definition of a successful ILC algorithm.

Definition 1. Consider a dynamic system with input u and output y. Let $\mathcal{Y}$ and $\mathcal{U}$ be the output and input function spaces respectively and let $r \in \mathcal{Y}$ be a desired reference trajectory from the system. Then an ILC algorithm is successful if, and only if, it constructs a sequence of control inputs $\{u_k(t)\}_{k\geq 0}$ which, when applied to the system or plant (under identical experimental conditions), produces an output sequence $\{y_k(t)\}_{k\geq 0}$ with the following properties of convergent learning

$$\lim_{k\to\infty} y_k = r, \quad \lim_{k\to\infty} u_k = u_\infty \tag{2}$$

Here convergence is interpreted in terms of the topologies assumed in $\mathcal{Y}$ and $\mathcal{U}$ respectively.

Note: This general description includes linear and nonlinear dynamics, continuous or discrete plants, and time-invariant or time-varying systems.

Now let the space of output signals $\mathcal{Y}$ be a real Hilbert space and $\mathcal{U}$ also be a real (and possibly distinct) Hilbert space of input functions. The respective inner products (denoted by $\langle\cdot,\cdot\rangle$) and norms $||\cdot||^2 = \langle\cdot,\cdot\rangle$ are indexed in a way that reflects the space if it is appropriate to the discussion.

The dynamics of the plant considered here are approximated by a linear model which in operator form can be written as

$$y = Gu + z_0 \tag{3}$$

where no loss of generality arises from setting $z_0 = 0$. Also it is clear that the ILC procedure, if convergent, solves the problem $r = Gu_\infty$ for u_∞ and, if G is invertible, the formal solution is just $u_\infty = G^{-1}r$. A basic premise of the ILC approach is that the direct inversion of G is regarded as an impractical solution because it requires exact knowledge of G and involves derivatives of r. This high-frequency gain characteristic would make such an approach sensitive to noise and other disturbances. Also it can be argued that inversion of the whole plant G is unnecessary as the solution only requires finding the pre-image of r under G.

The above problem is easily seen to be equivalent to finding the minimizing input u_∞ for the optimization problem

$$min_u\{||e||^2 : e = r - y, y = Gu\} \qquad (4)$$

The optimal error $||r - Gu_\infty||^2$ is a measure of how well the ILC algorithm has solved the inversion problem. It also represents the best that the system can do in tracking the signal r. The case of interest here is when the optimal error is zero, i.e. u_∞ is a solution of $r = Gu_\infty$. Also (4) is clearly a singular optimal control problem which by its very nature requires an iterative solution.

There are an infinity of potential iterative procedures for solving (4) and of these the gradient approach has the simplest form and has been extensively investigated in the ILC literature. A gradient based ILC algorithm has the form

$$u_{k+1} = u_k + \epsilon_{k+1}G^*e_k \qquad (5)$$

where $G^* : \mathcal{Y} \to \mathcal{U}$ is the adjoint operator to G, and ϵ_{k+1} is a step length to be chosen at each iteration. This general approach suffers from the need to choose this step length on each trial and the feedforward structure of the iteration takes no account of current trial effects - including disturbances and plant modeling errors.

Norm optimal ILC has the following two crucial properties relative to alternative gradient based algorithms

1. Automatic choice of step size.
2. Potential for improved robustness through the use of causal feedback of current trial data and feedforward of data from previous trials.

In particular, the ILC algorithms considered here compute the input on trial $k+1$ as the solution of the minimum norm optimization problem

$$u_{k+1} = \arg min_{u_{k+1}}\{J_{k+1}(u_{k+1})\} \qquad (6)$$

subject to

$$e_{k+1} = r - y_{k+1}, \; y_{k+1} = Gu_{k+1} \qquad (7)$$

where the performance index (or optimality criterion) used is

$$J_{k+1}(u_{k+1}) = ||e_{k+1}||^2_{\mathcal{Y}} + ||u_{k+1} - u_k||^2_{\mathcal{U}} \qquad (8)$$

The initial control $u_0 \in \mathcal{U}$ can be arbitrary but, in practice, will be a good first guess at the solution of the problem. The relative weighting of reducing the current trial error against minimizing the deviation in the control input signals used on successive passes can be absorbed into the definitions of the norms in $\mathcal{Y}$ and $\mathcal{U}$.

The benefits of this approach are immediate from the simple interlacing result

$$||e_{k+1}||^2 \leq J_{k+1}(u_{k+1}) \leq ||e_k||^2, \; \forall\, k \geq 0 \qquad (9)$$

which follows from optimality and the fact that the (non-optimal) choice of $u_{k+1} = u_k$ would lead to the relation $J_{k+1}(u_k) = ||e_k||^2$. This result states that the algorithm is a descent algorithm as the norm of the error is monotonically decreasing in k. Also equality holds if, and only if, $u_{k+1} = u_k$, i.e. when the algorithm has converged and no more input-updating takes place.

The controller on trial $k+1$ is given by

$$u_{k+1} = u_k + G^*e_{k+1}, \; \forall k \geq 0 \qquad (10)$$

This relationship, together with the error update relation

$$e_{k+1} = (I + GG^*)^{-1}e_k, \; \forall\, k \geq 0 \qquad (11)$$

and the recursive input update relation

$$u_{k+1} = (I + G^*G)^{-1}(u_k + G^*r), \; \forall\, k \geq 0 \qquad (12)$$

can be used to undertake a detailed analysis of the properties of this class of ILC laws (Amann *et al.*, 1996*b*).

Predictive optimal ILC (Amann *et al.*, 1998) extends the cost function to the form

$$J_{k+1,N}(u_{k+1}) = \sum_{i=1}^{N} \lambda^{i-1}(||e_{k+i}||^2_{\mathcal{Y}} + ||u_{k+i} - u_{k+i-1}||^2_{\mathcal{U}}) \qquad (13)$$

This criterion includes the error of the next N trials as well as the corresponding changes in the control input signals The weighting parameter $\lambda > 0$ determines the importance of more distant (future) errors and incremental inputs compared with the current ones. By including more future signals into the performance criterion, the algorithm becomes less 'short sighted'. The theory given above extends in a natural manner to this case and an obvious question to ask is: when does the extra (computational) cost become worthwhile?

3. ANALYSIS

The implementation of norm optimal and predictive norm optimal ILC schemes is a relatively straightforward exercise and is hence not detailed here. Instead, we proceed to use MATLAB based implementations to discuss and compare the performance of these approaches both stand alone and relative to alternative optimal control based algorithms. The plants used have the following transfer functions

$$G_1(s) = \frac{s+1}{s^2+5s+6} \qquad (14)$$

$$G_1(s) = \frac{s-1}{s^2+5s+6} \qquad (15)$$

Also we use a minimal state space realization of these processes with state space triple $\{A, B, C\}$, and assume that there are l inputs and m outputs.

In this paper, we will only consider the use of linear quadratic optimal control cost functions in the ILC setting. In particular, we will compare the optimal control based ILC of (Frueh and Pham, 2000) (and others) against norm optimal and predictive norm optimal control. The choice of input and output spaces is as follows

$$u \in \mathcal{U} = L_2^l[0,T]$$
$$(r, r(T)) \in \mathcal{Y} = L_2^m[0,T] \times \mathbb{R}^m \qquad (16)$$

Also the inner products in $\mathcal{Y}$ and $\mathcal{U}$ are defined as

$$\langle (y_1, z_1), (y_2, z_2) \rangle_{\mathcal{Y}} = \frac{1}{2} \int_{t=0}^{T} y_1^T(t) Q y_2(t)\, dt + \frac{1}{2} z_1^T F z_2$$
$$\langle u_1, u_2 \rangle_{\mathcal{U}} = \frac{1}{2} \int_{t=0}^{T} u_1^T(t) R u_2(t)\, dt \qquad (17)$$

where Q and R are symmetric positive definite matrices and F is a symmetric positive semi-definite matrix.

In all cases, the initial conditions are taken to be homogeneous without loss of generality since the plant response to non-zero initial conditions can be absorbed into $r(t)$ in a natural manner. The cost function in each case becomes, with the specified norms in $\mathcal{Y}$ and $\mathcal{U}$, a familiar linear quadratic performance criterion. For example, in the case of algorithms of the form considered by (Frueh and Pham, 2000) and others we have

$$J_{k+1} = \frac{1}{2} \int_{t=0}^{T} \{ e_{k+1}^T Q e_{k+1}(t) + u_{k+1}^T R u_{k+1}(t) \}\, dt \qquad (18)$$

In effect, the problem is now a combination of the well known optimal tracking problem (tracking of $r(t)$) and disturbance accommodation problem (regarding $u_k(t)$ as a known disturbance on trial $k+1$) from standard linear systems theory. The cost functions considered in (Frueh and Pham, 2000) and elsewhere result in the following solution algorithm

$$u_{k+1} = u_k + \delta u_{k+1}$$
$$\delta u_{k+1} = (R + P^T Q P)^{-1} (P^T Q e_k - R u_k) \qquad (19)$$

In the case of norm optimal control, standard optimal control theory now gives the solution as

$$\dot{\psi}_{k+1}(t) = -A^T \psi_{k+1}(t) - C^T Q e_{k+1}(t)$$
$$u_{k+1}(t) = u_k(t) + R^{-1} B^T \psi_{k+1}(t)$$
$$\psi_{k+1}(T) = C^T F e_{k+1}(T),\ t \in [0,T] \qquad (20)$$

This representation is non-causal (in the standard sense) but it can be transformed into a causal implementation as detailed next for the case of a relaxation factor α.

Transform the costate vector $\psi_{k+1}(t)$ using

$$\psi_{k+1}(t) = -K(t)\left[x_{k+1}(t) - \alpha x_k(t)\right] + \zeta_{k+1}(t) \qquad (21)$$

where the feedback gain matrix $K(t)$ satisfies the well known Riccati (matrix) differential equation

$$\dot{K}(t) = -A^T K(t) - K(t) A + K(t) B^T R^{-1} B^T K(t) - C^T Q C$$
$$K(T) = C^T F C \qquad (22)$$

Note that this last equation of independent of the inputs, states and outputs of the system and hence only needs to be computed once before the sequence of trials begin.

The predictive or 'feedforward' term $\zeta_{k+1}(t)$ needs to be computed on each trial using

$$\dot{\zeta}_{k+1}(t) = -(A - BR^{-1}B^T K)^T \zeta_{k+1}(t) - \alpha C^T Q e_k(t) + (1-\alpha) K B u_k(t) - (1-\alpha) C^T Q r(t) \qquad (23)$$

with terminal boundary condition

$$\zeta_{k+1}(T) = C^T F \left[\alpha e_k(T) + (1-\alpha) r(T)\right] \qquad (24)$$

The algorithm is now causal since (22) and (23) can be solved off-line by reverse time simulation using available previous trial data.

The following is the final form (with no relaxation factor) of the implementation algorithm for predictive optimal ILC (for complete details see (Amann *et al.*, 1998))

$$\begin{bmatrix} u_{k+1} \\ u_{k+2} \\ \vdots \\ u_{k+N} \end{bmatrix} = \begin{bmatrix} u_k \\ u_k \\ \vdots \\ u_k \end{bmatrix} - R_N^{-1} B_N^T (K(t) \times \left(\begin{bmatrix} x_{k+1} \\ x_{k+2} \\ \vdots \\ x_{k+N} \end{bmatrix} - \begin{bmatrix} x_k \\ x_k \\ \vdots \\ x_k \end{bmatrix} \right) - \xi_{k+1,N}(t)) \qquad (25)$$

$$\dot{K} = A_N^T K - K A_N +$$

$$KB_NR_N^{-1}B_N^TK - C_N^TQC_N,$$

$$K(T) = C_N^T(T)FC_N(T) \quad (26)$$

$$\dot{\xi}_{k+1,N}(t) = -\left(A_N - B_NR_N^{-1}B_N^TK\right)^T \xi_{k+1,N}(t) - C_N^TQ_N \begin{bmatrix} e_k(t) \\ e_k(t) \\ \vdots \\ e_k(t) \end{bmatrix} \quad (27)$$

with terminal condition

$$\xi_{k+1,N}(T) = C_N^TF_N \times \left[e_k^T(T) \dots e_k^T(T)\right]^T \quad (28)$$

where A_N is a block diagonal matrix with each diagonal entry equal to the state matrix A, and the matrices B_N and C_N constructed in an identical manner using the matrices B and C respectively. For the precise form of the weighting matrices Q_N, R_N and F_N, see again (Amann *et al.*, 1998). In the remainder of this paper, we apply these algorithms to the two plants defined by (14) and (15) respectively, where in all simulations $T = 6$, and (as required) $R = 0.1$, $Q = 10$, $F = 0$, $N = 5$. Also $r(t) = \sin t$, $u_0(t) = 0$.

Figure 1 shows the result of applying an optimal control algorithm of the form (19) to the minimum phase plant defined by (14). It has been reported many times in the literature that this algorithm is capable of very good performance. Our experience shows that in some cases it exhibits slow convergence and involves a matrix inversion which could result in a high computational cost. Also it is easy to demonstrate that this algorithm performs poorly when the plant is non-minimum phase (i.e. here when applied to (15)). Note also that in some cases the control signals required can be very large to the extent that they can be outside the effective operating ranges of many actuators.

The norm optimal algorithm does not involve matrix inversion and, because the cost function penalizes the change in the control input signals between successive trials, the resulting signals demanded are generally much smaller than with an algorithm of the form (19). Figure 2 shows the results of applying norm optimal ILC to the non-minimum phase plant (15). Overall, we have found that norm optimal control is more effective against non-minimum phase dynamics but still the performance can be rather poor.

Turning to the use of predictive ILC, Figure 3 shows the results of applying this control scheme to the minimum phase plant (14). Here we found that this scheme produces much faster convergence but at the expense of higher computational cost. Also the performance of such schemes is significantly dependent on the choice of the extra design parameters, i.e. the prediction horizon N and the weighting factor λ. An indication of the influence of the latter factor can be seen by comparing the results of Figure 4, where the value of λ is set at 10, with those of Figure 5 where λ has been reduced to 5. Another advantage of predictive ILC is that it can give much better performance when applied to non-minimum phase processes - see Figure 4 which gives the results of applying this algorithm to the plant defined by (15) - but note that this could come at the expense of 'very strong' control action. If such control action is unacceptable then it is still essentially an open research question of how to design effective ILC control schemes in such cases.

4. CONCLUSIONS

This paper has undertaken a detailed comparative study of three competing classes of ILC schemes for linear plants based on optimal control based design philosophies. The major conclusion is that those based on the norm optimal approach (including augmentation by prediction) have significant advantages over alternatives. A major reason for this is that in the norm optimal approach it is the difference in the control signals between successive trials which is penalized as opposed to the current trial input alone in the alternatives. The most significant additional factor to be considered before using predictive ILC is the extra computational cost but (in general terms) this should not be excessive unless the plant has a very large number of states, inputs, and outputs.

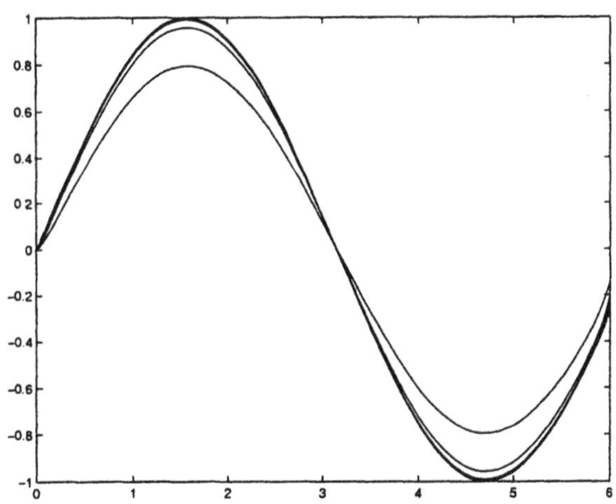

Fig. 1. ILC algorithm applied to the minimum phase plant.

In all simulation experiments reported here, the continuous-time versions of the algorithms were used. For experimental verification or applications studies, however, this may not be good practice due to high computation costs and many numerical problems that often can be avoided through the use of a discrete time version of a given algorithm (Amann, 1996). The discrete time version of the norm-optimal ILC algorithm considered here can be found in (Amann *et al.*, 1996*a*).

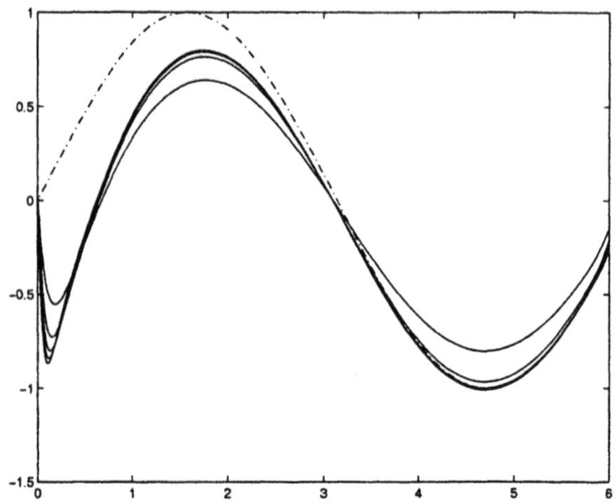

Fig. 2. ILC algorithm applied to the non-minimum phase plant.

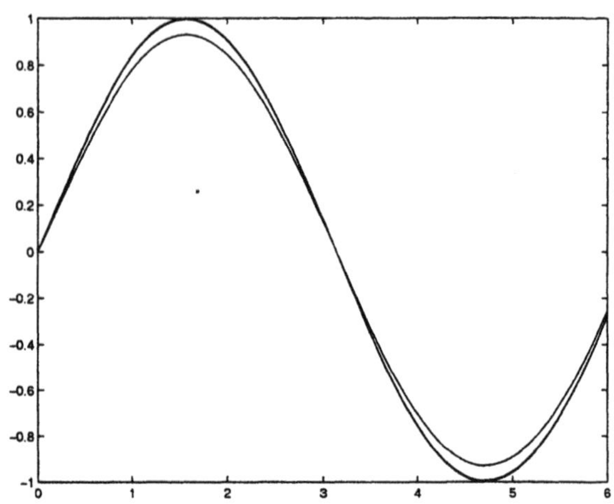

Fig. 3. Predictive ILC algorithm applied to the minimum phase plant.

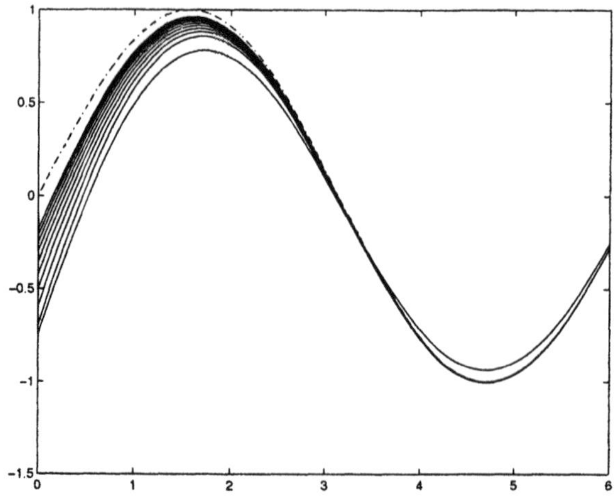

Fig. 4. Predictive ILC algorithm applied to non-minimum phase plant.

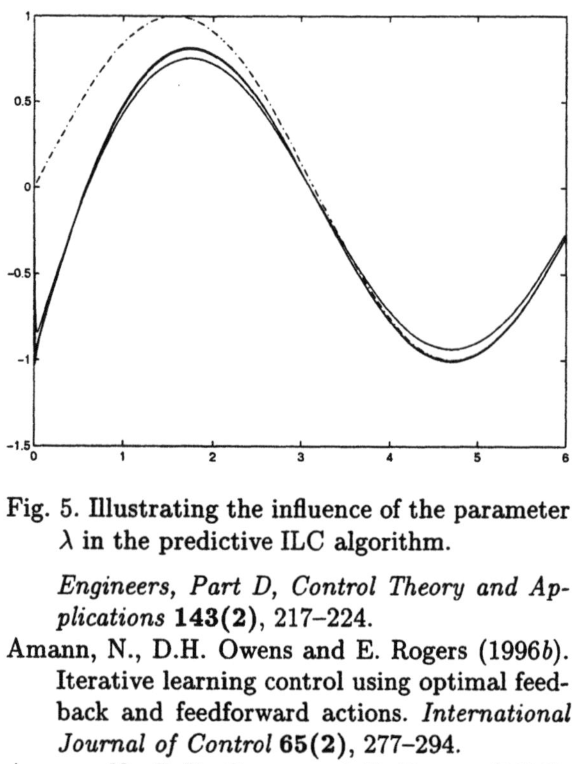

Fig. 5. Illustrating the influence of the parameter λ in the predictive ILC algorithm.

5. REFERENCES

Amann, N. (1996). Optimal Algorithms for Iterative Learning Control. *PhD Thesis, University of Exeter, UK.*

Amann, N., D.H. Owens and E. Rogers (1996*a*). Iterative learning control for discrete–time systems with exponential rate of convergence. *Proceeedings of The Institution of Electrical Engineers, Part D, Control Theory and Applications* **143(2)**, 217–224.

Amann, N., D.H. Owens and E. Rogers (1996*b*). Iterative learning control using optimal feedback and feedforward actions. *International Journal of Control* **65(2)**, 277–294.

Amann, N., D.H. Owens and E. Rogers (1998). Predictive optimal iterative learning control. *International Journal of Control* **69(2)**, 203–226.

Arimoto, S., S. Kawamura and F. Miyazaki (1984). Bettering of robots by learning. *Journal of Robotic Systems* **1**, 123–140.

Frueh, J.A. and M.Q. Pham (2000). Linear quadratic optimal learning control (LQL). *International Journal of Control* **70(10)**, 832–839.

Moore, K. (1993). *Iterative Learning Control for Deterministic Systems.* Springer–Verlag. London.

Rogers, E. and D.H. Owens (1992). *Stability Analysis for Linear Repetitive Processes, LNCIS Series, Vol 175.* Springer–Verlag. Berlin.

VIRTUAL REFERENCE FEEDBACK TUNING: A NEW FRAMEWORK FOR DATA-BASED DESIGN OF PID AND LINEAR CONTROLLERS

A. Lecchini [*,1], **M.C. Campi** [**,2] and **S.M. Savaresi** [***]

[*] *CESAME - Université Catholique de Louvain*
Avenue Georgés Lemaitre 4, B-1348 Louvain-la-Neuve, Belgium
lecchini@auto.ucl.ac.be
[**] *Dip. di Elettronica per l'Automazione - Università di Brescia*
Via Branze 38, 25123 Brescia, Italy
campi@ing.unibs.it
[***] *Dip. di Elettronica ed Informazione - Politecnico di Milano*
Piazza Leonardo da Vinci 32, 20133 Milano, Italy
savaresi@elet.polimi.it

Abstract: in this paper a new framework for the data-based design of feedback controllers is presented. The main features are that the controller is directly estimated from data (no model identification of the plant is needed) and that only a single set of data generated by the plant is used with no need for specific experiments or iterations.

1. INTRODUCTION

The Virtual Reference Feedback Tuning (VRFT) represents a new approach to the problem of designing a controller for a plant, whose transfer function $P(z)$ is unknown, on the basis of a set of I/O data generated by the plant. The control problem addressed by VRFT can be stated as follows.

Problem formulation

It is assumed that the plant is a *linear* SISO discrete-time dynamical system described by the rational transfer function $P(z)$. Such a transfer function is *unknown*. A set of I/O data, collected during an experiment on the plant, will be used for the design of the controller. The control problem addressed herein is a one-degree-of-freedom model-reference control problem (the VRFT has been extended also to a two-degree-of-freedom setting - see (Lecchini *et al.*, 2001)). Namely, given a reference model $M(z)$, and a class of parameterized controllers $\{C(z;\theta)\}$ the objective is the selection of the parameter vector $\bar{\theta}$ which minimizes the following model-reference criterion:

$$J_{MR}(\theta)=\left\|\left(\frac{P(z)C(z;\theta)}{1+P(z)C(z;\theta)}-M(z)\right)W(z)\right\|_2^2 \quad (1)$$

where $W(z)$ is a user-chosen weighting function. The attention is restricted to linear in the parameter controllers, namely the controller class takes the form $C(z;\theta) = \beta^T(z)\theta$, where $\beta(z) = [\beta_1(z)\,\beta_2(z)\cdots\beta_n(z)]^T$ is a vector of linear discrete-time transfer functions and $\theta = [\vartheta_1\,\vartheta_2\cdots\vartheta_n]^T \in \mathbf{R}^n$ is the n-dimensional vector of parameters (among others, PID controllers belong to this class).

Background

Since it is assumed that the transfer function $P(z)$ is not known, the controller selection cannot be performed by a direct minimization of the deign criterion (1). The goal is instead to perform an indirect minimization on the basis of data. This problem has been the object of several research efforts under different approaches. In the *indirect* approach, a model of the plant is identified from data and then the identified model, say $\hat{P}(z)$, is substituted to $P(z)$ in the design criterion. The so obtained criterion is then minimized in order to obtain the controller. An alternative is constituted by the *direct* approach where the data are used directly for the design of the controller without

[1] The European Commission is herewith acknowledged for its financial support in part to the research reported on in this paper. The support is provided via the Program Training and Mobility of Researchers (TMR) and Project System Identification (ERB FMRX CT98 0206) to the European Research Network System Identification (ERNSI).
[2] The research has been partly supported by MURST under the project "New techniques for the identification and adaptive control of industrial systems".

resorting to the identification of a model. In this contest a very interesting scheme is the Iterative Feedback Tuning (IFT) method ((Hjalmarsson *et al.*, 1998)). In (Hjalmarsson *et al.*, 1998) it has been shown that, for certain design criteria (e.g. $J_{MR}(\theta)$ defined in (1)), by injecting certain signals in the control system an estimate of the gradient of the design criterion can be obtained. The controller tuning is then performed by updating the controller parameters through an iterative gradient-based minimization scheme. In the long run the designed controller is expected to lie in the neighborhood of a local minimum of the design criterion.

Results presented in this paper

The purpose of this paper is to describe a new approach to the problem of minimizing (1) on the basis of data. As a starting point, it will be shown that a controller can be estimated by minimizing a quadratic criterion constructed on the basis of a *single* set of I/O data through the classical least-quare identification procedure. This criterion will be indicated as $J_{VR}(\theta)$. Then, it will be shown that, if $\{C(z;\theta)\}$ is sufficiently large to contain the ideal controller $C_0(z)$ defined as

$$C_0(z) := \frac{1}{P(z)}\frac{M(z)}{1-M(z)} \qquad (2)$$

(which realizes perfect matching between the closed-loop transfer function and the reference model $M(z)$) the minimum of $J_{VR}(\theta)$ corresponds to $C_0(z)$; whereas, if $C_0(z) \notin \{C(z;\theta)\}$ then the minimum point of $J_{VR}(\theta)$ is a sensible approximation of the global minimum point of $J_{MR}(\theta)$. (Notice that $C_0(z) \notin \{C(z;\theta)\}$ is the most interesting case since, in general, the ideal controller $C_0(z)$ should be a high-order controller, even more so it should result to be not proper, therefore it is not expected to belong to the class of available controllers $\{C(z;\theta)\}$.)

It is worth noticing that VRFT addresses a problem somehow different from iterative schemes. As a matter of fact, iterative schemes usually call for a sequence of experiments on the true plant; moreover, these experiments are ad-hoc, which means that they require the implementation of the last designed controller and, possibly, the use of specific inputs. Instead, we assume that a single set of data points is available (which should not be collected with an ad-hoc experiment), and our goal is to optimally exploit this information in the controller selection.

A comment on stability

The stability of the designed closed-loop system is not discussed here. Let us remark that when the plant is *unknown*, the stability conditions cannot be directly assessed. In this case, one should perform some a-posteriori evaluation of the closed-loop stability based on available data (see e.g. (Campi *et al.*, 2001; Bombois *et al.*, 1999)).

Outline of the paper

In Section 2 the design criterion $J_{VR}(\theta)$ is introduced and analyzed. For the sake of exposition clarity, in this section it is assumed that the data are not corrupted by noise. The presence of noise is discussed in Section 3. The final Section 4 is devoted to an illustrative numerical example.

2. THE DATA-BASED DESIGN CRITERION

We start by introducing the basic idea behind the Virtual Reference framework (see also (Guardabassi and Savaresi, 2000; Savaresi and Guardabassi, 1997; Campi *et al.*, 2000)).

The Virtual Reference idea

Let us assume that a set $\{u(t), y(t)\}_{t=1,..,N}$ of I/O data has been collected from a noise-free experiment on the plant. Given the measured $y(t)$ (i.e. the actual signal measured at the output of the plant), consider a reference $\bar{r}(t)$ such that $M(z)\bar{r}(t) = y(t)$. Such a reference is called "virtual" because it does not exist in reality and in fact it was not used in the generation of $y(t)$. Notice that $y(t)$ is the desired output of the closed-loop system when the reference signal is $\bar{r}(t)$. Then, compute the corresponding *tracking error* $e(t) = \bar{r}(t) - y(t)$. Even though plant $P(z)$ is not known, we know that when $P(z)$ is fed by $u(t)$ (the actually measured input signal), it generates $y(t)$ as an output. Therefore, a good controller, at least in the condition when the reference signal is the virtual reference $\bar{r}(t)$, is one that generates $u(t)$ when fed by $e(t)$. The idea is then to search for such a controller. What is important to note here is that, since both signals $u(t)$ and $e(t)$ are known, this task reduces to the *identification problem* of describing the dynamical relationship between $e(t)$ and $u(t)$.

The following simple algorithm implements the above idea. In the algorithm, we include also a possible filtering action whose choice will be discussed later on.

Design algorithm

Given the reference model $M(z)$, the family of available controllers $\{C(z;\theta)\}$ and the set of data $\{u(t), y(t)\}_{t=1,..,N}$, do the following:

1. Calculate:
   - a virtual reference $\bar{r}(t)$ such that $y(t) = M(z)\bar{r}(t)$, and
   - the corresponding tracking error $e(t) = \bar{r}(t) - y(t)$.
2. Filter the signals $e(t)$ and $u(t)$ with a suitable filter $L(z)$:

   $$e_L(t) = L(z)e(t), \qquad u_L(t) = L(z)u(t).$$
3. Select the controller parameter vector $\hat{\theta}_N$ that minimizes the following design criterion

   $$J_{VR}^N(\theta) = \frac{1}{N}\sum_{t=1}^{N}\left(u_L(t) - C(z;\theta)e_L(t)\right)^2. \quad (3)$$

Notice that when $C(z;\theta) = \beta^T(z)\theta$ (i.e. the controller depends linearly on the parameter vector), the parameter vector $\hat{\theta}_N$ can be easily obtained as:

$$\hat{\theta}_N = \left[\frac{1}{N}\sum_{t=1}^{N} \varphi_L(t)\varphi_L(t)^T\right]^{-1} \left[\frac{1}{N}\sum_{t=1}^{N} \varphi_L(t)u_L(t)\right]. \quad (4)$$

where $\varphi_L(t) = \beta(z)e_L(t)$.

Analysis of the design criterion

The analysis of the design criterion $J_{VR}^N(\theta)$ is based on asymptotic results.

Under the hypothesis of ergodicity of the involved signals, using the Parseval theorem (see (Ljung, 1999)) and the definition of $C_0(z)$ the following asymptotic (as $N \to \infty$) frequency domain representation of $J_{VR}^N(\theta)$ can be written:

$$J_{VR}(\theta) = \frac{1}{2\pi}\int_{-\pi}^{\pi} |P|^2 \frac{|C(\theta) - C_0|^2}{|1+PC_0|^2}\frac{|L|^2}{|M|^2}\Phi_u d\omega. \quad (5)$$

where $\Phi_u(\omega)$ is the power spectral density of $u(t)$ (throughout we drop the argument $e^{j\omega}$ of transfer functions). $J_{VR}(\theta)$ is the *asymptotic* counterpart of $J_{VR}^N(\theta)$, namely the criterion to which $J_{VR}^N(\theta)$ tends as the number of available data goes to infinity. Accordingly, as $N \longrightarrow \infty$, the minimum $\hat{\theta}_N$ of $J_{VR}^N(\theta)$ will converge to the minimum of $J_{VR}(\theta)$, say $\hat{\theta}$. In the following, for analysis purposes, $J_{VR}(\theta)$ will be used extensively in place of $J_{VR}^N(\theta)$. Consider now the criterion $J_{MR}(\theta)$. Using the definition of 2-norm of a discrete-time linear transfer function and the definition of the ideal controller $C_0(z)$, $J_{MR}(\theta)$ can be given the following form:

$$J_{MR}(\theta) = \frac{1}{2\pi}\int_{-\pi}^{\pi} |P|^2 \frac{|C(\theta) - C_0|^2}{|1+PC(\theta)|^2}\frac{|W|^2}{|1+PC_0|^2} d\omega. \quad (6)$$

The minimum point of $J_{MR}(\theta)$ is indicated as $\bar{\theta}$. Notice that, if $C_0(z) \in \{C(z;\theta)\}$, then the minimum of $J_{VR}(\theta)$ coincides with the minimum of $J_{MR}(\theta)$, whatever $L(z)$, $W(z)$, $M(z)$ and $P(z)$ are.

Now suppose that $C_0(z) \notin \{C(z;\theta)\}$. We note that if the following identity holds

$$|L|^2 = \frac{|M|^2|W|^2}{\Phi_u|1+PC(\theta)|^2} \quad (7)$$

then $J_{VR}(\theta) = J_{MR}(\theta)$. As a consequence, $\hat{\theta} = \bar{\theta}$. However, the choice of the filter $L(z)$ suggested by equation (7) is clearly not feasible since $L(z)$ in (7) depends on $P(z)$, which is not known, and θ. Here, the following choice of the filter is instead proposed:

$$|L|^2 = |1-M|^2|M|^2|W|^2\frac{1}{\Phi_u} = \frac{|M|^2|W|^2}{|1+PC_0|^2}\frac{1}{\Phi_u}. \quad (8)$$

Notice that all quantities in the right-hand-side of equations (8) are known and therefore $L(z)$ can be actually computed (the only $\Phi_u(\omega)$ may be considered known only in certain situations, when the input signal has been selected by the designer; otherwise $\Phi_u(\omega)$ can be estimated using many different techniques, among which a high-order AR or ARX model (Ljung, 1999), or a high-order state-space model (Van Overschee and De Moor, 1994)). Moreover, notice that choice (8) corresponds to substitute $|1 + PC(\theta)|^2$ with $|1 + PC_0|^2$ in equation (7), which appears to be a sensible selection since we expect that $|1 + PC(\theta)|^2 \approx |1 + PC_0|^2$ for $\theta = \bar{\theta}$, where $\bar{\theta}$ is the minimum of $J_{MR}(\theta)$. In Proposition 1 below, we show that choice (8) is in fact optimal in a certain sense. Before stating this result, some notations must be preliminary settled. Set

$$\Delta C(z) = C_0(z) - \beta^T(z)\bar{\theta}$$

$$\beta^+(z) = [\beta_1(z) \quad \beta_2(z) \quad \ldots \quad \beta_n(z) \quad \Delta C(z)]^T$$

$$\theta^+ = [\vartheta_1 \quad \vartheta_2 \quad \ldots \quad \vartheta_n \quad \vartheta_{n+1}]^T$$

($\bar{\theta}$ being the parameter vector which minimizes $J_{MR}(\theta)$). Then define an extended family of controllers as follows:

$$C^+(z;\theta^+) = \beta^+(z)^T\theta^+.$$

Finally, consider the extended performance index

$$J_{MR}^+(\theta^+) = \left\| \left(\frac{P(z)C^+(z;\theta^+)}{1+P(z)C^+(z;\theta^+)} - M(z)\right) W(z) \right\|_2^2.$$

The second order Taylor expansion around its global minimizer $\bar{\theta}^+$ is denoted by $\bar{J}_{MR}^+(\theta^+)$, namely:

$$J_{MR}^+(\theta^+) = \bar{J}_{MR}^+(\theta^+) + o(\|\theta^+ - \bar{\theta}^+\|_2^2).$$

Using the above definitions, Proposition 1 can now be stated.

Proposition 1. The parameter vector $\bar{\theta}$ which minimizes the performance index $J_{MR}(\theta)$, and the parameter vector $\hat{\theta}$ which minimizes the performance index $J_{VR}(\theta)$ when $L(z)$ is selected as in (8) fulfill the following relationships:

$$\bar{\theta} = \arg\min_{\theta} J_{MR}^+([\theta^T \quad 0]^T).$$

$$\hat{\theta} = \arg\min_{\theta} \bar{J}_{MR}^+([\theta^T \quad 0]^T).$$

proof: see (Lecchini, 2001) □

The above result is interesting since it provides a formal relationship between the parameter vector $\hat{\theta}$ obtained using the Virtual Reference approach (in the special case when $L(z)$ is selected according to (8)), and the "optimal" parameter vector $\bar{\theta}$, which minimizes the original performance index $J_{MR}(\theta)$. Based on this result, we conclude that if the transfer function $\Delta C(z)$ plays a marginal role in determining $C_0(z)$, namely the family of controllers $\{C(z;\theta)\}$ is only slightly under-parameterized given a certain reference model, then $C(z;\hat{\theta})$ is a good approximation to $C(z;\bar{\theta})$

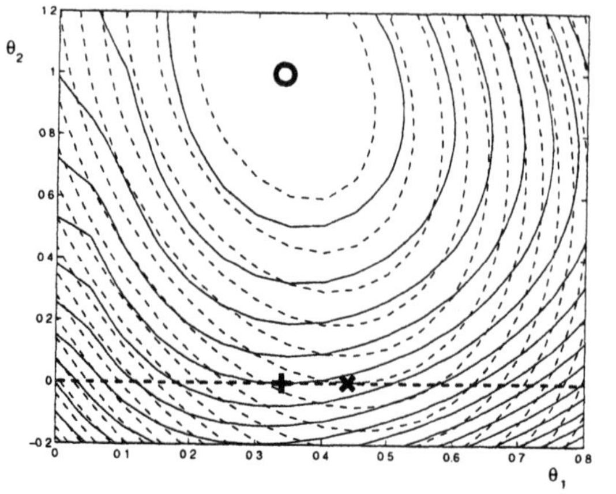

Fig. 1. Contour plots of $J^+_{MR}(\theta^+)$ (continuous line) and $\bar{J}^+_{MR}(\theta^+)$ (dashed line); (**O**)=$\bar{\theta}^+$; (**X**)=$[\hat{\theta} \;\; 0]$; (+)=$[\bar{\theta} \;\; 0]$.

since $J^+_{MR}(\theta^+)$ is well approximated in a neighborhood of its minimum by its second order expansion $\bar{J}^+_{MR}(\theta^+)$.

Illustrative example

Consider a plant and a reference model, characterized by the following transfer functions

$$P(z) = z^{-1}\frac{1}{1-0.6z^{-1}}, \quad M(z) = z^{-1}\frac{0.6}{1-0.4z^{-1}},$$

and assume that the following class of controllers is given:

$$C(z;\theta) = \frac{\theta}{1-z^{-1}}.$$

Notice that $C(z;\theta)$ can be rewritten as $C(z;\theta) = \beta(z)^T\theta$, where $\beta(z) = 1/(1-z^{-1})$. The ideal controller which exactly solves the model-matching problem is given by

$$C_0(z) = \frac{0.6 - 0.36z^{-1}}{1-z^{-1}}$$

(note that $C_0(z) \notin \{C(z;\theta)\}$). The global minimum of the model-reference criterion $J_{MR}(\theta)$ corresponds to $\bar{\theta} = 0.34$, whereas the global minimum of the criterion $J_{VR}(\theta)$ (when the filter is chosen as in (8)) corresponds to $\hat{\theta} = 0.44$. Correspondingly, $J_{MR}(\bar{\theta}) = 1.35$, and $J_{MR}(\hat{\theta}) = 1.45$.

Using the definitions introduced above, the extended controller class is given by $C^+(z;\theta^+) = \beta^+(z)^T\theta^+$, where

$$\beta^+(z)=\left[\frac{1}{1-z^{-1}} \quad \frac{0.26-0.36z^{-1}}{1-z^{-1}}\right]^T, \; \theta^+=[\vartheta_1 \;\; \vartheta_2]^T,$$

and $C_0(z) = \beta^+(z)^T\bar{\theta}^+$, with $\bar{\theta}^+ = [\bar{\theta}^T \;\; 1]^T = [0.34 \;\; 1]$. A graphical interpretation of the results is given in Fig.(1) where the contour plots of $J^+_{MR}(\theta^+)$ and $\bar{J}^+_{MR}(\theta^+)$ have been displayed. As stated in Proposition 1, it is apparent that $\bar{\theta}$ is the minimum of the extended performance index restricted to $\vartheta_2 = 0$, whereas $\hat{\theta}$ is the minimum of the quadratic approximation around $\bar{\theta}^+$ of the extended performance index restricted to $\vartheta_2 = 0$.

3. THE USE OF NOISY DATA

It has been shown that the controller can be estimated, on the basis of a set of noise-free data. Now let us assume that a set $\{u(t), \tilde{y}(t)\}_{t=1,..,N}$ of data has been collected from an open-loop noisy experiment. These data are described as:

$$\tilde{y}(t) = P(z)u(t) + d(t); \tag{9}$$

Throughout, we make the assumption that the signal $u(t)$ and the noise signal $d(t)$ are ergodic and uncorrelated signals (open-loop configuration).

If one applies the design algorithm of Section 2 to the set $\{u(t), \tilde{y}(t)\}_{t=1,..,N}$ of noisy data then he obtains a perturbed version, say $\tilde{\theta}_N$, of the right parameters vector $\hat{\theta}_N$. This perturbed estimate is given by:

$$\tilde{\theta}_N = \left[\frac{1}{N}\sum_{t=1}^{N}\tilde{\varphi}_L(t)\tilde{\varphi}_L(t)^T\right]^{-1}\left[\frac{1}{N}\sum_{t=1}^{N}\tilde{\varphi}_L(t)u_L(t)\right] \tag{10}$$

where $\tilde{\varphi}_L(t)$ is the regression vector constructed from $\tilde{y}(t)$, namely:

$$\tilde{\varphi}_L(t) = \beta(z)L(z)(M(z)^{-1} - 1)\tilde{y}(t).$$

In the limit, we have:

$$\lim_{N\to\infty}\tilde{\theta}_N \neq \hat{\theta} = \lim_{N\to\infty}\hat{\theta}_N. \tag{11}$$

The reason for (11) is that in (10) the noise affects the regression vector $\tilde{\varphi}(t)$. In the following, we propose the use of an Instrumental Variable method to counteract the effect of noise ((Ljung, 1999)). In this way, the results given in previous section can be saved for the case in which the data are corrupted by noise.

Instrumental Variables

The idea behind the Instrumental Variable method is to use a generic regression vector $\zeta(t)$ in place of $\tilde{\varphi}_L(t)$ in (10) as follows:

$$\hat{\theta}^{IV}_N = \left[\frac{1}{N}\sum_{t=1}^{N}\zeta(t)\tilde{\varphi}_L(t)^T\right]^{-1}\left[\frac{1}{N}\sum_{t=1}^{N}\zeta(t)u_L(t)\right] \tag{12}$$

The elements of $\zeta(t)$ are called instrumental variables. The basic requirements on the instrumental variables are:

$$E[\zeta(t)\varphi_L(t)^T] > 0 \tag{13}$$

$$E[\zeta(t)d(t)] = 0 \tag{14}$$

If (13-14), hold then the limit estimate $\hat{\theta}^{IV} = \lim_{N\to\infty}\hat{\theta}^{IV}_N$ does not depend on the noise. We can say that the noise is "correlated out" by the instrumental variables. However, the quality of the estimate depends on the choice of the instrumental variables. The more $\zeta(t)$ is close to $\varphi_L(t)$, the more $\hat{\theta}^{IV}$ is close to $\hat{\theta}$. Some appropriate choices are discussed below.

Choice of the instrumental variables

We propose two different choices for the instrumental variables. The first guarantees that asymptotically $\hat{\theta}^{IV} = \hat{\theta}$. However an additional experiment on the plant is required. The second does not guarantee that asymptotically $\hat{\theta}^{IV} = \hat{\theta}$ but the residual error is expected to be small. This choice does not require an additional experiment on the plant but calls for additional computational effort since it requires a model fitting step. The proposed choices are as follows:

- **Repeated experiment**

  Given the data $\{u(t), \tilde{y}(t)\}_{t=1,..,N}$, repeat the experiment on the plant using the same input $\{u(t)\}_{t=1,..,N}$ and collect the corresponding output sequence $\{\tilde{y}(t)'\}_{t=1,..,N}$. Then construct the instrumental variables as:

  $$\zeta(t) = \beta(z)L(z)(M(z)^{-1} - 1)\tilde{y}'(t) \quad (15)$$

  Notice that $\{\tilde{y}(t)'\}_{t=1,..,N}$ will be different from $\{\tilde{y}(t)\}_{t=1,..,N}$ since the two sequences are affected by two different realizations of the noise coming from two distinct experiments. It is reasonable to assume that the noise signals in the two experiments are uncorrelated, therefore (14) is satisfied.

- **Identification of the plant**

  Identify a model $\hat{P}(z)$ of the plant from the set of data $\{u(t), \tilde{y}(t)\}_{t=1,..,N}$ and generate the simulated output $\hat{y}(t) = \hat{P}(z)u(t)$. Then construct the instrumental variables as:

  $$\zeta(t) = \beta(z)L(z)(M(z)^{-1} - 1)\hat{y}(t) \quad (16)$$

  The identification of the plant is a standard open-loop identification problem. The model $\hat{P}(z)$ can be estimated using different techniques - see (Ljung, 1999; Van Overschee and De Moor, 1994).

It can be shown that, performing the choice (15), the effect of the noise can be perfectly removed as $N \to \infty$. However, this is possible only if two independent experiments characterized by the same input signal can be made on the plant. When only a single set of data is available, the choice (16) can be resorted to. In general, this choice does not deliver an estimate that asymptotically tends to $\hat{\theta}$ because of possible inaccuracy of the estimated $\hat{P}(z)$. Notice that, since the model $\hat{P}(z)$ is used only as a simulator of the plant, there are no strict limitations to its order (because it is not directly involved in the design of the controller). As a consequence, high-order models can be used to achieve high accuracy.

Closed-loop noisy data

The VRFT method can also be successfully applied to a set of noisy data collected in closed-loop (i.e. with a feedback controller connected to the plant) - see (Lecchini, 2001). An extended discussion on the use of closed-loop data goes beyond the scope of this paper.

4. A NUMERICAL EXAMPLE

We conclude the paper with an illustrative numerical example. The input-output dynamic behaviour of the plant can be described by the discrete-time linear transfer function $P(z)$ given by:

$$P(z) = z^{-3}B(z)/A(z)$$

$$A(z) = 1 - 1.4183z^{-1} + 1.5893z^{-2} - 1.316z^{-3} + .8864z^{-4}$$

$$B(z) = .28261 + .50666z^{-1}$$

The above transfer function $P(z)$ is the discrete-time model (using a sampling time $T_s = 0.05s$) of the system described in (Landau *et al.*, 1995). The control objective is expressed by the following second order reference model:

$$M(z) = z^{-3}\frac{(1-\alpha)^2}{(1-\alpha z^{-1})^2}, \quad \alpha = e^{-T_s\bar{\omega}}, \quad \bar{\omega} = 10,$$

where $\bar{\omega}$ is the desired bandwidth. The magnitude Bode plots of $P(z)$ and $M(z)$ are shown in Fig.(2). The weighting factor is $W(z) = 1$. The class of controllers is:

$$C(z;\theta) = \frac{\vartheta_0 + \vartheta_1 z^{-1} + \vartheta_2 z^{-2} + \vartheta_3 z^{-3} + \vartheta_4 z^{-4} + \vartheta_5 z^{-5}}{1 - z^{-1}}.$$

In order to compute $\hat{\theta}_N$ via the VRFT method, a set of data have been obtained by feeding $P(z)$ in open loop with $N = 512$ samples of a zero-mean Gaussian white noise ($\Phi_u = 0.01$). In the following, we will present three different VRFT design cases. The first two cases aim to illustrate the effect of a bad/good shaping of the filter $L(z)$. Specifically, in Case 1 the trivial filter $L(z) = 1$ is used, whereas in Case 2 $L(z)$ is designed as proposed in Section 2. The third design case is instead characterized by the presence of noise.

Case 1: $L(z) = 1$.

We obtain $J_{MR}(\hat{\theta}_N^1) = 0.232$. The magnitude Bode plot and step response of the corresponding closed-loop transfer function are shown in Fig.(3). Apparently, the control system has a behavior which remarkably differs from that of $M(z)$.

Case 2: $L(z) = M(z)(1 - M(z))$

In this case we obtain $J_{MR}(\hat{\theta}_N^2) = 0.0343$. The magnitude Bode plot and step response of the achieved closed-loop transfer function are shown in Fig.(4). Note that the so-obtained control system tracks almost perfectly the behaviour of the reference model. It is interesting to compare the performance of the controller $C(z;\hat{\theta}_N^2)$ obtained via VRFT with the optimal controller $C(z;\bar{\theta})$, for which $J_{MR}(\bar{\theta}) = 0.0340$. Since $J_{MR}(\hat{\theta}_N^2) = 0.0343$ it is apparent that the sub-optimality of $C(z;\hat{\theta}_N^2)$ is negligible.

Case 3: $L(z) = M(z)(1 - M(z))$ - noisy data

In this case the output signal has been corrupted

by a zero mean white disturbance such that the signal to noise ratio is $SNR = 10$. The filter $L(z)$ has been chosen as in Case 2. The controller parameter vector $\hat{\theta}_N^3$ is estimated without paying attention to the presence of noise. The step response of the achieved control system is shown in Fig.(5.a). Notice how the performance degrades dramatically with respect to the noise-free case (Case 2). The influence of noise can be counteracted via the use of instrumental variables as explained in Section 3. The parameter vector $\hat{\theta}_N^4$ estimated as in (12) performing the choice (16) - by estimating a 4th order ARX model of the plant - gives the step response shown in Fig.(5.b).

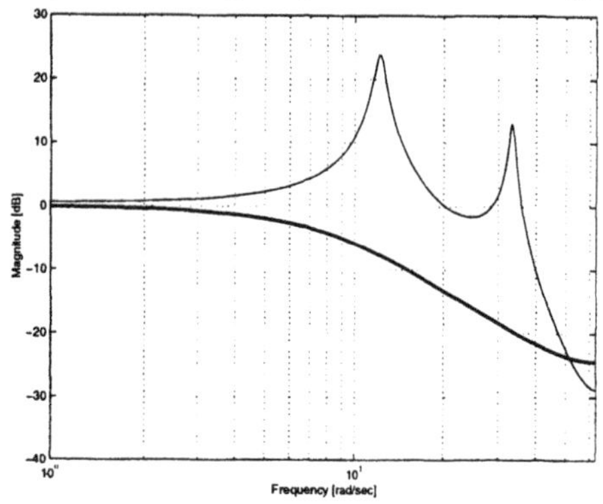

Fig. 2. Magnitude bode plots. The plant (thin line) and the reference-model (bold line).

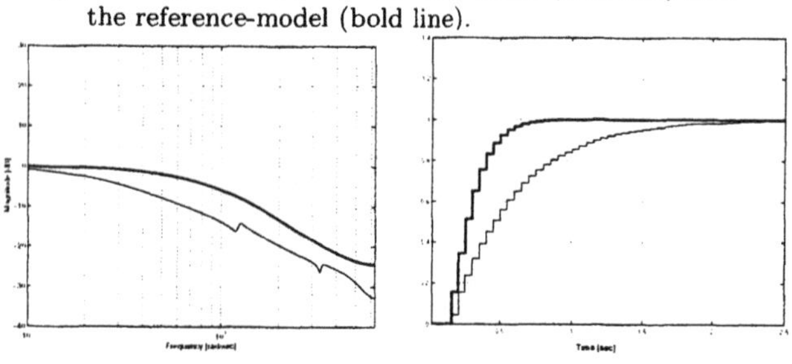

(a) (b)

Fig. 3. **(a)** Magnitude Bode plots: the control system achieved by $\hat{\theta}_N^1$ (thin line) and the reference-model (bold line); **(b)** Step responses: the control system achieved by $\hat{\theta}_N^1$ (thin line) and the reference-model (bold line);

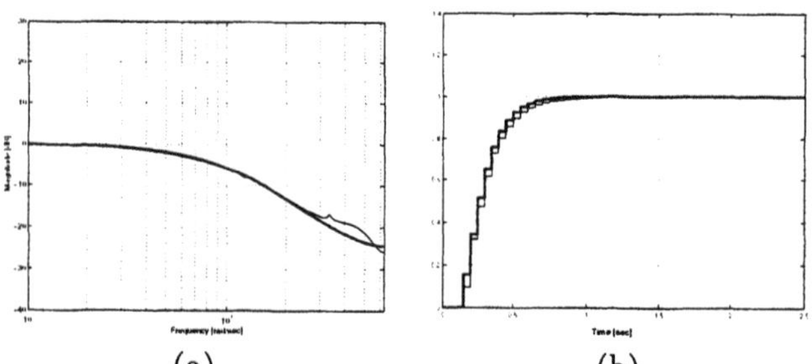

(a) (b)

Fig. 4. **(a)** Magnitude Bode plots: the control system achieved by $\hat{\theta}_N^2$ (thin line) and the reference-model (bold line); **(b)** Step responses: the control system achieved by $\hat{\theta}_N^2$ (thin line) and the reference-model (bold line);

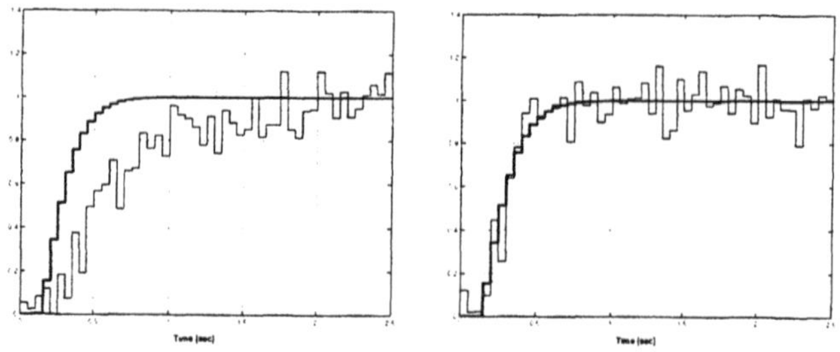

(a) (b)

Fig. 5. Step responses: **(a)** the control system achieved by $\hat{\theta}_N^3$ (thin line) and the reference-model (bold line); **(b)** the control system achieved by $\hat{\theta}_N^4$ (thin line) and the reference-model (bold line).

5. REFERENCES

Bombois, X., M. Gevers and G. Scorletti (1999). Controller validation based on an identified model. In: *38th IEEE Conf. on Decision and Control.* Phoenix, Arizona.

Campi, M.C., A. Lecchini and Savaresi S.M. (2001). Controller validation: a probabilistic approach based on prediction error identification. *Submitted.*

Campi, M.C., A. Lecchini and S.M. Savaresi (2000). Virtual Reference Feedback Tuning (VRFT): a new direct approach to the design of feedback controllers. In: *Proc. of the 39th IEEE Conference on Decision and Control.* Sydney, Australia.

Guardabassi, G.O. and S.M. Savaresi (2000). Virtual Reference Direct Design Method: an off-line approach to data-based control system design. *IEEE Trans. Automatic Control* **45**(5), 954–959.

Hjalmarsson, H., M. Gevers, S. Gunnarson and O. Lequin (1998). Iterative Feedback Tuning: theory and applications. *IEEE Control Systems* (August), 26–41.

Landau, I.D., D. Rey, A. Karimi, A. Voda and A. Franco (1995). A flexible transmission system as a benchmark for robust digital control. *European Journal of Control* **1**(2), 77–96.

Lecchini, A. (2001). *Virtual Reference Feedback Tuning: a new direct data-based method for the design of feedback controllers.* PhD Thesis - Università di Brescia.

Lecchini, A., M.C. Campi and S.M. Savaresi (2001). Virtual Reference Feedback Tuning for two degree of freedom controllers. In: *European Control Conference - ECC2001.* Porto, Portugal.

Ljung, L. (1999). *System Identification: theory for the user.* Prentice Hall.

Savaresi, S.M. and G.O. Guardabassi (1997). Approximate feedback linearization of discrete time non-linear systems using virtual input direct design. *System & Control Letters* **32**, 63–67.

Van den Hof, P.M.J. and R. Schrama (1995). Identification and Control - closed loop issues. *Automatica* **31**(12), 1751–1770.

Van Overschee, P. and B. De Moor (1994). N4SID: Subspace algorithms for the identification of combined deterministic-stochastic systems. *Automatica* **30**(1), 75–93.

EFFICIENT LINEAR MIMO ADAPTIVE INVERSE CONTROL

Gregory L. Plett [1]

University of Colorado at Colorado Springs,
Department of Electrical and Computer Engineering
1420 Austin Bluffs Parkway, P.O. Box 7150, Colorado Springs, CO
80933–7150 USA, `glp@eas.uccs.edu`

Abstract: Adaptive inverse control is an automatic control-system design method which "learns" over time how to control a particular dynamic system. This simple controller design approach can achieve very precise control of minimum- and nonminimum-phase systems, even with poor a priori knowledge of the plant dynamics. A problem with adaptive inverse control in its current state is that the learning algorithms for MIMO and nonlinear systems are very slow. This paper addresses some results concerning a new way to perform adaptive inverse control of linear MIMO systems which learns very quickly. An example is presented to demonstrate the dramatic speedup. *Copyright © 2001 IFAC*

Keywords: Adaptive control, Disturbance rejection, Efficient algorithms, Linear control systems, MIMO, Recursive least squares, Stochastic control, Wiener filters

1. INTRODUCTION

Adaptive inverse control is an automatic control-system design method which learns how to control a specific plant (Plett, n.d.; Plett, 1998; Widrow *et al.*, 1998; Widrow and Plett, 1997; Widrow and Walach, 1996; Widrow and Plett, 1996*a*; Widrow and Plett, 1996*b*; Widrow and Plett, 1996*c*). Adaptive filtering methods are used throughout. Invisible to the user, a three-part process is used internally. First, an adaptive plant model $\widehat{P}$ learns the dynamics of the plant. Secondly, an adaptive feedforward controller C learns to control the dynamics of the plant. Thirdly, an adaptive feedback disturbance canceler X learns to cancel disturbances which affect the plant. These processes may proceed concurrently. A block diagram of an adaptive inverse control system is shown in Fig. 1.

Methods are known to quickly adapt controllers for linear SISO plants (Widrow and Walach, 1996), but known methods for adapting controllers for linear MIMO and nonlinear plants learn very slowly (Plett, n.d.; Plett, 1998; Widrow and Walach, 1996). This paper presents some results regarding a training method for linear MIMO adaptive inverse control that learns about as quickly as methods for linear SISO control.

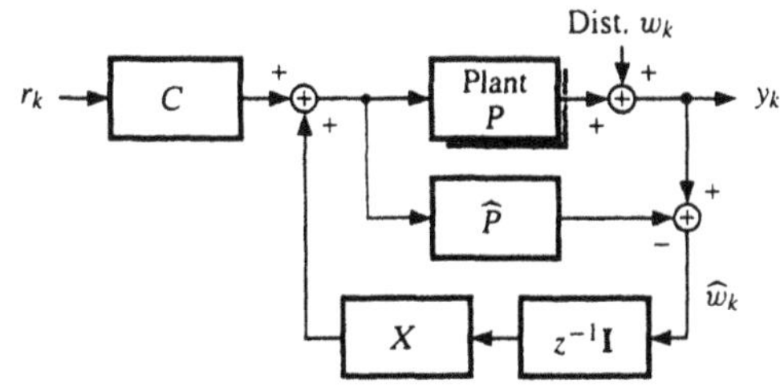

Fig. 1. Adaptive inverse control system block diagram.

2. ADAPTIVE DIGITAL FILTERING

Adaptive inverse control relies on adaptive digital filtering methods. Adaptive filters have an input, an output, and a "special input" called the desired response. The desired response d_k specifies the output we wish the filter to have. It is used to calculate an error signal e_k, which in turn is used to modify the internal parameters of the filter in such a way that the filter "learns" to

[1] Partially supported by a William Grigsby Sencenbaugh grant.

perform a certain function. Often, the trick to applying adaptive filtering to a specific application is finding a way to generate an appropriate d_k signal.

For the sake of the present discussion we limit ourselves to *finite impulse response* (FIR) filters. The filter output is computed as a weighted sum of its current and N previous inputs $y_k = \mathbf{W}X_k$, where the column-vector $X_k = [x_k^T, x_{k-1}^T \ldots x_{k-N}^T]^T$ and $\mathbf{W}$ is the *weight matrix* of the filter.

Linear SISO filters have a single input and a single output at each time instant; for $\mathbf{W}$ to represent a SISO system, x_k and y_k must be scalars. $\mathbf{W}$ is then a single row, and its components are the impulse response of the filter. Linear MIMO filters have (possibly) many inputs and outputs at each time instant so x_k and y_k are (column) vector signals. $\mathbf{W}$ then has many rows, and the individual impulse responses are interleaved in each row.

The weights in $\mathbf{W}$ may be adapted in a variety of ways in order that the output of the filter learns to closely match the desired output. For example, the weights may be updated by a gradient-descent optimization procedure such as $\Delta\mathbf{W} = 2\mu(d_k - y_k)X_k^T$, where μ is a small positive *learning constant*. This rule, commonly known as the matrix-LMS algorithm is well described in several textbooks (Haykin, 1996; Widrow and Stearns, 1985). Other, more sophisticated, algorithms, such as matrix-RLS (see the algorithm in Alg. 1) usually converge more quickly, and are very popular. For a good overview, see Glentis *et al.* (1999).

Algorithm 1 Matrix-RLS algorithm.

$$\pi(k) = X(k)^T \Phi_{xx}^{-1}(k-1)$$
$$r(k) = \frac{1}{\lambda + \pi(k)X(k)}$$
$$K(k) = r(k)\pi(k)$$
$$\zeta(k) = d(k) - \mathbf{W}(k-1)X(k)$$
$$\mathbf{W}(k) = \mathbf{W}(k-1) + \zeta(k)K(k)$$
$$\Phi_{xx}^{-1}(k) = \frac{1}{\lambda}\text{tri}\left\{\Phi_{xx}^{-1}(k-1) - \pi(k)^T K(k)\right\},$$

where the tri{} operator takes the upper or lower triangular part of a matrix and replicates it in the lower or upper part to preserve symmetry. λ is a forgetting constant, and is set slightly less than 1. $X(k)$ is the tap-delay-line input to the filter at time k, such that the filter output is $\mathbf{W}(k)X(k)$. $\Phi_{xx}^{-1}(k)$ is initialized to a diagonal matrix with large (order of 10,000) entries.

3. CONTROLLING LINEAR SISO PLANTS

We now consider adapting the feedforward controller C. (We shall restrict our development to apply only to stable plants. If the plant of interest is unstable, conventional feedback should be applied to stabilize it. Then the combination of the plant and its feedback stabilizer can be regarded as an equivalent stable plant.) The goal is to make the dynamics of the controlled system PC approximate the fixed filter M as closely as possible, where M is a user-specified *reference model.* The input reference signal r_k is filtered through M to create a desired response d_k for the plant output. The measured plant output is compared with the desired plant output to create a system error signal $e_k^{(sys)} = d_k - y_k$. We wish to adapt C to minimize the mean-squared system error.

We note that if the plant is minimum-phase, that is, has all of its poles and zeros inside the unit circle in the z-plane, then the inverse will be stable with all of its poles inside the unit circle. If the plant is nonminimum-phase, then some of the poles of the inverse will be outside the unit circle. According to the theory of two-sided z-transforms, the inverse will then either be unstable *or* noncausal. Since minimizing system error will not lead to an unstable solution—which would have unbounded system error—the algorithm will attempt to match a noncausal solution. This will result in very poor control unless the reference model has built-in latency. The longer the latency, the better we can approximate a delayed version of a noncausal inverse with a causal controller. A typical design latency is the transport delay of the plant.

3.1 *Linear SISO Feedforward Control:*

The challenge when attempting to adapt C is in generating its desired response signal. We notice that the system desired response is available at the output of the plant, and not at the output of the adaptive controller. A variety of solutions are known when the plant is a linear SISO system (Widrow and Walach, 1996). One simple and very effective method takes advantage of the commutability of transfer functions of linear SISO systems. That is, $P(z)C(z) = C(z)P(z)$. The block diagram of Fig. 2(a) may then be used to adapt the controller. A random signal n_k is filtered by a digital copy [2] of the plant model $\widehat{P}(z)$ and then by the controller $C(z)$. It is also filtered by the reference model $M(z)$. Since the output of the reference model is the desired output of the cascade $C(z)\widehat{P}(z)$, it is used as the desired response signal for $C(z)$. A filter whose weights are a digital copy of $C(z)$ is used as the feedforward controller.

This method is unbiased by zero-mean disturbance, but may be affected by plant modeling errors. More sophisticated methods are available to adapt controllers and which are not affected by plant modeling errors (Widrow and Walach, 1996). If the RLS algorithm is used to adapt $C(z)$, convergence occurs

[2] The weights of the digital copy are identical to the weights of the adaptive plant model, although the input to both filters is different and hence the outputs of the two filters are different. Note that alternative adaptive methods such as MRAC by Åström and Wittenmark (1995) do not require a digital copy of a plant model. For a good discussion comparing and contrasting STR/MRAC with AIC, see Appendix C of Widrow and Walach (1996).

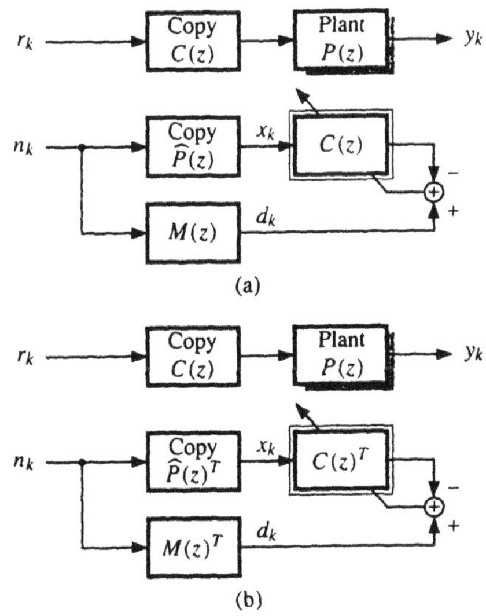

Fig. 2. Adapting a linear controller $C(z)$. (a) For a SISO linear plant; (b) For a MIMO linear plant. The plant modeling process has been omitted for clarity.

within twice as many iterations as there are taps in the controller filter C (Haykin, 1996).

3.2 *Linear SISO Disturbance Canceling*

The dynamic response of the system may now be controlled but plant disturbance, however, is not yet rejected. This can be accomplished via another special adaptive filter called the disturbance canceler X. In order to avoid bias in the adaptive plant model due to the disturbance, a special architecture must be used when adapting $\widehat{P}$ and X simultaneously. The solution is shown in Fig. 3. It can be shown that the plant model will adapt to the correct solution using this scheme (Plett, n.d.; Plett, 1998) so long as the disturbance is is zero-mean and uncorrelated with u_k.

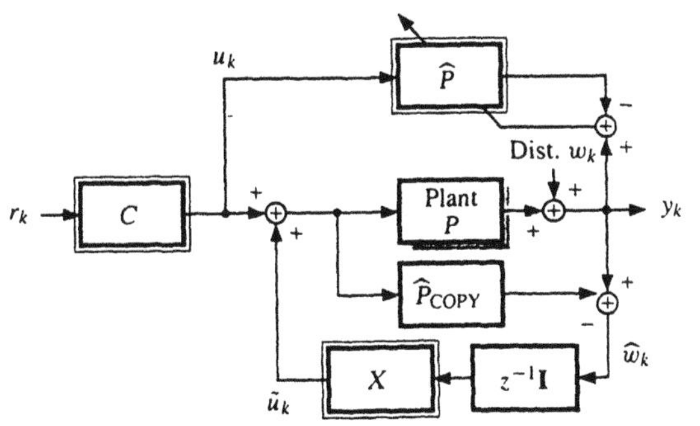

Fig. 3. Correct on-line adaptive plant modeling in conjunction with disturbance canceling for linear plants. The circuitry for adapting C has been omitted for clarity.

When considering the linear SISO case, we would like to adapt the disturbance canceler X such that $z^{-1}P(z)X(z) = -1$. This would entirely cancel the disturbance, but result in a non-causal X. We can still use this formula to adapt $X(z)$, as shown in Fig. 4(a), but by the adaptive method, $X(z)$ will adapt to the optimal causal solution. [3] As when adapting a controller, we take advantage of the fact that linear systems commute in order to generate an adaptation signal for X.

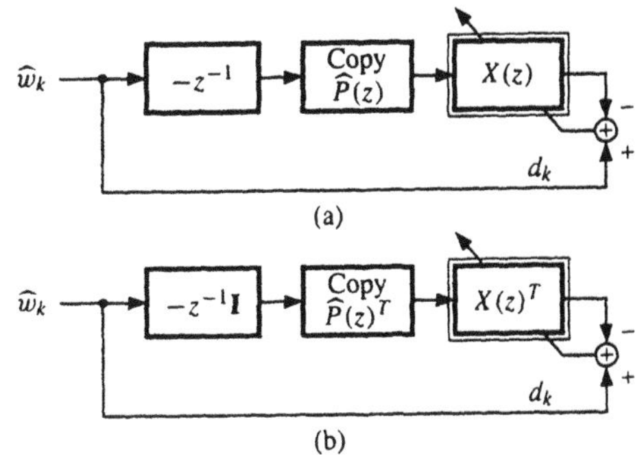

Fig. 4. Architectures for training disturbance canceler X; (a) for linear SISO systems; (b) for linear MIMO systems. The actual disturbance canceler is a digital copy of X.

4. THE NOVEL APPROACH TO LINEAR MIMO ADAPTIVE INVERSE CONTROL

4.1 *Feedforward Control*

While linear SISO systems have transfer functions, linear MIMO systems have transfer function matrices, denoted for example as $[P(z)]$. Matrix multiplication is not in general commutative; therefore, $[P(z)][C(z)] \neq [C(z)][P(z)]$ and we can not use the same simple method as with the SISO system. A number of ways have been proposed to adapt a controller for a linear MIMO system (Plett, n.d.; Plett, 1998; Widrow *et al.*, 1998; Widrow and Plett, 1997; Widrow and Walach, 1996). These methods tend to make inefficient use of the available data and are slow. Here, we present a very simple and fast method to adapt a linear MIMO controller (Plett, 2000). It uses the fact that $[P(z)][C(z)] = [C(z)]^T[P(z)]^T$. The block diagram in Fig. 2(b) may then be used to adapt the controller. The entire operation depends on being able to take the "transpose" of an adaptive filter representing a transfer function matrix. These filters are actually stored as impulse-response matrices, and the transpose operation is simply a re-organization of the components of the impulse-response matrices. The mechanics of taking a filter transpose are discussed in the next section. As with the SISO linear case, convergence using the RLS algorithm occurs within twice as many iterations as there are taps in the longest impulse response in the

[3] We note that this solution amounts to a predictor which predicts a future sample of the disturbance, cascaded with a plant inverse. Since optimal prediction is in general a nonlinear operation, the ideal disturbance canceler is a nonlinear adaptive filter even if the plant is linear!

MIMO controller filter C. This is an improvement of many orders of magnitude when compared with the other known cited methods.

4.2 *Transpose of Transfer Functions*

In order to train the controller, notice that we need the transposed filter $[\hat{P}(z)]^T$. The weight matrix for this filter is not the same as $\mathbf{W}_{\hat{P}}^T$. To find the correct weight matrix for an arbitrary transposed filter $[H(z)]^T$ consider first the weight matrix for some arbitrary filter $[H(z)]$. For the sake of example we will assume $[H(z)]$ has three inputs, two outputs and $N = 1$ so that the impulse response is two samples long. Then, the H-matrix has entries as shown in Fig. 5.

$[h_{11}]_0$	$[h_{12}]_0$	$[h_{13}]_0$	$[h_{11}]_1$	$[h_{12}]_1$	$[h_{13}]_1$
$[h_{21}]_0$	$[h_{22}]_0$	$[h_{23}]_0$	$[h_{21}]_1$	$[h_{22}]_1$	$[h_{23}]_1$

Fig. 5. Weight matrix for filter $[H(z)]$.

There are six impulse responses embedded in this matrix: h_{11}, h_{12}, h_{13}, h_{21}, h_{22} and h_{23}. When we take the z-transform and make the transfer-function matrix for this filter, we get

$$[H(z)] = \begin{bmatrix} H_{11}(z) & H_{12}(z) & H_{13}(z) \\ H_{21}(z) & H_{22}(z) & H_{23}(z) \end{bmatrix}$$

which has transpose

$$[H(z)]^T = \begin{bmatrix} H_{11}(z) & H_{21}(z) \\ H_{12}(z) & H_{22}(z) \\ H_{13}(z) & H_{23}(z) \end{bmatrix}.$$

So, the transpose of the weight filter has entries as shown in Fig. 6. It may be computed in Matlab, for example, with a sequence of matrix `reshape` and `permute` commands.

$[h_{11}]_0$	$[h_{21}]_0$	$[h_{11}]_1$	$[h_{21}]_1$
$[h_{12}]_0$	$[h_{22}]_0$	$[h_{12}]_1$	$[h_{22}]_1$
$[h_{13}]_0$	$[h_{23}]_0$	$[h_{13}]_1$	$[h_{23}]_1$

Fig. 6. Weight matrix for filter $[H(z)]^T$.

4.3 *Wiener Solution for Controller*

A property of linear adaptive filtering is that the solution to which the algorithm converges may be verified mathematically. This solution is known as the *Wiener* solution. There are three cases to consider when computing the Wiener solution for the controller weight matrix when the plant is linear MIMO: (1) the plant has more outputs than inputs; (2) the plant has fewer outputs than inputs; (3) the plant has an equal number of outputs and inputs. The details of the Wiener solution are in Plett (2000).

More outputs than inputs: If the plant has more outputs than inputs, then $\widehat{P}(z)^T\widehat{P}(z)$ is generally invertible. The controller converges to

$$C^{(opt)}(z) = \left[\widehat{P}(z)^T\widehat{P}(z)\right]^{-1}\widehat{P}(z)^T M(z),$$

which is the pseudo-inverse of $\widehat{P}(z)$ in cascade with the reference model. This is the desired solution.

More inputs than outputs: If the plant has more inputs than outputs, then $\widehat{P}(z)\widehat{P}(z)^T$ is generally invertible. One possible solution for $C(z)$ is

$$C^{(opt)}(z) = \widehat{P}(z)^T\left[\widehat{P}(z)\widehat{P}(z)^T\right]^{-1} M(z).$$

This is the minimum-norm solution. Generally, when there are more inputs than outputs there will be many solutions to $C(z)$ which all achieve zero error. The minimum-norm solution is one of these, and uses the least amount of control effort (in a mean-squared sense). It is not a guaranteed solution.

Equal number of inputs and outputs: If the plant has an equal number of inputs and outputs, and if the plant is invertible, then both solutions become

$$C^{(opt)}(z) = \widehat{P}(z)^{-1}M(z).$$

We note that all three Wiener solutions are based on the inverse of the plant model and not on the inverse of the plant. So, the efficacy of this method is dependent on an accurate plant model.

4.4 *Disturbance Canceling*

We can adapt a disturbance canceler for a linear MIMO system using the same equation as for the SISO case, and adopting the transpose method used when adapting C for the MIMO system. The block diagram of the adaptation method is shown in Fig. 4(b).

4.5 *Speed of Convergence*

There are two convergence issues to deal with. First, convergence of the plant model, and then convergence of the controller/disturbance canceler. The plant model needs to converge before either the controller or disturbance canceler can converge. Here we assume that matrix-RLS is used to adapt the two filters.

According to Haykin, RLS converges in two times as many iterations as there are taps in the FIR filter (Haykin, 1996). So, for example, if a matrix-FIR filter has five taps per sub-filter, convergence is achieved in about ten iterations. Similarly, convergence of the controller or disturbance canceler occurs in about twice as many iterations as there are taps in each filter. However, since the controller- and canceler adaptation process is done offline, it can be done quickly in the background and entire system convergence occurs can occur in about twice as many iterations as there are taps in the plant-model FIR filter.

5. EXAMPLE

Aspects of flight control for a Boeing 747 aircraft were selected to demonstrate linear, MIMO control. The dynamics of the airplane have been approximated by a linear model around an equilibrium point. In the case at hand, the equilibrium "point" is level flight at 40,000 ft and a nominal forward speed of Mach 0.8 (774 ft/sec). For further detail regarding particulars of this example, see Plett (1998). Here, three cases are considered. All simulations consider the system to have two inputs (rudder angle and aileron angle). Simulations have one through three outputs (yaw-rate, roll-rate, bank-angle).

The discrete-time state-space model of the plant has

$$A_d = \begin{bmatrix} 0.8876 & -0.3081 & 0.0415 & 0.0198 \\ 0.2020 & 0.3973 & -0.0046 & 0.0024 \\ -1.2515 & 0.5106 & 0.7617 & -0.0139 \\ -0.3313 & 0.1510 & 0.4407 & 0.9976 \end{bmatrix}$$

$$B_d = \begin{bmatrix} 0.4806 & -0.0013 \\ -1.5809 & 0.3887 \\ 0.0599 & 4.8390 \\ 0.0390 & 1.2585 \end{bmatrix},$$

with a sampling period of 0.5 *sec*. The output matrix was chosen to be the first one, two, or three rows of

$$C_d = \begin{bmatrix} 0 & 1 & 0 & 0 \\ 0 & 0 & 0 & 1 \\ 0 & 0 & 1 & 0 \end{bmatrix},$$

depending on whether the plant had one, two or three outputs, and $D_d = 0$. The reference input was filtered white noise, and disturbance was nonlinearly filtered white noise, added directly to the state. The reference model was a unit delay: $M(z) = z^{-1}\mathbf{I}$.

Tracking results for converged controllers (in the absence of disturbance) are presented in Fig. 7. Tracking is essentially perfect when there are more plant inputs than outputs, and is also very good when there are an equal number of plant inputs and outputs. When there are more plant outputs than inputs, it becomes impossible to get perfect tracking in general, and the controller adapts to find the solution which minimizes the mean-square output error.

Note: simulations in Plett (1998) considered two-input two-output control. The controller, trained with the BPTM method, converged in about 10^8 iterations. Learning curves for the present method and the three simulation cases are presented in Fig. 8. We see convergence in all cases within 200–400 iterations. Training time, to a similar level of steady-state error, is

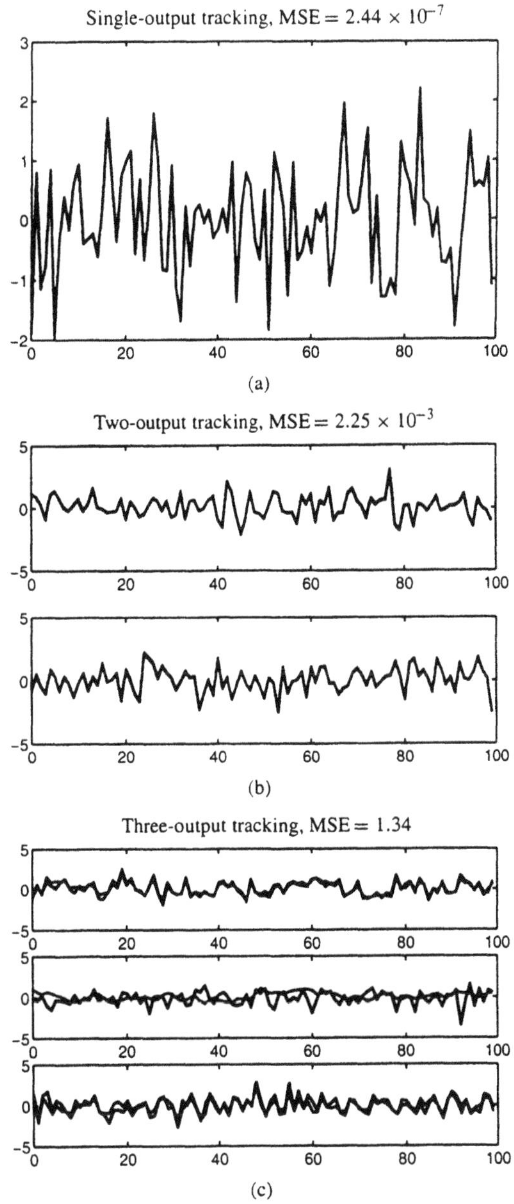

Fig. 7. Tracking performance for three trained controllers. The plant had one through three outputs in (a) through (c), respectively. The dark line is the desired plant output; the light line is the actual plant output.

improved by several orders of magnitude. As a point of reference, the plant model had 61 taps in each impulse response, and the controller had 71 taps in each impulse response.

Disturbance canceling was tested for the MIMO system, and results are plotted in Fig. 9. The figure shows the squared-norm of the system error plotted versus iteration. Disturbance was present at all times, and the (trained) disturbance canceler was turned "on" at time 1000. As can be seen, the disturbance canceler removes essentially all of the disturbance, resulting in near-perfect tracking even in the presence of disturbance.

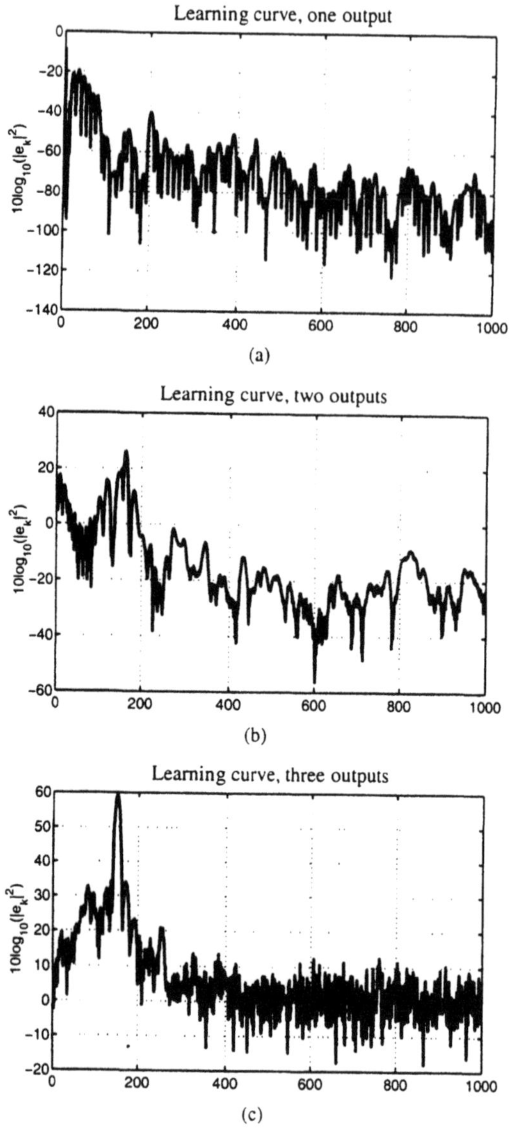

Fig. 8. Learning curves training the three controllers (in dB). The plant had one through three outputs in (a) through (c), respectively. The horizontal axis is the number of training iterations.

6. CONCLUSIONS

A new method is presented in this paper for adapting the controller and disturbance canceler in an adaptive-inverse-control system for linear MIMO plants. The method works very well and is very fast. Simulation results indicate orders-of-magnitude speed improvement over known methods. The method works for MIMO systems with equal or unequal numbers of inputs and outputs.

REFERENCES

Åström, K. J. and B. Wittenmark (1995). *Adaptive Control*. second ed.. Addison-Wesley. Reading, MA.

Glentis, G., K Berberidis and Sergios Theodoridis (1999). Efficient least squares adaptive algo-

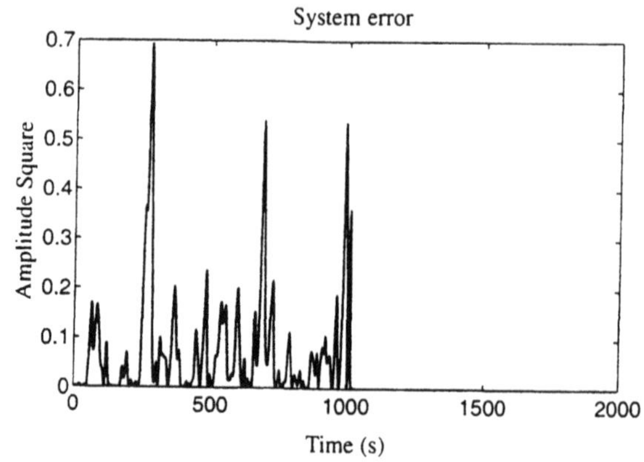

Fig. 9. Disturbance canceling for MIMO system.

rithms for FIR transversal filtering. *IEEE Signal Processing Magazine* **16**(4), 13–41.

Haykin, S. (1996). *Adaptive Filter Theory*. third ed.. Prentice Hall. Upper Saddle River, NJ.

Plett, G. L. (1998). Adaptive Inverse Control of Plants with Disturbances. PhD thesis. Stanford University. Stanford, CA 94305.

Plett, G. L. (2000). Some results concerning fast linear MIMO adaptive inverse control. Technical Report EAS_ECE_2000_11. ECE Department, University of Colorado at Colorado Springs. P.O. Box 7150, Colorado Springs, CO 80933–7150.

Plett, G. L. (n.d.). Adaptive inverse control of unmodeled stable SISO and MIMO linear systems. *International Journal of Adaptive Control and Signal Processing*. (in press).

Widrow, B. and E. Walach (1996). *Adaptive Inverse Control*. Prentice Hall P T R. Upper Saddle River, NJ.

Widrow, B. and G. L. Plett (1996*a*). Adaptive inverse control based on linear and nonlinear adaptive filtering. In: *Proceedings of International Workshop on Neural Networks for Identification, Control, Robotics and Signal/Image Processing*. (Venice, Italy: August 1996). pp. 30–38.

Widrow, B. and G. L. Plett (1996*b*). Adaptive inverse control based on linear and nonlinear adaptive filtering. In: *Proceedings of the World Congress on Neural Networks*. (San Diego, CA: September 1996). pp. 620–27.

Widrow, B. and G. L. Plett (1996*c*). 'Intelligent' adaptive inverse control. In: *Proceedings of IFAC*. (San Francisco, CA: July 1996). pp. 104–105.

Widrow, B. and G. L. Plett (1997). Nonlinear adaptive inverse control. In: *Proceedings of the 36th IEEE Conference on Decision and Control*. Vol. 2. (San Diego, CA: December 1997). pp. 1032–1037.

Widrow, B. and S. D. Stearns (1985). *Adaptive Signal Processing*. Prentice-Hall. Englewood Cliffs, NJ.

Widrow, B., G. L. Plett, E. Ferreira and M. Lamego (1998). Adaptive inverse control based on nonlinear adaptive filtering. In: *Proceedings of 5th IFAC Workshop on Algorithms and Architectures for Real-Time Control AARTC'98*. (Cancun, MX: April 1998). pp. 247–252. (invited paper).

NORM OPTIMAL ITERATIVE LEARNING CONTROL APPLIED TO CHAIN CONVEYOR SYSTEMS

T. Al-Towaim * P. Lewin * E. Rogers *

* *Dept. of Electronics and Computer Science, University of Southampton, Southampton SO17 1BJ UK*

Abstract: Norm optimal iterative learning control is an optimization-based approach where, in contrast to other such approaches, a term in the cost function is a measure of the difference between the control vectors used on successive trials. This approach has very well defined theoretical properties and is predicted to give superior performance relative to alternatives when applied to physical processes. In this paper, we describe on-going work towards the implementation of controllers designed using this approach to chain conveyor systems with the eventual aim of examining/verifying performance in 'real world' applications. *Copyright © 2001 IFAC*

Keywords: iterative learning control, optimal algorithms.

1. INTRODUCTION

Iterative learning control (ILC) is a technique to control systems operating in a repetitive mode with the additional requirement that a specified output trajectory $r(t)$ defined over a finite interval $[0, T]$ is followed to high precision. There are numerous examples of such systems including robot manipulators that are required to repeat a given task to high precision, chemical batch processes or, more generally, the class of tracking systems. Here we will consider another application area - chain conveyor systems.

Motivated by human learning, the basic of idea of ILC is to use information from previous executions of the task in order to improve performance from trial to trial in the sense that the tracking error is sequentially reduced ((Arimoto *et al.*, 1984; Moore, 1993)). Typical ILC algorithms construct the input to the plant on a given trial from the input used on the previous trial plus an additive increment that is typically a function of past values of the observed output error, i.e. the difference between achieved and desired plant output. The objective of constructing a sequence of input functions $\{u_k(t)\}_k$, $t \in [0, T]$, such that the performance as the task is repeated is gradually improving, can be refined to a convergence condition on the input and error:

$$\lim_{k\to\infty} ||e_k|| = 0, \quad \lim_{k\to\infty} ||u_k - u_\infty|| = 0 \quad (1)$$

where $e_k(t)$ is the error on trial k, i.e. the difference between $r(t)$ and the system output $y_k(t)$, and $u_k(t)$ is the input to the system on this trial.

The above definition of convergent learning is a stability problem on an infinite-dimensional two-dimensional (2D)-product space. As such it places the analysis of ILC schemes firmly outside the scope of traditional control theory. In particular, ILC must be studied in the context of fixed-point problems or, more precisely, linear repetitive processes ((Rogers and Owens, 1992; Rogers *et al.*, 2001)).

Since the ILC concept was proposed, a very large number of approaches have been considered but here we consider so-called norm optimal ILC which can be realized in terms of current trial mechanisms combined with feedforward of previous trial data. The algorithm is based on splitting the two-dimensional dynamics into two separate one-dimensional problems. This is done by introducing a performance criterion as the basis of specifying the control input to be used on each trial.

Norm optimal approach has a mature theoretical basis (Amann *et al.*, 1996) and the objective of this paper is to report results from a research programme whose ultimate goal is experimental application of ILC schemes designed using this approach to a chain conveyor rig to assess performance in 'real world applications - both stand alone and in comparison with alternatives. We begin in the next section by giving the required results from the norm optimal ILC theory.

2. BACKGROUND

The following is the formal definition of a successful ILC algorithm.

Definition 1. Consider a dynamic system with input u and output y. Let $\mathcal{Y}$ and $\mathcal{U}$ be the output and input function spaces respectively and let $r \in \mathcal{Y}$ be a desired reference trajectory from the system. Then an ILC algorithm is successful if, and only if, it constructs a sequence of control inputs $\{u_k\}_{k\geq 0}$ which, when applied to the system or plant (under identical experimental conditions), produces an output sequence $\{y_k\}_{k\geq 0}$ with the following properties of convergent learning:

$$\lim_{k\to\infty} y_k = r, \quad \lim_{k\to\infty} u_k = u_\infty \tag{2}$$

Here convergence is interpreted in terms of the topologies assumed in $\mathcal{Y}$ and $\mathcal{U}$ respectively.

Note: This general description includes linear and nonlinear dynamics, continuous or discrete plants, and time-invariant or time-varying systems.

Now let the space of output signals $\mathcal{Y}$ be a real Hilbert space and $\mathcal{U}$ also be a real (and possibly distinct) Hilbert space of input functions. Then the respective inner products (denoted by $\langle\cdot,\cdot\rangle$) and norms $||\cdot||^2 = \langle\cdot,\cdot\rangle$ are indexed in a way that reflects the space if it is appropriate to the discussion.

The dynamics of the plant considered here are approximated by a linear model which in operator form can be written as

$$y = Gu + z_0 \tag{3}$$

where no loss of generality arises from setting $z_0 = 0$. Also it is clear that the ILC procedure, if convergent, solves the problem $r = Gu_\infty$ for u_∞ and, if G is invertible, the formal solution is just $u_\infty = G^{-1}r$. A basic premise of the ILC approach is that the direct inversion of G is regarded as an impractical solution because it requires exact knowledge of G and involves derivatives of r. This high-frequency gain characteristic would make the approach sensitive to noise and other disturbances. Also it can be argued that inversion of the whole plant G is unnecessary as the solution only requires finding the pre-image of r under G.

The above problem is easily seen to be equivalent to finding the minimizing input u_∞ for the optimization problem

$$min_u\{||e||^2 \ : \ e = r - y, y = Gu\} \tag{4}$$

The optimal error $||r - Gu_\infty||^2$ is a measure of how well the ILC algorithm has solved the inversion problem. It also represents the best that the system can do in tracking the signal r. The case of interest here is when the optimal error is zero, i.e. u_∞ is a solution of $r = Gu_\infty$. Also (4) is clearly a singular optimal control problem which by its very nature requires an iterative solution.

There are an infinity of potential iterative procedures for solving (4) and of these the gradient approach has the simplest form and has been extensively investigated in the ILC literature. A gradient based ILC algorithm has the form

$$u_{k+1} = u_k + \epsilon_{k+1} G^* e_k \tag{5}$$

where $G^* : \mathcal{Y} \to \mathcal{U}$ is the adjoint operator to G, and ϵ_{k+1} is a step length to be chosen at each iteration. This general approach suffers from the need to choose this step length on each trial and the feedforward structure of the iteration takes no account of current trial effects - including disturbances and plant modeling errors.

Norm optimal ILC has the following two crucial properties relative to the gradient based algorithms discussed above.
1. Automatic choice of step size.
2. Potential for improved robustness through the use of causal feedback of current trial data and feedforward of data from previous trials.
In particular, the class of ILC algorithms considered here compute the input on trial $k+1$ as the solution of the minimum norm optimization problem

$$u_{k+1} = \arg min_{u_{k+1}}\{J_{k+1}(u_{k+1})\} \tag{6}$$

subject to

$$e_{k+1} = r - y_{k+1},\ y_{k+1} = Gu_{k+1} \tag{7}$$

where the performance index (or optimality criterion) used is

$$J_{k+1}(u_{k+1}) = ||e_{k+1}||^2_{\mathcal{Y}} + ||u_{k+1} - u_k||^2_{\mathcal{U}} \quad (8)$$

The initial control $u_0 \in \mathcal{U}$ can be arbitrary but, in practice, will be a good first guess at the solution of the problem. Also the relative weighting of reducing the current trial error against minimizing the deviation in the control input signals used on successive passes can be absorbed into the definitions of the norms in $\mathcal{Y}$ and $\mathcal{U}$.

The benefits of this approach are immediate from the simple interlacing result

$$||e_{k+1}||^2 \leq J_{k+1}(u_{k+1}) \leq ||e_k||^2, \; \forall \, k \geq 0 \quad (9)$$

which follows from optimality and the fact that the (non-optimal) choice of $u_{k+1} = u_k$ would lead to the relation $J_{k+1}(u_k) = ||e_k||^2$. This result states that the algorithm is a descent algorithm as the norm of the error is monotonically decreasing in k and also equality holds if, and only if, $u_{k+1} = u_k$, i.e. when the algorithm has converged and no more input-updating takes place.

The controller on trial $k + 1$ is given by

$$u_{k+1} = u_k + G^* e_{k+1}, \; \forall k \geq 0 \quad (10)$$

This relationship, together with the error update relation

$$e_{k+1} = (I + GG^*)^{-1} e_k, \; \forall \, k \geq 0 \quad (11)$$

and the recursive input update relation

$$u_{k+1} = (I + G^*G)^{-1}(u_k + G^* r), \; \forall \, k \geq 0 \quad (12)$$

can be used to undertake a detailed analysis of the (theoretical) properties of this class of ILC laws (Amann *et al.*, 1996).

In this paper, we will only consider the special case of $J_{k+1}(u_{k+1})$ defined as follows applied to a linear time invariant differential plant model with state space matrices (A, B, C) (state, input and output respectively)

$$\begin{aligned} J_{k+1}(u_{k+1}) = \frac{1}{2} \int_0^T & \{ e^T_{k+1}(t) Q e_{k+1}(t) \\ & + H^T R H \} dt \\ & + \frac{1}{2} e^T_{k+1}(T) F e_{k+1}(T) \end{aligned} \quad (13)$$

where

$$H = u_{k+1}(t) - u_k(t) \quad (14)$$

and the symmetric matrices $Q, R,$ and F satisfy the normal linear quadratic optimal control assumptions. Standard optimal control theory now gives the solution as

$$\begin{aligned} \dot{\psi}_{k+1}(t) &= -A^T \psi_{k+1}(t) - C^T Q e_{k+1}(t) \\ u_{k+1}(t) &= u_k(t) + R^{-1} B^T \psi_{k+1}(t) \\ \psi_{k+1}(T) &= C^T F e_{k+1}(T), \; t \in [0, T] \end{aligned} \quad (15)$$

This representation is non-causal (in the standard sense) but it can be transformed into a causal implementation as detailed next for the case of a relaxation factor α.

Transform the costate vector $\psi_{k+1}(t)$ using

$$\begin{aligned} \psi_{k+1}(t) = & -K(t) \left[x_{k+1}(t) - \alpha x_k(t) \right] \\ & + \zeta_{k+1}(t) \end{aligned} \quad (16)$$

where the feedback gain matrix $K(t)$ satisfies the well known Riccati (matrix) differential equation

$$\begin{aligned} \dot{K}(t) &= -A^T K(t) - K(t) A \\ & \quad + K(t) B^T R^{-1} B^T K(t) - C^T Q C \\ K(T) &= C^T F C \end{aligned} \quad (17)$$

Note that this last equation of independent of the inputs, states and outputs of the system and hence only needs to be computed once before the sequence of trials begin.

The predictive or 'feedforward' term $\zeta_{k+1}(t)$ needs to be computed on each trial using

$$\begin{aligned} \dot{\zeta}_{k+1}(t) = & -(A - BR^{-1} B^T K)^T \zeta_{k+1}(t) \\ & - \alpha C^T Q e_k(t) \\ & + (1 - \alpha) K B u_k(t) \\ & - (1 - \alpha) C^T Q r(t) \end{aligned} \quad (18)$$

with terminal boundary condition

$$\zeta_{k+1}(T) = C^T F \left[\alpha e_k(T) + (1 - \alpha) r(T) \right] \quad (19)$$

The algorithm is now causal since (18) and (19) can be solved off-line by reverse time simulation using available previous trial data.

Next we describe the target application area - chain conveyor systems and give the simulation based performance results against which the results from experimental implementation will eventually be compared.

3. CHAIN CONVEYOR SYSTEMS

Chain conveyors are extremely simple in construction as they generally consist of parallel lengths of chain that are connected at regular intervals to form a conveying surface. The use of chain conveyors is of particular importance to the food manufacturing industry, with the twin requirement of high standards of hygiene and low overheads. As chain conveyors have no sensitive components, and most if not all are metal, cleaning with high-pressure steam to ensure hygiene does not pose a problem. The use of standard components with no precision engineering makes them inexpensive- a very important factor in the highly competitive world of food manufacture.

Previous work [6] has applied three term (or PID) ILC schemes to this conveyor structure both in simulation and experiment. The major conclusion was that this is a highly relevant application area for ILC. Despite this, achievable performance was limited by the PID structure in certain cases. Hence the decision was made to apply norm optimal based ILC schemes outlined in the previous section to chain conveyor systems. The eventual goal is to assess the performance of such schemes in 'real world' operation - both stand alone and comparatively. In the remainder of this paper, we describe the chain conveyor system to be used, its mathematical modeling and configuration for the actual implementation of control action, and the design of the candidate ILC schemes.

The System.

The chain conveyor systems considered in this work have two possible operational modes - indexing and synchronization. When operating in an indexing mode the conveyor moves one item at a time under a dispenser. The dispenser remains stationary and product is dispensed when the conveyor comes to rest. This motion is then repeated for the next item. In synchronization mode the conveyor moves at constant velocity and the dispenser moves back and forth. Product is dispensed when the position and velocity of the dispenser are synchronized to that of item on the overall performance of the conveyor. The system is measured by its accuracy combined with rate of throughput and reliability. Each requirement introduces difficulties. Accuracy will degrade with time due to component wear. Commonly this is overcome by regular manual re-calibration of the system.

High rates of throughput imply large accelerations. These produce large electrical and mechanical stresses in the system components that increase wear and reduces accuracy. Ultimately high stresses will cause the premature failure of components, reducing reliability and overall throughput. It is therefore necessary to ensure that the controller demand does not require the actuator to perform outside of the manufacturers specifications. As described in (Barton *et al.*, 2000) the system has many problems that a PID controller cant deal with it at high enough accuracy to meet typical performance requirements.

The conveyor, see Figure 1, is constructed from a 3m long framework of right angle steel section

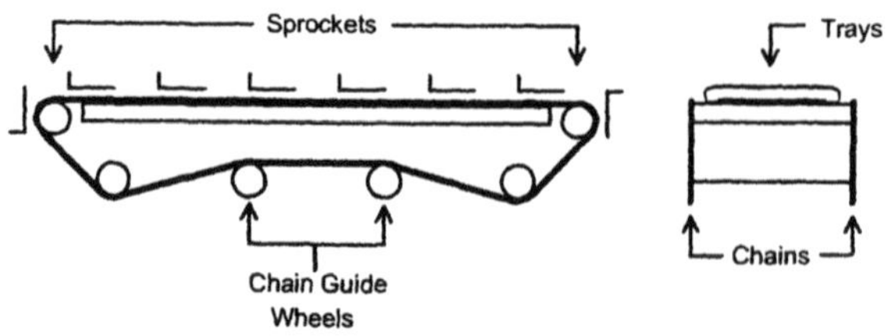

Fig. 1. Schematic of Conveyor Construction.

and consists of two parallel strings of 0.5 pitch steel roller chain. At 300 mm intervals there is an aluminum plate supported on a rod that is pinned through a bushing on each chain. A standard squirrel cage induction motor supplied by a variable voltage variable frequency (VVVF) inverter, that is delta connected to a 3 phase pulse width modulated (PWM) inverter drives the conveyor through a timing belt drive with a 5:2 reduction ratio. The induction motor is oversized for the mechanical load to ensure that the actuator would not limit system performance. A 500 pulse per revolution optical encoder, measured on the motor shaft with differential outputs, provides position feedback. Processing using a DEVA 004 motion control card increases the resolution to the equivalent of 2000 pulses per revolution.

The dispenser, see Figure 2, consists of a trolley that moves linearly above the conveyor. The trolley is an open frame, as this allows dispensing systems and instruments to be exchanged when required. A long belt supplies the linear motion, rotary motion being provided by an identical induction motor/belt drive system as for the conveyor.

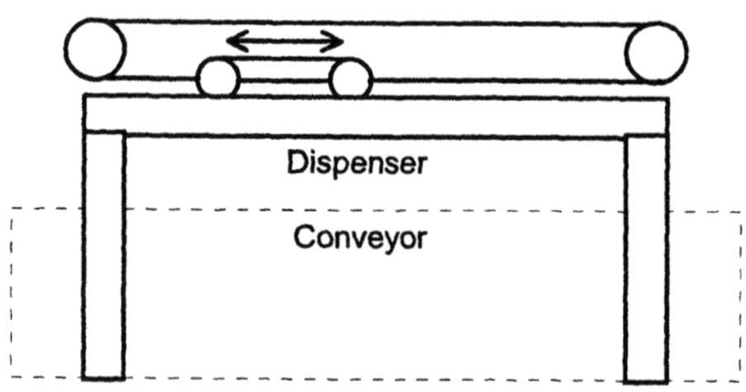

Fig. 2. Schematic of Conveyor Construction.

Frequency-Domain Model.

The models of the conveyor and the dispenser used are linear approximations, which were developed for simulation purposes. These were obtained by driving the conveyor and the dispenser with a variable frequency sinusoid, provided by a dc motor drive and recording the frequency response. The motor velocity was measured by using a tachometer, and then scaled to give a response relating input voltage to output velocity in counts/seconds. From the resulting Bode plots linear approximations were derived for the conveyor and dispenser respectively as

$$G_{\text{conv}}(s) = \frac{615.06 \times 10^6}{(s^2 + 49s + 35^2)(s^2 + 54s + 180^2)} \tag{20}$$

$$G_{\text{disp}}(s) = \frac{6.47 \times 10^6}{(s + 35)(s^2 + 99s + 110^2)} \tag{21}$$

Control Implementation.

A PC controls the system that includes the DEVA interface card. The card has two 14-bit Digital to Analogue Converters ($D\backslash A$) for speed demand output and two optional-isolated digital outputs for axes enabling. A programmable interrupt controller is provided to produce regularly timed processor interrupts suitable for running discrete controllers. As the inverters are unable to accept a -10V to +10V speed demand, a positive analogue speed signal is provided, with a single line of the parallel port linked to the direction setting pin of the inverter to provide direction control. A program, written in C, provides a user interface to the hardware and also implements the controller.

System Simulation.

In order to begin the evaluation of the performance of norm optimal ILC designs in this area, a simulation of the system operating in synchronous mode has been constructed in MATLAB/SIMULINK. Also a range of controllers have been designed. For example, Figure 3 shows a sample design (for the dispenser transfer function) where the reference signal used is the same as that used in the previous work (Barton *et al.*, 2000).

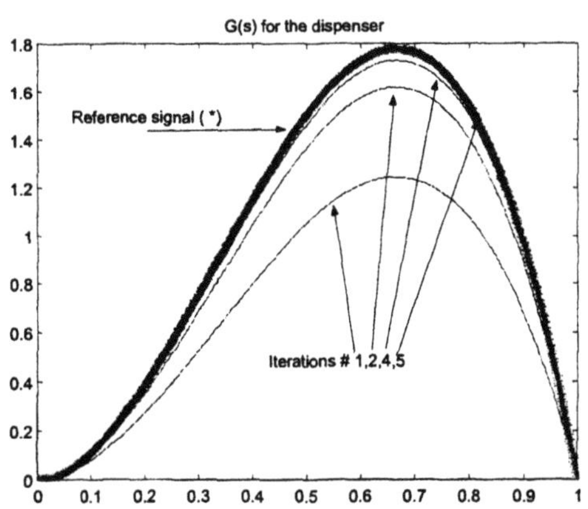

Fig. 3. Illustrative Design.

4. CONCLUSIONS

The ultimate goal of the research programme on which this paper is based is to evaluate the performance (both stand alone and comparative terms) of norm optimal based ILC schemes in 'real world' operating domain. Here we have shown that norm optimal control can indeed be applied to this case with highly promising simulation-based results. We have also described the basic experimental setup which is currently being used to experimentally assess the performance of these controllers. Results from this latter activity will be reported in due course.

5. REFERENCES

Amann, N., D.H. Owens and E. Rogers (1996). Iterative learning control using optimal feedback and feedforward actions. *International Journal of Control* **65(2)**, 277–294.

Arimoto, S., S. Kawamura and F. Miyazaki (1984). Bettering of robots by learning. *Journal of Robotic Systems* **1**, 123–140.

Barton, A.D., P.L. Lewin and D.J. Brown (2000). Practical implementation of a real-time iterative learning controller. *International Journal of Control* **73(10)**, 992–999.

Moore, K. (1993). *Iterative Learning Control for Deterministic Systems.* Springer–Verlag. London.

Rogers, E. and D.H. Owens (1992). *Stability Analysis for Linear Repetitive Processes, LNCIS Series, Vol 175.* Springer–Verlag. Berlin.

Rogers, E., K. Galkowski and D.H. Owens (2001). *Control Systems Theory and Applications for Linear Repetitive Processes, LNCIS Series, To appear.* Springer–Verlag. Berlin.

RBFN MODEL ADAPTATION BASED ON ORTHOGONAL DECOMPOSITION

D.L.Yu and J.B. Gomm

Control Systems Research Group, School of Engineering
Liverpool John Moores University, Byrom Street, Liverpool L3 3AF, U.K.
E-mail: **d.yu@livjm.ac.uk**

Abstract: An algorithm is developed in this paper to adapt a radial basis function network model on-line. This includes updating the output layer weights as well as the number of centres by pruning the centres contributing least to the network output and adding new centres when the modelling error is greater than a given criterion. The adaptation is based on the manipulation of a triangular matrix obtained in orthogonal decomposition of the hidden layer output matrix. The algorithm can be applied to modelling time varying dynamics of an unknown non-linear process and therefore is particularly useful in adaptive control. Simulations on a multivariable process demonstrate the effectiveness of the method. *Copyright © 2001 IFAC*

Keywords: radial basis function networks, model adaptation, time varying processes, adaptive control, non-linear systems.

1. INTRODUCTION

Neural network modelling and control of non-linear dynamic systems have been intensively researched in recent years. Several neural network control strategies that have been studied, including model reference, inverse model and internal model control, are reviewed in Hunt, *et al.* (1992). These approaches are mainly based on fixed neural network models which are trained off-line, in their implementation. However, it is realised that some industrial processes, especially chemical and biotechnological processes, have time-varying dynamics in their nature, or are subject to a time-varying environment. Thus, robust control of these plants could not be maintained using fixed neural network models. One strategy to cope with process time-variations is to develop adaptive neural models and a control scheme based on these models.

In various kinds of neural networks, radial basis function networks (RBFNs) are found to be suitable to on-line adaptation because their linear feature of the weights relating to the network output allows linear optimisation methods, e.g. least squares, to be used. Algorithms for recursive adaptation of the parameters (centres and weights) of an RBFN with a fixed structure have been investigated for process identification (Chen, *et al.*, 1992). However, only updating the weights and/or centre positions restricts the full utilisation of RBFNs. Adaptation of both the structure and weights would enable RBFNs to model time-varying and severely non-linear processes with reasonable network size.

Some off-line adaptation techniques have been proposed for RBFNs to achieve an optimal structure (number and location of centres) for a certain data set (e.g. Chen, *et al.*, 1991; Karayiannis and Mi, 1997). However, the structure will not be optimal when the model is used on-line and the process dynamics or environment changes. Therefore, on-line adaptation of both structure and parameters of an RBFN for use with adaptive control is needed.

Orthogonal decomposition is a well known numerically robust technique to eliminate ill-conditioning effects in solving least squares problems. Chen, *et al.* (1991) showed how batch orthogonal least squares (OLS) could be used to develop RBFNs by sequentially selecting centres based on evaluating an error reduction ratio in a forward selection algorithm. Yu, *et al.* (1997) developed a recursive OLS (ROLS) algorithm for training output layer weights in multi-input multi-output (MIMO) RBFNs. Results showed that the ROLS algorithm can be applied to a large training data set with a low requirement on computer memory, and can eliminate ill-conditioning in the data set, so that the accuracy of the RBFN model is greatly improved. The batch RBFN training method of Chen, *et al.* (1991) was extended by Luo, *et al.*

(1996), where a forward centre selection algorithm, using the same error reduction ratio, based on ROLS was used for on-line structure and parameter adaptation of RBFNs. More recently, Gomm and Yu (2000a) derived both forward and backward centre selection algorithms for the MIMO case within the ROLS framework. The algorithms were applied in off-line network training and, in the application reported, smaller RBFNs with acceptable accuracy were obtained with backward selection. This work was extended in (Gomm and Yu, 2000b), where an on-line algorithm is developed. In this algorithm, a fixed centre set with associated weights is formed off-line and includes sufficient centres to model the system dynamics in the entire workspace. Then, in each sample period, the centres having least contribution are identified one by one using the algorithm in Gomm and Yu (2000a) until the tolerance is reached. The remaining centres form an effective centre set which is used to produce the network output. In this way, the effective centre set almost always has much less centres than the fixed centre set, so that the prediction is more accurate. The inconvenience in this method is that a large fixed centre set needs a large amount of computations, while a small fixed centre set is difficult to model the plant dynamics in the entire workspace.

This paper describes an RBFN model adaptation algorithm with both centres and parameters updated on-line. The weights are updated by re-triangularising the R matrix using the Givens rotation as in Gomm and Yu (2000a). The centres are pruned using the backward selection method and new centres added on-line when the modelling error is greater than a pre-specified criterion and the new data is far from the nearest centre. The operation and performance of the algorithm is demonstrated by on-line prediction of a non-linear system and the simulation results are displayed.

2. ROLS TRAINING OF RBFN

This section reviews briefly how the ROLS algorithm is used to update RBFN weights on-line, which forms the basis for developing the centre adaptation. For the details of the training, readers are refereed to Yu, *et al.* (1997).

A standard RBFN performs a non-linear mapping $x \in \Re^m \to \hat{y} \in \Re^p$ via the non-linear basis functions and a linear output transformation,

$$\hat{y}^T(t) = \phi^T(t)W \tag{1}$$

where $\hat{y}(t)$ and $x(t)$ are the network output and input vectors at sample time t respectively, $W \in \Re^{n_h \times p}$ is the weighting matrix with element w_{ij} denoting the weight connecting the ith hidden node output to the jth network output, $\phi(t) \in \Re^{n_h}$ is the hidden layer output vector with the ith element $\phi_i(t)$. In this work the weighting matrix, W, is computed to minimise the error, $e(t) = y(t) - \hat{y}(t)$ where $y(t)$ is the process output and therefore is the target network output, using an ROLS algorithm.

The least squares problem is formed as follows. Considering (1) for a set of N input-output training data, we have

$$Y = \hat{Y} + E = \Phi W + E \tag{2}$$

where $Y \in \Re^{N \times p}$ is the desired output matrix, $\hat{Y} \in \Re^{N \times p}$ is the neural network output matrix, $\Phi \in \Re^{N \times n_h}$ is the hidden layer output matrix, $E \in \Re^{N \times p}$ is the error matrix and

$Y^T = [y(1), \cdots, y(N)]$, $\hat{Y}^T = [\hat{y}(1), \cdots, \hat{y}(N)]$
$\Phi^T = [\phi(1), \cdots, \phi(N)]$, $E^T = [e(1), \cdots, e(N)]$.

Equation (2) can be solved for $W(t)$ using the recursive MIMO least squares algorithm to minimise the following time-varying cost function,

$$J(t) = \left\| \begin{bmatrix} \lambda Y(t-1) \\ \cdots \\ y^T(t) \end{bmatrix} - \begin{bmatrix} \lambda \Phi(t-1) \\ \cdots \\ \phi^T(t) \end{bmatrix} W(t) \right\|_F \tag{3}$$

where the F-norm of a matrix is defined as $\|A\|_F^2 = trace(A^T A)$ and $\lambda < 1$ is used to introduce exponential forgetting to the past data. It has been shown (Bobrow and Murray, 1993; Yu, *et al.*, 1997) that minimising (3) is equivalent to minimising the following cost function,

$$J(t) = \left\| \begin{bmatrix} \lambda \hat{Y}(t-1) \\ \cdots \\ y^T(t) \end{bmatrix} - \begin{bmatrix} \lambda R(t-1) \\ \cdots \\ \phi^T(t) \end{bmatrix} W(t) \right\|_F \tag{4}$$

where R, an $n_h \times n_h$ upper triangular matrix, and $\hat{Y}$ are computed by an orthogonal decomposition as follows,

$$\begin{bmatrix} \lambda R(t-1) \\ \cdots \\ \phi^T(t) \end{bmatrix} = Q(t) \begin{bmatrix} R(t) \\ \cdots \\ 0 \end{bmatrix},$$

$$\begin{bmatrix} \hat{Y}(t) \\ \cdots \\ \eta^T(t) \end{bmatrix} = Q^T(t) \begin{bmatrix} \lambda \hat{Y}(t-1) \\ \cdots \\ y^T(t) \end{bmatrix} \tag{5}$$

where Q is an orthogonal matrix. Combining (4) and (5) and considering that the F-norm is preserved by orthogonal transformations, the following revised cost function is obtained,

$$J(t)=\left\| \begin{bmatrix} \hat{Y}(t)-R(t)W(t) \\ \cdots \\ \eta^T(t) \end{bmatrix} \right\|_F \quad (6)$$

which gives the straightforward optimal solution of $W(t)$ to be solved from

$$R(t)W(t)=\hat{Y}(t) \quad (7)$$

and leaves the residual at stage t as $\left\|\eta^T(t)\right\|_F$.

Since $R(t)$ is an upper triangular matrix therefore $W(t)$ can be easily solved from (7) by backward substitution. The decomposition in (5) can be achieved efficiently by applying Givens rotations to an extended matrix to obtain the following transformation (Bobrow and Murray, 1993),

$$\begin{bmatrix} \lambda R(t-1) & \lambda\hat{Y}(t-1) \\ \phi^T(t) & y^T(t) \end{bmatrix} \to \begin{bmatrix} R(t) & \hat{Y}(t) \\ 0 & \eta^T(t) \end{bmatrix} \quad (8)$$

The procedure of the ROLS algorithm is therefore the following: at stage t, calculate $R(t)$ and $\hat{Y}(t)$ according to (8), then solve $W(t)$ in (7). Initial values for $R(t)$ and $\hat{Y}(t)$ can be assigned as $R(0)=\alpha I$ and $\hat{Y}(0)=0$, where α is a small positive number.

3. STRUCTURE ADAPTATION

Structure adaptation includes on-line adaptation of the RBFN centres, i.e. to prune the centres having least contribution to the network output at each stage and to add new centres when new data falls into the area that is not covered by the existing centres. In this mode, the structure of the RBFN always remains the minimum size with a required modelling accuracy, which is believed to give the best generalisation or prediction. Thus, the adaptive model will be desirable in model-based on-line control.

3.1 Prune a centre

The additional modelling error caused by removing a particular centre, or equivalently a hidden node, can be calculated by re-optimising the remaining R matrix when the corresponding column in R is removed (Gomm and Yu, 2000a). This is also realised using the orthogonal decomposition,

$$R'=Q_j\begin{bmatrix} R_j \\ \cdots \\ 0 \end{bmatrix}, \quad \begin{bmatrix} \hat{Y}_j \\ \cdots \\ \eta_j^T \end{bmatrix}=Q_j^T\hat{Y}(t) \quad (9)$$

where R' is $R(t)$ with jth column removed. Then the cost function will become

$$J_j=\left\| Q_j\begin{bmatrix} \hat{Y}_j \\ \cdots \\ \eta_j^T \end{bmatrix}-Q_j\begin{bmatrix} R_j \\ \cdots \\ 0 \end{bmatrix}W_j \right\|_F=\left\| \begin{bmatrix} \hat{Y}_j-R_jW_j \\ \cdots \\ \eta_j^T \end{bmatrix} \right\|_F \quad (10)$$

where W_j is the new weight matrix with the jth centre removed. Hence, the increase in the residual caused by removing the jth centre is $\left\|\eta_j^T\right\|_F$. Similar to the ROLS algorithm, the orthogonal decomposition in (9) is computed efficiently by applying the Givens rotations to obtain the transform

$$\begin{bmatrix} R' & \hat{Y} \end{bmatrix} \to \begin{bmatrix} R_j & \hat{Y}_j \\ 0 & \eta_j^T \end{bmatrix} \quad (11)$$

Therefore, the procedure to remove a centre at stage t is, calculate the residual $\left\|\eta_j^T\right\|_F$ for each centre and actually remove the centre with the least residual. After removing the jth centre, the optimal weighting matrix is calculated from

$$R_jW_j=\hat{Y}_j \quad (12)$$

3.2 Add a centre

In the framework of orthogonal training, adding a centre can be realised by adding a column in the R matrix. In this study, the Gaussian non-linear basis function is adopted. When it is detected that a new centre is needed, the new data at this stage is used as the new centre, $c(t)=x(t)$. Therefore, the hidden layer output with the new centre, $\varphi_{n_h+1}(t)$ can be consequently computed. The new R matrix is constructed by adding a new column following the last one,

$$R_{new}=\begin{bmatrix} R_{old} & v_{(n_h+1)\times 1} \\ 0_{1\times n_h} & \end{bmatrix} \quad (13)$$

where v is the vector added to the R matrix for the new centre. v can be given the initial value of the R matrix, i.e.

$$v=\alpha[0 \ \cdots \ 0 \ 1]^T \quad (14)$$

The row vector in $\hat{Y}$ corresponding to the row vector $0_{1\times n_h}$ in R_{new} in (13) is also given the initial value of $\hat{Y}$, i.e. $0_{1\times p}$. Thus, to update the weights of the network with a new centre added, the Givens rotation is applied to the extended matrix to obtain the transformation,

$$\begin{bmatrix} \lambda R(t-1) & 0_{n_h\times 1} & \lambda\hat{Y}(t-1) \\ 0_{1\times n_h} & \alpha & 0_{1\times p} \\ \phi^T(t) & \varphi_{n_h+1}(t) & y^T(t) \end{bmatrix} \to \begin{bmatrix} R_{new}(t) & \hat{Y}_{new}(t) \\ 0 & \eta^T(t) \end{bmatrix} \quad (15)$$

where R_{new} and $\hat{Y}_{new}$ have one higher dimension than R and $\hat{Y}$ respectively.

The above assignment to the new column vector in R and the new row vector in $\hat{Y}$ using the initial value at $t = 0$ is natural and the value in these vectors will be built up with the arrival of new data. However, as these initial values have not been trained by the past data, the model with the added centre cannot predict output on-line as precisely as it does when it is trained for a short period. In this consideration, the on-line training data is stored in a form of moving window. When a new centre needs to be added, a new R matrix is obtained by applying the orthogonal decomposition to the stored past data window with the forgetting factor multiplied. Although the computing amount in this way is increased, the prediction accuracy is much improved for a number of samples immediately after the new centre is added. This can be described as follows,

$$\begin{bmatrix} \lambda^{\sigma-1} & & & \\ & \lambda^{\sigma-2} & & \\ & & \ddots & \\ & & & 1 \end{bmatrix} \begin{bmatrix} \phi^T(t-\sigma+1) \\ \phi^T(t-\sigma+2) \\ \vdots \\ \phi^T(t) \end{bmatrix} = Q \begin{bmatrix} R_{new} \\ 0 \end{bmatrix} \quad (16)$$

where σ is the length of the data window and can be chosen according to the forgetting factor λ, Q is an orthogonal matrix. Equation (16) can be easily realised using the QR decomposition algorithm, while an efficient calculation can be recursively realised applying the Givens rotation to the Φ matrix to save memory, as Q is not needed.

When a new centre is added, the width associated to the new centre in the Gaussian non-linear basis function used in this study needs to be designed. The *p-nearest-neighbour* rule used most frequently in construction of an RBFN (Moody and Darken, 1989)

$$\mu_l = \sqrt{\frac{1}{p} \sum_{i=1}^{p} \|c_l - c_i\|^2} \quad (17)$$

is adopted here and $p = 3$ is chosen.

3.3 Procedure and criteria

In each sample period the on-line adaptation of the RBFN including pruning and adding centres, updating the extended R matrix and updating weights, is done in the following procedure with the given criteria.

Step 1. A set of data with σ samples is used to obtain the initial values of matrix R and W off-line, with a pre-specified number of centres chosen using the *k-means* method.

Step 2. In each sample time, a centre is pruned if the minimum pruning error index is less than a pre-specified criterion

$$\left\| \eta_j^T \right\|_F < e_{pru} \quad (18)$$

The corresponding row in weight matrix W is also removed to form a new W.

Step 3. The RBFN model is used to predict the plant output for one step ahead and to form a prediction error moving window with length l. If the mean-squared-error (MSE) is greater than a pre-specified criterion

$$\frac{1}{l} \sum_{i=0}^{l-1} \|y(t+i) - \hat{y}(t+i)\|^2 > e_{add} \quad (19)$$

a centre will be added. The new centre uses the new data $x(t)$ and the corresponding width is calculated according to (17).

Step 4. Update matrix R and W using new data by applying Givens rotation to the extended R matrix as shown in (8).

The initial R and W matrices are obtained by off-line training using a set of data, instead of being chosen randomly, because the model is required to give quite precise prediction even at the beginning of the process when it is used in on-line adaptive control.

4. AN EXAMPLE

To demonstrate the operation and performance of the proposed method, the adaptive RBFN is used to on-line predict the system output for one step ahead. The system is chosen from examples in (Narendra and Parthasarathy, 1990),

$$y(t+1) = \frac{y(t)}{1 + y(t)^2} + u(t)^3 \quad (19)$$

which is non-linear and with first-order dynamics. The input is generated for 150 samples of which the first half has a constant value 0.6 superimposed by a random value uniformly distributed between 0 and 0.2, and the last half has a constant value 1.0 with the same random values. This is to enable the system output to fall into two clearly separated areas to test the adaptation ability of the algorithm.

The RBFN is chosen having two inputs, one output and 25 initial hidden layer nodes (centres). The Gaussian function is chosen as the radial basis function and initial weights are trained using the first 25 input data with the centres chosen as these input data. In this simulation, the error index for pruning a centre is $e_{pru} = 0.001$, the error index for adding a centre is $e_{add} = 0.01$, the length of the moving window for adding a centre is $l = 2$, the forgetting factor is $\lambda = 0.98$ and the data length for retraining the R and W matrices when the centre is added is $\sigma = 50$. When the sample time is greater than 50, the RBFN starts to adapt its structure and parameters according to the new data. The performance of the adaptive RBFN is displayed in the upper graph in Fig.1, and the centre on-line adaptation is shown in the lower graph in Fig.1.

It can be clearly seen that the centres are pruned before the step change of the input to minimise the size of the network structure until the prediction error reaches the error threshold e_{pru}. When the system input and correspondingly the network input has a step change, the RBFN adds several centres in the new area to enable accurate prediction for the new data, while old centres that do not have significant contribution currently are pruned. After a few steps the network adapted its centres and associated weights so that the prediction accuracy is recovered. The initial centre number is 25 and the final centre number is only 10. These are displayed in the lower graph in Fig.2 to show the movement of centres from the old area to the new area. The length of the moving window is chosen as $l = 2$. A long window will delay the adaptation when the input has a step change with large amplitude, while a short window cannot avoid the mis-addition of centres caused by data noise. Simulation of a two-input two-output system modelled by the adaptive RBFN has also been done but is not demonstrated here due to the limited space. The performance of the same RBFN with weight adaptation but the number of centres fixed is displayed in the upper graph in Fig.2 for comparison. It is demonstrated that if the RBFN is not adapted in the structure, the plant output in the new area cannot be predicted.

5. CONCLUSION

An on-line adaptation algorithm for RBFN is developed based on the recursive orthogonal least squares training. The novelty of the algorithm is to prune centres according to the error index generated in ROLS training and to add centres by modifying the extended R matrix. The algorithm is dedicated to neural model-based adaptive control where on-line adaptation without severe degradation of the prediction accuracy is required. The simulation on a non-linear dynamic system demonstrates the performance and effectiveness of the method.

ACKOWLEDGEMENT:

The project is supported by the U.K. EPSRC with the grant No. GR/N18697.

REFERENCES

Bobrow, J.E. and W. Murray (1993), An algorithm for RLS identification of parameters that vary quickly with time, *IEEE Trans. On Automatic Control*, Vol.38, No.2, pp. 35-254.

Chen, S., C.F.N. Cowan and P.M. Grant (1991), Orthogonal least squares learning algorithm for radial basis function networks, *IEEE Trans. On Neural Networks*, Vol.2, pp. 302-309.

Chen, S., S.A. Billings and P.M. Grant (1992), Recursive hybrid algorithm for non-linear system identification using radial basis function networks, *Int. J. Control*, Vol.55, pp. 1051-1070.

Gomm, J.B. and D.L. Yu (2000a), Selecting radial basis function network centres with recursive orthogonal least squares training, *IEEE Trans. on Neural Networks*, Vol. 11, No.3, pp. 306-314.

Gomm, J.B. and D.L. Yu (2000b), Backward centre selection method for on-line adaptation of RBF network models, *Proc. IFAC Symposium System Identification SYSID'2000*, California, USA, June.

Hunt, K.J., D. Sarbaro, R. Zbikowsky and P.J. Gawthrop (1992), Neural network for control systems – a survey, *Automatica*, Vol.28, No.6, pp. 1083-1112.

Karayiannis, N.B. and G.W. Mi (1997), Growing radial basis networks: merging supervised and unsupervised learning with network growth techniques, *IEEE Trans. on Neural Networks*, Vol.8, No.6, pp. 1492-1506.

Luo, W., M.N. Karim, A.J. Morris and E.B. Martin (1996), Control relevant identification of a pH waste water neutralisation process using adaptive radial basis function networks, *Computers Chem. Engng*. Vol.20, pp.1017-1022.

Narendra, K.S. and K. Parthasarathy (1990), Identification and control of dynamical systems using neural networks, *IEEE Trans. on neural networks*, Vol.1, No.1, pp. 4-27.

Yu, D.L., J.B. Gomm and D. Williams (1997), A recursive orthogonal least squares algorithm for training RBF networks, *Neural Processing Letters*, Vol.5, No.3, pp. 167-176.

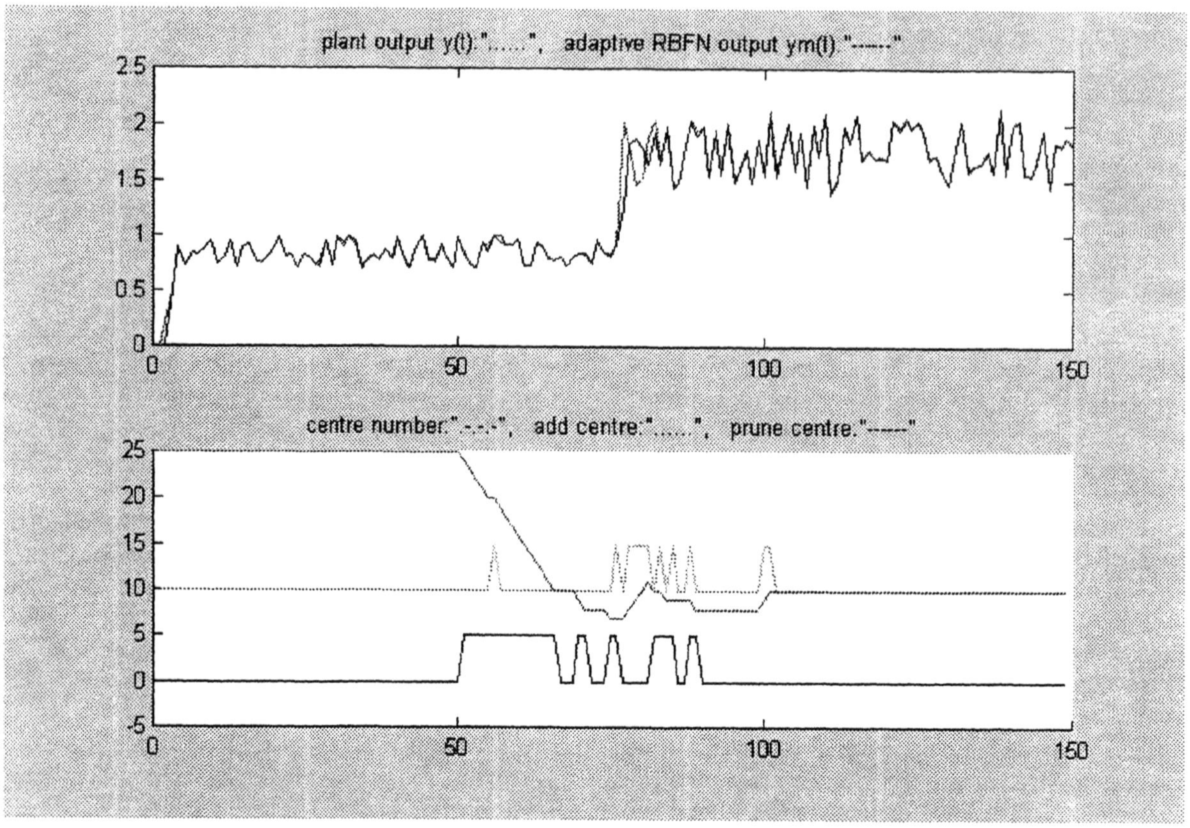

Fig.1 Top: the plant output and adaptive RBFN prediction, bottom: Centre adaptation.

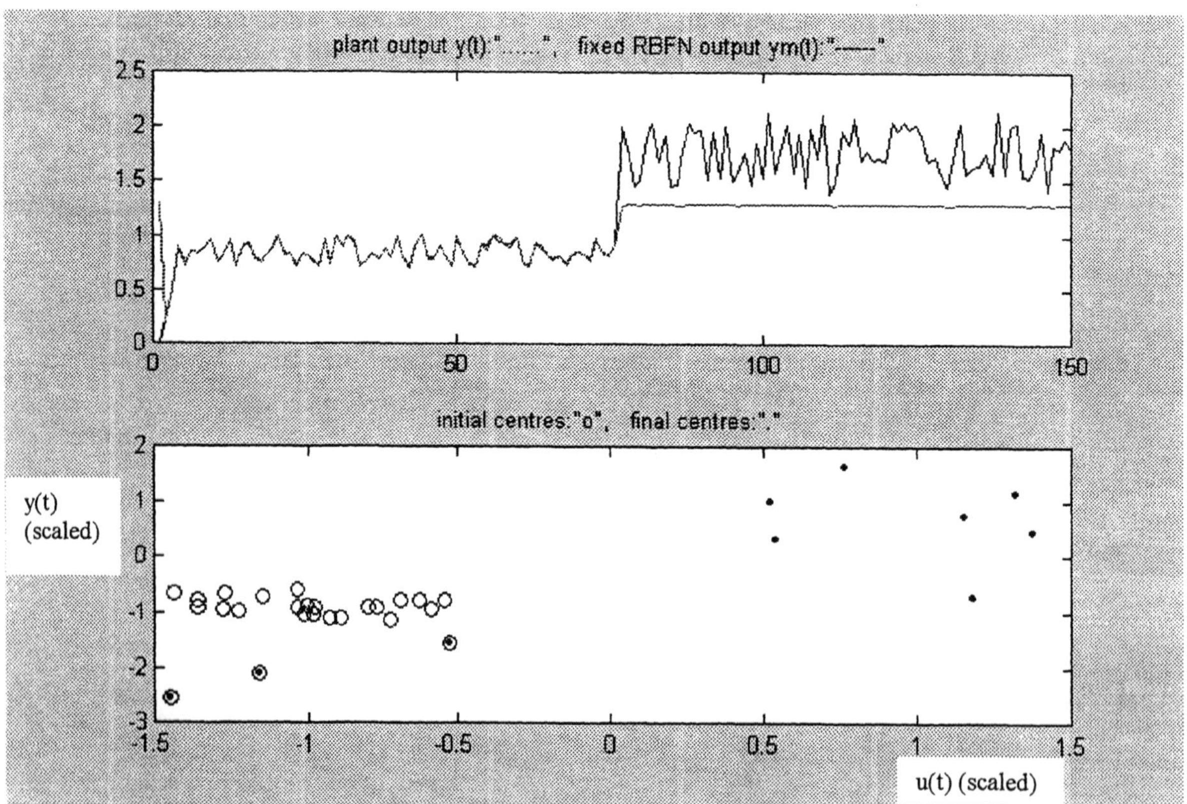

Fig.2. Top: prediction with fixed centre RBFN, Bottom: centre locations for adaptive RBFN

APPROXIMATING NETWORKS FOR THE SOLUTION OF T-STAGE STOCHASTIC OPTIMAL CONTROL PROBLEMS [1]

**Marco Baglietto * Cristiano Cervellera *
Thomas Parisini ** Marcello Sanguineti *
Riccardo Zoppoli ***

* *Dept. of Communications, Computer and System Sciences DIST–University of Genova, Via Opera Pia 13, 16145 Genova, Italy*
** *Dept. of Electronic Engineering and Information Sciences Politecnico di Milano, Piazza Leonardo da Vinci 32, 20133 Milano, Italy*

Abstract: Approximate solution of a general T–stage stochastic optimal control problem is considered. It is known that discretizing uniformly the state components may lead dynamic programming to incur the "curse of dimensionality". *Approximating networks*, i.e., linear combinations of parametrized basis functions provided with density properties in suitable normed spaces, are then defined and used in two approximate methods (examples of such networks are one hidden layered feedforward neural networks, Radial Basis Functions, etc.). The first one consists in approximating the optimal cost-to-go functions in dynamic programming. The second method reduces the original functional optimization problem to a nonlinear programming one that is solved by means of stochastic approximation. Approximating networks of suitable types benefit by the property that, for the approximation of members of some classes of smooth functions, the number of parameters to be optimized and the number of samples to be used increase moderately with the dimensions of the arguments of the functions. The two methods are tested and compared in a test problem involving a 10-dimension state vector. *Copyright © 2001 IFAC*

Keywords: Neural Networks, Dynamic Programming, Stochastic Approximation, Optimal Control, Applied Neural Control

1. INTRODUCTION

A general form of T-stage stochastic optimal control problem is addressed. The discrete-time dynamic system is given by

$$x_{t+1} = f(x_t, u_t, \xi_t), \quad t = 0, 1, \ldots, T-1, \quad (1)$$

where $x_t \in \Re^n$ is the state vector ($x_0 = \hat{x}$ is a given initial state), $u_t \in U_t \subseteq \Re^m$ is the control vector (U is the set of admissible controls), and $\xi_t \in \Re^q$ is a random noise. The state is assumed to be perfectly measurable and the stochastic vectors ξ_t, $t = 0, \ldots, T-1$, to be characterized by a known probability density function. The cost function is given by

$$\mathcal{J} = \sum_{t=0}^{T-1} h(x_t, u_t, \xi_t) + h_T(x_T). \quad (2)$$

[1] The work was supported in part by the Italian Ministry for the University and Research (MURST) Project "Identification and control of industrial systems".

The following optimal control problem is stated:

Problem P. *Find the sequence of optimal control functions* $u_t^\circ = \mu_t^\circ(x_t),\ t = 0,1,\ldots,T-1$, *that minimize the expected cost* $J(x_0) = \mathrm{E}_{\xi_0,\ldots,\xi_{T-1}}(\mathcal{J})$. □

It is common knowledge that Problem P can be solved analytically by dynamic programming (DP) only if suitable hypotheses are verified (typically, if Problem P is stated in the LQ framework and if the random vectors are mutually independent). In the general case, one has to look for approximate solutions. This is usually done by discretizing the state space. In such a way, the functional equation that defines the DP procedure has to be solved only in correspondence of a finite number of state values. Given a fixed number d of discretization levels for each component of the state vector, the use of a full uniform grid makes the total number of discretization points grow as d^n. This exponential growth leads to the "curse of dimensionality" (Bellman, 1957) and limits the applicability of DP to small dimensions.

The efforts to cope with the curse of dimensionality have followed two main directions: reduction of the complexity of the problem by using simpler models, and use of smart approximators for the cost-to-go functions in order to reduce the number of grid points. As for the first method, the aggregation/disaggregation of grid points to diminish the dimension of the state space (see, for example, (Turgeon, 1981) in the reservoir literature), and the partition of the original problem into smaller separable problems (see (Archibald *et al.*, 1997) and the references therein) can be cited. The major drawback of these methods lies in their lack of generality, since they are highly dependent on the particular problem they are applied to (in the aforementioned works, reservoir operation). As for the second method, it is possible to trace an evolution starting from polynomial approximation (see, for example, (Bellman *et al.*, 1963)), and continuing with cubic Hermite polynomials (Foufoula-Georgiou and Kitanidis, 1988) and spline interpolation (Johnson *et al.*, 1993). In (Johnson *et al.*, 1993) the use of cubic splines, instead of tensor-product linear interpolants, has allowed to solve a water supply problem up to dimension five. Although these strategies permit the use of fewer grid points to obtain a given precision in the approximation of the optimal cost-to-go functions, they still need a uniform discretization of the state space, thus leaving the curse of dimensionality eventually unavoided and their use still impracticable for large values of n.

Recently, an approximating architecture using multivariate adaptive regression splines and based on a statistical perspective has been proposed. This approach leads to a discretization of the state space different from the uniform grid. Only the values of the cost-to-go functions corresponding to the points belonging to a subset of the full uniform grid and with complexity growing as $O(n^3)$ have to be computed. This polynomial growth allows to deal (for the first time) with high values of n without any need of introducing a simpler model. Such an efficient choice of the discretization points has enabled the solution of an inventory forecasting problem with up to nine state variables (Chen *et al.*, 1999).

In this paper two approaches to solve Problem P in its general form and in high-dimensional settings are proposed. Both of them are based on the use of approximating networks, defined in Section 2. In Sections 3 and 4 the two methods of approximate solution are introduced. Some results from approximation theory and statistical learning, like those discussed in Section 5, justify the advantageous performances of the two proposed approaches in complex applications, such as the example presented in Section 6.

2. APPROXIMATING NETWORKS

Let us consider functions $\gamma(x)\colon \Re^n \mapsto \Re^s$. The proposed methods rely on approximating the unknown functions by means of fixed–structure functions $\gamma_\nu(x, w_\nu)$ of the form

$$\gamma_\nu(x, w_\nu) = \mathrm{col}\left[\sum_{i=1}^{\nu} c_{ij}\,\varphi(x,\kappa_i),\ j = 1,\ldots,s\right] \quad (3)$$

where $\varphi(\cdot,\cdot)$ is a given basis function, $c_{ij} \in \Re$ and the components of the vectors $\kappa_i \in \Re^k$ are parameters to be determined. w_ν is defined as $w_\nu \triangleq \mathrm{col}\,(c_{ij}, \kappa_i, i = 1,\ldots,\nu,\ j = 1,\ldots,s)$. Then $w_\nu \in \Re^{\mathcal{N}(\nu)}$, where $\mathcal{N}(\nu) = \nu(k+s)$. Following (Zoppoli *et al.*, 2001), let the set of functions to be approximated be a linear space H. $A_\nu \subset H$ is defined as the parametrized set of all functions of the form (3), that is, $A_\nu \triangleq \{\gamma_\nu(x, w_\nu) \in H,\ w_\nu \in \Re^{\mathcal{N}(\nu)}\}$. The sequence $\{A_\nu\}_{\nu=1}^{\infty}$ has the infinite nested structure $A_1 \subset A_2 \subset \cdots \subset A_\nu \subset \cdots$. As the distance between the functions γ and γ_ν has to be measured, the space H is equipped with a norm $\|\cdot\|$. Thus we let $\mathcal{H} \triangleq (H, \|\cdot\|)$ (Zoppoli *et al.*, 2001). In the approximation methods that are going to be presented, the following assumption plays a basic role.

A1. The functions $\gamma_\nu(x, w_\nu)$ are such that the set $\bigcup_{\nu=1}^{+\infty} A_\nu$ is dense in $\mathcal{H}$.

Functions of the form (3), which verify Assumption A1, benefit by the density property in the space $\mathcal{H}$ and are defined as *$\mathcal{H}$-approximating networks ($\mathcal{H}$-ANs)*. Feedforward neural networks with one hidden layer and linear output activation functions, linear combinations of sinusoidal functions with variable frequencies, and linear combinations of radial basis functions of the Gaussian type are examples of $\mathcal{H}$–ANs in the space $\mathcal{C}(K, \Re^s)$ of continuous functions on a compact set K with the supremum norm and in the space $\mathcal{L}_2(K, \Re^s)$ of square-integrable functions on K with the $\mathcal{L}_2$ norm. Hence, such approximators are both $\mathcal{C}$– and $\mathcal{L}_2$–ANs. The proofs of the related density properties can be found in several papers (see, for example, (Leshno *et al.*, 1993) and (Park and Sandberg, 1991)). *Linear $\mathcal{H}$-approximating networks* are those in which the basis functions φ_i do not contain the parameter vectors κ_i (in the case of linear ANs, the basis functions have to be specified by the subscript i, this is unnecessary for the nonlinear networks). Examples of $\mathcal{C}$– and $\mathcal{L}_2$–linear ANs are classical linear approximators such as algebraic and trigonometric polynomials.

The dimension $\mathcal{N}(\nu)$ of the parameter vector of an AN (or simply ν, as $\mathcal{N}(\nu)$ is linear in ν) will be considered as a "measure" of the complexity of the network. The term "network" is borrowed from the parlance of neural networks and it has been chosen since functions of the form (3) have the same structure as the feedforward neural networks with one hidden layer and linear output activation units. As the $\mathcal{H}$–ANs (or, briefly, the ANs) that will be presented in the next sections (in practice feedforward neural networks but possibly even other networks) benefit by both the $\mathcal{C}$- and the $\mathcal{L}_2$-density properties (and also by other density properties), in the following simply the term "approximating networks" will be used.

3. APPROXIMATE MINIMIZATION OF THE OPTIMAL COST-TO-GO FUNCTIONS IN THE DP PROCEDURE (METHOD A)

The first approximate method proposed here to solve problem P (that will be called *Method A*) is based on the DP procedure. What follows details the approach presented in (Baglietto *et al.*, 2000). Let us ignore for the moment how the state space has been discretized and just assume that, at each stage t, L discretized points $x_t^{(l)}$, $l = 1, \ldots, L$ have been selected (in order to simplify the notation, we suppose the number L to be the same at all stages). Let $X_{tL} \triangleq \left\{x_t^{(l)},\ l = 1, \ldots, L\right\}$. Then it is possible to write:

$$\tilde{J}^{\circ}_{T-1}(x^{(l)}_{T-1}) \triangleq J^{\circ}_{T-1}(x^{(l)}_{T-1}) = \min_{u_{T-1}\in U_{T-1}} \mathop{\mathrm{E}}_{\xi_{T-1}} \{h(x^{(l)}_{T-1}, u_{T-1}, \xi_{T-1}) + h_T[f(x^{(l)}_{T-1}, u_{T-1}, \xi_{T-1})]\},$$
$$x^{(l)}_{T-1} \in X_{T-1,L} \quad \text{(4a)}$$

$$\tilde{J}^{\circ}_t(x^{(l)}_t) = \min_{u_t\in U_t} \mathop{\mathrm{E}}_{\xi_t}\{h(x_t, u_t, \xi_t) + \hat{J}[f(x^{(l)}_t, u_t, \xi_t), w^{J\circ}_{t+1})]\},$$
$$t = T-2, \ldots, 0;\ x^{(l)}_t \in X_{t,L} \quad \text{(4b)}$$

(the $\xi_0, \ldots, \xi_{N-1}$ are assumed to be mutually independent). $\tilde{J}^{\circ}_t(x_t)$ is an approximation of the optimal cost-to-go $J^{\circ}_t(x_t)$ (at stage $T-1$, this approximation is not needed (see (4a)). $\hat{J}(x^{(l)}_t, w^{J\circ}_{t\nu_t})$ is an approximation of $J^{\circ}_t(x_t)$ derived by constraining this cost to take on the structure of an AN, that is,

$$\hat{J}_{\nu_t}(x_t, w^J_{t\nu_t}) = \sum_{i=1}^{\nu_t} c_i(t)\varphi\left[x_t, \kappa_i(t)\right],$$
$$t = T-1, \ldots, 1.$$

The ANs are assumed to maintain the same structure stage after stage. To simplify the notation, in (4b) and in the following the subscript ν_t has been dropped. The parameter vector $w^{J\circ}_t$ is obtained on the basis of the values of $\tilde{J}^{\circ}_t(x^{(l)}_t)$ by using some approximation criterion like the least-squares one, i.e.,

$$w^{J\circ}_t = \arg\min_{w^J_t} \sum_{l=1}^{L} \left[\tilde{J}^{\circ}_t(x^{(l)}_t) - \hat{J}(x^{(l)}_t, w^J_t)\right]^2,$$
$$t = T-1, \ldots, 1. \quad (5)$$

Of course, there is the need to assume that the cost functions to be approximated are sufficiently "smooth" in the sets of the admissible states, so that such functions belong to suitable subsets of linear spaces with the norm $\|\cdot\|$, hence, the spaces $\mathcal{H}_t \triangleq (\mathcal{H}_t, \|\cdot\|)$.

There are two ways of determining the optimal controls on-line: reoptimization and approximation of the optimal control functions by ANs.

1. *Reoptimization.* Using this technique the computation of the optimal control $\tilde{u}^{\circ}_t$ is performed on line by exploiting the values of the cost-to-go $\hat{J}(x_t, w^{J\circ}_t)$ stored in memory, that is,

$$\tilde{u}^{\circ}_t = \arg\min_{u_t\in U_t} \mathop{\mathrm{E}}_{\xi_t}\left\{ h(x_t, u_t, \xi_t) + \hat{J}\left[f(x_t, u_t, \xi_t), w^{J\circ}_{t+1}\right]\right\},$$
$$t = 0, 1, \ldots, T-1.$$

2. *Approximation of the optimal control functions by ANs.* The optimal control functions $\mu^{\circ}_t(x_t)$ are approximated by another family of ANs, denoted

by $\hat{\mu}(x_t, w_t^u)$, $t = 0, 1, \ldots, T-1$. Such a second family of ANs is derived as follows. While applying Eqs. (4), the L pairs $\left[x_t^{(l)}, \tilde{\mu}_t^{(l)}(x_t^{(l)})\right]$ are obtained at each stage. Then the optimal vectors $w_t^{u\circ}$ can be computed by minimizing (as in (5)) the approximation error $\sum_{l=1}^{L} ||\tilde{\mu}_t^{\circ}(x_t^{(l)}) - \hat{\mu}(x_t^{(l)}, w_t^u)||^2$, for $t = T-1, \ldots, 1$. Again, the optimal control functions μ_t°, $t = 0, 1, \ldots, T-1$, are assumed to belong to normed linear spaces in which the ANs are dense.

By using the second method instead of reoptimization, it is possible to obtain the important advantage of performing all computations off line. Indeed, reoptimization requires on-line computations whereas the use of ANs permits the on-line generation of the control vectors "almost instantaneously". The idea of approximating the optimal cost-to-go functions by approximating networks having the structure of neural approximators is not new. Bertsekas and Tsitsiklis in (Bertsekas and Tsitsiklis, 1996) have presented, under the name of "neuro-dynamic programming", several methods that involve the use of neural networks in order to deal with high-dimensional DP, though the attention (as for most of the whole "neuro-dynamic" literature) is by far more focused on the case of infinite-horizon than on T-stage optimal control problems. In Section 5, the use of ANs (either for the use of DP or for Method B presented in Section 4) is justified with recent results in the field of nonlinear approximation.

3.1 *On the choice of the sample sets X_{tL}*

In order for Method A to work, a uniformly good approximation of the optimal cost-to-go function over the sets of the feasible states is needed. In fact, an even slightly bad approximation of the $(t+1)$-th optimal cost-to-go function in some regions of its domain could lead the minimization algorithm to find a wrong minimum, at stage t, during the "exploration" of the $(t+1)$-th cost-to-go function. In other words, for the choice of a good training set, the various feasible sets should be represented without priviledging some regions over others, which would be unjustified in absence of prior information about the cost-to-go function. There are many possibilities of uniform sampling of the feasibe sets. For example, the following ones can be cited.

(i) A "full uniform grid". Each component of the state space is discretized by a fixed number of values. This is the simplest choice, and the first used since the introduction of the DP algorithm. As said before, its main drawback lies in the curse of dimensionality.

(ii) A random sequence of independent and identically distributed (i.i.d.) points drawn with an uniform probability distribution. In this way the dimension of the training set does not depend "structurally" on n (as in the full uniform grid case), but such a dependence is related to the accuracy of the approximation (in fact, it can be expected that functions with higher dimension n require larger training sets in order to guarantee the same level of accuracy). The main drawback lies in the fact that the original "deterministic" problem (approximation of a fixed unknown function) is transformed into a "statistical" one. In other words, a probability measure on the generation of the samples, is arbitrarily introduced (even if it is the most "reasonable" probability). In this way all the available theoretical results (see, for example, (Barron, 1994; Vapnik, 1995)) are applied to a problem that is not statistical in nature. Anyway, by using an uniform distribution such results can be helpful to give reasonable hints about the required complexity of the ANs and the sizes of the training sets.

(iii) A deterministic sequence of points, which tries to "spread" them in the most uniform way. Examples are Orthogonal Arrays, Latin Hypercubes, "quasi-Montecarlo" sequences (see, for example, (Niederreiter, 1992; Chen *et al.*, 1999)). In this case the sequences are deterministic, then they enable one to obtain "deterministic" results. Furthermore, generally the resulting training sets are more "uniformly spread" than the ones generated by a random uniform distribution (see (Niederreiter, 1992)). However, there are very few available theoretical results that can help find the complexity of the ANs and the required number of points in order to have a good accuracy of the approximation.

4. REDUCTION OF THE FUNCTIONAL PROBLEM P TO A NONLINEAR PROGRAMMING ONE (METHOD B)

Method A may not be an easy tool to handle, for it may require cumbersome numerical procedures. Therefore, another kind of approximate technique to solve Problem P (which will be called *Method B*) is considered. Such a technique consists in assigning the control functions $u_t = \mu_t(x_t)$ a fixed structure $\hat{u}_t = \hat{\mu}(x_t, w_t^u)$, $t = 0, 1, \ldots, T-1$, where $\hat{\mu}(\cdot, \cdot)$ is an AN and w_t^u is the vector of parameters to be determined. Even if the ANs introduced in this section differ from the ones used for Method A, the same notation is kept. Clearly, the control scheme gives rise to a chain of T approximating networks, each followed by the dynamic system. Now, by substituting the control functions $\hat{\mu}(x_t, w_t^u)$ into the state equation (1) and in (2), the cost function assumes the form $\mathcal{J}(w, \xi_0, \ldots, \xi_{T-1})$, where $w \triangleq \mathrm{col}(w_t,\ t =$

$0, 1, \ldots, T-1$). Then the vector w° that minimizes the expected cost $\mathrm{E}_{\xi}\left[\mathcal{J}(w, \xi)\right]$ has to be found, where $\xi \triangleq \mathrm{col}\,(\xi_0, \ldots, \xi_{T-1})$. It follows that the *functional optimization* Problem P has been approximated by a *nonlinear programming problem* that can be solved by using some gradient descent method. Provided that the numbers ν_t of basis functions in each AN are sufficiently large, the new problem can approximate Problem P to any degree of accuracy. The concept of *epi-convergence* is useful to state correctly the meaning of the term "approximation" of Problem P by a sequence of nonlinear programming problems parametrized by ν_t (see (Zoppoli *et al.*, 2001)). Unfortunately, due to the general statement of Problem P, it is generally impossible to express the average cost $\mathrm{E}_{\xi}[\mathcal{J}(w, \xi)]$ in explicit form. This leads to the computation of the "*realization*" $\nabla_w \mathcal{J}[w(k), \xi(k)]$ instead of $\nabla_w \mathrm{E}_{\xi}[\mathcal{J}(w, \xi)]$, thus giving rise to the *stochastic approximation* algorithm

$$w(k+1) = w(k) - \alpha(k) \nabla_w \mathcal{J}[w(k), \xi(k)] \qquad k = 0, 1, \ldots \quad (6)$$

The sequence $\{\xi(k), k = 0, 1, \ldots\}$ is generated by randomly selecting the vectors $\xi_0(k), \ldots, \xi_{T-1}(k)$ according to their probability density function. As it is well known, the step-size $\alpha(k)$ must suitably decrease to ensure convergence. In the example given in Section 6, it is $\alpha(k) = c_1 / (c_2 + k)$, $c_1, c_2 > 0$. To implement algorithm (6) at each iteration step k, the components of the gradient $\nabla_w \mathcal{J}[w(k), \xi(k)]$ have to be computed, i.e, the partial derivatives $\dfrac{\partial}{\partial w_t^i} \mathcal{J}[w(k), \xi(k)]$, for $t = 0, 1, \ldots, T-1$ and for $i = 1, \ldots, \mathcal{N}(\nu_t)$. w_t^i denotes the i–th component of the vector w_t. It is possible to write: $\dfrac{\partial \mathcal{J}}{\partial w_t^i} = \dfrac{\partial \mathcal{J}}{\partial u_t} \dfrac{\partial \hat{\mu}(x_t, w_t)}{\partial w_t^i}$, and then, by using some algebra, it is easy to see that $\dfrac{\partial \mathcal{J}}{\partial u_t}$ is given by

$$\frac{\partial \mathcal{J}}{\partial u_t} = \frac{\partial}{\partial u_t} h(x_t, u_t, \xi_t) + \lambda_{t+1}^{\top} \frac{\partial}{\partial u_t} f(x_t, u_t, \xi_t), \qquad t = 0, 1, \ldots, T-1$$

where $\lambda_t^{\top} \triangleq \dfrac{\partial \mathcal{J}}{\partial x_t}$. Moreover, λ_t can be computed as

$$\lambda_t^{\top} = \frac{\partial}{\partial x_t} h(x_t, u_t, \xi_t) + \lambda_{t+1}^{\top} \frac{\partial}{\partial x_t} f(x_t, u_t, \xi_t) + \frac{\partial \mathcal{J}}{\partial y_t}, \qquad t = T-1, \ldots, 0 \quad (7)$$

$$\lambda_T^{\top} = \frac{\partial}{\partial x_T} h_T(x_T).$$

In (7) y_t is the actual input to the AN at stage t (see, for example, (Parisini and Zoppoli, 1994) for a more detailed description of the algorithm). It is worth noting that (7) is the classical adjoint equation of T–stage optimal control theory, with the addition of one term (the third) to take into account the introduction of the parametrized feedback control law. Note also that the structure of (7) does not depend on the type of AN used to implement the control law. Furthemore, notice that Method B, unlike Method A, does not require the random vectors $\xi_0, \ldots, \xi_{T-1}$ to be mutually independent.

Method B has turned out to be simple (it does not require the construction of any grid in the state space), and very effective in solving optimal control problems with a large number of state components (see, for example, (Baglietto *et al.*, 2001) and (Parisini and Zoppoli, 1998)). Although the authors are not yet able to motivate analytically such powerful capabilities, there are good reasons to believe that the properties discussed in Section 5 can be in force not only to approximate functions belonging to suitable spaces, but also to solve approximately functional optimization problems by using a "moderate" number of free parameters and of samples, provided that the optimal solutions are characterized by appropriate smoothness properties.

5. ON THE CURSE OF DIMENSIONALITY IN APPROXIMATE OPTIMIZATION

If Method A is used, an effective methodology for deriving an approximate solution of problem P should require a limited number of parametrized basis functions and a limited number of samples ([state - optimal cost-to-go functions] and possibly [state - optimal controls] values) to obtain a good approximation of the optimal cost-to-go and control functions.

If Method B is used, a limited number of basis functions and of samples of stochastic variables should be required to obtain good approximations of the control functions. In particular, there is the need to avoid an unacceptably fast growth of the number of basis functions and samples with the dimension n of the state vector in order not to incur, if possible, the curse of dimensionality.

Method A and Method B share the following aspects: (i) the approximation of unknown functions by approximating networks, (ii) the use of sample sets in order to determine the values of the parameters. In this section, the target function $\gamma(x)$, defined on a compact set $B \subset \Re^n$, will be used as a unified notation representing, each time, the optimal t-th cost-to-go or the optimal t-th control function.

Now suppose to have a set Σ_P of P samples $[x^{(p)}, y^{(p)} = \gamma(x^{(p)})]$, $p = 1, \ldots, P$. The number P is called *sample complexity*. The quantity that can be minimized by using an approximating network with complexity ν and a set Σ_P of P samples is the so-called *empirical risk* $\frac{1}{P}\sum_{p=1}^{P} \|\gamma(x^{(p)}) - \gamma_{\nu,P}(x^{(p)}, w_\nu, \Sigma_P)\|^2$. Let

$$\gamma^\circ_{\nu P} \triangleq \gamma_{\nu P}(x, w^\circ_\nu, \Sigma_P) \triangleq \operatorname*{argmin}_{w_\nu \in \Re^{\mathcal{N}(\nu)}} \frac{1}{P}\sum_{p=1}^{P} \|\gamma(x^{(p)}) - \gamma_\nu(x^{(p)}, w_\nu)\|^2. \quad (8)$$

If, ideally, there was an infinite number of samples, it would be possible to obtain $\gamma^\circ_\nu \triangleq \gamma_\nu(x, w^\circ_\nu)$, $w^\circ_\nu \triangleq \arg \min_{w_\nu \in \Re^{\mathcal{N}(\nu)}} \mathrm{E}(\|\gamma(x) - \gamma_\nu(x, w_\nu)\|^2) = \arg \min_{w_\nu \in \Re^{\mathcal{N}(\nu)}} \int_B \|\gamma(x) - \gamma_\nu(x, w_\nu)\|^2 \rho(x) dx$,
where ρ is a probability measure on B. The quantity of interest to evaluate the effectiveness of the approximation is the distance between γ and $\gamma^\circ_{\nu P}$, called *generalization error* and defined as $\mathcal{E}_{\nu P}(\gamma, w_\nu, \Sigma_P) \triangleq \mathrm{E}(\|\gamma - \gamma^\circ_{\nu P}\|^2)$. By using the triangle inequality, the following can be written $\|\gamma - \gamma^\circ_{\nu P}\| \leq \|\gamma - \gamma^\circ_\nu\| + \|\gamma^\circ_\nu - \gamma^\circ_{\nu P}\|$, which evidences the two factors contributing to the generalization error. The term $\|\gamma - \gamma^\circ_\nu\|^2$, called *approximation error*, is due to the impossibility of perfectly representing a function belonging to an infinite-dimensional space by a function from a finite-dimensional space. The term $\mathrm{E}(\|\gamma^\circ_\nu - \gamma^\circ_{\nu P}\|^2)$, known as *estimation error*, is due to the fact that a limited number of samples contains insufficient information about the target function.

Now some analytical results are presented concerning the relationship among generalization error, computational complexity ν, sample complexity P and dimension n, for a random extraction of the set Σ_P according to the probability density ρ (see 3.1, case (ii)) and for an approximating network corresponding to feedforward sigmoidal neural networks. Such a network structure has been chosen for its wide and successful diffusion in applications. However, the results that are going to be presented hold in a similar form for wider classes of nonlinear approximating networks (see, for example, (Niyogi and Girosi, 1996) for radial basis function networks). To avoid burdening the notation and without losing generality, the case of a scalar function, i.e., $s = 1$ in (3), is considered. A feedforward neural network with one hidden layer is expressed as:

$$\gamma_\nu(x, w_\nu) = \sum_{i=1}^{\nu} c_i\, \sigma(x^\top \alpha_i + \beta_i), \quad (9)$$

where $\sigma(\cdot)$ is a sigmoidal function, i.e., a bounded measurable function on the real line such that $\lim_{z\to+\infty} \sigma(z) = 1$, $\lim_{z\to-\infty} \sigma(z) = 0$. For functions $\gamma : \Re^n \to \Re$ with a Fourier transform γ_F, let $C_\gamma \triangleq \int_{\Re^n} \|\omega\|_1 \, |\gamma_F(\omega)| \, d\omega$ (see (Barron, 1993)), where $\|\omega\|_1 \triangleq \sum_{i=1}^{n} |\omega^i|$, with ω^i denoting the i–th component of the vector ω, and $\Gamma \triangleq \{\gamma\colon \Re^n \to \Re \ \text{ such that } \ C_\gamma < \infty\}$. The following theorem is stated for an integral squared error on the set $B \subseteq [-1,1]^n$, but this is not a major limitation, since other bounded domains may be considered (see (Barron, 1993) for details).

Theorem 1(Barron, 1994). *Let A_ν be the set of sigmoidal feedforward neural networks with ν hidden units, and let $\gamma^\circ_{\nu P} \triangleq \gamma^\circ_{\nu P}(x, w^\circ_\nu, \Sigma_P)$, $w^\circ_\nu \triangleq \arg \min_{w_\nu \in \Re^{\mathcal{N}(\nu)}} \frac{1}{P}\sum_{p=1}^{P} \|\gamma(x^{(p)}) - \gamma_\nu(x^{(p)}, w_\nu)\|^2$. Moreover, let $\gamma \in \Gamma$, $B \subseteq [-1,1]^n$, and suitable assumptions on the parameters vector be satisfied (see (Barron, 1994)). Then*

$$\mathcal{E}_{\nu P}(\gamma, w_\nu, \Sigma_P) \leq O\left(\frac{C_\gamma^2}{\nu}\right) + O\left(\frac{\nu n}{P} \ln P\right).$$

A number of important remarks can be done about this result.

- To apply Theorem 1 to the proposed methods for the approximate solution of Problem P, the optimal cost-to-go functions or each component of the optimal controls must belong to the class Γ. However, this is not a major limitation, since Γ contains a large variety of functions of interest in applications (see (Barron, 1993)).
- For a fixed dimension n, the approximation error, as expected, goes to zero when $\nu \to \infty$. The estimation error goes to zero when ν, $P \to \infty$ only if $\frac{\nu}{P} \ln P \to 0$, i.e., only when the sample complexity ν grows slower than $P/\ln P$.
- For functions with C_γ showing a moderate growth with n, the bound $O(C_\gamma^2/\nu)$ allows to avoid the curse of dimensionality with respect to the computational complexity.
- If we let $\nu \sim C_\gamma \left(\frac{P}{n \ln P}\right)^{1/2}$, the upper bound on the generalization error is

$$O\left(C_\gamma \left(\frac{n}{P} \ln P\right)^{1/2}\right). \quad (10)$$

This implies that, for functions with C_γ showing a moderate growth with n, also the curse of dimensionality with respect to the sample complexity is avoided.
- With respect to the choice in the previous point, it must be pointed out that the number of units ν cannot generally be selected as a function of C_γ,

which is not known a priori. However, the same bound (10) can be obtained when the computational complexity is optimized by using one of the techniques described in (Barron, 1994).

It is interesting to compare the nonlinear ANs with the linear ones. By ignoring the logarithmic factor, for a given C_γ the bound (10) provides a rate of convergence $O\left(\frac{n}{P}\right)^{1/2}$ for the generalization error. In contrast, the rate of convergence by using linear approximating networks (e.g., fixed-knots splines and polynomials) for functions belonging to classical smoothness classes, such as Sobolev spaces of order s, is $O\left(\frac{1}{P}\right)^{2s/(2s+n)}$ (see, for example, (Nussbaum, 1986)). The dependence of the exponent on the dimension n implies that, in order to keep the generalization error below a fixed threshold, the smoothness s has to grow with the dimension n. In other words, to avoid the curse of dimensionality with respect to the sample complexity, the class of functions to be approximated has to be constrained more and more as the dimension increases. However, it must be considered that also the condition of finiteness of C_γ, used to define the class Γ, represents a smoothness constraint, although in a less evident form. As noticed in (Girosi *et al.*, 1995), functions in Γ become more and more constrained as the dimension n grows. In other words, the effect of the dimension reveals itself in a different way on Γ with respect to Sobolev spaces.

Moreover, in the bound (10) the dimension n is somehow "hidden" in C_γ. Such a bound allows to avoid the curse of dimensionality with respect to the samples complexity only for classes of functions with C_γ not exponentially large in n. The same problem arises if one remembers that the bound (10) corresponds to a computational complexity $\nu \sim C_f\left(\frac{P}{n \ln P}\right)^{1/2}$, which is characterized by the curse of dimensionality for exponentially large values of C_γ. However, a wide variety of functions of interest for which C_γ shows a moderate growth with the dimension is reported in (Barron, 1993).

6. A NUMERICAL COMPARISON BETWEEN METHOD A AND METHOD B

In order to test and compare Method A and Method B, a very classic example of application is considered: the optimal control of a system of reservoirs on a finite horizon of time. This problem has been faced many times in literature, very often by using some techniques related to DP ((Foufoula-Georgiou and Kitanidis, 1988), (Johnson *et al.*, 1993), (Yakowitz, 1982)), for which it has become a sort of benchmark. For the simulations in this paper, a 10-reservoir system with T=5 depicted in Fig.1 [2] has been chosen. The system is the same presented (for the deterministic case) in (Yakowitz, 1982). The state equation has the form:

$$x_{t+1}^j = \min(x_t^j - u_t^j + \sum_{h \in I_j^+} u_t^h + \xi_t^j, x_{max}^j)$$
$$t = 0, 1, \ldots, 4; \quad j = 1, 2, \ldots, 10$$
$$0 \le u_t^j \le x_t^j + \sum_{h \in I_j^+} u_t^h + \xi_{min}^j,$$

where x_t^j is the amount of water in the j-th reservoir at the beginning of the t-th period, u_t^j is the amount of water from the j-th reservoir released during the t-th period (planned at the beginning of the period), I_j^+ is the set of indexes relative to the reservoirs which release water directly into the j-th one, and ξ_t^j is the natural inflow of water into the j-th reservoir during the t-th period. The inflows are stochastic, all with probability density functions that are uniform between $\xi_{min} = \text{col}(1, 1, 1, 0, 0.7, 0.7, 0, 1.5, 0, 0)$ and $\xi_{max} = \text{col}(1.5, 1.5, 1.5, 0, 1.2, 1.2, 0, 2, 0, 0)$: they simulate the sum of deterministic minimum river inflows and stochastic rain inflows. The random variables ξ_t^j are mutually independent. The presence of a floodway for each reservoir is assumed, so that the state vector is never larger than $x_{max} = \text{col}(10, 10, 10, 10, 10, 10, 10, 10, 18, 25)$. The benefit function to be maximized is given by

$$\mathcal{J} = \sum_{t=0}^{4} (c_p u_t + c_f u_t^{10}) - 100 \sum_{j=1}^{10} \max\left(0, x_{min}^j - x_5^j\right)^2, \quad (11)$$

where $c_p = \text{col}(1.1, 1, 1, 1.2, 1.1, 1, 1.2, 1, 1.2, 2.5)$ describes the benefits for the release of water from each reservoir (for example, power generation) and $c_f = 1.9$ is the coefficient corresponding to an irrigation benefit, supposed to come only from the 10-th reservoir. J contains a penalty function to induce the condition $x_5 \ge x_{min}$, where $x_{min} = \text{col}(3, 3, 3, 3, 3, 3, 3, 3, 5, 6)$.

For what concerns Method A, ANs having the structure of feedforward neural networks have been used to approximate the optimal cost-to-go functions $\tilde{J}_t^o(x_t)$, $t = 1, 2, \ldots, 4$. Each network has $\nu = 40$ neural units, and it has been trained by using $L = 1800$ sample pairs [state vector - optimal cost-to-go function].

As for Method B, feedforward neural networks have been used again, this time to approxi-

[2] The numerical comparison has been drawn from (Baglietto *et al.*, 2000)

mate the optimal control functions $\mu_t^\circ(x_t)$, $t = 1, 2, \ldots, 4$. Each network has $\nu = 40$ neural units. Since Method B utilizes unconstrained optimization techniques, suitable differentiable penalty functions have been added to the benefit function (11) in order to implement the various constraints.

The results obtained from both methods have been tested by computing a mean benefit over different sequences of the inflows, randomly extracted according to their probability density functions, each one for $x_0 = x_{min}$. The mean benefit for Method A has turned out to be 282, while Method B has given a benefit of 285.

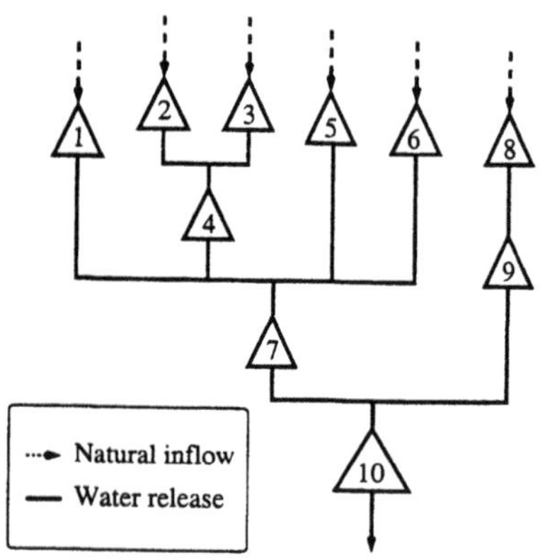

Fig. 1. Ten-reservoir system configuration.

7. REFERENCES

Archibald, T.W., K.I.M. McKinnon and L.C. Thomas (1997). An aggregate stochastic dynamic programming model of multireservoir systems. *Water Resour. Res.* **33**, 333–340.

Baglietto, M., C. Cervellera, T. Parisini, M. Sanguineti and R. Zoppoli (2000). Neural approximators, dynamic programming and stochastic approximation. *Proc. 19th Amer. Contr. Conf.* pp. 3304–3308.

Baglietto, M., T. Parisini and R. Zoppoli (2001). Distributed–information neural control: the case of dynamic routing in traffic networks. *IEEE Trans. on Neural Networks (to appear).*

Barron, A.R. (1993). Universal approximation bounds for superpositions of a sigmoidal function. *IEEE Trans. on Information Theory* **39**, 930–945.

Barron, A.R. (1994). Approximation and estimation bounds for artificial neural networks. *Machine Learning* **14**, 115–133.

Bellman, R. (1957). *Dynamic Programming.* Princeton University Press. Princeton.

Bellman, R., R. Kalaba and B. Kotkin (1963). Polynomial approximation - a new computational technique in dynamic programming. *Math. Comp.* **17**, 155–161.

Bertsekas, D.P. and J.N. Tsitsiklis (1996). *Neuro-Dynamic Programming.* Athena Scientific. Belmont.

Chen, V.C.P., D. Ruppert and C.A. Shoemaker (1999). Applying experimental design and regression splines to high-dimensional continuous-state stochastic dynamic programming. *Oper. Res.* **47**, 38–53.

Foufoula-Georgiou, E. and P.K. Kitanidis (1988). Gradient dynamic programming for stochastic optimal control of multidimensional water resources systems. *Water Resour. Res.* **24**, 1345–1359.

Girosi, F., M. Jones and T. Poggio (1995). Regularization theory and neural networks architectures. *Neural Computation* **7**, 219–269.

Johnson, S.A., J.R. Stedinger, C. Shoemaker, Y. Li and J.A. Tejada-Guibert (1993). Numerical solution of continuous-state dynamic programs using linear and spline interpolation. *Oper. Res.* **41**, 484–500.

Leshno, M., V. Ya, A. Pinkus and S. Schocken (1993). Multilayer feedforward networks with a nonpolynomial activation function can approximate any function. *Neural Networks* **6**, 861–867.

Niederreiter, H. (1992). *Random Number Generation and Quasi-Monte Carlo Methods.* SIAM. Philadelphia.

Niyogi, P. and F. Girosi (1996). On the relationship between generalization error, hypothesis complexity, and sample complexity for radial basis functions. *Neural Computation* **8**, 819–842.

Nussbaum, M. (1986). On nonparametric estimation of a regression function that is smooth in a domain on $\Re^k$. *Theory of Probability and its Applications* **31**, 118–125.

Parisini, T. and R. Zoppoli (1994). Neural networks for feedback feedforward nonlinear control systems. *IEEE Trans. on Neural Networks* **5**, 436–449.

Parisini, T. and R. Zoppoli (1998). Neural approximations for infinite-horizon optimal control of nonlinear stochastic systems. *IEEE Trans. on Neural Networks* **9**, 1388–1408.

Park, J. and I. W. Sandberg (1991). Universal approximation using radial–basis–function networks. *Neural Computation* **3**, 246–257.

Turgeon, A. (1981). A decomposition method for the long-term scheduling of reservoirs in series. *Water Resour. Res.* **17**, 1565–1570.

Vapnik, V. N. (1995). *The Nature of Statistical Learning Theory.* Springer-Verlag. New York.

Yakowitz, S. (1982). Dynamic programming applications in water resources. *Water Resour. Res.* **18**, 673–696.

Zoppoli, R., T. Parisini and M. Sanguineti (2001). Approximating networks and extended Ritz method for the solution of functional optimization problems. *J. of Optim. Theory and Appl. (to appear).*

NEURAL NETWORK BASED MIN-MAX PREDICTIVE CONTROL. APPLICATION TO A HEAT EXCHANGER

D.R. Ramírez M.R. Arahal E.F. Camacho

Departamento de Ingeniería de Sistemas y Automática,
Universidad de Sevilla
Escuela Superior de Ingenieros
Camino de los Descubrimientos s/n 41092 Sevilla
Phone: +34 954487357 Fax: +34 954487340
email: {danirr, arahal, eduardo}@cartuja.us.es

Abstract: Min-max model predictive controllers (MMMPC) have been proposed for the control of linear plants subject to bounded uncertainties. The implementation of MMMPC suffers a large computational burden due to the numerical optimization problem that has to be solved at every sampling time. This fact severely limits the class of processes in which this control is suitable. In this paper the use of a Neural Network (NN) to approximate the solution of the min-max problem is proposed. The number of inputs of the NN is determined by the order and time delay of the model together with the control horizon. For large time delays the number of inputs can be prohibitive. A modification to the basic formulation is proposed in order to avoid this later problem. Simulation and experimental results are given using a heat exchanger.

Keywords: Predictive control, Minimax techniques, Neural networks, Uncertain linear systems, Process control.

1. INTRODUCTION

Mathematical models and specially control models, which have to be kept simple, can only describe the dynamics of the process in an approximately way. Most of the approaches used for modeling uncertainties assume that there is a family of models and that the plant can be described exactly by one of the models belonging to the family. The approach considered here is global uncertainties. In this way, uncertainties will be considered to affect the 1-step ahead prediction equation, i.e. the uncertainties will affect the prediction capability of the model. The global uncertainties approach can be found in (Camacho and Bordons, 1999). A robust adaptive Min-Max MPC controller has been applied to a solar power plant in (Camacho and Berenguel, 1997).

Min-Max control techniques have in common a great computational burden which limits the range of process to which can be applied to those with slow dynamics. Neural Networks have been used with success in Model Predictive Control (Arahal *et al.*, 1998). In this paper we present a way to implement Min-Max Predictive Controllers for linear plants using a Neural min-max solver.

Another difficult arises when the plant has a large dead time. In this case we have to take into account the effect of past values of the control signal in the min-max optimization. We show here how to overcome this problem without loosing the robustness properties of Min-Max MPC and with only a gain of conservativeness in the control law.

The paper is organized as follows: section 2 presents the basic Min-Max MPC with bounded global uncertainties algorithm. Next the Neural min-max solver for plants without large delays is introduced. Section 3 is devoted to the Min-Max MPC with bounded global uncertainties algorithm for plants with large delays. Section 5 shows the application of the NN based MMMPC to a heat exchanger and section 6 presents the conclusions.

2. MIN-MAX MPC WITH GLOBAL UNCERTAINTIES

The objective of MPC control is to compute the future control sequence $u(t), u(t+1), \ldots, u(t+N-1)$ in such a way that the optimal j-step ahead predictions $y(t+j|t)$ are driven close to the set point sequence $w(t), w(t+1), \ldots, w(t+N-1)$ for the prediction horizon. The way in which the system will approach the desired trajectories will be indicated by a cost function J which depends on the presents and future control signals and uncertainties.

When bounded uncertainties are considered explicitly, it would seem that a more robust control would be obtained if the controller tried to minimize the objective function J for the worst situation. That is, by solving the following min-max problem:

$$\min_{u \in U} \max_{\theta \in \Theta} J(\theta, u) = \min_{u \in U} J^*(u) \qquad (1)$$

where θ represents the sequence of future uncertainties and:

$$J^*(u) = \max_{\theta \in \Theta} J(\theta, u) \qquad (2)$$

being $\Theta = \{\theta / \underline{\theta} \leq \theta \leq \overline{\theta}\}$. The function to be minimized $J^*(u)$ is the maximum of the objective function $J(\theta, u)$ which measures how well the process output follows the reference trajectories.

Min-Max MPC can be formulated either in state space or using input-output description. The latter is used here since process with large dead time are easy to describe using transfer function representations.

The most usual form of $J(\theta, u)$ is a quadratic criterion:

$$J(\theta, u) = \sum_{j=N_1}^{N_2} (y_{k+j|k} - w_{k+j})^2 + \lambda \sum_{j=1}^{N_u} (\Delta u_{k+j-1})^2 \qquad (3)$$

where: $\Delta = 1 - z^{-1}$,N_1 and N_2 define the beginning and end of the cost horizon, N_u is the control horizon and $y_{k+j|k}$ is the prediction output. Other types of objective functions have been used in literature. In (Campo and Morari, 1987) a $\infty - \infty$ norm is used while in (Allwright, 1994) a $1 - \infty$ norm is proposed.

When a global uncertainties approach is used the way of modeling the uncertainties is to assume that all modeling errors are globalized in a vector of parameters, such that the plant can be described by the following family of models:

$$\hat{y}(t+1) = \hat{f}(y(t), u(t)) + \theta(t) \qquad (4)$$

In (Camacho and Bordons, 1999) is shown that global uncertainties can be related to other types of uncertainties. When the model considered in (4) is linear it is common to choose a CARIMA model as prediction model (Camacho and Bordons, 1999).

The error concept present in a CARIMA model can be extended to consider explicitly the disturbances in the prediction model yielding:

$$\tilde{A}(z^{-1})y(t) = z^{-d}B(z^{-1})\Delta u(t-1) + \theta(t)$$

where $\tilde{A}(z^{-1}) = \Delta A(z^{-1})$ and $\theta(t) \in \Theta$. This prediction model will be referred as an integrated uncertainties prediction model ((Camacho and Bordons, 1999)).

It is easy to see that the prediction model is an affine function of $\theta(t)$, which in turn leads to a function $J(\theta, u)$ that is convex in the hypercube of $\theta(t)$. That implies that the maximum of $J(\theta, u)$ is reached on one of the vertices of Θ. Also the maximum is unique and it can be easily seen that function $J^*(u)$ is a piecewise quadratic function of u (Camacho and Bordons, 1999).

Using an integrated uncertainties prediction model allows to reject step disturbances but produces a continuos grow of the uncertainty band (Camacho and Bordons, 1999) and (Ramírez and Camacho, 2000). To have stable uncertainty bands while keeping the step response rejection the following constraints to optimization have to be added (Ramírez and Camacho, 2000):

$$[\underline{\Theta} \ldots \underline{\Theta}]^T \leq T\,[\theta(t+1) \ldots \theta(t+N_2)]^T \leq [\overline{\Theta} \ldots \overline{\Theta}]^T$$

where T is a lower triangular matrix of appropriate dimension.

3. WORST CASE PREDICTION OF DELAYED OUTPUTS MIN-MAX MPC

The strategy used here to apply the Min-Max MPC to process with large dead time splits the problem in two stages:

- Estimation of process output after the dead time. This is a prediction problem.
- Min-Max MPC control law calculation for the process after the dead time, taking into account the uncertainty in the estimation of the process output up to the dead time. This is a control design problem.

This strategy applied for a first order CARIMA prediction model can be formulated as the following steps:

- First, compute the bounds for $y(k+d)$ and $y(k+d-1)$:

$$\begin{aligned}\overline{\theta}'_{k+d-1} &= \max_{\theta}(y_{k+d-1} - \hat{y}_{k+d-1})\\ \underline{\theta}'_{k+d-1} &= \min_{\theta}(y_{k+d-1} - \hat{y}_{k+d-1})\\ \overline{\theta}'_{k+d} &= \max_{\theta}(y_{k+d} - \hat{y}_{k+d})\\ \underline{\theta}'_{k+d} &= \min_{\theta}(y_{k+d} - \hat{y}_{k+d})\end{aligned}$$

  where $\hat{y}_{k+d-1}$ y $\hat{y}_{k+d}$ are nominal values for y_{k+d-1} and y_{k+d}.
- Second, obtain the control law solving the min-max problem for the nominal plant without dead time using the following functional:

$$f(y_{k+d}, y_{k+d-1}) = \min_{\Delta u} \max_{\theta'_{k+d}, \theta} J(\Delta u, \theta'_{k+d}, \theta)$$

  taking into account the uncertainties in the plant output for $y(k+d)$ and $y(k+d-1)$, which have to be reflected in the future output predictions. Thus, the future output predicted will be:

$$\begin{aligned}y_{k+d+1} &= (1+a)(y_{k+d} + \theta'_{k+d}) - a(y_{k+d-1}\\ &\quad + \theta'_{k+d-1}) + \theta_{k+d+1}\\ y_{k+d+2} &= (1+a)y_{k+d+1} - a(y_{k+d} + \theta'_{k+d})\\ &\quad + \theta_{k+d+2}\\ &\ldots\end{aligned}$$

It can be seen that the prediction model is an affine function of $\theta'_{k+d-1}, \theta'_{k+d}$ as well of θ. This implies that the maximum of J for $\theta'_{k+d-1}, \theta'_{k+d}$ will be reach in one of the vertexes of the hypercube $\{\theta'_{k+d-1}, \theta'_{k+d}\}$. The computational burden increment is the the same to an increment of two in the prediction horizon.

This strategy is more conservative than a traditional Min-Max MPC in the sense that only the maximum and minimum values for $y(k+d)$ and $y(k+d-1)$ are considered as starting points to the min-max optimization. To illustrate this point consider a process with a dead time of $d = 6$. In a traditional Min-Max MPC there is 32 different possible values for y_{k+d-1} and 64 for y_{k+d}, which will be used as starting point for the rest of the optimization. With the strategy presented here only two possible values for each of them are considered, which yields 4 combinations of the most extreme values of y_{k+d-1} and y_{k+d}. In addition two of the combinations are unrealistic, because there is no way to get the plant outputs equal to the nominal values of y_{k+d-1} and y_{k+d} plus $\{\underline{\theta}'_{k+d-1}, \overline{\theta}'_{k+d}\}$ or $\{\overline{\theta}'_{k+d-1}, \underline{\theta}'_{k+d}\}$. Finally, two sources of uncertainty are considered in the prediction of the y_{k+d+1} and y_{k+d+2}, i.e. the global uncertainty $\theta(t)$ and the bounds for y_{k+d-1} and y_{k+d}.

To avoid this excess of conservatism only the combinations $\{\underline{\theta}'_{k+d-1}, \underline{\theta}'_{k+d}\}$ and $\{\overline{\theta}'_{k+d-1}, \overline{\theta}'_{k+d}\}$ are considered in the optimization which has the added benefit of halving the amount of vertexes to be considered to compute the max part of the min-max problem.

The amount of time required to compute the control sequence can be further reduced by considering that for linear prediction models the superposition principle holds, hence the part of the output prediction due to the uncertainty $\theta(t)$ can be separated from the part due to control actions. It can be easily seen that the bounds values $\{\underline{\theta}'_{k+d-1}, \underline{\theta}'_{k+d}\}$ and $\{\overline{\theta}'_{k+d-1}, \overline{\theta}'_{k+d}\}$ are independent of the values of $\{y_{k-1}, y_k\}$ and are also independent of the past and future control actions. This considerations imply that the bounds for $y(k+d)$ and $y(k+d-1)$ have only to be computed once.

Figure 1 illustrate the behaviour of a Min-Max MPC controller using this strategy compared with a traditional Min-Max MPC controller and a conventional MPC controller, using the same values for control and prediction horizons and control effort weighting. For the Min-Max MPC controllers the bounds for the uncertainty were $\underline{\theta} = \underline{\Theta} = -0.5$ and $\overline{\theta} = \overline{\Theta} = 0.5$. The faster response over the traditional Min-Max MPC is due to the greater conservativeness of considering only the maximum and minimum values of y_{k+d-1} and y_{k+d}.

Finally, it is noteworthy that this scheme is suitable for other implementations of Min-Max MPC even for the numerical one, with the benefit a reduced time of computation of the min-max solution.

4. NEURAL MIN-MAX SOLVER

Artificial Neural Networks have been used with success in predictive control and identification (Arahal *et al.*, 1998), especially when non linear systems are considered. When linear systems without taking into account uncertainties are considered there is no need to use a non linear estimation method like Neural Networks because

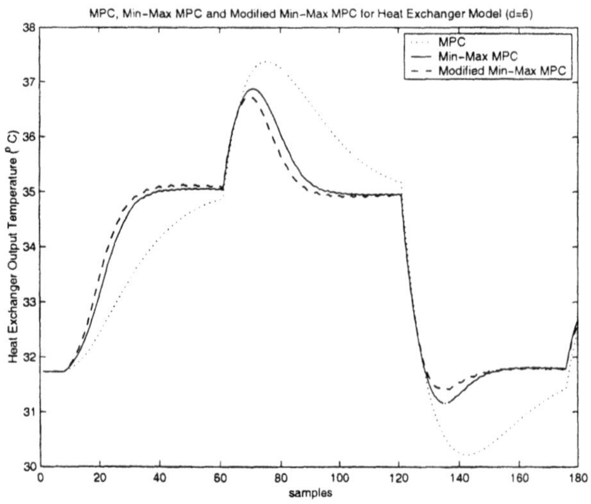

Fig. 1. Output of a MPC, Min-Max MPC and Min-Max MPC modified controllers applied to the plant $y_{k+1} = 0.9048y_k + 0.0128u_{k-6}$.

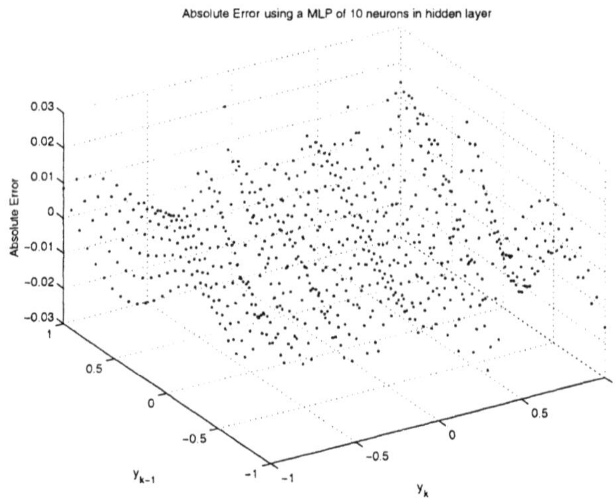

Fig. 2. Approximation error using a MLP with 10 neurons in hidden layer for a controller with $N_u = 4, N_2 = 8\lambda = 0.5, \overline{\theta} = 0.1, \underline{\theta} = -0.1$

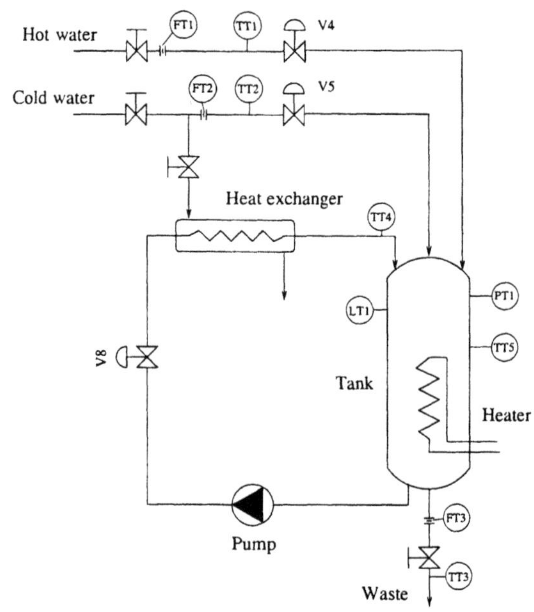

Fig. 3. Diagram of the Pilot Plant

efficient solutions exist in the form of closed formulae (for the unconstrained case) or mathematical programming (constrained case, see (Camacho and Bordons, 1999)). However if uncertainties are taking into account there is no efficient numerical methods to handle problems like (1). In this case Neural Networks can be a good choice to apply this strategy to process with reasonably fast dynamics.

The solution of the min-max problem can be precomputed for a number useful situations. A NN can be used to approximate function J^* in (2) with mild assumptions about the smoothness of the approximated mapping. If this is done properly, then solving the min-max problem would only amount to just query the NN for an output $u = NN(\mathbf{x})$ to current input $\mathbf{x}$ where $\mathbf{x}$ is the process state and set point. This allows the Min-Max MPC use in real time in a larger class of process.

Here we concentrate in a neural min-max solver for a first order plant using an integrated uncertainties prediction model:

$$y_{k+1} = (1+a)y_k - ay_{k-1} + b\Delta u_k + \theta \quad (5)$$

The function min-max to be approximated, which will take the set point and the actual and past values of process output as arguments, depends on the parameters of the plant as well of the parameters of the controller.

Figure 2 shows the approximation error using a Multi Layer Perceptron (MLP) with 10 neurons in the hidden layer when plant model parameters are $a = b = 0.25$ and controller parameters are $N_u = 4, N_2 = 8\lambda = 0.5, \overline{\theta} = 0.1, \underline{\theta} = -0.1$ and the set point is 0.25. A set of patterns distributed on an uniform grid with a spacing of 0.05, making a total of 1681 patterns, was built. The MLP was trained using the Levenberg-Marquardt method and a final sum of squared errors of 0.105 was obtained. The trained MLP was used with a different set of patterns built with a spacing of 0.08, and the errors obtained are shown in figure 2. The magnitude of errors is small enough to allow its real time use as shown for the heat exchanger in section 5.

5. APPLICATION TO A HEAT EXCHANGER

The controller presented in sections 4 and 3 has been applied to a heat exchanger which is part of a pilot plant. A diagram of the pilot plant which shows its main elements as well as the localization of the various instruments is given in Figure 3.

The main elements are:

- *Feed circuit.* The plant has two input pipes, a cold water one (which temperature is regulated by a cooling plant) and a hot water one (at about 70 ^oC). The temperatures and the flows of the inputs are measured with thermocouples and orifice plates respectively, with controlled pneumatic valves for regulating the input flows.
- *Tank.* It has a height of 1 m and an interior diameter of 20 cm, it is thermically insulated, and with an approximate volume of 31 l. In its interior there is a 15 kW electric resistor for heating, also an overflow pipe, an output pipe and another one for recirculating the water through the exchanger.
- *Recirculation circuit.* The hot water in the tank can be cooled by entering cold water through the cooling circuit. This circuit is composed of a centrifugal pump that circulates the hot water from the bottom of the tank through a tube bundle heat exchanger returning at a lower temperature at its top.

The plant control elements are connected to a PMC10 unit operated under the ORSI Automazione CUBE control software. The PMC10 architecture allows to implement control algorithms programmed in a PC using CUBE's ITER II language. However the execution time is restricted to 100 milliseconds and taking into account that PMC10 CPU is an old Intel 8086 it is clear that the min-max problem cannot be solved numerically in the PMC10. So it is a suitable scenario to the neural network implementation described in this paper.

Several predictive control strategies has been applied to the control loop $V8$ - $TT4$. In (Camacho and Bordons, 1999) a precomputed linear GPC is applied. On the other hand nonlinear predictive control has been applied to this control loop in (Ramírez *et al.*, 1999) (in this work control actions were calculated in a PC, rather than the PMC10 unit).

A first order lineal model for the transfer function from $V8$ to $TT4$ has been obtained by step response. The initial conditions were $V8 = 50\%, TT_4 = 31.73°C$, $TT5$ controlled with a PID around $50°C$, constant tank level of 76.8% and the set point for the cooling plant to get TT_2 around $23.3°C$. A step in the aperture of valve $V8$ from 50% to 70% yields the following model:

$$G(s) = \frac{0.135}{6s+1}e^{-6s} \tag{6}$$

A CARIMA model is obtained sampling model 6 with a sampling time of 1 second (giving a delay d of 6) and the controllers parameters have been chosen to be $N_u = 3, N_1 = 7, N_2 = N_\theta = 12, \lambda = 3.0, \overline{\theta} = 0.2, \underline{\theta} = -0.2$. To train the NN a set of patterns were generated for set point values of 32^oC, 35^oC and 37^oC. A total of 36963 patterns were generated for the training set and 6627 for the test set. A MLP with 10 neurons in the hidden layer was trained using the Levenberg-Marquardt rule until the average error in both sets were under 1.5%.

The resulting controller was applied to the heat exchanger and some of the experimental results are depicted in figures 4 to 5.

Figure 4 shows a set point tracking experiment. It is noteworthy that the set point is higher than the operating point considered for modeling the heat exchanger dynamics. The noisy output is due to the variations in the tank temperature (TT5) which is regulated by a local PI. Also it is noteworthy that the V8 aperture is not affected by fast TT5 changes, on the other hand V8 aperture follows TT2 changes and TT5 slow changes.

Disturbance rejection is illustrated in figure 5. In this case the manual valve of cold water was closed for 15 and 20 seconds, causing the temperature to greatly deviate from the set point. The controller reacts by closing the recirculation valve V8 to low the temperature by having less hot water to be cooled. Opening again the valve is another disturbance and the controller has to open again the V8 valve to bring the temperature TT4 to the desired value.

Finally a third experiment depicting a slow variation of plant dynamics is shown in figure 5. In this case, the tank temperature (TT5) was raised from the nominal point (50^oC) to a point around 70^oC. It is clear that the controller is able to keep the output close to the set point as soon as the tank temperature reaches the new operating point. Meanwhile a small offset is observed as expected, because the uncertainty grows as a ramp and the integrated uncertainties MMMPC is formulated to reject step disturbances.

6. CONCLUSIONS

A Neural Min-Max MPC controller has been proposed and applied to a pilot plant. Problems related to process delay have been tackled and discussed taking into account real time implementation requirements. The suitability of a NN approach has been discussed and brought to practice. Further investigations will be needed to include constraints and adaptive skills in the control strategy.

7. REFERENCES

Allwright, J.C. (1994). *Advances in Model-Based Predictive Control.* Chap. On min-max

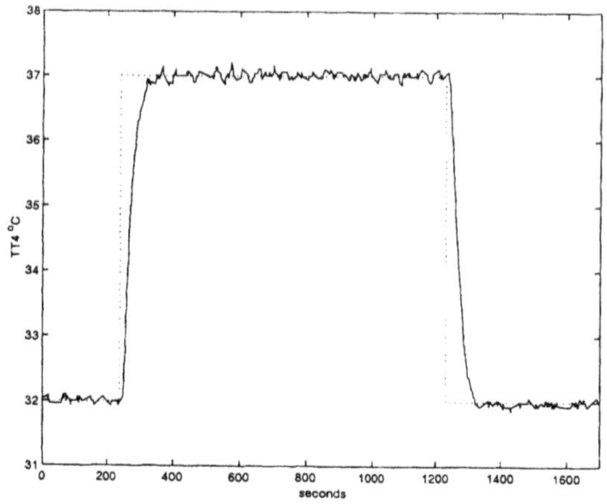

(a)

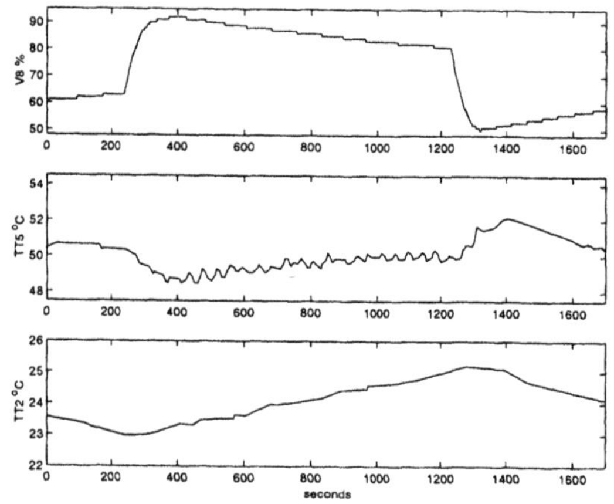

(b)

Fig. 4. Set point tracking experiment: a) set point and heat exchanger output temperature b) Valve V8 aperture, tank temperature TT5 and cold water temperature TT2

Model-Based Predictive Control. Oxford University Press.

Arahal, M.R., M. Berenguel and E.F. Camacho (1998). Neural identification applied to predictive control of a solar plant. *Control Engineering Practice* **6**, 333–344.

Camacho, E. F. and Manuel Berenguel (1997). Robust adaptive model predictive control of a solar plant with bounded uncertainties. *International Journal of Adaptive and Signal Processing* **11**, 311–325.

Camacho, E.F. and C. Bordons (1999). *Model Predictive Control.* Springer-Verlag.

Campo, P.J. and M. Morari (1987). Robust model predictive control. In: *Proc. American Control Conference.* pp. 1021–1026.

Ramírez, D.R. and E.F. Camacho (2000). Properties of min-max mpc with global uncertainties. Internal Report, Dept. Ingeniería de Sistemas y Automática, Universidad de Sevilla.

Ramírez, D.R., D. Limón-Marruedo, J. Gómez-Ortega and E.F. Camacho (1999). Model based predictive control using genetic algorithms. Application to a pilot plant. In: *Proc. European Control Conference, ECC'99.*

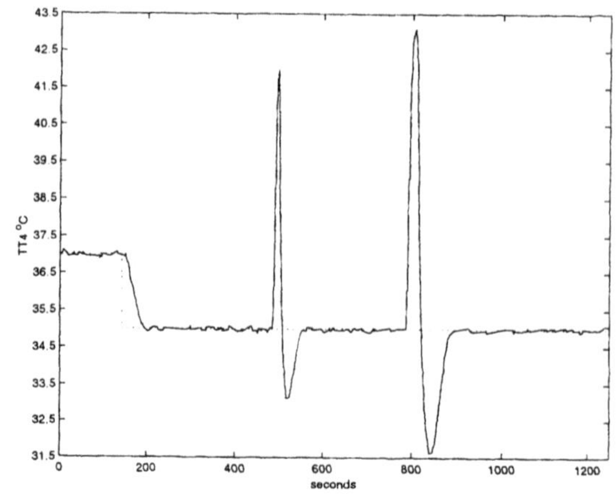

(a)

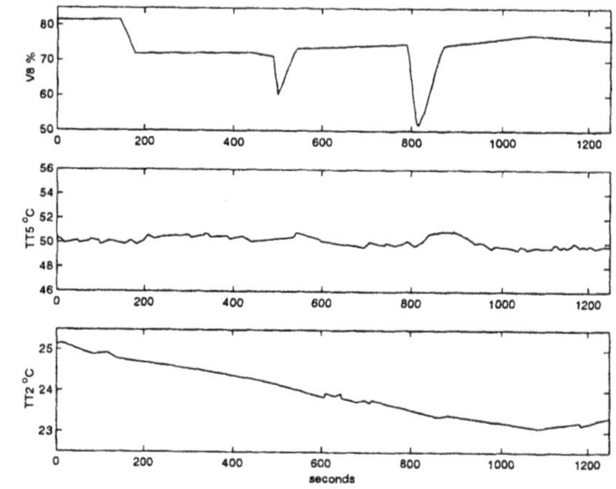

(b)

Fig. 5. Disturbance rejection experiment: a) set point and heat exchanger output temperature b) Valve V8 aperture, tank temperature TT5 and cold water temperature TT2

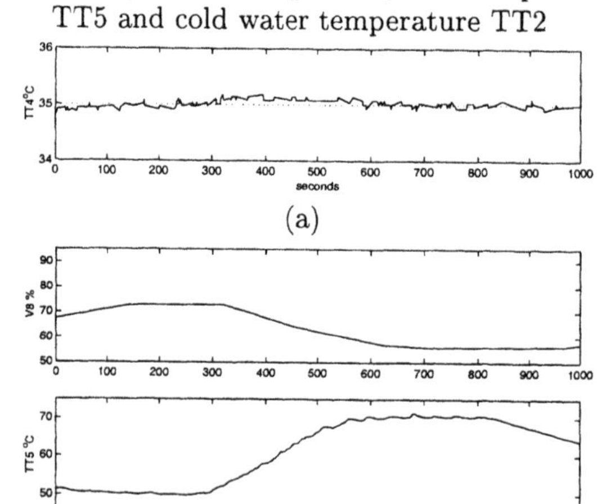

(a)

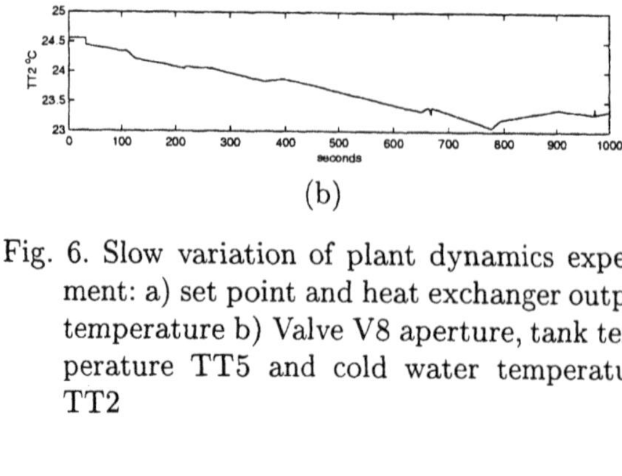

(b)

Fig. 6. Slow variation of plant dynamics experiment: a) set point and heat exchanger output temperature b) Valve V8 aperture, tank temperature TT5 and cold water temperature TT2

A DESIGN OF NEURAL-NET BASED GENERALIZED MINIMUM VARIANCE CONTROLLER

Yoshihiro Ohnishi* Toru Yamamoto Sigeru Omatu*****

* *Department of Electrical Engineering, Kure National College of Technology, 2-2-11 Aga-Minimi, Kure, Hiroshima, 737-8506, Japan, +81-823-73-8474,* onishi@kure-nct.ac.jp
** *Department of Information Science and Technology Education, Graduate School of Education, Hiroshima University, 1-1-1 Kagamiyama, Higashihiroshima, Hiroshima, 739-8524, Japan, +81-824-24-7160,* yama@hiroshima-u.ac.jp
*** *Department of Computer and Systems Sciences , Graduate School of Engineering, Osaka Prefecture University, 1-1 Gakuentyo, Sakai, Osaka, 599-8531, Japan, +81-722-54-9278,* omatu@cs.osakafu-u.ac.jp

Abstract: Much research on designing self-tuning control systems for linear systems have been proposed by using the least squares parameter identification method. However, it is difficult to employ the algorithm for the nonlinear systems except for the case where the unknown parameters are linearly combined with nonlinear terms. In this paper, a parameter estimation scheme for nonlinear systems is proposed by using a neural network. Furthermore, a design method of the control system is derived by minimizing a cost function of the generalized minimum variance control. This control input is calculated by using the estimated parameters. Finally, the effectiveness of the proposed scheme is numerically evaluated. *Copyright© 2001 IFAC*

Keywords: Self-Tuning Control, Neural Network, System Identification, Generalized Minimum Variance Control

1. INTRODUCTION

There are lots of works on designing self-tuning controllers (STCs) for linear systems (Åström, K.J.et al. 1977; Clarke and Gawthrop 1975; Clarke and Gawthrop 1979; Clarke *et al.*,1987;) Wellstead *et al.*, 1991). Most STCs are designed based on estimated parameters by using the least squares algorithm. However, it is difficult to employ the least squares algorithm for the nonlinear systems except for the case where the unknown parameters are linearly combined with nonlinear terms. Therefore, it is impossible to design the STCs for such systems. On the other hand, the effectiveness of neural networks is discussed for nonlinear systems(Narendra and Parthasarathy, 1990; Haykin, 1994). Especially, some control schemes by using the neural network have been proposed(Rumelhart *et al.*, 1987; Gupta and Rao, 1993). The conventional neural-net based control schemes can be classified into the two groups. The one is that control input is directly calculated by the neural network. As the control input is given by the output of the neural network, it is easy to implement for the controlled object. However, it is difficult to describe the structure of the controller as the transfer function. The other is that control parameters are calculated by the neural network, and this control input can be calculated by these control parameters. Although this scheme makes easy to grasp the physical meanings of control parameters, the properties of controlled object can not be directly understood.

In this paper, a generalized minimum variance controller (GMVC)(Clarke and Gawthrop, 1979) by using a neural network is proposed. According to the newly proposed control scheme, system parameters are firstly estimated by the neural network. The control law is derived by minimizing a GMVC cost function. Then, the control input is calculated by using the parameters which are estimated by the neural network. By this procedure, the explicit STC based on GMVC can be designed. Finally, the behavior of the newly proposed control scheme is investigated on some numerical simulation examples.

2. CONTROLLER DESIGN

2.1 *System description*

Consider a single-input and single-output system described as the following nonlinear discrete-time model:

$$\begin{aligned} y(t) = G\{&y(t-1), y(t-2), \dots, y(t-n), \\ &u(t-k-2), u(t-k-3), \dots \\ &\dots, u(t-k-m), \alpha_1, \alpha_2, \dots, \alpha_l\} \\ &+\sum_{i=1}^{p} c_i u^i(t-k-1), \end{aligned} \tag{1}$$

where $u(t)$ and $y(t)$ denote the control input signal and the corresponding output signal, respectively. Moreover, α_1,α_2, ...,α_l and c_i mean system parameters. Furthermore, k denotes the time-delay, and G denotes a nonlinear function. In order to simplify, the absolute values of α_1,α_2, ...,α_l and c_i in eqn.(1) are less than 1.

Suppose the following assumptions on the system (1).

Assumptions

[A.1] The structure of the nonlinear function G is known.
[A.2] The sytem parameters may be unknown.
[A.3] The time-delay k is known.

Note that there are some cases where eqn.(1) can not be described as the inner product of system parameter vector and date vector, it is difficult to apply the least squares algorithm for the parameter estimation.

Then, a parameter estimation scheme by using the neural network is proposed. First, the following identification model is introduced, which has the same structure as eqn.(1):

$$\begin{aligned} \hat{y}(t) = \hat{G}\{&y(t-1), y(t-2), \dots, y(t-n) \\ &u(t-k-2), u(t-k-3), \dots \\ &\dots, u(t-k-m), \hat{\alpha}_1, \hat{\alpha}_2, \dots, \hat{\alpha}_l\} \\ &+\sum_{i=1}^{p} \hat{c}_i u^i(t-k-1) \end{aligned} \tag{2}$$

In this equation, $\hat{\alpha}_1, \hat{\alpha}_2, \dots, \hat{\alpha}_l$ and $\hat{c}_i$ means the estimated values of $\alpha_1, \alpha_2, \dots, \alpha_l$ and c_i, respectively.

In this paper, the neural network as shown in Fig.1 is constructed for the parameter estimation. Note that every unit included in the output layer of the neural network corresponds to the estimated parameter.

In Fig.1, ○ at the hidden and the output layers mean the neuron shown in Fig.2.

In Fig.2, W_{ji}, net_j and O_j mean the weighing factor between the neuron i and the neuron j, the summation of the input to the neuron j and the output from the neuron j, respectively. And $f(\cdot)$ is the sigmoidal function used by

$$f(x) := \frac{2}{1+e^{-ax}} - 1. \tag{3}$$

Since it is assumed that the absolute value of the system parameters are under 1.0, the sigmoidal function described as eqn.(3) can be adopted. The output range of the sigmoidal function (3) should be tuned in proportion to the absolute value in the case where the absolute value in order 1.0.

The weighting factors included in the neural network are updated based on the back-propagation method whose cost function E_n is given by:

$$E_n = \frac{1}{2}\{y(t) - \hat{y}(t)\}^2. \tag{4}$$

The update rule of W_{kj} can be derived by the following procedure. First, differenciating E_n with W_{kj} yields

$$\frac{\partial E_n}{\partial W_{kj}} = \frac{\partial E_n}{\partial net_k}\frac{\partial net_k}{\partial W_{kj}} = -\delta_k O_j. \tag{5}$$

Furthermore, δ_k is given by

$$\delta_k := -\frac{\partial E_n}{\partial net_k} = e(t)\frac{\partial \hat{y}(t)}{\partial O_k}\frac{\partial O_k}{\partial net_k}. \tag{6}$$

On the other hand, the update rule of W_{ji} can be derived as follows. Diffrenciating E_n with W_{ji} yields

$$\frac{\partial E_n}{\partial W_{ji}} = \frac{\partial E_n}{\partial net_j}\frac{\partial net_j}{\partial W_{ji}} = -\delta_j O_i, \tag{7}$$

where

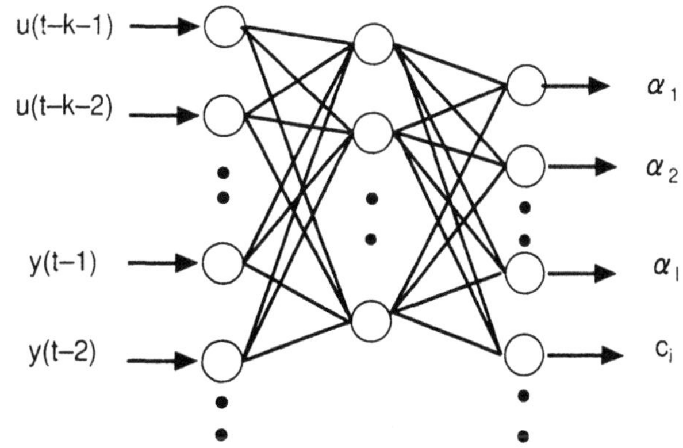

Fig. 1. Schematic figure of a neural network.

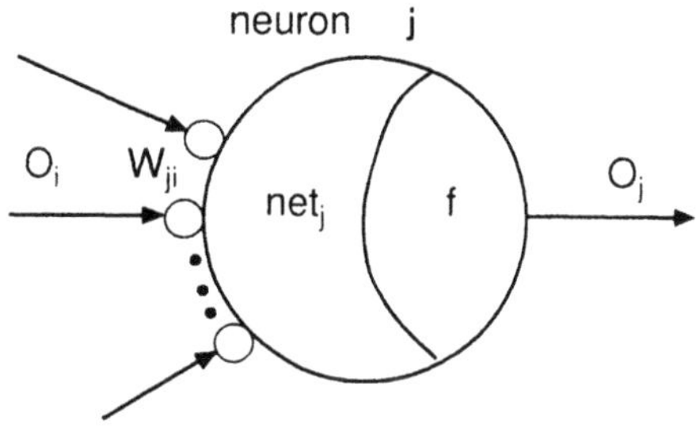

Fig. 2. Mathematical model of a neuron.

$$\delta_j := -\frac{\partial E_n}{\partial net_j} = \sum_k \delta_k W_{kj} O_j (1 - O_j). \quad (8)$$

Thus, update rules of W_{kj} and W_{ji} are given by

$$W_{kj}(t+1) = W_{kj}(t) + \eta\delta_k O_k + \alpha\Delta W_{kj}(t) (9)$$

$$W_{ji}(t+1) = W_{ji}(t) + \eta\delta_j O_j + \alpha\Delta W_{ji}(t) (10)$$

Furthermore, η and α denote the learning rate and the momentum rate, respectively.

The system identification can be realized via the above procedure. Note that according to the proposed scheme, a priori information about the system jacobian is not necessary.

2.2 *Controller Design*

The STC scheme using the parameters which are estimated by the neural network, is discussed in this section,.

Consider a GMVC criterion of the form(Clarke and Gawthrop, 1979):

$$J = E[\{P(z^{-1})y(t+k+1) \quad +\lambda\Delta u(t) \\ -P(1)r(t)\}^2] \quad (11)$$

where the reference signal $r(t)$ is given by piecewise constants. Furthermore, λ is the weighting factor for the control input, and $P(z^{-1})$ is the weighting polynomial given by the following equation:

$$P(z^{-1}) = 1 + p_1 z^{-1} + \ldots + p_n z^{-n}. \quad (12)$$

The system output $y(t+k+1)$ can be rewritten by

$$y(t+k+1) := \hat{y}(t+k+1|t) + \varepsilon(t+k+1), \quad (13)$$

where $\varepsilon(t+k+1)$ denotes the prediction error. $\varepsilon(t+k+1)$ approachs to zero in the case where the parameter estimation by the neural network works well.

Furthermore, $\hat{y}(t+k+1|t)$ is given by

$$\begin{aligned}\hat{y}(t+k+1|t) = \; & \hat{G}\{\hat{y}(t+k), \hat{y}(t+k-1), \ldots, \\ & \hat{y}(t+1), y(t), y(t-1), \ldots, \\ & y(t-n+k+1), \ldots, u(t-1), \\ & u(t-2), \ldots, u(t-m+1), \hat{\alpha}_1, \\ & \hat{\alpha}_2, \ldots, \hat{\alpha}_l\} + \sum_{i=1}^{p} \hat{c}_i u^i(t). \quad (14)\end{aligned}$$

In order to simplify, this equation is rewritten by

$$\begin{aligned}\hat{G}(\cdot|t) = \; & \hat{G}\{\hat{y}(t+k), \hat{y}(t+k-1), \ldots, \hat{y}(t+1), \\ & y(t), y(t-1), \ldots, y(t-n+k+1), \ldots \\ & u(t-1), u(t-2), \ldots, u(t-m+1), \\ & \hat{\alpha}_1, \hat{\alpha}_2, \ldots, \hat{\alpha}_l\}. \end{aligned} \quad (15)$$

In eqn.(15), $\hat{y}(t+k), \ldots, \hat{y}(t+1)$ can be calculated by using the estimated values by the neural network.

By minimizing the cost function J, the following control input can be obtained:

$$P(z^{-1})\{\sum_{i=1}^{p} \hat{c}_i(t)u^i(t) + \hat{G}(\cdot|t)\} + \lambda\Delta u(t) \\ -P(1)r(t) = 0. \quad (16)$$

Next, eqn.(16) can be rewritten by

$$\sum_{i=1}^{p} \hat{c}_i(t)u^i(t) + \lambda u(t) = \psi(t), \quad (17)$$

where

$$\psi(t) := P(1)r(t) - P'(z^{-1})\sum_{i=1}^{p} \hat{c}_i(t)u^i(t) \\ -P(z^{-1})\hat{G}(\cdot|t) + \lambda u(t-1) \quad (18)$$

$$P'(z^{-1}) = P(z^{-1}) - 1. \quad (19)$$

By solving eqn.(17), the control input which minimizes the cost function J can be obtained. As there are p solutions for eqn.(17), the minimum absolute value among real solutions is chosen as the actual control input. If all solutions are imaginary numbers, the previous input $u(t-1)$ is adopted instead of $u(t)$.

3. SIMULATION EXAMPLE

In order to investigate the behavior of the proposed control scheme, two numerical simulation examples are illustrated in this section.

3.1 *Case 1*

The system model considered by Narendra is given by the following equation(Narendra and Parthasarathy, 1990):

$$y(t) = \frac{0.5y(t-1)}{1+0.3y^2(t-1)} + 0.6u^2(t-2). \quad (20)$$

Next, the following model was used for the system identification:

$$\hat{y}(t) = \frac{\hat{\alpha}_2 y(t-1)}{1+\hat{\alpha}_1 y^2(t-1)} + \hat{\alpha}_3 u^2(t-2). \quad (21)$$

First, random input signals and the corresponding output signals during 100 steps were prepared for the learning of the neural network. Next, the unknown parameters in eqn.(21) were estimated by the neural network. The parameters of the neural network were set as

$$a = 1.0, \eta = 0.01, \alpha = 0.001. \quad (22)$$

$\partial\hat{y}(t)/\partial O_k$ in eqn.(6) were obtained from eqn.(21) as follows:

$$\left.\begin{aligned} \frac{\partial\hat{y}(t)}{\partial O_k(1)} &= -\frac{O_k(1)y^3(t-1)}{1+O_k(1)y^2(t-1)} \\ &\quad \cdot\{1+O_k(1)y^2(t-1)\} \\ \frac{\partial\hat{y}(t)}{\partial O_k(2)} &= \frac{y(t-1)}{1+O_k(1)y^2(t-1)} \\ \frac{\partial\hat{y}(t)}{\partial O_k(3)} &= u^2(t-2) \end{aligned}\right\} \quad (23)$$

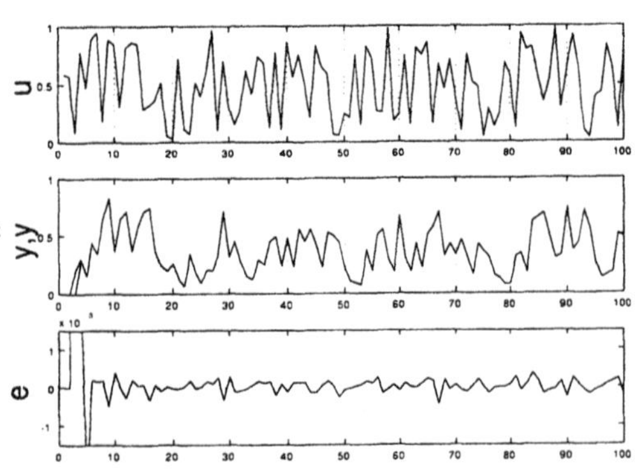

Fig. 3. The estimation result by the proposed estimator for the case 1.

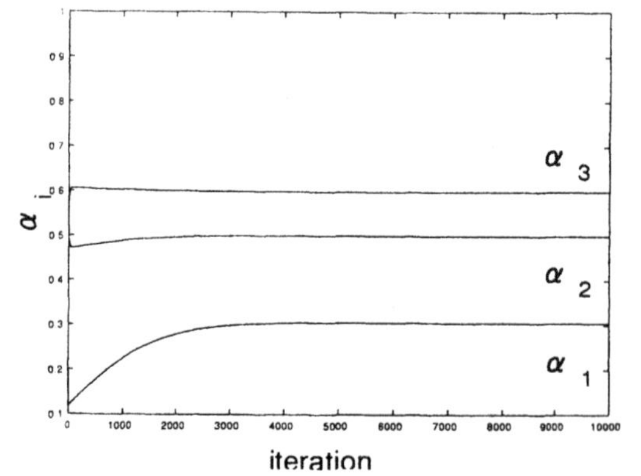

Fig. 4. The trajectories of estimated parameters for the case 1.

The number of units included in the input layer and the hidden layer of the neural network were set as 8 and 12. The learning was carried on 10000 iterations.

Fig.4 shows the input signal, the output signal, the prediction output signal and the prediction error signal. And, Fig.5 shows the trajectories of estimated parameters $\hat{\alpha}_1$, $\hat{\alpha}_2$ and $\hat{\alpha}_3$ during the learning iteration.

Next, the control system is designed by the following procedure. From eqn.(21), the prediction value $\hat{y}(t+2|t)$ is obtained:

$$\hat{y}(t+2|t) = \frac{\hat{\alpha}_2 y(t+1)}{1+\hat{\alpha}_1 y^2(t+1)} + \hat{\alpha}_3 u^2(t). \quad (24)$$

The control input which minimizes the cost function eqn.(11) can be obtained by solving the following equation.

$$\begin{aligned} \{\hat{\alpha}_3 u^2(t) + \lambda u(t)\} &= P(1)r(t) - P'(z^{-1})\hat{\alpha}_3 u^2(t) \\ &\quad -P(z^{-1})\frac{\hat{\alpha}_2 y(t+1)}{1+\hat{\alpha}_1 y^2(t+1)} \\ &\quad +\lambda u(t-1). \end{aligned} \quad (25)$$

This equation has the future value $y(t+1)$. Therefore, the prediction value $\hat{y}(t+1|t)$ is given by eqn.(21).

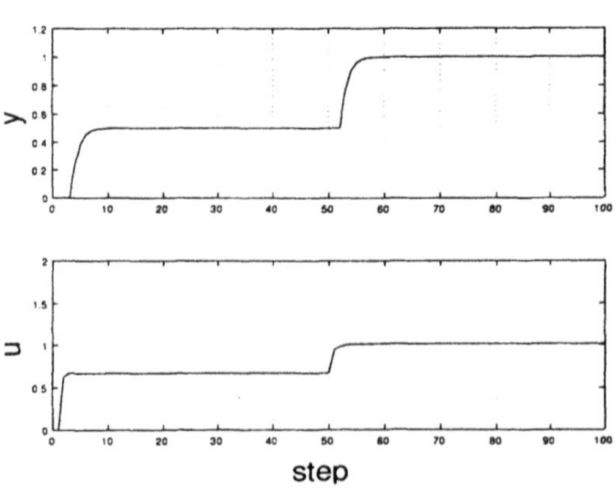

Fig. 5. The control results by the proposed control scheme for the case 1.

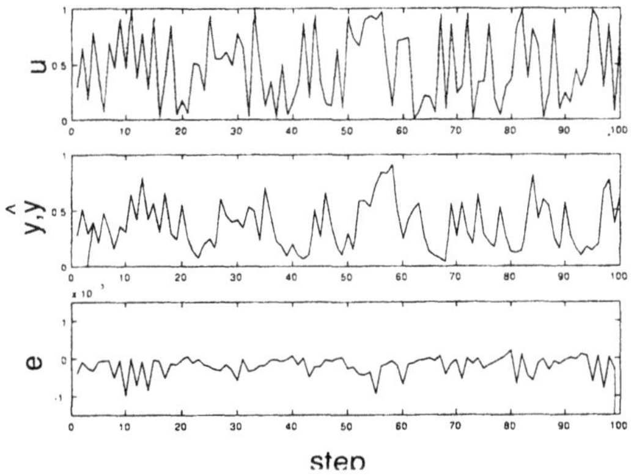

Fig. 6. The estimation result by employing other input signals for the case 1.

$$\hat{y}(t+1|t) = \frac{\hat{\alpha}_2 y(t)}{1+\hat{\alpha}_1 y^2(t)} + \hat{\alpha}_3 u^2(t-1). \quad (26)$$

After enough learning, $\hat{y}(t+1|t)$ is closed to $y(t+1)$. Therefore, the control input can be calculated.

After these preparation, this simulation example was carried out. The user-specified parameter λ was set as $\lambda = 0.1$, and the user-specified polynomial $P(z^{-1})$ was designed as

$$P(z^{-1}) = 1 - 0.4z^{-1}. \quad (27)$$

On the other hand, the generalization ability of the neural network was considered. First, other inputs were applied to the identification model, and the corresponding outputs were obtained. Next, the learning was carried out. Fig.7 shows the input, the output, the prediction output and the prediction error signals. Furthermore, Fig.8 shows estimated parameters $\hat{\alpha}_1$, $\hat{\alpha}_2$ and $\hat{\alpha}_3$

From Figs.4, 5 and 6, it is clear that the proposed control scheme works well with enough training. On the other hand, from Figs.7 and 8, the neural network has enough generalization ability.

3.2 *Case 2*

The system model considered by Pham is given by the following equation(Pham and Liu, 1995):

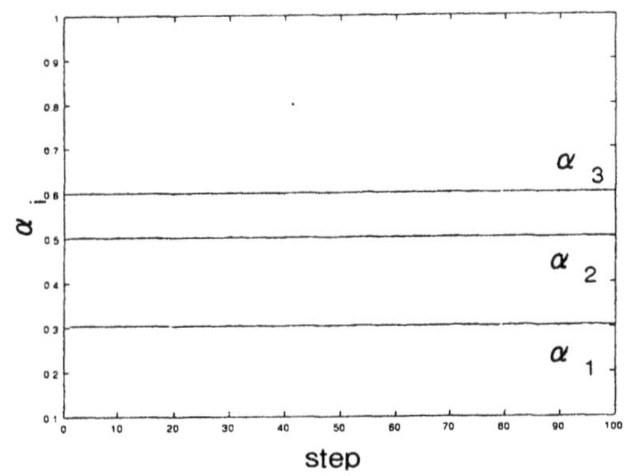

Fig. 7. The trajectories of estimated parameters corresponding to Fig.7.

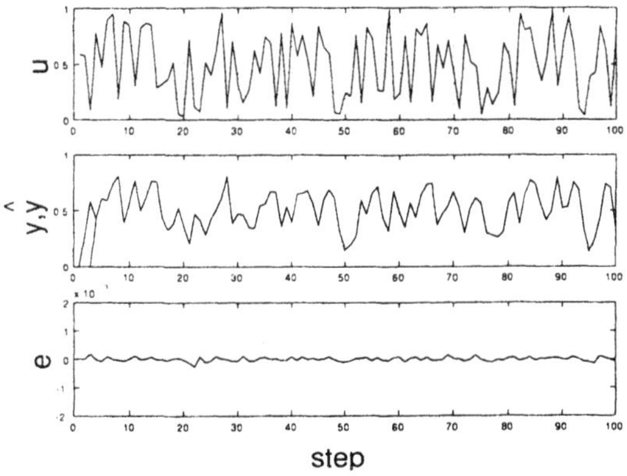

Fig. 8. The estimation results by the proposed estimator for the case 2.

$$y(t) = \frac{y(t-1)}{0.9+y^2(t-1)} - 0.3y(t-1) + 0.5u(t-2). \quad (28)$$

Next, the identification model was derived corresponding to eqn.(28) as follows:

$$\hat{y}(t) = \frac{y(t-1)}{\hat{\alpha}_1+y^2(t-1)} - \hat{\alpha}_2 y(t-1) + \hat{\alpha}_3 u(t-2). \quad (29)$$

The learning of the nural network was carried out by using 100 input-output data. The parameters used in the training were set as

$$a = 1.0, \eta = 0.001, \alpha = 0.001. \quad (30)$$

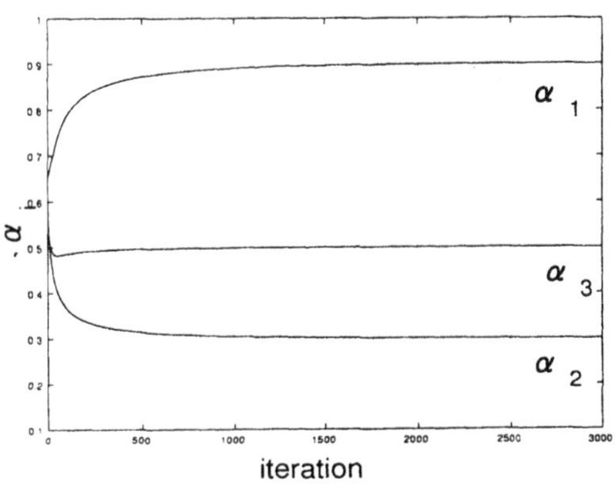

Fig. 9. The trajectories of estimated parameters for the case 2.

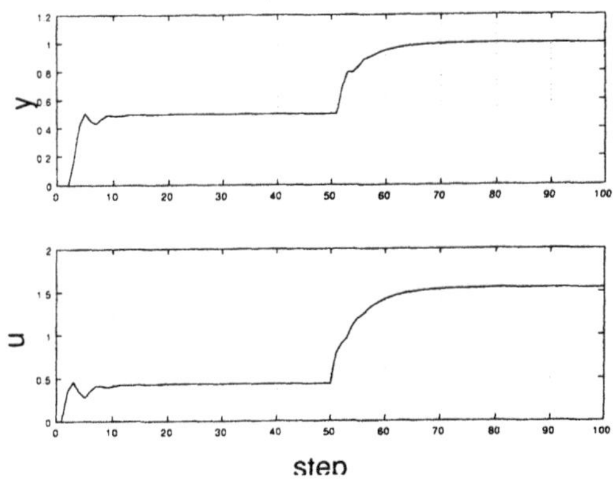

Fig. 10. The control results by the proposed control scheme for the case 2.

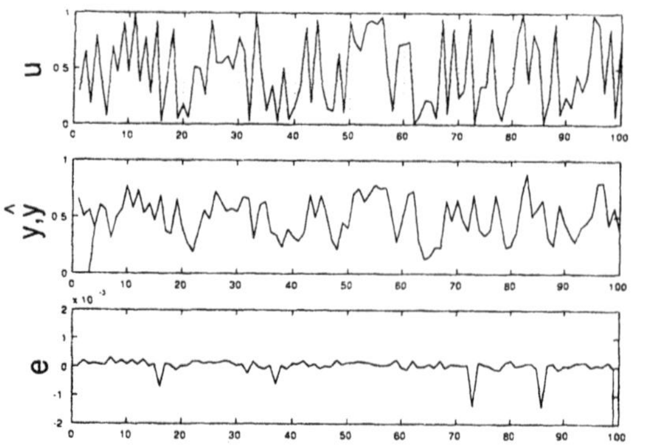

Fig. 11. The estimation result by employing other input signals for the case 2.

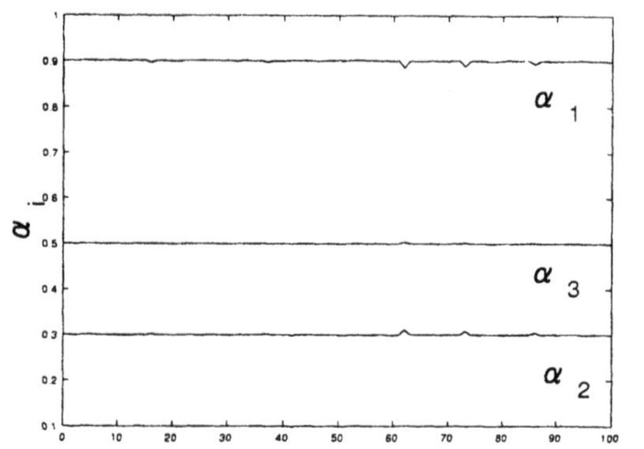

Fig. 12. The trajectories of estimated parameters corresponding to Fig.12.

$\partial\hat{y}(t)/\partial O_k$ in eqn.(6) were given by

$$\left.\begin{aligned}\frac{\partial\hat{y}(t)}{\partial O_k(1)} &= -\frac{y(t-1)}{\{O_k(1)+y^2(t-1)\}^2}\\ \frac{\partial\hat{y}(t)}{\partial O_k(2)} &= -y(t-2)\\ \frac{\partial\hat{y}(t)}{\partial O_k(3)} &= u(t-1).\end{aligned}\right\} \quad (31)$$

The number of units included in the inputs layers and the hidden layers of the neural network were set as 6 and 7. This learning was carried on 10000 iterations.

Fig.9 shows the input, the output, the prediction output and the prediction error. And, Fig.10 shows estimated parameters $\hat{\alpha}_1$, $\hat{\alpha}_2$ and $\hat{\alpha}_3$ during the learning iteration. Furthermore, Fig.11 shows control output and input signals. The user-specified parameter λ and polynomial $P(z^{-1})$ are designed as in case 1.

On the other hand, the generalization ability of neural network was investigated by the same way as case 1. Fig.12 shows the input, the output, the prediction output and the prediction error signals. Furthermore, Fig.13 shows estimated parameters $\hat{\alpha}_1$, $\hat{\alpha}_2$ and $\hat{\alpha}_3$.

From these results, it is clear that the neural network has also enough generalization ability for the case 2.

4. CONCLUSIONS

In this paper, a design method of GMVC using the neural network was considered. The control inputs are calculated based on the estimates which are parameters by using the neural network. Furthermore, some simulation examples have been illustrated to show the effectiveness of the newly proposed control scheme.

The main features of the proposed control scheme are summarized as follows.

- The control law can be employed for the non-linear system by using the neural network.
- The control input can be derived by minimizing the GMVC criterion.

The application of the newly proposed caontrol scheme for real system is currently considered.

5. REFERENCES

Åström, K.J.et al. (1977) Theory and Applications of Self-Tunig Regulator, In :*Automatica,* **Vol.13, No.5,457~476**

Clarke. D.W. and P.J.Gawthrop (1975) Self-tuning controller *IEE Proc.*, Vol.122D, No.9, pp.929-934

Clarke, D.W., and P.J.Gawthrop (1979) Self-tuning control:*IEE Proc.* , Vol.126D, No.6, pp.633-640

Clarke, D.W., C.Mohtadi and P.S.Tuffs (1987) Generalized predictive control ; *Automatica,* Vol.23, No.2, pp.137-160

Gupta, M.M. and D.H.Rao, (1993) Neuro-Control Systems Theory and Applications, IEEE Press

Haykin, S.(1994). Neural Networks. *Macmillan College Publishing,* New York,

Narendra, K.S. and K.Parthasarathy (1990) Identification and Control of Dynamical Systems Using Neural Networks. *IEEE Trans.* Neural Networks, **1**,No.1, 4

Rumelhart, D.E.,J.L.McClelland and the PDP Reseach Group (1987) Parallel Distributed Processing. The MIT Press

Pham, Duc T. and Liu, Xing. (1995) Neural Networks for Identification, Prediction and Control. Springer-Verlag,

Wellstead, P.E. and M.B.Zarrop (1991) *Self-Tuning Systems.* John Wiley & Sons Ltd.

SAFE SWITCHING IN MULTI-CONTROLLER IMPLEMENTATION

Brian D. O. Anderson† Thomas S Brinsmead†
Daniel Liberzon‡ A. Stephen Morse§

† *RSISE, Aust. Nat. Uni., ACT 0200, Australia*
‡ *CSL, Uni. Illinois U-C, IL 61801, USA*
§ *Dept. Elec. Eng., Yale Uni., CT 06520-208267, USA*

Abstract: Multiple Model Adaptive Control involves a supervisor switching among one of a finite number of controllers as more is learnt about the plant. Safe Adaptive Control is concerned with ensuring that when the controller is changed in an adaptive control algorithm, the frozen plant-controller combination is never (closed loop) unstable. A controller scheme is proposed that combines these principles and involves a frequency-dependent performance measure based on the Vinnicombe metric. Safe switching is guaranteed to the extent which closed loop transfer function identification is accurate. *Copyright ©2001 IFAC*
Keywords- Adaptive Control, Switching Algorithms, Stability Properties.

1. INTRODUCTION

This paper [1] combines two concepts in adaptive control, namely, Multiple Model Adaptive Control (MMAC) and Safe Adaptive Control. **Multiple Model Adaptive Control** (Anderson et al., 2000; Hespanha et al., 2001; Hespanha and Morse, 1999; Morse, 1996, 1998; Pait and Kassab, 2001) postulates that an unknown true plant is within a small uncertainty ball of one (or more) members of a given finite set of nominal plant models.

A controller that gives satisfactory performance is designed for each nominal model and the associated uncertainty ball, giving a finite set of alternative nominal controllers. A "high level" supervisor then switches among the controllers from the finite controller set based on the available observations. The overall objective is to converge to the best controller for the true unknown plant after some finite time. If the true plant coincides with one of the nominal plants, there is an obvious candidate for a good controller. The notion of the "best" controller, however, may be ambiguous. Note the following remarks.

(a) The determination of the finite set of nominal plants is addressed in Anderson et al. (2000).
(b) One can regard the supervisor's first task as plant identification, or alternatively, testing a set of hypotheses that the true plant lies in the uncertainty ball around each nominal plant model. That the plant is in a closed loop and that the controller may be changed in the future complicate this identification.
(c) After convergence to one of the nominal controllers further controller tuning for performance improvement is possible (Narendra and Balakrishnan, 1997).

One possible supervisory control architecture from Morse (1996) is depicted in Figure 1. There are m nominal models, $P_1, \cdots, P_m$, for an unknown plant $\bar{P}$, each associated with nominal controllers $C_1, \cdots, C_m$. A multiestimator is driven by the unknown plant input u and output y, with m outputs $y_1, \cdots, y_m$ with the property that if $\bar{P} = P_i$, then $y = y_i$ (after decay of initial condition effects, in the absence of noise, provided all signals are bounded). At time t, the controller $C_{\sigma(t)}$ is implemented. In Morse (1996) the switching logic relies on *monitoring signals*:

$$\mu_i = \int_0^t e^{-\lambda(t-\tau)}(y_i - y)^2 d\tau. \quad (1)$$

At each time t, the controller C_σ is implemented with $\sigma(t) \in \{1, 2, \dots, m\}$ taken as $\arg\min_{i \in I} \mu_i(t)$. A dwell-time or hysteresis may be imposed to slow down the switching.

If $\bar{P}$ coincides exactly with one of the nominal plants P_I, then, even if the switching process produces unstable signals, provided the exogenous signals are persistently exciting, then the monitoring signals μ_i will remain non-zero for $i \neq I$ and C_I will eventually be selected (Morse, 1996). This work investigates other possibilities for the monitoring signals, which are more compatible a safe switching objective.

[1] The authors acknowledge funding of this research by the US Army Research Office, Far East, the Office of Naval Research, Washington, grant number N00014-97-1-0946.

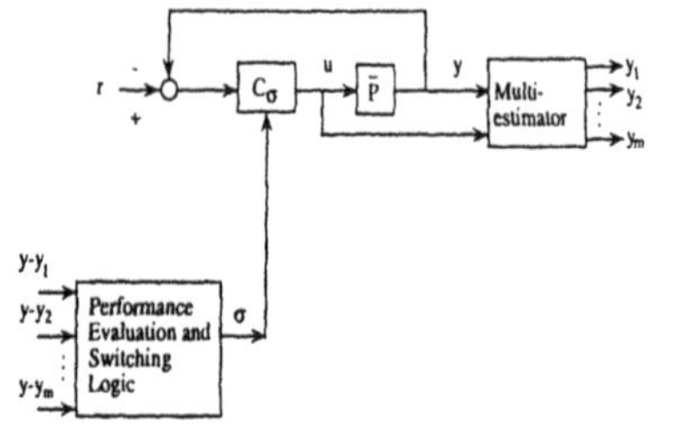

Fig. 1. Outline of multimodel adaptive controller, with $\sigma \in \{1, \cdots, m\}$

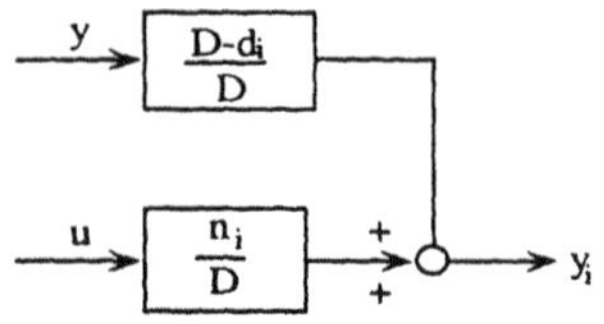

Fig. 2. Constituent part of multiestimator

The controller may switch before the monitoring signal has converged since this enables the earlier improvement of performance. Of course, if $\bar{P}$ is (slowly) time-varying, the controller may never converge.

For such a supervisory structure, the theoretical issues are: the multiestimator design; the boundedness of various signals; and the convergence of $\sigma(t)$ in finite time. One must also consider where $\bar{P}$ does not coincide exactly with any of the nominal plants.

Note that the concept of accuracy of plant approximation only makes sense with respect to the particular controller attached to the plant (Anderson and Gevers, 1998; Lee et al., 1995). The index associated with the "best" nominal model therefore depends upon the currently implemented controller and may not coincide with the index for the "best" controller.

It is hence important to make a careful choice of metric to measure the "closeness" of P_I and $\bar{P}$. We make extensive use of the δ_ν metric (Vinnicombe, 1993) which is related to, but less conservative in a well-defined sense (see Chapter 4 of Vinnicombe (2001)) than, the gap metric (Georgiou and Smith, 1990). Robust stability analysis based on the δ_ν metric relies on a small-gain argument for which more traditional small-gain results are a special case (Anderson and Brinsmead, 2001).

Many adaptive control algorithms, use an explicit or implicit identified model of the plant to design a controller. Hence (a) the true plant normally differs from the model used for controller design and (b) the controller undergoes changes. Such change is potentially dangerous, since even if the original closed loop appears stable the new (frozen) closed loop may be unstable. Of course this results in input signal excitation leading to more accurate identification and a different controller— the mechanism by which many adaptive control schemes guarantee that all signals remain bounded. This is of questionable utility if the "frozen" controller-plant combination is sometimes unstable. There are even algorithms for which all signals are bounded, but the bound is arbitrarily large!

Safe Adaptive Control refers to adaptive control algorithms in which any new controller is guaranteed *a priori* to yield a stable frozen closed loop when combined with the only partially known plant (Lee et al., 1995). In Safe Adaptive Control, controller changes will generally be small (slow). This requires that the set of controllers C_i and thus the set P_i of nominal plant models, is reasonably dense.

2. THE ESTIMATION SUPERVISOR

2.1 *Assumptions*

We derive some properties of the arrangement of Figure 1, investigating the situation where the switching supervisor is disconnected and the controller remains fixed. In order to enable better understanding, we make some assumptions, some of which will be relaxed later.

(A1) The reference signal r is stationary with a wide band spectrum;
(A2) No noise is present;
(A3) The true plant $\bar{P}$ is linear and time-invariant;
(A4) The controller C_σ is linear and time-invariant (and is not switched);
(A5) The nominal plants P_i have transfer functions $P_i = n_i/d_i$ with n_i and d_i coprime polynomials. The part of the multiestimator linking $[y, u]$ to y_i is depicted in Figure 2 where D is a stable polynomial.

2.2 *Interpretation of performance evaluation*

The transfer function: r to $(y_i - y)$ is (Figure 1)

$$W_i = \left(\frac{n_i}{D} - \frac{d_i}{D}\bar{P}\right)\frac{C}{1+\bar{P}C} \qquad (2)$$

If $(\bar{P}, C)$ is stable and r has power spectrum $\Phi_{rr}(\omega)$, the spectrum of $(y_i - y)$ is

$$\Phi_i(\omega) = \left|\frac{n_i}{D} - \frac{d_i}{D}\bar{P}\right|^2 \left|\frac{C}{1+\bar{P}C}\right|^2 \Phi_{rr}(\omega), \qquad (3)$$

and $\int_0^t (y_i - y)^2 d\tau \approx t\int_0^\infty \Phi_i(\omega)d\omega$ so that μ_i in equation (1) will behave like an integrated spectrum, that is, a measure of the error between P_i and $\bar{P}$, frequency weighted by Φ_{rr} and a C dependent term, and integrated over ω. Point-wise, rather than integral, measures of frequency domain quantities, however, may be more useful (see below).

Since for a stable closed loop the μ_i are bounded and for the unstable case they grow exponentially fast, this property could be used to distinguish whether the closed loop is stable. Detecting instability, while indicating that the currently connected C is unacceptable, is unhelpful for suggesting an appropriate replacement. If the closed loop is unstable, then the signals $(y_i - y)$ will grow exponentially and μ_i will be dominated by the transfer function error between P_i and $\bar{P}$ at only a single point in the (right half) complex plane. In such a case, the μ_i measures are likely to be of limited utility.

In the framework of Morse (1996), μ_i is a *scalar* measure associated with each plant (and thus controller) possibility. In principle, however, investigating the frequency content of $(y_i - y)$ allows for a more sophisticated metric.

2.3 *Relaxation of Certain Assumptions*

Under the above restrictive assumptions, we can identify the transfer function from r to $(y_i - y)$ if the $[\bar{P}, C]$ loop is stable. Assumption (A1) can be relaxed (see Section 6), and (A2) can be relaxed to permit noise. Assumption (A3) can be relaxed to allow time-variation, on a scale much slower than the time-scale for identification, and (A4) can be relaxed to permit contoller switching, provided that transient effects are allowed to decay. (A state-shared controller parameterisation ensuring bumpless transfer has been proposed by Morse (1998).)

The term "identification" means finding a transfer function estimate which is understood to have associated error. The error may be described probabilistically, or by hard bounds, depending on the noise models and identification method. "Identification" is not necessarily error free, nor even extremely accurate (which may be possible even in noise, given sufficient identification time). We merely need an estimate that is sufficiently accurate that robust stability and performance analysis can be performed.

2.4 *Performance Evaluation using Transfer Functions*

We make the following key observation.

Lemma 1. Adopt assumptions A1-A6. Let $W_i(j\omega)$ be the transfer function from r to $(y-y_i)$ in equation (2), for the scheme in Figure 1. Let $\kappa(P_i(j\omega), \bar{P}(j\omega))$ be the chordal distance (Vinnicombe, 1993) between P_i and $\bar{P}$ at $s = j\omega$, and let $T(\bar{P}(j\omega), C(j\omega))$ be the generalised sensitivity matrix of $(\bar{P}, C)$. Then $|W_i(j\omega)| =$

$$\kappa(P_i(j\omega), \bar{P}(j\omega))\bar{\sigma}\left[T(\bar{P}, C)\right] \frac{|C|}{\sqrt{1+|C|^2}} \left|\frac{D_i}{D}\right|, \quad (4)$$

where $D_i^* D_i = n_i^* n_i + d_i^* d_i$ for stable polynomial D_i. The proof is omitted for brevity. See Anderson et al. (2001).

The error signal $|W_i(j\omega)|$ in (4) is proportional to the spectrum of the input shaped by $\kappa(P_i, \bar{P})\bar{\sigma}[T(\bar{P}, C)]$. This quantity is critical for possibly allowing us to guarantee that C stabilises P_i, given that it stabilises $\bar{P}$. In a sense, the best P_i, *given the condition that C will be retained* (and a winding number condition involving P_i and $\bar{P}$), is the one that keeps $|W_i||D|\sqrt{1+|C|^2}/|D||C|$ small across the whole spectrum.

Although the quantities μ_i provide a measure of an integrated version of $|W_i|^2$ weighted by $\Phi_{rr}(\omega)$, the above argument indicates that if (approximate) frequency-wise identification is possible, then a Vinnicombe distance criterion may aid choice of the best P_i. Robust performance, as well as robust stability may also be considered.

An alternative multiestimator construction modestly simplifies the above at the expense of realisation complexity. By replacing D in Figure 2 with D_i one has the slightly simpler $W_i = \kappa(P_i, P)\bar{\sigma}[T(\bar{P}, C)]\frac{|C|}{\sqrt{1+|C|^2}}$ instead of (4). Of course, we cannot expect to identify the W_i exactly, but only to within some error bound. Further, the W_i may be changing slowly, due to changing $\bar{P}$. Also, even though quantities like $|W_i|$ measure the approximation error between P_i and $\bar{P}$, given a particular controller C, rather than measuring the best C_i to use for $\bar{P}$, we still use the $|W_i|$ for controller selection. We now focus on the safe switching issue.

2.5 *Changing the Controller*

The goal of the estimator-based supervisor is not to find which P_i is closest to the true plant $\bar{P}$ with the current controller, but what would make a better controller for $\bar{P}$ than the current one. We could select P_i corresponding to the smallest $\kappa(P_i, \bar{P})\bar{\sigma}\{T[(\bar{P}, C)]\}$ or an integrated version thereof, and then hope that C_i makes a good, and certainly stabilising, controller for $\bar{P}$. In order to *guarantee* that C_i stabilises $\bar{P}$, that is, to ensure safe adaptive control a sufficient condition (Vinnicombe, 1993), given the hypothesis that $(\bar{P}, C)$ is stable, is that $\kappa(C, C_i)\bar{\sigma}[T(\bar{P}, C)] < 1 \;\; \forall\omega$. In the following section we show how to verify this inequality on the basis of available data.

3. SAFE SWITCHING

We note the following lemma.

Lemma 2. (Vinnicombe, 1993) Suppose (P_1, C_1) is stable and

$$\kappa[P_1(j\omega), P_2(j\omega)]\bar{\sigma}\{T(P_1(j\omega), C_1(j\omega))\} < 1, \forall\omega. (5)$$

Then (P_2, C_1) is stable if and only if

$$\text{wno}(1 + P_2^* P_1) + \eta(P_1) - \tilde{\eta}(P_2) = 0. \; (6)$$

If $\kappa[C_1(j\omega), C_2(j\omega)]\bar{\sigma}\{T[P_1(j\omega), C_1(j\omega)]\} < 1, \; (7)$

then (P_1, C_2) is stable if and only if

$$\text{wno}(I + C_2^* C_1) + \eta(C_1) - \tilde{\eta}(C_2) = 0. \quad (8)$$

If the controller C_j is currently connected to $\bar{P}$ and we believe that it would be preferable to use C_I, then to be assured of safety, we would like to check

$$\kappa(C_I, C_j)\bar{\sigma}\left\{[T(\bar{P}, C_j)]\right\} < 1 \text{ for all } \omega. \quad (9)$$

Provided a winding number condition involving C_I, C_j holds in addition to (9), then by Lemma 2 C_I is certainly stabilising.

Since the C_i are all known, $\kappa(C_I, C_j)$ is also known. Thus, in order to check (9) we need only to evaluate $\bar{\sigma}\left\{[T(\bar{P}, C_j)]\right\}$. Although the performance estimator is not configured to yield $\bar{\sigma}[T(\bar{P}, C_j)]$ directly, even if it were to operate as a transfer function identifier, the following lemma and its corollaries, variants of the small gain theorem, specialised so that the terms can be estimated, allow us to calculate an overbound for this quantity.

Lemma 3. Suppose (P_1, C_1) is stable. If equations (5) and (6) hold, then at each ω,

$$\bar{\sigma}[T(P_2, C_1)] \leq \frac{\bar{\sigma}[T(P_1,C_1)]}{1-\kappa(P_1,P_2)\bar{\sigma}[T(P_1,C_1)]}. \quad (10)$$

If equations (7) and (8) hold then $\bar{\sigma}[T(P_1, C_2)] \leq \frac{\bar{\sigma}[T(P_1,C_2)]}{1-\kappa(C_1,C_2)\bar{\sigma}[T(P_1,C_2)]}$.

Lemma 4. Let $\bar{P}$ be the true plant and let $\{P_1, \cdots, P_m\}$ and $\{C_1, \cdots, C_m\}$ be the collection of nominal plants and controllers. Suppose that C_j stabilises $\bar{P}$ and P_i, and that for all ω

$$\frac{1}{2} > \kappa(P_i, \bar{P})\bar{\sigma}[T(\bar{P}, C_j)]. \text{ Then} \quad (11)$$

$$\bar{\sigma}[T(\bar{P},C_j)] \leq \bar{\sigma}[T(P_i,C_j)]\{1+\kappa(P_i,\bar{P})\bar{\sigma}[T(\bar{P},C_j)]\} \quad (12)$$

The proofs are omitted for brevity.

The Lemma may be used in the following way. With C_j connected to $\bar{P}$ stable, one uses the multiestimator signals to (approximately) identify $\kappa(P_i, \bar{P})\bar{\sigma}[T(\bar{P}, C_j)]$ as a function of frequency for each P_i which is stabilised by C_j (see Section 2.4). Equation (12) then overbounds $\bar{\sigma}[T(\bar{P}, C_j)]$.

A better bounds is given by defining, at each ω: $\bar{\sigma}[T(\bar{P}, C_j)] \leq \bar{T}_j :\stackrel{\text{def}}{=}$

$$\min_{i \in \Im_j} \bar{\sigma}[T(P_i, C_j)] \left\{1 + \kappa(P_i, \bar{P})\bar{\sigma}[T(\bar{P}, C_j)]\right\} \quad (13)$$

where $\Im_j \subset \{I, \cdots, m\}$ satisfies $i \in \Im_j$ if P_i is stabilised by C_j and equation (11) holds.

We can estimate the right hand side of equation (13) albeit with some error. By Lemma 2 a sufficient condition that C_I will stabilise $\bar{P}$ is that both $\text{wno}(1 + C_I^* C_j) + \eta(C_I) - \tilde{\eta}(C_j) = 0$ and (9) hold. In the light of (13) this leads to the following more conservative sufficient condition for stability, involving quantities that we can estimate.

Corollary 1. Given that $[\bar{P}, C_j]$ is stable, then a sufficient condition for $[\bar{P}, C_I]$ to be stable is that $\text{wno}(1 + C_I^* C_j) + \eta(C_I) - \tilde{\eta}(C_j) = 0$ and $\kappa(C_I, C_j)\bar{T}_j < 1$.

Notice that if the finite set $\{P_1, P_2, \ldots, P_m\}$ is dense enough in the full uncertainty set of possible unknown plants, then (11) will be straightforwardly satisfied for some i and with fixed j, the quantity $\kappa(C_i, C_j)$ will be small for some i.

If for some particular $i = I$, there holds

$$\kappa(P_I, \bar{P})\bar{\sigma}[T(\bar{P}, C_j)] < \kappa(P_i, \bar{P})\bar{\sigma}[T(\bar{P}, C_j)]$$

for all $i \neq I$, then it would be logical to hypothesise that $\bar{P}$ is best modelled by P_I, when C_j is attached. It would then be natural to check both whether $\kappa(C_I, C_j,)\bar{T}_j < 1$, together with the winding number condition for C_I and C_j, in order to determine whether C_I can be safely implemented.

Of course, we cannot, in practice, expect measurements on the closed loop system to yield exact values of $|W_i|$ at each frequency. It is also well-known (Ljung, 1987), that in the presence of stochastic noise, any parameter estimate is subject to variance which increases with the number of unknown parameters and decreases with the quantity of available identification data. Reduced identification variance and hence faster identification times can be achieved by reduction of the number of unknown parameters, but in general, only at the expense of increased identification bias.

If exact values of W_i were available, we could identify $\bar{P}$ and test each C_i with $\bar{P}$. However, exact values are not necessary, since robust stability is ensured merely by the satisfaction of particular inequalities. In the presence of norm bounded noise, however it is in principle possible to give hard error bounds on the identification error and hence give a hard guarantee of safe switching.

In the example of Section 5, we achieve satisfactory operation of the safe switching algorithm even though the identification of $\bar{T}_j$ is subject to noise, and with the checking of the inequality condition of Corollary 1 at only a finite number of discrete frequency points.

4. ALTERNATIVE METHOD TO SAFETY

Alternative methods to guarantee safe switching, based on different *a priori* assumptions, exist. As explained in Section 2, we (imperfectly) identify the frequency response $\bar{V}_i :\stackrel{\text{def}}{=} (P_i - \bar{P}) \frac{C_j}{1+\bar{P}C_j}$. We will make use of the following inequality:

$$\left|\frac{\bar{P}C_j}{1+\bar{P}C_j}\right| \leq \left|\frac{P_iC_j}{1+P_iC_j}\right| + \left|\frac{(P_i-\bar{P})C_j}{1+\bar{P}C_j}\right| \left(1+\left|\frac{P_iC_j}{1+P_iC_j}\right|\right) (14)$$

which is derived from the observations $\frac{\bar{P}C_j}{1+\bar{P}C_j} = \frac{P_iC_j}{1+\bar{P}C_j} + \frac{(\bar{P}-P_i)C_j}{1+\bar{P}C_j}$ and

$\frac{P_iC_j}{1+\bar{P}C_j} = \frac{P_iC_j}{1+P_iC_j}\left[1+\frac{(P_i-\bar{P})C_j}{1+\bar{P}C_j}\right]$. The following lemma is a variant of the small gain theorem.

Lemma 5. Suppose that $\bar{P}$ and P_i are each stabilised by C_j. Then $\bar{P}$ is stabilised by $C_k = C_j(I+\Delta)$ if the transfer functions C_k and C_j have the same number of right half plane poles and $|\Delta|\left|\frac{\bar{P}C_j}{1+\bar{P}C_j}\right| < 1, \quad \forall\omega$. A sufficient condition for the above is that

$$|\Delta|\left[\left|\frac{P_iC_j}{1+P_iC_j}\right|+\left|\frac{(P_i-\bar{P})C_j}{1+\bar{P}C_j}\right|\left(1+\left|\frac{P_iC_j}{1+P_iC_j}\right|\right)\right] < 1. \quad (15)$$

This lemma can be applied as follows. Using some identification algorithm, we estimate the frequency responses $\bar{V}_i$. All other quantities in (15) are known. We can check condition (15), with estimated quantities replacing true quantities. If C_k and C_j have the same number of right half plane poles and (15) holds, then C_k is safe. We consider all the P_i for which (15) holds subject to (P_i, C_j) being stable, and perform a minimisation over i frequency by frequency, thus obtaining a less conservative sufficiency condition.

Whether this scheme or that of Section 3 will give better results will depend on the particular problem. Depending on the controller pole distribution one or other of the two schemes can be used.

5. EXAMPLE

In order to demonstrate the method, the controller switching scheme with safety, as described in Section 3, was implemented in `Matlab`. The plant to be controlled was chosen as $\bar{P} = \frac{1.2(\frac{1}{2}s+1)(-\frac{1}{4}s+1)}{(\frac{2}{3}s+1)(\frac{1}{3}s+1)(\frac{1}{10}s+1)}$ with 441 plant models used for the multiple model set, each with varying DC gains and non-minimum-phase zero location. The controllers for each model were designed using discrete time Q-synthesis (equivalent to pole placement), with the designed bandwidth dependent on the nominal non-minimum-phase zero location.

To maintain persistent excitation, the reference signal was filtered white noise. For the estimation of W_i in equation (4), a standard recursive least squares (Ljung, 1987) algorithm was employed in order to directly identify a vector of discrete time ARMA (auto-regressive, moving average) parameters, and the multiestimator equation (2) was used as the monitoring signal to suggest controller switchings.

The minimum controller switching time was $\tau_S = 1$. At any time after the minimum controller switching time, the controller suggested by the supervisor corresponded to the minimum μ_i. In order to check safety, the inequality condition of Corollary 1 was investigated for a finite number of frequency points, with of course the *estimated* transfer functions $\bar{T}_j$ used in place of the actual values. For comparison,

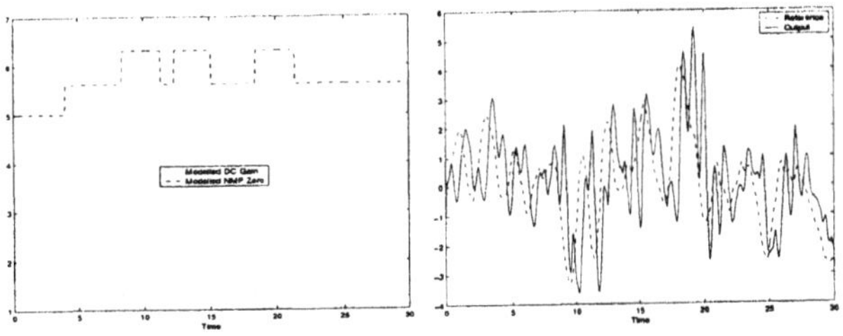

Fig. 3. (Safe) Controller Switching

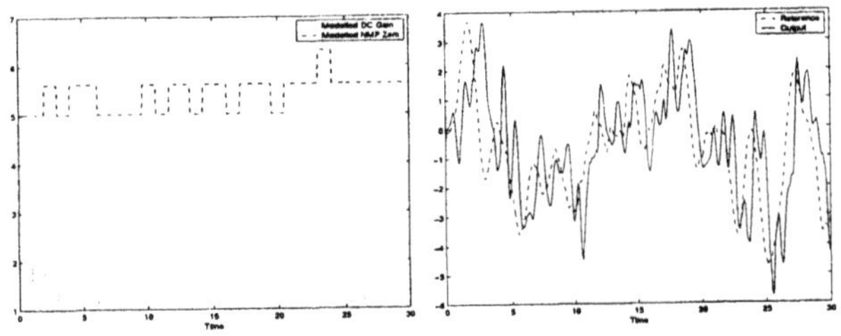

Fig. 4. Controller Switching (without safety)

simulations of the algorithm without the safety checking property were also conducted. The full specifications of the (somewhat arbitrary) choices of parameters and design methods used for the example appear in Anderson et al. (2001).

5.1 *Results*

The process was simulated and data recorded a total of sixteen (16) times. With the safety property enforced, the allowed controller switchings were much less frequent. A typical controller switching trajectory is shown in Figure 3. For the purposes of interpreting the switching graph, the lines correspond to two modeled parameters: DC gain and non-minimum phase zero location. The corresponding trajectory showing the output lagging the reference also appears in Figure 3. In all the simulation runs with the safety checking property, the routine did not once allow the switching in of any controller which was destabilising, although it did allow controllers with quite poor performance.

For simulations run without safety checking, more frequent controller switching was allowed. A typical controller switching trajectory and output trajectory are shown in Figure 4. Most of the simulations that were run without safety checking exhibited some periods of poor performance. In addition to resulting in poor performance, on a number of occasions, the implemented controller resulted in a destabilised closed loop, although only for a few periods τ_S. See Figure 5 for a simulation with a destabilising controller between 2 and 4 seconds (note the y-axis scale).

6. FURTHER ISSUES

Requirements on r: In the absence of more *a priori* information about $\bar{P}$, it is necessary to assume that r is a wideband signal. Excitation over a wide range of frequencies is required to enable checking the safe switching condition. In general, r needs to

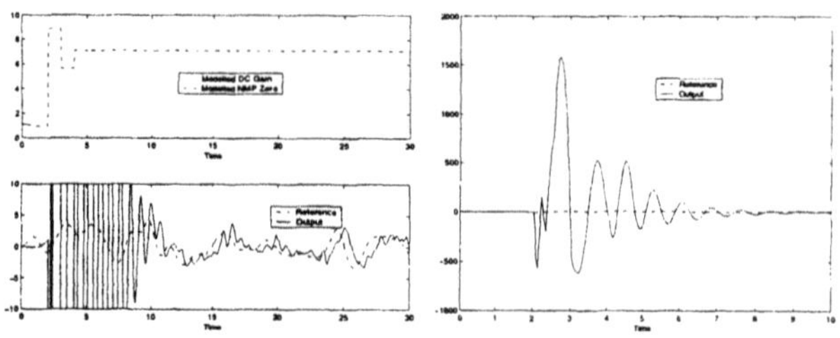

Fig. 5. Temporary Instability (without safety)

be such that a combination of experimental data and *a priori* information is sufficient to assure the satisfaction of the stability conditions.

If $\bar{P}$ is known to be a rational transfer function with known degree k, excitation at k complex frequencies is sufficient to determine $\bar{P}$ at all frequencies. Alternatively, if $\bar{P}$ is unknown up to some frequency Ω_l say, but for $|\omega| > \Omega_l$, one knows that $|\bar{P}| \leq (\omega^2+\Omega_l^2)^{-\frac{1}{2}}$ or if we know that $\bar{P} = \bar{P}_o(I+\Delta)$ for some unknown rational $\bar{P}_o$ of known degree, and some Δ lying within certain frequency-dependent bounds, then the requirement that r be a wideband signal may be somewhat relaxed.

Speed of switching: The switching speed will be limited in order to secure accurate identification for two reasons: switching transients must settle down and noise may be present.

Effect of noise: Although identification in noise can never be exact, we merely need assurance that certain inequalities involving transfer functions are fulfilled. For any given degree of confidence, noise will increase the identification time and thus limits the switching speed.

7. CONCLUSION

This paper has presented some methods to ensure that controller adaptation in Multiple Model Adaptive Control proceeds cautiously and safely, in the spirit of Lee et al. (1995). Vinnicombe metric properties ensure that no controller switchings result in a (frozen) unstable closed loop. A sufficient condition for guaranteeing safety can be checked by available measurements requiring an *estimate* of a particular transfer function. Given absolute or probabilistic confidence estimates on the identification method, it would be possible to give a quantitative absolute or probabilistic guarantees on switching safety. In addition, although the stability guarantee in principle requires checking a condition over an infinite number of frequencies only a finite number can be checked in practice. *A priori* bounds on *complexity* (Vinnicombe, 2001) of the true plant would enable this issue to be addressed rigourously.

A demonstration of the safe switching multiple model algorithm was presented. This showed that the safe switching algorithm resulted in a conservative switching regime which indeed maintained (frozen) closed loop stability, whereas the same supervisor without the safety check occasionally implemented a destabilising controller.

Open research issues include methods: to detect an unstable closed loop; to quickly find a stabilising controller the event that a destabilising controller is switched either initially or during operation; and to find an appropriate switching hysteresis time. This last issue involves a trade-off of fast supervisor response against identification confidence.

References

B. D. O. Anderson and M. Gevers. Fundamental problems in adaptive control. In Dorothée Normand-Cyrot, editor, *Perspectives in Control, Theory and Applications.* Springer, 1998.

B. D. O. Anderson and T. S. Brinsmead. Connections between ν-gap, additive, multiplicative and inverse multiplicative robust stability conditions. *IEEE Conf. Decision & Control,* 2001. submitted.

B. D. O. Anderson, T. S. Brinsmead, F. De Bruyne, J .P Hespanha, D. Liberzon, and A. S. Morse. Multiple model adaptive control, part 1: Finite coverings. *Int. JRNC,* 19:909–927, 2000.

B. D. O. Anderson, T. S. Brinsmead, J .P Hespanha, D. Liberzon, and A. S. Morse. Multiple model adaptive control with safe switching. *Int. J. Adapt. Cont. & Sig. Proc.,* 2001. to appear.

T. T. Georgiou and M. C. Smith. Optimal robustness in the gap metric. *IEEE Trans. Auto. Control,* 35(6):673–687, June 1990.

J. P. Hespanha, D. Liberzon, A. S. Morse, B. D. O. Anderson, T. S. Brinsmead, and F. De Bruyne. Multiple model adaptive control, part 2: Switching. *Int. J. Rob. & Nonl. C.,* 11:479–496, 2001.

J. P. Hespanha and A. Stephen Morse. Stability of switched systems with average dwell time. *Proc. 38th IEEE CDC,* 2655–2660, 1999.

W. S. Lee, B. D. O. Anderson, I. M. Y. Mareels, and R. L. Kosut. On some key issues in the windsurfer approach to adaptive robust control. *Automatica,* 31(11):1619–1636, 1995.

L. Ljung. *System Identification, A Theory for the User.* Prentice-Hall Int., NJ, 1987.

A. S. Morse. Control using logic-based switching. In Alberto Isidori, editor, *Trends in Control: A European Perspective,* 69–113. Springer, 1998.

A. S. Morse. Supervisory control of families of linear set-point controllers— part 1: Exact matching. *IEEE TAC,* 41(10):1413–1431, Oct. 1996.

K. S. Narendra and J. Balakrishnan. Adaptive control using multiple models. *IEEE Trans. Auto. Control,* 42(2):171–187, Feb. 1997.

F. M. Pait and F. Kassab. On a class of switched, robustly stable, adaptive systems. *Int. JACSP,* 2001. to appear.

G. Vinnicombe. Frequency domain uncertainty and the graph topology. *IEEE Trans. Auto. Control,* 38(9):1371–1383, Sep. 1993.

G. Vinnicombe. *Uncertainty and Feedback: $\mathcal{H}_\infty$ Uncertainty and the ν-gap metric.*to appear, 2001.

MULTIPLE MODEL ADAPTIVE ESTIMATION FOR MULTIUSER DETECTION IN CDMA COMMUNICATION

M.H. Jaward * V. Kadirkamanathan * S.G. Fabri **

* *Department of Automatic Control and Systems Engineering*
The University of Sheffield, Mappin Street
Sheffield S1 3JD, United Kingdom
** *Department of Electrical Power and Control Engineering*
University of Malta, Msida MSD06, Malta

Abstract: This paper introduces a blind adaptive multiuser detection scheme for code division multiple acces (CDMA) communication, based on the interacting multiple model algorithm. The system estimates the received power and input data sequence which are corrupted by multiple access interference (MAI) and noise. A hybrid system is used to model the synchronous code division multiple access, and multiuser detection is implemented using the interacting multiple model (IMM) technique. The performance of the IMM based multiuser detector is studied and simulations are used to compare with the decorrelating decision feedback detector (DDFD) and decision-feedback blind adaptive multiuser detector (DFBD).

Keywords: Adaptation, Adaptive equalization, Communication channels, Communication systems, Estimation

1. INTRODUCTION

In a code division multiple access (CDMA) system, several users share a common channel by modulating antipodally a set of signature waveforms. The received signal consists of a superposition of signals from all users plus additive noise. The synchronous CDMA applies in several cases (Varanasi and Aazhang, 1991) and provides a simple model which can be generalized to the asynchronous case and to channels with fading and multipath. Although signature waveforms are designed with low cross correlations, the users' signals are not orthogonal, leading to multiple access interference (MAI) at the receiver. If the powers from each user are widely varying, signature waveform design alone is not sufficient to suppress the MAI. This leads to the near far problem in the receiver.

The conventional receiver consists of a bank of single user filters matched to the signature waveforms followed by quantizers (Verdu, 1994). It is reliable if the powers of the users are not very different and if cross correlations are very low. These conditions are very difficult to meet in practice. The optimum receiver (Verdu, 1986) achieves low error probability at the expense of very high complexity. When the number of users is high, as it is often the case, suboptimum receivers with lower complexity are needed. The decorrelating detector (Verdu, 1994) is a simple receiver which easily outperforms the conventional receiver and does not require the estimation of user powers. But the inversion performed by the detector enhances the system noise, deteriorating the system performance. Another linear receiver is the minimum mean square error equalizer (Xie *et al.*, 1990) which retains some MAI in exchange for further noise reduction.

There are many nonlinear techniques for multiuser detection. In decision-feedback (DF) detectors (Duel-Hallen, 1993), all symbols except the symbol of interest would have been estimated and the effect of future symbols is eliminated by triangular matrix factorization. Mulgrew (1996), Aazhang *et al.* (1992) and Mitra and Poor (1994) use neural networks for multiuser detection in CDMA systems. Koulakiotis and Aghvami (2000) give an excellent survey on various multiuser detection schemes.

Most multiuser receivers assume *a priori* knowledge of each user's received energy, channel coefficient and timing information. But in a mobile environment, the amplitude (power) of the received signal varies with time due to the motion of mobile receivers and due to changes in the multipath propagation. Detectors which can adaptively estimate the unknown parameters, like the received amplitude, are thus required. Several blind adaptive multiuser detectors have been proposed for multiuser detection without a training sequence (Honig and Madhow, 1995; Ping *et al.*, 2000). These detectors do not assume the system parameters and still retain an acceptable performance level. Here we introduce a multiple model based approach to this communication problem, following our previous work (Fabri and Kadirkamanathan, 1998) in adaptive control.

The paper is organized as follows: In Section 2 we define our signal model and multiuser receivers. In Section 3, we review the IMM algorithm. We derive our IMM-based multiuser detector in Section 4. Simulation studies are presented in Section 5 and finally some conclusions are drawn in Section 6.

2. PROBLEM FORMULATION

Consider an antipodal K-user synchronous CDMA system. Each user is assigned with normalized signature waveforms $s_k(t)$, $k = 1, \cdots, K$ where waveforms are linearly independent and restricted to the symbol duration T such that $\int_0^T s_k^2(t)dt = 1$. For the synchronous case, it is sufficient to consider one single transmission. Let $\mathbf{b} = (b_1, \cdots, b_K)$; $b_k \in \{-1, +1\}$ denote the input vector in which symbols are equiprobable, *i.i.d* and independent of each user's symbols. We assume that the input energies w_k for each user are unknown but assume the input energies to be non-decreasing, i.e., $w_1 \geq w_2 \geq \cdots \geq w_K$ as in Duel-Hallen (1993). When a bank of matched filters matched to the set of signature waveforms is followed by samplers at time T, the resulting output is given by

$$\mathbf{y} = \mathbf{RWb} + \mathbf{n} \tag{1}$$

where $W = \{w_1, \cdots, w_K\}$, $\mathbf{n}$ is a zero mean Gaussian vector with variance $\sigma^2\mathbf{R}$ and $\mathbf{R}$ is the cross-correlation matrix of all s_k (Verdu, 1994). We can obtain a white noise model by decomposing the cross correlation matrix $\mathbf{R}$ such as $\mathbf{R} = \mathbf{F}^T \times \mathbf{F}$, where $\mathbf{F}$ is a lower triangular matrix. Then multiplying (1) by $(\mathbf{F}^T)^{-1}$, we can obtain the output of the noise whitening filter:

$$\tilde{\mathbf{y}} = \mathbf{FWb} + \tilde{\mathbf{n}} \tag{2}$$

where $\tilde{\mathbf{n}}$ is white noise with variance $\sigma^2\mathbf{I}$ ($\mathbf{I}$ is a $K \times K$ identity matrix). The objective here is to detect the input data vector $\mathbf{b}$ given the output vector $\tilde{\mathbf{y}}$. The complexity of the optimum detector (Verdu, 1994) increases exponentially with the increasing number of users. This leads us to use suboptimal receivers for multiuser detection.

If the transmitted user symbols $\mathbf{b}$ were known *a priori*, then the estimation of user powers would be a conventional linear Gaussian problem, and could be solved using the Kalman filter. But in our case, with $\mathbf{b}$ unknown, the model represented by (1) is dependent on a Markov switching process generated by the symbol sequences of K users. This suggests that if we consider our model as a hybrid system (Tugnait, 1982), we can find the input sequence together with the unknown parameters (power or amplitude in this case). We already used this approach in estimating the channel coefficients and equalization of single user systems (Jaward and Kadirkamanathan, 2001). This paper attempts to extend those ideas to multiuser detection problems. We propose to solve the problem using the interacting multiple model algorithm (IMM) (Bar-Shalom and Li, 1993; Bar-Shalom *et al.*, 1989), one of several methods available for hybrid systems.

3. THE INTERACTING MULTIPLE MODEL ALGORITHM

The received signal can be modeled as a linear hybrid system. The simplest linear hybrid system is described by a set of linear models as

$$\theta(t) = F_j(t-1)\theta(t-1) + G_j(t-1)w_j(t-1) \tag{3}$$

$$y(t) = H_j(t)\theta(t) + v_j(t); \qquad \forall\, j \in S \tag{4}$$

where $S = \{1, \ldots, r\}$ is the set of possible modes. This model can handle changes in the system structure as well as in the noise statistics but in our case, as will be seen, changes occur only in the observation matrix, $H_j(t)$. The mode at time t, $m(t)$ is assumed to be among the possible r modes. The mode transitions are governed by a first-order homogeneous Markov chain

$$P\{m_j(t+1)|m_i(t)\} = \pi_{ij}; \qquad \forall\, j \in S \quad (5)$$

where π_{ij} is the transition probability from mode i to mode j and $m_i(t)$ symbolizes that the mode at time t is i. Let $m_i^s(t-1)$ denote the sth mode history up to and including time $t-1$, with mode i active at time $t-1$. Then the mode probabilities are given by

$$\begin{aligned} P\{m_j(t)|Y^t\} &= \sum_s P\{m_j(t)|Y^t, m_i(t-1)\} \quad (6) \\ &\times P\{m_i^s(t-1)|Y^t\} \\ &= \sum_s \pi_{ij} P\{m_i^s(t-1)|Y^t\} \end{aligned}$$

where both summations are over all possible mode histories and $P\{m_i^s(t-1)|Y^t\}$ can be calculated using Bayes' formula. The number of possible mode histories through $(t-1)$ is r^{t-1}, since r number of modes are possible at each time. Hence the number of possible histories grows exponentially with time. To overcome this problem, fixed memory algorithms like Generalized Pseudo Bayes (GPB) (Bar-Shalom and Li, 1993) and IMM algorithms (Bar-Shalom and Li, 1993; Bar-Shalom *et al.*, 1989) were introduced.

The IMM algorithm is a recursive algorithm consisting of four major steps: interaction (mixing), filtering, mode update and combination. In each cycle, the initial condition for the filter matched to a certain mode is obtained by interacting (mixing) the state estimates of all filters at the previous time under the assumption that this particular mode is in effect at the current time. This is followed by a prediction and update step, performed in parallel for each mode. The mode probability update is performed next for each mode. Then a combination of the updated state estimates for all filters yields the state estimate. The mode probability acts as the weighting in the interaction step and in the combination of states and covariances. Implementation details are given by Bar-Shalom and Li (1993) and in Section 4 of this paper.

4. IMM ALGORITHM FOR MULTIUSER DETECTION

The IMM algorithm can be used for sequence estimation by assuming that the mode in effect is due to the one of transmitted symbols. For binary transmission, the received signal comes from one of two symbols and hence the number of modes, r is equal to 2. We need r number of filters in parallel at a time, as only one of r modes can be in effect. We define the state vector θ (input energies) for user k as

$$\theta_k(t) = [\ \sqrt{w_1(t)}\ \dots \sqrt{w_K(t)}\ ]. \qquad (7)$$

Let M be the set of symbols transmitted. Due to the lower triangular matrix $\mathbf{F}$ at the output of the whitening filter (2), the user signals are partially decorrelated. We can observe that the first user is MAI-free. Multiuser interference for the second user is $F_{21}\sqrt{w_1}b_1$. Thus the observation matrix for the user k at time t is $H_j(t) = [F_{11}\hat{b}_1 \cdots F_{kk}b_k(j)\ 0 \cdots 0]$ where $b_k(j)$ denotes the symbol corresponding to mode j. A random walk model is used in Equation (3) to model the unknown parameters.

Hence the hybrid system simplifies to,

$$\theta_k(t) = \theta_k(t-1) + w(t-1) \qquad (8)$$
$$\tilde{y}_k(t) = H_j(t)\theta_k(t) + \tilde{n}_k(t) \qquad \forall\, j \in S \quad (9)$$

where $\tilde{y}_k(t)$ is the whitening filter output for the user k, and $\tilde{n}_k(t)$ is the noise component corresponding to user k.

The algorithm can now be summarised as follows: For each user k,

- Calculate the probability that the symbol corresponding to mode i was in effect at time $t-1$, given that the mode j was in effect at t conditioned on past received data Y^{t-1}, for all $i, j = 1, \dots, r$. Here symbols corresponding to i and j take on values in M and the calculation is done for all r values in M.

$$\begin{aligned} &\mu_{i|j}(t-1|t-1) = P(m_i(t-1)|m_j(t), Y^{t-1}) \\ &= \frac{1}{c}P(m_j(t)|m_i(t-1), Y^{t-1}) \times \\ &\qquad\qquad P(m_i(t-1)|Y^{t-1}) \\ &= \frac{1}{c}\pi_{ij}\mu_i(t-1); \quad i,j = 1,..,r \end{aligned}$$

where c is the normalizing constant and $\mu_i(t-1)$ is the posterior symbol probability at time $t-1$.

- Compute the mixed initial condition for the filter matched to each symbol at time t using $\hat{\theta}^i(t-1|t-1)$, the state estimate of the Kalman filter matched to mode i at time $t-1$. The initial state estimate is given by

$$\begin{aligned} &\hat{\theta}^{0j}(t-1|t-1) = \qquad (10) \\ &\sum_{i=1}^{r} \hat{\theta}^i(t-1|t-1)\mu_{i|j}(t-1|t-1); \\ &\forall\, j = 1,..,r \end{aligned}$$

and the corresponding covariance matrix is

$$\begin{aligned} &P^{0j}(t-1|t-1) = \qquad (11) \\ &\sum_{i=1}^{r} \mu_{i|j}(t-1|t-1)\{P^i(t-1|t-1) \\ &+(\hat{\theta}^i(t-1|t-1) - \hat{\theta}^{0j}(t-1|t-1)) \end{aligned}$$

$(\hat{\theta}^i(t-1|t-1) - \hat{\theta}^{0j}(t-1|t-1))'\};$
$\forall\, j = 1,..,r$

where $\hat{\theta}^j(t|t)$ and $P^j(t|t)$ are the state estimate and covariance of the state estimate at time t.

- The above two estimates, (10) and (11), are used as inputs to two Kalman filters matched to the two symbols to obtain $\hat{\theta}^j(t|t)$ and $P^j(t|t)$, details of which can be found in Candy (1987). The likelihood corresponding to the two filters,

$$\Lambda_j(t) = p(y(t)|m_j(t), \hat{\theta}^{0j}(t-1|t-1), P^{0j}(t-1|t-1)) \tag{12}$$

are also computed. For Gaussian noise this reduces to

$$\Lambda_j(t) \propto \exp(-0.5\; e(t)R_e(t)^{-1}e(t)) \tag{13}$$

where $e(t)$ and $R_e(t)$ are respectively the innovation and innovation variance given by the Kalman filter.

- The probability that the mode j ($j = 1,..,r$) is in effect, is updated as follows

$$\mu_j(t) = P(m_j(t)|Y^t) = \frac{1}{\bar{c}}\Lambda_j(t)\sum_{i=1}^{r}\pi_{ij}\mu_i(t-1); \quad \forall\, j = 1,..,r \tag{14}$$

where $\bar{c}$ is the normalization constant. The mode at time t can now be estimated as

$$m(t) = \arg\max_j \mu_j(t) \tag{15}$$

Thus the symbol at t is the symbol corresponding to mode $m(t)$.

- Finally calculate the parameter estimates using the following mixture equation

$$\hat{\theta}_k = \sum_{j=1}^{r}\hat{\theta}^j(t|t)\mu_j(t) \tag{16}$$

The proposed algorithm successively decodes symbols of each user as in the decorrelating decision-feedback detector (DDFD) (Duel-Hallen, 1993). However, the proposed algorithm enables us to adaptively estimate the unknown parameters without a training sequence in addition to the user symbol sequence estimation.

5. SIMULATION RESULTS

The performance of the proposed algorithm was examined using simulations and compared to the DDFD and decision-feedback blind adaptive detector (DFBD) of Ping *et al.* (2000). The transmitted sequence was an independent and identically distributed (*i.i.d.*) binary sequence. The transition probability matrix, π_{ij} is given by

$$\pi_{ij} = \begin{bmatrix} 0.5 & 0.5 \\ 0.5 & 0.5 \end{bmatrix} \tag{17}$$

The simulations were carried out for 10000 symbols and repeated 20 times for different values of signal to noise ratio, to obtain an average value of bit error rate (BER).

Example 1. The first example is a four user synchronous CDMA system based on a set of signature waveforms derived from gold sequences of length seven (Varanasi and Aazhang, 1991; Duel-Hallen, 1993; Ping *et al.*, 2000). We assume four users are active with a cross-correlation matrix of the signature waveform,

$$\mathbf{R} = \frac{1}{7}\begin{bmatrix} 7 & -1 & 3 & 3 \\ -1 & 7 & -1 & 3 \\ 3 & -1 & 7 & -1 \\ 3 & 3 & -1 & 7 \end{bmatrix} \tag{18}$$

All users are assumed to have equal received energies. Figures 1 and 2 shows the probability of error of all users versus the signal-to-noise ratio (SNR) in dB. The performance of the DDFD detector is also shown for comparison. As seen from the graphs, the performance of all users is slightly better in the proposed scheme. This is partially due to the fact that in DDFD, although MAI is eliminated for user 1, no active effort is used to remove the additive noise. Whereas in the proposed scheme, an active effort is used to suppress both noise and MAI. This same example is simulated by Ping *et al.* (2000) where the adaptive version of the DDFD detector, called DFBD, is introduced. We note that there is no significant difference in the bit error performance of both detectors (DDFD and DFBD). On the other hand, despite being blind adaptive, the proposed algorithm can outperform both DDFD and DFBD multiuser detectors. The convergence of the algorithm is measured by checking the convergence of the parameters. They are found to converge within 800 time steps for our simulations. The results are not reported here due to space constraints.

Example 2. In this example, taken from Ping *et al.* (2000), we assume the amplitude of the received signal varies slowly according to the Rician fading process. The other assumptions are identical to the ones by Ping *et al.* (2000). The slowly varying fading signal is realized by sending the in-phase and quadrature components of a zero mean Gaussian process through two second-order Butterworth low-pass filters. If X_{kc} and X_{ks} are

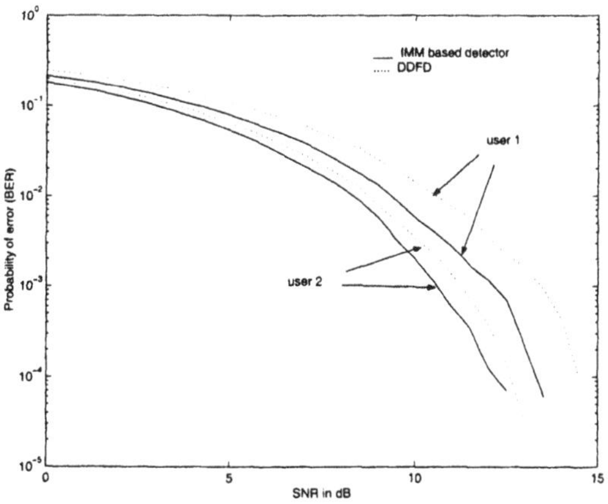

Fig. 1. Probability of symbol error (BER) for IMM based multiuser detector and DDFD

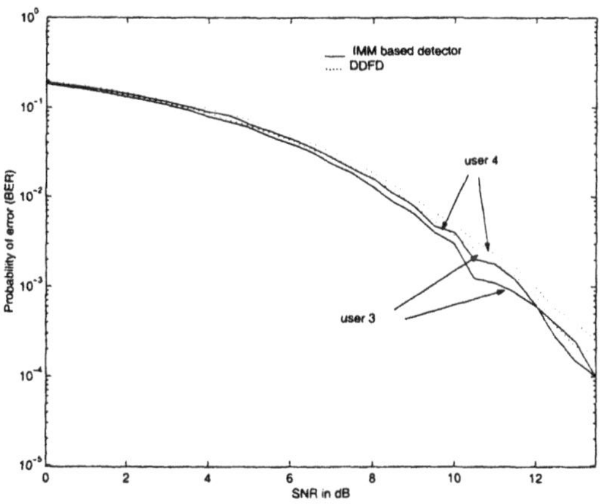

Fig. 2. Probability of symbol error (BER) for IMM based multiuser detector and DDFD

the in-phase and quadrature outputs with zero mean and variance σ_k^2, then the envelope of each signal at time t is given by,

$$\sqrt{w_k(t)} = \sqrt{(A_k + X_{kc}(t))^2 + X_{ks}^2}; \quad k = 1, \cdots, K$$

where A_k is a constant to represent the line-of-sight (LOS) component in the received signals. When no line of sight component exists, this becomes the Raleigh channel. Rician channels are characterized by the factor given by $K_c = A_k^2/2\sigma_k^2$.

We consider a four-user channel with the same cross correlation as in Example 1. The simulation parameters are identical to the ones used by Ping *et al.* (2000). The BER performance of the proposed algorithm for user 1 is compared with DDFD in Figure 3 for values of $K_c = 0,\ 10,\ 20$. The BER performance is significantly better in our IMM-based multiuser detector for all values of K_c. Comparing these results to the ones obtained by Ping *et al.* (2000) for DDFD and DFBD (where DDFD performs better), we can deduce that the IMM-based detector offers bit error rate advantages. In addition, the algorithm quickly

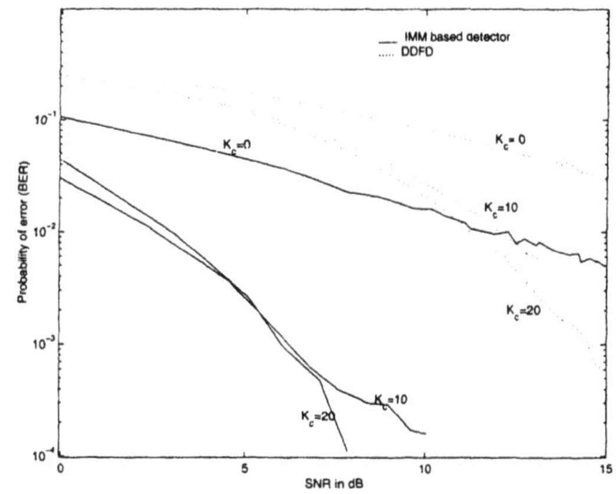

Fig. 3. Probability of symbol error (BER) for IMM based multiuser detector and DDFD (Rician channel)

converges to the time varying parameters and hence can be used also for rapidly fading channels.

6. CONCLUSIONS

A multiple model based blind adaptive multiuser detector has been proposed and the performance is compared with the DDFD and DFBD detectors. The proposed multiuser detector models the modes due to received bits as a set of models with parameters corresponding to amplitude and time delay of a CDMA system. Although we analysed only the synchronous operation, this can be extended to asynchronous CDMA by incorporating time delays as additional states in the system.

7. REFERENCES

Aazhang, B., B.P. Paris and G.C. Orsak (1992). Neural networks for multi-user detection in code-division multiple-access communication. *IEEE Trans. Commun.* **40**(7), 1212–1222.

Bar-Shalom, Y. and X.R. Li (1993). *Estimation and Tracking: Principles, Techniques and Software.* Artech House. Boston, MA.

Bar-Shalom, Y., K.C. Chang and H.A.P. Blom (1989). Tracking a maneuvreing target using input estimation versus the interacting multiple model algorithm. *IEEE Transactions on Aerospace and Electronic Systems* **25**(2), 296–300.

Candy, J.V. (1987). *Signal Processing: The model-based approach.* McGraw-Hill Book Company.

Duel-Hallen, A. (1993). Decorrelating decsion-feedback multiuser detector for synchronous code-division multiple-access channel. *IEEE Trans. Commun.* **41**(2), 285–290.

Fabri, S.G. and V. Kadirkamanathan (1998). A self-organised multiple model approach for neural-adaptive control of jump nonlinear systems. In: *Preprints of the IFAC Symposium on Adaptive Systems in Control and Signal Processing.* pp. 147–152.

Honig, M. and U. Madhow (1995). Blind adaptive multiuser detection. *IEEE Trans. Information Theory* **41**(4), 944–960.

Jaward, M.H. and V. Kadirkamanathan (2001). Interacting multiple models for single-user channel estimation and equalization. In: *Proc. of IEEE ICASSP 2001.*

Koulakiotis, D. and A.H. Aghvami (2000). Data detection techniques for DS/CDMA mobile systems: A review. *IEEE Personal Commun.*

Mitra, U. and H.V. Poor (1994). Neural network techniques for adaptive multiuser demodulation. *IEEE JSAC* **12**(9), 1460–1470.

Mulgrew, B. (1996). Applying radial basis functions. *IEEE Signal Processing Mag.* pp. 50–62.

Ping, H., T. T. Tjhung and L.K. Rasmussen (2000). Decision-feedback blind adaptive multiuser detector for synchronous CDMA system. *IEEE Transactions on Vehicular Technology* **49**(1), 159–166.

Tugnait, J.K. (1982). Detection and Estimation for Abruptly Changing Systems. *Automatica* **18**(5), 607–615.

Varanasi, M.K. and B. Aazhang (1991). Near-optimum detection in synchronous code-division multiple access systems. *IEEE Trans. Commun.* **39**, 725–736.

Verdu, S. (1986). Minimum probability of error asynchronous gaussian multiple-access channels. *IEEE Trans. Inform. Theory* **32**, 85–96.

Verdu, S. (1994). Adaptive multiuser detection. In: *Proc. of the 1994 International Symposium on Spread Spectrum Technology and Applications.* pp. 43–50.

Xie, Z., R.T. Short and C.K. Rushforth (1990). A family of suboptimum detectors for coherent multiuser communications. *IEEE Journal on Selected Areas of Commun.* **8**(4), 683–690.

GUARANTEED CLOSED LOOP PRECISION IN MULTIPLE MODEL BASED CONTROL

Josè Diez*, Fabio Previdi†

* Departamento de Ingenieria de Sistemas y Automatica
Universidad Politecnica de Valencia
P.O.Box: 22012 - 46071 Valencia, Spain
Ph.: +34-96-3877007, ext. 75794, Fax: +34-96-3879579
E-mail: jldiez@isa.upv.es

† Dipartimento di Elettronica & Informazione
Politecnico di Milano
via G. Ponzio 34/5 20133 Milano
Ph. +39-0223993467, Fax +39-0223993412
E-mail: previdi@elet.polimi.it

Abstract. In this paper, gain scheduled controllers designed on the basis of multiple modelling are considered. Specifically, Local Controller Networks are designed by interpolation of a set of linear controllers. The local linear controllers are obtained by Minimum Variance design based on a Local Model Network, i.e., a model of a nonlinear plant obtained by interpolation of linear models that guarantee a good approximation of the plant, at least locally, in the operating regime space. In particular, robustness issues are considered, based on the consideration that it is well known that the characteristics of the interpolating functions play a crucial role. This paper shows necessary qualitative conditions for designing robust control systems based on the interpolating function shape and rate of change. Additionally, under certain flatness conditions for those functions (well fulfilled by the so called super-Gaussian family), the upper bound for closed loop error can be precisely estimated and strictly connected to the interpolating function family features. *Copyright © 2001 IFAC*

Keywords. Multiple Model based control, Local Model Networks, Local Controller Networks, Gain Scheduling

1. INTRODUCTION

This paper is focused on the properties of nonlinear adaptive controllers designed on the basis of nonlinear time-varying models of a given system. In particular, the *Local Model Network* (LMN) family will be considered, i.e., nonlinear systems obtained by convex combinations of linear (or affine) local models. The basic idea behind LMN modelling is the partitioning of the operating regime space into a number of sets. Then, on each set, *local models* are estimated, which can well approximate the plant behaviour on the given set. Finally, local model interpolation is performed, by using convex coefficients depending on a given set of scheduling variables, usually past measurements of input and output. Specifically, interpolation is done by means of *validity functions*, i.e., smooth dynamic maps whose output is used for a linear weighted combination of the local model outputs.

This model set is very appealing for its full approximation capability (see Johansen and Foss, 1993; an interesting application can be found in Hunt and Gollee, 1997) and also because of the availability of software tools for identification (Johansen *et al.*, 1998). Furthermore, it is conceivable to design gain scheduling nonlinear controllers, i.e., *Local Controller Networks* (LCN), exploiting the peculiar model structure. In fact, it is straightforward to design a set of local linear controllers based on the local model set and then to interpolate them using the same convex coefficients of the model (Hunt and Johansen, 1997).

The use of controllers based on LMN is substantially different from the common gain scheduling approach. In fact, in classical gain scheduling design, the first step is the selection of a set of equilibrium points for the given plant. Then, often on the basis of an available nonlinear model, a corresponding set of linear (tangent) models is derived and finally the design of linear parametric controllers can be performed. After that, the controller parameters are interpolated depending on a scheduling variable or some external time-varying parameters (see, for instance, Shamma and Athans, 1990; Rugh and Shamma, 2000; Apkarian and Adams, 1995). The LMN approach goes further since it can incorporate local models which are not linear tangent to the nonlinear process in a steady state condition. So, at least in principle, specific models can be included, devoted to describe the "transient dynamics", i.e., the system behaviour while moving between different equilibria (Johansen, 1994). Moreover, since the interpolation variables are given by parametrized function of past measurements of input and output, also the dynamic behaviour of the scheduling variable can be estimated directly from data together with the local model parameters.

Unfortunately, controllers designed on the basis of a given LMN often suffer from robustness deficiency, mainly due to the difficulty of having "slowly varying" scheduling variables, when they are estimated in a "blind" way directly from data. A fundamental study on the closed loop precision of a specific LCN designed according to polynomial methods is presented in (Hunt and Johansen, 1997). In that paper, a bound is given for the closed loop error due to the specific model and controller parametrization.

In the present work, the application of LCN structure for non-linear dynamic control problems is going to be revisited, focusing on the effects of model approximation capability on the closed loop precision. It will be proved that all existing necessary "qualitative" conditions for designing a robust closed-loop stable system, such as fine partitions and slowly varying scheduling, can be reduced to the study of the shape and the rate of change of the validity functions.

As a consequence, a new (lower) bound for closed loop precision is computed and exploiting its explicit dependence on the validity function features, a new validity function family is proposed, specifically the generalized Gaussian (or super-Gaussian) functions.

The paper is organized as follows: in Section 2 the Linear LMN family is introduced. Section 3 is devoted to the description of a specific LMN based control strategy, i.e. a LCN based on Generalized Minimum Variance local controllers. Then, in Section 4 the closed loop precision properties are investigated and a precision bound is calculated and related to the validity function characteristics. In Section 5, results from a simple simulated example are shown and, finally, in Section 6, discussion of the results and future work are presented.

2. LOCAL MODEL NETWORKS

In this Section a particular family of nonlinear systems built using Local Models, the so called *Local Model Network* (LMN) (Johansen and Foss, 1993; Hunt and Johansen, 1997; Previdi, 1999), will be outlined. This model family is of great interest since well established techniques for parameter estimation (Johansen, et al., 1997) and the design of adaptive controllers based on it are available (Hunt and Johansen, 1997; Previdi, 1999). In particular, a controller is designed for each local model according to some design criterion. Then, a nonlinear controller is obtained by interpolation of the local controllers in the same way as the local models are scheduled, i.e., by weighting the local control actions using the validity functions.

First of all, let's recall some basic definitions on LMN.

Definition 1

Consider a family of real valued smooth functions $W = \{w_i(\cdot)\}_{1=1,\dots,M}$ where $w_i(\cdot): \Re^n \to \Re$ so that:

$$\text{(i)} \quad 0 \le w_i(\phi) \le 1, \ \forall i = 1,\dots,M \text{ and } \forall \phi \in \Re^n$$

$$\text{(ii)} \quad \sum_{i=1}^{M} w_i(\phi) = 1, \ \forall \phi \in \Re^n \tag{1}$$

The set W is a family of *validity functions.*

□

Definition 2

A SISO (*Single Input Single Output*) Local Model Network is a model whose dynamics is given by the following input/output equation

$$y(k) = \sum_{i=1}^{M} w_i(\phi(k-1)) f_i(\psi(k-1)) \tag{2}$$

where

- $\{f_i(\cdot)\}_{i=1,\dots,M}$ is a family of *local models* $f_i(\cdot): \Re^{\dim(\psi(k-1))} \to \Re$;
- $W = \{w_i(\cdot)\}_{1=1,\dots,M}$ is a family of *validity functions* $w_i(\cdot): \Re^{\dim(\phi(k-1))} \to \Re$
- The vector $\psi(k-1) \in \Psi \subseteq \Re^{\dim(\psi(k-1))}$ is the *local regressor*, i.e., it is a collection of past measurements of input and output:

$$\begin{aligned}\psi(k-1) = [&y(k-1),\dots,y(k-n_a),\\ &u(k-1),\dots,u(k-n_b)]^T\end{aligned} \tag{3}$$

- Similarly, the vector $\phi(k-1) \in \Phi \subseteq \Re^{\dim(\phi(k-1))}$ is the *validity regressor*, i.e., a collection of past measurements of input and output (and possibly of some other exogenous variable):

$$\begin{aligned}\phi(k-1) = [&y(k-1),\dots,y(k-n_y),\\ &u(k-1),\dots,u(k-n_u)]^T\end{aligned} \tag{4}$$

□

So, for each $k \in Z^+$, the current value of the vector $\phi(k-1) \in \Re^{n_y+n_u}$ is used by the validity functions to define the contribution of the output of each local model to the global output of the LMN. Therefore, the vector $\phi(\cdot)$ must be chosen so that it carries information about the current operating regime of the plant. The region $\Phi \subseteq \Re^{n_y+n_u}$ where the validity functions are significantly different from zero is often referred to as the *operating regime space* of the plant. Similarly the region $\Psi \subseteq \Re^{n_a+n_b}$ is called the *input space*.

Remark 1

As is evidenced by Eq. (2), each validity function is associated with a corresponding local model. So, a validity function is usually chosen to be very close to one in the region where the corresponding local model is a good approximation of the plant and very close to zero far from that region. To this aim, a proper choice is given by Gaussian functions, provided that normalization is performed in order to preserve the property of Eq. (1). □

Now, take the following structure for the local models:

$$y_i(k) \equiv f_i(\psi(k-1)) = L_i(\psi(k-1)) + d_i, \quad \text{for } i = 1,\dots,M \qquad (5)$$

where $y_i(k)$ is the output of the ith local model; $L_i(\cdot): \Psi \to \Re$ is a generic linear operator and $d_i \in \Re$.

Definition 3

A LMN built using affine local models is an *Affine Local Model Network* (ALMN). □

Now, consider the specific case when the local linear operator of Eq. (5) is given by an ARX (*AutoRegressive eXogenous*) model:

$$L_i(\psi(k-1)) = -\sum_{j=1}^{n_a} a_{ij} y(k-j) + \sum_{j=1}^{n_b} b_{ij} u(k-j) \equiv$$
$$\equiv \left(1 - A_i\left(q^{-1}\right)\right) y(k) + B_i\left(q^{-1}\right) u(k) \equiv \theta_i^T \psi(k-1) \quad \text{for } i = 1,\dots,M \qquad (6)$$

where:

- $A_i\left(q^{-1}\right) = 1 + a_{i1} q^{-1} + \dots + a_{in_a} q^{-n_a}$ and q^{-1} is the backward shift operator;
- $B_i\left(q^{-1}\right) = 1 + b_{i1} q^{-1} + \dots + b_{in_b} q^{-n_b}$;
- $\theta_i = \left[-a_{i1},\dots,-a_{in_a}, b_{i1},\dots,b_{in_b}\right]^T$ is the parameter vector. A delay of n_k time steps between input and output can be easily introduced by forcing to zero b_{ij}, for $j = 1,\dots,n_k$ and for all *i*.

Using Eq. (6) into Eq. (5) and then into the model definition (2) we finally obtain a parametric explicit expression for the model used in this paper.

Finally, consider a finite set of points in the input space, i.e., $\left\{\psi_i^*\right\}_{i=1,\dots,M}$ with $\psi_i^* \in \Re^{n_a+n_b}$. So, each ψ_i^* is in the form:

$$\psi_i^* = \left[\left(y_{i1}^*,\dots,y_{in_a}^*; u_{i1}^*,\dots,u_{in_b}^*\right)\right]^T \qquad (7)$$

Note that ψ_i^* need not be an equilibrium point and observe that, by using Eqs. (5), (6) and (7), the offset term d_i must depend on ψ_i^* since it fulfils the following condition:

$$d_i = \sum_{j=1}^{n_a} a_{ij} y_{ij}^* - \sum_{j=1}^{n_b} b_{ij} u_{ij}^* \qquad (8)$$

3. LOCAL CONTROLLER NETWORKS

In this Section, a specific controller structure is obtained exploiting the peculiar form of the model presented in the previous Section. In particular, for each local affine model $\left\{A_i\left(q^{-1}\right), B_i\left(q^{-1}\right), d_i\right\}_{i=1,\dots,M}$ defined by Eqs. (5), (6), (7), (8) a *Generalised Minimum Variance Controller* (GMVC) is designed following the traditional linear approaches (Goodwin and Sin, 1984). The structure of the local linear controllers is defined by the control action:

$$u_i(k) = \frac{1}{T_i(q^{-1})}\left[S_i(q^{-1}) r(k) - R_i(q^{-1}) y(k)\right] \qquad (9)$$

where integral action has been included in order to remove the effect of the offset term d_i, as it is considered as a static disturbance. In this way, for each local linear model a corresponding local linear controller $\left\{R_i\left(q^{-1}\right), S_i\left(q^{-1}\right), T_i\left(q^{-1}\right)\right\}_{i=1,\dots,M}$ can be designed, developing a local control action $\left\{u_i(k)\right\}_{i=1,\dots,M}$. A global control action to be applied to the plant can be obtained by interpolation of the local control actions, using the validity functions $w_i(\phi)$ employed in LMN:

$$u(k) = \sum_{i=1}^{M} u_i(k) w_i(\phi((k-1))) \qquad (10)$$

4. CLOSED LOOP PROPERTIES OF LOCAL CONTROLLER NETWORKS

In this section the closed loop equations derived in (Hunt and Johansen, 1997) will be used for the evaluation of the closed loop precision in the specific case described in the previous Sections 2 and 3.

The closed loop formulation of the control system can be obtained starting from the expressions of the global model and controller:

$$A\left(q^{-1}, \phi(k-1)\right) y(k) = B\left(q^{-1}, \phi(k-1)\right) u(k) + d(\phi(k-1)) + e(k) \qquad (11)$$

$$T\left(q^{-1}, \phi(k-1)\right) u(k) = S\left(q^{-1}, \phi(k-1)\right) r(k) - R\left(q^{-1}, \phi(k-1)\right) y(k) \qquad (12)$$

In the previous Eqs. (11), (12) the notation $P\left(q^{-1}, \phi(k-1)\right)$ indicates a *time-varying polynomial operator* (since it depends on the current value of the operating regime vector) defined as:

$$P\left(q^{-1}, \phi(k-1)\right) = \sum_{i=1}^{M} P_i\left(q^{-1}\right) w_i(\phi(k-1)) \qquad (13)$$

where $P_i\left(q^{-1}\right)$ indicates the usual time invariant polynomial in the backward shift operator. The properties of these operators are investigated in (Hunt and Johansen, 1997) and in (Previdi, 1999) where the algebraic properties of the corresponding non abelian *-algebra of operators are described.

In order to obtain a compact form for the closed loop equation, pre-multiply Eq. (11) by $T\left(q^{-1},\phi(k-1)\right)$ and Eq. (12) by $B\left(q^{-1},\phi(k-1)\right)$. The resulting equations can be merged observing that

$$\begin{aligned}&B\left(q^{-1},\phi(k-1)\right)T\left(q^{-1},\phi(k-1)\right)u(k)=\\&=T\left(q^{-1},\phi(k-1)\right)B\left(q^{-1},\phi(k-1)\right)u(k)+\\&-\Delta\left(T,B,q^{-1},\phi(k-1)\right)u(k)+\Delta\left(B,T,q^{-1},\phi(k-1)\right)u(k)+\\&-\delta\left(T,B,q^{-1},\phi(k-1)\right)u(k)+\delta\left(B,T,q^{-1},\phi(k-1)\right)u(k)\end{aligned}\tag{14}$$

The extra-terms $\Delta(\cdot)$ and $\delta(\cdot)$ are due to the non commutativity of the time-varying polynomial operators. Now, compacting all these undesired terms in $\varepsilon(\cdot)$ it is possible to write a single equation for the closed loop system

$$\begin{aligned}(TA+BR)\left(q^{-1},\phi(k-1)\right)y(k)=(BS)\left(q^{-1},\phi(k-1)\right)r(k)+\\+T\left(q^{-1},\phi(k-1)\right)e(k)+\varepsilon(k)\end{aligned}\tag{15}$$

Remark 2

The term $T\left(q^{-1},\phi(k-1)\right)d(\phi(k-1))$ coming from Eq. (11) has been dropped, supposing that the integrator compensates for its effect and so, it is not relevant to the static closed loop analysis.

□

From Eq. (15) it is clear that the closed loop precision can be improved if the term $\varepsilon(\cdot)$ is kept small. Note that $\varepsilon(\cdot)$ also depends on the validity functions. So, it is conceivable to look for conditions on the validity functions so that $\varepsilon(\cdot)$ is bounded by a bound as lower as possible.

To this aim it is necessary to explicitly evaluate $\varepsilon(\cdot)$ and, more specifically, the terms $\Delta(\cdot)$ and $\delta(\cdot)$. Following (Hunt and Johansen, 1997) we have:

$$\begin{aligned}&\Delta\left(T,B,q^{-1},\phi(k-1)\right)u(k)=\\&=\sum_{i=1}^{M}\sum_{h=1}^{M}w_i(\phi(k-1))w_h(\phi(k-1))T\left(q^{-1}\right)\times\\&\times\left[B_h\left(q^{-1}\right)-B_i\left(q^{-1}\right)\right]u(k)\end{aligned}\tag{16}$$

$$\begin{aligned}&\delta\left(T,B,q^{-1},\phi(k-1)\right)u(k)=\\&=\sum_{i=1}^{M}\sum_{j=1}^{n_t}\sum_{h=1}^{M}\sum_{l=1}^{n_b}t_{ij}b_{hl}w_i(\phi(k-1))[w_h(\phi(k-j-1))+\\&-w_h(\phi(k-1))]q^{-j-l}u(k)\end{aligned}\tag{17}$$

where $\{t_{ij}\}_{\substack{i=1,\dots,M\\ j=1,\dots,n_t}}$ are the parameters of the $T_i\left(q^{-1}\right)$ polynomials in the local controllers, being n_t their order and $\{b_{hl}\}_{\substack{h=1,\dots,M\\ l=1,\dots,n_b}}$ are the parameters of the $B_h\left(q^{-1}\right)$ polynomials in the local models, being n_b their order.

By looking for upper limits for $|\Delta(\cdot)|$ and $|\delta(\cdot)|$ an upper bound for $|\varepsilon(\cdot)|$ is proposed in (Hunt and Johansen, 1997):

$$\begin{aligned}|\varepsilon(k)|\le\varepsilon_I(k)=(\alpha+\gamma)K(|y(k)|+\ldots+|y(k-n)|+\\+|r(k)|+\ldots+|r(k-n)|+1)\end{aligned}\tag{18}$$

where

- $\alpha=\max\limits_{i=1,\dots,M}\left(\max\limits_{h\in N_i}\left\|\psi_i^*-\psi_h^*\right\|_2\right)$ (19)
- $\gamma=\max\limits_{i=1,\dots,M}\left(\sup\limits_{k\ge0}\left|w_i(\phi(k-1))-w_i(\phi(k))\right|\right)$ (20)
- $K\in\Re$; N_i is the index set of the validity functions that are nearest neighbours to the ith one.

Now, it is worth to do some qualitative comments on Eqs. (16-20). First, notice that, since the product $w_i(\cdot)w_h(\cdot)$ in Eq. (16) is significantly different from zero only when i and h indicate neighbour validity functions. So, the $\Delta(\cdot)$ term will be small if neighbour models have similar $B(\cdot)$ polynomial coefficients. Then, observe that $\delta(\cdot)$ is small when the validity functions are smooth, so that the difference term in Eq. (17) is small.

Specifically, the $\Delta(\cdot)$ terms are strictly dependent on the number of operating regimes (i.e., on the number of local models) that are used: the higher is the number of local models, the closer are the operating points $\{\psi_i^*\}_{i=1,\dots,M}$ and the smaller is $|\Delta(\cdot)|$. This effect is dominated by the value of α.

Similarly, the $\delta(\cdot)$ terms depend on the rate of change of the values assumed by a validity function: if the $\{w_i(\phi(\cdot))\}_{i=1,\dots,M}$ are slowly time-varying, the $|\delta(\cdot)|$ can be kept small. This effect is dominated by the value of γ.

Simulations indicate that some precision improvements could be obtained by acting on the shape of the validity functions. So, in the following, new bounds for $|\Delta(\cdot)|$ and $|\delta(\cdot)|$ will be calculated aiming to make a direct connection with the properties of the validity functions.

First of all note that

$$\left|b_{hj}-b_{ij}\right|\le\left|\frac{\partial^2 f}{\partial u^2(k-j)}\left(\psi^*\right)\right|\max_{i=1,\dots,M}\left(\max_{h\in N_i}\left\|\psi_i^*-\psi_h^*\right\|_2\right)\tag{21}$$

for $j=1,\dots,n_b$; $i,h=1,\dots,M$.

In Eq. (21) $\psi^* = \text{vec}\left(\left\{\psi_i^*\right\}_{i=1,\ldots,M}\right)$ and $f(\cdot)$ is the unknown nonlinear map to be approximated by using Eq. (2).

Then, in Eq. (21) the term containing the second derivative of $f(\cdot)$ can be approximated provided that the LMN is a good approximation of the system, so that we can suppose $f(\cdot) \cong \hat{f}(\cdot)$, using directly Eq. (2) in the calculations. Then, we can write, for $j = 1,\ldots,\max(n_b, n_u)$,

$$\frac{\partial^2 f(\psi(k-1))}{\partial u^2(k-j)} \cong \frac{\partial^2 \hat{f}(\psi(k-1))}{\partial u^2(k-j)} =$$

$$= \frac{\partial^2}{\partial u^2(k-j)}\left[\sum_{i=1}^{M} w_i(\phi(k-1)) f_i(\psi(k-1))\right] =$$

$$= \sum_{i=1}^{M}\left[w_i(\phi(k-1))\frac{\partial^2 f_i(\psi(k-1))}{\partial u^2(k-j)} + \right.$$

$$\left. + 2\frac{\partial w_i(\phi(k-1))}{\partial u(k-j)}\frac{\partial f_i(\psi(k-1))}{\partial u(k-j)} + f_i(\psi(k-1))\frac{\partial^2 w_i(\phi(k-1))}{\partial u^2(k-j)}\right] \tag{22}$$

In Eq. (22), the first term is zero, because of the linearity of the local models; the second term is only depending on the slope of the validity functions, since the first derivative of the local model is a constant. Moreover, in practice, it is often $n_u < n_b$, and so few terms from Eq. (22) enter Eq. (21).

So, using Eq. (22), it is possible to find a proper bound to be used in Eq. (21):

$$\max_{j=1,\ldots,n_b}\left|\frac{\partial^2 f}{\partial u^2(k-j)}(\psi^*)\right| \le$$

$$\le \max_{j=1,\ldots,n_b}\left[\max_{i=1,\ldots,M}\left|2b_{ij}\frac{\partial w_i(\phi(k-1))}{\partial u(k-j)}\right|_{\psi^*} + \right.$$

$$\left. + \max_{i=1,\ldots,M}\left|f_i(\psi^*)\frac{\partial^2 w_i(\phi(k-1))}{\partial u^2(k-j)}\right|_{\psi^*}\right] \tag{23}$$

So, through Eq. (21), (23) it is possible to directly relate the bound on $|\Delta(\cdot)|$ to the values of the first and second derivatives of the validity functions evaluated in the operating regimes $\left\{\psi_i^*\right\}_{i=1,\ldots,M}$. This fact puts in evidence that, not only the $\delta(\cdot)$ terms are related to the time-varying behaviour of the system. Specifically, Eq. (23) evidences that it is strongly advisable that the validity functions are "flat" close to the operating regime values. Moreover, from simulations it can be said that the bound $\varepsilon_{II}(\cdot)$ computed using the previous equations is pointwise lower than $\varepsilon_I(\cdot)$, i.e., it is lower for all $k \in Z^+$.

5. SIMULATION EXAMPLE

In this Section, a simple example will be outlined to test the results obtained in the previous Section. Only the noise-free case will be considered, in order to avoid any other effect different from validity function shape.

Two local models are considered ($M = 2$) of the form described in Eq. (6):

$$\begin{cases} A_1(q^{-1}) = 1 - q^{-1} + 0.09q^{-2} \\ B_1(q^{-1}) = 1 - 0.2q^{-1} \\ d_1 = 0 \\ \phi_1^* = (0,0) \end{cases}$$

$$\begin{cases} A_2(q^{-1}) = 1 - 0.9q^{-1} + 0.005q^{-2} \\ B_2(q^{-1}) = 1 - 0.5q^{-1} \\ d_2 = 2 \\ \phi_2^* = (19.0476,0) \end{cases}$$

where ϕ_i^* is the validity function centre. The operating regime space Φ is two-dimensional and the validity function regressor is $\phi(k-1) = [y(k-1), u(k-1)]^T$.

Super-Gaussian validity functions have been chosen:

$$w_i(\phi(\cdot)) = \frac{\exp\left[-\frac{1}{2}\left(\prod_n (\phi - \phi_i^*)\right)^T \Sigma_i^{-n}\left(\prod_n (\phi - \phi_i^*)\right)\right]}{\sum_{j=1}^{M}\exp\left[-\frac{1}{2}\left(\prod_n (\phi - \phi_j^*)\right)^T \Sigma_i^{-n}\left(\prod_n (\phi - \phi_j^*)\right)\right]} \tag{24}$$

where:

- the operator $\prod_n(\cdot)$ applied to a vector x is the n-times elementwise product of the vector x by itself;
- $n = 2,4,6,\ldots$ indicates the super-Gaussian order (only even values are allowed; note that for $n = 2$ the ordinary Gaussian case is obtained);
- Σ_i is a positive two-dimensional diagonal matrix indicating the validity functions width.

The choice of a fourth order super-Gaussian validity functions with $\Sigma_i = diag(18;18)$, leads to the shape of validity function $w_2(\cdot)$ shown in Fig. 1. This result fulfils the flatness requirements in the neighbourhood of ϕ_2^*. The same can be easily verified for $w_1(\cdot)$, which is just $w_2(\phi(\cdot)) = 1 - w_1(\phi(\cdot))$.

A LCN has been designed according to the strategy described in Sect. 3

In Fig. 2 the tracking of a sinusoidal reference signal is shown. The reference is

$$r(k) = 10\,\mathrm{sen}\left(\frac{2\pi}{100}k\right) + \frac{19.0476}{2} \tag{25}$$

This reference signal spans the range between ϕ_1^* and ϕ_2^*, so allowing the investigation of the control system properties when both the validity functions are different from their limit values (close to 0 and 1).

A comparison between $\varepsilon_I(k)$ and $\varepsilon_{II}(k)$ while the control system is tracking the reference of Eq. (25) is shown in Fig. 3, where it is evident that the estimated bound is lower than the one reported in (Hunt and Johansen, 1997).

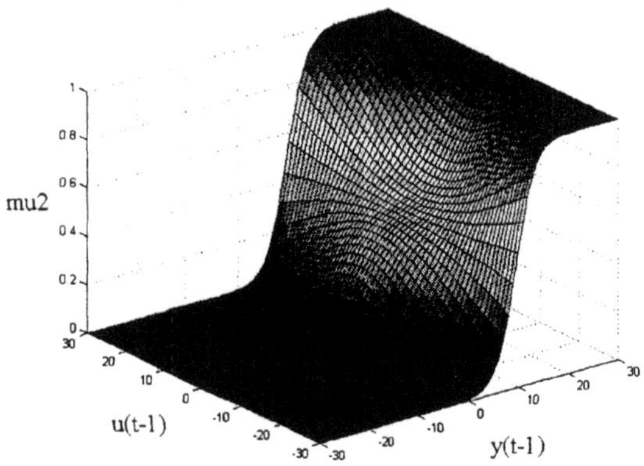

Fig. 1 - Validity function w_2.

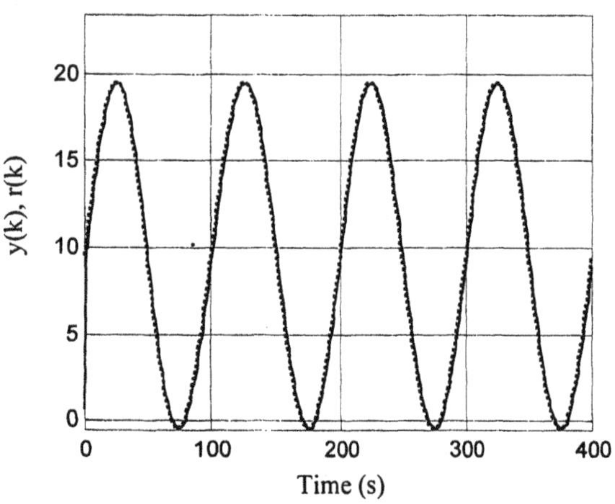

Fig. 2 - Tracking experiment of the sinusoidal reference $r(k)$ (dotted) of Eq. (25).

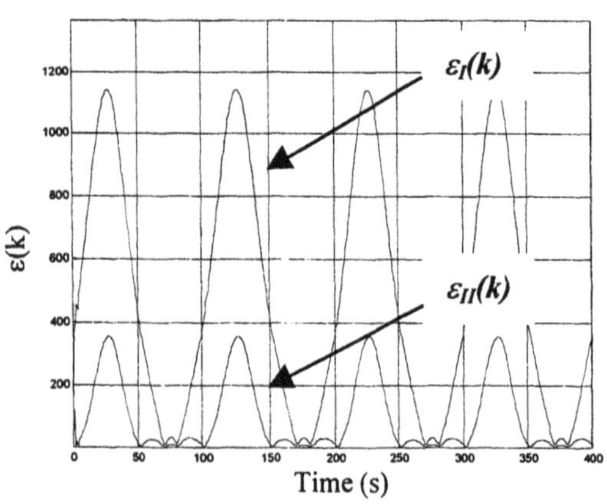

Fig. 3 - Comparison between $\varepsilon_I(k)$ and $\varepsilon_{II}(k)$ during the considered tracking experiment.

6. CONCLUSIONS

The application of LCN for the design of nonlinear controllers is a well known strategy. Unfortunately, to assess robustness issues is in general very difficult, due to the dependence of the closed loop precision from a number of different factors. Among these, it is well known that the characteristics of the validity functions play a crucial role. Moreover, it could happen that the closed loop systems have very poor robustness properties even though very precise LMN have been estimated.

Previous works related the design of stable LCN structures with a set of rules of thumb related to the granularity of the partition of the operating regime and the requirement of slow time-variations for the scheduling variables.

This paper shows that all the necessary qualitative conditions for designing robust closed-loop stable systems can be summarised and connected to the validity function shape and rate of change. Additionally, under certain flatness conditions for the validity functions (well fulfilled by super-Gaussian family), the upper bound for closed loop error can be lower and estimated more precisely.

7. REFERENCES

Apkarian P. and P. Gahinet (1995). A convex characterization of gain-scheduled controllers. *IEEE Trans. on Automatic Control*, **40**, 853-864.

Gollee H., K.J. Hunt (1997). Nonlinear modelling and control of electrically stimulated muscle: a local model network approach. *Int. J. of Control*, **68**, 6, 1259-1288.

Hunt K.J. and T.A. Johansen (1997). Design and analysis of gain-scheduled control using local controller networks. *Int. J. of Control*, **66**, 5, 619-651.

Johansen T.A. (1994). *Operating Regime based Process Modeling and Identification.* Dr. Ing. Thesis, Department of Engineering Cybernetics, Norwegian Institute of Technology, Trondheim, Norway, http://www.itk.unit.no/ansatte/Johansen_Tor.Arne/dring_taj.ps.gz

Johansen T.A., B.A. Foss (1993). Constructing NARMAX models using ARMAX models. *Int. J. of Control*, **58**, 1125-1153.

Johansen T.A., K.J. Hunt and H. Fritz (1998). A software environment for gain scheduled controller design. *IEEE Control System Mag.*, **18**, 48-60.

Previdi F. (1999). *Identification and Control using Local Linear Models*. Ph.D. Thesis, Politecnico di Milano, Italy.

Rugh W.J. and J.S. Shamma (2000). Research on gain scheduling. *Automatica*, **36**, 1401-1425.

Shamma J.S. and M. Athans (1990). Analysis of gain scheduled control for nonlinear plants. *IEEE Trans. on Automatic Control*, **35**, 898-907

LOCAL LINEAR MODELING FOR CONTROL OF BATCH PROCESSES

Dennis Bonné[†] [*] and Sten Bay Jørgensen[‡] [*]

[*] *Department of Chemical Engineering*
Technical University of Denmark
DK-2800 Lyngby, Denmark
[†] *E-mail : db@kt.dtu.dk*
[‡] *E-mail : sbj@kt.dtu.dk*

Abstract: This paper contains of four contributions. An existing modeling framework for batch processes, which consists of a set of Finite Impulse Response (FIR) models, has been extended with the effect of initial conditions on batch evolution. It has been shown that an analogous modeling framework can be derived from a set of AutoRegressive models with eXogenous (ARX) inputs. The ARX models are superior to the FIR models with respect to parameterization and thus with respect to indentifiability as well as for capturing multivariable correlations. Based on the presented modeling framework, an existing Iterative Learning Control (ILC) algorithm has been extended with the effect of initial conditions and noisy measurements. The ILC algorithm is set up in a Model Predictive Control (MPC) framework. The presented tools and algorithms are demonstrated on simulated fed-batch yeast fermentations. *Copyright ©2001 IFAC*

Keywords: Modeling batch processes, Control of batch processes, Learning control, Identification, Fermentation.

1. INTRODUCTION

Batch processes are experiencing a renaissance as products-on-command and first-to-marked strategies impel the need for flexible and specialized production methods. Furthermore, industries such as food, biochemical, and pharmaceutical depend on the confinement of faults and contaminations to single batches. This renaissance propels the need for modeling and control tools, which can facilitate optimal and reliable operation of batch processes. However, the traditional linear modeling and control tools are inadequate, when applied to the often highly nonlinear and time-varying behavior of batch processes.

In this contribution, it is proposed to model batch processes with sets of simple linear models identified from sparse and noisy operation data. Subsequently, a model predictive control framework is proposed, which will improve the tracking performance as knowledge is collected from batch to batch. Finally, the proposed tools are demonstrated on yeast fermentations and conclusions are given.

2. LOCAL MODELING

Most often the complex and time-varying dynamic behavior of batch processes can be approximated with a set of local linear models, each of which describe characteristic operation regimes, time-varying limitations, and/or transitions. The local modeling approach taken in the present paper adapts onto a batch to batch modeling framework presented by Lee *et al.* (2000). In the framework presented here the model set may be set up using

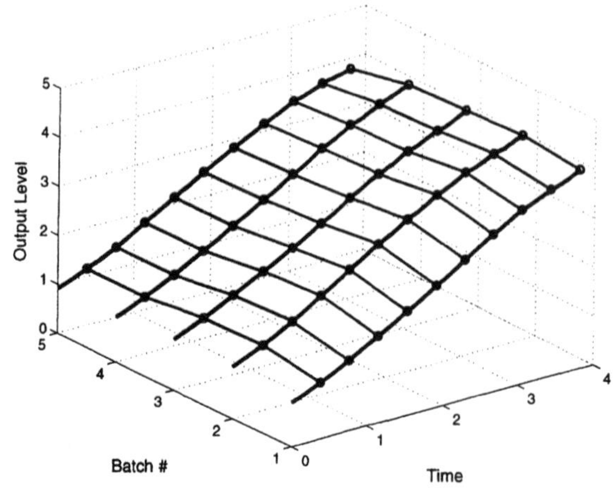

Fig. 1. The three dimensions of batch data, i.e., output level, time, and batch index. The rings symbolize sample times.

either Finite Impulse Response (FIR) (Lee *et al.*, 2000) models or AutoRegressive models with eXogenous inputs (ARX) (Gregersen, 2001).

Because of the periodic nature and finite horizon of batch processes it is possible to collect sufficient data from every sample time to approximate each sample time with one local model — see figure 1. As the operation of a batch run progresses, different inputs and outputs are in use depending on the current operation mode. Consequently, it seems reasonable to define the following variables:

- Input variable, $u(t) \in \mathbb{R}^{n_u(t)}$.
- Output variable, $y(t) \in \mathbb{R}^{n_y(t)}$, measured on-line.
- Disturbance variable, $d(t) \in \mathbb{R}^{n_d(t)}$.
- Quality variable, $q \in \mathbb{R}^{n_q}$, which is the end-product quality measured off-line.

Let N be the batch length(/number of samples) and define the input, output, and disturbance profiles respectively as:

$$\begin{aligned} \boldsymbol{u} &\equiv \left[u^T(0)\ u^T(1)\ \dots\ u^T(N-1)\right]^T \\ \boldsymbol{y} &\equiv \left[y^T(1)\ y^T(2)\ \dots\ y^T(N)\right]^T \\ \boldsymbol{d} &\equiv \left[d^T(1)\ d^T(2)\ \dots\ d^T(N)\right]^T, \end{aligned} \qquad (1)$$

and let $y(0)$ be the initial conditions. Note, not all initial conditions are measurable and/or physically meaningful (e.g. off-gas measurements).

The output profile is in general related to the input profile, initial conditions, and disturbance profile by a nonlinear algebraic model:

$$\boldsymbol{y} = \mathcal{M}(\boldsymbol{u}, y(0), \boldsymbol{d}) = \mathcal{N}(\boldsymbol{u}, \boldsymbol{d}) + \mathcal{N}^0(y(0)), \quad (2)$$

for which it is assumed that the effect of initial the conditions can be separated from that of the input and disturbance profiles. Let $[\bar{y}(0)^T \quad \bar{\boldsymbol{y}}^T]^T$ be the desired (and preferably optimal) output reference profile, $\bar{\boldsymbol{u}}$ the nominal input profile, and let k be the batch index. Then the output error profile $\boldsymbol{e}$ is defined as:

$$\begin{aligned} \boldsymbol{e} \equiv& \bar{\boldsymbol{y}} - \boldsymbol{y} \\ =& \mathcal{N}(\bar{\boldsymbol{u}}, \boldsymbol{0}) - \mathcal{N}(\boldsymbol{u}, \boldsymbol{d}) \\ &+ \mathcal{N}^0(\bar{y}(0)) - \mathcal{N}^0(y(0)). \end{aligned} \qquad (3)$$

By linearizing the output error profile given in (3) around $\bar{\boldsymbol{u}}$ and $\bar{y}(0)$, $\boldsymbol{e}$ is obtained as:

$$\boldsymbol{e} = \boldsymbol{G}(\bar{\boldsymbol{u}} - \boldsymbol{u}) + \boldsymbol{G}^0(\bar{y}(0) - y(0)) + \boldsymbol{r}, \quad (4)$$

where $\boldsymbol{G}$ and $\boldsymbol{G}^0$ are linear systems approximations, and $\boldsymbol{r}$ is a profile of residual terms due to bias in the input profile, the effect of $\boldsymbol{d}$ and the modeling errors due linear approximations. As $\boldsymbol{r}$ contain contributions from both batch wise persistent disturbances, such as input bias, modeling bias, and erroneous sensor readings, as well as from random disturbances, which occur with no batch wise correlation, $\boldsymbol{r}$ may be modeled with a random walk model with respect to batch index k:

$$\begin{aligned} \boldsymbol{x}_{k+1} &= \boldsymbol{x}_k + \boldsymbol{w}_k, \\ \boldsymbol{r}_k &= \boldsymbol{x}_k + \boldsymbol{v}_k, \end{aligned} \qquad (5)$$

where $\boldsymbol{w}_k$ represents batch wise persistent disturbances and $\boldsymbol{v}_k$ represents batch wise non-persistent disturbances. Then, defining the part of the output error profile that will reappear in the next batch run as:

$$\tilde{\boldsymbol{e}}_k \equiv \boldsymbol{G}(\bar{\boldsymbol{u}} - \boldsymbol{u}_k) + \boldsymbol{G}^0(\bar{y}(0) - y_k(0)) + \boldsymbol{x}_k, \quad (6)$$

and recognizing from (5) and (4) that:

$$\boldsymbol{e}_k = \boldsymbol{G}(\bar{\boldsymbol{u}} - \boldsymbol{u}_k) + \boldsymbol{G}^0(\bar{y}(0) - y(0)) + \boldsymbol{x}_k + \boldsymbol{v}_k, \quad (7)$$

it is easily seen that:

$$\begin{aligned} \tilde{\boldsymbol{e}}_{k+1} &= \tilde{\boldsymbol{e}}_k - \boldsymbol{G}\Delta\boldsymbol{u}_{k+1} - \boldsymbol{G}^0\Delta y_{k+1}(0) + \boldsymbol{w}_k \\ \boldsymbol{e}_k &= \tilde{\boldsymbol{e}}_k + \boldsymbol{v}_k, \end{aligned} \qquad (8)$$

where $\Delta\boldsymbol{u}_{k+1} = \boldsymbol{u}_{k+1} - \boldsymbol{u}_k$ and $\Delta y_{k+1}(0) = y_{k+1}(0) - y_k(0)$. Both $\boldsymbol{w}_k$ and $\boldsymbol{v}_k$ are assumed zero-mean i.i.d. (independent and identically distributed) sequences with respect to k.

Clearly, $\boldsymbol{G}$ is a set of FIR models and it has the following structure with N local and causal models:

$$\begin{aligned} \boldsymbol{G} &= \begin{bmatrix} g_{1,0} & 0 & \cdots & 0 \\ g_{2,0} & g_{2,1} & \cdots & 0 \\ \vdots & \vdots & \ddots & \vdots \\ g_{N,0} & g_{N,1} & \cdots & g_{N,N-1} \end{bmatrix} \\ &= \left[\boldsymbol{G}_0\ \boldsymbol{G}_1\ \cdots\ \boldsymbol{G}_{N-1}\right]. \end{aligned} \qquad (9)$$

However, eq. (8) may also be derived from a set of ARX models. Using an ARX model the output at time t may be correlated to past outputs and input as:

$$\bar{y}(t) - y(t) = -\theta_{t,t-1}\Big(\bar{y}(t-1) - y(t-1)\Big) - \dots$$
$$- \theta_{t,t-n_\Theta(t)}\Big(\bar{y}(t-n_\Theta(t))$$
$$- y(t-n_\Theta(t))\Big)$$
$$+ \phi_{t,t-1}\Big(\bar{u}(t-1) - u(t-1)\Big) + \dots$$
$$+ \phi_{t,t-n_\Phi(t)}\Big(\bar{u}(t-n_\Phi(t))$$
$$- u(t-n_\Phi(t))\Big) + \breve{r}(t), \qquad (10)$$

where $n_\Theta^y(t), n_\Phi^y(t) \in [1,\dots,t]$ are the model orders. Thus, by defining an additional output profile as:

$$\boldsymbol{y}_0 \equiv \left[\, y^T(0)\ y^T(1)\ \dots\ y^T(N-1)\,\right]^T, \qquad (11)$$

and defining its deviation from the reference profile $\bar{\boldsymbol{y}}_0$ as:

$$\boldsymbol{e}_0 \equiv \bar{\boldsymbol{y}}_0 - \boldsymbol{y}_0, \qquad (12)$$

the relation between the input profile and the output profile can be described with a set of ARX models:

$$\boldsymbol{e} = \boldsymbol{\Theta}\boldsymbol{e}_0 + \boldsymbol{\Phi}(\bar{\boldsymbol{u}} - \boldsymbol{u}) + \breve{\boldsymbol{r}}. \qquad (13)$$

Let $\breve{\boldsymbol{\Theta}}$ and $\acute{\boldsymbol{\Theta}}$ be partitions of $\boldsymbol{\Theta}$ corresponding to $\left[\, y^T(1)\ y^T(2)\ \dots\ y^T(N-1)\,\right]^T$ and $y(0)$, respectively. Then eq. (13) can be reformulated as:

$$\boldsymbol{e} = \boldsymbol{\Gamma}\acute{\boldsymbol{\Theta}}(\bar{y}(0) - y(0)) + \boldsymbol{\Gamma}\boldsymbol{\Phi}(\bar{\boldsymbol{u}} - \boldsymbol{u}) + \boldsymbol{r}, \qquad (14)$$

where $\boldsymbol{\Gamma}$ is given as:

$$\boldsymbol{\Gamma} = \Big(\boldsymbol{I} - \begin{bmatrix}\breve{\boldsymbol{\Theta}} & \boldsymbol{0}\end{bmatrix}\Big)^{-1}, \qquad (15)$$

and $\boldsymbol{r} = \boldsymbol{\Gamma}\breve{\boldsymbol{r}}$ is given by (5). It should then be straightforward to realize that $\boldsymbol{\Gamma\Phi} = \boldsymbol{G}$ and $\boldsymbol{\Gamma}\acute{\boldsymbol{\Theta}} = \boldsymbol{G}^0$ assuming that the model orders are correct. Note, the common model structure is significant as the FIR model set often will have considerably more parameters than the ARX model set. The parameterization of FIR models versus ARX models will be discussed in more detail in section 3. More importantly, contrary to the uncorrelated FIR models, in the ARX model setup the multivariable correlations between the outputs may be captured by the model.

Applying the same methodology as above, the deviation from the desired end-product quality may be correlated to the error and input profiles as:

$$e_k^q \equiv \bar{q} - q_k = A^y\boldsymbol{e}_k + A^q(\bar{\boldsymbol{u}} - \boldsymbol{u}_k) + r^q, \qquad (16)$$

and included in the error profile model (8) as:

$$\begin{bmatrix}\tilde{\boldsymbol{e}}_{k+1}\\ \tilde{e}_{k+1}^q\end{bmatrix} = \begin{bmatrix}\tilde{\boldsymbol{e}}_k\\ \tilde{e}_k^q\end{bmatrix} - \begin{bmatrix}\boldsymbol{G}\\ A^y\boldsymbol{G} + A^q\end{bmatrix}\Delta\boldsymbol{u}_{k+1}$$
$$- \begin{bmatrix}\boldsymbol{G}^0\\ A^y\boldsymbol{G}^0\end{bmatrix}\Delta y_{k+1}(0) \qquad (17)$$
$$+ \begin{bmatrix}\boldsymbol{w}_k\\ A^y\boldsymbol{w}_k + w_k^q\end{bmatrix}$$
$$\begin{bmatrix}\boldsymbol{e}_k\\ e_k^q\end{bmatrix} = \begin{bmatrix}\tilde{\boldsymbol{e}}_k\\ \tilde{e}_k^q\end{bmatrix} + \begin{bmatrix}\boldsymbol{v}_k\\ A^y\boldsymbol{v}_k + v_k^q\end{bmatrix},$$

As eq.'s (8) and (17) have the same structure, eq. (8) will be assumed to include the end-product quality for the sake of simplicity in the remaining part of the present paper.

To enable usage of a Kalman filter for estimation of at-line or off-line measured initial conditions, the initial conditions are modeled with a simple batch wise random walk model:

$$y_{k+1}(0) = y_k(0) + v_k(0)$$
$$z_k(0) = y_k(0) + \xi_k(0),$$

where v_k^0 is assumed a batch wise white disturbance term, $z_k(0)$ is the measurement of $y_k(0)$, and $\xi_k(0)$ is assumed batch wise white measurement noise. Note, it is assumed that there are no batch wise persistent perturbations of the initial conditions.

3. IDENTIFICATION

One major drawback of the proposed modeling approach is the immense dimensionality of the resulting set of models. E.g., assuming causality a process with one output and one input sampled 100 times during a batch will be described by 5150 parameters using FIR models versus 198 model orders and assuming all model orders set equal to three then 594 parameters using ARX models. In practice, this immense dimensionality will render any identification problem ill-conditioned.

Fortunately, there is some structural information as well as some desired model properties that may be imposed onto the resulting set of models $\boldsymbol{G}$:

(1) By causality the upper triangular is **0**.
(2) The time variation can reasonably be assumed smooth ($\boldsymbol{L}_1$), i.e., the diagonal elements in $\boldsymbol{G}$ will vary in a smooth manner.
(3) The impulse responses can reasonably be assumed smooth ($\boldsymbol{L}_2$), i.e., the column elements in $\boldsymbol{G}$ will vary smoothly.

Regularization exploits such *a priory* knowledge, reducing the amount of data needed, by introducing constraints or penalties in order to constrain excessive degrees of freedom. Unfortunately, regularization *will* result in biased estimates, i.e., there will be a tradeoff between variance and bias. The optimal tradeoff is found by selecting the regularization weights λ_i that minimize an average prediction error over the cross-validation batches.

The regularized system is formulated as a regression:

$$\boldsymbol{b}_{reg} = \boldsymbol{A}_{reg}\varphi, \tag{18}$$

and the unknown model parameters φ are estimated by solving (18) in a Least Squares (LS) sense. The regularized system is set up from the un-biased system and the additional penalties:

$$\begin{bmatrix} \boldsymbol{b} \\ \boldsymbol{0} \\ \boldsymbol{0} \end{bmatrix} = \begin{bmatrix} \boldsymbol{A} \\ \lambda_1 \boldsymbol{L}_1 \\ \lambda_2 \boldsymbol{L}_2 \end{bmatrix} \varphi, \tag{19}$$

The un-biased system is set up from N_k batch data sets:

$$\boldsymbol{A} = \begin{bmatrix} \boldsymbol{A}_1 \\ \vdots \\ \boldsymbol{A}_{N_k} \end{bmatrix}, \quad \boldsymbol{b} = \begin{bmatrix} \boldsymbol{b}_1 \\ \vdots \\ \boldsymbol{b}_{N_k} \end{bmatrix}, \tag{20}$$

where $\boldsymbol{b}_k$ is the system outputs and $\boldsymbol{A}_k$ is the system inputs, initial conditions, and (for ARX models) outputs.

In literature, several approaches to selection of scalar optimal regularization weights have been presented (Hansen, 1996). However, seemingly no work has been reported with vectorial regularization weights based on desired model properties.

Furthermore, usage of high quality data for the identification will decrease estimation variance. The quality of batch identification input data can be determined by statistical comparison to a *normal* reference model obtained through *multiway analysis*. For further discussion of quality or normal batch identification data the reader is referred to Gregersen and Jørgensen (1999), Bro (1998) and Nomikos (1996).

4. MODEL PREDICTIVE CONTROL

Based on the above described batch transition model, Model Predictive Control (MPC) can be applied for control of the system outputs as well as the end-product quality. Model predictive control has been comprehensively covered in literature and most recently by Mayne *et al.* (2000) and Rao and Rawlings (2000). Lee *et al.* (2000) have presented an Iterative Learning Control (ILC) algorithm in a model predictive control framework, which offers batch to batch self-improving MPC. The block diagram of this ILC-MPC control algorithm with the extensions introduced in the present paper, i.e., initial conditions and measurement noise, is shown in figure 2.

From equation (8) the following periodic, Linear Time-Varying (LTV) state-space model is obtained for control design:

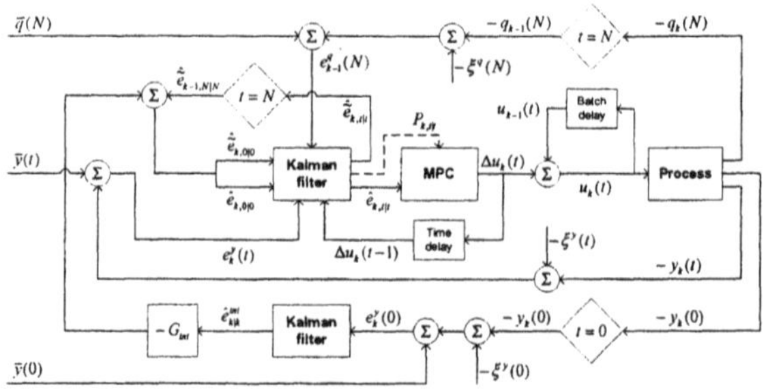

Fig. 2. Block diagram of an Iterative Learning batch Control (ILC) algorithm in a Model Predictive Control (MPC) framework with combined output trajectory tracking and end-product quality control.

$$\begin{aligned} \begin{bmatrix} \tilde{\boldsymbol{e}}_{k,t} \\ \boldsymbol{e}_{k,t} \end{bmatrix} &= \begin{bmatrix} \tilde{\boldsymbol{e}}_{k,t-1} \\ \boldsymbol{e}_{k,t-1} \end{bmatrix} - \begin{bmatrix} \boldsymbol{G}_{t-1} \\ \boldsymbol{G}_{t-1} \end{bmatrix} \Delta u_k(t-1) \\ z_k(t) &= \begin{bmatrix} \boldsymbol{0} & \boldsymbol{H}_t \end{bmatrix} \begin{bmatrix} \tilde{\boldsymbol{e}}_{k,t} \\ \boldsymbol{e}_{k,t} \end{bmatrix} + \xi(t), \end{aligned} \tag{21}$$

for $t = 1, 2, \ldots, N$, where $z_k(t)$ is the measurement of the output error $e_k(t)$, and $\xi(t)$ is assumed zero-mean i.i.d. measurement noise. At the start of a batch the states in (21) satisfy:

$$\begin{aligned} \tilde{\boldsymbol{e}}_{k,0} &= \tilde{\boldsymbol{e}}_{k-1,N} - \boldsymbol{G}^0 \Delta y_k(0) + \boldsymbol{w}_{k-1}, \\ \boldsymbol{e}_{k,0} &= \tilde{\boldsymbol{e}}_{k,0} + \boldsymbol{v}_k, \end{aligned} \tag{22}$$

Note that the structure of $\boldsymbol{H}_t$ and $e_k(t)$ depend on which measurements are available at sample t.

Let $\hat{\boldsymbol{e}}_{k,t+m|t} = \hat{\boldsymbol{e}}_{k,t|t} - \boldsymbol{G}_t^m \Delta \boldsymbol{u}_{k,t}^m$ be the optimal estimate of the error profile of the kth batch given measurements up to time t and m future control moves. The optimal input movements are then determined by employing the standard quadratic objective function known from conventional MPC control.

$$\begin{aligned} \min_{\Delta \boldsymbol{u}_{k,t}^m} & \left[\frac{1}{2} \{ \hat{\boldsymbol{e}}_{k,t+m|t}^T \boldsymbol{Q}_t \hat{\boldsymbol{e}}_{k,t+m|t} \right. \\ & \left. + \Delta \boldsymbol{u}_{k,t}^{mT} \boldsymbol{R}_t \Delta \boldsymbol{u}_{k,t}^m \} \right], \\ s.t. \quad & \hat{\boldsymbol{e}}_{k,t+m|t} = \hat{\boldsymbol{e}}_{k,t|t} - \boldsymbol{G}_t^m \Delta \boldsymbol{u}_{k,t}^m, \end{aligned} \tag{23}$$

where $\boldsymbol{Q}_t$ and $\boldsymbol{R}_t$ are weighting matrices. The analytical solution to (23) is given by:

$$\Delta \boldsymbol{u}_{k,t}^m = [\boldsymbol{G}_t^{mT} \boldsymbol{Q}_t \boldsymbol{G}_t^m + \boldsymbol{R}_t]^{-1} \boldsymbol{G}_t^{mT} \boldsymbol{Q}_t \hat{\boldsymbol{e}}_{k,t|t}, \tag{24}$$

where only $\Delta u_k(t)$ from $\Delta \boldsymbol{u}_{k,t}^m$ is implemented. In practice, (23) should be extended to:

$$\begin{aligned} \min_{\Delta \boldsymbol{u}_{k,t}^m, \epsilon_{k,t}} & \left[\frac{1}{2} \{ \Delta \boldsymbol{u}_{k,t}^{mT} \mathcal{Q}_t \Delta \boldsymbol{u}_{k,t}^m + 2 \mathcal{R}_t^T \Delta \boldsymbol{u}_{k,t}^m \right. \\ & \left. + \epsilon_{k,t}^T \boldsymbol{S}_t \epsilon_{k,t} \} \right], \\ s.t. \quad & \mathcal{A}_t \Delta \boldsymbol{u}_{k,t}^m \leq \mathcal{B}_t(\epsilon_{k,t}), \\ & \epsilon_{k,t} \geq \boldsymbol{0}, \end{aligned} \tag{25}$$

where $\mathcal{Q}_t = \boldsymbol{G}_t^{mT} \boldsymbol{Q}_t \boldsymbol{G}_t^m + \boldsymbol{R}_t$ and $\mathcal{R}_t^T = -\hat{\boldsymbol{e}}_{k,t|t}^T \boldsymbol{Q}_t \boldsymbol{G}_t^m$. By proper configuration of $\mathcal{A}$ and $\mathcal{B}$ both the outputs and the inputs may be constrained. The *constraint softening* slack variable

ϵ insures feasible solutions and $\mathcal{S}_t$ is the corresponding weighting matrix.

The presented control algorithm utilizes the updated control error profile as initial condition for the next batch. Furthermore, it computes the input profile as a change from the previous batch based on the control error prediction. This use of information from previous batches provides the control algorithm with an integral action along the batch index (Lee *et al.*, 2000). Through this integral action the control algorithm is expected to asymptotically reject the effect of the disturbances represented by $\boldsymbol{w}$, i.e., input bias, modeling bias, erroneous sensor readings, etc. Lee *et al.* (2000) have proven convergence for a batch MPC controller based on state-space model (21). This proof can be extended to hold for this algorithm also.

5. APPLICATION TO A FERMENTER

To validate the presented modeling and control framework a set of FIR models was identified on data from yeast (*Saccharomyces ceravisiae*) producing fed-batch fermentations simulated with a biochemically structured model (Lei *et al.*, 2001).

As output variables the ethanol concentration (EtOH), Carbon dioxide Evolution Rate (CER), Oxygen Uptake Rate (OUR), and Respiratory Quotient (RQ) were chosen and the feed rate was chosen as control variable. In fact, RQ $\equiv$ CER/OUR, but here RQ was modelled as a measurement to see if its dynamics could be captured with the suggested modeling framework. White measurement noise was added to all inputs and outputs: EtOH $\pm3\%$, OUR $\pm2\%$, CER $\pm2\%$, RQ $\pm2\%$, feed rate $\pm1\%$. The level of measurement noise was based on laboratory experiments and industrial data. It was assumed that there were no time delays.

The model set was identified on data from 10 normal batch runs and cross-validated on 3 normal data sets. The identification was regularized to ensure smooth model parameter evolutions and smooth impulse responses. The regularization weights were optimized by trial and error, minimizing the mean prediction error from pure simulation model cross-validation. The optimal regularization weights are shown in table 1 along with their respective mean prediction fits. As demonstrated in figure 3, the identified model set possesses reasonable predictive capabilities.

From figure 4 it may be evidenced that the time evolution of the model parameters is indeed nonlinear and smooth as desired. The nonlinearity of the time evolution of the model parameters confirms the need for multiple linear models. The smooth time evolution of the model parameters ensures a continuous model set suitable for control applications. The impulse responses (columns in the model set) shown in figure 5 are less smoothened compared to the diagonal elements in the model (figure 4), which corresponds to the chosen regularization weights — see table 1.

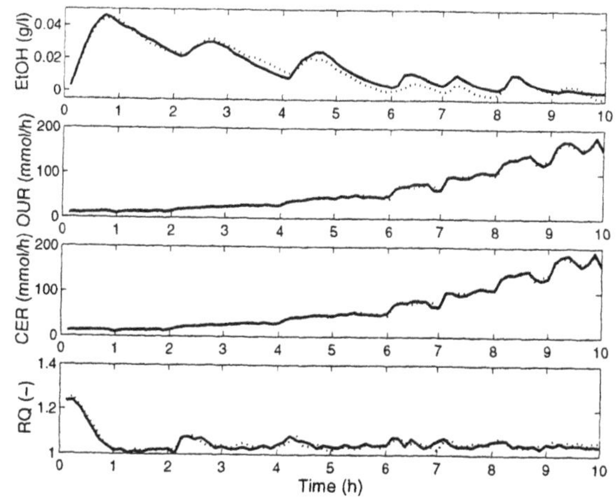

Fig. 3. Validation of noisy FIR model set identification. True output, solid line; predicted output, dotted line.

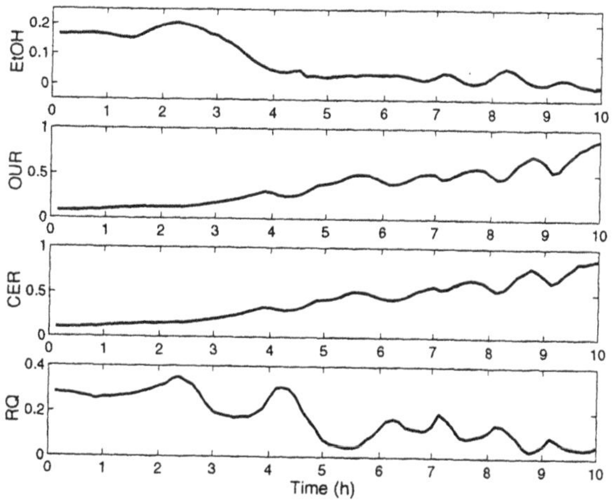

Fig. 4. Time evolution of the first impulse response coefficient $g_{t,t-1}$.

Based on the identified FIR model set the reference profiles of the four outputs were tracked for further model validation. In the control simulation scenario the initial conditions were randomly perturbed by $\pm10\%$ in each batch run and the feed concentration was perturbed by a persistent 10% reduction in all batch runs.

Table 1. Optimal regularization weights and relative prediction fit for noisy identification.

Output variable	λ_1	λ_2	*fit*
ΔEtOH	1.18	0.05	1.5708×10^{-2}
ΔOUR	0.05	0.01	2.4224×10^{-2}
ΔCER	0.05	0.01	2.0479×10^{-2}
ΔRQ	0.26	0.02	2.3902×10^{-2}

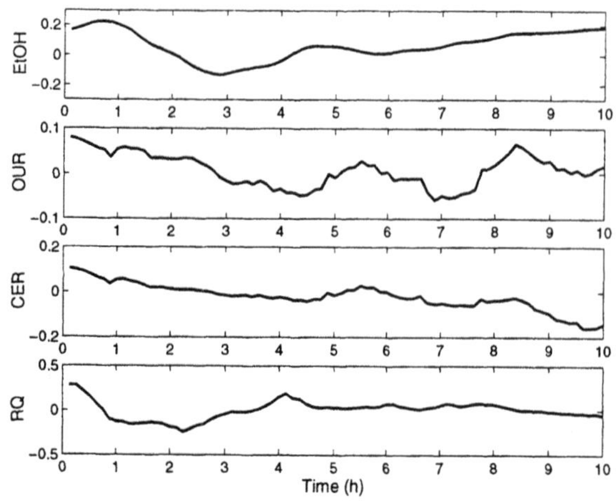

Fig. 5. Impulse response from the first input, i.e., $g_{t,0}$.

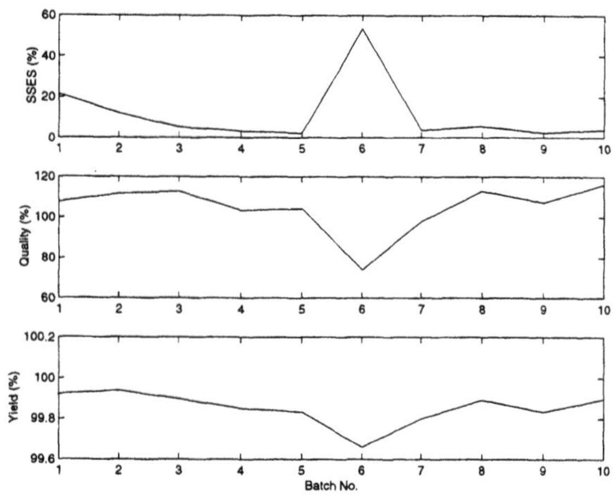

Fig. 6. Summed squared error sequence, quality, and yield evolution in noisy reference tracking of EtOH, OUR, CER, and RQ. No control corresponds to 100%.

In figure 6 it can be observed that the controller rejected approximately 80% of the Summed Squared Error Sequence (SSES) in the first batch run and that after having been trained on the first three batch runs more than 90% of SSES with exception of batch run no. 6. The sluggish performance in batch run no. 6 was due to a particularly intractable disturbance direction, i.e., the yeast had been starved before seeding. However, the controller still rejected approximately 40% of SSES and the performance in the subsequent batch run was not affected. In addition to SSES, also the end-product quality and the product yield are shown in figure 6, and it can be observed that the end-product quality can be improved by output reference tracking alone and that the product yield may endure a minute decrease, which is the cost of high tracking performance.

6. CONCLUSION

In the present paper, it has been demonstrated how multiple local, linear models may be applied when modeling nonlinear and time-varying batch processes. A tool for identification of the resulting high dimensional model from sparse and noisy data has been presented. The identification tool will integrate both physiological constraints and desired model properties providing a tradeoff between model parameter variance, model bias, and desired model properties. The resulting model sets have been set up in a model predictive control framework, which offers batch-to-batch adapting operational reliability with respect to both output reference tracking and end-product quality control. Finally, the potential of the presented tools has been demonstrated through application of ILC control to fed-batch yeast fermentations.

REFERENCES

Bro, Rasmus (1998). *Multi-way Analysis in the Food Industry.* Academish Proefschrift. Universiteit van Amsterdam.

Gregersen, Lars (2001). Monitoring and Fault Diagnosis of Fermentation Processes. PhD thesis. Technical University of Denmark.

Gregersen, Lars and Sten Bay Jørgensen (1999). Supervision of fed-batch fermentations. *Chemical Engineering Journal* **75**, 69–76.

Hansen, Per Christian (1996). *Rank-Deficient and Discrete Ill-Posed Problems.* Polyteknisk Forlag. Lyngby.

Lee, Jay H., Kwang S. Lee and Won C. Kim (2000). Model-based iterative learning control with a quadratic criterion for time-varying linear systems. *Automatica* **36**(5), 641–657.

Lei, F., M. Rotbøll and S. B. Jørgensen (2001). A Biochemically Structured Model for *Saccharomyces cerrevisiae.* Accepted for Journal of Biotechnology.

Mayne, D. Q., J. B. Rawlings, C. V. Rao and P. O. M. Scokaert (2000). Constrained model predictive control: Stability and optimality. *Automatica* **36**, 789–814.

Nomikos, Paul (1996). Detection and diagnosis of abnormal batch operations based on multi-way principal component analysis. *ISA Transactions* **35**(3), 259–266.

Rao, Christopher V. and James B. Rawlings (2000). Linear programming and model predictive control. *Journal of Process Control* **10**, 283–289.

AN EXTENSION OF SELF-TUNING GMVC BASED ON STATE-SPACE APPROACH

Akira Yanou * , Akira Inoue * and Yoichi Hirashima *

* *Department of Systems Engineering, Faculty of Engineering, Okayama University, Okayama, 700-8530, Japan*
Phone +81-86-251-8233, Fax +81-86-251-8233
E-mail aki@suri.sys.okayama-u.ac.jp
inoue@suri.sys.okayama-u.ac.jp
hira@suri.sys.okayama-u.ac.jp

Abstract: This paper proposes an extension of self-tuning Generalized Minimum Variance Control (GMVC) based on state-space approach by using coprime factorization approach. The extended controller has a new design parameter which is chosen to stabilize the controller without changing the closed-loop poles, that is, the controller is a strongly stable controller. *Copyright © 2001 IFAC*

Keywords: minimum variance control, self-tuning control, coprime factorization, state space, adaptive control

1. INTRODUCTION

To control plants with uncertainty, Generalized Minimum Variance Control (GMVC) is widely applied in industry. GMVC is first proposed by Clarke and others(Clarke *et al.*, 1979). Using a generalized output, GMVC can be applied to a wider class of plants such as, unstable plants or non-minimum phase plants.

In designing GMVC, the generalized output is usually selected for the closed-loop system of the GMVC to be stable. Then the controller is designed to minimize the variance of the generalized output. And the poles of the controller can not be designed independently to the poles of the closed-loop system.

To design both of the poles of the controller and the poles of the closed-loop system independently, the authors have proposed a new GMVC design method for single-input single-output (SISO) systems(Inoue *et al.*, 1999*b*). The method have new design parameters introduced by using coprime factorization approach(Vidyasagar, 1985), In the method, the poles of the controller are designed by selecting the newly introduced parameters and are chosen independently to the poles of the closed-loop systems(Inoue *et al.*, 1999*a*; Inoue *et al.*, 2000*b*). The method to design a stable GMVC controller with stable closed-loop is applied to design a self-tuning PID controller(Sato *et al.*, 2000).

The method uses transfer functions to describe the plant and the controllers and based on polynomial algebraic approach. And it is difficult to apply the method to multi-input multi-output (MIMO) plants having different numbers of inputs and outputs. To extend the method for MIMO cases it is easier to use a method given by state-space approach than a method based on polynomial algebraic approach.

Although there are many design methods of Model Predictive Control (MPC) in the state-space approach(Ordys and Clarke, 1993; Rossiter *et al.*, 1998; Inoue *et al.*, 2000*a*), there are only a few methods of GMVC in the state-space approach(Masuda *et al.*, 2000).

Hence this paper extends the self-tuning GMVC design method in the state-space approach. The method also has new design parameters by using coprime factorization approach and Youla Parametrization(Vidyasagar, 1985). By choosing newly introduced parameters, the controller poles can be designed without changing the closed-loop poles, that is, the strongly stable system in the state-space approach can be designed. Although the method in this paper is for single-input single-output systems, the method can be extended to a method for multi-input multi-output (MIMO) systems straightforwardly.

Notations: z^{-1} denotes backward shift operator: $z^{-1}y(t) = y(t-1)$. A polynomial function and a rational function are distinguished by $A[z^{-1}]$ and $A(z^{-1})$.

2. SELF-TUNING GMVC IN THE STATE-SPACE APPROACH

In this section a self-tuning GMVC controller is constructed by the state-space approach.

Consider a single-input single-output system given by a difference equation

$$A[z^{-1}]y(t) = z^{-k_m}B[z^{-1}]u(t) + \xi(t) \qquad (1)$$
$$t = 0,\ 1,\ 2\cdots$$

where $u(t)$ is the input, $y(t)$ is the output, k_m is the time delay, $\xi(t)$ is a white Gaussian noise with zero mean. $A[z^{-1}]$ and $B[z^{-1}]$ are polynomials;

$$A[z^{-1}] = 1 + a_1z^{-1} + \cdots + a_nz^{-n}$$
$$B[z^{-1}] = b_0 + b_1z^{-1} + \cdots + b_{n-1}z^{-(n-1)}$$

On the system (1), the following assumptions are assumed;

[1.] The degrees n and m of $A[z^{-1}]$ and $B[z^{-1}]$ and the time delay k_m are known.

[2.] The coefficients $a_1, \cdots, a_n, b_0, \cdots, b_m$ of $A[z^{-1}]$, $B[z^{-1}]$ are unknown.

[3.] The polynomials $A[z^{-1}]$ and $B[z^{-1}]$ are coprime.

To design the control law, first the coefficients of $A[z^{-1}]$ and $B[z^{-1}]$ are assumed to be known and a structure of the control law is given, then the self-tuning controller is obtained by using the adaptive observer.

The control objective is that the output $y(t)$ has a desirable response to the reference input $r(t)$. To accomplish this objective, the generalized minimum variance control given by Clarke and others(Clarke *et al.*, 1979) minimizes variance $J = E[\Phi^2(t+k_m)]$ of a generalized output $\Phi(t+k_m)$;

$$\Phi(t+k_m) = P[z^{-1}]y(t+k_m) + Q[z^{-1}]u(t) - R[z^{-1}]r(t) \qquad (2)$$

where the objective function J is averaged over the noise $\xi(t), \xi(t-1), \cdots$ and for simplicity the design polynomials $P[z^{-1}]$, $Q[z^{-1}]$ and $R[z^{-1}]$ are assumed to be 0th-order;

$$P[z^{-1}] = 1,\ Q[z^{-1}] = q_0,\ R[z^{-1}] = r_0 \qquad (3)$$

To construct GMVC law in state-space approach, the equation (1) is expressed in the state equation in observable canonical form;

$$\boldsymbol{x}(t+1) = \boldsymbol{A}_p\boldsymbol{x}(t) + \boldsymbol{b}_pu(t) - \boldsymbol{a}_p\xi(t) \qquad (4)$$
$$y(t) = \boldsymbol{c}_p{}^{\mathrm{T}}\boldsymbol{x}(t) + \xi(t) \qquad (5)$$

$$\boldsymbol{A}_p = \left[-\boldsymbol{a}_p \quad \begin{matrix} \boldsymbol{I}_{n+k_m-2} \\ \boldsymbol{0}_{1\times(n+k_m-2)} \end{matrix} \right]$$
$$\boldsymbol{a}_p = [a_1,\ \cdots,\ a_n\ \boldsymbol{0}_{1\times(k_m-1)}]^{\mathrm{T}}$$
$$\boldsymbol{b}_p = [\boldsymbol{0}_{1\times(k_m-1)}\ b_0,\ \cdots,\ b_{n-1}]^{\mathrm{T}}$$
$$\boldsymbol{c}_p = [1,\ \boldsymbol{0}_{1\times(n+k_m-2)}]$$

From (4) $y(t+k_m)$ is given by

$$y(t+k_m) = \boldsymbol{c}_p^{\mathrm{T}}\boldsymbol{A}_p^{k_m}\boldsymbol{x}(t) + b_0u(t) + \sum_{i=0}^{k_m-1}\boldsymbol{c}_p^{\mathrm{T}}\boldsymbol{A}_p^{k_m-1-i}\boldsymbol{c}_p\xi(t+i) + \xi(t+k_m) \qquad (6)$$

A prediction $\hat{y}(t+k_m|t)$ of the output is defined by eliminating the future values of noise from $y(t+k_m)$.

$$\hat{y}(t+k_m|t) = \boldsymbol{c}_p^{\mathrm{T}}\boldsymbol{A}_p^{k_m}\boldsymbol{x}(t) + b_0u(t) \qquad (7)$$

Then the k_m-ahead optimally predicted value $\hat{\Phi}(t+k_m|t)$ of the generalized output $\Phi(t+k_m)$ is given by

$$\hat{\Phi}(t+k_m|t) = \boldsymbol{c}_p^{\mathrm{T}}\boldsymbol{A}_p^{k_m}\boldsymbol{x}(t) + (b_0+q_0)u(t) - r_0r(t) \qquad (8)$$

Since the predicted value $\hat{\Phi}(t+k_m|t)$ and the noise term in $\Phi(t+k_m)$ have no correlation with each other, the control $u(t)$ to minimize the variance J is obtained by choosing $u(t)$ to make $\hat{\Phi}(t+k_m|t) = 0$;

$$u(t) = \frac{1}{b_0+q_0}(r_0r(t) - \boldsymbol{c}_p^{\mathrm{T}}\boldsymbol{A}_p^{k_m}\boldsymbol{x}(t)) \qquad (9)$$

Then a self-tuning GMVC is given by, first, identifying values of parameters in $\hat{\boldsymbol{A}}_p(t)$, $\hat{\boldsymbol{b}}_p(t)$ and estimating value $\hat{\boldsymbol{x}}(t)$ of $\boldsymbol{x}(t)$ from an adaptive observer, then, replacing the parameters $\boldsymbol{A}_p$, $\boldsymbol{b}_p$ and the state $\boldsymbol{x}(t)$ by the identified and estimated values.

To construct the adaptive observer(Ioannou and Sun, 1996), the plant equation (4) is reformed to the following equations;

$$\boldsymbol{x}(t+1) = \boldsymbol{F}\boldsymbol{x}(t) + (-\boldsymbol{a}_p + \boldsymbol{f})y(t) + \boldsymbol{b}_p u(t) \quad (10)$$

$$y(t) = \boldsymbol{c}_p^{\mathrm{T}}\boldsymbol{x}(t) \quad (11)$$

$$\boldsymbol{F} = \begin{bmatrix} -\boldsymbol{f} & \begin{matrix} \boldsymbol{I}_{n+k_m-2} \\ \boldsymbol{0}_{1\times(n+k_m-2)} \end{matrix} \end{bmatrix}$$

$$\boldsymbol{f} = [f_1, \cdots, f_n \; \boldsymbol{0}_{1\times(k_m-1)}]^{\mathrm{T}}$$

where $\boldsymbol{F}$ is chosen as a stable matrix and its eigenvalues are the roots of the following polynomial.

$$f[z^{-1}] = 1 + f_1 z^{-1} + \cdots + f_n z^{-n}$$

$\boldsymbol{R}_1(t)$, $\boldsymbol{R}_2(t) \in R^{(n+k_m-1)\times(n+k_m-1)}$ are defined as

$$\boldsymbol{R}_1(t+1) = \boldsymbol{F}\boldsymbol{R}_1(t) + \boldsymbol{I}y(t), \; \boldsymbol{R}_1(0) = 0$$

$$\boldsymbol{R}_2(t+1) = \boldsymbol{F}\boldsymbol{R}_2(t) + \boldsymbol{I}u(t), \; \boldsymbol{R}_2(0) = 0$$

Using $\boldsymbol{R}_1(t), \boldsymbol{R}_2(t)$, (10) and (11) are expressed by the following equations;

$$\boldsymbol{x}(t) = \boldsymbol{R}(t)\boldsymbol{\theta} \quad (12)$$

$$y(t) = \boldsymbol{\zeta}^{\mathrm{T}}(t)\boldsymbol{\theta} \quad (13)$$

$$\boldsymbol{R}(t) = [\boldsymbol{R}_1(t) \;\; \boldsymbol{R}_2(t)]$$

$$\boldsymbol{\zeta}^{\mathrm{T}}(t) = [\boldsymbol{c}_p{}^{\mathrm{T}}\boldsymbol{R}_1(t) \;\; \boldsymbol{c}_p{}^{\mathrm{T}}\boldsymbol{R}_2(t)]$$

$$\boldsymbol{\theta}^{\mathrm{T}} = [(-\boldsymbol{a}_p + \boldsymbol{f})^{\mathrm{T}} \;\; \boldsymbol{b}_p{}^{\mathrm{T}}]$$

where the terms with the initial state value $\boldsymbol{x}(0)$ are removed from equations (12) and (13), since $\boldsymbol{x}(0)$ decreases exponentially.

Replacing the plant parameter vector $\boldsymbol{\theta}$ with the identified value vector $\hat{\boldsymbol{\theta}}(t)$ in equations (12) and (13), the adaptive observer is obtained as

$$\hat{\boldsymbol{x}}(t) = \boldsymbol{R}(t)\hat{\boldsymbol{\theta}}(t) \quad (14)$$

$$\hat{y}(t) = \boldsymbol{\zeta}^{\mathrm{T}}(t)\hat{\boldsymbol{\theta}}(t) \quad (15)$$

The parameter identification law is given by

$$\hat{\boldsymbol{\theta}}(t) = \hat{\boldsymbol{\theta}}(t-1) - \boldsymbol{\Gamma}(t-1)\boldsymbol{\zeta}(t)\varepsilon(t)$$

$$\varepsilon(t) = \frac{\boldsymbol{\zeta}^{\mathrm{T}}(t)\hat{\boldsymbol{\theta}}(t-1) - y(t)}{1 + \boldsymbol{\zeta}^{\mathrm{T}}(t)\boldsymbol{\Gamma}(t-1)\boldsymbol{\zeta}(t)}$$

$$\boldsymbol{\Gamma}(t) = \boldsymbol{\Gamma}(t-1) - \frac{\boldsymbol{\Gamma}(t-1)\boldsymbol{\zeta}(t)\boldsymbol{\zeta}^{\mathrm{T}}(t)\boldsymbol{\Gamma}(t-1)}{\lambda + \boldsymbol{\zeta}^{\mathrm{T}}(t)\boldsymbol{\Gamma}(t-1)\boldsymbol{\zeta}(t)}$$

$$0 < \lambda \leq 1, \; \boldsymbol{\Gamma}(0) = \alpha\boldsymbol{I}, \; 0 < \alpha < \infty$$

3. SELF-TUNING GMVC USING COPRIME FACTORIZATION APPROACH

In order to include new design parameters in the next section, this section gives the coprime factorization of the self-tuning GMVC law in the state-space approach. In this section, it is assumed that $\hat{\boldsymbol{A}}_p(t)$ and $\hat{\boldsymbol{b}}_p(t)$ converge on true values.

The coprime factorization approach uses the family of stable rational functions:

$$RH_\infty = \{G(z^{-1}) = \frac{G_n[z^{-1}]}{G_d[z^{-1}]}, \; G_d[z^{-1}] : \text{stable polynomial}\}$$

<u>Remark</u> Since $G_d[z^{-1}] = 1$ is stable polynomial, the rational function whose denominator is equal to 1, that is, the polynomials belong to RH_∞.

The transfer function (4) is expressed by a ratio of rational functions in RH_∞,

$$G(z^{-1}) = N(z^{-1})D^{-1}(z^{-1}) = \boldsymbol{c}_p^{\mathrm{T}}(z\boldsymbol{I} - \boldsymbol{A}_p)^{-1}\boldsymbol{b}_p \quad (16)$$

$$N(z^{-1}), \; D(z^{-1}) \in RH_\infty$$

where $N(z^{-1})$ and $D(z^{-1})$ are right-coprime factorization.

Functions $X(z^{-1}), Y(z^{-1}) \in RH_\infty$ denote the solutions of the following Bezout equations;

$$X(z^{-1})N(z^{-1}) + Y(z^{-1})D(z^{-1}) = 1 \quad (17)$$

Since it is assumed that the plant parameters converge on true values, the adaptive observer (14) has the following form;

$$\hat{\boldsymbol{x}}(t) = (z\boldsymbol{I} - \boldsymbol{F})^{-1}(-\boldsymbol{a}_p + \boldsymbol{f})y(t) + (z\boldsymbol{I} - \boldsymbol{F})^{-1}\boldsymbol{b}_p u(t) \quad (18)$$

A coprime factorization $N(z^{-1})$ and $D(z^{-1})$ and a set of solutions of Bezout equation (17), $X(z^{-1})$ and $Y(z^{-1})$ in state-space form are obtained by using a state-feedback controller (9) and an observer (18)(Nett *et al.*, 1984);

$$X(z^{-1}) = \boldsymbol{L}(z\boldsymbol{I} - \boldsymbol{F})^{-1}(-\boldsymbol{a}_p + \boldsymbol{f}) \quad (19)$$

$$Y(z^{-1}) = 1 + \boldsymbol{L}(z\boldsymbol{I} - \boldsymbol{F})^{-1}\boldsymbol{b}_p \quad (20)$$

$$N(z^{-1}) = \boldsymbol{c}_p^{\mathrm{T}}(z\boldsymbol{I} - \boldsymbol{A}_p + \boldsymbol{b}_p\boldsymbol{L})^{-1}\boldsymbol{b}_p \quad (21)$$

$$D(z^{-1}) = 1 - \boldsymbol{L}(z\boldsymbol{I} - \boldsymbol{A}_p + \boldsymbol{b}_p\boldsymbol{L})^{-1}\boldsymbol{b}_p \quad (22)$$

$$K(z^{-1}) = \frac{r_0}{b_0 + q_0}, \quad \boldsymbol{L} = \frac{1}{b_0 + q_0}\boldsymbol{c}_p^{\mathrm{T}}\boldsymbol{A}_p^{k_m}$$

Using the solutions of the equations, all the stabilizing controllers are given in Youla Parametrization by

$$u(t) = \frac{K(z^{-1})}{Y(z^{-1})}r(t) - \frac{X(z^{-1})}{Y(z^{-1})}y(t)$$
$$+U(z^{-1})(\hat{y}(t) - y(t)) \qquad (23)$$

where $U(z^{-1}) \in RH_\infty$ is a design parameter.

4. AN EXTENSION OF SELF-TUNING GMVC IN STATE-SPACE APPROACH USING YOULA PARAMETRIZATION

This section extends the self-tuning GMVC law (9) in the state space approach in order to include a new design parameter by using Youla Parametrization (23).

The self-tuning GMVC law (9) can be extended by Youla Parametrization (23),

$$u(t) = \frac{1}{b_0 + q_0}(r_0 r(t) - c_p^T A_p^{k_m} \hat{x}(t))$$
$$+U(z^{-1})(c_p^T \hat{x}(t) - y(t)) \qquad (24)$$

where $U(z^{-1})$ becomes a newly introduced design parameter and $\hat{x}(t)$ is the estimated value of $x(t)$ obtained by the adaptive observer.

the controller in Youla Parametrization (23) has the following equivalent form;

$$u(t) = C_1(z^{-1})r(t) - C_2(z^{-1})y(t) \qquad (25)$$
$$C_1(z^{-1}) = (Y(z^{-1}) - U(z^{-1})\widetilde{N}(z^{-1}))^{-1}K(z^{-1})$$
$$C_2(z^{-1}) = (Y(z^{-1}) - U(z^{-1})\widetilde{N}(z^{-1}))^{-1}$$
$$\cdot(X(z^{-1}) + U(z^{-1})\widetilde{D}(z^{-1}))$$

where $\widetilde{N}(z^{-1})$ and $\widetilde{D}(z^{-1})$ are the left-coprime factorization of the plant equation (4);

$$\widetilde{N}(z^{-1}) = c_p(zI - F)^{-1}b_p$$
$$\widetilde{D}(z^{-1}) = 1 - c_p(zI - F)^{-1}(-a_p + f)$$

The closed-loop characteristic function from $r(t)$ to $y(t)$ by (24) is given by

$$|zI - A_p + b_p L| = 0 \qquad (26)$$

which is equal to the one by (9). So the closed-loop poles from $r(t)$ to $y(t)$ are independent of $U(z^{-1})$.

The controller's characteristic function in (25) is given by

$$1 + L(zI - F)^{-1}b_p - U(z^{-1})c_p(zI - F)^{-1}b_p = 0 \qquad (27)$$

So the controller's poles can be changed by using a newly introduced parameter $U(z^{-1})$.

By selecting a new design parameter $U(z^{-1})$ introduced in (24), a strongly stable system can be constructed with the closed-loop characteristics given by (26).

5. EXAMPLES

Consider a system and a generalized output

$$A[z^{-1}] = 1 + 0.7z^{-1} + 0.9z^{-2}$$
$$B[z^{-1}] = 0.5 - 1.5z^{-1}, \; k_m = 1$$
$$\Phi(t + k_m) = y(t+1) + 0.5u(t) - r_0 r(t)$$
$$P[z^{-1}] = 1, \; Q[z^{-1}] = 0.5$$

and the noise $\xi(t)$ is a white Gaussian noise with zero mean and the variance 0.1^2. In order to make the steady state gain equal to 1, r_0 is calculated at each sampling period by the following equation $r_0 = \hat{r}_0(t)$;

$$\hat{r}_0(t) = \frac{\hat{b}_0(t) + q_0}{c_p^T(I - \hat{A}_p(t) + \hat{b}_p(t)\hat{L}(t))^{-1}\hat{b}_p(t)}$$

The adaptive observer's poles are assumed to -0.2 and 0.1 and the nominal values of the plant parameters are set to be equal to 0.9 of the true values.

λ and $\Gamma(0)$ in the parameter identification law are set to be:

$$\lambda = 1, \; \Gamma(0) = I$$

The reference input $r(t)$ is assumed to be a rectangular wave with amplitude 1.0 and period of 80 steps and the simulation steps are 200.

The closed-loop's poles are $|0.575 \pm 0.3455i| < 1$ and stable. But the absolute values of poles of controller (9) are $|z| = 1.07$ and unstable, so this controller is not strongly stable.

Selecting the design parameter in the extended self-tuning GMVC law (24) as $U(z^{-1}) = -0.15$, the absolute values of the controller are $|z| = 0.96$ and are improved to be stable, and the poles of the closed-loop system are same as the poles by the controller (9), that is, the proposed controller (24) is strongly stable.

Simulation results are shown in Figure 1 for the case of using the controller (9) and Figure 2 with the strongly stable controller (24) proposed in this paper. In both figures the top figures show the output with a solid line and the reference input with a dotted line. And the bottom figures show the inputs.

From these figures, it is shown that, the poles of the controller are improved to be stable by using newly introduced parameter $U(z^{-1})$ and the response of the closed-loop system shown in Figure 2 is not affected by the change of the controller and is the same as the response by the controller without $U(z^{-1})$ shown in Figure 1.

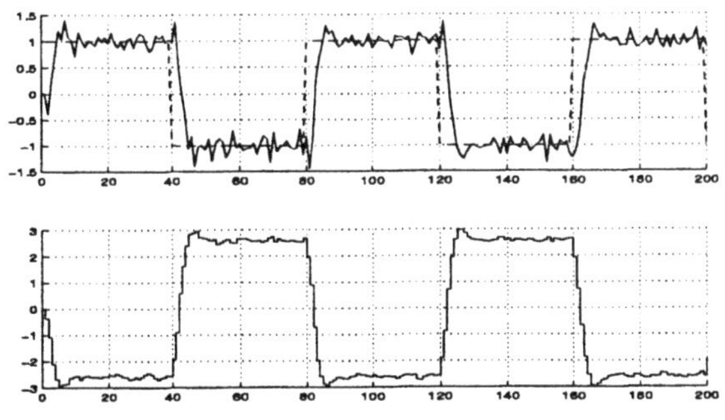

Fig. 1. simulation result by (9)

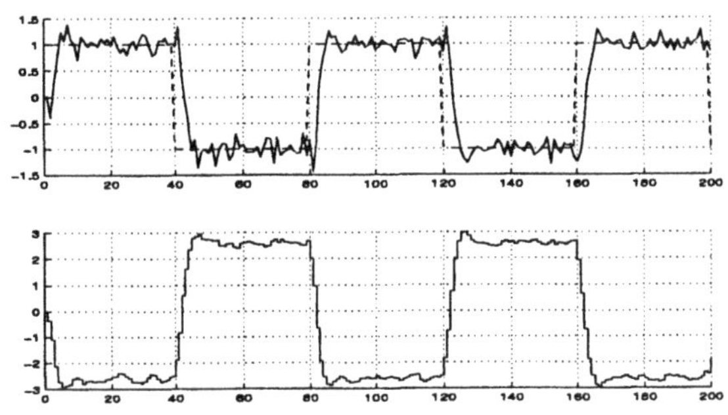

Fig. 2. simulation result by (24)

6. CONCLUSION

In this paper, the self-tuning GMVC in the state-space approach is extended to an extended self-tuning GMVC including the new design parameter by using coprime factorization in state-space form and comparing the most general two-degree-of-freedom compensator in Youla Parametrization form. Without changing the response of output to reference input, the poles of the compensator can be changed by selecting the newly introduced design parameter.

By designing the poles of the controller, a strongly stable controller is obtained.

The extension to multi-input multi-output systems of the proposed method and the design of the controller for pole assignments using the newly introduced parameter in this paper are future works.

7. REFERENCES

Clarke, D. W., M. A. D. Phil and P. J. Gawthrop (1979). Self-tuning control. *Proc. IEE* **126**, 633–640.

Inoue, A., A. Yanou and Y. Hirashima (1999*a*). A design of a strongly stable generalized predictive control using coprime factorization approach. *Proceedings of the 1999 American Control Conference* pp. 652–656.

Inoue, A., A. Yanou and Y. Hirashima (1999*b*). A design of a strongly stable self-tuning controller using coprime factorization approach. *Preprints of the 14th IFAC World Congress* **C**, 211–216.

Inoue, A., A. Yanou, T. Sato and Y. Hirashima (2000*a*). A coprime factorization of a generalized predictive controller in state space model. *Preprints of the 43st Japan Joint Automatic Control Conference* pp. 7–8.

Inoue, A., A. Yanou, T. Sato and Y. Hirashima (2000*b*). An extension of generalized minimum variance control for multi-input multi-output systems using coprime factorization approach. *Proceedings of the 2000 American Control Conference* pp. 4184–4188.

Ioannou, P. A. and J. Sun (1996). *Robust Adaptive Control.* Prentice Hall. New Jersey.

Masuda, S., A. Inoue and Y. Hirashima (2000). A design scheme for a self-tuning two-degrees-of-freedom generalized minimum variance control based on a state-space approach. *Preprints of 20th SICE Symposium on Adaptive Control* pp. 63–66.

Nett, C. N., C. A. Jacobson and M. J. Balas (1984). A connection between state-space and doubly coprime fractional representations. *IEEE Trans. Automat. Contr.* **AC-29**, 831–832.

Ordys, A. W. and D. W. Clarke (1993). A state-space description for gpc controllers. *Int. J. System Sci.* **24**, 1727–1744.

Rossiter, J. A., B. Kouvaritakis and M. J. Rice (1998). A numerically robust state-space approach to stable-predictive control strategies. *Automatica* **34**, 65–73.

Sato, T., A. Inoue, T. Yamamoto and Sirish. L. Shar (2000). Self-tuning pid controllers based on the strongly stable generalized minimum variance control law. *IFAC Digital Control: Past, Present and Future of PID Control* pp. 511–516.

Vidyasagar, M. (1985). *Control System Synthesis;A factorization Approach.* MIT Press. Cambridge.

Overparametrization in Predictive Adaptive Control: Experimental results and an Entropy Interpretation

J. M. Lemos

INESC-ID/IST, R. Alves Redol 9, 1000-029 Lisboa, Portugal, jlml@inesc.pt

Abstract: This paper is concerned with the links between predictive adaptive control algorithms and Statistic Learning theory. In particular, the MUSMAR controller is considered. Through the review of industrial scale plant control examples (arc welding, superheated steam and a solar collector field) it is shown that this algorithm applies to widelly different situations, outperforming standard industrial controllers. Using Saridi's Maximum Entropy formulation of Optimal and Adaptive Control, relationships between performance and certain aspects of controller structure, observed both in simulations and experiments, are shown to hold. The entropy approach also sheds light on the benefic effect of using more parameters than one would strictly need (referred as "redundant modeling") would the plant order be perfectly known.

Keywords: Adaptation and Learning in process control, Adaptive control, predictive control, entropy, process control.

1 Introduction

Adaptive controllers relying on the minimization of an extended receding horizon quadratic cost are known, since the 80'ies, to possess a number of properties which render them adequate for applications. GPC [1] and MUSMAR [3] are two well known examples. In these algorithms, the plant dynamics is described by a set of predictive models, whose parameters are estimated on-line. These estimates are then used for the control synthesis through the minimisation of a quadratic cost.

If the selected controller structure perfectly matches plant dynamics (ideally assumed linear and of ARMAX type), only one predictive model has to be directly identified from plant i/o data, the others being computed from it. This is the approach followed in GPC [1]. If, however, there is a mismatch between the actual plant and the assumed model structure (as unavoidable in applications), *separately* estimation of the various predictors provides more information to the controller, thereby contributing to its improved tuning. This is the approach followed in MUSMAR [3], being referred as "redundant modeling". In both cases, the loss yielded often presents a U-shapped dependency on the horizon T.

This paper aims at providing a link between predictive adaptive control algorithms, when applied to uncertain plants, and the methods of Statistic Learning Theory [11]. In this framework, known relationships between performance and controller structure (*e. g.* the length of the horizon T, assumed model orders or modeling redundancy or no redundancy) are explained, at least in general terms. The emphasis is on the MUSMAR algorithm, but at least some conclusions apply as well to other controllers based on the minimisation of quadratic costs.

First, through the review of industrial scale plant control examples, it is shown that, if properly configured, the same algorithm applies to widelly different situations (in physical terms), yielding a performance which is clearly better than the one achieved by standard industrial means. In all these examples, despite the level of uncertain being very high, the control algorithm is able to "resolve" this high level of uncertainty by extracting the adequate information pattern from the available observations of plant i/o data (and without explicit concern to the specific physic relations of the variables in each plant).

Second, the relation between performance and controller structure is established, providing a general framework for results known from both experiments and simulations. In particular, the above mentioned U-shapped dependency of the observed loss on the horizon length is explained. This is done by relating predictive adaptive controllers with the Entropy approach to Optimal and Adaptive Control presented in [7, 10]. The resulting trade-of ressembles the MDL principle known for modeling [11].

The paper is organised as follows: Section 1 (this section) introduces and formulates the objective of the paper. Section 2 briefly reviews the basic MUSMAR controller. Section 3 reviews 3 process control applications (arc welding, superheated steam and a solar collector field) controlled with the same algorithm. Section 4 establishes the link with the Entropy approach to control and Section 5 draws conclusions.

2 The MUSMAR algorithm

The MUSMAR algorithm is described in the available literature [3], as well as its properties [5]. It relies on the minimization of the multistep quadratic cost function:

$$J(T) = E\left[\frac{1}{T}\sum_{i=0}^{T-1} y^2(t+i+1) + \rho u^2(t+i) \mid I(t)\right] \quad (1)$$

where $E[\cdot|I^t]$ denotes the mean conditioned on the observations I^t made up to time t, T is an integer hereafter referred as the "prediction horizon" and $\rho \geq 0$ is a penalty in the manipulated variable effort. The variables u and y denote the plant manipulated variable.

For the sake of minimizing (1), the plant is described by the set of predictive models:

$$\begin{aligned} \hat{y}(t+i \mid t) &= \theta_i u(t) + \Psi_i' s(t) \\ \hat{u}(t+i-1 \mid t) &= \mu_{i-1} u(t) + \Phi_{i-1}' s(t) \\ i &= 1, \ldots, T \end{aligned} \quad (2)$$

where $\hat{y}(t+i \mid t)$ and $\hat{u}(t+i-1 \mid t)$ are predictors in least squares sense, given I^t of, respectively $y(t+i)$ and $u(t+i-1)$, and $s(t)$ is the so called pseudostate defined by:

$$s(t) = [y(t) \ldots y(t-n_a+1)\, u(t-1) \ldots u(t-n_b)]' \quad (3)$$

The vector $s(t)$ is called "pseudostate" [5] because, even in the presence of correlated noise, although it is not a state, it is a sufficient statistic for computing the control. The entries of $s(t)$ define the structure of the controller. In addition to samples of the input and output variables, samples of other variables such as accessible disturbances may also be included. In order to provide a feedforward effect from the reference to track, samples of this variable are also included in $s(t)$.

The coefficients θ_i, μ_{i-1} and vectors Ψ_i, Φ_{i-1} in equation (2) are parameters to be estimated online. For each i, $i = 1, \ldots, T$, this set of parameters describes the dynamic behavior of the process output and manipulated variable from $t+1$ to $t+T$, assuming a constant feedback of the pseudostate over the prediction horizon. It is remarked that this set contains more parameters than what would be strictly needed, were the plant linear and n_a and n_b correctly selected according to its order. In order to tackle the above mentioned overparametrization problem, the identification algorithm used was recursive least squares with directional forgetting factor [4].

Minimization of (1) assuming (2) yields the control law:

$$u(t) = F's(t) + \eta(t) \quad (4)$$

with the vector of optimal gains given by:

$$F = -\frac{\sum_{i=1}^{T} \theta_i \Psi_i + \rho \sum_{i=1}^{T-1} \mu_i \Phi_i}{\sum_{i=1}^{T} \theta_i^2 + \rho\left(1 + \sum_{i=1}^{T-1} \mu_i^2\right)} \quad (5)$$

and η a low power dither noise injected in order to fulfil a persistency of excitation condition.

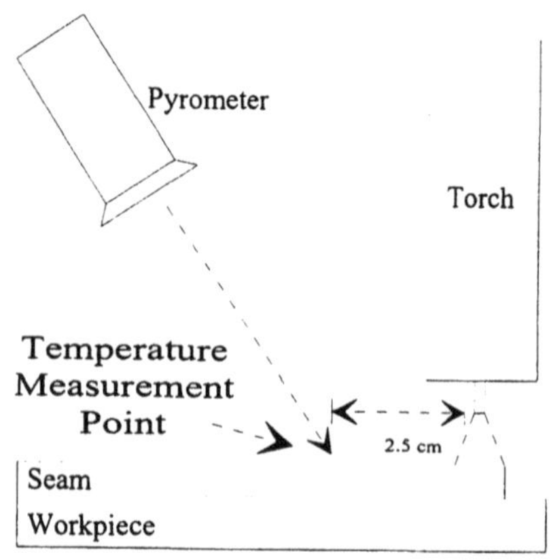

Figure 1: (f1) Trailing centerline temperature measurement.

It is remarked that the above algorithm is very general. Since it relies only on plant i/o data, it may be applied to plants whose order is underestimated and even to slightly nonlinear plants. Of course, this does not mean that the vector F of controller gains will converge for sure to a value which stabilizes the closed loop. However, it has been proved [5] that the only possible convergence points are, for T large enough, close approximations to the local minima of the underlying quadratic cost, constrained to the *a priori* chosen structure of s. Furthermore, extensive simulations in the presence of plant unmodelled dynamics show that F will actually converge. Next section reports significant experimental results in three industrial scale plants.

3 Experimental examples

3.1 Arc welding cooling rate

In this example, trailing centerline temperature control in the gas metal arc welding process is considered. The temperature measurement is achieved by using an infrared pyrometer focused on the trailing weld bead, 2.5 cm behind the tip of the electrode, so that the cooling rate is actually measured. The manipulated variable is the welding voltage. Wire feed speed is adjusted in conformity with the imposed voltage to ensure a stable electric arc. More details may be found in [6].

For the purpose of characterising the difficulties facing control system synthesis, incremental ARMAX models were identified using open loop experimental plant data obtained around several operation points as defined by the input voltage (28 V, 30 V, 32 V, 34 V and 36 V) and using different workpiece thicknesses (6 mm and 12 mm). The frequency response diagrams of the local ARMAX models obtained using standard system identification software are presented in figure 2. These results show the great variability and thereby uncertainty associated with the process, which motivates the consideration of adaptive control. It should also be mentioned that, due to heat diffu-

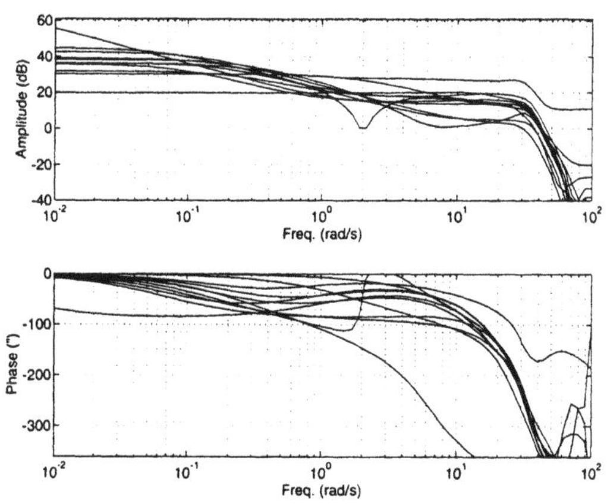

Figure 2: (f2) Bode diagrams of experimental models on a $6\,mm$ and $12\,mm$ plates.

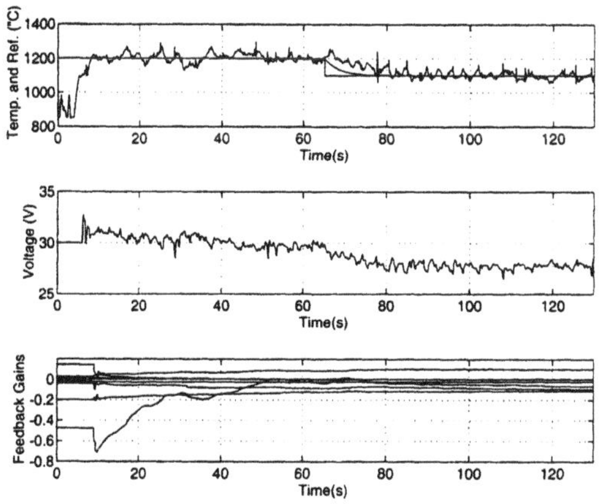

Figure 3: (f3) Experimental plant behavior under closed loop control. $12\,mm$ plate, $F_s = 3\,Hz$, $\rho = 2000$, $T = 10$, $n_a = 3$, $n_b = 3$, $n_g = 1$, $n_v = 1$.

sion, this is a distributed parameter, and therefore infinite dimensional, process.

The adjustment of the controller parameters must take into account the compromise between the control signal (voltage) oscillation and temperature setpoint tracking and requires the basic MUSMAR algorithm to be complemented with integral action. A complete discussion based on experiments is found in [6]. Figure (3) reports an experiment with the MUSMAR based control system. These results show that the temperature reference was tracked and the weld seam has stayed reasonably regular.

In general, it was found that a trade-off held for the selection of model order parameters and the horizon T. It should also be remarked that the standard self-tuning corresponding to $T = 1$ could not stabilize the closed loop. Furthermore, in the great majority of the tests performed, the PID controller also yielded an unstable closed-loop. A stable closed-loop has also never been achieved with pole-placement designed on the basis of the *a priori* identified

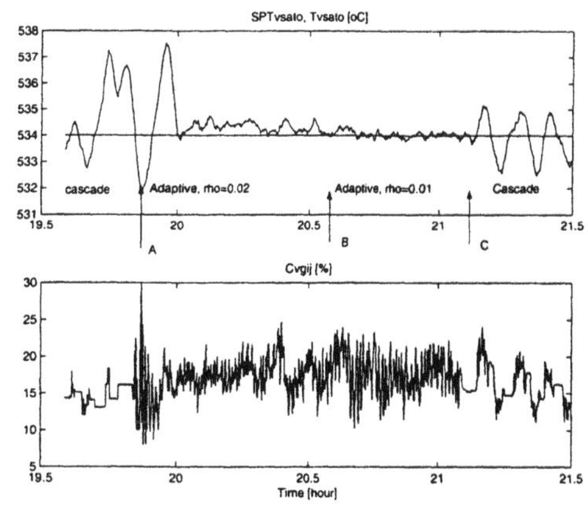

Figure 4: (f5) Steam temperature: Set-point tracking with different values of ρ and comparison with standard cascade control (temperature and manipulated variable).

models.

3.2 Superheated steam temperature

The second example is concerned with the steam super heating subsystem of a boiler, yielding loads between $120\,t/h$ and $150\,t/h$. More details, as well as experimental results may be found in [9]. The structure of the process is as follows: The steam coming from the boiler drum passes through the low temperature superheater and receives a spray water injection before passing through the high temperature superheater (HTSH) to the steam collector. From the collector, the steam is extracted for use, either by the turbine or by industrial users who induce frequent and unexpected load changes in their normal mode of operation. The process variable to be controlled is $y = T_{vsato}$, the steam temperature at the output of HTSH. The manipulated variable is $u = C_{vgij}$, the command of the spray water valve, which influences the spray water flow, C_{gij}. When the spray water valve opens, the flow C_{gij} increases and the steam temperature decreases.

The measure T_{vsati} of steam temperature after spray water injection and before the HTSH is available. The action of the manipulated variable in T_{vsati} is fast. The influence of a change of T_{vsati} in T_{vsato} is much slower. This dynamics is influenced by the load imposed on the system, the super heated steam flow, C_{vsato}. Both C_{vsato} and the inflow of air to the furnace, C_{afrn}, disturb the system, their measurements being used by the controller.

For comparison purposes, a PID cascade controller with feed-forward is considered as a base line. This is the standard controller in use and has been the subject of carefully tuning by plant's staff.

Fig. 4 documents the performance of MUSMAR in comparison with the standard cascade control. The experiment starts with standard cascade control. At $19.85\,h$ (point A in Fig. 4), the adaptive controller is turned on, the initial value of ρ being 0.02. The estimation algorithm

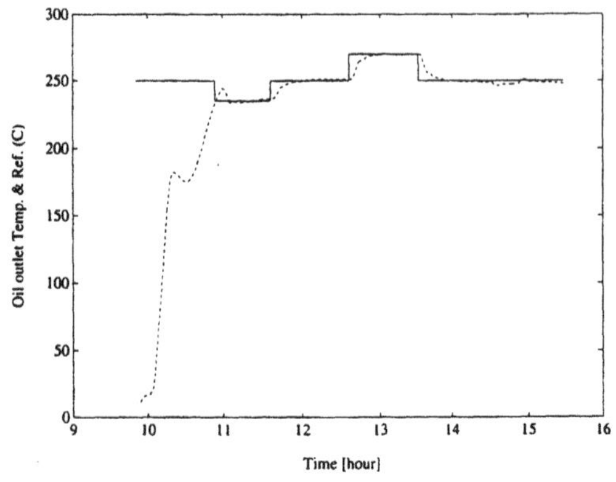

Figure 5: (f6) Solar collector field: Outlet oil temperature and reference.

is initialized with an high value of the entries of the covariance matrix. After an initial learning transient in which the trace of the covariance matrix P is reduced, the gains are adjusted so that a noticeable reduction in temperature fluctuations is observed. At $20.7\,h$ (point B in Fig. 4), ρ is reduced to 0.01. This entails tighter set-point tracking but results in increased variations of the manipulated variable. When the standard controller is brought once more to action, steam temperature fluctuations raise again.

Again this is an highly uncertain process. Its dynamics is altered unpredictably either due to the way the other plant variables are operated or due to random factors such as dust deposition affecting heat transfer coefficients. It also depends on the demand of steam flow.

3.3 Solar collector field temperature

The last example deals with outlet oil temperature control in a distributed collector solar field. The field considered is the ACUREX field of Plataforma Solar de Almeria, located in the South of Spain, and has been the subject of several studies on the application of Adaptive Control. See [2, 8] for a review, as well as further details of the application of the MUSMAR algorithm.

The oil to be heated in the field is extracted from the bottom of a storage tank and pumped through a pipe which passes at the focus of the solar collectors. The manipulated variable is oil flow and the process variable to control is the average of temperatures measured at the outlet of ten parallel collector loops. This is also a distributed parameter system whose parameters (*e. g.* mirror reflectivity) may change in an unpredictably way. It is also a nonlinear system.

Fig. 5 shows the time evolution of the process variable as well as of its reference, since start-up in a day with no clouds. Up to $11\,h$ the manipulated variable was selected manually, MUSMAR being turned on at this time.

4 Entropy interpretation

Consider the system described by

$$x(k+1) = Ax(k) + Bu(k) + w(k) \tag{6}$$

$$k = t,\ t+1,\ \ldots \quad x(t) \quad given$$

where k stands for discrete time, u is the manipulated variable, $x \epsilon \Omega_x$ is a state, A and B are real matrices and w is a random white process. Associate with this system the cost

$$J(U_t) = \sum_{k=t+1}^{t+T} R((x_k, u_k - 1)$$

where

$$U_t = [u_{t+T-1} \ \cdots \ u_t]'$$

is a sequence of control values defined over the optimization horizon of length T. Denote by U^* the sequence minimizing $J(U)$, for each t.

For applying the maximum entropy principle (MEP) in order to find U^* define, over the set Ω of admissible controls, a probability density function (pdf) $p(U)$, which expresses the uncertainty of selecting the proper U. Define the entropy of this pdf by

$$H(U, p) = -\int_{\Omega_x} p(U(x))dx \tag{7}$$

According to the MEP, $p(U)$ is selected such as to maximize $H(U, p)$ subject to the constraints

$$\int_{\Omega_x} p(U)dx = 1 \quad and \quad E[J(U)] = M \tag{8}$$

with $M = J(U^*)$ when $U = U^*$. The following theorem is shown in [7]:

Theorem: A necessary and suficient condition for U^* to minimize J subject to (6) is that U^* minimizes the entropy $H(U, p)$, where $p(u)$ is the maximum entropy pdf.

□

In other words, the sequence U^* minimizing J and such that (6) is satisfied, is given by

$$U^* = \underset{U}{\min}\ \underset{p}{\max}\ H(U, p)$$

This theorem establishes the equivalence of the optimal feedback control problem with the minimization (with respect to the sequence of controls U) of the worst case entropy (*i. e.* the entropy corresponding to the maximum entropy density). Hereafter, $H(U)$ is to be understood as the entropy sequence U when the pdf $p(U)$ is of maximum entropy and satisfies (8).

4.1 Overparametrization and entropy

Having recalled the result which establishes the equivalence between cost minimization and (maximum) entropy minimization, the relation between performance and modeling is now considered. The discussion refers to an adaptive structure (such as MUSMAR) comprising two blocks: The *identifier*, which estimates the parameters of a set of predictive models (defined by (2) in the case of MUSMAR), and the *controller* (defined by (4,5) in the case of MUSMAR), which uses these estimates for minimising an extended horizon cost.

Assume that the control synthesis relies on two predictive models, and let Θ_1 and Θ_2 denote, respectively, the parameter vectors defining them. For instance, in the case of MUSMAR with $\rho = 0$, $\Theta_1 = [\theta_1\ \psi_1']'$ and $\Theta_2 = [\theta_2\ \psi_2']'$. As shown by a modification of the results in [7, 10], the optimal cost (entropy) may be decomposed as

$$H(U^*) = H(U^*|\Theta_1, \Theta_2) + H(\Theta_2|\Theta_1) + H(\Theta_1) - H(\Theta_1, \Theta_2|U^*) \quad (9)$$

The term on the left side is invariant with respect to the type of adaptive control algorithm used and represents the minimum cost $J(U^*) = H(U^*)$ yielded by the optimal control sequence. It is made up of two parts. The first part is the cost minimised by the adaptive controller, being given by

$$H(U^*|\Theta_1, \Theta_2) + H(\Theta_2|\Theta_1) + H(\Theta_1)$$

The term $H(U^*|\Theta_1, \Theta_2)$ represents the cost minimised by the controller, assuming the model parameters to be known. The sum

$$H(\Theta_2|\Theta_1) + H(\Theta_1)$$

is the term minimised by the parameter estimator (identifier entropy).

The second part in which $H(U^*)$ is decomposed is $-H(\Theta_1, \Theta_2|U^*)$. Its symmetric is called the *equivocation* [7]. The bigger the equivocation, the bigger the difference between the true cost and the cost "seen" by the adaptive controller. The above decomposition, due to [7] may be interpred as in fig. 6.

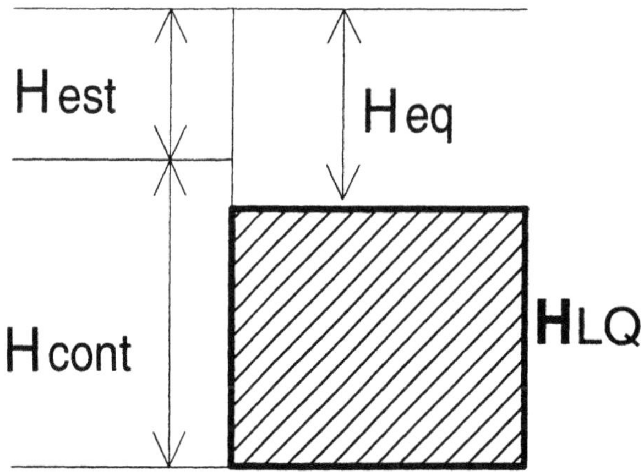

Figure 6: (f7) Relation, provided by the entropy approach, between the performances of adaptive and optimal controllers.

4.2 Extending the horizon

Using the above entropy framework, a comparison is made between the situations in which only one predictive model is used (the horizon is $T = 1$) and in which two predictive models are used ($T = 2$). For the situation in which $T = 1$, eq. (9) is replaced by

$$H(U^*) = H(U^*|\Theta_1) + H(\Theta_1) - H(\Theta_1|U^*) \quad (10)$$

Comparing (9) and (10) it is seen that, when $T = 2$, the identifier's entropy (contribution to the cost) is increased by $H(\Theta_2|\Theta_1)$. This reflects the fact that an extra set of parameters to estimate will render this task more difficult and less accurate. On the other way, assuming that the two steps ahead predictor provides additional information on system dynamics, the controller will be able to perform a better optimization with two predictors than just with one. This means that

$$H(U^*|\Theta_1, \Theta_2) < H(U^*|\Theta_1) \quad (11)$$

Therefore, when the horizon is extended, the performance yielded by the adaptive controller results from two conflicting effects:

- Increasing the horizon requires more parameters to be estimated. This degrades the identifier performance and, therefore, the performance attainable by the adaptive controller. In other words, the equivocation will be higher.
- On the other way, the controller cost will be lower, and so will be the equivocation.

These results have been established for $T = 1$ and $T = 2$, but they can be generalised to an arbitrary horizon. When the horizon increases, the second of the above aspects starts by dominating and the performance of the adaptive controller increases (*i. e.* the experimental loss decreases when the horizon increases). For longer values of the horizon, the estimator entropy increases more and more but, on the other way, the added value of additional predictors for control becomes relatively low (when a full complexity controller is used, this last effect is a consequence of the behaviour of the Riccati equation for the underlying LQ problem when the horizon increases; for reduced complexity controllers see [5]). Therefore, the overall performance will decrease. Thus, in a typical situation, the experimental loss will exhibit a U-shaped dependence on T. The eact form can only be computed from knowledge of the exact plant model.

4.3 Incorporating redundancy

Two cases for the estimation of model parameters are compared hereafter:

- The entries of Θ_2 (parameters of the two steps ahead predictor) are computed from Θ_1. This situation is denoted as NR (no redundancy);
- Both Θ_1 and Θ_2 are estimated directly from plant data. This situation is denoted R (redundancy).

The decomposition provided by eq. (9) is the basis for comparing these two situations. For the identifier entropy, it is expected that:

$$H_R(\Theta_2|\Theta_1) > H_{NR}(\Theta_2|\Theta_1) \qquad (12)$$

and, for the controller entropy (for situations in which there are mismatches between plant and its assumed structure):

$$H_R(U^*|\Theta_1, \Theta_2) < H_{NR}(U|\Theta_1, \Theta_2) \qquad (13)$$

since it is expected that the additional information brought by redundancy will yield better control decisions. Again, these factors are conflicting in what concerns their influence on the global performance. Although an exact computation can only be made on the basis of a given plant model, it is to be expected that redundancy will be benefic in the presence of unmodelled plant dynamics.

5 Conclusions

Although the original motivation for MUSMAR stemed from considerations related to Optimal Control and Estimation, this algorithm turns out to have a close relationship with Statistic Learning Theory. As shown in [5], MUSMAR provides a way of tuning a vector of controller gains such as to minimise an underlying cost, on the basis of plant measured data and with litle *a priori* concern to detailed plant structure. Actually, the experimental examples presented in this paper show that the same algorithm can be applied to different physical situations for which little *a priori* information is available.

A key aspects in this respect is overparametrization, achieved through the use of multiple predictive models and redundancy. Using the Maximum Entropy formulation of Optimal and Adaptive Control it was possible to assess, at least in general terms, the influence of choices made concerning the structure of the controller, *e. g.* the prediction horizon. It was shown that there are trade-offs which affect these choices depending on the controller and the estimator. These lead to a U-shapped dependency of global algorithm performance on the structure parameters, which is predicted by the entropy approach.

Acknowledgement

The experimental work concerning arc welding has been performed in cooperation with T. O. Santos and R. Caetano, concerning superheated steam temperature control, with P. O. Shirley and R. N. Silva and concerning oil temperature in the ACUREX field with F. Coito, R. N. Silva and L. M. Rato. Part of this work has been supported by the POSI program of the IIIrd EC Framework Program.

Appendix

Definition of Conditional entropy

Let x and y be two random variables. The conditional entropy $H(y|x)$ is given by

$$H(y|x) \triangleq -\int_{\Omega_x}[\int_{\Omega_y} p(y|x)\ln p(y|x)dy]p(x)dx$$

References

[1] Clarke, D., C. Mohtadi and P. S. Tuffs (1987). Generalized Predictive Control – Part I. The Basic Algorithm. *Automatica*, 23:137-148.

[2] Coito, F., J. M. Lemos, R. N. Silva and E. Mosca (1997). Adaptive Control of a solar energy plant: Exploiting accessible disturbances. *Int. J. Adaptive Control and Signal Proc.*, 11:327-342.

[3] Greco, C., G. Menga, E. Mosca and G. Zappa (1984). Performance Improvements of Self-tuning Controllers by Multistep Horizons: The MUSMAR Approach. *Automatica*, 20:681-699.

[4] Kulhavý, R. (1987). Restricted exponential forgetting in real-time identification, *Automatica*, 23(5):589-600.

[5] Mosca, E., G. Zappa. and J.M. Lemos (1989). Robustness of multipredictor adaptive regulators: MUSMAR *Automatica*, 25:521 - 529.

[6] Santos, T., R. Caetano, J. M. Lemos and F. Coito (2000). Multipredictive Adaptive Control of Arc Welding Trailing Centerline Temperature. *IEEE Trans. Control Syst. Techn.*, 8:159-169.

[7] Saridis, G. (1988). Entropy formulation of optimal and adaptive control. *IEEE Trans. Autom. Control*, 33:713-721.

[8] Silva, R. N., L. M. Rato, J. M. Lemos and F. Coito (1997). Cascade control of a distributed collector solar field. *J. Proc. Control*, 7:111-117.

[9] Silva, R. N., P. O. Shirley, J. M. Lemos and A. C. Gonçalves (2000). Adaptive regulation of superheated steam temperature: A case study in an industrial boiler. *Control Eng. Practice*, 8:1405-1415.

[10] Tsai, Y., F. Casiello and K. Loparo (1992). Discrete time entropy formulation of optimal and adaptive control problems. *IEEE Trans. Autom. Control*, 37:1083-1088.

[11] Vapnik, V. N. (1995). *The Nature of Statistical Learning Theory*. Springer.

ADAPTIVE CONTROL OF A TWO INPUT – TWO OUTPUT SYSTEM

Marek Kubalčík, Vladimír Bobál

Department of Control Theory, Institute of Information Technologies
Tomas Bata University
Nám. TGM 275, 762 72 Zlín, Czech Republic
Phone, Fax: +420 67 754 3217, E – mail:kubalcik@ft.utb.cz

Abstract: This paper presents the design and simulation of adaptive control for a two input-two output system together with the real-time control of a laboratory model using this design method. The synthesis is based on a polynomial approach. Decoupling, where the compensator is placed ahead of the system, suppresses the interactions between control loops. The results of the simulation and the real-time experiments are also given. *Copyright © 2001 IFAC*

Keywords: multivariable control, control algorithms, decoupling, compensators, adaptive control

1. INTRODUCTION

Many technological processes require that several variables relating to one system are controlled simultaneously. Each input may influence all system outputs. The design of a controller able to cope with such a system must be quite sophisticated. There are many different methods of controlling multivariable systems. Several of these use decentralized PID controllers (Luyben 1986), others apply single input-single-output (SISO) methods extended to cover multiple inputs (Chien *et al.* 1987). Here decoupling methods are used to transform the multivariable system into a series of independent SISO loops (Krishnawamy *et al.* 1991; Tade *et al.* 1986; Wittenmark *et al.* 1987; Skogestad and Postlethwaite 1996).

This paper is organized as follows: Section 2 presents the controlled model; Section 3 describes how feedback control without decoupling is designed; Section 4 describes two decoupling methods; Section 5 describes the system identification method; Section 6 gives the simulation results; Section 7 contains the experimental results; finally, Section 8 concludes the paper.

2. A DESCRIPTION OF A TWO INPUT – TWO OUTPUT SYSTEM

The internal structure of the the system is shown in Fig. 1

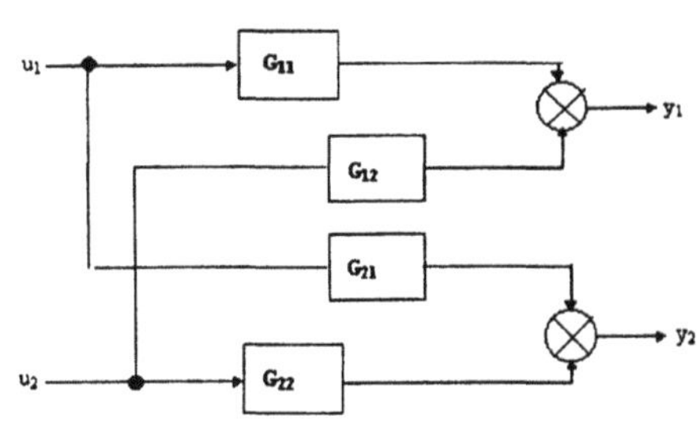

Fig. 1. A two input – two output system – the "P" structure

The transfer matrix of the system is

$$G = \begin{bmatrix} G_{11} & G_{12} \\ G_{21} & G_{22} \end{bmatrix} \tag{1}$$

It is possible to assume that the system is described by the matrix fraction

$$G\left(z^{-1}\right) = A^{-1}\left(z^{-1}\right)B\left(z^{-1}\right) = B_1\left(z^{-1}\right)A_1^{-1}\left(z^{-1}\right) \tag{2}$$

Where polynomial matrices $A \in R_{mm}[z^{-1}]$, $B \in R_{mn}[z^{-1}]$ are the left indivisible decomposition of matrix $G(z^{-1})$ and matrices $A_1 \in R_{mm}[z^{-1}]$, $B_1 \in R_{mn}[z^{-1}]$ are the right indivisible decomposition.

The matrices of the discrete model are

$$A(z^{-1}) = \begin{bmatrix} 1 + a_1 z^{-1} + a_2 z^{-2} & a_3 z^{-1} + a_4 z^{-2} \\ a_5 z^{-1} + a_6 z^{-2} & 1 + a_7 z^{-1} + a_8 z^{-2} \end{bmatrix} \quad (3)$$

$$B(z^{-1}) = \begin{bmatrix} b_1 z^{-1} + b_2 z^{-2} & b_3 z^{-1} + b_4 z^{-2} \\ b_5 z^{-1} + b_6 z^{-2} & b_7 z^{-1} + b_8 z^{-2} \end{bmatrix}$$

and the differential equations of the model are

$$\begin{aligned} y_1(k) = & -a_1 y_1(k-1) - a_2 y_1(k-2) - a_3 y_2(k-1) - \\ & - a_4 y_2(k-2) + b_1 u_1(k-1) + b_2 u_1(k-2) + \\ & + b_3 u_2(k-1) + b_4 u_2(k-2) \end{aligned} \quad (4)$$

$$\begin{aligned} y_2(k) = & -a_5 y_1(k-1) - a_6 y_1(k-2) - a_7 y_2(k-1) - \\ & - a_8 y_2(k-2) + b_5 u_1(k-1) + b_6 u_1(k-2) + \\ & + b_7 u_2(k-1) + b_8 u_2(k-2) \end{aligned}$$

3. DESIGNING FEEDBACK CONTROL

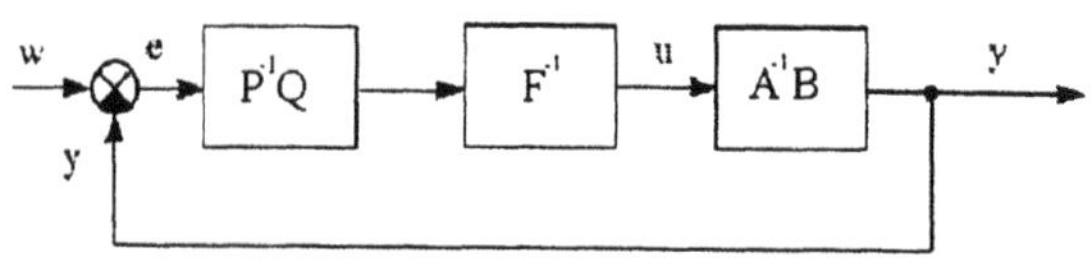

Fig. 2. Block diagram of the closed loop system

In the same way as the controlled system, the transfer matrix of the controller takes the form of matrix fraction

$$G(z^{-1}) = P^{-1}(z^{-1})Q(z^{-1}) = Q_1(z^{-1})P_1^{-1}(z^{-1}) \quad (5)$$

The matrix of an integrator for permanent zero control error is

$$F(z^{-1}) = \begin{bmatrix} 1 - z^{-1} & 0 \\ 0 & 1 - z^{-1} \end{bmatrix} \quad (6)$$

The control law apparent in the block diagram (operator z^{-1} will be omitted from some operations for the sake of simplification) has the form

$$U = F^{-1} Q_1 P_1^{-1} E \quad (7)$$

It is possible to derive the following equation for the system output

$$Y = A^{-1} B F^{-1} P^{-1} Q E = A^{-1} B F^{-1} P^{-1} Q (W - Y) \quad (8)$$

which can be modified to give

$$Y = P_1 (A F P_1 + B Q_1)^{-1} B Q_1 P_1^{-1} W \quad (9)$$

The closed loop system is stable when the following diophantine equation is satisfied

$$A F\, P_1 + B Q_1 = M \quad (10)$$

where $M(z^{-1}) \in R_{mm}[z^{-1}]$ is a stable diagonal polynomial matrix.

$$M(z^{-1}) = \begin{bmatrix} 1 + m_1 z^{-1} + m_2 z^{-2} + \\ + m_3 z^{-3} + m_4 z^{-4} & 0 \\ 0 & 1 + m_5 z^{-1} + m_6 z^{-2} + \\ & + m_7 z^{-3} + m_8 z^{-4} \end{bmatrix} \quad (11)$$

The roots of this polynomial matrix are the ruling factor in the behaviour of the closed loop system. They must be inside the unit circle if the system is to be stable.

The degree of the controller matrices polynomials depends on the internal properness of the closed loop. The structure of matrices P_1 and Q_1 was chosen so that the number of unknown controller parameters equals the number of algebraic equations resulting from the solution of the diophantine equations using the uncertain coefficients method.

$$P_1(z^{-1}) = \begin{bmatrix} 1 + p_1 z^{-1} & p_2 z^{-1} \\ p_3 z^{-1} & 1 + p_4 z^{-1} \end{bmatrix} \quad (12)$$

$$Q_1(z^{-1}) = \begin{bmatrix} q_1 + q_2 z^{-1} + q_3 z^{-2} & q_4 + q_5 z^{-1} + q_6 z^{-2} \\ q_7 + q_8 z^{-1} + q_9 z^{-2} & q_{10} + q_{11} z^{-1} + q_{12} z^{-2} \end{bmatrix}$$

The solution to the diophantine equation results in a set of sixteen algebraic equations with unknown controller parameters. The controller parameters are given by solving these equations.

4. DESIGNING DECOUPLING CONTROL USING COMPENSATORS

There are several ways to control multivariable systems with internal interactions. Some make use of decentralized PID controllers, whilst others are composed of a string of single input – single output methods.

One possibility is the serial insertion of a compensator ahead of the system (Krishnawamy *et al.* 1991; Peng 1990; Tade *et al.* 1986; Wittenmark *et al.* 1987). The aim here is to suppress of undesirable interactions between the input and output variables so that each input affects only one controlled variable.

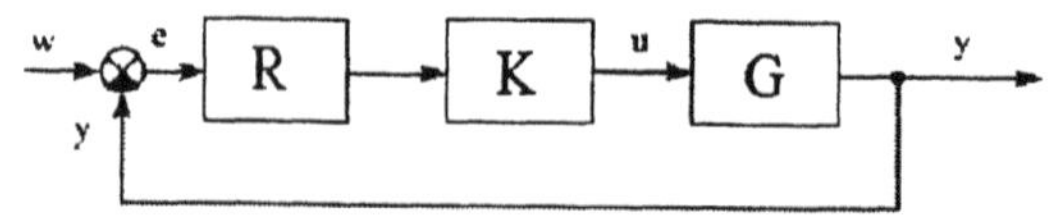

Fig. 3. A Closed loop system with compensator

The resulting transfer function H is then given by

$$H = KG \quad (13)$$

The decoupling conditions are fulfilled when matrix H is diagonal.

Several well – known compensators are given in (Krishnawamy *et al.* 1991; Peng 1990; Tade *et al.*

1986; Wittenmark *et al.* 1987). Control algorithms were derived for the model above with two compensators. These will be referred to as $\mathbf{C}_1$ and $\mathbf{C}_2$.

Compensator $\mathbf{C}_1$ is the inversion of the controlled system. Matrix $\boldsymbol{H}$ is,therefore, a unit matrix.

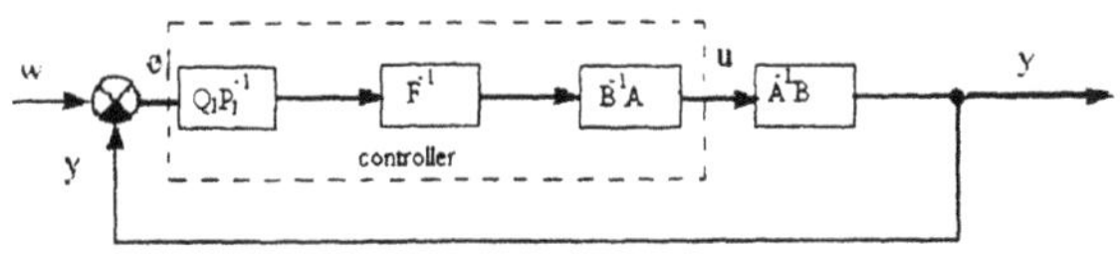

Fig. 4. The closed loop system with compensator $\mathbf{C}_1$

This block diagram leads to an equation for the system output which takes the form

$$Y = P_1(FP_1 + Q_1)^{-1} Q_1 P_1^{-1} W \tag{14}$$

The following equation must be satisfied if the closed loop system is to be stable

$$FP_1 + Q_1 = M \tag{15}$$

The structure of the polynomial matrices of the controller were chosen to suit physical demands.

$$P_1(z^{-1}) = \begin{bmatrix} 1 & 0 \\ 0 & 1 \end{bmatrix} \tag{16}$$

$$Q_1(z^{-1}) = \begin{bmatrix} q_1 z^{-1} & 0 \\ 0 & q_2 z^{-1} \end{bmatrix}$$

Consequently, matrix $\boldsymbol{M}$ was chosen to be

$$M(z^{-1}) = \begin{bmatrix} 1 + m_1 z^{-1} & 0 \\ 0 & 1 + m_2 z^{-1} \end{bmatrix} \tag{17}$$

The controller parameters are the result from the equation (15). The control law can be described by matrix equation

$$FU = B^{-1} A Q_1 P_1^{-1} E \tag{18}$$

Compensator $\mathbf{C}_2$ is adjugated matrix $\boldsymbol{B}$. When $\mathbf{C}_2$ was included in the design of the closed loop the model was simplified by considering matrix $\boldsymbol{A}$ as diagonal. The multiplication of matrix $\boldsymbol{B}$ and adjugated matrix $\boldsymbol{B}$ results in diagonal matrix $\boldsymbol{H}$. The determinants of matrix $\boldsymbol{B}$ represent the diagonal elements. When matrix $\boldsymbol{A}$ is nondiagonal, its inverted form must be placed ahead of the system in order to obtain diagonal matrix $\boldsymbol{H}$, otherwise it may increase the order of the controller and sophistication of the closed loop system. Although designed for a diagonal matrix, compensator $\mathbf{C}_2$ also improves the control process for non – diagonal matrix $\boldsymbol{A}$ in the controlled system. This is demonstrated in the simulation results.

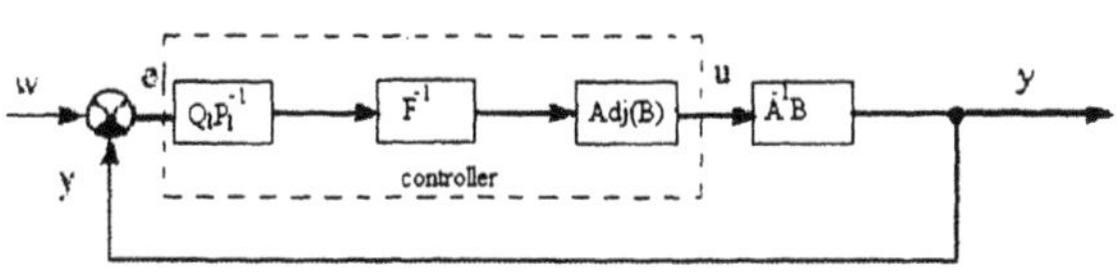

Fig. 5. The closed loop system with compensator $\mathbf{C}_2$

The equation for the system output as shown in this block diagram takes the form

$$Y = P_1(AFP_1 + B_v Q_1) B_v Q_1 P_1^{-1} W \tag{19}$$

where

$$B_v = B\,adj(B) = \begin{bmatrix} det(B) & 0 \\ 0 & det(B) \end{bmatrix} \tag{20}$$

To achieve stability in the closed loop system the following diophantine equation must be fulfilled

$$AFP_1 + B_v Q_1 = M \tag{21}$$

The controller polynomial matrices are chosen as shown below

$$P_1(z^{-1}) = \begin{bmatrix} 1 + p_1 z^{-1} & p_2 z^{-1} \\ p_3 z^{-1} & 1 + p_4 z^{-1} \end{bmatrix} \tag{22}$$

$$Q_1(z^{-1}) = \begin{bmatrix} q_1 + q_2 z^{-1} + q_3 z^{-2} & q_4 + q_5 z^{-1} + q_6 z^{-2} \\ q_7 + q_8 z^{-1} + q_9 z^{-2} & q_{10} + q_{11} z^{-1} + q_{12} z^{-2} \end{bmatrix}$$

and matrix $\boldsymbol{M}$ is

$$M(z^{-1}) = \begin{bmatrix} \begin{matrix} 1 + m_1 z^{-1} + \\ + m_2 z^{-2} + m_3 z^{-3} + \\ + m_4 z^{-4} + m_5 z^{-5} \end{matrix} & 0 \\ 0 & \begin{matrix} 1 + m_6 z^{-1} + \\ m_7 z^{-2} + m_8 z^{-3} + \\ + m_9 z^{-4} + m_{10} z^{-5} \end{matrix} \end{bmatrix} \tag{23}$$

Solving the diophantine equation defines a set of algebraic equations which we subsequently use to obtain the unknown controller parameters.

The control law is given by the block diagram

$$FU = adj(B) Q_1 P_1^{-1} E \tag{24}$$

5. IDENTIFICATION

The algorithms designed here were incorporated into an adaptive control system with recursive identification. The recursive least squares method proved effective for self-tuning controllers (Bittanti *et al.* 1990) and was used as the basis for our algorithm. For our two-variable example we considered the disintegration of identification into two independent parts.

The parameter vectors are completed as shown below:

$$\begin{aligned} \Theta_1^T(k) &= [a_1, a_2, a_3, a_4, b_1, b_2, b_3, b_4] \\ \Theta_2^T(k) &= [a_5, a_6, a_7, a_8, b_5, b_6, b_7, b_8] \end{aligned} \tag{25}$$

The data vector is

$$\begin{aligned} \phi^T(k-1) = [&-y_1(k-1), -y_1(k-2), -y_2(k-1), \\ &-y_2(k-2), u_1(k-1), u_1(k-2), u_2(k-1), u_2(k-2)] \end{aligned} \tag{26}$$

The parameter estimates are actualized using the recursive least squares method plus directional forgetting.

6. SIMULATION

Matlab + Simulink for Windows (The MathWork, Inc.) were used to create a program and diagrams to simulate and verify the algorithms. Verification by simulation was carried out on a range of systems with varying dynamics. The control of the model below is given here as our example.

$$A(z^{-1}) = \begin{bmatrix} 1+0{,}3z^{-1}+0{,}1z^{-2} & 0{,}1z^{-1}+0{,}2z^{-2} \\ 0{,}1z^{-1}+0{,}3z^{-2} & 1+0{,}3z^{-1}+0{,}1z^{-2} \end{bmatrix} \quad (27)$$

$$B(z^{-1}) = \begin{bmatrix} 0{,}1z^{-1}+0{,}4z^{-2} & 0{,}9z^{-1}+0{,}4z^{-2} \\ 0{,}6z^{-1}+0{,}2z^{-2} & 0{,}3z^{-1}+0{,}4z^{-2} \end{bmatrix}$$

Fig. 6 shows the system's step response

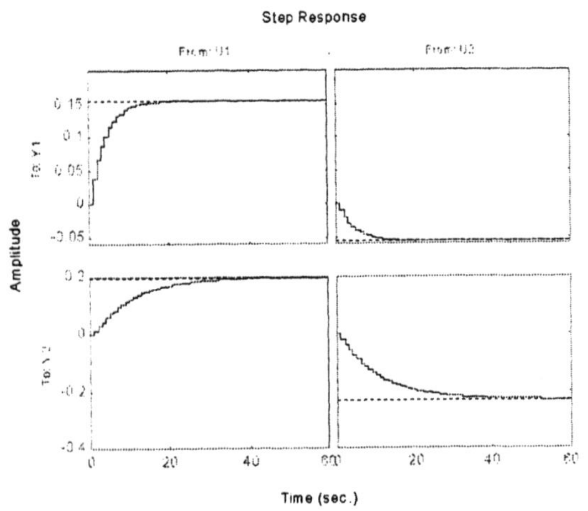

Fig. 6. The step response of the system

The right side control matrices are denoted as follows: without compensator - M_1, with compensator C_1 - M_2, and with compensator C_2 - M_3.

$$M_1(z^{-1}) = \begin{bmatrix} (1-0{,}1z^{-1})^4 & 0 \\ 0 & (1-0{,}1z^{-1})^4 \end{bmatrix} \quad (28)$$

$$M_2(z^{-1}) = \begin{bmatrix} 1-0{,}1z^{-1} & 0 \\ 0 & 1-0{,}1z^{-1} \end{bmatrix}$$

$$M_3(z^{-1}) = \begin{bmatrix} (1-0{,}1z^{-1})^5 & 0 \\ 0 & (1-0{,}1z^{-1})^5 \end{bmatrix}$$

The same initial conditions for system identification were used for all the types of adaptive control we tested. The initial parameter estimates were chosen to be

$$\Theta_1^T(0) = [0.1, 0.2, 0.3, 0.4, 0.1, 0.2, 0.3, 0.4] \quad (29)$$

$$\Theta_2^T(0) = [0.5, 0.6, 0.7, 0.8, 0.5, 0.6, 0.7, 0.8]$$

The results of simulation are shown in Figs 7 – 12 .

It is possible to draw several conclusions from the simulation results of the experiments on linear static systems. The basic requirement to ensure permanent zero control error was satisfied in all cases. The criteria on which we judge the quality of the control process are the overshoot on the controlled values and the speed with which zero control error is achieved. According to these criteria the controller incorporating compensator C_1 performed the best. However, this controller appears to be unsuited to adaptive control due to the size of the overshoot and the large numbers of process and controller outputs. The controller which uses compensator C_2 seems to work best in adaptive control. With regards to decoupling, it is clear that controllers with compensators greatly reduce interaction.

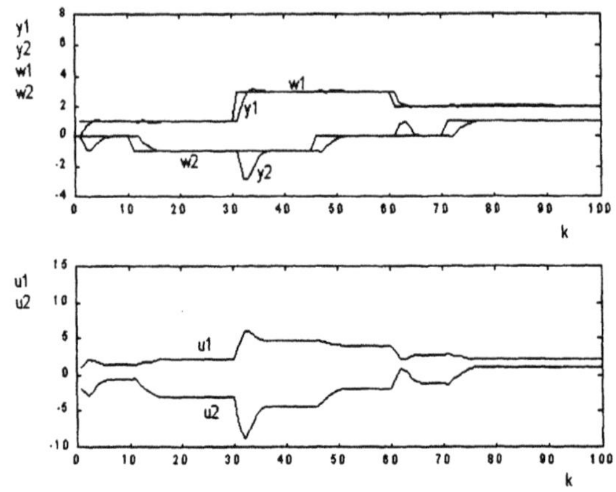

Fig. 7. Deterministic control without compensator

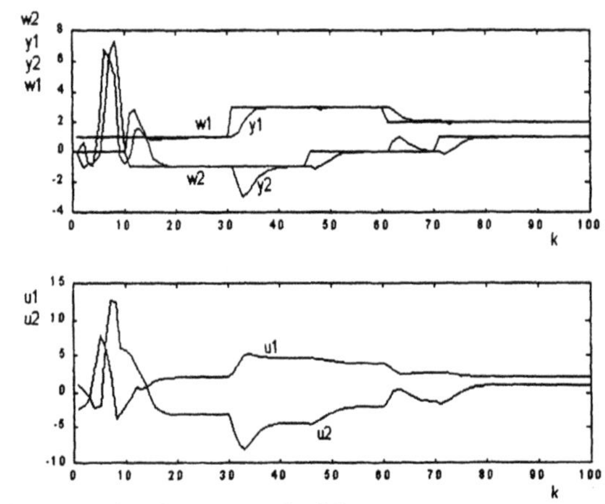

Fig. 8. Adaptive control without a compensator

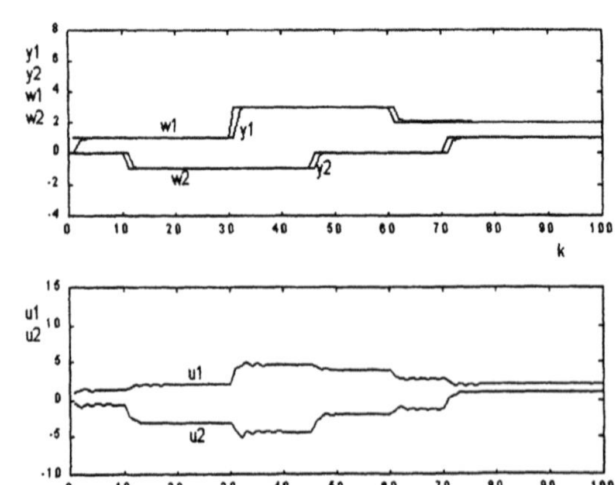

Fig. 9. Deterministic control with compensator C_1

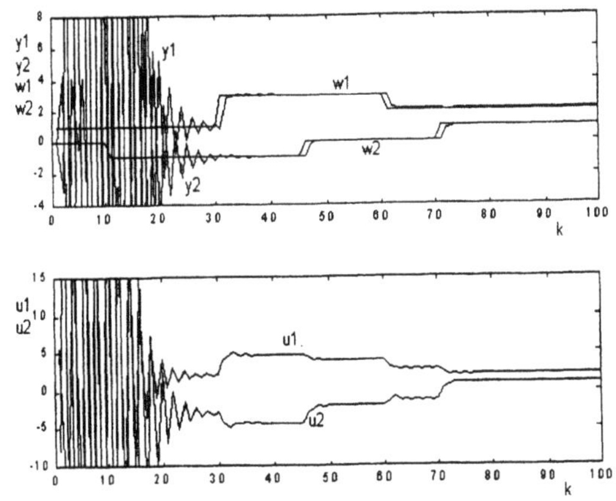

Fig. 10. Adaptive control with compensator $\mathbf{C_1}$

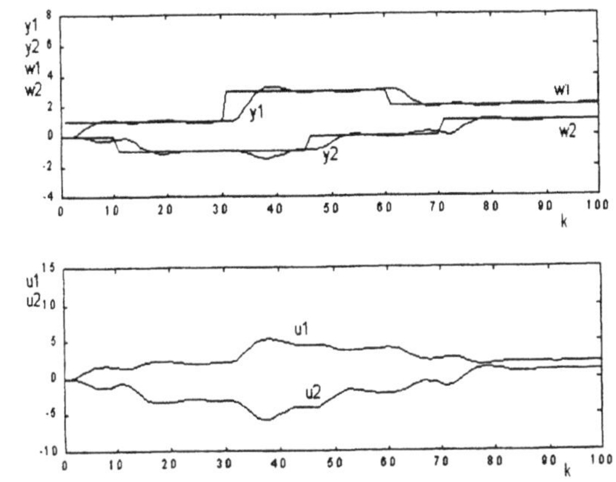

Fig. 11. Deterministic control with compensator $\mathbf{C_2}$

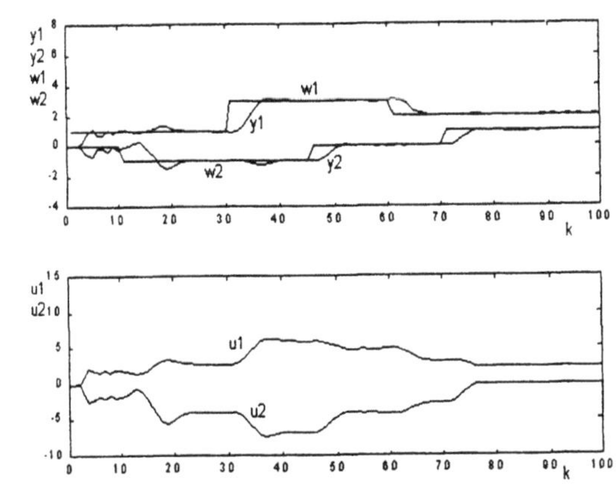

Fig. 12. Adaptive control with compensator $\mathbf{C_2}$

7. VERIFICATION – CONTROLLING A LABORATORY MODEL

Our department has experimental laboratory model CE 108 - couples drives apparatus. This apparatus, based on experience with authentic industrial control applications, was developed in cooperation with the University of Manchester and made by a British company, TecQuipment Ltd. It allows us to investigate the ever-present difficulty of controlling the tension and speed of material in a continuous process. The process may require the material speed and tension to be controlled to within defined limits. Examples of this occur in the paper-making industry, strip metal and wire manufacture and, indeed, any process where the product is manufactured in a continuous strip.

The industrial type material strip is replaced by a continuous flexible belt. The principle scheme of the model is shown in the Fig. (13). It consists of three pulleys, mounted on a vertical panel so that they form a triangle resting on its base. The two base pulleys are directly mounted on the shafts of two nominally identical servo motors and the apparatus is controlled by manipulating the drive torques to these servo motors. The third pulley, the jockey, is free to rotate and is mounted on a pivoted arm. The jockey pulley assembly, which simulates a material work station, is equipped with a special sensor and tension measuring equipment. It is the jockey pulley speed and tension which form the principle system outputs. The belt tension is measured indirectly by monitoring the angular deflection of the pivoted tension arm to which the jockey pulley is attached.

The manipulated variables are the inputs to the servo motors and the controlled variables are the tension and speed at the work station. There are interactions between the control loops.

The task was to apply the methods we designed for the adaptive control of a model representing a non-linear system with variable parameters which is, therefore, impossible to control deterministically.

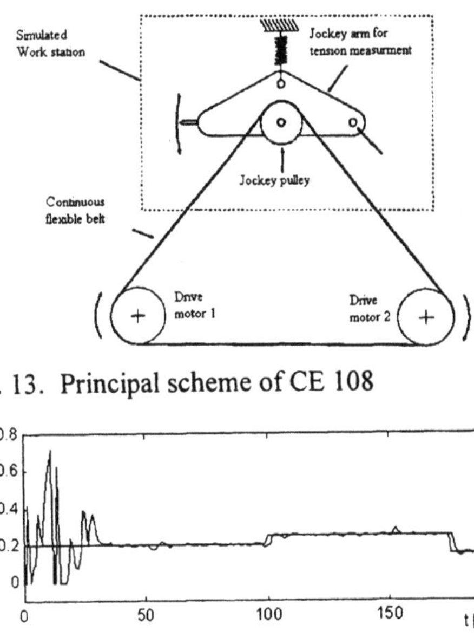

Fig. 13. Principal scheme of CE 108

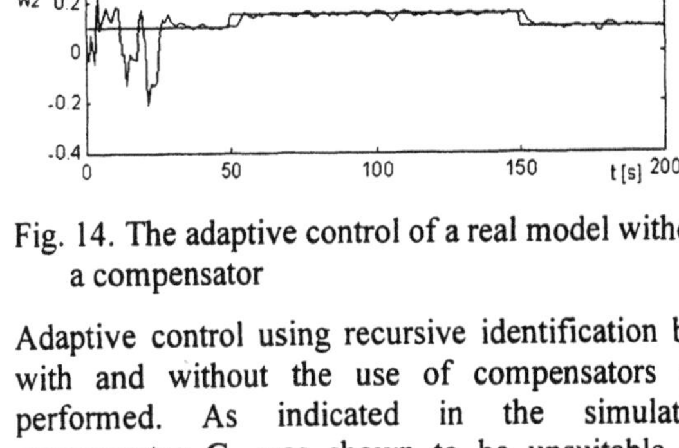

Fig. 14. The adaptive control of a real model without a compensator

Adaptive control using recursive identification both with and without the use of compensators was performed. As indicated in the simulation, compensator $\mathbf{C_1}$ was shown to be unsuitable and control broke down. The other two methods gave satisfactory results. The time responses of the control for both cases are shown in Fig. 14, Fig. 15, Fig. 16 and Fig. 17. The figures demonstrate that control with a compensator reduces interaction. The controlled variable y_1 is the speed and the controlled variable y_2 is the tension.

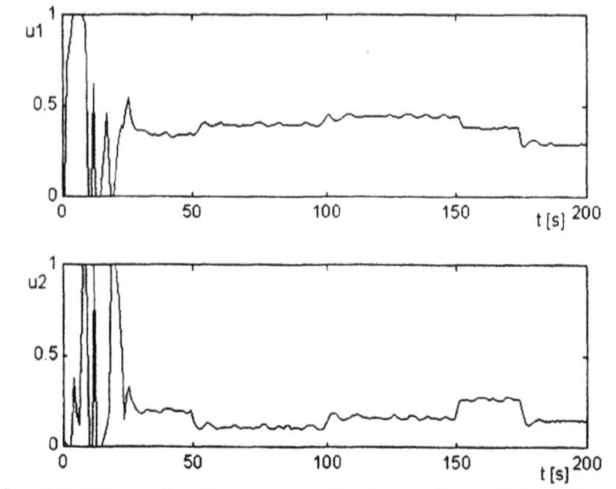

Fig. 15. The adaptive control of a real model without a compensator – controller output

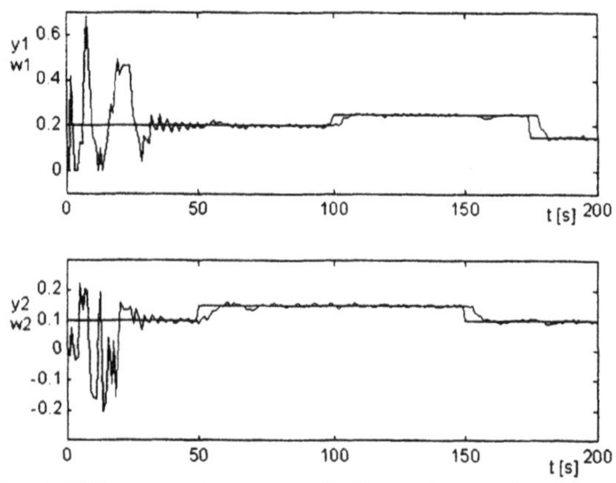

Fig. 16 The adaptive control of a real model using compensator $\mathbf{C}_2$

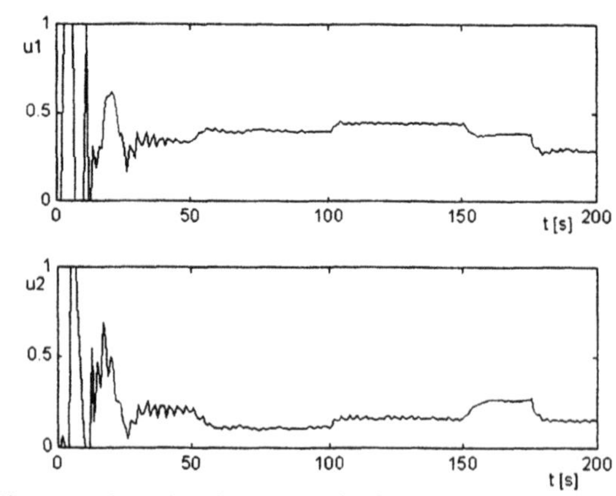

Fig. 17. The adaptive control of a real model using compensator $\mathbf{C}_2$– controller output

8. CONCLUSIONS

The adaptive control of a two-variable system based on polynomial theory was designed. Decoupling problems were solved by the use of compensators. The designs were simulated and used to control a laboratory model. The simulation results proved that these methods are suitable for the control of linear systems. The control tests on the laboratory model gave satisfactory results despite the fact that the non-linear dynamics were described by a linear model.

ACKNOWLEDGMENTS

The work has been supported by the Grant Agency of the Czech Republic under grants No. 102/99/1292 and 102/01/P013 and by the Ministry of Education of the Czech Republic under grant No. MSM 281100001. This support is very gratefully acknowledged.

REFERENCES

Bittanti, S., Bolzern, P., Campi, M. (1990). Convergence and Exponential Convergence of Identification Algorithms with Directional Forgetting Factor. *Automatica*, 26, 5, 929 – 932.

Chien, I.L., Seborg, D.E., Mellichamp, D. A. (1987). Self-Tuning Control with Decoupling. *AIChE* J., 33, 7, 1079 – 1088.

Krishnawamy, P.R. et al. (1991). Reference System Decoupling for Multivariable Control. *Ind. Eng. Chem. Res.*, 30, 662-670.

Luyben W. L. (1986). Simple Method for Tuning SISO Controllers in Multivariable Systems. *Ing. Eng. Chem. Process Des. Dev.*, 25, 654 – 660.

Peng, Y. (1990). A General Decoupling Precompensator for Linear Multivariable Systems with Application to Adaptive Control. *IEEE Trans. Aut. Control*, AC-35, 3, 344-348.

Skogestad, S., Postlethwaite, J. (1996) *Multivariable Fedback Control - Analysis and Design.* New York, J. Willey.

Tade, M.O., Bayoumi, M.M., Bacon, D.W. (1986). Adaptive Decoupling of a Class of Multivariable Dynamic Systems Using Output Feedback. *IEE Proc. Pt.D*, 133, 6, 265-275.

Wittenmark, B., Middelton, R. H., Goodwin, G.C. (1987). Adaptive Decoupling of Multivariable Systems. *Int. J. Control*, 46, 6, 1993-2009.

ADAPTIVE CONTROL USING CONTROLLERS OF RESTRICTED STRUCTURE

M.J. Grimble and P. Martin

Industrial Control Centre, EEE Department, University Of Strathclyde, Glasgow, Scotland

Abstract: A novel approach to adaptive control is presented, based on a multiple-model LQG optimal control problem. The control law calculation is for a restricted structure controller, in this case a modified-PID form. The restricted structure algorithm given enables a simplified cost-function to be optimised numerically. The $n+1$ models are obtained from a non-linear system description linearised at n appropriate operating points, plus one on-line linear model identified with recursive least squares, thus providing the adaptation. The main advantage of the approach is the cautious nature of the solution. Each new update of the controller involves averaging the cost across fixed and present models, providing cautious robust adaptive control action. The method is applied to a non-linear dynamic ship-positioning simulation, and results are given. *Copyright © 2001 IFAC*

Keywords: Adaptive control, Multiple models, Optimisation, Polynomial methods

1. INTRODUCTION

An original adaptive control technique is suggested, which utilises low-order optimal controllers of a pre-specified structure and multiple-model descriptions of a non-linear system. The algorithm depends upon recent advances in restricted structure optimal control, Grimble (1999a), and in control of multiple-model systems, Grimble (1999b), first combined in Grimble (2000a).

The strategy followed is to minimize an LQG criterion over $n+1$ separate linear models in such a way that the controller is of the desired structure and is causal. A simple analytic solution cannot be obtained, as in the case where the controller structure is unconstrained. However, a relatively straightforward direct optimisation problem can be established which provides the desired solution.

A non-linear system can often be represented by a set of linear models, or these models may arise due to uncertainty in system description. An adaptive controller is constructed by simply estimating a new plant model at each sample time, and adding this to the set of pre-determined linearised models at a time interval greater than or equal to the sampling period. Each new model estimate replaces the previous one, so avoiding any dependence on earlier identified models.

The method has some relation to self-tuning control, Astrom and Wittenmark (1973), since the control laws and plant model estimates are separate computations. The contrast with traditional self-tuning control is clear, however, as in existing self-tuning controllers the present control law depends upon the current plant model estimates only. In the new approach the current controller depends both upon the initial assumed plant models and the subsequent model estimates, providing a greater degree of robustness.

In this paper, a non-linear dynamic ship positioning simulation is utilised to examine the success of the proposed technique.

2. SYSTEM MODEL

The system shown in Fig. 1 is assumed to be linear, discrete-time, and single-input single-output. The external white noise sources drive colouring filters that represent the reference, $W_r(z^{-1})$, and disturbance subsystems, $W_d(z^{-1})$. The system equations become:

Output: $y(t) = d(t) + Wu(t)$ (1)

Input disturbance: $d(t) = W_d \xi(t)$ (2)

Reference: $r(t) = W_r \zeta(t)$ (3)

Tracking error: $e(t) = r(t) - y(t)$ (4)

Control signal: $u(t) = C_0(r(t) - y(t))$ (5)

The system transfer functions are all assumed to be functions of the delay operator z^{-1} in the time-domain, or the inverse of the z-transform complex number in the complex frequency domain. For notational simplicity, the z^{-1} arguments in $W(z^{-1})$, and the disturbance and reference models, are omitted.

2.1. *Assumptions.*

(i) The white noise sources, ξ *and* ζ, are zero mean and mutually statistically independent. The covariances for these signals are without loss of generality taken to be unity.

(ii) The plant, W, is free of unstable hidden modes and the reference, W_r, and disturbance, W_d, subsystems are asymptotically stable.

2.2. Signals and Sensitivity.

The following expressions may easily be derived for the output, error and control signals:

$$y(t) = WC_0(1 + WC_0)^{-1} r(t) + (1 + WC_0)^{-1} d(t) \quad (6)$$

$$e(t) = r(t) - y(t) \quad (7)$$

$$e(t) = r(t) - WC_0(1+WC_0)^{-1} r(t) - (1+WC_0)^{-1} d(t) \quad (8)$$

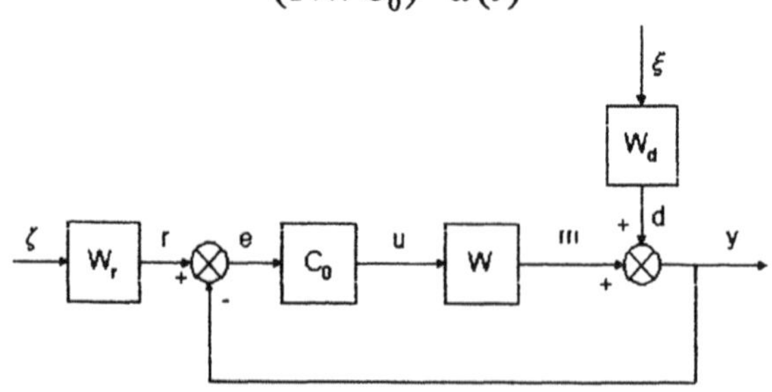

Fig.1. Closed-Loop System

$$e(t) = r(t) - WC_0(1+WC_0)^{-1} r(t) - (1+WC_0)^{-1} d(t) \quad (8)$$

$$u(t) = C_0(1+WC_0)^{-1}(r(t) - d(t)) \quad (9)$$

These equations include the following sensitivity operators:

Sensitivity: $S = (1 + WC_0)^{-1}$ (10)

Complementary sensitivity: $T = 1 - S = WC_0 S$ (11)

Control sensitivity: $M = C_0 S = C_0(1+WC_0)^{-1}$ (12)

2.3. Polynomial system description

The system shown in Fig. 1 may be represented in polynomial form, where the system transfer-functions are written as:

Plant: $W = A^{-1}B$ (13)

Reference generator: $W_r = A^{-1}E$ (14)

Input disturbance: $W_d = A^{-1}C_d$ (15)

The various polynomials are not necessarily coprime but the system elements are assumed to be free of unstable hidden modes.

3. LQG CRITERION AND RESTRICTED STRUCTURE CONTROLLER

The LQG cost-function for a single-model controller, from Grimble and Johnson (1988), is defined as:

$$J = \frac{1}{2\pi j} \oint_{|z|=1} \{Q_c \Phi_{ee} + R_c \Phi_{uu}\} \frac{dz}{z} \quad (18)$$

where Q_c, R_c represent dynamic weighting elements, acting on the spectra of the error $e(t)$ and feedback control $u(t)$ signals. The R_c weighting term is assumed to be positive definite and Q_c is assumed to be positive-semidefinite on the unit-circle of the z-plane. The weightings can be chosen as in Grimble (1994) and can be written in polynomial form as:

$$Q_c = Q_n / (A_q^* A_q) \text{ and } R_c = R_n / (A_r^* A_r) \quad (19)$$

where the adjoint of an operator $A(z^{-1})$ is denoted by $A^*(z^{-1})$ or simply A^*.

In the multiple model case, let J_j denote the value of (18) for the *j*th system model, and let the probability of occurrence of this event be denoted by

$p_j=\Pr(J=J_j)$. Then the total cost index to be minimised can be written as:

$$\bar{J}=\sum_{j=1}^{n+1} p_j J_j \qquad (20)$$

The overall problem will then be to minimize the above criterion with the controller chosen to have a specified structure:

Modified PID:

$C_0(z^{-1})=k_P+k_I/(1-z^{-1})+k_D(1-z^{-1})/(1-\alpha z^{-1})$

The optimisation must be performed subject to the parameters which are not fixed, i.e. the three gains k_P, k_I, and k_D in this case.

The assumption must be made that a stabilising control law exists for the assumed controller structure. The optimal solution is also required to be causal. The solution to this problem may be obtained using the results of the following theorem.

Theorem 3.1: Restricted Structure Single-Degree of Freedom LQG Problem for Single-Model Case

Consider the LQG criterion in (18), where the weightings may be expressed as the ratio of polynomials in (19). The LQG controller, of restricted structure, to minimize the cost function (18), for the system of section 2, may be calculated from the solution of a simpler, more direct, optimisation problem.

Compute the strictly Schur spectral factors D_f, D_c using:

$$D_f D_f^* = EE^* + C_d C_d^* \qquad (21)$$

$$D_c^* D_c = B^* A_r^* Q_n A_r B + A^* A_q^* R_n A_q A \qquad (22)$$

Diophantine equations:

Calculate (G_0, H_0, F_0), with F_0 of smallest degree:

$$D_c^* G_0 z^{-g} + F_0 A A_q = B^* A_r^* Q_n D_f z^{-g} \qquad (23)$$

$$D_c^* H_0 z^{-g} - F_0 B A_r = A^* A_q^* R_n D_f z^{-g} \qquad (24)$$

The optimal controller $C_0 = C_{0n} C_{0d}^{-1}$ must be chosen to minimize the causal component in the cost-function term:

$$J_{csl}=\frac{1}{2\pi j}\oint_{|z|=1} \{T_1^+ T_1^{+*}\} dz/z$$

where

$$T_1^+ = \frac{C_{0n} H_0 A_q - C_{0d} G_0 A_r}{A_w (A C_{0d} + B C_{0n})} \qquad (25)$$

For the unconstrained optimal solution, Grimble (2001), the minimum is achieved when $T_1^+ = 0$ and the minimum of J_{csl} (denoted J_{cslmin}) is zero. If the controller has a specified limited structure, the minimum of the cost term J_{cslmin} will be non-zero.

4. PARAMETRIC OPTIMISATION PROBLEM

The computation of the optimal feedback controller C_0 for a single model reduces to minimization of the term J_{csl}:

$$J_{csl} = \frac{1}{2\pi j}\oint_{|z|=1} T_1^+ T_1^{+*} \frac{dz}{z}$$

$$= \frac{T}{2\pi}\int_0^{2\pi/T} T_1^+(e^{-j\omega T}) T_1^+(e^{j\omega T}) d\omega$$

where T denotes the sampling period and $\theta=\omega T$.

It is clear from (25) that T_l^+ can be written in the form:

$$T_1^+=(C_{0n}L_1-C_{0d}L_2)/(C_{0n}L_3+C_{0d}L_4) \qquad (26)$$

where $C_0 = C_{0n} / C_{0d}$ has a specified structure.

Assume that C_0 has a modified PID structure so that

$$C_{0n} = k_P(1-z^{-1})(1-\alpha z^{-1}) + k_I(1-\alpha z^{-1}) + k_D(1-z^{-1})^2 \qquad (27)$$

and

$$C_{0d} = (1-z^{-1})(1-\alpha z^{-1}) \qquad (28)$$

The numerator includes the unknown gains to be optimised and the denominator is known in this case.

Let the superscripts r and i denote the real and imaginary parts of a complex function, so that:

$$C_{0n}=C_{0n}^r+jC_{0n}^i \text{ and } C_{0d}=C_{0d}^r+jC_{0d}^i \qquad (29)$$

The numerator term may be split into complex frequency dependent components, through comparison with (27):

$$C_{0n} = k_P\alpha_0(z^{-1}) + k_I\alpha_1(z^{-1}) + k_D\alpha_2(z^{-1}) = C_{0n}^r + jC_{0n}^i$$

where

$$\alpha_0 = (1-z^{-1})(1-\alpha z^{-1}),\quad \alpha_1 = (1-\alpha z^{-1}),\quad \alpha_2 = (1-z^{-1})^2 \qquad (30)$$

and hence,

$$C_{0n}^r = k_P\alpha_0^r + k_I\alpha_1^r + k_D\alpha_2^r$$

and

$$C_{0n}^i = k_P\alpha_0^i + k_I\alpha_1^i + k_D\alpha_2^i \qquad (31)$$

In deriving the following expressions, account is taken of the fact that the solution of the optimisation problem is to be found by iteration. See Luenberger

(1969). The denominator term in T_1^+ will therefore be assumed known, and the minimisation will be performed on the numerator terms. Thus let,

$$T_1^+ = C_{0n} L_{n1} - C_{0d} L_{n2}$$

where L_{n1} and L_{n2} are defined as:

$$L_{n1} = L_1 / (C_{0n} L_3 + C_{0d} L_4)$$

and

$$L_{n2} = L_2 / (C_{0n} L_3 + C_{0d} L_4) \qquad (32)$$

Substituting from (29)

$$\begin{aligned} T_1^+ &= (C_{0n}^r + jC_{0n}^i)(L_{n1}^r + jL_{n1}^i) \\ &\quad -(C_{0d}^r + jC_{0d}^i)(L_{n2}^r + jL_{n2}^i) \\ &= C_{0n}^r L_{n1}^r - C_{0n}^i L_{n1}^i - C_{0d}^r L_{n2}^r + C_{0d}^i L_{n2}^i \\ &\quad + j(C_{0n}^i L_{n1}^r + C_{0n}^r L_{n1}^i - C_{0d}^r L_{n2}^i - C_{0d}^i L_{n2}^r) \end{aligned}$$

and after substitution from (31), obtain:

$$\begin{aligned} T_1^+ &= k_P\big((\alpha_0^r L_{n1}^r - \alpha_0^i L_{n1}^i) + j(\alpha_0^i L_{n1}^r + \alpha_0^r L_{n1}^i)\big) \\ &\quad + k_I\big((\alpha_1^r L_{n1}^r - \alpha_1^i L_{n1}^i) + j(\alpha_1^i L_{n1}^r + \alpha_1^r L_{n1}^i)\big) \\ &\quad + k_D\big((\alpha_2^r L_{n1}^r - \alpha_2^i L_{n1}^i) + j(\alpha_2^i L_{n1}^r + \alpha_2^r L_{n1}^i)\big) \\ &\quad -\big(C_{0d}^r L_{n2}^r - C_{0d}^i L_{n2}^i\big) - j\big(C_{0d}^r L_{n2}^i + C_{0d}^i L_{n2}^r\big) \end{aligned}$$

The real and imaginary part of T_1^+ may therefore be written as $T_1^+ = T_1^{+r} + jT_1^{+i}$, and it follows that:

$$|T_1^+|^2 = T_1^+(e^{-j\omega T}) T_1^+(e^{j\omega T}) = (T_1^{+r})^2 + (T_1^{+i})^2$$

A vector form of the above equations that will enable the optimisation to be performed for the single-model case, with respect to the unknown gains k_P, k_I, k_D is:

$$\begin{bmatrix} T_1^{+r} \\ T_1^{+i} \end{bmatrix} = F \begin{bmatrix} k_P \\ k_I \\ k_D \end{bmatrix} - L = Fx - L$$

where

$$F = \begin{bmatrix} (\alpha_0^r L_{n1}^r - \alpha_0^i L_{n1}^i) & (\alpha_0^i L_{n1}^r + \alpha_0^r L_{n1}^i) \\ (\alpha_1^r L_{n1}^r - \alpha_1^i L_{n1}^i) & (\alpha_1^i L_{n1}^r + \alpha_1^r L_{n1}^i) \\ (\alpha_2^r L_{n1}^r - \alpha_2^i L_{n1}^i) & (\alpha_2^i L_{n1}^r + \alpha_2^r L_{n1}^i) \end{bmatrix}^T$$

and

$$L = \begin{bmatrix} C_{0d}^r L_{n2}^r - C_{0d}^i L_{n2}^i \\ C_{0d}^r L_{n2}^i + C_{0d}^i L_{n2}^r \end{bmatrix}$$

The cost-function can be optimised directly, but a simple iterative solution can be obtained if the integral is approximated by a summation with a sufficient number of frequency points $\{\omega_1, \omega_2, ..., \omega_k, ..., \omega_N\}$. The optimisation can then be performed by minimising the sum of squares at each of the frequency points as used by Yukitomo *et al.* (1998). The maximum frequency can be taken as the foldover frequency. The minimization of the cost term J_0 is then required, where

$$\begin{aligned} J_0 &= \sum_{k=1}^{N} (Fx-L)^T (Fx-L) \\ &= (b-Ax)^T (b-Ax) \end{aligned} \qquad (33)$$

and

$$A = \begin{bmatrix} F(\omega_1 T) \\ \vdots \\ F(\omega_N T) \end{bmatrix}, b = \begin{bmatrix} L(\omega_1 T) \\ \vdots \\ L(\omega_N T) \end{bmatrix}, x = \begin{bmatrix} k_P \\ k_I \\ k_D \end{bmatrix} \qquad (34)$$

The multiple-model solution can therefore be found by minimising the weighted sum of the costs of each individual system model, J_{0j}, as below:

$$\begin{aligned} \overline{J_0} &= \sum_{j=1}^{n+1} p_j J_{0j} \\ &= \sum_{j=1}^{n+1} p_j (b_j - A_j x)^T (b_j - A_j x) \\ &= (\underline{b} - \underline{A}x)^T \underline{P} (\underline{b} - \underline{A}x) \end{aligned} \qquad (35)$$

where

$$\underline{b} = \begin{bmatrix} b_1 \\ b_2 \\ \vdots \\ b_n \end{bmatrix}, \quad \underline{A} = \begin{bmatrix} A_1 \\ A_2 \\ \vdots \\ A_n \end{bmatrix} \text{ and } \underline{P} = diag\{p_1, \cdots, p_n\}$$

Assuming the matrix $\underline{A}^T \underline{P} \underline{A}$ is not singular the least squares optimal solution, Noble (1969), follows as:

$$x = (\underline{A}^T \underline{P} \underline{A})^{-1} \underline{A}^T \underline{P} \underline{b} \qquad (36)$$

Since in practice, the denominator of the term T_1^+ is a function of these gains, iteration is required to successively perform the gain calculation and the calculation of this denominator term. The convexity of the solution is easy to establish when simple parameterisations are employed for this linear system, quadratic cost problem.

4.1 Iterative Solution

By adding an identification algorithm, the following successive approximation algorithm, Luenberger (1969), can be used to compute the restricted structure LQG adaptive controller as described in Grimble (2000b)

5. ADAPTIVE DYNAMIC SHIP POSITIONING USING MULTIPLE MODELS

The example to be dealt with in this paper is one dimension of a 3DOF non-linear ship model used for dynamic ship positioning. Four second-order models are used to represent the ship at various operating points, where $j \in \{1,2,3,4\}$ is the model index for the transfer functions and polynomials given in equations (13) - (15), restated below:

$$W_j = \frac{b_{0_j} z^{-1} + b_{1_j} z^{-2}}{1 + a_{0_j} z^{-1} + a_{1_j} z^{-2}} \qquad (37)$$

The reference transfer function is:

$$W_{r_j} = \frac{0.1}{1 + a_{0_j} z^{-1} + a_{1_j} z^{-2}} \qquad (38)$$

The disturbance transfer function is:

$$W_{d_j} = \frac{10}{1 + a_{0_j} z^{-1} + a_{1_j} z^{-2}} \qquad (39)$$

The error weighting in the LQG cost function is:

$$Q_c = \frac{10}{(z-1)(z^{-1}-1)} \qquad (40)$$

The control weighting in the LQG cost function is:

$$R_c = \frac{10^{-4}(z-1)(z^{-1}-1)}{(z-0.5)(z^{-1}-0.5)} \qquad (41)$$

In equations (37)-(39), the B and A polynomials are time-invariant for $j = \{1,2,3\}$, and vary from one sample time to the next for $j = \{4\}$. In the latter case, the parameter identification algorithm is Recursive Least Squares with a forgetting factor $\lambda = 0.95$. The sample time used is 1 second, and the adaptation occurs every 4 seconds, due to the long time constant of the ship. Figure 2 is a schematic of the overall control scheme.

The simulation was executed with a filtered step reference input to the longitudinal ship direction under adaptive control. See Figure 6. Sinusoids with 400s period were applied to the lateral and rotational degrees of freedom in order to excite the ship nonlinearities. Wind, wave, and current disturbances were also employed to give realistic operating characteristics.

To establish the properties of the control scheme under investigation, it was necessary to run the

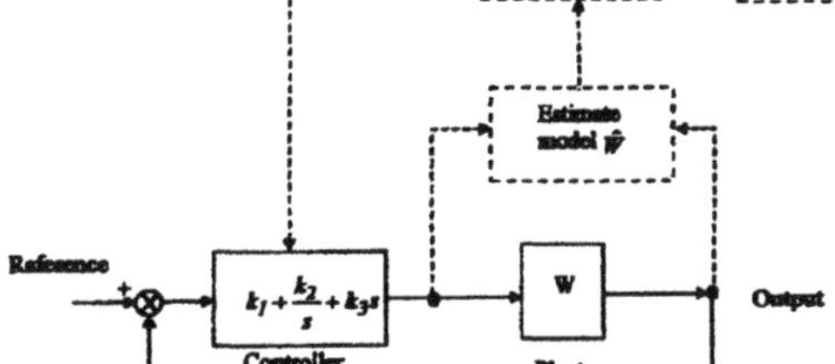

Fig.2. Adaptive Control Scheme

simulation with an assortment of weighting values for probability p_j in equation (20). Starting with zero weight on the identified model and increasing by 0.1 for each simulation run, with the remaining weight shared equally among the fixed models, it was possible to see the effect of the adaptation. ..,

By evaluating equation (18) in the discrete time domain for each probability weighting, the graph in Figure 3 was produced, depicting output error cost versus identified-model weighting. Figure 3 shows that error variance is highest for the pure self-tuning solution ($p_4 = 1$), and lowest for the fixed+adaptive case with weighting of 0.5.

To further illustrate the controller's performance, note the 400s near-periodicity of Figures 4 and 5, as the system and controller parameters are able to follow the ship non-linearity previously mentioned.

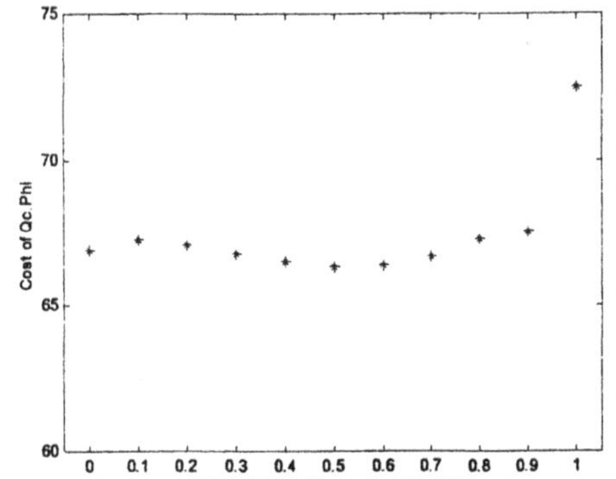

Fig.3. Cost versus Identified Model Weight

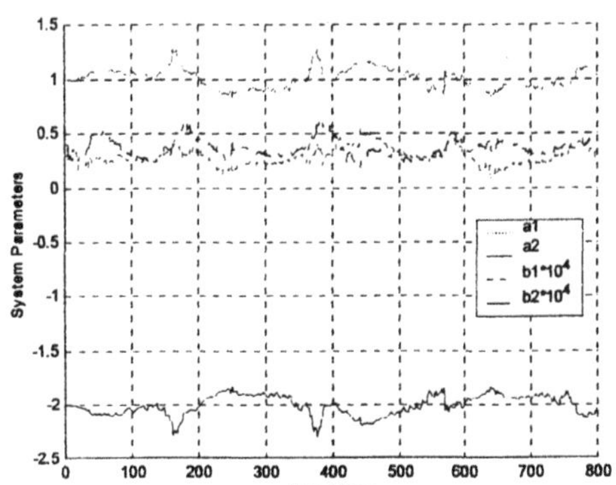

Fig.4. System Parameters

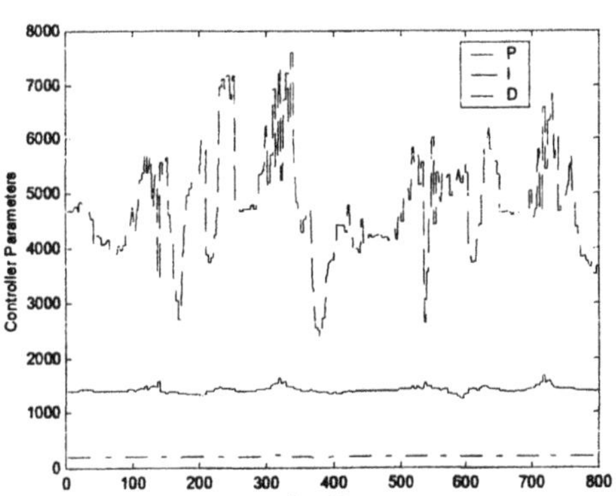

Fig.5. Control Parameters

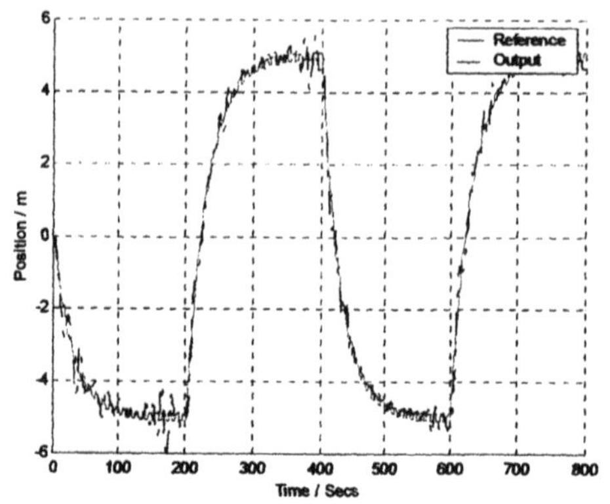

Fig.6. Longitudinal Ship Position

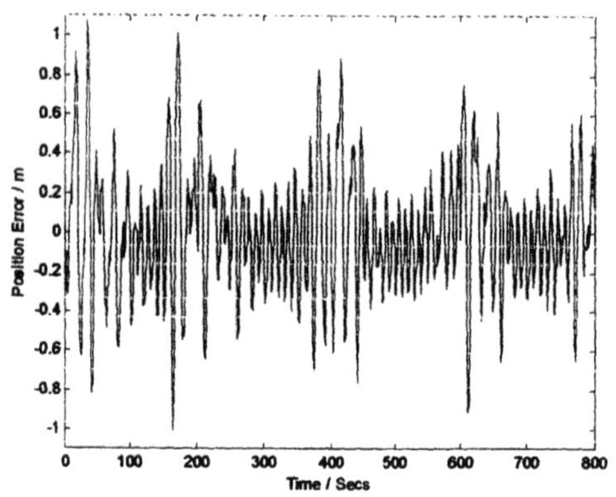

Fig.7. Longitudinal Ship Position Error

Figure 6 depicts the reference-following capability of the fixed+adaptive controller, whilst Figure 7 portrays the error between the two quantities in Figure 6. Figures 4 to 7 were acquired from the 0.5 probability weighting case.

6. CONCLUSIONS

In this paper, a novel adaptive control technique has been presented. The advantage of this method is the combination of the benefits of self-tuning and multiple-model restricted-structure optimal controller designs into one scheme. A self-tuner is able to adapt to changing system parameters at the expense of possible instability. A multiple-model optimal controller gives greater assurance of stability over a wide range of operating points with the expense of conservative performance.

A multiple-model adaptive controller is intermediate to these two schemes. It provides a certain amount of confidence in stability, due to the weighted effect of fixed known models in the optimisation, plus a performance enhancement due to the incorporation of system identification knowledge from one sample point to the next. The restricted structure of the control law provides simplicity of implementation, and transparency of the solution to those acquainted with much-used classical control laws.

To further extend and bring rigour to this work, an investigation of the convergence of the restricted structure algorithm would be desirable. Also, a criterion for the robustness of a given multiple-model problem would be beneficial.

ACKNOWLEDGEMENTS

The authors are grateful to the EPSRC for their general support on the Platform Grant, Number GR/R04683/01

REFERENCES

Astrom, K.J. and B. Wittenmark (1973). On self-tuning regulators. *Automatica*, **Vol. 9**, pp. 185-199.

Grimble, M.J. (2001). *Industrial Control Systems Design*. John Wiley, Chichester, 2001.

Grimble, M.J. (2000a). Restricted Structure Adaptive Control Including Series 2 DOF tracking. ICC Report 167.

Grimble, M.J. (2000b). Restricted Structure LQG Optimal Control. *Proc. IEE*, **Vol. 147**, No.2, March, pp. 185-195.

Grimble, M.J. (1999a). Restricted structure feedforward and feedback stochastic optimal control. *IEEE CDC Conference*, Phoenix, Arizona.

Grimble, M.J. (1999b). Robust control of multiple model systems and application to hot strip rolling. ICC Report 155.

Grimble, M.J. (1994). *Robust industrial control*, Prentice Hall, Hemel Hempstead.

Grimble, M.J. and M.A. Johnson (1988). *Optimal multivariable control and estimation theory: theory and applications, Vols. I and II*. John Wiley, London.

Luenberger, D. (1969). *Optimisation by vector space methods*. John Wiley, New York.

Noble, B. (1969). *Applied Linear Algebra*. Prentice Hall, Englewood Cliffs.

Yukitomo, M., I. Yutaka, S. Hino, K. Takahashi and K. Nagata (1998). A new PID controller tuning system and it's application to a flue gas temperature control in a gas turbine power plant. IEEE CCA. Trieste, Italy.

APPLICATIONS OF STATISTICAL-LEARNING METHODS IN SYSTEMS AND CONTROL

Marco Ariola* Chaouki T. Abdallah**
Vladimir Koltchinskii***

* *Dipartimento di Informatica e Sistemistica, Università degli Studi di Napoli Federico II, Via Claudio 21, 80125, Napoli, Italy*
** *Department of Electrical & Computer Engineering, University of Tennessee, Knoxville, TN 37996-2100, USA*
*** *Department of Mathematics and Statistics, The University of New Mexico, Albuquerque, NM 87131, USA*

Abstract: This paper shows an application of a statistical learning method to the design of a fixed-order controller for a flexible beam. Making use of a randomized algorithm proposed by the authors, a third-order controller is designed and its performance are compared with those achieved by controllers previously proposed by other authors. *Copyright © 2001 IFAC*

Keywords: Statistical Learning, Robust Control, Fixed-order Controllers.

1. INTRODUCTION

It has been recently shown that many control problems are too difficult to admit analytic solutions (Blondel, 1994; Blondel and Tsitsiklis, 1997). These controls problems include fixed-order controller design (Syrmos *et al.*, 1997), robust multiobjective control (Abdallah *et al.*, 1999), hybrid systems control (Lygeros *et al.*, 1999), time-delay systems (Koltchinskii *et al.*, 1999), and others (Jirstrand, 1996). Even though such questions may be too difficult to answer analytically, or may not be answerable exactly given a reasonable amount of computational resources, researchers have shown that we can "approximately" answer these questions "most of the time", and have "high confidence" in the correctness of the answers. This is the gist of the probabilistic methods, which many authors have recently proposed in control analysis and design. These methods build on the standard Monte Carlo approach (with justifications based on Chernoff Bounds, Hoeffding Inequality, and other elementary probabilistic tools (Vidyasagar, 1997*b*)) with ideas advanced during the 1960s and on the theory of empirical processes and statistical learning developed in the 1970s and 1980s. In control theory, some of the original (Monte Carlo) ideas have already been used in Tempo *et al.* (1997), Khargonekar and Tikku (1996), to solve robust analysis problems while Vidyasagar (Vidyasagar, 1998) and the authors (Koltchinskii *et al.*, 2000) used learning theory to solve robust design problems. The basic idea of using learning methods in such problems is to convert a highly complex (potentially undecidable) design problem into a probabilistic optimization framework. This framework is further reduced to a sequence of testable analysis problems whose outcome is a probabilistic solution to the original design problem. Such design problems may include the congestion control of high-speed communications network, the design of power control algorithms in spread-spectrum communications and others.

Unfortunately, and as acknowledged by the various authors, probabilistic and learning methods, while more efficient than gridding techniques (which suffer from the curse of dimensionality), still require a large number of samples in order to guarantee accurate designs. This has precluded their use in a practical design setting. As an example, in Vidyasagar (1998), the author calculates that more than 2 million samples are needed in order to probabilistically guarantee a certain performance level in a robust control design problem. On the other hand, it was conjectured and verified experimentally that much smaller bounds on the number of samples (tens of thousands instead of millions) may be sufficient to guarantee a certain level of performance (Vidyasagar, 1998). In fact, in Vidyasagar (1998), the author uses 200 samples instead of the millions implied by his bounds, while acknowledging that, in this way, the theoretical guarantees of accuracy and confidence no longer hold. In Koltchinskii *et al.* (2000) the authors, invoking different versions of *bootstrap sequential learning* algorithms, found that a smaller bound on the number of samples can still guarantee the desired level of accuracy and confidence.

This paper is organized as follows. In Section 2 we give some backgrounds on randomized algorithms and their purpose. Then in Section 3, using the algorithm proposed in (Koltchinskii *et al.*, 2000), we design a fixed-order controller for a flexible beam. Our conclusion are drawn in Section 4.

2. BACKGROUNDS

To fix ideas about randomized algorithms in controller design, let us consider an uncertain plant $G(s, X)$, where X indicates the unknown plant parameters, and that we want to find some controller parameters Y of controller $C(s, Y)$ such that a perfomance index is minimized. What is usually done in the worst-case approach is to minimize a function in the form

$$\sup_X [g(X, Y)],$$

whereas in the probabilistic approach we attempt to minimize the expected value

$$\mathbb{E}_X [g(X, Y)],$$

where $g(X, Y)$ is a performance measure. One limitation of this approach is that in practice, we do not have the necessary information to calculate $\mathbb{E}_X [g(X, Y)]$ since all we have are sample plants and compensators. Moreover, how do we find minimize $\mathbb{E}_X [g(X, Y)]$ when all we have are the values of $g(X, Y)$ at sample points? In Vidyasagar (1997*a*), the empirical mean of $g(X, Y)$ is used instead of $\mathbb{E}_X [g(X, Y)]$ for a given Y

$$\frac{1}{n} \sum_{j=1}^{n} g(X_j, Y), \tag{1}$$

which then leaves us with two questions:

(1) Will (1) be a good approximation of $\mathbb{E}_X [g(X, Y)]$ uniformly in Y as n increases?
(2) How to minimize the empirical mean (1) with respect to Y?

It turns out that the first question (concerning the accuracy of empirical approximation) has been studied thoroughly in the theory of empirical processes and statistical learning. Minimization of a function defined by equation (1) is a case of empirical risk minimization. In the context of control problems, Vidyasagar (Vidyasagar, 1997*a*; Vidyasagar, 1998) used for this purpose a simple Monte Carlo minimization technique based on simulation of random points Y_i (independently of X_j) and minimizing the empirical risk over a finite set of such points. We also used this approach in Koltchinskii *et al.* (2000), but we emphasized there that such optimization technique becomes imprecise in high dimensional problems. This motivates the development of more efficient algorithms of empirical risk minimization that are at the same time self-bounding (i.e. they provide sharp enough data-driven bounds on the accuracy of empirical approximation).

In the following example we use the algorithm proposed in Koltchinskii *et al.* (2000) to design a robust controller. The control problem we consider was originally presented in Doyle *et al.* (1992). What we will do is to solve the robust control problem using our randomized algorithm and then to interpret the result using classical analysis tools.

3. APPLICATION TO THE DESIGN OF A FIXED-ORDER CONTROLLER

3.1 *Controller design*

Let us consider the feedback system in Figure 1. The plant $G(s)$ is a simplified model of a flexible beam. The input is the voltage to a power amplifier while the output is the tip deflection of the beam. The transfer function $G(s)$ is given by

$$G(s) = \frac{-6.4750s^2 + 4.0302s + 175.7700}{s(5s^3 + 3.5682s^2 + 139.5021s + 0.0929)}. \tag{2}$$

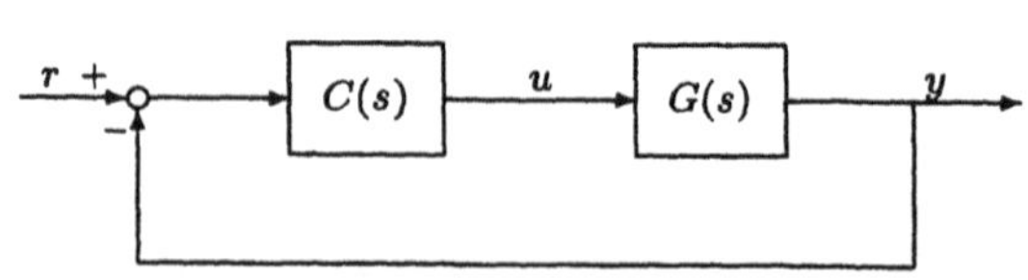

Fig. 1. The closed-loop system

In Doyle *et al.* (1992) the author designed a controller $C(s)$ considering some time domain specifications and an amplitude constraint on the plant input. In order to use the $\mathcal{H}_\infty$ approach, some appropriate weighting functions $W_1(s)$ and $W_3(s)$ were chosen and the problem was reformulated as the minimization of the mixed-sensitivity norm (see Figure 2)

$$\left\| \begin{pmatrix} W_1 S \\ W_3 C S \end{pmatrix} \right\|_\infty \tag{3}$$

where S is the sensitivity function $S = (1 + GC)^{-1}$. After an iterative procedure, the weighting functions W_1 and W_3 turned out to be

$$W_1(s) = 0.8 \frac{s^2 + 1.2s + 1}{(s + 0.001)(s + 1.2)(0.001s + 1)}$$

$$W_3(s) = \frac{s + \frac{1}{3}}{s + \frac{100}{3}}.$$

The suboptimal controller found in Doyle *et al.* (1992) achieves an $\mathcal{H}_\infty$ norm of the mixed sensitivity (3) of about 0.938 and has order eight, the same order of the augmented plant (plant plus weighting function).

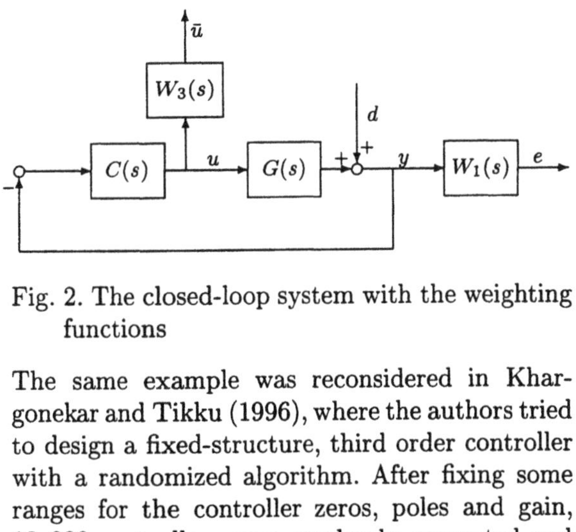

Fig. 2. The closed-loop system with the weighting functions

The same example was reconsidered in Khargonekar and Tikku (1996), where the authors tried to design a fixed-structure, third order controller with a randomized algorithm. After fixing some ranges for the controller zeros, poles and gain, 10,000 controllers were randomly generated and the controller minimizing the norm (3) was chosen. This controller resulted to achieve an $\mathcal{H}_\infty$ norm of the mixed sensitivity (3) of about 1.02, not far from the optimal one.

Now let us re-consider the beam transfer function (2) where we introduce uncertainties in the gain, the damping factor, and the natural frequency of the two complex poles which model the first flexible mode

$$\hat{G}(s) = K \frac{-6.4750s^2 + 4.0302s + 175.7700}{5s(s + 6.66 \cdot 10^{-4})(s^2 + 2\zeta\omega_n s + \omega_n^2)} \tag{4}$$

where the original plant $G(s)$ is recovered letting $K = 1$, $\zeta = 0.0675$, $\omega_n = 5.28$.

Rewriting (4) as

$$\hat{G}(s) = K \frac{-6.4750s^2 + 4.0302s + 175.7700}{5s(s + 6.66 \cdot 10^{-4})(s^2 + As + B)} \tag{5}$$

we consider the following intervals for the coefficients K, A and B

$$K \in (0.5, 1.5),\ A \in (0, 1.4259),\ B \in (19.53, 36.27), \tag{6}$$

which result in the following ranges for the original parameters ζ and ω_n

$$\zeta \in (0, 0.1184), \quad \omega_n \in (4.4176, 6.0201).$$

What we intend to find is a third-order controller which minimizes the mixed sensitivity norm (3) over all the uncertain plants (4), guaranteeing at the same time some nominal performance

$$\min \left\| \begin{pmatrix} W_1(1 + \hat{G}C)^{-1} \\ W_3 C(1 + \hat{G}C)^{-1} \end{pmatrix} \right\|_\infty \text{ subject to}$$
$$\left\| \begin{pmatrix} W_1(1 + GC)^{-1} \\ W_3 C(1 + GC)^{-1} \end{pmatrix} \right\|_\infty < 2 \tag{7}$$

A possible *classical* way to approach the problem could be to use the μ-synthesis to design a controller and then reduce its order hoping that its optimal properties would be retained. Hereafter we will try a randomized approach based on our algorithm. As in Khargonekar and Tikku (1996), we look for a stable, minimum-phase, strictly proper, third order controller. Its coefficients are randomly generated using the algorithm proposed in Khargonekar and Tikku (1996).

Next, we reformulate our original problem. We assume that the plant uncertain parameters have uniform distribution in the intervals (6). We denote by $X \in \mathcal{X} \subset \mathbb{R}^3$ the vector of the random coefficients of the plant $X = [K \ \ A \ \ B]$, and by $Y \in \mathcal{Y} \subset \mathbb{R}^6$ the the vector of the random coefficients of the controller. Let us introduce a cost function

$$\Psi(Y) = \max\{\psi_1(Y), \psi_2(Y)\} \tag{8}$$

where

$$\psi_1(Y) = \begin{cases} 0 \text{ if } \left\| \begin{pmatrix} W_1(1 + G(s)C(s,Y))^{-1} \\ W_3 C(1 + G(s)C(s,Y))^{-1} \end{pmatrix} \right\|_\infty < 2 \\ 1 \text{ otherwise} \end{cases} \tag{9}$$

and $\psi_2(Y) = E\left(\zeta(X, Y)\right)$ with

$$\zeta(X, Y) = \begin{cases} 1 \text{ if } (1 + \hat{G}(s, X)C(s, Y))^{-1} \text{ is unstable} \\ \left\| \begin{pmatrix} W_1(1 + \hat{G}(s, X)C(s, Y))^{-1} \\ W_3 C(1 + \hat{G}(s, X)C(s, Y))^{-1} \end{pmatrix} \right\|_\infty \text{ otherwise} \end{cases}$$

Now we are ready to apply the randomized algorithm proven in Koltchinskii *et al.* (2000) to find a probably approximate near minimum of $\Psi(Y)$ with confidence $1 - \delta$, level α and accuracy ε (Vidyasagar, 1997*a*).

Algorithm 1. (Controller Design).

Step 1. Fix the values for δ, α and ε.

Step 2. Choose n controllers with random uniformly distributed coefficients $Y_1, \ldots, Y_n \in \mathcal{Y}$ where (we indicate by $\lfloor \cdot \rfloor$ the floor operator)

$$n = \left\lfloor \frac{\log(2/\delta)}{\log[1/(1-\alpha)]} \right\rfloor$$

Evaluate for these controllers the function $\psi_1(Y)$ (9) and discard those controllers for which $\psi_1(Y) = 1$. Let $\hat{n}$ be the number of the remaining controllers.

Step 3. Choose m plants $\hat{G}(s, X)$ generating random parameters $X_1, \ldots, X_m \in \mathcal{X}$ with uniform distribution, where

$$m = 2^k \left\{ \left\lfloor \frac{100}{\varepsilon^2} \log\left(\frac{8}{\delta}\right) \right\rfloor + 1 \right\}$$

Step 4. Evaluate the stopping variable

$$\gamma = \max_{1 \le j \le \hat{n}} \left| \frac{1}{m} \sum_{i=1}^{m} r_i \zeta(X_i, Y_j) \right|$$

where r_i are *Rademacher* random variables, i.e. independent identically distributed random variables taking values $+1$ and -1 with probability 1/2 each. If $\gamma \le \varepsilon/5$, go to Step 5. If $\gamma > \varepsilon/5$, add m more independent plants to the plant samples, set $m := 2m$ and repeat Step 4.

Step 5. Choose the controller which minimizes the function

$$\frac{1}{m} \sum_{i=1}^{m} \zeta(X_i, \cdot)$$

This is the statistical optimal controller in the sense defined above.

◇

In our simulation, we chose $\varepsilon = 0.1$, $\delta = 0.05$ and $\alpha = 0.001$. Therefore n evaluated to 3,688 and m initially evaluated to 50,753. Out of the n controllers, only $\hat{n} = 15$ gave $\psi_1 = 0$; the others were discarded. One iteration was sufficient. The *best* controller is

$$C(s) = \frac{0.5781s^2 + 0.6602s + 0.1198}{0.2012s^3 + 0.9687s^2 + 11.7181s + 1}. \tag{10}$$

and the corresponding value of the cost function (8) is $\Psi = 0.600$.

3.2 *Analysis of the designed controller*

Now let us go back to our original intent. We wanted to design a *good* third order controller attempting to minimize the norm in (7). So how *good* is our controller? To answer this question, we have analyzed the *nominal performance*, *robust stability* and *robust performance* of our closed-loop system. The tools we used have become classical control tools; for a detailed discussion see e.g. Zhou and Doyle (1998) and the references therein.

To analyze the nominal performance, it is sufficient to evaluate the mixed sensitivity norm (3). This norm evaluates to about 1.08, which is not far from the values achieved by the $\mathcal{H}_\infty$ controller proposed in Doyle *et al.* (1992) and by the random controller proposed in Khargonekar and Tikku (1996).

The analysis of both robust stability and robust performance can be carried out using the μ-analysis. To this aim we need to write our uncertain system (4) using LFTs. First of all let us rewrite (5) as

$$\hat{G}(s) = K \frac{as^2 + bs + c}{5s(s+d)(s^2 + As + B)}$$

where

$$a = -6.4750, \qquad b = 4.0302,$$
$$c = 175.77, \qquad d = 6.66 \cdot 10^{-4}.$$

Let the uncertain parameters K, A and B have the expressions

$$K = \bar{K}(1 + \alpha_K \delta_K) \qquad A = \bar{A}(1 + \alpha_A \delta_A),$$
$$B = \bar{B}(1 + \alpha_B \delta_B),$$

with

$$\bar{K} = 1 \qquad \alpha_K = 0.5$$
$$\bar{A} = 0.7130 \qquad \alpha_A = 1$$
$$\bar{B} = 27.90 \qquad \alpha_B = 0.3.$$

In this way we have normalized the uncertainties

$$-1 < \delta_K, \delta_A, \delta_B < 1$$

and the nominal plant (2) is recovered simply letting

$$\delta_K = 0, \quad \delta_A = 0, \quad \delta_B = 0.$$

Now, letting

$$\bar{G} = -a\bar{K}/5$$
$$\hat{u} = \begin{pmatrix} u_A & u_B & u_K & u \end{pmatrix}^T$$
$$\hat{y} = \begin{pmatrix} y_A & y_B & y_K & y \end{pmatrix}^T$$

we can write our system isolating the uncertain terms as

$$\dot{x} = \begin{pmatrix} 0 & 0 & 0 & 0 \\ 1 & 0 & 0 & -\bar{B}d \\ 0 & 1 & 0 & -\bar{B} - \bar{A}d \\ 0 & 0 & 1 & -\bar{A} - d \end{pmatrix} x + \begin{pmatrix} 0 & 0 & c & \bar{G}c \\ 0 & d & b & \bar{G}b \\ d & 1 & 1 & \bar{G} \\ 1 & 0 & 0 & 0 \end{pmatrix} \hat{u}$$

$$\hat{y} = \begin{pmatrix} 0 & 0 & 0 & -\alpha_A \bar{A} \\ 0 & 0 & 0 & -\alpha_B \bar{B} \\ 0 & 0 & 0 & 0 \\ 0 & 0 & 0 & 1 \end{pmatrix} x + \begin{pmatrix} 0 & 0 & 0 & 0 \\ 0 & 0 & 0 & 0 \\ 0 & 0 & 0 & \bar{G}\alpha_G \\ 0 & 0 & 0 & 0 \end{pmatrix} \hat{u} \tag{11}$$

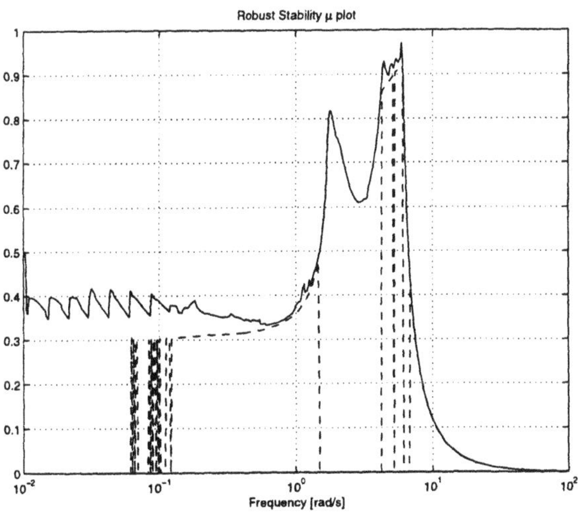

Fig. 3. The lower and upper bounds for $\mu_{\mathbf{\Delta}}(G_c(s))$

The input u_A, u_B and u_K and the outputs y_A, y_B and y_K are fictitious inputs and outputs used to model the uncertainties

$$\begin{pmatrix} u_A \\ u_B \\ u_K \end{pmatrix} = \begin{pmatrix} \delta_A & 0 & 0 \\ 0 & \delta_B & 0 \\ 0 & 0 & \delta_K \end{pmatrix} \begin{pmatrix} y_A \\ y_B \\ y_K \end{pmatrix}.$$

Let now $u = -C(s)y$ in (11), with $C(s)$ given by (10), and denote by $G_c(s)$ the resulting closed-loop system

$$\begin{pmatrix} y_A \\ y_B \\ y_K \end{pmatrix} = G_c(s) \begin{pmatrix} u_A \\ u_B \\ u_K \end{pmatrix}$$

Consider the block structure

$$\mathbf{\Delta} := \left\{ \begin{pmatrix} \delta_1 & 0 & 0 \\ 0 & \delta_2 & 0 \\ 0 & 0 & \delta_3 \end{pmatrix}, \delta_i \in \mathbb{R} \right\}$$

We evaluated by means of the *μ-Analysis and Synthesis Toolbox* (Balas *et al.*, 1998) the real structured singular value $\mu_{\mathbf{\Delta}}(G_c(s))$. Its plot is shown in Figure 3. Since its peak value is approximately equal to 0.97, our system (11) is guaranteed to be stable for all $\Delta \in \mathbf{\Delta}$ with $\|\Delta\|_2 < 1/0.97 \approx 1.03$. Anyway this result may be conservative, since the lower and upper μ bounds are significantly different. For the sake of comparison, we evaluated $\mu_{\mathbf{\Delta}}(G_c(s))$ for the controller proposed in (Doyle *et al.*, 1992), which however had *not* been designed to be robust. The peak value of $\mu_{\mathbf{\Delta}}(G_c(s))$ is approximately equal to 1.04.

For the evaluation of the closed-loop performance let us refer to Figure 4 (see also Figure 2). We denote by $T(G_P, \Delta)$ the transfer matrix between d and $(e\ \bar{u})^T$ (it is indeed the transfer matrix whose norm we minimized in (7)). Define an augmented block structure

$$\mathbf{\Delta}_P := \left\{ \begin{pmatrix} \Delta & 0 \\ 0 & \Delta_r \end{pmatrix} : \Delta \in \mathbf{\Delta},\ \Delta_r \in \mathbb{C}^{1\times 2} \right\}$$

We evaluated the structured singular value $\mu_{\mathbf{\Delta}_P}(G_P(s))$. Its plot is shown in Figure 5. Since its peak value

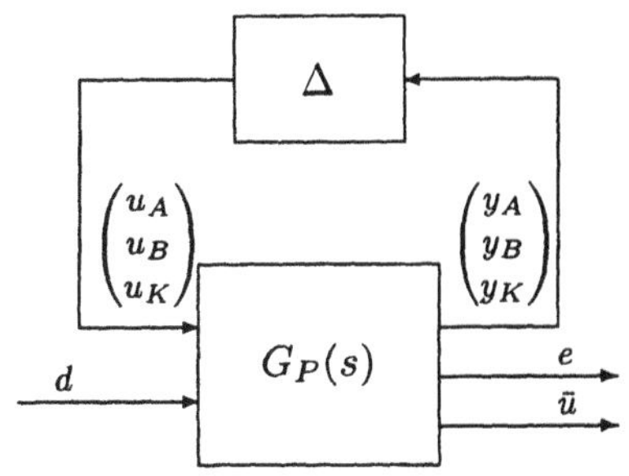

Fig. 4. The closed-loop for the evaluation of robust perfomance

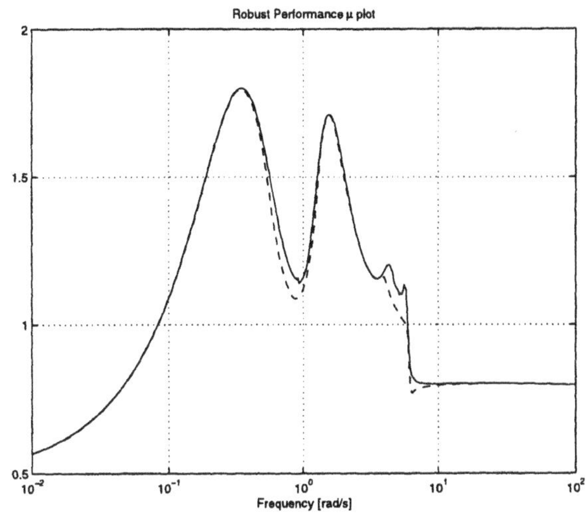

Fig. 5. The lower and upper bounds for $\mu_{\mathbf{\Delta}_P}(G(s))$

is approximately equal to 1.80, $\|T(G_P, \Delta)\|_\infty \leq 1.80$ for all $\Delta \in \mathbf{\Delta}$ with $\|\Delta\|_2 < 1/1.80 \approx 0.556$. We evaluated $\mu_{\mathbf{\Delta}_P}(G_c(s))$ also for the controller proposed in (Doyle *et al.*, 1992). The peak value of $\mu_{\mathbf{\Delta}_P}(G_c(s))$ in this case is approximately equal to 1.60.

4. CONCLUSIONS

In this paper we have shown through an example how a combination of statistical methods and classical theoretical results can be a good way of approaching *difficult* control design problems. A third-order controller for a flexible beam has been designed using a randomized algorithm; then its performance has been evaluated by means of μ-analysis.

5. REFERENCES

Abdallah, C. T., M. Ariola, P. Dorato and V. Koltchinskii (1999). Quantified Inequalities and Robust Control. In: *Robustness in Identification and Control* (A. Garulli, A. Tesi and A. Vicino, Eds.). pp. 373–390. Number 245 In: *Lecture Notes in Control and Information Sciences*. Springer Verlag.

Balas, G. J., J. C. Doyle, K. Glover, A. Packard and R. Smith (1998). *μ-Analysis and Synthesis Toolbox*. The MathWorks, Inc.. Natick, MA. User's Guide *Version 3*.

Blondel, V. (1994). *Simultaneous Stabilization of Linear Systems*. 1st ed.. Springer Verlag. London.

Blondel, V. and J. Tsitsiklis (1997). NP-Hardness of Some Linear Control Design Problems. *SIAM Journal of Control and Optimization* **35**, 2118–2127.

Doyle, J. C., B. A. Francis and A. R. Tannenbaum (1992). *Feedback Control Theory*. Macmillan Publishing Company. New York.

Jirstrand, M. (1996). Algebraic Methods for Modeling and Design in Control. PhD thesis. Linköping University. Sweden.

Khargonekar, P. and A. Tikku (1996). Randomized Algorithms for Robust Control Analysis and Synthesis have Polynomial Complexity. In: *Proceedings IEEE Conference on Decision and Control*. Kobe, Japan. pp. 3470–3475.

Koltchinskii, V., C. T. Abdallah, M. Ariola, P. Dorato and D. Panchenko (2000). Improved Sample Complexity Estimates for Statistical Learning Control of Uncertain Systems. *IEEE Transactions on Automatic Control* **45**(12), 2383–2388.

Koltchinskii, V., M. Ariola, C. T. Abdallah and P. Dorato (1999). Statistical Controller Design for the Linear Benchmark Problem. In: *Proceedings IEEE Conference on Decision and Control*. Phoenix, AZ. pp. 507–509.

Lygeros, J., J. Tomlin and S. Sastry (1999). Controllers for Reachability Specifications for Hybrid Systems. *Automatica* **35**(3), 349–370.

Syrmos, V., C. T. Abdallah, P. Dorato and K. Grigoriadis (1997). Static Output Feedback: A Survey. *Automatica* **33**(2), 125–137.

Tempo, R., E. Bai and F. Dabbene (1997). Probabilistic Robustness Analysis: Explicit Bounds for the Minimum Number of Samples. *Systems & Control Letters* **30**, 237–242.

Vidyasagar, M. (1997*a*). Statistical Learning Theory and Its Applications to Randomized Algorithms for Robust Controller Synthesis. In: *ECC97*. Vol. Plenary Lectures and Mini-Courses. Brussels, Belgium. pp. 162–190.

Vidyasagar, M. (1997*b*). *A Theory of Learning and Generalization*. Springer Verlag. New York.

Vidyasagar, M. (1998). Statistical Learning Theory and Randomized Algorithms for Control. *IEEE Control Systems Magazine* **18**(6), 69–85.

Zhou, K. and J. C. Doyle (1998). *Essentials of Robust Control*. Prentice Hall. Upper Saddle River, NJ.

A NEW SYSTEM IDENTIFICATION METHOD BASED ON SUPPORT VECTOR MACHINES

Shuichi Adachi and Tomonori Ogawa

Department of Electrical and Electronic Engineering
Utsunomiya University
7-1-2 Yoto, Utsunomiya, 321-8585, Japan
Phone/Fax : +81 28 689 6125
e-mail : adachis@cc.utsunomiya-u.ac.jp

Abstract: Support Vector Machines (SVM) have become a subject of intensive study in statistical learning theory. They have been applied to successfully to classification problems and recently extended to regression problems. Support vector machines for regression problems is called Support Vector Regression (SVR). In this paper, a brief introduction to SVR is presented and then a new system identification method based on SVR is proposed for linear in parameter models. The effectiveness of the proposed method is examined through numerical examples.

Keywords: system identification, statistical learning theory, support vector machines, regularization, robust estimation.

1. INTRODUCTION

Support Vector Machines (SVM) proposed by V.N. Vapnik early in 90's have become a subject of intensive study in statistical learning theory. They have been applied to successfully to classification problems and recently extended to regression problems. SVM for regression problems is called Support Vector Regression (SVR) (Vapnik, 1998, 1999; Scholköpf, 1999).

In this paper, a brief introduction to SVR is presented and then a new system identification method based on SVR for linear regression models such as Finite Impulse Response (FIR) and Auto-Regressive with eXtra input (ARX) models (Ljung, 1999) is proposed. It will be shown through numerical examples that the proposed method is robust for data with outliers and is effective for identification problem with band-limited input signals. Finally, SVR is applied to actual data which is measured from rolling mill. It will be shown that SVR has a good learning performance for finite data samples.

2. SUPPORT VECTOR REGRESSION

The regression problem is to estimate a parameter vector $\boldsymbol{w} = [w_1, w_2, \dots, w_m]^T$ of an unknown function $f(\boldsymbol{x}_i, \boldsymbol{w})$ based on the observed data y_i:

$$y_i = f(\boldsymbol{x}_i, \boldsymbol{w}) + e_i, \qquad i = 1, 2, ..., N \tag{1}$$

where $\{\boldsymbol{x}_i\}$ is the input data, e_i is an error and N is the number of data. The pair $(\boldsymbol{x}_i, y_i)$ is called training data set, that is, input-output data. In order to approximate the regression, a function which is linear in parameters $\{w_i\}$ is used:

$$f(\boldsymbol{x}_i, \boldsymbol{w}) = \sum_{j=1}^{m} w_j \boldsymbol{g}_j(\boldsymbol{x}_i) \tag{2}$$

where $\boldsymbol{g}_j(\boldsymbol{x}_i)$ denotes a set of nonlinear transformation chosen *a priori*.

A key idea in generalization of SVM from the classification problems to the regression problems is an adoption of a new type of loss function, *i.e.*, so-called ε-insensitive or Vapnik's loss function defined by

$$L_\varepsilon(y_i, f(\boldsymbol{x}_i, \boldsymbol{w})) = \begin{cases} 0, & |y_i - f(\boldsymbol{x}_i, \boldsymbol{w})| \le \varepsilon \\ |y_i - f(\boldsymbol{x}_i, \boldsymbol{w})| - \varepsilon, & \text{otherwise} \end{cases} \tag{3}$$

where a parameter ε controls the width of the insensitive zone and it determines the number of support vectors.

The loss function is suitable for robust regression (Huber, 1981), *i.e.*, under more relaxed assumptions on the error e_i than the conventional least-squares loss function which is appropriate only for Gaussian noise. If the density of the noise is a symmetric convex function with continuous second derivatives, then minimizing Vapnik's loss function with $\varepsilon = 0$ provides the best mini-max approximation to the regression.

Based on the Structural Risk Minimization (SRM) principle (Vapnik, 1999), a model structure which provides minimal risk values in

$$\mathcal{S}_k = \left\{ \sum_{j=1}^{m} w_j g_j(\boldsymbol{x}) : (\boldsymbol{w} \cdot \boldsymbol{w}) \le c_k \right\} \tag{4}$$

is selected where c_k is a constant which is selected *a priori.*

Now, the SVR problem is summarized as follows.

SVR Problem (Primal problem): Find the parameter vector $\boldsymbol{w}$ that minimizes the empirical risk functional

$$R_{emp}(\boldsymbol{w}) = \frac{1}{N} \sum_{i=1}^{N} L_\varepsilon \left(y_i, \sum_{j=1}^{m} w_j g_j(\boldsymbol{x}_i) \right) \tag{5}$$

subject to constraint

$$(\boldsymbol{w} \cdot \boldsymbol{w}) \le c_k \tag{6}$$

given the training data $(\boldsymbol{x}_i, y_i)$, $i = 1, ..., N$. □

By introducing slack variables ξ_i, ξ_i', $i = 1, ..., N$ and a regularization constant C, the SVR problem can be transformed to the following dual problem.

Dual Problem: Find the parameters α_i^* and β_i^*, $i = 1, ..., N$ that maximize the functional

$$\begin{aligned} Q(\alpha, \beta) &= -\varepsilon \sum_{i=1}^{N} (\alpha_i + \beta_i) + \sum_{i=1}^{N} y_i(\alpha_i - \beta_i) \\ &\quad -\frac{1}{2} \sum_{i,j=1}^{N} (\alpha_i - \beta_i)(\alpha_j - \beta_j) H(\boldsymbol{x}_i, \boldsymbol{x}_j) \end{aligned} \tag{7}$$

subject to constraint

$$\sum_{i=1}^{N} \alpha_i = \sum_{i=1}^{N} \beta_i, \; 0 \le \alpha_i, \beta_i \le C, \quad i = 1, ..., N \tag{8}$$

given the training data $(\boldsymbol{x}_i, y_i)$, $i = 1, ..., N$, an inner product kernel H, an insensitive zone ε , and a regularization parameter C. □

The dual problem can be solved via standard quadratic programming. Once solutions α_i^* and β_i^*, $i = 1, ..., N$ are calculated, the regression function is constructed in terms of these values as

$$f(\boldsymbol{x}) = \sum_{i=1}^{N} \gamma_i^* H(\boldsymbol{x}_i, \boldsymbol{x}) \tag{9}$$

where

$$\gamma_i^* = \alpha_i^* - \beta_i^*. \tag{10}$$

In the regression problem, both ε and C which play an important role for noise and smoothness of the function, respectively, have to be selected *a priori.* However, systematic selection of these parameters has not been proposed.

3. A NEW SYSTEM IDENTIFICATION METHOD BASED ON SVR

In this section, a new system identification method based on SVR is proposed for linear regression models such as FIR and ARX models (Ljung, 1999). Since a nonlinear kernel is usually utilized as an inner kernel H in Eq.(7), an explicit expression for parameter $\boldsymbol{w}$ could not be derived. On the other hand, a linear kernel is utilized in the proposed method, so an explicit expression for $\boldsymbol{w}$ can be formulated.

First, an FIR model

$$y_i = \sum_{j=0}^{m} w_j x_{i-j} + e_i, \quad i = 1, 2, \ldots, N \tag{11}$$

is considered where $\{x_i\}$ is the input data, $\{y_i\}$ is the output data, $\{w_j\}$ is the impulse response of the system, and $\{e_i\}$ is a white noise uncorrelated with the input data. The purpose of the system identification problem considered here is to estimate the impulse response based on finite input-output data.

As a basis function for SVR in Eq.(2),

$$\boldsymbol{g}_k(\boldsymbol{x}) = [x_{1-k}, \ldots, x_{N-k}]^T, \; k = 0, 1, \ldots, m \tag{12}$$

is utilized whose elements are time-shifted input data. Then, the output prediction in the primal problem can be written as

$$\widehat{\boldsymbol{y}}_N = \boldsymbol{G}_N \widehat{\boldsymbol{w}} \tag{13}$$

where

$$\begin{aligned} \widehat{\boldsymbol{y}}_N &= [\hat{y}_1, \hat{y}_2, ..., \hat{y}_N]^T, \\ \widehat{\boldsymbol{w}} &= [\hat{w}_0, \hat{w}_1, ..., \hat{w}_m]^T, \\ \boldsymbol{G}_N &= [\boldsymbol{g}_0(\boldsymbol{x}), \boldsymbol{g}_1(\boldsymbol{x}), \ldots, \boldsymbol{g}_m(\boldsymbol{x})]. \end{aligned}$$

On the other hand, the prediction output in the dual problem can be written as

$$\widehat{\boldsymbol{y}}_N = \boldsymbol{G}_N \boldsymbol{G}_N^T \boldsymbol{\gamma} \tag{14}$$

where $\boldsymbol{\gamma} = [\gamma_1, \gamma_2, ..., \gamma_N]^T$ is calculated by Eq.(10).

From Eqs.(13) and (14), estimates of the unknown parameter can be calculated by

$$\widehat{\boldsymbol{w}} = \boldsymbol{G}_N^T \boldsymbol{\gamma}. \tag{15}$$

In general, the purpose of SVR in the statistical learning theory is to predict output data precisely, so there was little interest in the expression of the parameter estimates such as Eq.(15). From control systems point of view, however, parameters of the system are very important to design a controller.

Next, an ARX model which is written as

$$y_i = -\sum_{j=1}^{m} a_j y_{i-j} + \sum_{j=1}^{m} b_j x_{i-j} + e_i, \; i = 1, \ldots, N \quad (16)$$

is considered. Since the ARX model is also linear in unknown parameters, $\boldsymbol{\theta} = [a_1, \ldots, a_m, b_1, \ldots, b_m]^T$, the result for the FIR model can be easily extended, so

$$\widehat{\boldsymbol{\theta}} = \boldsymbol{G}_N^T \boldsymbol{\gamma} \quad (17)$$

is derived where

$$\boldsymbol{G}_N^T = \begin{bmatrix} y_1 & y_2 & \cdots & y_N \\ y_0 & y_1 & \cdots & y_{N-1} \\ \vdots & \vdots & & \vdots \\ y_{1-m} & y_{2-m} & \cdots & y_{N-m} \\ x_1 & x_2 & \cdots & x_N \\ x_0 & x_1 & \cdots & x_{N-1} \\ \vdots & \vdots & & \vdots \\ x_{1-m} & x_{2-m} & \cdots & x_{N-m} \end{bmatrix}. \quad (18)$$

Moreover, the proposed system identification method can be extended to nonlinear system identification problems as long as the corresponding nonlinear model is linear in parameters such as a discrete-time bilinear model. Details of these results are omitted.

4. NUMERICAL EXAMPLES

4.1 ARX model identification

First, an ARX model identification problem is considered. The parameter vector $\boldsymbol{\theta}$ was assumed as

$$\boldsymbol{\theta} = [a_1, a_2, b_1, b_2]^T = [-1.60, 0.680, 0.0364, 0.0437]^T.$$

As an identification input, a PRBS (Pseudo Random Binary Signal) was applied, and the output was corrupted by a white noise $N(0, 0.03^2)$ where number of data was 50.

The output prediction result is shown in Fig.1. The insensitive and regularization parameters were selected by comparing Mean Square Error (MSE) shown in Fig.2, as a result $\varepsilon = 0.022$ and $C = 48.5$ were chosen. Number of the corresponding support vector was 24. The parameter estimates by the proposed method ($\hat{\boldsymbol{\theta}}_{SVR}$) and the conventional Least Squares Method (LSM) ($\hat{\boldsymbol{\theta}}_{LSM}$) were

$$\hat{\boldsymbol{\theta}}_{SVR} = [-1.60, 0.681, 0.0386, 0.0496]^T,$$
$$\hat{\boldsymbol{\theta}}_{LSM} = [-1.59, 0.679, 0.0379, 0.0492]^T.$$

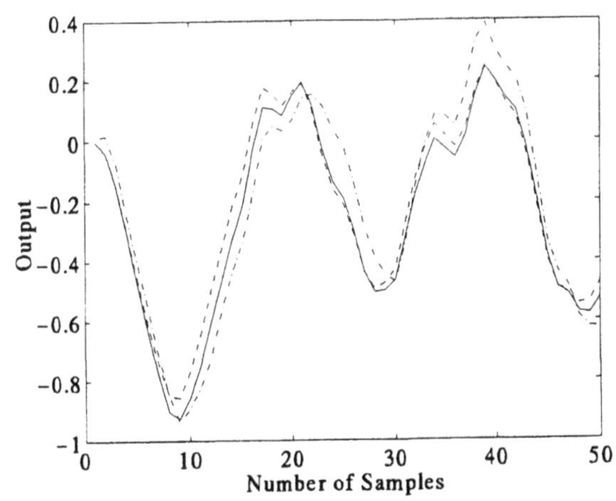

Fig.1: Comparison between output predictions by the proposed method (solid line) and the conventional LSM (dashed line), where dashdot line represents the measured output.

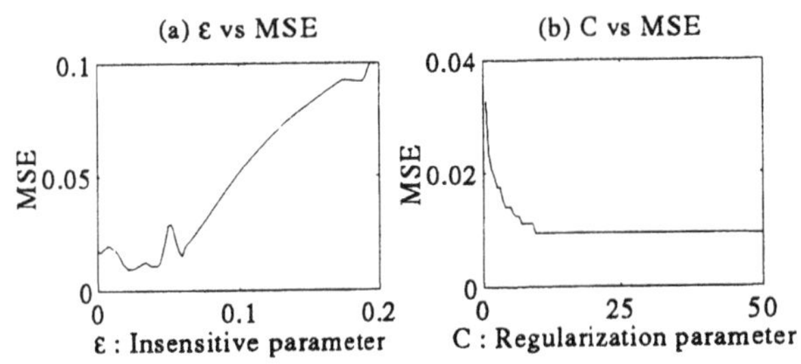

Fig.2: Selection of insensitive and regularization parameters.

The prediction accuracy and parameter estimates of the proposed method are almost the same as the LSM.

4.2 FIR model identification with an outlier

An FIR model whose impulse response is given by

$$\{w_j\} = \{1, 0.82, 0.67, 0.55, 0.45, 0.37, 0.3, 0.25, 0.2, 0.17, 0.14, 0.11, 0.09\}$$

is considered. A PRBS was applied as an identification input and the output was corrupted by a white noise $N(0, 0.1^2)$. Moreover, an outlier was contained in output data at 18th point. The input and output data are shown in Fig.3.

The identification results are shown in Figs.4 and 5. In this numerical experiment, $\varepsilon = 0.1$ and $C = 1.4$ were selected. In Fig.4, an output prediction by the proposed method (solid line) and measured output (dashdot line) are compared for the validation data. The output prediction by the LSM is also shown by a dotted line. It is clear that the proposed method has an excellent prediction ability. On the other hand, the LSM has a poor prediction performance due to the outlier.

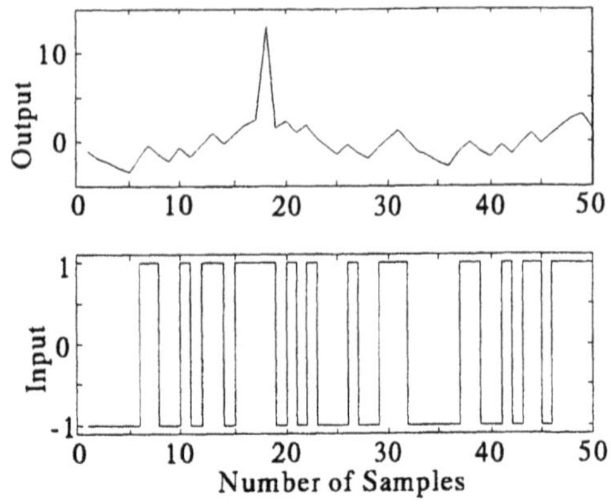

Fig.3: Input-output data with outlier.

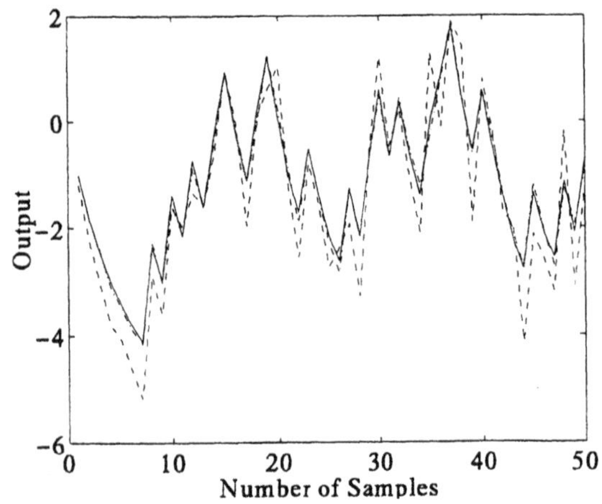

Fig.4: Comparison between output predictions by the proposed method (solid line) and the conventional LSM (dashed line), where dashdot line represents the measured output.

Fig.5 shows the true impulse response and estimated one. The impulse response of the proposed (○) and the true one (+) are almost the same, however, the impulse response by the LSM (□) fluctuates around the true values. From this example, it is shown that the proposed method is robust against outlier.

4.3 System identification with band-limited input signal

It is well known that white noise is one of the best input signals for system identification experiments, because it contains a lot of frequency components. However, white noise can not be applied in many practical situations, since it fluctuates stably operated plants. In such cases, system identification with band-limited, that is, smooth input signal, is required. It was reported that the conventional LSM has bad convergence property for the band-limited input signal, so system identification based on regularization was proposed (Sano, *et al.*, 1988).

Now, the proposed identification method based on SVR has regularization effect, so it is expected that it is effective for band-limited input data. To examine this, the following numerical example is presented.

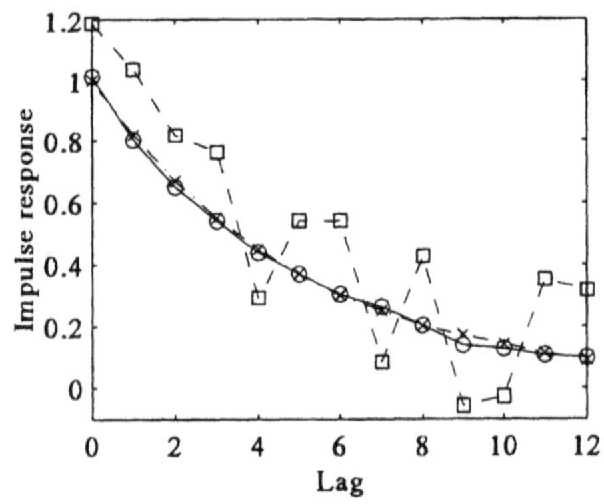

Fig.5: Comparison of impulse responses : true value (+), estimated by SVR (○) and estimated by LSM (*).

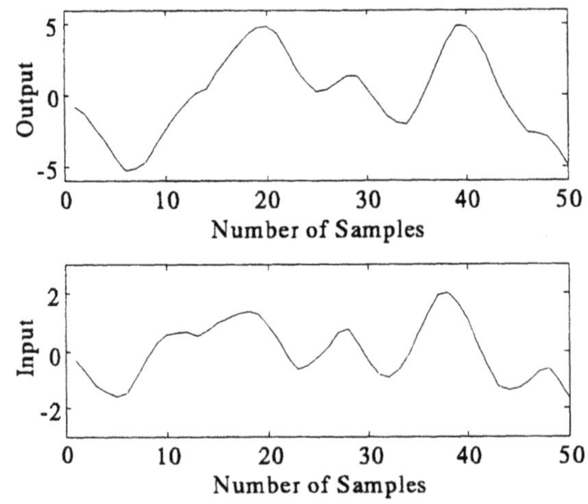

Fig.6: Band-limited input signal (lower) and corresponding output signal (upper).

A band-limited input signal shown in Fig.6 (lower) was employed for system identification. From the power spectrum shown in Fig.7, it is clear that the signal consists of low-frequency components. The system to be identified was the same as the FIR model in Section 4.2. The output of the system which is shown in Fig.6 (upper) was corrupted by a white noise $N(0, 0.1^2)$ where number of data was 50.

The identification result is shown in Fig.8. In this example, $\varepsilon = 0.04$ and $C = 0.3$ were used for the proposed method. It is clear that the proposed method is effective even when the input is band-limited signal. On the other hand, estimates by the LSM fluctuates seriously around the true values.

5. APPLICATION OF SVR TO ROLLING-MILL DATA

In order to examine the performance of SVR, it is applied to actual data which is measured from a hot

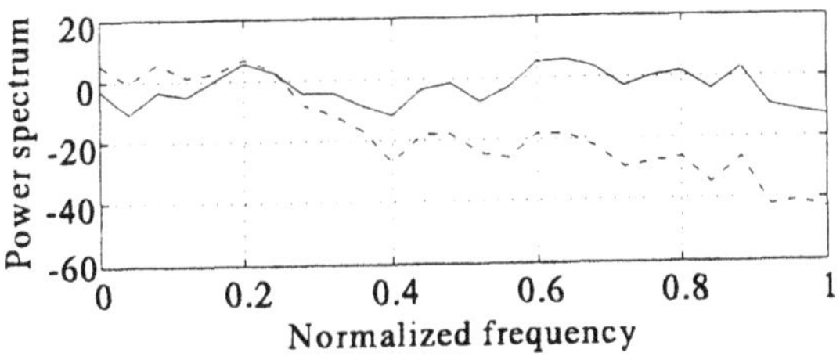

Fig.7: Power spectrum of the band-limited signal (dashed line) and white noise (solid line).

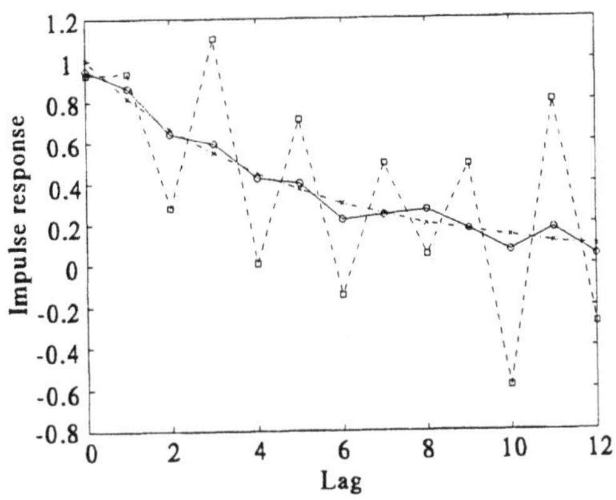

Fig.8: Comparison of impulse response : true value (×), estimated by SVR (○) and estimated by LSM (□).

strip rolling mill in Japan. The flowchart of the roughing rolling method is shown in Fig.9. The measured data consists of the width of the rolled steel strip after horizontal rolling. It contains 561 instances and each instance includes 7 attributes.

An experimental nonlinear model, which is called a set-up model, for the width of the rolled strip after horizontal rolling (W_O)

$$W_O = W_E + \Delta W_H + \Delta W_D \tag{19}$$

was proposed (Shibata, et al. 1981) where

$$\begin{aligned}
\Delta W_H &= W_E\left\{\left(\frac{H_I}{H_O}\right)^{C_H} - 1\right\}, \\
\Delta W_D &= C_D\left(W_I - W_E\right), \\
C_H &= \exp\left\{b_0\left(\frac{W_E}{H_I}\right)^{b_1}\right. \\
&\qquad \left.\cdot\left(\frac{W_E}{L_D}\right)^{b_2\left(\frac{W_E}{H_I}\right)}\left(\frac{H_I}{R_H}\right)^{b_3\left(\frac{W_E}{H_I}\right)}\right\}, \\
C_D &= \exp\left\{a_0\left(\frac{W_I - W_E}{W_I}\right)^{a_1}\right. \\
&\qquad \left.\cdot\left(\frac{H_I}{R_E}\right)^{a_2}\left(\frac{R_E}{W_I}\right)^{a_3}\left(\frac{W_I}{W_E}\right)^{a_4}\right\}.
\end{aligned}$$

W_E and W_I are the width of the rolled strip after edging and before rolling, respectively, H_I and H_O

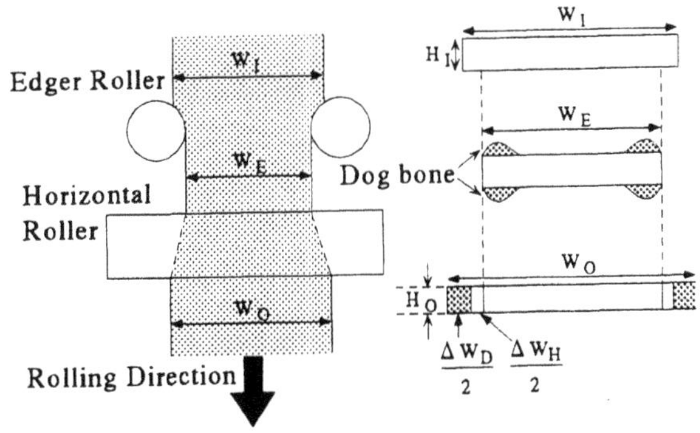

Fig.9: Illustration of roughing rolling method.

are the thickness of the rolled strip before rolling and after horizontal rolling, respectively, and R_E and R_H are the diameter of the edger and horizontal roller, respectively.

The regression problem considered here is to predict W_O based on input variables W_I, W_E, H_I and H_O. To solve the static regression problem Eq.(19), model parameters $a_0 \sim a_4$ and $b_0 \sim b_3$ have to be estimated based on the measured data. However, the model Eq.(19) is very complicated, so satisfactory prediction model could not be obtained by the conventional method.

In this paper, SVR is applied to learn the set-up model. A linear kernel is employed as an inner kernel H and three kinds of candidates for basis

- Case 1 : $\{W_I, W_E, H_I, H_O\}$
- Case 2 : $\left\{W_I, W_E, H_I, H_O, W_E\dfrac{H_I}{H_O}\right\}$
- Case 3 :
$$\left\{W_I, W_E, H_I, H_O, W_E\frac{H_I}{H_O}, W_I^2, W_E^2, W_I W_E\right\}$$

are utilized. For example, a model corresponds to Case 1 is

$$W_O = w_1 W_I + w_2 W_E + w_3 H_I + w_4 H_O$$

which is the most simple regression model, and a model corresponds to Case 3 is

$$\begin{aligned}
W_O \;=\; & w_1 W_I + w_2 W_E + w_3 H_I + w_4 H_O \\
& + w_5 W_E\frac{H_I}{H_O} + w_6 W_I^2 + w_7 W_E^2 + w_8 W_I W_E
\end{aligned}$$

It is noticed that the basis for SVR such as H_I/H_O is selected based on the experimental model Eq.(19).

Now, SVR is applied to actual data. The input data W_I, W_E, H_I and H_O are shown in Fig.10 and output data W_O is shown in Fig.11 where all data is normalized. For a model building, 50 data from instance 401 to 450 is used.

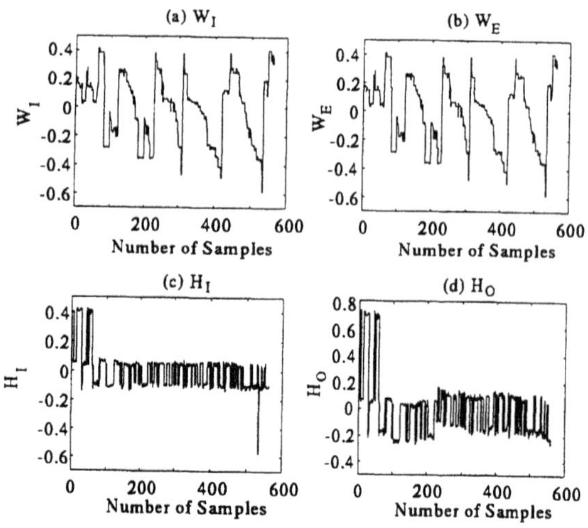

Fig.10: Input data: W_I, W_E, H_I, H_O.

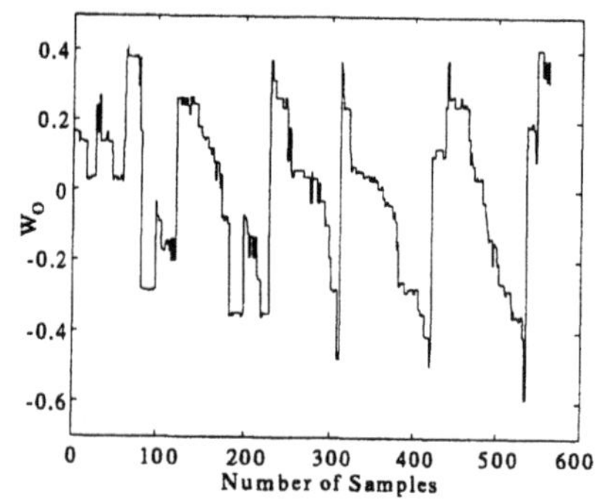

Fig.11: Output data : W_O.

Table 1 summarizes the learning results of SVR for Case 1 to 3. Especially, the learning results of Case 3 is shown in Fig.12. It is clear that appropriate selection of basis improves its learning performance. The important point of this example is to select the basis based on the experimental physical insight.

6. CONCLUSIONS

In this paper, SVR has been briefly introduced and a new system identification method based on SVR has been proposed. It has been shown that the proposed identification method has several advantages, that is, it is robust against outlier and is effective for band-limited input signal through numerical experiments. Also, SVR has been applied to the regression problem of the rolling mill data and it has a good learning performance.

REFERENCES

Huber, P.J. (1981). *Robust Statistics*, Wiley, New York.

Ljung, L. (1999). *System Identification - Theory for the User (2nd Edition)*, Prentice Hall PTR.

Table 1: Comparison of learning results for various basis

Case	ε	C	No. of SV	MSE	
				SVR	LSM
1	0.02	0.080	48	1.201	1.349
2	0.14	0.010	39	1.029	1.345
3	0.02	0.025	47	0.869	2.423

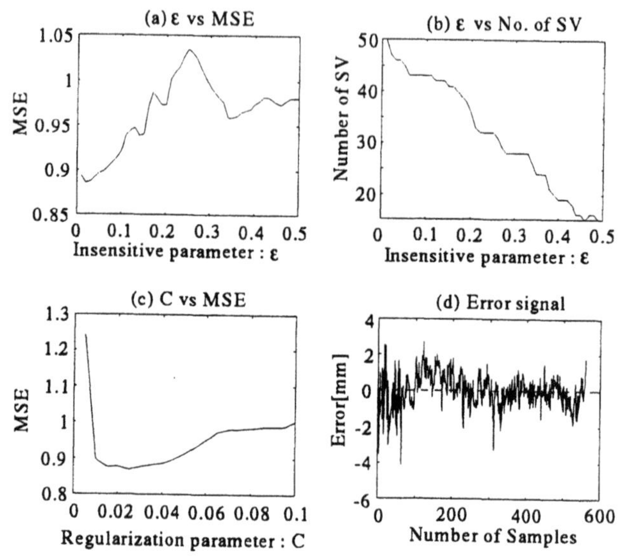

Fig.12: Learning result of Case 3.

Sano, A., S.Adachi and H.Ohmori (1988). Convergency improvement in least squares type of adaptive digital filter, Signal Processing IV : Theories and Applications, edited by J.L. Lacoume, pp.815–818, Elsevier Science Publishers B.V., North-Holland.

Shibata,T., et al. (1981). Edger set-up model at roughing train in hot strip mill, Iron and Steel, Vol.15, pp.2509–2514.

Scholköpf, B. et al. ed. (1999). *Advances in Kernel Methods*, The MIT Press.

Vapnik, V.N. (1998). *Statistical Learning Theory*, J. Willey & Sons.

Vapnik, V.N. (1999). *The Nature of Statistical Learning Theory (2nd Edition)*, Springer-Verlag.

NON-PARAMETRIC HEAVY-TAILED DENSITY ESTIMATION AND CLASSIFICATION PROBLEM

Natalia M. Markovitch * Udo R. Krieger **

* *Institute of Control Sciences, Russian Academy of Sciences, Profsoyuznay 65, 117997 Moscow, Russia*
** *T-Nova Deutsche Telekom, Technologiezentrum, Am Kavalleriesand 3, D-64295 Darmstadt, Germany*

Abstract: This report is devoted to the non-parametric probability density function (PDF) estimation for distributions with heavy tails. It is proposed to transform the data to a compact interval. The PDF of the original data is obtained by the "back-transform" of the PDF estimate of the transformed data. The transformation adapted to the data is explored to keep the tail decay rate of the true PDF and to provide a stable estimation regarding the deviations of the tail index estimates. The relative risk of the misclassification is proposed as a measure of quality for the PDF estimates.

Keywords: Non-parametric density estimation, Transformations, Classifiers, Risk of classification

1. INTRODUCTION

In the last few years heavy-tailed (Sigman, 1999) distributed random variables (r.v.) have been accepted as realistic models for various phenomena: the length of transmitted files, on and off periods of transmission, the claim sizes of catastrophes in insurance. The analysis of measurements of WWW-traffic by statistical methods also shows that WWW characteristics are often heavy-tailed distributed.

Here, the non-parametric estimation of heavy-tailed PDFs is considered. For the finite and light-tailed distributions (i.e. without heavy tails) the histogram is a good estimate of the PDF. But if the distribution is heavy-tailed, the histogram provides an absolutely misleading estimate in the "tail" domain. The same is true for most of the common non-parametric PDF estimates such as kernel, projection and spline estimates. In general, they have sharp peaks at "outliers" and don't provide the right rate of the decay at infinity.

The necessity of non-parametric PDF estimates with a good behavior at the tail domain is obvious. It is especially significant if PDFs of many populations are compared. Classification (pattern recognition) is another problem which is nearly related to the comparison of densities. If one uses an empirical Bayesian classification algorithm then the observation may be classified by the comparison of the PDF estimates of each class. Since the object can arise in the tail domain as well as in the body a good behavior of the estimate tails is principal for the classification.

To get more exact estimates at the tails it was proposed in (Devroye and Györfi, 1985) to use a preliminary transformation of the data which compresses them to a compact interval. Then one can estimate the PDF of the compressed sample and stretch this estimate by the inverse transformation. Based on this idea 1) a transformation, which is adapted to empirical data, 2) PDF estimates to keep the rate of the tail decay at infinity, and 3) the ratio of the empirical mean risk of the misclassification by an empirical Bayesian classifier to the risk of the misclassification by a Bayesian classifier as a quality measure of the PDF estimates are proposed.

2. ESTIMATE CONSTRUCTION

2.1 *Preliminary transformation*

Let $X_1,\ldots, X_n$ be i.i.d. r.v.s with the CDF F and the PDF f. Let us consider some monotone increasing continuous function $T : R_+ \to [0,1]$ and a sample $Y_j = T(X_j), j = 1,\ldots,n$. The CDF of Y_j is

$$G(x) = \Pr\{Y_j < x\} = \Pr\{T(X_j) < x\} \\ = \Pr\{X_j < T^{-1}(x)\} = F(T^{-1}(x)) \quad (1)$$

and the PDF is

$$g(x) = G'(x) = f(T^{-1}(x))(T^{-1}(x))', \quad (2)$$

analogously, since $X_j = T^{-1}(Y_j)$,

$$f(x) = g(T(x))T'(x).$$

One can estimate $g(x)$ by $Y_1,\ldots,Y_n$ using one of the non-parametric methods. Denote this estimate by $\breve{g}_n(x)$. Then one can use

$$\breve{f}_n(x) = \breve{g}_n(T(x))T'(x) \quad (3)$$

as the estimate of f. In (Devroye and Györfi, 1985) the general theory of such estimates is considered. For kernel estimates with compact kernels the transformation to an isosceles triangular PDF $\phi^{\text{tr}}(x)$ on $[0,1]$ and for a histogram the transformation to a uniform PDF $\phi^{\text{un}}(x)$ are recommended. These $T(x)$ provide an optimal convergence rate in L_1. Since such $T(x)$ and, therefore, the distribution of $Y_j = T(X_j)$ depend on the unknown CDF F, it is impossible to obtain an absolute identity of g and ϕ^{tr} (or $\phi^{\text{un}}(x)$). Hence, Devroye et al. (1985) used instead of F some family of CDFs Ψ_γ which depends on the parameter γ and adapt γ to the sample. In (Wand, et al., 1991) this idea was extended, i.e. transformations not only to compactly supported PDFs are considered. A transformation from suitable parametric families is selected which provides the minimal mean-square error of the PDF estimation of a transformed r.v. by a kernel estimate. Here, as Ψ_γ some family of heavy-tailed distributions is proposed, where the tail index γ is estimated by Hill's estimate.

2.2 *Estimation algorithm*

The algorithm proceeds as follows:

- The tail index of X_j is estimated by the sample $\{X_1,\ldots,X_n\}$ using the Hill estimate $\hat{\gamma}_n$.
- The transformation $T = T_{\hat{\gamma}_n}$ is constructed such that if ξ has the guess CDF $\Psi_{\hat{\gamma}_n}$ then $T_{\hat{\gamma}_n}(\xi)$ has the target CDF Φ, i.e. a uniform or triangular one. (Here $\hat{\gamma}_n$ is considered as a fixed value).
- The transformed sample $Y_j = T_{\hat{\gamma}_n}(X_j), j = 1,\ldots,n$ is constructed.
- The PDF of $Y_1,\ldots,Y_n$ is estimated by some non-parametric estimate $\breve{g}_n(x)$.
- The PDF of X_j is estimated by (3).

2.3 *Analysis of the algorithm*

For the realization of the algorithm the guess family, the target distribution and the estimate $\breve{g}_n$ must be chosen. The guess family must be such that the transformation T and its derivative can be easily evaluated. If Ψ_γ is the guess and Φ is the target CDF, then by (1) $T(x) = \Phi^{-1}(\Psi_\gamma(x))$ and $T^{-1}(x) = \Psi_\gamma^{-1}(\Phi(x))$ hold. The generalized Pareto distribution

$$\Psi_\gamma(x) = \begin{cases} 1-(1+\gamma x)^{-1/\gamma}, & \text{if } x \geq 0 \\ 0, & \text{if } x < 0 \end{cases}$$

is assumed as a guess CDF. Evidently, the guess PDF is

$$\psi_\gamma(x) = \Psi_\gamma'(x) = (1+\gamma x)^{-(1/\gamma+1)} \quad (4)$$

Such an assumption is based on facts from extreme value theory, namely, that for a sufficiently high threshold u of the r.v. X, the distribution of $Y = X - u$ converges to the Generalized Pareto distribution (Pickands, 1975). As target PDFs the uniform PDF $\phi^{\text{un}}(x) = \mathbf{1}\{\mathbf{x} \in [\mathbf{0},\mathbf{1}]\}$ and the positive triangular PDF $\phi^{+\text{tr}}(x) = 2(1-x)\mathbf{1}\{\mathbf{x} \in [\mathbf{0},\mathbf{1}]\}$ are considered. The corresponding CDFs are

$$\Phi^{\text{un}}(x) = x\mathbf{1}\{\mathbf{x} \in [\mathbf{0},\mathbf{1}]\} + \mathbf{1}\{\mathbf{x} > \mathbf{1}\},$$
$$\Phi^{+\text{tr}}(x) = (2x - x^2)\mathbf{1}\{\mathbf{x} \in [\mathbf{0},\mathbf{1}]\} + \mathbf{1}\{\mathbf{x} > \mathbf{1}\}.$$

Two questions are clarified here:

- What transformation ensures a more stable estimation algorithm with respect to the deviations of a tail index estimate?
- What tail behavior is ensured by the inverse transformation?

Let the target be $\phi = \phi^{\text{un}}, \Phi = \Phi^{\text{un}}$. Then

$$T_\gamma(x) = (1-(1+\gamma x)^{-1/\gamma})\mathbf{1}\{\mathbf{x} \in [\mathbf{0},\mathbf{1}]\},$$
$$T_\gamma'(x) = (1+\gamma x)^{-(1/\gamma+1)}\mathbf{1}\{\mathbf{x} \in [\mathbf{0},\mathbf{1}]\},$$
$$T_\gamma^{-1}(x) = ((1-x)^{-\gamma} - 1)/\gamma,$$
$$(T_\gamma^{-1}(x))' = (1-x)^{-\gamma-1}, x \in [0,1).$$

Suppose that the true PDF of X is

$$f(x) = \ell(x)\psi_\alpha(x), \quad (5)$$

where $\ell(x)$ is a bounded function with

$$\lim_{x\to\infty} \ell(x) = \ell_\infty < \infty$$

(the tail index of this PDF is α). Then by (2) the PDF of $Y = T_\gamma(X)$ is

$$g(x) = \left(1 + ((1-x)^{-\gamma} - 1)\alpha/\gamma\right)^{-(1/\alpha+1)} \cdot (1-x)^{-\gamma-1}\ell\left(((1-x)^{-\gamma} - 1)/\gamma\right)$$

for $x \in [0,1)$. This PDF will be estimated by $\breve{g}_n$ for some value of γ. Hence, its behavior must be good at its support $x \in [0,1)$. If x is close to zero, then $g(x)$ is bounded. Let us consider $g(x)$ for $x \uparrow 1$. Note that in this case $\ell(T_\gamma^{-1}(x)) \to \ell_\infty < \infty$, since $T_\gamma^{-1}(x) \to \infty$. Then

$$g(x) \simeq \ell_\infty \left(1 + ((1-x)^{-\gamma} - 1)\alpha/\gamma\right)^{-(1/\alpha+1)} \cdot (1-x)^{-\gamma-1} \text{ as } x \uparrow 1.$$

If $\alpha = \gamma$ then $g(x) \simeq \ell_\infty$, i.e. $g(x)$ behaves as a uniform PDF in the neighborhood of 1 (up to an insignificant multiplier ℓ_∞). Let $\alpha \neq \gamma$. For $x \uparrow 1$ and any constant C it holds $(1-x)^{-\gamma} + C \sim (1-x)^{-\gamma}$. Hence,

$$g(x) \simeq \ell_\infty (\alpha/\gamma)^{-(1/\alpha+1)} (1-x)^{\gamma/\alpha - 1}.$$

If γ overestimates the true tail index α, i.e. $\gamma/\alpha - 1 > 0$, then PDF $g(x)$ behaves nice in the neighborhood of 1 (while it is not like the uniform distribution). But if $\gamma < \alpha$ then $g(x) \to \infty$ for $x \uparrow 1$ and the performance of PDF estimates is bad since they are appointed for the estimation of finite PDFs.

Suppose now that the target PDF is a triangular one. Then

$$\Phi^{-1}(x) = (\Phi^{+tr})^{-1}(x) = 1 - \sqrt{1-x},$$
$$T_\gamma(x) = 1 - (1+\gamma x)^{-1/(2\gamma)},$$
$$T_\gamma^{-1}(x) = ((1-x)^{-2\gamma} - 1)/\gamma,$$
$$T_\gamma'(x) = 0.5(1+\gamma x)^{-1/(2\gamma)-1},$$
$$(T_\gamma^{-1}(x))' = 2(1-x)^{-2\gamma-1}.$$

If the PDF of X has the form (5) then the PDF of $Y = T_\gamma(X)$ is

$$g(x) \simeq \ell_\infty (\alpha/\gamma)^{-(1/\alpha+1)} (1-x)^{2\gamma/\alpha - 1},$$

as $x \uparrow 1$. For $\gamma = \alpha$ the PDF $g(x) \to 0$ as $x \uparrow 1$ and the rate of its decay is the same as for the triangular PDF ($\sim C(1-x)$). If $\gamma \neq \alpha$ and $\gamma > \alpha/2$ then $g(x)$ tends to 0 for $x \uparrow 1$. The rate of this decay is not $\sim C(1-x)$, but the PDF estimates will work well in this case. These problems can arise only if $\gamma < \alpha/2$. If α is estimated by a consistent estimate (e.g. by Hill's one) such a rough underestimation will be a rare event. Really, the probability to get γ, which is far from α tends to 0 as the sample size $n \to \infty$. The non-asymptotic confidence interval

$$I_\epsilon = [\hat{\gamma}_n(1 + y_\epsilon k^{-0.5})^{-1}; \hat{\gamma}_n(1 - y_\epsilon k^{-0.5})^{-1}]$$

of level $1-\epsilon$ for the index α has been proved in (Novak, 1998). Here, $\Phi(-y_\epsilon) = (\epsilon/2 - c_* k^{-0.5})_+$ and $C_* < 0.8$.

Suppose now that $g(x)$ is estimated by some estimate $\breve{g}_n$. What value will attain the tail index of the PDF estimate $\breve{f}_n$ defined by (3)?

Let $\breve{g}_n$ be a histogram estimate. Since for both transformations $T_\gamma(x) \to 1$ as $x \to \infty$ follows, then by (3) $\breve{f}_n(x) \simeq \breve{g}_n(1)T_\gamma'(x)$ as $x \to \infty$ holds. Note that $\breve{g}_n(1)$ is the height of the right-most bar of the histogram and $\breve{g}_n(1) > 0$ holds a.s. for large n if the support of X is R_+.

If the target PDF is a uniform one, it means that $\breve{f}_n(x) \simeq \breve{g}_n(1)(1+\gamma x)^{-(1/\gamma+1)}$, i.e. the estimate has the tail index $\hat{\gamma}_n$ which has been chosen for the guess PDF. For the kernel estimate with a compactly supported kernel (e.g., Epanechnikov's kernel: $K(x) = 0.75(1-x^2), |x| \leq 1$) $\breve{g}_n(x)$ depends on the smoothing parameter h and $\breve{g}_n(1,h) \sim O(h^{-1})$ for sufficiently large $h > 1 - T_\gamma(x_{(n)})$, $\breve{g}_n(1,h) = 0$ for other h. For a such kernel estimate $\breve{f}_n(x,h) \simeq \breve{g}_n(1,h)(1+\gamma x)^{-(1/\gamma+1)}$ holds. Here, boundary effects of a kernel estimate applied to the PDF with a compact support are observed due to the truncation of a kernel near the boundaries.

If the target PDF is triangular then

$$\breve{f}_n(x) \simeq 0.5\breve{g}_n(1)(1+\gamma x)^{-(1/(2\gamma)+1)}.$$

holds. In this case the tail index is two times larger than the true one. The same is true for a kernel estimate with $\breve{g}_n(1)$ replaced by $\breve{g}_n(1,h)$. To remove this effect a smoothed histogram (or a polygram, i.e. a histogram with a variable bin width (Tarasenko, 1968)) may be used. For such $\breve{g}_n$ and for $x \simeq 1$ $\breve{g}_n(x) = C_n(1-x)$ holds, where C_n is the slope of the line which connects the center of the top of the right-most histogram bar with the point $(1,0)$. Then for the triangular target PDF

$$\breve{f}_n(x) \simeq 0.5C_n(1 - (1 - (1+\gamma x)^{-1/(2\gamma)})) \cdot (1+\gamma x)^{-(1+1/(2\gamma))} \simeq 0.5C_n(1+\gamma x)^{-(1+1/\gamma)},$$

holds, i.e. in this case the tail index of the estimate coincides with its estimate for the guess PDF. The polygram is preferable to the histogram in this case. Really, a density g with $g(x) \to 0$ as $x \to 1$ is estimated. If the bin lengths are equal, then very few observations fall into the right-most bin. Hence, the histogram is unstable at the tail. The polygram adapts the bin widths to the data and works better. Figure 1 shows the poor behavior of

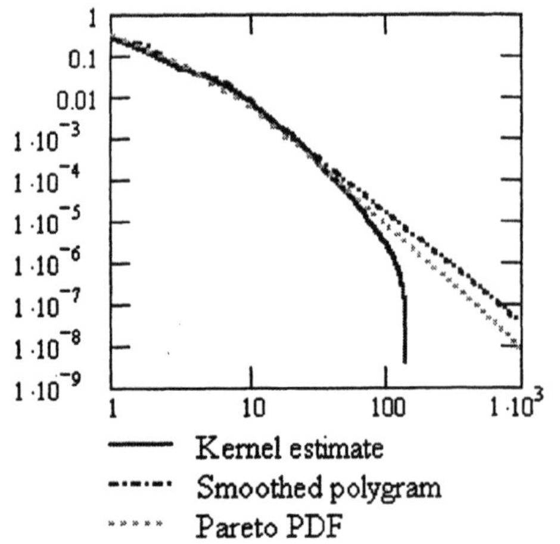

Fig. 1. Comparison of the tail estimates

the kernel estimate with Epanechnikov's kernel at the tail and its vanishing for large x in comparison to the smoothed polygram for a Pareto PDF with $\gamma = 0.5$. The estimates have been obtained using the transformation T_γ to a triangular PDF with $\gamma = 0.59$. The sample size is $n = 100$, the parameter $h = 0.1 < 1 - T_\gamma(x_{(n)})$, the number of bins for a polygram is 16.
Summarizing one may conclude that the best combination of the algorithm parameters among the considered ones is the triangular PDF as a target PDF and a smoothed polygram (or a kernel estimate with compactly supported kernel) as the PDF estimate on [0,1].

3. CLASSIFICATION PROBLEM AND QUALITY OF THE PDF ESTIMATION

Here, the problem of a quality measure of the PDF estimate is discussed. Normally, an expectation of a divergence in some functional space such as L_1, L_2, C is considered. The consistency of PDF estimates in the Kullback metric, in the total variation or χ^2-divergence is often considered, too. From the practical viewpoint the ability of an estimate to solve a specific applied problem is a much more important characteristic than its deviation from the estimated PDF in some sense. Here, the probability of the classification (pattern recognition) error is defined as a measure of the quality of the PDF estimate.
Suppose that the observed object O belongs to one of M different populations P_k, $k = \{1, \dots, M\}$. The number of this population is unknown. Let $p_k = \Pr\{O \in P_k\}$ be an a priori probability of a population and some random characteristic (feature) $X \in R_+$ of O is observed. Denote the PDF of X if $O \in P_k$ as $f_k(x)$. The aim is to estimate the number k of the population to which O belongs by the observation X. The solution of this problem is a classifier, i.e. a function $\eta : R_+ \to \{1, \dots, M\}$ which assigns a value to the characteristic $X = x$ corresponding to the estimated number $k \in \{1, \dots, M\}$ of the population.
Suppose that the penalty of the mistake $q_k(x)$ depends on the number of the true population k and the value of the characteristic x. The risk of the misclassification is the mean of the loss

$$L(\eta) = E\sum_{k=1}^{M} \mathbf{1}\{\mathbf{O} \in \mathbf{P_k}, \eta(\mathbf{X}) \neq \mathbf{k}\}\mathbf{q_k}(\mathbf{x})$$

$$= \sum_{k=1}^{M} p_k \int_{\eta(X)\neq k} q_k(x) f_k(x)dx \qquad (6)$$

The minimal risk is attained at the Bayesian classifier

$$\eta_B(x) = k \quad \text{if} \quad p_k q_k(x) f_k(x) \geq p_i q_i(x) f_i(x) \quad \text{for all} \quad i \neq k$$

Normally, the true f_k and p_k are unknown, but there are some estimates $\hat{f}_k, \hat{p}_k$ of them. The empirical Bayesian estimate is used

$$\eta_{EB}(x) = k \quad \text{if} \quad \hat{p}_k q_k(x) \hat{f}_k(x) \geq \hat{p}_i q_i(x) \hat{f}_i(x) \quad \text{for all} \quad i \neq k$$

Obviously, $L(\eta_{EB}) > L(\eta_B)$. Moreover $L(\eta_{EB})$ defined by (6) is a random variable since η_{EB} is a random function which depends on the data by which the estimates $\hat{f}_n$ were evaluated.
The penalties q_i are defined by the loss of the missclassification in a real classification problem. If the tail behavior is not important, then one can define $q_k(x) = 1$ for all k. Then, since the observations in the tail domain are rare, the improvement of the classification at the tail provides a negligible decrease of the risk of the misclassification. In problems where the tail behavior is important (like the WWW-traffic analysis), $q_k(x)$ should be larger on the tail and less on the body, i.e. $q_k(x) \to \infty$ as $x \to \infty$. The Bayesian approach can be applied only if the condition holds:

$$\int_0^\infty q_k(x) f_k(x)dx < \infty \qquad (7)$$

For the PDF estimation the classification means that M different PDFs are estimated and used in the empirical Bayesian classifier for objects with these PDFs. However, a classifier η_{EB} is not sensitive with respect to the accuracy of the PDF estimates, i.e. different estimates classifying the objects to the same classes may have the same value $L(\eta_{EB})$. Hence, the estimate of $L(\eta)$

$$\widetilde{L(\eta)} = \sum_{k=1}^{M} \hat{p}_k \int_{\eta(X)\neq k} q_k(x)\hat{f}_k(x)dx$$

may be used which reflects the quality of the PDF estimator. $\widetilde{L(\eta_{EB})}$ is the risk of Bayesian classifier when $\hat{f}_i(x)$ and $\hat{p}_i$, $i = 1, \dots, M$ are the true PDFs and a priori probabilities of classes, respectively. Therefore, $\widetilde{L(\eta_{EB})}$ fluctuates around the minimal Bayesian risk. The more accurate the PDF estimates are the more close is $\widetilde{L(\eta_{EB})}$ to $L(\eta_B)$. Note, that $L(\eta)$ is invariant regarding any "one-to-one" transformation with the properties mentioned above.

4. SIMULATION STUDY

The closeness of $\widetilde{L(\eta_{EB})}$ to $L(\eta_B)$ is compared for a kernel estimate

$$\hat{f}_n^K(x) = \frac{1}{nh}\sum_{j=1}^{n} K\left(\frac{x - X_j}{h}\right),$$

where h is a smoothing parameter, $K(x)$ is a kernel function and a smoothed polygram

$$f_{L,n}(t) = L((n+1)\lambda(\Delta_{rL}))^{-1} \text{ for } \quad t \in \Delta_{rL}.$$

Here $\lambda(\Delta)$ is the length of Δ, $\lambda(\Delta_{rk}) \to 0$, $L = o(n)$, the number of observations inside each interval Δ_{rL} is less or equal to L. The fixed transformation $T(x) = (2/\pi)\arctan(x)$ and the transformation T_γ adapted to the empirical data (see section 2.3) are applied to both estimates. The obtained estimates are compared for different distributions by means of a Monte Carlo study. Couples of populations with PDFs $f_1(x)$ and $f_2(x)$ were considered. Samples of the known PDFs of a Pareto distribution (4) with the extreme value index $\gamma = \{0.3, 1.2\}$, and of a Burr distribution Burr(γ, ρ) with $\gamma = 1, \rho = -1$ and $\gamma = 4, \rho = -2$, $f(x) = \lambda\tau x^{\tau-1}(1 + x^\tau)^{-\lambda-1}$, where the extreme value index γ and a parameter ρ are defined as $\gamma = 1/(\lambda\tau), \rho = -1/\lambda$ have been generated. The mixed couples Burr(1,-1)-Pareto(1.2) are also considered.

The mean penalty of the misclassification for two classes is

$$L(\eta) = p_1 \int_0^\infty q_1(x)f_1(x)\mathbf{1}\{\eta(\mathbf{x}) = \mathbf{2}\}\mathbf{dx}$$
$$+ p_2 \int_0^\infty q_2(x)f_2(x)\mathbf{1}\{\eta(\mathbf{x}) = \mathbf{1}\}\mathbf{dx}.$$

Here, $\eta : R_+ \to \{1, 2\}$ is the classifier. Since for known PDFs one can calculate an exact Bayesian risk of the misclassification $L(\eta_B)$ (the best possible), then as a measure of the relative accuracy of the PDF estimate one can consider

Table 1. Loss of the misclassification

PDFs	n	$J(g_{EB}) \cdot 10^3$			
		Pl	PlF	Ke	KeF
Pareto	50	21	87	84	86
	100	34	46	11	1
	500	24	43	3.365	5.029
Burr	50	45	29	41	63
	100	46	51	24	58
	500	42	55	13	15
Burr-Pareto	50	133	279	201	242
	100	102	209	121	186
	500	89	156	41	131

$$J(\eta_{EB}) = |\frac{1/m\sum_{i=1}^{m} L_i(\widetilde{\eta_{EB}})}{L(\eta_B)} - 1|,$$

where

$$\widetilde{L(\eta_{EB})} = \hat{p}_1 \int_0^\infty q_1(x)\hat{f}_1(x)\mathbf{1}\{\eta_{\mathbf{EB}}(\mathbf{x}) = \mathbf{2}\}\mathbf{dx}$$
$$+ \hat{p}_2 \int_0^\infty q_2(x)\hat{f}_2(x)\mathbf{1}\{\eta_{\mathbf{EB}}(\mathbf{x}) = \mathbf{1}\}\mathbf{dx}.$$

The mean is taken over m samples of a Monte Carlo study. The worse is the PDF estimate the higher is $J(\eta_{EB})$. As proportions $p_1 = p_2 = 0.5$ were taken.

Penalties $q_k(x)$ are defined in such a manner that the switching of a Bayesian classifier from one population to another is in the tail domain, i.e. for some sufficiently large x $q_i f_i(x) = q_j f_j(x), i \neq j$. It was made to reject the situation where some $q_i f_i(x)$ dominates the other for tail points. Otherwise the empirical Bayesian classifier will classify such points as the points from one class and the effect of the improvement of the accuracy of the PDF estimation will be negligible. Penalty functions $q_1(x) = (d_1 + x)^\delta$ and $q_2(x) = (d_2 + x)^\delta$ have been used, where $d_1, d_2 > 0$, $\delta = 0.5\min\{1/\gamma_1, 1/\gamma_2\}$. Here γ_1 and γ_2 are the estimates of tail indexes of the population couple. They ensure the convergence of (7). The constants d_k for two populations ($k = 1, 2$) are selected from the assumption that switching of the Bayesian criterion occurs at the $(1 - \alpha)-$quantile of an arbitrary population, i.e. $q_1(x)f_1(x)$ and $q_2(x)f_2(x)$ cross each other in the point $x^* = F_1^{-1}(1 - \alpha)$ ($F_1(x)$ is the CDF of one of the population). For the Generalized Pareto distribution

$$x^* = (\alpha^{-\gamma_1} - 1)/\gamma_1$$

holds. The values d_1 and d_2 are defined by the equation:

$$(d_1 + x^*)^\delta f_1(x^*) = (d_2 + x^*)^\delta f_2(x^*),$$

i.e. $d_1 = (d_2 + x^*)\left(f_2(x^*)/f_1(x^*)\right)^{1/\delta} - x^*$,

$$d_2 > x^* \left((f_2(x^*)/f_1(x^*))^{-1/\delta} - 1\right).$$

The quantile α is selected which provides positive values of d_1, d_2 and similar values of the penalties $q_1(x)$ and $q_2(x)$.
In the Table 1, $J(\eta_{EB})$ is shown for two types of transformed smoothed polygrams (Pl and PlF) and transformed kernel estimates with Epanechnikov's kernel (Ke and KeF). Namely, Pl and Ke are calculated by means of the transformation T_γ, but PlF and KeF are calculated by means of the fixed transformation $T(x)$. A smoothing parameter $h = \sigma n^{-1/5}$ is selected (σ^2 is the dispersion of the transformed data), which performs well with Epanechnikov's kernel (Devroye and Györfi, 1985). 10, 16, 40 were chosen as the number of quantile intervals in the polygram for the sample sizes 50, 100 and 500, respectively. The latter corresponds to the selection of the number of points inside each interval of a polygram by the discrepancy method $L = [(\sqrt{n}/2 + 1/n)(n + 1)]$, $[r]$ denotes the integer part of r (Markovitch and Krieger, 2000). The number of Monte Carlo repetition times is $m = 25$. Penalty functions corresponding to the quantile $\alpha = 0.25$ have been used.
Hill's estimate for the tail index with a bootstrap selection of the smoothing parameter k is used (Caers and Van Dyck, 1999).
The simulation study shows that:

- The transformation adapted to the tail index provides better estimates than the fixed transformation despite of the rough estimation of the tail indexes.
- A kernel estimate is worse than a polygram for relatively small samples and better for large samples.

5. CONCLUSIONS

In the paper the non-parametric estimation of a heavy-tailed PDF is considered. The compression of the data to a finite interval in such a way that the PDF of the new r.v. will be convenient for the estimation, the restoration of this PDF and an inverse transformation to get a reliable and accurate PDF estimate are explored. Since the true PDF is unknown, it is impossible to transform the r.v. to the r.v. with an exact (preliminary known) finite PDF. The transformation adapted to the data is constructed under the assumption that the PDF is Generalized Pareto and that its parameter (a tail index) is estimated by Hill's estimate. Such a transformation may provide a better estimation of the heavy-tailed PDFs than a fixed transformation (independent on the data), especially if Hill's estimate ensures a quite accurate estimate. The transformation to a triangular PDF and a smoothed polygram (or a kernel estimate) are established as the best combination for the estimation. It allows: 1) to get a stable estimation of the PDF regarding small perturbations of the transformation due to the rough tail index estimation; 2) to keep the tail decay (of the true PDF) after the inverse transformation. In contrast to a polygram, a kernel estimate does not ensure a correct tail decay rate. Besides, a kernel estimate restores the PDF of a transformed r.v. poor on the boundaries of the finite interval due to the truncation of the kernel. In this respect, a polygram is preferable, especially for quite limited sample sizes and a tail estimation. For sufficiently large samples kernel estimates restore the PDF body better.
As a quality measure of a heavy-tailed PDF estimate the relative risk of the misclassification is proposed. It is investigated for different estimates and heavy-tailed PDFs by a Monte Carlo study.

REFERENCES

Caers, J. and J. Van Dyck (1999). Nonparametric tail estimation using a double bootstrap method. *Computational Statistics & Data Analysis*, **29**, 191–211.

Devroye, L. and L. Györfi (1985). *Nonparametric density estimation: The L_1 view*, Wiley, New York.

Markovich, N.M. and U.R. Krieger (2000). Nonparametric estimation of long-tailed density functions and its application to the analysis of World Wide Web traffic. *Performance Evaluation*, **42**, 205-222.

Novak, S.Y. (1998). Berry-Esseen inequalities for a ratio of sums of random variables. *University of Sussex: Research Report No 98/07 SMS/CSSM.*

Pickands, J. (1975). Statistical inference using extreme order statistics. *Annals of Statistics*, **3**, 119–131.

Sigman, K. (1999). Appendix: A primer on heavy-tailed distributions. *Queueing Systems*, **33**, 261-275.

Tarasenko, F.P.(1968). On the evaluation of an unknown probability density function, the direct estimation of the entropy from independent observations of a continuous random variable and the distribution-free test of goodness-of-fit In: *Proceedings IEEE*, Vol.56, pp. 2052–2053.

Wand, M.P., J.S. Marron and D. Ruppert (1991). Transformations in density estimation. *Journal of American Statistical. Association, Theory and Methods*, **86(414)**, 343–353.

TORQUES GENERATION FOR DRIVING AN INSECT-LIKE ROBOT BY EVOLUTIONARY LEARNING APPROACH

A. Pedrocchi, C. Valentini, G. Ferrigno

Dept. Bioengineering Politecnico di Milano P.zza L. da Vinci, 33- 20133 Milan (Italy)

Abstract

Aim of the present work is to design and develop a system of torques generation for a mobile six-legged robot, based on biological inspiration, namely on studies on the stick insect (Carausius morosus). Virtual learning by simulation of kinematics and dynamic is here proposed as test bed of solutions before building the robot. Results confirm the goodness of the evolution learning approach in the design of the hexapod robot, showing good performances very close to nature solutions and efficient in terms of adaptation to environment constraints changes. *Copyright ©2001 IFAC*

Keywords: biocybernetics, adaptation, learning algorithms, genetic algorithms

1. INTRODUCTION

Aim of the present work is to design and develop a system of torques generation for a mobile six-legged robot, based on biological inspiration, namely on studies on the stick insect (Carausius morosus).
The biological approach in the engineering design is not easy: the number of parameters is usually much greater than the number of constraints and at the same time it's important to not reduce the potential flexibility of the system. This implies the optimization among different solutions, which is in animals achieved by a learning and adaptation process. Indeed, intrinsic aim of our biological approach is to provide the robot of a learning ability in order to successfully adapt to different environments. This way the robot shall be intelligent and autonomous. Virtual learning by simulation of kinematics and dynamic is here proposed as test bed of solutions before building the robot.
Results confirm the goodness of the evolution learning approach in the design of the hexapod robot, showing good performances very close to nature solutions and efficient in terms of adaptation to environment constraints changes.

2. PHYSICAL DESIGN

The most relevant constraint for the design of the kinematics of the robot is stability both static and dynamic, where static stability accounts for equilibrium maintenance and dynamic stability takes into account requirements of uniformity and smoothness of walking. Locomotion by legs represents the best solution for adaptation to different and uneven support surfaces, as

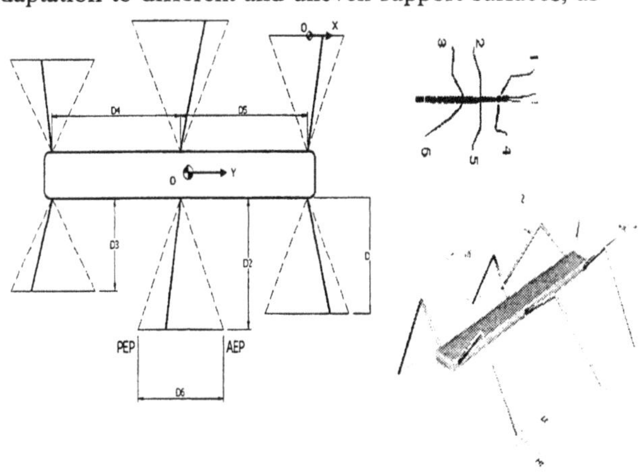

figure 1 Hexapode scheme plant, numeration of the legs and 3D visualization. Measures are in cm: D1=18, D2=20,D3=15, D4=D5=D6=24.

nature teaches. The number of legs is the result of a trade off between posture stability along with movement smoothness and control complexity. Six legs assure static and dynamic stability, good flexibility and the resulting computational effort is not too big. Again keeping nature as teacher, the number of joints for each leg is three, in order to allow walking on steep support.

3. PAWS KINEMATICS COORDINATION

The first problem to define kinematics concerns the definition of the mutual interactions between legs movements, i.e. movement coordination.
For this purpose, neurological investigations and experiments on insect walking, namely on stick insect (Carausius morosus) have been studied (Cruse et al.,).
The non-smooth movement of each leg is defined by two phases: the *power stroke*, when the support is present, and the *return stroke*, the flying phase of the leg. The alternation of those phases is called gait. A specific gait describes the coordinated movement of legs, defining the time profile and geometry of each leg end effector (paw).
Two assumptions drive the kinematics design:

1. each paw moves in straight path , parallel to the body axis
2. the body moves only parallel to itself (2D forward movement)

Each paw moves along OX axis, defining two points of transitions between the power and return strokes called *Anterior Extreme Position* (AEP) and *Posterior Extreme Position* (PEP). The leg during the power stroke achieves the PEP, here it receives the stimulus to raise and fly forward, 'return stroke', until it reaches the AEP, when it lays on the surface again.

The kinematics of each leg is described by two variables (x_i, s_i), where x_i is the coordinate of the distal point of the i-th leg with respect to the reference OX and $s_i \in \{0,1\}$ represents the status ($s_i=0$ return stroke, $s_i=1$ power stroke).

The coordination mechanisms control the transitions AEP and PEP, considering the influence of ipsilateral anterior and posterior and controlateral same leg. Four mechanisms are proposed in literature [2] for describing such coordination: three of them modify PEP and one AEP.

With respect to the nominal PEP, PEP_0, PEP can move forward and backward, without exceeding a minimal constraint PEP_{min}, based on geometric considerations, for AEP, the condition is $AEP_0 \geq AEP(t)$, $\forall t$.
In table 1 the complete expressions of the four mechanisms of coordination is reported, hereby we describe only the correspondent effect.

Mechanism 1 represents the effect of the omolateral posterior leg (if it exists) and of the controlateral leg on the PEP: the raising point can be delayed to wait for the close legs to end their return stroke and thus assure stability.

Mechanism 2 represents the effect due to the status of the immediate posterior leg (if it exists) on the PEP: the PEP can be anticipated if the immediate posterior leg can support for stability.
Mechanism 3 considers that the PEP can move forward if stability is assured by the immediate anterior leg and the controlateral leg.

Mechanism 4 describes the variations on the AEP, requested to avoid interference when the PEP of the anterior leg moved backward.

Considering all the mechanism the kinematics of each leg can be expressed as follows:

$$x_d(t+\Delta t) = x_d(t) - V\cdot\Delta t\cdot s_d(t) + \Delta t\cdot[1-s_d(t)] \qquad (1)$$

$$\begin{aligned} s_d(t+\Delta t) &= 1, && \text{if } x_d(t+\Delta t) > AEP_d \text{ and } s_d(t)=0 \\ s_d(t+\Delta t) &= 0, && \text{if } x_d(t+\Delta t) < PEP_d \text{ and } s_d(t)=1 \\ s_d(t+\Delta t) &= s_d(t), && \text{otherwise} \end{aligned} \qquad (2)$$

$$\begin{aligned} AEP_d(t+\Delta t) &= AEP_0 + \Delta^4_{d,e}(t)\,, && e\in N^4_d \\ PEP_d(t) &= PEP_0 + \sum_{m=1}^{3} \Delta^m_{d,e}(t) && e\in N^m_d \end{aligned} \qquad (3)$$

V indicates the ratio between the velocity of the power stroke and the return stroke velocity, equations are thus normalized by return stroke velocity (V_r). V_r also represents the body progress velocity, under the assumptions that all the legs move at the same velocity and the PEP_0 and AEP_0 are independent from velocity. The total number of parameters in the kinematics equation is 58, on the basis of symmetries they can reduce to 11.
The identification of those parameters depends on the environment constraints and defines the capability of adaptation of the robot.

4. KINEMATICS PARAMETERS IDENTIFICATION: LEARNING PHASE

The coordination between the legs has to result in a stable and flexible walking; stable walking, first of all, means that the acceleration of the center of mass of the robot has to be maintained in the convex polygon defined by the points of contact with the supporting surface.
The trajectory of the robot body is determined once the legs initial positions '$\bar{q}$', velocities 'V' and the set of parameters '$\bar{p}$'are known, and thus stability can be verified.

In order to find the best set of parameters a genetic approach has been chosen so to not impose any a priori rule [3]. The occurrence of loss of stability $I(\bar{p},V)$ represents the function to be minimized for best fitness to the environment. A set of 100 initial configurations, all with the constraints of initial velocity zero and uniform acceleration to get to V, are used for each individual (set of parameters) and the mean of the

table 1 Mechanisms of coordination between the paws of the stick insect. In the left column the variations Δ^j to the d-th leg parameters AEP_d and PEP_d due to the e-th leg is defined, where the e-th leg has to belong to N_d^j group (right column). The way each mechanism influences the motus law of the paws is summarized in the text.

CORRECTIONS	LEGS
$\Delta^1_{d,e}(t+\Delta t) = [s_e(t)-1]\cdot P^1_{d,e}$	$N^1_d = \begin{cases} (d+1, d+3), & \text{if } d=1,2 \\ (d+3), & \text{if } d=3 \\ (d+1, d-3), & \text{if } d=4,5 \\ (d-3), & \text{if } d=6 \end{cases}$
$\Delta^2_{d,e}(t+\Delta t) = \begin{cases} (AEP_0-PEP_{min})\cdot P^2_{d,e}, & \text{if } s_e(t)=1,\ s_e(t-\Delta t)=0 \\ 0, & \text{otherwise} \end{cases}$	$N^2_d = \begin{cases} (d+1), & \text{if } d=1,2,4,5 \\ \{0\}, & \text{if } d=3,6 \end{cases}$
$P^{ntr}_{d,e} = P^{3,1}_{d,e} + P^{3,4}_{d,e}\cdot V_r$ $\Lambda^3_{d,e}(t+\Delta t) = \begin{cases} P^{3,2}_{d,e}\cdot s_e(t)\cdot [P^{ntr}_{d,e}-x_e(t)], & \text{if } x_e(t) \le P^{ntr}_{d,e} \\ P^{3,3}_{d,e}\cdot s_e(t)\cdot [P^{ntr}_{d,e}-x_e(t)], & \text{if } x_e(t) > P^{ntr}_{d,e} \end{cases}$ With the limit: $\Delta^3_{d,e}(t+\Delta t) = \begin{cases} \Lambda^3_{d,e}(t+\Delta t) & \text{if } \Lambda^3_{d,e}(t+\Delta t) < (AEP_0-PEP_0)\cdot 0.2 \\ (AEP_0-PEP_0)\cdot 0.2, & \text{if } \Lambda^3_{d,e}(t+\Delta t) \ge (AEP_0-PEP_0)\cdot 0.2 \end{cases}$	$N^3_d = \begin{cases} (d+3), & \text{if } d=1 \\ (d-1, d+3), & \text{if } d=2,3 \\ (d-3), & \text{if } d=4 \\ (d-3, d-3), & \text{if } d=5,6 \end{cases}$
$\Delta^4_{d,e}(t+\Delta t) = \begin{cases} [PEP_{min}-AEP(t)]\cdot P^4_{d,e}, & \text{if } y_d(t) \ge y_e(t) \\ 0, & \text{otherwise} \end{cases}$	$N^4_d = \begin{cases} (d-1), & \text{if } d=2,3,5,6 \\ \{0\}, & \text{if } d=1,4 \end{cases}$

instability over the all initial configurations represents the fitness of that individual at velocity V.

$$I_N(\bar{p},V) = \frac{\sum_{i=1}^{N} d^t(\bar{q}^i,\bar{p},V)}{N} \tag{4}$$

Fitness to the environment is then calculated as the mean over all the allowed velocities (from 0.1 to 1, every 0,1), so to produce a set of parameters optimized over all the available velocities

$$\eta(\bar{p}) = \frac{\sum_{V=0.1}^{1} I_N(\bar{p},V)}{10} \tag{5}$$

Every population W^n is composed by L individuals $\overline{W}_i^n$, i=1,..., L (L=10), each representing and array of $\bar{p}$: the computation of $I_N(\bar{p},V)$, with N=100, allows to associate to each individual a fitness $\eta(\bar{p})$.

Initial population is randomly build, including the '0-individual'. Each individual has a probability to survive in the next generation depending on its performance $\eta(\bar{p})$. Survived individuals from one generation to the next are mutated so to assure exploration in the whole domain. Learning process lasts for 100 generations.

Results on genetic learning are reported in figure 2, performances of the best individual at all walking velocities after 50 (triangles) and 100 (squares) generations and in case of random parameters (circles). Learning process is able to strongly reduce the instability, by 100 generations no strong improvements are shown thus the process is ended.

5. DRIVING TORQUES CALCULATION

The complete kinematics of the joints for each leg is then calculated using Denavit-Hartenberg approach. End effectors laws of movement have been defined imposing the paws horizontal motions derived by the coordination mechanisms and that the paws vertical motions follow a smooth bell shape trajectory, avoiding high accelerations.

For robot driving, we need to derive from kinematics the torques to be applied at each joint. D'Alembert equations are adopted with boundary conditions about Coulomb forces and directions of the constraints, but infinite solutions are still available. Legs are all equal, they consist of two cylinders: the femur (10cm lenght and ∅=0,8cm) and the tibia (20cm lenght and ∅=0,6). The inertial characteristics are calculated imposing the robot to be steel (P_s=7,8 kg/dm^3). To choose the best solution a typical biological inspired approach is used: we hypothesized that the developed torques at all the joints has to be the minimal, indicating a strategy of muscular effort save.

6. RESULTS

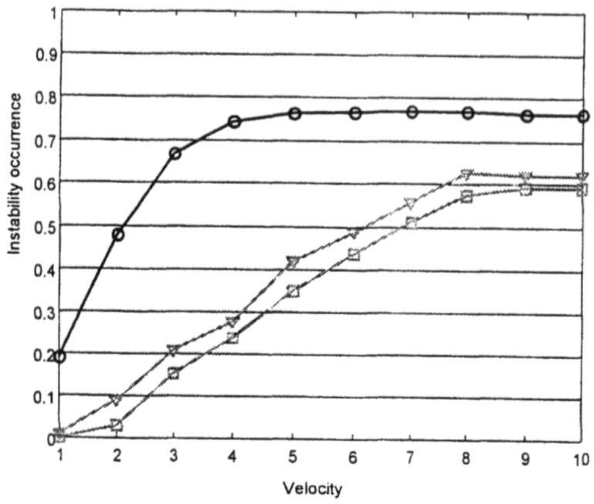

P^1_i	= 11.99	P^1_c	= 5.99
P^2	= 0		
$P^{3,1}_i$	= -9,31	$P^{3,1}_c$	= -2.63
$P^{3,2}_i$	= 6.23	$P^{3,2}_c$	= 0.08
$P^{3,3}_i$	= 0.08		
$P^{3,4}_i$	= 16.06	$P^{3,4}_c$	= 0
P^4_i	= 1		

Figure 2 –Upper panel: Performances (instability occurrence) of the best individual of after 50 generations (triangles) and after 100 generations (squares) compared with random coordinated kinematics (circles). Lower panel: resulting set of parameters after 100 generations

The mutual interactions resulting from the learning process for the definition of the coordination parameters are very interesting with respect to nature solutions. At low velocity, a wave oscillation characterized the coordination, similar pattern is observed in biology and indicated as 'wave gait'. Increasing velocity, strategies of synchronization emerge, and two or three legs make the return strokes simultaneously, analogously with real insects.

In figure 3 examples of gait cycles are reported at low velocity and at high velocity.

Complete kinematics at all the joints have been calculated by Denavit-Hartenberg equations and resulting profiles are feasible, producing a smooth forward movement of the body.

Driving torques, calculated by inverse dynamics imposing proper inertial characteristics and a strategy of energy save, are feasible (2-7 Nm) and meaningful values of the torques at all the joints have been found confirming the goodness of the adopted approach.

In order to evaluate the capability of adaptation to environment changes, some realistic scenarios have been simulated: sliding, lesion and slow down of one paw. In all those simulations the kinematics of the other legs changed, resulting in a successful adaptation to the different constraints, confirming the flexibility of the robot (figure 4).

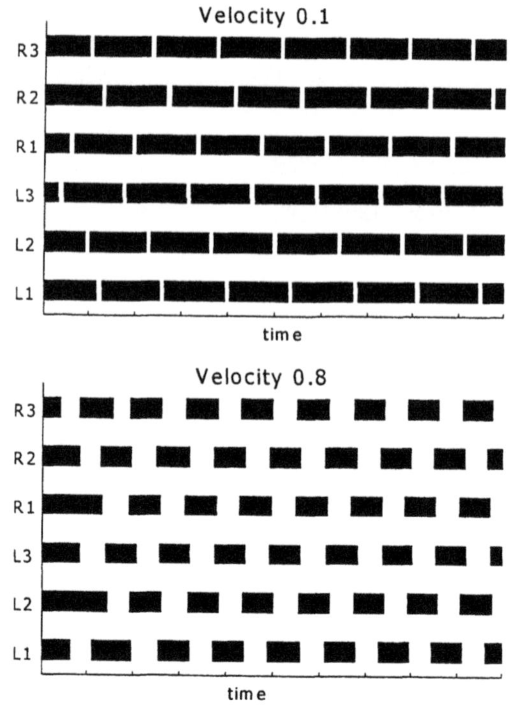

Figure 3 – Examples of gait cycles at low velocity (upper) and at high velocity (lower). Each line corresponds to one leg (R1 right anterior, R2: right central and so on). Black rectangles indicate power stroke phase, while white rectangles indicate return phase. In the low velocity case, wave swing is recognized R3, R2, R1, L3,L2, L1; while in case of high velocity, a synchronization of groups of legs emerges: (R3,L2,R1) and (L3,R2, L1). In addition, an initial non periodic gait is then substituted by fixed cycles as the environment characteristics are known.

7. DISCUSSION AND CONCLUSIONS

Biology and neurophysiology studies on the stick insect have been here adopted for designing kinematics and dynamics characteristics of a hexapod robot. In addition, the project has been defined with the specific aim of achieving the typical flexibility and adaptation capability of insects walking. In nature, these features are the results of a learning and evolutionary process; analogously, the design optimization have been performed by genetic approach based on typical physiological and biological criteria.

It is important to notice that learning is performed without explicit rules: there don't exist specific corrections in case of instability, it is the interaction with the environment that gives the performance of the individual, this way the learning process adopted can be defined by trials and errors, as in insects life.

The results confirm the similarity of the robot to real insect behaviors, particularly in coordination strategies and in capability to react to changes in the environment.

In conclusion, the study demonstrates the power of a genetic and biological-drive approach to intelligent robot design; furthermore the virtual learning here adopted represents a very good test bed for robot verification, before hardware tests.

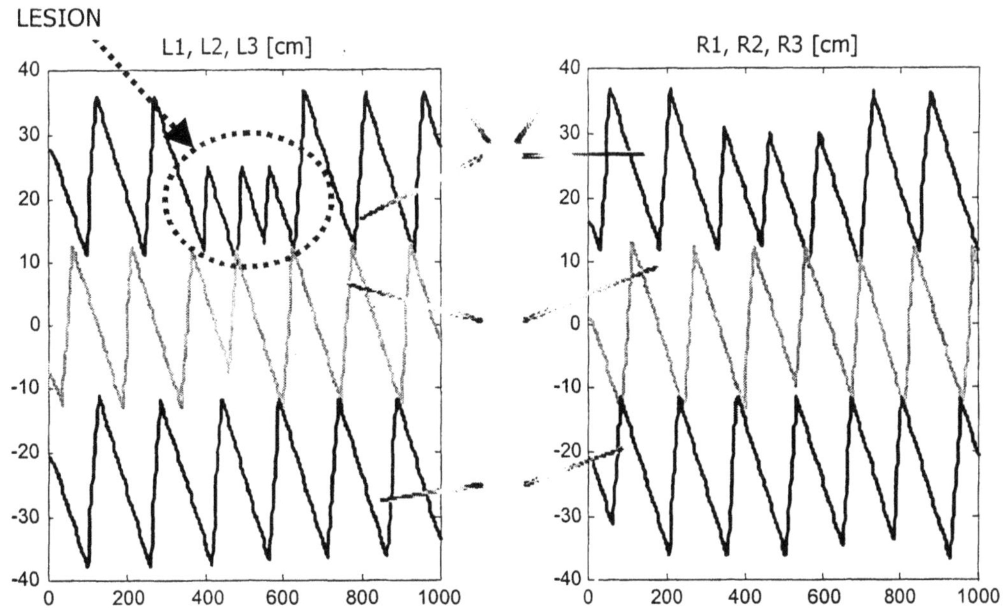

Figure 4 – Simulation of the legs kinematics in case of temporary lesion at the left anterior leg (L1). The more close legs kinematics are changed to assure a correct walking and the lesion is not visible at the level of the far legs, uindicating that the complex walking proceed correctly.

REFERENCES

H. Cruse, J. Dean, T. Kindermann, J. Schmitz, M. Schumm (1998) Simulation of complex movements using artificial neural networks *Verlag der Zeitschrift fur Naturforschung 53c: 628-38*

G.S. Cymbalayuk, R.M. Borisyuk, U. Müeller-Wilm, H. Cruse (1998) Oscillatory network controlling six-legged locomotion. Optimization of model parameters, *Neural Networks 11: 1149-1460*

J. Dean (1991) A model of leg coordination in stick insects Carausuis morosus *Biol. Cybern 64: 403-11*

S. Kimura, M. Yano, H. Shimizu (1994) A self-organizing model of walking patterns of insects, *Biol. Cybern. 70(6): 505-12*

M. Michalewicz (1994) Genetic algorithms + data structures = evolution programs, *AI Series, Springer–Verlag, New York*

ADAPTATION AND LEARNING IN NEURAL NETWORKS MULTIPLE MODELS BASED CONTROL OF MOBILE ROBOTS

A. D'Amico G. Ippoliti S. Longhi

Dipartimento di Elettronica e Automatica,
Università di Ancona,
Via Brecce Bianche, 60131 Ancona, Italy,
E-mail: {g.ippoliti, s.longhi}@ee.unian.it .

Abstract: The paper proposes a multiple models based control scheme for the solution of the tracking problem for mobile robots. The proposed method utilizes multiple models of the robot for its identification in an adaptive and learning control framework. Radial Basis Function Networks (RBFN) are considered for the multiple models. The training of the nets and the control performances analysis have been done in a real experimental setup. The proposed solutions are implemented on a PC-based control architecture for the real-time control of the LabMate mobile base. The experimental results are satisfactory in terms of tracking errors and computational efforts and show the improvement in the tracking performance when the proposed methodology is used for tracking tasks in dynamical uncertain environments.

Keywords: Multiple models control, Mobile robots, Neural networks.

1. INTRODUCTION

The point stabilization and the tracking trajectory problems for mobile robots have been given great attention by automatic control researchers in recent literature (see e.g. (Canudas de Wit *et al.*, 1993; Corradini *et al.*, 1999; Corradini *et al.*, 2000*c*; Fierro and Lewis, 1997; Jiang and Nijmeijer, 1997; Kolmanovsky and McClamroch, 1995; Yang and Kim, 1999)). As pointed out in (Fierro and Lewis, 1997; Kolmanovsky and McClamroch, 1995), although an extensive theoretical framework has been developed in this field, relatively few results have been presented about the robustness issue, i.e. the problem of control of nonholonomic systems when there are model uncertainties (Kolmanovsky and McClamroch, 1995).

The experimental validation of any controller using commercially available vehicles imposes the use of the kinematic model, since the robot is usually controlled setting its linear and angular velocities. However this experimental setup based on the kinematic model, hinders to take into account the parameter variations in the design of the control algorithm. In fact, the kinematic model does not contain information on vehicle wheels dimensions, robot mass and inertia.

In order to provide the control schemes with some degree of robustness, still preserving the possibility of its experimental testing, Neural Networks (NN) based controllers have been proposed in the past (Corradini *et al.*, 2000*a*; Corradini *et al.*, 2000*b*) with satisfactory results. The nonlinear approximation capabilities of the considered Neural Networks (NN) have been used for modeling the kinematic behaviour of the vehicle and the dynamic effects introduced by the wheels rolling and by the relative strong linear and angular accelerations imposed by path planners.

In this paper a neural network based control is proposed which makes use of multiple models of the mobile robot in an adaptive and learning control framework. This approach situates in the rapidly emerging area of multiple models logic-based switching control (Baldini *et al.*, 1999; Chen and Narendra, 2000; Ippoliti and Longhi, 2001*a*; Ippoliti and Longhi, 2001*b*; Morse, 1996; Narendra and Xiang, 2000; Narendra *et al.*, 1995). A switching strategy between these models determines the best possible control input at any given instant to improve the overall tracking performances with respect to single model adaptive RBFN based control and fixed RBFN based control scheme. In this work a learning and adaptive control scheme is proposed for achieving fast and accurate output response in the presence of large uncertainties. Such scheme leads to switching to the appropriate controller and eventually tuning such a controller to improve the performance.

The approach here recalled has been specialized to the control of a mobile robot. In particular, radial basis function networks have been used for the identification and control in order to avoid the non linear optimization techniques used in the learning algorithm of the Multilayer Neural Networks (MNN) considered in (Corradini *et al.*, 2000*a*; Corradini *et al.*, 2000*b*) and the related problems of local minima. The experimental results using RBFN control schemes (D'Amico *et al.*, 2001) have shown that RBFN are a possible alternative to control schemes based on MNN (Corradini *et al.*, 2000*a*; Corradini *et al.*, 2000*b*). Moreover these nets are useful for the identification and control of complex nonlinear dynamical systems in the presence of uncertainty because the particular structure of RBFN allows their applicability to adaptive and adaptive-learning control schemes.

Robots usually works in different operative environments which assume different forms depending on the applications. The robot model usually incorporates the motion dynamics and features of the environment as external disturbances, parametric variations or unmodeled phenomena. In this context a multiplicity of models can be used to identify different environments and to determine the necessary control actions (Chen and Narendra, 2000; Narendra and Xiang, 2000; Narendra *et al.*, 1995).

The experimental testing of the proposed NN multiple models control schemes has been performed on the LabMate, available at the Robotics Lab of the University of Ancona and provided with an automatic navigation system. The experimental results show a noticeable improvements of the vehicle performances with respect to a single model adaptive RBFN based control scheme and a fixed RBFN based controller confirming that the theoretical concepts on switching, learning and tuning developed in (Narendra *et al.*, 1995) can be applied to achieve significant improvement in the control of mobile robots.

The paper is organized as follows. The radial basis function networks used for implementing the multiple models control scheme, and their learning algorithm are recalled in Section 2. The details of the multiple models control scheme based on the inversion of the mobile robot kinematic model via neural networks is described in Section 3, while experimental results are reported in Section 4. The paper ends with some comments on the performances of the proposed controllers with respect to that obtained with single model adaptive RBFN based control schemes and with fixed RBFN based controllers.

2. LEARNING ALGORITHM FOR RADIAL BASIS FUNCTION NETWORKS

2.1 Radial Basis Function Networks

A RBFN with n inputs and a scalar output implements a mapping $f_r : \mathbb{R}^n \to \mathbb{R}$ according to

$$\hat{y} = f_r(x) = \lambda_0 + \sum_{i=1}^{n_r} \lambda_i \phi(\|x - c_i\|) \quad (1)$$

where $x \in \mathbb{R}^n$ is the input vector, $\phi(\cdot)$ is a given function from $\mathbb{R}^+$ to $\mathbb{R}$, $\|\cdot\|$ denotes the Euclidean norm, λ_i, $i = 0, 1, \cdots, n_r$ are the weights or parameters, $c_i \in \mathbb{R}^n$, $i = 1, 2, \cdots, n_r$, are the radial basis functions centers, and n_r is the number of centers (Chen *et al.*, 1991; Girosi and Poggio, 1990; Warwick *et al.*, 1996).

In the estimation of the output y of a dynamical system by RBFN, the dynamics can be taken into account through the input vector x, that is composed by dynamical system input and output samples acquired in a finite set of past time instants (Hunt *et al.*, 1992; Warwick *et al.*, 1996). Therefore the network input vector is formed as follows

$$\begin{aligned} x(k) = [&y(k-1)\; y(k-2)\; \cdots\; y(k-n_y) \\ &u(k-1)\; u(k-2)\; \cdots\; u(k-n_u)], \end{aligned} \quad (2)$$

where $u(\cdot)$ and $y(\cdot)$ are respectively the dynamical system input and output samples and n_y, n_u are positive integers to be chosen. Then, the network output $\hat{y}$ is an estimation of the plant output y, at time instant k.

In the RBFN the functional form $\phi(\cdot)$ and the centers c_i are suitably fixed. By providing a set of the network input $x(k)$ and the corresponding desired plant output $y(k)$ for $k = 1, 2, \cdots, N$, the values of the weights λ_i, $i = 0, 1, \cdots, n_r$ can be determined using the linear Least Squared (LS) method. The choices of $\phi(\cdot)$ and c_i must be carefully considered in order the RBFN closely matches the performances of the multilayer neural networks. Theoretical investigation and practical results show that the choice of the non-linearity $\phi(\cdot)$, a function of the distance d_i between the current input x and the centre c_i, does not significantly influence the performance of the RBFN. Therefore, the following guassian function is considered:

$$\phi(d_i) = \exp\left(-d_i^2/\beta^2\right), \ i = 1, 2, \cdots, n_r \quad (3)$$

where $d_i = \|x - c_i\|$ and the real constant β is simply scaling or "width" parameter.

The centers c_i, $i = 1, 2, \cdots, n_r$ are generally chosen from the data set $\{x(k)\}_{k=1}^N$. The Orthogonal Least Squares (OLS) algorithm (Chen *et al.*, 1991) is an efficient method for selecting centers from the data set obtaining adeguate and parsimonious RBFN thus reducing computational complexity and numerical ill-conditioning. In this work the centers have been selected using the OLS algorithm operating in a forward regression way until a network having adeguate performance is obtained (Chen *et al.*, 1991).

2.2 K-Means Clustering Algorithm

The Orthogonal Least Squares (OLS) method is employed as a forward regression procedure (Chen *et al.*, 1991) to select a suitable set of centers from a large initial set of candidates.

In the present study the initial set of centers is obtained using the k-means unsupervised clustering algorithms (Hush and Horne, 1993) that starting from a reasonable high number of centers randomly chosen in the input space, moves them in the most significative regions of the input space.

The k-means clustering algorithm is given by the sequential execution of the following steps:

Step 1 Initialize the cluster centers c_j, $j = 1, 2, \cdots, n_c$. This centers are randomly chosen from the input data set $\{x(k)\}_{k=1}^N$.

Step 2 Group all the inputs of the data set $\{x(k)\}_{k=1}^N$ in the sets Ψ_j, $j = 1, 2, \cdots, n_c$, where Ψ_j is the set of input vectors $x(\cdot)$ closest to the cluster center c_j, i.e. $\Psi_j := \{x(\cdot) \in \{x(k)\}_{k=1}^N | \ \|x(k) - c_j\| = \min_{i=1}^N \|x(k) - c_i\|\}$.

Step 3 For all $j = 1, 2, \cdots, n_c$ do $c_j = \frac{1}{M_j} \sum_{x(k)\in\Psi_j} x(k)$

where M_j is the number of input vectors $x(k) \in \Psi_j$.

Step 4 If there is no change in the cluster assignments of step 3 from one iteration to the next the algorithm is stopped otherwise go back to step 2.

For a graphic representation of the clustering algorithm see Figure 1 where panels a and b respectively represent *Step 2* and *Step 3* of the algorithm.

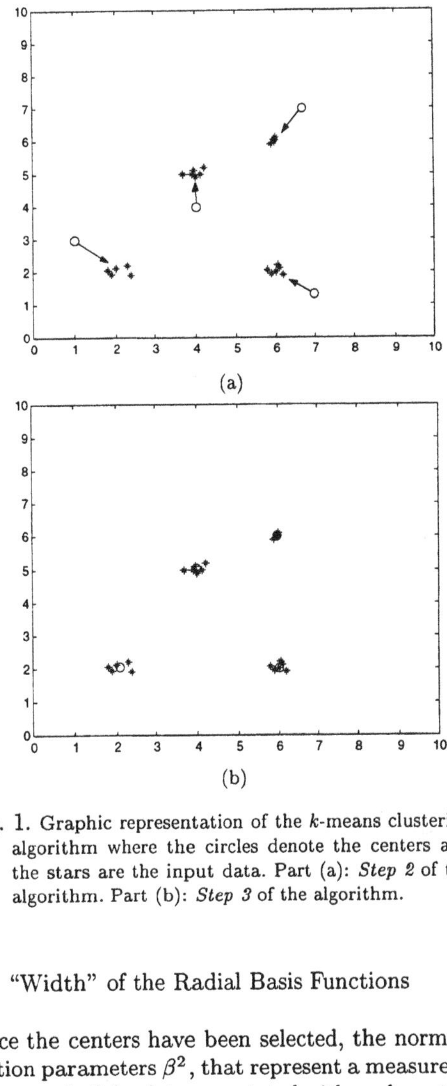

Fig. 1. Graphic representation of the k-means clustering algorithm where the circles denote the centers and the stars are the input data. Part (a): *Step 2* of the algorithm. Part (b): *Step 3* of the algorithm.

2.3 "Width" of the Radial Basis Functions

Once the centers have been selected, the normalization parameters β^2, that represent a measure of the spread of the data associated with each centre, have to be determined. No rigorous method exists to calculate these parameters.

In this work an Akaike-type criteria (Chen *et al.*, 1991) is proposed for choosing the normalization parameters obtaining a good compromise between estimation accuracy and network complexity. Akaike-Type criteria which compromise between the performance and the number of parameters has the following form:

$$AICT(\chi) = N\log(\sigma_\varepsilon^2) + M_s\chi \quad (4)$$

where χ is the critical value of the chi-squared distribution with one degree of freedom and for a given level of significance, N is the number of inputs, σ_ε^2 is the variance of the residuals, and M_s is the number of significant regressors.

2.4 Adaptive Neural Networks

The RBFN is a special two-layer network which is linear in the parameters λ_i, $i = 0, 1, \cdots, n_r$ by fixing all radial basis functions centers and nonlinearities in the hidden layer. The second layer implements a linear combiner of the outputs of the hidden layer where the weights $\lambda_i, i = 0, 1, \cdots, n_r$ of this linear combiner are determined using standard least squares algorithms.

This particular structure of RBFN allows their applicability to adaptive control schemes (Warwick *et al.*, 1996; D'Amico *et al.*, 2001), where the weights of the network second layer are on-line adapted making use of plant input and output signals. This adaptation allows the control scheme to cope with parameters variations and external disturbances.

2.5 Autonomous Learning Strategy

The structure of the RBFN allows also an on-line adaptive clustering by the update of the number of the radial basis function centers and of the weights of the second layer. The proposed on-line adaptive clustering algorithm, for each new input vector, checks if the outputs of the first RBFN layer are significantly different from zero. If the outputs of the hidden layer are under a proper fixed threshold it means that the new input vector is outside the input space covered by the RBFN. Therefore the algorithm adds a new center to the RBFN to cover the new input vector and update the weight of the second layer with the LS method.

3. MULTIPLE MODELS CONTROL FOR MOBILE ROBOTS

In this work the use of multiple models in an adaptive and learning control architecture is proposed for the control of mobile robots. The main idea is to generate a control vector making use of the model which closely approximates the mobile robot at each instant. This leads to a switching control scheme which improve the performance of the tracking error.

3.1 Radial Basis Function Neural Networks Control for Mobile Robots

The discrete-time model of the dynamics of an unicycle mobile robot tracking errors has the following form (Corradini *et al.*, 2000*a*; Corradini *et al.*, 2000*b*):

$$\begin{aligned}
\varepsilon_1(k+1) &= \varepsilon_1(k) + f_1(\varepsilon_2)\omega(k) + f_2(\varepsilon_3, v_r) - \\
&\quad T_c v(k) \\
\varepsilon_2(k+1) &= \varepsilon_2(k) + f_3(\varepsilon_1)\omega(k) + f_4(\varepsilon_3, v_r) \\
\varepsilon_3(k+1) &= \varepsilon_3(k) + f_5(\omega_r) - T_c\omega(k) \qquad (5)
\end{aligned}$$

where $\varepsilon_i(\cdot)$, $i = 1, 2, 3$ are the tracking errors, the argument k denotes the variable evaluated in $t = kT_c$ where T_c is the sampling time, $v(\cdot)$ and $\omega(\cdot)$ are respectively the linear and angular robot velocities assumed constant over the time-interval $[kT_c, (k+1)T_c)$, $v_r(\cdot)$ and $\omega_r(\cdot)$ are respectively the reference linear and angular robot velocities evaluated over the time-interval $[kT_c, (k+1)T_c)$ and f_i, $i = 1, \cdots, 5$, are nonlinear functions depending on $\varepsilon_i(\cdot)$, $i = 1, 2, 3$.

The most immediate application of neural networks for controlling the mobile vehicle directly originates from NN theoretical ability to approximate arbitrary nonlinear mappings (Girosi and Poggio, 1990). Exploiting this feature, the inverse model of the vehicle kinematic system can be obtained using a number of different schemes (see e.g. (Hunt *et al.*, 1992). In the following a control scheme for mobile robots based on RBFN is proposed.

Consider the following control laws respectively for the linear and angular velocities of the mobile robot:

$$\begin{aligned}
v(k) &= \frac{1}{T_c}(k_1\varepsilon_1(k) + f_1(\varepsilon_2)\omega(k) + f_2(\varepsilon_3, v_r)) \\
\omega(k) &= \frac{1}{T_c}(f_5(\omega_r) + k_2\varepsilon_2(k) + k_3\varepsilon_3(k)) \qquad (6)
\end{aligned}$$

where k_1, k_2 and k_3 are the control gains. Under the assumption of $v_r(k)$ or $\omega_r(k)$ different to zero (Canudas de Wit *et al.*, 1993), these control gains are chosen for imposing the local asymptotic stability of the equilibrium point ($\varepsilon_1 = \varepsilon_2 = \varepsilon_3 = 0$) of the closed loop system described by (5) and (6).

For the considered control laws (6) a RBFN-based implementation has been proposed. The overall control scheme is shown in Figure 2. In particular the term $f_1(\varepsilon_2)$ in (6) is substitute with its rectangular approximation expressed by:

$$f_1(\varepsilon_2) := \int_{kT_c}^{(k+1)T_c} \varepsilon_2(\tau)d\tau \simeq T_c\,\varepsilon_2(k). \qquad (7)$$

The term $f_2(\varepsilon_3, v_r)$ is easily estimated using the neural network NN1, off-line trained using experimental data according to the relationship given by the first row of (5). The term $f_5(\omega_r)$ is estimated using the neural network NN2, off-line trained using experimental data, according to the relationship given by the third row of (5).

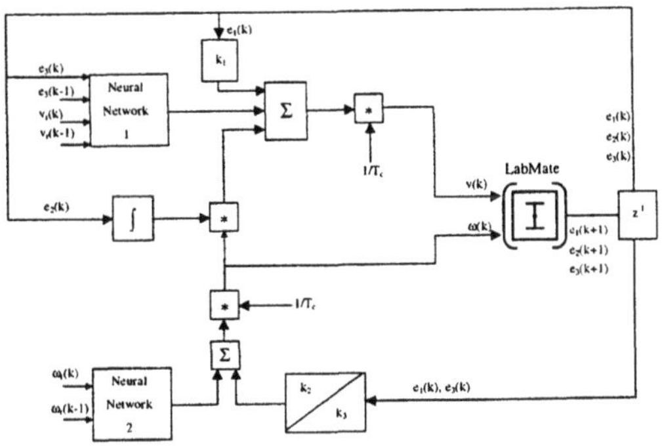

Fig. 2. RBFN-based Control Scheme.

3.2 Multiple Models Framework for Mobile Robots Control

Consider a set of L models $\{M_j\}_{j=1}^{L}$ that have the same form as the dynamics of the mobile robot tracking errors (5):

$$\begin{aligned} \varepsilon_1^j(k+1) &= \varepsilon_1(k) + f_1^j(\varepsilon_2)\omega(k) + f_2^j(\varepsilon_3, v_r) - \\ &\quad T_c v(k) \\ \varepsilon_2^j(k+1) &= \varepsilon_2(k) + f_3^j(\varepsilon_1)\omega(k) + f_4^j(\varepsilon_3, v_r) \\ \varepsilon_3^j(k+1) &= \varepsilon_3(k) + f_5^j(\omega_r) - T_c\omega(k) \end{aligned} \quad (8)$$

with $j = 1, 2, \cdots, L$.

Moreover, a set of L regulators $\{R_j\}_{j=1}^{L}$ is considered, where the controller R_j is designed for controlling the corresponding model M_j, $j = 1, 2, \cdots, L$. In the proposed multiple models control scheme at every time instant k only one controller of the set $\{R_j\}_{j=1}^{L}$ is applied to the plant and a parallel architecture of L models $\{M_j\}_{j=1}^{L}$, forced by the same control input of the plant, yields the outputs $y_j(k) := [\varepsilon_1^j(k)\ \varepsilon_2^j(k)\ \varepsilon_3^j(k)]^T$, $j = 1, 2, \cdots, L$.

With this control structure, the performances of the different controllers can not be directly evaluated in parallel because at any time instant only one control input can be chosen. Therefore, the output responses of the multiple models are used for an estimation of the controllers performances. Defining the prediction errors:

$$e_j(k) := y(k) - y_j(k), \ j = 1, 2, \cdots, L, \quad (9)$$

where $y(k) := [\varepsilon_1(k)\ \varepsilon_2(k)\ \varepsilon_3(k)]^T$ the following performance indices are considered:

$$J_j(k) := a e_j^T(k) e_j(k) + b \sum_{\tau=k_0}^{k-1} l^{(k-\tau-1)} e_j^T(\tau) e_j(\tau),$$

$$j = 1, 2, \cdots, N, \quad k \geq 0, \ k \in \mathbb{Z}, \quad (10)$$

where the design parameters a and b are positive constants, the module of l is lesser then one and k_0 is the initial time instant.

In this multiple models control scheme, at every time instant the performance indices are monitored in order to deduce the controller to be applied to the plant i.e. the controller corresponding to the model with the minimum index is used (switched) for controlling the plant.

Different control strategies can be incorporated into this multiple models framework. In this work the multiple models control scheme incorporates three different controllers.

(i) The first controller is a fixed RBFN based control system described by equations (5) and (6), where the neural networks have been trained with data acquired in a set of trajectories chosen with different shapes, curvature radii and tracking velocities as described in Subsection 3.1. The measures of the position and orientation of the mobile robot and of the control efforts applied along the set of trajectories have been used for nets training.
(ii) The second controller is a single model adaptive RBFN based controller implemented as described in Subsection 2.4. The overall control scheme of this solution is the same as Figure 2 but the terms $f_2(\varepsilon_3, v_r)$, and $f_5(\omega_r)$ are estimated respectively with the neural networks NN1 and NN2 where the weight of the second layers are on-line tuned making use of the data acquired from the mobile robot.
(iii) The last controller is an adaptive and learning RBFN based controller implemented as described in Subsection 2.5. The overall control scheme of this solution is the same as Figure 2 but the terms $f_2(\varepsilon_3, v_r)$, and $f_5(\omega_r)$ are estimated respectively with the neural networks NN1 and NN2 where the number of centers of the first layer and the weight of the second layer are on-line updated making use of the acquired data from the mobile robot.

4. EXPERIMENTAL RESULTS

The experimental testing of the proposed controller has been performed on the LabMate mobile robot, available at the Robotics Lab of the University of Ancona and provided with an automatic navigation system (Conte *et al.*, 1996). The navigation system is based on the hybrid structure (Angeloni *et al.*, 1996; Jetto *et al.*, 1999) composed by the motion planning module, the localization and control module, the perception module and the obstacle avoidance module. All these modules of the navigation system are implemented on a PC based architecture. Figure 3 shows the LabMate mobile base equipped with the PC based architecture.

Fig. 3. LabMate mobile robot.

With this experimental setup the real-time implementations of the developed multiple models RBFN based controllers have been tested in the control module of the mobile robot. The training phases of the RBFN have been performed with data acquired in a set of trajectories. The measures of the position and orientation of the mobile robot and of the control efforts applied along the set of trajectories have been used.

A set of 15 trajectories, chosen with different shapes, curvature radii and tracking velocities, have been used for the training phase. The hidden layer of the nets used in the proposed control schemes contains from 6 to 9 neurons.

The real-time multiple models RBFN based controller has been implemented on a PC 486 DX2 with PC 104 bus installed on the vehicle (see Figure 3). A sampling time of 0.4 s has been used, in order to guarantee a maximum vehicle velocity of 0.3 m/s.

Performances of the proposed multiple models RBFN based control scheme have been tested in many different environmental conditions and different planned trajectories with satisfactory results.

Samples of the performed experimental tests are reported in the following. The performances of the proposed multiple models RBFN based controller with the trajectories of Figures 4a and 5a are reported. The proposed multiple models RBFN based controller has been tested with constant disturbances acting on the control inputs of the mobile robot (Canudas de Wit and Khennouf, 1995). The constant disturbances start at the 50th sample for all the reported trajectories and finish at the 200th sample for the trajectories of Figure 4a and at the 140th sample for the trajectory of Figure 5a. These performances are compared with that obtained by a single model adaptive RBFN based control scheme and by a fixed RBFN based controller.

Figure 4 (panels b–d) and Figure 5 (panels b–d) show, respectively for the trajectories of Figures 4a and 5a, the position tracking errors (panels b, c) and orientation tracking error (panel d) obtained with the RBFN-based control schemes. The continuous lines represent the tracking errors of the proposed multiple models RBFN based controller, the dotted lines represent the tracking errors of the single model adaptive RBFN based control scheme and the dashed lines represent the tracking errors of the fixed RBFN based controller. The applied sequence of regulators R_j $j = 1, 2, 3$ of the multiple models RBFN based control scheme, respectively for the trajectories of Figures 4a and 5a, is displayed in Figures 4e and 5e. Note that the considered multiple models control structure switches on the adaptive and adaptive-learning control systems only in the presence of constant disturbance inputs, switching back to the fixed controller when the disturbances end. This fact shows that the fixed RBFN based controller is suitable for undisturbed trajectories. In fact, it is based on neural networks trained on a large set of different undisturbed trajectories, and therefore this nets have the ability to generalize when presented with undisturbed trajectories not appearing in the training set.

The investigated multiple models RBFN-based controller is satisfactory in terms of tracking errors and computational efforts and shows an improvement of the control performances with respect to single model adaptive RBFN based control schemes and fixed RBFN based controllers in the case of constant disturbance inputs on the vehicle.

5. CONCLUSIONS

In this paper a multiple models RBFN based approach for the tracking problem of a mobile vehicle is analyzed. A multiple models RBFN-based controller is developed and compared to a single model adaptive RBFN based control scheme and to a fixed RBFN based controller by experimental tests performed on the mobile base LabMate.

The proposed approach is an application of a general learning control methodology which is based on the use of multiple models, switching, and tuning (Narendra *et al.*, 1995).

Fixed, adaptive and adaptive-learning RBFN based control schemes are integrated into this multiple models framework. The control algorithm makes use of multiple models of the mobile robot and of a switching scheme which determines the specific control input that drives the robot actuators. Moreover the proposed adaptive-learning control algorithms is able to introduce new models during robot tasks.

The experimental tests were carried out for different environmental conditions and confirm that the proposed multiple models based approach improves the overall tracking performance as compared to the performance of a single model adaptive RBFN based controller and of a fixed RBFN based control scheme.

The neural networks have been trained on experimental data acquired from the mobile base LabMate. Different training trajectories have been considered. The neural networks tracking algorithms have been implemented on the PC based real-time control architecture installed on the vehicle.

6. REFERENCES

Angeloni, A., T. Leo, S. Longhi and R. Zulli (1996). Real time collision avoidance for mobile robots. In: *Proceedings of the 6th Symposium on Measurement and Control in Robotics (ISMCR 96)*. Brussels, Belgium. pp. 239–244.

Baldini, M., M.L. Corradini, L. Jetto and S. Longhi (1999). A multiple-model based approach for the intelligent control of underwater remotely operated vehicles. In: *Proceedings of the 14th Triennial World Congress of IFAC*. Vol. Q. Beijing, P.R. China. pp. 19–24.

Canudas de Wit, C., H. Khennouf, C. Samson and O. J. Sordalen (1993). Nonlinear control design for mobile robots. In: *Recent Trends in Mobile Robots* (Y. F. Zheng, Ed.). pp. 121–156. World Scientific. Singapore.

Canudas de Wit, C. and H. Khennouf (1995). Quasi-continuous stabilizing controllers for nonholonomic systems: design and robustness considerations. In: *Proceedings of the 3rd European Control Conference*. Rome, Italy. pp. 2630–2635.

Chen, L. and K. S. Narendra (2000). Nonlinear adaptive control using neural networks and multiple models. In: *Proceedings of the 2000 American Control Conference*. Vol. 6. pp. 4199–4203.

Chen, S., C. F. N. Cowan and P. M Grant (1991). Orthogonal least squares learning algorithm for radial basis function networks. *IEEE Transactions on Neural Networks* **2**(2), 302–309.

Conte, G., S. Longhi and R. Zulli (1996). Motion planning for unicycle and car-like robots. *International Journal of Systems Science* **27**(8), 791–798.

Corradini, M. L., T. Leo and G. Orlando (1999). The tracking problem for a mobile robot: a sliding mode controller for the dynamical model. In: *Proceedings of the 1999 European Control Conference (ECC99)*. Karlsruhe, Germany.

Corradini, M. L., G. Ippoliti, S. Longhi and S. Michelini (2000*a*). Neural networks inverse model approach for the tracking problem of mobile robot. In: *Proceedings of the 9th International Workshop on Robotics in Alpe-Adria-Danube Region (RAAD 2000)*. Maribor, Slovenia. pp. 17–22.

Corradini, M. L., G. Ippoliti, T. Leo and S. Longhi (2000*b*). The tracking problem of mobile robots: experimental results using a neural network approach. In: *Proceedings of the 2000 World Automation Congress (WAC 2000)*. Maui, Hawaii, USA.

Corradini, M. L., T. Leo and G. Orlando (2000*c*). Robust control of a mobile robot subjected to skidding effects: a sliding mode controller using the kinematic model and its experimental testing. In: *Proceedings of the 6th IFAC Symposium on Robot Control (SYROCO 00)*. Vienna, Austria.

D'Amico, A., G. Ippoliti and S. Longhi (2001). A radial basis function networks approach for the tracking problem of mobile robots. In: *Proceedings of the 2001 IEEE/ASME International Conference on Advanced Intelligent Mechatronics (AIM '01)*. Como, Italy.

Fierro, R. and F. L. Lewis (1997). Robust practical point stabilization of a nonholonomic robot using neural networks. *Journal of Intelligent and Robotic Systems* **20**, 295–317.

Girosi, F. and T. Poggio (1990). Neural networks and the best approximation property. *Biol. Cybernetics* **63**, 169–176.

Hunt, K. J., D. Sbarbaro, R. Zbikowski and P. J. Gawthrop (1992). Neural networks for control systems - a survey. *Automatica* **28**(6), 1083–1112.

Hush, D. R. and B. G. Horne (1993). Progress in supervised neural networks. *IEEE Signal Processing Magazine* **10**, 8–39.

Ippoliti, G. and S. Longhi (2001*a*). Hybrid control with switching on fixed and adaptive models for the intelligent control of underwater remotely operated vehicles. In: *9th Mediterranean Conference on Control and Automation (MED01)*. Dubrovnik, Croatia.

Ippoliti, G. and S. Longhi (2001*b*). Minimum variance multiple models adaptive control: Stability analysis and simulation tests. In: *Proceedings of the European Control Conference (ECC 2001)*. Seminario de Vilar, Porto, Portugal.

Jetto, L., S. Longhi and G. Venturini (1999). Development and experimental validation of an adaptive extended kalman filter for the localization of mobile robots. *IEEE Trans. on Robotics & Automation* **15**, 219–229.

Jiang, Z. P. and H. Nijmeijer (1997). Tracking control of mobile robots: a case study in backstepping. *Automatica* **33**, 1393–1399.

Kolmanovsky, I. and N. H. McClamroch (1995). Developments in nonholonomic control problems. *IEEE Control Systems* **15**, 20–36.

Morse, A.S. (1996). Supervisory control of families of linear set-point controllers - part 1st: Exact matching. *IEEE Trans. Autom. Contr.* **41**, 1413–1431.

Narendra, K. S., J. Balakrishnan and M. K. Ciliz (1995). Adaptation and learning using multiple models, switching, and tuning. *IEEE Control Systems Magazine* **15**(3), 37–51.

Narendra, K. S. and C. Xiang (2000). Adaptive control of discrete-time systems using multiple models. *IEEE Transactions on Automatic Control* **45**(9), 1669–1686.

Warwick, K., C. Kambhampati, P. Parks and J. Mason (1996). Dynamic systems in neural networks. In: *Neural Network Engineering in Dynamic Control Systems* (K. Hunt, G. Irwin and K. Warwick, Eds.). pp. 27–41. Springer-Verlag. London, Great Britain.

Yang, J. M. and J. H. Kim (1999). Sliding mode motion control of nonholonomic mobile robots. *IEEE Control Systems* **19**(2), 15–23.

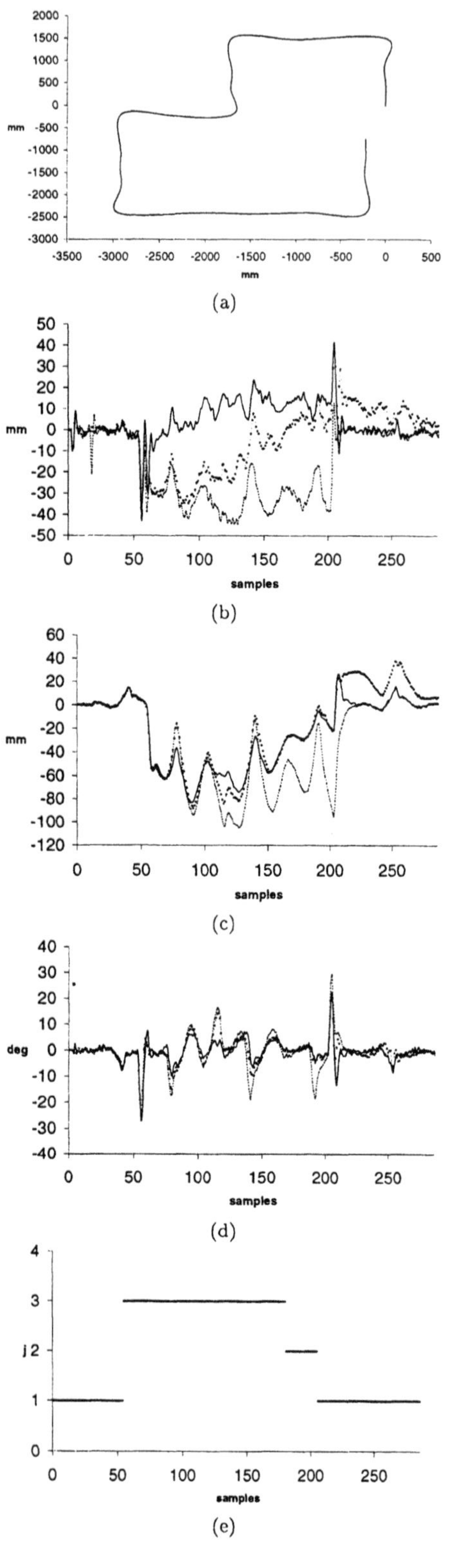

Fig. 4. Part (a): planned trajectory. Parts (b), (c), (d): tracking errors $\varepsilon_1(k)$, $\varepsilon_2(k)$ and $\varepsilon_3(k)$ respectively. This tracking errors have been obtained with the multiple models RBFN based control scheme (continuous line), with the single model adaptive RBFN based control scheme (dotted line) and with the fixed RBFN based controller (dashed line). Part (e): index j of the applied sequence of regulators R_j, $j = 1, 2, 3$ of the multiple models RBFN based control scheme.

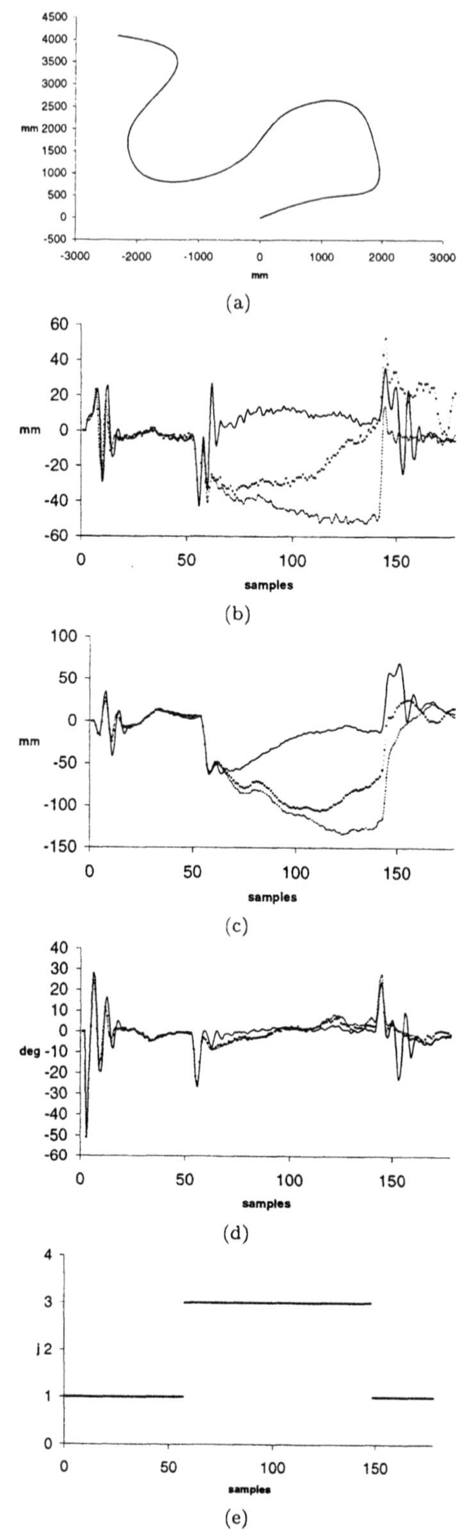

Fig. 5. Part (a): planned trajectory. Parts (b), (c), (d): tracking errors $\varepsilon_1(k)$, $\varepsilon_2(k)$ and $\varepsilon_3(k)$ respectively. This tracking errors have been obtained with the multiple models RBFN based control scheme (continuous line), with the single model adaptive RBFN based control scheme (dotted line) and with the fixed RBFN based controller (dashed line). Part (e): index j of the applied sequence of regulators R_j, $j = 1, 2, 3$ of the multiple models RBFN based control scheme.

SELF-TUNING TWO-DEGREE-OF-FREEDOM PID COMPENSATOR BASED ON TWO-DEGREE-OF-FREEDOM GENERALIZED MINIMUM VARIANCE CONTROL

Takao Sato * Akira Inoue * and Yoichi Hirashima *

* *Department of Systems Engineering, Faculty of Engineering, Okayama University, Okayama, 700–8530, Japan. Phone and Fax: +81-86-251-8233, E-mail: inoue@suri.sys.okayama-u.ac.jp*

Abstract: This paper proposes a new design scheme for a self-tuning PID controller. Expressions to calculate PID parameters from the identified plant coefficients are given. These expressions are derived by approximating the Two-Degree-of-Freedom PID controller to the Two-Degree-of-Freedom generalized minimum variance control (GMVC) law. Since the proposed self-tuning PID control law is Two-Degree-of-Freedom, reference response and disturbance response can be designed independently. Using a numerical example, the control performance of the proposed method is compared with that of another self-tuning PID controller.

Keywords: PID controllers, self-tuning control, minimum variance control, feedforward-compensation

1. INTRODUCTION

PID controllers are the most widely used control laws in industry. This is because that PID controllers have simple structures and their control parameters are intuitively understandable to practical engineers. However, it is rather difficult to find 'optimal' PID controller parameter values in implementing such controllers for practical industrial processes. Hence, many papers have been devoted to automatic tuning or adaptive selection of PID parameters.

There are two main classes of methods to design PID parameters. The first is to design the parameters to satisfy given control specifications, such as phase and amplitude margins (Hägglund(1995)). Yu(1998) has given an auto-tuning method combining an identification scheme with relay feedback. Wang, et al(1999) have proposed a PID tuning method based on a frequency domain approach. The second class of automatic PID tuning technique is based on the comparison and approximation of a PID controllers to controllers designed by other methods. The methods to be approximated are a pole-placement scheme (Tjokro and Shah(1985)), a generalized predictive control (Miller et al(1995)) and a generalized minimum variance controllers (GMVC) (Cameron and Seborg(1983) and Yamamoto et al(1998)).

Among the methods to be approximated, the GMVC was based on the work of Clarke and Gawthrop in 1979 and can also be interpreted as a pole-placement controller. Since then, it has been widely applied in industry. The self-tuning PID controller given by Yamamoto et al(1998) based on GMVC was applied to a temperature control system of an industrial-scale polybutene reactor and obtained successful control results (Yamamoto et al(1998)). Yamamoto et al(1999) have extended the self-tuning PID controller to a self-tuning pole-assignment controller with a PID structure. To approximate better, Sato et al(2000) have proposed a self-tuning PID controller with time-varying proportional gain based on the strongly stable GMVC (Inoue et al(1999)). However, because these methods are designed in the one-degree-of-freedom (abbreviated as "1DOF") PID control, both of reference response and disturbance response are decided simultaneously and can not be designed independently. On the other hand, in the two-degree-of-freedom("2DOF") PID control, each of responses can be designed independently, so the performance of 2DOF control system is higher than that of 1DOF system.

Therefore in this paper we will propose a design method of 2DOF PID controller based on the 2DOF GMVC(Takahashi et al(1998)) having a structure of a conventional GMVC plus a feedforward compensator.

The numerator of feedforward compensator in 2DOF GMVC must be approximated by the second-order numerator of the feedforward compensator in 2DOF PID controller. Generally the numerator of feedforward compensator in 2DOF GMVC is higher-order than 2. This paper proposes a method approximating the higher-order numerator of feedforward compensator in 2DOF GMVC by a second-order numerator by choosing the most significant roots of the numerator.

If the process is given by a second-order plus dead-time model, then the feedforward compensator of 2DOF GMVC law has a second-order numerator and is approximated exactly by a 2DOF PID controller. Hence, the obtained 2DOF PID controller will provide the same control performance as 2DOF GMVC. Since most chemical processes are overdamped and can be sufficiently well approximated by a second-order process with a time-delay, it is expected that the proposed 2DOF PID controller should yield good control results in most cases. Therefore, in this paper, the plant to be controlled is assumed to have a second-order numerator and dead-time.

Notations: z^{-1} denotes backward shift operator; $z^{-1}y(t) = y(t-1)$ and Δ is the difference operator $\Delta = 1 - z^{-1}$.

2. PLANT MODEL AND PID CONTROLLER

Consider a plant described by the following discrete-time ARIMAX model:

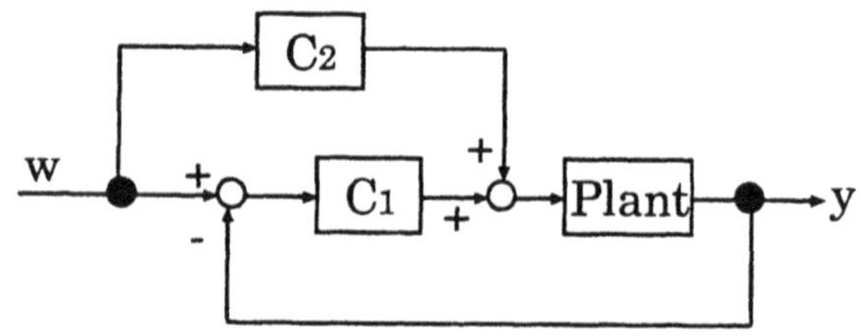

Fig. 1. Two-Degree-of-Freedom PID

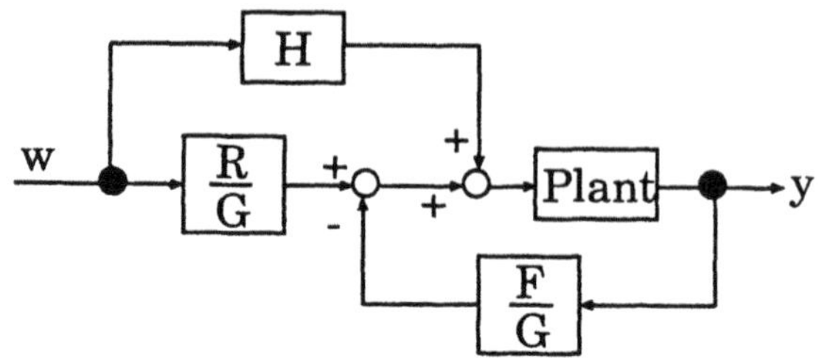

Fig. 2. Two-Degree-of-Freedom GMVC

$$A(z^{-1})y(t) = z^{-k_m}B(z^{-1})u(t-1) + \frac{\xi(t)}{\Delta} \quad (1)$$

$$A(z^{-1}) = 1 + a_1 z^{-1} + a_2 z^{-2} \quad (2)$$

$$B(z^{-1}) = b_0 + b_1 z^{-1} + \cdots + b_m z^{-m} \quad (3)$$

and $u(t)$ is the input, $y(t)$ is the output, k_m is dead time and $\xi(t)$ is Gaussian white noise.

The discrete-time 1DOF PID control law has its structure given by:

$$\Delta u(t) = C_1(z^{-1})(w(t) - y(t)) \quad (4)$$

$$C_1(z^{-1}) = k_c\left\{\Delta + \frac{T_s}{T_I} + \frac{T_D}{T_s}\Delta^2\right\} \quad (5)$$

where $w(t)$ is the reference input to be followed by the output $y(t)$. Parameters k_c, T_I and T_D are the proportional gain, the reset time and the derivative time respectively. The sampling time is denoted by T_s. The structure of 2DOF PID control law is given by 1DOF PID plus feedforward compensator, that is:

$$\Delta u(t) = C_1(z^{-1})(w(t) - y(t)) + C_2(z^{-1})w(t) \quad (6)$$

$$C_2(z^{-1}) = -k_c\left\{\alpha\Delta + \beta\frac{T_D}{T_s}\Delta^2\right\} \quad (7)$$

and is shown in Fig.1, where α and β are 2DOF parameters. The tuning problem is to design suitable values of parameters k_c, T_I, T_D, α and β for plant (1). This paper solves the design problem by approximating the controller (6) to 2DOF GMVC.

3. TWO-DEGREE-OF-FREEDOM GMVC

2DOF GMVC is given by a conventional 1DOF GMVC law plus a feedforward compensator that improves the reference response from reference input to system output. In this section, the schemes of 1DOF and 2DOF GMVC control law are derived. These methods will be approximated by PID controllers in the next section.

GMVC law(Clarke and Gawthrop, 1979) is derived to minimize the following variance of a generalized output:

$$J = E[\{P(z^{-1})y(t+k_m+1) + Q(z^{-1})\Delta u(t) - R(z^{-1})w(t)\}^2] \qquad (8)$$

where $P(z^{-1})$, $Q(z^{-1})$ and $R(z^{-1})$ are design polynomials.

The controller to minimize the covariance (8) is given by

$$\Delta u(t) = \frac{R(z^{-1})w(t) - F(z^{-1})y(t)}{G(z^{-1})} \qquad (9)$$

where polynomials $E(z^{-1})$ and $F(z^{-1})$ are the solutions of the Diophantine equation

$$P(z^{-1}) = \Delta A(z^{-1})E(z^{-1}) + z^{-(k_m+1)}F(z^{-1}) \qquad (10)$$

and $G(z^{-1})$ and $R(z^{-1})$ are obtained by the following equation,

$$G(z^{-1}) = E(z^{-1})B(z^{-1}) + Q(z^{-1}) \qquad (11)$$

$$R(z^{-1}) = F(z^{-1}) \qquad (12)$$

Substituting (9) into plant (1), closed-loop transfer function is obtained by the following equation,

$$y(t) = \frac{z^{-(k_m+1)}B(z^{-1})R(z^{-1})}{T(z^{-1})}w(t) + \frac{G(z^{-1})}{T(z^{-1})}\xi(t) \qquad (13)$$

where the characteristic polynomial $T(z^{-1})$ of closed-loop system is

$$T(z^{-1}) = P(z^{-1})B(z^{-1}) + Q(z^{-1})\Delta A(z^{-1}) \qquad (14)$$

GMVC given by (9) is 2DOF because it has feedforward compensator $R(z^{-1})/G(z^{-1})$ and feedback one $F(z^{-1})/G(z^{-1})$. But PID controller to approximate has the form of (6). And GMVC to be approximated must have the same structure. Therefore $R(z^{-1}) = F(z^{-1})$ is needed, then GMVC(9) has only one degree of freedom. Hence in this paper GMVC(9) is called 1DOF GMVC. We will consider another GMVC which is obtained by adding a feedforward compensator to GMVC(9) and it will be called as 2DOF GMVC.

The 2DOF GMVC law with feedforward compensator $H(z^{-1})$ which improves reference response is obtained by the following equation and shown in Fig.2,

$$\Delta u(t) = \left\{\frac{R(z^{-1})}{G(z^{-1})} + H(z^{-1})\right\}w(t) - \frac{F(z^{-1})}{G(z^{-1})}y(t) \qquad (15)$$

where $H(z^{-1})$ is designed such that the transfer function from reference input to output follows the discrete-time reference model and improves characteristics of tracking reference. Substituting 2DOF GMVC law (15) into plant (1), closed-loop transfer function is obtained by the following equation,

$$y(t) = \frac{G(z^{-1})}{T(z^{-1})}\xi(t) \qquad (16)$$
$$+ \frac{z^{-(k_m+1)}B(z^{-1})(R(z^{-1}) + G(z^{-1})H(z^{-1}))}{T(z^{-1})}w(t)$$

In the (13), once the polynomials $P(z^{-1}), Q(z^{-1})$ and $R(z^{-1})$ are designed, then both reference response and disturbance response are decided simultaneously therefore each responses can't be designed independently. On the other hand, in the (17) disturbance response is equal to (13), but reference response by (15) can be improved by selecting the additional feedforward compensator.

The feedforward compensator $H(z^{-1})$ is designed so that the closed-loop transfer function from reference input to system output matches discrete-time reference model $z^{-(k_m+1)}G_m(z^{-1})$. Then $H(z^{-1})$ is obtained by following equation.

$$H(z^{-1}) = \frac{G_m(z^{-1})T(z^{-1}) - B(z^{-1})R(z^{-1})}{B(z^{-1})G(z^{-1})} \qquad (17)$$

4. SELF-TUNING PID CONTROLLER

In this section, we derive a self-tuning 2DOF PID controller to approximate 2DOF GMVC law (15). 2DOF PID control law (6) is rewritten as the following equation.

$$\Delta u(t) = (C_1(z^{-1}) + C_2(z^{-1}))w(t) - C_1(z^{-1})y(t) \qquad (18)$$

To obtain an approximating 2DOF PID controller (18) to 2DOF GMVC law (15), $C_1(z^{-1})$ and $C_2(z^{-1})$ are selected to satisfy the following relation approximately,

$$C_1(z^{-1}) \simeq \frac{F(z^{-1})}{G(z^{-1})} \qquad (19)$$

$$C_2(z^{-1}) \simeq H(z^{-1}) \qquad (20)$$

The self-tuning controller given by Yamamoto et al(1998) is derived by replacing polynomial $G(z^{-1})$ by steady state gain $G(1)$ in the equation

(19). And also they restrict the plant to be of order 2 so that the numerator $F(z^{-1})$ of the right-hand side of the equation (19) has the same order (2) as the order of the left-hand side of (19), $C_1(z^{-1})$. In this paper also, $G(z^{-1})$ is replaced with steady state gain $\nu = G(1)$.

To make the equation (20) hold, the polynomial $H(z^{-1})$ should be approximated to be of second order because the order of $C_2(z^{-1})$ is 2. The reduction of the order of the polynomial $H(z^{-1})$ to 2 is obtained in the following way; first calculate roots of the equation with the numerator of $H(z^{-1}) = 0$ and choose the roots λ_{max} by having the largest absolute values $|\lambda_{max}|$, and considering the condition that closed-loop steady gain is equal to 1: $\tilde{h}(1) = 0$, then the approximated 2nd-order polynomial $\tilde{h}(z^{-1})$ is

$$\tilde{h}(z^{-1}) = H(1)\frac{(1-z^{-1})(\lambda_{max}-z^{-1})}{(\lambda_{max}-1)} \quad (21)$$

$$= \tilde{h}_0 + \tilde{h}_1 z^{-1} + \tilde{h}_2 z^{-2} \quad (22)$$

From equations (5) and (19) the parameters of PID controller are given by

$$\begin{aligned} k_c &= -\frac{1}{\nu}(f_1 + 2f_2) \\ T_I &= -\frac{f_1 + 2f_2}{f_0 + f_1 + f_2} T_s \\ T_D &= -\frac{f_2}{f_1 + 2f_2} T_s \end{aligned} \quad (23)$$

From (7) and (20) and using coefficients of $\tilde{h}(z^{-1})$, 2DOF parameters are given by

$$\alpha = -\frac{2\tilde{h}_0 + \tilde{h}_1}{k_c}, \qquad \beta = -\frac{\tilde{h}_2}{k_c T_D} T_s \quad (24)$$

If the coefficients $a_1, a_2, b_0, b_1, \cdots, b_m$ of the plant are known, then the parameters of the PID controller can be calculated through equations (17),(21),(23) and (24). Since in this paper the case with the unknown coefficients is considered, the recursive least squares identification law (Goodwin (1984)) is used to obtain the estimated values $\hat{a}_1(t)$, $\hat{a}_2(t)$, $\hat{b}_0(t)$, $\hat{b}_1(t)$, $\cdots$, $\hat{b}_m(t)$ of the unknown parameters a_1, a_2, b_0, b_1, $\cdots$, b_m.

5. AN EXAMPLE

In this section, the self-tuning PID controller is designed for the following plant

$$\begin{aligned} &(1 - 0.981z^{-1} + 0.069z^{-2})y(t) \\ &= z^{-2}(0.029 + 0.035z^{-1})u(t) \end{aligned} \quad (25)$$

To design the controller, the GMVC law is first designed, that is, the controller to minimize the variance $E[\Phi(t+2)^2]$ of the generalized output:

$$\begin{aligned} \Phi(t+2) &= (1 - 1.399z^{-1} + 0.518z^{-2})y(t+2) \\ &\quad + 0.2\Delta u(t) - R(z^{-1})w(t) \end{aligned} \quad (26)$$

where $R(z^{-1})$ is given by equation (12). In the simulations, these polynomials $F(z^{-1})$ and $G(z^{-1})$ are unknown because the plant parameters are unknown. These polynomials are obtained by the following procedure: first, identified plant parameters $\hat{a}_1(t)$,$\hat{a}_2(t)$,$\hat{b}_0(t)$,$\cdots$,$\hat{b}_3(t)$ are calculated; second, $\hat{E}(z^{-1})$, $\hat{F}(z^{-1})$ are calculated from equation (10). Reference model(Yamamoto, Tshihara, Omtu and Kitamori, 1992) is given as

$$G_m(z^{-1}) = \frac{19(0.029 + 0.035z^{-1})}{(1 - 0.74z^{-1} + 0.14z^{-2})(1 + 1.5z^{-1} + 0.56z^{-2})} \quad (27)$$

with sampling time: $\tau = 1.0$, rise time: $\sigma = 2.0$ and polynomial of shaping response trajectory: $V(z^{-1}) = 1 + 1.5z^{-1} + 0.56z^{-2}$. Feedforward compensator $H(z^{-1})$ is calculated by equation (17) and

$$H(z^{-1}) = \frac{H_{num}(z^{-1})}{H_{den}(z^{-1})} \quad (28)$$

$$\begin{aligned} H_{num}(z^{-1}) &= 3.38 - 8.44z^{-1} + 3.77z^{-2} \\ &\quad + 0.29z^{-3} - 0.076z^{-4} \end{aligned} \quad (29)$$

$$\begin{aligned} H_{den}(z^{-1}) &= 0.23 + 0.23z^{-1} - 0.016z^{-2} \\ &\quad - 0.018z^{-3} + 0.0075z^{-4} - 0.013z^{-5} \\ &\quad - 0.0016z^{-6} + 0.0016z^{-7} \end{aligned} \quad (30)$$

The order of $H(z^{-1})$ is greater than 2, then the approximated $\tilde{h}(z^{-1}) = 2.91 - 8.40z^{-1} + 5.49z^{-2}$ is derived from (21). Finally 2DOF PID parameters are obtained from equations (23) and (24) and are $k_c = 1.54$, $T_I = 4.42$, $T_D = 0.081$, $\alpha = 1.67$ and $\beta = -43.7$.

Simulations are conducted under the conditions that simulation steps are 200, step disturbance with amplitude 0.1 is added at 100 steps and the recursive least square identification law having reset with the forgetting factor 0.99 is used. The initial value of the estimated covariance matrix is $100I$ and the initial values of identified coefficients is the nominal values which are multiplied true values of (25) by 1.5.

In Fig.3 ~ Fig.6, simulation result of the proposed method is given. To compare with this result, the result obtained by self-tuning 1DOF PID controller designed by Yamamoto et al(1998) is shown in Fig.7 ~ FIg.9. Fig.3 and Fig.7 show simulated outputs of each methods, Fig.4 and Fig.8 inputs, Fig.5 and Fig.9 the calculated values of PID gains k_c, T_I and T_D and Fig.6 shows the calculated values of 2DOF PID parameters α and β.

The result by the proposed method shows no overshoot at the initial stage of the identification and is better than the results by a method by Yamamoto et al without deteriorating the response to disturbance. When a disturbance is added to plant, the performance of proposed 2DOF PID is similar to 1DOF PID.

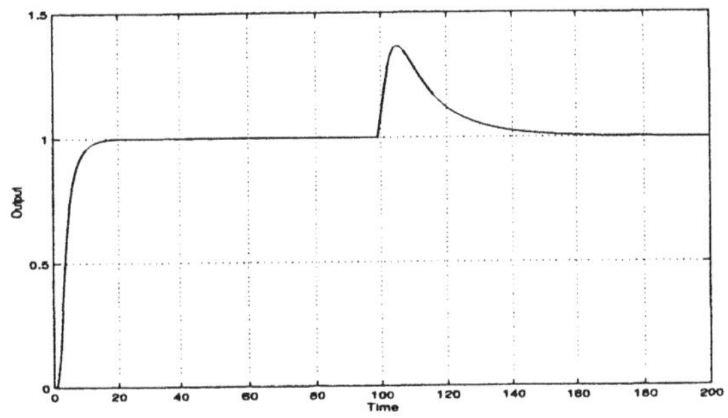

Fig. 3. Simulated Output: 2DOF

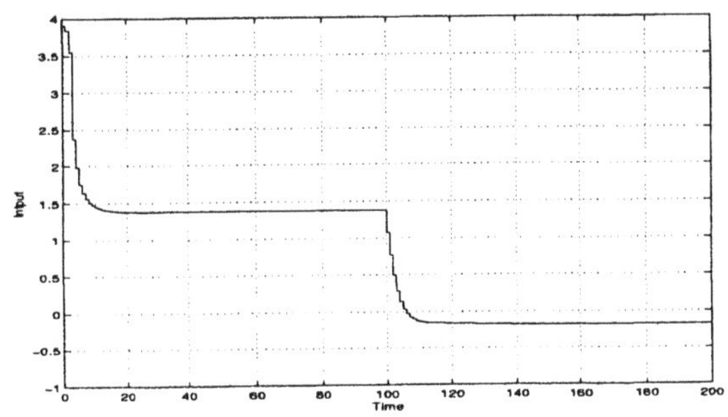

Fig. 4. Simulated Input: 2DOF

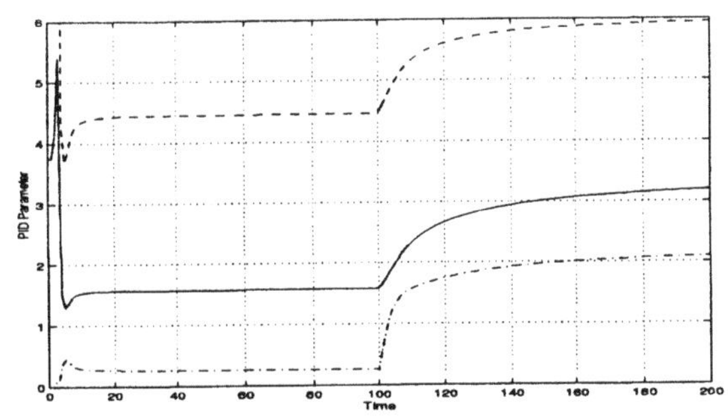

Fig. 5. PID parameters: 2DOF

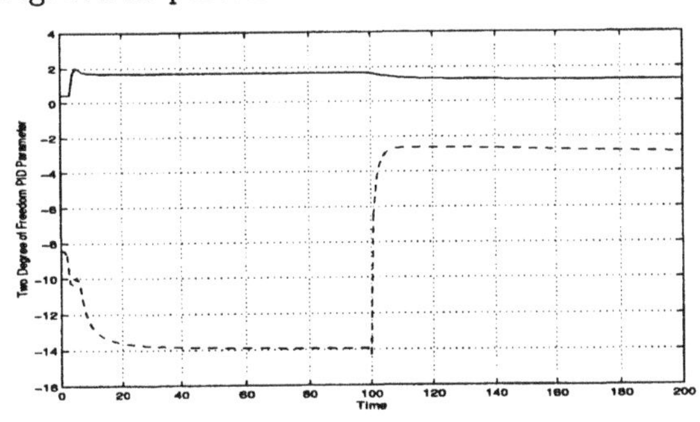

Fig. 6. 2DOF PID parameters: 2DOF

6. CONCLUSION

In this paper, a design scheme of a self-tuning PID controller is proposed. Expressions to calculate

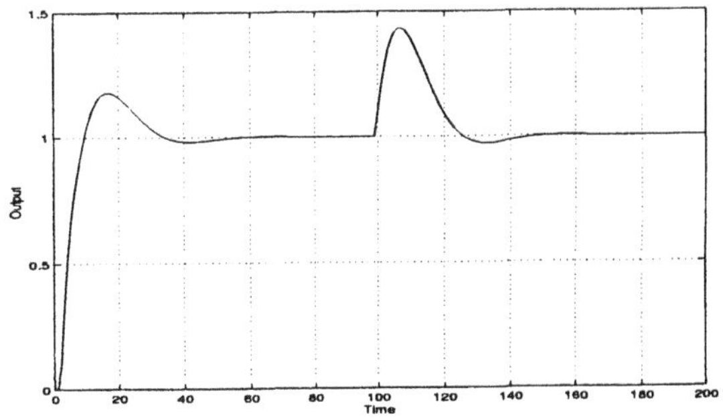

Fig. 7. Simulated Output: 1DOF

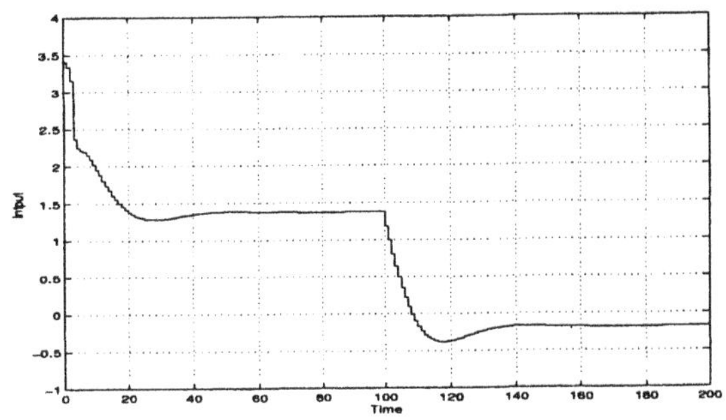

Fig. 8. Simulated Input: 1DOF

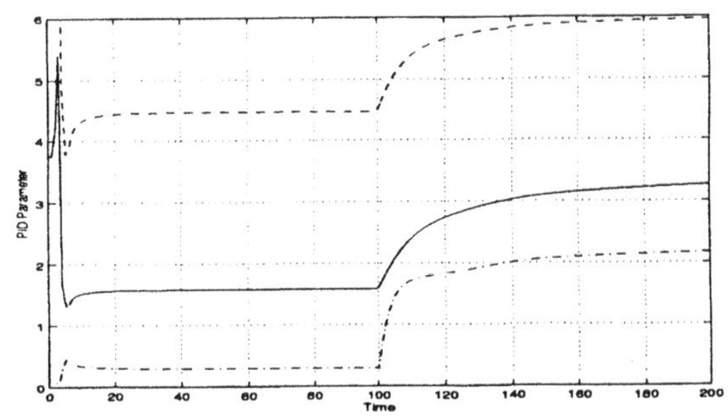

Fig. 9. PID parameters: 1DOF

PID parameters from the identified coefficients are given. These expressions are derived by approximating the 2DOF PID controller to the 2DOF GMVC law.

This controller law can improve reference response and disturbance response simulataneouly. The proposed self-tuning PID controller includes a parameter identification algorithm.

A numerical example is given to show the effectiveness of the proposed scheme and is compared with a self-tuning 1DOF PID controller that approximates a 1DOF GMVC.

7. REFERENCES

Cameron, F. and D. E. Seborg (1983). A Self-Tuning Controller with a PID Structure. *Int. I. Control.* Vol.38, No.2, 401–417.

Chien, K. L., I. A. Hrones and J. B. Reswick (1952). On the Automatic Control of Generalized Passive Systems; *Trans on ASME*, Vol.74, 175–185.

Clarke, D. W. and P. J. Gawthrop (1979). Self-Tuning Control. *IEE Proc.* Vol.126D, No.6. 633–640.

Goodwin, G. C. and K. S. Sin (1984). *Adaptive Filtering Prediction and Control*, Prentice-Hall.

Hägglund, T. and K. J. Astrom (1995). Automatic Tuning of PID Controllers. *The Control Handbook.* CRC Press, 817–826.

Inoue, A., A. Yanou and Y. Hirashima(1999). A Design of a Strongly Stable Self-Tuning Controller Using Coprime Factorization Approach. *14th IFAC World Congress*, Beijin, Vol.C, 211–216.

Miller, R. M., K. E. Kwok, S. L. Shah and R. K. Wood (1995). Development of a Stochastic Predictive PID Controller. *Proc. American Control Conference*, Seattle. 4204–4208.

Sato, T., A. Inoue, T .Yamamoto and S. L. Shah(2000). Self-Tuning PID Controllers Based on the Strongly Stable Generalized Minimum Variance Control Law. *IFAC Digital Control: Past, Present and Future of PID Control*, Terrassa, 511–516.

Takahashi, K., Y. Nakauchi and Y. Mori(1998). Two-Degree-of-Freedom Generalized Minimum Control. *The Society of Instrument and Control Engineers*, Japan. Vol.34, No.6, 510–515. (in Japanese)

Tjokro, S. and S. L. Shah (1985). Adaptive PID Control. *Proc. of American Control Conference*, Boston. 1528–1534.

Vidyasagar, M. (1985). Control System Synthesis. *A Factrization Approach*, The MIT Press.

Wang, Q. G., T. H. Lee, H. W. Fung, Q. Bi and Y. Zhang (1999). PID Tuning for Improved Performance. *IEEE Trans on Control Systems Tech.* Vol. 7, No. 4, 457–465.

Yamamoto, T. H. Ishihara, S. Omatu and T. Kitamori (1992). A Construction of Multivariable Self-Tuning Controller with Two-Degree-of-Freedom PID Structure for Discrete-Time System. *The Society of Instrument and Control Engineers*, Japan. Vol. 28, No. 4, 484–491. (in Japanese)

Yamamoto, T. and M. Kaneda (1998). A Design of a Self-Tuning PID Controllers Based on the Generalized Minimum Variance Control Law. *The Institute of System, Control and Information Engineers*, Japan. Vol. 11, No. 1, 1–9. (in Japanese)

Yamamoto, T., K. Fujii and M. Kaneda (1998). Design and Implementation of a Self-Tuning PID Controller. *Preprints of IFAC Workshop on Adaptive Control and Signal Processing.* Glasgow. 58–63.

Yamamoto, T., A. Inoue and S. L. Shah(1999). Generalized Minimum Variance Self-Tuning Pole-Assignment Controller with a PID Structure. *Pro. of 1999 IEEE Int Conf. on Control Applications*, Hawaii. 125–130.

Yu, C. C. (1998). Auto-tuning of PID Controllers. *Advances in Industrial Control.* Springer.

RELAY BASED GAIN AND PHASE MARGINS ITERATIVE PID CONTROLLER DESIGN

Gustavo Henrique Machado de Arruda
Péricles Rezende Barros

Departamento de Engenharia Elétrica,
Universidade Federal da Paraíba,
Cx.P. 10105, Campina Grande, PB - BRAZIL.
E-mail: arruda@dee.ufpb.br, prbarros@dee.ufpb.br

Abstract: In this paper it is addressed the problem of achieving gain and phase margin specifications on a closed loop system with a PID controller. The nonlinear design problem is solved iteratively using relay feedback experiments to estimate the gain and phase margin at each iteration. The relay tests are fast and easy to perform, which makes the procedure suitable for industrial applications. The iterative scheme is illustrated with simulation examples using desired gain and phase margin specifications. *Copyright © 2001 IFAC*

Keywords: Relay tests; PID controller tuning; gain and phase margins; Iterative design.

1. INTRODUCTION

The gain and phase margins are well known robustness and performance measurements from the classical control theory. The tuning of PID controllers based on such specifications requires the solution of a set of nonlinear equations, and has been recently studied in Ho et al. (1995), Ho et al. (1998), Kiong et al. (1999) and references therein.

A simple solution to the gain and phase margin based PID controller design is presented in Åström and Hägglund (1995). The idea generalizes the well known frequency domain method presented in Ziegler and Nichols (1942) by moving a single point of the Nyquist curve of the process to any desired location in the complex plane. The method, however, does not yield an exact solution. The critical point of the process is moved to another location, and it is expected the resulting loop transfer function to have the desired gain and phase margins.

Under the assumption that the process is first order plus time delay, an analytical solution to the nonlinear design problem is presented in Ho et al. (1995) and Ho et al. (1998). The controller parameters are calculated by solving the set of nonlinear equations using approximations. Alternatively, an exact solution is derived from a graphical approach in Kiong et al. (1999), but the full knowledge of the process frequency response is required.

The gain margin of a process can be easily estimated with the relay test presented in Åström and Hägglund (1995). Recently, in de Arruda and Barros (2000), a new relay feedback structure is presented, allowing the estimation of the frequency where a desired transfer function achieves a selected gain. The relay test was applied to a closed loop system during normal operation and its use for estimation of the loop transfer function and sensitivity function was studied.

In this paper it is presented a new combination of sequential relay tests for achieving gain and phase margins specifications in a closed loop system. The nonlinear problem is solved in an iterative manner, and two relay tests are used in each iteration. The advantage of this procedure is that no assumptions on process model structure is required, neither complete knowledge of its frequency response.

2. NOMENCLATURE

Consider the closed loop shown in Fig. (1). The unknown plant transfer function is given by $G(s)$ while the PID controller in the standard form is given by

$$C(s) = K_p\left(1 + \frac{1}{T_i s} + T_d s\right) = K_p \frac{s + 1/T_i + s^2 T_d}{s} . \tag{1}$$

The closed loop transfer function from reference $y_r(t)$ to output $y(t)$ is

$$T(s) = \frac{Y(s)}{Y_r(s)} = \frac{G(s)C(s)}{1 + G(s)C(s)} , \tag{2}$$

which is also known as the Complementary Sensitivity Function and

$$L(s) = G(s)C(s) \tag{3}$$

is the Loop Transfer Function.

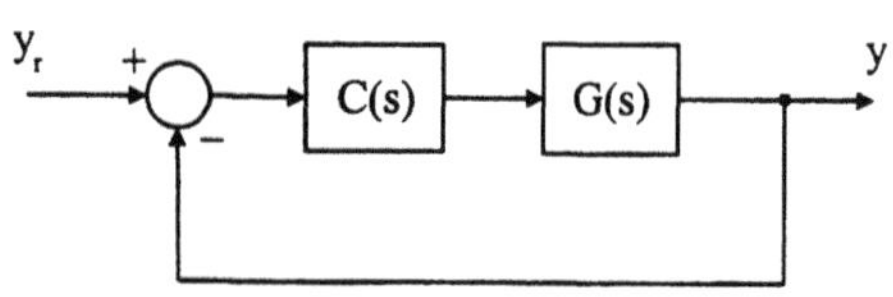

Fig. 1. Closed loop system.

The Gain Margin (GM) and Phase Margin (PM) of a closed loop are defined as

$$GM = \frac{1}{|L(j\omega_u)|} \quad \text{and} \quad PM = \pi + \angle L(j\omega_g) , \tag{4}$$

where ω_u and ω_g are obtained from

$$\angle L(j\omega_u) = -\pi \quad \text{and} \quad |L(j\omega_g)| = 1 . \tag{5}$$

Note that GM represents the maximum gain increase allowed on loop gain $|L(j\omega)|$ such that the closed loop system remains stable. Similarly PM represents the maximum amount of phase lag that can be added to $L(j\omega)$ at ω_g before the closed loop becomes unstable. Typical choices for GM ranges from 2 to 3, and for PM are within 30° to 60°, depending on the characteristics of the process dynamics.

The problem posed in this paper is to find the PID controllers parameters using relay tests, such that a closed loop system with specified gain and phase margins is obtained.

3. RELAY BASED IDENTIFICATION

3.1 *Gain Margin Estimation*

The estimation of the critical point using a relay test is presented in Åström and Hägglund (1995) and has became a standard in autotune procedures. The idea is to replace the controller in Fig. 1 by a relay with amplitude d. For most types of processes, this configuration leads to limit cycle operation and an estimate of the critical gain of the process is obtained from

$$\hat{K}_u = \frac{4d}{\pi a} , \tag{6}$$

where a is the amplitude of the oscillation at the process output. This procedure is named here the *standard relay test.*

It can be shown (Schei, 1994) that if this relay test is applied to a closed loop system, $T(s)$, the limit cycle occurs at the critical frequency of the loop transfer function, i.e.,

$$L(j\omega) = G(j\omega)C(j\omega) \cong \frac{m}{1-m}, \tag{7}$$

where

$$m = -\frac{\pi a}{4d} , \tag{8}$$

and a is the process output amplitude in closed loop. In this paper, this relay test will be referred as the *standard relay test on closed loop.* An estimate of the gain margin of the control system in Fig. 1 is then obtained with the standard relay test on closed loop, and calculated as

$$\hat{GM} = \left|\frac{1-m}{m}\right| = 1 + \frac{4d}{\pi a} , \tag{9}$$

and the critical frequency estimate, $\hat{\omega}_u$, is obtained from the frequency of the limit cycle.

3.2 *Phase Margin Estimation*

A basic procedure for the estimation of a general frequency point of a given transfer function using a relay feedback is presented in de Arruda and Barros (2000). The feedback structure applied for loop transfer function estimation is presented in Fig. 2. The conditions of the limit cycle operation are defined by the following proposition.

Proposition 1. Consider the closed loop system shown in Fig (2). Assume that for a stable closed loop $T(s)$ and a real positive number r, the transfer function

$$F(s) = \frac{2}{r} \frac{T(s)}{T(s)\left(\frac{1-r}{r}\right) + 1} - 1 \tag{10}$$

is also stable. Then if a limit cycle is present it oscillates at a frequency ω_o such that

$$|L(j\omega_o)| \approx r . \tag{11}$$

Proof. See de Arruda and Barros (2000). ■

This procedure can be seen as a generalization of the feedback structure presented in Schei (1994), and is referred here as the *frequency response relay*

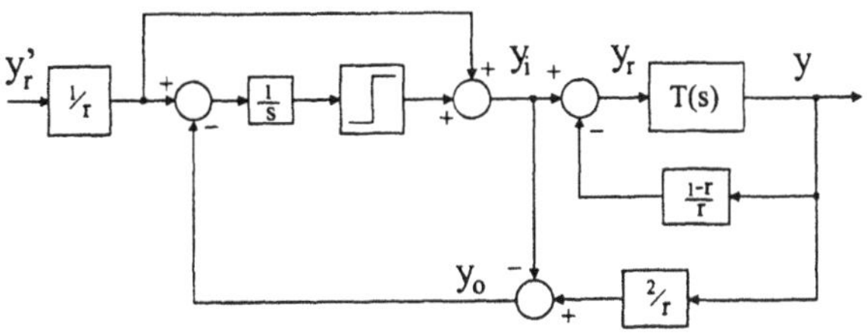

Fig. 2. Relay closed loop experiment for loop transfer function estimation.

test. In that work, it was used $r = 1$, and a limit cycle was obtained at the loop gain crossover frequency. Here it will be used to estimate the phase angle at this frequency.

To estimate the phase margin of the feedback system in Fig. 1, it is necessary to obtain the phase lag of $L(j\omega)$ at crossover frequency ω_g. That can be accomplished using discrete Fourier techniques to estimate the magnitude and phase at that particular frequency (see de Arruda and Barros (2000)). The procedure is briefly described in the sequel. Collect N samples of the closed loop input and output over one period of the oscillation during the frequency response relay test with $r = 1$. Estimate the crossover frequency, $\hat{\omega}_g$, measuring the frequency of the oscillation, and obtain the estimate of the loop transfer function using

$$\hat{L}\left(e^{j\hat{\omega}_g}\right) = \frac{Y_N(\hat{\omega}_g)}{Yr_N(\hat{\omega}_g) - Y_N(\hat{\omega}_g)} . \tag{12}$$

where Y_N and Yr_N are the N-point discrete Fourier transform of the output, $y(t)$, and the input reference, $y_r(t)$, respectively. The phase margin estimate of the control system in Fig. 1 is obtained from

$$\hat{PM} = \pi + \angle\hat{L}\left(e^{j\hat{\omega}_g}\right) . \tag{13}$$

4. PID CONTROLLER ITERATIVE TUNING

Consider the closed loop system shown in Fig. 1 and defined gain and phase margin specifications given by A_m and ϕ_m respectively. For the desired specifications the controller parameters must be designed to satisfy the following set of equations:

$$\angle G(j\omega_u) C(j\omega_u) = -\pi , \tag{14}$$

$$|G(j\omega_u) C(j\omega_u)| = \frac{1}{A_m} , \tag{15}$$

$$|G(j\omega_g) C(j\omega_g)| = 1 , \tag{16}$$

$$\angle G(j\omega_g) C(j\omega_g) = -\pi + \phi_m . \tag{17}$$

For the PID controller given by Eq. (1), the solution is obtained solving the set of equations (14) to (17) for K_p, T_i, T_d, ω_u and ω_g, i.e., a total of five unknowns and four equations. Thus another equation must be introduced in order to obtain a finite number of solutions to the design problem.

As pointed out in Åström and Hägglund (1995), a common procedure is to specify that the ratio of the parameters T_i and T_d is constant,

$$T_d = \alpha T_i . \tag{18}$$

A typical choice is $\alpha = 0.25$.

The solution of the design equations is not a trivial task because of the non-linearity of the problem. A procedure was presented in Kiong et al. (1999) for numerically calculating the controller parameters for both PI and PID controllers, but it requires the knowledge of the process frequency response, $G(j\omega)$. It is also not guaranteed that such a solution exists, as pointed out by the authors.

In this paper, instead, an iterative algorithm is used which requires the estimation of the frequencies ω_u and ω_g and the gain and phase margins at each iteration. Observe that these frequencies are the solutions to Eqs. (14) and (16), and the estimates can be obtained using the relay tests presented. The following two step design approach can be used to tune the PID controller, where a superscript k denotes the current iteration.

I. ***Initial Conditions:*** *Start from an initial controller $C^0(s)$, with K_p^0, T_i^0 and T_d^0 (in this step it is assumed that the system is operating in closed loop and $k = 1$).*

II. ***Gain Margin Estimation:*** *use the* standard relay test on closed loop *and estimate the critical frequency, $\hat{\omega}_u^k$, and the gain margin, $\hat{GM}^k$, according to the procedure presented in section 3.1.*

III. ***Stop criterion:*** *If $\left|\hat{GM}^k - A_m\right| \leq \varepsilon_1$ and this is not the first iteration, stop the procedure and make $C(s) = C^k(s)$. Else, continue.*

IV. ***Controller Redesign for Gain Margin:*** *The controller gain can now be calculated for achieving the gain margin A_m using Eq. (15). That is, with the current estimated gain margin, $\hat{GM}^k$, and the estimated critical frequency, $\hat{\omega}_u^k$, compute the controller proportional gain, $\bar{K}_p^{k+1}$, from*

$$\bar{K}_p^{k+1} = \frac{K_p^k \hat{GM}^k}{A_m} . \tag{19}$$

Now the new intermediate controller is

$$\bar{C}^{k+1}(s) = \bar{K}_p^{k+1}\left(\frac{s + 1/T_i^k + s^2 T_d^k}{s}\right) . \tag{20}$$

which can be applied immediately because the closed loop is stable by construction.

V. ***Phase Margin Estimation:*** *Use the* frequency response relay test *with $r = 1$ and estimate $\hat{\omega}_g^k$ and $\hat{PM}^k$ according to the procedure presented in section 3.2.*

*VI. **Stop criterion:** If $\left|\hat{PM}^k - \phi_m\right| \le \varepsilon_2$, then stop the procedure and make $C(s) = \bar{C}^{k+1}(s)$. Else, continue.*

*VII. **Controller Redesign for Phase Margin:** This step is separated into two parts:*

***A.** Determine T_i^{k+1} and T_d^{k+1} such that Eq. (17) and (18) are both satisfied. This is accomplished by solving the following system using $\hat{\omega}_g^k$ and $\hat{PM}^k$ obtained from the previous step*

$$\frac{1}{T_i^{k+1}} - \left(\hat{\omega}_g^k\right)^2 T_d^{k+1} = \frac{\hat{\omega}_g^k}{\beta}, \qquad (21)$$

$$T_i^{k+1} = 4T_d^{k+1}, \qquad (22)$$

where

$$\beta = \tan\left[\phi_m - \hat{PM}^k + \tan^{-1}\left(\frac{\hat{\omega}_g^k}{1/T_i^k - \left(\hat{\omega}_g^k\right)^2 T_d^k}\right)\right], \qquad (23)$$

which has two solutions, but only one is positive and therefore one valid solution is obtained. The phase contribution from the PID controller ranges from $-90°$ to $90°$, and this information must be used in order to avoid invalid values of T_i^{k+1} and T_d^{k+1}. Since $\angle G\left(j\hat{\omega}_g^k\right) + \angle C^{k+1/2}\left(j\hat{\omega}_g^k\right) = -180° + \phi_m$, then the following condition must be satisfied in the above steps

$$-270° + \phi_m < \angle G\left(j\hat{\omega}_g^k\right) < -90° + \phi_m, \qquad (24)$$

If Eq. (24) is not satisfied, stop the iteration.

***B.** Now, update the controller proportional gain K_p^{k+1} such that the loop gain at the frequency $\hat{\omega}_g^k$ is equal to one,*

$$K_p^{k+1} = \bar{K}_p^{k+1} \frac{\sqrt{\left[1/T_i^k - \left(\hat{\omega}_g^k\right)^2 T_d^k\right]^2 + \left(\hat{\omega}_g^k\right)^2}}{\sqrt{\left[1/T_i^{k+1} - \left(\hat{\omega}_g^k\right)^2 T_d^{k+1}\right]^2 + \left(\hat{\omega}_g^k\right)^2}}. \qquad (25)$$

The controller at the end of the iteration is finally given by

$$C^{k+1} = K_p^{k+1}\left(\frac{s + 1/T_i^{k+1} + s^2 T_d^{k+1}}{s}\right). \qquad (26)$$

Note that, in this last step, stability of the closed loop is not easily verified as in step IV.

VIII. Increment k and go to step 2.

In steps III and VI, it is considered that ε_1 and ε_2 are small tolerances. The final controller is the last one applied to the closed loop, which is obtained after one of these two steps. Note that both specifications A_m and ϕ_m are satisfied in this case, except if an invalid value of T_i is calculated in step $VII(A)$. Extensive simulations using the exact values of ω_g and ω_u at each iteration instead of the estimates obtained with the relay tests were performed. The results point out that the algorithm converges to the same solution obtained using the method in Kiong et al. (1999), whenever such a solution exists.

5. SIMULATION EXAMPLES

To illustrate the capabilities of the proposed method, two simulation examples are presented in this section.

5.1 *High Order Dynamics*

Consider the eighth order process given by

$$G(s) = \frac{1}{(s+1)^8}.$$

The initial controller is calculated from Ziegler-Nichols tuning (see Åström and Hägglund (1995)), using the critical point information from a relay test. The controller parameters are

$$K_p^0 = 1.1191, \; T_i^0 = 7.5839 \text{ and } T_d^0 = 1.8960,$$

and the gain and phase margins obtained with this controller are

$$GM_0 = 1.8262 \text{ and } PM_0 = 74.236°.$$

It is desired a gain margin of $A_m = 2.5$ and a phase margin of $\phi_m = 60°$. The exact controller parameters that yield the desired specifications are

$$K_p = 0.7636, \; T_i = 4.9345 \text{ and } T_d = 1.2336.$$

It should be clear that this controller can not be obtained without fully knowing the process frequency response.

Now the iteration scheme is performed. Using the same specifications for gain and phase margins, $\varepsilon_1 = 0.01$ and $\varepsilon_2 = 0.5°$, the controller is obtained after 5 iterations at the step VI of the algorithm. The obtained parameters are

$$K_p^5 = 0.5497, \; T_i^5 = 3.7181 \text{ and } T_d^5 = 0.9295,$$

and the gain and phase margins calculated with the process transfer function for this controller are

$$GM^5 = 2.7074 \text{ and } PM^5 = 53.94°$$

Although the controllers are quite different, the gain and phase margins achieved using the above parameters are close to the desired specifications. The difference is justified by the error in the approximation of the describing function of the relay, which yields a bias in the estimated gain and phase margins. This can be seen in Table 1, where both the estimated and the exact gain and phase margins achieved at steps III and VI of the algorithm at each iteration are shown.

Table 1. Calculated and estimated gain and phase margins at each iteration.

	Exact		Estimated	
Iter.	GM	PM	$\hat{GM}$	$\hat{PM}$
1	1.8262	84.0670	1.7984	85.8171
2	2.8999	50.0629	2.6731	54.8893
3	3.0934	51.2836	2.8508	56.6256
4	2.8598	53.1755	2.6389	58.9497
5	2.7389	53.9442	2.5291	59.5478

The controller parameters at the end of each iteration are shown in Table 2. Note that $T_i^5 = T_i^4$ and $T_d^5 = T_d^4$ because the iteration scheme has stopped at step VI of the algorithm.

Table 2. Controller parameters at the end of each iteration.

Iter.	K_p	T_i	T_d
0	1.1191	7.5839	1.8960
1	0.2514	2.1547	0.5387
2	0.3737	3.0310	0.7578
3	0.4946	3.5583	0.8896
4	0.5434	3.7181	0.9295
5	0.5497	3.7181	0.9295

Finally, the closed loop output for a step reference is shown in Fig. 3. Note that the difference between the exact solution and the one obtained with the iterative procedure is acceptable.

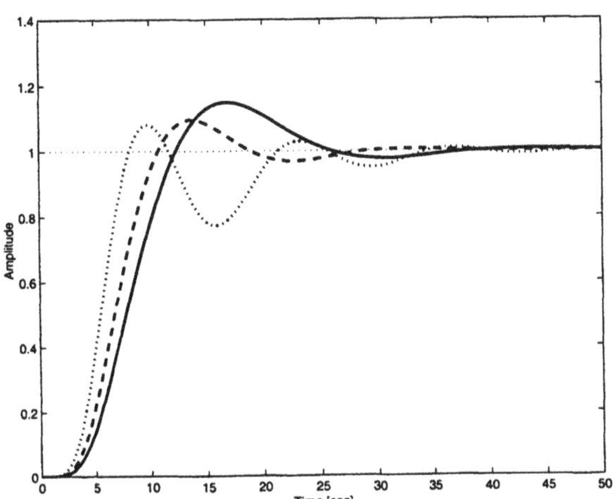

Fig. 3. Step response for obtained closed loop systems: initial controller (dotted), exact solution (dashed) and using the proposed iterative algorithm (continuous).

5.2 *Second Order Dynamics with Transport Delay*

Consider now the process

$$G(s) = \frac{\omega_o^2}{s^2 + 2\xi\omega_o s + \omega_o^2} e^{-\tau s} ,$$

with $\omega_o = 2$, $\xi = 0.8$ and $\tau = 1$. The controller parameters obtained from the Ziegler-Nichols frequency response table and the critical point estimate are

$$K_p^0 = 0.8293, \; T_i^0 = 1.7666 \text{ and } T_d^0 = 0.4416, \tag{27}$$

and the corresponding gain and phase margins are

$$GM_0 = 1.6926 \text{ and } PM_0 = 84.249° .$$

The desired gain and phase margins for this example are $A_m = 2$ and $\phi_m = 60°$. The exact controller parameters that meet these specifications are

$$K_p = 0.6957 \, , \; T_i = 1.1064 \text{ and } T_d = 0.2766 \, . \tag{28}$$

The exact and the estimated gain and phase margins at steps III and VI are shown in Table 3. The controller parameters obtained at the end of each iteration are shown in Table 4. The tolerances in this example are chosen as $\varepsilon_1 = 0.01$ and $\varepsilon_2 = 0.1°$.

Table 3. Calculated and estimated gain and phase margins at each iteration.

	Exact		Estimated	
Iter.	GM	PM	$\hat{GM}$	$\hat{PM}$
1	1.6926	87.6369°	1.5168	88.954°
2	2.4913	46.1098°	2.3141	50.580°
3	2.7929	47.0348°	2.5659	51.665°
4	2.4559	52.3244°	2.2056	58.889°
5	2.2417	stopped	1.999	stopped

Table 4. Controller parameters at the end of each iteration.

Iter.	K_p	T_i	T_d
0	0.8293	1.7666	0.4416
1	0.0163	0.0429	0.0107
2	0.2027	0.4636	0.1159
3	0.4180	0.7618	0.1905
4	0.4826	0.8008	0.2002

A comparison between the step responses of the obtained closed loop systems is shown in Fig. 4.

6. CONCLUSIONS

In this paper an iterative procedure combining two relay experiments is applied to PID controller tuning based on gain and phase margins specifications. The relay tests are used to estimate the critical point and the gain crossover point of the loop transfer function of a closed loop system. This information is used to solve the nonlinear design problem in an iterative manner. The advantage is that no assumptions on process model structure is required, neither complete knowledge of its frequency response. Simulation examples are used to illustrate the procedure. The results point out that with a few iteration steps it is possible to design a PID controller such that the desired gain and phase margins are approximately achieved.

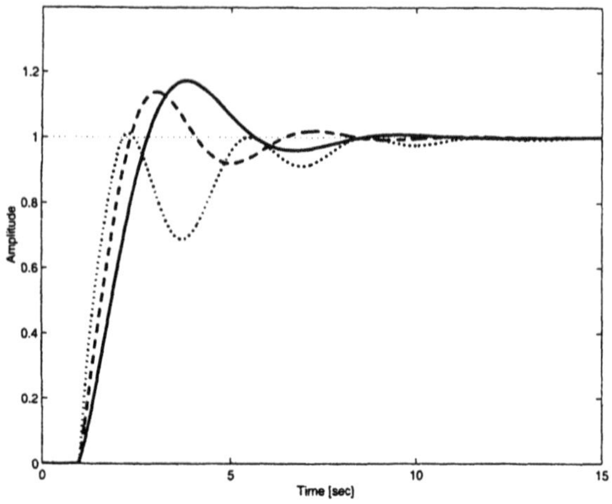

Fig. 4. Step response for obtained closed loop systems: initial controller (dotted), exact solution (dashed) and using the proposed iterative algorithm (continuous).

7. REFERENCES

Åström, K. J. and Hägglund, T. (1995). *PID Controllers: Theory, Design and Tuning*, 2nd edn, Instrument Society of America, Research Triangle Park, North Carolina.

de Arruda, G. H. M. and Barros, P. R. (2000). Relay based closed loop transfer function estimation, ACC, American Control Conference, Chicago, USA.

Ho, W. K., Hang, C. C. and Cao, L. S. (1995). Tuning of PID controllers based on gain and phase margin specifications, *Automatica* **31**(3): 497–502.

Ho, W. K., Lim, K. W. and Xu, W. (1998). Optimal gain and phase margin tuning for PID controllers, *Automatica* **34**(8): 1009–1014.

Kiong, T. K., Qing-Guo, W., Chieh, H. C. and Hägglund, T. J. (1999). *Advances in PID Control*, Springer-Verlag, London.

Schei, T. S. (1994). Automatic tuning of PID controllers based on transfer function estimation, *Automatica* **30**(12): 1983–1989.

Ziegler, J. G. and Nichols, N. B. (1942). Optimum settings for automatic controllers, *Transactions of the ASME* **42**: 759–768.

A DESIGN OF EVOLUTIONARY TUNING PID CONTROLLERS

Toru Yamamoto * Michiyo Suzuki * Yasue Mitsukura **

* *Department of Technology and Information Education,*
Graduate School of Education, Hiroshima University
1-1-1 Kagamiyama, Higashi-Hiroshima, 739-8524, Japan
Tel: +81-824-24-7160 Fax: +81-824-22-7171
{yama / suzuki}@hiroshima-u.ac.jp

** *Department of Information Science and Intelligent Systems,*
Faculty of Engineering, University of Tokushima
2-1 Minami-josanjima, Tokushima, 770-8506, Japan
Tel: +81-88-656-7488 Fax: +81-88-656-7488
mitsu@is.tokushima-u.ac.jp

Abstract: PID control schemes have been widely used in most process control systems represented by chemical processes for a long time. However, it is still a very important problem how to determine or tune the PID parameters, because these parameters have a great influence on the stability and the performance of the control system. On the other hand, in the last twenty years, a genetic algorithm(GA) is attracted as one method which gives us suitable answers for search, optimization and machine learning problems. In this paper, a new evolutuinary tuning algorithm of PID parameters is proposed, in which the search area of PID parameters is sharply reduced by considering an effective parameters' area from the viewpoint of the control engineering. Finally, the newly proposed scheme is numerically evaluated on a simulation example. *Copyright © 2001 IFAC*

Keywords: PID control, Genetic algorithms, Process control, Minimum variance control, Intelligent control

1. INTRODUCTION

Lots of studies on advanced complex control schemes have been carried out vigorously due to the marvelous development of the computer technology, However, the conventional PID control scheme has been widely used in most process industries. Since it plays a great role in determining the behavior of the control system how to tune the PID parameters, a lot of tuning methods have been studied up to now(Zieglar and Nichols, 1942; Chien, *et al.*, 1952; Åström and Hägglund, 1988; Åström, *et al.*, 1992). The author also has proposed a tuning method(Yamamoto and Kaneda, 1998) of PID parameters from the viewpoint of the self-tuning control scheme based on the relationship between the PID control law and the generalized minimum variance control(GMVC) law(Clarke and Gawth-rop,1979), and has illustrated the effectiveness of the proposed method by applying it to a chemical process(Fujii, *et al.*, 1997). However, the proposed scheme includes some user-specified parameters, and it was left how to choose these parameters.

On the other hand, the genetic algorithm(GA) which simulates the mechanism of heredity or evolution in the living things, has attracted our attentions as a method which gives us fairly quickly suitable or suboptimal answers for search, optimization or machine learning problems(Goldberg, 1992). Moreover, some searching methods of PID parameters using GA have also been reported(Porter and Jones, 1992; Jones, *et al.*, 1996; Omatu and Deris, 1996). However, since three PID parameters are directly searched in those methods, it costs fairly long time to get a suitable set of these parameters due to the wide search area.

In this paper, a new tuning scheme of PID parameters is proposed, in which the user-specified parameters included in the tuning method of PID parameters are sought using GA, and are transformed into the corresponding PID parameters. That is, it means that a suitable set of the user-specified parameters in the performance criterion of the GMVC law is firstly sought by using GA, and then a set of PID parameters is calculated using the above set as a mediator. Therefore, according to the newly proposed PID tuning scheme, since search areas of the user-specified parameters are sharply reduced, the time to get a suitable set of PID parameters can be drastically shorten. This paper is organized as follows. First, the descriptive model of the controlled object is derived, and the previously proposed tuning scheme of PID parameters is explained in brief. Next, a new tuning scheme of PID parameters as mentioned above is proposed. Finally, the effectiveness of the proposed scheme is evaluated on a numerical simulation example.

2. PROBLEM FORMULATION

2.1 *System description*

Let z^{-1} be the backward shift operator, and the following discrete-time model describes the process:

$$\alpha(z^{-1})y(t) = z^{-k}\beta(z^{-1})u(t-1) + f(u,y) + \eta(t)/\Delta \qquad (1)$$

where

$$\left.\begin{aligned} \alpha(z^{-1}) &= 1 + \alpha_1 z^{-1} + ... + \alpha_{n_1} z^{-n_1} \\ \beta(z^{-1}) &= \beta_0 + \beta_1 z^{-1} + ... + \beta_{n_2} z^{-n_2}. \end{aligned}\right\} \qquad (2)$$

Moreover, $u(t)$, $y(t)$ and $\eta(t)$ are the input, the output and the modeling error in (1) with zero mean and covariance σ_η^2, respectively. k denotes the dead time and $f(u,y)$ implies the nonlinear components. Furthermore, the system model (1) is called as the 'full model', because it explains the behavior of the system as exactly as possible. In this paper, it is assumed that the full model (1) can be given. If it is not so, the full model may be replaced by an appropriate model, for example, a neural network which simulates approximately the behavior of the controlled object. It is used only for calculating the fitness function in GA as shown in the section 4.3.

On the other hand, consider the second-order linear model which approximates the full model (1) by the following equation:

$$A(z^{-1})y(t) = z^{-k}B(z^{-1})u(t-1) + \xi(t)/\Delta \qquad (3)$$

where

$$\left.\begin{aligned} A(z^{-1}) &= 1 + a_1 z^{-1} + a_2 z^{-2} \\ B(z^{-1}) &= b_0 + b_1 z^{-1} + ... + b_m z^{-m}. \end{aligned}\right\} \qquad (4)$$

$\xi(t)$ denotes the modeling error in (3) with zero mean and covariance σ_ξ^2. In general, since the control law based on the full model (1) seems to be complex and very difficult in implementation, the second-order linear model (3) is often used in designing the process control system. Therefore, the model (3) is called as a 'design-oriented model'. Note that a_2 is set to 0 in (4) if the controlled object is approximately expressed as a first-order system. Moreover, the degree of $B(z^{-1})$, m, should be set large enough in order to cope with the higher components which can not be approximated as the second-order linear system.

2.2 *Brief review of PID control law*

Consider the velocity-type PID control law(Cameron and Seborg, 1983) described as

$$\Delta u(t) = \frac{k_c \cdot T_s}{T_I} e(t) - k_c \Delta y(t) - \frac{k_c \cdot T_D}{T_s} \Delta^2 y(t) \qquad (5)$$

where

$$e(t) := w(t) - y(t). \qquad (6)$$

Δ denotes the differencing operator defined as $\Delta := 1 - z^{-1}$. $w(t)$ is the reference signal and is given by piecewise constant components. k_c , T_I and T_D are the proportional gain, the reset time and the derivative time, respectively. Moreover, T_s denotes the sampling interval. For convenience, let $L(z^{-1})$ be

$$L(z^{-1}) := k_c\{\Delta + \frac{T_s}{T_I} + \frac{T_D}{T_s}\Delta^2\} \qquad (7)$$

then, (5) can be rewritten by

$$L(z^{-1})y(t) + \Delta u(t) - L(1)w(t) = 0. \qquad (8)$$

Since it plays a great role in determining the behavior of the closed-loop system how to tune the PID parameters in (5) or (8), a lot of works for PID tuning schemes have been reported up to now. The author also has proposed a new tuning scheme of PID parameters based on the relation between the PID control and the GMVC laws(Yamamoto and Kaneda, 1998). Then, the previously proposed PID parameter tuning scheme is briefly explained in the following section.

3. PID PARAMETER TUNING

Consider the following cost function to derive a GMVC law:

$$J = E[\{P(z^{-1})y(t+k+1) + \lambda\Delta u(t) - P(1)w(t)\}^2]. \quad (9)$$

λ in (9) is the weighting factor with respect to the control input, and $P(z^{-1})$ is the user-specified polynomial given by:

$$P(z^{-1}) = 1 + p_1 z^{-1} + p_2 z^{-2} \quad (10)$$

$$\left.\begin{aligned} p_1 &= -2e^{-\frac{\rho}{2\mu}}\cos\frac{\sqrt{4\mu-1}}{2\mu}\rho \\ p_2 &= e^{-\frac{\rho}{\mu}} \\ \rho &= \frac{T_s}{\sigma} \\ \mu &= 0.25(1-\delta) + 0.51\delta. \end{aligned}\right\} \quad (11)$$

σ and δ in (11) are parameters related to the rise-time and the damping property, respectively(Yamamoto, *et al.*, 1992). The control input minimizing the cost function (9) is given by the following equation(Clarke and Gawthrop, 1979):

$$F(z^{-1})y(t) + \{E(z^{-1})B(z^{-1}) + \lambda\}\Delta u(t) - P(1)w(t) = 0, \quad (12)$$

where $F(z^{-1})$ and $E(z^{-1})$ are obtained by solving the following Diophantine equation:

$$P(z^{-1}) = \Delta A(z^{-1})E(z^{-1}) + z^{-(k+1)}F(z^{-1}) \quad (13)$$

$$\left.\begin{aligned} E(z^{-1}) &= 1 + e_1 z^{-1} + \cdots + e_k z^{-k} \\ F(z^{-1}) &= f_0 + f_1 z^{-1} + f_2 z^{-2}. \end{aligned}\right\} \quad (14)$$

In order to obtain a control law with a PID structure, consider the following equation in which $E(z^{-1})B(z^{-1})$ is replaced by the static gain $E(1)B(1)$:

$$F(z^{-1})\ y(t) + \{E(1)B(1) + \lambda\}\Delta u(t) - P(1)w(t) = 0. \quad (15)$$

Here, ν is defined as

$$\nu := B(1)E(1) + \lambda, \quad (16)$$

and then (15) can be rewritten as

$$\frac{F(z^{-1})}{\nu}y(t) + \Delta u(t) - \frac{P(1)}{\nu}w(t) = 0. \quad (17)$$

Furthermore, if the following relation is satisfied:

$$L(z^{-1}) = \frac{F(z^{-1})}{\nu}, \quad (18)$$

(17) becomes identical to (8). Because the relation $P(1) = F(1)$ can be immediately obtained from (13). Therefore, if the parameters, λ, σ and δ are suitably chosen, the PID parameters can be calculated based on (8) and (18) as follows:

$$\left.\begin{aligned} k_c &= -\frac{1}{\nu}(f_1 + f_2) \\ T_I &= -\frac{f_1 + 2f_2}{f_0 + f_1 + f_2}T_s \\ T_D &= -\frac{f_2}{f_1 + 2f_2}T_s. \end{aligned}\right\} \quad (19)$$

However, it was left how to choose these user-specified parameters. Then, a new scheme in which a suitable set of them is sought by using the GA, is considered in the following section.

4. EVOLUTIONARY TUNING OF PID PARAMETERS USING GA

4.1 *Summary of evolutionary tuning of PID parameters*

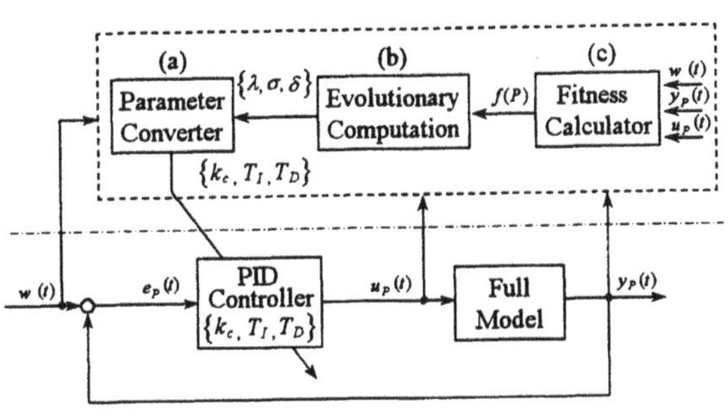

Fig. 1. Schematic diagram of the evolutionary computation of PID parameters using the GA.

Fig.1 shows the schematic diagram of the evolutionary computation of PID parameters. In Fig.1, the under part of the dash-dotted line composes a usual feedback control system which has the full model and the PID controller with parameters $\{k_c, T_I, T_D\}$. Here, the full model is what expresses the behavior of the controlled object as exactly as possible, as mentioned in the section 2.1.

On the other hand, the upper part of the dash-dotted line is the tuning part of PID parameters every one generation in GA. It is composed of (a) the parameter conversion part in which the user-specified parameters $\{\lambda, \sigma, \delta\}$ sought by GA are converted to the corresponding PID parameters, (b) the evolutionary computation part in which a suitable set of the user-specified parameters are chosen based on the fitness values, and (c) the calculation part of the fitness function for each individual.

The features of the proposed method can be summarized as follows.

1) As mentioned in the section 3, PID parameters are not directly sought by GA, but they are calculated using the user-specified parameters $\{\lambda, \sigma, \delta\}$ included in the performance criterion of the GMVC law, which are sought by GA. Here, since the search areas of the parameters $\{\lambda, \sigma, \delta\}$ are restricted by taking the control performance of the closed-loop system into consideration, the time to get a suitable set of PID parameters can be drastically shorten.
2) By means of selecting the fitness function appropriately, the response of the closed-loop system can be shaped fairly freely. For example, the rise property, the overshoot and the fluctuation of the control input can be specified fairly in detail.
3) The full model does not have to be an analytical formulation, but it may be only the model whose input-output property is closely similar to that of the controlled object in a sense. Therefore, as mentioned in the section 2, various neural network models can be utilized, and it is possible to take any nonlinear property in the model freely.

Here, consider more precisely the tuning part of PID parameters (a), (b) and (c) mentioned above.

4.2 *Parameter converter*

The calculation flow of the parameter conversion is as follows.

$$\xrightarrow{\text{GA}} \{\lambda, \sigma, \delta\} \xrightarrow{\text{eqn.(11)}} \{\lambda, p_1, p_2\} \longrightarrow$$
$$\xrightarrow{\text{eqns.(13) and (14)}} \{\lambda, e_1, \cdots, e_k, f_0, f_1, f_2\} \longrightarrow$$
$$\xrightarrow{\text{eqn.(16)}} \{\nu, f_0, f_1, f_2\} \xrightarrow{\text{eqn.(19)}} \{k_c, T_I, T_D\}$$

The significant search spans of the parameters λ, σ and δ in the control engineering sense can be considered as follows.

On λ, investigating the stability of the closed-loop system using an appropriate approach, for example, the Bode diagram, the search span of λ can be determined so as to give a suitable gain or phase margin. On σ, since it is said that it is desirable to set it between 1/3 and 1/2 of the system's time constant, it is suggested that the practical search span of σ had better to be set a little larger than the above span. Furthermore, on δ, it is known that a desirable response of the closed-loop system can be obtained when δ is set between 0.0 and 2.0(Yamamoto, *et al.*, 1992). Therefore, the practical search span of δ also had better to be a little larger than the above span.

From these control engineering restrictions, the seek span of the corresponding PID parameters can be limited to fairly small area in practice.

4.3 *Evolutionary computation*

The evolutionary computation of the parameters $\{\lambda, \sigma, \delta\}$ is discussed in this section. These parameters are calculated by using a real-coded GA(Ono, *et al.*).

First, the user-specified parameters λ, σ and δ are arranged as cells included in the l–th string $s_l(n_3)(l = 1, 2, \cdots, n_4; n_3 = 1, 2, 3)$, that is, $s_l(1) = \lambda$, $s_l(2) = \sigma$ and $s_l(3) = \delta$. Furthermore, n_4 denotes the number of population. Therefore, $s_l(\cdot)$ denotes the l–th individual. After the above preparation, a real-coded GA is employed, which is explained as follows.

i) **Initialization** Determine the search spans of the user-specified parameters by taking account of the control engineering sense, and product the initial individuals with random real coded within the search spans.
ii) **Selection** Calculate the fitness value $f(l)(l = 1, 2, \cdots, n_4)$ corresponding to the individual s_l, and select γ percent individuals with superior fitness values. Here, the fitness function $f(\cdot)$ can be freely designed by taking account of the desired control performance, which is discussed in the following section. γ percent superior individuals are saved in the next generation.
iii) **Crossover** The $(100-\gamma)$ percent remains are generated by the crossover. Choose two individuals s_a and s_b among γ percent superiors. Then, the new individual s_c is generated by

$$s_c(n_3) = \frac{I(X) + Y}{2}, \tag{20}$$

where

$$X := \max\{f(s_a(n_3)), f(s_b(n_3))\} \tag{21}$$

$$Y := \frac{s_a(n_3) + s_b(n_3)}{2}. \tag{22}$$

$I(X)$ denotes the individual with the fitness value X, that is, $I(\max\{f(s_a(n_3)), f(s_b(n_3))\})$ means the individual with the superior fitness value within $s_a(n_3)$ and $s_b(n_3)$. Note that this procedure is employed for every cell $n_3(= 1, 2, 3)$ included in s_a and s_b.

iv) **Mutation** ω percent cells for all new individuals are replaced with

$$s_l(n_3) \leftarrow s_{\max}(n_3) - s_l(n_3), \tag{23}$$

where $s_{\max}(n_3)$ denotes the maximum value of the search span.

The procedure from ii) thru iv) is repeated for generations which are designed in advance.

4.4 *Fitness function*

As mentioned above, the response of the closed-loop system can be shaped fairly freely by using a suitable fitness function. For example, consider the following fitness function:

$$f(P) = \sum_{t=1}^{N} \{y_m(t) - y_P(t)\}^2. \tag{24}$$

$f(P)$ corresponds to the quadratic performance criterion which is often used designing the control systems[1]. $y_m(t)$ denotes the reference model output excited by the reference signal $w(t)$. $y_P(t)$ is the output by the p-th individual when the PID controller is employed for the full model, whose PID parameters correspond to each individual in the real-coded GA.

Usually, a kind of quadratic performance criteria is mainly used so that it may be convenient to obtain a control law analytically. However, in parameter tuning by the GA, it is a feature that the fitness function equivalent to the performance criterion can be freely chosen for the control purpose as mentioned above.

5. SIMULATION EXAMPLE

In this section, the newly proposed scheme is evaluated on a numerical simulation example.

Consider the following nonlinear system expressed by a Hammerstein model:

$$\begin{aligned} y(t) = {} & 0.8y(t-1) - 0.3y(t-2) + 0.4x(t-1) \\ & +0.2x(t-2) + \eta(t)/\Delta, \end{aligned} \tag{25}$$

where $x(t-1)$ is defined by

$$x(t-1) := u(t-1) + 0.5u^2(t-1). \tag{26}$$

$\xi(t)$ denotes a white Gaussian noise with zero mean and covariance 0.01. Suppose that the full model is also given by (25) and (26). Carrying out the parameter estimation by a recursive least squares method using the input-output data of (25) and (26), the design-oriented model was obtained as follows:

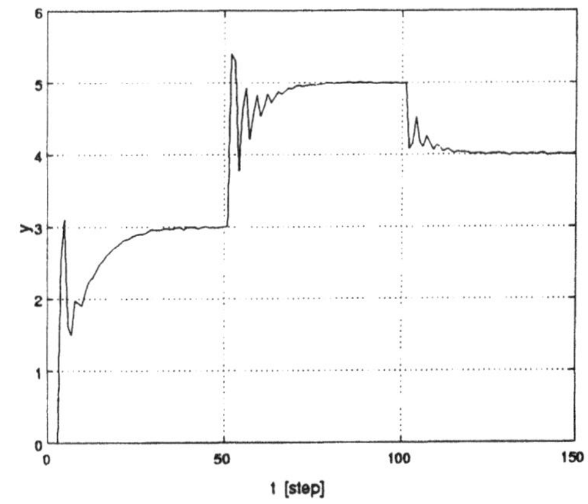

Fig. 2. Control result by using CHR type PID control scheme.

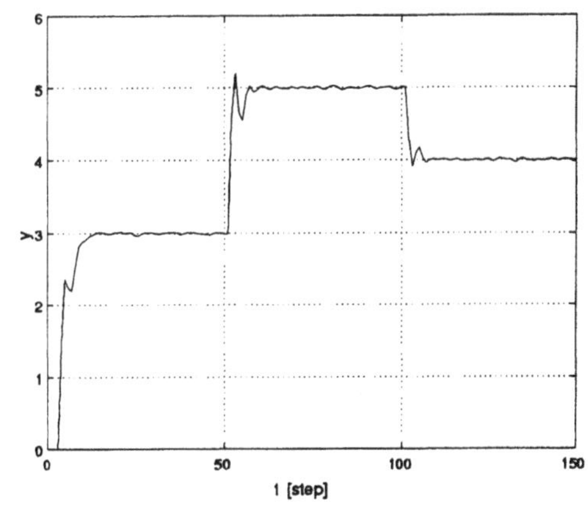

Fig. 3. Control result by using the proposed PID control scheme.

$$\begin{aligned} y(t) = {} & 1.11y(t-1) - 0.47y(t-2) + 1.58u(t-1) \\ & +0.26u(t-2) - 0.47u(t-3) + \xi(t)/\Delta. \end{aligned} \tag{27}$$

First, for comparison purposes, the control result by Chien, Hrones & Reswick(CHR) method(1952), which is well-known as a PID parameter tuning method, is shown in Fig.2. Here, PID parameters were chosen as $k_c = 0.36$, $T_I = 4.07$ and $T_D = 1.25$. Fig.2 illustrates that the control result is oscillatory in the transient state, and the settling time becomes large.

Next, the newly proposed scheme was employed, and the control result is shown in Fig.3, where the strategy of the real-coded GA is summerized as follows.

- Individual: The string is composed of λ, σ and δ corded by real numbers, whose initial values were randomly generated. The number of the population was set to 100.
- Generation: The generation number was set to 50.

[1] Taking meaning of fitness into consideration, the inverse of the cost function is usually used as a fitness function in GA. However, in order to make the relation to the control theory clear, (24) is defined as a fitness function in this paper.

- Fitness function: The fitness function was given by (24), where $N = 150$.
- Selection: The conventional Elite strategy (Goldberg, 1992) was employed, that is, 20 percent superior individuals were saved in the next generation.
- Crossover: The 80 percent remains were generated by the crossover. The crossover was performed by (20), (21) and (22).
- Mutation: The ratio of the mutation was set to 30 percent. The mutation was carried out by (23).
- Seek spans: $0.1 \leq \lambda \leq 3.0$, $1.0 \leq \sigma \leq 10.0$, $0.0 \leq \delta \leq 2.0$

Moreover, the reference model was designed as:

$$y_m(t) = \frac{z^{-1}0.40}{1 - 0.74z^{-1} + 0.14z^{-2}} w(t). \quad (28)$$

PID parameters calculated using λ, σ and δ which were sought by GA, were the following: $k_c = 0.20$, $T_I = 1.06$ and $T_D = 1.15$, and the sampling interval T_s was set to 1.0[sec].

From Fig.3, it is clear that the oscillation is considerably removed, and that the newly proposed scheme works well.

6. CONCLUSION

In this paper, a new evolutionary tuning scheme of PID parameters has been proposed. Moreover, the effectiveness of the newly proposed scheme has been verified by the numarical simulation example. The procedure of the newly proposed method is summerized as follows.

1) A suitable set of the user-specified parameters in the performance criterion of the GMVC law is firstly sought by the GA.
2) PID parameters are calculated by using the suitable set of the user-specified parameters.

Therefore, since the seek spans of the user-specified parameters are restricted to a fairly small area from the viewpoint of the control engineering restrictions, the time required to seek a suitable set of the parameters can be drastically shortend. Now, in most design-oriented models, there exist various modeling errors and uncertainties due to the change of the operating conditions. The control scheme taking the system uncertainties into consideration will have to be investigated in near future. For example, it may be possible to utilize the proposed scheme in designing the self-tuning and/or robust control systems. Especially, since the time to seek parameters in the proposed scheme is very short, it seems to be possible to tune PID parameters in an on-line manner.

REFERENCES

Åström, K.J. and T.Hägglund, (1988). *AutomaticTuning of PID Controllers*, Instrument Society of America.

Åström, K.J., C.C.Hang, P.Persson and W.K.Ho (1992). Towards Intelligent PID Control, *Automatica*, **28**, 1–9.

Cameron, F. and D.E.Seborg (1983). A Self-Tuning Controller with a PID Structure, *Int.J. of Control*, **38**, 401–417.

Chien, K.L., J.A.Hrones and J.B.Reswick (1952). On the Automatic Control of Generalized Passive Systems, *Trans. ASME*, **74**, 175–185.

Clarke, D.W. and P.J.Gawthrop (1979). Self-Tuning Control, *Proc. IEE*, **126D**, 633–640.

Fujii, K., T.Yamamoto and M.Kaneda (1997). Self-Tuning Control of a Polybutene Process System, *Proc. of IFAC Symposium on Advanced Control of Chemical Process*, Banff, 107–112.

Goldberg, D.E. (1992). *Genetic Algorithm in Search, Optimization & Machine Learning*, Addison-Wesley.

Jones, A.H., N.Ajlouni and M.Uzam (1996). On-Line Frequency Domain Identification and Genetic Tuning of PID Controllers, *Proc. of IEEE Conference on Emerging Tech nologies and Factory Automation*, Hawaii, 261–266.

Omatu, S. and S.Deris (1996). Stabilization of Inverted Pendulum by the Genetic Algorithm, *Proc. of IEEE International Conference on Evolutionary Computation*, Nagoya, 700–705.

Ono, I., M.Yamamura and H.Kita (2000). Real-Coded Genetic Algorithms and Their Applications (in Japanese), *J.of Japanese Society for Artificial Intelligence*, **15**, 259–266.

Porter, B. and A.H.Jones (1992). Genetic Tuning of Digital PID Controllers, *Electronics Letter*, **28**, 843–844.

Yamamoto, T., H.Ishihara, S.Omatu and T.Kitamori (1992). Self-Tuning Controller with Two-Degree-of-Freedom PID Structure and Its Application, *Preprints of IFAC Symposium on Adaptive Systems in Control and Signal Processing*, Grenoble, 527–532.

Yamamoto, T., K.Fujii and M.Kaneda (1998). Design and Implementation of a Self-Tuning PID Controller, *Preprints of IFAC Workshop on Adaptive Control and Signal Processing*, Glasgow, 58–63.

Ziegler, J.G. and N.B.Nichols (1942). Optimum Settings for Automatic Controllers, *Trans. ASME*, **64**, 759–768.

A DESIGN OF MULTILOOP PREDICTIVE SELF-TUNING PID CONTROLLERS

Masaru Katayama * **Toru Yamamoto** * **Yasuhiro Mada** *
Sirish L. Shah **

* *Department of Technology and Information Education, Graduate School of Education, Hiroshima University, 739-8524, Japan. Tel: +81-824-24-7160, Fax: +81-824-22-7171, E-mail:yama@hiroshima-u.ac.jp*
** *Department of Chemical and Materials Engineering, University of Alberta, Edmonton, Alberta, T6G 2G6, Canada. Tel: +1-403-492-5162, Fax: +1-403-437-6879, E-mail:sirish.shah@ualberta.ca*

Abstract: In this paper, a design scheme of multiloop predictive self-tuning PID control systems is proposed for multivariable systems. The proposed scheme firstly uses a static pre-compensator as an approximately decoupling device, in order to roughly decouple the controlled object. Furthermore, by regarding the approximately decoupled system as a series of single-input single-output subsystems, a single-input single-output PID controller is designed for each subsystem. The PID parameters are calculated on-line based on the relationship between the PID control and the generalized predictive control laws. *Copyright © 2001 IFAC*

Keywords: PID control, Self-tuning control, Generalized predictive control, Multivariable control systems, Decoupling pre-compensators, Process control

1. INTRODUCTION

Self-tuning control schemes (Åström, 1983; Clarke and Gawthop, 1979; Clarke, *et al.*, 1987) are useful for systems with unknown or slowly time-varying parameters, and represent a class of advanced control algorithms. On the other hand, PID (Ziegler and Nichols, 1942; Chein *et al.*, 1972) control schemes have been widely used for a large class of process control systems. However, most process control systems in industries have usually unknown time-delays, and also are multivariable systems with strong interactions. Therefore, it is necessary to consider the choice of PID parameters for such systems.

In this paper, a design scheme of multiloop predictive self-tuning PID control systems is proposed for multivariable systems with uncertain time-delays. The proposed scheme utilizes a static pre-compensator which is well known as an approximate decoupling device based on the inverse Nyquist method of Rosenbrock (Rosenbrock, 1970), to approximately decouple the controlled object. The static pre-compensator is designed of by an on-line estimator. The $p \times p$ pre-compensated system is then controlled via 'p' self-tuning PID controllers whose parameters are adjusted by a second identifier placed around the pre-compensated plant. The PID parameters are calculated on-line based on the relationship between the PID control and the generalized predictive control(GPC) (Clarke, *et al.*, 1987) laws. It is well known that GPC is useful for systems with uncertain time-delays. This paper is organized as follows. First, a descriptive model for the controlled object is considered. Next, a design of pre-compensators is discussed, and a multiloop predictive self-tuning PID control system is developed. Finally, the effectiveness of the newly proposed control scheme is evaluated on a simulation example.

2. SYSTEM DESCRIPTION

The process mentioned above can be expressed by the following p-input and p-output multivariable system:

$$A(z^{-1})y(t) = DB(z^{-1})u(t-1) + \frac{\xi(t)}{\Delta} \quad (1)$$

where $u(t)$ and $y(t)$ are respectively the input vector and the output vector with p-elements, *i.e.*,

$$\left.\begin{array}{l} u(t) = [u_1(t), u_2(t), \cdots, u_p(t)]^T \\ y(t) = [y_1(t), y_2(t), \cdots, y_p(t)]^T, \end{array}\right\} \quad (2)$$

and $\xi(t)$ denotes a white Gaussian noise vector. $A(z^{-1})$ is a diagonal polynomial matrix given by

$$A(z^{-1}) = \text{diag}\{A_1(z^{-1}), \cdots, A_p(z^{-1})\}. \quad (3)$$

Although most process systems are usually high-order, they can be approximated time-delay plus second-order lag. Therefore, $A_i(z^{-1})$ as considered in this study is a second-order polynomial:

$$A_i(z^{-1}) = 1 + a_{i,1}z^{-1} + a_{i,2}z^{-2} \quad (4)$$
$$(i = 1, 2, \cdots, p).$$

$B(z^{-1})$ is given by the following polynomial matrix with full elements:

$$B(z^{-1}) = \begin{bmatrix} B_{11}(z^{-1}) & B_{12}(z^{-1}) & \cdots & B_{1p}(z^{-1}) \\ B_{21}(z^{-1}) & B_{22}(z^{-1}) & \cdots & B_{2p}(z^{-1}) \\ \vdots & \vdots & \ddots & \vdots \\ B_{p1}(z^{-1}) & B_{p2}(z^{-1}) & \cdots & B_{pp}(z^{-1}) \end{bmatrix}$$
$$= B_0 + B_1 z^{-1} + \cdots + B_m z^{-m}, \quad (5)$$

where the coefficient matrix B_l is defined as

$$B_k = \begin{bmatrix} b_{k,11} & b_{k,12} & \cdots & b_{k,1p} \\ b_{k,21} & b_{k,22} & \cdots & b_{k,2p} \\ \vdots & \vdots & \ddots & \vdots \\ b_{k,p1} & b_{k,p2} & \cdots & b_{k,pp} \end{bmatrix}. \quad (6)$$

All interactions are considered in the polynomial matrix $B(z^{-1})$. Moreover, D is the time-delay matrix of the form:

$$D = \text{diag}\{z^{-k_{m_1}}, z^{-k_{m_2}}, \cdots, z^{-k_{m_p}}\}, \quad (7)$$

where k_{m_i} denotes the minimum estimated value of the time-delays for the i-th row. If the true time-delays, k_i, are known exactly in advance, it is better to set

$$k_{m_i} = k_i \ (:= \min_{l=1,\cdots,p} \{k_{i,l}\}), \quad (8)$$

where $k_{i,l}$ denotes the time-delay between the i-th output and the j-th input signals. On the other hand, k_{m_i} is set to 0 in the case where *a priori* information about time-delays is not available. For the system (1), we make the following assumptions.

[A.1] The polynominal matrix $A(z^{-1})$ is stable.

[A.2] The degree of $B(z^{-1})$, m is known, and the following relationship is satisfied:

$$k_{m_i} \le k_i \le m \quad (9)$$

[A.3] $\det B(1) \neq 0$.

3. MULTILOOP PREDICTIVE SELF-TUNING PID CONTROLLER

The block diagram of the proposed multiloop predictive self-tuning PID control system is shown in Fig.1. This control system has two estimators, Estimator 1 and Estimator 2 as shown in Fig.1. Each estimator plays the designated role of on-line tuning of the pre-compensator and PID parameters, respectively.

3.1 *Pre-compensator*

In designing multivariable control systems, it is important to first remove or compensate for interactions. The simplest pre-compensator (Yamamoto and Shah, 1998) is employed, which is given by the inverse static gain H of the form:

$$H = B^{-1}(1)A(1) = \begin{bmatrix} h_{11} & h_{12} & \cdots & h_{1p} \\ h_{21} & h_{22} & \cdots & h_{2p} \\ \vdots & \vdots & \ddots & \vdots \\ h_{p1} & h_{p2} & \cdots & h_{pp} \end{bmatrix}. \quad (10)$$

That is, H is given by the inverse of the static gain matrix of the controlled object. In fact, the pre-compensator could be designed as $B^{-1}(z^{-1})A(z^{-1})$ in order to exactly decouple the system. However, this would be needlessly cumbersome for realization purposes and require the assumption that $B^{-1}(z^{-1})$ is asymptotically stable. Therefore, a simple static pre-compensator as given by eqn.(10) is employed in this paper.

Then, the augmented system constructed by eqns.(1) and (10) is described as

$$A(z^{-1})y(t) = DB(z^{-1})Hv(t-1) + \frac{\xi(t)}{\Delta}, \quad (11)$$

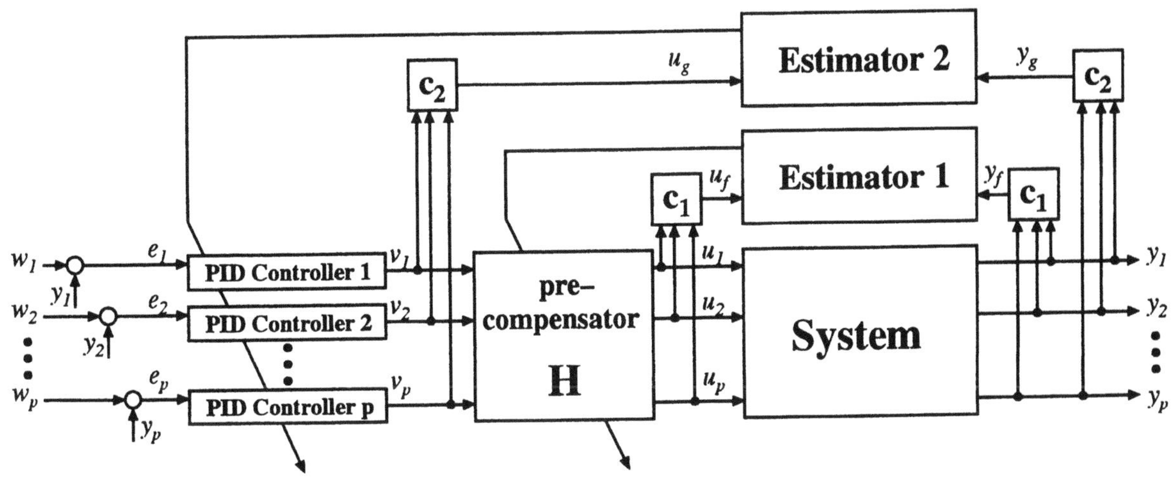

Fig. 1. Block diagram of the proposed multiloop predictive self-tuning PID control system.

where $v(t)$ denotes the input vector to the augmented system. By regarding the augmented system (11) as an approximately decoupled system, the following model for each diagonal element can be obtained:

$$A_i(z^{-1})y_i(t) = z^{-k_{m_i}}\bar{B}_i(z^{-1})v_i(t-1) + \frac{\xi_i(t)}{\Delta}, \quad (12)$$

where $A_i(z^{-1})$ is given by eqn.(4), and $\bar{B}_i(z^{-1})$ is defined as:

$$\bar{B}_i(z^{-1}) := \sum_{l=1}^{p} B_{il}(z^{-1})h_{li} = \bar{b}_{i,0} + \bar{b}_{i,1}z^{-1} + \cdots + \bar{b}_{i,m}z^{-m}. (13)$$

3.2 *Multiloop PID controller design*

Next, the design scheme of PID controllers is considered for the augmented system given by eqn.(12). The digital PID control law to be considered in this paper, is described as

$$\Delta v_i(t) = \frac{k_{c_i}T_s}{T_{I_i}}e_i(t) - k_{c_i}\left(\Delta + \frac{T_{D_i}}{T_s}\Delta^2\right)y_i(t)(14)$$

where Δ denotes the differencing operator defined as $\Delta := 1 - z^{-1}$; $e_i(t)$ denotes the control error signal given by

$$e_i(t) := w_i(t) - y_i(t), \quad (15)$$

and k_{c_i}, T_{I_i} and T_{D_i} are the proportional gain, the reset time and the derivative time, respectively. Furthermore, T_s denotes the sampling interval. For convenience, let $L_i(z^{-1})$ be

$$L_i(z^{-1}) := k_{c_i}\left\{\left(1 + \frac{T_s}{T_{I_i}} + \frac{T_{D_i}}{T_s}\right) - \left(1 + \frac{2T_{D_i}}{T_s}\right)z^{-1} + \frac{T_{D_i}}{T_s}z^{-2}\right\} (16)$$

then, eqn.(14) can rewritten as

$$L_i(z^{-1})y_i(z^{-1}) + \Delta v_i(t) - L_i(1)w_i(t) = 0. (17)$$

The tuning of the control parameters in PID control laws (14) or (17), is important, since the performance of the control system strongly depends on them. For systems with unknown parameters and unknown time-delays, however, it is difficult to find the "optimal" PID parameters. On the other hand, the GPC scheme is useful for such systems. Therefore, a self-tuning PID control algorithm based on the relationship between PID control and the GPC laws, is derived below.

3.3 *PID tuning*

Consider the following cost function to derive a GPC law(Clarke, *et al.*, 1987):

$$J_i = E\left[\sum_{j=k_{m_i}+1}^{N_i+k_{m_i}} \{y_i(t+j) - w_i(t)\}^2 + \sum_{j=1}^{N_i} \lambda_i(j)\{\Delta v_i(t+j-1)\}^2\right], \quad (18)$$

where $\lambda_i(j)$ is the control weighting factor and N_i is the prediction horizon. The control law based on minimizing the cost function (18) is given by

$$\sum_{j=k_{m_i}+1}^{N_i+k_{m_i}} p_{i,j}F_{i,j}(z^{-1})y_i(t) + \left\{1 + z^{-1}\sum_{j=1}^{N_i} p_{i,j}S_{i,j}(z^{-1})\right\}\Delta v_i(t) - \sum_{j=1}^{N_i} p_{i,j}w_i(t) = 0, \quad (19)$$

where $F_{i,j}(z^{-1})$ is calculated by solving the following Diophantine equation:

$$1 = \Delta A_i(z^{-1})E_{i,j}(z^{-1}) + z^{-j}F_{i,j}(z^{-1}), \quad (20)$$

where

$$\left.\begin{array}{l} E_{i,j}(z^{-1}) = 1 + e_{i,1}z^{-1} + \cdots + e_{i,j-1}z^{-(j-1)} \\ F_{i,j}(z^{-1}) = f_{i,0}^{j} + f_{i,1}^{j}z^{-1} + f_{i,2}^{j}z^{-2}. \end{array}\right\}$$

Furthermore, $R_{i,j}(z^{-1})$ is calculated by

$$E_{i,j}(z^{-1})B_i(z^{-1}) = R_{i,j}(z^{-1}) + z^{-j}S_{i,j}(z^{-1}), \quad (21)$$

where $R_{i,j}(z^{-1})$ is a polynomial of the form:

$$R_{i,j}(z^{-1}) = r_{i,0} + \cdots + r_{i,j-1}z^{-(j-1)}. \quad (22)$$

Moreover, $P_i(z^{-1})$ is defined by

$$P_i(z^{-1}) := \sum_{j=1}^{N_i} p_{i,N_i+k_{m_i}-j+1}z^{-(j-1)}, \quad (23)$$

where $p_{i,j}$ is calculated by

$$[p_{i,1}, p_{i,2}, \cdots, p_{i,N_i+k_{m_i}}] := [1, 0, \cdots, 0] \cdot (R_i^T R_i + \Lambda)^{-1} R_i^T. \quad (24)$$

In eqn.(24), the matrix R_i which consists of coefficient of $R_{i,j}(z^{-1})$ is defined as

$$R_i := \begin{bmatrix} r_{i,0} & & & 0 \\ r_{i,1} & r_{i,0} & & \\ \vdots & \vdots & \ddots & \\ r_{i,N_i+k_{m_i}-1} & r_{i,N_i+k_{m_i}-2} & \cdots & r_{i,0} \end{bmatrix}, \quad (25)$$

and Λ is as follows:

$$\Lambda := \text{diag}\{\lambda_i(j)\} \quad (j = 1, 2, \cdots, N_i). \quad (26)$$

Next, based on the relationship between the PID control and the GPC laws, a tuning method of PID parameters is derived. In order to obtain a control law with a PID structure, the following equation with $S_{i,j}(z^{-1})$ replaced by the static gain $S_{i,j}(1)$ is considered:

$$\sum_{j=k_{m_i}+1}^{N_i+k_{m_i}} p_{i,j}F_{i,j}(z^{-1})y_i(t) + \left\{1 + \sum_{j=1}^{N_i} S_{i,j}(1)\right\}\Delta v_i(t) - \sum_{j=1}^{N_i} p_{i,j}w_i(t) = 0. \quad (27)$$

Here, X_i is defined as

$$X_i = 1 + \sum_{j=1}^{N_i} S_{i,j}(1). \quad (28)$$

Furthermore, if the following relation is satisfied:

$$L_i(z^{-1}) = \frac{1}{X_i}\sum_{j=k_{m_i}+1}^{N_i+k_{m_i}} p_{i,j}F_{i,j}(z^{-1}) = \frac{1}{X_i}(\tilde{f}_{i,0} + \tilde{f}_{i,1}z^{-1} + \tilde{f}_{i,2}z^{-2}) \quad (29)$$

as in eqn.(27), then eqn.(27) becomes identical to eqn.(17). Therefore, based on eqns.(16) and (29), the PID parameters can be calculated as follows:

$$\left.\begin{array}{rcl} k_{c_i} & = & -\frac{1}{X_i}(\tilde{f}_{i,1} + 2\tilde{f}_{i,2}) \\ T_{I_i} & = & -\frac{\tilde{f}_{i,1} + 2\tilde{f}_{i,2}}{\tilde{f}_{i,0} + \tilde{f}_{i,1} + \tilde{f}_{i,2}}T_s \\ T_{D_i} & = & -\frac{\tilde{f}_{i,2}}{\tilde{f}_{i,1} + 2\tilde{f}_{i,2}}T_s. \end{array}\right\} \quad (30)$$

3.4 *Multiloop predictive self-tuning controller*

Based on the control scheme discussed above, a multiloop predictive self-tuning PID control is designed in this section. This control system has two estimators. Each estimator plays the designated role of on-line tuning of the pre-compensator and PID parameters, respectively.

In estimator 1, unknown parameters concerned with the static gain are estimated by using the following recursive least squares(RLS) algorithm:

$$\hat{\theta}_{1,i}(t) = \hat{\theta}_{1,i}(t-1) + \frac{\Gamma_{1,i}(t-1)\psi_{1,i}(t-1)}{1 + \psi_{1,i}^T(t-1)\Gamma_{1,i}(t-1)\psi_{1,i}(t-1)}\varepsilon_{1,i}(t), \quad (31)$$

$$\Gamma_{1,i}(t) = \Gamma_{1,t}(t-1) - \frac{\Gamma_{1,i}(t-1)\psi_{1,i}(t-1)\psi_{1,i}^T(t-1)\Gamma_{1,t}(t-1)}{1 + \psi_{1,i}^T(t-1)\Gamma_{1,i}(t-1)\psi_{1,i}(t-1)} \quad (32)$$

$$\varepsilon_{1,i} = \Delta y_{f_i}(t) - \hat{\theta}_{1,i}^T(t-1)\psi_{1,i}(t-1). \quad (33)$$

$\hat{\theta}_{1,i}(t)$ and $\psi_{1,i}(t-1)$ are unknown parameter and data vectors, respectively, *i.e.*,

$$\hat{\theta}_{1,t}(t) = [\hat{a}_{i,1}(t), \hat{a}_{i,2}(t), \hat{b}_{0,i1}(t), \hat{b}_{0,i2}(t), \cdots, \hat{b}_{0,ip}(t), \cdots, \hat{b}_{m,i1}(t), \hat{b}_{m,i2}(t), \cdots, \hat{b}_{m,ip}(t)]^T \quad (34)$$

$$\psi_{1,t}(t-1) = [-\Delta y_{f_i}(t-1), -\Delta y_{f_i}(t-2), \Delta u_f^T(t-k_{m_i}-1), \cdots, \Delta u_f^T(t-k_{m_i}-m-1)]^T. \quad (35)$$

In eqn.(35), the subscript 'f' denotes filtered value of the inputs and the outputs. Since the pre-compensator H given by eqn.(10) is supposed to capture the static dynamics and gain of this system, the following low-pass data pre-filter $C_1(z^{-1})$ is used for the purpose of improving the reliability of the static gain calculation:

$$C_1(z^{-1}) := \frac{1-c_1}{1-c_1 z^{-1}}. \tag{36}$$

It is well known that the filter plays an important role in the identification of real systems. The filtered values $y_{f_i}(t)$ and $u_{f_i}(t)$ included in (33) and (35) are given by

$$\left.\begin{array}{l} y_{f_i}(t) := C_1(z^{-1})y_i(t) \\ u_{f_i}(t) := C_1(z^{-1})u_i(t). \end{array}\right\} \tag{37}$$

Based on the on-line estimates included in $\hat{\theta}_{1,i}(t)$, the pre-compensator H is calculated.

In estimator 2, the unknown parameters included in the augmented system (12) are estimated via the following RLS algorithm:

$$\begin{aligned}\hat{\theta}_{2,i}(t) &= \hat{\theta}_{2,i}(t-1) \\ &+ \frac{\Gamma_{2,i}(t-1)\psi_{2,i}(t-1)}{1+\psi_{2,i}^T(t-1)\Gamma_{2,i}(t-1)\psi_{2,i}(t-1)}\varepsilon_{2,i}(t),\end{aligned} \tag{38}$$

$$\begin{aligned}\Gamma_{2,i}(t) &= \Gamma_{2,t}(t-1) \\ &- \frac{\Gamma_{2,i}(t-1)\psi_{2,i}(t-1)\psi_{2,i}^T(t-1)\Gamma_{2,t}(t-1)}{1+\psi_{2,i}^T(t-1)\Gamma_{2,i}(t-1)\psi_{2,i}(t-1)}\end{aligned} \tag{39}$$

$$\varepsilon_{2,i} = \Delta y_{f_i}(t) - \hat{\theta}_{2,i}^T(t-1)\psi_{2,i}(t-1). \tag{40}$$

$\hat{\theta}_{2,i}(t)$ and $\psi_{2,i}(t-1)$ are given by

$$\begin{aligned}\hat{\theta}_{2,t}(t) = [&\hat{a}_{i,1}(t), \hat{a}_{i,2}(t), \hat{\bar{b}}_{i,0}(t) \\ &, \hat{\bar{b}}_{i,1}(t), \cdots, \hat{\bar{b}}_{i,m}(t)]^T\end{aligned} \tag{41}$$

$$\begin{aligned}\psi_{2,t}(t-1) = [&-\Delta y_{g_i}(t-1), -\Delta y_{g_i}(t-2), \\ &\Delta v_{g_i}^T(t-k_{m_i}-1), \cdots, \Delta v_{g_i}^T(t-k_{m_i}-m-1)]^T.\end{aligned} \tag{42}$$

For the purpose of improving reliability of the parameter estimation and once again emphasizing estimation accuracy at lower frequencies, the following estimator filter $C_2(z^{-1})$ is also utilized.

$$C_2(z^{-1}) := \frac{1-c_2}{1-c_2 z^{-1}} \tag{43}$$

Therefore, y_{g_1} and v_{g_i} included in (40) and (42) are given by

$$\left.\begin{array}{l} y_{g_i}(t) := C_2(z^{-1})y_i(t) \\ v_{g_i}(t) := C_2(z^{-1})v_i(t). \end{array}\right\} \tag{44}$$

By solving the Diophantine equation based on estimates included in $\hat{\theta}_{2,i}(t)$, and calculating eqns.(28) and (30), PID parameters can be obtained. The proposed multiloop predictive self-tuning PID(ST-PID) control algorithm is then realized via the following step.

Multiloop Predictive ST-PID control algorithm

1. Choose N_i and $\lambda_i(j)$.
2. Design the estimator filters $C_1(z^{-1})$ and $C_2(z^{-1})$.
3. Estimate $\hat{\theta}_{1,i}(t)$ by using the RLS algorithm in eqns.(31)-(35).
4. Calculate the pre-compensator H based on eqn.(10).
5. Estimate $\hat{\theta}_{2,i}(t)$ by using the RLS algorithm in eqns.(38)-(42).
6. Solve the Diophantine eqns. (20) and (21).
7. Calculate X_i based on eqn.(28).
8. Calculate PID parameters based on eqn.(30)
9. Calculate the control input vector $v(t)$ based on eqn.(14)
10. Update t and return to step 3.

4. NUMERICAL EVALUATION

In this section, the effectiveness of the proposed design scheme is evaluated on the following simulation example:

$$\begin{aligned}\boldsymbol{y}(t) = &\begin{bmatrix} 0.9 & 0 \\ 0 & 0.8 \end{bmatrix} \boldsymbol{y}(t-1) - \begin{bmatrix} 0.2 & 0 \\ 0 & 0.1 \end{bmatrix} \boldsymbol{y}(t-2) \\ &+ \begin{bmatrix} 0.6 & 0.5 \\ 0.4 & 0.5 \end{bmatrix} \boldsymbol{u}(t-3) + \begin{bmatrix} 0.3 & 0.4 \\ 0.3 & 0.3 \end{bmatrix} \boldsymbol{u}(t-4).\end{aligned} \tag{45}$$

First, for the purpose of investigating the interaction of the controlled object (45), a fixed PID control scheme without the pre-compensator was employed, whose PID parameters were tuned based on the Chien, Hrones & Reswick(CHR) scheme(Chein, *et al.*, 1972). The control results are shown in Fig.2, where $k_{c_1} = 0.16$, $k_{c_2} = 0.18$, $T_{I_1} = 10.61$, $T_{I_2} = 10.73$. Owing to the interactions and the time-delays, it is shown that the control responses are very oscillatory.

Next, the proposed multiloop predictive self-tuning PID control scheme was employed. The control results are summarized in Fig.3, where $N_1 = 12$, $N_2 = 12$, $\lambda_1 = 5.0$, $\lambda_2 = 2.0$, $c_1 = 0.85$ and $c_2 = 0.4$, respectively. The results of the estimator to compute the PID parameters are shown in Fig.4. From Fig.3 and Fig.4, it is found that satisfactory control performance can be obtained, and it is clear that the proposed control scheme works well.

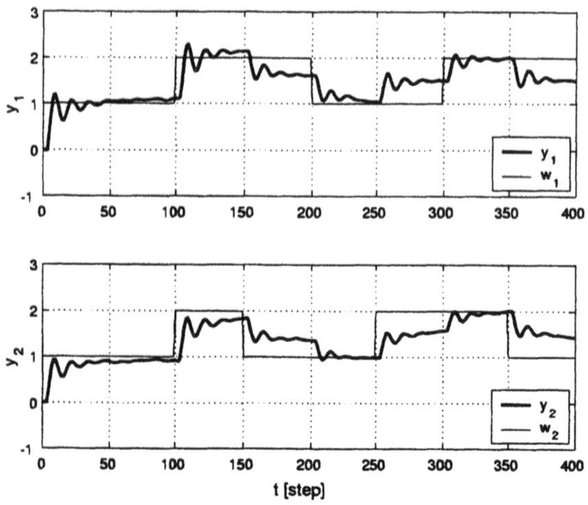

Fig. 2. Control result by using the PID controller without the pre-compensator.

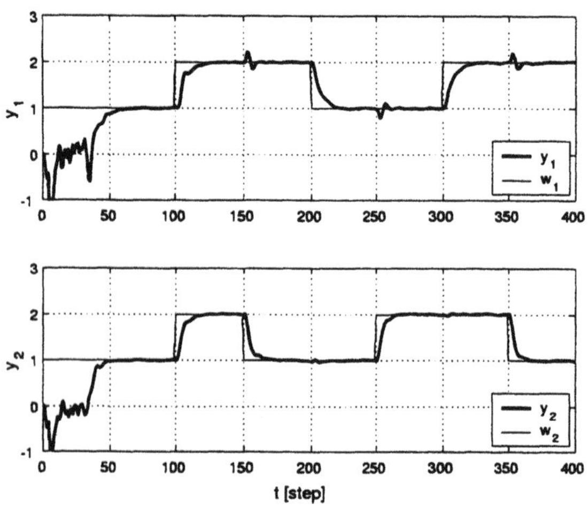

Fig. 3. Control result by using the newly proposed control scheme.

5. CONCLUSION

In this paper, a design scheme of multiloop predictive self-tuning PID control systems has been proposed for multivariable systems. The proposed scheme utilizes a static pre-compensator as an approximately decoupling device, in order to roughly decouple the controlled object. Furthermore, by regarding the roughly decoupled system as a series of single-input single-output subsystems, a single-input single-output self-tuning PID control system based on the relationship between the PID control and the GPC laws, is designed for each subsystem. According to this procedure, each PID control system can be easily designed. The effectiveness of the proposed scheme are illustrated by application to a simulation example. The applications of the newly proposed scheme to real systems are currently in progress.

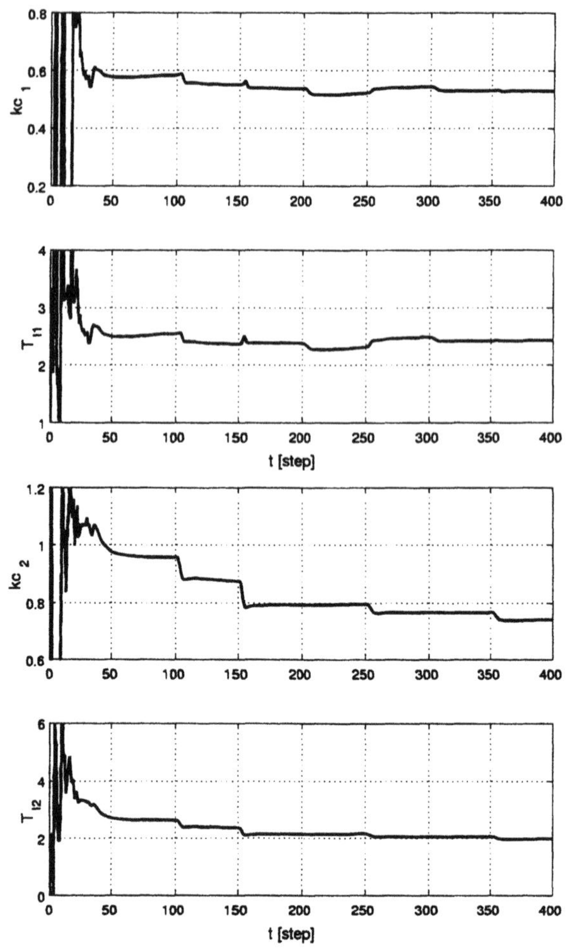

Fig. 4. Controller parameter trajectories for the control law used in Fig.3.

6. REFERENCES

Åström, K.J. (1983). Theory and Applications of Adaptive Control -A Survey. *Automatica*. **19-5**, 471–486.

Chein, K.L., J.A. Horones, and J.B. Reswick (1972). On the Automatic Control of Generalized Passive System. *Trans. ASME*. **74**, 175–185.

Clarke, D.W., and P.J. Gawthrop (1979). Self-Tuning Control. *Proc. IEEE*. **126D-6**, 633–640.

Clarke, D.W., C. Mohtadi, and P.S. Tuffs (1987). Generalized Predictive Control. *Automatica*. **23-6**, 859–875.

Rosenbrock, H.H. (1970). State-Space and Multivariable Theory. *Wiley-Interscience*.

Yamamoto, T. and S.L. Shah (1998). Multivariable Self-Tuning PID Controller. *Proc. of IEEE Conference on Control Applications*. 1230–1234. Trieste.

Ziegler, J.G., and N.B. Gawthrop (1942). Optimum Setting for Automatic Controllers. *Trans. ASME*. **64-8**, 759–768.

SELF-TUNING POLE-ASSIGNMENT PD SWAY CONTROL OF A LIFTED LOAD FOR A MOBILE CRANE

Kazuo Kawada* Hiroyuki Sogo* Toru Yamamoto**

** Department of Electro-Mechanical Systems Engineering, Takamatsu National College of Technology, 355 Chokushi, Takamatsu, Kagawa 761-8058, Japan*
*** Department of Technology and Information Education, Graduate School of Education, Hiroshima University, 1-1-1 Kagamiyama, Higashi-Hiroshima, Hiroshima 739-8524, Japan*
TEL:+81-824-24-7160 FAX:+81-824-22-7171
E-mail:yama@hiroshima-u.ac.jp

Abstract: In the sway control system of a lifted load for a mobile crane, control parameters must be tuned corresponding to the rope length, the jib length and the jib angle in order to obtain the good control performance. In this paper, a new self-tuning PD control scheme is proposed, whose PD parameters are self-tuned corresponding to the change of these parameters. The effectiveness of the newly proposed control scheme is experimentally evaluated on the sway control system of a lifted load for a mobile crane.

Keywords: self-tuning control, PID control, adaptive control, pole assignment, mechanical systems

1. INTRODUCTION

In recent years, many complicated control algorithms can be developed owing to the rapid progress of digital computers, and lots of advanced and intelligent control schemes have been considered. On the other hand, PID control(Ziegler and Nichols, 1942; Chien *et al.*, 1952; Suda, 1992) schemes have been widely employed for real control systems. The reasons are summarized as follows:

1. PID controllers have simple control structures.
2. It is relatively easy to understand the physical meanings of the control parameters.
3. It is possible to maintain and tune PID parameters based on operators' experience.

However, it is difficult to tune PID parameters in the case where system dynamics vary due to the change of movement and environment conditions. For example, in case of the sway control system of a lifted load for a mobile crane to be considered in this work, the system parameters and the equilibrium states are changed by the rope length, the jib length and the jib angle. Then, it is necessary to prepare PID parameters corresponding to the system parameters and equilibrium states, whereas it is difficult to mount the huge combination of PID parameters in real systems.

In this paper, a new self-tuning PD controller design scheme is proposed, and applied for a sway control system of a lifted load for a mobile crane. Then, PD controller which is removed the integral control action from PID controller is employed, because the jib angular acceleration is utilized as the control input. The state feedback control(Tada *et al.*, 1997a, 1997b) and fuzzy control(Hino *et al.*, 1998) have been considered for the sway control system of a lifted load for a mobile crane up to now. However, to the best of our knowledge, the sway control system of a lifted load for a mobile crane is not considered in an adaptive manner. According to the newly proposed control scheme, PD parameters are adjusted adequately corresponding to the rope length, the jib length and the jib angle. This paper is organized as follows. First, the physical model for the sway of a lifted load by the jib derricking motion for a mobile crane is derived. Next, a self-tuning PD control algorithm is considered based on the pole-assignment control scheme. Finally, the newly proposed scheme is experimentally evaluated on the sway control system of a lifted load for a mobile crane.

2. MODELING OF MOBILE CRANE SYSTEM

2.1 *Physical model*

The simulation model of a mobile crane system to be considered in this paper, is shown in **Fig.1.** The control objective in this paper is to regulate

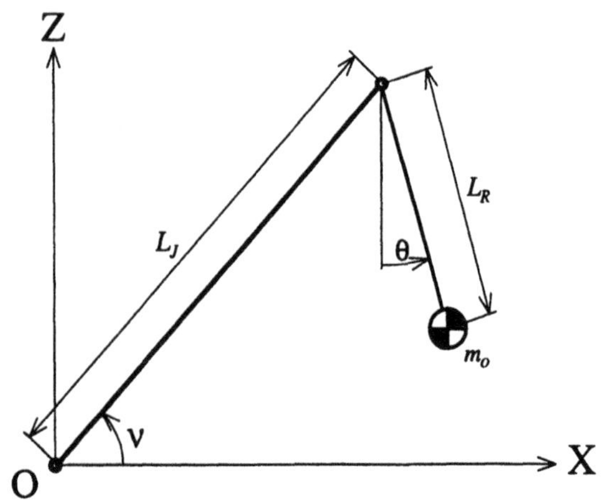

Fig. 1. Simulation model of a mobile crane system.

the sway of a lifted load and the the jib angle. The revolving super structure of a mobile crane, which affects the behavior of a lifted load, is modeled. In order to derive the differential equation, suppose the following assumptions:

1. Lifted load is one particle.
2. Jib and rope are rigid.
3. Mobile crane system has no mechanical loss.
4. Jib angle and sway angle of the lifted load can be measured by sensors.

The differential equation based on the jib angle ν and the sway angle of the lifted load θ, is derived for examining the behavior of the lifted load during the jib derricking motion. In Fig.1, m_o is the mass of a lifted load; L_J is the jib length and L_R is the rope length. The sway angle of the lifted load θ and the jib angle ν, are controlled by manipulating the jib angular acceleration $\ddot{\nu}$. The angle of the lifted load is given by

$$\left.\begin{aligned} X_o &= L_J \cos\nu + L_R \sin\theta \\ Z_o &= L_J \sin\nu - L_R \cos\theta, \end{aligned}\right\} \quad (1)$$

where X_o and Z_o denote positions of the horizontal direction and of the vertical direction from the origin O to m_o, respectively.

The velocity of the lifted load can be derived by differentiating (1) with respect to the time t. Then, the kinetic and the potential energy of the lifted load can be obtained by

$$\left.\begin{aligned} T &= \frac{1}{2} m_o (\dot{X}_o^2 + \dot{Z}_o^2) \\ V &= m_o g Z_o, \end{aligned}\right\} \quad (2)$$

where g is the gravitational acceleration. Substitute (2) into the following Lagrange's equation of motion:

$$\frac{d}{dt}\left(\frac{\partial T}{\partial \dot{q}}\right) - \frac{\partial T}{\partial q} + \frac{\partial V}{\partial q} = Q \quad (3)$$

$$\left.\begin{aligned} q &= \theta \\ Q &= 0, \end{aligned}\right\} \quad (4)$$

where q and $\dot{q}$ denote the generalized coordinates, and the differentiate q with respect to the time t, respectively. Q denotes the external force for each generalized coordinate. Then, the following relationship can be obtained:

$$A_1 \ddot{\theta} = A_2 + A_3 \ddot{\nu} \quad (5)$$

$$\left.\begin{aligned} A_1 &:= L_R \\ A_2 &:= -g \sin\theta - 2\dot{L}_R \dot{\theta} + L_J \dot{\nu}^2 \cos(\nu - \theta) \\ A_3 &:= L_J \sin(\nu - \theta). \end{aligned}\right\} \quad (6)$$

The following assumptions are introduced in order to linearize the non-linear terms in (5).

1. The sway of the lifted load is controlled by the jib angular acceleration $\ddot{\nu}$ around the desired jib angle ν_d.
2. The sway angle of the lifted load θ is so small.
3. The jib angular velocity $\dot{\nu}$ and the sway angular velocity of the lifted load $\dot{\theta}$ are so small.

Then (5) can be simplified to

$$L_R \ddot{\theta} = -g\theta + L_J \ddot{\nu} \sin\nu_d. \quad (7)$$

Moreover, the jib angle, the jib angular velocity and the jib angular acceleration are related to the following equation:

$$\ddot{\nu} = \frac{d}{dt}\dot{\nu} = \frac{d^2}{dt^2}\nu. \tag{8}$$

2.2 *Experimental Model*

The scale model is shown in **Fig.2**, which is prepared for experimentally evaluation. The jib angle and the sway angle of the scale model are controlled by manipulating the jib angular acceleration. A DC motor as an actuator and two rotary encoders as angle sensors, are equipped with the scale model.

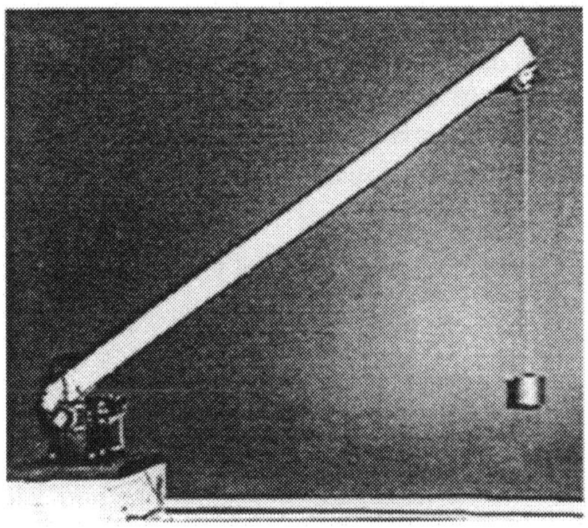

Fig. 2. Experimental scale model.

3. SELF-TUNING PD CONTROLLER

3.1 *Control system*

Based on the physical model (7) and (8), the controlled object is described as the following state space model:

$$\left.\begin{aligned} \dot{\mathbf{x}}(t) &= \mathbf{A}\mathbf{x}(t) + \mathbf{b}u(t) \\ \mathbf{y}(t) &= \mathbf{C}\mathbf{x}(t), \end{aligned}\right\} \tag{9}$$

where $\mathbf{A}$ denotes the system matrix, which depends on the rope length L_R. $\mathbf{b}$ denotes the control vector, which depends on the rope length L_R, the jib length L_J and the desired jib angle ν_d. $\mathbf{C}$ denotes the output matrix. These matrices and vector are mathematically given by

$$\left.\begin{aligned} \mathbf{A} &= \begin{bmatrix} 0\,0\,0 & 0 \\ 1\,0\,0 & 0 \\ 0\,0\,0 & -\dfrac{g}{L_R} \\ 0\,0\,1 & 0 \end{bmatrix} \\ \mathbf{b} &= \left[1\;0\;\frac{L_J}{L_R}\sin\nu_d\;0\right]^T \\ \mathbf{C} &= \begin{bmatrix} 0\,1\,0\,0 \\ 0\,0\,0\,1 \end{bmatrix}. \end{aligned}\right\} \tag{10}$$

$\mathbf{x}(t)$, $u(t)$ and $\mathbf{y}(t)$ denote the state variable vector, the control input and the output vector, respectively, which are mathematically given by

$$\left.\begin{aligned} \mathbf{x}(t) &= \left[\,\dot{\nu}(t)\;\nu(t)-\nu_d\;\dot{\theta}(t)\;\theta(t)\,\right]^T \\ u(t) &= \ddot{\nu}(t) \\ \mathbf{y}(t) &= \left[\,\nu(t)-\nu_d\;\theta(t)\,\right]^T. \end{aligned}\right\} \tag{11}$$

3.2 *Pole-assignment controller design*

The pole-assignment is realized by the following state feedback control law:

$$\begin{aligned} u(t) &= -\mathbf{k}^T\mathbf{x}(t) \\ &= -k_1\dot{\nu}(t) - k_2\{\nu(t)-\nu_d\} \\ &\qquad - k_3\dot{\theta}(t) - k_4\theta(t). \end{aligned} \tag{12}$$

Then, the following closed-loop characteristic polynomial $T(s)$, can be obtained:

$$T(s) = \det\{s\mathbf{I} - (\mathbf{A} - \mathbf{b}\mathbf{k}^T)\}. \tag{13}$$

Therefore, by designing the desired characteristic polynomial $T^*(s)$ and matching the coefficients in the relationship $T^*(s) = T(s)$, the pole-assignment control system can be constructed, where $T^*(s)$ is the user-specified polynomial[7] given by

$$T^*(s) := s^4 + t_1s^3 + t_2s^2 + t_3s + t_4, \tag{14}$$

where

$$\left.\begin{aligned} t_1 &= \frac{5}{\sigma} \\ t_2 &= \frac{50}{3\sigma^2} \\ t_3 &= \frac{100}{3\sigma^3} \\ t_4 &= \frac{100}{3\sigma^4}. \end{aligned}\right\} \tag{15}$$

These parameters $t_i (i = 1, 2, 3, 4)$ give us the response minimizing the ITAE(Integral of Time squared Absolute Error) criterion(Shigemasa *et al.*, 1983). In (15), σ denotes the rise-time, and is set by taking account of the regulation properties.

3.3 *PD control law*

(12) can be rewritten by

$$u(t) = -k_{C1}\{(\nu(t) - \nu_d) + T_{D1}\dot{\nu}(t)\} - k_{C2}\{\theta(t) + T_{D2}\dot{\theta}(t)\}, \quad (16)$$

where

$$\left.\begin{aligned} k_{C1} &= k_2 \\ T_{D1} &= \frac{k_1}{k_2} \\ k_{C2} &= k_4 \\ T_{D2} &= \frac{k_2}{k_4}. \end{aligned}\right\} \quad (17)$$

By rearranging the control law (12) as the PD control law (16), it is easy to grasp the physical meaning of the control parameters. Furthermore, PD parameters can be expressed as the following equations by using system parameters L_R, L_J and ν_d, and user-specified parameters $t_i (i = 1, 2, 3, 4)$:

$$\left.\begin{aligned} k_{C1} &= \frac{L_R}{g} t_4 \\ T_{D1} &= \frac{t_3}{t_4} = \sigma \\ k_{C2} &= -\frac{g^2 - gL_R t_2 + L_R^2 t_4}{gL_J \sin \nu_d} \\ T_{D2} &= \frac{L_R(L_R t_3 - g t_1)}{g^2 - gL_R t_2 + L_R^2 t_4}. \end{aligned}\right\} \quad (18)$$

From (18), it is clear that PD parameters strongly depend on the rope length L_R, the jib length L_J and the desired jib angle ν_d.

3.4 *Self-tuning PD controller*

Here, the rope length L_R and the jib length L_J are changed during the jib derricking motion, and suppose that these parameters can not be practically measured. Then, these parameters are estimated on-line.

First, the state space model(9)-(11) are rewritten by the following the transfer functions:

$$G_1(s) = \frac{1}{s^2} \quad (19)$$

$$G_2(s) = \frac{c}{s^2 + d^2}. \quad (20)$$

where c and d in (20) are defined as

$$\left.\begin{aligned} c &= \frac{L_J}{L_R} \sin \nu_d \\ d &= \sqrt{\frac{g}{L_R}}. \end{aligned}\right\} \quad (21)$$

Then, $G_1(s)$ and $G_2(s)$ denote the transfer functions of the jib angle system and the sway angle system, respectively. Discretizing (19) and (20) with zero-order holder yields the following discrete-time transfer functions:

$$G_1^*(z^{-1}) = \frac{T_S^2}{2} \frac{z^{-1}(1 + z^{-1})}{(1 - z^{-1})^2} \quad (22)$$

$$G_2^*(z^{-1}) = \frac{c}{d^2} \frac{\{1 - \cos(d \cdot T_S)\}(1 + z^{-1})}{1 - 2\cos(d \cdot T_S)z^{-1} + z^{-2}}, \quad (23)$$

where T_S denotes the sampling interval. Note that $G_1^*(z^{-1})$ does not change even if L_R and L_J are changed, but $G_2^*(z^{-1})$ does. Therefore, the recursive least squares method is employed for (23) in order to estimate L_R and L_J. Thus, (23) is rewritten by the discrete-time model:

$$\tilde{y}_2(k) = -a_1 \tilde{y}_2(k-1) + \tilde{y}_2(k-2) + b_0\{\tilde{u}(k-1) + \tilde{u}(k-2)\}, \quad (24)$$

where $\tilde{y}_2(k)$ and $\tilde{u}(k)$ respectively denote the sway angle and the jib angular acceleration in discrete-time description. And a_1 and b_0 are estimated by using the following least squares method with a forgetting factor λ:

$$\hat{\theta}(k) = \hat{\theta}(k-1) + \frac{\Gamma(k-1)\phi(k-1)}{1 + \phi^T(k-1)\Gamma(k-1)\phi(k-1)} \epsilon(k) \quad (25)$$

$$\Gamma(k) = \frac{1}{\lambda}\{\Gamma(k-1) - \frac{\Gamma(k-1)\phi(k-1)\phi^T(k-1)\Gamma(k-1)}{\lambda + \phi^T(k-1)\Gamma(k-1)\phi(k-1)}\} \quad (26)$$

where $\epsilon(k)$ denotes the prediction error defined as

$$\epsilon(k) = \tilde{y}_2(k) + \tilde{y}_2(k-2) - \hat{\theta}(k-1)\phi(k-1), \quad (27)$$

and $\hat{\theta}(k)$ and $\phi(k)$ denote the unknown parameter vector and the corresponding data vector, respectively, which are given by

$$\hat{\theta}(k) = [\hat{a}_1(k) \quad \hat{b}_0(k)]^T \quad (28)$$

$$\phi(k) = [-\tilde{y}_2(k-1) \quad \tilde{u}(k-1) + \tilde{u}(k-2)]^T. \quad (29)$$

Then, L_R and L_J are calculated by using estimates $\hat{\theta}(k)$, that is, the estimated values $\hat{L}_R(k)$ and $\hat{L}_J(k)$ can be obtained from the relations (21) and (23) as follows:.

$$\left.\begin{aligned} \hat{L}_R(k) &= \frac{gT_S^2}{\{\cos^{-1}\dfrac{-\hat{a}_1(k)}{2}\}^2} \\ \hat{L}_J(k) &= \frac{\hat{b}_0(k)}{\sin \nu_d}\{1 + \frac{\hat{a}_1(k)}{2}\}. \end{aligned}\right\} \quad (30)$$

PD parameters are adjusted corresponding to $\hat{L}_R(k)$, $\hat{L}_J(k)$ and ν_d by using (18). Therefore, the self-tuning PD controller can be designed.

4. SWAY CONTROL OF LIFTED LOAD

4.1 *Strategy of sway control*

The proposed self-tuning PD control scheme is experimentally evaluated on a scale model as shown in Fig.2. The strategy of the sway control of the lifted load during the jib derricking motion is summarized as the following three steps.

1. The jib derricking motion from the initial jib angle to the desired jib angle is generated by the jib angular acceleration pattern which is divided into three sections as shown in **Fig.3**. Note that the jib angular acceleration pattern shown in Fig.3 is utilized in the case where the jib goes down. Therefore, section 1 and section 3 are replaced each other in the case where the jib goes up.
2. In the acceleration section(section 3), L_R and L_J are estimated by obtaining $\ddot{\nu}$ and θ.
3. The jib angle ν in which the acceleration pattern(section 3) is finished, is determined as the desired jib angle ν_d, and the self-tuning PD sway control is started.

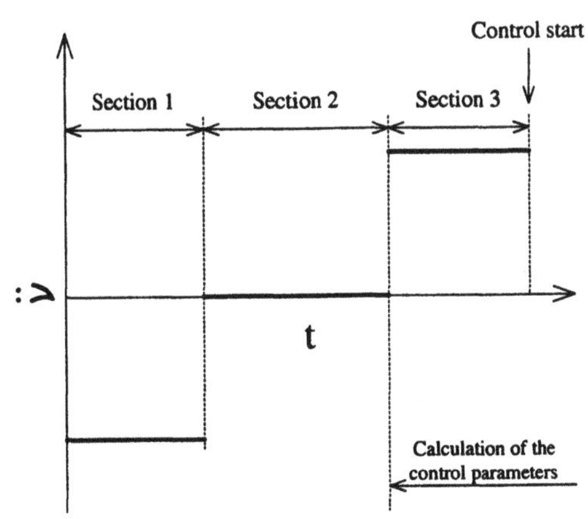

Fig. 3. Jib angular acceleration pattern.

It is pointed out that if the periods of the decelration section(section 1) and the acceleration section(section 3) are respectively equal to the natural period of the rope, the sway angle of the lifted load can be regulated(Mita and Kanai, 1978; Ohto and Yasunobu, 1997; Tanuzumi *et al.*, 1995).

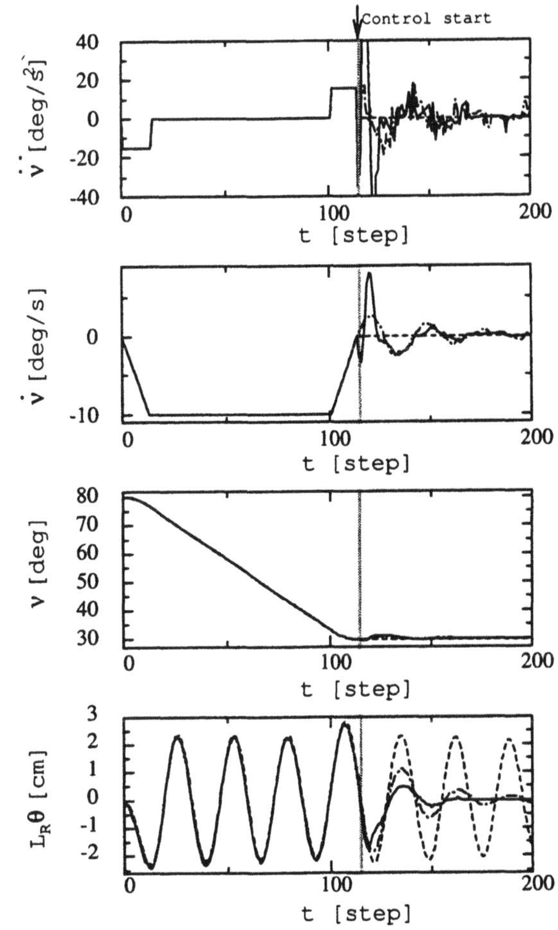

Fig. 4. Control parameters trajectories corresponding to Fig.4.

However, in the case where the length of the rope becomes long, the period of the jib derricking motion also becomes large, and so it is not good for the practical use. Therefore, in this paper, the periods of the deceleration section(section 1) and acceleration section(section 3) are set as a half of the natural period of the rope.

4.2 *Experimental results*

The effectiveness of the proposed self-tuning PD control scheme is experimentally evaluated on the sway control. The control result by using the newly proposed self-tuning PD controller is shown in **Fig.4**, where L_J = 0.9 [m] and L_R = 0.45 [m]. And also, the jib angle ν is changed from 80 [deg] thru 30 [deg]. Figure 4 shows the behavior of the jib angular acceleration $\ddot{\nu}$, the jib angular velocity $\dot{\nu}$, the jib angle ν and sway width of the lifted load, $L_R\theta$. In Fig.4, the solid line shows the proposed self-tuning PD control result in which σ and λ set as 0.6 and 0.98, respectively; the broken line shows the free response without the sway control; and the dashed and dotted line shows the PD control result whose PD parameters are

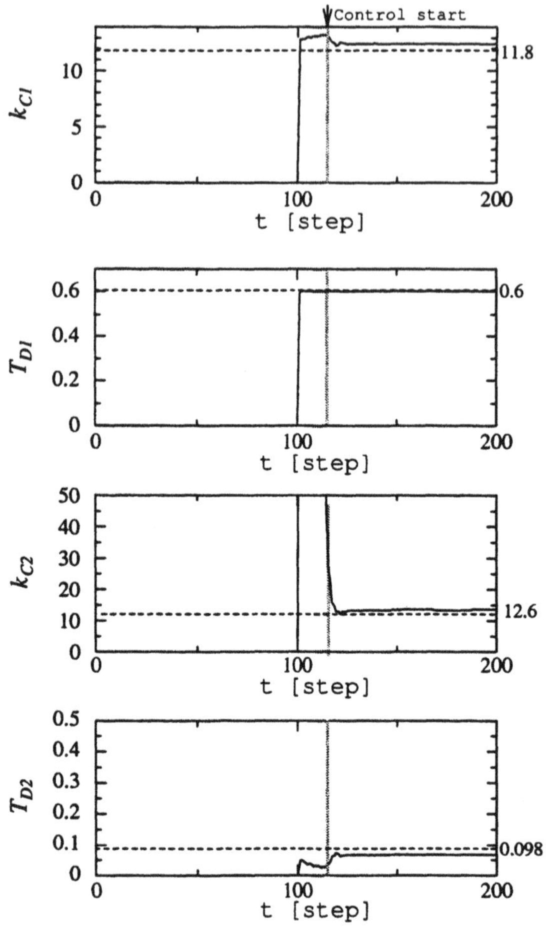

Fig. 5. Experimental result by the proposed self-tuning PD controller, where $L_J = 0.9$ [m] and $L_R = 0.45$.

determined around $\nu_d = 80$ [deg](initial jib angle). It is clear that the proposed control scheme is experimentally effective. Furthemore, the trajectories of control parameters are shown in **Fig.5.** The control parameters are adequately estimated.

From these control results, the effectiveness of the newly proposed self-tuning PD control algorithm is illustrated through the experiment on the sway control of the lifted load for the mobile crane.

5. CONCLUSION

In this paper, a new self-tuning PD controller design scheme has been proposed, whose algorithm is derived based on the pole-assignment control scheme. The features of the proposed control scheme are summerized as follows:

- Control parameters are given by the proportional gain and derivative time.
- The sway angle can be regulated even if the rope length, the jib length and the jib angle are unknown.
- The regulation properties can be adjusted by choosing the user-specified parameter in the pole-assignment scheme.

The effectiveness of the newly proposed self-tuning PD control scheme has been experimentally evaluated on the sway control system of a lifted load for a mobile crane.

6. REFERENCES

Chien, K.L., Hrones, J.A. and Reswick, J.B. (1952). On the automatic control of generalized passive systems. *Trans. of American Society Mechanical Engineers* **74**, 175–185.

Hino, J., Fujita, K. and Yoshimura, T. (1998). Fuzzy control for a load of truck-crane in elevation and swing motion(in Japanese) *Trans. of Japan Society Mechanical Engineers C* **64**, 106–112.

Mita, T. and Kanai, Y. (1978). Optimal control of the crane using the maximum speed of the trolley(in Japanese) *Trans. of Society of Instrument and Control Engineers* **15**, 833–838.

Ohto, M. and Yasunobu, S. (1997). A rotary crane method based on a skillfuroperator's knowledge(in Japanese) *Trans. of Society of Instrument and Control Engineers* **33**, 923–929.

Shigemasa, T., Takagi, T., Ichikawa, Y. and Kitamori, T. (1997). A practical reference model for control system design(in Japanese) *Trans. of Society of Instrument and Control Engineers* **19**, 592–594.

Suda, N. (1992). PID control(in Japanese) *Asakura Publishing Co.Ltd.*

Tada, H., Inoue, Y., Ohshima, M. and Yoshimura, Y. (1997a). Studies on sway control systems for rotary cranes(Development of a sway angle sensor and sway control of a scale-down model during swing operation)(in Japanese) *Trans. of Japan Society Mechanical Engineers C* **63**, 47–54.

Tada, H., Inoue, Y., Ohshima, M. and Yoshimura, Y. (1997b). Studies on sway control systems for rotary cranes(Sway control of a scale-down model during boom hoisting operations)(in Japanese) *Trans. of Japan Society Mechanical Engineers C* **63**, 55–62.

Tanizumi, K., Yamamoto, K., Sakai, T., Yoshimura, T. and Hino, J. (1995). Modeling of dynamic behavior and control of truck cranes(Velocity and sway control of truck cranes with hydraulic system in elevating operation)(in Japanese) *Trans. of Japan Society Mechanical Engineers C* **61**, 55–63.

Ziegler, J.G. and Nichols, N.B. (1942). Optimum settings for automatic controllers *Trans. of American Society Mechanical Engineers* **64**, 759–768.

ADAPTIVE CONTROL OF THE POWER FACTOR PRECOMPENSATOR: A COMPARATIVE STUDY

Georgia Kaliora *,[1] **Alessandro Astolfi** *,**

* *Department of Electrical and Electronic Engineering*
Imperial College of Science, Technology and Medicine
London, SW7 2BT, United Kingdom
Tel: +44 20 75946295
Fax: +44 20 75946282
E-mail: {g.kaliora,a.astolfi}@ic.ac.uk
** *Dipartimento di Elettronica e Informazione*
Politecnico di Milano,
Piazza Leonardo da Vinci 32, 20132 Milano, Italy

Abstract: An adaptive control scheme for the *power factor precompensator* circuit is proposed. The adaptive feature of the design is based on a simple Hopfield network architecture and relies on least squares approximation. Comparisons with existing results are also presented. *Copyright ©2001 IFAC*

Keywords: Power electronics, adaptive control, recurrent neural networks.

1. INTRODUCTION

In this paper we are interested in the problem of performance enhancement of unity power factor AC–to–DC converters by means of adaptive control. Such devices are also known as Power Factor Precompensators (PFP) and are commonly used as a first step in a power source. The control objective is to achieve efficient load voltage regulation to a desired level, while preserving a nearly unit power factor at the input mains of the converter. After some simplifying assumptions, see *e.g.* (Escobar *et al.*, 2000), the model describing the average behavior of a *full bridge boost type PFP* is described by the equations

$$\begin{aligned} L\dot{x}_1 &= -ux_2 + v_i \\ C\dot{x}_2 &= ux_1 - R_{inv}x_2, \end{aligned} \tag{1}$$

where x_1 and x_2 are the input inductor current and the output capacitor voltage variables respectively, v_i is the voltage of the AC-line source and R_{inv} is the inverse of the output resistance ($R_{inv} = \frac{1}{R}$). The signal $u \in [-1, 1]$ represents the duty ratio of a PWM circuit which generates the sequence of switching positions and acts as the control signal. The output resistance represents the load of the power line, so it is unknown and varying in a random way within some known bounds. However, the knowledge of the exact value of this parameter in the controller implementation would enhance the performance of the control action.

In (Escobar *et al.*, 2000) the regulation problem is solved with the dynamic controller

$$\begin{aligned} L\dot{x}_{1d} &= -ux_{2d} + v_i + R_1(x_1 - x_{1d}) \\ C\dot{x}_{2d} &= ux_{1d} - \hat{\theta}x_{2d} + R_2^{-1}(x_2 - x_{2d}), \end{aligned} \tag{2}$$

where $x_{1d} = I_d \sin(\omega t)$, with I_d and ω known, R_1, R_2^{-1} controller gains and the control u is obtained solving the first equation. This controller is designed with the Passivity Based Control methodology (Ortega *et al.*, 1998) and achieves *total energy shaping.* $\hat{\theta}$ represents an estimation

[1] Supported by the TMR Network, NACO 2 and the State Scholarships Foundation of Greece.

of the unknown parameter $R_{inv} \in [R_{inv}^{min}, R_{inv}^{max}]$ and, in the adaptive version of the controller (2) is given by the dynamic law

$$\dot{\hat{\theta}} = \gamma(\hat{\theta} - R_{inv}^{max})(\hat{\theta} - R_{inv}^{min})x_{2d}(x_2 - x_{2d}). \quad (3)$$

In this paper, we propose another adaptive law for the update of $\hat{\theta}$. This will perform a *least squares curve fitting* and at the same time will be realized by a recurrent neural network. The new adaptive law is implemented with the controller (2), but would also be suitable to collaborate with any other controller, as it is not driven by the controller states.

The paper is organized as follows. In Section 2 we discuss the architecture of the proposed family of adaptive laws, for the general case of a nonlinear, linearly parameterized systems. In Section 3 we present our controller for the specific PFP problem highlighted above, and in Section 4 we present our simulation results, comparing the performance of the new adaptive scheme to the performance of (2)–(3). In Section 5 we summarize our results and discuss future work.

2. A HOPFIELD NETWORK AS A PARAMETERS ESTIMATOR

Recurrent neural networks and adaptive recurrent networks can be very relevant - and similar - to many control applications. The dynamical character of recurrent networks, in which the state of the neurons depends not only on the input of the network, but also on the previous states, can very well be connected with issues of system stabilization and control. From early times single layered recurrent networks have been used as point attractors, in a sense that they could converge to a stable point from any starting state.

The Hopfield network is a one layered recurrent network. The output of each neuron is fed back to all the others (most of the times, no self feedback is considered), and the exogenous inputs are only part of the input of each neuron. In this "closed-loop" architecture, it is able to store information in a dynamically stable configuration (Haykin, 1994). Hopfield networks have been most commonly used as Content Addressable Memories for pattern recognition and restoration purposes.

However, in the paper (Chu *et al.*, 1990) a Hopfield network is utilized to implement a least-squares estimation for a parameter identification problem. The proposed scheme employes no training or learning algorithm. The weights of the network vary over time, driven by the state of the system, while the history of the unknown parameters is described by the states of the neurons. The architecture of the network is also straightforward, since each neuron is assigned to the estimation of each parameter.

In this paper we illustrate how this technique can be used for online estimation of unknown parameters, that are needed for the control of the plant, *i.e.* for a closed loop operation. In other words, the Hopfield network is incorporated in the setup that controls the system on-line, taking care of the adaptation of the controller's parameters. However, the architecture is such, that minimum interference with the well behaved controller is imposed. The aim is not to redesign a controller, in the perspective of adaptive control, but enhance an existing controller with an extra feature, adaptation to parameters changes.

In the remaining of this section we present a method for the derivation of a dynamic law, through which identification of the unknown parameters of a control system can be achieved. The method is inspired by (Chu *et al.*, 1990), where it was applied for the identification of the matrices A and B of a linear system. Here, we generalize the applicability of the concept to nonlinear systems. However, we still require that the system equations are linear with respect to the vector of the unknown parameters.

Consider a nonlinear system with state $x \in \mathbb{R}^n$ and control $u \in \mathbb{R}$, linearly parameterized by a vector of unknown parameters, $\theta \in \mathbb{R}^m$. We consider that the full state x is composed by the state of the system and the state of the controller. Suppose that if θ is known, there exists a known stabilizing controller for the system, $u = k(x)$, and that the closed loop system (or part of it, see Remark 3) can be written as

$$\begin{aligned} \dot{x} &= \Psi(x,u)\theta + \Psi_o(x,u) \\ u &= k(x) \end{aligned} \quad (4)$$

where $\Psi(x,u)$ and $\Psi_o(x,u)$ are $n \times m$ and $n \times 1$ matrices, respectively.

In the case of unknown θ, we consider an estimation of θ denoted by $\hat{\theta}$ and define the estimation error and an objective function as

$$e = \dot{x} - \Psi(x,u)\hat{\theta} - \Psi_o(x,u), \quad (5)$$

$$J = \frac{1}{t}\int_0^t \frac{1}{2}e^T(\tau)e(\tau)d\tau \quad (6)$$

Since J is quadratic in $\hat{\theta}$, a natural choice for an adaptation law for $\hat{\theta}$ is

$$\frac{d\hat{\theta}}{dt} = -h(\hat{\theta})\frac{\partial J}{\partial \hat{\theta}}, \quad (7)$$

where $h(\hat{\theta})$ could be a diagonal matrix of positive non-zero functions that act as filtering or weighting gains to the adaptation. After simple calculations (7) gives

$$\begin{aligned}\frac{d\hat{\theta}}{dt} &= h(\hat{\theta})\left[-\left(\frac{1}{t}\int_0^t \Psi^T(x,u)\Psi(x,u)d\tau\right)\hat{\theta}\right]\\ &+ h(\hat{\theta})\left[\frac{1}{t}\int_0^t \Psi^T(x,u)\left(\dot{x}-\Psi_o(x,u)\right)d\tau\right]\\ &= h(\hat{\theta})\left[W(t)\hat{\theta}+I(t)\right],\end{aligned} \tag{8}$$

where $W(t)$ and $I(t)$ are matrices whose elements are defined by the corresponding integrals.

On the other hand, we know that the dynamical behavior of a one-layered recurrent (Hopfield) network is described by an equation of the form, see (Khalil, 1996; Haykin, 1994),

$$\dot{y} = h(y)[Wy + I],$$

where y is considered to be the vector of the outputs of the neurons, W is the matrix of the feedback synaptic weights, and I is a vector representing the external inputs to the neurons. If the output and the input of each neuron are related through the activation function ϕ, where most typically, $\phi(s) = \tanh(s)$,

$$y_j = \phi_j(v_j),$$

then $h(y)$ is defined as the matrix

$$h(y) = diag\{h_i(y)\} = diag\{\left.\frac{d\phi_i}{dv_i}\right|_{v_i=\phi_i^{-1}(y_i)}\} \tag{9}$$

for $i = 1, 2, \ldots, n$. Due to (8), it is clear that by choosing $h(\hat{\theta})$ as in (9), we assign to the adaptation of $\hat{\theta}$ the dynamical behavior of a Hopfield network. The network that performs the estimation will then have as many neurons as the dimension of $\hat{\theta}$, each one with activation function ϕ_i, while the feedback interconnections will be characterized by the matrix W and the external inputs by the vector I. $\hat{\theta}$ will then be the vector of the outputs of the neurons and, as such, will always be bounded.

The above procedure mimics the building of a least squares regressor. The neural-networks specifics are introduced through the particular choice of the functions $h_i(\hat{\theta}_i)$.

Remark 1. In general (8) requires measurement of $\dot{x}$. However in some special cases, the integral

$$\int_0^t \Psi^T(x(\tau), u(\tau))\dot{x}(\tau)d\tau$$

can be analytically computed. In others, the *dirty derivative* can be used, or numerical approximations can be used. This issue is currently under investigation.

Remark 2. The equilibrium for the new states $\hat{\theta}$ has been naturally assigned. Indeed, notice that

$$\frac{\partial J}{\partial \hat{\theta}} = -W(t)(\hat{\theta} - \theta),$$

where $W(t)$ is by definition non zero. Also, the use of $h(\hat{\theta})$ introduces no other equilibria, since this function is always non zero.

Remark 3. The representation (4) implies that all state vector is present in x. This is not always necessary. In fact, some state space equations, may not depend on $\hat{\theta}$, in which case they do not need to be involved in the construction of the estimation error and objective function. Also it is true, that, regardless of the size of x, e and J are scalar functions, and during the derivation of $\frac{dJ}{d\hat{\theta}}$, J is treated as a mapping from $\mathbb{R}^m$ to $\mathbb{R}$.

3. ADAPTIVE CONTROL OF THE POWER FACTOR PRECOMPENSATOR CIRCUIT

In this section we illustrate the use of an adaptive law derived following the ideas in Section 2. The network will be required to estimate the unknown output load resistance of the Power Factor Precompensator (PFP) circuit, presented in Section 1. We still use the controller (2), replacing the adaptive law (3) for the estimation of the unknown parameter with the Hopfield Network like estimator. Since there is only one unknown parameter, we only need one computing unit, one neuron, so the term "network" is probably unnecessary.

For the specific problem both states are considered to be measurable which makes the estimator design unconstrained.

Following the procedure in Section 2, it is not difficult to see that the Hopfield network like adaptation law for $\hat{\theta}$, that will be used in (2) to control (1), is

$$\dot{\hat{\theta}} = h(\hat{\theta})\left[W\hat{\theta} + I\right], \tag{10}$$

with

$$\begin{aligned} W &= -\frac{1}{t}\int_0^t x_2^2(\tau)d\tau,\\ I &= -\frac{1}{t}\int_0^t \left[C\dot{x}_2x_2 - x_1x_2\right]d\tau \end{aligned} \tag{11}$$

and

$$h(\hat{\theta}) = \left.\frac{d\phi}{dv}\right|_{v=\phi^{-1}(\hat{\theta})} . \qquad (12)$$

The activation function is chosen in a way that the upper and lower limit imposed on the neuron output, *i.e* $\hat{\theta}$, coincide with the known upper and lower limit of the real parameter R_{inv}. In particular,

$$\hat{\theta} = \phi(v) = a + b \tanh(\frac{v}{b}) \qquad (13)$$

hence

$$\begin{aligned} \frac{d\phi}{dv} &= \frac{2e^{\frac{-v}{b}}}{\left(1+e^{\frac{-v}{b}}\right)^2} \\ v = \phi^{-1}(\hat{\theta}) &= \frac{b}{2}\ln\left(\frac{\hat{\theta}-a+b}{a+b-\hat{\theta}}\right) \end{aligned} \qquad (14)$$

with a and b both positive numbers depending on R_{inv}^{min} and R_{inv}^{max}. Thus, $h(\hat{\theta})$ is explicitly defined.

The estimator described employs a simple structure and is driven only by x and u, unlike the estimator (3) which uses x_{2d} as well. This means that the estimator (10) might be used even when the system is controlled by a different controller.

To react to continuous changes of R_{inv} we introduce a "forgetting" action. Namely, we set

$$\begin{aligned} W &= -\frac{1}{t}\int_0^t e^{-\lambda(t-\tau)}x_2^2(\tau)d\tau \quad \text{and} \\ I &= -\frac{1}{t}\int_0^t e^{-\lambda(t-\tau)}\left[C\dot{x}_2x_2 - ux_1x_2\right]d\tau, \end{aligned} \qquad (15)$$

where λ is the forgetting factor, that is kept constant. In this way the information stored in the integrals is weighted, and more recent events are given more value. With this modification the estimator is able to follow even rapid changes in R_{inv}.

Note that the term containing $\dot{x}_2$ can be replaced by

$$\int_0^t e^{\lambda\tau}Cx_2\dot{x}_2d\tau = e^{\lambda t}\left.\frac{Cx_2^2}{2}\right|_0^t - \int_0^t \frac{\lambda C}{2}e^{\lambda\tau}x_2^2d\tau$$

The asymptotic properties of the adaptive law (10) when used in closed loop with (1)–(2) can be established by means of the Lyapunov function

$$\begin{aligned} H = &\frac{L}{2}(x_1 - x_{1d})^2 + \frac{C}{2}(x_2 - x_{2d})^2 \\ &+ \frac{1}{2}(\hat{\theta} - R_{inv})^2 \end{aligned} \qquad (16)$$

For the details of the proof see (Kaliora, 2000).

4. SIMULATION RESULTS

Simulation results of the overall control and adaptation scheme are presented in this section. In Figure 1 we see the behavior of the system, when $\hat{\theta}$ is initially considerably different from the real value R_{inv}, and the circuit is required to track a step change in the output voltage. The performance of the estimator (10) is compared with (3) used in the paper (Escobar *et al.*, 2000). We see that the results are comparable.

Next, we examine the reaction of the estimator in the case of continuous and abrupt changes of the real value of the parameter. In Figure 2 we consider a large step change in R_{inv}, and in Figure 3 a series of step changes. Finally, in Figure 4, we test the system against a sinusoidal change of R_{inv}.

In all cases, we use an initial value for $\hat{\theta}$ which is sufficiently different from the real value of the parameter R_{inv} at time $t = 0$. The new adaptive law manages to track the correct value almost instantly in all cases.

5. CONCLUSIONS AND FUTURE WORK

In this paper we presented a method for the design of a parameters estimator to be used as a part of an adaptive control scheme. The method is applicable to nonlinear systems, linearly parameterized, and exploits the properties and architecture of Hopfield neural networks, while performing a least-squares approximation task. The method has been tested on a example from power electronics, the PFP circuit, where the estimation of the output load resistance is needed. Comparisons of our results to those previously presented in the literature have been presented. For the future, we plan to test and further develop the method to systems exhibiting larger dependance on unknown parameters.

6. REFERENCES

Chu, S. R., R. Shoureshi and M. Tenorio (1990). Neural networks for system identification. *IEEE Control Systems magazine.*

Escobar, G., D. Chevreau, R. Ortega and E. Mendes (2000). An adaptive passivity-based controller for a unity power factor rectifier using a full bridge boost circuit: Stability analysis and experimental results. *Proceedings of the IEEE International Symposium on Industrial Electronics, ISIE'99- Bled, Slovenia* **2**, 451–456.

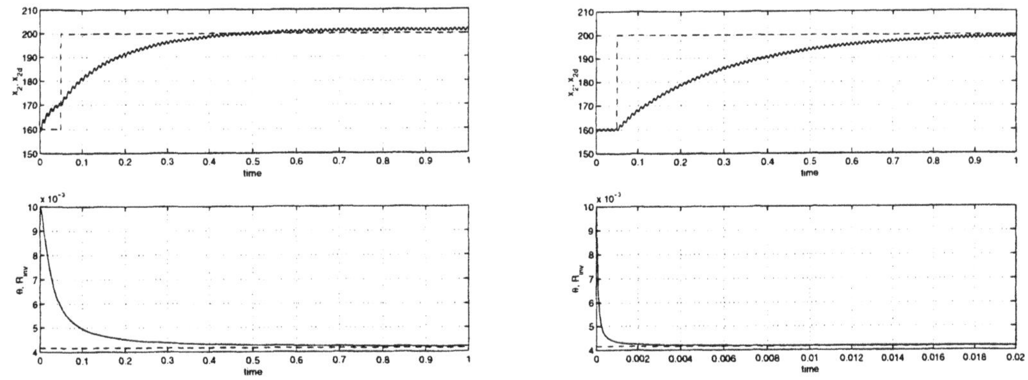

Fig. 1. Performance of the estimator proposed in (Escocar *et al.*, 2000) (graphs on the left), against the estimator proposed in this work (graphs on the right). Output voltage (graphs on the top), and $\hat{\theta}$ against the real value of R_{inv} (graphs on the bottom, solid line for $\hat{\theta}$, dashed line for R_{inv}). Notice the different time scale on the last graph on the right.

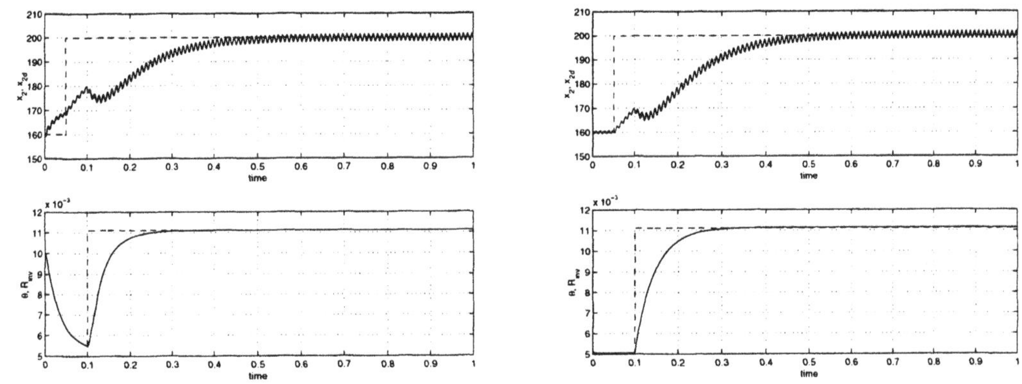

Fig. 2. Tracking of the output voltage with simultaneous step change of R_{inv}, with the estimator proposed in (Escocar *et al.*, 2000) (graphs on the left), and the estimator proposed in this work (graphs on the right).

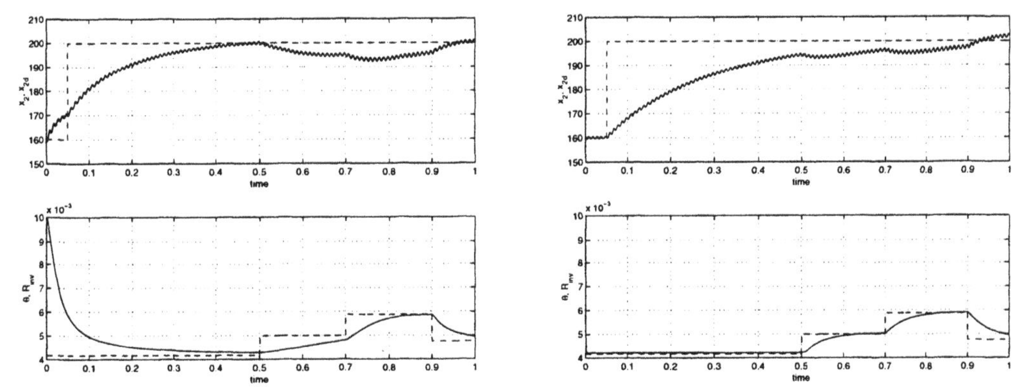

Fig. 3. Tracking of the output voltage with simultaneous step changes of R_{inv}, with the estimator proposed in (Escocar *et al.*, 2000) (graphs on the left), and the estimator proposed in this work (graphs on the right).

Haykin, S. (1994). *Neural Networks, A Comprehensive Foundation.* Macmillan College Publishing Company, Inc.

Kaliora, G. (2000). Robust and adaptive control of nonlinear systems with bounded controls. *M.Phil to Ph.D. Transfer Report, Imperial College, London.*

Khalil, H. (1996). *Nonlinear Systems.* Prentice-Hall.

Ortega, R., A. Loria, P. J. Nicklasson and H. Sira-Ramirez (1998). *Passivity-based Control of Euler-Lagrance Systems.* Springer–Verlag.

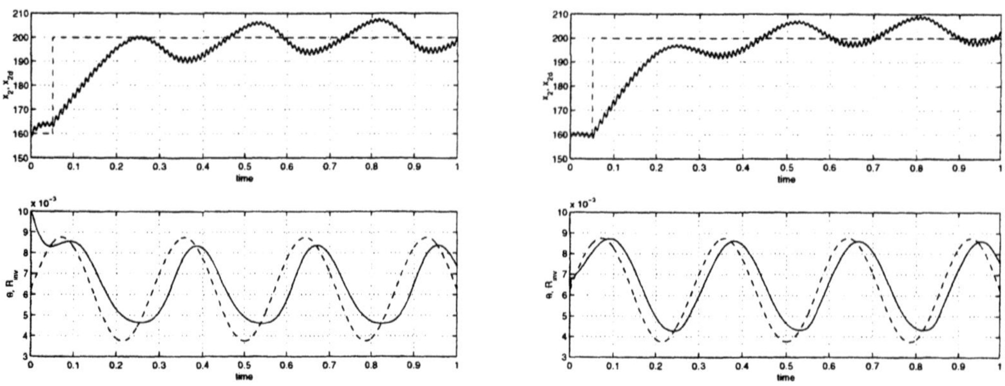

Fig. 4. Tracking of the output voltage with simultaneous sinusoidal change of R_{inv}, with the estimator proposed in recent literature (graphs on the left), and the estimator proposed in this work (graphs on the right).

ADAPTIVE COMPENSATION OF NONLINEARITY IN HIGH POWER AMPLIFIER BY SUPPORT VECTOR MACHINE

Toshiyuki Eda, Takanori Ito, Hiromitsu Ohmori, Akira Sano *

* *Keio University, 3-14-1 Hiyoshi, Yokohama 223-8522, Japan*
sano@sd.keio.ac.jp

Abstract. This paper is concerned with a new adaptive predistortion method for compensation of linear and nonlinear distortions of a high power amplifier (HPA), which is frequently used in the orthogonal frequency division multiplexing (OFDM) communication scheme and mobile communication systems. The proposed nonlinear method is based on the support vector machine (SVM), which can construct an inverse system of the nonlinear amplifier through learning using a given number of training samples. An on-line adaptive algorithm for updating the parameters of SVM is also investigated to reduce its computational burden of the batch-processed optimization. By comparing with a case without predistortion and ordinary table-lookup scheme, it is shown that the proposed data-based approach can improve the suppression of the power spectrum of the amplifier output outside the input frequency band caused by the nonlinear distortions. *Copyright ©2001 IFAC*

Keywords. Nonlinear system, system identification, support vector regression, inverse system, high power amplifier, nonlinearity compensation

1. INTRODUCTION

This paper is concerned with an adaptive predistortion technique which is based on the support vector machine (SVM) approach (Vapnik, 1998) to compensate for unknown nonlinearities of high power amplifier (HPA) used in the orthogonal frequency division multiplexing (OFDM) communication systems. The technique allows bandwidth efficient modulation schemes to be used with nonlinear HPAs with small degradation in bit error rate (BER) performance relative to a purely linear transmitter system. In addition, the technique adapts to changes in the nonlinearity of the HPA which could be caused by changes to the input power, carrier frequency, aging, temperature, or component replacement.

Several other techniques have been proposed to this problem. The most obvious ways are to use a linear class A of HPA or to operate a nonlinear HPA sufficiently far from saturation to obtain near linear performance. These techniques tend to require larger, heavier, more expensive, and less efficient HPAs than necessary. Other predistortion techniques include analog feedback, analog feedforward, and digital table-lookup mapping. The analog techniques have been limited by narrow operating bandwidth, extreme sensitivity of HPA variations, instabilities, and in some techniques the requirement for additional RF amplifiers. The digital mapping techniques have been limited by the massive amount of storage required for a sufficiently accurate mapping to be stored, and by degradation caused by discontinuity of the table-based mapping (Saleh and Salz, 1983; Cavers, 1990; Matsuoka *et al.*, 1997). In addition, inverse modeling approach of a nonlinear PHA has been proposed by using a Volterra series analysis (Nojima and Okamoto, 1981; Lazzarin and Sarti, 1994; Kang and Youn, 1997; Okada and Komaki, 1998; Horiguchi and Takagi, 1999; Cheng and Power, 1999), and digital techniques based on neural networks have also been developed (Watkins and Tummla, 1995; Shin and Hahn, 1996). The proposed SVM-based approach adopting a polynomial type of kernel is related to these polynomial series modeling, however, the SVM method is a data-based learning approach using a finite number of training samples.

The purpose of this paper is to present a predistortion scheme based on the SVM which is one of powerful tools for constructing a nonlinear input-output mapping on the basis of a given number of training samples. The obtained SVM functions as an inverse nonlinear model which the nonlinear HPA follows. To determine the parameters in the SVM from the training samples, we have to solve an optimization problem subject to inequality constraints. Ordinary SVM approaches adopt a batch-processing type of calculations, however, it is not suitable for on-line signal processing. Then we present a recursive adaptive algorithm for updating the parameters of the SVM on an instant-by-instant basis. By comparing with a case without predistortion and ordinary table-lookup scheme, it is shown that the proposed data-based approach can suppress the power spectrum of the amplifier output outside the input frequency band caused by the nonlinear distortions.

2. SUPPORT VECTOR MACHINES AND REGRESSION

The SVM approach has been successfully extended to nonlinear regression problem (Vapnik, 1998). Let the training input-output data be given by $(\boldsymbol{x}_1, y_1)$, $(\boldsymbol{x}_2, y_2)$, $\cdots$, $(\boldsymbol{x}_n, y_n)$. The purpose is to obtain a regression function as

$$y = f(\boldsymbol{x}, \boldsymbol{\theta}) = \langle \boldsymbol{w}, \boldsymbol{x} \rangle + b \tag{1}$$

which provides the model output predicting an actual output. By introducing an ε-insensitive loss function denoted by

$$L_{1,\varepsilon}(y, f(\boldsymbol{x},, \boldsymbol{w})) = \max\left(0, |y - f(\boldsymbol{x}, \boldsymbol{w})| - \varepsilon\right) \tag{2}$$

we can take the structual risk function given by

$$\min_{\boldsymbol{w}} \left\{ \frac{1}{2} \|\boldsymbol{w}\|^2 + C \sum_{i=1}^{n} L_{1,\varepsilon}(y_i, f(\boldsymbol{x}_i, \boldsymbol{w})) \right\} \tag{3}$$

where the first term corresponds to the model complexity and the second term indicates the empirical risk due to the regression error. The above minimization can be described as the primary problem as

$$\min \quad \frac{1}{2} \|\boldsymbol{w}\|^2 + C \sum_{i=1}^{n} \left(\xi_i + \hat{\xi}_i\right)$$

$$\text{subject to} \quad \begin{cases} y_i - \langle \boldsymbol{w}, \boldsymbol{x}_i \rangle - b \leq \varepsilon + \xi_i \\ \langle \boldsymbol{w}, \boldsymbol{x}_i \rangle + b - y_i \leq \varepsilon + \hat{\xi}_i \\ \xi_i \geq 0, \quad \hat{\xi}_i \geq 0, \quad i = 1, \cdots, n \end{cases} \tag{4}$$

where ξ_i and $\hat{\xi}_i$ are the non-negative sluck variables.

By introducing the following Lagrangian

$$L = \frac{1}{2} \|\boldsymbol{w}\|^2 + C \sum_{i=1}^{n} \left(\xi_i + \hat{\xi}_i\right) - \sum_{i=1}^{n} \left(\eta_i \xi_i + \hat{\eta}_i \hat{\xi}_i\right)$$
$$- \sum_{i=1}^{n} \alpha_i (\varepsilon + \xi_i - y_i + \langle \boldsymbol{w}, \boldsymbol{x}_i \rangle + b)$$
$$- \sum_{i=1}^{n} \hat{\alpha}_i (\varepsilon + \hat{\xi}_i + y_i - \langle \boldsymbol{w}, \boldsymbol{x}_i \rangle - b)$$

we can transform the primary problem into the dual problem stated below

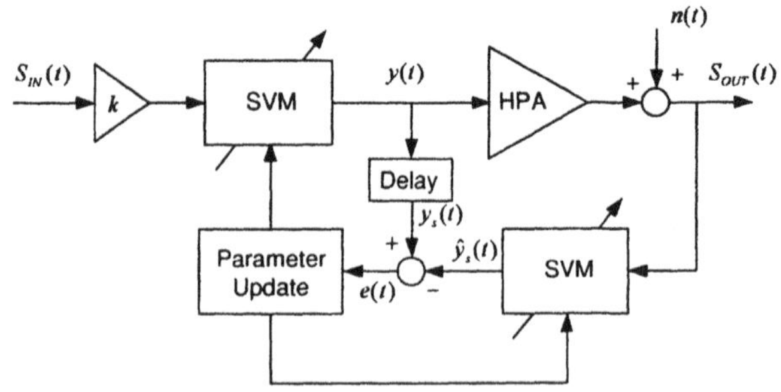

Fig. 1. Adaptive predistortion of HPA nonlinearity by SVM

$$\max_{\boldsymbol{\alpha}} \; W(\boldsymbol{\alpha}) = -\frac{1}{2} \sum_{i=1}^{n} \sum_{j=1}^{n} (\alpha_i - \hat{\alpha}_i)(\alpha_j - \hat{\alpha}_j) \langle \boldsymbol{x}_i, \boldsymbol{x}_j \rangle$$
$$-\varepsilon \sum_{i=1}^{n} (\alpha_i + \hat{\alpha}_i) + \sum_{i=1}^{n} y_i (\alpha_i - \hat{\alpha}_i)$$
$$\text{subject to} \quad \begin{cases} \sum_{i=1}^{n} (\alpha_i - \hat{\alpha}_i) = 0 \\ 0 \leq \alpha_i \leq C, \quad 0 \leq \hat{\alpha}_i \leq C \end{cases} \tag{5}$$

For the purpose of extension to the nonlinear regression problem, a mapping $\boldsymbol{z}(\boldsymbol{x})$ from the input vectors $\boldsymbol{x} \in \mathcal{X}$ to a high-dimensional feature space $\mathcal{Z}$ is introduced, (Vapnik, 1998) as

$$\begin{aligned} y_t &= f(\boldsymbol{x}_t, \boldsymbol{\theta}) \\ &= \langle \boldsymbol{w}, \boldsymbol{z}_t \rangle + b = \sum_{i=1}^{n-1} (\alpha_i - \hat{\alpha}_i) \langle \boldsymbol{z}_i, \boldsymbol{z}_t \rangle + b \\ &= \sum_{i=1}^{n-1} (\alpha_i - \hat{\alpha}_i) K(\boldsymbol{x}_i, \boldsymbol{x}_t) + b \end{aligned} \tag{6}$$

where $K(\boldsymbol{x}_i, \boldsymbol{x}_t)$ is introduced to estimate the inner product $\langle \boldsymbol{z}_i, \boldsymbol{z}_t \rangle$ in the feature space implicitly without performing the nonlinear mapping from the input space $\boldsymbol{x} \in \mathcal{X}$ into feature space $\boldsymbol{z} \in \mathcal{Z}$ (Vapnik, 1998), where a variety of kernels have been proposed, for instance, radial-basis functions and polynomial functions.

3. SVM-BASED PREDISTORTION OF HPA

3.1 *Inverse Modeling by Support Vector Machine*

Fig.1 gives a schematic diagram for adaptive predistortion of a nonlinear HPA by constructing an inverse model by SVM obtained from the training input-output data set. The amplifier ordinarily includes uncertain memoryless saturation type of nonlinearity and uncertain low-pass dynamics in input and output parts of the amplifier. k is a nominal gain, and $n(t)$ is a zero-mean white noise. The aim is to construct a nonlinear input-output inverse model for the HPA with the above uncertainties by taking the SVM-based approach using a finite number of training samples, so that $S_{\text{out}} = k \cdot S_{\text{in}}$ can be realized.

Let the $n-1$ training data samples be denoted by

$$\boldsymbol{x}_s(iT) = [S_{\mathrm{OUT}}(iT), \ldots, S_{\mathrm{OUT}}((i-m+1)T)] \quad (7)$$

$$y_s(iT) = y((i-d)T) \qquad \text{for } i = 1, \ldots, n-1 \quad (8)$$

where T is a sampling interval, d is a delay, m is the dimension of the input vector $\boldsymbol{x}_s(iT)$. T will be omitted for simplicity of notation.

Fig.1 shows an input-output matching scheme, in which the SVM is adjusted so as to minimize the magnitude of input errors $e(i) = y_s(i) - \hat{y}_s(i)$ in principle, in order to construct the inverse system of HPA. In the case, we consider the following minimization problem from (3):

$$\min_{w} \left\{ \frac{1}{2} \|\boldsymbol{w}\|^2 + C \sum_{i=1}^{n} L_{1,\varepsilon}(y_s(i), f(\boldsymbol{x}_s(i), \boldsymbol{w})) \right\} \quad (9)$$

An output-matching scheme can also be adopted as an alternative, in which the SVM is constructed so as to minimize the magnitude of errors $\bar{e}(i) = S_{IN}(i - d) - S_{OUT}$.

3.2 *Off-Line Algorithm for SVM-Based Predistortion*

The SVM approach can be formulated as a convex problem by (5) and (6). In the nonlinearity predistortion case, in which the nonlinearity of amplifiers can be expressed by an odd degree of polynomial function of input signal, we can decide that $b = 0$ in (6). Furthermore, $\alpha_i - \hat{\alpha}_i$ can be replaced by one variable, for instance, β_i, so we also define β_i as α_i again. When the data set consisting of $n-1$ input-output training samples given by (7) and (8), from the above consideration and the Kuhn-Tucker optimality conditions, we can show that the dual problem (5) can be rewritten in the next alternative expression as

$$\max_{\boldsymbol{\alpha}_{n-1}^{(n-1)}} W(\boldsymbol{\alpha}_{n-1}^{(n-1)}) = \max_{\boldsymbol{\alpha}_{n-1}^{(n-1)}} \sum_{i=1}^{n-1} \alpha_i^{(n-1)} y_s(i)$$
$$-\frac{1}{2} \sum_{i,j=1}^{n-1} \alpha_i^{(n-1)} \alpha_j^{(n-1)} K(\boldsymbol{x}_s(i), \boldsymbol{x}_s(j))$$
$$-\varepsilon \sum_{i=1}^{n-1} |\alpha_i^{(n-1)}| \quad (10)$$

subject to $0 \le |\alpha_i^{(n-1)}| \le C$ for $i = 1, \ldots, n-1$

Some of $\{\alpha_i^{(n-1)}\}$ are zero, but some of them take non-zero values which correspond to the support vectors $\{\boldsymbol{x}_s(i)\}$. By solving the above problem, the output $\hat{y}_s(t)$ of the optimal SVM has an expression form as

$$\hat{y}_s(t) = \sum_{i=1}^{n-1} \alpha_i^{(n-1)} K(\boldsymbol{x}_s(i), \boldsymbol{x}_s(t)) \quad (11)$$

where $K(\boldsymbol{x}_s(i), \boldsymbol{x}_s(t))$ is a kernel function satisfying the Mercer's conditions. As described later, we adopt a polynomial type of kernel as $K(\boldsymbol{x}_s(i), \boldsymbol{x}_s(t))$, which seems suitable to description of an inverse model of the Volterra series systems.

3.3 *Adaptive Algorithm for SVM-Based Predistortion*

The SVM-based algorithm given in the previous section is not suitable for on-line calculation. We present an adaptive algorithm for updating the parameters in the SVM on every instant when new input-output data is obtained.

It is assumed that the SVM parameters $\{\alpha_i^{(n-1)}\}$ are given by solving the minimization problem (10) using the $n-1$ input-output training samples. When new input-output data are added at n-th time instant, the issue is how to update and calculate the new SVM parameters which can maximize the performance:

$$\max_{\boldsymbol{\alpha}_n^{(n)}} W(\boldsymbol{\alpha}_n^{(n)}) = \max_{\boldsymbol{\alpha}_n^{(n)}} \sum_{i=1}^{n} \alpha_i^{(n)} y_s(i)$$
$$-\frac{1}{2} \sum_{i,j=1}^{n} \alpha_i^{(n)} \alpha_j^{(n)} K(\boldsymbol{x}_s(i), \boldsymbol{x}_s(j)) - \varepsilon \sum_{i=1}^{n} |\alpha_i^{(n)}|$$

$$\text{subject to} \quad 0 \le |\alpha_i^{(n)}| \le C \quad \text{for } i = 1, \ldots, n \quad (12)$$

We show an adaptive scheme for updating the old parameters $\alpha_1^{(n-1)}, \alpha_2^{(n-1)}, \cdots, \alpha_{n-1}^{(n-1)}$ to the new parameters $\alpha_1^{(n)}, \alpha_2^{(n)}, \cdots, \alpha_n^{(n)}$. The algorithm consists of two steps: In the first step, we calculate $\{\alpha_n^{(n-1)}\}$ first, so that the criterion $W(\boldsymbol{\alpha}_n^{(n-1)})$ can be maximized with respect to only $\alpha_n^{(n-1)}$, on the condition that other $\alpha_1^{(n-1)}, \alpha_2^{(n-1)}, \cdots, \alpha_{n-1}^{(n-1)}$ are remained fixed at the values previously obtained. Its solution is given as follows:

$$\frac{\partial W(\boldsymbol{\alpha}_n^{(n-1)})}{\partial \alpha_n^{(n-1)}} = e^{(n-1)}(n) - \alpha_n^{(n-1)} K(\boldsymbol{x}_s(n), \boldsymbol{x}_s(n))$$
$$-\varepsilon \mathrm{sign}(\alpha_n^{(n-1)}) = 0 \quad (13)$$

where

$$e^{(n-1)}(n) = y_s(n) - \hat{y}_s^{(n-1)}(n) \quad (14)$$

$$\hat{y}_s^{(n-1)}(n) = \sum_{i=1}^{n-1} \alpha_i^{(n-1)} K(\boldsymbol{x}_s(i), \boldsymbol{x}_s(n)) \quad (15)$$

Then it follows from (13) that

$$\alpha_n^{(n-1)} = \min \left[C, \frac{||e^{(n-1)}(n)| - \varepsilon|}{K(\boldsymbol{x}_s(n), \boldsymbol{x}_s(n))} \right] \mathrm{sign}(e^{(n-1)}(n)) \quad (16)$$

Now we can have all the parameters $\{\alpha_i^{(n-1)}$: for $i = 1, 2, \cdots, n\}$, however, they should further be updated by the new data set $\{y_s(n), \boldsymbol{x}_s(n)\}$ already obtained at n-th instant.

In the next step, we update $\alpha_i^{(n-1)}$ to new parameters $\alpha_i^{(n)}$ for $i = 1, \cdots, n$ as

$$\alpha_i^{(n)} = \alpha_i^{(n-1)} + \Delta\alpha_i^{(n)} = \alpha_i^{(n-1)} + \eta_i \frac{\partial W(\boldsymbol{\alpha}_n^{(n-1)})}{\partial \alpha_i^{(n-1)}} \quad (17)$$

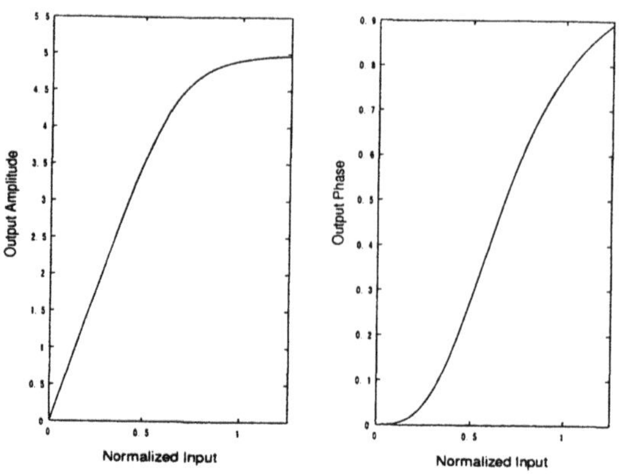

Fig. 2. Amplifier characteristics (*left*: Gain, *right*: Phase)

where

$$\frac{\partial W(\boldsymbol{\alpha}_n^{(n-1)})}{\partial \alpha_i^{(n-1)}} = e^{(n)}(i) - \varepsilon \text{sign}(\alpha_i^{(n-1)}) \qquad (18)$$

$$e^{(n)}(i) = e^{(n-1)}(i) - \alpha_n^{(n-1)} K(\boldsymbol{x}_s(n), \boldsymbol{x}_s(i)) \quad (19)$$

where $e^{(0)}(i) = y_s(i)$.

The optimal step size η_i can also be determined so that the criterion $W(\boldsymbol{\alpha}_n^{(n-1)} + \Delta\boldsymbol{\alpha}_n^{(n)})$ can be maximized, then we have

$$\eta_i = \frac{1}{K(\boldsymbol{x}_s(i), \boldsymbol{x}_s(i))} \quad \text{for } i = 1, \ldots, n \quad (20)$$

Thus the updated parameters given by the second step is expressed as

$$\alpha_i^{(n)} = \alpha_i^{(n-1)} + \frac{1}{K(\boldsymbol{x}_s(i), \boldsymbol{x}_s(i))} \cdot \left[e^{(n-1)}(i) - \alpha_n^{(n-1)} K(\boldsymbol{x}_s(n), \boldsymbol{x}_s(i)) - \varepsilon \text{sign}(\alpha_i^{(n-1)}) \right]$$
$$i = 1, \ldots, n \qquad (21)$$

Thus the adaptive algorithm executed at every instant when the sample data is given can be summarized as

Step 1: Let $n = 1$, $\alpha_1^{(0)} = 0$ and $\hat{y}_s^{(0)}(1) = 0$.
Step 2: Obtain the new data $\boldsymbol{x}_s(nT), y_s(nT)$, as (7) and (8).
Step 3: Calculate $e^{(n-1)}(n)$ by using (14). If $e^{(n-1)}(n) \leq \varepsilon$, set $\alpha_n^{(n)} = 0$ and $n = n + 1$, and go to Step 2, otherwise go to Step 4.
Step 4: Calculate the parameters $\alpha_n^{(n-1)}$ by using (16).
Step 5: Calculate the perturbations for all parameters $\Delta\boldsymbol{\alpha}_n^{(n-1)}$ by using (17), (18) and (20), and update them, although the parameters $\alpha_i^{(n-1)} = 0$ are not updated.
Step 6: Calculate the criterion $W(\boldsymbol{\alpha}_n^{(n)})$ from (12). If $W(\boldsymbol{\alpha}_n^{(n)}) \geq W(\boldsymbol{\alpha}_{n-1}^{(n-1)})$, then let $n = n + 1$ and go to Step 2, otherwise do not update the parameters and set $\alpha_n = 0$ and $n = n + 1$ and go to Step 2.

4. SIMULATION RESULTS

In this simulation, we investigate the effectiveness of the proposed SVM-based approach to compensation of linear and nonlinear distortions of HPA, which is frequently used in OFDM transmission systems. The OFDM is one of the multi-carrier modulations, and its base-band signal form is expressed by

$$s(t) = \sum_{n=0}^{N-1} \{a_n \cos(2\pi n f_0 t) - b_n \sin(2\pi n f_0 t)\} \quad (22)$$

where f_0 is a fundamental carrier frequency and $T = 1/f_0$ is a sampling interval. $(a_n,\ b_n)$ represent the data symbols to be transmitted, and 64QAM scheme is adopted. Thus, since the signal has widely changeable amplitudes dependently of transmitted symbols, the nonlinearity compensation is essentially needed to improve the distortion.

Let the nonlinear amplifier have the gain and phase characteristics given by

$$\text{Gain:} \quad A(x) = \frac{|x|}{(1 + |x|^{2b})^{1/2b}} \qquad (23)$$

$$\text{Phase:} \quad \Phi(x) = \frac{\pi}{3} \frac{|x|^2}{1 + |x|^2} \qquad (24)$$

where x is the amplitude of the input to HPA, and b is a degree denoting the saturation of the amplifier. We will consider two cases in which HPA has no dynamics and HPA has dynamics.

(a) *Case of nonlinear amplifier without dynamics*

We consider the nonlinear amplifier given by (23) and (24) with $p = 3$. Its input-output nonlinearity without dynamics is shown in Fig.2. The setup for simulation is summarized as follows: $S_{IN}(t)$ is 64QAM-OFDM signal, the number of carriers is $N = 1405$, and the FFT size is 2048. The size of the regression vector $\boldsymbol{x}_s(i)$ is chosen $m = 2$, and the delay is chosen $d = 1$. The kernel is selected as the polynomial type given by $K(\boldsymbol{x}_1, \boldsymbol{x}_2) = (a{\boldsymbol{x}_1}^T \boldsymbol{x}_2 + 1)^p$ where $a = 10^{-6}$ and $p = 3$. The constant parameters in the design of the SVM are chosen as $C = 0.072$ and $\varepsilon = 10^{-10}$. These constants are determined as follows: The constant C relating with regularization can be chosen from a maximum value among actually calculated α_i by changing C from larger to smaller one. The constant a in the kernel can be given from the magnitude of the inner product of the support vectors and input vectors. The appropriate constants m, d and p can be determined so that the sum of the squares error $e^2(t)$ may be minimized. It will be noticed that m and d are smaller than the case with dynamics in HPA.

The predistortion performance is evaluated by the reduction of the output power spectra outside the transmission frequency band caused by high nonlinearity of HPA. The results are compared to a case with no predistortion scheme (see Fig.3) and a case with an ordinary table-lookup scheme (see Fig.4). The proposed SVM based method can im-

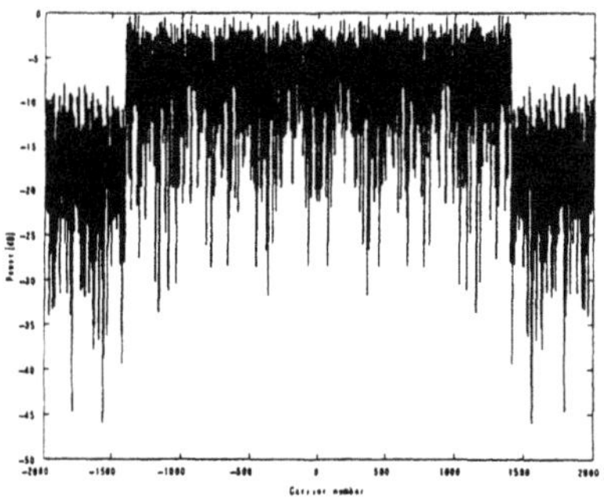

Fig. 3. Output power spectrum: No predistortion

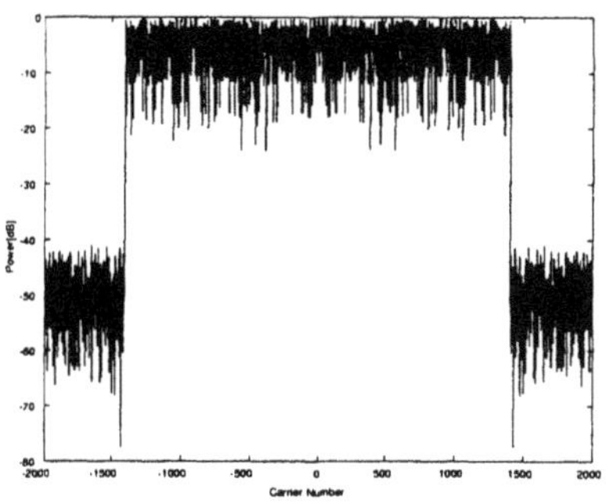

Fig. 4. Output power spectrum: Predistortion by ordinary table lookup scheme

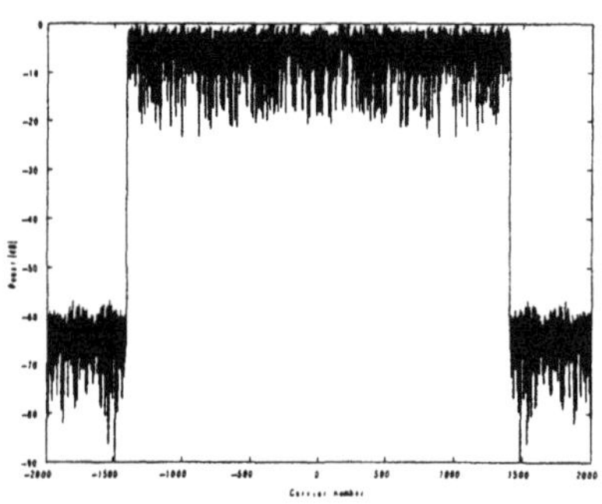

Fig. 5. Output power spectrum: Proposed SVM

prove the predistortion by 50dB from the no predistortion case and by 18dB from the table lookup scheme (see Fig.5).

(b) *Case of nonlinear amplifier with dynamics*

In this simulation, it is assumed that the nonlinear amplifier has dynamics in its input and output parts as shown in Fig.6, and they are given by linear lowpass filters $A(j\omega)$ and $B(j\omega)$ respectively, for instance, expressed as

$$A(j\omega) = \frac{4.4381}{1 + 0.001118j\omega}, \quad B(j\omega) = \frac{1.16}{1 + 0.007437j\omega}$$

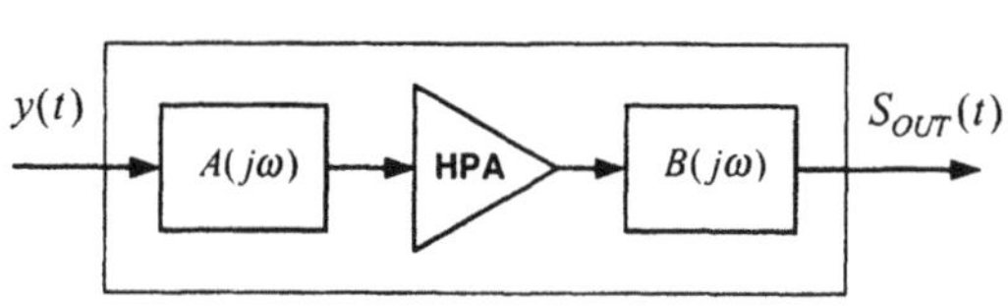

Fig. 6. Nonlinear structure of HPA with dynamics

The setup for simulation in the case of amplifier with the above unknown dynamics is summarized as follows: The transmitted signal $S_{IN}(t)$ is the 64QAM-OFDM signal same as the former simulations. The number of carriers, the size of FFT are also the same. The delay is chosen as $d = 3$, the dimension of the regressor vector is chosen as $m = 6$. The kernel is given by $K(\boldsymbol{x}_1, \boldsymbol{x}_2) = (a{\boldsymbol{x}_1}^T \boldsymbol{x}_2 + 1)^p$, where $a = 5 \times 10^{-3}$ and $p = 7$. The constants in design for the SVM are chosen by $C = 0.1$ and $\varepsilon = 10^{-9}$. These constants needed in the SVM approach can also be determined similarly as in the case (a).

The predistortion performance is also evaluated by the reduction of the output power spectra outside the transmission frequency band. The results are compared to a case with no predistortion (see Fig.7) and a case by an ordinary table-lookup scheme (see Fig.8). The table-lookup scheme can hardly compensate uncertain linear dynamics included in HPA. The proposed SVM based method can improve the predistortion by 30dB from the no predistortion case and 20dB from the table lookup scheme (see Fig.9). It should be remarked that the appropriate values of d, m and p given from minimization of the squares error are larger than those in the case (a) with no dynamics. However we can use the same structure of the SVM predistortor. The recovered 64-QAM symbols are plotted in Fig.10 obtained in a case without predistortion, in Fig.11 obtained by the table-lookup scheme considering no linear dynamics, and in Fig.12 obtained by the proposed adaptive SVM based predistortion. The detection performance can also be improved as well as the leakage of output power spectra.

5. CONCLUSIONS

The new adaptive predistortion algorithm has been presented on the basis of the data-based nonlinear input-output modeling using the SVM. By setting some constant parameters appropriately chosen, the obtained adaptive SVM algorithm has very simple structure compared to ordinary table lookup schemes and mathematical modeling approach.

REFERENCES

Cavers, J.K. (1990). Amplifier linearization using a digital predistortor with fast adaptation and low memory requirements. *IEEE Trans. Vehicular Tech.* **89**, 374–382.

Cheng, C.H. and E.J. Power (1999). A reconsideration of the p-th order inverse predistorter. *1999 IEEE 49th Vehicular Technology Conference*, 1501–1504.

Horiguchi, K., K. Yamauchi, K. Mori M. Nakayama Y. Ikeda and T. Takagi (1999). A distortion analysis method for fet amplifiers using novel frequency-dependent complex power series model. *IEICE Trans. Electron.* **E82-C**, 738–743.

Kang, H.W., Y.S. Cho and D.H. Youn (1997). An efficient adaptive predistorter for nonlinear high

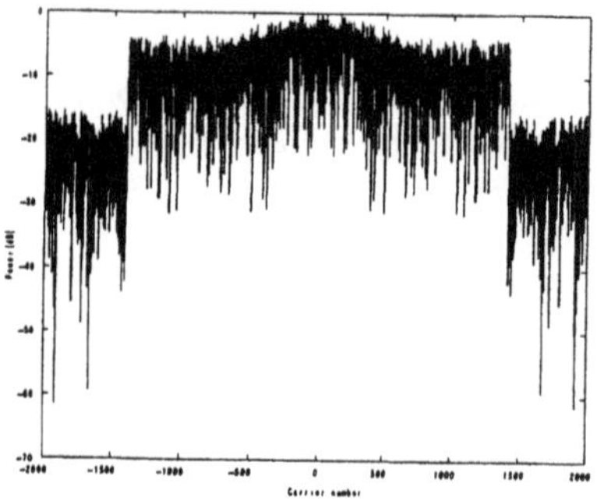

Fig. 7. Output power spectrum: No predistortion

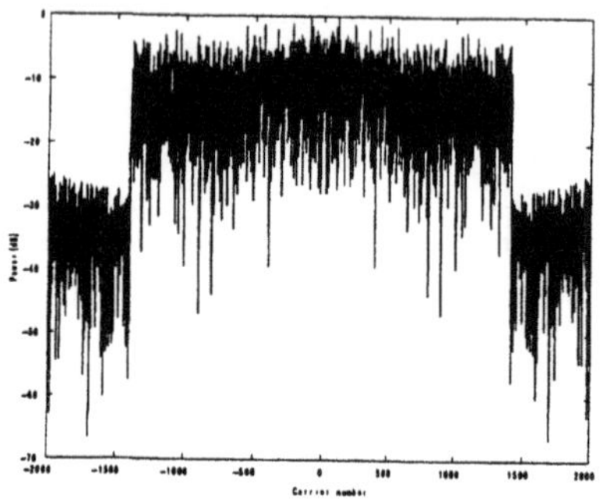

Fig. 8. Output power spectrum: Predistortion by ordinary table lookup scheme

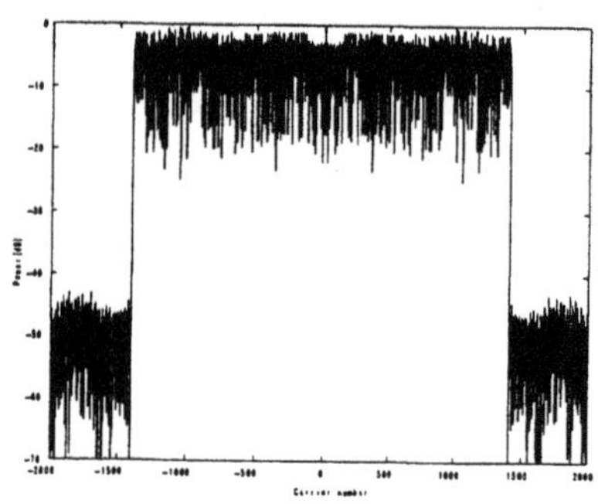

Fig. 9. Output power spectrum: Proposed SVM

power amplifier in satellite communication. *Proc. 1997 IEEE International Symposium on Circuits and Systems*, 2288–2291.

Lazzarin, G., S. Pupolin and A. Sarti (1994). Nonlinearity compensation in digital radio systems. *Proc. IEE* **134A**, 525–546.

Matsuoka, T., M. Orihashi, M. Sagara, H. Ikeda and K. Misaiz (1997). Compensation of nonlinear distortion during transmission based on the adaptive predistortion method. *IEICE Trans. Electron.* **E80-C**, 782–787.

Nojima, T. and Y. Okamoto (1981). Analysis and compensation of twt nolinearities based on complex power series representation. *Trans. IECE* **J64-B**, 1449–1456.

Okada, M., H. Nishijima and S. Komaki (1998). A maximum likelihood decision based nonlinear distortion compensator for multi-carrier modulated signals. *IEICE Trans. Commun.* **E81-B**, 737–744.

Saleh, A.A.M. and J. Salz (1983). Adaptive lin-

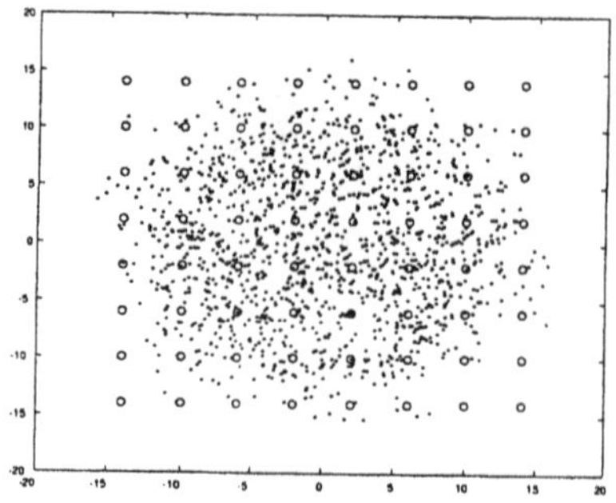

Fig. 10. De-mapped signal of 64QAM: No predistortion

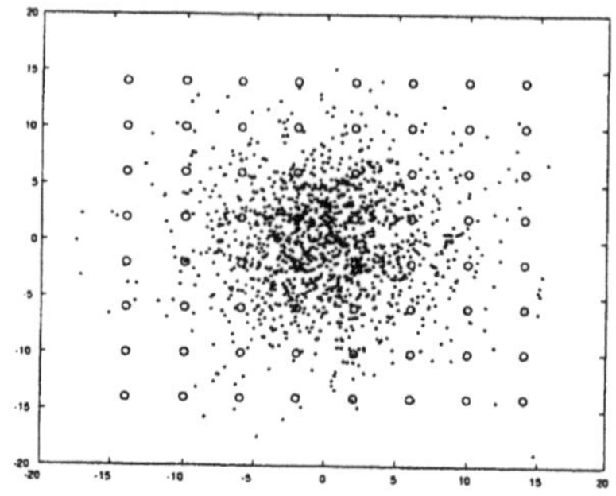

Fig. 11. De-mapped signal of 64QAM: Predistortion by table lookup scheme considering no linear dynamics

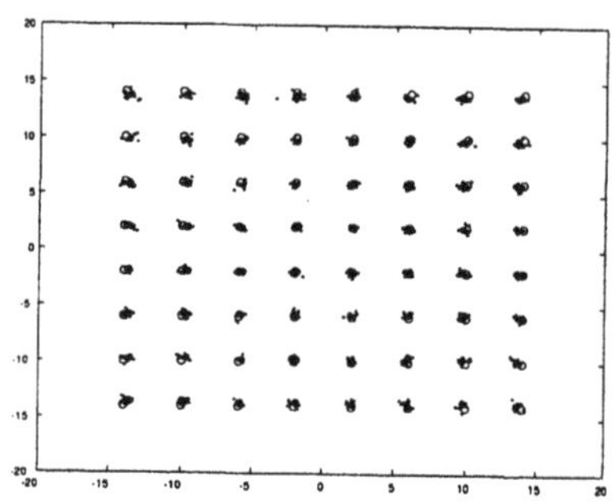

Fig. 12. De-mapped signal of 64QAM: Proposed SVM

earization of power amplifiers in digital radio systems. *The Bell System Technical Journal* **62**, 1019–1034.

Scholkop, B., C.J.C. Burges and A. J. Smola (1998). *Advances in Kernel Methods: Support Vector Learning.* MIT Press. Boston.

Shin, Y., D.M. Yoon, Y.S. Lim and H.S. Hahn (1996). Equalization of nonlinear digital satellite communication channels using a complex radial basis function network. *Proc. IASTED Conf. Artificail Intelligence, Expert Systems, and Neural Networks.*

Vapnik, V. (1998). *Statistical Learning Theory.* John Wiley & Sons. Boston.

Watkins, B.E., R. North and M. Tummla (1995). Neural network based adaptive predistortion for the linearization of nonlinear rf amplifiers. *Proc. MILCOM-95*, 145–149.

AN HEURISTIC LEARNING MODELLING STRATEGY FOR NONLINEAR SYSTEMS

Giovanni L. Santosuosso *

* *Dipartimento di Ingegneria Elettronica,*
Università di Roma Tor Vergata,
via di Tor Vergata 110,
00133 Rome, Italy. e-mail: santosuosso@ing.uniroma2.it

Abstract: We propose a new heuristic learning algorithm that we call atomic metaphor optimization strategy (AMOS), retaining similarities with some aspects of both simulated annealing and genetic algorithms, following recent advances in the literature. The new strategy is parallel in nature, like genetic algorithms, but does not generate "offsprings" between parents, and avoids a mechanism to select among all generated points the new parents. It maintains the random evolution of new candidate points of simulated annealing without a decreasing acceptance probability performance. We apply this algorithm to the on-line modeling of continuous-time nonlinear systems via a neural approximation of the system dynamics, with a nonlinearly parametrized multilayer neural network. *Copyright © 2001 IFAC*

Keywords: Reinforcement Learning, Genetic Algorithms, Simulated Annealing, Nonlinear Adaptive Observer, Neural Networks.

1. INTRODUCTION

The application of heuristic optimization methods to intelligent systems modelling and control problems has attracted considerable interest in recent years. The main feature of these optimization strategies is the capability to yield "feasible" sub-optimal solutions when searching for a global optimum in a crispy environment, escaping local minima without gradient knowledge. Two important algorithms developed in this context are simulated annealing (SA) and genetic algorithms (GA); in SA, that was first proposed by Kirkpatrik *et al.* in 1983, a candidate solution is randomly generated, and also a new solution in the neighborhood of the first one is produced. If the new one is found to be better then the current point then it is unconditionally accepted to be the next current point; if not, it can either accepted or rejected with a certain probability. To begin with, the probability of accepting a worse point is kept high. This reduces the chance of getting trapped in a local optimum. As the number of iterations increases, this probability is reduced according to a given distribution.

In GA, proposed by Holland (1975), a population of randomly generated candidate points is generated. By combining "parent" candidate points a number of new "offsprings" points are generated, and the best among all the points are selected as new "parent" candidate points. Recently, hybrid SA-GA algorithms have been proposed in discrete combinatorial optimization problems as in Lin *et al.* (1998), Mantawy, *et al.* (1999). The key advantages of both heuristic optimal search techniques is the capability to find rapidly a relatively "good" solution to a problem, escaping local minima, but their limitation is poor exact convergence to the real optimal solution, when the algorithm proceeds. The implementation of GA or SA is further restricted by the fact that these optimization algorithms have been originally developed in the computer science community in a discrete time context, while many control or identification problems arising in systems science are inherently continuous time.

A different on-line modeling approach, consists in the use of neural networks with linear parametrization (see Anderson *et al.*, 1986, Marino and Tomei 1995, Rovithakis and Christodoulou 2000). Their fundamental advantage is the capability to guarantee with Lyapunov stability arguments that if the approximating error tends to zero, then it can be constructed an adaptive observer that provides an estimation error of the neural network parameters that also tends deterministically to zero. A second positive feature of the above estimation strategy is that parameter estimates converge to their true values with exponential velocity. The real drawback is in the so called "curse of dimensionality" which is inherent in linearly parametrized approximating functions. A better approximation can be obtained by using neural networks with nonlinear parametrization. However, adaptive continuous-time learning algorithms hardly secure the convergence properties in the nonlinear setting. The reason lies in the fact that adaptive observation learning strategy in Anderson *et al.*, 1986, Marino, and Tomei 1995, Rovithakis and Christodoulou 2000 hides an error gradient descending approach, and for nonlinear parametrization neural networks may exist many local error minima. In conclusion, both heuristic optimal search and deterministic gradient descending learning strategies have advantages and limitations, and this strongly motivates us to propose a strategy that combines the advantages to operate in continuous time setting with the capabilities of discrete time heuristic search algorithms.

In this note we construct a continuous time modelling observer with the following features:

- We propose a new continuous time heuristic learning modeling algorithm retaining similarities with some aspects of both simulated annealing and genetic algorithms, by introducing however a fairly different data structure, that we call atomic metaphor optimization strategy (AMOS). The new algorithm is parallel in nature, like GA, but does not generate "offsprings" between parents, and avoids a selection mechanism among all generated points. It maintains the random evolution of new candidate points of SA without a decreasing acceptance probability performance.
- We describe a continuous time version of AMOS optimization strategy for the on line modelling of a nonlinear system whose dynamics are approximated with a nonlinear neural network.

2. PROBLEM STATEMENT

In this note we address the problem of on-line identification of a nonlinear system

$$\dot{\zeta} = f(\zeta, u(t)) + g(\zeta, u(t)) \qquad (1)$$

with state $\zeta \in \Re^n$, control input $u \in \Re^m$ and $f(\cdot,\cdot) : \Re^n \times \Re^m \to \Re^n$ $g(\cdot,\cdot) : \Re^n \times \Re^m \to \Re^n$ smooth drift vector fields. In the following we assume that both ζ and u are available for measurement, the mapping $g(\cdot,\cdot)$ is known, but the function $f(\cdot,\cdot)$ is unknown. We assume that $f(\cdot,\cdot)$ admits a suitable neural approximation; notice however that the term "neural networks" can be used in a very general sense to define a class of parameter dependent functions that exhibit approximation capabilities of given mappings for a suitably complex individual neural network and by a proper choice of the parameters in the net. Neural modeling has attracted a considerable interest in the research community in the last decade, and there are many classes of approximating networks that are available in the literature.

In the context of on-line modeling, there have been concentrated efforts towards the design and analysis of learning algorithms that are based on the Lyapunov stability theory, considering classes of approximating functions with linear parametrization as in Anderson *et al.*, 1986, Marino and Tomei 1995, Rovithakis and Christodoulou 2000. To be more specific, assume that the unknown function is approximated with a linear parametrization neural network

$$f(\zeta, u(t)) = \Psi(\zeta, u(t))\theta + \varepsilon(t) \qquad (2)$$

where $\Psi(\cdot,\cdot)$ is a $(n \times p)$ matrix of known functions, $\theta \in \Re^p$ is a parameter vector and $\varepsilon(t)$ is the approximation error. The linear parametrization (2) includes polynomial approximations (see Rovithakis and Christodoulou 2000), or multidimensional Fourier expansions. The fundamental advantage of using neural networks with linear parametrization is the capability to guarantee with Lyapunov stability arguments that if the approximating error $\varepsilon(t)$ tends to zero, then it can be constructed an adaptive observer that provides an estimation error of the neural network parameters θ that also tends to zero. The adaptive observer's structure is (see Marino *et al.* 2001)

$$\begin{aligned} \dot{\hat{\zeta}} &= K(\zeta - \hat{\zeta}) + \Psi(\zeta, u(t))\hat{\theta} + g(\zeta, u(t)) \qquad (3) \\ \dot{\hat{\theta}} &= \Pr oj[\Gamma\Psi^T(\zeta, u(t))(\zeta - \hat{\zeta}), \hat{\theta}], \end{aligned}$$

with state $\left(\hat{\zeta}, \hat{\theta}\right) \in \Re^n \times \Re^p$, K and Γ are symmetric positive definite matrices of suitable dimension, $\mathrm{Pr}\,oj[\cdot]$ is the projection operator as introduced in Pomet and Praly 1992. A better approximation can be obtained by using neural networks with nonlinear parametrization, i.e. by setting

$$f(\zeta, u(t)) = \Phi(\zeta, u(t), w) + \varepsilon(t), \qquad (4)$$

where now the vector $w \in \Re^q$ appears nonlinearly in the network equations. The neural network class most commonly used in control systems is the feedforward multilayer neural network. It has been shown that even with one single layer feedforward neural networks can approximate any continuous function uniformly over a compact domain, (see Hornik *et al.* 1989; Cybenco *et al.* 1989). However, adaptive continuous-time learning algorithms like (3) for the linear parametrization case that secure deterministic convergence properties with Lyapunov function geometric arguments are not available in general, for nonlinear neural networks.

We state in a formal definition the problem we address in this paper.

Definition 1. Consider the system

$$\dot{\zeta} = g(\zeta, u(t)) + \Phi(\zeta, u(t), w) + \varepsilon(t). \qquad (5)$$

A class of **pseudo-ISS adaptive modelling observers** for system (5) is a set of r dimensional continuous time systems, parametrized in the integer number $r \in \mathcal{N}$, described by

$$\dot{\chi} = \alpha_1(\chi, \zeta, u) \;\; \chi \in \Re^r$$
$$\dot{\hat{w}}_C = \alpha_2(\chi, \zeta, u) \;\; \hat{w}_C \in \Re^q$$

where $\alpha_i(.)$ $i = 1, 2$ are suitable functions, driven by the inputs $\zeta(t)$, $u(t)$, such that for any $\chi(0) \in \Re^r$ $\chi(t)$ is globally bounded, and by setting $\Delta(t) = [w - \hat{w}_C(t)]^T$ the inequality

$$\|\Delta(t)\|^2 \leq \eta\left(\|\Delta(0)\|^2, r, t\right) + \gamma\left(r, \sup_{0 \leq \tau \leq t} \|\varepsilon(\tau)\|\right) \qquad (6)$$

holds with suitable functions $\eta(\cdot,\cdot)$ and $\gamma(\cdot)$ with $\gamma(r, 0) = 0$, for all $t \geq 0$, and all $r \in \mathcal{N}$, that exhibit the following property:

i) $\lim_{\substack{r \to \infty \\ t \to \infty}} \eta\left(\|\Delta(0)\|^2, r, t\right) = 0$, for all $\Delta(0) \in \mathcal{B}(\mathcal{R})$, where $\mathcal{B}(\mathcal{R}) = \{\Delta \in \Re^q : \|\Delta\| \leq \mathcal{R}\}$.

A word is in order to explain the features of the modelling estimator definition above. The observer's error dynamics evolve according to an heuristic optimal search strategy, and inequality (6) simply states that the neural network parameter estimates are attracted in a compact set. Property **i)** can be interpreted in the following way: if the dimension of the observer in the class tends to infinity, then parameter estimates are attracted in an open ball whose size is a function of $\sup_{0 \leq \tau \leq t} \|\varepsilon(\tau)\|$ and if $\varepsilon(\tau) \equiv 0$, then the neural network parameters converge to the optimum. An infinite dimensional observer is not implementable of course, but the meaning of **i)** is that the greater is the dimension of the observer, the better is the approximation.

3. AMOS : ATOMIC METAPHOR OPTIMIZATION STRATEGY

Let $J(w) : \Re^q \longrightarrow \Re^+$ be a non-negative real-valued fitness function. Consider a static optimization problem as the retrieval of the minimum of $J(w)$ on a compact domain $\mathcal{D} \subset \Re^q$. We introduce an algorithm whose data structure retain some vague similarities with the structure of an atom (AMOS). In an atom of an element there is a nucleus -a central bulk- composed of neutrons and protons, and a set of regions around it on which the electrons move. The learning strategy that we propose there is still a point in the search space conceptually similar to an atom bulk, and sets points in the search space with the role of electrons confined in suitable regions around it. We are more specific as follows.

- A "center" point $w_C \in \Re^q$ is introduced. This is a point in the search space that turns out to be the minimum fitness point when the algorithm proceeds.
- There is a number of sets of points of the search space $\Pi = \{P_1, \ldots, P_\mu\}$ where $P_i = \{w_{i1,\ldots}w_{i\nu}\}$ and $w_{ij} \in \Re^q$, $i = 1, \ldots \mu$, $j = 1, \ldots \nu$, that constitute a population of possible "explorer" optimal solutions, like GA algorithms.. The population Π is grouped in the subsets P_i since the elements of each subgroup will evolve according to a different criterion. To be more specific, at each subset of the population is associated a region of attraction $\mathcal{RA}_i(w_C)$ centered in w_C, for instance a ball centered in w_C and radius R_i, $i = 1, \ldots \mu$, $\mathcal{B}_i(w_C, R_i) = \{w : \|w - w_C\| \leq R_i\}$. Every point $w_{ij} \in P_i$, will be attracted in $\mathcal{RA}_i(w_C)$ if $w_{ij} \notin \mathcal{RA}_i(w_C)$, for $i = 1, \ldots \mu$, $j = 1, \ldots \nu$.
- A "best explorer" $w_B \in \Re^n$ is determined at each algorithm step, that is the point $w_B \in \Pi$ such that $J(w_B) = \min_{w \in \Pi} J(w)$.

The structure of the algorithm is as follows:

- *Initialize the population* $\Pi(0)$, $x_C(0)$ *randomly selecting the search points in* $\mathcal{D}$.
- *set* $k = 0$.
- *While* ($k \leq N$) *DO:(*
  - ▷ *find* $w_B(k) \in \Pi$ *as* $J(w_B(k)) = \min_{x \in \Pi(k)} J(w)$.
  - ▷ *If* $J(w_B(k)) < J(w_C(k))$ *then set* $w_C(k+1)$ *so that* $\|w_C(k+1) - w_B(k)\| < \|w_C(k) - w_B(k)\|$, *otherwise set* $w_C(k+1) = w_C(k)$.
  - ▷ *For all the subclasses* P_i *in the population DO:(*
    For all the points in the subclass P_i *DO:(*
    if $w_{ij}(k) \notin \mathcal{RA}_i(w_C)$,
    then set $w_{ij}(k+1)$ *so that* $\|w_{ij}(k+1) - w_C(k+1)\| < \|w_{ij}(k) - w_C(k)\|$,
    otherwise set $w_{ij}(k+1)$ *according to a different strategy* $S_i(k+1)$*))))*

We notice that in the algorithm structure above two issues have to be specified. The first one is the specific strategy for the center w_C to approach w_B if $J(w_B) < J(w_C)$, and for a point in the population to approach w_C if it is out of the region $\mathcal{RA}_i(w_C)$. However, in a practical algorithm implementation w_C can be set to converge asymptotically to w_B if $J(w_B) < J(w_C)$, the points in the population $w_{ij} \in \Pi$ are attracted inside the respective region $\mathcal{RA}_i(w_C)$. The second and more critical issue is the evolution of a point $w_{ij} \in \Pi$ in the population when it is inside the region $\mathcal{RA}_i(w_C)$; if it is imposed a random motion "sufficiently rich", inside $\mathcal{RA}_i(w_C)$, then the explorer element in the population will eventually discover a minimum inside the ball, and w_C will tend to it.

3.1 *AMOS implementation for continuous time modeling*

The optimization strategy introduced above can be adapted to the problem addressed in this note. To this purpose, we have to define a continuous time dynamical behavior for elements in the population whose coordinates in the search state space are $\hat{w}_{ij}$, $i = 1, \ldots \mu$, $j = 1, \ldots \nu$, and for the center with coordinates $\hat{w}_C$.

Consider an element of the population that belongs to the class $\hat{w}_{ij} \in P_i$, associated to a region of attraction consisting in a ball centered in $\hat{w}_C$ and radius R_i, $i = 1, \ldots \mu$, $\mathcal{B}_i(\hat{w}_C, R_i) = \{w : \|w - \hat{w}_C\| \leq R_i\}$. The element $\hat{w}_{ij} \in P_i$, is associated with two continuous time dynamical systems; $\sum_{1ij}$ that characterizes the time trajectory of $\hat{w}_{ij}$, and $\sum_{2ij}$ with the role of extracting a performance measure of the trial solution $\hat{w}_{ij}$. A possible implementation of system $\sum_{1ij}$ is

$$\begin{cases} \dfrac{d\hat{w}_{ij}(t)}{dt} = -q_{ij}(t) \\ if \ \|\hat{w}_{ij} - \hat{w}_C\| > R_i - \rho_i \\ \text{and } \beta_{ij}^T(t)\,(\hat{w}_{ij} - \hat{w}_C) \geq 0 \end{cases} \qquad (7)$$
$$\begin{cases} \dfrac{d\hat{w}_{ij}(t)}{dt} = -\beta_{ij}(t) \text{ otherwise,} \end{cases}$$

where

$$q_{ij}(t) = \left[1 - p(\hat{w}_{ij}) \frac{\left[\frac{\partial p(\hat{\theta})}{\partial t}\right]^T \left[\frac{\partial p(\hat{\theta})}{\partial t}\right]}{\left\|\frac{\partial p(\hat{\theta})}{\partial t}\right\|^2}\right] \beta_{ij}(t)$$

$$p(\hat{w}_{ij}) = \frac{\|\hat{w}_{ij} - \hat{w}_C\| - (R_i - \rho_i)^2}{\rho_i^2 + 2\rho_i (R_i - \rho_i)}$$

and ρ_i is an positive scalar, such that $\rho_i < R_i$. The vectors $\beta_{ij}(t) \in \Re^q$ are important parameters in the modeling strategy, and are chosen by the designer, with the constraint that are continuous time random vector signals, uncorrelated one with each other. Note that $\frac{d\hat{w}_{ij}(t)}{dt} = Proj\left[\beta_{ij}(t), (\hat{w}_{ij} - \hat{w}_C)\right]$ as in Pomet, and Praly 1992, and adapted to our case. In particular if x_C is constant and $\|\hat{w}_{ij}(t_0) - \hat{w}_C(t_0)\| < R_i$, then for all $t \geq t_0$

$$\begin{aligned} \|\hat{w}_{ij}(t) - \hat{w}_C(t)\| &\leq \quad R_i \\ \left\|\frac{d\hat{w}_{ij}(t)}{dt}\right\| &\leq \|\beta_{ij}(t)\|. \end{aligned}$$

If $\sup_{t \in [t_0, \infty)} \|\beta_{ij}(t)\|$ is sufficiently greater than $\sup_{t \in [t_0, \infty)} \left\|\frac{d\hat{w}_C(t)}{dt}\right\|$ then the element $\hat{w}_{ij}(t)$ is attracted in the ball $\mathcal{B}_i(\hat{w}_C, R_i)$ and moves randomly inside it. We also assume the control input $u(t)$ to be sufficiently "rich" in a period of time T so that the couple $(\zeta(\tau) \times u(\tau))$ is sufficiently "dense" in compact set $(\mathcal{B}_\zeta \times \mathcal{B}_u) \subset (\Re^{\nu_1} \times \Re^{\nu_2})$ for $t \leq \tau \leq t + T$. An implementation of $\sum_{2ij}$ can be

$$\begin{aligned} &\dot{\hat{\zeta}}_{ij} = K(\zeta - \hat{\zeta}_{ij}) + \Phi(\zeta, u(t), \hat{w}_{ij}(t)) \\ &+ g(\zeta, u(t)) \\ &J(\hat{w}_{ij}(t), t) = AEF\left(\left\|\zeta - \hat{\zeta}_{ij}\right\|^2\right) \end{aligned} \qquad (8)$$

and $AEF(\cdot)$ is an average extracting filter of the quantity $\left\|\zeta - \hat{\zeta}_{ij}\right\|^2$ over time T. Notice that if we consider a fixed point in the search space $\bar{w} \in \Re^q$ such that $f(\zeta, u(t)) = \Phi(\zeta, u(t), \bar{w}) + \varepsilon(t)$, and $\sup_{t \in [t_0, \infty)} \|\varepsilon(t)\| \to 0$, then also $\lim_{t \to \infty} \left\|\zeta - \hat{\zeta}_{ij}\right\|^2 \to 0$. Therefore, if time T is sufficiently long to describe the performance of a fixed point in the search space $\bar{w} \in \Re^q$, and $\sup_{t \in [t, T)} \left\|\frac{d\hat{w}_{ij}(t)}{dt}\right\|$ is kept sufficiently "small", then for suitable parameters the variable $J(\hat{w}_{ij}(t), t)$ gives an idea of the performance of $\hat{w}_{ij}(t)$.

The center has a performance indicator identical to $\sum_{2ij}$ with $\hat{w}_C$ replacing $\hat{w}_{ij}(t)$ and yielding $J(\hat{w}_C, t)$. The dynamics of the center turn out to be

$$\dot{\hat{w}}_C = \delta\left[J(\hat{w}_B, t), J(\hat{w}_C, t)\right](\hat{w}_B(t) - \hat{w}_C(t)) \quad (9)$$

where $\hat{w}_B(t) \in \Pi$ is the element of the population such that

$$J(\hat{w}_B(t), t) = \min_{\substack{i\in[1,\mu] \\ j\in[1,\nu]}} J(\hat{w}_{ij}, t).$$

and $\delta(a, b)$ is a suitable smooth function such that $\delta(a, b) = 0$ if $a \geq b$ and $\delta(a, b) > 0$ if $a < b$.

4. A SIMPLE SIMULATION EXAMPLE

We present the on line modelling simulation of the elementary "benchmark" scalar nonlinear system

$$\dot{\zeta} = -\zeta + f(\zeta) + u(t)$$

with state $\zeta \in \Re$, control input $u \in \Re$, where $u(t) = 150\sin(11.1t)$, and

$$\begin{aligned} f(\zeta) = &(2 + 4\,|\zeta - 4| + 4\,|1.5\zeta + 6| \\ &+ 8\,|1.5\zeta + 6|\,|\zeta - 4|)^{-1} \\ &\cdot (139 + 198\,|\zeta - 4| + 86\,|1.5\zeta + 6| \\ &12\,|1.5\zeta + 6|\,|\zeta - 4| + 14\zeta \\ &+ 48\zeta\,|\zeta - 4| - 20\zeta\,|1.5\zeta + 6|) \end{aligned}$$

is a function to be modelled. We consider a neural approximation of $f(\zeta)$ by introducing the sigmoidal function $\sigma(x) = \frac{1}{2}\left[\frac{1}{\pi}\arctan(x) + 1\right]$and defining the simple neural approximation

$$\Phi(\zeta, w) = w_3\sigma(w_2\zeta + w_1) + w_6\sigma(w_5\zeta + w_4).$$

The goal of AMOS modelling strategy is to find suitable parameters for the parameter vector $w = [w_1, w_2, \ldots w_6]^T$, in order to minimize the quantity $|f(\zeta) - \Phi(\zeta, w)|$ in the compact set $-10 \leq \zeta \leq 10$. We have chosen a population of 300 elements, that belong to 6 different classes, P_1, P_2, P_3, P_4, P_5, P_6; in each class there are 50 elements. The "explorers" elements in the class P_1, are attracted in a ball centered $\hat{w}_C$ and radius $R_1 = 10$, and when inside evolve randomly with an update law $\beta_{1j}(t) \in \Re^6$ with $j = 1, 2, \ldots, 50$, whose entries are sinusoidal functions in the class $0.25\sin(\omega_1 t + \varphi) + 0.25\cos(\omega_2 t + \varphi)$ and ω_1 , ω_2, φ are different positive real numbers for each entry and each element in the class, that are randomly chosen in the range $1 \leq \omega_1, \omega_2, \varphi, \leq 1.3$. The elements in classes P_2, ... P_5, evolve in a similar way, except the respective radii of attraction are $R_2 = 7$, $R_3 = 5$, $R_4 = 2.5$, $R_5 = 1$. The elements of the class P_6 evolve in a ball of radius $R_6 = 1$, and the entries of $\beta_{6j}(t) \in \Re^6$ are functions in the class $2.5\sin(\omega_1 t + \varphi) + 2.5\cos(\omega_2 t + \varphi)$. The center $\hat{w}_C$ basically moves towards the best element in the population if it has a better fitness the fitness of the center itself, otherwise is fixed. The algorithm has been simulated for $t_F = 600$ seconds, and $\hat{w}_C(t_F)$ turns out to be $\hat{w}_C(t_F) = (11.17, 2.94, 8.45, -5.63, 1.75, -5.44)^T$.

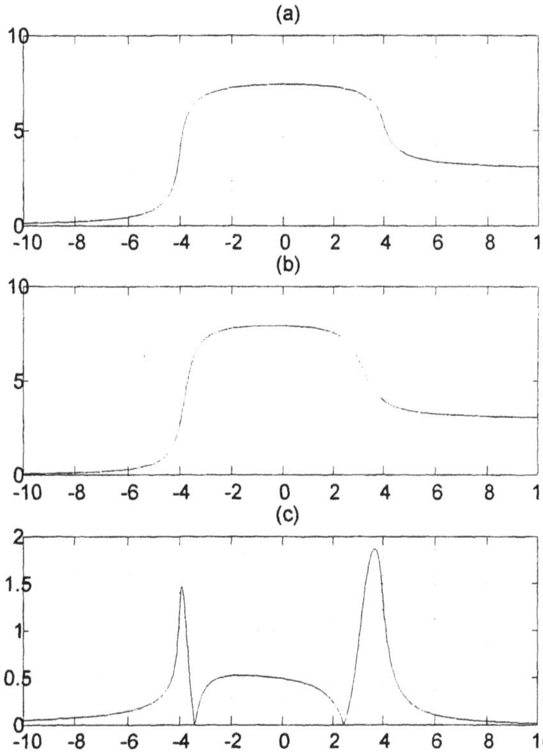

Fig. 1: (a): Plot of $f(\zeta)$; (b): plot of $\Phi(\zeta, \hat{w}_c(t_F))$ and (c) plot of $|\Phi(\zeta, \hat{w}_c(t_F)) - f(\zeta)|$.

Figure 1 shows for $-10 \leq \zeta \leq 10$, respectively the plots of function $f(\zeta)$, of the neural approximation $\Phi(\zeta, \hat{w}_c(t_F))$, and of the modelling error $|f(\zeta) - \Phi(\zeta, \hat{w}_c(t_F))|$. Figure 2 shows the index, between 1 and 300, of the element $\hat{w}_B$ in the entire population, plotted with time $0 \leq t \leq 600$; in fact the elements of class P_1, are numbered in the simulation between 1 and 50, the elements of class P_2, between 50 and 100, et cetera, up to the elements of P_6 whose index is between 250 and 300. This plot shows implicitly which class P_i, is capable to express in each instant of the simulation the best "explorer" element. Figure 3 shows the plot of the 6 entries of the center $\hat{w}_c(t)$ for time $0 \leq t \leq 600$. The algorithm makes use of a Runge-Kutta 6-th order simulation, with fixed step equal to 0.01. Notice that class P_1 never does, and as long as the learning algorithm proceeds the elements in classes with smaller orbits around the center $\hat{w}_c$ take the lead in expressing a best "explorer" element.

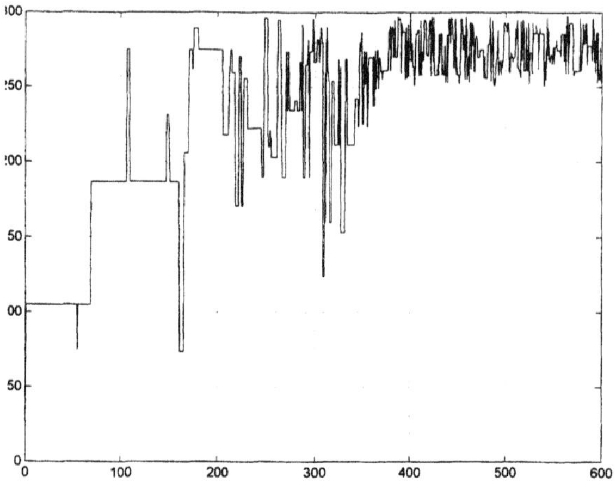

Fig. 2. Plot of index of best element with respect to time.

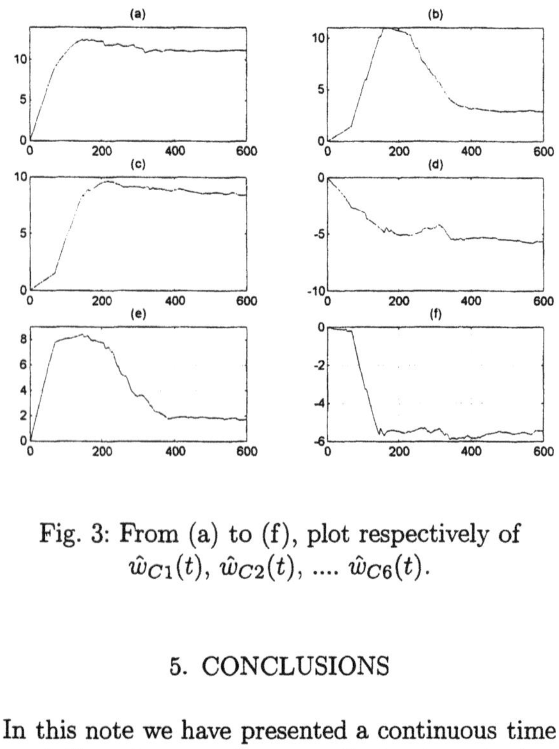

Fig. 3: From (a) to (f), plot respectively of $\hat{w}_{C1}(t)$, $\hat{w}_{C2}(t)$, ... $\hat{w}_{C6}(t)$.

5. CONCLUSIONS

In this note we have presented a continuous time modelling observer that we call atomic metaphor optimization strategy (AMOS). This learning procedure can be fast at the beginning of the learning phase, but slow in the final approach to a global optimum. Like many heuristic optimal search strategy, if we do not make further assumptions on the nature of the random dynamic behavior $\beta_{ij}(t) \in \Re^q$ in (7) of the elements in the population and on the structure of the nonlinear function to be modelled, then it is not trivial to prove that AMOS solves the problem addressed in Definition 1. However on real simulations the proposed modelling strategy is capable to closely approximate a sub-optimal neural approximation of a nonlinear function, as the example we presented has shown. Future research direction will focus on the closer integration with the continuous time adaptive observer algorithm based on the Lyapunov stability theory, developed for classes of approximating functions with linear parametrization.

Acknowledgments. The author wish to thank Prof. Riccardo Marino for helpful discussions and suggestions regarding this work.

REFERENCES

Anderson, B., R. Bitmead, C. Johnson, P. Kokotovic, R. Kosut, I. Mareels, L. Praly, and B. Riedle (1986). Stability of Adaptive systems: Passivity and Averaging Analysis. Cambridge, MA: MIT Press.

Cybenco C., (1989). Approximations by Superimposition of Sigmoidal Function. Mathematics of Control, Signals and Systems., **2**. pp. 303-314, 1989.

Holland, J. H. (1975). Adaptation in Natural and Artificial Systems. Univ of Michigan Press.

Hornik K.K.M., Stinchombe M., White H., (1989). "Multilayer Feedforward Networks are Universal Approximators". Neural Networks, **2**, pp. 359-366.

Lin, C-T., Jou, C-P, and Lin, C-J. (1998). GA-Based Reinforcement Learning for Neural Networks. Int. J. of systems Sc. **29**, pp.233-247.

Khalil, H.K. (1996). Nonlinear Systems. Prentice Hall, Upper Saddle River, (NJ), 2-nd edition.

Mantawy, (1999). A. H., Abdel-Magid, Selim S.Z. "Integrating Genetic algorithms, Tabu search, and Simulated Annealing for the Unit Commitment Problem". IEEE Trans. on Power Syst. **14**, Aug.

Marino, R. and P. Tomei. (1992). Global Adaptive Observers for Nonlinear Systems via Filtered Transformations. IEEE Trans. Automat. Contr., **37**, pp.1239-1245, Aug.

Marino, R., and P. Tomei (1995). Nonlinear Control Design - Geometric, Adaptive and Robust. Prentice Hall, Hemel Hempstead.

Marino, R. Santosuosso, G. and P. Tomei (2001). Robust Adaptive Observers for Nonlinear Systems with Bounded Disturbances. IEEE Trans. Automat. Contr.,**46**, N.6, June.

Rovithakis, G. A. and M. A. Christodoulou (2000). Adaptive Control with Recurrent High Order Neural Networks Springer Verlag, Berlin.

Sastry, S. and M. Bodson (1989). Adaptive Control- Stability, Convergence and Robustness. Prentice Hall, Englewood Cliffs, NJ..

Pomet, J.B. and L. Praly (1992). Adaptive Nonlinear Regulation: Estimation from the Lyapunov Equation. IEEE Trans. Automat. Contr., **37**, pp.729-740, June.

Adaptive predictive fault tolerant control of super-heated steam temperature in an industrial boiler*

R. V. Dionísio[a,†] J. M. Lemos[b] P. O. Shirley[c]

[a,b,c]INESC-ID, R. Alves Redol 9, 1000-029 Lisboa, Portugal
[a]ESTSetúbal/IPS [b]IST [c]UA

Abstract: This paper describes a fault tolerant control TC) system consisting of a fault detection and isolation DI) scheme and a reconfigurable control (RC) strategy ving on the use of an adaptive predictive controller. The in goal is to allow that a certain plant continues to work n under the occurrence of actuator or sensor faults. e considered faults are additive and of single occurrence, responding to abrupt transitions. The FTC system is plied to a continuous non-linear physical based model super-heated steam temperature in an industrial boiler. nulation results are presented. *Copyright © 2001 IFAC*

Keywords: Adaptive predictive control; Discrete mod- Generalized predictive controller; Fault tolerant control; ılt detection and isolation; Parity space; Additive faults.

Introduction

C relies on the combined interplay between FDI and on- RC. More important than the integration of these two ks, which can be solved by a more or less complex logic cuit, are the robustness issue of the residual generation in presence of unknown plant disturbances and the control onfiguration strategy to choose. Unfortunately, despite ny efforts performed in the RC and in the FDI areas, se investments have not resulted in a great number of pa- s integrating both in real applications. This paper aims providing a contribution in this sense, in the line of [6].

The main idea is to use an adaptive predictive controller ether with an online RC strategy, able to compensate for ametric and additive faults either in the actuator or in the sor.

An adaptive predictive controller has the in-built advan- e of naturally reacting to parametric faults (adaptation), these are seen as plant dynamic modifiers, and the in-built antage of naturally fitting in a parity space method for idual generation (prediction) through the use of plant I/O dictors.

*Part of this work has been supported by the POSI program under IIIrd EC Framework Program.

†The work presented in this paper was prepared under a 3 years ching leave given by ESTSetúbal/IPS. Corresponding author: Tel.: 51.21.3100322. E-mail address: rvilela@est.ips.pt (R. V. Dionísio).

The FTC system is tested with a continuous non-linear physical based model of super-heated steam temperature in an industrial boiler (from now on referred as continuous plant model), which is modelled by a discrete linear time-invariant system (from now on referred as discrete plant model). A fully detailed description of the continuous plant model and its validation against real plant data can be found in [7].

2 Adaptive predictive control

2.1 Discrete plant model description

Consider a discrete SISO plant described by the following difference equation:

$$y(t) + a_1 y(t-1) + \dots + a_{n_a} y(t-n_a) = b_0 u(t-1) + \dots + b_{n_b} u(t-n_b-1) \quad (1)$$

with $a_{n_a}.b_{n_b} \neq 0$, $b_0 \neq 0$, $n_a > n_b$ and t being the discrete time. Define the $A^*(q^{-1})$ and $B^*(q^{-1})$ polynomials in the backward shift operator q^{-1}:

$$A^*(q^{-1}) := 1 + \sum_{i=1}^{n_a} a_i q^{-i} \quad B^*(q^{-1}) := \sum_{i=0}^{n_b} b_i q^{-i} \quad (2)$$

A^* and B^* being coprime. Define the (non-minimal) state vector $s(t)$:

$$s(t) := [y(t)\ y(t-1)\ \dots\ y(t-n_a+1)\ u(t-1)\ \dots\ u(t-n_b)]' \in \Re^{n_a+n_b} \quad (3)$$

Write (1) in the state space form:

$$s(t+1) = \Phi s(t) + \Gamma u(t) \quad y(t) = H s(t) \quad (4)$$

with:

$$\Phi_{(n_a+n_b)\times(n_a+n_b)} := \quad (5)$$
$$\begin{bmatrix} -a_1 \ -a_2 \ \dots \ -a_{n_a-1} \ -a_{n_a} \ b_1 \ b_2 \ \dots \ b_{n_b-1} \ b_{n_b} \\ I_{(n_a-1)\times(n_a-1)} \quad O_{(n_a-1)\times 1} \quad O_{(n_a-1)\times n_b} \\ O_{1\times(n_a+n_b)} \\ O_{(n_b-1)\times n_a} \quad I_{(n_b-1)\times(n_b-1)} \quad O_{(n_b-1)\times 1} \end{bmatrix}$$

$$\Gamma_{(n_a+n_b)\times 1} := [b_0\ 0\ \dots\ 0\ 1\ 0\ \dots\ 0]' \quad (6)$$

$$H_{1\times(n_a+n_b)} := [1\ 0\ \dots\ 0] \quad (7)$$

where the '1' in Γ is in the $n_a + 1th$ position.

Note that Φ is singular though the pair (Φ, Γ) is reachable and the pair (H, Φ) is reconstructible (polynomials A^* and B^* are coprime).

2.2 Multipredictive models

Rewrite (4) considering the existence of an exogenous signal $d(t)$ which models the unknown additive disturbances acting on the plant:

$$\begin{aligned} s(t+1) &= \Phi s(t) + \Gamma u(t) + Gd(t) \\ y(t) &= Hs(t) \end{aligned} \tag{8}$$

with $G_{(n_a+n_b)\times 1} := [1\ 0\ ...\ 0]'$.

Predictive control relies on the optimization of multistep costs defined over an extended time horizon T. The plant dynamics is described over this horizon by a set of predictive models obtained from (8). Forward iterating (8) up to time $t+i$, the following equation is obtained:

$$y(t+i) = \pi_i' s(t) + \sum_{j=0}^{i-1} w_{i-j-1} u(t+j) + \epsilon_i(t) \tag{9}$$

with:

$$w_j := H\Phi^{j-1}\Gamma \qquad \pi_j' := H\Phi^j \qquad ; j = 1, ..., i \tag{10}$$

$$\epsilon_i(t) = \sum_{j=0}^{i-1} \Phi^{i-j-1} Gd(t+j) \tag{11}$$

Assuming $\epsilon_i(t)$ to be orthogonal to the other terms of (9), the following multipredictive model holds:

$$Y_{t+1}^{t+T} \approx WU_t^{t+T-1} + \Pi' s(t) \tag{12}$$

with:

$$Y_{t+1}^{t+T} := \begin{bmatrix} y(t+1) \\ \vdots \\ y(t+T) \end{bmatrix} \quad U_t^{t+T-1} := \begin{bmatrix} u(t) \\ \vdots \\ u(t+T-1) \end{bmatrix} \tag{13}$$

$$W := \begin{bmatrix} w_1 & \cdots & \cdots & O \\ w_2 & w_1 & \cdots & O \\ \vdots & \vdots & \ddots & \vdots \\ w_T & \cdots & w_2 & w_1 \end{bmatrix} \qquad \Pi := [\pi_1 | \cdots | \pi_T] \tag{14}$$

2.3 GPC control law

The GPC is a receding-horizon LQ control law, minimizing the receding horizon multistep quadratic cost, [5]:

$$J_T = E\left\{ \frac{1}{T} \sum_{p=1}^{T} \left[\tilde{y}(t+p)^2 + \rho u^2(t+p-1) \right] | \mathcal{O}^t \right\} \tag{15}$$

where $\tilde{y}(t) = y(t) - y^*(t)$ is the tracking error, $E\{.|\mathcal{O}^t\}$ is the means conditioned on the observations up to time t, $\mathcal{O}^t$, and $y^*(t)$ is the reference to track, considering the mult predictive model (12). The parameter ρ represents a weig in the control effort, being a design variable.

The GPC family of algorithms for the regulation case defined by:

$$u(t) = F's(t) \qquad (1$$

with the vector of controller gains F given by:

$$F'_{1\times(n_a+n_b)} = -\bar{e}_1' M^{-1} W' \Pi' \qquad (1$$

with $\bar{e}_1' := [1\ 0\ ...\ 0]$ and $M := \rho I + W'W$.

The GPC is rendered adaptive through online paramet estimation of the predictive models. This allows the co troller to recalculate the gains in order to tackle changes plant dynamics and providing some ability to deal with par metric faults.

As recently proposed by [2] a directional forgetting recu sive least-squares algorithm based on the decomposition the information matrix (RLS-DFDIM) $R(t)$ is used to est mate the discrete plant parameters.

3 Fault tolerant control

A solution for estimation and online compensation of act ator and sensor additive faults, which is adequate for int gration with GPC for yielding a reconfigurable controller presented.

Consider the regulation scheme presented in Fig.1. Th GPC is initially designed to behave as a regulator for th tracking error integral $z(t)$ and considering $y^*(t) = 0$. The it is transformed in a controller for the variable $y(t)$ throug the introduction of $y^*(t) \neq 0$. This means that the GPC designed considering the plant in series with an integrato the new pseudo-state being:

$$\begin{aligned} \tilde{s}(t) := [z(t)\ z(t-1)\ ...\ z(t-n_a) \\ u(t-1)\ ...\ u(t-n_b-1)]' \in \Re^{n_a+n_b+2} \end{aligned} \qquad (1$$

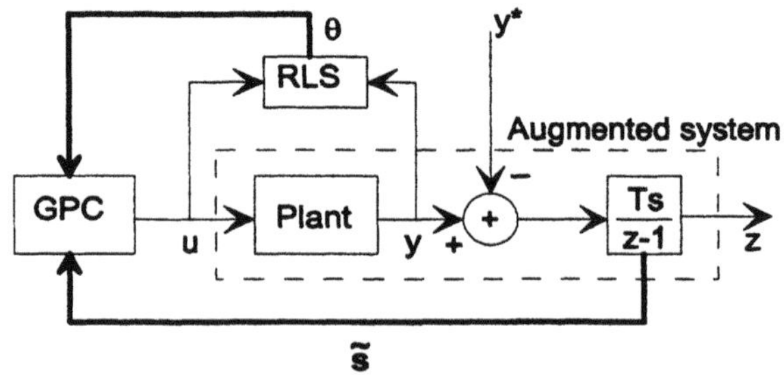

Figure 1: Regulation scheme for the continuous plant mod in series with an integrator (T_s is the sampling period).

.1 Actuator fault estimation

n actuator additive fault $f_a(t)$ can be described in the I/O presentation as:

$$A^*(q^{-1})y(t) = B^*(q^{-1})\left[u(t-1) + f_a(t-1)\right] \tag{19}$$

ccording to [6], but adopting a different plant description, t us start from (4) and consider the augmented state space cluding the tracking error integral, and a generic actuator ult:

$$\begin{bmatrix} s(t+1) \\ z(t+1) \end{bmatrix} = \begin{bmatrix} \Phi & O \\ T_s H & 1 \end{bmatrix} \begin{bmatrix} s(t) \\ z(t) \end{bmatrix} + \begin{bmatrix} \Gamma \\ O \end{bmatrix} u(t) + \begin{bmatrix} O \\ -T_s \end{bmatrix} y^*(t) + \begin{bmatrix} F^a \\ O \end{bmatrix} \xi_a(t) \tag{20}$$

$$y(t) = [H \quad O] \begin{bmatrix} s(t) \\ z(t) \end{bmatrix} \tag{21}$$

ith F^a the actuator fault matrix, and $\xi_a(t)$ the actuator ult time vector:

$$F^a = \begin{bmatrix} b_0 \; b_1 \; \dots \; b_{n_b-1} \; b_{n_b} \\ O_{(n_a+n_b-1)\times(n_b+1)} \end{bmatrix} \tag{22}$$

$$\xi_a(t) = [f_a(t) \; \dots \; f_a(t-n_a+n_b)]' \tag{23}$$

ewrite now (20) and (21) in the following way:

$$\bar{E}^a \bar{S}^a(t+1) = \bar{\Phi}^a \bar{S}^a(t) + \bar{\Gamma}^a \bar{U}(t) + \bar{G}^a y^*(t) \tag{24}$$

ith:

$$\bar{E}^a = \begin{bmatrix} I_{(n_a+n_b)\times(n_a+n_b)} & O_{(n_a+n_b)\times 1} & -F^a \\ O_{1\times(n_a+n_b)} & 1 & O_{1\times(n_b+1)} \\ H & O & O_{1\times(n_b+1)} \end{bmatrix} \tag{25}$$

$$\bar{\Phi}^a = \begin{bmatrix} \Phi & O_{(n_a+n_b)\times 1} & O_{(n_a+n_b)\times(n_b+1)} \\ T_s H & 1 & O_{1\times(n_b+1)} \\ O_{1\times(n_a+n_b)} & O & O_{1\times(n_b+1)} \end{bmatrix} \tag{26}$$

$$\bar{\Gamma}^a = \begin{bmatrix} \Gamma & O_{(n_a+n_b)\times 1} \\ O & O \\ O & 1 \end{bmatrix} \quad \bar{G}^a = \begin{bmatrix} O_{(n_a+n_b)\times 1} \\ -T_s \\ O \end{bmatrix} \tag{27}$$

$$\bar{S}^a(t) = \begin{bmatrix} s(t) \\ z(t) \\ \hat{\xi}_a(t-1) \end{bmatrix} \quad \bar{U}(t) = \begin{bmatrix} u(t) \\ y(t+1) \end{bmatrix} \tag{28}$$

general, $\bar{E}^a$ is a long matrix (more columns than rows; nder-specified problem). This means that to solve (24) in der to $\hat{\xi}_a(t)$ (estimated actuator fault time vector) one must ft multiply its second member by the right pseudo-inverse $\bar{E}^a$, $\bar{E}^{a+} = \bar{E}'^a(\bar{E}^a\bar{E}'^a)^{-1}$. Doing so, from all infinite sotions, choose the one with minimum euclidean norm. Only in the particular case of $n_b = 0$, $\bar{E}^a$ becomes square, and if it is non-singular, it is possible to compute the exact estimate of $\xi_a(t)$. Due to this particular state space construction the solution has the form:

$$\hat{\xi}_a(t) = [\hat{f}_a(t) \; \frac{b_1}{b_0}\hat{f}_a(t) \; \dots \; \frac{b_{n_b}}{b_0}\hat{f}_a(t)]' \tag{29}$$

with $\hat{f}_a(t) = f(t, s(t), u(t), y(t+1))$ the estimated actuator fault.

3.2 Actuator fault compensation

An actuator additive fault may be seen as a disturbance in the control action. In this sense, it is possible to compensate this loss by adding the signal $u_{ac}(t)$ to (16) and replacing $s(t)$ by $\tilde{s}(t)$:

$$u(t) = F'\tilde{s}(t) + u_{ac}(t) \tag{30}$$

Rewriting the first line of (20) for the system in closed loop and making use of (29):

$$s(t+1) = \Phi s(t) + \Gamma(F'\tilde{s}(t) + u_{ac}(t)) + F^a\hat{\xi}_a(t) \tag{31}$$

After some calculations it is found that, in order to cancel the fault term, it must be:

$$u_{ac}(t) = -\frac{\sum_{i=0}^{n_b}\left(\frac{b_i^2}{b_0}\right)}{B^*(q^{-1})}\hat{f}_a(t) \tag{32}$$

From this last equation it is concluded that if the identified discrete plant is non-minimum phase, then the FTC system will be unstable. One way to overcome this situation is by increasing the sampling time, [1], which is not always a good idea since one may lose control capabilities. Another way is to do only static compensation through the use of the static gain of (32). Furthermore, since $\hat{f}_a(t)$ depends on $y(t+1)$, $u_{ac}(t)$ should be delayed one sample period.

3.3 Sensor fault estimation

A sensor additive fault $f_s(t)$ can be described in an I/O representation as:

$$A^*(q^{-1})y(t) = B^*(q^{-1})u(t-1) + A^*(q^{-1})f_s(t) \tag{33}$$

Also in this case it is possible to obtain expressions similar to (20), (21) and (24) - (28), as it was done for the actuator fault estimation, with F^s the sensor fault matrix and $\xi_s(t)$ the sensor fault time vector:

$$F^s = \begin{bmatrix} 1 \; a_1 \; \dots \; a_{n_a-1} \; a_{n_a} \\ O_{(n_a+n_b-1)\times(n_a+1)} \end{bmatrix} \tag{34}$$

$$\xi_s(t) = [f_s(t+1) \; \dots \; f_s(t-n_a+1)]' \tag{35}$$

The discussion about the existence of the solution $\hat{\xi}_s(t)$ (estimated sensor fault time vector) is identical to the one of $\hat{\xi}_a(t)$. One remark: in order for $\bar{E}^s$ to be square one should have $n_a = 0$, which corresponds to a static system. Only an approximate estimate of $\xi_s(t)$ will be obtained. The solution has the form:

$$\hat{\xi}_s(t) = [\hat{f}_s(t+1)\ a_1\hat{f}_s(t+1)\ \ldots\ a_{n_a}\hat{f}_s(t+1)]' \tag{36}$$

with $\hat{f}_s(t) = f(t, s(t), u(t), y(t+1))$ the estimated sensor fault.

3.4 Sensor fault compensation

A sensor additive fault causes an over-reacted control action by the actuator, trying to compensate the sensor offset. To compensate this 'exaggerated' controller reaction one should subtract $u_{sc}(t)$ from (16). Proceeding as before for the actuator, yields:

$$u_{sc}(t) = -\frac{\left(1+\sum_{i=1}^{n_a} a_i^2\right)}{B^*(q^{-1})}\hat{f}_s(t+1) \tag{37}$$

Now, instead of acting directly on the control signal, and considering that the identified plant description is given by (1), the sensor signal may be compensated by adding to $y(t)$:

$$y_{sc}(t) = -q^{(-n_a+n_b+1)}\frac{\left(1+\sum_{i=1}^{n_a} a_i^2\right)}{A^*(q^{-1})}\hat{f}_s(t+1) \tag{38}$$

If the identified discrete plant is stable, then the FTC system will also be stable and it will always be possible to do dynamic compensation. Also here, since $\hat{f}_s(t)$ depends on $y(t+1)$, $u_{sc}(t)$ should be delayed one sample period.

4 Fault detection and isolation

FDI is the first level of FTC and can be described as based on three main tasks: (i) residual generation, (ii) residual testing and (iii) decision making. Related with the first task is the robust residual generation in the presence of unknown disturbances, noise and model errors. This means that one should be able to generate residuals sensitive to faults and as insensitive as possible to disturbances, noise and model errors.

4.1 Residual generation

Two scalar residuals (one for each type of fault) were generated using the parity space method [3, 4] (backward iterating the state space corresponding to the superposition of (19) and (33) added with $d(t)$ over a time window of size σ):

$$\begin{aligned} r_i(t) &= \omega_i'\left[Y_{t-\sigma}^t - KU_{t-\sigma}^t\right] = \\ &\omega_i'\left[Js(t-\sigma) + L^F P_{t-\sigma}^t + L^D D_{t-\sigma}^t\right] \end{aligned} \tag{39}$$

(using the same notation as in (13)), with ω the parity vecto and:

$$p(t) = \begin{bmatrix} \xi_a(t) \\ \xi_s(t) \end{bmatrix} \quad J_{(\sigma+1)\times(n_a+n_b)} = \begin{bmatrix} H \\ H\Phi \\ H\Phi^2 \\ \vdots \\ H\Phi^\sigma \end{bmatrix} \tag{40}$$

$$K_{(\sigma+1)\times(\sigma+1)} = \begin{bmatrix} O & O & \cdots & O & O \\ H\Gamma & O & \cdots & O & O \\ H\Phi\Gamma & H\Gamma & \cdots & O & O \\ \vdots & \vdots & \ddots & \vdots & \vdots \\ H\Phi^{\sigma-1}\Gamma & H\Phi^{\sigma-2}\Gamma & & H\Gamma & O \end{bmatrix} \tag{41}$$

$$L^F_{(\sigma+1)\times(\sigma+1)(n_a+n_b+2)} = \begin{bmatrix} O & O & \cdots & O & O \\ HE^F & O & \cdots & O & O \\ H\Phi E^F & HE^F & \cdots & O & O \\ \vdots & \vdots & \ddots & \vdots & \vdots \\ H\Phi^{\sigma-1}E^F & H\Phi^{\sigma-2}E^F & \cdots & HE^F & O \end{bmatrix} \tag{42}$$

$$L^D_{(\sigma+1)\times(\sigma+1)} = \begin{bmatrix} O & O & \cdots & O & O \\ HE^D & O & \cdots & O & O \\ H\Phi E^D & HE^D & \cdots & O & O \\ \vdots & \vdots & \ddots & \vdots & \vdots \\ H\Phi^{\sigma-1}E^D & H\Phi^{\sigma-2}E^D & \cdots & HE^D & O \end{bmatrix} \tag{43}$$

$$E^F = [F^a \quad F^s] \qquad E^D = G \tag{44}$$

Comparing (40) and (41) with (10) and (14), and if $\sigma = T$ it is easily verified that:

$$J = \begin{bmatrix} H \\ \Pi' \end{bmatrix} \qquad K = \begin{bmatrix} O & O \\ W & O \end{bmatrix} \tag{45}$$

Considering the state decoupling condition $\omega_i' J = O'$, an defining $z_{F_i}' := \omega_i' L^F$ and $z_{D_i}' := \omega_i' L^D$, the following equa tion must be solved:

$$\omega_i'\left[J\ L^F\ L^D\right] = \left[O'\ z_{F_i}'\ z_{D_i}'\right] \tag{46}$$

where z_{F_i}' and z_{D_i}' are design vectors. One should bear in mind that the z_{F_i}' used for the generation of the residua associated with the actuator fault is distinct from the on associated with the sensor fault, and that one also desire residual robustness against unknown disturbances ($z_{D_i}' \approx 0$)

4.2 Residual testing

A single observation of the two scalar residuals allows a sim ple detection strategy [4]. This consists in testing each ob servation of each scalar residual individually and declarin ult when the test fires. In order to perform the test, density function $f(r_i)$ for a single observation of a scalar dual $r_i(t)$ is needed. This is assumed to be:

$$f(r_i) = \frac{1}{\sqrt{(2\pi)}\sigma_{r_i}} \exp(-\frac{r_i^2}{2\sigma_{r_i}^2}) \quad (47)$$

re $\sigma_{r_i}^2$ is the variance of the residual. The test thresh- η is given by $\int_{-\eta}^{\eta} f(r_i)dr = 1 - \alpha$ (where α is the false m rate, a design parameter). The test to perform is: *"if $|r_i(t)| < \eta$, no fault ("0") is declared; if $|r_i(t)| \geqslant \eta$, fault ") is declared".*

Fault identification

two residuals are designed so that each one responds only one fault and both are as insensible as possible to itive disturbances, in order to obtain a set of structured duals [4], as pointed out by the following structure matrix:

	$f_a(t)$	$f_s(t)$	$d(t)$
$r_1(t)$	1	0	0
$r_2(t)$	0	1	0

Simulation results

e results presented in this section were obtained by simulation in a detailed physical based model [7].

Continuous plant description

e plant considered is the steam super-heating subsystem a boiler (Barreiro thermoelectric power plant of CPPE). .2 gives a simplified overview of the subsystem referring main variables. The steam coming from the boiler drum sses through the low-temperature super-heater (LTSH) d receives a spray water injection before passing through high-temperature super-heater (HTSH) to the steam collector. From the collector, the steam is extracted for use, her by the turbine or by industrial clients. The process iable to be controlled is $y = T_{vsato}$, the steam temperae at the output of the HTSH. The manipulated variable is $= C_{vgij}$, the command of the spray water valve, which inences the spray water flow, C_{gij}. The additive disturbance $d = C_{vsato}$, the steam flow requested by industrial clients. The overall plant dynamics is quite slow ($\tau \simeq 230s$) and anges due to the influence of the load imposed on the sysn, the super-heated steam flow, C_{vsato}, which affects the namic behavior between T_{vsati} and T_{vsato}.

2 Discrete plant model identification and validation

sampled model was identified in open loop with a discrete linear time-invariant SISO model, with 4 poles, 3 zes and a unit delay, at the equilibrium point defined by $C_{vgij}(0) = 30\%$ and $T_{vsato}(0) = 535.15^oC$. The sampling period was $T_s = 5s$, [8], and the input signal was a band-limited white noise with variance of 3.3% of $C_{vgij}(0)$, superimposed to the constant input $C_{vgij}(0)$. The RLS-DFDIM algorithm was running in parallel, with a forgetting factor $\lambda = 0.98$, estimating the plant model parameters. The identified discrete plant model is of non-minimum phase having a static incremental negative gain. With this set of parameters and a new collection of data, the discrete linear model was validated, Fig.3.

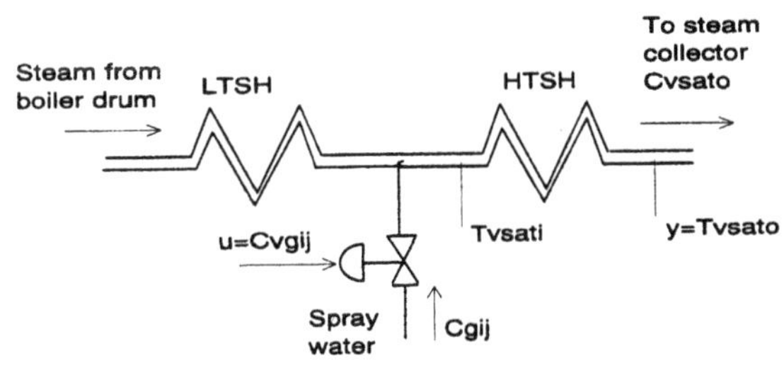

Figure 2: Plant schematic view.

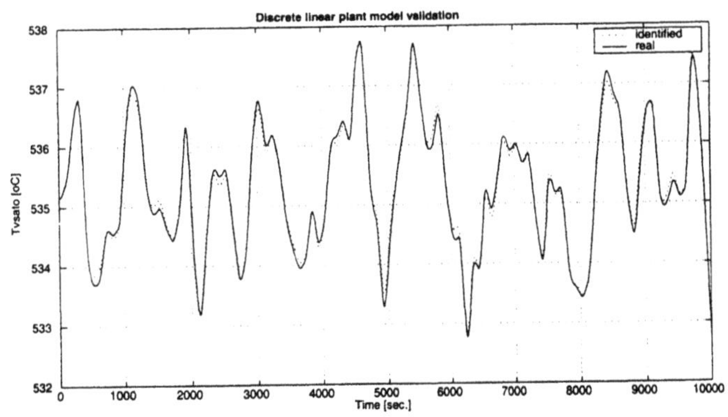

Figure 3: Continuous plant model T_{vsato} data (solid) vs. identified discrete plant model T_{vsato} data (dashed).

5.3 Integrated RC and FDI strategy

Actuator and sensor additive faults were generated (one at a time) in order to allow the comparison between the adaptive predictive controller alone (without fault compensation) and the adaptive predictive controller with fault compensation. The faults considered are of the type "abrupt" (step) and occurred at $t = 500s$. The GPC parameters were $T = 15$ and $\rho = 1$. Real field data was used during simulation as additive disturbance C_{vsato}. The reference $y^*(t)$ was filtered with a low-pass filter in order to avoid significant overshoots in the controlled variable T_{vsato} whenever the former incrementaly changes from -0.3^oC at $t = 2500s$ to 0.4^oC at $t = 4500s$.

Residuals were generated according to the structure matrix defined earlier. Residual $r_1(t)$ was designed to be insensitive to disturbance $d(t)$ and to fault $f_s(t)$ and residual $r_2(t)$ was designed to be insensitive to disturbance $d(t)$ and to fault $f_a(t)$. Although detection of both faults was achieved in 3 to 4 sample periods, isolation was not possible because both residuals performed in the same way. In part, this happens because when we try to solve (46), we can only find a solution in the least squares sense (the problem is over-specified) and almost never the state decoupling condition of the parity space is met, corrupting the residuals structure. Nevertheless, it was decided to present FTC results for actuator and sensor faults. Other residual generation approaches involving more sensor signals (e.g., the steam temperature after spray water injection and before the HTSH, T_{vsati}) are currently being investigated.

5.3.1 Actuator

The simulated fault has an amplitude of 25% (corresponding to an actuator loss of acting). From Fig.4 (a detail of the full simulation) one can see that the system with FTC does not degrade the undershoot and reduces significantly the overshoot of the controlled variable T_{vsato}, as well as the overall amplitude oscillation when compared with the system without FTC. Also, the settling time is much shorter for the system with FTC. The fault compensation also works well if new changes on the reference occur, and the system performance is not degraded.

5.3.2 Sensor

The simulated fault has an amplitude of -1^oC (corresponding to a sensor negative offset). From Fig.5 one can see that the system with FTC is able to compensate the sensor offset, although introducing some small oscillation during the transition response of the controlled variable T_{vsato}.

6 Conclusions

A fault tolerant control (FTC) system based on the integration of an adaptive predictive controller and a fault detection and isolation (FDI) algorithm has been presented. Both GPC and the parity space method were selected since they naturally fit. The overall algorithm has been demonstrated by simulations on a detailed physical based model of the superheated steam temperature in an industrial boiler.

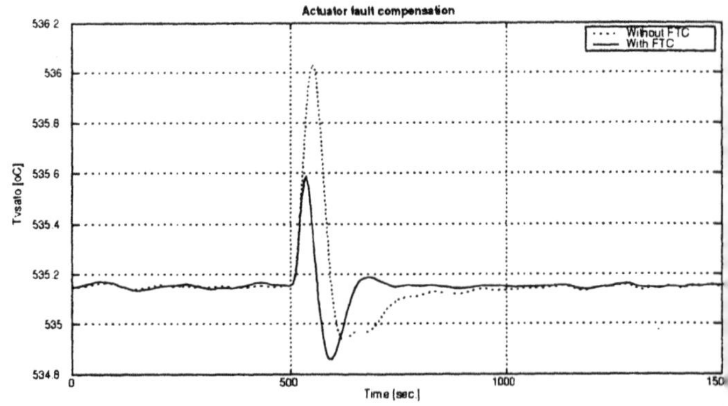

Figure 4: System output T_{vsato} without (dashed) and wit (solid) FTC. Fault occuring at $t = 500s$. Detail of the fu simulation.

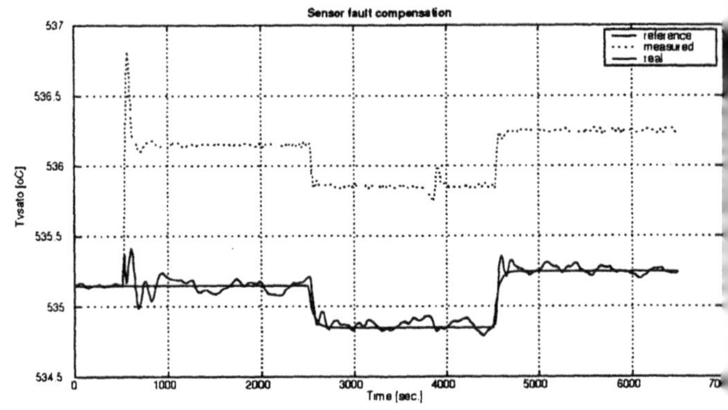

Figure 5: System output T_{vsato} measured (dashed) and re (solid) FTC. Fault occuring at $t = 500s$.

References

[1] Åström, K. J., Hagander, P., Sternby, J. (1980). Analytical Redundancy and the design of robust failure detection systems. *IEEE Transactions on automatic control*, AC-19, pp. 716-722.

[2] Cao, L., Schwartz, H. (2000). A directional forgetting algorithm based on the decomposition of the information matrix. *Automatica*, 36, 1725-1731.

[3] Chow, E. Y., Willsky, A. S. (1984). Analytical Redu dancy and the design of robust failure detection system *IEEE Transactions on automatic control*, vol. 29, no. pp. 603-614.

[4] Gertler, J. J. (1998). Fault detection and diagnosis in e gineering systems. Marcel Dekker.

[5] Mosca, E. (1995). Optimal, predictive, and adaptive co trol. Prentice Hall.

[6] Noura, H., Sauter, D., Hamelin, F., Theilliol, D. (200 Fault tolerant control in dynamic systems: application a winding machine. *IEEE Control Systems*, vol. 20, no. pp. 33-49.

[7] Silva, R. N., Shirley, P. O., Gonçalves, A. C., Lemos, J. (1998). Model of a power plant superheating system. *P ceedings Control 98*, Coimbra, Portugal, (pp. 471-476)

[8] Silva, R. N., Shirley, P. O., Lemos, J. M., Gonçalv A. C. (2000). Adaptive regulation of super-heated ste temperature: a case study in an industrial boiler. *Cont Engineering Practice*, 8, 1405-1415.

Copyright © IFAC Adaptation and Learning in Control and
Signal Processing, Cernobbio-Como, Italy, 2001

REALIZATION IN MATLAB-SIGLAB ENVIRONMENT OF A REAL TIME ESTIMATOR OF THE DAMPING OF THE ELECTROMECHANICAL MODE OF THE ALTERNATOR-NETWORK SYSTEM

Claudio Luigi Brasca - Aldo Danelli
CESI Spa
Via Rubattino,54 – 20134 Milano -ITALY
Email : claudio.brasca@cesi.it ---danelli@cesi.it

Keywords - Adaptive voltage control, Real-time Damping Estimators –Prototype of Identifiers

Abstract

The paper present the realization, in Matlab-Siglab environment ,of the prototype of a real-time estimator of the electromechanical mode using only the measure of active power, which is perturbed by the noise of the network and of the power plant.
This real-time estimator is essential to decide when it is necessary to activate the adaptation of the additional feedbacks gains to guarantee a predefined margin of stability

1 Introduction

The most critical problem in the voltage control of a turboalternator is that to guarantee an enough large stability margin to the electromechanical mode with the changing of the operation point during the normal conditions (active and reactive power, voltage).
This need led to the introduction of some additional feedbacks from the active power and from the speed.
In order to obtain a pre-fixed value for this damping is necessary to adapt in real-time the additional feedbacks gains. This adaptation is strictly necessary everytimes the damping becomes near to zero.

2 The low damping of electromechanical oscillations of the alternator-network system

2.1. The trend toward the instability of the electromechanical mode.

First of all it considered a turbo-alternator connected to the electrical transmission network, this is a non linear system which is in operation with an angular speed equal to the speed of the grid. In this work the linearised system around the operating point is taken into consideration, so the stability analysis for small variations is developed.
In brief the model of the alternator connected to the electrical transmission grid can be describe by the powers and kinetic equations of balance and by the electrical relationship between the generator electric power P_e and the electric angle δ .
The first equation gives the link between the angular speed of the alternator and the difference between the mechanical power absorbed by the alternator and the active electric power sent to the grid.

The second equation gives the difference between the angular speed of the alternator and the angular speed of the network in function of the load angle, that is the position of the rotor in respect of a reference with the angular speed of the network.

(1)

$$P_m - P_e = J\Omega \frac{d\Omega}{dt} \approx J\Omega_s \frac{d\Omega}{dt}$$

$$\Omega - \Omega_s = \frac{d\delta}{dt}$$

where :

Pm = mechanical power
Pe = electric power
J = Inertia of alternator-turbines
Ω = rotation speed of the rotor
Ω s = speed of synchronism of the rference
δ = angolar position of the rotor with respect to reference which have speed equal to Ω s

The link between the mechanical power P_m and the load angle δ is then of second degree, in addition there is a feedback between the two variables for non linear relationships which take into account the electromechanical equation of the alternator, the network and of the excitation control system, ect.
The schematic representation of this is in Fig.1.

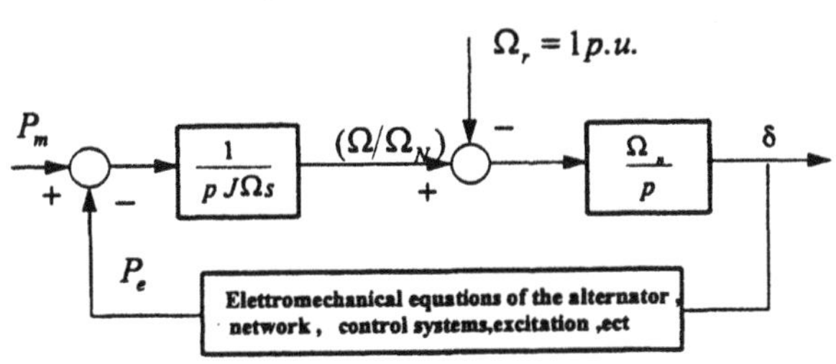

Fig.1 – The electromechanical loop

If the system is linearised, making some important simplifications:

a- only one generator is connected to the grid by a line with reactance x_e, constant ,
b- the grid is so strong that it is not influenced by the operating conditions of the generator. Its voltage V_r is constant in module and phase.

a- At the first step the effects of the Automatic Voltage Regulator (AVR) are not considered, this means the only contribute taken into account is that of the feedback from the electric angle and mechanical power, which is represented by the transfer function Gδp(s) (generator in manual operation)

The result of the linearisation is shown in Figure 2.

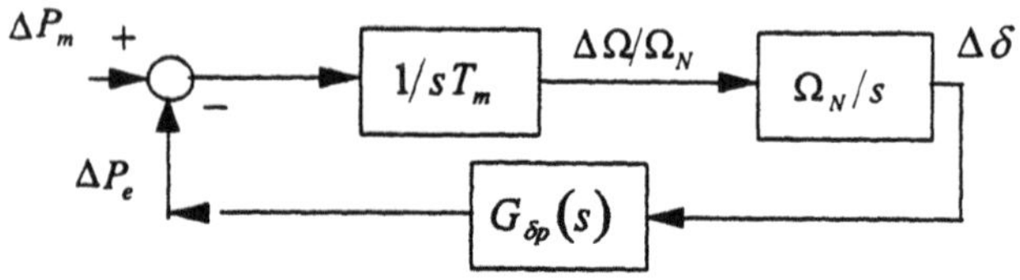

Fig.2 – The linearized electromechanical loop

In such a case the electromechanical loop has two imaginary poles, with angular speed :

$$\omega_0 = \sqrt{\frac{\Omega_N k}{T_m}} \qquad (2)$$

$$\text{where } k = \left.\frac{\partial P_e}{\partial \delta}\right|_{P_e = P_e^0} = \frac{e_i V_r}{x_i + x_e}\cos\delta$$

k is the coefficient obtained substituting the operational reactances of the alternator with their approximations at the synchronism speed.

With this situation should seem impossible to stabilise the system under control for its trend to oscillate around the equilibrium operating point. However in the reality the two poles are very close to the imaginary axe, but with a negative real part. So the damping is very low and so close to zero and ω_0 is still a good approximation of the natural frequency of oscillation.

The oscillation mode associated at this couple of complex pole is "more visible" on the electromechanical variables , as the load angle and the electric power , meanwhile it is less visible on the electro-magnetic variables , as the voltage and the magnetic fluxes.

For these consideration the oscillation associated with the this mode are called "electromechanical oscillation".

2.2 The additional stabilizing feedbacks

Till now it was been considered only the elettomechanical loop without the Automatic Voltage Regulator , this means without the feedback loop from the generator voltage V to the voltage reference node, see Figure 3. The voltage V is strictly dependant from the electric power P_e and from the load angle δ, the voltage loop is strictly correlated with the electromechanical one.

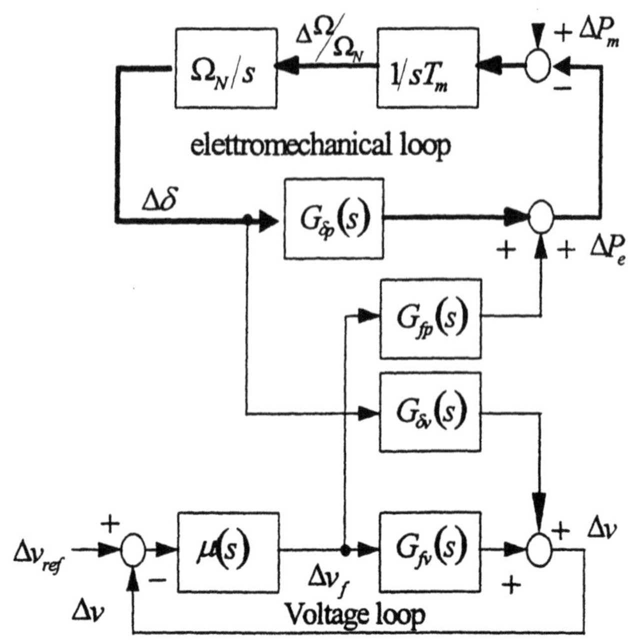

Fig.3 – The linearised electromechanical and voltage loops

It is possible to prove that the electromagnetic variables of the voltage loop are characterized by a set of poles : real or complex with a very high damping, and so they are less critical for the stability of the generator than the electromechanical ones. The influence of the voltage loop on the electromechanical one can be schematised by an additional block in parallel on the feedback loop, see Figure 4. The gain of this additional block is always positive, so it pushes the system toward the instability decreasing the stability margin of the controlled system.

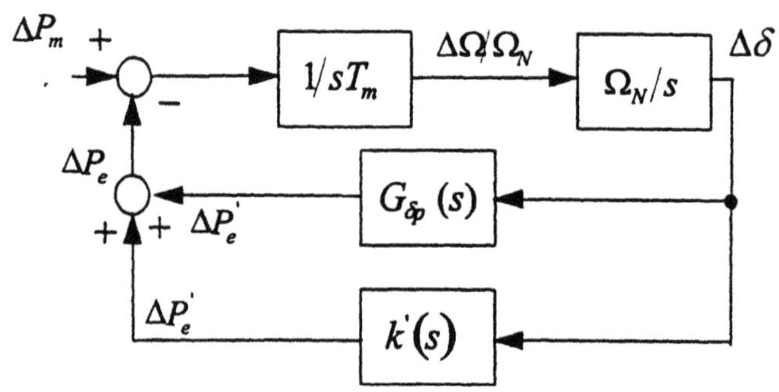

Figure 4 – Effect of the Automatic Voltage regulator on the electromechanical loop

In order to resolve this critical situation in the past were introduced some additional stabilising feedbacks, from the speed ω and from the electric power Pe, see the figure 5. The scope was not only to recover the destabilising effect of the AVR, but also to have an increasing in the damping of the system under control.

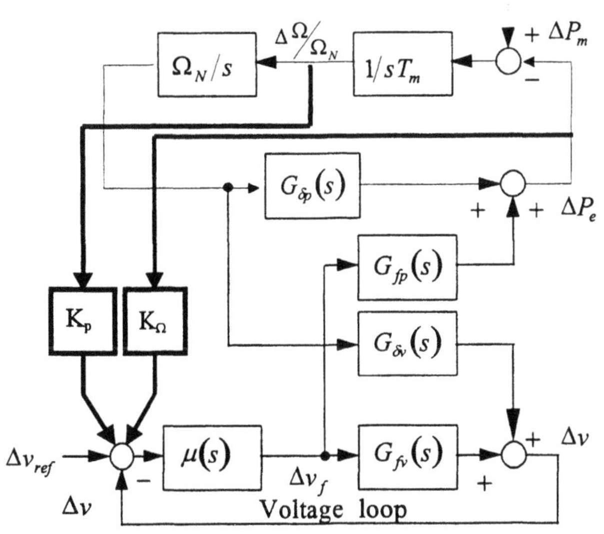

Figure 5 – The additional stabilising feedbacks

From a qualitative view point the additional feedbacks increase the correlation between the two loops under consideration, improving the stability of the electromechanical loop and decreasing the damping of the voltage one. This trade-off between the two loops is put in evidence in the figure 6 , where it is shown the movement of the two poles changing the values of the gains of the additional feedbacks.

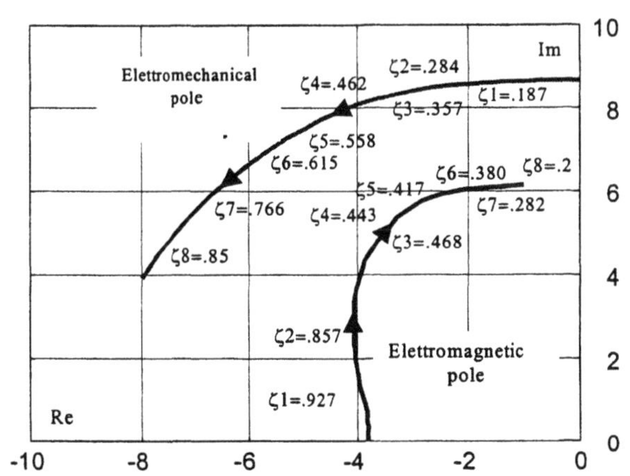

Figure 6 – Placement of electromechanical and electromagnetic poles changing the gains Kp and Kω

3 Adaptive Voltage Control of the generator

The most critical problem in the voltage control of a turboalternator is that to guarantee an enough large stability margin to the electromechanical mode with the changing of the operation point during the normal conditions (active and reactive power, voltage).
This need led to the introduction of some additional feedbacks from the active power and from the speed.
In order to obtain a pre-fixed value for this electromechanical damping is necessary to adapt in real-time the additional feedbacks gains. This adaptation is strictly necessary everytimes the damping becomes near to zero.
An electric power plant normally has N groups connected to the same bus-bur, it represents a dynamic system with one oscillatory mode in which the different generators oscillate together , with amplitudes almost equal, in respect to the external network and it has N-1 oscillatory modes in which the generators oscillate with opposite phase.
The need to identify the characteristics of the electromechanical mode of the whole power station toward the network lead to send the same perturbation signal to each Automatic Voltage Regulator of the groups and to have only one adaptive regulator for the power station. The need to have an adequate electromechanical damping without to have too negative consequences on dynamic performances of the voltage loop, suggests to fix a reasonable value for the damping ,for instance 0.2, to be kept in all the realistic operating points of the power station.This by acceptable values of the gains Kp and Kω of the additional feedbacks, to avoid unacceptable disturbance on the voltage.

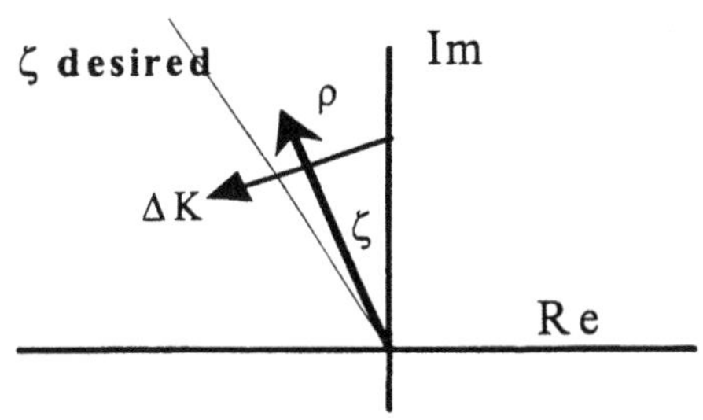

Fig. 7 – The shift of the electromechanical pole to have the desired damping

The optimisation criterion of the values of the additional feedbacks gains becomes that to find the gains which put the electromechanical pole on the semi-line of the complex plane which corresponds to the desired damping.
Nowdays the two gains are tuned on the basis of history data and once defined and chosen are fixed parameters , which seldom are modified. Eventually the tuning can be made taking into account the different amount of the power
required in the various period of the year, different hour of the day.

4 Estimators of the electromechanical mode

4.1 Traditional estimators

The estimation of the electromechanical mode, that is of its natural frequency ω_o and of the damping factor ζ, is constituted by the problem to estimate the poles of a second order looped system
On the basis of the knowledge of the electromechanical mode and of the residues is possible to adapt the gains

Kp and Kω of the additional feedbacks to kept the damping factor ζ, inside an acceptable range around the desired value ζd. This approach is shown in Brasca C.L.,1993.The traditional methods to make this estimation are the following two:

- The RLS (Recursive Least Squares) injects a White Noise on the voltage reference of the generators . From the electric power it is possible to estimate the electromechanical mode and the residues on the power with an analysis in the time domain , see reference Brasca 1994;

- The response to a step on the voltage reference gives the opportunity to estimate ω and the damping factor ζ on the basis of the over-oscillations of the electric power in the time domain

The main disadvantage of these two methods is that both need a perturbation on the voltage reference and so a consequent disturbance on the voltage and on the electric power. This fact makes these methods not very pleasant for the power plant. the network voltage (for changes in the loads, changes on absorption, etc).However in the alternator-network system is naturally present a background noise, due to the sum of noise of mechanical nature and that due to the noise on the voltage of the network.
Using this noise naturally present on the plant is possible to make an estimation of the electromechanical mode making an analysis in frequency of the electric power measured at the bus-bur.

4.2 Description of an estimator based on the natural noise of the power plant

This paragraph describes the algorithm used to estimate the damping factor and the natural speed of the electromechanical poles without any additional perturbation injected on the power plant.

Some experimental considerations lead to put in evidence that the mechanical noise influences only the frequencies under 1 Hz. The noise due to voltage network can be represented by a white noise sent in input to a LP-low pass filter of the first order and it influences the frequencies till 5 Hz (($T_{Vr} \approx 0{,}03$ s):

$$v_r = G_{v_r}(s)\,e_{v_r} = \frac{\mu_{v_r}}{\left(1 + sT_{v_r}\right)} e_{v_r} \qquad (3)\text{-}(4)$$

where $e_{v_r} \approx WN(0,1)$;

The electromechanical poles are excited only by the voltage network noise : substituting typical data in the equations which determine the poles it is possible to observe that their frequency is between 1 and 2 Hz.

$$\omega_0 = \sqrt{\frac{\Omega_N k}{T_m}}$$

Tm = Time constant of generator to start = 7.5 s

$$k = \left.\frac{\partial P_e}{\partial \delta}\right|_0 = \left.\frac{\partial}{\partial \delta}\left(\frac{e_i V_r}{x} \sin\delta\right)\right|_0 = \left.\frac{e_i V_r}{x}\cos\delta\right|_0 \approx 2$$

essendo :

$$e_i \approx V_r \approx 1, \delta \approx 0$$

$x = x_i + x_e \approx 0.5$ p.u. internal reactance of the generator and external reactance

Substituting

$$\omega_0 = \sqrt{\frac{\Omega_N k}{T_m}} = \sqrt{\frac{2\pi\, 50\; 2}{7.5}} \cong 10\,[\text{rad/s}]$$

$$f_0 \approx 1.6\,\text{Hz}$$

(5)-(10)

Now it is necessary to consider deeper the block diagram of the transfer function Gvr(s) between the voltage network noise ΔVr and the electric power ΔPe.
From the alternator and control system equations it is possible to prove that it is approximated by an equation of third order , and in particular :

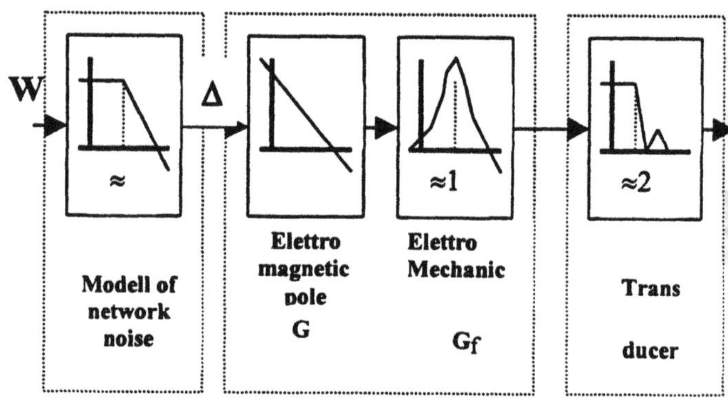

Simplified aternator model

Figure 8– Contributes of the different parts to the spectrum of the elettric power Pe

$$G_{Vr} = G_e(s)\; G_f(s) \qquad (11)$$

$$G_e(s) = \sqrt{k}\,\frac{\dfrac{s}{\Omega_N}}{\left(\dfrac{s}{\Omega_N}\right)^2 + 2\zeta\dfrac{s}{\Omega_N} + 1} \qquad (12)$$

$$G_f(s) = s\,\frac{1 - sT_1}{1 + sT_2} \qquad (13)$$

Where Ge(s) e Gf(s) are the contributions of the electromechanical loop and of the electromagnetic loop. This equation is obtained through some zero-pole "almost cancellations " at frequencies very close, which not affect the model at the frequencies of interest.The electromagnetic mode can be considered known, because T1 and T2 can be computed from the standard data and from the identification tests normally given by the alternator constructor. The good correspondence between the model and the real system is put in evidence in figure 9, which shows the power spectrum of the electric power computed from data collected in a power plant: it is very clear the peak due to the electromechanical mode.

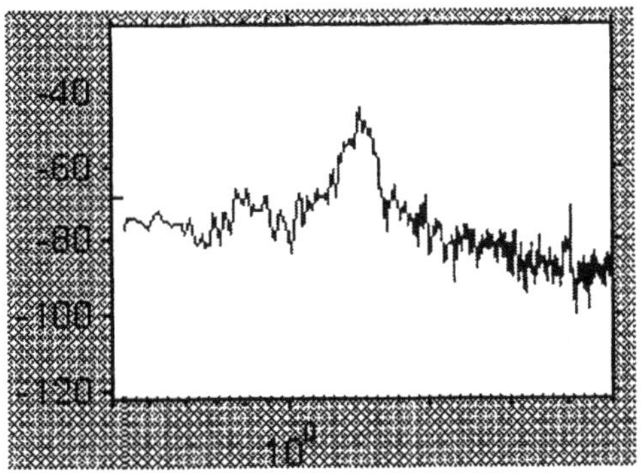

Figure 9 – Power spectrum of the electric power computed from data collected in Turbigo power plant

4.3 The estimation algorithm

The estimation algorithm is an estimator of the first order. It uses the power spectrum estimated , first of all to estimate the natural frequency of the electromechanical mode as the argument of the point of maximum between the two frequencies ω ω_2.
The gain is estimate by linear regression in the bi-logarithmic plane where the asintote is a stright line which decrease of two order of magnitude every decade in frequency. Using the weighting function:

$$w_i = \frac{\log\left(1 + \frac{\Delta\omega}{\omega_i}\right)}{\sum \log\left(1 + \frac{\Delta\omega}{\omega_i}\right)} \qquad (14)$$

where $\Delta\omega$ is the piece of the spectrum which is approximated by the stright line between $\omega 2$ and $\omega 3$, it is possible ,using the bi-logarithmic plane and calling A the stright line which approximate the asintote , to compute the cost function to be minimized:

$$\min J(k) = \sum w_i\left(\log S_e(\omega_i) - \log A(\omega_i)\right)^2 =$$
$$= \sum w_i\left(\log S_e(\omega_i) - \log k + 2\log\left(\frac{\omega_i}{\omega}\right)\right)^2$$

from which : (15)

$$\frac{\partial J}{\partial(\log k)} = 0 \Rightarrow \log\hat{k} = \mathbf{W}^T\mathbf{S}_e + 2(\mathbf{W}^T\mathbf{F} - \log\omega_n) \qquad (16)$$

$$\mathbf{W} = \frac{\mathbf{P}}{\|\mathbf{P}\|_1}$$
$$\mathbf{P} = \begin{bmatrix} \log\left(1 + \frac{\Delta\omega}{\omega_1}\right) \\ \vdots \\ \log\left(1 + \frac{\Delta\omega}{\omega_m}\right) \end{bmatrix} \qquad (17)$$
$$\mathbf{S}_e = \begin{bmatrix} \log S_e(\omega_1) \\ \vdots \\ \log S_e(\omega_m) \end{bmatrix}$$
$$\mathbf{F} = \begin{bmatrix} \log \omega_1 \\ \vdots \\ \log \omega_m \end{bmatrix}$$

Once the value of the gain and of the natural frequency it can be found the estimation of the damping :

$$\hat{\zeta} = \frac{1}{2}\sqrt{\frac{A(\hat{\omega}_n)}{S_e(\hat{\omega}_n)}} = \frac{1}{2}\sqrt{\frac{\hat{k}}{S_e(\hat{\omega}_n)}} \qquad (18)$$

A correction parameter k to take into account particular situations has been introduced.

5 Implementation in matlab siglab environment of the real-time estimation algorithm

5.1 The choice of the DSP of SigLab

In the previous paragraphs has been introduced an algorithm able to put in evidence the trend of the electromechanical mode toward the instability, based on the use of the natural network noise . The first implementation of this estimation algorithm was made in Matlab and it was off-line. The choice to use the DSP of Siglab in connection with Matlab was due to the need to have,with reduced time of development,a prototype of the real-time version of the algorithm to verify its behaviour with the real time simulator of the alternator-network system and after to the power plant. Using the Matlab-Siglab environment to implement the real-time estimator of the damping electromechanical mode it has been obtained a customised virtual instrument.
The tests of this real time estimator were been made using the digital real-time simulator STAR of the alternator-network system, see reference Tagliabue,, and the real signal registered on the field in different power stations(Turbigo, Edolo, Piacenza, etc,).
The on-line estimator must plays different tasks :

1. Acquisition of the electric power signal Pe from the power plant
2. Transformation by FFT of the data to obtain the power spectrum Sp of the electric power Pe
3. Prefiltering of the signal in frequency domain, in order to eliminate the effect of the electromagnetic mode, so putting in evidence only the electromechanical mode, which is that of interest
4. Estimation of the peak and of the right asintote of the spectrum

5. Estimation of the frequency Ωn and of the damping factor ζ of the mode
6. Filtering of the damping factor values to reduce the variance .

The next step was the implementation of all these functions. The DSP Siglab can operate only together with the PC , at which gives the monitoring functions.

One of the main feature of Siglab is that to have "Virtual Instruments ", these are SW which give easily the most important acquisition and computing functions.

It was been necessary to create a customized virtual instrument because it is essential to make acquisition, elaboration, filtering,etc.

So it was developed a virtual instrument in which meanwhile Siglab makes acquisition and after the FFT of the electric power signal the PC in Matlab environment makes the pre-filtering of the spectrum, implements the algorithm to estimate the damping factor , and finally make the filtering at the output. Figure 10 gives a schematisation of these tasks.

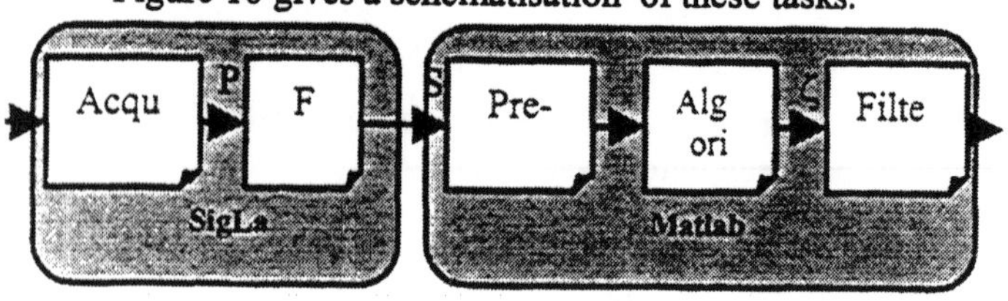

Figure 10 – Operation of the real-time estimator

In this case the acquisition speed is not critical because the transducer give new data every 20ms, so with a frequency of 50 Hz. The frequencies of acquisition of Siglab are prefixed, this does not give any problem to the implementation of the estimation algorithm.

Siglab makes a samplig at 51.2Hz of the input signal, after an analogue anti aliasing filtering .After this treatment the signal is filtered by a digital low pass filter ,which eliminates the frequencies higher 20Hz, finally the signal is reduced to have the sampling frequency of 51.2 Hz. On the basis of all the previous considerations it is possible to conclude that the spectrum of the electric signal is not significantlymodified in the frequencies under 20Hz, which are those of interest for us .

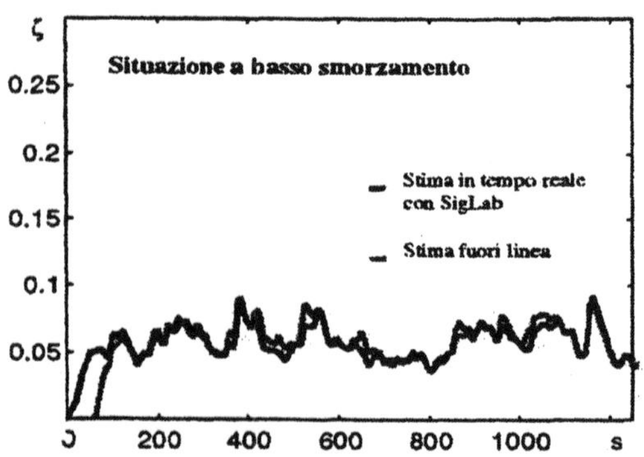

Figure 11 – Comparison between the estimation of damping off-line and on-line

The time to compute the spectrum of the electric power is critical. Some experimental considerations had put in evidence that a good estimation needs at least 1024 samples, that is acquisition time equal to 20s. Having 20s available for all the counts necessary to compute the spectrum the implementation uses Siglab for the acquisition and for the FFT leaving all the other tasks to the PC in Matlab.

To verify the good behaviour of the estimator two type of tests were been done.

-one makes a comparison between the estimation of the damping obtained off-line and that made on-line, an example is in Figure 11.These tests were very good.

-another type of tests use the digital real time simulator of the alternator-network system to check if the online estimator is able to follow changes in the real damping. Figure 12 gives one case, as in this one all the other tests give good results.

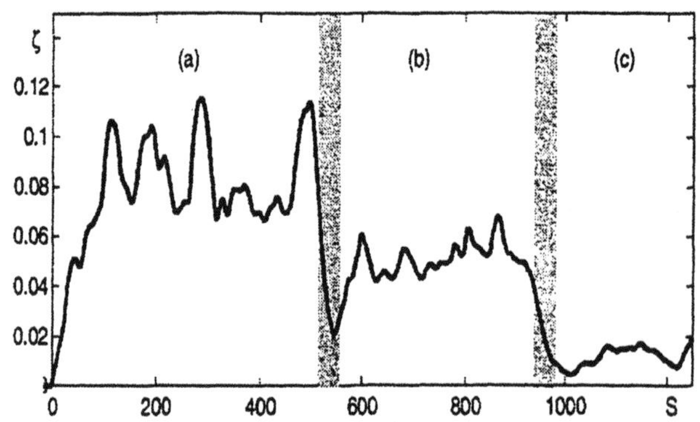

punto di lavoro	t scalino s	P p.u.	Q p.u.	Xr p.u.	ζ	ζ scalino
(a)	900	0.9	0.375	0.3	0.083	0.085
(b)	550	0.9	0	0.3	0.058	0.045
(c)	0	0.9	-0.35	0.3	0.018	0.012

Figure 12 – Estimated damping factor moving from under-excitation to the over-excitation

6 References

Brasca, C.; Arcidiacono, V.; Corsi, S.: An Adaptive Excitation Controller for Synchronous Generators: studies and experimental results at a power station, *2 nd IEEE Conf. on Control Appl.*, Vancouver, B.C., pp. 305--318, 1993.

Brasca, C.L. Grimble, M.J., 1998, *Limited authority self-tuning control and applications*, IEEE CCA Trieste, Italy.

Grimble, M.J., M.A. Johnson,V. Arcidiacono and C.L. Brasca, 1999, *Self –Tuning control and industrial application :the Limited authority control concept* , International Journal Of Adaptive Control and Signals Processing, Vol 13, Pp521-536 .

Savaresi S.M., G.O. Guardabassi, C. Brasca, S. Corsi, and M. Pozzi (1997). Improving the Dynamics of a Synchronous Generator by Nonlinear Stabilizers: a Virtual Reference Direct Design Approach. Proceedings of the *6th IEEE Conference on Control Applications*, Hartford, CT, pp.571-575.

ADAPTIVE TRACKING CONTROL OF NONLINEAR POWER CONVERTERS

Enric Fossas Colet *Alan S. I. Zinober **

* *Institut d'Organització i Control*
Universitat Politecnica de Catalunya
Avda. Diagonal 647, 11a. Pl.
08028 Barcelona
Spain
E-mail: fossas@ioc.upc.es
** *Department of Applied Mathematics*
The University of Sheffield
Sheffield S10 2TN
UK
E-mail: a.zinober@sheffield.ac.uk

Abstract: Tracking control of basic nonlinear power converters is studied for systems with *time-varying* parameters. The switching control laws accomplish indirect asymptotic sinusoidal voltage tracking. The exposition emphasizes the use of Fourier expansions in solving an ordinary differential equation of the Abel type. In this paper adaptation schemes are studied. *Copyright © 2001 IFAC*

Keywords: Adaptive control systems, nonlinear control, converters, parameter estimation, discontinuous control.

1. INTRODUCTION

Switched mode power converters lie at the heart of DC power supplies, bringing the advantages of high efficiency and low mass. The converters can be represented mathematically as nonlinear time-varying dynamical systems. They can be modelled as variable structure systems because of the abrupt topological changes that the circuit, commanded by a discontinuous control action, undergoes. They constitute a natural field of application of nonlinear control techniques and (Venkataramanan *et al.*, 1985) and (Sira-Ramírez, 1987) have applied sliding control techniques to the regulation problem in basic dc-to-dc power converters. Other important contributions include (Karsenti *et al.*, 1995), (Fossas-Colet and Biel, 1996), (Karsenti and Rouchon, 1997) and (Rios-Bolívar *et al.*, 1998).

Tracking problems in nonlinear converters are studied in different practical situations; for instance, uninterruptable power systems (UPS). The main nonlinear converters considered in these cases are the buck-boost and the Čuk converters. We describe here the tracking control of a boost and a buck-boost converter using sliding mode control techniques. The boost converter has been considered for simplicity and as an academic example; the buck-boost converter can be considered as a practical example.

Direct regulation/tracking control of an output voltage for the boost and buck-boost converters results in a non-minimum phase system and therefore an unstable controller. However, as Sira-Ramírez shown in (Sira-Ramírez, 1987), both converters can be indirectly controlled in regulation/tracking tasks. In previous studies the knowledge of the parameters of the plant has been assumed. Here we consider the case of time-varying unknown parameters. Fundamental to the tracking of a periodic function is an ordinary differential equation of Abel type, relating the state variables and the Fourier approximation of the periodic solution.

The paper is organised as follows: Section 2 is devoted to the tracking problem for basic nonlinear converters. In Section 3 the problem is defined. Sections 4 and 5 present solutions providing two estimates and a procedure to optimise the state transients. Conclusions are summarised in Section 6. The MATLAB m-files used are listed in the Appendix.

2. THE TRACKING PROBLEM IN 2D NON-LINEAR CONVERTERS

Basic 2D nonlinear converters can be modelled as bilinear systems. The usual state variables are the current through the inductor and the voltage in the capacitor. The problem considered here deals with the voltage tracking the periodic signal $A + B\sin(\omega t)$. The plant can be described by the dimensionless system

$$\frac{dx_1}{dt} = (u-1)x_2 + 1 \qquad (1)$$

$$\frac{dx_2}{dt} = (1-u)x_1 - \gamma(x_2 - \delta) \qquad (2)$$

where $\gamma \in [0,1]$ is a piecewise constant parameter and $\delta \in \{0,1\}$ is $\delta = 0$ for the boost converter and $\delta = 1$ for the buck-boost. The dimensionless variables are related by

$$x_1 = \frac{1}{V_g}\sqrt{\frac{L}{C}}i_L$$

$$x_2 = \begin{cases} \frac{1}{V_g}v_C & \text{boost converter} \\ 1 - \frac{1}{V_g}v_C & \text{buck-boost converter} \end{cases}$$

$$t = \frac{\tau}{\sqrt{LC}}$$

$$\gamma = \frac{1}{R}\sqrt{\frac{L}{C}}$$

where L is the inductance, C the capacitance and R the load. V_g is the input voltage, i_L the current through the inductor and v_C the voltage across the capacitor.

From (1) and (2) one gets

$$x_1\left(\frac{dx_1}{dt} - 1\right) = -x_2\left(\frac{dx_2}{dt} - \gamma(x_2 - \delta)\right) \quad (3)$$

a differential equation that can be rewritten as an Abel ODE. This equation is unstable if it is considered as an ODE in the variable x_1, and stable if it is considered as an ODE in the variable x_2. This is so because the plant is minimum phase when the variable x_1 is considered as output, and non-minimum phase when the variable x_2 is considered. However, in the latter case, it can be proved that if x_2 is periodic, then there exists a bounded x_1-solution of equation (3) which can be analytically approximated by a truncated Fourier series $a_0 + b_1\sin(\omega t + \phi)$, and numerically computed by reverse-time integration.

Proposition 1. (Olm-Miras and Fossas-Colet, 1999), (Zinober *et al.*, 1998) Let

$$k(t) = \alpha_0 + \sum_{l=1}^{\infty}(\alpha_l\cos(l\omega t) + \beta_l\sin(l\omega t)) \quad (4)$$

be a solution of (3) and

$$A_0 + \sum_{l=1}^{\infty}(A_l\cos(l\omega t) + B_l\sin(l\omega t))$$

the Fourier expansion of the right hand-side of equation (3)

$$x_{2d}\left(\frac{dx_{2d}}{dt} + \gamma(x_{2d} - \delta)\right)$$

Then

$$A_0 = \alpha_0$$

$$A_1 = \alpha_1 - \alpha_0\omega\beta_1 + \frac{\omega}{2}\sum_{n=1}^{\infty}(\beta_n\alpha_{n+1} - \alpha_n\beta_{n+1})$$

$$B_1 = \beta_1 + \alpha_0\omega\alpha_1 + \frac{\omega}{2}\sum_{n=1}^{\infty}(\alpha_n\alpha_{n+1} + \beta_n\beta_{n+1})$$

$$A_l = \alpha_l - l\alpha_0\omega\beta_l - \frac{\omega}{2}\sum_{n=1}^{l-1}n(\alpha_n\beta_{l-n} + \beta_n\alpha_{l-n}) + \frac{l\omega}{2}\sum_{n=1}^{\infty}(\beta_n\alpha_{n+l} - \alpha_n\beta_{n+l})$$

$$B_l = \beta_l + l\alpha_0\omega\alpha_l - \frac{\omega}{2}\sum_{n=1}^{l-1}n(\beta_n\beta_{l-n} - \alpha_n\alpha_{l-n}) + \frac{l\omega}{2}\sum_{n=1}^{\infty}(\alpha_n\alpha_{n+l} + \beta_n\beta_{n+l})$$

■

Sliding mode control based on the time-dependent sliding surface $S(x,t) := x_1 - k(t) = 0$, can now be designed. Using standard techniques regarding the stability of periodic orbits, it can be proved that the steady-state behaviour is given by $x_2 = x_{2d}$, i.e. the error $|x_2 - x_{2d}|$ in the ideal sliding dynamics tends asymptotically to zero.

Proposition 2. Let x_{2d} be a periodic signal to be tracked and $k(t)$ the periodic solution of the Abel equation (3). Assume

$$-1 \leq \frac{1}{x_{2d}}\left(\frac{dk(t)}{dt} - 1\right) \leq 0 \qquad (5)$$

then the control policy defined by the surface

$$S(x,t) := x_1 - k(t) = 0 \qquad (6)$$

and the control law

$$u(x,t) = \begin{cases} 0 & \text{if } x_2 S(x,t) > 0 \\ 1 & \text{if } x_2 S(x,t) < 0 \end{cases} \qquad (7)$$

provides us with sliding motion on the surface $S(x,t) = 0$, and the absolute error $|x_2 - x_{2d}|$ in the ideal sliding dynamics tends asymptotically to zero.

Proof:

$$\begin{aligned}\frac{d}{dt}\left(S^2(x,t)/2\right) &= (x_1(t) - k(t))(u(t) - u_{eq}(t))x_2(t) \\ &= S(x,t)(u(t) - u_{eq}(t))x_2(t)\end{aligned}$$

Moreover, equation (5) is equivalent to $0 \leq u_{eq} \leq 1$, hence the control policy defined in (6) and (7) makes $0.5S^2(x,t)$ a Lyapunov function. ■

Remark 3. Inequalities (5) for x_{2d} and $k(t)$ are the conditions to be satisfied in order for sliding motion to exist.

In practice, we use the sliding surface $S(x,t) := x_1 - k_a(t) = 0$, where $k_a(t)$ is the truncated approximating Fourier series

$$k_a(t) = \alpha_0 + \sum_{l=1}^{m} (\beta_l \sin(l\omega t + \phi_l))$$

Following (Sira-Ramírez, 1989),

$$\begin{aligned}u = \max\left(0, \min\left(1, 1 + \frac{1}{x_{2d}}\left(\frac{dk_a(t)}{dt} - 1\right)\right.\right. \\ \left.\left. + \mu * \text{sgn}(S(x,t))\right)\right)\end{aligned}$$

is considered instead of the bang-bang control in simulations. The system is sampled np times per period and a zero-order hold is applied to the control u. Then, the system is linear in each sampling period.

Example 4. For $x_{2d} = A + B\sin(\omega t)$, the periodic solution $k(t)$ of (3) can be approximated by a first harmonic Fourier expansion $k_a(t) = \alpha_0 + \beta_1(\sin(\omega t + \phi_1)$ where

$$\alpha_0 = \frac{1}{2}\left(2A^2 - 2A\delta + B^2\right)\gamma \qquad (8)$$

$$\beta_1 = B\sqrt{\frac{\omega^2 A^2 + \gamma^2(2A - \delta)^2}{\alpha_0^2\omega_0 + 1}} \qquad (9)$$

$$\tan(\phi_1) = \frac{(1 + 2\gamma\alpha_0)\omega A - \delta\alpha_0\gamma\omega}{A(2\gamma - \alpha_0\omega^2) - \delta\gamma} \qquad (10)$$

On the other hand, system (1)-(2) is static feedback linearizable, or flat. The flat output is

$$y = x_1^2 + x_2^2$$

So, trajectory-planning or tracking problems can be solved presuming the parameter γ is known since it appears in the diffeomorphism relating the state variables (x_1, x_2) and the linearized new variables $(y_1, dy_1/dt)$. It is worth to note that parameter γ models the load which, in turn, can suffer perturbations.

3. PROBLEM STATEMENT

Let $X_2(t)$ be the target periodic x_2 reference and $X_1(t)$ the corresponding x_1 reference derived from equations (8)-(10). Suppose that the actual parameter γ is a step function taking values γ_1 and γ_2. We need to answer:

(1) How does one estimate γ?
(2) How can one modify $X_1(t)$ in order to get a minimum x_2-error in the transient state?

4. ESTIMATION OF γ

Assuming that γ is constant and the steady-state is reached, the FFT applied to $x_2(t)$ yields $x_2(t) \simeq \hat{A} + \hat{B}\sin(\omega t + \hat{\psi})$, and γ can be derived by comparing these coefficients with the nominal values $(A, B, 0)$. Namely,

$$\hat{\gamma} = \frac{A^2 - A\delta + B^2/2}{\hat{A}^2 - \hat{A}\delta + \hat{B}^2/2}\gamma_{nom} \qquad (11)$$

with γ_{nom} the nominal (or expected) value of γ.

One should note that on-line determination of a_0, b_1 and ϕ_1 may produce oscillations in the system transients that can lead to instability. In the case of a change in the load with the initial values of a_0, b_1 and ϕ_1 retained, the FFT procedure detects the changes with great accuracy. Thus, one can proceed with the computation of new values for a_0, b_1 and ϕ_1 if $|\hat{\gamma} - \gamma_{nom}| > \tau_0$, for a certain threshold τ_0.

By integrating equation (2) between consecutive samples we have

$$(x_{f_2} - x_{in_2}) = (1-u)\int_0^s x_1(t)dt + \gamma\int_0^s (x_2(t) - \delta)dt$$

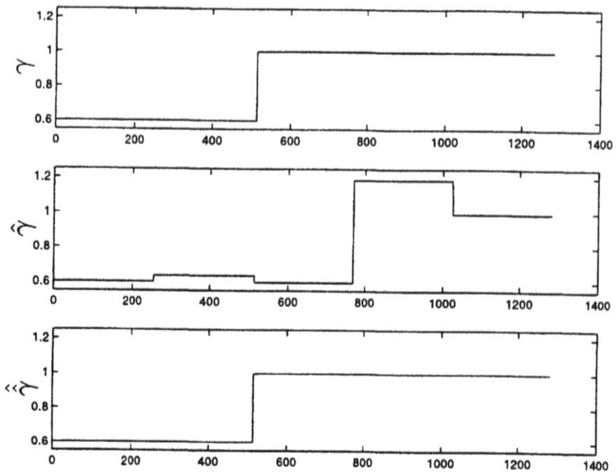

Fig. 1. Load variation $\gamma(t)$ and estimates $\hat{\gamma}$ and $\hat{\hat{\gamma}}$

and, by numerical approximation of the integrals

$$(x_{f_2} - x_{in_2}) = (1-u)s\frac{x_{f_1} + x_{in_1}}{2} + \\ + \gamma s \frac{x_{f_2} + x_{in_2} - 2\delta}{2}$$

Thus, solving for γ one obtains using equations (1)-(2) and the sampled values

$$\hat{\hat{\gamma}} = \frac{-(x_{f_2} - x_{in_2}) + s(1-u)(x_{f_1} + x_{in_1})/2}{s(x_{in_2} + x_{f_2} - 2\delta)/2}$$

where s is the sampling period. This approach leads to an excellent estimate.

Figure 4 shows from top to bottom the load variation $\gamma(t)$ and the estimates $\hat{\gamma}$, $\hat{\hat{\gamma}}$.

5. OPTIMIZATION OF THE TRANSIENT

Let x_0 be initial conditions defined by

$$x_0 = [a_0 + b_1 \sin(\phi_1), A]$$

where a_0, b_1 and ϕ_1 are derived from $\gamma = \gamma_1$. Let us study the system for $t \in [0, 2\pi/\omega]$ and γ defined by

$$\gamma(t) = \begin{cases} \gamma_1 & \text{if } t = 0 \\ \gamma_2 & \text{if } t > 0 \end{cases}$$

The system is sampled n_p times and the function

$$MQE = \sum_{k=1}^{n_p} (x_2[k] - x_{ref_2}[k])^2 / n_p$$

is considered as the cost function to be minimized.

In a first approach the interval $[0, 2\pi/\omega]$ is divided into $2, 3, \ldots, m+1$ subintervals and auxiliary loads $g_1, g_2, g_3, \ldots, g_{m+1} = \gamma_2$ are considered in the definition of $X_{1k} = a_{0k} + b_{1k}\sin(\omega t + \phi_k)$. The optimal values for $g_1, \ldots, g_m$ are found using the MATLAB function FMINSEARCH [1], a minimization routine based on the Nelder-Mead algorithm. The initial conditions for $g_1, \ldots, g_m$ are obtained from evaluating function MQE in a square net of mesh 0.2 defined on $[0., 1.4]^m$. Some results for $m = 2$ corresponding six step changes in the load can be read in table 1, the information is organised as follows:

- step change is in the first column,
- in second column there is the MQE value when no auxiliary load is considered,
- the minimum error obtained after taking into account two auxiliary loads defined by a net of mesh 0.2 in the square $[0.0, 1.4]^2$ is given in column three,
- finally, the auxiliary loads obtained through FMINS starting at the previous auxiliary loads are given in column fourth.
- auxiliary load values, if any, are in odd rows, the values the function MQE takes are in even rows.

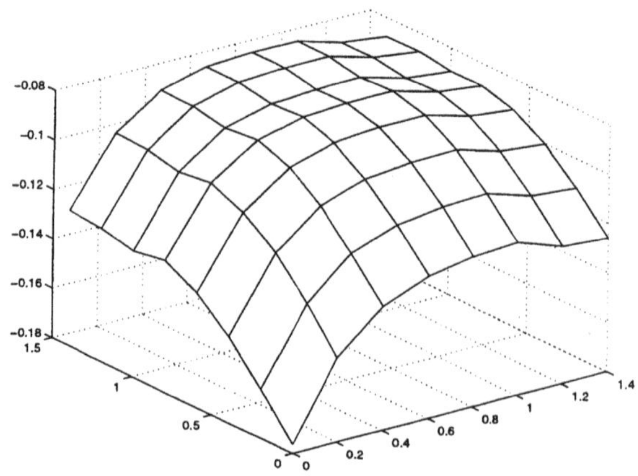

Fig. 2. Plot of $-MQE$ values on the net of mesh 0.1 defined in $[0., 1.4]^2$

As is shown in the table, appropriate auxiliary loads for a step change in the load from γ_1 to γ_2 are given by $g_1 = \gamma_2 + k_1(\gamma_2 - \gamma_1)$, $g_2 = \gamma_2 - k_2(\gamma_2 - \gamma_1)$ where k_1 and k_2 are positive.

A 3D plot of the values the function $-MQE$ [2] takes in the mesh when a step change from 0.6 to 1.0 happens in the load is depicted in Figure 2. The minimum of MQE on the net is 0.0802 and is achieved in (1.2, 1.0). Then, the function starting at [1.2, 1.0] gives [1.31, 0.97] and $MQE(1.31, 0.97) = 0.0798$. If no auxiliary loads are considered, MQE takes the value 0.0826.

In another approach the appropriate transient X_1 is derived from the numerical integration of equation (3) in reverse time. The right hand side of the equation is obtained from $X_2(t) =$

[1] FMINSEARCH Multidimensional unconstrained nonlinear minimization (Nelder-Mead).

[2] For sake of clarity in the picture, function $-MQE$ is depicted instead MQE

Table 1. Auxiliary loads

step ch.	load MQE	aux. loads min. on mesh	aux. loads FMINS
0.2 − 1.0	1.0	[1.4, 0.8]	[1.33, 0.85]
	0.33	0.3263	0.3261
1.0 − 0.2	0.2	[0.0, 0.2]	[0.00, 0.29]
	1.49	1.4420	1.4379
0.2 − 0.6	0.6	[1.0, 0.4]	[1.04, 0.41]
	0.14	0.1312	0.1311
0.6 − 0.2	0.2	[0.2, 0.2]	[0.09, 0.22]
	0.34	0.3374	0.3293
0.6 − 1.0	1.0	[1.2, 1.0]	[1.31, 0.97]
	0.08	0.0802	0.0798
1.0 − 0.6	0.6	[0.2, 0.4]	[0.25, 0.40]
	0.14	0.1323	0.1318

$A + B\sin(\omega t)$ and $\gamma(t)$. The initial condition is $x_1(T) = X_1(T)$, with $T = 2\pi/\omega$.

The numerical results of integrating the Abel equation in reverse time are in perfect agreement with the Fourier approximation $X_1 = a_0 + b_1\sin(\omega t + \phi_1)$. There is a problem arising from the reverse time aspects. The x_1-transient and the change in the load appear in reverse order depending whether one is considering time advancing forwards or backwards, i.e. in one of the cases the transient appears before of the change of the load. Therefore, the results obtained by numerical integration do not inform us precisely what X_1 to use. The Abel equation can be integrated in reverse time using Maple in order to get an approximate solution depending the values of the load γ_1 and γ_2.

6. CONCLUSIONS

Tracking control of basic nonlinear power converters have been studied for systems with *time-varying* parameters. The switching control laws accomplish indirect asymptotic sinusoidal voltage tracking whilst Fourier expansions are used to solve an ordinary differential equation of the Abel type. Different estimation approaches have been tried and the stability of the state variables during the parameter estimate transient has been ensured.

7. REFERENCES

Fossas-Colet, E. and D. Biel (1996). A sliding mode approach to robust generation on dc-to-dc converters. *Proceedings of the IEEE Conference on Decision and Control, Kobe* pp. 4010–4012.

Karsenti, L. and P. Rouchon (1997). A tracking controller-observer scheme for dc-to-dc converters. *Proceedings of the European Control Conference.*

Karsenti, L., Sira-Ramírez H. and F. Lammabhi-Lagarrigue (1995). Backstepping adaptive pwm stabilization of dc-to-dc converters towards minimum or non-minimum phase equilibria. *Proceedings of the IEEE Conference on Decision and Control, New Orleans.*

Olm-Miras, J.M. and E. Fossas-Colet (1999). *Abel Equations and Asymptotic Tracking in Power Converters.* Internal report IOC, Universitat Politècnica de Catalunya. Barcelona.

Rios-Bolívar, E.M. and A.S.I. Zinober (1998). Dynamic adaptive regulation of nonlinear continuous processes. In: *Theory and Practice of Control and Systems* (G. Conte A. Tornambe and A.M. Perdon, Eds.). World Scientific. Singapore.

Sira-Ramírez, H. (1987). Sliding motions in bilinear switched networks. *IEEE Trans. on Circuits and Systems* **34**, 919–933.

Sira-Ramírez, H. (1989). A geometric approach to pulse-width-modulated control in nonlinear dynamical systems. *IEEE Trans. on Aut. Control* **34**, 184–187.

Venkataramanan, R., A. Sabanovič and S. Čuk (1985). Sliding mode control of dc-to-dc converters. *Proceedings of the International conference on Industrial Electronics Control and Instrumentation* pp. 251–258.

Zinober, ASI, E. Fossas-Colet, J. Scarratt and D. Biel (1998). Two sliding mode approaches to the control of a boost system. In: *Theory and Practice of Control and Systems* (G. Conte A. Tornambe and A.M. Perdon, Eds.). pp. 343–348. World Scientific. Singapore.

APPENDIX

Two m-files describing the averaged dynamical system and the m-function to be minimized are listed below.

BCD.M

function xdot=bcd(t,x,flag,delta,u);

This program describes the averaged model of the two-dimensional basic converters boost and buck-boost. The duty- cycle is used as a constant input during the sample-period. There is an external function g_load(t), which provides parameter gamma vs. time

mu is a constant. It must be negative (u=ueq + mu*sign(s)), nc is the number of cycles. It sould be an integer. In this case mu=-0.02 and nc=200.

% System equations

```
xdot1=1+(u-1)*x(2);
xdot2=(1-u)*x(1)-gload(t)*x(2)+...
delta*gload(t);
xdot=[xdot1;xdot2];
```

JEY.M

function outs2=J(gg,delta,np,initg,finalg);

Returns the Mean Quadratic Error $\sum(x2(k) - xr2(k))^2/np$.

gg should be a vector of parameters. The load varies in a discrete set taking the values given by [gg finalg] on subperiods of the same length.

delta=0 boost, delta=1 buck-boost

np=number of samples (np=200 in this case)

initg= the initial value of gamma. It will provide initial conditions for the plant.

finalg= the final value of gamma.

```
A=2;
B=0.5;
omega=0.2513;
mu=-0.5;
g=[gg finalg];
sp=2*pi/(omega*np);
l=length(g);

% Initial conditions
hatgam=initg;
a0=0.5*(2*A*A-2*A*delta+B*B)*hatgam;
b1=B*sqrt(((omega*A)^2+hatgam^2*...
(2*A-delta)^2)/(a0^2*omega^2+1));
a_num=((1+2*hatgam*a0)*A-delta*...
a0*hatgam)*omega;
a_den=A*(2*hatgam-a0*omega^2)-...
delta*hatgam;
phi1=atan2(a_num,a_den);
tin=0;
x2rd=A+B*sin(omega*tin);
x1rd=a0+b1*sin(omega*tin+phi1);
xr=[x1rd x2rd];
x=[x1rd x2rd];
ins=0;
s2=0;

% processing
for n=1:l,
hatgam=g(n);
a0=0.5*(2*A*A-2*A*delta+B*B)*hatgam;
b1=B*sqrt(((omega*A)^2+hatgam^2*...
(2*A-delta)^2)/(a0^2*omega^2+1));
a_num=((1+2*hatgam*a0)*A-delta*...
a0*hatgam)*omega;
a_den=A*(2*hatgam-a0*omega^2)-...
delta*hatgam;
phi1=atan2(a_num,a_den);
% hg=[hg; hatgam];
for i=1:np/l,
inx=x;
xr=[a0+b1*sin(omega*tin+phi1) ...
A+B*sin(omega*tin)];
dx1ref=omega*b1*cos(omega*...
tin+phi1);
s=x(1)-xr(1);
var_d=max(0,min(1,1+((dx1ref-1)/...
xr(2))+mu*sign(s)));
[T auxx]=ode23('bcd',[tin tin+sp],...
x, [], delta, var_d);
nn=length(auxx);
x=auxx(nn,:);
tin=tin+sp;
s2=s2+(x(2)-xr(2))^2;
end;
end;
outs2=s2/np;
```

TIDEA: A CONTROL SYSTEM FINE TUNING SOFTWARE TOOL APPLIED TO A LARGE POWER STATION

Francesco Pretolani, George Oluwande*, Giambattista Garbossa**, Megan Boardman***

**CESI S.p.A. Milan (Italy)
*Innogy Plc Swindon (UK)

Abstract: TIDEA is a Windows NT application that has been developed for the automatic tuning of the parameters of PID regulators on industrial plants. This paper reviews the TIDEA package and a trial application at Didcot "A" Power Station (Innogy Plc, UK). *Copyright © 2001 IFAC.*

Keywords: PID controllers, identification algorithms, autotuners, controlling power stations, industrial control

1. INTRODUCTION

Regulation control systems are aimed at ensuring stable and efficient running of industrial plants and minimising the effects of external disturbances. The correct performance of the regulation systems tends to deteriorate with time, due to process changes and the degradation of plant components between maintenance intervals. This makes the parameters of the control loop (e.g. gain, time constants) no longer adequate, and requires intervention by control technicians.

The need is felt in process industry for packages to identify quickly and automatically the appropriate controller settings. With this aim, an EU project is in progress, under the Information Society Technologies Programme 1999, Framework 5. This project, named TIDE, is aimed at fostering the introduction in the European market of TIDEA (Time Delay Estimation Algorithm).

TIDEA is a package developed and tested by CESI, Italy, for tuning regulation loops based on dynamic identification of the process parameters. The package features:

- derivation of the optimal set of regulator parameters (gains, time constants),
- capability to link to the actual control system for the subsequent tuning of the regulator,
- the ability to evaluate the degradation of the dynamic parameters of the systems, and
- the facility to archive results with standard format reporting.

Currently the package has been used by CESI to service a few units of plants operated by their mother company ENEL (Bolis and Villa, 1994), the largest Italian utility.

The aims of TIDE are to test and validate TIDEA, to revise and strengthen the package, and to disseminate the resulting product in the process automation market. It includes a trial application of the TIDEA package on the Didcot "A" Power Station, owned and operated by Innogy Plc, UK.

The specific technical objectives of this paper are as follows:

- to describe TIDEA and its core algorithms,
- to discuss the trial application of TIDEA on Didcot "A" Power Station,
- to evaluate the benefit of the trial application to Didcot, and
- to evaluate the product performance and ease-of-use for a subsequent re-engineering of the package.

2. TIDEA DESCRIPTION

TIDEA is a Windows NT application that has been developed for the calculation of the optimal parameters of Proportional, Integral and Derivative

(PID) regulators. It allows the tuning of a single regulation loop at a time.

The TIDEA algorithm is the end result of a joint research project with Polytechnic of Milan (Ferretti et al., 1991; Leva et al., 1994; Ferretti et al., 1995 and 1996). It consists of:

- a process identification algorithm developing a process model that focuses in particular on the time delay estimate, and
- a PID tuning algorithm operating on the basis of the identified model using the classical control theory.

The precise delay estimate can be used in the model, to consider those phase shifts present in a real process but not attributable to a physical delay. The process to be controlled is identified through the acquisition of output Y(t) and input U(t) values of the process to be controlled (see Fig. 1). During data acquisition the overall system must be stimulated by means of small step changes to the set point S(t) up and down around its nominal value in a closed loop mode.

Although the TIDEA algorithms could acquire data and identify the process model in true real time, an off-line mode passing through an ASCII file is preferred so as to decouple the data acquisition from the identification and tuning phases.

The TIDEA recursive identification algorithm requires an initialization in which the process's general dynamic characteristics are roughly given. The user must select the type of dynamic model to use (1 pole, 2 poles or 1 pole plus integrator, see Table 1). They must also provide a rough quantitative characteristic of the response to the step (gain, rise time, etc.).

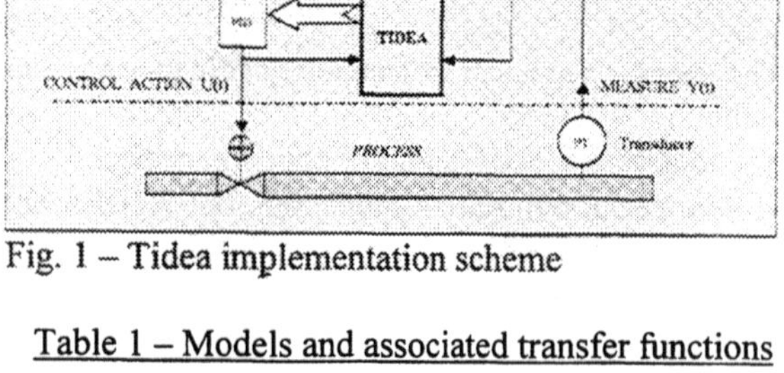

Fig. 1 – Tidea implementation scheme

Table 1 – Models and associated transfer functions

1-POLE	$G(s) = RO * \frac{e^{-sTAU}}{s + A2}$
1-POLE + INTEGRAL	$G(s) = RO * \frac{e^{-sTAU}}{s*(s + A2)}$
2-POLES	$G(s) = RO * \frac{e^{-sTAU}}{s^2 + A1*s + A2}$

Various industrial PID regulators are taken into account: H&B (Contronic), ABB (Simphony), Elsag Bailey (Infi 90), Toshiba (EC310), and Moores Product Company (APACS).

The PID parameter calculation method is based on the classic synthesis technique in the frequency domain. The regulator parameters are determined based on the two concepts of φ "phase margin" and ω_n "critical frequency" of the closed-loop transfer function of the overall system consisting of the process plus the regulation. However, these two indices are too complex for industrial plant technicians to apply and thus two other more intuitive parameters that are characteristic of the controlled system's response over time have been chosen: the desired "overshoot" and "rise time" of the response to the step of the closed loop system.

3. TIDEA ALGORITHM

The most important part of TIDEA is the identification algorithm: the system being analyzed is described using a first or second-order transfer function that includes a time delay. The critical point of the algorithm is the delay estimate since it introduces a non-linear identification problem.

The algorithm utilizes the following basic theoretical steps:

1. A generic transfer function *G(s)* can be converted, under certain hypotheses, into an "all-pole" transfer function, i.e.:

$$G(s) \cong \frac{e^{-\tau s}}{D_R(s)}$$

2. Given an "all-pole" transfer function, it is demonstrated that its digitalization with sampling time h, setting $\tau = h \cdot (D+E)$ with integer D and $E \in (-1, 1)$, becomes:

$$H(z) = z^{-(d+1)} \frac{b_2 z^2 + b_1 z + b_0}{A_R(z)}$$

whose zeroes z_1 and z_2 are related to the time delay with the formulas:

$$\frac{1}{1-z_1} + \frac{1}{1-z_2} = 1 - \varepsilon$$

$$\tau = h \cdot \left(d + \varepsilon + \frac{n-1}{2} \right)$$

where n is the order of the polynomial $D_R(s)$. The values of ε and d represent the "equivalent" delays, respectively fractional and integer, which generally do not coincide, due to the approximation procedure, with the "real" delays E and D.

3. Once the fractional delay ε has been estimated, a check is performed to determine if it belongs to the confidence range (-0.8, 0.8). If this does not

occur, it may indicate an incorrect initial estimate of the integer delay d and thus a relative increase or decrease.

All this must be preceded by an estimate algorithm that, given the output values of process y and the control variable u as input values, generates an estimate of the discrete time transfer function $H(z)$. For this purpose a recursive estimate algorithm based on minimum squares (RLS) has been used. From a general viewpoint, the algorithm's block diagram can be represented in Fig. 2.

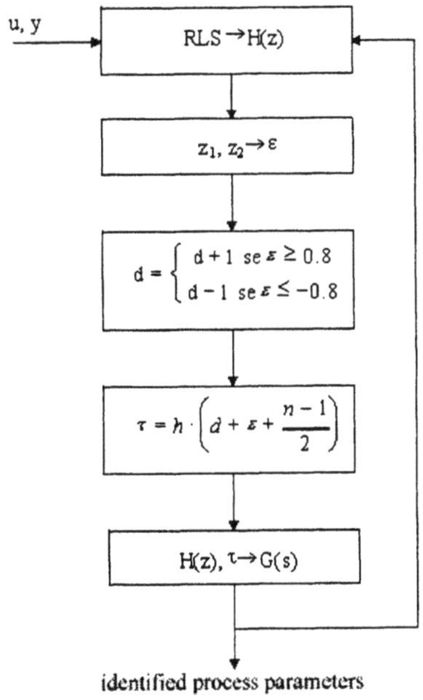

Fig. 2 – Identification algorithm block diagram

4. TIDEA PLANT TRIAL AT DIDCOT "A" POWER STATION

Didcot "A" Power Station is owned by Innogy Plc, a leading energy utility in the UK. The station consists of 4 x 500MW units, with 3 of these units capable of dual-firing on coal or gas, whilst the 4th unit is currently capable of coal-firing only. The station is in the middle of refurbishing the existing control systems on each of the units, with the replacement of the existing proprietary control system Swepspeed by a control system called APACS from Moore Products (a subsidiary of Siemens). Currently only two of the units (units 3 and 4) have had their control systems replaced.

Also as part of the refurbishment, a new man-machine interface (MMI) has been installed called the Advanced Plant Management System (APMS). APMS is a software package which has been specified by National Power (which demerged into Innogy Plc and International Power Plc) and developed by Syseca Ltd. It was developed so as to enable the integration of existing and future control systems and act as a platform for the installation of strategic packages, i.e. expert systems shell G2 and RapidMVC, a multivariable predictive control package. APMS has been installed on all the four units at the station.

TIDEA was installed on the APMS Application Server. The plant data required by TIDEA for its identification and control parameters determination was logged from APACS on the APMS host and the resulting ASCII file(s) transferred to the Applications server.

4.1 Selected Control Loops

Unit 3 at Didcot "A" Power Station was chosen as the unit on which TIDEA would be installed for its trial application. The three control loops to which TIDEA was applied were: Primary Air (PA) differential pressure, mill feeder speed and mill outlet temperature. The reasons why these specific loops were considered for the trial are:

- Improved tuning of the mill/combustion control loops (PA and mill feeder) could result in better combustion.
- Improved tuning of the temperature control would result in improved combustion efficiency from better milling performance. Better mill outlet temperature control would also help to minimise risk of fire in mills from excessive temperatures.

There are 8 mills ("A" - "H") per unit, although at full load typically no more than 6 mills are in operation. Tuning was carried out on all of these mills (with the exception of C-mill), for each of the 3 control loops on the mills.

5. TEST RESULTS

5.1 Feeder Speed Loops

The feeder speed loop is used to control the amount of coal that is sent into the mill. Its set-point is dependent on an air/fuel ratio relationship to the PA differential pressure set-point.

After installation of TIDEA in February, an initial trial of the software was applied to the A-mill feeder speed control. TIDEA revealed that a consistent improvement could be made to this particular feeder speed loop. For this mill, it recommended reducing the integral action time by about 86%, and to reduce the proportional gain term by about 63%. To build up confidence, the new controller terms identified by TIDEA were not implemented in one go, but gradually built up to in a series of small changes to the existing terms, as this was felt to be prudent (see Fig. 3).

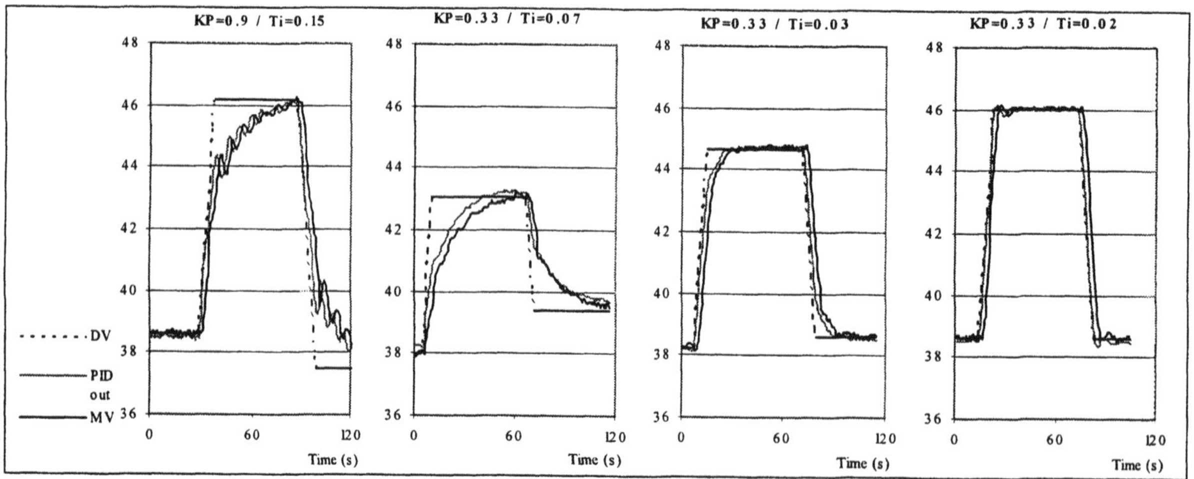

Fig. 3 - Effect of gradually entering Tidea suggested parameters on A-mill feeder speed loop, from the untuned loop response at the left, to the Tidea tuned loop at the right.

Table 2, below, summarises the results of the feeder loops for the 7 mills identified. In general, the table suggests that the current tuning parameters were conservative, and that a performance improvement could be gained without causing a loss in stability.

Table 2 - Feeder Loop Tuning Results

Loop	Old KP	New KP	Old Ti (min)	New Ti (min)
Feeder A	0.9	0.33	0.15	0.02
Feeder B	0.8	0.33	0.15	0.037
Feeder D	0.85	0.4	0.15	0.035
Feeder E	0.8	0.55	0.15	0.047
Feeder F	0.8	0.5	0.15	0.04
Feeder G	0.72	0.37	0.15	0.05
Feeder H	0.9	0.71	0.15	0.05

5.2 PA Differential Pressure Loops

The mill PA differential pressure control loops are used to control the amount of fuel that gets conveyed from the mills into the boiler. TIDEA's recommended changes to the PA differential pressure loop were consistent in terms of the gains, in that on average it doubled the existing proportional gain for all of the mills.

The results for the integral time were more mixed, with the recommended settings increasing the values for each mill by between 25 - 150%.

Table 3 summarises the changes in the controller terms from the existing to the TIDEA recommended settings. The tuning results for G-mill are shown in Figures 4-7.

Table 3 – PA Diff Pressure Loop Tuning Results

Loop	Old KP	New KP	Old TI (min)	New TI (min)
PA Diff. A	10	39.38	0.2	0.36
PA Diff. B	10	25.6	0.2	0.41
PA Diff. D	15.75	16.6	0.2	0.25
PA Diff. E	10	19.67	0.2	0.25
PA Diff. F	10.37	17.8	0.2	0.28
PA Diff. G	10	23.8	0.2	0.43
PA Diff. H	16.2	34.7	0.2	0.49

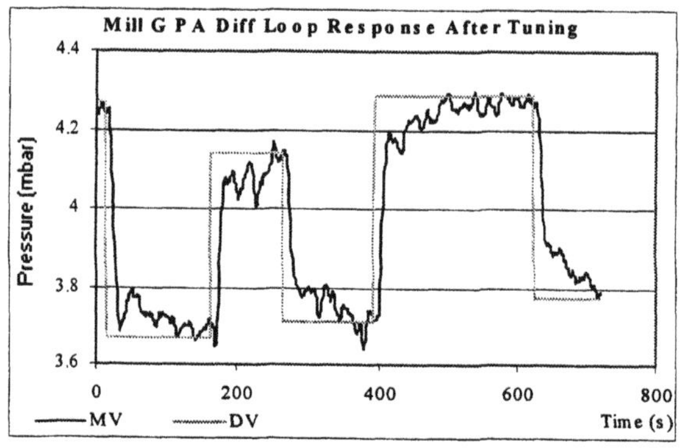

Fig. 4 - G Mill PA Differential Pressure Loop tuning results

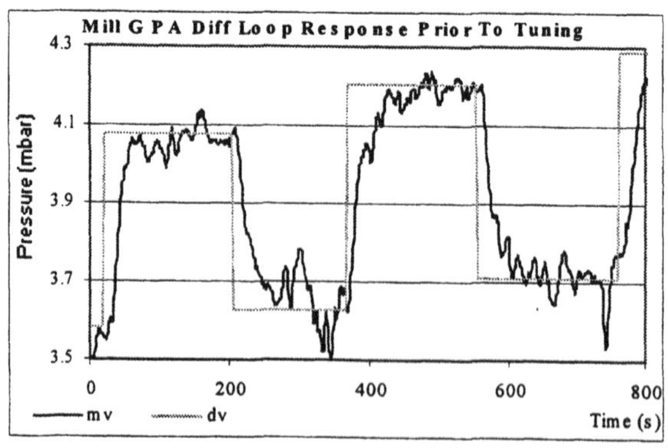

Fig. 5 - G Mill PA Differential Pressure Loop performance prior to tuning

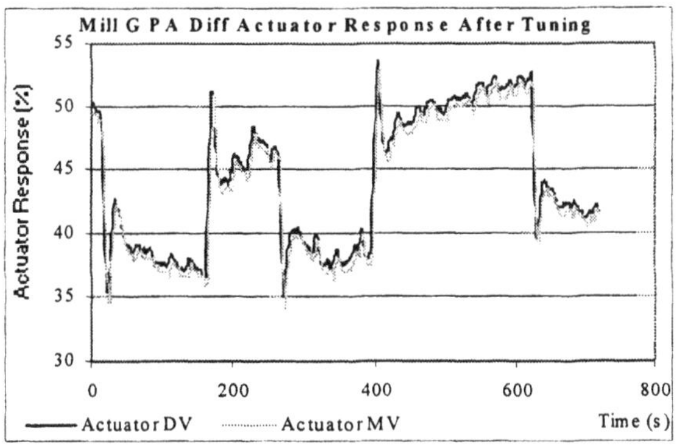

Fig. 6 - G Mill PA Differential Pressure Loop actuator response after tuning

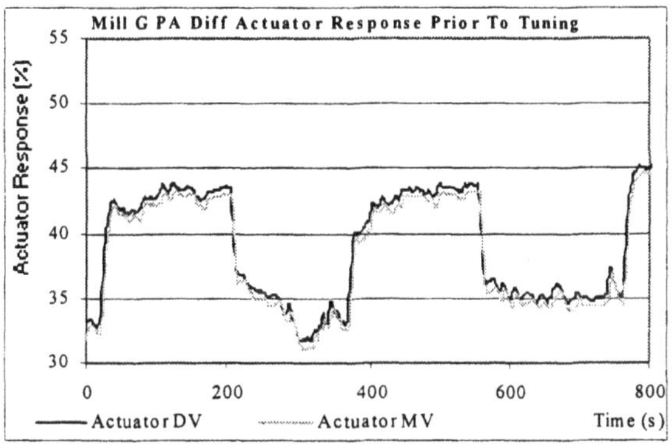

Fig. 7 - G Mill PA Differential Pressure Loop actuator response prior to tuning

5.3 Mill Outlet Temperature Loops

The mill outlet temperature control loops are used to control the "dryness" of the product of the mills. The temperature control loop is the slowest of the three mill control loops that were being reviewed.

It was difficult to get sufficiently clean data from the step tests that were conducted on the mills for the identification stage of the tuning. The problem encountered during the step responses was the amount of external disturbances impacting on the temperature loops. This meant that ultimately only two loops could successfully be modelled and new controller terms established by TIDEA (mills A and B). Currently only the B-mill results have been tried, as A-mill has been off on repairs.

The results for B-mill are shown in Figures 8-11.

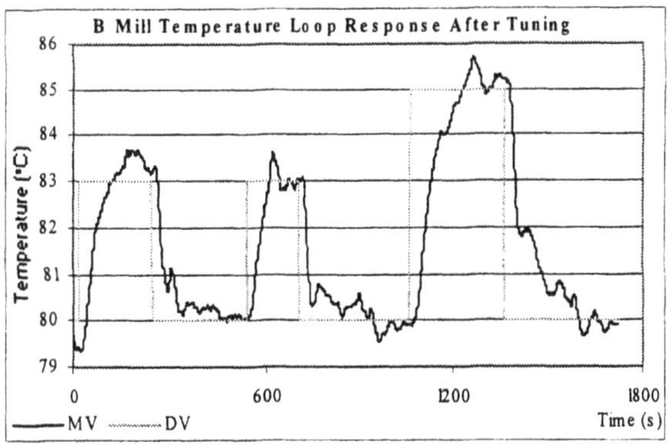

Fig. 8 - B Mill Temperature Loop tuning results

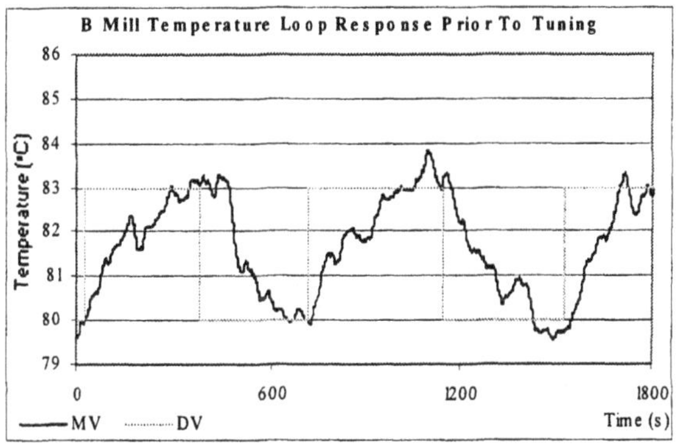

Fig. 9 - B Mill Temperature Loop performance prior to tuning

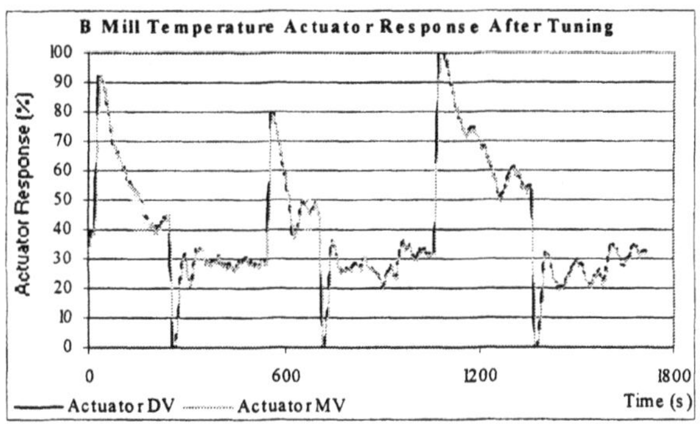

Fig. 10 - B Mill Temperature Loop actuator response after tuning

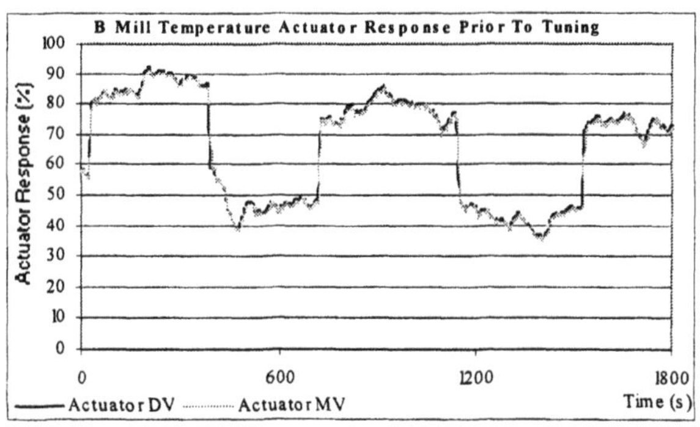

Fig. 11 - B Mill Temperature Loop actuator response prior to tuning

6. CONCLUSIONS

TIDEA has demonstrated the benefits to be gained from having a control loop-tuning package on a power plant. In general it appears to show that the current controller terms are conservative and, therefore, the loops tend to run in a sub-optimal way.

As for the benefits in terms of improved combustion, evaluations are still in progress at Didcot Power Station, requiring long term trials.

An overall economic benefits evaluation will be performed as well taking into consideration the fuel savings and the improvement of the plant dynamic performances.

The following issues came up during the TIDEA evaluation and are areas for improvement and research:

- TIDEA needs to improve its user documentation for the end user, to help improve understanding of its functionality.
- The robustness of the identification package needs to be strengthened, as was highlighted by the difficulties in identifying the mill outlet temperature dynamics.
- The need to balance the optimal tuning of a loop in isolation with compromising the controller terms to suit the needs of a higher (or supervisory) controller. This was the case with the PA differential pressure controllers, where it was necessary to back away from the TIDEA generated terms due to their impact on a supervisory load controller that was on the unit.
- The need to balance the derived optimal tuning terms against actuator wear and tear, as the tradeoff required might result in a more conservative set of parameters being used by the control engineer.

In addition the following development activities are ongoing:

- An archiving feature based on standard database utility will be implemented.
- An OPC client to make easy the connection and the data collection from modern DCS will be developed.

7. REFERENCES

Bolis, W., G.W. Villa (1994): “A Boiler/Turbine Control System Tuning Method”, AEI (Italian Electrotechnical Association) Annual Meeting, Bologna (I)

Ferretti, G., C. Maffezzoni, R. Scattolini (1991): "Recursive Estimation of Time Delay for Sampled Systems", *Automatica*, Vol. 27, n.4, 653-661.

Ferretti, G., C. Maffezzoni, R. Scattolini (1995): "The Recursive Estimation of Time Delay in Sampled-Data Control Systems", *Control and Dynamic Systems* (C.Leondes ed.), Vol.73, pp. 159-206.

Ferretti, G., C. Maffezzoni, R. Scattolini (1996): "On the Identifiability of Time Delay with Least Squares Methods", *Automatica*, Vol.32, n.3, pp.449-453.

Leva, A., C. Maffezzoni, R. Scattolini (1994): "Self-Tuning PI-PID Regulators for Stable Systems with Varying Delay", *Automatica*, Vol.30,n.7, pp.1171-1183.

MATLAB-TOOLBOX FOR CAD OF SIMPLE SELF-TUNING CONTROLLERS

Vladimír Bobál*, Josef Böhm, Petr Chalupa***

**Department of Control Theory, Institute of Information Technologies,*
Tomas Bata University in Zlín, Nám. TGM 275, 762 72 Zlín, Czech Republic,
tel., fax:+420 67 754 3217, E-mail: bobal@ft.utb.cz

***Institute of Information Theory and Automation*
Academy of Sciences of the Czech Republic
Pod vodárenskou věží 4, 182 08, Prague 8, Czech Republic
E-mail: bohm@utia.cas.cz

Abstract: The contribution presents a MATLAB-Toolbox for design, simulation and verification of single input single output (SISO) discrete self-tuning controllers. The proposed adaptive controllers who are included into a Toolbox can be divided into two parts. The first part covers PID adaptive algorithms using traditional Ziegler-Nichols method for the setting of the controller parameters, the second part of described controllers is based on the pole placement design. The MATLAB-Toolbox is very successfully used in Adaptive Control Course in education practice for design, simulation and verification of self - tuning control systems in real - time conditions. It is suitable for design and verification of the industrial controllers, too. *Copyright © 2001 IFAC.*

Key words: Self-tuning control; ARX model; recursive least squares; PID control; pole assignment; MATLAB–Toolbox.

1. INTRODUCTION

Adaptive control methods have been developing significantly over the past tree decades. The aim of this research was to solve the problem of designing a controller for systems where the characteristics are not completely known or vary. Different approaches were proposed and utilized. One successful approach is represented by self-tuning controllers (STC). The main idea of an STC is based on the combination of a recursive identification procedure and a selected controller synthesis. Different branches of the STC differ mainly in the method of the controller design while the identification recursive least squares method (RLSM) applied to a regression (ARX) model forms a standard.

Despite intensive research activity, there is a lack of controllers which are able to cope with practical requirements. Industrial applications on a large scale are still missing, however, many realized cases were successful. It is evident that until to now application has required a control engineer skilled in specific technology as well as versed in modern control methods.

Also the aim of this contribution is to bridge the gap between theory and practice and to present some simple controller algorithms in a form acceptable for industrial users. The theoretical background not only these simple algorithms, also algorithms based on polynomial solutions of controller synthesis and controllers what are derived from the use of minimization of linear quadratic criterion, are given in Bobál, *et al.* (1999).

All the explicit self-tuning controllers who are included into MATLAB-Toolbox have been algorithmically modified in the form of mathematical relations or as flow diagrams so as to make them easy to program and apply. Some are original algorithms based on a modified Ziegler-Nichols criterion, others have been culled from publications and adapted to make them more accessible to the user.

2. THEORETICAL BACKGROUND

2.1 Recursive identification

In the identification part of the designed controller algorithms the regression (ARX) model of the following form

$$y(k) = \Theta^T(k)\phi(k-1) + n(k) \tag{1}$$

is used, where

$$\Theta^T(k) = [a_1\ a_2\ \ldots\ a_{na}\ b_1\ b_2\ \ldots\ b_{nb}] \tag{2}$$

is the vector of the parameter estimates and

$$\phi^T(k-1) = [-y(k-1)\ -y(k-2)\ \ldots\ -y(k-na)\quad u(k-1)\ u(k-2)\ \ldots\ u(k-nb)] \tag{3}$$

is the regression vector ($y(k)$ is the process output variable, $u(k)$ is the controller output variable). The non-measurable random component $n(k)$ is assumed to have zero mean value $E[n(k)] = 0$ and constant covariance (dispersion) $R = E[n^2(k)]$. For calculating of the parameter estimates (2) is utilized the recursive least squares method, where adaptation is supported by directional forgetting (Kulhavý, 1987).

2.2 Algorithms of digital Ziegler-Nichols PID controllers

The PID controllers are still widely used in industry. These types of controllers are more convenient for users owing to their simplicity of implementation, which is generally well known. Provided the controller parameters are well chosen they can control a considerable part of continuous technological processes. To get a digital version of the PID controller, it is necessary to discretize the integral and derivative component of the continuous-time controller. For discretizing the integral component we usually employ the forward rectangular method (FRM), backward rectangular method (BRM) or trapezoidal method (TRM). The derivative component is mostly replaced by the 1st order difference (two-point difference). For practical use the recurrent control algorithms which compute the actual value of the controller output $u(k)$ from the previous value $u(k\text{-}1)$ and from compensation increment seem to be suitable

$$u(k) = q_0 e(k) + q_1 e(k-1) + q_2 e(k-2) + u(k-1) \tag{4}$$

where q_0, q_1, q_2 are the controller parameters and T_0 is the sampling period.

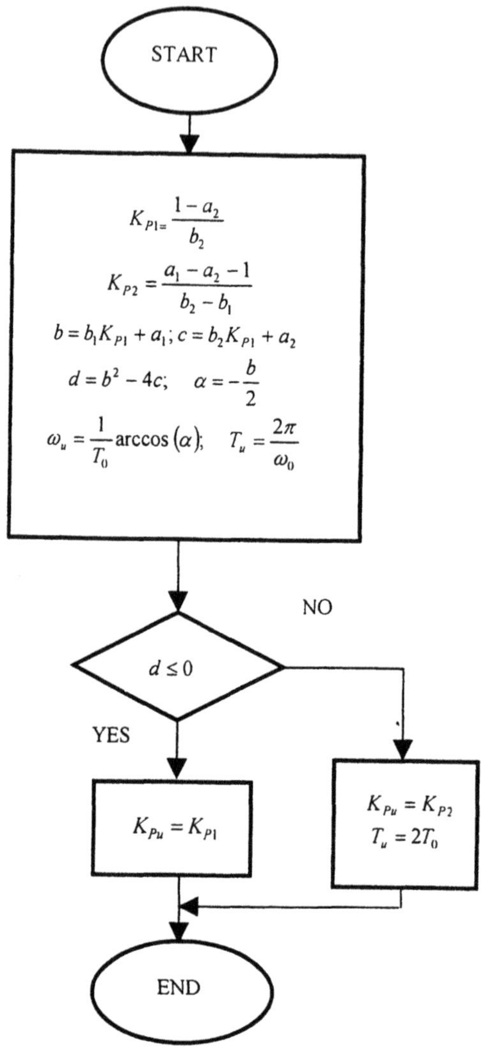

Fig. 1 Flow diagram for the calculation of control parameters for a second-order system.

It is subsequently possible to derive further variants of digital PID controllers. For example the PID controller with a filter constant in the D-part (Isermann, 1989), Takahashi's PID controller (Takahashi, *et al.*, 1971), the PID controller based on the ideas of Åström, *et al.* (1992), Dahlin's PID controller (Dahlin, 1968) or the PID controller designed by Bányász and Keviczky (1993).

First group of proposed self-tuning PID controllers is based on the classical Ziegler and Nichols 1942) method. In this well-known approach the parameters of the controller are calculated from the ultimate

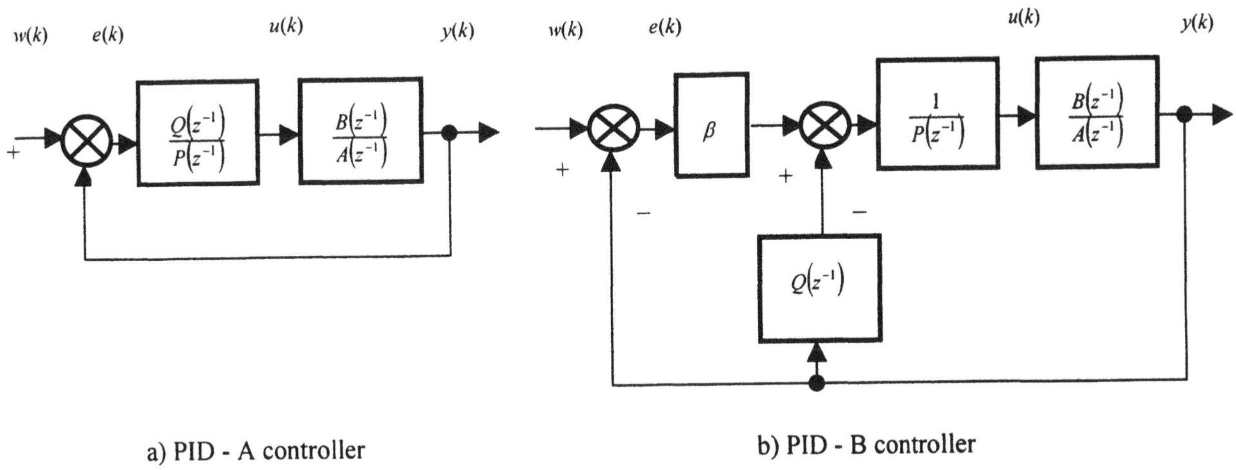

Fig. 2 Block diagrams of control loops with pole placement PID controllers

(critical) gain K_{pu} and the ultimate period of oscillations T_u of the closed loop system. The analytical expressions for computing of these critical parameters are derived in Bobál, *et al.* (1999a) and Bobál, *et al.* (1999b). The flow diagram for the computing of the critical parameters for the second- is presented in Fig. 1 and and for the third-order model is introduced in Bobál, *et al.* (1999b).

2.3 Pole placement PID controllers

A controller based on the pole placement method in a closed feedback control loop is designed to stabilise the closed control loop whilst the characteristic polynomial should have a previously determined pole. Digital PID controllers is possible can be expressed in the form of a discrete transfer function

$$G_R(z)=\frac{U(z)}{E(z)}=\frac{Q(z^{-1})}{P(z^{-1})}=\frac{q_0+q_1z^{-1}+q_2z^{-2}}{(1-z^{-1})(1+\gamma z^{-1}]} \quad (5)$$

If the process is given by the transfer function

$$G_P(z)=\frac{Y(z)}{U(z)}=\frac{B(z^{-1})}{A(z^{-1})} \quad (6)$$

If the process is given by the transfer function

$$G_P(z)=\frac{Y(z)}{U(z)}=\frac{B(z^{-1})}{A(z^{-1})} \quad (7)$$

with the polynomials $A(z^{-1})$, $B(z^{-1})$ for degree $n = 2$ then the characteristic polynomial of the closed-loop system with a PID-A controller (see Fig. 2a)) is in the form

$$A(z^{-1})P(z^{-1})+B(z^{-1})Q(z^{-1})=D(z^{-1}) \quad (8)$$

The pole placement of the characteristic polynomial (8) determines the dynamic behaviour of the closed - loop system. The characteristic polynomial $D(z^{-1})$ can be specified by different methods. It is usually described by the dominant poles for a second order continuous model (controller PID A-1)

$$D(s)=s^2+2\xi\omega_n s+\omega_n^2 \quad (9)$$

The dominant poles are given by the desired damping factor ξ and the natural frequency ω_n of the closed-loop.

A PID controller with the characteristic polynomial (controller PID A-2)

$$D(z)=(z-\alpha)^2[z-(\alpha+j\omega)][z-(\alpha-j\omega)] \quad (10)$$

is proposed in Bobál, *et al.* (1999a). Characteristic polynomial (10) has a double real pole $z_{1,2}=\alpha$ within interval $0 \le \alpha < 1$ and a pair of complex conjugated poles $z_{3,4}=\alpha \pm j\omega$, where $\alpha^2+\omega^2<1$. The parameter α influences the speed of the control-loop transient response and influences controller output changes, too. By changing parameter ω it is possible to influence the desired overshoot.

The structure of the control loop with controller PID B is shown in Fig. 2b). The characteristic polynomial has in this case form

$$A(z^{-1})P(z^{-1})+B(z^{-1})[Q'(z^{-1})+\beta]=D(z^{-1}) \quad (11)$$

where polynomial $P(z^{-1})$ has the same form as in the transfer function (5) and second polynomial in equation (11) is given

$$Q'(z^{-1})=(1-z^{-1})(q_0'-q_2'z^{-1}) \quad (13)$$

Using equation (8) it is possible to derived controller A-2, using equation (11) PID B-2.

Table 1 Survey of individual controllers

Contr. No. / m-file	Controller algorithm	Input parameters / Controller type
1 zn2fr.m zn3fr.m	$u(k)=q_0e(k)+q_1e(k-1)+q_2e(k-2)+u(k-1)$ $q_0=K_P\left(1+\frac{T_D}{T_0}\right)$; $q_1=-K_P\left(1-\frac{T_0}{T_I}+2\frac{T_D}{T_0}\right)$; $q_2=K_P\frac{T_D}{T_0}$	K_{Pu}; T_u; T_0 PID - forward rectangular method
2 zn2br.m zn3br.m	$u(k)=q_0e(k)+q_1e(k-1)+q_2e(k-2)+u(k-1)$ $q_0=K_P\left(1+\frac{T_0}{T_I}+\frac{T_D}{T_0}\right)$; $q_1=-K_P\left(1+2\frac{T_D}{T_0}\right)$; $q_2=K_P\frac{T_D}{T_0}$	K_{Pu}; T_u; T_0 PID - backward rectangular method
3 zn2tr.m zn3tr.m	$u(k)=q_0e(k)+q_1e(k-1)+q_2e(k-2)+u(k-1)$ $q_0=K_P\left(1+\frac{T_0}{2T_I}+\frac{T_D}{T_0}\right)$; $q_1=-K_P\left(1-\frac{T_0}{2T_I}+\frac{2T_D}{T_0}\right)$; $q_2=K_P\frac{T_D}{T_0}$	K_{Pu}; T_u; T_0 PID - trapezoidal rectangular method
4 zn2fpd.m zn3fpd.m	$u(k)=K_P\left\{e(k)-e(k-1)+\frac{T_0}{T_I}e(k)+\frac{T_D}{6T_0}\left[e(k)+2e(k-1)-6e(k-2)+2e(k-3)+e(k-4)\right]\right\}+$ $+u(k-1)$	K_{Pu}; T_u; T_0 PID - backward rectangular method, replacing derivation by a four- point difference
5 zn2fd.m zn3fd.m	$u(k)=p_1u(k-1)+p_2u(k-2)+q_0e(k)+q_1e(k-1)+q_2e(k-2)$ $p_1=\frac{-4\frac{T_f}{T_0}}{\frac{2T_f}{T_0}+1}$; $p_2=\frac{\frac{2T_f}{T_0}-1}{\frac{2T_f}{T_0}+1}$; $q_0=\frac{K_P+2K_P\frac{T_f+T_D}{T_0}+\frac{K_PT_0}{2T_I}(\frac{2T_f}{T_0}+1)}{\frac{2T_f}{T_0}+1}$ $q_1=\frac{\frac{K_PT_0}{2T_I}-4K_P\frac{T_f+T_D}{T_0}}{\frac{2T_f}{T_0}+1}$; $q_2=\frac{\frac{T_f}{T_0}(2K_P-\frac{K_PT_0}{T_I})+2\frac{K_PT_D}{T_0}+\frac{K_PT_0}{2T_I}-K_P}{\frac{2T_f}{T_0}+1}$; $T_f=\frac{T_D}{\alpha}$; $\alpha\in\langle 3;20\rangle$	K_{Pu}; T_u; T_0; T_f PID - filtration of D-component, Tustin approximation
6 zn2tak.m zn3tak.m	$u(k)=K_R[y(k-1)-y(k)]+K_I[w(k)-y(k)]+K_D[2y(k-1)-y(k-2)-y(k)]+$ $+u(k-1)$ $K_R=0.6K_{Pu}-\frac{K_I}{2}$; $K_I=\frac{1.2K_{Pu}T_0}{T_u}$; $K_D=\frac{3K_{Pu}T_u}{40T_0}$	K_{Pu}; T_u; T_0 PID - Takahashi's controller
7 zn2sic.m zn3sic.m	$u(k)=u_P(k)+u_I(k)+u_D(k)$ $u_P(k)=K_Pe(k)$; $u_I(k)=K_P\frac{T_0}{T_I}e(k)+u_I(k-1)$; $u_D(k)=K_P\frac{T_D}{T_0}[e(k)-e(k-1)]$	K_{Pu}; T_u; T_0 PID - component form
8 zn2ast.m zn3ast.m	$u(k)=u_{PI}(k)+u_D(k)$ $u_{PI}(k)=K_P[y(k-1)-y(k)]+\frac{K_PT_0}{2T_I}[e(k)+e(k-1)]+\beta K_P[w(k)-w(k-1)]+u_{PI}(k-1)$ $u_D(k)=K_P\frac{T_D\alpha}{T_D+T_0\alpha}[y(k-1)-y(k)]+\frac{T_D}{T_D+T_0\alpha}u_D(k-1)$	K_{Pu}; T_u; T_0 PID - Åström's controller
9 ba2.m	$u(k)=q_0e(k)+q_1e(k-1)+q_2e(k-2)+u(k-1)$ $q_0=\frac{k_I}{b_0}$; $q_1=q_0a_1=\frac{k_I}{b_0}a_1$; $q_2=q_0a_2=\frac{k_I}{b_0}a_2$ $k_I=\frac{1}{2d-1}$ $(\gamma=0)$; $k_I=\frac{1}{2d(1+\gamma)(1-\gamma)}$ $(\gamma>0)$; $\gamma=\frac{b_1}{b_0}$	d - time delay PID - Bányász - Keviczky's controller
10 da2.m	$u(k)=K_P\left\{e(k)-e(k-1)+\frac{T_0}{T_I}e(k)+\frac{T_D}{T_0}[e(k)-2e(k-1)+e(k-2)]\right\}+u(k-1)$ $K_P=-\frac{(a_1+2a_2)Q}{b_1}$; $T_I=-\frac{T_0}{\frac{1}{a_1+2a_2}+1+\frac{T_D}{T_0}}$ $T_D=\frac{T_0a_2Q}{K_Pb_1}$; $Q=1-e^{-\frac{T_0}{B}}$	B - adjustment factor T_0 PID - Dahlin's controller
11 pp2a_1.m	$u(k)=q_0e(k)+q_1e(k-1)+q_2e(k-2)+(1-\gamma)u(k-1)+\gamma u(k-2)$ $q_0=\frac{1}{b_1}(d_1+1-a_1-\gamma)$; $q_1=\frac{a_2}{b_2}-q_2(\frac{b_1}{b_2}-\frac{a_1}{a_2}+1)$; $q_2=-\frac{s_1}{r_1}$; $\gamma=q_2\frac{b_2}{a_2}$ $s_1=a_2\left[(b_1+b_2)(a_1b_2-a_2b_1)+b_2(b_1d_2-b_2d_1-b_2)\right]$; $r_1=(b_1+b_2)(a_1b_1b_2+a_2b_1^2+b_2^2)$	ω_o - natural frequency ξ - damping factor PID A-1 pole placement controller

12 pp2a_2.m	$u(k)=q_0e(k)+q_1e(k-1)+q_2e(k-2)+(1-\gamma)u(k-1)+\gamma u(k-2)$ $q_0=\frac{r_2-r_3}{r_1}$; $q_1=-\frac{r_4+r_5}{r_1}$; $q_2=\frac{x_4+\gamma a_2}{b_2}$; $\gamma=\frac{r_6}{r_1}$ $r_1=(b_1+b_2)(a_1b_1b_2+a_2b_1^2+b_2^2)$; $r_2=x_1(b_1+b_2)(a_1b_2-a_2b_1)$ $r_3=b_1^2x_4-b_2\left[b_1x_3-b_2(x_1+x_2)\right]$; $r_4=a_1\left[b_1^2x_4+b_2^2x_1-b_1b_2(x_2+x_3)\right]$ $r_5=(b_1+b_2)\left[a_2(b_1x_2-b_2x_1)-b_1x_4+b_2x_3\right]$; $r_6=b_1(b_1^2x_4-b_1b_2x_3+b_2^2x_2)-b_2^3x_1$	α, ω - real and imaginary component of the pole PID A-2 pole placement controller
13 pp2b_1.m	$u(k)=-\left[(q_0'+\beta)y(k)-(q_0'+q_2')y(k-1)+q_2'y(k-2)\right]-$ $-(\gamma-1)u(k-1)+\gamma u(k-2)+\beta w(k)$ $q_0'=q_2'(\frac{b_1}{b_2}-\frac{a_1}{a_2})-\frac{a_2}{b_2}$; $q_2'=\frac{s_1}{r_1}$; $\gamma=q_2'\frac{b_2}{a_2}$; $\beta=\frac{1}{b_1}(d_1+1-a_1-\gamma-b_1q_0')$ $s_1=a_2\left\{b_2\left[a_1(b_1+b_2)+b_1(d_2-a_2)-b_2(d_1+1)\right]-a_2b_1^2\right\}$ $r_1=(b_1+b_2)(a_1b_1b_2-a_2b_1^2-b_2^2)$	ω_n - natural frequency ξ - damping factor PID B-1 pole placement controller
14 pp2b_2.m	$u(k)=-\left[(q_0'+\beta)y(k)-(q_0'+q_2')y(k-1)+q_2'y(k-2)\right]-$ $-(\gamma-1)u(k-1)+\gamma u(k-2)+\beta w(k)$ $q_0'=-\frac{r_2-r_3+r_4}{r_1}$; $q_2'=\frac{r_6+r_7}{r_1}$; $\gamma=\frac{r_5}{r_1}$; $\beta=\frac{x_1+x_2-x_3+x_4}{b_1+b_2}$ $r_1=(b_1+b_2)(a_1b_1b_2-a_2b_1^2-b_2^2)$; $r_2=a_1b_2\left[b_1(x_2+x_3+x_4)-b_2x_1\right]$ $r_3=a_2b_1\left[b_2x_1-b_1(x_2-x_3+x_4)\right]$; $r_4=(b_1+b_2)\left[b_1x_4+b_2(-x_3-x_4)\right]$ $r_5=b_1(b_1^2x_4-b_1b_2x_3+b_2^2x_2)-b_2^3x_1$; $r_6=b_1^2(-a_2x_3+a_1x_4-a_2x_4)$ $r_7=b_2\left[b_1(a_1x_4+a_2x_2-x_4)-b_2(a_2x_1+x_4)\right]$	α, ω - real and imaginary component of the pole PID B-2 pole placement controller
15 zn2pd.m	$u(k)=K_P\left\{w(k)-y(k)+\frac{T_D}{T_0}\left[y(k-1)-y(k)\right]\right\}$ $K_P=0.4K_{Pu}$; $T_D=\frac{T_U}{20}$	K_{Pu}; T_u; T_0 PD controller
16 zn2pi.m	$u(k)=q_0e(k)+q_1e(k-1)+u(k-1)$ $q_0=K_P\left(1+\frac{T_0}{2T_I}\right)$; $q_1=-K_P\left(1-\frac{T_0}{2T_I}\right)$	K_{Pu}; T_u; T_0 PI controller
17 mv2.m	$u(k)=\frac{1}{q}\left[a_1y(k-1)+a_2y(k-2)-b_1u(k-1)-b_2u(k-2)+w(k)\right]+u(k-1)$	q - penalisation factor minimum variance controller

Notes:
m-file name: **xx(x)a(yyyy).m**

xx(x) - controller type, **n** = 2 or 3 - process model order, **(yyyy)** - controller further details

Controllers number 1,2,3,4,6,7,8,16 and 17:

$K_P=0.6K_{Pu}$; $T_I=0.5T_u$; $T_D=0.125T_u$

Controllers number 11 and 13:

$d_1=-2\exp(-\xi\omega_nT_0)\cos(\omega_nT_0\sqrt{1-\xi^2})$ for $\xi\le 1$, $d_1=-2\exp(-\xi\omega_nT_0)\cosh(\omega_nT_0\sqrt{1-\xi^2})$ for $\xi>1$,

$d_2=\exp(-2\xi\omega_nT_0)$

Controllers number 12 and 14:

$x_1=c+1-a_1$; $x_2=d+a_1-a_2$; $x_3=f+a_2$; $x_4=g$

$c=-4\alpha$; $d=6\alpha^2+\omega^2$; $f=-2\alpha(2\alpha^2+\omega^2)$; $g=\alpha^2(\alpha^2+\omega^2)$

3. TOOLBOX DESCRIPTION

The toolbox has two different forms. It can be considered as a SIMULINK block library of adaptive controllers. In this case it is a completely open system based on SIMULINK features. The second form is a closed one, controlled by a GUI (Bobál, *et all.*, 1999, Bobál and Böhm, 2000). Here all actions are guided by prepared graphical windows that checks any action and leads the user step by step to correct settings. This version is typically efficient in education.

These two versions differ in the upper level appearance, however, use the same or slightly different lower level functions. The SIMULINK oriented part uses only SIMULINK features. The choice of a specific controller, setting its parameters connection to the system, use of additional blocks etc. all must be done by user. Compared with a GUI form this offers, on one side, large flexibility, on the other side, it requires some knowledge and

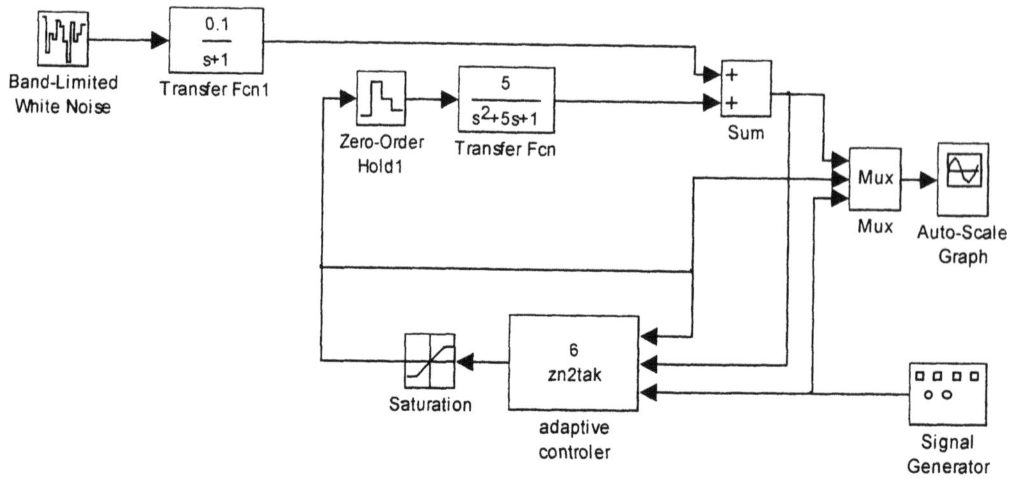

Fig. 3 Typical SIMULINK scheme of self-tuning controller

responsibility to avoid misuse of the controller.

The survey of the individual controllers with name of the m-files and requisite input parameters is contained in Table 1. The detailed description of all controller algorithms with the recursive identification procedure is introduced in Bobál, *et al.* (1999a). This Toolbox is suitable for real-time control using the acquisition card ADVANTECH PCL-812 PG with Extended Real Time Toolbox (see User's Manual, 1996), too.

5. CONCLUSIONS

The Toolbox is widely used in the education, laboratory training and some practical applications. Its structure can address various users. The example of a typical SIMULINK scheme of adaptive controller is shown in Fig. 3.

The MATLAB - Toolbox is used not only for simulation of designed self-tuning controllers but it is used for the control a number of laboratory equipments in practical courses. PID controllers have been implemented e. g. for control of the coupled tanks apparatus, ball and plate model, electrical analog models.

The purpose of this contribution was to arrange the brief informative User's Manual about MATLAB-Toolbox.

ACKNOWLEDGEMENTS

This work was supported in part by the Grant Agency of the Czech Republic under grants No.102/99/1292, 102/00/0526 and in part by the Ministry of Education of the Czech Republic under grant MSM 281100001.

REFERENCES

Åström, K. J., C. Hang, P. Pearson and W. K. Ho (1992). Towards intelligent PID control, *Automatica*, **28**, 1-9.

Bányász, Cs. and L. Keviczky (1993). Design of adaptive PID regulators based on recursive estimation of the process parameters. *Journal of Process Control*, **3**, 1993, 53-59.

Bobál, V., J. Böhm, J. Prokop and J. Fessl (1999a). *Practical Aspects of Self-Tuning Controllers: Algorithms and Implementation.* VUTIUM Press, Brno University of Technology, Brno, (in Czech).

Bobál, V., J. Böhm, and R. Prokop (1999b). Practical aspects of self-tuning controllers. *International Journal of Adaptive Control and Signal Processing*, **13**, 1999, 671-690.

Dahlin, D. B. (1968). Designing and tuning digital controllers. Inst. Cont. Syst., 42, 1968, 77-83.

Isermann, R., *Digital Control Systems*, Springer Verlag, Berlin, 1989.

Kulhavý, R. (1987). Restricted exponential forgetting in real time identification. *Automatica*, **23**, 586-600.

Takahashi, Y., C. Chan and D. Auslander (1971). Parametereinstellung bei linearen DDC-Algorithmen, *Regelungstech. Prozessdatenver.*, **19**, 237-284.

User's Manual (1996). *Real Time Toolbox for Use with MATLAB.* Humusoft s. r. o. Praha.

Ziegler, J. G. and N. B. Nichols (1942). Optimum settings for automatic controllers. *Trans. ASME*, **64**, 759-768.

FEEDBACK-LOOP MONITORING FOR CONTROLLER FALSIFICATION

Tommaso Agnoloni * David Angeli * Edoardo Mosca *

* *Dip. Sistemi e Informatica, Università di Firenze*
Via di S. Marta 3, 50139 Firenze (Italy)
email: {agnoloni,angeli,mosca}@dsi.unifi.it

Abstract: The paper tackles the controller falsification problem in the framework of switching supervisory control. A comparison is carried out between two approaches: one is based on on-line testing candidate Lyapunov functions, the other on a suitable statistic which allows to detect divergence drifts in the currently operating loop.

Keywords: Fault tolerant control; Switching supervisory control; Lyapunov methods; Statistical approaches to fault detection.

1. INTRODUCTION

A well-estabilished approach to the control problem of uncertain plants is the use of adaptation. However, conventional continuous adaptation is not always capable of performing satisfactorily. This may be particularly true whenever the plant switches among different modes of operation and/or if not enough excitation is present. Under such circumstances undesirable transients typically arise because of slow adaptation. In recent years, *switching supervisory control* (SSC) has emerged as an alternative approach for tackling the problem with the appealing feature of resembling an adaptive version of classic gain-scheduling control, (Morse, 1996), (Narendra and Balakrishnan, 1997). As a matter of fact, SSC aims at extending gain-scheduling control to cases where the supervisor has no full information on the current dynamical behaviour of the plant to be controlled. A typical situation is the one where only records of past plant I/O data are available in order to let the supervisor decide whether the current controller is adequate, and, in the negative, select another candidate controller. This is in fact similar to what happens in adaptive control wherein I/O data are used so as to choose in real-time a controller from a continuum of possible feedback laws. If the unknown plant belongs to a certain pre-specified set of models, the idea is to have a set of candidate controllers $\mathcal{C}_k$, usually finite, so that each plant model performs "satisfactorily" when controlled by at least one of the $\mathcal{C}_k$'s. Then, a suitably designed supervisory unit takes care of orchestrating the switching among the candidate controllers so as to ensure closed loop stability and possibly performance in the face of uncertainty or even in the presence of a sufficiently slow drift of the uncertain parameters. Within SSC the fundamental problems to be tackled are the following two: controller falsification, and inference of candidate loop behaviour. *Controller falsification* (Safonov and Tsao, 1997), (Kosut, 1998) deals with the problem of finding criteria according to which the supervisor decides whether the acting controller is adequate or not. This is a binary decision problem: if the controller is falsified, the supervisor replaces the formerly acting controller with a different one selected amongst the available collection of candidate controllers. In order to rule out even temporarily the use in the loop of unsatisfactory controllers, such a selection should be made without directly checking the behaviour of all candidate controllers via their effective use in the control system. *Inference of candidate loop behaviour* (Mosca and Agnoloni, 2001) deals with the problem of inferring the behaviour of the potential loop made up by a candidate controller connected in feedback to the uncertain plant on the grounds of data taken from the currently operating loop.

With particular attention to its application to SSC, the objective of this paper is to focus on the issue of closed-loop performance monitoring and controller falsification. To this end, two different approaches will be described and compared. One is based on on-line testing candidate Lyapunov functions, the other on a suitable statistic which allows to detect the occurrence of divergence drifts in the currently operating loop.

2. UNCERTAIN FEEDBACK SYSTEM

Let the plant be a discrete-time SISO linear dynamic system $\mathcal{P}(\gamma)$ given in a finite-difference representation

$$\mathcal{P}(\gamma) \;:\; A_\gamma(d)y(t) = B_\gamma(d)\delta u(t) + n(t) \quad (1)$$

where $y(t)$ is the plant output, $\delta u(t)$ the input-increment, $n(t)$ an exogenous disturbance, $A_\gamma(d)$ and $B_\gamma(d)$ two polynomials

$$\begin{aligned} A_\gamma(d) &= \Delta(d)a_\gamma(d) = 1 + a_1(\gamma)d + \ldots + a_{n_a}(\gamma)d^{n_a} \\ \Delta(d) &= 1 - d \qquad\qquad (2) \\ B_\gamma(d) &= b_1(\gamma)d + \ldots + b_{n_b}(\gamma)d^{n_b} \end{aligned}$$

where d denotes the unit backward-shift operator, $dy(t) = y(t-1)$, and γ an uncertain parameter vector taking values in a compact set $\Gamma \subset \mathbb{R}^{n_\gamma}$. Along with the family of plants (1), consider a one-degree of freedom linear controller

$$\mathcal{C} \;:\; R(d)\delta u(t) = -S(d)[y(t) - r(t)] \quad (3)$$

The feedback control system (1) and (3) is denoted by $(\mathcal{P}(\gamma)/\mathcal{C})$. Define $\mathcal{S} = \{\gamma \in \Gamma : (\mathcal{P}(\gamma)/\mathcal{C})$ is stable$\}$, as the ***stability set***, and $\mathcal{D} \subseteq \mathcal{S}$ where some performance specifications are met.

Our aim is to provide criteria and related tests by which to detect divergence and even possibly performance degradation. In the present note two different approaches will be described. One is based on on-line testing candidate Lyapunov functions, the other on a suitable statistic in the form of a ratio of closed-loop variables.

3. LYAPUNOV-BASED FALSIFICATION

The central assumption in this approach is that performance specifications be given, for a tuned-loop, in terms of Lyapunov dissipation inequalities. Specifically, each controller is designed in order to fulfill in a neighborhood Γ_i, $\Gamma_i \subset \Gamma$, of some nominal value γ_i° of the uncertain parameter a suitable matrix inequality. Since such inequalities are usually formulated in the state-space domain, it is convenient to introduce a matrix representation of the system. In particular, we adopt the following choice for the plant state $\chi(t) \doteq [y(t), y(t-1), \ldots, y(t-n_a+1), \delta u(t-1), \delta u(t-2) \ldots \delta u(t-n_b+1)]'$, whose time-update can be represented as follows

$$\begin{aligned} \chi(t+1) &= \Phi_\gamma \chi(t) + G_\gamma \delta u(t) + Ln(t) \\ y(t) &= H\chi(t) \end{aligned} \qquad (4)$$

where Φ_γ, G_γ, L and H are defined in the obvious way. Notice that this choice of coordinates, though corresponding to a non-minimal realization, allows to interpret dynamic output feedback controllers in (3) as static state-feedback ones for the extended system (4). In particular we denote by $\Phi_\gamma^{cl} := \Phi_\gamma + G_\gamma F$, where F is the state feedback corresponding to controller (3). Accordingly, we may define $\mathcal{D}$ as the subset of Γ where matrix inequalities of the following type are fulfilled for some $P = P' > 0$:

$$\Phi_\gamma^{cl'} P \Phi_\gamma^{cl} \leq \lambda P$$

$$\begin{bmatrix} \Phi_\gamma^{cl'} P \Phi_\gamma^{cl} - P + H'H & \Phi_\gamma^{cl'} PL \\ L'P\Phi_\gamma^{cl} & L'PL - \sigma^2 I \end{bmatrix} < 0. \quad (5)$$

In (5) the first LMI enforces an upper-bound ($\lambda^{1/2}$) to the spectral radius of the closed-loop system. The second LMI instead guarantees an r.m.s.-gain from n to y smaller than σ. Of course, since the plant parameters are not apriori known we cannot check directly the validity of (5). Nevertheless, we still have the possibility, on the basis of I/O data, of falsifying a controller, viz. of detecting violations of such inequalities. To this end, defining $V(\chi) = \chi' P \chi$, and using (4) and the first LMI in (5), we get

$$V(\chi(t+1)) < \lambda V(\chi(t)) + n(t)'L'PLn(t) + 2n(t)'L'P\Phi_\gamma^{cl}\chi(t) \quad (6)$$

In a somewhat analogous way, the second LMI reads along trajectories of (4)

$$V(\chi(t+1)) - V(\chi(t)) < -y^2(t) + \sigma^2 n^2(t) \quad (7)$$

If the disturbances were accessible, one could falsify a controller just by checking validity of inequalities (6) and/or (7). In general this is not the case. Therefore, one possibility is to assume that $|n(t)| \leq N$ so that suitable upper-bounds to the right hand side of equations (6) and (7) can be computed. For instance, a falsification test at time t, based on (7), could read:

- if $V(\chi(t)) - V(\chi(t-1)) \geq -y^2(t-1) + \sigma^2 N^2$ then falsify controller. (8)

Notice that satisfaction of (8), even over a semi-infinite horizon, does not guarantee stability of the closed-loop system; however, as it is made precise in the following lemma, it ensures for the particular trajectory under consideration that the system is compatible with the performance of the nominal loop under the action of some disturbance (not necessarily the actual one) within the a priori known bounds.

Lemma. Assume that a controller in the loop is never falsified, viz. for all $t \geq 0$ condition (8) is satisfied. Then, the following average power bound holds:

$$\lim_{n \to +\infty} \frac{1}{n} \sum_{t=0}^{n} y^2(t) \leq \sigma^2 N^2. \quad (9)$$

4. STATISTICAL FALSIFICATION

Here the idea is to detect divergence drifts in the feedback loop by comparing a ratio of appropriate closed-loop variables against a suitably chosen threshold.

Let $r(t)$ be a stepwise reference, $\tilde{y}(t) := y(t) - r(t)$ and $\varphi(t) := [\tilde{y}(t) \ldots \tilde{y}(t - n_a + 1) \quad -\delta u(t) \ldots - \delta u(t - n_b + 1)]'$.

Denoted $\gamma^\circ \in \mathcal{D}$ as the "nominal" value of the uncertain parameter in (1), the one-step ahead prediction error of $\tilde{y}(t)$ based on γ° is given by

$$\varepsilon(t) = A_{\gamma^\circ}(d)\tilde{y}(t) - B_{\gamma^\circ}(d)\delta u(t) \quad (10)$$

We then introduce the statistical test

$$g(t) := \frac{\|\varphi(t)\|}{\max|\varepsilon^t|} \begin{matrix}(\leq)\\(>)\end{matrix} T \quad (11)$$

where $\max|\varepsilon^t| := \max_{t_0 \leq k \leq t} |\varepsilon(k)|$, in order to discriminate between the two mutually exclusive hypothesis $h_{\bar{F}} := \{\gamma \in \mathcal{D}\}$ and $h_F := \{\gamma \in \mathcal{S}^c\}$, being t_0 the time at which C is switched on in feedback with $\mathcal{P}(\gamma)$. The threshold T should be chosen, provided that this is possible, in such a way that

$$P_D(T) := \mathbb{P}\{g(t) > T, \text{ some } t \in [t_0, \infty) | h_F\} = 1 \quad (12)$$

$$P_{FA}(T) := \mathbb{P}\{g(t) > T | h_{\bar{F}}\} \simeq 0 \quad (13)$$

where $\mathbb{P}\{Q\}$ stands for probability of the event Q, and P_D and P_{FA} are to be interpreted as probability of detection of faults and, respectively, probability of false alarms. In (Mosca and Agnoloni, 2000), assuming $n(t) = A(d)w(t)$ in (1), $w(t)$ i.i.d. (white) sensor noise, conditions are provided under which (12) and (13) are satisfied. However, whiteness of $w(t)$ is a limitative assumption which can be removed by appropriately filtering the variables entering the falsification test, provided that the noise colour, or at least an estimate of the latter, be known.

The choice of the ratio (11) is based on a result provided by a suitable version of the small gain stability lemma, see (Mosca and Agnoloni, 2000). It ensures that, if $\|\varphi(t)\|$ diverges, the statistic $g(t)$ must be greater than a number depending on the size of Γ. Then, if for values of γ in the stability set, the ratio $g(t)$ takes on values smaller than the above mentioned number, one can use in (11) a threshold T under which (12) and (13) can be both satisfied.

One of the important features of the test (11) is that, being expressed as a ratio of closed-loop variables, it turns out to be insensitive to the noise intensity. This allows one to choose the threshold, irrespective of the noise intensity.

Suitable bounds for the threshold can be computed in an analytical way or, for less conservative results, via simulations whereby the test variable $g_\gamma(t)$ is computed for values of γ varying over the whole uncertainty set. For practical applications the ratio $g(t)$ should then be computed from closed-loop I/O pairs and prediction errors collected on-line over a time window of length T_w in order to allow γ time-variations. Thus, (11) practically becomes

$$g(t) := \frac{\|\varphi(t)\|}{\max_{t-T_w \leq k \leq t} |\varepsilon(k)|} \begin{matrix}(\leq)\\(>)\end{matrix} T \quad (14)$$

5. AN EXAMPLE: DOUBLE CART WITH ELASTIC COUPLING

In order to clarify the concepts introduced so far, let us consider an application of the two described falsification tests in the context of Switching Supervisory Control. The plant, depicted in Fig. 1, consists of two carts moving along a straight horizontal line. The two carts are coupled via an elastic force proportional to the deformation of the spring. Discretizing the model with sampling time

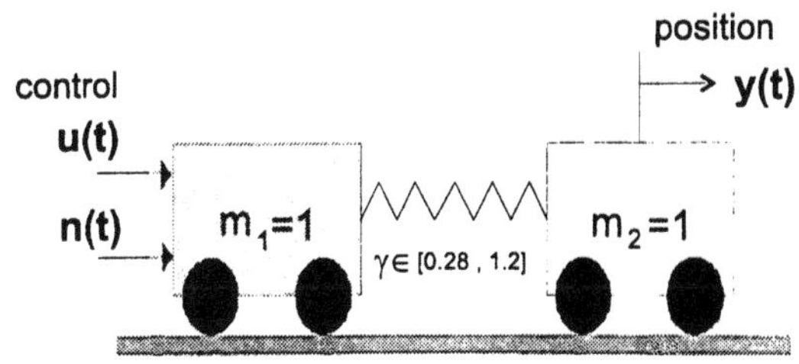

Fig. 1. Plant

T and taking y as the output of the system we obtain the following discrete time transfer function from the input force u to y

$$y = \frac{\gamma T^4 d}{(d-1)^2[d^2 + (2\gamma T^2 - 2)d + 1]} u \quad (15)$$

The elasticity constant γ plays the role of an uncertain parameter and is only known to belong to the interval $[0.28, 1.2]$. The size of this interval prevents one from finding a single stabilizing linear controller with adequate performance. Our aim is to design a supervisory logic in order to switch between a certain number of dynamic controllers according to past input-output data. (See Fig. 2). In the simulations reported hereafter we let

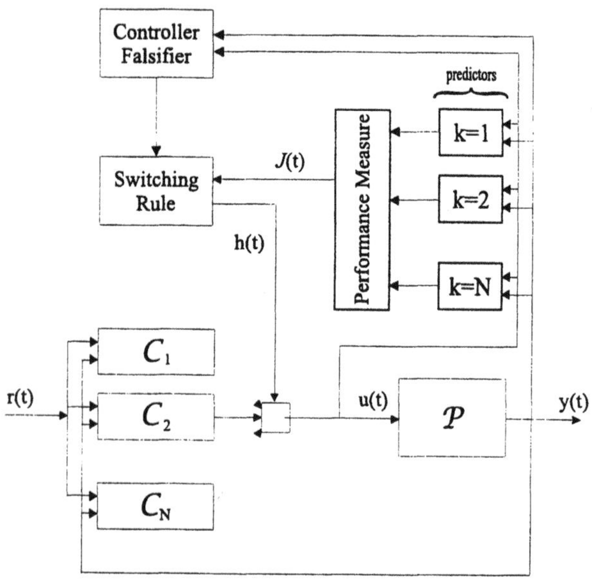

Fig. 2. SSC system with controller falsification

$T = 0.1$ *sec.* and all variables expressed in MKS units. The synthesis of the candidate controllers was carried out by choosing 4 nominal values of the stifness parameter, namely $\gamma_1 = 0.3$, $\gamma_2 = 0.5$, $\gamma_3 = 0.7$, $\gamma_4 = 1$ and designing for each of them a "tuned" weighted $\mathcal{H}_\infty$ mixed sensitivity controller. This design criterion is particularly suited for the adopted statistical approach as pointed out in (Mosca and Agnoloni, 2000). In order to have a common ground for comparison we employed this same set of controllers in the experimental set-up of the Lyapunov-based approach. Using a prescribed set of controllers as a bank of candidate stabilizers for a Switching Supervisory Logic based on the statistical approach suffices that the plant be stabilized by at least one of the controllers for each γ in the uncertainty range. The Lyapunov method instead needs a stronger covering condition. Namely a set of Lyapunov functions must be known such that, for each γ in the uncertainty range, there exists a controller and a Lyapunov function which is decreasing as in (7) for that particular value of the uncertain parameter. For this check, linearity of the plant realization with respect to unknown parameters is essential. Under such an assumption it is easier to design a polytopic covering of the uncertainty set and this allows to check the covering condition just by solving a finite number of LMIs, each one corresponding to suitable Lyapunov inequalities at the vertices of the prescribed polytopes. The following subintervals S_i were adopted: $S_1 : [0.28, 0.45]$, $S_2 : [0.45, 0.58]$, $S_3 : [0.58, 0.82]$, $S_4 : [0.82, 1.2]$. It turned out that both covering conditions were satisfied by the above controllers. However, in general, better results can be achieved if the synthesis method is tailored to the falsification logic. In order to verify robustness of the method we performed simulations in the presence of exogenous input disturbances. This is particularly suited for the Lyapunov based approach, since it is essentially a state-space method (Angeli and Mosca), which assumes exact knowledge of the present state (in this case a vector of I/O data) and allows for additive process disturbances of apriori known magnitude. The statistical approach, instead, due to the peculiar synthesis method, is known to behave correctly when the equivalent disturbance at the plant output is white or at least of known colour. The latter information allows one to recast the problem as in the white output disturbance case through a whitening filtering. Whenever a controller is falsified, the next controller for the closed-loop is chosen among the three remaining candidates on the basis of performance signals generated through the filtering of prediction errors as in (Mosca and Agnoloni, 2001). In the statistical method, once a falsification takes place, a new test can be performed only after enough time (dwell-time) has elapsed to let the statistic discard past data and let the effect of initial conditions decay. On the other hand the Lyapunov falsification method operates on the basis of istantaneous test of functions thus allowing faster recovery in case of wrong commutation. This implies that in the latter case a rougher inference logic can be tolerated.

The simulations were run with γ piecewise constant as can be seen in Fig. 3(b) and 4(b); in the same figure we plot the nominal value of γ corresponding to the controller acting in the loop. A white uniformly distributed input noise of suitable amplitude was taken into account and only roughly known upperbounds of it were used in order to tune the Lyapunov based algorithm as shown in Fig. 3(d). Fig. 4(d) refers to the filtered falsification signal, plotted against its threshold. A quick review of the merits of the two approaches is detailed in Table 1.

	Lyapunov-based falsification	Statistical falsification
Uncertainty covering	common Lyapunov functions: large number of controllers	robust stability regions: small number of controllers
Falsification test	increment of Lyapunov functions	ratio between closed-loop variables
Noise	known bounded amplitude	known colour and unknown amplitude
Performance	no dwell-time insensitive to noise colour sensitive to noise amplitude	dwell-time insensitive to high freq. noise and noise amplitude sensitive to low freq. noise
Uncertainty	plant model linear in the parameter	any

Table 1. Comparison of the two approaches.

6. CONCLUSIONS

Controller falsification is a fundamental problem in SSC where one of the two main supervisor tasks is to decide whether the acting controller is adequate for the current operating conditions or it should be discarded and replaced by a different candidate controller.

To this end two different approaches were proposed and compared: a falsification test based on a worst-case evaluation of Lyapunov functions variations and a statistical test based on the ratio of closed loop variables. Such tests can be easily embodied in a supervisory logic for SSC. Their effectiveness was tested on a benchmark example where the algorithm was shown to be prompt to detect incipient divergence due to plant parameter variations even in the absence of any reference change and in the only presence of measurement noise.

ACKNOWLEDGEMENTS

This work was partially supported by the Italian Ministry for University and Research in Science and Technology (MURST) under the Project "Fault detection and diagnosis, supervision and control reconfiguration in industrial process automation".

References

Angeli, D. and E. Mosca, "Lyapunov based supervisory control of uncertain nonlinear systems", submitted to *IEEE Trans. on Automatic Control.*

Kosut, R.L. (1998). "Uncertainty model unfalsification for robust adaptive control", *Proc. IFAC Workshop on Adaptive Systems in Control and Signal Processing*, pp. 1-11, Glasgow, U.K.

Kulkarni, S.R. and P. J. Ramadge (1996). "Model and controller selection policies based on output prediction errors", *IEEE Trans. on Automatic Control*, Vol. 41, pp. 1594-1604.

Kwakernaak, H. (1991). "The polynomial approach to H_∞-optimal regulation", in E. Mosca and L. Pandolfi (Eds.): *H_∞-Control Theory.* Lecture Notes in Mathematics n. 1496, Springer, pp. 141-221.

Morse, A. S. (1996). "Supervisory control of families of linear set-point controllers- part 1: Exact matching", *IEEE Trans. on Automatic Control*, Vol. 41, pp. 1413-1431.

Morse, A. S. (1998). "Control using logic based switching", in A. Isidori editor, *Trends in control: a European Perspective*, pp. 69-113.

Mosca, E. and T. Agnoloni (2000). "Controller falsification for self-scheduling control of uncertain noisy plants", *Proc. of the 14th Mathematical Theory of Networks and Systems*, Perpignan, France.

Mosca, E. and T. Agnoloni (2001). "Inference of candidate loop performance and data filtering for switching supervisory control", *Automatica.* Vol. 37, pp. 527-534.

Narendra, K.S. and J. Balakrishnan (1997). "Adaptive control using multiple model", *IEEE Trans. on Automatic Control*, Vol. 42, pp. 171-187.

Pait, F.M. and F. Kassab Jr. (1999). "On a class of switched, robustly stable, adaptive systems", *Proc. of the 38th Conference on Decision and Control*, Phoenix, AZ, U.S.A.

Safonov, M.G. and T. C. Tsao (1997). "The unfalsified control concept and learning", *IEEE Trans. on Automatic Control*, Vol. 42, pp. 843-847.

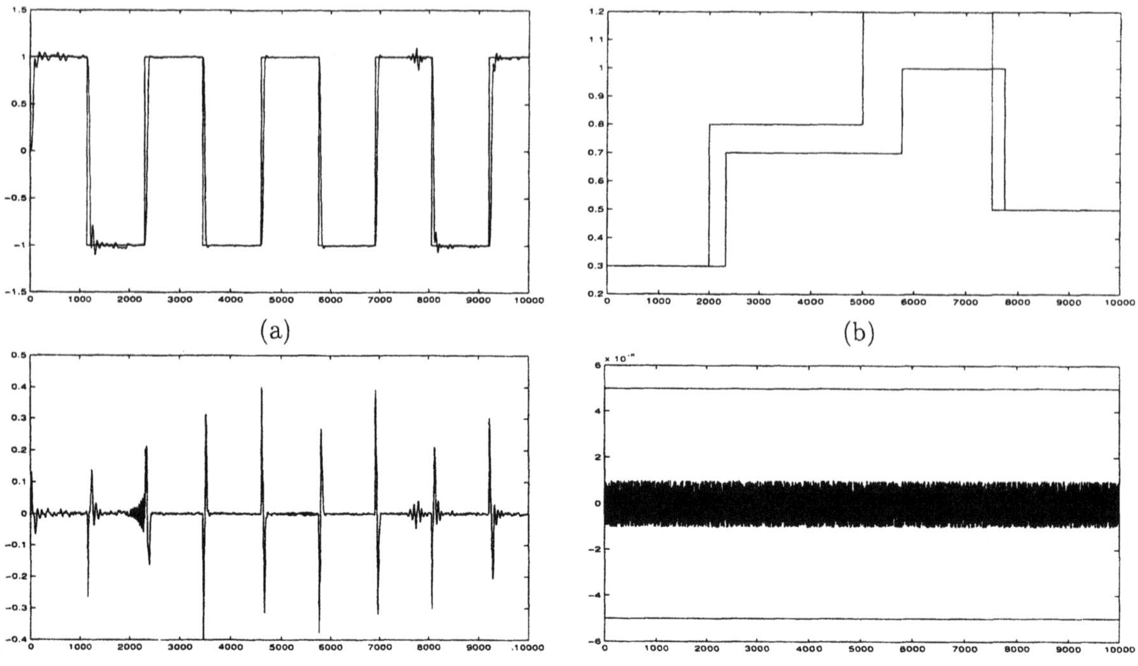

Fig. 3. Lyapunov falsification - (a) Position of cart No. 2; (b) Parameter variation and controller selection; (c) Input signal; (d) Noise and upperbound.

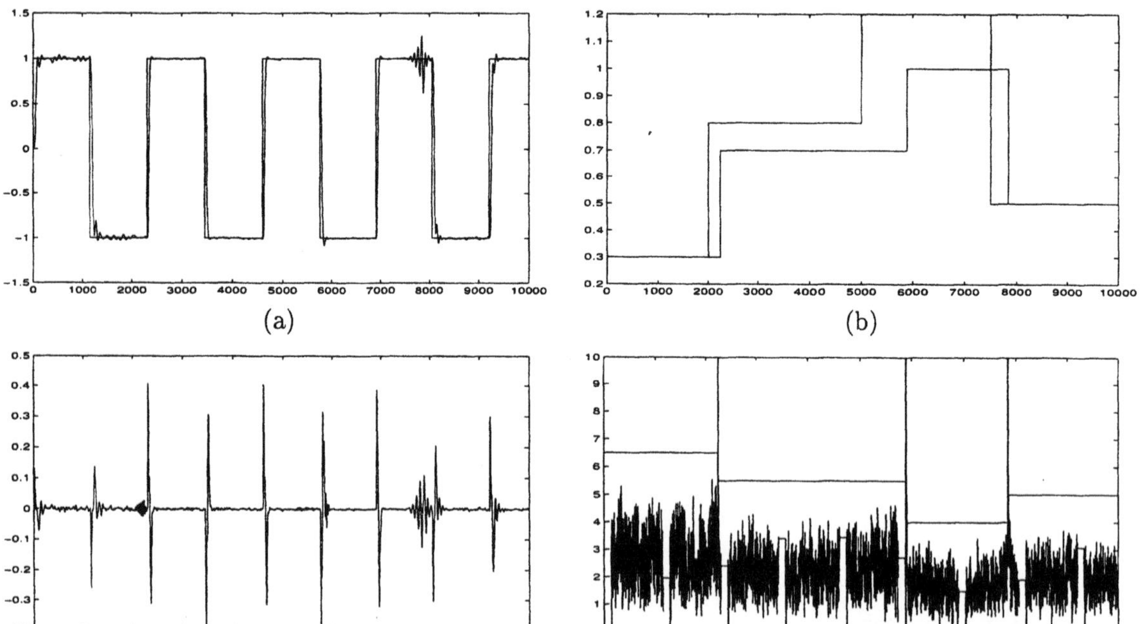

Fig. 4. Statistical falsification - (a) Position of cart No. 2; (b) Parameter variation and controller selection; (c) Input signal; (d) Falsification signal and thresholds.

DESIGN METHOD FOR ROBUST SUPERVISOR CONTROLLER

Toshihiko Nakajima *Hiromitsu Ohmori ** Akira Sano **

* *School of Electric Engineering, Keio University*
** *Department of System Design Engineering, Keio University*

Abstract: We suggest a method to design supervisor which guarantees robust stability for linear time-invariant systems in the presence of the uncertainty via output feedback. The system is a composite of continuous-time uncertain plant and switched controller. The controller is defined by a collection of given controllers whose state is shared. Furthermore the proposed supervisor controller determines the mode of the controller to guarantee robust stability of the total closed-loop system. A sufficient condition of robust stability is given in terms of an algebraic Riccati equation and bilinear matrix inequalities(BMI). *Copyright © 2001 IFAC*

Keywords: supervisory control, switching, robustness, bilinaer, hybrid

1. INTRODUCTION

Recently the effectiveness of supervisory controller is pointed out, which is capable of switching into feedback of a controller from a family of candidate controllers. One advantage of supervisory control is that it can achieve closed-loop performance objectives that cannot be achieved using classical linear or nonlinear controllers (J.A.De Dona and G.C.Goodwin, 1999), (N.H.Mcclamroch and I.Kolmanovsky, 2000). Moreover it is effective for real-time changes in plant dynamics (A.S.Morse, 1996), (A.S.Morse, 1992), (Edoardo Mosca, 1999), where the nominal plant is a member of a known class of admissible set. In some situations the choice of linear or nonlinear controllers available to the designer is limited and the design task is to use the available set of controllers.

In this paper, we consider such a situation and output feedback robust stabilization problem for a class of uncertain linear time-invariant systems. The controller is selected from a set of given linear controllers whose state is shared.

We are motivated mainly by (A.V.Savkin and R.J.Evans, 1999). Compared with it, the contributions of this paper are that solving BMI for designing the switching-law is used, and the controller is allowed to have dynamics. Then, a sufficient condition for robust stability is given in terms of the existence of solutions of an algebraic Riccati equation for an observer and BMI. In order to verify the effectiveness of the proposed switching system, numerical simulations are carried out.

2. PROBLEM STATEMENT

We consider the linear time-invariant uncertain system $\boldsymbol{P}$ defined on the infinite time interval $[0, \infty)$,

$$\boldsymbol{P}: \begin{cases} \dot{\boldsymbol{x}}_p(t) = \boldsymbol{A}\boldsymbol{x}_p(t) + \boldsymbol{B}_1\boldsymbol{\xi}(t) + \boldsymbol{B}_2\boldsymbol{u}(t), \\ \boldsymbol{z}(t) = \boldsymbol{C}_1\boldsymbol{x}_p(t), \\ \boldsymbol{y}(t) = \boldsymbol{C}_2\boldsymbol{x}_p(t) + \boldsymbol{v}(t), \end{cases} \tag{1}$$

where $\boldsymbol{x}_p(\cdot) \in \Re^{n_p}$ is the state of the plant, $\boldsymbol{\xi}(\cdot) \in \Re^{p}, \boldsymbol{v}(\cdot) \in \Re^{\ell}$ are uncertainty inputs, $\boldsymbol{u}(\cdot) \in \Re^{h}$ is the control input, $\boldsymbol{z}(\cdot) \in \Re^{m}$ is the performance output, and $\boldsymbol{y}(\cdot) \in \Re^{\ell}$ is the measured output, $\boldsymbol{A} \in \Re^{n_p \times n_p}, \boldsymbol{B}_1 \in \Re^{n_p \times p}, \boldsymbol{B}_2 \in \Re^{n_p \times h}, \boldsymbol{C}_1 \in$

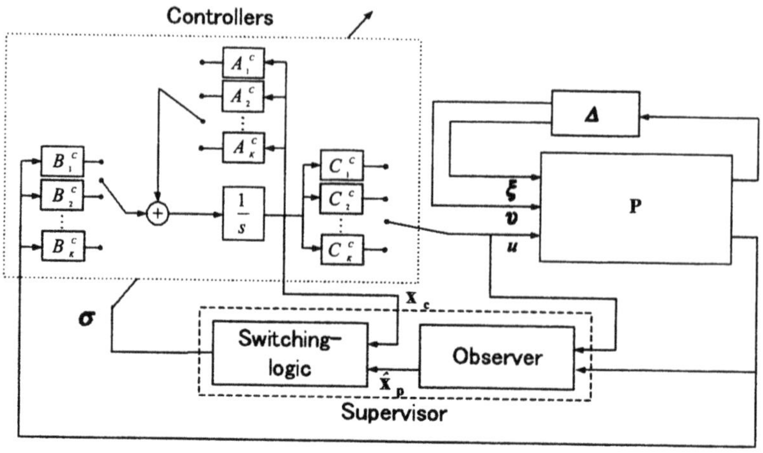

Fig. 1. Supervisory control system

$\Re^{m\times n_p}$, and $C_2 \in \Re^{\ell\times n_p}$ are given constant matrices.

Assumption 1. P admits a realization (1) satisfying,
(i) (C_2, A) are detectable,
(ii) (A, B_2) are stabilizable.

Suppose we have a collection of given linear output feedback controllers $\{K_i\}$ whose state is shared. The design task is to decide the switching-time and the controller to switch. Then the controller is,

$$K_\sigma : \begin{cases} \dot{x}_c(t) = A^c_{\sigma(t)} x_c(t) + B^c_{\sigma(t)} y(t), \\ u(t) = C^c_{\sigma(t)} x_c(t), \\ \sigma(\cdot) \in M = \{1, 2, \cdots, K\}, \end{cases} \tag{2}$$

where $x_c(\cdot) \in \Re^{n_c}$ is the state of the controller, $A^c_i \in \Re^{n_c\times n_c}, B^c_i \in \Re^{n_c\times\ell}, C^c_i \in \Re^{h\times n_c}$, $i = 1, 2, \cdots, K$ are given constant matrices, $\sigma(t)$ is the switch index which is active, and M is the set of symbols. Note that it is allowed for controllers to own dynamics against the paper (A.V.Savkin and R.J.Evans, 1999).

We will consider uncertainty inputs $\xi(\cdot), v(\cdot)$ such that the following Integral Quadratic Constraint holds,

Assumption 2. (Integral Quadratic Constraint) For any locally square integrable control input $u(\cdot)$ for which any solution to Eq.(1) with $[u(\cdot), \xi(\cdot), v(\cdot)]$ exists on $[0, \infty)$ and there exists a constant $d \geq 0$ such that,

$$\int_0^t (\|\xi(t)\|^2 + \|v(t)\|^2)dt \leq d + \int_0^t \|z(t)\|^2 dt, \quad t \in [0, \infty). \tag{3}$$

In this paper $\|\cdot\|$ denotes the standard Euclidean norm, $L_2[0,t)$ denotes the Hilbert space of square integrable vector valued functions defined on $[0,t)$ where $t \in (0, \infty]$.

We now introduce a corresponding notion of robust stability for the uncertain system Eqs. (1)-(3) . Here, we define the composite state

$$\bar{x}(t) = \begin{bmatrix} \hat{x}_p(t) \\ x_c(t) \end{bmatrix}. \tag{4}$$

Definition 1. The system Eqs. (1)-(3) is said to be stable, if there exists a constant $c > 0$ and a function $\hat{V}(\bar{x}) : \Re^{n_p+n_c} \to \Re$ such that $\hat{V}(\bar{x}(0)) \geq 0, \hat{V}(\mathbf{0}) = 0$, such that the following inequality holds.

$$\int_0^\infty (\|x_p(t)\|^2 + \|\xi(t)\|^2 + \|v(t)\|^2)dt \leq \hat{V}(\bar{x}(0)) + c\,[\|\hat{x}_p(0) - x_p(0)\|^2 + d\,], \tag{5}$$

where $\hat{x}_p(0)$ is an expected value of initial condition $x_p(0)$.

3. H_∞ OBSERVER

For $\epsilon > 0$, we involve the following algebraic Riccati equation

$$AP + PA^T + P[C_1^T C_1 + \epsilon I - C_2^T C_2]P$$

$$+B_1B_1^T = \mathbf{0}. \quad (6)$$

The H_∞ observer is described as

$$\dot{\hat{x}}_p(t) = [A + P[C_1^T C_1 + \epsilon I - C_2^T C_2]\,]\hat{x}_p(t) + PC_2^T y(t) + B_2 u(t). \quad (7)$$

The following lemma plays an important rule to design supervisor.

Lemma 1. (A.V.Savkin and R.J.Evans, 1999) Let $\hat{x}_p(0)$ be a given vector, $\epsilon > 0$ and $\tilde{V}(\bar{x}(0))$ be given constants. Suppose that a matrix $P > 0$ is a solution for the algebraic Riccati equation (ARE) (6). Then if the condition

$$\int_0^{t_N} (\|C_1\hat{x}_p(t)\|^2 + \epsilon\|\hat{x}_p(t)\|^2 - \|C_2\hat{x}_p(t) - y(t)\|^2)dt \leq \tilde{V}(\bar{x}(0)), \quad (8)$$

holds, the following condition holds for all solutions to system (1), with any inputs $\xi(\cdot)$ and $v(\cdot)$ belong to $L_2[0, t_N]$.

$$\int_0^{t_N} (\|z(t)\|^2 + \epsilon\|x_p(t)\|^2 - \|\xi(t)\|^2 - \|v(t)\|^2)dt \leq (x_p(0) - \hat{x}_p(0))^T P^{-1}(x_p(0) - \hat{x}_p(0)) + \tilde{V}(\bar{x}(0)). \quad (9)$$

If condition (9) holds, by combining it with the Integral Quadratic Constraint (3), we obtain condition of Definition 1, so the total closed system will be stable.

Next, for each $i \in M$, we define matrices

$$\bar{A}_i = \begin{bmatrix} A + P[C_1^T C_1 + \epsilon I - C_2^T C_2] & B_2 C_i^c \\ \mathbf{0} & A_i^c \end{bmatrix}, \quad (10)$$

$$\bar{B}_i = \begin{bmatrix} PC_2^T \\ B_i^c \end{bmatrix}, \quad (11)$$

$$\bar{C}_1 = \begin{bmatrix} C_1^T C_1 + \epsilon I & \mathbf{0} \\ \mathbf{0} & \mathbf{0} \end{bmatrix}^{\frac{1}{2}}, \quad \bar{C}_2 = \begin{bmatrix} C_2 & \mathbf{0} \end{bmatrix}. (12)$$

Using these definitions enable us to obtain the differential equation

$$\dot{\bar{x}}(t) = \bar{A}_{\sigma(t)}\bar{x}(t) + \bar{B}_{\sigma(t)}y(t). \quad (13)$$

Then, we can rewrite the inequality (8) as follows

$$\int_0^{t_N} (\|\bar{C}_1\bar{x}(t)\|^2 - \|\bar{C}_2\bar{x}(t) - y(t)\|^2)dt \leq \tilde{V}(\bar{x}(0)). \quad (14)$$

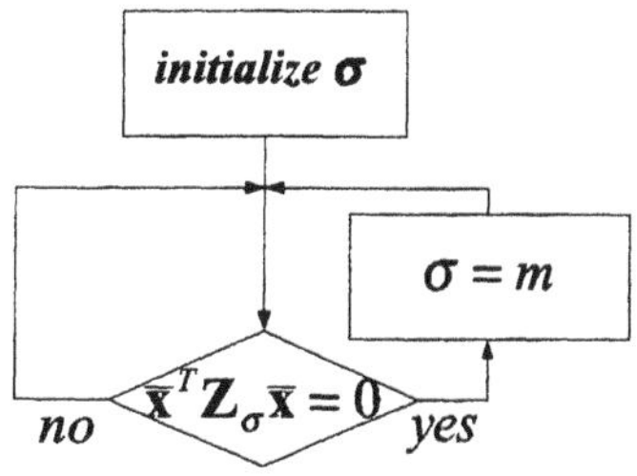

Fig. 2. Switching-law

So our object will be achieved by designing switching-law such that condition (14) holds for the system (13).

4. SWITCHING-LAW AND SUPERVISORY CONTROL

Now we consider the following BMI such that $X = X^T, Z_i = Z_i^T, \mu_i > 0, \tau_i > 0$ and $\gamma_i > 0$, for $\forall i \in M$,

$$\begin{bmatrix} \tilde{A}_i^T X_i + X_i \tilde{A}_i + \bar{C}_1^T \bar{C}_1 + Z_i & X_i \bar{B}_i \\ \bar{B}_i^T X_i & -I \end{bmatrix} < \mathbf{0}, (15)$$

$$X_i - \gamma_i Z_i > \mathbf{0}, (16)$$

$$\tau_1 Z_1 + \tau_2 Z_2 + \cdots + \tau_k Z_K > \mathbf{0}, (17)$$

where

$$\tilde{A}_i = \bar{A}_i + \bar{B}_i \bar{C}_2, \; X_i = X - \mu_i Z_i. \quad (18)$$

Then, we propose the following switching-law,

$$\sigma(t) = \begin{cases} \sigma(t^-) & \text{if } \; \bar{x}(t)^T Z_{\sigma(t^-)}\bar{x}(t) > 0, \\ m(t) & \text{else if } \; \bar{x}(t)^T Z_{\sigma(t^-)}\bar{x}(t) = 0. \end{cases} \quad (19)$$

$$m(t) = \{m \in M \mid \bar{x}(t)^T Z_m \bar{x}(t) = \max_{i \in M} \bar{x}(t)^T Z_i \bar{x}(t)\}. \quad (20)$$

Figure 2 shows the computation diagram of the switching-law Eq. (19).When $\bar{x}(t)^T Z_{\sigma(t^-)}\bar{x}(t)$ is 0, the switch index $\sigma(t)$ is set equal to the value $m(t)$. The next switch index $m(t)$ is defined by Eq. (20), which is the switch index that maximizes the value $\bar{x}(t)^T Z_i \bar{x}(t)$, where Z_i is calculated by BMI (15)-(17). Figure 3 shows an example of the switching surface and the state trajectory. The gray region implies the region $\{\bar{x} : \bar{x}^T Z_1 \bar{x} > 0\} \cup \{\bar{x} : \bar{x}^T Z_2 \bar{x} > 0\}$. The border $\bar{x}^T Z_1 \bar{x} = 0$ is the surface where switching $(1 \rightarrow 2)$ occurs, and $\bar{x}^T Z_2 \bar{x} = 0$ is the surface where switching $(2 \rightarrow 1)$ occurs. Because the two switching surfaces are defferent from each other, this switching-law is robust about some swiching-delay.

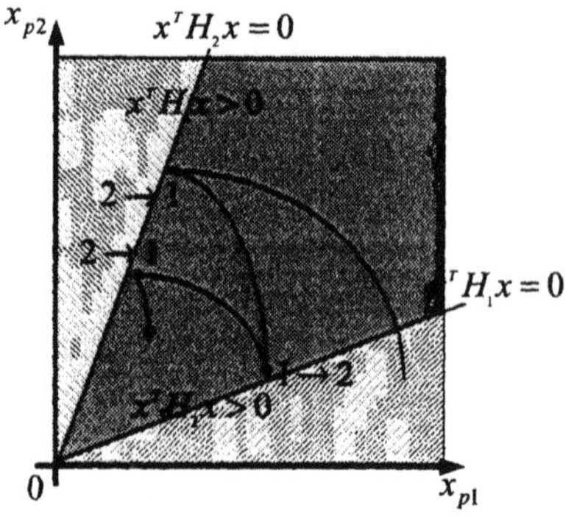

Fig. 3. State trajectory

5. STABILITY ANALYSIS OF TOTAL CLOSED LOOP SYSTEM

In the sequel, the following theorem is obtained.

Theorem 1. Consider the system Eqs. (1)-(3) supervisory controlled by Eqs. (7), (19). Suppose that there exist $\epsilon > 0$, $\boldsymbol{X} = \boldsymbol{X}^T$, $\boldsymbol{Z}_i = \boldsymbol{Z}_i^T, \mu_i > 0, \tau_i > 0$ and $\gamma_i > 0$ such that the following conditions hold:
(i) There exits a solution $\boldsymbol{P} > 0$ to the algebraic Riccati equation Eq. (6) such that the matrix $\boldsymbol{A} + \boldsymbol{P}[\boldsymbol{C}_1^T \boldsymbol{C}_1 + \epsilon \boldsymbol{I} - \boldsymbol{C}_2^T \boldsymbol{C}_2]$ is stable.
(ii) $\forall i \in \boldsymbol{M}$, BMI Eqs. (15)-(17) hold.
Then, the uncertain system Eqs. (1)-(3) can be stabilized by the supervisor Eqs. (7), (19) in the sense of Definition 1.

Now, we give a lemma that is used in this section.

Lemma 2. (S-procedure). Let $\boldsymbol{\Theta}, \boldsymbol{S}_1, \boldsymbol{S}_2, \cdots, \boldsymbol{S}_m \in \Re^{n\times n}$ be symmetric matrices. Then, $\boldsymbol{\zeta}^T \boldsymbol{\Theta} \boldsymbol{\zeta} < 0$, forall $\boldsymbol{\zeta} \neq \boldsymbol{0}$ such that $\boldsymbol{\zeta}^T \boldsymbol{S}_i \boldsymbol{\zeta} \le 0\ (i = 1, 2, \cdots, m)$, if there exist scalars $\tau_i > 0$ such that $\boldsymbol{\Theta} - \sum_{i=1}^{m} \tau_i \boldsymbol{S}_i < 0$.

The derived switching sequence would be of the form

$$S = (\sigma(t_0), t_0), (\sigma(t_1), t_1), \cdots, (\sigma(t_n), t_n), \cdots, \tag{21}$$

which means the controller switches from $\sigma(t_{i-1})$ to $\sigma(t_i)$ at time t_i.

Proof 1. (Theorem1). Suppose that the conditions (i),(ii) hold. Then, we get

$$\tilde{\boldsymbol{A}}_i^T \boldsymbol{X}_i + \boldsymbol{X}_i \tilde{\boldsymbol{A}}_i + \boldsymbol{X}_i \bar{\boldsymbol{B}}_i \bar{\boldsymbol{B}}_i^T \boldsymbol{X}_i + \bar{\boldsymbol{C}}_1^T \bar{\boldsymbol{C}}_1 + \boldsymbol{Z}_i < \boldsymbol{0},\ \forall i \in \boldsymbol{M}. \tag{22}$$

The switching-law (19) implies $\bar{\boldsymbol{x}}(t)^T \boldsymbol{Z}_{\sigma(t)} \bar{\boldsymbol{x}}(t) \ge 0$, and applying Lemma2 (S-procedure) for (15), it follows

$$\bar{\boldsymbol{x}}(t)^T (\tilde{\boldsymbol{A}}_{\sigma(t)}^T \boldsymbol{X}_{\sigma(t)} + \boldsymbol{X}_{\sigma(t)} \tilde{\boldsymbol{A}}_{\sigma(t)} + \boldsymbol{X}_{\sigma(t)} \bar{\boldsymbol{B}}_{\sigma(t)} \bar{\boldsymbol{B}}_{\sigma(t)}^T \boldsymbol{X}_{\sigma(t)} + \bar{\boldsymbol{C}}_1^T \bar{\boldsymbol{C}}_1) \bar{\boldsymbol{x}}(t) \le 0. \tag{23}$$

Here, we define matrix $\boldsymbol{H}_i$ as follows

$$-\boldsymbol{H}_i = \tilde{\boldsymbol{A}}_i^T \boldsymbol{X}_i + \boldsymbol{X}_i \tilde{\boldsymbol{A}}_i + \boldsymbol{X}_i \bar{\boldsymbol{B}}_i \bar{\boldsymbol{B}}_i^T \boldsymbol{X}_i + \bar{\boldsymbol{C}}_1^T \bar{\boldsymbol{C}}_1. \tag{24}$$

Moreover, we consider the following Riccati differential equation

$$-\dot{\boldsymbol{X}}_i(t) = \tilde{\boldsymbol{A}}_i^T \boldsymbol{X}_i(t) + \boldsymbol{X}_i(t) \tilde{\boldsymbol{A}}_i + \boldsymbol{X}_i(t) \bar{\boldsymbol{B}}_i \bar{\boldsymbol{B}}_i^T \boldsymbol{X}_i(t) + \bar{\boldsymbol{C}}_1^T \bar{\boldsymbol{C}}_1 + \boldsymbol{H}_i = \boldsymbol{0}, \tag{25}$$

it follows

$$-\bar{\boldsymbol{x}}(t_{i+1}) \boldsymbol{X}_j(t_{i+1}) \bar{\boldsymbol{x}}(t_{i+1}) + \bar{\boldsymbol{x}}(t_i) \boldsymbol{X}_j(t_i) \bar{\boldsymbol{x}}(t_i) + \int_{t_i}^{t_{i+1}} (\dot{\bar{\boldsymbol{x}}}(t) \boldsymbol{X}_j(t) \bar{\boldsymbol{x}}(t) + \bar{\boldsymbol{x}}(t) \dot{\boldsymbol{X}}_j(t) \bar{\boldsymbol{x}}(t) + \bar{\boldsymbol{x}}(t) \boldsymbol{X}_j(t) \dot{\bar{\boldsymbol{x}}}(t)) dt = 0. \tag{26}$$

Adding $\int_{t_i}^{t_{i+1}} (\|\bar{\boldsymbol{C}}_1 \bar{\boldsymbol{x}}(t)\|^2 - \|\bar{\boldsymbol{C}}_2 \bar{\boldsymbol{x}}(t) - \boldsymbol{y}(t)\|^2)\, dt$ to Eq.(26), and combining it with Eqs.(23) and (25), we get the following

$$\begin{aligned}
&\int_{t_i}^{t_{i+1}} (\ \|\bar{\boldsymbol{C}}_1 \bar{\boldsymbol{x}}(t)\|^2 - \|\bar{\boldsymbol{C}}_2 \bar{\boldsymbol{x}}(t) - \boldsymbol{y}(t)\|^2) dt \\
&\le \bar{\boldsymbol{x}}(t_i)^T \boldsymbol{X}_{\sigma(t_i)} \bar{\boldsymbol{x}}(t_i) - \bar{\boldsymbol{x}}(t_{i+1})^T \boldsymbol{X}_{\sigma(t_i)} \bar{\boldsymbol{x}}(t_{i+1}) \\
&\quad - \int_{t_i}^{t_{i+1}} (\|[\bar{\boldsymbol{C}}_2 + \bar{\boldsymbol{B}}_{\sigma(t_i)}^T \boldsymbol{X}_{\sigma(t_i)}] \bar{\boldsymbol{x}}(t) - \boldsymbol{y}(t)\|^2) dt.
\end{aligned} \tag{27}$$

Applying inequality Eq. (27) for $i = 0, 1, 2, \cdots, N$, we obtain it

$$\begin{aligned}
&\sup\ [\int_{t_0}^{t_N} (\|\bar{\boldsymbol{C}}_1 \bar{\boldsymbol{x}}(t)\|^2 - \|\bar{\boldsymbol{C}}_2 \bar{\boldsymbol{x}}(t) - \boldsymbol{y}(t)\|^2) dt] \\
&\le \bar{\boldsymbol{x}}(t_0)^T \boldsymbol{X}_{\sigma(t_0)} \bar{\boldsymbol{x}}(t_0) - \bar{\boldsymbol{x}}(t_1)^T \boldsymbol{X}_{\sigma(t_0)} \bar{\boldsymbol{x}}(t_1) \\
&+ \bar{\boldsymbol{x}}(t_1)^T \boldsymbol{X}_{\sigma(t_1)} \bar{\boldsymbol{x}}(t_1) - \bar{\boldsymbol{x}}(t_2)^T \boldsymbol{X}_{\sigma(t_1)} \bar{\boldsymbol{x}}(t_2) \\
&+ \cdots \\
&+ \bar{\boldsymbol{x}}(t_{N-1})^T \boldsymbol{X}_{\sigma(t_{N-1})} \bar{\boldsymbol{x}}(t_{N-1}) \\
&- \bar{\boldsymbol{x}}(t_N)^T \boldsymbol{X}_{\sigma(t_{N-1})} \bar{\boldsymbol{x}}(t_N).
\end{aligned} \tag{28}$$

From the definition of $\boldsymbol{X}_i$ in Eq.(18), it follows

$$-\bar{\boldsymbol{x}}(t_{i+1})^T \boldsymbol{X}_{\sigma(t_i)} \bar{\boldsymbol{x}}(t_{i+1}) + \bar{\boldsymbol{x}}(t_{i+1})^T \boldsymbol{X}_{\sigma(t_{i+1})} \bar{\boldsymbol{x}}(t_{i+1})$$

$$= \bar{x}(t_{i+1})^T(\mu_{\sigma(t_i)}Z_{\sigma(t_i)} - \mu_{\sigma(t_{i+1})}Z_{\sigma(t_{i+1})})\bar{x}(t_{i+1})$$
$$= -\bar{x}(t_{i+1})^T\mu_{\sigma(t_{i+1})}Z_{\sigma(t_{i+1})}\bar{x}(t_{i+1})$$
$$\leq 0. \tag{29}$$

Combining Eq. (29) with Eq. (28), it follows

$$\sup\ [\int_{t_0}^{t_N}(\|\bar{C}_1\bar{x}(t)\|^2 - \|\bar{C}_2\bar{x}(t) - y(t)\|^2)dt]$$
$$\leq \bar{x}(t_0)^T X_{\sigma(t_0)}\bar{x}(t_0) - \bar{x}(t_N)^T X_{\sigma(t_{N-1})}\bar{x}(t_N), \tag{30}$$

Eq.(16) implies $\bar{x}(t)^T X_{\sigma(t)}\bar{x}(t) \geq 0$. Then, we get

$$\sup\ [\int_{t_0}^{t_N}(\|\bar{C}_1\bar{x}(t)\|^2 - \|\bar{C}_2\bar{x}(t) - y(t)\|^2)dt]$$
$$\leq \bar{x}(t_0)^T X_{\sigma(t_0)}\bar{x}(t_0). \tag{31}$$

Combining this inequality with the Integral Quadratic Constraint (3), we obtain the condition of (5). This completes the proof of this theorem.

6. NUMERICAL SIMULATIONS

Consider the uncertain system (1)-(3) with the coefficients

$$A = \begin{bmatrix} 0 & 1 \\ -1.25 & 1 \end{bmatrix}, B_1 = \begin{bmatrix} 0 \\ 1 \end{bmatrix}, B_2 = \begin{bmatrix} 0 \\ 1 \end{bmatrix},$$
$$C_1 = \begin{bmatrix} 0.1 & 0 \end{bmatrix}, C_2 = \begin{bmatrix} 1 & -2 \end{bmatrix}.$$

This plant will be controlled by (2) with the coefficients,

$$A_1^c = \begin{bmatrix} -1 & 2 \\ 3 & -7 \end{bmatrix}, B_1^c = \begin{bmatrix} 1 \\ -1 \end{bmatrix}, C_1^c = \begin{bmatrix} 3 & -6 \end{bmatrix},$$
$$A_2^c = \begin{bmatrix} -1.5 & 3 \\ -1 & -0.5 \end{bmatrix}, B_2^c = \begin{bmatrix} 1 \\ -1 \end{bmatrix}, C_2^c = \begin{bmatrix} -1 & 2 \end{bmatrix}.$$

Note that this plant can't be stabilized even if we use whichever in these two controllers.
We apply Theorem1 to design a stabilizing supervisor. We numerically solve the Riccati equation (6) by setting $\epsilon = 0.01$, then we get

$$P = \begin{bmatrix} 2.155 & 0.537 \\ 0.537 & 0.857 \end{bmatrix} > 0$$

From the $P > 0$, it follows the H_∞ observer (7).
We solve BMI (15),(16),(17), and we have

$$X = \begin{bmatrix} 0.892 & -0.521 & -0.328 & 0.874 \\ -0.521 & 4.116 & 1.468 & -3.919 \\ -0.328 & 1.468 & 0.805 & -1.305 \\ 0.874 & -3.919 & -1.305 & 4.360 \end{bmatrix},$$

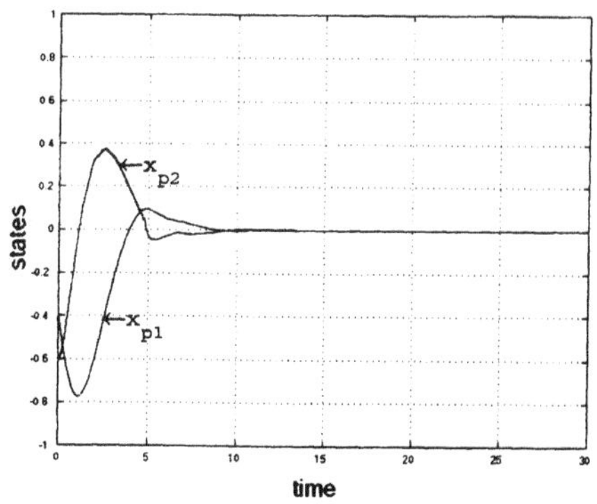

Fig. 4. time vs states (Case.1)

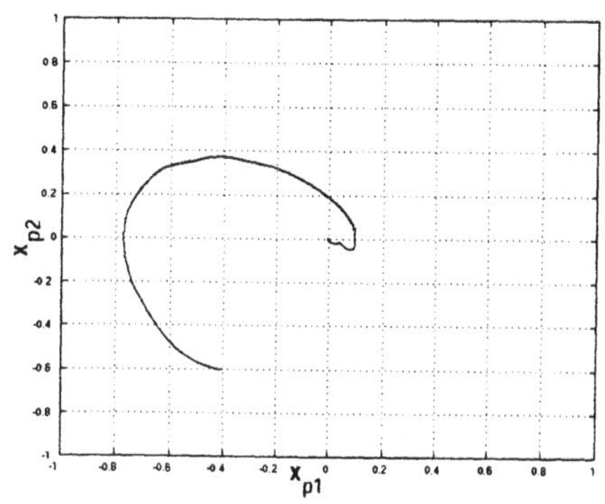

Fig. 5. x_{p1} vs x_{p2} (Case.1)

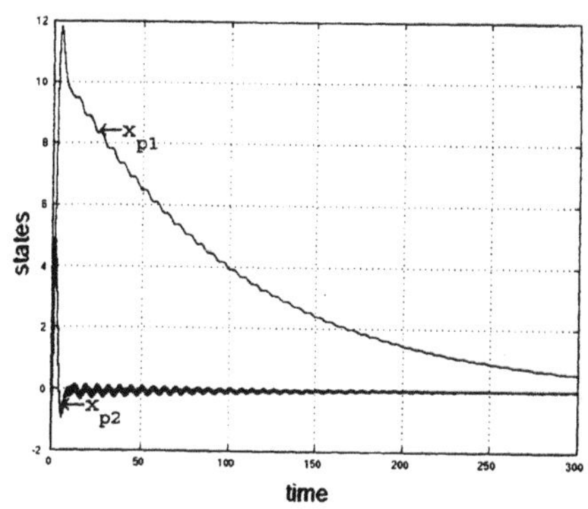

Fig. 6. time vs states (Case.2)

$$Z_1 = \begin{bmatrix} -0.043 & 0.501 & -0.462 & -0.193 \\ 0.501 & -2.270 & -1.429 & 7.536 \\ -0.462 & -1.429 & -1.193 & 2.460 \\ -0.193 & 7.536 & 2.460 & -8.896 \end{bmatrix},$$
$$Z_2 = \begin{bmatrix} 0.174 & -0.601 & 0.448 & 0.253 \\ -0.601 & 2.827 & 1.697 & -8.804 \\ 0.448 & 1.697 & 1.415 & -2.858 \\ 0.253 & -8.804 & -2.858 & 10.484 \end{bmatrix},$$

$$\mu_1 = 5.152 \times 10^{-3}, \mu_2 = 2.881 \times 10^{-3},$$
$$\tau_1 = 4.40, \tau_2 = 3.91,$$
$$\gamma_1 = 5.22 \times 10^{-2}, \gamma_2 = 11.11 \times 10^{-2}.$$

Here we take initial condition $x_p(0) = [-0.4\ -0.6]^T$. We present the simulation results as follows,

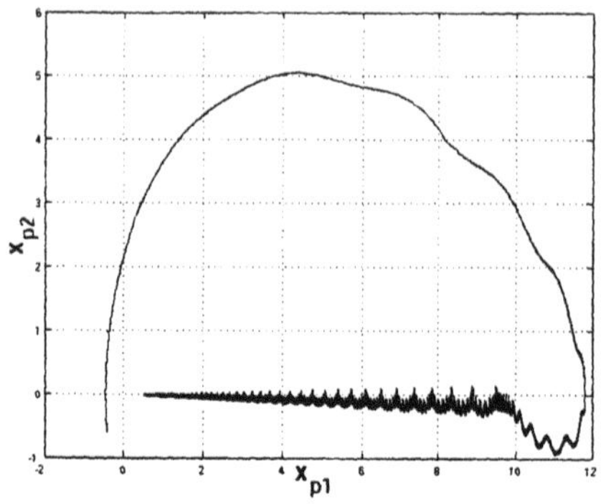

Fig. 7. x_{p1} **vs** x_{p2} **(Case.2)**

- Case.1. $\xi(t) = \sin(10t)z(t)$, $v(t) = 0.1\cos(t)z(t)$ (Fig.4 and Fig.5).
- Case.2. $\xi(t) = \sin(10t)z(t) + 3e^{-0.01t}$, $v(t) = 0.1\cos(t)z(t) + 3e^{-0.01t}$(Fig.6 and Fig.7).

7. CONCLUSIONS

We proposed a design method of supervisor that robustly stabilizes a class of uncertain system. A sufficient condition for stabilizability is described. The result is given in terms of the existence of suitable solutions to an algebraic Riccati equation of the H_∞ filter and BMI. Compared with existing result, the main contribution of this paper is that we proposed solving BMI, which get it simply to design switching-law.

8. REFERENCES

A.S.Morse (1992). Applications of hysteresis switching in parameter adaptive control. *IEEE Trans.Automat.Contr* **vol.37, no.9**, 1343–1354.

A.S.Morse (1996). Supervisory control of families of linear set-point controllers -part1:exact matching. *IEEE Trans.Automat.Contr* **vol.41, no.10**, 1413–1431.

A.V.Savkin, E.Skafidas and R.J.Evans (1999). Robust output feed back stabilizability via controller switching. *Automatica* **35**, 69–74.

Edoardo Mosca, Tommaso Agnoloni (1999). Adaptive gain scheduling control:inference of feedback-loop behavior. *Proceeding of 14th IFAC* pp. 313–318.

J.A.De Dona, S.O.Reza Moheimani and G.C.Goodwin (1999). Allowing for over-saturation in robust switching control of a class of uncertain systems. *Proceedings 38th CDC* pp. 3053–3058.

N.H.Mcclamroch and I.Kolmanovsky (2000). Performance benefits of hybrid control design for linear and nonlinear systems. *Proceedings IEEE* **vol.88, no.7**, 1083–1096.

FAULT DETECTION AND ISOLATION APPLIED TO THE SUPERVISION OF ADAPTIVE CONTROL SYSTEMS: A NEURAL NETWORK BASED SCHEME

Fernando Valles Barajas *
Ricardo A. Ramírez Mendoza **

* *Center for Artificial Intelligence and Department of Mechatronics and Automation, ITESM Campus Monterrey, Monterrey, Nuevo León Av. Eugenio Garza Sada No. 2501, C.P. 64849 Tel. +52 8358-2000, México*
f_valles@hotmail.com
** *Department of Mechatronics and Automation, ITESM Campus Monterrey, Monterrey, Nuevo León Av. Eugenio Garza Sada No. 2501, C.P. 64849 Tel. +52 8358-2000, México*
raramire@campus.mty.itesm.mx

Abstract: This paper presents a novel scheme to implement a supervisor for an indirect adaptive control system. The underlying supervisory is performed by using neural networks that heavily borrow from the artificial intelligence culture. Intensive simulations are carried out to emphasize the performance of the proposed supervisor. An experimental study case is examined for demonstrating the applicability of this scheme. *Copyright © 2001 IFAC*

Keywords: Neural Networks, Adaptive Control Systems, Supervision and Fault Detection and Isolation (FDI)

1. INTRODUCTION

It has been pointed out, e.g. see (Frank, 1996; Gertler, 1991), that the industrial applications of feedback control system require an adequate supervision to maintain the required performance. This performance decreases because the faults, that can appear in the process, see, e.g., (IFAC, 1992; IFAC, 1994). Malfunctions of process equipment and instrumentation increase the operating costs of any plant. Even, more serious are the consequences of a gross accident due to faulty design of operation (Himmelblau, 1978).

For this reason, the supervision of the industrial process has become a very important issue in the industry. Usually the supervision is made by the operators. The lack of this scheme is that the effectiveness will depend on the skills of the operator. One of the tools that has been applied to the supervision of control systems is Fault Detection and Isolation (FDI), see, e.g., (Gertler, 1991; R.J. Patton, 1991). As pointed out in (Gertler, 1991), a FDI system is composed by two main parts. The first one is properly, the fault detection, by means of checking procedure of some abnormal condition in the process. The second one is related with the isolation of the fault, this concerning the localization of the fault, and finally the fault identification, it estimate the size of the fault. The figure 1 shows the stages of a FDI system, where $\bar{u}(k)$ and $\bar{y}(k)$ represent respectively the input vector and the output vector.

In the other hand, adaptive control is a technique that is used to control time-varying-process. Figure 2 shows a diagram of adaptive control system. One of the key elements of an adaptive control sys-

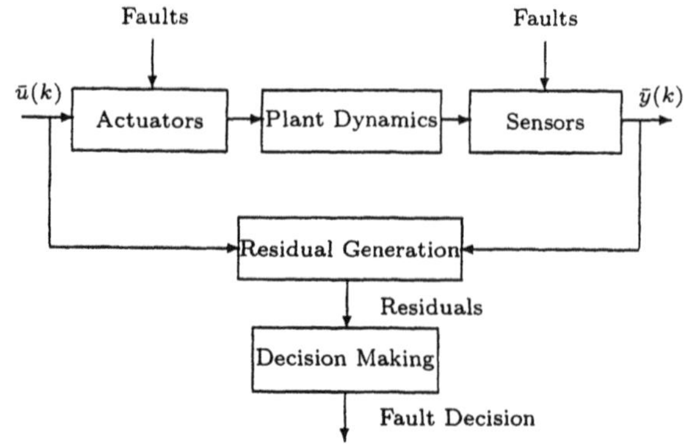

Fig. 1. Stages of a FDI system

tem is the Parametric Adaptive Algorithm (PAA) block. This block estimates the parameters of the process based of measurements of the input and output. Afterwards, the parameters estimated by the PAA are sent to the control design block, which based on the performance index and on the parameters estimated by the PAA, will decide the structure and the values of the parameters of the controller, $G_c(z^{-1})$.

This kind of control technique has some faults that are inherent from its design. These faults cause the deterioration of the performance of the entire system. The faults involved in an adaptive control systems are: fail to achieve the persistent excitation condition, disturbances, noise, set-point changes, oscillations, stop of the adaptation, parameter estimates out of the valid zone or inadmissible, signal saturation and bumpless transfer (mode transition), see, e.g.,(Aström and Witternmark, 1995; Hägglund and Aström, 2000).

The adaptive control problem represents a chal-

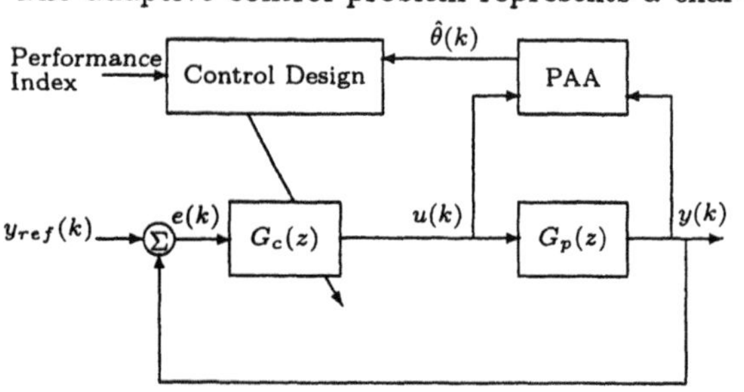

Fig. 2. Adaptive control system

lenging opportunity to investigate advanced FDI techniques that have reached a reasonable level of maturity. This is mainly a result of many years of effort devoted to the understanding of the FDI techniques and control theory as well as the tremendous progress in the computer technology that makes the implementation of supervised control systems easier and cheaper.

Supervision for adaptive control has been poorly studied during the last ten years. In what follows, some of most relevant works are described.

In (Hägglund and Aström, 2000), the authors remark the importance of having a supervision block to ensure the integrity of the closed loop performance and robustness issues. In (Linkens and Mahfouf, 1994), a proposal for the supervision of generalized predictive control anesthesia system is developed, the authors define the faults in three different kinds, sensor failures, actuator failures and algorithmic failures. The detection of the faults is made by using binary logic.

In the present paper, the aim of the authors is to investigate the applicability of the FDI techniques for the design of a supervisor scheme to an adaptive control system. The supervisor scheme proposed in this paper, which will be called the NN-FDI scheme, has a supervisor block and a backup system that works when the performance of the adaptive control system (2) does not achieve a prescribed specification. This backup system is based on FDI techniques and is in charge of the detection and isolation of any parametric variation that can occur in the system.
The residual generator block was implemented taking the difference between the reference variable and the process variable. Two neural networks, which are FDI tools, were used to build the decision making block.

A realistic experimental framework involving a level process was used to evaluate this scheme. This experimental process was developed at the Department of Mechatronics and Automation at the Monterrey Institute of Technology, to investigate adaptive control problems. The results show that the scheme proposed by the authors performs well under certain conditions. The involved experimental evaluations were carried out using the Labwindows [1] industrial software package.

The paper is organized as follows: in this section the motivation to investigate the supervision of adaptive control systems has been give. The NN-FDI scheme for supervision is presented in Section 2. Section 3 presents the simulations and experimental results. Some concluding remarks are mentioned in the last Section.

2. NN-FDI SCHEME

Figure 3 shows the general scheme of the NN-FDI scheme for fault detection and isolation of an adaptive control system. Two new elements have been added to the adaptive control system of figure 2: the supervisor block, which orchestras the entire system of the figure 3, and the FDI backup system of the adaptive control system, which

[1] ©National Instruments

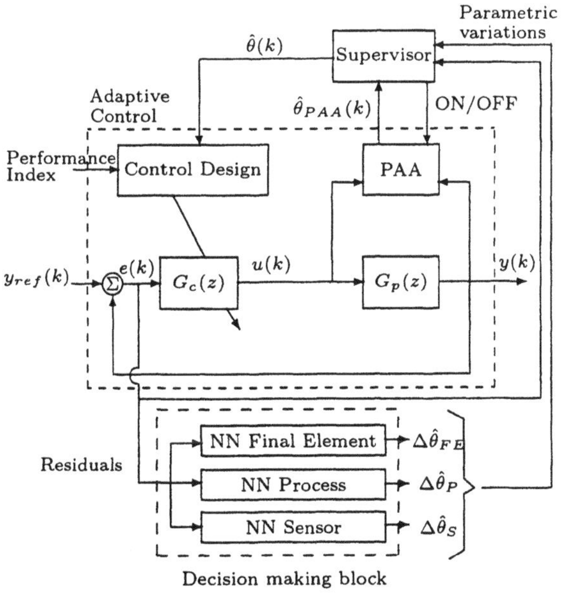

Fig. 3. NN-FDI Scheme

informs to the supervisor block when a parametric variation occurs in the process. The FDI system is formed by the residual generator block, which in this case is a simple difference between $y_{ref}(k)$ and $y(k)$, and the decision making block, which is formed by three neural networks, see figure 3.
In what follows, the function of this two new elements is explained.

2.1 *The supervisor block*

The supervisor block receives information of the system from three different sources: the PAA, the FDI backup system and the closed loop error $e(k)$. The PAA and the FDI backup system are two sources that give information related with the dynamic of the process; this information is tagged in figure 3 as $\hat{\theta}_{PAA}(k)$, which is the estimated parameter vector obtained by the PAA, and the parametric variations of the process, which are the output of the FDI backup system. The third source of information of the process is the closed-loop error $e(k)$; this variable is used by the supervisor block to calculated the Integral Absolute Error (IAE), which is taken as a measurement of the performance of the adaptive control system. When the performance of the adaptive control system, reflexed by the IAE, is under a prescribed value, the supervisor will turn off (see figure 3) the PAA and will start to take into account the information sending by the FDI backup system. Another case when the Supervisor turn off the PAA is when systematically the PAA obtains parameters out of the valid zone. The output of the supervisor $\hat{\theta}(k)$ corresponds to the best parameters of the process that the supervisor can be found based on the parameters estimated by the PAA $\hat{\theta}_{PAA}(k)$, and the parametric variations of the process estimated by the FDI backup system. An explanation of the supervisor block has been given in this section in the following section a description of the FDI backup system will be described.

2.2 *The FDI backup system*

When the supervisor establishes that the adaptive control system does not have a good level of performance because of the IAE reaches some undesirable value or because the PAA systematically obtains parameters out of the valid zone, the FDI backup system will be in charge to inform to the supervisor of any change in the parameters of the process.
The backup system proposed in this research is based on the FDI philosophy. As it was explained in the introduction the two main parts of a FDI system are the residual generator block and the decision making block. In what follow, how these two blocks are built using the proposed approach is explained.

2.2.1. *Residual generation block*

To generate the residuals the following methodology has to be followed:

(1) Firstly, a the controller is tuned taking the nominal plant of the process. This controller can be; PID, RST, GPC or another controller.
(2) A parametric variation of the process (which is divided in the final element, the actual process and the sensor) are simulated, keeping the controller with the parameters obtained using the nominal plant.
(3) Later a step input in the reference, $y_{ref}(k)$, is applied.
(4) The difference between process variable, $y(k)$, and the reference, $y_{ref}(k)$ is taken as the residuals, as it is shown in figure 3.
(5) Steps 2, 3 and 4 are repeated for every possible parametric variation in the process.

This section has described the form to build the residual generator using the proposed scheme by the authors of this paper. The building of the decision making block is explained in the next section.

2.2.2. *Decision making block*

To build the decision making block the following aspects must be considered:

(1) The decision making block is built by using the three neural networks that appear in figure 3.
(2) Each NN is associated to an element of the process (the final element, the process itself

or the sensor) and learns the faults of its associated element. The reason for use three NN's instead of one NN is to obtain a faster training.

(3) Each NN receives as inputs the closed-loop error $e(k)$, and gives as outputs the parametric variations $\Delta\hat{\theta}_x$, of its associated element. Where $\Delta\hat{\theta}_x$ represents: the changes of the parameters in the final element $\Delta\hat{\theta}_{FE}$, in the process $\Delta\hat{\theta}_P$ or in the sensor $\Delta\hat{\theta}_S$, with respect to their nominal model.

(4) The number of the inputs of each NN depends on the time constant τ, of its associated element and the number of outputs of each NN depends on the number of parameters that form the model of its associated element; for example, the NN associated with the final element will have 3 outputs if the final element is represented as a first order process plus dead time (FOPDT). The reason for this is that a FOPDT can be described by using three parameters: the dead time t_o, the constant time τ, and the static gain K, as is shown in Eq. (1).

$$G_p(s) = \frac{K}{\tau s + 1} e^{-t_o s} \qquad (1)$$

Once the FDI backup system has been built using the above method, it can by used together with the supervisor block to detect any parametric variation that can appear in the process. The description of the system working with the FDI backup system and the supervisor block is as follows: The supervisor is always checking the value of the IAE, when this variable reaches some prescribed value, this will be an indication of a possible parametric variation in the process. In order to know if a parametric variation has occurred, a step input is applied in the reference $y_{ref}(k)$, and then the pattern of the generated error $e(k)$, is taken by the FDI system, which based on the patterns "stored" in the NN's that form the decision making block, will isolate the parametric variation that occurred in the process.
In this section, an explanation of the NN-FDI scheme to supervise an adaptive control system has been given. In the following section, an experimental evaluation is presented to demonstrate the applicability of this proposal in a realistic environment involving a level process.

3. EXPERIMENTAL EVALUATION

An experimental evaluation was carried out in an industrial process using the scheme proposed by the authors. The process to be controlled is a pilot process represented in figure 4. The characteristics of this process are explained in what follows:

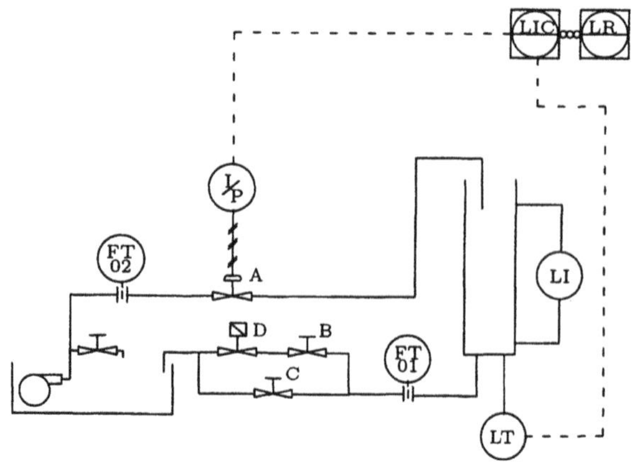

Fig. 4. Experimental Process

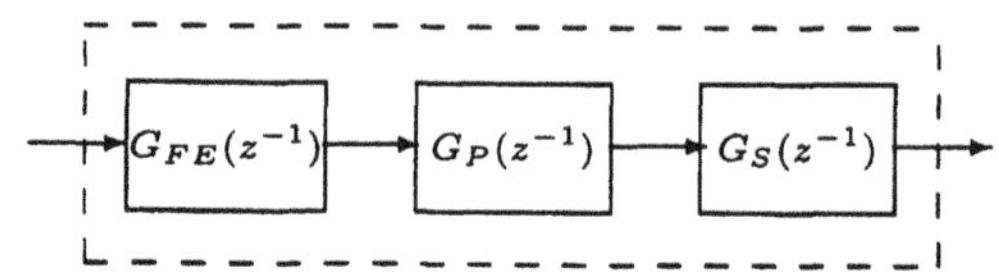

Fig. 5. Block diagram of the entire system

- the objective of the control system is to maintain the level of the tank in a desirable value.
- the pump is always sending water to the tank in a constant velocity.
- the valve A, which is the final element of this control system, opens with pressure. The amount of water that the tank receives can be regulated by opening or closing this valve.
- faults in the system are generated by changing the position of the manual valves B and C.
- the solenoids valve D also serves to generate faults.
- the controller, LIC (Level Indicator Controller), is a RST controller; see (Landau *et al.*, 1998) for further information about this controller.

The dynamic characteristics of the process are:

- for the purpose of detecting additive and multiplicative faults, the final element (valve A) the process and the sensor are modelled as a first order processes with dead time (FOPDT), see figure 5, where $G_{FE}(z^{-1})$, $G_P(z^{-1})$ and $G_S(z^{-1})$ represent respectively the transfer function of the final element, the transfer function of the process and the transfer function of the sensor.
- for the purpose of control, the final element (valve A) the process and the sensor LT (Level Transmitter) are modelled as a single first order process plus dead time (FOPDT). This structure is representative for most of the industrial process (Pradeep B. Deshpande, 1988).
- this system is nonlinear and time varying, in the sense that its dynamic depends on the

operation zone that the system is working and on the time.

To test the reliability of the proposed scheme the final element, the process and the sensor were represented as a FOPDT. For the the process and final element all the parameters were monitored (K, τ, and θ) and for the sensor only the static gain K was monitored. Two NN were used to build the decision making block, one of them was associated with the final element, which presented 3 outputs, and the other one was associated with the process and the sensor, presented 4 outputs; 3 outputs to monitor the variables of the process and one output to monitor the K of the sensor. In what follows the results that were obtained when the NN-FDI scheme was used to supervise the level process are described.

When the FDI backup system was working, the neural networks performed well most of the time a parametric variation occurred in the process. In particular, figure 6 shows some of the patterns of the error that was learned by the neural network assigned to the process, when the τ of the process was varied and figure 7 shows a similar graph when the K of the process was varied. In these figures; Δ_τ is the difference between the nominal τ of the process and the current τ of the process and Δ_K is the difference between the nominal K of the process and the current K of the process.

Three particular cases that the NN assigned to

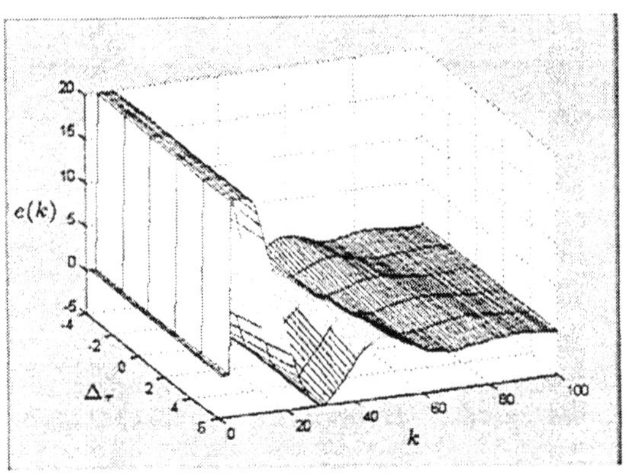

Fig. 6. Error behavior for some variations of the time constant τ of the process

the process, identified successfully the parametric variation that occurred in the process are presented in figure 8, 9 and 10. Figure 8 shows the response of the system when an increase in the K of the process of 60.0% occurred. In the figure 9 a decrease in the τ of the process of 42.0% happened. Figure 10 shows the response of the system when a decrease in the *to* of the process of 20.0% occurred.
These parametric variations were correctly identified by the FDI backup system, given to the

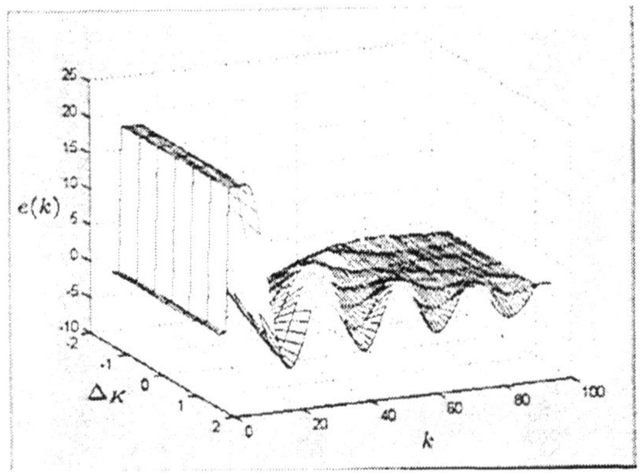

Fig. 7. Error behavior for some variations of the gain K of the process

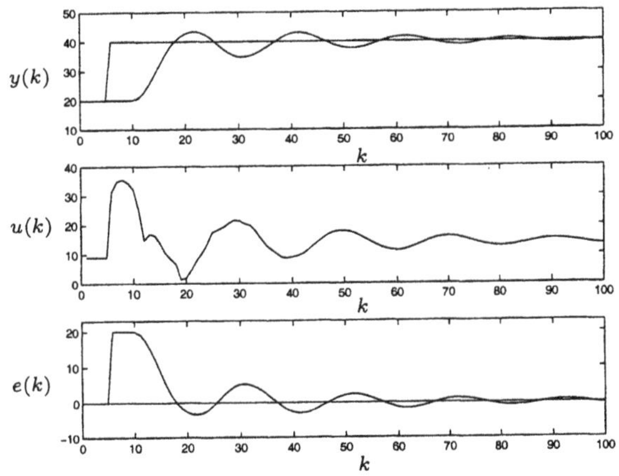

Fig. 8. Output variable, control variable and error of the system for mismatch conditions $\Delta_K = 60.0\%$ in the process.

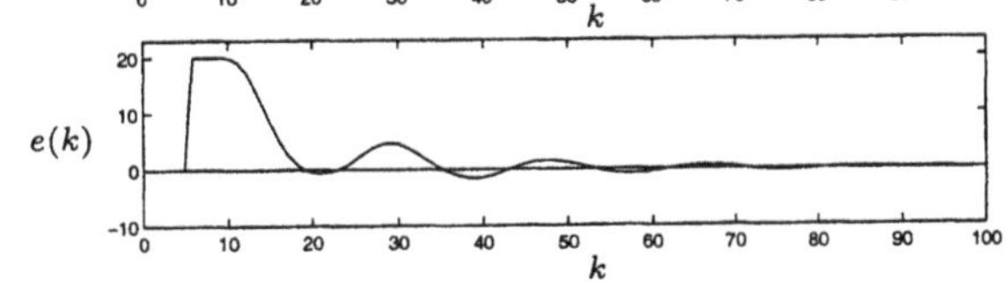

Fig. 9. Output variable, control variable and error of the system for mismatch conditions $\Delta_\tau = -42.0\%$ in the process.

supervisor support to retune the controller. Figure 11 shows the response of the system after the controller was retuned when the parametric variation in the K of 60.0% was detected. As it can be seen the response of the system has a better performance that the response of the system of figure

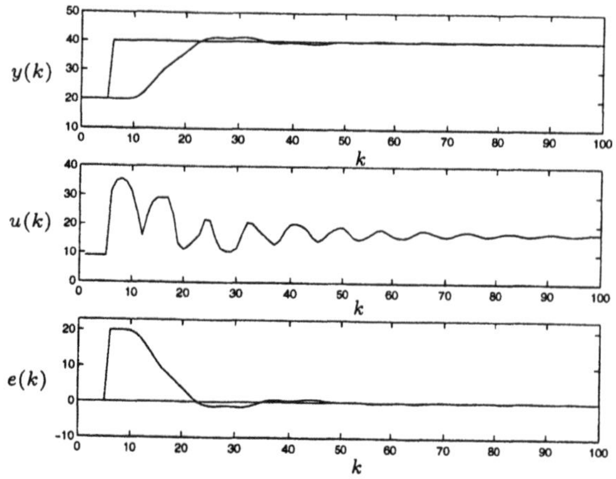

Fig. 10. Output variable, control variable and error of the system for mismatch condition $\Delta_{to} = -20.0\%$ in the process.

8. Similar results were found when the controller was retuned, after the variations of the t_o and τ of the process occurred.

The cases where the neural networks gave bad information to the supervisor were when for some combinations of the K of the final element $G_{FE}(z^{-1})$, the process $G_P(z^{-1})$, and the sensor $G_S(z^{-1})$, the total K of the system was the same. In these situations the patterns of error given to the neural network were the same. Because of this the neural networks were not enable to find out where the parametric variation occurred.

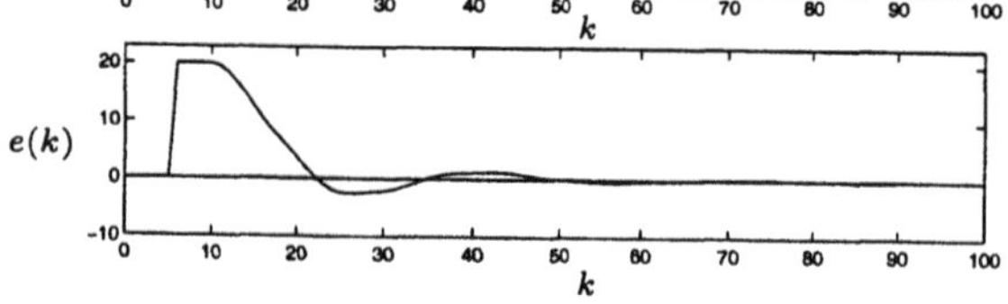

Fig. 11. Output variable, control variable and error of the system after the controller was retuned

4. CONCLUSIONS AND FURTHER RESEARCH

FDI are techniques that have shown a good potential to supervise control process. Adaptive control systems have some problems that are inherent from its design. Because of these problems, adaptive control systems must be supervised in order to maintain a good level in the performance of the system.

In this paper the authors have shown that FDI techniques can be applied to the supervision of adaptive control systems. In particular the potential of the neural networks was investigated.

A disadvantage of this scheme is that for some combination of the static gains of the blocks of the system, the neural networks were not capable of giving a reliable information. Further research by the authors will be focus on the building of a supervisor of the adaptive control systems that will more robust against parametric variations that the scheme presented in this paper.

5. REFERENCES

Aström, K. and K.J. Witternmark (1995). *Adaptive Control.* Addison-Wesley. USA.

Frank, P.M. (1996). Robust model-based fault detection in dynamic systems. In: *Summer School Continuous system supervision.* pp. 275–287. LAG INPG-CNRS. Grenoble, France, September 2-6.

Gertler, J. (1991). Analytical redundancy methods in fault detection and isolation. IFAC. Baden–Baden.

Hägglund, T. and K. Aström (2000). Supervision of adaptive control algorithms. *Automatica.*

Himmelblau, David M. (1978). *Fault detection and diagnosis in chemical and petrochemical process.* Elsevier Scientific Publishing Company. U.S.A.

IFAC (1992). *Fault detection, supervision and ,safety for technical processes.* Pergamon. Baden Baden,Germany.

IFAC (1994). *Fault detection, supervision and safety for technical processes.* Pergamon. Espoo,Finland.

Landau, I.D., R. Lozano and M. M'Saad (1998). *Adaptive control.* Springer-Verlag London Ltd. New York.

Linkens, D.A. and M. Mahfouf (1994). Supervisory generalised predictive control and fauls detection for multivariable anaesthesia. *IEE Proc. Control Theory.*

Pradeep B. Deshpande, Raymond H. Ash (1988). *Computer process control, with advanced control applications.* Instrument Society of America. U.S.A.

R.J. Patton, J. Chen (1991). Parity space approach to model-based fault diagnosis. a tutorial survery and some results. IFAC. Baden–Baden.

SLIDING MODE CONTROL WITH ADAPTIVE DECREASE OF CHATTERING

Ryszard Gessing

Politechnika Śląska Instytut Automatyki,
ul. Akademicka 16, 44-101 Gliwice, Poland,
fax: +4832 372127, email: gessing@ia.gliwice.edu.pl

Abstract: The idea of the system with sliding mode control and adaptation of the switched values of the relay output is researched. The observation that the sliding mode control is equivalent to the bang-bang relay control of the sliding surface signal on the level zero is exploited. The slopes of the increasing and decreasing parts of the sliding surface signal are used for independent adaptation of both the switched values of the relay output, respectively. The adaptation causes that the difference between both the values is decreased and the mean value tend to the required value of the control. Owing to this the chattering effect is decreased, significantly and the system remains very robust. Simplified model of the system is proposed and simulations are performed.

Keywords: Sliding mode control; robust control; relay control.

1. INTRODUCTION

The control systems which uses sliding mode technique have now good theoretical elaboration (e.g. Slotine and Li (1991), Utkin (1992)) as well as successful practical applications (e.g. commonly used voltage stabilisation of car alternators). This kind of systems works well both with linear and non-linear plants.

It is well known, that the systems with sliding mode control are very robust so they work well even in the case of large and rapid parameter changes. However with the switching action of the relay there is connected the so called chattering effect (Åström, 1989) which sometimes is not accepted by users and/or by actuators. Therefore chattering decrease is interesting from application point of view.

The relay control and sliding mode technique is also exploited in self oscillating adaptive systems. There was some trials of decreasing the chattering effect in these systems by means of applying different amplitudes of the relay output for different absolute values of the control error (Åström, 1989).

In the present paper the system with sliding mode control and adaptation of both the switched amplitudes of the relay output, for non-linear plants is considered. The independent adaptation of both the amplitudes $Max = \beta^+$ and $Min = \beta^-$ causes that the difference $(\beta^+ - \beta^-)$ is decreased and both the amplitudes are symmetrically placed with respect to required average value of the control. Owing to this the chattering effect is decreased, significantly and the system remains very robust with respect to large and rapid parameter changes.

The idea researched first by Gessing (1997) was then applied to the plant with delay in (Gessing, 1999). In the present paper the case of nonlinear plants is considered and the simplified model of adaptation used in the system is described.

The contribution of the paper is partly in applying independent adaptation of both the switched amplitudes of the relay output in the system with nonlinear plant and sliding mode control and partly in creating the simplified model of adaptation.

2. THE CHOICE OF THE SWITCHED AMPLITUDES OF THE RELAY

Assume that a single input single output plant is described by the state space model in the form

$$\dot{x} = f(x,u) \tag{1}$$

$$y = g(x) \tag{2}$$

where x is n-dimensional state, u, y are scalar input and output, while $f(x,u)$ and $g(x)$ are vector and scalar functions of the mentioned arguments, respectively. Assume that the plant is stable.

Both the switched values of the relay outputs are denoted by β^+ and β^-. The goal of the preliminary design is the determination of β_{mx} and β_{min} for which the system will work for the disturbances and initial states occurring in the system.

Let

$$\begin{aligned} \dot{y} &= \frac{\partial g}{\partial x} f(x,u) = g^1(x) \\ \ddot{y} &= \frac{\partial g^1}{\partial x} f(x,u) = g^2(x) \\ &\ldots\ldots\ldots\ldots\ldots\ldots \\ y^{(d-1)} &= \frac{\partial g^{d-2}}{\partial x} f(x,u) = g^{d-1}(x) \\ y^{(d)} &= \frac{\partial g^{d-1}}{\partial x} f(x,u) = g^d(x,u) \end{aligned} \tag{3}$$

Note, that the introduced notation $g^j(x)$, $j = 1, 2, ..., d$ shows that the derivatives $y^{(j)}(t)$, $j = 1, 2, ..., d-1$ are independent directly of u, while $y^{(d)}(t)$ is dependent, i.e.

$$\frac{\partial g^j}{\partial u} = 0 \quad j = 0, 1, 2, ..., d-1 \tag{4}$$

and

$$\frac{\partial g^d}{\partial u} \neq 0 \tag{5}$$

Assume that relations (4), (5) are fulfilled for any x occurring in the system.

Let w be the set point of the closed-loop (CL) system and $e = w - y$, be the control error. Let $a_0, a_1, ..., a_{d-2}$ be the coefficients chosen so that the equation

$$e^* = a_0 e^{(d-1)} + a_1 e^{(d-2)} + ... + e = 0 \tag{6}$$

has stable transients of a good quality. The equation (6) describes the sliding surface.

Let us denote

$$w^* = a_0 w^{(d-1)} + a_1 w^{(d-2)} + ... + w \tag{7}$$

$$y^* = a_0 y^{(d-1)} + a_1 y^{(d-2)} + ... + y \tag{8}$$

$$e^* = w^* - y^* \tag{9}$$

From (3)-(9) it results

$$\dot{e}^* = \dot{w}^* - g^*(x,u) \tag{10}$$

$$g^*(x,u) = a_0 g^d + a_1 g^{d-1} + ... + g^1 \tag{11}$$

By means of the proper choice of the two values β_{mx} and β_{min} the following bang-bang relay control law can be created

$$\begin{aligned} u &= \beta_{mx}, \quad if\ e^* > 0 \\ u &= \beta_{min}, \quad if\ e^* < 0 \end{aligned} \tag{12}$$

This control law may implement the sliding mode control if the two switched values of the relay β_{mx} and β_{min} fulfil the inequalities

$$\begin{aligned} \dot{e}^* &= \dot{w}^* - g^*(x, \beta_{mx}) < 0 \\ \dot{e}^* &= \dot{w}^* - g^*(x, \beta_{min}) > 0 \end{aligned} \tag{13}$$

The inequalities (19) should be fulfilled for any possible values of $\dot{w}^*$, x occurring in the system.

For any bounded set of the states x and velocities $\dot{w}^*$ some appropriate values of $= \beta_{mx}$ and $= \beta_{min}$ may be chosen so that the inequalities (13) are fulfilled. Of course, this is possible for functions $g^*(x,u)$ which for any bounded state x take any positive or negative value for appropriately chosen u. The choice of the values β_{mx}, β_{min} assuring (13) is possible for sufficiently smooth variations of the excitations.

3. THE IDEA OF CHATTERING DECREASE

One from the possible choices is $\beta_{mx} = \bar{\beta}$ and $\beta_{min} = -\bar{\beta}$ where $\bar{\beta}$ is sufficiently large value. Larger value $\bar{\beta}$ – faster variations of the output y are possible and bigger variations of the disturbance may be compensated. However with the switching appearing in the sliding mode control the effect of chattering is related, which is stronger when $\bar{\beta}$ is larger. Some times users or actuators do not accept to strong chattering effects, especially if they are lasting to long time. Therefore, there arises the idea of adaptation of the values of β^+ and β^- to the actual system work conditions (expressed by required output variations, parameter changes or appearing actually disturbances). This idea makes it possible to create the control system in which the identification of the plant model (usually performed in adaptive systems) is not needed. The identification is replaced by the current adaptation of the values β^+, β^- so that the difference $(\beta^+ - \beta^-)$ is sufficiently small; the latter is increased when it is needed. In this system the chattering effect is reduced significantly. On the other hand the sliding mode control appearing in the system causes that it is very robust even in the case of large and fast parameter changes.

One can note that the considered system with sliding mode control works as bang-bang relay control of the variable e^* on the level zero. The successive switchings of the relay causes characteristic oscillations of the variable e^* shown in Fig. 1. These oscillations contain information about actual values of β^+ and β^-. Namely, there is a dependence between maximum positive slope $\dot{e}^*_{mx}$ in the interval (t_1, t_2) of increase and the value of β^-, as well as the minimum (negative) slope $\dot{e}^*_{min}$ in the interval (t_2, t_3) of decrease and the value of β^+. Smaller values of β^- causes greater values of the maximum slope $\dot{e}^*_{mx}$ and greater value of β^+ causes smaller value of the minimum slope $\dot{e}^*_{min}$.

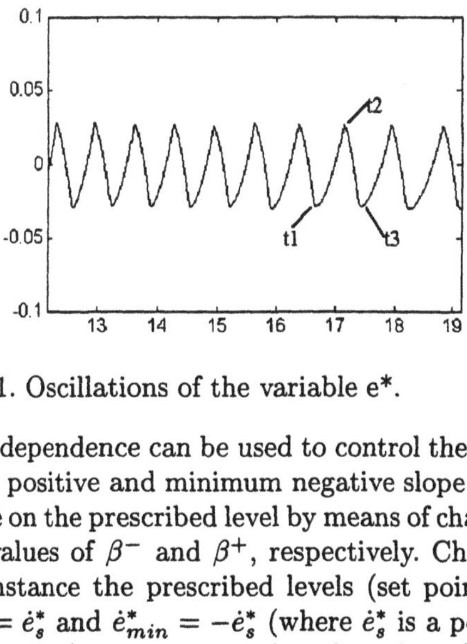

Fig. 1. Oscillations of the variable e*.

This dependence can be used to control the maximum positive and minimum negative slope of the curve on the prescribed level by means of changing the values of β^- and β^+, respectively. Choosing for instance the prescribed levels (set points) of $\dot{e}^*_{mx} = \dot{e}^*_s$ and $\dot{e}^*_{min} = -\dot{e}^*_s$ (where $\dot{e}^*_s$ is a positive small value) we can cause that the values β^- and β^+ are mutually close and symmetrically placed with respect to the required value of the control u.

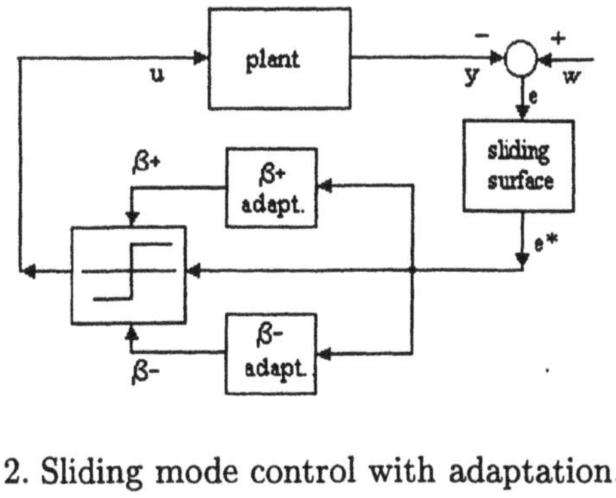

Fig. 2. Sliding mode control with adaptation.

The block diagram of the system, which implements this idea, is shown in Fig. 2. The system implements sliding mode control and contains additionally two channels for independent adaptation of the values of β^+ and β^-.

4. ADAPTATION OF THE SWITCHED AMPLITUDES

Now the simplified model of the amplitudes adaptation will be described. Assume that sliding mode control stabilises the plant output y on the level $w = const.$ Let β_y denotes the constant value of the input u which in steady state assures that $y = w$. Since during the sliding mode control with adapted both the amplitudes, all the variables of the plant are varying in-significantly around the constant values, then the plant may be linearized. The linearized model of the plant (1), (2) takes the form

$$\dot{x} = \frac{\partial \bar{f}}{\partial x} x + \frac{\partial \bar{f}}{\partial u} u \tag{14}$$

$$y = \frac{\partial \bar{g}}{\partial x} u \tag{15}$$

The bar over the partial derivatives denotes that they are evaluated for the constant values of x and $u = \beta_y$, which correspond to steady state value of y. The model (14), (15) is linear and may be described by the transfer function (TF)

$$G(s) = \frac{Y(s)}{U(s)} = \frac{b_0 s^m + b_1 s^{m-1} + \ldots + b_m}{s^n + a_1 s^{n-1} + \ldots + a_n} \tag{16}$$

where $U(s)$, $Y(s)$ are the Laplace transforms of the input $u(t)$ and output $y(t)$ of the plant; n, m are the degrees of denominator and numerator of the TF (16), $n > m$; $d = n - m$ is the relative degree of $G(s)$. Assume that the model (16) is stable and minimum phase.

From (16) and (8) it results

$$G^*(s) = \frac{Y^*(s)}{U(s)} = \frac{\bar{b}_0 s^{n-1} + \bar{b}_1 s^{n-2} + \ldots + \bar{b}_{n-1}}{s^n + a_1 s^{n-1} + \ldots + a_n} \tag{17}$$

where $Y^*(s)$ is the Laplace transform of $y^*(t)$ and the numerator polynomial results from performing multiplication

$$\begin{aligned} &\bar{b}_0 s^{n-1} + \bar{b}_1 s^{n-2} + \ldots + \bar{b}_{n-1} = \\ &(b_0 s^m + b_1 s^{m-1} + \ldots + b_m)(c_0 s^{d-1} + c_1 s^{d-2} + \ldots + 1) \end{aligned} \tag{18}$$

If for $t < 0$ the output takes the constant value y, $u = \beta_y$, and at $t = 0$ it appears the stepwise change by $\Delta\beta^+ = \beta^+ - \beta_y$ or by $\Delta\beta^- = \beta^- - \beta_y$, then from the initial value theorem of the Laplace transform applied to (17) it results

$$\begin{aligned} &\dot{y}^*(0^+) = \bar{b}_0 \Delta\beta^+ = b_0 c_0 \Delta\beta^+ \\ &\text{or } \dot{y}^*(0^+) = b_0 c_0 \Delta\beta^- \end{aligned} \tag{19}$$

where 0^+ denotes the right hand side limit at zero. Then if $\beta^+ > 0$ and $\beta^- < 0$ then for constant w, from (9) and (19) it results

$$\begin{aligned} \dot{e}^* &= -b_0 c_0 \Delta\beta_{mx} < 0 \quad \text{if} \quad e^* > 0 \\ \dot{e}^* &= -b_0 c_0 \Delta\beta_{min} > 0 \quad \text{if} \quad e^* < 0 \end{aligned} \tag{20}$$

where $\Delta\beta_{mx} = \beta_{mx} - \beta_y$, $\Delta\beta_{min} = \beta_{min} - \beta_y$.

Let β_n^+ and β_{n+1}^+ be the values of β^+ in the successive intervals of decrease and β_n^- and β_{n+1}^- the values of β^- in the successive intervals of increase of the variable e^* (see Fig. 1). The adaptation of the values β_n^+ and β_n^- will be realised in accordance with the following algorithms

$$\beta_{n+1}^+ = \beta_n^+ + c(\dot{e}_{min}^* + \dot{e}_s^*) \tag{21}$$

$$\beta_{n+1}^- = \beta_n^- + c(\dot{e}_{mx}^* - \dot{e}_s^*) \tag{22}$$

where $\dot{e}_{min}^*$ is the minimal measured value of $\dot{e}^*$ in the n-th interval of decrease and $\dot{e}_{mx}^*$ is the maximal measured value of $\dot{e}^*$ in the n-th interval of increase; $(-\dot{e}_s^*)$ and $\dot{e}_s^*$ be the reference values for the variables $\dot{e}_{min}^*$ and $\dot{e}_{mx}^*$, respectively, c is the gain coefficient.

For given constant y denote

$$\Delta\beta_n^+ = \beta_n^+ - \beta_y, \quad \Delta\beta_n^- = \beta_n^- - \beta_y \tag{23}$$

From (20) it results

$$\dot{e}_{mx}^* = -b_0 c_0 \Delta\beta_n^-, \quad \dot{e}_{min}^* = -b_0 c_0 \Delta\beta_n^+ \tag{24}$$

Subtracting from both sides of equations (21), (22) the term β_y and accounting (23) and (24) we obtain

$$\Delta\beta_{n+1}^+ = \Delta\beta_n^+ + c(\dot{e}_s^* - b_0 c_0 \Delta\beta_n^+) \tag{25}$$

$$\Delta\beta_{n+1}^- = \Delta\beta_n^- + c(-\dot{e}_s^* - b_0 c_0 \Delta\beta_n^-) \tag{26}$$

The equations (25), (26) describe the successive changes of the values $\Delta\beta_n^+$ and $\Delta\beta_n^-$. Note the successive intervals of the increase and decrease of e^* are varying, therefore the equations (25), (26) describe the control in an unconventional discrete-time linear systems with varying sampling intervals.

The gain c should be chosen so that the systems (25) and (26) are stable. The characteristic equation for the system (25) is

$$z - (1 - c b_0 c_0) = 0 \tag{27}$$

The system is of the first order and the stability condition is

$$0 < c b_0 c_0 < 2 \tag{28}$$

However to assure aperiodic decay of the value $\Delta\beta_n^+$ the condition (28) should be replaced by

$$0 < c b_0 c_0) < 1 \tag{29}$$

It is easy to note that for the equation (26) the conditions (28) and (29) are also valid.

Then the sliding mode control in the main channel of the system from Fig. 2 is realised in the continuous-time relay system, while the adaptation of β^+ and β^- is realised by means of the discrete-time control algorithms.

In implementation of the sliding mode control the signal e^* is related to the error y by means of the formula (9). Then the successive differentiation of y must be performed. The differentiation gains the noises. Therefore, it is better to use in the place of the ideal differentiation described by the TF s the real differentiation described by

$$G_d(s) = \frac{k_d s}{s + k_d} \tag{30}$$

where k_d is sufficiently large gain (e.g. $k_d = 100$). Owing to the inertia appearing in (30) the high frequency noise is not gained.

It is worthwhile to realize that implementation of the sliding mode control is possible in the case of not to high d (e.g $d = 2, 3$). For higher d some difficulties connected with higher order differentiation can appear.

5. EXAMPLE

Consider the plant described by the equations

$$\begin{aligned} \dot{x}_1 &= -2x_1 + k2(1 + |u|)u \\ \dot{x}_2 &= x_1 = \frac{1}{1 + |x_2|} x_2, \quad y = x_2 \end{aligned} \tag{31}$$

where k is the gain which will be varried during the simulations. One can note that for the plant (31) we have $d = 2$. Let $c_0 = T, c_1 = 1$, then the dependence (6) takes the form

$$\dot{e}^* = T\dot{e} + e \tag{32}$$

where T is the time constant of the transients appearing when the sliding mode control is realized.

The SIMULINK block diagram of the system with the plant (31) and sliding mode relay control together with adaptation of the values β^+ and β^- is shown in Fig.3. The derivative $\dot{e}$ is obtained by means of element (30) with $k_d = 100$. The relay with hysteresis H is realized by means of two elements: "Blacklash" and "Sign". Some small hysteresis makes it possible to obtain more expressive oscillations bringing the information about values of $\beta+$ and $\beta-$.

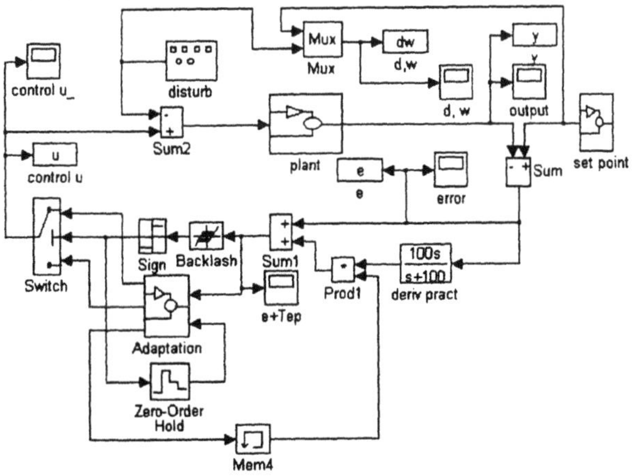

Fig. 3. System with sliding mode control and adaptation of β^+ and β^- (SIMULINK).

The adaptation of the values β^+ and β^- realizes the block "adaptation" which works in the discrete-time with sampling period h. The period h should be several times smaller than the minimal switching period of the relay. The main part of this block is the MATLAB function *slidad22* described by the program shown in Fig. 4. In this program the notation of some variables or parameters is different form that used in the paper. Namely $ep = \dot{e}^*$, $dep = \dot{e}^*_s$, $bpl = \beta^+$, $bml = \beta^-$, $p_1(t) = sign(e^*(t-h))$, $eplp = \dot{e}^*_{mx}$, $eplm = \dot{e}^*_{min}$, $epl(t) = \dot{e}^*(t-h)$, $p = H(e^*)$ where $H(e^*)$ denotes the characteristic of the relay with hysteresis; $t1$–instant of the last switching; dt–time interval assuring that the correction of β^+, β^- is made when the sliding control appears; bmx–maximum value of β^+ ($-bmx$ is the minimum value of β^-); $e = e^*$, de–the threshold of the error e: for $e < de$ the adaptation β^+, β^- is possible; the same notation with indexes 1 or 2 e.g. $bp1$ and $bp2$ means that $bp1(t) = bp2(t-h)$; c–denotes a positive coefficient in the dependencies establishing new value of $\beta^+(\beta^-)$ in accordance with (21), (22). The formulas (21) and (22) are used in the conditions determined by the program.

```
function b=slidad22(u)

e=u(1); ep=u(2); t=u(3); dt=u(4); dep=u(5); c=u(6);
bmx=u(7); bpl=u(8); bml=u(9); pl=u(10); tl=u(11);
eplp=u(12); eplm=u(13); epl=u(14); p=u(15);
de=u(16); T=u(17);
bp2=bpl; bm2=bml; t2=tl; ep2p=eplp; ep2m=eplm;

if abs(e)>de
 T2=T;
 bp2=bmx; bm2=-bmx;
else
 T2=T;
 if sign(ep)<sign(epl)
  t3=t;
  if t3-tl<dt
   bp2=bpl+c*(eplm+dep); ep2m=0;
  end
 end
 if sign(ep)>sign(epl)
  t3=t;
  if t3-tl<dt
   bm2=bml+c*(eplp-dep); ep2p=0;
  end
 end
 if sign(ep)==sign(epl)
  if sign(ep)==1
   if ep>eplp
    ep2p=ep;
   end
  end
  if sign(ep)==-1
   if ep<eplm
    ep2m=ep;
   end
  end
 end
end
if p~=pl
 t2=t;
end
b=[bp2,bm2,t2,ep2p,ep2m,T2];
```

Fig. 4. Algorithm of adaptation of the relay output.

6. RESULTS OF SIMULATIONS

The results of simulations shown in Fig.5 were obtained for the following values of the parameters: $k = 2$, $dt = 0.5$, $dep = 0.05$, $c = 0.8$, $bmx = 4$, $de = 0.05$, $T = 0.2$, $h = 0.01$, $H = 2 \times 0.005$. These parameters chosen experimentally assured the best work of the system for the assumed excitations. The disturbance d added to the input u and the set point w have stepwise variations shown in Fig. 5. The stepwise change of the disturbance d causes a rapid increase of the difference $(\beta^+ - \beta^-)$, which then is fast decreasing; owing to the adaptive action both the values β^+ and β^- become close to the required value of the control signal and the chattering effect is decreased, significantly. The stepwise change of the disturbance d gives un-notable effect on the output signal y and a fast decaying small peak (of order 0.01) in the error signal e. The stepwise change of the set point w causes a good fast output response without overshot. After that, the control u in some period of time takes the constant value β_{min} (or β_{mx}); then the sliding mode control appears with decreasing difference $(\beta^+ - \beta-)$ causing the decrease of chattering effect.

The results of simulations shown in Fig. 6 were obtained for the following values of the parameters: $dt = 0.5$, $dep = 0.1$, $c = 1$, $bmx = 4$, $de = 0.05$, $T = 0.2$, $h = 0.005$, $H = 2 \times 0.005$. The values of the parameters were chosen experimentally to assure the best work of the system for the assumed work conditions. The set point w is varied in accordance with the sinusoidal function $w(t) =$

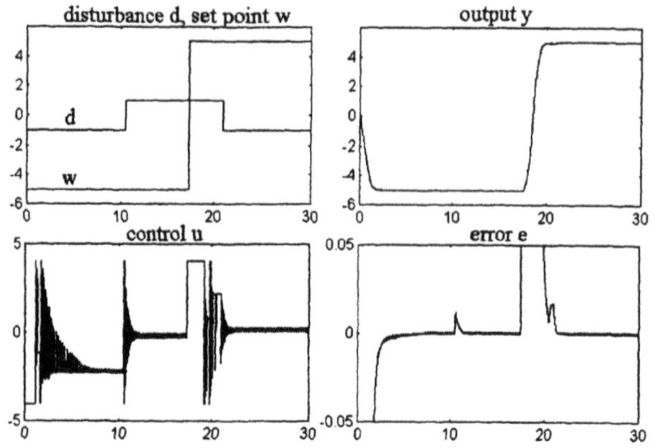

Fig. 5. Simulations results for stepwise change of d and w.

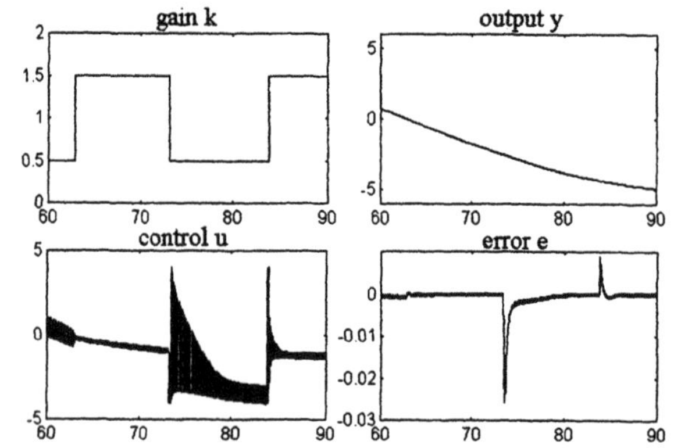

Fig. 6. Simulations results for stepwise change of k and sinusoidal w.

$5sin(0.05t)$. The stepwise variation of the gain k from the value 0.5 to 1.5 and vice versa appears. The stepwise change of the gain k gives usually a rapid increase of the difference $(\beta^+ - \beta^-)$ to the value $(\beta_{mx} - \beta min)$; then the adaptive action causes the exponential decrease of the difference and of the chattering effect. Note, that in the intervals of higher gain the difference $(\beta^+ - \beta^-)$ is smaller and in the intervals of smaller gain the difference is higher. For the first stepwise change of the gain there is no increase of the difference to maximal value, but it appears a decrease. This is caused by the fact that the stepwise change of k appears when y is approximately equal to zero. Note that the stepwise change of the gain gives unnotable effect on the output y and the relatively small peaks in the error e.

From the performed simulations it results that the system is very robust, even in the case of nonlinear plant and large and rapid parameter changes; during the relatively small velocity of the required mean value of the control, the chattering effect is reduced, significantly.

7. CONCLUSIONS

The proposed system with sliding mode control and adaptive action of the relay output remains all good properties of the system which uses sliding mode technique. Especially, it is very robust even in the case of such large and rapid parameter changes which create great difficulties for other adaptive control systems.

Owing to the adaptive action, the switched outputs β^+ and β^- of the relay are suited to the required value of the control u, so that the chattering effect connected with the sliding mode control is decreased, significantly.

The simplified model of the adaptive action has been created making it possible to choose the value of the parameter c assuring the stability. The instability of the adaptive action is critical because then the system doesn't work correctly. However the final values of the parameters were chosen experimentally taking into account the following properties: smaller $\dot{e}_s$ gives smaller difference $(\beta^+ - \beta^-)$; higher switching frequency (may by obtained by smaller H) creates possibility of tracking the required varying control without increasing the difference $(\beta^+ - \beta^-)$. To high c gives instability and is dangerous.

Generally if the adaptive action is stable the system works well and the temporary increase of the difference $(\beta^+ - \beta^-)$ influences the output y in-significantly.

ACKNOWLEDGMENT

The paper was partially supported by the Science Research Committee (KBN), grant No. 8 T11A 012 19.

8. REFERENCES

Åström, K.J. and B. Wittenmark (1989). *Adaptive Control.* Addison-Wesley, NY.

Gessing, R. (1997). Sliding Mode Control with Decreased Chattering Effect. *Preprints of the 7-th IFAC Symposium on Computer Aided Control System Design CACSD'97.* Gent, Belgium, 28-30 April 1997, pp.91-96.

Gessing, R. (1999). Sliding Mode Control with Decreased Chattering for Plants with Delay. *Proceedings of the 1999 IEEE Hong Kong Symposium on Robotics and Control HK-SRC'99.* Hong Kong, July 2-3 1999, Vol.II, pp. 465-470.

Slotine, J. J. E. and W. Li (1991). Applied Nonlinear Control, Englewood Cliffs, Prentice Hall, NJ.

Utkin, V. I. (1992) *Sliding Modes in Control and Optimization.* Springer Verlag, New York.

SLIDING MODE CONTROL FOR SYSTEMS WITH UNKNOWN CONTROL DIRECTION [1]

Giorgio Bartolini * **Antonella Ferrara** **
Luisa Giacomini **

* *Department of Electrical and Electronic Engineering University of Cagliari - Piazza d' Armi - 09123 Cagliari Italy - Tel. +39-070-675.5869 - fax +39-070-675.5900 E-mail: giob@{diee.unica.it, dist.unige.it}*
** *Department of Computer Engineering and Systems Science University of Pavia - Via Ferrata 1 - 27100 Pavia - Italy*
Tel. +39(0382)505757 - fax +39(0382)505373
E-mail: ferrara@conpro.unipv.it

Abstract: A novel control scheme is proposed to solve the problem of stabilizing nonlinear uncertain systems with unknown sign of the control gain and potentially unbounded nonlinearities. If the drift term $f(x)$ has a growth rate such that the overall system has a finite escape time t^*, only a guess of a lower bound, different from zero, of t^* is necessary to design the control law. The properties of the proposed algorithm are compared with those of a well-known adaptive control procedure when applied to the same class of systems.

Keywords: Adaptive control, uncertain high-frequency gain, sliding-mode control.

1. INTRODUCTION

In the present paper, the problem of controlling systems with unknown control gain, i.e., generic nonlinear systems $\dot{x} = f(x) + g(x)u$, with the nonlinear drift term $f(x)$ and $g(x)$ unknown but bounded by unknown functions. Moreover, the sign of $g(x)$ is assumed to to be constant but unknown, is addressed. Moreover,

From a practical viewpoint, the motivation of the interest towards this class of systems is due to the fact that there are systems, such as those of mechanical type, for which it is sensible to regard the drift term and the function multiplying the control signal in their differential equation as bounded quantities (since, for instance, the working space of the system is bounded and the actuator speeds are saturated). Yet, the use of the upperbounds of such quantities, determined on the basis of the physical limit values, directly in the control law produces extremely conservative control actions, with, in particular, excessive amplitudes. If, in turn, the control space is a subset of the working space, there exist upperbounds smaller than those computable relying on the limit values, though difficult to be determined. Thus, the design of a peculiar control strategy capable of dominating the uncertain drift term, and unsensitive to the control direction, whichever they are, is undoubtedly valuable in many practical applications.

Morse in (Morse, 1995) raised a famous conjecture, i.e. that the class of one-dimensional linear systems $\dot{x} = x + bu$, where $b \neq 0$ is unknown

[1] Partially supported by the contract IST 10107-PROTECTOR of the European Community, and by MURST Project "Identification and control of industrial systems".

with unknown sign, is not globally stabilizable via an adaptive control law. In (Nussbaum, 1983) this conjecture has been disproved and a global adaptive stabilizer has been designed. The control law is obtained introducing a control gain that is a suitable nonlinear function of an adaptive gain. This law allows one to automatically increase the gain of the control signal while dynamically changing its sign.

Starting from linear systems (Willems and Byrnes, 1984), Nussbaum-based controllers have been applied to first–order nonlinear systems (Mårtensson, 1990), and to nonlinear systems of any order with triangular structure (Mårtensson and Polderman, 1993; Ryan, 1994; Xudong, 2001), up to the design of an universal stabilizer (Ryan, 1995), that applies to the class of systems we addressed at the beginning.

However, as Willems and Byrnes (Willems and Byrnes, 1984) pointed out, the high gain feature of these procedures are restricting their interest to a pure theoretical level. Indeed, it is not surprising that a closer look to the transient behaviour of the controlled system reveals the presence of high peaking phenomena.

Willems and Byrnes (Willems and Byrnes, 1984) argued that the verification of Morse's conjecture would depend on the allowed smoothness of the control law: i.e., a discontinuous control might solve the problem. However it is well known that sliding mode control (Utkin, 1992) requires the knowledge of the sign of the control gain as well. In this view, the aim pursued in this paper is twofold: find a control for the specified class of system free from high gain peaking, and design a sliding mode control independent of the control direction. Note that the same problem is solved in different way in (Drakunov, 1993).

The new proposal presented in this paper is a two phase controller. First no control input is applied and the magnitude of the drift term is identified through a sliding mode observer module. Then, the control is applied with amplitude equal to the identified one scaled by the lower bound of the absolute value of the function pre-multiplying the control vector and sign equal to an initial guess. With the knowledge that the nonlinearities are all dominated, the correct sign of the control gain is identified from the subsequent behaviour of the closed-loop system.

2. UNIVERSAL ADAPTIVE STABILIZER

Given the system

$$\begin{cases} \dot{x} = f(x) + bd(u) \\ y = e(x) \end{cases} \tag{1}$$

with $f(x)$ and b unknown, $f \in C^1(\mathbb{R})$, $f(0) = 0$, $x \in \mathbb{R}$, $u \in \mathbb{R}$, a stabilizing control law independent from the knowledge of the sign of b is sought. In the literature, an adaptive control was proposed as a possible solution, the so–called *universal controller* (Ryan, 1994). With respect to previous control schemes (see for example (Mårtensson, 1990)), that controller applies to a broader class of systems, since it does not require the knowledge of the rate of growth of the unknown functions involved. For the reader's convenience, let us briefly recall the design principle of the universal controller in the simplified hypotheses that $d(u) = u$ and $e(x) = cx$ (i.e., linear sensors and actuators). Further, necessary hypotheses are that the system has relative-degree one, i.e. $cb \neq 0$, and that $f(x)$ is a positive homogeneous function of degree k, i.e. given $\rho \in \mathbb{R}$, $f(\rho x)) = \rho^k f(x)$. More specifically, in (Ryan, 1995), the control law

$$u(t) = \Lambda(\lambda(t))e^{|y(t)|}y(t) \tag{2}$$

$$\dot{\lambda} = \alpha e^{|y(t)|}y^2(t), \ \lambda(0) = \lambda^0 \tag{3}$$

is designed for the considered system, where α is a real constant and $\Lambda : \mathbb{R} \to \mathbb{R}$ is any continuous function with the Nussbaum properties

$$\lim_{\lambda \to \infty} \sup \frac{1}{\lambda} \int^{\lambda} \Lambda(\kappa) d\kappa = +\infty \tag{4}$$

$$\lim_{\lambda \to \infty} \inf \frac{1}{\lambda} \int^{\lambda} \Lambda(\kappa) d\kappa = -\infty \tag{5}$$

In (Ryan, 1995), Theorem 3 states that given system (1), controlled by (2), and (3), and it being $(x(\cdot), \lambda(\cdot)) : [0, \omega) \to \mathbb{R} \times \mathbb{R}$ the maximal solution of the resulting initial-value problem, then (i) $\omega = \infty$; (ii) $lim_{t\to\infty}\lambda(t)$ exists and is finite; (iii) $x(t) \to 0$ as $t \to \infty$.

Relying on the various hypotheses previously mentioned it can be shown that, along the solutions of the closed loop system (1), (2), (3), the inequality

$$x\dot{x} \leq \mu |x|^p + cb\alpha\Lambda\dot{\lambda} \tag{6}$$

holds $\forall t \in [0, \omega)$, where μ is a suitable constant, $p = k+1$, and $[0, \omega)$ is the maximal interval of existence of the solution $(x(\cdot), \lambda(\cdot))$ of the feedback control system. From integration it follows that, for all $t, \tau \in [0, \omega)$, $t \geq \tau$

$$0 \leq \frac{1}{2}x^2(t) \leq \frac{1}{2}x^2(\tau) \tag{7}$$
$$+\mu_2[\lambda(t) - \lambda(\tau)] + cb\alpha \int_{\lambda(\tau)}^{\lambda(t)} \Lambda(\theta) d\theta$$

where μ_1, μ_2 are suitable constants. At this point a proof based on contradiction argument is used.

By (3), $\dot{\lambda} \geq 0$, i.e. $\lambda(t)$ is a monotone non decreasing function. Let us suppose that $\lambda(t)$ is unbounded. Then, there exists τ such that $\lambda(\tau) \geq 1$ and $\lambda(t) \geq \lambda(\tau)$, $\forall t \geq \tau$. Dividing the above disequality by $\lambda(t)$, it yields

$$0 \leq \text{constant} + \lim_{\lambda \to \infty} \inf \frac{cb\alpha}{\lambda} \int_{\lambda(\tau)}^{\lambda} \Lambda(\theta) d\theta \quad (8)$$

which contradicts the Nussbaum properties of $\Lambda(\lambda)$ (one of (4) and (5), depending on the sign of b). It follows that $\lambda(t)$ is a bounded function. From boundedness and monoticity of λ, it follows that $x(t)$ is L^p. Since $\dot{x}(t)$ is bounded as well, by virtue of Barbalat's Lemma, $x(t) \to 0$ as $t \to \infty$.

2.1 *Peaking*

In (Ryan, 1995) it is suggested that a Nussbaum function of the form $\Lambda(\lambda) = e^{\lambda^2} cos(\lambda)$ suffices in most of the applications. Let us write the closed loop system with respect to the output variable

$$\dot{y} = cf(x) + cb\alpha y e^{|y|} e^{\lambda^2} cos(\lambda) \quad (9)$$
$$\dot{\lambda} = e^{|y|} y^2 \quad (10)$$

Since $x = 0$ when $y = 0$, and $f(0) = 0$, this system has one equilibrium point in $(y, \lambda) = (0, \lambda^*)$ that is reached for $t \to \infty$. The exact value of λ^* is not determined by the presented method, but it is guaranteed that it is bounded. Let us suppose that after a time t_1, $y \leq \varepsilon$, i.e. x enters a small residual interval B_ε centered on $x = 0$ and $\lambda = \lambda^* - \Delta_\varepsilon$, $\Delta_\varepsilon \in \mathbb{R}_+$. The situation is such that

$$sign(bcos(\lambda^* - \Delta_\varepsilon)) = -1 \quad (11)$$

On the other hand $cos(\lambda) = 0$ for $\lambda = (2k+1)\frac{\pi}{2}$, $k \in \mathbb{N}$. It is possible that it exists $\bar{k} \in \mathbb{N}$ such that $(2\bar{k}+1) \in [\lambda^* - \Delta_\varepsilon, \lambda^*]$, that means that it surely exists a time t_2, $t_1 < t_2 < \infty$ such that $cos(\lambda)$ changes its sign. As a result the system reaches an unstable state and the adaptation of λ restarts. This behavioural scheme can repeat itself more than once and the final value of λ can be quite large.

Let us now suppose that the system is affected by a bounded perturbation, such that $f(x) = \bar{f}(x) + \delta(t)$ (i.e. $f(0) \neq 0$). In this case system (9), (10) does not have an equilibrium point. In fact $\dot{\lambda} = 0$ only for $y = 0$, that implies $\dot{y} \neq 0$. Clearly λ is unbounded and the peaking phenomenon will be even more serious. Indeed, in (Ryan, 1994; Blanchini and Ryan, 1999), the adaptive law has been modified to be applied to a class of perturbed nonlinear systems. The considered class contains the class of homogeneous functions and allows the presence of bounded disturbances. To avoid peaking, (3) is replaced by

$$\dot{\lambda} = d_\varepsilon(y)[|y| + \phi(y)] \quad (12)$$

where

$$d_\varepsilon(y) = \begin{cases} |y| - \varepsilon, & |y| \geq \varepsilon \\ 0, & |y| < \varepsilon \end{cases} \quad (13)$$

and $\phi(y)$ is a known function such that $f(x) \leq \gamma(\| x \| + \phi(y))$, with γ suitable real constant. In the presented hypotheses $\phi(y) = ye^{|y|}$. Obviously, the controlled variable y converges to the compact interval $[-\varepsilon, \varepsilon] \subset \mathbb{R}$.

3. A SLIDING MODE CONTROL ALGORITHM

The dynamic system (1), with the assumption introduced in Section 2 and generalizing the unknown gain b to be a function of x, $g(x)$, results in

$$\begin{cases} \dot{x} = f(x) + g(x)u \\ y = cx \end{cases} \quad (14)$$

where $f(x)$ possesses an unknown time-varying bound such that $|f(x)| < \bar{k} + e^{|y|}$ and $g_1 \leq |g(x)| \leq g_2$, g_1, g_2 known with positive sign. The control algorithm here proposed can be divided into three steps

(1) find, in finite time, an overestimate of the drift term $f(x)$;
(2) identify the sign of the control gain, and, at the same time, guarantee, that the system trajectories are contained in Ω, bounded neighborhood of the origin of the state space;
(3) adapt the drift estimate such that it is an overestimate of the drift term for all x in Ω.

The Nussbaum functions are replaced, in our proposal, by a decision-based process.

3.1 *Step 1.*

An overestimate of the drift term $f(x)$ is obtained setting $u = 0$ in (14), and introducing a simple observer module of the form

$$\dot{w} = \alpha e^{|w|} sign(y - w) \quad (15)$$

When the error signal $\varepsilon = y - w$ reaches the manifold $\varepsilon = 0$, say at time $t = \bar{t}$, this control phase ends. Due to stability reasons, the observer sub-system will remain in sliding mode here after.

Proposition 1. Given the system

$$\begin{cases} \dot{x} = f(x), & y = cx \\ \dot{\varepsilon} = cf(x) - \alpha e^{|w|} sign\varepsilon & w = y - \varepsilon \end{cases} \quad (16)$$

where $|f(x)| < \overline{k} + e^{|y|}$, ε goes to zero in a finite time.

Proof: Defined $B_\epsilon = \{x \in \mathbb{R} \mid x < \epsilon\}$, $\exists\, \epsilon \in \mathbb{R}$ such that $\forall x \notin B_\epsilon$, $|f(x)| < \alpha e^{|y|}$. Then,

$$\varepsilon\dot{\varepsilon} = \varepsilon(cf(x) - \alpha e^{|w|} sign\varepsilon) \leq -\gamma|\varepsilon| \quad (17)$$

for $w \notin B_\epsilon$. That is, $\varepsilon = 0$ is reached in finite time as soon as w grows outside B_ε. △

This first step is proved to be accomplishable even in presence of a finite escape time system (14 with $u = 0$, i.e. $\dot{x} = f(x)$), if the finite escape time of the system observer (15) is smaller than that of the system one, as proved hereafter.

Proposition 2. Let $\dot{x} = f(x)$, where $f(x)$ is any function such that it is possible to define, a-priori, a minimal existence interval $[0, t^*]$. Then, with reference to systems (14), (15), there exists an α such that a sliding mode on the sliding manifold $\varepsilon = 0$, is established at any prespecified $\tilde{t} = \beta t^*$, with $\beta < 1$.

Proof: The result is a consequence of the fact that $w = \pm ln(1 - \alpha t)$. Indeed, (15) has finite escape time $\frac{1}{\alpha}$. If α is chosen such that $\frac{1}{\alpha} < t^*$ the time for which $\alpha e^{|w|}$ becomes greater than $|f(x)|$ is surely less than $\frac{1}{\alpha}$, i.e., less than t^*. △

3.2 *Step 2.*

From $\tilde{t}$ on ,the quantity $\alpha e^{|w|} sign(\varepsilon)$ is Equivalent, in the sense of Filippov (Filippov, 1988), to $f(x)$ and its actual modulus can be used as an upper bound of the drift term. Thus, at time $t = \tilde{t}$, letting $\tilde{y} = y(\tilde{t})$, the control

$$\begin{aligned} u = &-\frac{\alpha}{g_1} e^{|w|} \rho(t) sign(y) sign(\tilde{y} \\ &+ \Delta sign(\tilde{y}) - y) \quad (18) \\ \dot{w} = &\ \alpha e^{|w|} sign(\varepsilon) + g_2 |u|\, sign(\varepsilon) \quad (19) \\ \rho(t) \in &\ \{-1, 1\},\ \rho(\tilde{t}) = 1 \quad (20) \end{aligned}$$

is applied, where Δ is a suitable design constant, and $\rho(t)$ is a gain, switching between 1 and -1 with a logic still to be defined.

Two situations can arise within the time interval $(\tilde{t}, t^*]$:

a) a sliding mode arises on $y = \tilde{y} + \Delta sign(\tilde{y})$(this happens if $\rho \cdot sign(g(x)) = -1$),
b) a sliding mode arises on $y = \overline{y} < \tilde{y}$ (this happens if $\rho \cdot sign(g(x)) = 1$).

If, at time $t = \overline{t}$, Case (a) is verified, $\rho(\overline{t}^+)$ is set to $-\rho(\overline{t}^-)$, and y becomes a non increasing time function as in Case (b). Case (b) can be divided into two sub-cases: (i) $\overline{y} = 0$ and (ii) $\overline{y} \neq 0$.

Proposition 3. At time $t = \tilde{t}$, one between $S_1 = y = 0$ and $S_2 = y - (\tilde{y} + \Delta sign(\tilde{y})) = 0$ is an attractive manifold.

Proof. Let us suppose that the system is such that only sub-case (i) of case (b) may happen, i.e. $\overline{y} = 0$. Taking into account S_1 and S_2 previously defined, one has

$$\dot{S}_1 = \dot{S}_2 = f(x) + (\rho g(x))\frac{\alpha}{g_1} e^{|w|} sign S_1 sign S_2 \quad (21)$$

Again, two cases are possible: $\tilde{y} > 0$ and $\tilde{y} < 0$.
First case: $\tilde{y} > 0$.
In this case, in a neighborhood of $\tilde{y}$, $signS_1 signS_2 = signS_1 = signS_2 = 1$; in a neighborhood of $S_1 = 0$, $signS_1 signS_2 = signS_1$; finally, in a neighborhood of $S_2 = 0$, $signS_1 signS_2 = signS_2$. For $\rho g(x) > 0$, starting at $\tilde{y}$ and given the dominance of the control gain term,

$$\begin{aligned} S_1\dot{S}_1 &= S_1(f(x) - \gamma(x, w) signS_1) \\ &\leq -\overline{\gamma}|S_1| < 0 \end{aligned} \quad (22)$$

while

$$\begin{aligned} 0 < S_2\dot{S}_2 &= S_2(f(x) + \gamma(x, w) signS_2) \\ &< 2\overline{\gamma}|S_2| \end{aligned} \quad (23)$$

That is, S_1 is attractive and S_2 is not.
For $\rho g(x) < 0$, starting at $\tilde{y}$

$$\begin{aligned} 0 < S_1\dot{S}_1 &= S_1(f(x) + \gamma(x, w) signS_1) \\ &\leq 2\overline{\gamma}|S_1| \end{aligned} \quad (24)$$

while

$$\begin{aligned} S_2\dot{S}_2 &= S_2(f(x) - \gamma(x, w) signS_2) \\ &< -\overline{\gamma}|S_2| \end{aligned} \quad (25)$$

That is, S_2 is attractive and S_1 is not.

Second case: $\tilde{y} < 0$.
For $\rho g(x) > 0$, starting at $\tilde{y}$ and given the dominance of the control gain term,

$$\begin{aligned} 0 < S_1\dot{S}_1 &= S_1(f(x) + \gamma(x, w) signS_1) \\ &\leq 2\overline{\gamma}|S_1| \end{aligned} \quad (26)$$

while

$$\begin{aligned} S_2\dot{S}_2 &= S_2(f(x) - \gamma(x, w) signS_2) \\ &< -\overline{\gamma}|S_2| \end{aligned} \quad (27)$$

That is, S_2 is attractive.
For $\rho g(x) < 0$, starting at $\tilde{y}$

$$\begin{aligned} S_1\dot{S}_1 &= S_1(f(x) - \gamma(x, w) signS_1) \\ &\leq -\overline{\gamma}|S_1| < 0 \end{aligned} \quad (28)$$

while

$$0 < S_2 \dot{S}_2 = S_2 (f(x) + \gamma(x, w) sign S_2) < 2\overline{\gamma} |S_2| \quad (29)$$

That is, S_1 is attractive. Note that for $x(\tilde{t}) < 0$, the sign of the gain ρg should be positive due to the presence of the term $sign S_2$, which is negative in a neighborhood of $S_1 = 0$.

Because $\tilde{y}$ is contained in the compact interval determined by the two manifolds $y = 0$ and $y = \tilde{y} + \Delta sign(\tilde{y})$, y (and, as a consequence, x, since $y = cx$) remains bounded by a fixed value when the control gain has an 'incorrect' sign.

3.3 *Step 3.*

The fact that the control gain, in modulus, keep on dominating $f(x)$ after the closure of the control loop is guaranteed by the term $g_2 |u| sign \varepsilon$ in (19) that dominates $|g(x)|u$.

In sub-case (ii) of the previous section, a further adaptation is required. The convergence of y to zero can be ensured if the control amplitude is greater than the maximum value of $f(x)$ in a closed ball centered at the origin. This property can be achieved in finite time by an integral adaptation gain $\dot{K} = \gamma\sqrt{|y|}$ multiplying the control signal u in eq.(18).

4. COMMENTS ON THE GROWTH RATE OF THE OBSERVER MODULE

In Section 3 it has been proved that the observer module (15) will catch the system trajectory even in presence of a finite escape time, provided that the finite escape time of the system observer is smaller than the system one. Under this assumption, convergence of the error ε to zero is assured with two possible modalities: (1) $\varepsilon = 0$ and w is decreasing (i.e. $w = y \to 0$) or (2) $\varepsilon = 0$ and y is increasing (i.e. $w = y \to \infty$). In this second case, that, hopefully, will last for a very short time (i.e. until the control with the right sign is eventually applied), it is clear that the quantity $e^{|w|}$ will outgrow the term it is estimating, namely $f(x)$. This implies that, potentially, the observer subsystem is affected by high-gain switchings, and, the control gain will be unnecessarily high. This justifies a further modification of eq. (15)

$$\dot{w} = \alpha e^{|w|} sign(\varepsilon) + K sign(\varepsilon) \quad (30)$$

$$\dot{K} = e^{|w|} \varepsilon \quad (31)$$

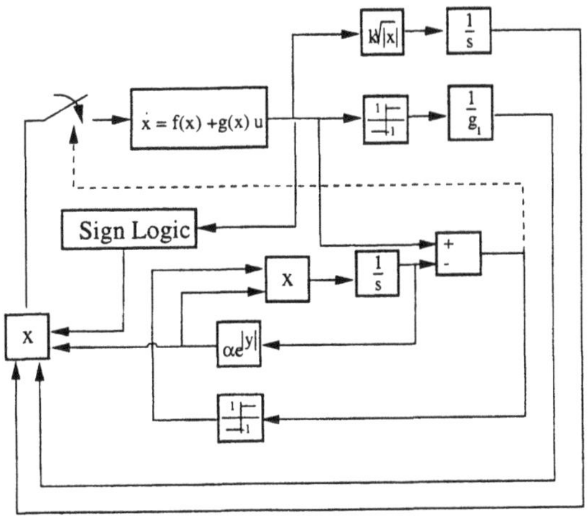

Fig. 1. The proposed control scheme.

5. SOME REMARKS

The overall control scheme is shown in figure 1. It can be described as

$$\begin{cases} \dot{x} &= f(x) + g(x)u, \; y = cx \\ u &= -\dfrac{\alpha}{g_1} e^{|w|} \rho(t) sign(y) sign(\tilde{y} \\ & \quad + \Delta sign(\tilde{y}) - y) \\ \dot{w} &= \alpha e^{|w|} sign(\varepsilon) + g_2 |u| \; sign(\varepsilon) \\ \varepsilon &= y - w \end{cases} \quad (32)$$

where ρ is generated by the following logic:

```
ρ = 0;
while (true){
        if (in sliding on ε = 0){
          ρ = 1;
          break;
        }
}
while (true){
        if (in sliding on y = ỹ + Δsign(ỹ)){
          ρ = −1;
          break;
        }
}
```

Due to the nature of the control law, only one peak is present during transients and its amplitude, usually quite small, can be further reduced through the design parameters of the controller.

Note that it is possible to manage also more general cases similar to that dealt with by Ryan in which the actuator is more complex than that here considered.

6. AN EXAMPLE

Consider the system

$$\dot{x} = x^2 - 50\cos(t) + gu \quad (33)$$

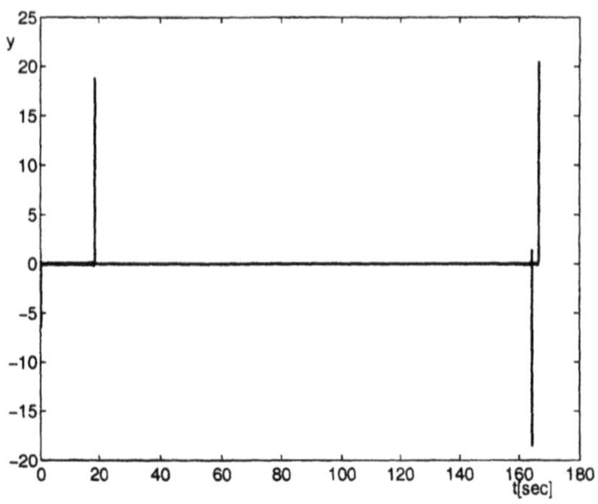

Fig. 2. The output quantity when the *universal controller* is applied.

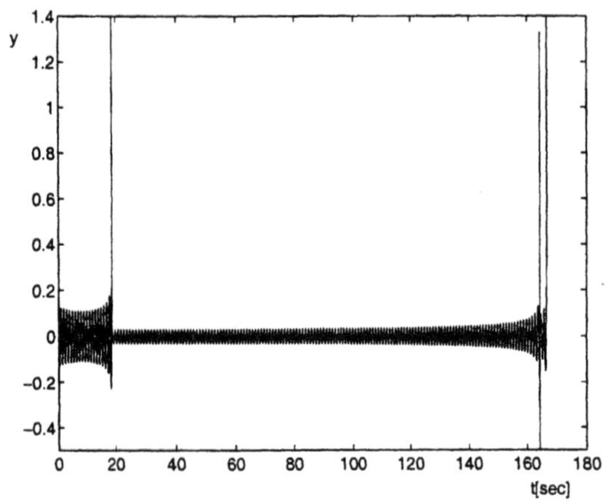

Fig. 3. Zoom of figure 2.

where $g \in \{-1, 1\}$. Note that, the drift term is made by an homogeneous part and a bounded non-homogenous one, that is the control scheme in (Ryan, 1995) should not work properly. In figure 2 (figure 3 shows an enlargement about the time axis) the results obtained are shown when the *universal stabilizer* is applied, when eq. (10) is used[2]. When eq. (12) is applied the system goes unstable after a few seconds. In figure 4 is shown the output variable trajectory with the algorithm proposed in this paper.

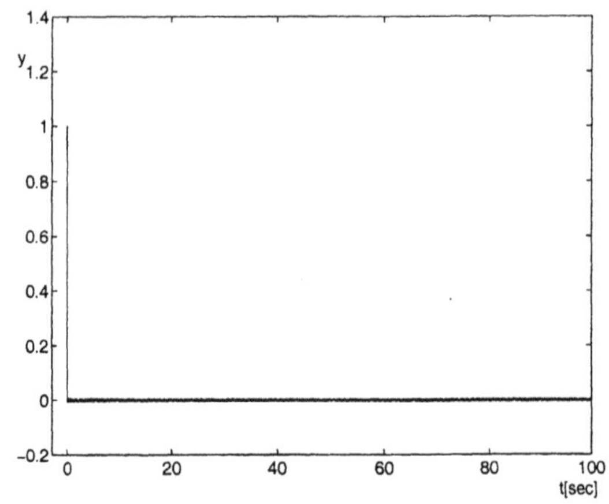

Fig. 4. The output quantity when the proposed algorithm is applied.

7. REFERENCES

Blanchini, F. and E.P. Ryan (1999). A razumikhin-type lemma for functional differential equations with application to adaptive control. *Automatica* **35**, 809–818.

Drakunov, S. (1993). Sliding modes control of the systems with uncertain direction control vector. *Proc. 32rd Conf. on Decision and Control, San Antonio, Texas* pp. 2477–2478.

Filippov, A.F. (1988). *Differential equations with discontinuous righthand sides.* Vol. ?. Kluwer Academic Press. Dordrecht. NL.

Morse, A.S. (1995). Control using logic-based switching. In: *Trends in Control. An European Perspective.* pp. 89–113. Springer Verlag. Berlin.

Mårtensson, B. (1990). Remarks on adaptive stabilization of first order non-linear systems. *Systems & Control Letters* **14**, 1–7.

Mårtensson, B. and J.W. Polderman (1993). Correction and simplification to 'the order of any stabilizing regulator is sufficient information for the adaptive stabilization'. *Systems & Control Letters* **20**, 1–7.

Nussbaum, R.D. (1983). Some remarks on a conjecture in parameter adaptive control. *Systems & Control Letters* **3**, 243–246.

Ryan, E.P. (1994). A nonlinear universal servomechanism. *IEEE Trans. on Automatic Control* **AC-39**, 753–761.

Ryan, E.P. (1995). Universal stabilization of a class of nonlinear systems with homogeneous vector fields. *Systems & Control Letters* **26**, 177–184.

Utkin, V.I. (1992). *Sliding modes in control and optimization.* Vol. ?. Springer Verlag. Berlin.

Willems, J.C. and C.I. Byrnes (1984). Global adaptive stabilization in the absence of information on the sign of the high frequency gain. In: *Lecture Notes on Computer and Information Science.* Vol. 62. pp. 49–57. Springer Verlag. New York.

Xudong, Y. (2001). Adaptive nonlinear output-feedback control with unknown high-frequency gain sign. *IEEE Trans. on Automatic Control* **AC-46**, 112–115.

[2] Tests with different integration algorithm have been made to verify that the peaks are not due to a numerical integration problem.

Copyright © IFAC Adaptation and Learning in Control and
Signal Processing, Cernobbio-Como, Italy, 2001

TURBO-LEARNING: A NOVEL APPLICATION TO MIMO WIRELESS COMMUNICATIONS

Simon Haykin and Mathini Sellathurai

McMaster University
Hamilton, Ontario, Canada
haykin@mcmaster.ca

Abstract: This paper briefly reviews the turbo-learning principle, thereby setting the stage for the description of a novel multiple-input, multiple-cutput (MIMO) wireless communication system. This new system, called Turbo-BLAST, builds on two key principles:

• The Turbo-learning principle
• The Bell-Labs Space-Time Layered (BLAST) architecture

Experimental results are presented, demonstrating the high-spectral efficiency of Turbo-BLAST.

A distinctive feature of the Turbo-BLAST receiver is its rapidly converging iterative nature. Indeed, it may be described as a nonlinear stochastic sampled-data feedback system. The paper identifies the development of a mathematical framework for convergence analysis of the receiver as an open research problem.

Keywords: Communication system. Convergence problem. Feedback. Information. Likelihood function. Markov model.

1. INTRODUCTION

In 1993, Berrou et al [1] developed the revolutionary iterative "turbo" receiver for decoding two-dimensional product-like codes. Two properties constitute the hallmark of turbo codes:

- The error performance of the turbo decoder improves with the number of iterations of the decoding algorithm.
- The turbo decoder is capable of closely approaching the Shannon theoretical limit of channel capacity in a computationally feasible manner.

What is even more profound is the fact that the "turbo-learning principle" has been successfully applied not only to channel decoding, but also channel equalization, coded modulation, multi-user detection, joint-source and channel decoding [2]. In this paper, we describe another novel application of the turbo-learning principle to wireless communications using multiple antennas at both the transmitting and receiving ends of the wireless channel.

2. BASIC IDEAS OF TURBO LEARNING

2.1 The Turbo-learning Principle

The design of a digital communication system, rooted in turbo learning, is based on the *maximum a posteriori probability (MAP) criterion.* Let s denote a message symbol transmitted over a communication channel. Let $\mathbf{x}$ denote the received signal whose conditional probability density function (pdf), given that s was transmitted, namely, $f(\mathbf{x}|s)$, is governed by a Markov model. According to Bayes' rule, the probability that symbol s was transmitted, given $\mathbf{x}$, is

$$P(s|\mathbf{x}) = \frac{f(\mathbf{x}|s)P(s)}{f(\mathbf{x})}$$

$$= \frac{f(\mathbf{x}|s)P(s)}{\sum_s f(\mathbf{x}|s)P(s)} \quad (1)$$

where $P(s)$ is the a priori probability of symbol s being sent. Equation (1) shows that the computation of the a posteriori probability $P(s|\mathbf{x})$ depends on the a priori probabilities $P(s)$ of all the symbols that constitute the source alphabet. The particular value of s that maximizes $P(s|\mathbf{x})$ yields the MAP estimate of the transmitted symbols.

The turbo-learning principle may now be formally stated as follows [2]:

- Compute iterative MAP estimates of the transmitted message symbols using successively refined a priori distributions $P_i(s)$, where the subscript i refers to a particular iteration in this computation.
- For the calculation of $P_i(s)$, use all statistically independent information (i.e., set of sufficient statistics) that is already available to the receiver at iteration i.

For example, in a serially concatenated turbo code, $P_i(s)$ may be obtained by using the a posteriori probabilities computed for the decoding of an outer code at the previous iteration i - 1.

Let the symbol 0 and 1 in a binary alphabet be respectively denoted by the levels -1 and +1. Then, using Eq. (1) we may define the *a posteriori log-likelihood ratio* as

$$L(s|\mathbf{x}) = \log\left(\frac{P(s=+1|\mathbf{x})}{P(s=-1|\mathbf{x})}\right)$$

$$= \log\left(\frac{f(\mathbf{x}|s=+1)}{f(\mathbf{x}|s=-1)}\right) + \log\left(\frac{P(s=+1)}{P(s=-1)}\right) \quad (2)$$

whose sign defines the hard decision on which binary symbol was transmitted and its magnitude defines the reliability of the decision. The first term on the right-hand side of Eq. (2) represents the incremental information about symbol s that is gained by the decoder as a result of processing the received signal vector $\mathbf{x}$. The second term of the equation is the *a priori log-likelihood ratio.*

2.2 Turbo Encoding

The channel encoder of a turbo code involves two encoders joined together by means of an interleaver; the encoders can be concatenated serially or in parallel, as illustrated in Fig. 1. An interleaver is an input-output mapping device that permutes the ordering of the sequence of symbols from a fixed alphabet in a completely deterministic manner; that is, it takes the symbols at the input and produces identical symbols at the output but in different order. The interleaver can be of many types, of which the periodic and pseudo random are two. Turbo codes use a pseudo-random interleaver, which operates only on the systematic (message) bits.

There are two reasons for the use of an interleaver in a turbo code:

- To tie together errors that are easily made in one half of the turbo code to errors that are exceptionally unlikely to occur in the other half. This is indeed the main reason why the turbo code performs better than a traditional code.
- To provide a robust performance with respect to mismatched decoding, which is a problem that arises when the channel statistics are not known or have been incorrectly specified.

In the parallel concatenated turbo encoder of Fig. 1(a), two systematic convolutional encoders are concatenated in parallel. A block of k message bits is fed directly to encoder 1; for encoder 2 the same block of k message bits is permuted first before being applied to it. (Permuting is applied to the parity-check bits only if the channel has memory.) The code rates of the two encoders do not have to be equal, but for best decoding performance the recommendation is to choose $R_1 \leq R_2$, where R_1 and R_2 are respectively the code rates of encoders 1 and 2. (The code rate of a block code is defined as the ratio of the number of message bits in each block to the number of bits in the codeword.)

Alternatively, we may use serially concatenated convolutional encoders by concatenating two systematic convolutional encoders as shown in Fig. 1(b). A block of k message bits is encoded directly by the outer encoder, then the interleaver permutes the encoded codewords before passing them onto the inner encoder.

The term "systematic" used in describing the convolutional encoders that constitute the two versions of turbo codes shown in Fig. 1 refers to the fact that in both cases the incoming message bits are transmitted in unaltered form.

2.3 Turbo Decoding

Our interest in this paper is in serially concatenated turbo codes. Figure 2 shows the block diagram of the *turbo decoder* for the turbo encoder of Fig. 1(b).

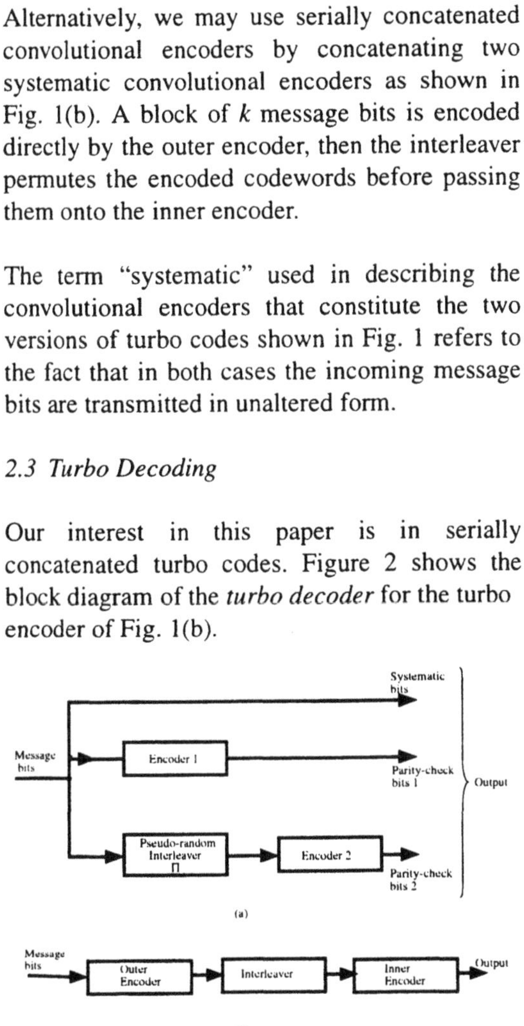

Figure 1: (a) Parallel concatenated channel encoder; (b) serial concatenated channel encoder

Central to the operation of a turbo receiver exemplified by that of Fig. 2 is the notion of extrinsic information, the most convenient representation for which is as a log-likelihood ratio; see Eq. (2). Formally, *extrinsic information* generated by a decoding stage in the receiver for a set of bits is defined as the difference between two quantities: (1) the log-likelihood ratio computed at the output of that decoding stage, and (2) the *intrinsic information* (also expressed as a log-likelihood ratio) applied to its input by the other decoding stage in the receiver. In effect, extrinsic information is the incremental information gained by exploiting the dependencies that exist between a message bit of interest and incoming raw data (received over the channel) processed by the decoder in question.

As shown in Fig. 2, the turbo receiver is in the form of a feedback loop consisting of an inner decoder and an outer decoder separated by an interleaver (Π) and de-interleaver (Π^{-1}). The receiver operates in an iterative manner in accordance with the turbo principle: Assuming that the message-bit estimates computed by the inner decoder are accurate, the outer decoder is then supplied with reliable information about the message bits. Hence, by having the outer decoder reinitiate the decoding process, even more accurate estimates of the message bits are passed onto the inner decoder, so the iterative estimation of message bits goes on.

In specific terms, the receiver performs the following computations:

1. The inner decoder generates soft estimates of the message bits conditioned on the inner code constraint. It computes the extrinsic information of the message and parity-check bits. This extrinsic information is applied to the outer decoder as the intrinsic information of its encoded bits. However, before application to the outer decoder, the extrinsic information is reordered to compensate for the pseudo-random interleaving introduced in the turbo encoder. During the first iteration, the input labeled "channel output" is initialized with soft outputs consisting of the log-likelihood ratios of the binary symbols received from the channel. The second input labeled "intrinsic information" is initialized to zero since no a priori information about the incoming symbols is available at the start of the computation.

2. The outer decoder computes the log-likelihood ratios of both message and parity-

check bits subject to the outer code constraints, thereby determining the extrinsic information of all those bits. The output of the outer decoder provides the intrinsic information to the inner decoder after re-ordering to conform to the random interleaving introduced in the turbo encoder. The input labeled "0" is always initialized to zero on the assumption that the binary symbols generated by the original source of information at the transmitter input are equally likely.

Both the inner and outer decoders in Fig. 2 are soft-input, soft-output (SISO) processing units, with the result that the iterative turbo receiver of Fig. 2 may be viewed as a *nonlinear stochastic sampled-data feedback system.*

Another point of interest is that the decoders in the iterative receiver of Fig. 2 update the log-likelihood ratios of both the message as well as parity-check bits subject to the respective code constraints. In contrast, the iterative receiver for the parallel-concatenated turbo encoder of Fig. 1(a) updates only the log-likelihood ratios of the message bits.

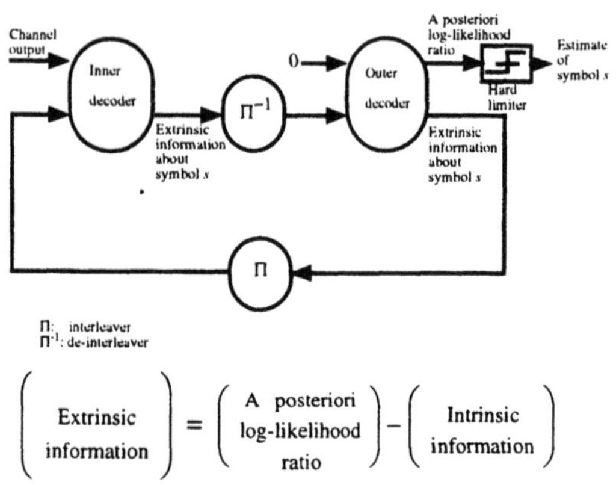

Figure 2: Turbo decoder for serially concatenated codes

3. BLAST ARCHITECTURE

With the brief introduction to turbo learning presented in Section 2, we are now ready to focus our attention in the rest of the paper on its application to wireless communications, using space-time processing.

Future-generation wireless communications will focus on high-throughput data transmission in order to achieve a rate of tens of Mbits/sec in local area wireless networks in office complexes and from building to building. High-throughput data transmission over band-limited wireless channels demands novel multiplexing techniques that offer increased channel capacity and spectral efficiency. Spatial multiplexing is the most popular choice since it exploits spatial diversity to increase the single-user capacity, while minimizing the use of both time and frequency. In particular, the multiple-input and multiple-output (MIMO) communications structures, popularized as *Bell Labs Space-Time (BLAST) architectures,* have received considerable attention recently as they have the potential to provide very high data-rate communication over wireless channels without increasing the total transmit power.

The communication system of interest consists of multiple antennas at both ends of the wireless communication channel with n_T transmit antennas and n_R receive antennas, as illustrated in Figure 3. In this system, information-bearing signals are demultiplexed into multiple substreams, and an array of antennas is used to launch the substreams simultaneously, using the same channel bandwidth and with the total transmitted power always held constant. At the receiving end, the transmitted signals are picked up by a receiving antenna array. Each antenna receives all of the transmitted signals as one superimposed signal, modified by the channel dynamics. Even though the signals are transmitted in the same frequency band, the signals from the different transmit antennas are located at different points in space, and each signal is scattered differently. Hopefully the received signals at each receive antenna element still contain useful information about the transmitted signal for it to be estimated reliably. Since BLAST does not require additional spectrum resources to transmit parallel substreams, that is, each antenna operates in a co-channel manner, the BLAST architecture is spectrally efficient. However, the spatial multiplexing and simultaneous use of the same portion of the spectrum leads to *co-antenna interference,* which is the major source of channel impairment in the BLAST architecture.

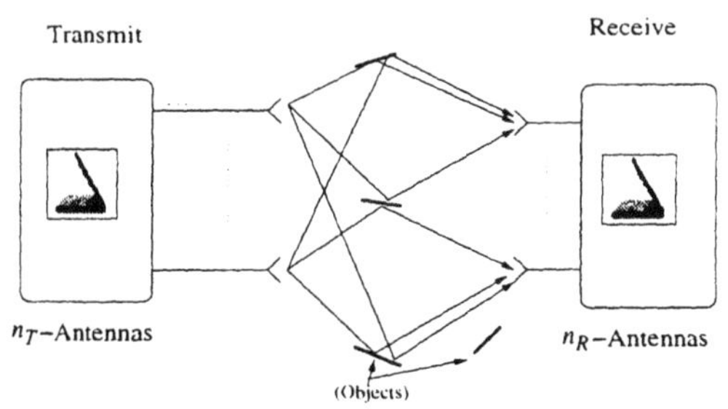

Figure 3: BLAST system

Optimal codecs (coders/decoders) for MIMO schemes are multi-dimensional and can be found through exhaustive search methods as in the maximum likelihood literature. However, the search complexity increases exponentially with the number of transmit antennas, the number of bits per modulation symbol and the burst size, which makes the problem NP hard. In practice, we are therefore compelled to seek a sub-optimal design for the receiver.

In a seminal paper [3] on space-time processing for wireless communications, Foschini raised the following fundamental question:

> *Can one construct a MIMO system whose capacity scales linearly with the number of transmit antennas, using building blocks of n_T separately coded one-dimensional subsystems of equal capacity?*

The motivation for raising this question is that the system can be designed with the well- developed 1-D codec technology with lower complexity interference- cancellation receivers.

It turns out that the answer to Foschini's question is an emphatic yes, provided that the right technology is used for the design of both the transmitter and the receiver, as described in Section 4.

3.1 Diagonal-BLAST

The first version of BLAST, proposed by Foschini [3], used a novel diagonal layered space-time architecture, hence the terminology *Diagonal-BLAST* or *D-BLAST*. The D-BLAST architecture can attain the channel capacity limits by using one-dimensional (1-D) codec technology. The innovative feature of the D-BLAST transmitter is the space-time encoding structure constructed with n_T diagonal layering 1-D coded subsystems of equal capacity, which permit the decoding complexity to grow linearly with the number of transmit antennas. However, with this architecture requiring the use of diagonal layering, the resulting space-time use of resources at the start and end of each data burst is wasteful, particularly for a practical burst length of few hundred symbols. Here, we should note that, in practice, the use of a short packet size is desirable in wireless communications for two practical reasons:

- Long packets require channel tracking inside a packet since the wireless channel varies with time.
- Wireless communication is usually delay-limited.

3.2 Vertical-BLAST

The second major version of BLAST, namely, *Vertical-BLAST*, described in [4], overcomes the practical limitation of D-BLAST just mentioned. Indeed, Vertical-BLAST, or V-BLAST, was the first practical BLAST configuration demonstrated in the literature. In V-BLAST, every antenna transmits its own independent stream of data, using a simple vector encoding and linear decoding structure. This architecture can achieve up to fifty percent channel capacity with no channel coding.

The major difference between D-BLAST and V-BLAST transmitters is in the coding scheme. D-BLAST uses a diagonal layering space-time code, whereas in V-BLAST a vector encoding process is used. The vector encoding process is simply a demultiplexing operation followed by independent bit-to-symbol mapping of each substream. Although no space-time coding can be employed in V-BLAST, a conventional 1-D channel coding of the individual substreams may certainly be applied.

Unfortunately, V-BLAST does not use any transmit diversity naturally provided in the MIMO system and therefore achieves a much lower information capacity than D-BLAST.

4. TURBO-BLAST

Turbo-BLAST (T-BLAST), described in [5-10], built on the combination of BLAST and the turbo principle, provides a practical answer to Foschini's fundamental question as described here.

T-BLAST attains near Shannon capacity of MIMO systems by exploiting the following ideas:

- The Bell-Labs Layered Space Time (BLAST) architecture.
- A novel space-time coding scheme based on the use of 1-D channel codes and space-time interleaving that are designed off-line [9].
- Sub-optimal turbo-like receiver that performs joint decoding of the proposed space-time codes and co-antenna interference-cancellation in an iterative manner and, most important, simple fashion [5-8].

Simply put, T-BLAST combines the benefits of the BLAST architecture and the turbo-learning principle in a layered space-time coding design. The structure of the layered space-time encoder and the turbo-like operation allows Turbo-BLAST to realize the full benefits of BLAST in a computationally feasible manner. Furthermore, it offers the following unique features:

- High information capacity (efficiency)
- Low bit-error rate (reliability)

The encoding process involved in the T-BLAST architecture, having n_T transmitting and n_R receiving antenna elements, is as follows:

- De-multiplexing of the user information bits into n_T substreams $\{\mathbf{b}_k\}_{k=1}^{n_T}$ of equal data rate.

- Independent block-encoding of each data substream, which uses the same predetermined linear block forward error correction (FEC) code, defined by

$$\mathbf{C}=[\mathbf{b}_1\mathbf{G}, \mathbf{b}_2\mathbf{G}, \cdots, \mathbf{b}_{n_T}\mathbf{G}]=[\mathbf{c}_1, \mathbf{c}_2, \cdots, \mathbf{c}_{n_T}] \quad (3)$$

  where $\mathbf{G}$ is the K x L binary code generator, and the $\{\mathbf{b}_k\}_{k=1}^{n_T}$ are K-dimensional information sequences, and the $\{\mathbf{c}_k\}_{k=1}^{n_T}$ are L-dimensional code sequences.
- The encoded substreams are bit-interleaved using an off-line designed space-time random permuter (interleaver), denoted by Π. We use $\{\tilde{\mathbf{c}}_k\}_{k=1}^{n_T}$ to denote the permuted substreams, written in matrix form as

$$\tilde{\mathbf{C}} = \Pi(\mathbf{C}) \quad (4)$$

- The space-time interleaved substreams are independently mapped onto M-ary PSK symbols $\{\mathbf{a}_k\}_{k=1}^{n_T}$, written in matrix form as

$$\mathbf{A} = f(\tilde{\mathbf{C}}) \quad (5)$$

  where f is the mapper.

The combined use of block codes and interleaving provides the basis for the *random block codes*, namely, parallel and serially concatenated turbo codes. Using this principle, for MIMO systems, T-BLAST provides turbo space-time codes by concatenating block codes and space-time interleavers. In theory, it is possible to model the proposed turbo space-time codes as a single Markov process, and a trellis can be formed to include the effect of interleaving. Such a trellis representation is extremely complex and does not lend itself to feasible decoding algorithms. Instead, T-BLAST uses a practical sub-optimum detection scheme based on iterative turbo detection principles. The inter-substream coding proposed as independent encoding and space-time interleaving can be viewed as a serially concatenated code, as illustrated in Fig. 4.

- *Inner code*: n_T parallel channel codes
- *Outer code*: time-varying matrix channel

The inner and outer encoders are separated by n_T parallel interleavers. The concatenated code can be decoded using a low-complexity iterative receiver similar to the iterative schemes proposed for serially concatenated turbo codes.

In the iterative decoding scheme, we separate the optimal decoding problem into two stages and exchange extrinsic information learned from one stage to the other stage iteratively until the receiver converges. The two decoding stages are:

- *Inner decoder*: SISO minimum mean-square error (MMSE) detectors
- *Outer decoders*: A set of n_T parallel SISO channel decoders.

The detector and decoder stages are separated by space-time interleavers and de-interleavers. As mentioned in Subsection 2.2, the interleavers and de-interleavers are used to compensate for the interleaving operation used in the transmitter and

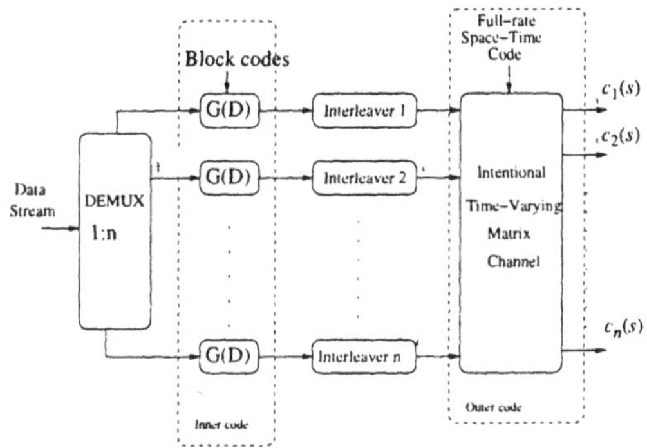

Figure 4: Turbo-BLAST: Serially concatenated transmitter

to decorrelate the correlated outputs before feeding them to the next decoding stage. The iterative receiver produces new and better estimates of the transmitted symbols at each iteration and repeats the information exchange process for a relatively small number of iterations to improve the estimation of transmitted message bits.

Figure 5 shows a block diagram of the turbo receiver, which has three learning strategies:

Outer decoder:

(i) Soft-interference cancellation using short training sequence supplied by the transmitter within each packet (burst) of data transmission.

Inner decoder:

(ii) Implicit form of supervision provided by the forward error-correction (FEC) codes.

(iii) Re-estimation of the channel matrix by using a training sequence provided by the complete set of decoded message bits at each iteration.

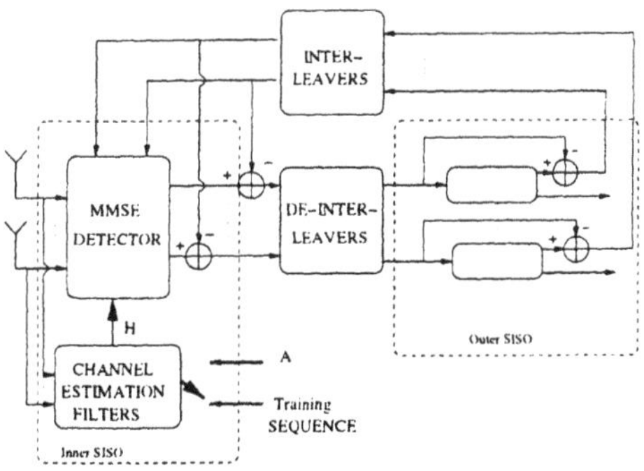

Figure 5: Turbo-BLAST: Iterative receiver

The property that underlines the hallmark of T-BLAST is summarized here:

- The error performance of T-BLAST improves with the number of iterations of the decoding algorithm. This is achieved by feeding extrinsic information from the output of one decoding stage as intrinsic information to the input of the next decoding stage, which permits the iterative decoding process to take its natural course in response to the received noisy signal and known channel code constrainst. Most important, T-BLAST is capable of achieving a performance significantly better than V-BLAST in the course of 2 to 4 iterations.

5. SPECTRAL EFFICIENCY

Figure 6 illustrates the capacity margins of the various BLAST schemes for $n_T = n_R = 8$. The capacity of T-BLAST falls between the capacity upper bounds of D-BLAST (or the channel capacity) and V-BLAST. Figure 6 also includes the orthogonal space-time codes (Orthogonal STC), namely, the well-known Tarokh-Seshadri-Calderbank space-time codes [11]. These latter codes have poor bandwidth efficiency since they use a form of repetition coding, but the codes are designed in such a way that the code matrix maximizes the transmit-diversity gain.

6. EXPERIMENTS

This section describes the performance of QPSK modulated T-BLAST and the coded V-BLAST using indoor real-channel measurements. In the transmit end, each substream of 100 information bits is independently encoded using a rate 1/2

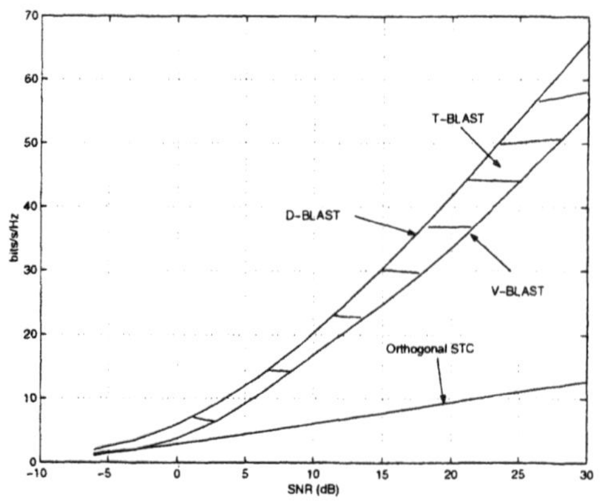

Figure 6: The 10% outage capacity of BLAST systems with $n_T = n_R = 8$

convolutional code generator (7,5) and then interleaved using space-time interleavers. We refer to horizontal coded V-BLAST when each of the substreams is provided with an amount of channel coding equal to that used in T-BLAST. Note that V-BLAST does not use any space-time coding or iterative decoding.

The goal of this experiment is to illustrate the superior performance of T-BLAST over V-BLAST when equal numbers of transmit and receive antenna elements are used in a BLAST configuration. In Figure 7, we compare the bit error rate performance of the following schemes:

- Scheme 1: T-BLAST with $n_T = n_R = 8$,
- Scheme 2: Coded V-BLAST with $n_T = n_R = 8$,
- Scheme 3: V-BLAST with $n_T = n_R = 8$,
- Scheme 4: V-BLAST with $n_T = 4$, $n_R = 8$.

As expected the performance of uncoded V-BLAST with $n_T = n_R = 8$ is inferior to all of the other schemes, whereas a significant gain (2-3dB) is achieved by the T-BLAST scheme over the coded V-BLAST with $n_T = n_R = 8$ and uncoded V-BLAST with $n_T = 4$, $n_R = 8$. Moreover, the performance of uncoded V-BLAST with $n_T = 4$ and $n_R = 8$ outperforms that of the coded V-BLAST with 8 transmit and receive antennas for signal-to-noise ratios SNR between -3dB to 5dB.

The complexity of the iterative receiver used in T-BLAST increases with the number of iterations. However, most of the iterative gain is achieved within 3 to 4 iterations. To confirm this, in Fig. 8, the bit error rate (BER) versus the number of iterations at SNR = 4dB is plotted. Naturally, the V-BLAST system uses hard decisions and is therefore independent of the number of iterations. In direct contrast, the performance of T-BLAST, which uses soft decisions, improves with iterations and approaches steady state in about 7 iterations, but performance gain after 3 iterations is, for all practical purposes, minimal.

7. OPEN RESEARCH PROBLEM: CONVERGENCE ANALYSIS

As mentioned in Section 2, the Turbo-BLAST receiver is a *nonlinear stochastic sampled-data feedback system* with extrinsic information providing the error signal that propagates around the feedback loop. What is remarkable about the T-BLAST receiver, a system rooted in space-time processing, are the following observations derived from the experiments reported in Section 6 and many others reported elsewhere [5-10].

(i) In terms of information capacity, the receiver converges close to the Shannon channel capacity in a relatively small number of iterations, in the order of 6 to 8.

(ii) The receiver is robust in that it is capable of delivering a good performance despite wide variations in environmental conditions, extending from indoor to outdoor wireless environments.

Undoubtedly, the presence of *global feedback* around the receiver, consisting of two soft-input, soft-output decoders separated by an interleaver and deinterleaver, is responsible for these highly desirable properties.

Feedback is an integral part of practically every-control system in operation. By contrast, the use of global feedback around a communication receiver is not that common, but nevertheless it has a history of its own. In this context, we may-mention the *FM frequency demodulator with feedback* described in [12], the motivation for which was to extend the *threshold* with respect to the standard frequency discriminator (i.e., one without feedback) for the demodulation of frequency-modulated signals in noise.

In any event, we presently lack a mathematical

framework for a *convergence analysis* of the Turbo-BLAST receiver, the availability of which would enable a designer to do two things:

(i) Prediction of the asymptotic performance of the receiver, which is attained when the receiver has converged to a fixed point.

(ii) Evaluation of the effect of varying the bit energy-to-noise spectral density (E_b/N_0) ratio on the asymptotic performance of the receiver.

Extending what we know about the noise behavior of a frequency demodulator with feedback, we conjecture that Turbo-BLAST exhibits a threshold phenomenon of its own:

> Once the (E_b/N_0) ratio drops below a certain critical value, the bit error rate of the receiver begins to deteriorate sharply.

We base this conjecture on the thought that once the inner and outer decoders of the turbo receiver start producing unreliable estimates of the transmitted message bits, the quality of the extrinsic information exchanged between these two decoders deteriorates rapidly from one iteration of the receiver to the next. In such a situation, the feedback acting around the receiver begins to work against it, causing a sharp increase in the bit error rate of the receiver.

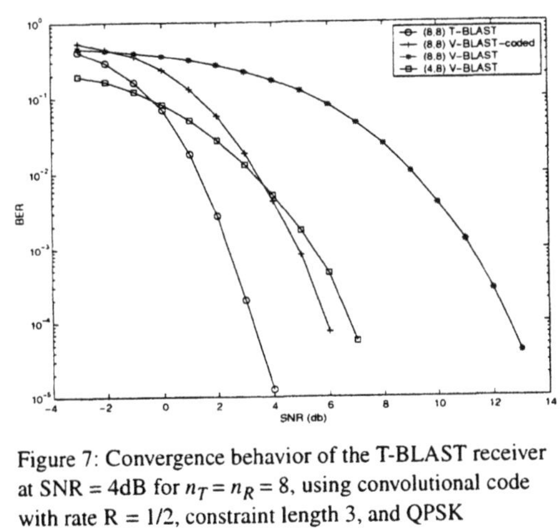

Figure 7: Convergence behavior of the T-BLAST receiver at SNR = 4dB for $n_T = n_R = 8$, using convolutional code with rate R = 1/2, constraint length 3, and QPSK modulation.

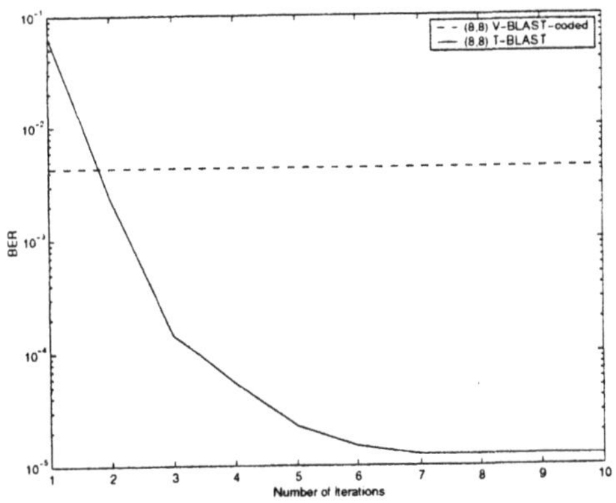

Figure 8: T-BLAST vs V-BLAST using real-channel measurements for $n_T = n_R = 8$, convolutional code with rate $R = 1/2$ and constraint length 3, and QPSK modulation.

8. CONCLUSIONS

In this paper, we described a novel MIMO wireless communication system called Turbo-BLAST, which uses a random space-time block encoder and an iterative receiver that is turbo-like in operation. Most important, Turbo-BLAST is capable of achieving a high spectral efficiency in the course of 3 to 4 iterations of the receiver.

The new transceiver (transmitter/receiver) is not only computationally efficient, but it also yields a probability of error performance that is orders of magnitude smaller than the corresponding coded Vertical-BLAST. This performance has been confirmed using computer simulations as well as tests with real-life data for both indoor and outdoor wireless communication environments.

REFERENCES

[1] C. Berrou, A. Glavieux, and P. Thitimajshima, "Near Shannon limit error-correcting coding and decoding turbo codes", ICC, pp. 1064-1070, 1993.

[2] J. Hagenaur, "The turbo principle: Tutorial introduction and state of the art", International Symposium on Turbo Codes, Brest, France, pp. 1-11, 1997.

[3] J. G. Foschini, "Layered space-time architecture for wireless communication in a fading environment when using multi element antennas", *Bell Labs Technical Journal*, Autumn 1996.

[4] G. D. Golden, J. G. Foschini, R. A. Valenzuela, and P. W. Wolniansky,

"Detection algorithm and initial laboratory results using V-BLAST space-time communication architecture", *Electronics Letters*, Val. 35, pp. 14-15, January 1999.

[5] M. Sellathurai and S. Haykin, "Turbo-BLAST for high speed wireless communications", *Proceedings of wireless communications and network conf.*, Chicago, IL, September 2000.

[6] M. Sellathurai and S. Haykin, "A nonlinear iterative beamforming technique for wireless communications", *33rd ASILOMAR conf. on signals, systems, and computers*, Pacific Grove, CA, Val. 2, pp. 957-961, November, 1999.

[7] M. Sellathurai and S. Haykin, "Turbo-BLAST: A novel technique for multi-transmit multi-receive wireless communications", *Multiaccess, Mobility, and Teletraffic for Wireless Communications*, vol. 5, pp. 13-24, Kulwer Academic Publishers, ISBN 0-7923-7275-1.

[8] M. Sellathurai and S. Haykin, "A simplified diagonal BLAST architecture with iterative parallel-interference cancellation", *IEEE Int. Conf. on Communications (ICC)*, Helsinki, Finland, June 2001.

[9] M. Sellathurai and S. Haykin, "Random space-time codes with iterative decoders for BLAST architectures", the *IEEE Int. Symposium on Information Theory (ISIT)*, June 2001.

[10] M. Sellathurai, *TURBO-BLAST: A Novel Technique for Multi-Transmit and Multi-Receive Wireless Communications*, Ph.D. Thesis, McMaster University, April 2001.

[11] V. Tarokh, N. Seshadri, and A. R. Calderbank, "Space-time codes for high data rate wireless communications: performance criterion and code construction", *IEEE Trans. on Information Theory*, vol. 44, pp. 744-765, March 1998.

[12] J.G. Chafee, "The application of negative feedback to frequency-modulation systems:, Bell System Technical Journal, vol. 18, pp. 404-437, 1939.

NEW TIME-VARYING LMS AND RLS ALGORITHMS WITH APPLICATION TO DIRECTION-OF-ARRIVAL TRACKING

Keisuke Yanagihara, Hiromitsu Ohmori, Akira Sano* Koichi Hidaka**

** Keio University, 3-14-1 Hiyoshi, Yokohama 223-8522, Japan*
*** Tokyo Metropolitan College of Aeronautical Engineering, Tokyo, Japan*

Abstract. New adaptive identification algorithms are proposed for a time-varying linear system in which the system parameters change as piecewise polynomial function of time. By appropriately incorporating the internal model corresponding to the polynomial degree into the parameter adjustment dynamics of the new algorithm, its BIBO stability can be assured by the aid of the small gain theorem and modified passivity theorem. Its effectiveness of the proposed LMS and RLS types of adaptive algorithms is validated by comparing to ordinary approaches in adaptive direction-of-arrival (DOA) estimation for fast moving targets. *Copyright ©2001 IFAC*

Keywords. Adaptive identification, time varying system, adaptive filter, internal model principle, passivity, tracking, direction-of-arrival

1. INTRODUCTION

Adaptive filtering is needed to deal with uncertain changes in dynamical systems and becomes now one of important tools in the area of digital signal processing and adaptive control (Haykin, 1996; Tsakalis and Ioannou, 1993; Niedzwiecki, 2000). Almost all adaptive filtering algorithms have been derived on the assumption that the system parameters are unknown but constant. In order to treat with its time-varying situations, appropriate selection of a step size, forgetting factor or sliding window interval plays a very important role in the LMS or RLS algorithm respectively. However, their choice should be done by compromise between the tracking ability and steady-state performance (Eleftheriou and Falconer, 1986).

One of typical method for modeling time-varying parameters is to expand these parameters by basis functions or polynomial functions of time (Tsatsanis and Giannakis, 1996). However, these parameterization causes increase of number of parameters to be identified and then needs much more convergence time. The extended Kalman filter approach is also an efficient but approximate method to estimate both states and parameters of dynamical systems (Guo, 1990).

The purpose of the present paper is to propose new adaptive identification algorithms for a time-varying linear system in which the system parameters change as piecewise polynomial function of time. By including the internal model corresponding to the polynomial degree in the parameter adjusting dynamics of the proposed algorithm, its BIBO stability can be assured by using the small gain theorem and modified passivity theorem. One of the advantages of the proposed approach is that it is not necessary to identify each parameter of the polynomial time function, but only use prior information on a maximum degree of the polynomial. Its effectiveness of the proposed LMS and RLS types of algorithms is validated by comparing to ordinary approaches in (DOA) estimation for fast moving target.

2. ADAPTIVE IDENTIFICATION PROBLEM

In order to mention the principle of the time-varying adaptive algorithm, we first consider a time-varying FIR system expressed by the input-output relation as

$$x(n) = \sum_{k=0}^{M} u(n-k)h(k,n) = \boldsymbol{u}^T(n)\boldsymbol{h}(n) \quad (1)$$

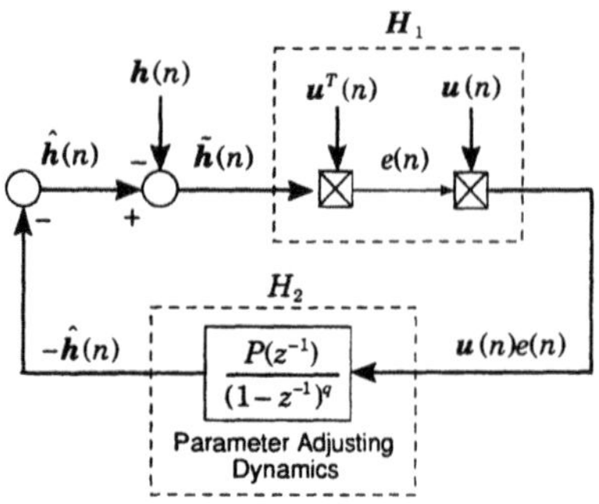

Fig. 1. New stucture of time-varying LMS algorithm

where $u(n)$ is an accessible input signal, $x(n)$ is a received signal. $h(k,n)$ is a time-varying impulse response of the FIR system. $\boldsymbol{h}(n)$ and $\boldsymbol{u}(n)$ are the impulse response vector and input signal vector defined by $\boldsymbol{h}(n) = [h(0,n),\ h(1,n),\ \cdots\ ,h(M,n)]^T$ and $\boldsymbol{u}(n) = [u(n),\ u(n-1),\ \cdots\ ,u(n-M)]^T$ respectively.

It is assumed in the following discussion that

(A1) An upper bound of the impulse response length is assumed to be known *a priori* as $M+1$.

(A2) The coefficients $\{h(i,n)\}$ of the FIR system are bounded and change as a piecewise polynomial function of time, where the maximum order is known as $q-1$.

(A3) The input $u(n)$ is a bounded white random signal.

3. TIME-VARYING PARAMETER IDENTIFICA- TION BY INTERNAL MODEL PRINCIPLE

3.1 *New LMS Algorithm for Time-varying System*

A new LMS algorithm for identifying the time-varying parameter vector $\boldsymbol{h}(n)$ is described as

$$\hat{\boldsymbol{h}}(n) = -\frac{P(z^{-1})}{(1-z^{-1})^q}[\boldsymbol{u}(n)e(n)] \quad (2)$$

$$e(n) = \boldsymbol{u}^T(n)\hat{\boldsymbol{h}}(n) - x(n) = \boldsymbol{u}^T(n)\tilde{\boldsymbol{h}}(n) \quad (3)$$

where $\hat{x}(n)$ is the model output, and $\hat{\boldsymbol{h}}(n) = (\hat{h}(0,n),\ \cdots\ ,\ \hat{h}(M,n))^T$ is an identified model parameter vector, and $\boldsymbol{u}(n)$ is the input vector. $P(z^{-1})$ is the numerator polynomial:

$$P(z^{-1}) = p_0 + p_1 z^{-1} + \ \cdots\ + p_r z^{-r} \quad (4)$$

where $r \leq q$ and $p_0 \neq 0$. Fig.1 shows that the adaptive algorithm is a closed-loop system consisting of the parameter adjusting dynamics (2) and the error system (3). Then, the unknown time-varying $\boldsymbol{h}(n)$ can be regarded as an external input added into the closed-loop system in Fig.1. If the true parameters $\boldsymbol{h}(n)$ change piecewise as a polynomial function of time, for instance, a first-order function of time, second-order function of time, or so, by incorporating the corresponding internal model into the closed-loop, such as $1/(1-z^{-1})^2$, $1/(1-z^{-1})^3$, respectively, we can achieve the BIBO stability of the adaptive algorithm. The coefficients of $P(z^{-1})$ should be determined so that the closed-loop system can be stabilized by using the strictly output passivity of $\boldsymbol{H}_1$ and the small gain theorem, as described later.

The error $e(n)$ should also be rewritten by substituting (2) into (3), as

$$e(n) = \frac{\boldsymbol{u}^T(n)(1-(1-z^{-1})^q)[\hat{\boldsymbol{h}}(n)]}{1+p_0\boldsymbol{u}^T(n)\boldsymbol{u}(n)} - \frac{\boldsymbol{u}^T(n)(P(z^{-1})-p_0)[\boldsymbol{u}(n)e(n)] - x(n)}{1+p_0\boldsymbol{u}^T(n)\boldsymbol{u}(n)} \quad (5)$$

For instance, the proposed time-varying LMS algorithm in a case with $P(z^{-1}) = p_0 + p_1 z^{-1}$ and $q=2$ can be expressed by

$$\begin{aligned}\hat{\boldsymbol{h}}(n) &= -\frac{P(z^{-1})}{(1-z^{-1})^2}[\boldsymbol{u}(n)e(n)] \\ &= 2\hat{\boldsymbol{h}}(n-1) - \hat{\boldsymbol{h}}(n-2) \\ &\quad -[p_0\boldsymbol{u}(n)e(n) + p_1\boldsymbol{u}(n-1)e(n-1)]\end{aligned}$$

where

$$e(n) = \frac{\boldsymbol{u}^T(n)}{1+p_0\boldsymbol{u}^T(n)\boldsymbol{u}(n)}\Big[2\hat{\boldsymbol{h}}(n-1) - \hat{\boldsymbol{h}}(n-2) - p_1\boldsymbol{u}(n-1)e(n-1) - x(n)\Big] \quad (6)$$

It can be noticed that since $\hat{\boldsymbol{h}}(n)$ is not included in $e(n)$, the algorithm becomes causal. The number of identified parameters is the same as the size of $\boldsymbol{h}(n)$, and then it is independent of the degree of the polynomial time function.

3.2 *New RLS Algorithm for Time-varying System*

Similarly we can derive a new RLS algorithm for time-varying FIR systems, which is described by

$$\hat{\boldsymbol{h}}(n) = -\frac{P(z^{-1})}{(1-z^{-1})^q}[\boldsymbol{\Gamma}(n)\boldsymbol{u}(n)e(n)] \quad (7)$$

$$\boldsymbol{\Gamma}(n) = \boldsymbol{\Gamma}(n-1) + \frac{\boldsymbol{\Gamma}(n-1)\boldsymbol{u}(n)\boldsymbol{u}(n)^T\boldsymbol{\Gamma}(n-1)}{1+\boldsymbol{u}(n)^T\boldsymbol{\Gamma}(n-1)\boldsymbol{u}(n)} \quad (8)$$

$$e(n) = \frac{\boldsymbol{u}^T(n)\left(1-(1-z^{-1})^q\right)[\hat{\boldsymbol{h}}(n)]}{1+p_0\boldsymbol{u}^T(n)\boldsymbol{\Gamma}(n)\boldsymbol{u}(n)} - \frac{\boldsymbol{u}^T(n)\left(P(z^{-1})-p_0\right)[\boldsymbol{\Gamma}(n)\boldsymbol{u}(n)e(n)] + x(n)}{1+p_0\boldsymbol{u}^T(n)\boldsymbol{\Gamma}(n)\boldsymbol{u}(n)} \quad (9)$$

where the initial conditions are given by $\hat{\boldsymbol{h}}(0) = \boldsymbol{h}_0$, $\boldsymbol{\Gamma}(0) = \gamma\boldsymbol{I}$ where $\gamma > 0$ is a lrage positive number.

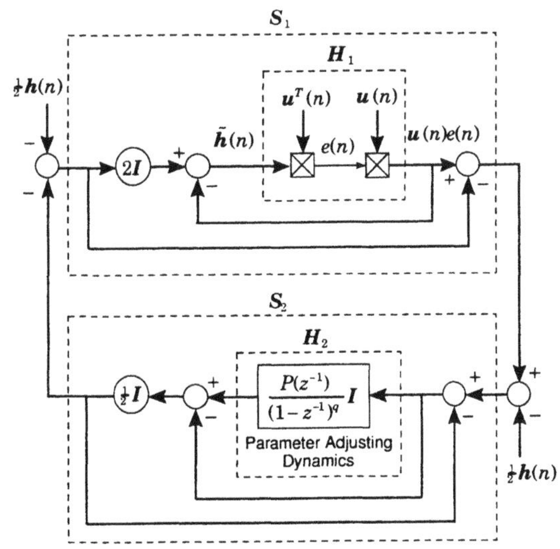

Fig. 2. Stability analysis by using loop transformation

4. STABILITY ANALYSIS OF PROPOSED LMS ALGORITHM

By taking the loop-transformation which preserves the closed-loop property to the original operators $\boldsymbol{H}_1$ and $\boldsymbol{H}_2$, the structure consisting of the operators $\boldsymbol{S}_1$ and $\boldsymbol{S}_2$ can be obtained as shown in Fig.2, where $\boldsymbol{H}_1$, $\boldsymbol{H}_2$, $\boldsymbol{S}_1$ and $\boldsymbol{S}_2$ are expressed by

$$\boldsymbol{H}_1 = \boldsymbol{u}(n)\boldsymbol{u}^T(n) \tag{10}$$

$$\boldsymbol{H}_2 = \frac{P(z^{-1})}{(1-z^{-1})^q}\boldsymbol{I} \tag{11}$$

$$\begin{aligned}\boldsymbol{S}_1 &= [\boldsymbol{H}_1 - \boldsymbol{I}][\boldsymbol{H}_1 + \boldsymbol{I}]^{-1}\\ &= [\boldsymbol{u}(n)\boldsymbol{u}^T(n) - \boldsymbol{I}][\boldsymbol{u}(n)\boldsymbol{u}^T(n) + \boldsymbol{I}]^{-1} \quad (12)\end{aligned}$$

$$\begin{aligned}\boldsymbol{S}_2 &= (H_2 - 1)(H_2 - 1)\boldsymbol{I}\\ &= \frac{P(z^{-1}) - (1-z^{-1})^q}{P(z^{-1}) + (1-z^{-1})^q}\boldsymbol{I} \quad (13)\end{aligned}$$

Let the input-output relation be denoted by $\boldsymbol{y}(n) = \boldsymbol{H}\boldsymbol{z}(n)$, where $\boldsymbol{z}(n) \in l_2$ and l_2 is a space such that $\|\boldsymbol{z}\| = \sum_{k=0}^{\infty} \boldsymbol{z}^T(k)\boldsymbol{z}(k) < \infty$ is bounded. An extended space l_{2e} is also introduced, where $\boldsymbol{z}_N(n)$ is obtained by truncating $\boldsymbol{z}(n)$ at an arbitrary time N, and is defined as $\boldsymbol{z}_N(n) = \boldsymbol{z}(n)$ (for $0 \le n \le N$), $\boldsymbol{z}_N(n) = 0$ (for $n > N$). Then let the inner product and norm be denoted by

$$\langle \boldsymbol{z}, \boldsymbol{y}\rangle \equiv \sum_{k=0}^{N} \boldsymbol{z}^T(k)\boldsymbol{y}(k)$$

$$\|\boldsymbol{y}\|^2 \equiv \sum_{k=0}^{N} \boldsymbol{y}^T(k)\boldsymbol{y}(k) = \sum_{k=0}^{N} \|\boldsymbol{y}(k)\|^2$$

An important property of the operator $\boldsymbol{H}$ is passivity (Anderson, 1972). Let the strict output passivity be defined as follows:
Definition: Let the operator be denoted by $\boldsymbol{y}(n) = \boldsymbol{H}\boldsymbol{z}(n)$. If it holds that $\langle \boldsymbol{z}, \boldsymbol{y}\rangle > \alpha\|\boldsymbol{y}\|^2$ for $\alpha > 0$, $\boldsymbol{H}$ is called strictly output passive.

The next lemma is known as the passivity theorem:
Lemma 1: (Anderson, 1972)(Åström and Wittenmark, 1995) The necessary and sufficient condition for that $\boldsymbol{H}$ is strictly output passive is that $\gamma(\boldsymbol{S}) < 1$, where $\boldsymbol{S}$ is an operator obtained by the transformation in Fig.2, and $\gamma(\boldsymbol{S})$ is an operator gain defined by

$$\gamma(\boldsymbol{S}) \equiv \sup_{\boldsymbol{v} \in X_e} \frac{\|\boldsymbol{S}\boldsymbol{v}\|}{\|\boldsymbol{v}\|} \tag{14}$$

where $X_e \equiv \{x(n) \in X,\ x_N(n) \in X,\ \exists N > 0,\ \|x_N\| < \infty\}$, and the norm is an arbitrary one and X_e is an extended space.

First, the next lemma is given by using the above preparations.
Lemma 2: If the vector norm of the input signal $\boldsymbol{u}(n)$ satisfies that

$$\|\boldsymbol{u}(n)\|^2 < 1/\varepsilon \tag{15}$$

for all n and $\varepsilon > 0$, then the operator $\boldsymbol{H}_1$ in Fig.2 with the input $\tilde{\boldsymbol{h}}(n)$ and the output $e(n)\boldsymbol{\psi}(n)$ is strictly output passive.
Proof: Let the input $\boldsymbol{z}$ and output $\boldsymbol{y}$ of $\boldsymbol{H}_1$ be $\boldsymbol{z}(n) = \tilde{\boldsymbol{h}}(n)$ and $\boldsymbol{y}(n) = e(n)\boldsymbol{u}(n)$ respectively. Then it can be confirmed that

$$\langle \boldsymbol{z}, \boldsymbol{y}\rangle - \varepsilon\langle \boldsymbol{y}, \boldsymbol{y}\rangle = \sum_{k=0}^{n} (1 - \varepsilon\|\boldsymbol{u}(k)\|^2)\, e^2(k) \tag{16}$$

From Lemma 1, the strict output passivity of $\boldsymbol{H}_1$ gives $\gamma(\boldsymbol{S}_1) < 1$, however, more detailed evaluation of $\gamma(\boldsymbol{S}_1)$ is needed to assure the stability of the closed-loop system. The main result of this paper is given next:
Theorem: The gain of $\boldsymbol{S}_1$ obtained by the loop transformation is given by

$$\gamma(\boldsymbol{S}_1) < 1/\sqrt{1+\varepsilon} < 1 \tag{17}$$

Proof: The operator $\boldsymbol{S}_1$ in Fig.2 has the input-output expression as

$$\boldsymbol{\eta}(n) = \boldsymbol{S}_1\boldsymbol{\zeta}(n) = [\boldsymbol{H}_1 - \boldsymbol{I}][\boldsymbol{H}_1 + \boldsymbol{I}]^{-1}\boldsymbol{\zeta}(n) \tag{18}$$

where $\boldsymbol{\zeta}(n) = \{\tilde{\boldsymbol{h}}(n) + e(n)\boldsymbol{\psi}(n)\}/2$ and $\{e(n)\boldsymbol{\psi}(n) - \tilde{\boldsymbol{h}}(n)\}/2$. Furthermore, let the state $\boldsymbol{\xi}(n)$ be defined by $\boldsymbol{\xi}(n) = [\boldsymbol{H}_1 + \boldsymbol{I}]^{-1}\boldsymbol{\zeta}(n)$, then it gives

$$\boldsymbol{\eta}(n) = [\boldsymbol{H}_1 - \boldsymbol{I}]\boldsymbol{\xi}(n) \tag{19}$$

$$\boldsymbol{\zeta}(n) = [\boldsymbol{H}_1 + \boldsymbol{I}]\boldsymbol{\xi}(n) \tag{20}$$

It follows that

$$\|\eta\|^2 = \langle H_1\xi, H_1\xi\rangle + \|\xi\|^2 - 2\langle H_1\xi, \xi\rangle$$
$$\|\zeta\|^2 = \langle H_1\xi, H_1\xi\rangle + \|\xi\|^2 + 2\langle H_1\xi, \xi\rangle$$

Then it leads to that

$$\|\eta\|^2 = \|\zeta\|^2 - 4\langle H_1\xi, \xi\rangle$$

As given in Lemma 2, since H_1 is strictly output passive, *i.e.* $\langle H_1\xi, \xi\rangle \geq \varepsilon\|H_1\xi\|^2$, it follows that

$$\begin{aligned}\frac{\|\eta\|^2}{\|\zeta\|^2} &= 1 - \frac{4\langle H_1\xi, \xi\rangle}{\|\zeta\|^2} \leq 1 - \frac{4\varepsilon\|H_1\xi\|^2}{\|\zeta\|^2} \\ &= 1 - \frac{\varepsilon\|\eta + \zeta\|^2}{\|\zeta\|^2} \\ &\leq 1 - \varepsilon\left(1 + \frac{\|\eta\|^2}{\|\zeta\|^2} + 2\frac{\langle \eta, \zeta\rangle}{\|\zeta\|^2}\right) \end{aligned} \quad (21)$$

Thus, we have

$$\frac{\|\eta\|^2}{\|\zeta\|^2} \leq \frac{1-\varepsilon}{1+\varepsilon} - \frac{2\varepsilon\langle \eta, \zeta\rangle}{1+\varepsilon} \quad (22)$$

By using the squared gain $\gamma(S_1)$ given by

$$\gamma^2(S_1) = \sup \frac{\|\eta\|^2}{\|\zeta\|^2} = \sup_{\|\zeta\|^2=1} \|\eta\|^2$$

we will evaluate a tight upper bound of $\gamma(S_1)$. For this purpose, we evaluate $\langle \eta, \zeta\rangle$ in the RHS of (22). Then it follows from (19) and (20) that

$$\begin{aligned} 2\langle \eta, \zeta\rangle &= \frac{1}{2}(\|\eta + \zeta\|^2 - \|\eta - \zeta\|^2) \\ &= \frac{1}{2}(\|(I + S_1)\zeta\|^2 - \|(I - S_1)\zeta\|^2) \end{aligned} \quad (23)$$

Then we use the following inequalities at each instant n:

$$\begin{aligned} \underline{\sigma}(I \pm S_1)\|\zeta(n)\|^2 &\leq \|(I \pm S_1)\zeta(n)\|^2 \\ &\leq \bar{\sigma}(I \pm S_1)\|\zeta(n)\|^2 \end{aligned} \quad (24)$$

Further it holds that $I - S_1 = 2[I + H_1]^{-1}$, $I + S_1 = 2[I + H_1]^{-1}H_1$, and H_1 is given in (10), the each maximum and minimum singular values are obtained by $\underline{\sigma}(I + S_1) = 0$, $\bar{\sigma}(I + S_1) = 2\|u(n)\|^2/(\|u(n)\|^2+1)$, $\underline{\sigma}(I - S_1) = 2/(1+\|u(n)\|^2)$, $\bar{\sigma}(I - S_1) = 2$. It is possible to calculate the maximum of $\gamma(S_1)$ in possible range, by evaluating the maximum magnitude with minus sign of the RHS of (23). This can be attained in a case with $\underline{\sigma}(I + S_1) - \bar{\sigma}(I - S_1) = -2$, which is fortunately independent of the actual input $u(n)$, then finally we can obtain

$$\gamma^2(S_1) < \frac{1-\varepsilon}{1+\varepsilon} + \frac{\varepsilon}{1+\varepsilon} = \frac{1}{1+\varepsilon}$$

Then, the proof is completed. □

By combining the above result and the small gain theorem, the stability of the proposed algorithm is assured by determining the polynomial function $P(z^{-1})$ so that the gain of S_2 satisfies that

$$\gamma(S_2) = \left\|\frac{P(z^{-1}) - (1 - z^{-1})^q}{P(z^{-1}) + (1 - z^{-1})^q}\right\|_\infty < \sqrt{1+\epsilon} \quad (25)$$

and the denominator of (13) should be stable.

Thus the design procedure will be summarized in the following steps:

S1: Decide the maximum order $q - 1$ of the polynomial time function of the time-varying parameters from the prior information.

S2: Determine a region of the parameters of $P(z^{-1})$ so that the denominator of (13) may be stable.

S3: Determine the coefficients of $P(z^{-1})$ so that the infinity norm of (25), and calculate ε such that the obtained minimum norm satisfies (25).

S4: Generate the input signal $u(n)$ such that (15) is satisfied at any time n.

5. APPLICATION TO DOA TRACKING

Consider a linear array of M uniformly spaced sensors, in which N mobile stations ($N \leq M$) emit narrow-band plane waves which impinge on the array from distinct directions $\theta_1, \cdots, \theta_N$. Let $x(n) = (x_1(n), \cdots, x_M(n))^T$ be the observation vector, $s(n) = (s_1(n), \cdots, s_N(n))^T$ be the signal vector, $v(n) = (v_1(n), \cdots, v_M(n))^T$ be the sensor noise vector. Thus the array output vector is described by

$$x(n) = A(\theta)s(n) + v(n), \quad n = 1, 2, \ldots \quad (26)$$

where let $A(\theta) = [a(\theta_1), \cdots, a(\theta_N)]$ be the array response matrix. $a(\theta_n) = (a_1(\theta_n), \cdots, a_M(\theta_n))^T$ and $a_m(\theta_n) = \exp(j2\pi d(m-1)\sin\theta_n/\lambda)$, where d is the intersensor spacing and λ is the signal carrier wavelength. It is assumed that (a) $s_n(n)$ is statistically independent of other signals, (b) the noise $v_m(n)$ is zero mean white Gaussian noise independent of the signals.

The MUSIC is one of the noise subspace approaches, and needs computation of eigenvalue decomposition of the observation covariance matrix $R_x = E[x(n)x^H(n)] = A(\theta)R_sA^H(\theta) + R_v$. By adopting the eigenvalue decomposition $R_x = U_s\Lambda U_s^H + U_n\Lambda_n U_n^H$, the DOA spectrum is given by

$$P_{MUSIC}(\theta) = 1/(M - |U_s^H a(\theta)|^2) \quad (27)$$

To perform this procedure, an approximate calculation was presented such that the next cost function is minimized (Yang, 1995):

$$J(W) = E\{\|x - WW^Hx\|^2\} \quad (28)$$

where $\boldsymbol{W}$ is an $M \times N$ matrix ($N \leq M$) defined by $\boldsymbol{W} = [\boldsymbol{w}_1 \quad \boldsymbol{w}_2 \cdots \boldsymbol{w}_N]$. If $\boldsymbol{W}^H\boldsymbol{W} = \boldsymbol{I}$, $\boldsymbol{W}\boldsymbol{W}^H = \boldsymbol{W}[\boldsymbol{W}^H\boldsymbol{W}]^{-1}\boldsymbol{W}^H$ is a projection matrix, and $J(\boldsymbol{W})$ has a minimum if $\boldsymbol{W}\boldsymbol{W}^H$ projects $\boldsymbol{x}$ into the signal subspace. The solution is given by $\boldsymbol{W}^o = \boldsymbol{U}_s\boldsymbol{Q}$, where $\boldsymbol{U}_s$ is a signal subspace and $\boldsymbol{Q}$ is an arbitrary unitary matrix. Then it follows that $\boldsymbol{W}^o\boldsymbol{W}^{oH} = \boldsymbol{U}_s\boldsymbol{U}_s^H$. Thus the MUSIC spectrum can be obtained by using $\boldsymbol{W}$.

Let the criterion (28) be denoted at the n-th time instant by

$$J(\boldsymbol{W}(n)) = E\{\|\boldsymbol{x}(n) - \boldsymbol{W}(n)\boldsymbol{W}^H(n)\boldsymbol{x}(n)\|^2\} \quad (29)$$

The minimization of (29) is executed in two steps: First the term $\boldsymbol{W}^H(n)\boldsymbol{x}(n)$ is approximately expressed by $\boldsymbol{y}(n) = \boldsymbol{W}^H(n-1)\boldsymbol{x}(n)$ using the former estimate $\boldsymbol{W}^H(n-1)$. Secondly the criterion (29) described by

$$J(\boldsymbol{W}(n)) = E\{\|\boldsymbol{x}(n) - \boldsymbol{W}(n)\boldsymbol{y}(n)\|^2\} \quad (30)$$

is minimized with respect to $\boldsymbol{W}(n)$. The proposed LMS and RLS time-varying adaptive algorithms are applicable to the minimization of (30), as summarized as follows:

The time-varying LMS adaptive algorithm given in (2) and (5) can be expressed by

$$\hat{\boldsymbol{W}}(n) = -\frac{P(z^{-1})}{(1-z^{-1})^q}\boldsymbol{E}(n) \quad (31)$$

$$\boldsymbol{E}(n) = \boldsymbol{y}^*(n) \otimes \boldsymbol{e}(n) \quad (32)$$

where $\boldsymbol{y}(n) = (y_1(n), \cdots, y_N(n))^T$. The error defined by $\boldsymbol{e}(n) = \hat{\boldsymbol{W}}(n)\boldsymbol{y}(n) - \boldsymbol{x}(n)$ can be calculated by

$$\boldsymbol{e}(n) = \frac{(1-(1-z^{-1})^q)[\hat{\boldsymbol{W}}(n)]\boldsymbol{y}(n)}{1 + p_0\boldsymbol{y}^H(n)\boldsymbol{y}(n)} - \frac{((P(z^{-1}) - p_0)[\boldsymbol{E}(n)]\boldsymbol{y}(n) - \boldsymbol{x}(n)}{1 + p_0\boldsymbol{y}^H(n)\boldsymbol{y}(n)} \quad (33)$$

where $\boldsymbol{y}(n) = \hat{\boldsymbol{W}}^H(n-1)\boldsymbol{x}(n)$.

Similarly, the time-varying RLS algorithm given in (7) to (9) can also be applied to the DOA tracking algorithm:

$$\hat{\boldsymbol{W}}(n) = -\frac{P(z^{-1})}{(1-z^{-1})^q}\left[\boldsymbol{e}(n)\boldsymbol{y}^H(n)\boldsymbol{\Gamma}(n)\right] \quad (34)$$

$$\boldsymbol{\Gamma}(n) = \boldsymbol{\Gamma}(n-1) - \frac{\boldsymbol{\Gamma}(n-1)\boldsymbol{y}(n)\boldsymbol{y}^H(n)\boldsymbol{\Gamma}(n-1)}{1 + \boldsymbol{y}^H(n)\boldsymbol{\Gamma}(n-1)\boldsymbol{y}(n)} \quad (35)$$

$$\boldsymbol{e}(n) = \frac{\{1-(1-z^{-1})^q\}\left[\hat{\boldsymbol{W}}(n)\right]\boldsymbol{y}(n)}{1 + p_0\boldsymbol{y}^H(n)\boldsymbol{\Gamma}(n)\boldsymbol{y}(n)} - \frac{\{P(z^{-1}) - p_0\}\left[\boldsymbol{e}(n)\boldsymbol{y}^T(n)\boldsymbol{\Gamma}(n)\right]\boldsymbol{y}(n) + \boldsymbol{x}(n)}{1 + p_0\boldsymbol{y}^H(n)\boldsymbol{\Gamma}(n)\boldsymbol{y}(n)} \quad (36)$$

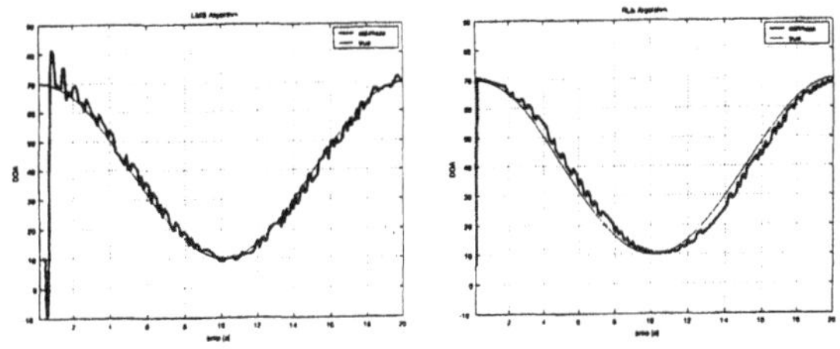

Fig. 3. DOA tracking performance for slowly moving target: (*Left*) Tracking performance of ordinary LMS algorithm, (*Right*) Tracking performance of ordinary RLS algorithm.

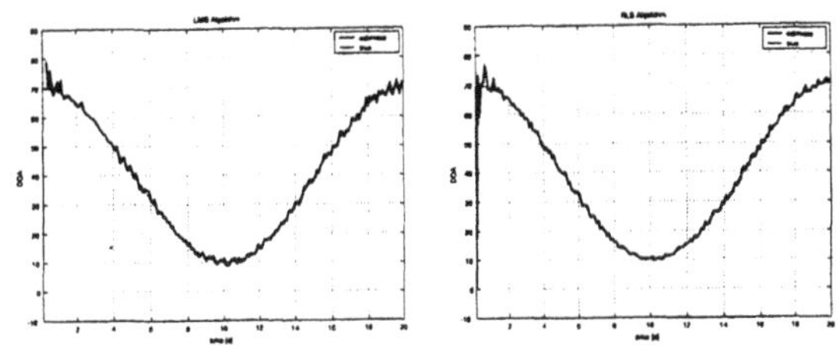

Fig. 4. DOA tracking performance for slow moving target: (*Left*) Tracking performance of proposed LMS algorithm with $q = 2$, (*Right*) Tracking performance of proposed RLS algorithm with $q = 2$.

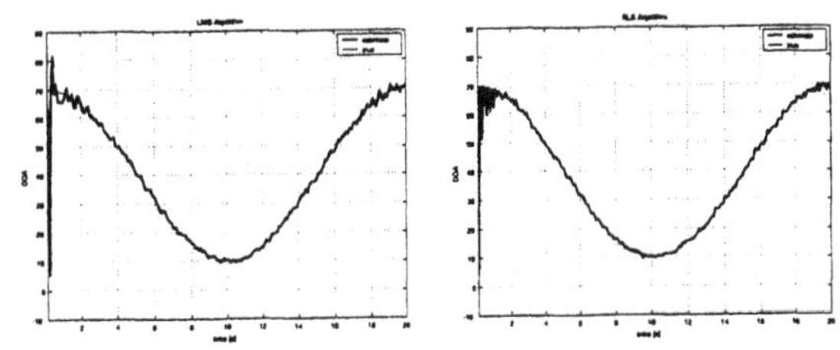

Fig. 5. DOA tracking performance for slow moving target: (*Left*) Tracking performance of proposed LMS algorithm with $q = 3$, (*Right*) Tracking performance of proposed RLS algorithm with $q = 3$.

where $\boldsymbol{y}(n) = \hat{\boldsymbol{W}}^H(n-1)\boldsymbol{x}(n)$.

The simulation setup is given as follows: the number of sensors is $M = 8$, the SNR is 20 dB at the sensor output. Figs.3, 4 and 5 show the estimated DOA obtained by the ordinary methods based on (Yang, 1995), the proposed time-varying subspace methods for $q = 2$ and $q = 3$, for slowly moving target. Similarly, in a case of faster moving target, the estimation results are also summarized in Figs.6, 7 and 8 respectively. It should be noticed that the ordinary LMS and RLS types of subspace methods should have different step size and forgetting factor to track the both slow and fast moving target, for instance, the step size should be chosen as $\mu = 0.04$ in the slow case and $\mu = 0.1$ in the fast case respectively, and the forgetting factor is chosen as $\beta = 0.8$ and $\beta = 0.6$ respectively. However, they cannot give good performance but some delay in the DOA estimation. On the other hand, the proposed time-varying LMS and RLS type of algorithms use the same polynomials $P(z^{-1})$ in the both slow and fast cases. The tracking performance of the proposed method can be improved by increasing the polynomial degree in the faster case.

6. CONCLUSIONS

The novel LMS and RLS adaptive algorithms for tracking time-varying parameters have been presented based on the internal model principle. The sufficient conditions for stability and convergence is clarified by applying the passivity and small gain theorems to the closed-loop system of the adaptive algorithms for time-varying systems. The effectiveness has been validated in the simulations of identification and DOA estimations.

REFERENCES

Anderson, B.D.O. (1972). The small-gain theorem, the passivity theorem and their equivalence. *Journal of the Franklin Institute* **298**, 105–115.

Åström, K.J. and B. Wittenmark (1995). *Adaptive Control*, Addison-Wesley Publishing Company.

Desoer, C.A. and M. Vidyasager (1975). *Feedback Systems: Input-Output Properties*, Academic Press, NY.

Eleftheriou, E. and Falconer, D. (1986). Tracking properties and steady-state performance of RLS adaptive filter algorithmsm. *IEEE Trans. Acoust. Speech, Signal Process.* **ASSP-34**, 1097–1109.

Guo, L. (1990). Estimating time-varying parameters by the Kalman filter based algorithm. *IEEE Trans. Automat. Contr.* **AC-35**, 141–147.

Haykin, S. (1975). *Adaptive Filter Theory, 3rd ed.*, Prentice-Hall, NJ.

Hidaka, K., H. Ohmori and A. Sano (1999). Adaptive control design for linear time-varying system based on internal model principle. *IEICE Trans. Fundamentals* **E82-A**, 1047–1054.

Katkovnik, V.Y. (1997). Identification of physical rapidly time-varying parameters. *Proc. SYSID*, 349–354.

Ljung, L. (1999). *System Identification: Theory for the User, 2nd ed.* Prentice-Hall, NJ.

Niedzwiecki, M. (2000). *Identification of Time-varying Processes*, John Wiley & Sons.

Saleh, A.A.M. and J. Salz (1993). *Linear Time-varying Systems*, Prentice-Hall, Englewood Cliffs, NJ.

Tsatsanis, M.K. and G.B. Giannakis (1996). Modeling and equalization of rapidly fading channels. *Int. J. Adaptive Contr. Signal Process* **10**, 1592–176.

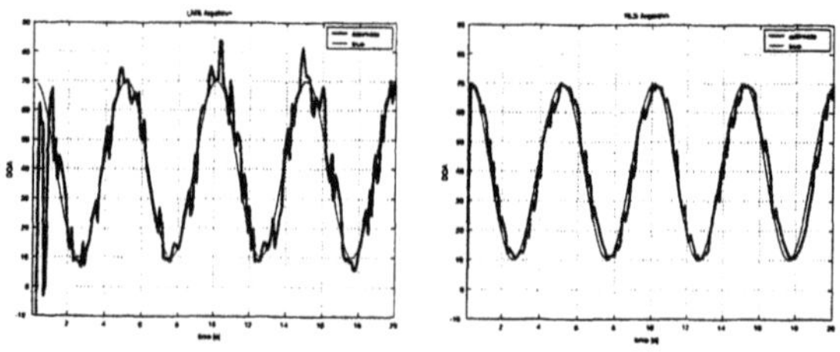

Fig. 6. DOA tracking performance for fast moving target: (*Left*) Tracking performance of ordinary LMS algorithm, (*Right*) Tracking performance of ordinary RLS algorithm.

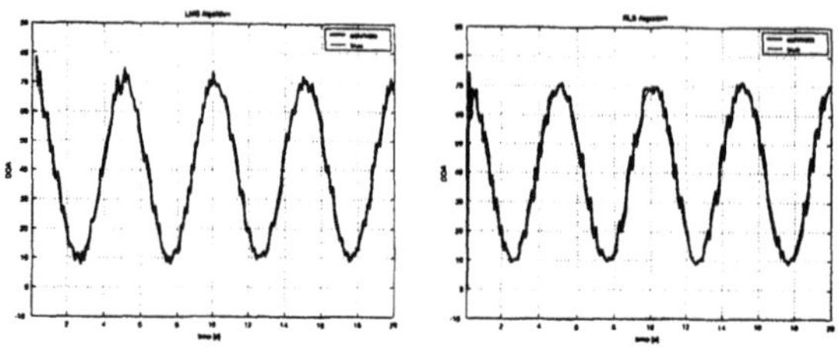

Fig. 7. DOA tracking performance for fast moving target: (*Left*) Tracking performance of proposed LMS algorithm with $q = 2$, (*Right*) Tracking performance of proposed RLS algorithm with $q = 2$.

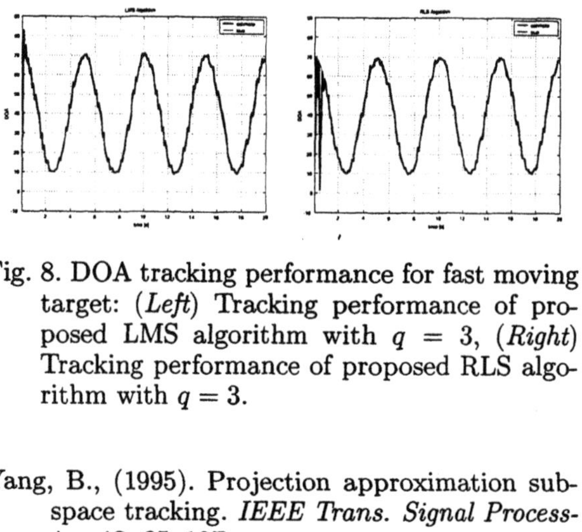

Fig. 8. DOA tracking performance for fast moving target: (*Left*) Tracking performance of proposed LMS algorithm with $q = 3$, (*Right*) Tracking performance of proposed RLS algorithm with $q = 3$.

Yang, B., (1995). Projection approximation subspace tracking. *IEEE Trans. Signal Processing* **43**, 95–107.

Ydstie, B.E. and R.W.H. Sargent (1986). Convergence and stability properties of an adaptive regulator with variable forgetting factor. *Automatica* **22**, 749-751.

THE ACCELERATING ADAPTIVE FILTERING ALGORITHM

Pablo E. Jojoa * Max Gerken ** Felipe M. Pait *,[1]**

* *pablo@lcs.poli.usp.br*
** *mgk@lcs.poli.usp.br*
*** *pait@lac.usp.br*
Universidade de São Paulo - EPUSP - PTC
Av. Prof. Luciano Gualberto, travessa 3, n. 158
05508-900 São Paulo SP - Brazil

Abstract: In a previous paper we introduced the Accelerating Adaptive Filtering algorithm which, at the cost of a moderate increase in computational complexity, presents a more favorable compromise between convergence speed and steady-state estimation error than the LMS or NLMS algorithms. In this paper we show that the AAF algorithm is of the quasi-Newton type and present a simplified version that is easier to adjust than its general form. We show the results of a first and second order moment analysis and obtain expressions for the transient behavior of the mean squared error and the misadjustment. Using these expressions, a simple comparison with the LMS algorithm clearly shows the advantages of the AAF algorithm. Simulation results are presented. *Copyright ©2001 IFAC*

Keywords: Adaptation, Adaptive Filters

1. INTRODUCTION

In a previous paper (Gerken, *et al.*, 2000) we introduced a fast, discrete-time adaptive filtering algorithm, derived from a continuous-time tuner used in adaptive control which sets the second derivative of the parameter estimates (Pait, 1998) . By means of simulations we showed that, for colored input signals, the proposed algorithm presents a more favorable compromise between convergence speed and steady-state estimation error than the LMS or NLMS algorithms, a property obtained at the cost of a moderate increase in computational complexity. We have also provided a deterministic convergence proof and presented some simulation results.

The Accelerating Adaptive Filtering (AAF) algorithm is based on a continuous time tuner that adjusts the second derivative, with respect to time, of the filter coefficients. Similarly, the popular LMS algorithm can be obtained from a continuous time tuner that adjusts the first derivative, with respect to time, of the filter coefficients. The LMS algorithm also uses an estimative of the first derivative, with respect to the filter coefficients, of a cost function, so one may suspect that the AAF algorithm will make use of an estimate of the second derivative with respect to the filter coefficients. Indeed, this suspicion is correct and in this paper we show that the AAF algorithm can be interpreted as a quasi-Newton method.

In its general form the AAF algorithm depends on four parameters, a scalar and three positive definite matrices, which must be chosen appropriately. On the one hand the extra parameters add to the freedom of filter's designer, on the other hand choosing them is a difficulty in the use of the AAF algorithm, a that motivation to obtain

[1] This work was supported by CNPq - Brasilian Research Council and FAPESP -São Paulo State Research Council.

versions that are more straightforward to adjust. In this paper we present a simplified version of the AAF algorithm and the results of a first and second order moments analysis of its filter coefficient errors. We then obtain expressions that characterize the transient behavior of the mean squared error and the misadjustment. Using these expressions a simple comparison with the LMS algorithm is presented, showing the advantages of using the AAF algorithm. For example, we show that for similar convergence speeds the AAF algorithm presents a lower misadjustment than the LMS algorithm. This misadjustment "attenuation" can be made larger for more ill-conditioned input signals. As a result from the performance comparison with the well known LMS algorithm, we also obtain some practical hints for choosing the parameters of the AAF algorithm. They are of interest for the adaptive filter design engineer.

The paper is organized as follows. First we present the simplified version of the AAF algorithm. Thereafter, we show that it is a quasi-Newton algorithm. Then we present some results of the first and second order moment analysis and a comparison with the LMS algorithm. We close the paper with simulation results.

2. THE AAF ALGORITHM: A SIMPLIFIED VERSION

Like other adaptive filtering algorithms, the AAF algorithm adjusts recursively an estimate $\mathbf{p}[n]$ of a coefficient vector $\mathbf{p}_o$ using a measured signal $d[n] = \mathbf{x}^T[n]\mathbf{p}_o + \eta[n]$, where $\mathbf{x}[n]$ is the input signal vector (regressor) and $\eta[n]$ is the measurement noise. The goal is to keep the *a posteriori* estimation error $e[n] = \mathbf{x}^T[n]\mathbf{p}[n] - d[n]$, the *a priori* estimation error $e_a[n]$ and the coefficient error $\delta\mathbf{p}[n] = \mathbf{p}[n] - \mathbf{p}_o$ as small as possible.

The original form of the AAF algorithm is given by (Gerken, 2000):

$$e_a[n] = \mathbf{x}^T[n]\mathbf{p}[n-1] - d[n] \quad (1)$$

$$\mathbf{G}[n] = \mathbf{I} + \alpha^2\mathbf{M}_1\mathbf{x}[n]\mathbf{x}^T[n] + +2\alpha\mathbf{M}_1\left(\mathbf{M}_2 + \mathbf{x}[n]\mathbf{x}^T[n]\mathbf{M}_1\mathbf{M}_3\right) \quad (2)$$

$$\mathbf{q}[n] = (\mathbf{G}[n])^{-1}\left(\mathbf{q}[n-1] - \alpha\mathbf{M}_1\mathbf{x}[n]e_a[n]\right) \quad (3)$$

$$\mathbf{p}[n] = \mathbf{p}[n-1] + \alpha\mathbf{q}[n]. \quad (4)$$

In the preceding equations α is a positive scalar, $\mathbf{I}$ is the identity matrix and $\mathbf{M}_i$, $i = 1,2,3$, are symmetric positive definite matrices that must satisfy the following inequalities to guarantee stability:

$$4\mathbf{M}_1\mathbf{M}_3\mathbf{M}_1 \geq \mathbf{M}_2^{-1} \quad (5)$$

$$\mathbf{M}_2\mathbf{M}_1\mathbf{M}_3 + \mathbf{M}_3\mathbf{M}_1\mathbf{M}_2 \geq \mathbf{M}_1^{-1}/2. \quad (6)$$

In Gerken (2000), we have shown that the update (3) of $\mathbf{q}[n]$ can be done as follows:

$$\mathbf{C}[n] = \frac{\alpha e_a[n] + \mathbf{x}^T[n]\mathbf{B}\mathbf{M}_1^{-1}\mathbf{q}[n-1]}{1 + \mathbf{x}^T[n]\mathbf{B}\mathbf{x}[n]}\mathbf{M}_1$$

$$\mathbf{q}[n] = \mathbf{A}\left(\mathbf{q}[n-1] - \mathbf{C}[n]\mathbf{x}[n]\right),$$

with

$$\mathbf{A} = (\mathbf{I} + 2\alpha\mathbf{M}_1\mathbf{M}_2)^{-1},$$

$$\mathbf{B} = \alpha\left(\mathbf{I} + 2\alpha\mathbf{M}_1\mathbf{M}_3\right)\mathbf{A}\mathbf{M}_1.$$

Motivated by the fact that fastest convergence is obtained at the upper bound of (5) and (6), we introduce the relations $\mathbf{M}_1\mathbf{M}_2 = \mathbf{I}/(2\gamma)$ and $\mathbf{M}_1\mathbf{M}_3 = \gamma\mathbf{I}/2$. Thus, for a positive scalar γ and a positive definite matrix $\mathbf{M}_1$ the upper bounds of (5) and (6) are reached. The simplified version of the AAF algorithm is given below:

$$\begin{aligned} e_a[n] &= \mathbf{x}^T[n]\,\mathbf{p}[n-1] - d[n] \\ g[n] &= \frac{e_a[n] + \gamma\mathbf{x}^T[n]\,\mathbf{q}[n-1]}{1 + \alpha\gamma\mathbf{x}^T[n]\,\mathbf{M}_1\mathbf{x}[n]} \\ \mathbf{q}[n] &= \frac{\gamma}{\alpha+\gamma}\left(\mathbf{q}[n-1] - \alpha g[n]\,\mathbf{M}_1\mathbf{x}[n]\right) \\ \mathbf{p}[n] &= \mathbf{p}[n-1] + \alpha\mathbf{q}[n]. \end{aligned} \quad (7)$$

A further simplification can be introduced using $\mathbf{M}_1 = m_1\mathbf{I}$, with $\mathbf{I}$ being the identity matrix. We reinforce that the above algorithm will be stable for positive parameters α, γ and m_1 (or $\mathbf{M}_1$ positive definite).

3. THE AAF ALGORITHM AND THE NEWTON METHOD

The idea behind Newton's method is to locally approximate a function $f(\mathbf{p})$ to be minimized by a quadratic function, and find the minimizer of the approximation (Luenberger, 1989). Near $\mathbf{p}_{n-1} = \mathbf{p}[n-1]$ we can approximate $f(\mathbf{p})$ by the truncated Taylor series

$$\begin{aligned} f(\mathbf{p}) \simeq\; & f(\mathbf{p}_{n-1}) + \nabla f^T(n, \mathbf{p}_{n-1})(\mathbf{p} - \mathbf{p}_{n-1}) \\ & + \frac{1}{2}(\mathbf{p} - \mathbf{p}_{n-1})^T\mathbf{F}(n, \mathbf{p}_{n-1})(\mathbf{p} - \mathbf{p}_{n-1}), \end{aligned} \quad (8)$$

with $\nabla f(.,.)$ being the gradient vector defined as an M-dimensional column vector and $\mathbf{F}(.,.)$ being the Hessian matrix of $f(\mathbf{p})$. Now $\mathbf{p}[n]$ is set to the value which minimizes the right side of (8), namely:

$$\mathbf{p}_n = \mathbf{p}_{n-1} - \mathbf{F}^{-1}(n, \mathbf{p}_{n-1})\nabla f(n, \mathbf{p}_{n-1}). \quad (9)$$

It is illustrative to consider the minimization of the quadratic cost function $f(\mathbf{p}) = E\{e^2[n]\}/2$, with $E\{.\}$ being the expectation operator and $e[n] = \mathbf{x}^T[n]\mathbf{p} - d[n]$. Considering an approximation as in (8) one readily obtains:

$$\mathbf{F}(n, \mathbf{p}_{n-1}) = \mathbf{R}_x = E\left\{\mathbf{x}[n]\mathbf{x}^T[n]\right\} \quad (10)$$

$$\nabla f(n, \mathbf{p}_{n-1}) = \mathbf{R}_x\mathbf{p}[n-1] - E\{d[n]\mathbf{x}[n]\} \quad (11)$$

$$\mathbf{p}[n] = \mathbf{R}_x^{-1}E\{d[n]\mathbf{x}[n]\}.$$

Thus, in this case Newton's algorithm converges in only one step.

Although the convergence properties of Newton's method are very attractive, evaluation of the inverse Hessian matrix is impractical. The idea underlying quasi-Newton methods is to use approximations to the inverse in place of the true one. The first modification in direction of a quasi-Newton method is usually the introduction of a "step-size" parameter ρ_n:

$$\mathbf{p}_n = \mathbf{p}_{n-1} - \rho_n \mathbf{F}^{-1}(n, \mathbf{p}_{n-1}) \nabla f(n, \mathbf{p}_{n-1}). \tag{12}$$

We now show that the AAF algorithm can be put in the form (12) by using approximations to the inverse Hessian matrix and to the gradient vector. Combine expressions (3) and (4) to obtain

$$\begin{aligned}\mathbf{p}[n] &= \mathbf{p}[n-1] + \alpha(\mathbf{G}[n])^{-1}\{\mathbf{q}[n-1] \\ &\quad - \alpha \mathbf{M}_1 \mathbf{x}[n] e_a[n]\} \\ &= \mathbf{p}[n-1] - (\mathbf{G}[n])^{-1}\{\alpha^2 \mathbf{M}_1 \mathbf{x}[n] e_a[n] \\ &\quad - (\mathbf{p}[n-1] - \mathbf{p}[n-2])\}. \end{aligned} \tag{13}$$

Comparing (13) with (12) we immediately conclude that the AAF algorithm makes use of the following estimates:

$$\begin{aligned}\widehat{\nabla f}(n, \mathbf{p}_{n-1}) &= \beta(\alpha^2 \mathbf{M}_1 e_a[n] \mathbf{x}[n] \\ &\quad - \mathbf{p}[n-1] + \mathbf{p}[n-2]) \end{aligned} \tag{14}$$

$$\widehat{\mathbf{F}}(n, \mathbf{p}_{n-1}) = \beta \mathbf{G}[n], \tag{15}$$

with β being an appropriate positive constant. Considering the simplified version with $\mathbf{M}_1 = m_1 \mathbf{I}$ and $\beta = \gamma/(\alpha+\gamma)$ we obtain expressions that look more familiar:

$$\rho_n = \frac{\alpha}{\alpha+\gamma} \alpha \gamma m_1 \tag{16}$$

$$\begin{aligned}\widehat{\nabla f}(n, \mathbf{p}_{n-1}) &= e_a[n] \mathbf{x}[n] - \frac{1}{\alpha^2 m_1}(\mathbf{p}[n-1] \\ &\quad - \mathbf{p}[n-2]) \end{aligned} \tag{17}$$

$$\widehat{\mathbf{F}}(n, \mathbf{p}_{n-1}) = \mathbf{I} + \alpha \gamma m_1 \mathbf{x}[n] \mathbf{x}[n]^T. \tag{18}$$

Expression (17) is an instantaneous estimate of the true gradient $E\{e_a[n]\mathbf{x}[n]\}$ "corrected" by a term proportional to the previous coefficient increment, and (18) is a regularized instantaneous estimate of the true Hessian matrix $\mathbf{R}_x = E\{\mathbf{x}[n]\mathbf{x}^T[n]\}$. It is illustrative to observe that for the LMS algorithm we have $\rho_n = \mu$ (the step size), $\widehat{\nabla f}(n, \mathbf{p}_{n-1}) = e_a[n]\mathbf{x}[n]$ and $\widehat{\mathbf{F}}(n, \mathbf{p}_{n-1}) = \mathbf{I}$.

To illustrate the Newton character of the AAF algorithm we consider the simple example of a two-tap predictor filter excited by $\sin((5\pi/128)n + 0.4\pi)$ under the following conditions: white Gaussian measurement noise with variance 10^{-4} and initial coefficients -1.5 and -1.7. We considered three algorithms: LMS ($\mu = 0.03$), Newton ($\rho_n = 0.05$; see Eq. (10), (11) and (12)) and AAF ($\alpha = 0.125$, $m_1 = 0.09$, $\gamma = 15$ and $\gamma = 320$). Step-sizes and parameters of all algorithms, except Newton's, have been adjusted to produce the same coefficient standard deviation (0.002) after convergence. Fig. 1 shows the trajectories in the coefficient plane where the contour curves of the cost function are also represented. Two distinct behaviors of the AAF algorithm are shown, one similar to the LMS algorithm ($\gamma = 15$) and one similar to the Newton algorithm ($\gamma = 320$). To close this section we summarize the main results:

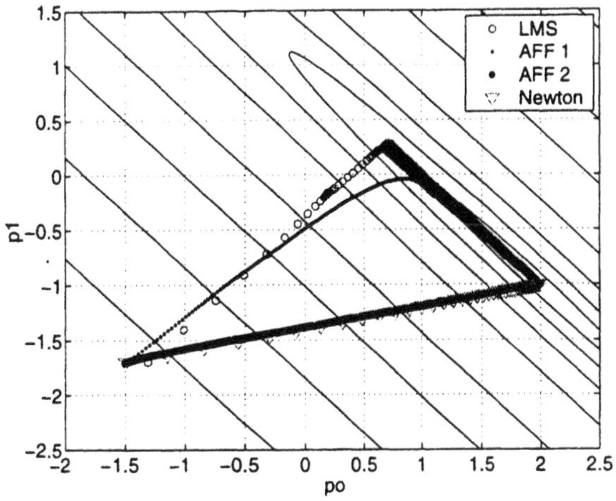

Fig. 1. Trajectories in the coefficient plane for a two-tap predictor.

(i) The AAF algorithm minimizes $E\{e^2[n]\}$.
(ii) At each step it considers a local approximation of the cost function as given in (8) using nontrivial estimates for the gradient vector and the Hessian matrix, given respectively by (14) and (15), or by (17) and (18) for the simplified version.
(iii) In the sense of (i) and (ii) the AAF algorithm can be classified as a quasi-Newton method.
(iv) As shown in Gerken, *et al.* (2000), the AAF algorithm is stable, which means that using the presented estimates of the gradient vector and the Hessian matrix guarantees convergence.

4. FIRST AND SECOND ORDER MOMENT ANALYSIS

Recall that $d[n] = \mathbf{x}^T[n]\mathbf{p}_o$ and $\delta\mathbf{p}[n] = \mathbf{p}[n] - \mathbf{p}_o$, so that

$$\begin{aligned} e_a[n] &= \mathbf{x}^T[n]\mathbf{p}[n-1] - d[n] + \eta[n] \\ &= \mathbf{x}^T \delta\mathbf{p}[n] + \eta[n], \end{aligned}$$

with $\eta[n]$ being the measurement noise. Introducing the variable $\mathbf{r}[n] = \delta\mathbf{p}[n] + \gamma\mathbf{q}[n]$ in equations (7) we obtain

$$\mathbf{r}[n] = \mathbf{\Gamma}[n]\mathbf{r}[n-1] + \alpha\gamma m_1 \mathbf{\Gamma}[n]\mathbf{x}[n]\eta[n] \tag{19}$$

$$\delta\mathbf{p}[n] = a\delta\mathbf{p}[n-1] + b\mathbf{r}[n], \tag{20}$$

where $\mathbf{\Gamma}[n] = \mathbf{I} - \dfrac{\alpha\gamma m_1 \mathbf{x}[n]\mathbf{x}^T[n]}{1 + \alpha\gamma m_1 \mathbf{x}^T[n]\mathbf{x}[n]}$ and $a = 1 - b = \gamma/(\alpha+\gamma)$. An analysis of the first and second order moments of $\mathbf{r}[n]$ and $\delta\mathbf{p}[n]$, similar to the usual moment analysis of the LMS algorithm (Solo and Kong, 1995; Haykin, 1996) , can be performed using the preceding equations. We make the following assumptions:

- $\mathbf{x}[n]$ is a zero mean Gaussian white noise vector with covariance matrix

$$\mathbf{R}_x = E\{\mathbf{x}[n]\mathbf{x}^T[n]\}.$$

- $\eta[n]$ is a zero mean Gaussian white noise with variance σ_η^2 and statistically independent from $\mathbf{x}[n]$.
- The order M of the adaptive filter is sufficiently high to make the variable $\mathbf{x}^T[n]\mathbf{x}[n]$ "slow" in comparison with $\mathbf{x}[n]$.

The last assumption enables us to apply the averaging property (Samsom, 1983) to obtain

$$\begin{aligned} E\{\mathbf{\Gamma}[n]\} &\approx \mathbf{I} - \frac{\alpha\gamma m_1 E\{\mathbf{x}[n]\mathbf{x}^T[n]\}}{1+\alpha\gamma m_1 E\{\mathbf{x}^T[n]\mathbf{x}[n]\}} \\ &= \mathbf{I} - \frac{\alpha\gamma m_1}{1+\alpha\gamma m_1 \mathrm{tr}\{\mathbf{R}_x\}}\mathbf{R}_x = \mathbf{I} - \mu_a\mathbf{R}_x, \end{aligned} \tag{21}$$

with tr{.} being the trace of a matrix. Under these conditions we are able to show that:

• The first order moments of $\mathbf{r}[n]$ and $\delta\mathbf{p}[n]$ satisfy

$$E\{\mathbf{r}[n]\} = (\mathbf{I} - \mu_a\mathbf{R}_x)\, E\{\mathbf{r}[n-1]\} \tag{22}$$

$$E\{\delta\mathbf{p}[n]\} = aE\{\delta\mathbf{p}[n-1]\} + bE\{\mathbf{r}[n]\} \tag{23}$$

• The covariance matrices $\mathcal{R}[n] = E\{\mathbf{r}[n]\mathbf{r}^T[n]\}$ and $\mathcal{P}[n] = E\left\{\delta\mathbf{p}[n]\delta\mathbf{p}^T[n]\right\}$ satisfy

$$\begin{aligned} \mathcal{R}[n] &= (\mathbf{I} - \mu_a\mathbf{R}_x)\,\mathcal{R}[n-1]\,(\mathbf{I} - \mu_a\mathbf{R}_x) \\ &\quad + \mu_a^2\mathbf{R}_x\sigma_\eta^2 \end{aligned} \tag{24}$$

$$\begin{aligned} \mathcal{P}[n] &= a^2\mathcal{P}[n-1] + b^2\mathcal{R}[n] \\ &\quad + abE\{\delta\mathbf{p}[n-1]\mathbf{r}^T[n]\} \\ &\quad + abE\left\{\mathbf{r}[n]\delta\mathbf{p}^T[n-1]\right\} \\ &\quad + \frac{2\alpha}{\gamma}a^2E\{\delta\mathbf{p}[n-1]\}\,E\{\mathbf{r}^T[n-1]\} \end{aligned} \tag{25}$$

• The mean square estimation error $E\{e_a^2[n]\}$ satisfies

$$E\{e_a^2[n]\} = \mathrm{tr}\{\mathbf{R}_x\mathcal{P}[n-1]\} + \sigma_\eta^2. \tag{26}$$

From Equations (22) to (26) we obtain, for $n \to \infty$, the trace of the steady state covariance matrix of the coefficient errors

$$\mathrm{tr}\{\mathcal{P}_\infty\} = \sigma_\eta^2\frac{\alpha}{\alpha+2\gamma}\sum_{k=1}^{M}\frac{\mu_a}{2-\mu_a\lambda_k} \tag{27}$$

and the steady-state mean square estimation error

$$E\{e_a^2\} = \sigma_\eta^2 + \underbrace{\sigma_\eta^2\frac{\alpha}{\alpha+2\gamma}\sum_{k=1}^{M}\frac{\mu_a\lambda_k}{2-\mu_a\lambda_k}}_{\mathcal{M}_a}, \tag{28}$$

with λ_k being the eigenvalues of $\mathbf{R}_x$. We observe that, for equal step-sizes ($\mu = \mu_a$), the misadjustment of the AAF is equal to the misadjustment of the LMS (see Haykin (1996)) attenuated by a factor $\alpha/(\alpha+2\gamma) < 1$. On the other hand, the LMS coefficient-error covariance matrix follows the same law shown in (24) for $\mathcal{R}[n]$. Therefore, as long as

$$\gamma/(\alpha+\gamma) < 1-\mu_a\lambda_{\min}, \quad \text{or} \quad \alpha/(\alpha+\gamma) > \mu_a\lambda_{\min},$$

$\mathcal{P}[n]$ in (25) has essentially the same convergence behavior as $\mathcal{R}[n]$ in (24). Thus, with this condition satisfied, we may expect the AAF algorithm to have approximately the same convergence speed as the LMS one with step-size $\mu = \mu_a$. Recalling that for the LMS algorithm $\mu\lambda_{\min} < 2\lambda_{\min}/\lambda_{\max}$, we conclude that the misadjustment attenuation of the AAF over the LMS algorithm can be increased proportionally to the ill-conditioning of the matrix $\mathbf{R}_x$, while keeping the convergence speed constant.

To verify to what extend the above reasoning is true we performed some simulations. We considered the identification of a FIR system with $M = 6$ coefficients equal to 1. The input signal was a zero mean colored Gaussian noise with variance σ_x^2 and the measurement noise was a zero mean Gaussian noise with variance σ_η^2.

In Figure 2 we have investigated the influence of $\gamma/(\alpha+\gamma)$ on the the convergence speed of the AAF algorithm. The corresponding values of the parameters are shown in Table 1. A comparison of the values of $\gamma/(\alpha+\gamma)$ and $1-\mu\lambda_{\min}$ and the corresponding mean square error curves confirm the expected behavior. In Table 2 we are

Table 1. Parameters of Figure 2.

m_1	γ	$\frac{\gamma}{\alpha+\gamma}$	$1-\mu\lambda_{min}$	Fig.2
60	0.01	0.242	0.987	AAF 1
6	0.1	0.762	0.987	AAF 2
0.6	1	0.970	0.987	AAF 3
$\sigma_x^2 = 4$, $\sigma_\eta^2 = 0.5$, $\alpha = 0.0313$, $\mu_a = \mu = 0.0129$				

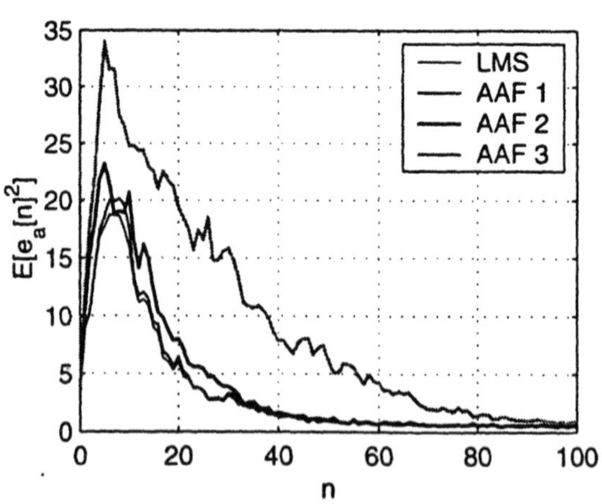

Fig. 2. Learning curves of AAF and LMS algorithms with parameters in Table 1. Averaged over 200 experiments.

able to compare measured misadjustments of the AAF and LMS algorithms and the theoretical misadjustment atenuation $\alpha/(\alpha + 2\gamma)$ given by Equation (28). We observe that with increasing α the achieved misadjustment attenuation approximates well the theoretical value, while for low values of α the achieved misadjustment attenuation is not as good as expected. Table 2 is shown here as an example of this effect, which has been also observed in more extensive simulations. To some extent the results presented in Table 2 are disappointing, since the achieved misadjustment reductions are not as large as predicted by expression (28). On the bright side, even for stepsizes μ for which the LMS algorithm's behavior is ugly, the AAF algorithm is stable and converges. An example is shown in Figure 3. The learning curves shown were obtained under the following conditions: $m_1 = 460$, $\alpha = 0.3125$, $\gamma = 0.1$, $\mu = 0.0416$, $\sigma_x^2 = 4$ and $\sigma_\eta^2 = 0.5$. The condition number of the correlation matrix $\mathbf{R}_x$ was $\lambda_{\max}/\lambda_{\min} \approx 6.5$. The obtained misadjustment of the AAF algorithm was $\mathcal{M}_a = 0.64$, while for LMS we measured $\mathcal{M}_a = 3.7 \cdot 10^6$.

Table 2. Misadjustment of AAF ($\mathcal{M}_a$) and LMS ($\mathcal{M}_{lms}$) algorithms. Averaged over 200 experiments.

α	$\alpha/(\alpha+2\gamma)$	$\mathcal{M}_a$	$\mathcal{M}_{lms}$	$\mu = \mu_a$
α_o	0.135	0.075	0.102	0.013
$2\alpha_o$	0.238	0.130	0.202	0.020
$4\alpha_o$	0.347	0.190	0.536	0.025
$\sigma_x^2 = 4$, $\sigma_\eta^2 = 0.5$, $\gamma = 0.01$, $m_1 = 6$, $\alpha_o = 0.0313$				

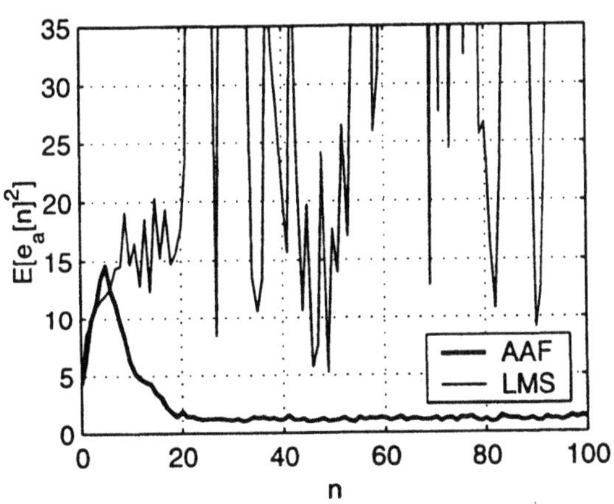

Fig. 3. Learning curves of AAF and LMS algorithms with $\mu = \mu_a$. Averaged over 200 experiments.

5. CONCLUSION

We presented a simplified version of the AAF algorithm that turns out to be a quasi-Newton algorithm, and is easier to adjust than its general form. By means of a first and second order analysis we obtained expressions for the mean square error and the misadjustment. To some extent simulations confirmed these results. Moreover, the robust convergence behavior of the proposed algorithm makes it a good choice for applications where ill conditioned signals are processed by high-order adaptive filters, for example acoustic echo cancellation.

REFERENCES

Gerken M., F.M. Pait and P.E. Jojoa (2000). An Adaptive Algorithm with Parameter Acceleration. *Proceedings of ICASSP'2000*, **I**, 17-20.

Haykin S. (1996). *Adaptive Filter Theory* Prentice Hall, 3. Ed.

Luenberger D.G. (1989). *Linear and Nonlinear Programming* Addison-Wesley. 2.Ed.

Pait F.M. (1998). A Tuner that Accelerates Parameters, *Systems & Control Letters*, **35**, 65-68.

Samson C. and V.U. Reddy (1983). Fixed point error analysis of the normalized ladder algorithms, *IEEE Trans. on ASSP*, **31**, 5, 1177-1191.

Solo V. and X. Kong (1995). *Adaptive Signal Processing Algorithms, Stability and Performance*, Prentice Hall.

ADAPTIVE NOTCH FILTER WITH GLOBAL STABILITY

Tatsuo Suzuki [*,1] **Kazuyasu Hamada** [**,2]

* *Mitsubishi Heavy Industries, 1200 Higashitanaka, Komaki, Aichi, JAPAN*
** *Gifu University, 1-1 Yanagido, Gifu, JAPAN*

Abstract: A new synthesis of adaptive notch filter is proposed for the system that is imposed sinusoidal signal with unknown time-varying amplitude and frequency. It is shown that the proposed adaptive notch filter guarantees the global stability. Some numerical examples are also given. *Copyright © 2001 IFAC*

Keywords: Adaptive filters, Notch filters, Gloval stability, Stability analysis, Sinusoidal signals

1. INTRODUCTION

In the control of mechanical system, it is important to remove unnecessary resonance phenomena.

To avoid the resonance, controller synthesis is usually conducted in the situation that the control signal's bandwidth and structural vibration mode are separated.

However, it is often difficult to do this in practice. In the case that controller's bandwidth and resonance frequencies are coupled, using a notch filter

$$G(\theta_*, s) = \frac{s^2 + \theta_*^2}{s^2 + 2\zeta\theta_* s + \theta_*^2} \qquad (1)$$

,where θ_*,ζ denote notch frequency, damping ratio, respectively, will be a good idea to reduce the control gain near the structural vibration mode only.

The notch frequency and the damping ratio are usually given by constants based on the identification of the structural vibration mode through some pre-structural test. However, in the case that the structural vibration mode will be often varying with time, notch filter with constant coefficients is not effective. Moreover, to estimate the time-varying structural vibration mode in advance is often very difficult. In such case, the adaptive notch filter will be required.

P.Regalia was the first who formulated the adaptive notch filter (Regalia, 1995). But his adaptive algorithm has no guarantee for the global stability.

Recently, L.Hsu et al. improved the Regalia's algorithm (Hsu *et al.*, 1999). Different from Regalia's algorithm, their algorithm is guaranteed the global stability by introducing the normalizing and the time scale changeing, but there are some restrictions for initial parameters and adaptive gains because their algorithm uses internal signals of the filter. This makes serious restrictions for real applications.

Moreover, these previous studies only dealt with the case that the system input has only one sinusoidal signal with unknown constant amplitude and frequency. In real applications, the amplitude and the frequency of the sinusoidal signal is often time-varying.

In this paper, a new algorithm that is based on the adaptive observer for estimating the frequency of

[1] E-mail: suzuki@twics.com
[2] E-mail: k-hamada@cc.gifu-u.ac.jp

the disturbance sinusoidal signal will be proposed. It is shown that the proposed algorithm guarantees the global stability. Numerical examples are also given.

2. NOTATION

The input-output formulation of multivariable systems is taken throughout this paper and the notation and terminology are standard (see, for example (Kosut and Friedlander, 1985; Desoer and Videyasagar, 1975)).

The input and output signals are assumed to be imbedded in either the normed functional space or its extensions as follows.

$$L_p^n = \{x : [0,\infty) \to \mathbf{R}^n \mid \|x\|_p < \infty\}$$
$$L_{pe}^n = \{x : [0,T) \to \mathbf{R}^n \mid \|x\|_{Tp} < \infty, \quad T < \infty\}$$

The respective norms $\|\cdot\|_p$, $\|\cdot\|_{Tp}$ are defined as follows.

$$\|x\|_p = \lim_{T\to\infty} \|x\|_{Tp}$$

$$\|x\|_{Tp} = \begin{cases} (\int_0^T \|x(t)\|^p \, dt)^{1/p}, & p \in [1,\infty) \\ \sup_t \|x(t)\|, \quad 0 \le t \le T, & p = \infty \end{cases}$$

where $\|\cdot\|$ is the Euclidean norm on $\mathbf{R}^n$. Also L_{2e}^n is an inner product space with the following inner product.

$$\langle x,y\rangle_T = \int_0^T x(t)'y(t)dt, \quad x,y \in L_{2e}^n$$

where $x(t)'$ denotes the tranpose of vector $x(t)$. In the case of $T \to \infty$, L_2^n is the inner product space with the inner product $\langle x,y\rangle = \lim_{T\to\infty}\langle x,y\rangle_T$.

Regulated function $V(\cdot) : \mathbf{R}_+ \to \mathbf{R}^{n\times m}$ is called Persistently Exciting $(V \in PE)$ if there exist some positive constant α_1, α_2 and δ such that

$$\alpha_1 I_n \le \int_s^{s+\delta} V(t)V'(t)dt \le \alpha_2 I_n, \quad \forall s \in \mathbf{R}_+$$

3. ADAPTIVE NOTCH FILTER AND ERROR SYSTEM

Consider the following system depicted in Fig.1.

$$\begin{cases} e = P(s)(w-u) - w \\ z = \frac{1}{s}P(s)(w-u) \\ u = z\hat{\theta}^2 \\ \hat{\theta} = \Omega(e, z, \hat{\theta}, \hat{\theta}(0)) \\ y = G(\hat{\theta}, s)w \end{cases} \tag{2}$$

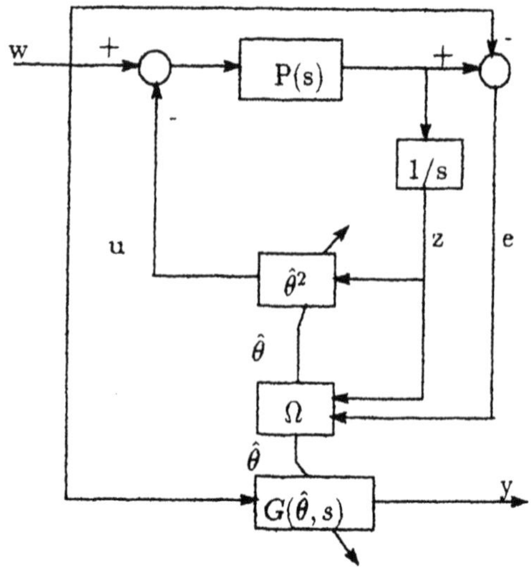

Fig. 1. Adaptive notch filter

where

$$P(s) = \frac{s}{s^2 + s + \theta_0^2} \tag{3}$$

and θ_0 is a lower limit value for adaptation. z is the regressor signal to produce the adaptive control signal u. w is the externel filter input signal. y is the filterd output signal. $\hat{\theta}$ is the estimeted unknown frequencies. $G(\hat{\theta}, s)$ is the notch filter represented as in (1) with fixed ζ .

Consider the following adaptive algorithm to estimate the unknown frequencies.

$$\dot{\hat{\theta}} = \gamma\hat{\theta}ze \tag{4}$$

Proposition 1. This algorithm has a good feature such that for the initial condition $\hat{\theta}(0) > 0$, $\hat{\theta}(t) > 0$ for $\forall t \ge 0$

Define the adaptive gain error to analyze the convergence of the parameter $\hat{\theta}(t)$ as follows.

$$\theta(t) = \hat{\theta}(t) - \theta_* \tag{5}$$

where θ_* is some constant value that is so called tuned gain .

Remark 2. θ_* is not always a true value because there does not exist the true value for some input signal w.

By virtue of Proposition 1, $\theta(t) > -\theta_*$ holds.

Define $v = z(2\theta_* + \theta)\theta$, and after some tedious calculations, the following error system is given.

$$\begin{bmatrix} e \\ z \end{bmatrix} = H(\theta_*) \begin{bmatrix} w \\ v \end{bmatrix} = \begin{bmatrix} H_{ew} & -H_{ev} \\ H_{zw} & -H_{zv} \end{bmatrix} \begin{bmatrix} w \\ v \end{bmatrix}$$

$$= \begin{bmatrix} -\dfrac{s^2+\theta_0^2+\theta_*^2}{s^2+s+\theta_0^2+\theta_*^2} & \dfrac{s}{s^2+s+\theta_0^2+\theta_*^2} \\ \dfrac{1}{s^2+s+\theta_0^2+\theta_*^2} & \dfrac{1}{s^2+s+\theta_0^2+\theta_*^2} \end{bmatrix} \begin{bmatrix} w \\ v \end{bmatrix} \tag{6}$$

When adaptation was completed (i.e. $\theta \to 0$, $\hat{\theta} \to \hat{\theta}_*$), tuned error signal e_* and tuned regressor signal z_* are represented as follows.

$$e_* = H_{ew}w, \quad z_* = H_{zw}w \tag{7}$$

Remark 3. If w includes only one sinusoidal signal that is $w(t) = k\sin(\omega t + \phi)$, then there exists θ_* such that $\theta_*^2 = \omega^2 - \theta_0^2$, and $e_* = 0$ holds. This is clear from the shape of H_{ew}.

Let S and S_0 denote the set of stable proper transfer functions and the set of stable strictly proper transfer functions, respectively. And define two subsets in strictly positive real (SPR) functions as follows.

Definition 4.

$$SPR_+ = \{G \in S | Re[G(j\omega)] \geq \mu, \quad \forall\omega \in \mathbf{R}, \quad \mu > 0\}$$

$$SPR_0 = \{G \in S_0 | Re[G(j\omega)] \geq \mu |G(j\omega)|^2, \quad \forall\omega \in \mathbf{R}, \quad \mu > 0\}$$

Proposition 5. $H_{ev} \in SPR_0$ for $\forall\theta_*$, $\theta_0 > 0$.

4. GLOBAL STABILLITY

In this section, the result of the global stability of the proposed adaptive notch filter is shown. The proof of the result is omitted by the limitation of the space. It is almost similar to (Kosut and Friedlander, 1985)

Assumption 6. For the error system (6), assume the followings.

(i) The system is well posed in the sense that the signals $e, v, z, \theta, \dot{\theta}$ that is generated by any input $w \in W$ are all in $L_{\infty e}$.
(ii) $H_{ev} \in SPR_+$.

Remark 7. $H_{zv} \in S_0$ for any θ_*. To achieve $H_{ev} \in SPR_+$, some technique are required (see, for example (Hamada and Suzuki, 1988)).

Under these assumptions, the following theorem holds.

Theorem 8. Suppose that the system (2) holds Assumption 6. Then

(i) If $(e_*, \dot{e}_*) \in L_2 \cap L_\infty$ ($\Rightarrow e_* \to 0$) and $(z_*, \dot{z}_*) \in L_\infty$, then
  (a) $(\theta, \ddot{\theta}) \in L_\infty$, $\dot{\theta} \in L_\infty$ and $\dot{\theta} \to 0$, $v \in L_2 \cap L_\infty$, $\dot{v} \in L_\infty$
  (b) $e \in L_2 \cap L_\infty$, $\dot{e} \in L_\infty$ and $e - e_* \to 0$
  (c) $v \in L_2 \cap L_\infty$, $\dot{v} \in L_\infty$ and $v \to 0$
  (d) $(z, \dot{z}) \in L_\infty$, $(z - z_*, \dot{z} - \dot{z}_*) \in L_2 \cap L_\infty$ and $z - z_* \to 0(\exp)$.
(ii) If $(e_*, \dot{e}_*) \in L_\infty$ and $(z_*, \dot{z}_*) \in L_\infty$, then
  (a) $z \in L_\infty$
  (b) If $z \in PE$, then $\theta, \dot{\theta}, e, \dot{e}, v, \dot{v}, \dot{z} \in L_\infty$.

Remark 9. $H_{zv} \in S_0$ for any θ_*. To achieve $H_{ev} \in SPR_+$, some technique are required (see, for example (Hamada and Suzuki, 1988)).

Corollary 10. If w includes only one sinusoidal signal (i.e.; $w(t) = k\sin(\omega t + \phi)$), then there exist θ_* such that $\theta_*^2 = \omega^2 - \theta_0^2$, and $\theta \to 0$.

5. NUMERICAL EXAMPLE

For numerical examples, assume that the input signal to adaptive notch filter is composed of $s(t)+n(t)$, where $s(t)$ and $n(t)$ are sensor signal and sinusoidal disturbances, respectively. The purpose of the adaptive notch filter is to estimate the frequency of n(t) and remove it from the input signal.

Throughout this section, the damping ratio of the notch filter (1) is fixed with $\zeta = 0.7$.

5.1 *One Sinusoidal Disturbance Noise Case*

Let $s(t)$ as a sinusoidal sensor signal with amplitude 0.5 and frequency 0.1Hz. And unknown disturbing noise $n(t)$ has a constant amplitude $k_* = 1.0$ and a time-varying frequency $f_*(t)(= \theta_*/2\pi)$ that sweeps from 30Hz ($t = 0sec$) to 50Hz ($t = 10sec$) and fixed after $t > 10sec$. Fig.2 depict the result of simulation with $\hat{\theta}(0) = 40Hz \times 2\pi$, $\gamma = 500$. From Fig.2, it is shown that $n(t)$ is reduced in about 3.0 sec and $s(t)$ is reproduced from the input signal $s(t) + n(t)$. And estimated frequency $f(t)$ traces the true frequency $f_*(t)$ of $n(t)$.

5.2 *Two Sinusoidal Disturbance Noises Case*

To remove two sinusoidal noises, two cascade adaptive notch filters are used. The sensor signal $s(t)$ is same as above example.

The disturbance noise $n(t)$ has

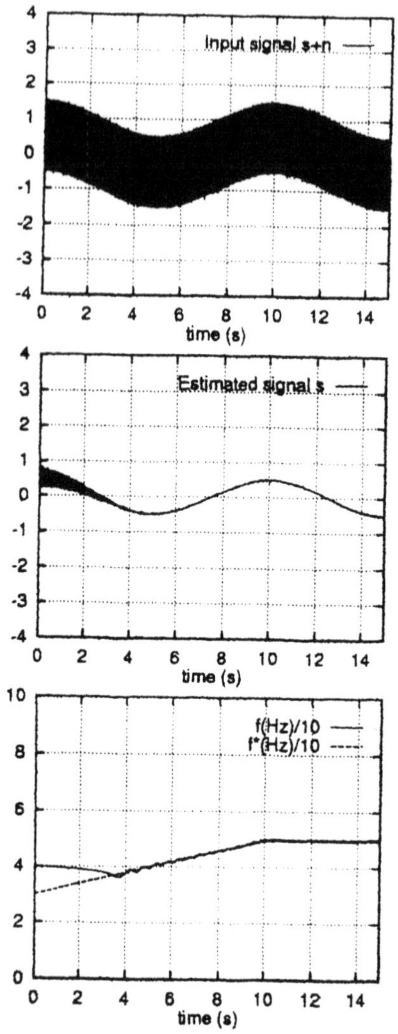

Fig. 2. One sinusoidal noise case

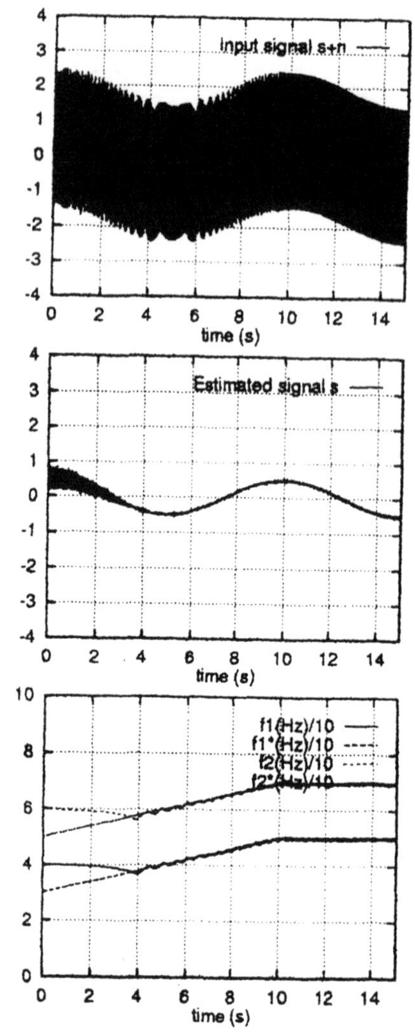

Fig. 3. Two sinusoidal noises case

- $n_1(t)$ with a constant amplitude $k_{1*}(t) = 1.0$, a time-varying frequency $f_{1*}(t)$ that sweeps from 30Hz ($t = 0sec$) to 60Hz ($t = 10sec$) and fixed after $t > 10sec$,
- And $n_2(t)$ with a constant amplitude $k_{2*}(t) = 1.0$, a time-varying frequency $f_{2*}(t)$ that sweeps from 50Hz ($t = 0sec$) to 70Hz ($t = 10sec$) and fixed after $t > 10sec$.

The simulation runs with $f_1(0) = 40Hz$ and $\gamma_1 = 500$ for the first notch filter, and $f_2(0) = 60Hz$ and $\gamma_2 = 2000$ for the second one.

Fig.3 depicts the result and it is shown that $n(t)$ is removed from the input signal $s(t) + n(t)$, and estimated frequencies $f_1(t)$, $f_2(t)$ track the true frequencies $f_{1*}(t)$, $f_{2*}(t)$, respectively.

6. CONCLUSION

In this paper, a new synthesis of adaptive notch filter with global stability is proposed.

On the contrary to the Hsu's adaptive algorithm, since the proposed adaptive algorithm has constant adaptive gain, the tuning procedure is rather simple. Moreover the proposed algorithm can be assigned the lower limit frequency of adaptation. These features are convenient to real applications.

In the numerical example, the noise cancellation problem is shown in the case that the filter input signal is composed of the sensor signal and the single or multiple disturbance noise with time-varying frequency.

7. REFERENCES

Desoer, C.A. and M. Videyasagar (1975). *Feedback Systems: Input-Output Properties.* Academic Press. New York.

Hamada, K. and M. Suzuki (1988). Robust adaptive algorithm with non-linear decay term. *Int. J. of Control* **47**, 665-680.

Hsu, L., R. Ortega and G. Damm (1999). A globally convergent frequency estimator. *IEEE Trans.* **AC44**, 698–713.

Kosut, R.L. and B. Friedlander (1985). Robust adaptive control: Conditions for global stability. *IEEE Trans.* **AC30**, 698–713.

Regalia, P. (1995). *IIR Filtering in Signal Processing and Control.* Marcel Dekker. New York.

ROBUST WIENER DESIGN OF ADAPTATION LAWS WITH CONSTANT GAINS

Mikael Sternad*, Lars Lindbom[1], and Anders Ahlén*

*Signals and Systems, Uppsala University, PO Box 528, SE-75120, Uppsala, Sweden.
[1]Ericsson Infotech, PO Box 1038, SE-65115 Karlstad
mikael.sternad@signal.uu.se (corresponding author), fax +46 18 555096

Abstract: Filters can be introduced into LMS-like adaptation algorithms to improve their tracking performance. This paper discusses the systematic model-based design of such filters. Parameter variations in coefficients of linear regression models are modeled as ARIMA-processes. The aim is to provide high performance filtering, prediction or fixed lag smoothing estimates for arbitrary lags. The properties of the time-varying parameters are in general not known exactly, so a robust design for a set of possible models will be of interest. We minimize the average tracking MSE, based on probabilistic descriptions of the model uncertainty. The method is based on a novel signal transformation that recasts the algorithm design into a robust Wiener filtering problem. The performance is illustrated on the tracking of mobile radio channels in IS-136 systems, based on a model of the time-variations affected by parametric uncertainty.

1. INTRODUCTION

Algorithms that estimate time-varying parameters of models and filters in linear regression form are important tools for signal processing, control and digital communication. Kalman estimators are the optimal linear algorithms when the statistics of the parameter variations are known. However, their complexity is often unacceptable in high-speed applications.

The required low complexity of channel estimators in mobile radio systems has motivated us to develop a class of adaptation laws that for channel tracking attain close to the optimal Kalman performance, at a computational complexity close to that of LMS algorithms (Lindbom 1995, Sternad *et.al.* 2000, Lindbom *et.al.* 2000a). The approach also opens up new ways of analyzing adaptation laws for fast variations (Ahlén *et.al.* 2000), that provide accurate estimates of the parameter error variance. These and related algorithms can, for example, be applied effectively on the fading 1900MHz channels of North American IS-136 mobile radio systems (Lindbom *et.al.* 2000b). Design of a related class of algorithms has been investigated by Benveniste *et.al.* (1990) for slowly varying parameters. Section 4 and 5 outline an iterative Wiener design that is effective also for fast variations.

A model-based Wiener design may become sensitive to the assumed model. We here propose a method for decreasing this sensitivity. Using tools from Sternad and Ahlén (1993) and Öhrn *et.al.* (1995), the design equations are in Section 6 modified to minimize the *average* tracking MSE, based on probabilistic descriptions of the uncertainty in the parameter models. In Section 7, time-varying mobile radio channels in IS-136 systems are estimated. The model of the time-variations is there affected by parametric uncertainty in the Doppler frequency and the robust design is performed by using an averaged covariance function.

Notation: Here, $R(q^{-1})$, $\boldsymbol{R}(q^{-1})$ and $\boldsymbol{\mathcal{R}}(q^{-1})$ denote polynomials, polynomial matrices and causal rational matrices, respectively. Conjugate matrices $\boldsymbol{P}_*(q)$ or $\boldsymbol{\mathcal{R}}_*(q)$ are obtained by conjugating complex coefficients, transposing and substituting the forward shift operator q for the backward shift operator q^{-1}.

2. OUTLINE OF THE PROBLEM

A sequence of measurement vectors $\{y_t\}$ of dimension $n_y|1$ is assumed available at the discrete time instants $t = 0, 1, 2, \ldots$ and to be generated by a linear regression

$$y_t = \varphi_t^* h_t + v_t \quad , \tag{1}$$

where v_t is a noise vector that is uncorrelated with the $n_y|n_h$ regression matrix φ_t^*. We assume the possibly complex-valued regressors to be known at time t and to be persistently exciting, so that

$$\mathbf{R} \triangleq \mathrm{E}\{\varphi_t \varphi_t^*\} \tag{2}$$

is nonsingular. The covariance matrix $\mathbf{R}$ will here be assumed time-invariant, but it can in practice be slowly varying. The time-varying parameter vector

$$h_t = (h_{0,t} \ \ldots \ h_{n_h-1,t})^T \tag{3}$$

is to be estimated, with the order n_h assumed known. Models describing the variation of h_t are sometimes called *hypermodels* (Benveniste *et.al.* 1990). We here use linear time-invariant stochastic models

$$h_t = \boldsymbol{\mathcal{H}}(q^{-1}) e_t \quad , \tag{4}$$

where e_t is white noise with covariance matrix $\mathbf{R}_e$ and where $\mathcal{H}(q^{-1})$ is an $n_h|n_h$ matrix of stable or marginally stable transfer operators. The model (4) is for now assumed known, but will in Section 6 be assumed uncertain. Denote the tracking error by

$$\tilde{h}_{t+k|t} \triangleq h_{t+k} - \hat{h}_{t+k|t} \tag{5}$$

where the estimate $\hat{h}_{t+k|t}$ may be obtained by filtering ($k=0$), prediction ($k>0$) or fixed lag smoothing ($k<0$). Kalman estimators, based on (1) and on state-space realizations of (4), are the linear estimators that minimize the tracking covariance matrix

$$\mathbf{P}_{k,t} \triangleq \mathrm{E}\{\tilde{h}_{t+k|t}\tilde{h}^*_{t+k|t}\} \ . \tag{6}$$

Since φ_t^* in (1) is time-varying, the Kalman gains will not converge as $t \to \infty$, so on-line Riccati updates are required.

We here consider a class of adaptation laws that avoid on-line Riccati updates. Instead, pre-designed linear time-invariant filters $\mathcal{M}_k(q^{-1})$ operate on the negative instantaneous gradient of $|\varepsilon_t|^2$ with respect to $\hat{h}_{t|t-1}$,

$$\varepsilon_t = y_t - \varphi_t^*\hat{h}_{t|t-1} \tag{7}$$
$$\hat{h}_{t+k|t} = \mathcal{M}_k(q^{-1})\varphi_t\varepsilon_t \ . \tag{8}$$

The LMS algorithm

$$\hat{h}_{t+1|t} = \frac{\mu}{1-q^{-1}}\mathbf{I}\,\varphi_t\varepsilon_t \ , \tag{9}$$

where $\mu > 0$ is a scalar gain, constitutes a simple special case of the structure (7),(8). The rational matrix $\mathcal{M}_k$ can be selected to asymptotically minimize (6) under various constraints and assumptions.

3. THE LOOP TRANSFORMATION

The algorithm (7),(8) can be expressed as a stable and causal filter, denoted the *learning filter* $\mathcal{L}_k(q^{-1})$, that operates on a signal vector

$$f_t \triangleq \varphi_t\varepsilon_t + \mathbf{R}\hat{h}_{t|t-1} \ . \tag{10}$$

Since (7),(8) give $\hat{h}_{t|t-1} = q^{-1}\mathcal{M}_1(q^{-1})\varphi_t\varepsilon_t$,

$$\varphi_t\varepsilon_t = (\mathbf{I} + q^{-1}\mathbf{R}\mathcal{M}_1(q^{-1}))^{-1}f_t \ .$$

Thus, we obtain

$$\hat{h}_{t+k|t} = \mathcal{M}_k(q^{-1})(\mathbf{I}+q^{-1}\mathbf{R}\mathcal{M}_1(q^{-1}))^{-1}f_t \triangleq \mathcal{L}_k(q^{-1})f_t \ . \tag{11}$$

By (7) and (1),

$$\varphi_t\varepsilon_t = \varphi_t\varphi_t^*\tilde{h}_{t|t-1} + \varphi_t v_t \ . \tag{12}$$

Adding and subtracting $\mathbf{R}\tilde{h}_{t|t-1}$ on the right-hand side of (12) gives

$$\varphi_t\varepsilon_t = \mathbf{R}h_t - \mathbf{R}\hat{h}_{t|t-1} + (\varphi_t\varphi_t^* - \mathbf{R})\tilde{h}_{t|t-1} + \varphi_t v_t \ . \tag{13}$$

We now define

$$Z_t \triangleq \varphi_t\varphi_t^* - \mathbf{R} \tag{14}$$
$$\eta_t \triangleq Z_t\tilde{h}_{t|t-1} + \varphi_t v_t \tag{15}$$

which we call the *autocorrelation matrix noise* and the *gradient noise*, respectively. The signal f_t which can be regarded as a fictitious measurement, can then by using (10), (13), (14) and (15), be expressed as

$$f_t = \mathbf{R}h_t + Z_t\tilde{h}_{t|t-1} + \varphi_t v_t = \mathbf{R}h_t + \eta_t \ , \tag{16}$$

see Fig. 1. The design of our adaptation law (7),(8) has now been transformed into a *Wiener filter design* for $\mathcal{L}_k(q^{-1})$, where η_t plays the role of noise, see Fig. 2.

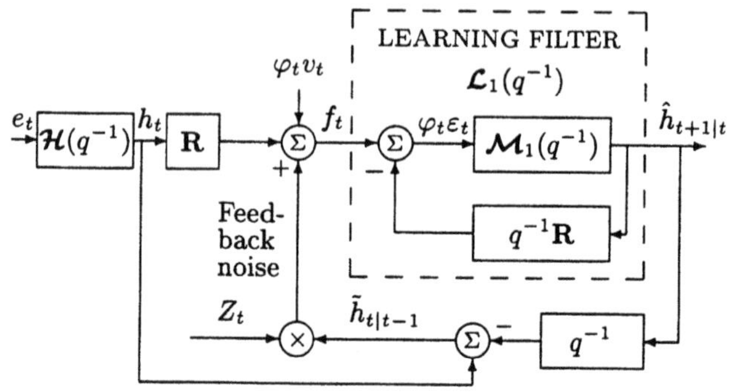

Figure 1: The adaptation algorithm (8) operates in closed loop. This loop can be decomposed into an inner time-invariant feedback of $\mathbf{R}\hat{h}_{t|t-1}$ and an outer time-varying loop via the feedback noise $Z_t\tilde{h}_{t|t-1}$.

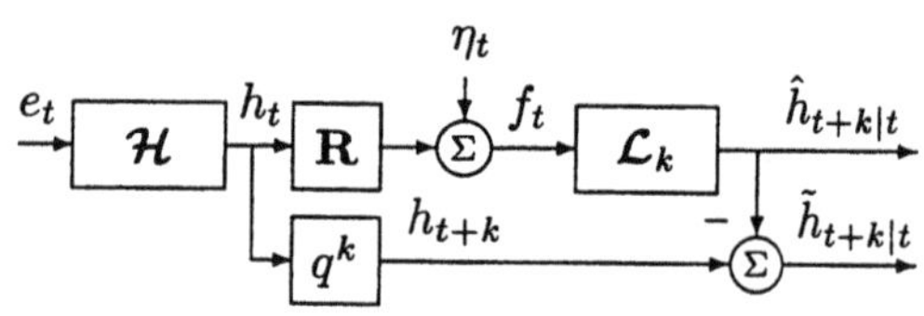

Figure 2: The filter design problem. The vector h_{t+k} is to be estimated from f_t, such that the steady state tracking error covariance matrix of the parameter error $\tilde{h}_{t+k|t}$ is minimized.

The gradient noise η_t is affected by the term $Z_t\tilde{h}_{t|t-1}$, here called the *feedback noise*. It is shown in Ahlén *et.al.* (2000) that the feedback noise is negligible either when h_t has small increments or when the noise v_t has high variance. Such situations are denoted "slow variations" (Ahlén *et.al.*, 2000; Macchi, 1995). The optimal learning filter will then operate in open loop, with $\eta_t \approx \varphi_t v_t$. Stability and convergence in MSE is then guaranteed by stability of the learning filter, which follows directly from a Wiener design. (While the learning filter $\mathcal{L}_k(q^{-1})$ must be stable, the filter $\mathcal{M}_1(q^{-1})$ need not be stable, since it works within the inner feedback loop of Fig. 1.) An iterative design must bee performed when $Z_t\tilde{h}_{t|t-1}$ cannot be neglected, see Section 5.

4. LEARNING FILTER OPTIMIZATION

The criterion (6) for $t \to \infty$ could be minimized directly by adjusting $\mathcal{L}_k(q^{-1})$, if $\mathcal{H}(q^{-1})$ in (4) and the properties of η_t were known exactly. Using a polynomial approach to Wiener filtering, the learning filter is here designed under the constraint of stability, and under the following assumptions.

Assumption A1: The sequence $\{\varphi_t^*\}$ is stationary and known, with $\mathbf{R}$ known and nonsingular □
Assumption A2: The gradient noise η_t is white and stationary with zero mean and known covariance matrix $\mathbf{R}_\eta$. The correlation of η_t with h_{t-i} and with $\hat{h}_{t-i|t-i-1}, i \geq 0$ is negligible. □
Assumption A3: The time-varying parameters are described by a known vector-ARIMA process

$$\boldsymbol{D}(q^{-1})h_t = \boldsymbol{C}(q^{-1})e_t \ , \tag{17}$$

with $\mathbf{R}_e = \mathrm{E}\, e_t e_t^*$ nonsingular, where $\boldsymbol{D}(q^{-1}) = D_u(q^{-1})\boldsymbol{D}_s(q^{-1})$. Moreover, $\boldsymbol{C}(q^{-1})$ and $\boldsymbol{D}_s(q^{-1})$ are monic and stably invertible, while the polynomial $D_u(q^{-1})$ has zeros on the unit circle □

Assumption A3 implies that e.g. random walks, integrated random walks and filtered random walk models can be considered, but that the unstable dynamics $D_u(q^{-1})$ must then affect all the elements of h_t. We can now present the optimal learning filter.

Theorem 1: Under Assumptions A1-A3, the stable and causal learning filter minimizing the asymptotic parameter covariance matrix (6) is

$$\hat{h}_{t+k|t} = \mathcal{L}_k^{opt} f_t = \boldsymbol{D}_s^{-1}\boldsymbol{Q}_k\beta^{-1}\boldsymbol{D}_s\mathbf{R}^{-1}f_t \ , \tag{18}$$

where the polynomial matrix $\beta(q^{-1})$ of dimension $n_h|n_h$ and degree $n_\beta = \max(n_C, n_D)$ is the stable left spectral factor obtained from

$$\beta\beta_* = \boldsymbol{C}\mathbf{R}_e\boldsymbol{C}_* + \boldsymbol{D}\mathbf{R}^{-1}\mathbf{R}_\eta\mathbf{R}^{-1}\boldsymbol{D}_* \ . \tag{19}$$

The unique solution to the Diophantine equation

$$q^k\boldsymbol{C}\mathbf{R}_e\boldsymbol{C}_* = \boldsymbol{Q}_k\beta_* + q\boldsymbol{D}\boldsymbol{L}_{k*} \tag{20}$$

provides polynomial matrices $\boldsymbol{Q}_k(q^{-1})$ and $\boldsymbol{L}_{k*}(q)$ of dimension $n_h|n_h$, with generic degrees

$$n_Q = \max(n_C - k, n_D - 1) \ , \ \ n_L = \max(n_C + k, n_\beta) - 1 \tag{21}$$

respectively. The estimation error $\tilde{h}_{t+k|t}$ will be stationary with finite covariance matrix and zero mean.
Proof: See Sternad *et.al.* (2000), where a generalization to colored gradient noise is also presented □

Under Assumptions A1-A3, the (generalized) innovations model of $f_t = \mathbf{R}h_t + \eta_t$ can be expressed as

$$f_t = \mathbf{R}\boldsymbol{D}^{-1}(q^{-1})\beta(q^{-1})\epsilon_t \tag{22}$$

where ϵ_t is the white zero mean innovation sequence with unit covariance matrix. By defining the signal

$$\bar{\epsilon}_t \triangleq \frac{1}{D_u(q^{-1})}\epsilon_t = \beta^{-1}(q^{-1})\boldsymbol{D}_s(q^{-1})\mathbf{R}^{-1}f_t \ , \tag{23}$$

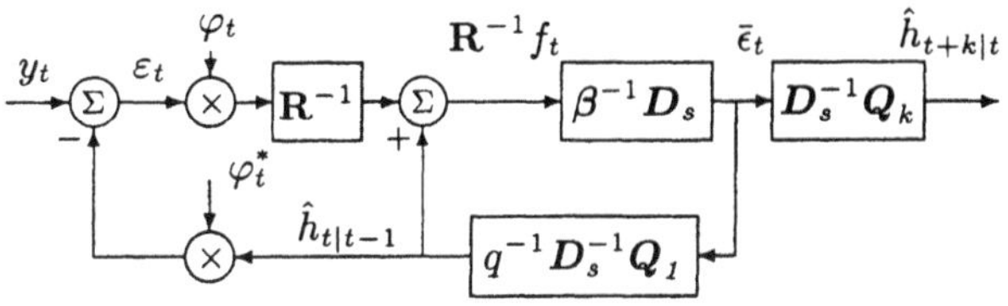

Figure 3: The Wiener optimized tracking algorithm.

the adaptation law (18) can be realized as in Fig. 3. The polynomial matrix $\boldsymbol{Q}_k(q^{-1})$ can be obtained from closed-form expressions, see Sternad *et.al.* (2000). In particular, $\boldsymbol{Q}_1(q^{-1}) = q(\beta(q^{-1}) - \boldsymbol{D}(q^{-1})\beta_0)$, with β_0 being the leading coefficient matrix of $\beta(q^{-1})$.

With this expression and (11),(18), the Wiener optimized filter matrix $\mathcal{M}_k(q^{-1})$ in (8) can for white gradient noise be shown to be given by

$$\mathcal{M}_k^{opt}(q^{-1}) = \boldsymbol{D}^{-1}(q^{-1})\boldsymbol{Q}_k(q^{-1})\beta_0^{-1}\mathbf{R}^{-1} \ . \tag{24}$$

The filter has the inverse regressor covariance matrix $\mathbf{R}^{-1}$ as a right factor. The optimized tracker can thus be seen as a generalization of the LMS-Newton adaptation law (Widrow and Stearns 1985).

5. ITERATIVE WIENER DESIGN

For slow time-variations, the feedback noise $Z_t\tilde{h}_{t|t-1}$ is negligible, so we may perform a one-shot design by assuming $\eta_t = \varphi_t v_t$. Otherwise, the properties of η_t depend on $\mathcal{L}_1(q^{-1})$ via (15). The multiplication by Z_t in (15), see also Fig. 1, acts as a scrambler. For FIR models (1) with white regressors, it will reduce the correlation between the feedback noise and $\tilde{h}_{t|t-1}$. Assumption A2 will then hold for white regressor elements and a design for fast parameter variations can be obtained iteratively. (See Sternad *et.al.* (2000) for a design example.)

1. Design a one-step predictor for slow variations, i.e. use $\mathbf{R}_\eta = \mathrm{E}\{\varphi_t v_t v_t^* \varphi_t^*\}$ to design $\mathcal{L}_1(q^{-1})$. Verify that the closed loop around $\mathcal{L}_1(q^{-1})$ of Fig. 1 is stable. If not, scale up $\mathbf{R}_\eta$ to decrease the gain of $\mathcal{L}_1(q^{-1})$.

2. Based on a simulation of φ_t, v_t, h_t and of $\hat{h}_{t|t-1}$, estimate $\mathbf{R}_\eta$ from $\hat{\eta}_t = \varphi_t\varepsilon_t - \mathbf{R}(h_t - \hat{h}_{t|t-1})$ (see (10),(16)), by using sample averages over $\hat{\eta}_t$.

3. Design a new estimator $\mathcal{L}_1(q^{-1})$.

Repeat 2. and 3. until the difference in consecutive $\hat{h}_{t+1|t}$ becomes small. Then, design $\mathcal{L}_k(q^{-1})$ for lag k.

It will be possible to find an initial stable solution under mild conditions. If $\mathcal{H}(q^{-1})$ is stable, then $\mathcal{L}_1(\omega) \to 0 \ \forall \ \omega$ when the assumed noise power is increased. If Z_t has bounded elements, then the small gain theorem implies that the closed loop of Fig. 1

can be stabilized by assuming a sufficiently high noise power in the design of $\mathcal{L}_1(q^{-1})$.

6. ROBUST WIENER DESIGN

The statistics of time-varying parameters will rarely be exactly known. Small uncertainties can be disregarded, but large uncertainties should be taken into account in a model-based design.

One can then use a gain scheduling approach or a minimax robust design, as investigated in Lindbom *et.al.* 2000b). Another alternative is to optimize the parameter error covariance *on average* over a set of possible dynamics for h_t. Such an approach to the design of robust Wiener filters was originally suggested by Speyer and Gustafsson (1975) The minimization of averaged quadratic criteria has been developed into a systematic design methodology in Sternad and Ahlén (1993), Öhrn *et.al.* (1995) and Öhrn (1996) and it can be applied directly here. Assume that a set of hypermodels is described by the probabilistic *extended design model*

$$h_t = (\mathcal{H}^o(q^{-1}) + \Delta\mathcal{H}(q^{-1}))\, e_t \ . \tag{25}$$

Here, $\mathcal{H}^o(q^{-1})$ is the nominal model, while the error model $\Delta\mathcal{H}(q^{-1})$ represents the set of possible deviations, described by a set of time-invariant transfer functions, parametrized by random coefficients.

Define the operation $\bar{\mathrm{E}}(\cdot)$ as an average over all random coefficients parameterizing the error model. We assume that

$$\bar{\mathrm{E}}\,(\Delta\mathcal{H}(q^{-1})) = 0$$

so the nominal model is defined as being the average model of the set. For a set of models (25), it will be possible to minimize the average of the asymptotic tracking error covariance matrix (6)

$$\bar{\mathbf{P}}_k \triangleq \bar{\mathrm{E}}\left(\lim_{t\to\infty} \mathbf{P}_{k,t}\right) \ , \tag{26}$$

if A3 is substituted by the following assumption:

Assumption A4: The set of hypermodels can be represented by

$$h_t = \frac{1}{\tilde{D}_u(q^{-1})}\,(\mathcal{H}_s^o(q^{-1}) + \Delta\mathcal{H}_s(q^{-1}))\, e_t \ , \tag{27}$$

where the known polynomial $\tilde{D}_u(z^{-1})$ has zeros on $|z| = 1$, $\mathcal{H}_s^o(q^{-1})$ is a known, stable and stably invertible rational matrix and $\Delta\mathcal{H}_s(q^{-1})$ is a set of stable random rational matrices that are independent of e_t, with zero mean and known second order statistics. The noise e_t is white, with nonsingular covariance matrix $\mathbf{R}_e = \mathrm{E}\{e_t e_t^*\}$ □

The known polynomial $\tilde{D}_u(z^{-1})$ must thus include all marginally stable factors of transfer function denominators appearing in the set (25). If the marginally stable modes are unknown, there exists no single learning filter which can provide a finite average covariance matrix $\bar{\mathbf{P}}_k$.

Under Assumptions A1,A2 and A4, a robust design can be obtained by a modification of Theorem 1.[1] The key modification of the derivation in Sternad *et.al.* (2000), is the use of an averaged measurement spectrum

$$\bar{\mathrm{E}}\,(\phi_f) \triangleq \bar{\mathrm{E}}\,(\mathbf{R}\,(\mathcal{H}^o + \Delta\mathcal{H})\,\mathbf{R}_e\,(\mathcal{H}^o + \Delta\mathcal{H})_*\,\mathbf{R}) + \mathbf{R}_\eta \tag{28}$$

Introduce the *averaged hypermodel* $\tilde{\boldsymbol{D}}^{-1}(q^{-1})\tilde{\boldsymbol{C}}(q^{-1})$, with $\tilde{\boldsymbol{D}}(q^{-1}) = \tilde{D}_u(q^{-1})\tilde{\boldsymbol{D}}_s(q^{-1})$ where $\tilde{\boldsymbol{D}}_s(q^{-1})$ and $\tilde{\boldsymbol{C}}(q^{-1})$ are assumed stable. It is a (generalized) innovation description, having spectral density equal to that of the average of the set (25),

$$\tilde{\boldsymbol{D}}^{-1}\tilde{\boldsymbol{C}}\tilde{\boldsymbol{C}}_*\tilde{\boldsymbol{D}}_*^{-1} \triangleq \bar{\mathrm{E}}\,(\mathcal{H}^o + \Delta\mathcal{H})\,\mathbf{R}_e\,(\mathcal{H}^o + \Delta\mathcal{H})_*$$
$$= \frac{1}{\tilde{D}_u}\,(\mathcal{H}_s^o\mathbf{R}_e\mathcal{H}_{s*}^o + \bar{\mathrm{E}}\,(\Delta\mathcal{H}_s\mathbf{R}_e\Delta\mathcal{H}_{s*}))\,\frac{1}{\tilde{D}_{u*}} \ . \tag{29}$$

By defining an averaged spectral factorization

$$\tilde{\beta}\tilde{\beta}_* = \tilde{\boldsymbol{C}}\tilde{\boldsymbol{C}}_* + \tilde{\boldsymbol{D}}\mathbf{R}^{-1}\mathbf{R}_\eta\mathbf{R}^{-1}\tilde{\boldsymbol{D}}_* \ , \tag{30}$$

the averaged measurement spectrum becomes

$$\bar{\mathrm{E}}\,(\phi_f) = \mathbf{R}\tilde{\boldsymbol{D}}^{-1}\tilde{\beta}\tilde{\beta}_*\tilde{\boldsymbol{D}}_*^{-1}\mathbf{R} \ .$$

The square polynomial matrix $\tilde{\beta}(q^{-1})$ is monic and $\det\tilde{\beta}(z^{-1})$ is stable since the right hand sides of (28)-(30) are nonsingular on the unit circle, due to the assumed non-singularity of $\mathbf{R}_e$ and stability of $\boldsymbol{C}(q^{-1})$. The Diophantine equation (20) is in a similar way modified by substituting $\tilde{\boldsymbol{C}}(q^{-1})\tilde{\boldsymbol{C}}_*(q)$ for $\boldsymbol{C}(q^{-1})\mathbf{R}_e\boldsymbol{C}_*(q)$ and $\tilde{\boldsymbol{D}}(q^{-1})$ for $\boldsymbol{D}(q^{-1})$:

$$q^k\tilde{\boldsymbol{C}}\tilde{\boldsymbol{C}}_* = \tilde{\boldsymbol{Q}}_k\tilde{\beta}_* + q\tilde{\boldsymbol{D}}\tilde{\boldsymbol{L}}_{k*} \ . \tag{31}$$

The averaged robust design is summarized as follows:

Theorem 2: For the model set (25), with specified second order moments, a learning filter can be designed under Assumptions A1, A2 and A4 which minimizes the average covariance matrix (26), by obtaining a polynomial spectral factor $\tilde{\beta}(q^{-1})$ from (30) and a polynomial matrix $\tilde{\boldsymbol{Q}}_k(q^{-1})$, together with $\tilde{\boldsymbol{L}}_{k*}(q)$, as the unique solution of the Diophantine equation (31). The robust learning filter is then given by

$$\mathcal{L}_k^{rob} = \tilde{\boldsymbol{D}}_s^{-1}\tilde{\boldsymbol{Q}}_k\tilde{\beta}^{-1}\tilde{\boldsymbol{D}}_s\mathbf{R}^{-1} \tag{32}$$

□

At frequencies where $\bar{\mathrm{E}}\,(\Delta\mathcal{H}(e^{-j\omega})\mathbf{R}_e\Delta\mathcal{H}_*(e^{j\omega}))$ is

[1] More general cases are covered by Theorem 4.1 in Öhrn (1996) which take into account also uncertainty in the transducer, here corresponding to $\mathbf{R}$, and in the noise model.

significant as compared to the nominal spectral density $\mathcal{H}^o(e^{-j\omega})\mathbf{R}_e\mathcal{H}^o_*(e^{j\omega})$, the averaged hypermodel (29) will have higher gain than the nominal hypermodel. At such frequencies, the principal gains of the robust learning filter (32) will be larger than the principal gains of the nominal filter (18), since the average signal-to-noise ratio is higher than the nominal SNR. Note that only second order moments

$$\bar{\mathrm{E}}\,(\Delta\mathcal{H}_s(q^{-1})\mathbf{R}_e\Delta\mathcal{H}_{s*}(q))$$

need to be specified, since the type of distribution, and higher order moments, will not affect the design. The required function on the right-hand side of (29) can be obtained by *averaging over hypermodels*: Draw n samples from the set of their stable parts $\{\mathcal{H}_{s_i} = \mathcal{H}^o_{s_i} + \Delta\mathcal{H}_{s_i}\}_{i=1}^n$. The corresponding spectra (or covariance functions) can then be averaged to obtain (28),(29).

7. EXAMPLE: TRACKING OF IS-136 CHANNELS WITH UNCERTAIN DOPPLER ESTIMATES

In IS-136 mobile radio systems, the symbol-spaced sampled baseband channels can be described by FIR filters with one or two time-varying taps

$$y_t = h_{0,t}u_t + h_{1,t}u_{t-1} + v_t = \varphi_t^* h_t + v_t \ , \qquad (33)$$

where y_t is the received scalar complex-valued signal while v_t is noise and co-channel interference. The transmitted symbols u_t are here assumed known, although they would in reality partly be estimated by the receiver. They have variance σ_u^2 and are mutually uncorrelated, so $\mathbf{R} = \sigma_u^2\mathbf{I}_2$ is known exactly. For a mobile terminal, the channel coefficients $h_{0,t}$ and $h_{1,t}$ will be subject to fading characterized by the maximum Doppler frequency $\omega_D = 2\pi v_o/\lambda$, where v_o denotes the speed of the mobile and λ is the carrier wavelength, which in the following is assumed to be 16cm (~1900 MHz). For the purpose of our investigation, we shall use Jakes' (1974) fading model, which assumes an infinite number of nearby scatterers and is parametrized by ω_D. When the vehicle velocity is constant, the channel coefficient vector $h_t = (h_{0,t}\ h_{1,t})^T$ will then be a stationary, complex circular Gaussian process with zero mean and covariance function

$$r_h(\ell) = \mathrm{E}\{h_t h_{t-\ell}^*\} = \mathbf{R}_h J_0(\Omega_D\,\ell) \quad \ell = 0, \pm 1 \cdots \ . \qquad (34)$$

Here, $\mathbf{R}_h \triangleq \mathrm{E}\{h_t h_t^*\}$, $J_0(\cdot)$ denotes the Bessel function of the first kind and zero order and $\Omega = \omega T$, $\Omega_D = \omega_D T$. The symbol time T is 41.15μs in IS-136. This yields the classical Rayleigh fading spectrum

$$\phi_h(\Omega) = \begin{cases} \frac{2}{\sqrt{\Omega_D^2 - \Omega^2}}\mathbf{R}_h & |\Omega| < \Omega_D \\ 0 & |\Omega| > \Omega_D \ . \end{cases} \qquad (35)$$

When Ω_D is known, the model (4), (17) can be adjusted to the autocorrelation function (34). Perfect adjustment would require models of infinite degree, but good performance can be obtained with simple models. In the following, third order autoregressive models (AR$_3$ models) $(1/D(q^{-1}))\mathbf{I}_2$ will be fitted to the relevant covariance function.

The maximum normalized Doppler frequency Ω_D can be estimated from data, but such estimates will be imperfect. We therefore here investigate the design of an algorithm for tracking h_t that is robustified against uncertainties in the Doppler estimate. The Doppler spectrum averaged with respect to the uncertain Ω_D constitutes our averaged model (29):

$$\tilde{\phi}_h(\Omega) \triangleq \mathrm{E}_{\Omega_D}(\phi_h(\Omega)) = \int_{-\pi}^{\pi} \phi_h(\Omega)p(\Omega_D)\,d\Omega_D \ . \qquad (36)$$

Here, $p(\Omega_D)$ denotes the assumed probability density function of the normalized maximum Doppler frequency Ω_D. When assuming Jakes model giving (34), the covariance function corresponding to (36) is

$$\tilde{r}_h(\ell) \triangleq \int_{-\pi}^{\pi} \mathbf{R}_h J_0(\Omega_D\,\ell)p(\Omega_D)\,d\Omega_D \ . \qquad (37)$$

In Figure 4, an element of the averaged covariance function (37) is displayed for a uniformly distributed probability density function, with different uncertainty regions. A wider uncertainty region will increase the damping of the averaged covariance function, yielding a spectrum with a less pronounced peak.

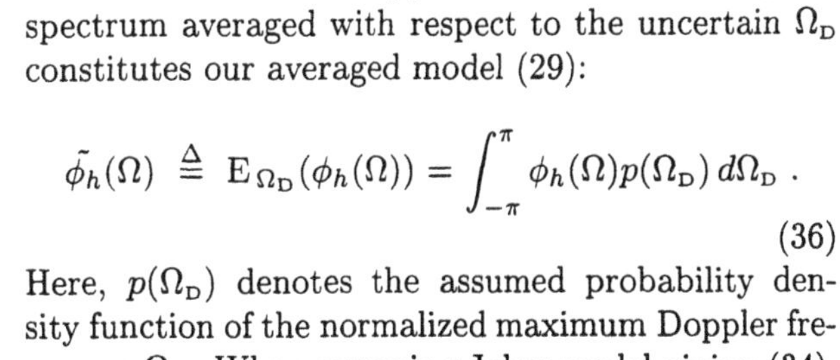

Figure 4: Auto-covariance function $r_h(\ell) = J_0(\Omega_D\ell)$ with $\Omega_D = 0.02$ (solid) and the averaged covariance function (37), with $\Omega_D \in U[0.01\ \ 0.03]$ (dotted) and $\Omega_D \in U[0.015\ \ 0.025]$ (dashed).

A (stable) averaged AR model can now be adjusted to $\tilde{r}_h(\ell)$ (Lindbom *et.al.* 2000b) and a robust tracking algorithm is then designed, as outlined in Section 6, using this model[2]. Assume the signal-to-noise ratio to be 15dB and let $\mathbf{R}_h = \mathbf{I}_2$. Use a nominal $\Omega_D^o = 0.02$, while the true Doppler frequency is varied. A nominal Wiener design is now performed by adjusting an AR$_3$ model to (34) and then using Theorem 1 once, to design an algorithm that minimizes the one-step prediction error. (Iterations are not needed in this case.) Then, robust designs are performed based on

[2]In the work Lin *et. al.* (1995), a windowed LS algorithm was designed to take uncertainties about the Doppler frequency into account. The averaged covariance function (37) was there utilized in the choice of the adaptation window.

AR_3 models adjusted to (37).

The effect of using the robust designs is presented in Figure 5, where it is compared to the nominal design. We also compare to a perfectly matched AR_3 model-based design, based on the correct Doppler frequency. To compute the one-step prediction tracking MSE tr $\mathbf{P}_1$, we use a novel analytical expression which is *exactly* valid for two-tap FIR channels with white inputs, and which gives very good approximations for higher order FIR models. It is derived in Ahlén *et.al.* (2000), and utilized extensively also in Lindbom *et.al.* (2000b). The averaged AR_3 model is matched to the covariance functions described by the dotted and the dashed lines in Figure 4, for lags < 200. The left hand diagram of Figure 5 reflects the performance when the uncertainty level of Ω_D is assumed moderate: $\Omega_D \in U[0.015\ \ 0.025]$, whereas the right hand diagram displays the performance for a larger uncertainty interval, $\Omega_D \in U[0.01\ \ 0.03]$.

When the Doppler frequency is known, the Wiener design (dash-dotted) is much superior to LMS tracking by (9) (dotted). Its performance is almost equal to that of a time-varying Kalman predictor designed for the same AR_3 model. For uncertain Ω_D the averaged robust design improves both the worst-case performance and the average performance significantly. The average MSE performance (area under solid line) is 33% higher with an averaged robust design, than for a known Ω_D (area under dash-dotted line). It would be 80% higher for the nominal design (dashed). The effect is significant also for the more moderate uncertainties (left-hand diagram).

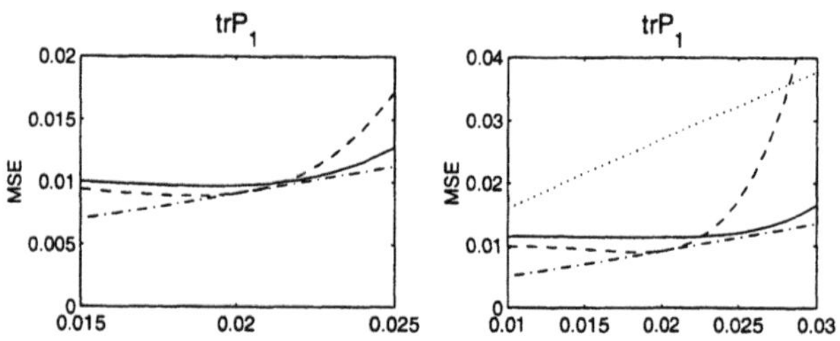

Figure 5: One-step prediction MSE performance of Wiener designs based on AR_3 fading models: Averaged robust (solid), nominal design for $\Omega_D = 0.02$ (dashed) and optimally matched to known model (dash-dotted) . The assumed uncertainty $\Omega_D \in U[0.015\ \ 0.025]$ (left) and $\Omega_D \in U[0.01\ \ 0.03]$ (right). Also shown is an LMS design tuned for a known Ω_D (upper dotted in right figure).

As a generalization of the design above, deviations from the idealized Jakes' model can be introduced. They can be regarded as unstructured uncertainty, which could also be incorporated in an averaged robust design, using methods described in Sternad and Ahlén (1993) and Öhrn (1996).

REFERENCES

Ahlén, A., L.Lindbom and M.Sternad (2000). Tracking of time-varying systems, Part II: Analysis of stability and performance of adaptation algorithms with time-invariant gains. Report R002, Signals and Systems, Uppsala Univ. www.signal.uu.se/Publications/abstracts/r002.html

Benveniste, A., M.Métivier and P.Priouret (1990). *Adaptive Algorithms and Stochastic Approximations.* Springer.

Jakes, W.C. (1974). Multipath Interference. In W. C. Jakes ed. *Microwave Mobile Communications.* Wiley, New York, NY. Reissued by IEEE.

Lindbom L. (1995). *A Wiener filtering approach to the design of tracking algorithms.* PhD Thesis, Dept. of Technology, Uppsala University, Sweden.

Lindbom L., M. Sternad and A. Ahlén (2000a). Tracking of time-varying mobile radio channels. Part I: The Wiener LMS algorithm. To appear in *IEEE Trans. Com.* www.signal.uu.se/Publications/abstracts/r003.html

Lindbom, L., A.Ahlén M.Sternad and M. Falkenström (2000b). Tracking of time-varying mobile radio channels. Part II: A case study. To appear in *IEEE Trans. Com..* www.signal.uu.se/Publications/abstracts/r004.html

Macchi, O. (1995). *Adaptive Processing: The Least Mean Squares Approach with Applications in Transmission.* Wiley, New York.

Lin, J., J. G. Proakis, F. Ling and H. Lev-Ari (1995). Optimal channel tracking of time-varying channels: A frequency domain approach for known and new algorithms. *IEEE Trans. Select. Areas Commun.*, **vol. 13**, 141-154.

Speyer J.L. and D. E. Gustafson (1975). An approximation method for estimation in linear systems with parameter uncertainty. *IEEE Transactions on Automatic Control*, **vol. 20**, pp. 354-359, 1975.

Sternad M. and A. Ahlén (1993). Robust filtering and feedforward control based on probabilistic descriptions of model errors. *Automatica*, **vol 29**, pp. 661-679.

Sternad, M., L.Lindbom, and A.Ahlén (2000). Tracking of time-varying systems. Part I: Wiener design of algorithms with constant gains. Submitted. www.signal.uu.se/Publications/abstracts/r001.html

Widrow B. and S. D. Stearns (1985) *Adaptive Signal Processing*, Prentice Hall, Englewood Cliffs, NJ.

Öhrn K., A. Ahlén and M. Sternad (1995). A probabilistic approach to multivariable robust filtering and open-loop control. *IEEE Trans. Aut. Contr.*, **vol. 40**, pp. 405-417.

Öhrn, K. (1996) *Design of Multivariable Cautious Wiener Filters.* PhD Thesis Dept. Tech, Uppsala Univ., Sweden.

ON BIAS COMPENSATED RECURSIVE LEAST-SQUARES ALGORITHM FOR FIR ADAPTIVE FILTERING

L-J JIA, C-Z JIN, K. WADA

*Department of Electrical and Electronic System Engineering,
Kyushu University, Fukuoka, 812-8581 JAPAN
Tel: +81-92-642-3958; Fax:+81-92-642-3939;
E-mail: {jia, czjin, wada} @dickie.ees.kyushu-u.ac.jp*

Abstract: It is known that recursive least squares (RLS) algorithm or least mean square (LMS) algorithm leads to biased FIR filter coefficients in the presence of input and output noise. In this paper, a new type of bias compensated recursive least squares (BCRLS) algorithm is proposed to produce consistent results for adaptive FIR filtering in the input and output noise case. The proposed algorithm introduces an auxiliary estimator to estimate unknown input noise variance. Owing to this, the bias of the RLS solution due to the input noise can be compensated to yield the consistent filter coefficients. Computational simulations indicate that the proposed algorithm is very efficient. *Copyright ©2001 IFAC*

Keywords: Linear systems, adaptive filters, recursive least squares algorithm, noisy systems, system identification, bias compensation

1. INTRODUCTION

Adaptive filtering, in which adjustable parameters are used, is widely applied in many fields such as control, signal processing and system identification (Haykin, 1991). Adaptive finite impulse response (FIR) filters are always used in adaptive filtering owing to their stability associated with adaptive infinite impulse response (IIR) filters (Haykin, 1991). The object of this paper is to investigate an efficient adaptive FIR filtering algorithm applied in system identification from measured input and output. Least mean squares (LMS) algorithm, which is based on the minimum mean squared error (MMSE) criterion, and recursive least squares (RLS) algorithm which is based on the least squares error criterion, are two widely used classes of adaptive filtering algorithms. It is known that if input measurement is noise-free, the LMS algorithm or the RLS algorithm can provide unbiased filter coefficients (Widrow and Stearns, 1989). However, this condition is not satisfied in some practical situations. When both input and output are corrupted by additive noise, the LMS solution and the RLS solution are biased.

Total least squares (TLS) method is an unbiased estimation method when both input and output are corrupted by noises. Recently, several TLS-based algorithms have been proposed for FIR adaptive filtering in this noise environment (Davila, 1994; Gao, Ahmad and Swamy, 1994; Feng, Bao and Jiao, 1998). More recently, under the assumption that the noise-free input is also a white Gaussian random process independent of input and output noise, the modified LMS (MLMS) (So. 1999) and the modified RLS (MRLS) (Feng, Bao, and Zhang, 2000) were proposed to give consistent FIR filtering coefficients with both input and output noise.

The bias compensated least squares (BCLS) method was proposed by Sagara and Wada (1977) and was further extended by Wada, Eguchi and Sagara (1990) to the input and output noise case. In this paper, an efficient algorithm which is derived based on the BCLS approach, is proposed to provide consistent coefficients for FIR adaptive filtering with both input and output noise. Theoretical analysis illustrates that the bias of RLS estimator is caused only by the input noise whereas the output noise does not affect the estimate. Then by compensating the bias of the RLS estimator with the estimated input noise variance, the proposed algorithm called bias compensated recursive least squares (BCRLS) algorithm is obtained.

The proposed algorithm consists of three major steps. Firstly, using the classical RLS algorithm the RLS estimates of adaptive FIR filter coefficients are obtained. The obtained estimates are biased and the next work is to find a solution to eliminate the bias of the RLS estimates. Secondly, as mentioned in above, the key is to obtain the estimate of the variance of input noise. In the proposed algorithm an auxiliary estimator which is similar to RLS estimator is introduced to estimate the input noise variance. Finally, by compensating the bias of the RLS estimator with the estimated input noise variance, consistent BCRLS estimate is obtained.

Computer simulations are carried on in two cases that noise-free input signal is white process and colored process. Simulation results illustrate that the proposed algorithm is consistent in both cases. It is shown that the proposed algorithm can be applied more widely than the other algorithms because the assumption of input signal is not strict. Comparison with several other algorithms is also done to show that the proposed algorithm is more accurate.

The paper is organized as follows. In the next section, the problem statement is presented and the RLS filter bias resulting from the presence of additive input noise is analyzed. In section 3, the BCRLS algorithm for FIR adaptive filtering in the presence of input and output noise is derived. The simulation results are presented in section 4 and finally section 5 summarizes the conclusion.

2. STATEMENT OF PROBLEM

Consider a system identification problem illustrated in Figure 1. An unknown FIR system $H(z)$ with L-point impulse response vector $\boldsymbol{h}$ has an input $s(k)$ and an output $d(k)$. $x(k)$ and $y(k)$ are measured input and output corrupted by additive noise respectively. Our task is to estimate the parameter of $H(z)$ using adaptive FIR filter $W(z)$ from mearured input $x(k)$ and output $y(k)$. Then the following relationships are given:

$$H(z^{-1}) = \sum_{i=0}^{L-1} h_i z^{-i} \tag{1}$$

$$W(z^{-1}) = \sum_{i=0}^{L-1} w_i z^{-i} \tag{2}$$

$$x(k) = s(k) + n_i(k) \tag{3}$$

$$y(k) = d(k) + n_o(k) \tag{4}$$

$$d(k) = \sum_{i=0}^{L-1} h_i s(k-i) \tag{5}$$

$$r(k) = \sum_{i=0}^{L-1} w_i x(k-i) \tag{6}$$

where input noise $n_i(k)$ and output noise $n_o(k)$ are assumed to be zero-mean white noise processes, independent of input $s(k)$ with variances σ_i^2 and σ_o^2 respectively.

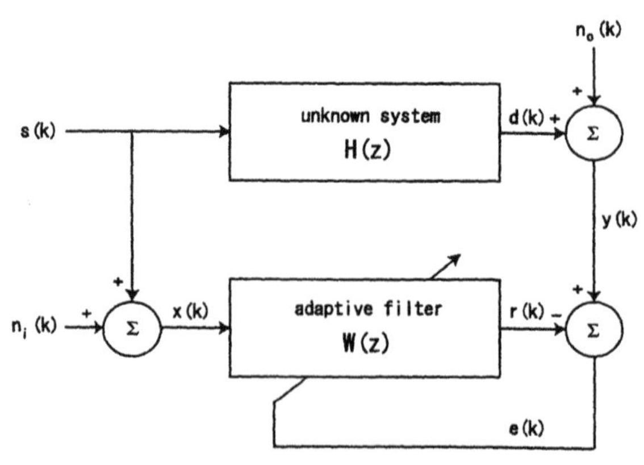

Figure 1: System block diagram for system identification

Let

$$\begin{aligned}
\boldsymbol{h} &= [h_0 h_1 \ldots h_{L-1}]^T \\
\boldsymbol{y}(k) &= [y(k)\, y(k-1) \ldots y(k-L+1)]^T \\
\boldsymbol{x}(k) &= [x(k)\, x(k-1) \ldots x(k-L+1)]^T \\
\boldsymbol{n}_i(k) &= [n_i(k)\, n_i(k-1) \ldots n_i(k-L+1)]^T \\
\boldsymbol{w} &= [w_0 w_1 \ldots w_{L-1}]^T,
\end{aligned}$$

then by equations (3) and (5), equation (4) can be written as:

$$y(k) = \boldsymbol{x}(k)^T \boldsymbol{h} + v(k) \tag{7}$$

where

$$v(k) = n_o(k) - \boldsymbol{n}_i(k)^T \boldsymbol{h}. \tag{8}$$

The error signal $e(k)$ is given by

$$\begin{aligned}
e(k) &= y(k) - r(k) \\
&= y(k) - \sum_{i=0}^{L-1} w_i x(k-i) \\
&= y(k) - \boldsymbol{x}(k)^T \boldsymbol{w}. \qquad (9)
\end{aligned}$$

Minimization of the cost function $\sum_{k=1}^{N} e(k)^2$ with respect to the impulse response vector $\boldsymbol{w}$ leads to LS estimate $\widehat{\boldsymbol{w}}_{LS}$ as follows

$$\widehat{\boldsymbol{w}}_{LS} = \left(\sum_{k=1}^{N} \boldsymbol{x}(k)\boldsymbol{x}(k)^T \right)^{-1} \sum_{k=1}^{N} \boldsymbol{x}(k) y(k). \tag{10}$$

The RLS estimate is obtained by

$$\widehat{\boldsymbol{w}}_N = \widehat{\boldsymbol{w}}_{N-1} + \frac{P_{N-1}\boldsymbol{x}(N)(y(N) - \boldsymbol{x}(N)^T \widehat{\boldsymbol{w}}_{N-1})}{1 + \boldsymbol{x}(N)^T P_{N-1} \boldsymbol{x}(N)}$$

$$P_N = P_{N-1} - \frac{P_{N-1}\boldsymbol{x}(N)\boldsymbol{x}(N)^T P_{N-1}}{1 + \boldsymbol{x}(N)^T P_{N-1}\boldsymbol{x}(N)} \tag{11}$$

where

$$P_N = \left(\sum_{k=1}^{N} \boldsymbol{x}(k)\boldsymbol{x}(k)^T \right)^{-1}. \tag{12}$$

Substituting equation (7) into equation (10), yields

$$\widehat{\boldsymbol{w}}_{LS} = \boldsymbol{h} + P_N \sum_{k=1}^{N} \boldsymbol{x}(k) v(k) \tag{13}$$

Taking the probability limit, we have

$$\operatorname*{plim}_{N\to\infty} \widehat{\boldsymbol{w}}_{LS} = \boldsymbol{h} - \sigma_i^2 R_{xx}^{-1} \boldsymbol{h} \tag{14}$$

where

$$R_{xx} = \operatorname*{plim}_{N\to\infty} \frac{1}{N} \sum_{k=1}^{N} \boldsymbol{x}(k)\boldsymbol{x}(k)^T = E[\boldsymbol{x}(k)\boldsymbol{x}(k)^T].$$

This equation indicates that the asymptotic bias of the LS estimate is

$$\boldsymbol{b} = -\sigma_i^2 R_{xx}^{-1} \boldsymbol{h}.$$

It is clear that in the case of both input and output noise, the RLS estimate is biased, and the bias of the RLS estimate is caused only by the input noise whereas the output noise does not affect the estimate.

3. BIAS COMPENSATED RECURSIVE LEAST SQUARES ALGORITHM

If we can compensate the RLS estimate $\widehat{\boldsymbol{w}}_{LS}$ for the asymptotic bias $\boldsymbol{b}$, the consisent estimate of $\boldsymbol{h}$ will be obtained.
Because the estimate of $\boldsymbol{b}$ is obtained by

$$\widehat{\boldsymbol{b}}_N = -\sigma_i^2 N P_N \widehat{\boldsymbol{h}}_{N-1}, \tag{15}$$

the BCLS estimate is given by

$$\widehat{\boldsymbol{h}}_N = \widehat{\boldsymbol{w}}_{LS} + N P_N \sigma_i^2 \widehat{\boldsymbol{h}}_{N-1}. \tag{16}$$

We will derive the BCRLS algorithm in the following.

3.1. Estimation of the input noise variance

We have learnt that the key point is to obtain the estimate of the input noise variance. Now we introduce an estimator $\widehat{\varphi}_N$ defined as:

$$\widehat{\varphi}_N = \left(\sum_{k=1}^{N} \boldsymbol{x}(k-1)\boldsymbol{x}(k-1)^T \right)^{-1} \sum_{k=1}^{N} \boldsymbol{x}(k-1) y(k) \tag{17}$$

The recursive algorithm for $\widehat{\varphi}_N$ is given by

$$\widehat{\varphi}_N = \widehat{\varphi}_{N-1} + \frac{Q_{N-1}\boldsymbol{x}(N-1)\left(y(N) - \boldsymbol{x}(N-1)^T \widehat{\varphi}_{N-1}\right)}{1 + \boldsymbol{x}(N-1)^T Q_{N-1} \boldsymbol{x}(N-1)}$$

$$Q_N = Q_{N-1} - \frac{Q_{N-1}\boldsymbol{x}(N-1)\boldsymbol{x}(N-1)^T Q_{N-1}}{1 + \boldsymbol{x}(N-1)^T Q_{N-1} \boldsymbol{x}(N-1)} \tag{18}$$

where

$$Q_N = \left(\sum_{k=1}^{N} \boldsymbol{x}(k-1)\boldsymbol{x}(k-1)^T \right)^{-1}. \tag{19}$$

Consider the LS error described as

$$\varepsilon(k) = y(k) - \boldsymbol{x}(k)^T \widehat{\boldsymbol{w}}_{LS}. \tag{20}$$

From equation (7), we have

$$\varepsilon(k) = (\boldsymbol{h} - \widehat{\boldsymbol{w}}_{LS})^T \boldsymbol{x}(k) + v(k). \tag{21}$$

Moreover, consider the auxiliary error for $\widehat{\varphi}_N$ described as

$$\xi(k) = y(k) - \boldsymbol{x}(k-1)^T \widehat{\varphi}_N. \tag{22}$$

From equations (17) and (22), we have

$$\sum_{k=1}^{N} \boldsymbol{x}(k-1)\xi(k) = \boldsymbol{0}. \tag{23}$$

Now define f_N as

$$f_N = \sum_{k=1}^{N} \varepsilon(k-1)\xi(k). \tag{24}$$

By equations (21), (22) and (23), f_N can be rewritten as

$$\begin{aligned} f_N &= \sum_{k=1}^{N} y(k-1)\xi(k) & (25)\\ &= \sum_{k=1}^{N} v(k-1)\xi(k). & (26)\end{aligned}$$

Thus, we have

$$\operatorname*{plim}_{N\to\infty} \frac{1}{N} f_N = \sigma_i^2 \boldsymbol{h}^T \operatorname*{plim}_{N\to\infty} \widehat{\varphi}_N. \tag{27}$$

According to the above equation, we can calculate the input noise variance estimate as follows:

$$\widehat{\sigma}_i^2 = \frac{f_N}{N\widehat{\boldsymbol{h}}^T \widehat{\varphi}_N} \tag{28}$$

It is shown that if we want to obtain the estimate of input noise variance $\widehat{\sigma}_i^2$, we should firstly calculate the value of f_N. We will derive the recursive algorithm of f_N in the next subsection.

3.2. The recursive algorithm of f_N

In order to derive the recursive form of f_N, firstly define matrix ψ_N as follows:

$$\psi_N = \sum_{k=1}^{N} \begin{bmatrix} -y(k-1) \\ \boldsymbol{x}(k-1) \end{bmatrix} \begin{bmatrix} -y(k) & \boldsymbol{x}(k-1)^T \end{bmatrix}. \tag{29}$$

From equation (17) and (25), we have

$$\psi_N \begin{bmatrix} 1 \\ \widehat{\varphi}_N \end{bmatrix} = \begin{bmatrix} -f_N \\ \boldsymbol{0} \end{bmatrix}. \tag{30}$$

Now let us introduce η_N and $\boldsymbol{s}_N$ defined by

$$\psi_N \begin{bmatrix} 0 \\ \boldsymbol{s}_N \end{bmatrix} = \begin{bmatrix} -\eta_N \\ \boldsymbol{x}(N-1) \end{bmatrix}. \tag{31}$$

It is shown that

$$\boldsymbol{s}_N^T \sum_{k=1}^{N} \boldsymbol{x}(k-1)y(k-1) = \eta_N, \tag{32}$$

$$\left[\sum_{k=1}^{N} \boldsymbol{x}(k-1)\boldsymbol{x}(k-1)^T\right] \boldsymbol{s}_N = \boldsymbol{x}(N-1). \tag{33}$$

Then it yields

$$\begin{aligned} \boldsymbol{s}_N &= Q_N \boldsymbol{x}(N-1) \\ &= \frac{Q_{N-1}\boldsymbol{x}(N-1)}{1+\boldsymbol{x}(N-1)^T Q_{N-1}\boldsymbol{x}(N-1)}. \end{aligned} \tag{34}$$

Introducing

$$\boldsymbol{\kappa}_N = \sum_{k=1}^{N} \boldsymbol{x}(k-1)y(k-1), \tag{35}$$

we have $\eta_N = \boldsymbol{\kappa}_N^T \boldsymbol{s}_N$. The recursive form of $\boldsymbol{\kappa}$ is

$$\boldsymbol{\kappa}_N = \boldsymbol{\kappa}_{N-1} + \boldsymbol{x}(N-1)y(N-1). \tag{36}$$

From the definition of ψ_N, we have

$$\psi_N = \psi_{N-1} + \begin{bmatrix} -y(N-1) \\ \boldsymbol{x}(N-1) \end{bmatrix} \begin{bmatrix} -y(N) & \boldsymbol{x}(N-1)^T \end{bmatrix}. \tag{37}$$

Post multiplying by $[1\ \widehat{\varphi}_{N-1}^T]^T$ and using equation (30) yield

$$\psi_N \begin{bmatrix} 1 \\ \widehat{\varphi}_{N-1} \end{bmatrix} = \begin{bmatrix} f_{N-1} \\ \boldsymbol{0} \end{bmatrix} - \begin{bmatrix} -y(N-1) \\ \boldsymbol{x}(N-1) \end{bmatrix} \left(y(N) - \boldsymbol{x}(N-1)^T \widehat{\varphi}_{N-1}\right). \tag{38}$$

Multiplying equation (31) by $y(N) - \boldsymbol{x}(N-1)^T \widehat{\varphi}_{N-1}$ and adding to the above equation, we can obtain

$$\psi_N \left(\begin{bmatrix} 1 \\ \widehat{\varphi}_{N-1} \end{bmatrix} + \begin{bmatrix} 0 \\ \boldsymbol{s}_N \end{bmatrix} \left(y(N) - \boldsymbol{x}(N-1)^T \widehat{\varphi}_{N-1}\right) \right) = \begin{bmatrix} f_{N-1} + (y(N-1) - \eta_N)\left(y(N) - \boldsymbol{x}(N-1)^T \widehat{\varphi}_{N-1}\right) \\ \boldsymbol{0} \end{bmatrix}. \tag{39}$$

Hence the recursive form of f_N is given by

$$\begin{aligned} f_N = f_{N-1} &+ \left(y(N-1) - \boldsymbol{\kappa}_N^T \boldsymbol{s}_N\right) \\ &\cdot \left(y(N) - \boldsymbol{x}(N-1)^T \widehat{\varphi}_{N-1}\right). \end{aligned} \tag{40}$$

3.3. BCRLS Algorithm

Based on the above discussions, the BCRLS algorithm is summarized as follows:

Step1.
Estimate $\boldsymbol{w}$ with noisy data by the conventional RLS algorithm:

$$\widehat{\boldsymbol{w}}_N = \widehat{\boldsymbol{w}}_{N-1} + \frac{P_{N-1}\boldsymbol{x}(N)\left(y(N) - \boldsymbol{x}(N)^T \widehat{\boldsymbol{w}}_{N-1}\right)}{1 + \boldsymbol{x}(N)^T P_{N-1}\boldsymbol{x}(N)},$$

$$P_N = P_{N-1} - \frac{P_{N-1}\boldsymbol{x}(N)\boldsymbol{x}(N)^T P_{N-1}}{1 + \boldsymbol{x}(N)^T P_{N-1}\boldsymbol{x}(N)}.$$

Step2.

Obtain the estimate of the input noise variance according to the following steps :
2.1 Calculate $\widehat{\varphi}_N$:

$$\widehat{\varphi}_N = \widehat{\varphi}_{N-1} + \frac{Q_{N-1}\boldsymbol{x}(N-1)\left(y(N) - \boldsymbol{x}(N-1)^T\widehat{\varphi}_{N-1}\right)}{1+\boldsymbol{x}(N-1)^TQ_{N-1}\boldsymbol{x}(N-1)}$$

$$Q_N = Q_{N-1} - \frac{Q_{N-1}\boldsymbol{x}(N-1)\boldsymbol{x}(N-1)^TQ_{N-1}}{1+\boldsymbol{x}(N-1)^TQ_{N-1}\boldsymbol{x}(N-1)}.$$

2.2 Calculate f_N:

$$\boldsymbol{\kappa}_N = \boldsymbol{\kappa}_{N-1} + \boldsymbol{x}(N-1)y(N-1),$$

$$\boldsymbol{s}_N = \frac{Q_{N-1}\boldsymbol{x}(N-1)}{1+\boldsymbol{x}(N-1)^TQ_{N-1}\boldsymbol{x}(N-1)},$$

$$f_N = f_{N-1} + \left(y(N-1) - \boldsymbol{\kappa}_N^T\boldsymbol{s}_N\right) \cdot \left(y(N) - \boldsymbol{x}(N-1)^T\widehat{\varphi}_{N-1}\right).$$

2.3 Compute the estimate of input noise variance

$$\widehat{\sigma}_i^2 = \frac{f_N}{N\widehat{\boldsymbol{h}}_{N-1}^T\widehat{\varphi}_N}.$$

Step3.

Calculate the BCRLS estimate via

$$\widehat{\boldsymbol{h}}_N = \widehat{\boldsymbol{w}}_N + N\widehat{\sigma}_i^2P_N\widehat{\boldsymbol{h}}_{N-1}.$$

Step4.

N=N+1 and return to Step 1.

Initial values at $N = 0$ are given as follows:

$$P_0 = \alpha\boldsymbol{I}, \widehat{\boldsymbol{w}}_0 = \boldsymbol{0}, Q_0 = \alpha\boldsymbol{I},$$

$$\widehat{\varphi}_0 = \boldsymbol{0}, f_0 = 0, \boldsymbol{\kappa}_0 = \boldsymbol{0}.$$

4. SIMULATION RESULTS

The unknown FIR system is characterized by $\boldsymbol{h} = [-0.3; -0.9; 0.8; -0.7; 0.6]$.
Computer simulations have been carried out to evaluate the impulse response estimation performance of the BCRLS algorithm in presence of input and output noise in the following two cases respectively.

Case1:
The noise-free input $s(k)$, input noise $n_i(k)$ and output noise $n_o(k)$ are independent white Gaussian random processes. The variance of $s(k)$ is fixed to be unity, and the input and output signal-to-noise ratios, $SNR_i = \sigma_s^2/\sigma_i^2$ and $SNR_o = \sigma_s^2/\sigma_o^2$, are chosen 5dB and 10dB.

Case2:
The noise-free input signal $s(k)$ is the AR process

$$s(k) + 0.9s(k-1) = g(k)$$

where $g(k)$ is a zero mean white Gaussian random variable. The variances of the input noise and output noise are σ_i^2=σ_o^2=0.25.

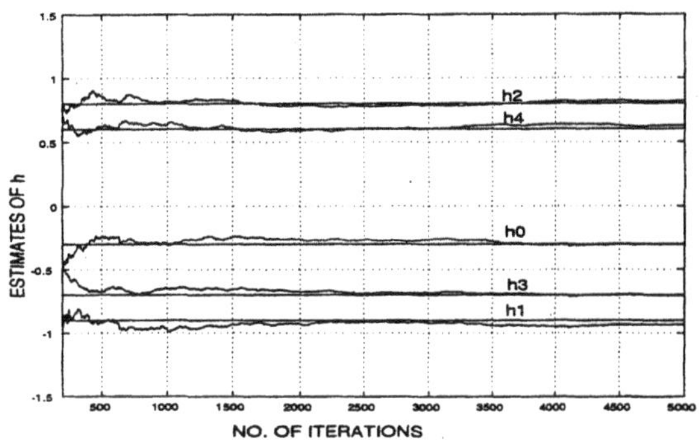

Figure 2: Parameter estimates in Case 1.

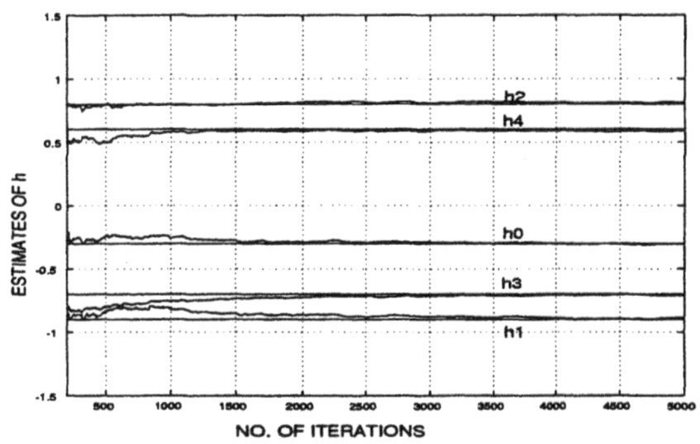

Figure 3: Parameter estimates in Case 2.

Fig.2 and Fig.3 display the results of adaptive FIR filter parameter estimates obtained by BCRLS algorithm in one trial for both cases of input signal. Fig.4 and Fig.5 indicate the mean squared errors (MSE) given by BCRLS algorithm in both cases and these results are averages of 50 independent trials with a data length of 5000. It is shown that the BCRLS algorithm can provide consistent FIR filter coefficients in the presence of input and output noise and the estimation accuracy of BCRLS algorithm is satisfactory in this case.

Comparisons are also made with the MLMS algorithm, the MRLS algorithm and the GAO's TLS

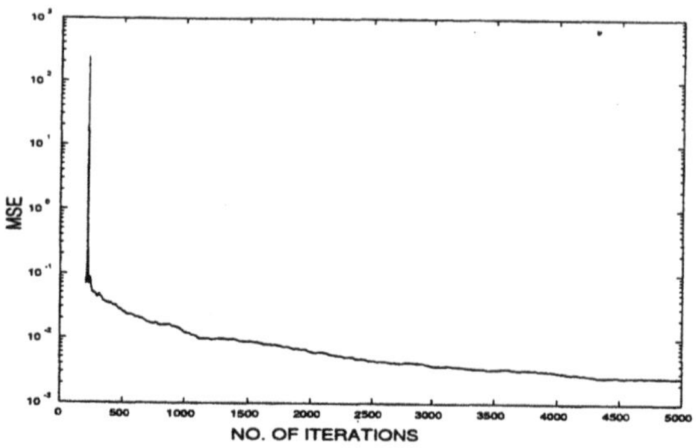

Figure 4: MSE of BCRLS in Case 1.

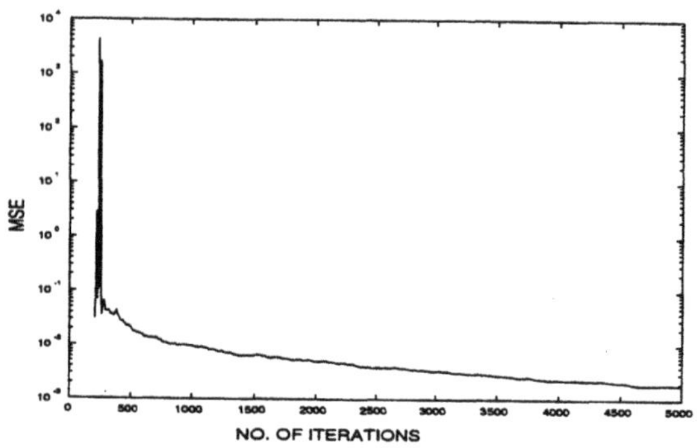

Figure 5: MSE of BCRLS in Case 2.

algorithm (Gao, *et al.*,1994). The MLMS algorithm and MRLS algorithm can give consistent estimate only in case 1 because the derivations of the two algorithms depend on the assumption that $s(k)$, $n_o(k)$ and $n_i(k)$ are independent white Gaussian random processes, while in Case 2, the two algorithms can not give the consistent estimate. Figure 6 illustrates the learning trajectories of MSE of the four algorithms. It is shown that the estimation accuracy of the BCRLS algorithm is superior to the other algorithms.

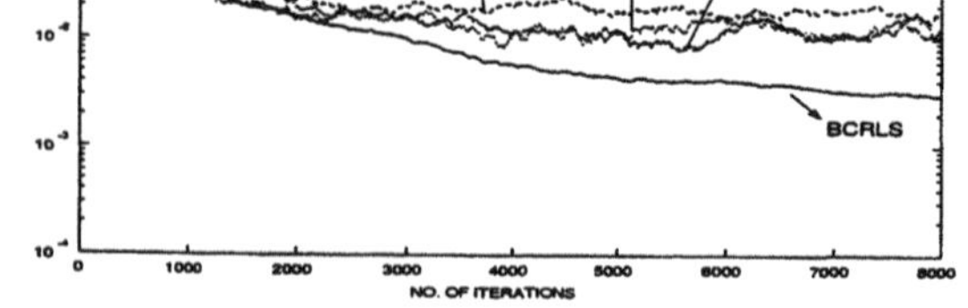

Figure 6: MSE of Four Algorithms in Case 1.

5. CONCLUSIONS

In this paper, we focus on the derivation of the BCRLS algorithm for consistent adaptive FIR filter coefficients in the presence of the input and output noise. It has been shown that when the input vector contains additive noise, the resulting FIR filter coefficients given by RLS algorithm are biased. The BCRLS algorithm, on the other hand, is shown to result in consistent FIR filter coefficients. The idea is to compensate the bias of RLS algorithm with the obtained estimate of input noise variance. By computer simulations, it is illustrated that the proposed BCRLS algorithm gives good results and it can be applied more widely than the other algorithms because the kind of input signal in the BCRLS algorithm is not strict and it's accuracy is better.

REFERENCES

Haykin, S.S. (1991). *Adaptive filter theory* Prentice-Hall, 2nd edn.

Widrow, B. and Stearns, S.D.(1989). *Adaptive signal processing* Prentice-Hall.

Sagara, S. and Wada, K. (1977). On-line modified least-squares parameter estimation of linear discrete dynamic systems, *Int.J.Control,* **25**, No.3, 329-343.

Wada, K., Eguchi, M., Sagara, S. (1990). Estimation of pulse transfer function via bias-compensated least-squares method in the prensence of input and output noise, *Systems Science*, 16, No.3, 57-70.

Davila, C.E.(1994). An efficient recursive total least squares algorithm for FIR adaptive filtering, *IEEE Trans. on SP*, **41**, No.2, 268-280.

Gao,K., Ahmad, M.O., and Swamy, M.N.S. (1994). A constrained anti-Hebbain learning algorithm for total least-squares estimation with applications to adaptive FIR and IIR filtering, *IEEE Trans., on CAS-II*, 41, No.11, 718-729.

Feng, D.Z., Bao, Z., and Jiao, L.C. (1998). Total least mean squares algorithm, *IEEE Trans., on SP*, **46**, No.8, 2122-2130.

So, H.C.(1999). Modified LMS algorithm for unbiased impulse response estimation in nonstationary noise, *Electron. Lett.*, , **35**, No.10, 791-792.

Feng, D.Z., Bao, Z. and Zhang, X.D. (2000). Modified RLS algorithm for unbiased estimation of FIR system with input and output noise, *Electronics Letters* , **36**, No.3, 273-274.

Longitudinal Attitude Control of a Small Aircraft Using Cascade LCN's

T. Caetano T. Loureiro J. M. Lemos†

†Corresponding author

INESC-ID/IST, Rua Alves Redol 9, 1000-029 Lisboa, Portugal,
Fax +351.21.314 58 43, email: jlml@inesc.pt.

Abstract: This paper is concerned with gain scheduling control of the longitudinal attitude of a small aircraft. The parameter variations of the attitude dynamics induced by total velocity changes are compensated by a cascade of local controller networks (LCN's) where pitch rate and pitch are fedback according to the so called Stability Augmentation System. The design procedure is presented and the influence of several options are discussed. It is shown that this scheme allows to achieve a substantially higher speed of response with respect to the situation in which only pitch is used. Simulations are performed with the aid of a detailed physically based model.

Keywords: Gain scheduling, Aerospace, Attitude Control, Computer Control.

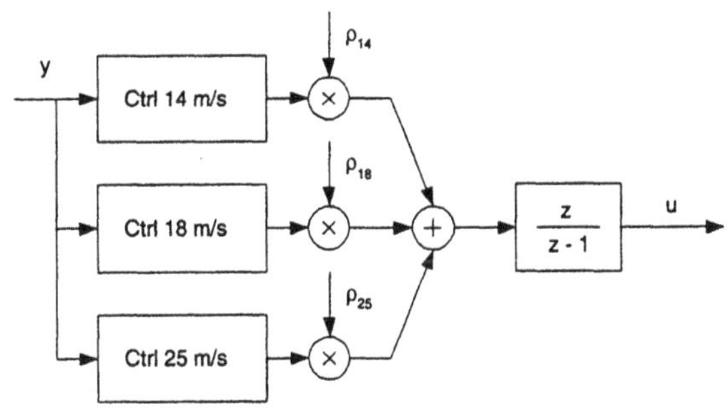

Figure 1: Local Controller Network with series integrator.

1 Introduction

Gain scheduling is by now a well defined field of research, with both significant applications and supporting theory available [10, 13]. It provides an interesting example of a topic born heuristically in the industrial field for which work from the "academic world" is providing solid ground.

Among other possibilities, Local Controller Networks (LCN) is a form of designing gain scheduled controllers which is receiving increasing attention in recent years. It provides a natural and simple methodology of controller design, applicable to a class of nonlinear systems, for which theory supported guidelines are available [6]. The LCN (fig. 1) is made of a bank of linear controllers (named "local controllers") whose outputs are multiplied by weights (named ρ_i in fig. 1) and then added. Each local controller matches the linearised dynamics of the process to be controlled, around each possible operating region. The weights are functions of (one or more) variables, named "indicator variables". The partition of the indicator variables space define the possible operating regions. For each local controller the corresponding weight is close to 1 when well inside the operating region matched by this controller, and close to 0 otherwise. When the operation takes place near the border of two operating regions, the weights are in between 0 and 1 and the control action actually applied to the plant results from the interpolation of the "adjacent" controllers.

Since aircrafts are nonlinear systems which, in general, may not be adequatelly controlled over the whole flight envelope with a single linear controller, they form a natural field of application for gain scheduling techniques [3, 7, 12, 15]. In [3] the Multiple Model Adaptive Control (MMAC) method, which results from an application of the Partitioning approach to Adaptive Control [9] has been considered in relation to the F-8C aircraft. It should be remarked that MMAC has some relation with the LCN method. There are, however, important differences. Among others, LCN (as described in [6] and used hereafter in this paper) relies on input/output models thereby avoiding the robustness problems which may arise from the inclusion in the control loop of a Kalman-Bucy state estimator. Furthermore, the weights of MMAC are adjusted on-line, depending on the observed matching between measured data and the bank of local models. While this provides an advantage for dealing with plant uncertainty, it poses an harder stability problem.

This paper is concerned with the longitudinal attitude control of a small aircraft using cascade LCN's. The aircraft considered has been developed by others [4, 5] with the aim of serving either as a RPV[1] or a UAV[2]. A detailed phisically based SIMULINK model has been used in this work, being described in [4, 5]. In this aircraft, velocity changes induce parameter variations of the longitudinal attitude dynamics which may destabilize the pitch control loop. The use of a cascade of LCN's where both pitch rate and pitch are fedback compensates this effect, allowing to achieve insensitivity of the pitch response with respect to the longitudinal velocity. The design procedure is presented and the influence of several options are discussed.

The contribution of this paper consists of the above case study of application of gain scheduling. It is shown that the cascade of LCN's allows to achieve a substantially higher speed of response with respect to the situation in which only pitch is used, while preserving reasonable uniform response over the whole flight envelope.

The paper is organised as follows: After the introduction (this section) in which related work is briefly reviewed, the problem is formulated (section 2) being shown that longitudinal velocity changes induce parameter variations of the attitude dynamics. The problem is then solved with just one LCN (section 3) and then with a cascade of LCN's (section 4). Comparisons show that a faster control system results in this later case. Section 5 draws conclusions.

2 Problem Formulation

The aim of the work reported herein is the design of a pitch controller for a small unmaned aircraft. The output variable is the pitch angle, θ, and the manipulated variable is the elevator deflection angle, δ_e. The pitch rate measurement, Q, is assumed to be available to the controller [3]. The controller is desired to be valid throughout the whole range of flight velocity. Ideally, the performance of the controlled aircraft should be constant over this flight envelope.

The performance of the controlled system is considered to be acceptable if the pitch loop is such that:

- The rise time is of the order of magnitude of 1 s;
- The relative overshoot is at most 10%;

[1] Unmanned Aircraft Vehicle
[2] Remotelly Piloted Vehicle
[3] The capital letter Q is used for pitch rate in replacement to q in order to avoid confusion with the forward shift operator.

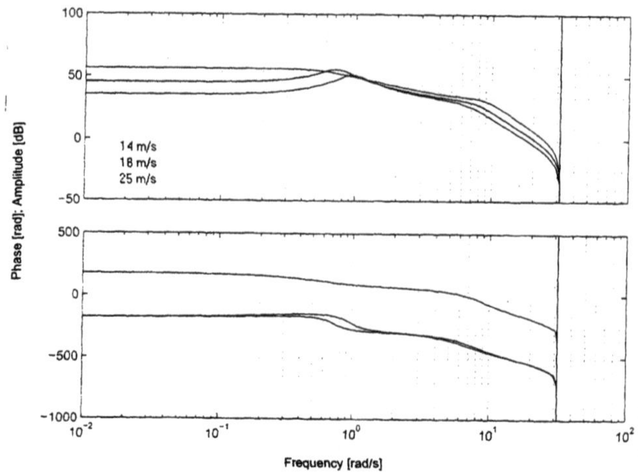

Figure 2: Frequency response of the local models at different total speeds.

- The movements demanded by the controller to the elevator are acceptable from a mechanical point of view.

For the sake of control design, local linear models of ARX type were identified by Least Squares. Here only one scheduling variable is considered, thereby simplifying the choice of the local models. More complicated situations may be dealt systematically by the methods described in [8]. The local models are defined by

$$A(q)y(t) = B(q)u(t) + e(t) \tag{1}$$

where A and B are polynomials in the forward shift operator q, y is the plant output (pitch angle increment with respect to a trimmed situation), u is the manipulated variable (elevator deflection angle increment) and e is a white noise gaussian random process. Following [11], a value of 4 was selected for the degree of A. Since the type of plant considered has no pure delay, the difference of degrees between A and B (equal to the pure delay of the system) was made equal to 1. The degree of B was thus made equal to 3.

According to the procedures of flight testing [1], the sampling rate should be made at least five times the one corresponding to the short period mode. In this case a sampling rate of 10 Hz was used, which is almost 10 times the frequency of the short period mode of the plane considered (given by 1.076 Hz). Another aspect to consider when selecting the sampling rate is the increase in bandwidth which may occur when the loop is closed. In this case it was found that the maximum frequency of the closed-loop poles (which corresponds to the observer poles) is 12 rad/s. Since this corresponds to about 1/5 of the sampling frequency selected above, the selected value was considered to be adequate.

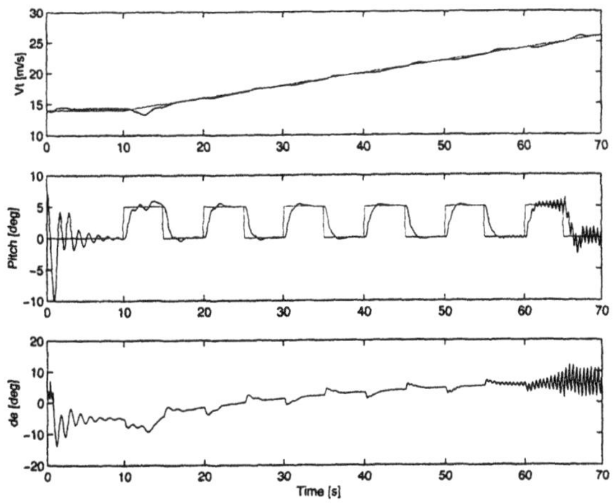

Figure 3: Pitch control in the presence of a total speed increase.

For performing the identification, data has been collected from simulated experiments conducted for horizontal flight at total velocities of $14\,m/s$, $18\,m/s$ and $25\,m/s$. The excitation inputs have been chosen as random signals superimposed to the trimming value corresponding to each velocity and having a significant spectral content over the frequency range of interest.

Fig. 2 shows the frequency response of the local models, *i.e.*, corresponding to the three above mentioned trimming total speeds. As it is apparent, these frequency responses are different. The models for $14\,m/s$ and $18\,m/s$ have a similar frequency response, differing mostly in the low frequency band gain. The model for $25\,m/s$ is noticeably different from the others. At this speed the phugoid mode is much more damped and the region of high gain around the phugoid frequency does not exist. At high frequencies, the shape of the response is similar for all models, but with a gain which is higher the greater the velocity. This is due to the fact that in these regimes the control surfaces have greater effect due to the dynamic pressure, which depends on the speed.

As a consequence of the above differences, when the pitch controller designed for the total speed of $18\,m/s$ is employed at different speeds, undesirable effects may occur. This is shown in fig. 3 where the pitch reference (middle) varies as a square wave when the total velocity (top) increases linearly. As expected, the best response is obtained when the velocity is around $18\,m/s$. For lower velocities the response degrades and overshoot increases. For higher values, ringing starts to appear and the closed loop becomes unstable.

The main issue considered in this paper is the design of a gain scheduling controller which prevents these effects from occurring.

3 Pitch control with LCN.

In order to achieve a constant behaviour of the pitch control loop over the whole velocity range, a LCN is designed. In this section, only the pitch is fedback. The classical solution where both pitch and pitch rate are used by the controller [11] will be considered in the following section (and shown to yield a faster rise time).

The LCN used is shown in fig. 1. It consists of 3 linear controllers designed on the basis of the linear incremental models valid for the velocities of $14\,m/s$, $18\,m/s$ and $25\,m/s$ and considered to be valid, respectively, for small, medium and high total velocities. The outputs of these controllers are multiplied by the weights ρ_{14}, ρ_{18} and ρ_{25}, added, and then passed through a common integrator to form the manipulated variable u actually applied to the elevators.

Each local controller is designed by a standard pole placement polynomial based procedure [2, 6] and has the structure

$$R(q)u(k) = T(q)u_c(k) - S(q)y(k) \tag{2}$$

where u_c is the reference to track and R, T and S are polynomials in the forward shift operator. The nomenclature is the same as in [2]. Since many pole placement problems had to be solved, the procedure was programmed based on the software package described in [14]. The following specifications were made:

- The specified discrete time poles are obtained by the zero order hold sampling of a pair of continuous complex conjugate poles with natural frequency $6\,rad/s$ and damping ratio 0.7 plus a stable real pole at $-3\,rad/s$. This specification stems from the controlled plane specifications.
- The observer polynomial has roots in 0.3. This choice results from a trial and error procedure (see also the considerations below).
- All zeros are cancelled.
- One integrator is forced in the design.

As already mentioned, the integrators of all local controllers are replaced by a common integrator located after the summation point. The presence of this integrator provides not only zero static position error but it also compensates for the off-sets due to different operating regimes (in this example no feedforward action as suggested in [6] was found to be needed). Furthermore, the common integrator

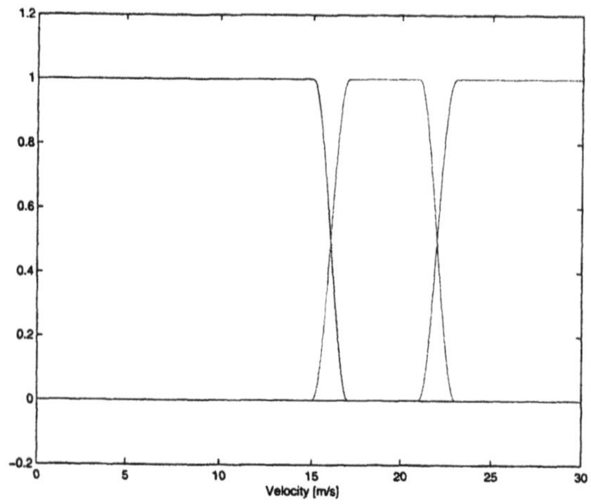

Figure 4: Weight functions.

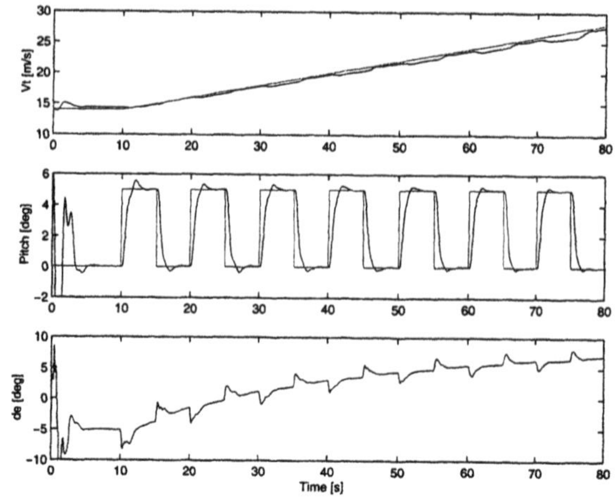

Figure 5: Pitch control with LCN (no pitch rate feedback).

smoths out eventual abrupt changes due to change of operating regime (although the weight functions are designed for avoiding this, as discussed below).

An important aspect is the selection of the observer polynomial. A too fast observer rises robust stability problems. On the other way, a too slow observer leads to an increase of sensitivity in the low frequency band. A trade-off must be made.

The weights are functions of the total speed, taking values between zero and one. In the range of velocities in which a local controller is valid, the corresponding weight should be close to one, having smaller values (eventually zero) otherwise.

As a consequence of the stability analysis presented in [6], the weights (and the corresponding operating regimes) must capture the nonlinearities and be such as to avoid interaction between operating regimes which are not adjacent.

Fig. 4 shows the weight functions which were chosen so as to respect these constraints. The transitions in the values of the weights were made according to a sin^2 function in order to yield a smooth behaviour of the controller when a change of operating regime takes place.

Fig. 5 shows the pitch response to a square wave reference when the velocity is made to vary as in fig. 3. The instability noticeable in fig. 3 has been eliminated by the LCN controller, resulting in more constant performance under the presence of total velocity changes. In fig. 5 the overshoot decreases with an increase in velocity, a behaviour which is consistent with the models identified, for which the phugoid mode is more damped at higher velocities.

Although the performance of the response yielded by LCN with only pitch feedback is acceptable, one would expect to achieve a faster response. This was however not possible to attain by just redesigning the local controllers. For that sake, a different control structure, in which access to the pitch rate is exploited, must be used.

4 Pitch control with cascade LCN's.

A classical solution for designing the pitch controller is the so called Stability Augmentation System (SAS) [11]. This structure consists of a cascade in which the inner loop controls the pitch rate Q by acting on the elevator and the outer loop controls pitch θ by manipulating reference to the inner loop.

Hereafter, the classical SAS structure is considered with a modification consisting in the inclusion of a LCN in each cascade loop. The design proceeds as follows: The open-loop dynamics between the elevator deflection and the pitch rate is first identified. On the basis of the model thereby obtained, a controller is designed by pole placement. With this loop closed, the dynamics between the pitch rate reference and the pitch is identified. The resulting model serves as a basis for controller design. This procedure takes place for the 3 total velocities considered.

According to the short period approximation [11], a *2nd* order model with one zero was identified for expressing the relation between the elevator deflection and the pitch rate. Although it only models the short period mode and does not take into account the phugoid mode, it was found appropriate for designing the pitch rate controller embedded in the SAS.

The dependence of pitch rate dynamics on total speed is shown in fig. 6. Both bandwidth and gain increase with total speed, which again motivates the use of a LCN for the inner loop.

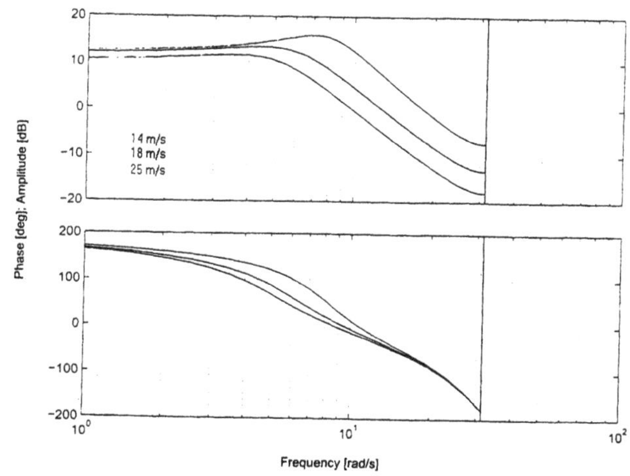

Figure 6: Frequency response of the models for pitch rate at different total speeds.

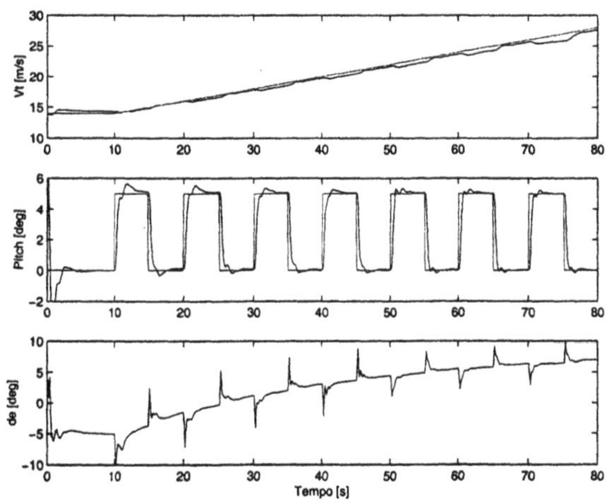

Figure 7: Pitch control with cascade LCN's.

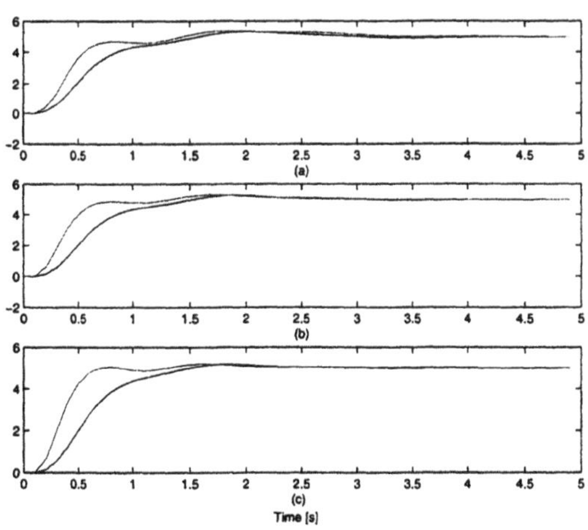

Figure 8: Comparison of the rise times with and without cascade control for: a) 14 m/s, b) 18 m/s, c) 25 m/s.

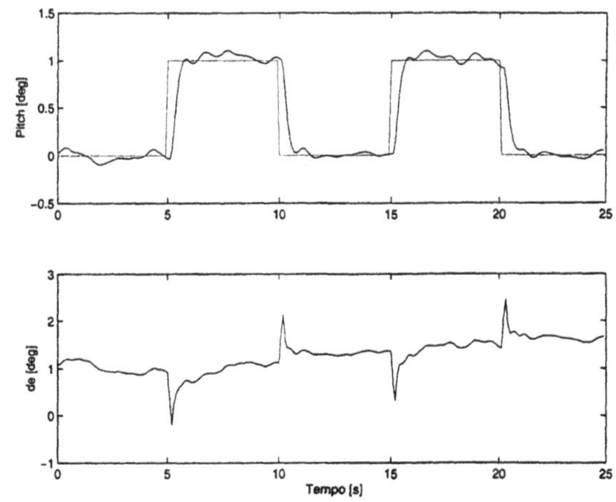

Figure 9: Response of cascade LCN's in the presence of disturbances.

Fig. 7 shows the resulting behaviour of the pitch controller with SAS and cascade LCN's. Comparing with fig. 5, it is seen that SAS results in a net improvement of rise time. See also fig. 8 for details. With SAS the control action becomes more "energetic" but the actuator movements demanded are still within acceptable mechanical limits.

Fig. 9 shows the response obtained at a speed of $18\,m/s$ under the disturbance of a random wind with a standard deviation of $5\,m/s$.

5 Conclusion

A case study on the design of LCN gain scheduling for longitudinal attitude control of a small unmaned aircraft has been presented. Results have been obtained by simulation in a detailed physically based model. Both single loop and SAS control systems were considered. The use of LCN yielded a reasonable constant performance when the total velocity varies over a wide range. Employing SAS results in a substantial improvement of rise time and implies a cascade of LCN's. The main aspects of controller design with LCN's and input/output models were highlighted so this paper provides a guideline for application of this method in other types of plants as well.

Acknowledgement

The authors wish to thank prof. J. R. Azinheira of Instituto Superior Técnico, Lisboa, for providing the SIMULINK model used for the simulations reported in this work. Part of the work reported in this paper was produced within the research project EQUIPA, PRAXIS/C/EEI/10081/1998.

References

[1] AGARDograph (1993) N 160, Vol. 1 (Issue 2), *Basic Principles of Flight Test Instrumentation Engineering.*

[2] Åström, K. J. and B. Wittenmark (1992). *Computer Controlled Systems.* Prentice Hall.

[3] Athans, M., D. Castañon, K.-P. Dunn, C. S. Greene, W. H. Lee, N. R. Sandell, Jr. and A. S. Willsky (1977). The stochastic control of the F-8C aircraft using a multiple model adaptive control (MMAC) method – Part I: Equilibrium Flight. *IEEE Trans. Autom. Control*, AC-22(5):768-780.

[4] Azinheira, J. R., P.M. Lourtie, J. P. Rente (1996). Robust Control and Guidance for a Low Cost Civilian RPV, *XIIth Bristol Int. Conf on RPV's*, Bristol, U. K.

[5] Azinheira, J. R., J. P. Rente and M. G. Kellet (1998). Longitudinal Auto-landing for a Low Cost UAV. *IAV'98*, Madrid, Spain.

[6] Hunt, K. J. and T. A. Johansen (1997). Design and Analysis of gain-scheduling control using local controller networks. *Int. J. Control*, 66(5):619-651.

[7] Hyde, R. A. and K. Glover (1993). The application of Scheduled H_∞ Controllers to a VSTOL Aircraft. *IEEE Trans. Autom. Control*, 38(7):1021-1039.

[8] Johansen, T. A. and B. A. Foss (1995). Identification of non-linear system structure and parameters using regime decomposition. *Automatica*, 31(2):321-326.

[9] Lainiotis, D. (1976). Partitioning: A Unifying Framework for Adaptive Systems, II: Control. *Proc. IEEE*, 64(8):1182-1198.

[10] Leith, D. J. and W. E. Leithead (2000). Survey of gain-scheduling analysis and design. *Int. J. Control*, 73(11):1001-1025.

[11] McLean, D. (1990). *Automatic Flight Control Systems.* Prentice Hall.

[12] Nichols, R. A., R. T. Reichert and W. J. Rugh (1993). Gain Scheduling for H-Infinity Controllers: A Flight Control Example. *IEEE Trans. Control Systems Technology*, 1(2):69-79.

[13] Rugh, W. and J. Shamma (2000). Research on Gain scheduling. *Automatica*, 36(10):1401-1425.

[14] Henrion, D., M. Hromcik, H. Kwakernaak, S. Pejchov, M. Sebek and R. Trijbos (1998). *Polynomial Toolbox Version 1.6 - Manual.* Available at http://www.utwente.nl/polbox.

[15] Wang, J. (1995). Scheduling of local nonlinear control laws by exogenous signals – an application to flight control. *Int. J. Control*, 61(4):759-782.

ON-LINE TUNING OF ONE-MASS MOTOR DRIVE SYSTEM BY SIMPLE ADAPTIVE CONTROL

Takaaki Sekiai *Takeshi Higashiyama *
Hiromitsu Ohmori * Akira Sano * Tsutomu Miyashita **
Hideyuki Nishida ** Yuji Todaka **

* *Department of System Design Engineering, Keio University*
** *Industrial Systems Division, Fuji Electric Corporation*

Abstract: A new auto-tuning algorithm for motor velocity control is presented to adjust the parameters in the feedforward and feedback PI controllers, according to identified physical parameters such as a motor time constant, viscous friction terms proportional to motor velocity and squared velocity, and Coulomb friction. The physical parameters are identified by the proposed adaptive algorithm which appropriately set up a reference model and use its state variables to adjust the feedforward controller parameters. The proposed scheme is robust to disturbance torque, compared to ordinary closed-loop identification using the motor output directly. Its effectiveness is investigated in velocity control experiment using an AC synchronous servo motor. *Copyright©2001 IFAC*

Keywords: nonliner control, adaptive control, PI controllers, motor control, friction

1. INTRODUCTION

Friction is a nonlinear phenomenon which sometimes degrades its control performance (Armstrong *et al.*, 1994). Fine tuning of the controller parameters is essentially needed in motor drive systems to attain high tracking performance and quick recovery response for stepwise torque disturbances (Higashiyama *et al.*, 2000). The aim of this paper is to present a new auto-tuning algorithm for motor velocity control to adjust the parameters in the feedforward and feedback PI controllers, according to identified physical parameters such as a motor time constant, viscous friction terms proportional to motor velocity and squared velocity, and Coulomb friction. These physical parameters are identified in a feedforward reference model which is appropriately designed and set up so that the simple adaptive control approach (Kaufman *et al.*, 1994) may be applied. Comparing to ordinary closed-loop identification directly using the input-output data of motor drive system, the proposed adaptive scheme is robust and stable to stepwise torque disturbances. Stability of the proposed algorithm is also investigated in cases of P and PI feedback controllers, and its effectiveness is also examined in velosity control of an AC synchronous servo motor.

2. TUNING ALGORITHM BASED ON SAC

2.1 *Case with P-Controller*

The state-space description of the controlled system illustrated in Fig.1 is given by

$$\dot{x}_p(t) = a_p x_p(t) + b_p u_p(t) + \boldsymbol{a}_\gamma{}^T \gamma(y_p(t)) \quad (1)$$

$$y_p(t) = c_p x_p(t) \quad (2)$$

where $a_p = -D_M/T_M$, $b_p = 1/T_M$, $c_p = 1$, $\boldsymbol{a}_\gamma = [-a/T_M \; -c/T_M]^T$, $\gamma(y_p(t)) = [|y_p(t)|y_p(t) \; tanh(\alpha y_p(t))]^T$. Physical parameters $T_M[s]$, D_M $[Nm/rad]$, y_p $[rad]$, u_p $[Nm]$ denote the motor time constant, viscous friction, motor speed and input torque, and $\alpha \gg 1$. A specific reference model is employed to identify uncertain motor drive system with frictions. The state-space de-

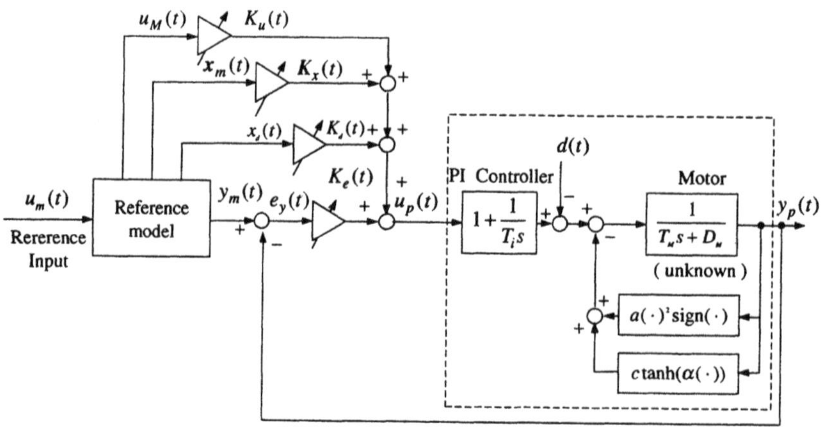

Fig. 1. Controller Structure consisting of feedforward and PI controller

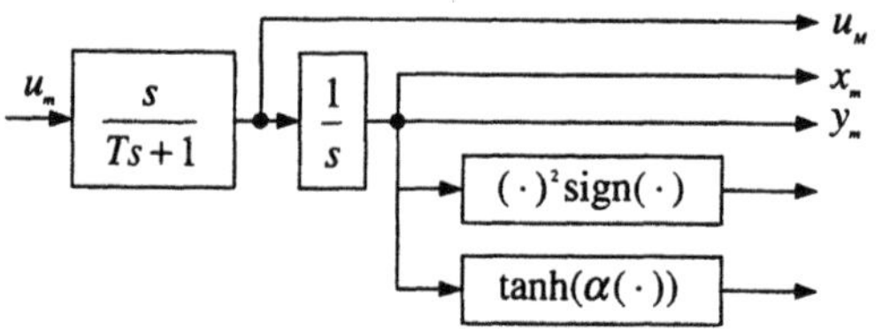

Fig. 2. Reference model in case of P control

scription of the reference model is also illustrated in Fig.2 and expressed as

$$\dot{x}_m(t) = a_m x_m(t) + b_m u_M(t) \tag{3}$$
$$y_m(t) = c_m x_m(t) \tag{4}$$

where $a_m = 0$, $b_m = 0$, $c_m = 1$, and T is a motor time constant.

Adaptive control input is given and its parameters are updated as follows:

$$u_p = \boldsymbol{K}^T(t)\boldsymbol{r}(t) \tag{5}$$
$$\boldsymbol{K}(t) = \left[\, K_e(t)\ K_x(t)\ \boldsymbol{K_d}(t)\ K_u(t) \,\right]^T$$
$$\boldsymbol{K_d}(t) = \left[\, K_{d1}(t)\ K_{d2}(t) \,\right]^T$$
$$\boldsymbol{r}(t) = \left[\, e_y(t)\ x_m(t)\ \boldsymbol{x_d}\ u_M(t) \,\right]^T$$
$$\boldsymbol{x_d}(t) = \left[\, |y_m(t)|y_m(t)\ tanh(\alpha y_m(t)) \,\right]^T$$
$$e_y(t) = y_m(t) - y_p(t)$$
$$\dot{\boldsymbol{K}}(t) = e_y(t)\Gamma\boldsymbol{r}(t)\ ,\ \Gamma = \Gamma^T > 0 \tag{6}$$

The adaptive control scheme with P-feedback controller has the following property.

Theorem 1. If the controll law (5) and parameter adaptation law (6) are applied to the controlled system (1) and (2), then it follows that $\lim_{t\to\infty} e_y(t) = 0$ and the adaptive parameters converge as

$$\lim_{t\to\infty} K_x(t) = D_M,\ \lim_{t\to\infty} K_u(t) = T_M$$
$$\lim_{t\to\infty} K_{d1}(t) = a,\ \lim_{t\to\infty} K_{d2}(t) = c \tag{7}$$

Proof: From the CGT property (Kaufman *et al.*, 1994), we assume that the ideal trajectory is a linear function of the state and input of the reference model as

$$x_p^*(t) = S_{11}x_m(t) + S_{12}u_M(t) \tag{8}$$
$$u_p^*(t) = S_{21}x_m(t) + S_{22}u_M(t) + \boldsymbol{S_\gamma}^T\gamma(y_m(t)) \tag{9}$$

Substituting (1) and (2) into (8) and (9) using $\dot{u}_M = 0$, we can obtain

$$\dot{x}_p^*(t) = [a_pS_{11} + b_pS_{21}]x_m(t) + [a_pS_{12} + b_pS_{22}]u_M(t)$$
$$+[b_p\boldsymbol{S_\gamma} + \boldsymbol{a_\gamma}]\gamma(y_m(t))$$
$$= a_mS_{11}x_m(t) + b_mS_{11}u_M(t)$$
$$y_m(t) = c_pS_{11}x_m(t) + c_pS_{12}u_M(t) = c_mx_m(t)$$

Thus it follows that

$$b_p\boldsymbol{S_\gamma} = -\boldsymbol{a_\gamma},\quad c_pS_{11} = c_m,\quad c_pS_{12} = 0$$
$$a_pS_{12} + b_pS_{22} = S_{11}b_m,\quad a_pS_{11} + b_pS_{21} = S_{11}$$

Thus the solution is given by $S_{11} = 1$, $S_{12} = 0$, $S_{21} = D_M$, $S_{22} = T_M$, $\boldsymbol{S_\gamma} = [a\ \ c]^T$. Then, the ideal input is also rewritten by

$$u_p^*(t) = S_{21}x_m(t) + S_{22}u_M(t) + \boldsymbol{S_\gamma}^T\gamma(y_m(t))$$
$$= D_Mx_m(t) + a|y_p(t)|y_p(t)$$
$$+c\tanh(\alpha y_m(t)) + T_Mu_M(t) \tag{10}$$

For stability analysis, from (1) to (4) the error system can be expressed by

$$\dot{e}_x(t) = a_p^*e_x(t) + b_p(\boldsymbol{K}^* - \boldsymbol{K}(t))\gamma(t)$$
$$+\boldsymbol{a_\gamma}^T[\gamma(y_m(t)) - \gamma(y_p(t))]$$
$$e_y(t) = c_pe_x(t) \tag{11}$$

where $e_x(t) = x_p^*(t) - x_p(t)$, $a_p^* = a_p - K_e^*b_pc_p$, $\boldsymbol{K}^* = [\,K_e^*\ \ S_{21}\ \ \boldsymbol{S_\gamma}^T\ \ S_{22}\,]^T$.

Since the linear part of (11) satisfies the ASPR property, it follows from the KYP lemma that there exist $P > 0$ and $Q \geq 0$ such that $a_p^*P + Pa_p^* = -Q$ and $Pb_p = c_p$. The solution is given by $P = T_M, Q = 2(D_M + K_e^*)$.

Now let a candidate of the Lyapunov function be

$$V(t) = Pe_x(t)^2$$
$$+(\boldsymbol{K}(t) - \boldsymbol{K}^*)^T\Gamma^{-1}(\boldsymbol{K}(t) - \boldsymbol{K}^*) \tag{12}$$

The derivative of $V(t)$ along (11) becomes

$$\dot{V}(t) = -Qe_x(t)^2 + 2e_x(t)P\boldsymbol{a}_\gamma[\gamma(y_m(t)) - \gamma(y_p(t))] \quad (13)$$

The second term of the RHS is calculated and is shown to be negative as shown below.

$$\begin{aligned} & 2e_x(t)P{\boldsymbol{a}_\gamma}^T[\gamma(y_m(t)) - \gamma(y_m(t))] \\ &= -2e_y(t)[a \quad c][\gamma(y_m(t)) - \gamma(y_p(t))] \\ &= -2ae_y(|y_m(t)|y_m(t) - |y_p(t)|y_p(t)) \\ &\quad -2ce_y(tanh(\alpha y_m(t)) - tanh(\alpha y_p(t))) \le 0 \end{aligned}$$

The above result and (13) leads to $\dot{V} \le 0$. Thus it gives that the parameter convergence and $\lim_{t\to\infty} e_x = 0$.

Thus, unlike ordinary closed-loop adaptive identification scheme using the actual input $u_p(t)$ and output $y_p(t)$ directly, the proposed approach sets up the specific reference model in which the reference output $y_m(t)$ and the error $e_y(t)$ are used and the adaptive identification can be done in the feedforward way. Still in the case, the above theorem assures the parameter convergence and stability.

2.2 *Case with PI Control*

The PI controller is employed to compensate for the closed-loop response to stepwise torque disturbances. The state space description of the controlled system with PI controller illustrated in Fig.1 is given by

$$\dot{\boldsymbol{x}}_p = \boldsymbol{A}_p\boldsymbol{x}_p + \boldsymbol{b}_p u_{p1} + \boldsymbol{A}_{\gamma p}\gamma(y_p) \quad (14)$$

$$y_p = \boldsymbol{c}_p^T\boldsymbol{x}_p \quad (16)$$

where

$$\boldsymbol{x}_p = \begin{bmatrix} x_{p1} \\ x_{p2} \end{bmatrix}, \gamma(y_p) = \begin{bmatrix} |y_p|y_p \\ tanh(\alpha y_p) \end{bmatrix}$$

$$\boldsymbol{A}_p = \begin{bmatrix} -D_M/T_M & 1/T_M \\ 0 & 0 \end{bmatrix}, \boldsymbol{b}_p = \begin{bmatrix} 1/T_M \\ 1/T_i \end{bmatrix}$$

$$\boldsymbol{c}_p^T = \begin{bmatrix} 1 & 0 \end{bmatrix}, \boldsymbol{A}_{\gamma p} = \begin{bmatrix} -a/T_M & -c/T_M \\ 0 & 0 \end{bmatrix}$$

The state space description of reference model illustrated in Fig.4 is

$$\dot{\boldsymbol{x}}_m = \boldsymbol{A}_m\boldsymbol{x}_m + \boldsymbol{b}_m u_M + \boldsymbol{A}_{\gamma m}\gamma(y_m) \quad (17)$$

$$y_m = \boldsymbol{c}_m^T\boldsymbol{x}_m \quad (18)$$

where

$$\boldsymbol{x}_m = \begin{bmatrix} x_{m1} \\ x_{m2} \\ x_{m3} \\ x_{m4} \end{bmatrix}, \gamma(y_m) = \begin{bmatrix} |y_m|y_m \\ tanh(\alpha y_m) \end{bmatrix}$$

$$\boldsymbol{A}_m = \begin{bmatrix} 0 & 0 & 0 & 0 \\ 1/T_i & 0 & 0 & 0 \\ 0 & 0 & 1/T_i & 0 \\ 0 & 0 & 0 & -1/T_i \end{bmatrix}, \boldsymbol{b}_m = \begin{bmatrix} 1 \\ 0 \\ 0 \\ 0 \end{bmatrix}$$

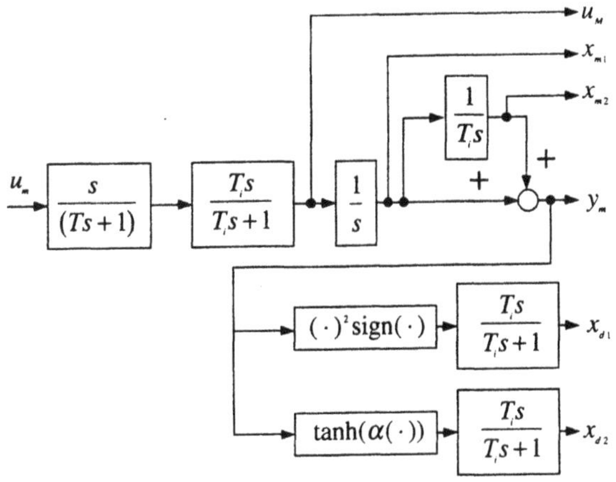

Fig. 3. Reference model in case with PI controller

$$\boldsymbol{c}_m^T = \begin{bmatrix} 1 & 1 & 0 & 0 \end{bmatrix}, \boldsymbol{A}_{\gamma m} = \begin{bmatrix} 0 & 0 \\ 0 & 0 \\ -1/T_i & 0 \\ 0 & -1/T_i \end{bmatrix}$$

The control law and adaptation law are given by

$$u_{p1} = \boldsymbol{K}^T\boldsymbol{r} \quad (19)$$

$$\boldsymbol{r} = \begin{bmatrix} e_y & x_{m1} & x_{m2} & x_{d1} & x_{d2} & u_M \end{bmatrix}^T$$

$$\boldsymbol{K} = \begin{bmatrix} K_e & K_{x1} & K_{x2} & K_{d1} & K_{d2} & K_u \end{bmatrix}^T$$

$$e_y = y_m - y_p$$

$$\dot{\boldsymbol{K}} = e_y\boldsymbol{\Gamma}\boldsymbol{r}, \boldsymbol{\Gamma} = \boldsymbol{\Gamma}^T \ge 0 \quad (20)$$

Theorem 2. If the control law (19) and adaptation law (20) are applied to the controlled system given by (14) and (15), where let $D_0(e_{x2})$ be defined by

$$D_0(e_{x2}) = \left\{ e_{x2} \middle| \theta_2/\beta < |e_{x2}| < \beta/\theta_1 \right\} \quad (21)$$

where β, θ_1, θ_2 are defined later, then the adaptive parameters have the convergence property as

$$\lim_{t\to\infty} K_{x1}(t) = D_M, \lim_{t\to\infty} K_{x2}(t) = 0$$

$$\lim_{t\to\infty} K_{d2}(t) = a, \lim_{t\to\infty} K_{d2}(t) = c$$

$$\lim_{t\to\infty} K_u(t) = T_M \quad (22)$$

Proof. Similarly as Theorem 1, we assume that the ideal trajectorie is a linear function of the states and input of the reference model as

$$\boldsymbol{x}_p^* = \boldsymbol{S}_{11}\boldsymbol{x}_m + \boldsymbol{S}_{12}u_M \quad (23)$$

$$u_p^* = \boldsymbol{S}_{21}^T\boldsymbol{x}_m + \boldsymbol{S}_{22}u_M + \boldsymbol{S}_\gamma^T\gamma(y_p^*) \quad (24)$$

$$\boldsymbol{A}_{\gamma p} + \boldsymbol{b}_p\boldsymbol{S}_\gamma = \boldsymbol{S}_{11}\boldsymbol{A}_{\gamma m}$$

$$\boldsymbol{A}_p\boldsymbol{S}_{11} + \boldsymbol{b}_p\boldsymbol{S}_{21}^T = \boldsymbol{S}_{11}\boldsymbol{A}_m$$

$$\boldsymbol{A}_p\boldsymbol{S}_{12} + \boldsymbol{b}_p\boldsymbol{S}_{22} = \boldsymbol{S}_{11}\boldsymbol{b}_m$$

$$\boldsymbol{c}_p^T\boldsymbol{S}_{11} = \boldsymbol{c}_m^T, \quad \boldsymbol{c}_p^T\boldsymbol{S}_{12} = 0$$

Thus, the solutions are given by

$$\boldsymbol{S}_{11} = \begin{bmatrix} 1 & 1 & 0 & 0 \\ T_M/T_i & D_M & -a & -c \end{bmatrix}, \boldsymbol{S}_{12} = \begin{bmatrix} 0 \\ 0 \end{bmatrix}$$

$$S_{21}^T = \begin{bmatrix} D_M & 0 & a & c \end{bmatrix}, S_{22} = T_M, S_\gamma = \begin{bmatrix} a & c \end{bmatrix}^T$$

Then the ideal input is also given by

$$u_p^* = D_M x_{m1}(t) + 0 x_{m2}(t) + a x_{d1}(t) + c x_{d2}(t) + T_M u_M(t) \quad (25)$$

The error system is expressed by (14) to (18) as

$$\dot{e}_x = \dot{x}_p^* - \dot{x}_p = A_p^* e_x + b_p (K^* - K)^T r + A_{\gamma p}[\gamma(y_m) - \gamma(y_p)]$$

$$e_y = y_m - y_p = c_p^T e_x \quad (26)$$

where $e_x = x_p^* - x_p$, $A_p^* = A_p - K_e^* b_p c_p^T$, $K^* = [K_e^* \ S_{21}^T \ S_\gamma^T \ S_{22}]$.

Since the linear part of (26) satisfies the ASPR property, there exist the matrices $P = P^T > 0$, $Q \geq 0$ such that $A_p^* P + P A_p^* = -Q$ and $P b_p = c_p$, where the elements are given by $P_{11} = T_M(1 - p/T_i)$, $P_{21} = P_{12} = p$, $P_{22} = -T_i p / T_M$, $Q_{11} = 2(D_M + K_e^* - D_M p/T_i)$, $Q_{12} = Q_{21} = D_M p/T_M - p/T_i - 1$, $Q_{22} = -2p/T_M$, and p is selected to satisfy $P = P^T > 0$, $Q \geq 0$.

Taking a candidate of Lyapunov function as

$$V = e_x^T P e_x + (K - K^*)^T \Gamma^{-1} (K - K^*) \quad (27)$$

then we have the derivative of $V(t)$ as

$$\dot{V} = -e_x Q e_x + 2 e_x^T P A_{\gamma p}[\gamma(y_m) - \gamma(y_p)] \quad (28)$$

Let $e_{x1} = y_m - y_p$ and $e_{x2} = x_2^* - x_{p2}$ and let a' and c' be such that

$$\| |y_m| y_m - |y_p| y_p \| \leq a' |e_{x1}|^2$$

$$\| tanh(\alpha y_m) - tanh(\alpha y_p) \| \leq c'$$

Then the second term can be rewritten by

$$\begin{aligned} & e_x^T P A_{\gamma p}[\gamma(y_m) - \gamma(y_p)] \\ &= -a(1 - p/Ti) e_{x1} (|y_m| y_m - |y_p| y_p) \\ &\quad - c(1 - p/Ti) e_{x1} (tanh(\alpha y_m) - tanh(\alpha y_p)) \\ &\quad - a(p/T_M) e_{x2} (|y_m| y_m - |y_p| y_p) \\ &\quad - c(p/T_M) e_{x2} (tanh(\alpha y_m) - tanh(\alpha y_p)) \\ &\leq \theta_1 |e_{x1}|^2 |e_{x2}| + \theta_2 |e_{x2}| \end{aligned} \quad (29)$$

where $\theta_1 = paa'/T_M, \theta_2 = pcc'/T_M$, and define $\beta = \lambda_{min}(Q)$, then

$$\begin{aligned} \dot{V} &\leq -\beta e_{x1}^2 - \beta e_{x2}^2 + \theta_1 |e_{x1}|^2 |e_{x2}| + \theta_2 |e_{x2}| \\ &= -e_{x1}^2 (\beta - \theta_1 |e_{x2}|) - e_{x2}^2 (\beta - \theta_2 / |e_{x2}|) \end{aligned} \quad (30)$$

Inside $D_0(e_{x2})$, $\dot{V}(t) < 0$ and therfore $V(t)$ decreases. The solution e_{x2} which starts in $D_0(e_{x2})$ remains inside $D_0(e_{x2})$.

2.3 *Extension to Two-Mass Motor Drive System*

The proposed approach can be extended to adaptive identification of two-mass motor drive system, in which the inertia of motor and load and the shaft conductance can be identified in the feedforward manner by taking a specific reference model. The its algorithm are omitted here for the sake of space.

3. DESIGN OF PI CONTROLLER FOR MOTOR DRIVE SYSTEM

The PI controller is employed to compensate for the closed-loop response to stepwise torque disturbances. If the motor parameters are known and no nonliner frictions exist, the PI parameters K_e and T_i can be designed as follows, where K_e corresponds to the proportional gain parameter of the PI controller. Then, the closed-loop dynamics consisting of the PI controller and the motor is given by a second-order linear transfer function with the denominator as

$$s^2 + \frac{T_i D_M + K_e T_i}{T_i T_M} s + \frac{K_e}{T_i T_M} \quad (31)$$

which is characterized by the natural frequency $\omega_n = \sqrt{K_e / T_i T_M}$ and the damping factor $\zeta = (K_e + D_M)\sqrt{T_i} / 2\sqrt{K_e T_M}$. If the cutoff frequency ω_c of the closed-loop system is specified as $\sqrt{\phi} \cdot \omega_n$ where ϕ is chosen as 2, for instance, the PI controller parameters can be given as $K_e = \omega_c T_M$ and $T_i = \frac{\phi}{\omega_c}$. Thus, the damping factor of the closed-loop system is given by $\zeta = \sqrt{\phi}(\omega_c T_M + D)/(2\omega_c T_M)$. Then the damping factor becomes approximately 0.7 if $D_M = 0$ and $\phi = 2$. Thus, if the motor parameters T_M and D_M are uncertain, the tuning of the PI controller parameters is needed.

4. DIGITAL IMPLEMENTATION OF AUTO-TUNING BASED ON SAC

Fig.4 summarizes the digital implementation of the proposed adaptive tuning based on the SAC scheme. To compensate the relative degree of the discrete-time model, ordinary approach adopted a PFC. The proposed approach employs the serial compensation as illustrated in Fig.4, in which the adaptive control is given by

$$u_{p1} = (\hat{\xi}(k) - \hat{\xi}(k-1)) + \hat{\xi}(k) \quad (32)$$

$$\hat{\xi}(k) = K^T(k) \varphi(k)$$

$$K_e(k) = (1 - \sigma) K_e(k-1) + \frac{e_y(k) \Gamma_e e_y(k)}{m^2(k) + \varphi^T(k) \varphi(k)}$$

$$K_{x1}(k) = K_{x1}(k-1) + \frac{e_y(k) \Gamma_{x1} x_{m1}(k)}{m^2(k) + \varphi^T(k) \varphi(k)}$$

$$K_u(k) = K_u(k-1) + \frac{e_y(k) \Gamma_u u_M(k)}{m^2(k) + \varphi^T(k) \varphi(k)}$$

$$K_{d1}(k) = K_{d1}(k-1) + \frac{e_y(k) \Gamma_{d1} x_{d1}(k)}{m^2(k) + \varphi^T(k) \varphi(k)}$$

$$K_{d2}(k) = K_{d2}(k-1) + \frac{e_y(k) \Gamma_{d2} x_{d2}(k)}{m^2(k) + \varphi^T(k) \varphi(k)}$$

where

$$\varphi(k) = \begin{bmatrix} e_y(k) & x_{m1}(k) & x_d(k) & u_m(k) \end{bmatrix}^T$$

$$K(k) = \begin{bmatrix} K_e(k) & K_{x1}(k) & K_d(k) & K_u(k) \end{bmatrix}^T$$

$$x_d = \begin{bmatrix} x_{d1} & x_{d2} \end{bmatrix}, K_d = \begin{bmatrix} K_{d1} & K_{d2} \end{bmatrix}$$

$$m(k+1) = \delta_0 m(k) + \delta_1 (|\hat{\xi}_1(k)| + |y_p(k)| + 1)$$

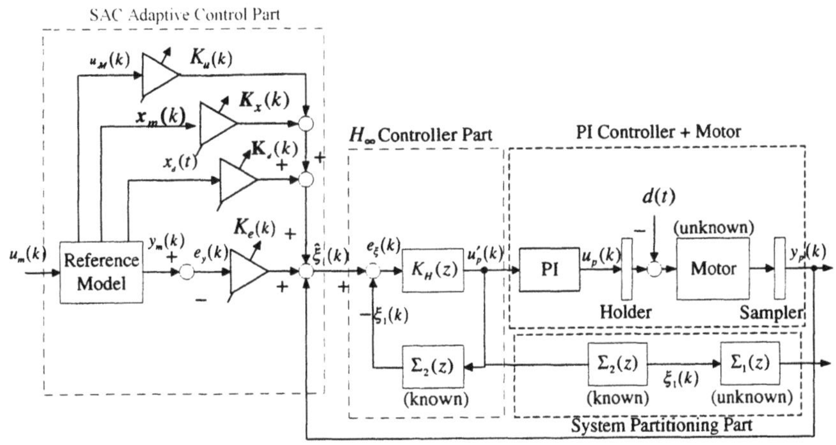

Fig. 4. Digital implementation of the proposed closed-loop tuning

where $m(0) \geq \delta_1/(1-\delta_0)$, $0 < \delta_0 < 1$, and $\delta_1 > 0$, where δ_0 should be chosen so that the poles of $W(z/\delta_0)$ may be stable, $W(z)$ is a transfer function of the error system.

5. EXPERIMENTAL RESULTS

The effectiveness of the proposed adaptive algorithm for auto-tuning is investigated in experiments using an AC synchronous servo-motor. The control purposes are the tracking ability to the reference velocity changes and the recovery performance to torque disturbances. The experimental setup of the control system using an AC servo motor is summarized as follows: The nominal values of the motor velocity, output power and motor torque are given by 3000 [rpm], 200 [W] and 0.64 [Nm] respectively. The velocity is measured by using the pulse encoder with resolution 4000 [pulses/rev]. Three cases of various inertia will be considered in the experiments. The time constant T_M can be identified in an open-loop manner by adding the M-sequence signals to the motor control unit. The obtained parameters were obtained as $T_M = 0.269, 0.317$ and 0.379[s]. These identified time constants will be used for comparing with the parameter estimates obtained by the proposed adaptive scheme. The sampling interval is set at 1ms. The cutoff frequency of the resultant closed-loop system is specified by $\omega_c = 100$[rad/s]. The reset time T_i of the PI controller is given by 0.02[s], which can keep the damping factor at $\zeta \cong 0.7$. As for the reference model, the time constant T of the first order system is chosen as $T = 0.05$[s].

In the experiments, we applied the proposed adaptive algorithm to the velocity control in which the reference velocity changes in the four modes (acceleration, positive constant, deceleration, negative constant) as shown in Fig.7. The identified parameters are plotted in Fig.5 in which the only linear viscous friction term D and other linear parameters are estimated, and in Fig.6 in which the nonlinear friction terms are also estimated. The estimation error of the motor time constant is indicated by $T_M - K_u(K)$ in Figs.5 and 6. Comparing these results it is shown that these estimation errors of T_M can be reduced by taking account of the nonlinear friction terms. T_M is a measeured value in the off-line way as mentioned. Fig.7 illustrates the torque disturbance rejection performance, when the stepwise torque disterbances (100 %) are added twice from 2.5s to 2.7s, and form 3.5s to 3.7s. The parameter estimates are not so much affected by these disturbances, while the quick recovery can be attained in the both cases with different time constants.

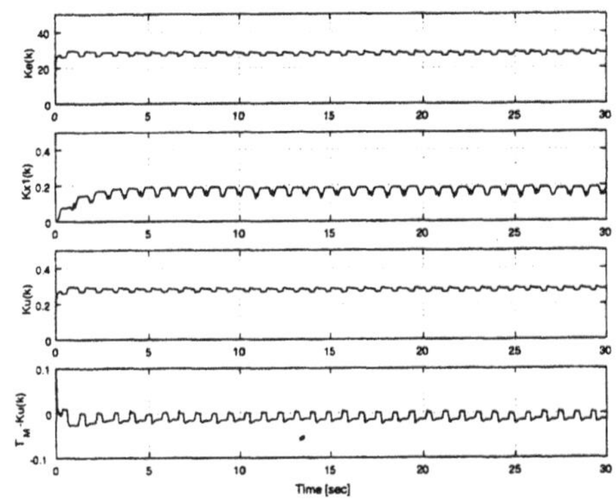

Case 1 : $T_M = 0.269$s

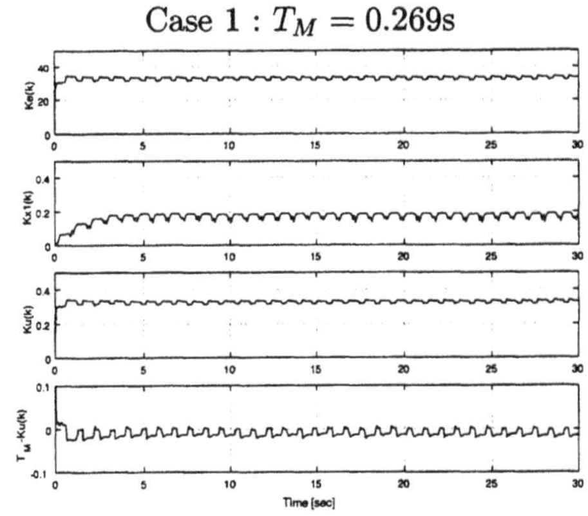

Case 2 : $T_M = 0.379$s

Fig. 5. Identified physical parameters and estimation error of motor time constant ($T_M - K_u(k)$)

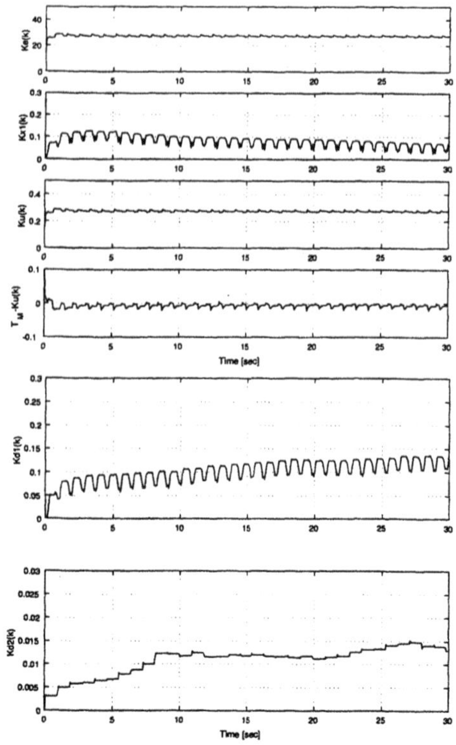

Case 1 : $T_M = 0.269$s

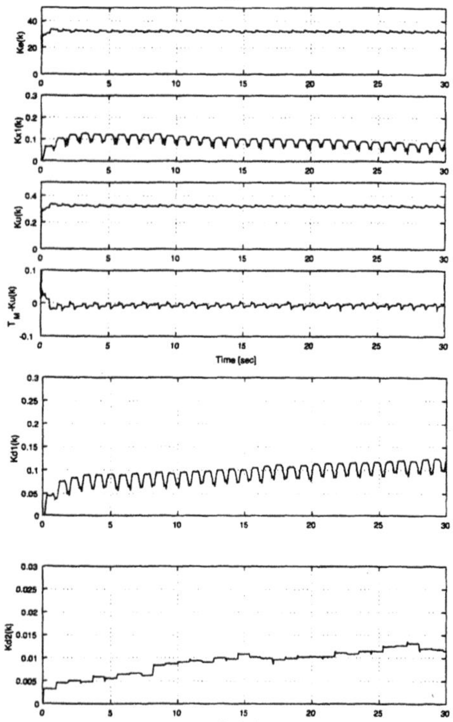

Case 2 : $T_M = 0.379$s

Fig. 6. Identified physical parameters and nonlinear friction terms, and estimation error of motor time constant $(T_M - K_u(k))$

6. CONCLUSION

The new adaptive algorithm for identifying the physical parameters of the motor drive system has been proposed to attain the auto-tuning of the feedforward and feedback controllers. Incorporating the physical structure such as motor time constant and various friction terms into the specific reference model, the adaptive algorithm can obtain the stability and robustness for stepwise torque disturbances.

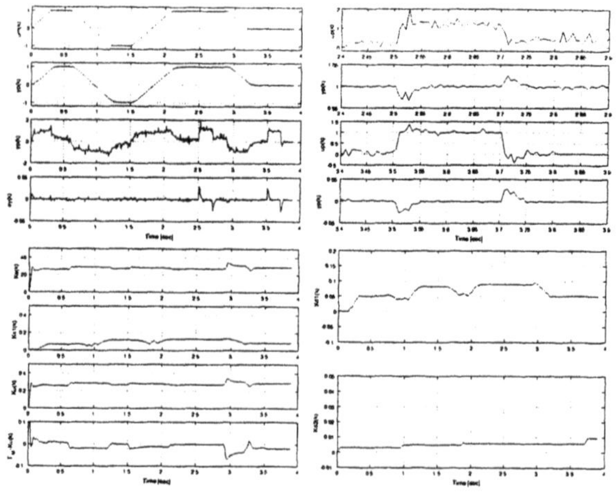

Case 1 : $T_M = 0.269$s

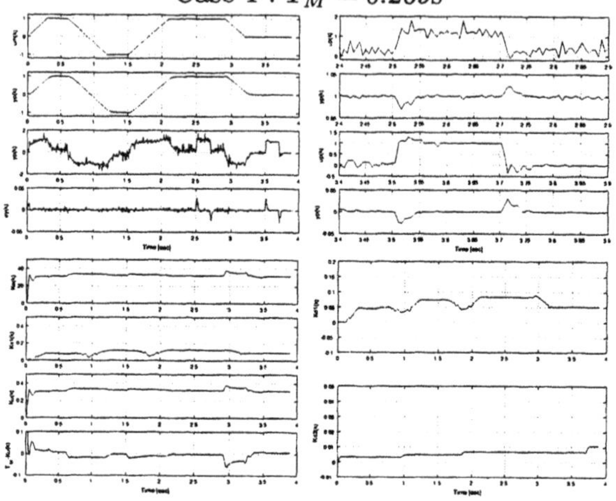

Case 2 : $T_M = 0.379$s

Fig. 7. Response profiles for torque disturbances

7. REFERENCES

Armstrong-Helouvry B., P. Dupont, and C. Canudas de Wit (1994). A survey of models, analysis tools and compensation methods for the control of machines with friction. *Automatica* **30-7** , 1083–1138.

Dosho, T., H. Ohmori and A. Sano (1997). New design method of simple adaptive controll for system with arbitrary relative degree. *Proc. the 36th IEEE Conference on Decision and Control* , 1914–1917.

Higashiyama, T., M. Manabu, H. Ohmori, A. Sano, H. Nishida, and Y. Todaka (2000). Auto-tuning of motor drive system by simple adaptive control approach. *Proc. The 2000 IEEE International Conference Control Applications* , 868–873, Anchorage, USA.

Ioannou, P.A. and Jing Sun (1996). *Robust Adaptive Control.* Prentice Hall PTR.

Kaufman, H., I. Bar-Kana and K. Sobel (1994). *Direct Adaptive Control Algorithms.* Springer-Verlag.

DIRECT ADAPTIVE PREDICTIVE CONTROL OF AN AUTOMATIC GEAR BOX

G. Ramond [1,2], D. Dumur [2], S. Perez [3], P. Boucher [2]

[1] *ESIEA F 94200 Ivry sur Seine Phone : +33 (0)1 46 71 87 98*
Fax : +33 (0)1 46 71 88 11 Email : ramond@esiea.fr
[2] *Supélec F 91192 Gif sur Yvette cedex Phone : +33 (0)1 69 85 13 75*
Fax : +33 (0)1 69 85 13 89 Email : didier.dumur@supelec.fr
[3] *ETSII, Universidad Politecnica de Valencia, E 46022 Valencia*

Abstract: Even if predictive control algorithms have proved to be efficient and flexible for industrial applications, they may have some difficulties to maintain acceptable performance for systems with time-varying parameters, without considering an adaptive framework. This paper presents a direct version of adaptive generalized predictive control (DAGPC), with a polynomial GPC controller and a recursive least squares-type strategy for the controller parameters on-line identification. An application of this structure to an engine drive with an automatic gear box, including a variable transmission with continuous variations of the reduction ratio, is finally developed.

Keywords: Adaptive control, predictive control, digital control, least-squares identification, engine modelling, continuously variable transmission.

1. INTRODUCTION

Predictive control algorithms based on Generalized Predictive Control (GPC) or even long range predictive control strategies have proved to be efficient, flexible and successful for industrial applications. However problems may appear trying to maintain acceptable performances for a plant with time-varying parameters. A solution is then to develop an adaptive version of the predictive controller for systems with parametric disturbances (Wang and Henriksen, 1992, 1994).

This paper presents a direct version of adaptive generalized predictive control. The GPC algorithm (Clarke, *et al.*, 1987) is rewritten in an adequate adaptive form and the controller parameters are updated with a recursive least-squares type strategy.

A first part reminds the basic definitions of the GPC algorithm and the elaboration of the control law under a polynomial RST form. A second part presents the adaptive structure with the identification strategy. An application of this structure to an engine drive with an automatic gear box, including a variable transmission with continuous variations of the reduction ratio, is finally developed. The model of this system is derived as a starting point from the well-known two-masses problem. In this study, only the engine speed is measured, the acceleration (variable to be controlled) remains not measurable. The aim is to control the acceleration in the presence of vibratory effect due to the shafts elasticity.

2. PREDICTIVE CONTROL

This part reminds the main developments of Generalized Predictive Control (Clarke, *et al.*, 1987), starting first with the definition of the prediction model, given by the polynomial relation:

$$A(q^{-1})y(t) = B(q^{-1})u(t-1) + \xi(t)/\Delta(q^{-1}) \quad (1)$$

u is the control signal applied to the system, y the output of the system, $\Delta(q^{-1}) = 1 - q^{-1}$ the difference operator, A and B polynomials in the backward shift operator q^{-1}, of respective order n_a and n_b, ξ an uncorrelated zero-mean random sequence.
From the previous model (1), a polynomial optimal predictor is designed under the following form:

$$\hat{y}(t+j) = \underbrace{F_j(q^{-1})y(t) + H_j(q^{-1})\Delta u(t-1)}_{\text{free response}} + \underbrace{G_j(q^{-1})\Delta u(t+j-1) + J_j(q^{-1})\xi(t+j)}_{\text{forced response}} \quad (2)$$

Unknown polynomials F_j, G_j, H_j and J_j are derived solving Diophantine equations, with unique solutions. In the further developments, the term related to future disturbances is taken equal to zero, corresponding to the best prediction of $\xi(t+j)$.
The criterion is a weighted sum of square predicted future errors and square control signal increments:

$$J = \sum_{j=N_1}^{N_2} (\hat{y}(t+j) - w(t+j))^2 + \lambda \sum_{j=1}^{N_u} \Delta u(t+j-1)^2 \quad (3)$$

With the matrix form of the predictor (2):

$$\hat{\mathbf{y}} = \mathbf{G}\tilde{\mathbf{u}} + \mathbf{if}\, y(t) + \mathbf{ih}\, \Delta u(t-1) \quad (4)$$

The minimization of relation (3) written in a matrix form provides the future control sequence:

$$\tilde{\mathbf{u}} = \mathbf{M}\left[\mathbf{w} - \mathbf{if}\, y(t) - \mathbf{ih}\, \Delta u(t-1)\right] \quad (5)$$

with:

$$\mathbf{w} = \left[w(t+N_1) \;\cdots\; w(t+N_2)\right]'$$

$$\mathbf{if} = \left[F_{N_1}(q^{-1}) \;\cdots\; F_{N_2}(q^{-1})\right]'$$

$$\mathbf{ih} = \left[H_{N_1}(q^{-1}) \;\cdots\; H_{N_2}(q^{-1})\right]'$$

$$\tilde{\mathbf{u}} = \left[\Delta u(t) \;\cdots\; \Delta u(t+N_u-1)\right]'$$

$$\hat{\mathbf{y}} = \left[\hat{y}(t+N_1) \;\cdots\; \hat{y}(t+N_2)\right]'$$

$$\mathbf{G} = \begin{bmatrix} g_{N_1}^{N_1} & g_{N_1-1}^{N_1} & \cdots & \cdots \\ g_{N_1+1}^{N_1+1} & g_{N_1}^{N_1+1} & \cdots & \cdots \\ \cdots & \cdots & \cdots & \cdots \\ g_{N_2}^{N_2} & g_{N_2-1}^{N_2} & \cdots & g_{N_2-N_u+1}^{N_2} \end{bmatrix}$$

$$\mathbf{Q} = \left[\mathbf{G}'\mathbf{G} + \lambda \mathbf{I}_{N_u}\right]^{-1}$$

$$\mathbf{M} = \mathbf{Q}\,\mathbf{G}' \quad N_u \times (N_2 - N_1 + 1) \text{ matrix}$$

Only the first value of Equation (5) is finally applied to the system according to the receding horizon strategy. The equivalent RST controller is computed through a difference equation (with $\mathbf{m}_1$ first line of the $\mathbf{M}$ matrix):

$$\Delta u(t) = T(q)\, w(t) - R(q^{-1})\, y(t) - S^*(q^{-1})\Delta u(t) \quad (6)$$

with:

$$S^*(q^{-1}) = \mathbf{m_1}\, \mathbf{ih}(q^{-1})q^{-1}$$

$$R(q^{-1}) = \mathbf{m_1}\, \mathbf{if}(q^{-1})$$

$$T(q) = \mathbf{m_1}\left[q^{N_1} \;\cdots\; q^{N_2}\right]$$

3. DIRECT ADAPTIVE GPC

In the time-varying parameters case, the previous controller (5) must be included within an adaptive structure. The classical indirect scheme approach (Aström and Wittenmark, 1989) is however time-consuming with predictive structures, because all predictors must be re-computed on-line for the design of the new RST controller. Consequently, it is proposed here an original approach, which **directly updates the controller parameters**. This strategy, which **avoids to compute on-line all predictors**, imposes to reformulate the controller under an adequate form, without changing the structure of the control law, then to develop an identification algorithm which updates the controller parameters.

3.1 Reformulation of the predictive law

The control law (5) is equivalent to the following reformulation (Ramond, *et al.*, 1998):

$$\mathbf{Mw} = \theta'\Phi(t) \quad (7)$$

With $\theta = \left[\mathbf{M\,if} \;\; \mathbf{I}_{N_u} \;\; \mathbf{M\,ih}\right]'$ the controller parameters matrix ($\mathbf{M\,if}$, $\mathbf{M\,ih}$ formed with the $\mathbf{M\,if}(q^{-1})$, $\mathbf{M\,ih}(q^{-1})$ coefficients) and:

$$\Phi(t) = \left[y(t)\cdots y(t-n_a)\,\tilde{\mathbf{u}}\,\Delta u(t-1)\cdots \Delta u(t-n_b)\right]' \quad (8)$$

called 'regressor' formed with the input and output measurements, of dimension $(n_a + n_b + N_u + 1)$.

3.2 Performance error

Considering the nominal model, the fixed GPC control law explicitly cancels the following performance error:

$$\mathbf{e}_f(t+N_2) = \mathbf{iP}(t+N_2) - \mathbf{iPw}(t+N_2) \quad (9)$$

with:

$$\mathbf{iP}(t+N_2) = \left[\mathbf{M} \;\; \lambda\mathbf{Q}\right]\begin{bmatrix}\hat{\mathbf{y}} \\ \tilde{\mathbf{u}}\end{bmatrix} = \mathbf{M}\,\hat{\mathbf{y}} + \lambda\mathbf{Q}\,\tilde{\mathbf{u}}$$

$$\mathbf{iPw}(t+N_2) = \left[\mathbf{M} \;\; \lambda\mathbf{Q}\right]\begin{bmatrix}\mathbf{w} \\ \mathbf{0}\end{bmatrix} = \mathbf{M\,w} = \theta'\Phi(t)$$

With these definitions, $\mathbf{iP}(t+N_2)$ is an indication of the measured performances and $\mathbf{iPw}(t+N_2)$ an evaluation of the expected performances, considering

the fact that the output vector $\hat{y}$ has to converge to the reference vector $\mathbf{w}$ and at the same time the control signal $\tilde{\mathbf{u}}$ has to tend to zero.
Including the RST structure and the performance error, the DAGPC algorithm is given Figure 1.

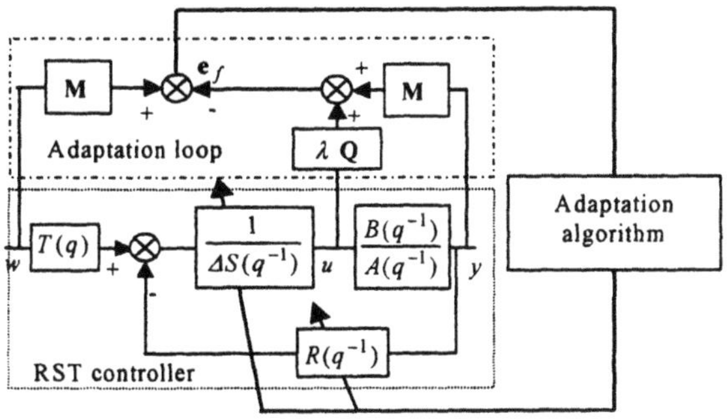

Fig. 1. Equivalent structure of the DAGPC.

From this, the objective in the adaptive case is to minimize this performance index $\Im$ at each step, in order to reach asymptotically and without plant parameter knowledge:

$$\lim_{t \to +\infty} \mathbf{e_f}(t+1) = 0 \qquad (10)$$

3.3 Recursive parameters estimation

In the time-varying parameters case, an estimated controller parameters matrix $\hat{\theta}(t)$ is considered to insure the minimization of the performance error. This one is updated according to the recursive least squares method (Isermann, *et al.*, 1992; Söderström and Stoica, 1989), as follows:

$$\hat{\theta}(t+N_2) = \hat{\theta}(t+N_2-1) + \Gamma(t)\Phi(t)\mathbf{e}_f(t+N_2)' \quad (11)$$

$$\Gamma(t) = \Gamma(t-1) - \frac{\Gamma(t-1)\Phi(t)\Phi(t)'\Gamma(t-1)}{1+\Phi(t)'\Gamma(t-1)\Phi(t)} \quad (12)$$

Furthermore, a conditional updating can be included (M'Saad, *et al.*, 1993). The method specificity has consisted since the beginning in the comparison between the measured performances index $\mathbf{iP}(t+N_2)$ and the expected performances index $\mathbf{iPw}(t+N_2)$. Within this context, the idea is to maintain the performances at the same level as they were when the algorithm started, which was supposed to be the nominal point. The adaptation must thus be set off considering the following conditional test:

$$\mathbf{iPw}(t+N_2) > \mathbf{iPw}_{\text{nom}} \qquad (13)$$

3.4 Design methodology

The first aspect of the algorithm is the design on the 'nominal' system at time $t=0$ of a 'fixed' RST GPC controller, which will initialize the whole procedure.

Then, at time t, the different steps are the following:

1. Measurement of the output value $y(t)$
2. Computation of the performance error (Eq. 9)
3. Computation of the controller parameters matrix (Eqs. 11 and 12)
4. Controller parameters update.
5. Application of the control value $u(t)$ to the system (Eq. 6)

4. APPLICATION TO AN AUTOMATIC GEAR BOX

The developed application, which is at the moment a crucial issue in the automotive field (Richard, *et al.*, 1999), considers an automatic gear box of a thermal engine, equipped with a variable transmission, considering important variations of the reduction ratio. The purpose is the realization of a torque controller, looking at the acceleration (as an image of the torque) as the manipulated variable, the main task being to avoid the jolts generated by the global structure. Another difficulty which must be taken into account in the structure is coming from the fact that the acceleration can not be measured, only information from the engine velocity is available. The modeling phase starts with the two-masses system described below, which is of most interest for systems with coupled inertia through particular shafts, such as steel-rolling plants (Hori, *et al.*, 1994).

4.1 System description

Consider the two-masses system of Figure 2, with one upper inertia (moment J_1) devoted to the engine, the other one (moment J_2) related to the load, both connected by a non-stiff coupling shaft (stiffness coefficient k) with small elasticity.

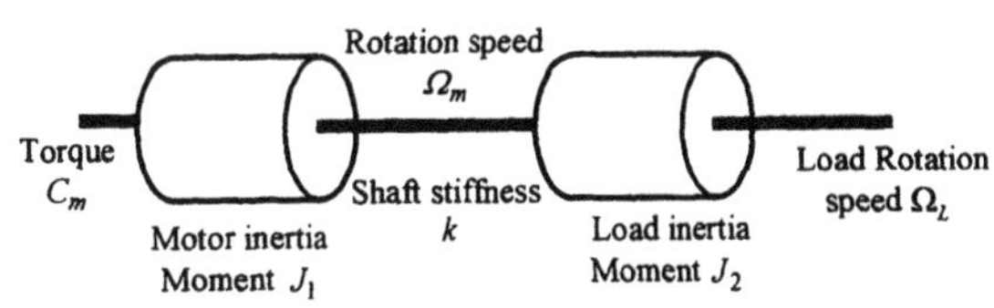

Fig. 2. General scheme of the two-masses system.

The model of the automatic gear box with variable transmission can be derived from this scheme considering some slight modifications. Introducing two rubbing coefficients $f_i|_{\{i=1,2\}}$ related to the moments of inertia $J_i|_{\{i=1,2\}}$, it appears as proposed on Figure 3, that a moment of inertia $J = N^2 J_1$, associated to its rubbing coefficient $f = N^2 f_1$, can be considered as the model of the engine through a gear box with a N reduction ratio. In the same way, a second moment of inertia $M_v = J_2/R_r^2$, with its rubbing coefficient $b = f_2/R_r^2$, can be considered as coming from the wheels moving the car, in further developments of radius R_r.

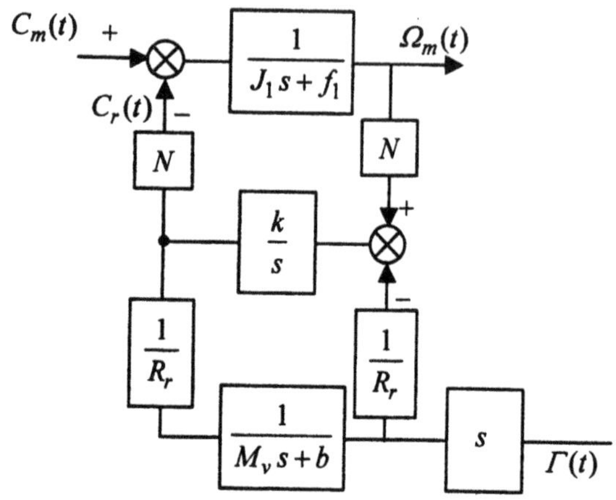

Fig. 3. Block diagram of the automatic gear box with continuous variable transmission.

The global relation between the torque and the acceleration is thus given by the equation:

$$H_{TA}(s) = \frac{\Gamma(s)}{C_m(s)} = \frac{kNR_r s}{D(s)}$$

$$\begin{aligned} D(s) &= J_1 M_v R_r^2 s^3 + (J_1 b R_r^2 + f_1 M_v R_r^2) s^2 \\ &+ (J_1 k + f_1 b R_r^2 + k N^2 R_r^2 M_v) s \\ &+ k f_1 + b k N^2 R_r^2 \end{aligned} \quad (14)$$

with Γ the vehicle acceleration, C_m the engine torque, s the Laplace transform, and the parameters values:

$$J = 0.2 \ \text{kg m}^2, M_v = 1200 \ \text{kg}, R_r = 0.3 \ \text{m},$$

$$k = 800, b = 50 \, \text{kg s}^{-1}, f = 0.28 \, \text{Nm s rad}^{-1}$$

Its characteristics depends on the reduction ratio N of the variable transmission. Figure 4 shows the open-loop acceleration response to a step input for different values of N. It clearly illustrates a very different behavior from $N = 0.1$ up to $N = 0.5$.

The objective of the control algorithm is to obtain the same performances whatever the reduction ratio may be, without overshoot, with as less oscillations as possible, and with a response time as short as possible.

Only the engine speed is measured, without information from the acceleration. In this case, the GPC controller is designed considering the architecture of Figure 5, with the velocity as the controlled variable and the a posteriori calculation of the acceleration.

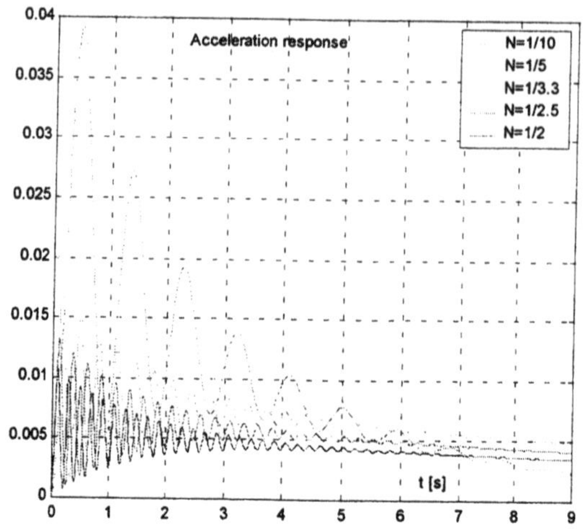

Fig. 4. Open-loop acceleration response to a step input.

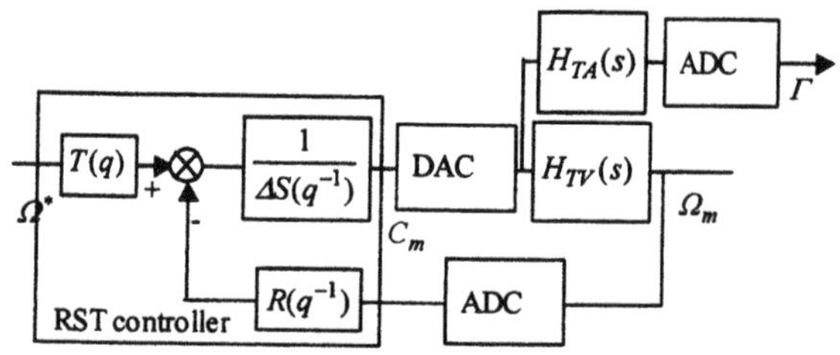

Fig. 5. Control architecture when the vehicle acceleration is unknown.

For this structure, the following transfer function must be taken into account for the control:

$$H_{TV}(s) = \frac{\Omega_m(s)}{C_m(s)} = \frac{M_v R_r^2 s^2 + b R_r^2 s + k}{D(s)} \quad (15)$$

Figure 6 gives the Bode diagram of the previous transfer function for different values of N. The different behaviors observed on this figure point out the interest for adaptive control.

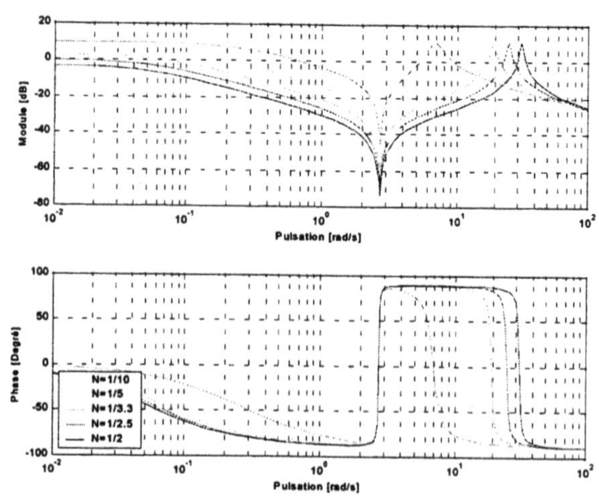

Fig. 6. Bode diagram of the transfer function between torque and engine velocity.

4.2 Simulation results

The adaptive strategy is applied to the gear box, considering the motor velocity setpoint of Figure 7.

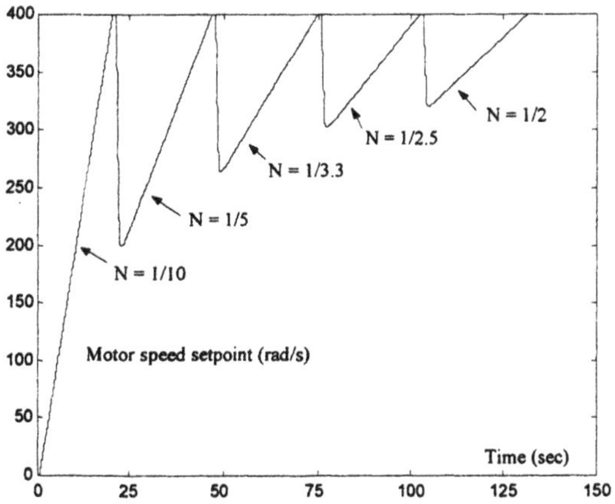

Fig. 7. Motor velocity setpoint.

This setpoint reproduces velocity variations from 0 to 4000 rpm, with five different values of the reduction ratio N (from 0.1 to 0.5).

With a sampling period T_e of 50 ms, a GPC controller, which ensures at time $t = 0$ stability and robust performances in terms of phase and gain margins and sensitivity functions (Boucher, *et al.*, 1993) is first elaborated for the reduction ratio $N = 0.5$ (nominal model) with the 'optimal' tuning:

$$N_1 = 1 \; ; \; N_2 = 30 \; ; \; N_u = 1 \; ; \; \lambda = 7 \, .$$

The DAGPC algorithm, starting from the previous initial GPC controller as nominal performance, has been applied to the gear box with the particular motor speed setpoint of Figure 7. The corresponding motor speed output and car acceleration are given on Figures 8 and 9. It has been assumed that during the change of the reduction ratio (transient from one value to an other one), the vehicle acceleration remains zero (friction is here neglected).

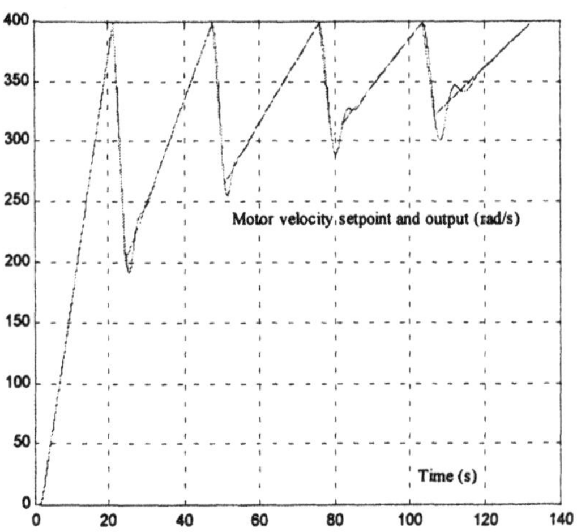

Fig. 8. Motor velocity output and setpoint (DAGPC).

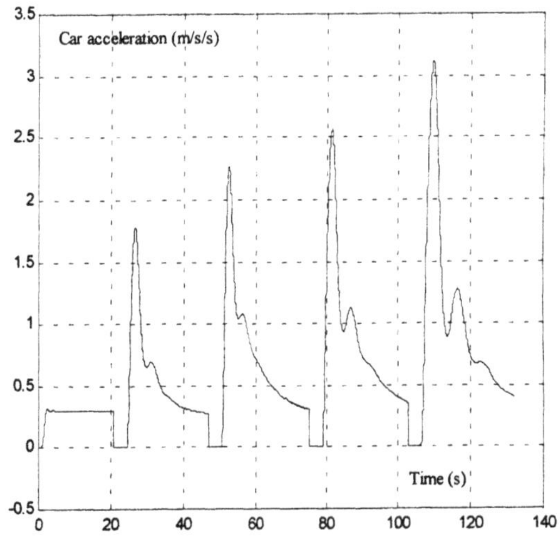

Fig. 9. Car acceleration (DAGPC).

For comparison purposes, the 'fixed' GPC algorithm has been applied to the gear box in the same conditions, providing results of Figures 10 and 11.

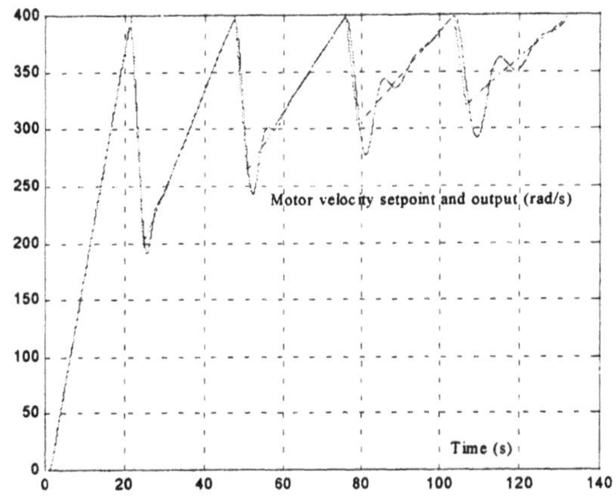

Fig. 10. Motor velocity output and setpoint (GPC).

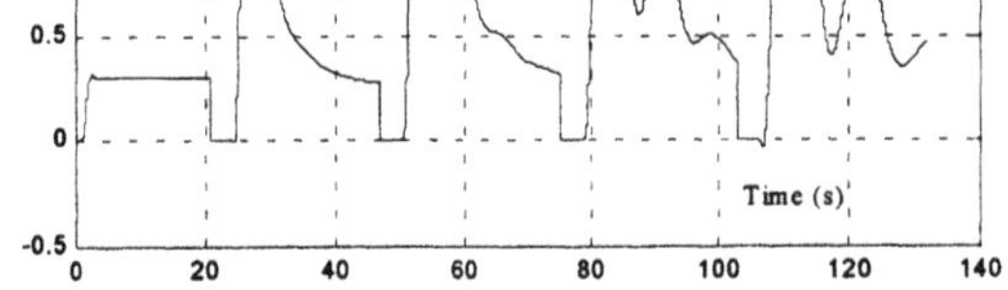

Fig. 11. Car acceleration (GPC).

These figures show the improvements of the adaptive structure, in terms of overshoot and oscillations, which enables to maintain nominal performance even with important variations of the reduction ratio.
Moreover, nominal performance can be maintained using only one adaptive controller, with only one

controller design at the beginning, which avoids the computation of multi 'fixed' controllers to reach the same level of performance.

For comparison purposes, Figure 12 shows the previous car acceleration of Figures 9 and 11 (DAGPC and GPC) on the same plot. Figure 13 also gives a comparison between the resulting motor torque with GPC or DAGPC.

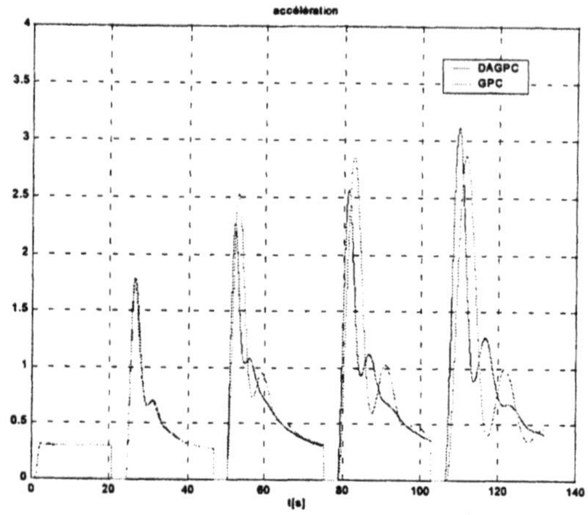

Fig. 12. Car acceleration, comparison between GPC (magenta) and DAGPC (blue).

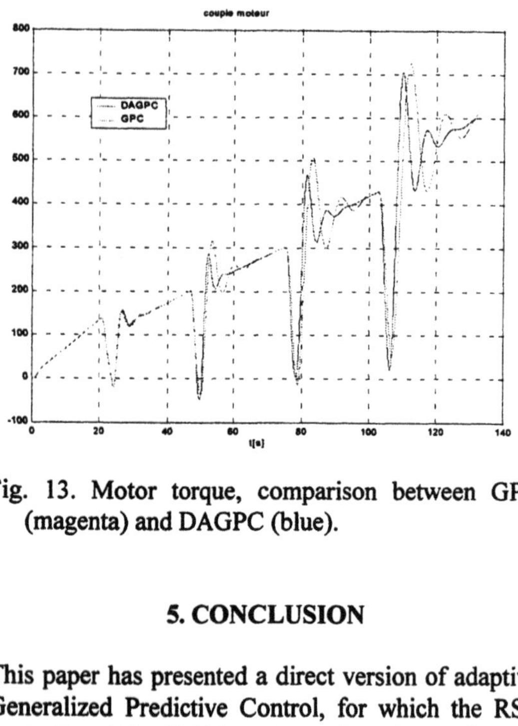

Fig. 13. Motor torque, comparison between GPC (magenta) and DAGPC (blue).

5. CONCLUSION

This paper has presented a direct version of adaptive Generalized Predictive Control, for which the RST controller of the GPC loop is rewritten in an adequate form for adaptation. The recursive adaptation loop has allowed to update the controller parameters vector according to a least square-type strategy, including a particular conditional updating test.

An application to an automatic gear box with a continuous variable transmission has been examined, with a model derived from the two-masses system study. It has proved that, even with important variations of the reduction ratio, the developed adaptive structure allows to maintain a high level of performances, in terms of rapidity and overshoot, cancellation of oscillations. Furthermore, it has been shown that this adaptive structure using only one initial controller design avoids the problem of multiple syntheses related to the 'fixed' controller structure case.

This adaptive control law appears to be very easy to implement, and considering the small number of identified parameters (due to small degrees of the R and S polynomials), it may be easily integrated in real-time context with small sampling periods.

6. REFERENCES

Aström, K. and B. Wittenmark, (1989). *Adaptive Control*, Addison-Wesley Editions.

Boucher, P., D. Dumur and S. Daumüller (1993). Autotuned Predictive Control. In: *Mutual Impact of Computing Power and Control Theory*, pp. 277-287.

Clarke, D.W., C. Mohtadi and P.S. Tuffs (1987). Generalized Predictive Control: Part I The Basic Algorithm, Part II: Extensions and Interpretation. In: *Automatica*, **Vol. 23:2**, pp. 137-160.

Hori, Y., H. Iseki and K. Sugiara (1994). Basic consideration of Vibration suppression and Disturbance Rejection Control of Multi-inertia System using SFLAC. In: *IEEE Transactions on Industry Applications*, **Vol. 30, n° 4**, pp. 889-896.

Isermann, R., K.-H. Lachmann and D. Matko (1992). *Adaptive control systems*, Prentice Hall.

M'Saad, M., L. Dugard and SH. Hammad (1993). A suitable generalized predictive adaptive controller case study: Control of a flexible arm. In: *Automatica*, **Vol. 29:3**, pp. 589-608.

Ramond, G., D. Dumur and P. Boucher (1998). Direct Adaptive Generalized Predictive Control for a Minimum Phase Plant with Parametric Disturbances. In: *Proceedings of the IEEE-SMC-IMACS Multiconference CESA'98*, **Vol. 1**, pp. 347-352, Hammamet, April.

Richard, S., P. Chevrel, P. de Larminat and B. Marguerie (1999). Polynomial Pole Placement revisited: application to active control of car longitudinal oscillations. In: *Proceedings of the European Control Conference ECC'99*, September.

Söderström, T. and P. Stoica (1989). *System identification*, Prentice Hall.

Wang, W. and R. Henriksen (1992). A direct adaptive generalized predictive controller and some of its global convergence properties. In: *Proceedings of the 4th IFAC Symposium on Adaptive Systems in Control and Signal Processing*, pp. 187-192.

Wang, W. and R. Henriksen (1994). A direct adaptive generalized predictive controller. In: *IEEE Transaction on Modeling Identification and Control*, **Vol. 15**, pp. 23-32.

A ROAD-ADAPTIVE LQG CONTROL FOR SEMI-ACTIVE SUSPENSION SYSTEMS

Hyun-Chul Sohn, Keum-Shik Hong,* Wan-Suk Yoo, and O-Kaung Lim

School of Mechanical Engineering, Pusan National University, 30 Changjeon-dong, Kumjeong-ku, Pusan 609-735, Korea. Tel: +82 (51) 510-2454, Fax: +82 (51) 514-0685, Emails: hcson@hyowon.pusan.ac.kr & kshong@hyowon.pusan.ac.kr
* The corresponding author

Abstract: In this paper, a road-adaptive LQG control for the semi-active Macpherson strut suspension system of hydraulic type is investigated. A new control-oriented model, which incorporates the rotational motion of the unsprung mass, is introduced. A semi-active suspension controller adapting to road variations is proposed. First, based on the extended least squares estimation algorithm, a LQG controller adapting to the estimated road characteristics is designed. Through the computer simulations, the performance of the proposed semi-active suspension is compared with that of a non-adaptive one. The results show better control performance of the proposed system over the compared one. *Copyright © 2001 IFAC*

Keywords: Macpherson strut suspension, semi-active suspension, road variation, identification, LQG control

1. INTRODUCTION

The roles of a suspension system are to support the vehicle weight, to isolate the vehicle body from road disturbances, and to maintain the traction force between the tire and the road surface. Suspension systems are classified into a passive system and an active system according to the existence of control input. The active suspension system can be further classified into two types: a semi-active system and a fully active system according to the control input generation mechanism. Semi-active suspension systems are getting more attention because of their low cost and competitive performance to fully active ones. In this paper, a road-adaptive LQG control for the semi-active Macpherson strut suspension system of hydraulic type is investigated.

The performance of a suspension system is characterized by the ride quality, the handling performance of vehicle, the size of the rattle space, and the dynamic tire force. The prime purpose of adopting an active/semi-active suspension system is to improve the ride quality and the handling performance of vehicle. For a fixed suspension spring constant, the better isolation of the sprung mass the road disturbances can be achieved with a soft damping by allowing a larger suspension deflection. However, the better road contact can be achieved with a hard damping by not allowing unnecessary suspension deflections. Therefore, the ride quality and the handling performance of vehicle are two conflicting criteria in the control system design of suspension systems.

In this paper, assuming the use of a single acceleration sensor for the quarter-car model, a road-adaptive LQG control is investigated: First, a new control-oriented model for the semi-active Macpherson suspension system is derived. Second, a control structure including the estimation of road conditions using the extended least squares estimation method is suggested. Third, a LQG control using estimated road profiles is developed. One final comment is that four sensors (three vertical sensors and one lateral sensor) for the entire vehicle control are used, but the issues beyond the quarter-car model are not discussed in this paper.

2. A QUARTER-CAR MODEL

A schematic diagram for controlling the Macpherson strut suspension system including its quarter-car model is shown in Fig. 1. The quarter-car model admits the rotational motion of the unsprung mass. For the brevity of this paper, the detailed assumptions made for this model are referred to Hong et al. (1999). If the mass of the control arm is neglected and the bushing is assumed to be a pin joint, then the degree-of-freedom becomes two: z_s and θ, where z_s is the vertical displacement of the sprung mass and θ is the angular displacement of the unsprung mass. In this section, the derivation in Hong et al. (1999) is modified to fit to the control problem of the semi-active suspension system with a continuously variable damper.

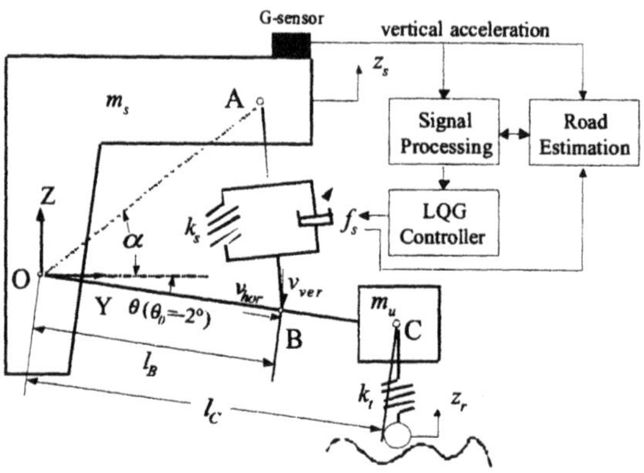

Fig. 1 A new model for the semi-active Macpherson suspension.

For the two generalized coordinates $q_1 \equiv z_s$ and $q_2 \equiv \theta$, the equations of motion become:

$$(m_s+m_u)\ddot{z}_s+m_u l_C\cos(\theta-\theta_0)\ddot{\theta}-m_u l_C\sin(\theta-\theta_0)\dot{\theta}^2 \\ +k_t\{z_s+l_C(\sin(\theta-\theta_0)-\sin(\theta_0))-z_r\}=0, \tag{1}$$

$$m_u l_C^2\ddot{\theta}+m_u l_C\cos(\theta-\theta_0)\ddot{z}_s+k_t l_C\cos(\theta-\theta_0)\{z_s \\ +l_C(\sin(\theta-\theta_0)-\sin(-\theta_0))-z_r\}-\frac{1}{2}k_s\sin(\alpha'-\theta) \\ \{b_l+\frac{d_l}{(c_l-d_l\cos(\alpha'-\theta)^{1/2})}\}=-l_B f_s, \tag{2}$$

where $a_l=l_A^2+l_B^2$, $b_l=2l_A l_B$, $\alpha'=\alpha+\theta_0$,

$c_l=a_l^2-a_l b_l\cos(\alpha')$ and $d_l=a_l b_l-b_l^2\cos(\alpha')$.

The parameters in Fig. 1 are collected in Table 1.

Table 1 Vehicle Parameters

Parameters	Descriptions	Values
m_s	Sprung mass	453Kg
m_u	Unsprung mass	36Kg
k_s	Coil spring constant	17,658 N/m
k_t	Tire spring constant	183,887 N/m
l_A	Distance from O to A	0.66 m
l_B	Distance from O to B	0.34 m
l_C	Control arm length	0.37 m
α	Angle of $\overline{OA}$	74°
θ_0	Initial angle of control arm	−2°

Now, let the state variables be $[x_1\ x_2\ x_3\ x_4]^T = [z_s\ \dot{z}_s\ \theta\ \dot{\theta}]^T$ and let the output be $y=\ddot{z}_s$. The linearization of (1)-(2) at an equilibrium point $x_e=(x_{1e},x_{2e},x_{3e},x_{4e})=(0,0,\theta_0,0)$ yields:

$$\dot{x}(t)=A_m x(t)+B_1 f_s+B_2 z_r(t), \quad x(0)=x_0 \tag{3}$$

$$y(t)=C_m x(t)+D_1 f_s+D_2 z_r, \tag{4}$$

where

$$A_m=\begin{bmatrix}0&1&0&0\\a_{21}&0&a_{23}&0\\0&0&0&1\\a_{41}&0&a_{43}&0\end{bmatrix}=\begin{bmatrix}0&1&0&0\\-0.49437&0&21.177&0\\0&0&0&1\\-13796&0&-5105.4&0\end{bmatrix},$$

$$B_1=\begin{bmatrix}0\\ \dfrac{l_B\cos(-\theta_0)}{m_s l_C+m_u l_C\sin^2(-\theta_0)}\\0\\ -\dfrac{(m_s+m_u)l_B}{m_s m_u l_C^2+m_u^2 l_C^2\sin^2(-\theta_0)}\end{bmatrix}, \quad B_2=\begin{bmatrix}0\\ \dfrac{k_t l_C\sin^2(-\theta_0)}{m_s l_C+m_u l_C\sin^2(-\theta_0)}\\0\\ \dfrac{m_s k_t l_C\cos(-\theta_0)}{m_s m_u l_C^2+m_u^2 l_C^2\sin^2(-\theta_0)}\end{bmatrix},$$

$$C_m=[a_{21}\ \ 0\ \ a_{23}\ \ 0],$$

$$D_1=\left[\frac{l_B\cos(-\theta_0)}{m_s l_C+m_u l_C\sin^2(-\theta_0)}\right],$$

$$D_2=\left[\frac{k_t l_C\sin^2(-\theta_0)}{m_s l_C+m_u l_C\sin^2(-\theta_0)}\right],$$

and see Hong et al. (1999) for the expressions of a_{21}, a_{23}, a_{41}, and a_{43}.

3. ROAD ADAPTATION

The two control objectives are the improvement of both the ride quality and the handling performance of vehicle. If fixed control gains are used, these two conflicting objectives cannot be achieved. However, by adapting road conditions, i.e., by changing controller gains for various road conditions, both objectives can be achieved (Gordon and Sharp, 1998). In this paper, as shown in Fig. 2, a LQG control with road adaptation, in the form of an output feedback control, is proposed.

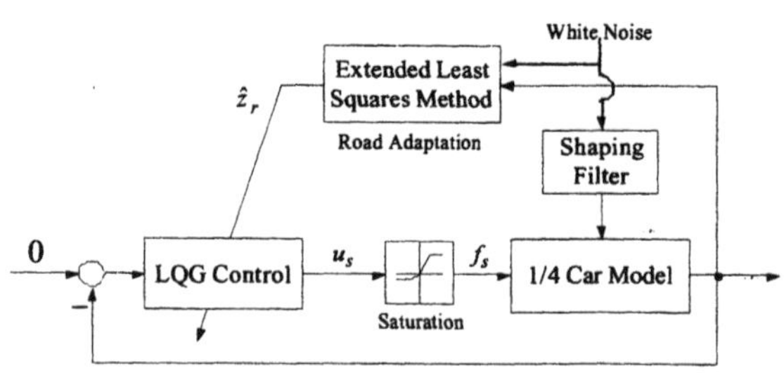

Fig. 2. Road adapting semi-active control system.

3.1 Road Profiles

The power spectral density of a road surface is assumed to be given approximately by the following analytical form (Jeong, 1990; Crolla and Abdel Hady, 1991)

$$S_v(v)=\frac{\sum_{j=0}^{m} b_{vj}v^{2j}}{\sum_{i=0}^{n} a_{vi}v^{2i}}, \tag{5}$$

where v is the spatial frequency[cycle/m] of road surface, and a_{vi}, b_{vj} are parameters of the shaping filter. When a vehicle is moving along the road with velocity V, the excitation frequency of the road input $\omega[rad/\sec]$ becomes $\omega = 2\pi vV$. The mean squared value of road surface roughness, that is the total area of the power spectral density function, does not change with the velocity of a vehicle. Therefore we have the following relation:

$$S_r(\omega)d\omega = S_v(v)dv, \tag{6}$$

where $S_r(\omega)$ represents the power spectral density of road input with respect to displacement excitation frequency. We can express the power spectral density of road input using the relation of $dv=(2\pi V)^{-1}d\omega$ as

$$S_r(\omega)=\frac{\sum_{j=0}^{m} b_{vj}(2\pi V)^{-2j-1}\omega^{2j}}{\sum_{i=0}^{n} a_{vi}(2\pi V)^{-2i}\omega^{2i}}. \tag{7}$$

We now determine the transfer function $G_r(j\omega)=z_r(t)/\varepsilon(t)$, where $\varepsilon(t)$ is the input white noise and $z_r(t)$ is the actual road profiles, of the shaping filter, so that the power spectral density of the output colored noise of the shaping filter is the same as that of the actual road surface. The relationship between the two power spectral densities is represented as

$$S_r(\omega)=|G_r(\omega)|^2 S_\varepsilon(\omega), \tag{8}$$

where $S_\varepsilon(\omega)$ is the normalized power spectral density function of white noise. Therefore, the transfer function of the shaping filter can be determines as

$$G_r(s)=\sqrt{(2\pi V)^{2n-2m-1}}\frac{\sqrt{b_{vm}}\prod_{j=0}^{m}(s-z_j)}{\sqrt{a_{vn}}\prod_{i=0}^{n}(s-p_i)} \tag{9}$$

where p_i and z_j are negative roots of the following equations:

$$\sum_{i=0}^{n}(-1)^i a_{vi}(2\pi V)^{-2i}s^{2i}=0, \tag{10}$$

$$\sum_{j=0}^{m}(-1)^j b_{vj}(2\pi V)^{-2j-1}s^{2j}=0. \tag{11}$$

By increasing the orders(m, n) in (5), the statistical properties of road input will be increased. However, the increase of the orders will increase the calculation time, which may deteriorate the overall control performance. In this paper, $m=1$, and $n=2$ are used.

3.2 Extended Least Square

The shaping filter of (9) can be rewritten in the following ARMA (auto-regressive moving averages) model with respect to input white noise $\varepsilon(t)$ by using the zero-order holder and z-transform.

$$G_r(z^{-1})=\frac{z_r(t)}{\varepsilon(t)}=\frac{1+b_{k_1}z^{-1}+b_{k_2}z^{-2}}{1+a_{k_1}z^{-1}+a_{k_2}z^{-2}}. \tag{12}$$

The ARMA model can be rewritten in the following regression form:

$$z_r(t)=\Phi^T(t-1)\theta+\varepsilon(t), \tag{13}$$

where θ is the parameter vector to be estimated and $\Phi(t-1)$ includes all known variables as

$$\theta=[a_{k_1}\quad a_{k_2}\quad b_{k_1}\quad b_{k_2}]^T, \tag{14}$$

$$\Phi(t-1)=[-z_r(t-1)\ \ -z_r(t-2)\ \ \varepsilon(t-1)\ \ \varepsilon(t-2)]^T. \tag{15}$$

The estimated output $\hat{z}_r(t)$ is denoted as

$$\hat{z}_r(t)=\Phi^T(t-1)\hat{\theta}(t). \tag{16}$$

The parameter vector $\hat{\theta}$ should be chosen to minimize the least squares loss function J_{LS} in terms of the estimated error $e_s(t)=z_r(t)-\hat{z}_r(t)$

$$J_{LS}(\hat{\theta},t)=\frac{1}{2}\sum_{i=1}^{t}\lambda^{t-i}\left\{z_r(i)-\Phi^T(i-1)\hat{\theta}(i)\right\}^2, \tag{17}$$

where λ, $0<\lambda\le 1$, is the forgetting factor. In this work, $\lambda=0.9$ is used. If $\Phi(t-1)$ has full rank, then the least squares estimate $\hat{\theta}(t)$ satisfies the recursive equations

$$\hat{\theta}(t)=\hat{\theta}(t-1)+L(t)[z_r(t)-\Phi^T(t)\hat{\theta}(t-1)], \tag{18}$$

$$L(t)=\frac{P(t-1)\Phi(t)}{\lambda+\Phi^T(t)P(t-1)\Phi(t)}, \tag{19}$$

$$P(t)=\frac{1}{\lambda}\{I-L(t)\Phi^T(t)\}P(t-1). \tag{20}$$

In this paper, the first two terms in (15) are obtained from the measured data $\ddot{z}_s$. The transfer function from $z_r(s)$ to $\ddot{z}_s(s)$ is derived from (3) and (4) as follows:

$$G_{rs}(s)\overset{\Delta}{=}\frac{\ddot{z}_s(s)}{z_r(s)}=C'(sI-A')^{-1}B_2+D_2$$

$$=\frac{0.5s^4+17212.4s^3+317325s^2-11.5s-212.6}{s^4+45.7s^3+509765s^2+17212.4s+317359}, \tag{21}$$

where $A'=A+B_1H$, $C'=C+D_1H$, (A, B_2, C, D_2) are from (3)-(4), and H satisfies $f_s = c_p\Delta\dot{l} \cong Hx$, where $H \overset{\Delta}{=} c_p\left[0\ \ 0\ \ \partial\Delta\dot{l}/\partial\theta\ \ \partial\Delta\dot{l}/\partial\dot{\theta}\right]_{(0,0,\theta_0,0)}$. It is noted that H has been introduced to avoid possible roots on the imaginary axis. In this work, $H=[0\ \ 0\ \ 0\ \ 614]$ has been used. Therefore, using the transfer function above, the road input can be estimated as follows:

$$z_r(s)=\frac{40\pi}{s+40\pi}\cdot G^{-1}{}_{rs}(s)\cdot\ddot{z}_s(s), \tag{22}$$

where a low-pass filter of cutoff frequency 20 Hz has been added to enhance the stability of the filter.

The last two terms in (15) are approximated by using (13) and (16) as follows:

$$\varepsilon(t)\cong\zeta(t)=z_r(t)-\Phi^T(t-1)\hat{\theta}(t). \tag{23}$$

Finally, $\Phi(t-1)$ takes the form as

$$\Phi(t-1)=[-z_r(t-1)\ \ -z_r(t-2)\ \ \zeta(t-1)\ \ \zeta(t-2)]^T. \tag{24}$$

4. A ROAD-ADAPTIVE LQG CONTROL

In this section, by incorporating the estimated road roughness in the performance index, a road adaptive LQG controller is designed. The control objective is achieved by leffing the semi-active suspension system to track the control force calculated in the LQG controller.

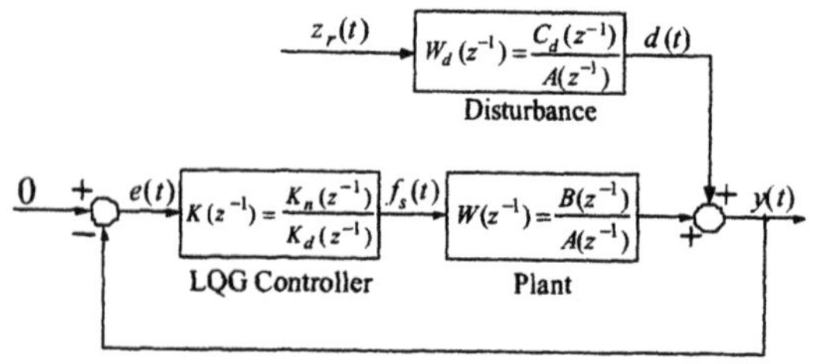

Fig. 3. Output feedback control system.

4.1 LQG Optimal Control

In Fig. 3, the transfer function $W(z^{-1})$ from f_s to y and the transfer function $W_d(z^{-1})$ from z_r to y are obtained by z-transforming (3)-(4) with sampling time 0.01sec.

$$W(z^{-1})=\frac{y(t)}{f_s(t)}\overset{\Delta}{=}\frac{B(z^{-1})}{A(z^{-1})}, \tag{25}$$

$$W_d(z^{-1})=\frac{y(t)}{z_r(t)}\overset{\Delta}{=}\frac{C_d(z^{-1})}{A(z^{-1})}. \tag{26}$$

It is noted that $W(z^{-1})$ must not contain unstable modes and $W_d(z^{-1})$ should be asymptotically stable. Also, the white noise $\varepsilon(t)$ is assumed to be $N(0,\sigma^2)$. Let $S(z^{-1})$ and $M(z^{-1})$ be the sensitivity function and the control sensitivity function, respectively, as follows:

$$S(z^{-1})=\frac{y(t)}{d(t)}=\frac{1}{1+W(z^{-1})K(z^{-1})}, \tag{27}$$

$$M(z^{-1})=-\frac{u_s(t)}{d(t)}=\frac{K(z^{-1})}{1+W(z^{-1})K(z^{-1})}. \tag{28}$$

The performance index is defined as follows:

$$J=\frac{1}{2\pi j}\oint_{|z|=1}X(z^{-1})\frac{dz}{z}. \tag{29}$$

$X(z^{-1})$ is expressed as

$$X(z^{-1})=Q_c\Phi_{ee}+R_c\Phi_{uu}+G_c\Phi_{ue}+G_c^*\Phi_{eu}. \tag{30}$$

where * denotes the adjoint of a polynomial; Q_c, R_c, and G_c are dynamic weightings as follows:

$$Q_c=\frac{B_q^*B_q}{A_q^*A_q},\quad R_c=\frac{B_r^*B_r}{A_r^*A_r},\quad G_c=\frac{B_q^*B_r}{A_q^*A_r}. \tag{31}$$

For the sake of brevity the arguments of shift operators are suppressed. In equation (30), Φ_{uu}, Φ_{ee}, and Φ_{ue} are power spectra of the control, error, and cross-weighting terms as

$$\Phi_{uu}=M\Phi_{dd}M^*, \tag{32}$$

$$\Phi_{ee}=S\Phi_{dd}S^*, \tag{33}$$

$$\Phi_{ue}=M^*\Phi_{dd}S, \tag{34}$$

where Φ_{dd} denotes the power spectrum of the disturbance as

$$\Phi_{dd}=Y_f^*Y_f=\left\{\frac{C_d}{A}\varepsilon(t)\right\}^*\left\{\frac{C_d}{A}\varepsilon(t)\right\}=\frac{C_d^*C_d}{A^*A}. \tag{35}$$

Substituting (32)-(34) and (35) into (30) yields:

$$X=Y_f^*[M^*(W^*Q_cW+R_c-W^*G_c-G_c^*W)M+Q_c -M^*W^*Q_c(z^{-1})-Q_cWM+MG_c+M^*G_c^*]Y_f(z^{-1}). \tag{36}$$

This equation suggests that the following spectral factors are defined:

$$Y_c^*Y_c=W^*Q_cW+R_c-W^*G_c-G_c^*W, \tag{37}$$

$$\Phi_h=W^*\Phi_{dd}Q_c-G_c^*\Phi_{dd}. \tag{38}$$

Hence, the substitution of (37) and (38) into (36) gives:

$$\begin{aligned}X&=Y_f^*[M^*(W^*Q_cW+R_c-W^*G_c-G_c^*W)M+Q_c\\&\quad-M^*W^*Q_c-Q_cWM+MG_c+M^*G_c^*]Y_f\\&=(Y_cMY_f-Y_c^{*-1}\Phi_hY_f^{*-1})^*(Y_cMY_f-Y_c^{*-1}\Phi_hY_f^{*-1})\\&\quad+Q_c\Phi_{dd}-Y_c^{*-1}\Phi_hY_f^{*-1}Y_f^{-1}\Phi_h^*Y_c^{-1}\end{aligned}. \tag{39}$$

The spectral factor $Y_c^*Y_c$ is defined by substituting equation (25) and (31) into (30).

$$Y_c^*Y_c=\frac{D_c^*D_c}{AA^*A_qA_q^*A_rA_r^*}. \tag{40}$$

Using (40), the terms of (39) are derived, respectively, as

$$Y_cMY_f=\frac{D_cC_dK_n}{AA_qA_r(AK_d+BK_n)}, \tag{41}$$

$$Y_c^{*-1}\Phi_hY_f^{*-1}=\frac{B_qC_d(A_r^*B^*B_q^*-A^*A_q^*B_r^*)}{D_c^*AA_q}. \tag{42}$$

4.2 Optimization

It is necessary to introduce several Diophantine equations to simplify the polynomial expression. The term in (42) can be written as a partial-fraction expansion, in terms of the solution G and F, of the following Diophantine equation:

$$D_c^*Gz^{-g}+FAA_q^*=B_qC_d(A_r^*B^*B_q^*-AA_q^*B_r^*)z^{-g}. \tag{43}$$

where g is the smallest value which ensures the terms of equation (43) to involve only polynomials in z^{-1}, whereas D_c^*, for example, is a polynomial in z.

Attention may now return to the optimization argument based on equation (39). The first term in (39) is expressed as

$$Y_cMY_f-Y_c^{*-1}\Phi_hY_f^{*-1}=\frac{D_cC_dK_n}{AA_qA_r(AK_d+BK_n)}-\frac{G}{AA_q}-\frac{Fz^g}{D_c^*}. \tag{44}$$

A second Diophantine equation is required to calculate $D_c C_d$ in (44). The equation follows in terms of the solution polynomials H and F, where F also arose in (43):

$$D_c^* A_r H z^{-g} - FBA_r A_q = (B_r B_r^* A^* A_q A_q^* - B^* B_q^* B_r A_q A_r^*) C_d z^{-g} . \tag{45}$$

The adding of (43) and (45), appropriately multiplied, gives:

$$D_c^* (GA_r B + HAA_r) z^{-g} = D_c^* D_c C_d z^{-g} . \tag{46}$$

By canceling the common D_c^* and z^{-g} in both sides of (46), $D_c C_d$ is obtained as

$$GA_r B + HAA_r = D_c C_d . \tag{47}$$

Substituting (47) into (44) gives:

$$Y_c M Y_f - Y_c^{*-1} \Phi_h Y_f^{*-1} = \frac{HK_n - GK_d}{A_q (AK_d + BK_n)} - \frac{Fz^g}{D_c^*} . \tag{48}$$

Thus, the expression (48) has been separated into a causal and non-causal terms. The only term in the performance index (29) which depends upon the controller K is the first causal term. It follows that the performance index is minimized when the first causal term is zero.

$$K = \frac{K_n}{K_d} = \frac{G}{H} . \tag{49}$$

In section 3.2, using the estimated parameters of the shaping filter, the weightings in (31) are designed as follows:

$$A_q = A_r = A_k , \; B_q = B_k , \; B_r = \rho B_k . \tag{50}$$

Hence, the desired control force is determined as

$$u_s(t) = -Ky(t) \tag{51}$$

where K is defined in (49).

4.3 Continuously Variable Damper

Even though the calculation of control input is carried out by (51), the generation of control input should be provided by a variable damper. Also, because of saturation, the actual control force from a CVD is limited as follows:

$$f_s = \begin{cases} f_s^* , & if \quad f_s^* \le u_s \\ u_s , & if \quad f_{s*} < u_s < f_s^* \\ f_{s*} & if \quad f_{s*} \ge u_s \end{cases} \tag{52}$$

where f_s^* and f_{s*} denote the maximum and the minimum damping forces available at a given relative velocity. With this control law, the actuator saturation would be unavoidable. Therefore, the control performance of a semi-active damper would be less perfect in severe road conditions.

The characteristics of a typical CVD are depicted in Fig. 4. The solid line in the middle section denotes the damping force characteristics of a typical passive damper. The dotted line of highest slope denotes the characteristics for 0 Ampere current input, which is the most hard case. The triangles overlaid on top of the dotted line represent the lookup table values of the maximum damping force. The dotted line of lowest slope denotes the minimum damping force characteristics for 1.6 Ampere current input. Again, the squares denote the lookup table values of the minimum damping force. The damping forces in the extension region (the first quarter in Fig. 4) are larger than those in the compression region (the third quarter).

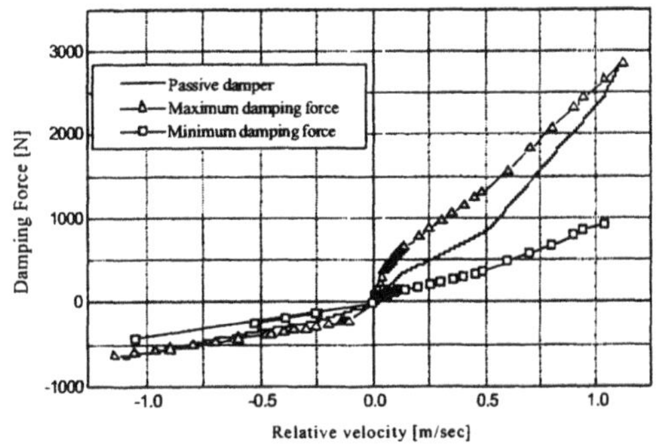

Fig. 4. Damping force characteristics of a typical continuously variable damper.

5. SIMULATIONS

Through computer simulations, the performance of the road-adaptive semi-active suspension system that takes into account of the road roughness Fig. 5 is compared to that of a conventional semi-active suspension system (skyhook control) in Fig. 6. The sampling time used during the system identification is 0.01sec, and the initial values used for $P(t-1)$ and λ are $1 \times 10^6 I$ and 0.9, respectively.

In Fig. 6, a vehicle is driven on a paved road for 10sec, a standard unpaved road for 10sec, and again on a paved road for 10sec, sequentially, at the speed of 60km/h. Fig. 6(a) compares the ride quality of the road-adaptive LQG control to that of a passive suspension system (c_p=2,000 Nsec/m). Also, Fig. 6(b) compares the ride quality of the road-adaptive LQG control to that of a conventional skyhook control. The root mean square(RMS) values for the road-adaptive LQG control, the conventional skyhook control, and the passive suspension system during the entire period are 3.20 m/s^2, 4.04 m/s^2 and 5.37 m/s^2, respectively. As can be noticed in Fig. 6(b), for the first 10sec, the vertical accelerations of the road-adaptive LQG control and the conventional skyhook control are similar, but the ride quality of the road-adaptive LQG control is superior to that of a conventional skyhook control when the vehicle reaches the unpaved road. During the unpaved road, the RMS values for the road-adaptive LQG control and the conventional skyhook control show much larger difference: 3.74 m/s^2 and 4.48 m/s^2, respectively.

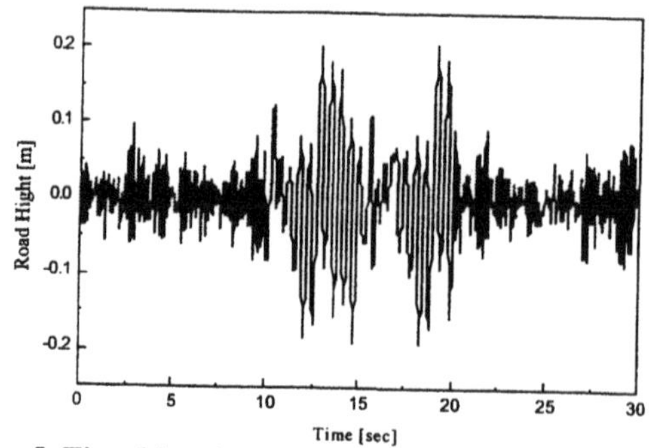

Fig. 5. Time histories of a typical road disturbance.

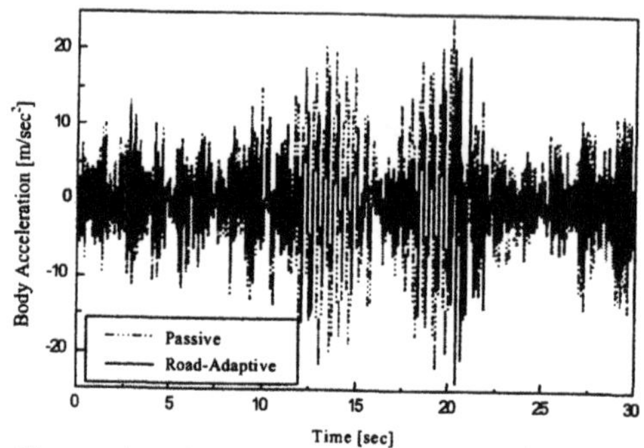

(a) Comparison between the passive damper and the road adaptive LQG control.

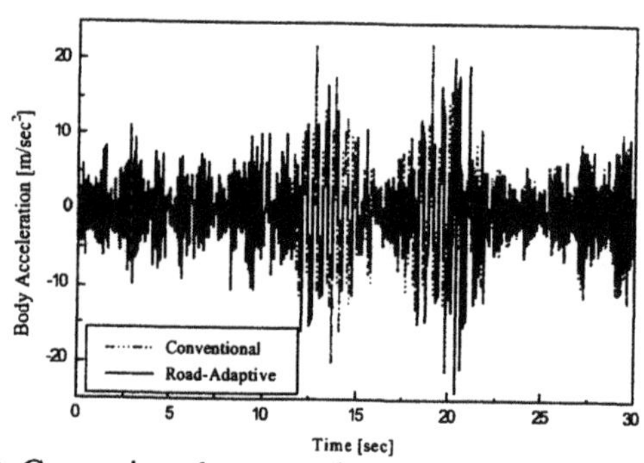

(b) Comparison between the conventional skyhook control and the road adaptive LQG control.

Fig. 6. Acceleration responses of the sprung mass.

6. CONCLUSIONS

In this paper, a new control-oriented model and a road-adaptive LQG control scheme for the semi-active Macpherson suspension system were investigated. The new model incorporates the rotational motion of the unsprung mass, giving a better description for the plant dynamics and keeping the degree-of-freedom of the plant model by two. Upon the requirement of using only one acceleration sensor, a LQG control with road adaptation was developed. The road estimation and the control variables estimation are based upon the linearized model developed. Because the control law design, the road estimation method, developed in this paper are not restricted to a particular suspension system, the entire strategy can be extended to any semi-active system including the ER damper and the MR damper.

ACKNOWLEDGMENT

This work was supported in part by the Basic Research Program of the Korea Science and Engineering Foundation under grant no. 2001-1-30400-006-2.

REFERENCES

Alleyne, A. and J.K. Hedrick (1995). Nonlinear Adaptive Control of Active Suspensions. *IEEE Transaction on Control Systems Technology*, **3(1)**, 94-101.

Besinger, F.H., D. Cebon and D.J. Cole (1995). Force Control of a Semi-Active Damper. *Vehicle System Dynamics*, **24**, 695-723.

Crolla, D.A. and M.B.A. Abdel Hady (1991). Active Suspension Control; Performance Comparisons using Control Laws Applied to a Full Vehicle Model. *Vehicle System Dynamics*, **20**, 107-120.

Gordon, T.J. and R.S. Sharp (1998). On Improving The Performance of Automotive Semi-Active Suspension Systems through Road Preview. *Journal of Sound and Vibration*, **217(1)**, 163-182.

Hac, A. (1995). Decentralized Control of Active Vehicle Suspension with Preview. *ASME Transactions, Journal of Dynamic Systems, Measurement, and Control*, **117(4)**, 478-483.

Hong, K.S., D.S. Jeon, W.S. Yoo, H. Sunwoo, S.Y. Shin, C.M. Kim and B.S. Park (1999). A New Model and an Optimal Pole-placement Control of the Macpherson Suspension System *SAE International Congress and Exposition*, **1999-01-1331**, 267-276.

Jeong, W.B. (1990). State Estimation of Road Surface and Vehicle System Using a Kalman Filter. *JSME International Journal Series III*, **33(4)**, 528-534.

Karnopp, D.C., M.J. Crosby and R.A. Harwood (1974). Vibration Control Using Semi-Active Force Generators. *ASME Journal of Engineering for Industry*, **96(2)**, 619-626.

Kim, H. and Y.S. Yoon (1995). Semi-Active Suspension with Preview Using a Frequency-Shaped Performance Index. *Vehicle System Dynamics*, **24**, 759-780.

Lin, J.S. and I. Kanellakopoulos (1997). Nonlinear Design of Active Suspensions. *IEEE Control System Magazine*, **17(3)**, 45-59.

Novak, M. and M. Valasek (1996). A New Concept of Semi-Active Control of Trucks Suspension. *Proc. of AVEC 96, International Symposium on Advanced Vehicle Control*, 141-151.

Robson, J.D. (1979). Road Surface Description and Vehicle Response. *Int. J. Vehicle Design*, **1(1)**, 25-35.

Sharp, R.S. and D.A. Crolla (1987). Road Vehicle Suspension System Design – a Review. *Vehicle System Dynamics*, **16**, 167-192.

SELF-TUNING SWING CONTROL OF A VARIABLE-LENGTH PENDULUM

Hiroyuki Sogo * Akira Matsumoto * Toru Yamamoto **

* *Department of Electro-Mechanical Systems Engineering, Takamatsu National College of Technology, 355 Chokushi, Takamatsu, Kagawa 761-8058, Japan*
** *Department of Technology and Information Education, Graduate School of Education, Hiroshima University, 1-1-1 Kagamiyama, Higashi-Hiroshima, Hiroshima 739-8524, Japan*

Abstract: In this paper, a new self-tuning swing control scheme of a variable-length pendulum is proposed, which is utilized the Coriolis' force. Using the newly propose control scheme, the pendulum can be controlled without knowing information about the system parameters and the phase angle information between the swing angle of the pendulum and its velocity. Furthermore, the algorithm considering disturbances for the swing angle direction of the pendulum is considered. The effectiveness of the newly proposed control scheme is experimentally evaluated on the swing control system of the variable-length pendulum. *Copyright© 2001 IFAC*

Keywords: Self-tuning regulators, Discrete digital dynamic control, Nonlinear systems, Dead zones, Disturbance

1. INTRODUCTION

It is well known that the variable-length pendulum(VLP) which is composed of rigid pendulum and sliding mass, is controlled by Coriolis' force. Since Coriolis' force is expressed by the outer product of rectilinear velocity with rotational velocity, this system can not be exactly described as a linear model around an equilibrium point. The swing control scheme of the VLP in which the energy based method is employed, has been reported(Yoshida *et al.*, 2000a; Yoshida *et al.*, 2000b). According to the energy based method, since control input is calculated based on the stability condition of the system which is derived based on the mechanical energy of the VLP system, it is necessary to grasp some system parameters and phase angle conditions in advance. On the other hand, the self-tuning regulator(Åström and Wittenmark, 1989; Wellstead and Zarrop, 1991) has been widely used for systems with unknown parameters.

In this paper, a design scheme of self-tuning regulators is proposed for the VLP system described as one-input two-output system. According to the proposed scheme, the regulator can be designed without a priori information with respect to system parameters and phase angle information between the swing angle of the pendulum and its velocity. However, it is well known that a self-tuning control method is easily affected by disturbances. Then, the parameter estimation algorithm considering disturbances is discussed in this paper. This paper is organized as follows. The equation of motion of the VLP system is first derived, followed by discretizing this model. Next, the self-tuning swing controller is designed, in which an improved scheme of control performance for disturbances is especially considered. Finally, the newly proposed scheme is experimentally evaluated on the swing control of the VLP system.

2. MODELING OF VLP SYSTEM

2.1 *Equation of motion*

Fig.1 shows the simulation model of the VLP. The VLP is composed of the rigid pendulum and the sliding mass. The sliding mass can be moved along the rigid pendulum. The rigid pendulum is controlled by this motion. J_O is the moment of inertia of the rigid pendulum; m_O is the mass of sliding mass; m_P is the mass of rigid pendulum; L_P is the distance from the rigid pendulum axis to its center of gravity; L_R is the initial position of the sliding mass; $\theta(t)$ is the angle of the rigid pendulum and $\delta_R(t)$ is the displacement of the sliding mass from the initial position.

The sliding mass is driven by a DC servomotor. Then $\delta_R(t)$ can be expressed by using the angle of the motor $\alpha_W(t)$ as follows:

$$\delta_R(t) = R_W G_R \alpha_W(t), \tag{1}$$

where G_R and R_W denote the reduction ratio of the gear head and radius of pinion gear, respectively. The equation of motion with respect to $\theta(t)$ is written as follows:

$$\begin{aligned}&\Big\{ J_O + m_O\{L_R + R_W G_R \alpha_W(t)\}^2\Big\}\ddot{\theta}(t)\\ &+ 2m_O R_W G_R\{L_R + R_W G_R\alpha_W(t)\}\dot{\alpha}_W(t)\dot{\theta}(t)\\ &+ \Big\{m_O\{L_R + R_W G_R \alpha_W(t)\} + m_P L_G\Big\}\\ &\times g\sin\theta(t) = 0. \end{aligned}\tag{2}$$

In this equipment, since the DC servomotor is controlled by a velocity control motor driver, $\dot{\alpha}_W(t)$ is expressed by using the input voltage of the motor driver $V_{in}(t)$ as follows:

$$\dot{\alpha}_W(t) = K_M V_{in}(t), \tag{3}$$

where K_M denotes the proportional constant of the motor driver. In this system, the control input is the voltage of the motor driver $V_{in}(t)$, and the system outputs are the angle of the rigid pendulum $\theta(t)$ and the angle of the motor $\alpha_W(t)$. That is, the control input $u(t)$ and system output vector $\boldsymbol{y}(t)$ are mathematically described as

$$u(t) = V_{in}(t) \tag{4}$$

$$\boldsymbol{y}(t) = \left[\,\theta(t)\ \ \alpha_W(t)\,\right]^T. \tag{5}$$

2.2 *Derivation of discrete model*

In Eq.(2), the following assumptions are introduced in order to derive the discrete-time model:

i) $L_R + R_W G_R \alpha_W(t) \fallingdotseq L_R$

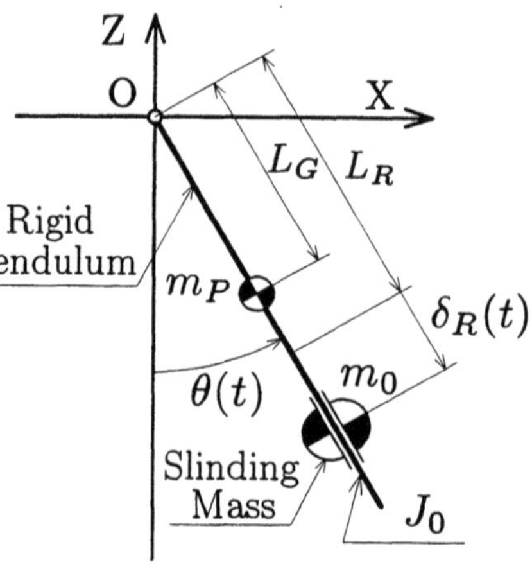

Fig. 1. Simulation model

ii) $\sin\theta(t) \fallingdotseq \theta(t)$.

Furthermore, velocity and acceleration are approximately discretized as follows:

$$\dot{\theta}(t) := \frac{\theta[k] - \theta[k-1]}{T_S} \tag{6}$$

$$\dot{\alpha}_W(t) := \frac{\alpha_W[k+1] - \alpha_W[k]}{T_S} \tag{7}$$

$$\begin{aligned}\ddot{\theta}(t) &:= \frac{\dot{\theta}(t+T_S) - \dot{\theta}(t)}{T_S}\\ &= \frac{\theta[k+1] - 2\theta[k] + \theta[k-1]}{T_S^2},\end{aligned}\tag{8}$$

where T_S denotes the sampling interval and k denotes the sampling number which has the relation $t{=}kT_S$. In Eqs.(6),(7) and (8), (t) and $[k]$ denote signals in the continuous-time and discrete-time forms, respectively. Moreover, note that $\ddot{\theta}(t)$ is approximated as the forward difference jointly with the backward difference approximation, as shown in Eq.(8). Using the data at before and behind one step which are the nearest at the current step k, yields the greatly accurate difference approximation.

Substituting Eqs.(6), (7) and (8) into Eqs.(2) and (3), and using the assumption (i) and (ii) yield

$$\begin{aligned}\theta[k+1] &= a_1\theta[k] + a_2\{\theta[k] - \theta[k-1]\}V_{in}[k]\\ &\quad -\theta[k-1]\end{aligned}\tag{9}$$

$$\alpha_W[k+1] = a_3 V_{in}[k] + \alpha_W[k], \tag{10}$$

where a_1, a_2 and a_3 are given by

$$\begin{aligned}a_1 &= 2 - \frac{\{m_O L_R + m_P L_G\}g}{J_O + m_O L_R^2}T_S^2\\ &= 2 - \omega_n^2 T_S^2\end{aligned}\tag{11}$$

$$a_2 = -\frac{2m_O L_R R_W G_R K_M}{J_O + m_O L_R^2}T_S \tag{12}$$

$$a_3 = K_M T_S. \tag{13}$$

In Eq.(11), ω_n is the natural angular frequency of the pendulum. Note that system parameters included in Eqs.(9) and (10) are only three parameters which are a_1, a_2 and a_3.

3. SELF-TUNING CONTROL OF VLP SYSTEM

3.1 *Parameter estimation*

The system parameters a_1, a_2 and a_3 are estimated by using the recursive least squares method(RLSM), where prediction error signals $\varepsilon_P[k]$ and $\varepsilon_W[k]$ are given by

$$\varepsilon_P[k] = \theta[k] - \left\{\hat{\Theta}_P^T[k-1]\varphi_P[k-1] - \theta[k-2]\right\} \quad (14)$$

$$\varepsilon_W[k] = \alpha_W[k] - \left\{\hat{\Theta}_W^T[k-1]\varphi_W[k-1] + \alpha_W[k-1]\right\}, \quad (15)$$

where the subscriptions P and W express signals with respect to the pendulum and the DC servomotor , respectively.

Furthermore, $\hat{\Theta}_i[k]$ denote the unknown parameter vector, and $\varphi_i[k-1]$ denote data vector, that is,

$$\hat{\Theta}_P[k] = \left[\, \hat{a}_1[k] \ \hat{a}_2[k] \,\right]^T \quad (16)$$

$$\hat{\Theta}_W[k] = \hat{a}_3[k] \quad (17)$$

$$\varphi_P[k-1] = \begin{bmatrix} \theta[k-1] \\ \{\theta[k-1] - \theta[k-2]\}V_{in}[k-1] \end{bmatrix} \quad (18)$$

$$\varphi_W[k-1] = V_{in}[k-1] \quad (19)$$

where the symbol "ˆ" means parameters estimated by the RLSM.

Then, the dead zone shown in Fig.2 is utilized, that is, $\Theta_i[k]$ is not updated in the case where the absolute value of $\varepsilon_i[k]$ is smaller than $\varepsilon_{i,cr}$ which is so called dead zone. $\hat{\Theta}_i[k]$ are updated by the following RLSM:

$$\hat{\Theta}_i[k] = \hat{\Theta}_i[k-1] + \frac{\Gamma_i[k-1]\varphi_i[k-1]}{\lambda + \varphi_i^T[k-1]\Gamma_i[k-1]\varphi_i[k-1]}\eta_i[k] \quad (20)$$

$$\Gamma_i[k] = \frac{1}{\lambda}\left\{\Gamma_i[k-1] - \frac{\Gamma_i[k-1]\varphi_i[k-1]\varphi_i^T[k-1]\Gamma_i[k-1]}{\lambda + \varphi_i^T[k-1]\Gamma_i[k-1]\varphi_i[k-1]}\right\} \quad (21)$$

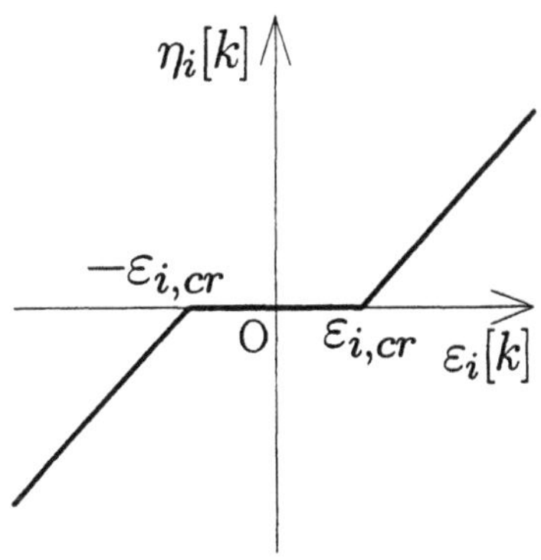

Fig. 2. Dead zone for $\varepsilon_i[k]$

$$\eta_i[k] = \begin{cases} \varepsilon_i[k] + \varepsilon_{i,cr} & (\varepsilon_i[k] < -\varepsilon_{i,cr}) \\ 0 & (|\varepsilon_i[k]| \le \varepsilon_{i,cr}) \\ \varepsilon_i[k] - \varepsilon_{i,cr} & (\varepsilon_i[k] > \varepsilon_{i,cr}). \end{cases} \quad (22)$$

3.2 *Linear quadratic regulator*

The linear quadratic regulator, is designed by minimizing the following quadratic cost function J (Clarke and Gawthrop, 1975):

$$J = E\left[\boldsymbol{y}^T[k+1]\boldsymbol{Q}\boldsymbol{y}[k+1] + \rho u^2[k]\right], \quad (23)$$

where $u[k]$ and $\boldsymbol{y}[k+1]$ denote the control input and the output vector as shown in Eq.(4) and Eq.(5). Moreover, $\boldsymbol{Q}$ and ρ are the user-specified parameters, where $\boldsymbol{Q}$ is defined as

$$\boldsymbol{Q} := \begin{bmatrix} q_\theta & 0 \\ 0 & q_\alpha \end{bmatrix}. \quad (24)$$

In order to cope with the stochastic noise in the VLP system, Eq.(23) is evaluated in the expectation form. The control input $V_{in}[k]$ which minimizes J, is derived as follows:

$$V_{in}[k] = -\Big\{q_\theta \hat{a}_2[k]\{y_1[k] - y_1[k-1]\} \times \{\hat{a}_1[k]y_1[k] - y_1[k-1]\} + q_\alpha \hat{a}_3[k]y_2[k]\Big\}/D \quad (25)$$

where D is expressed by

$$D = q_\theta \hat{a}_2^2[k]\{y_1[k] - y_1[k-1]\}^2 + q_\alpha \hat{a}_3[k]^2 + \rho. \quad (26)$$

The control input $V_{in}[k]$ becomes small in proportion to the amplitude of $\theta[k]$, when the fixed $\boldsymbol{Q}$ is employed for the VLP system. Consequently,

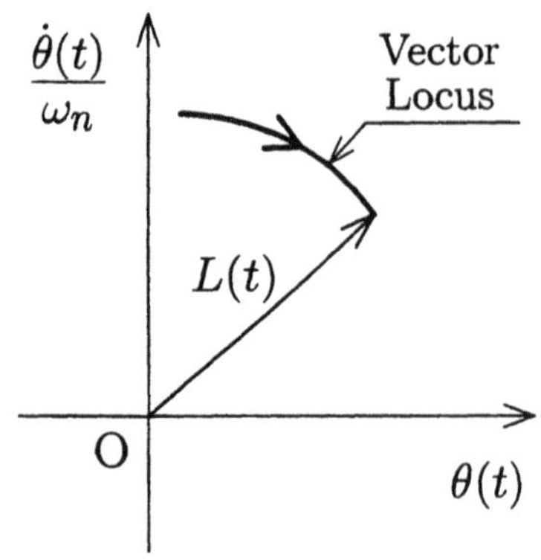

Fig. 3. Vector $\boldsymbol{L}(t)$ on the phase plane

the generated Coriolis' force also becomes small, and the swing control performance deteriorates. In this paper, q_θ is utilized as the variable weighting factor in order to prevent the degradation of the swing control performance. That is, q_θ is changed corresponding to the pendulum angle $\theta[k]$, and given by the following equation:

$$q_\theta = \frac{K_\theta}{\left\|\boldsymbol{L}(t)\right\|^2}, \tag{27}$$

where K_θ is the user-specified parameter, and $\left\|\boldsymbol{L}(t)\right\|$ means the norm of the vector $\boldsymbol{L}(t)$ as shown in Fig.3. Thus, $\left\|\boldsymbol{L}(t)\right\|^2$ is given by

$$\begin{aligned}\left\|\boldsymbol{L}(t)\right\|^2 &= \theta^2(t) + \left(\dot{\theta}(t)/\omega_n\right)^2 \\ &\cong \theta^2[k] + \frac{(\theta[k]-\theta[k-1])^2}{2-\hat{a}_1[k]}.\end{aligned} \tag{28}$$

By choosing these parameters suitably and calculating the control input based on estimates, the self-tuning regulator can be designed.

4. ALGORITHM CONSIDERING DISTURBANCES

Suppose that disturbances are put for the direction of $\theta[k]$. The flow chart of the proposed algorithm considering disturbances is shown in Fig.4. Since the prediction error $\varepsilon_P[k]$ becomes large in the case where disturbances are put, the difference of the prediction error, $\Delta\varepsilon_P[k]$ is defined for the purpose of judging the existing of disturbances as follows:

$$\Delta\varepsilon_P[k] = \Big|\varepsilon_P[k] - \varepsilon_P[k-1]\Big|. \tag{29}$$

If $\Delta\varepsilon_P[k]$ becomes larger than the threshold $\Delta\varepsilon_{P,cr}$ during last n steps, it is judged that disturbances are put. Then, the parameter estimation is stopped, and the parameters estimated at the previous step are utilized. The reason why the parameter estimation is stopped, is that estimates become considerably large owing to disturbances, and the control system falls in the unstable state. Here, n is set as $n = 4$ by trial and error in this work. The parameter estimation is re-started, when the difference $\Delta\varepsilon_P[k]$ becomes smaller than the threshold $\Delta\varepsilon_{P,cr}$ again.

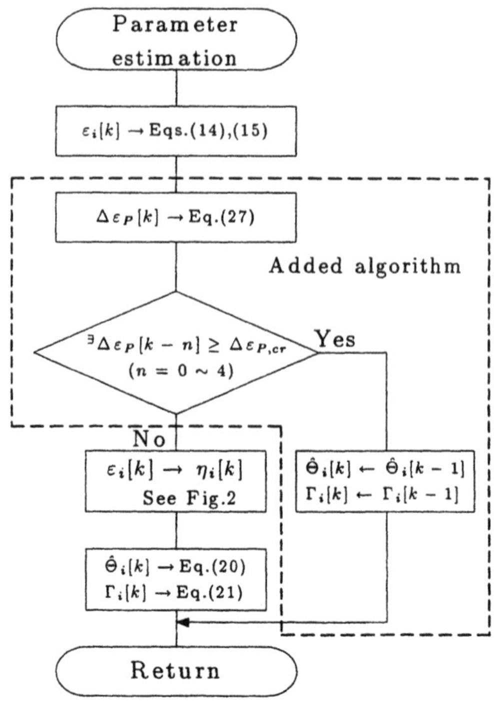

Fig. 4. Flow chart of the proposed algorithm considering disturbances

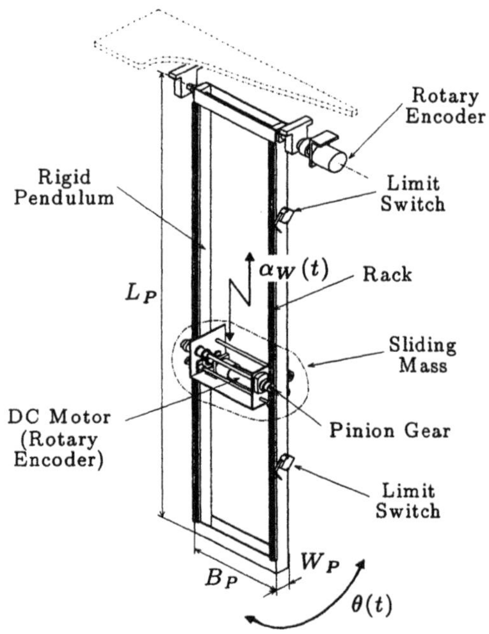

Fig. 5. Schematic figure of the experimental system

Table 1. Specifications of the experimental system

Symbol	Value	Symbol	Value
B_P	0.188(m)	L_P	1.0 (m)
G_R	0.192	R_W	0.01(m)
J_O	0.2613(kgm^2)	W_P	0.03 (m)
K_M	270(rad/sV)	m_O	0.870 (kg)
L_G	0.3742(m)	m_P	0.985 (kg)

5. EXPERIMENT WITH REAL MACHINE

5.1 *Experimental System*

The schematic figure of the experimental system is shown in Fig.5. The pendulum angle $\theta(t)$ is measured by the rotary encoder equipped at the shaft of the pendulum, and the rotation angle of the motor $\alpha_W(t)$ by the rotary encoder equipped at the DC servomotor. The specifications of the experimental system are summarized in Table 1.

The initial values of estimation parameters are set as all zero. The initial value of the covariance matrix in the RLSM, $\Gamma_i[0]$ is $100\boldsymbol{I}$ ($\boldsymbol{I}$:identity matrix), and the forgetting coefficient is set as λ=0.98. The sampling interval is set as T_S=0.1(sec). Then the user-specified parameters are designed as K_θ=35, q_α=0.005 and ρ=5e-06, respectively. The dead zones shown in Fig.2 are set as $\varepsilon_{P,cr}$=1e-3(rad) and $\varepsilon_{W,cr}$=1e-1(rad), respectively. Furthermore, the threshold to discriminating the existence of disturbances is set as $\Delta\varepsilon_{P,cr}$=2.5e-2(rad). The swing control is stopped when the extreme value of $\theta[k]$, *i.e.*, θ_{pol} drops within $|\theta_{pol}| \leq 0.02$ (rad).

The experimental procedure is shown as follows:

(1) The rigid body pendulum is the free vibration. (0~5step)
(2) Parameter estimation using PRB sequence (Shah and Cluett, 1991). (6~50step)
(3) Self-tuning swing control. (51step~)
(4) Stop the control if the local extreme value of $\theta[k]$ drops within the range determined in advance.

5.2 *Experimental Results and Discussion*

The newly proposed self-tuning regulator was employed for this system in order to verify the effectiveness. The experimental results are shown in Fig.6, where the initial values were set as $\theta[0]$=1.1(rad) and L_R=0.465(m). Fig.6 (a) indicates the value of $\theta[k]$, (b) the value of $\alpha_W[k]$, and (c) the value of $V_{in}[k]$. In Fig.6 (a), the solid line designates the control result by using the proposed self-tuning regulator, and the broken line without control. During the first 50 steps, in order to improve the reliability of estimates, the control

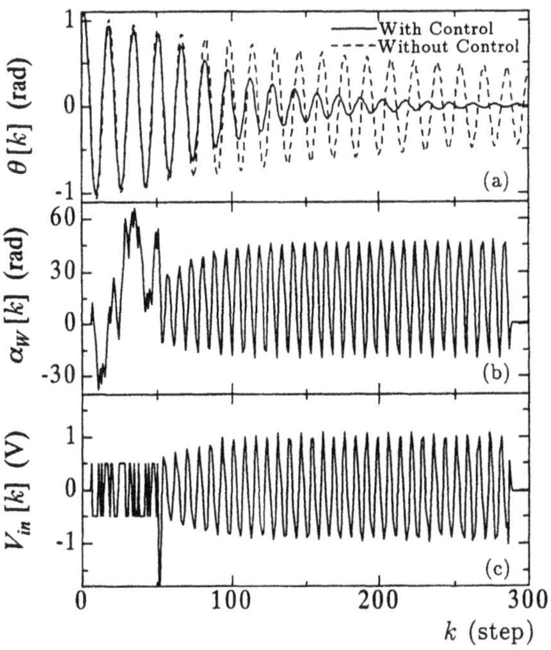

Fig. 6. Control results by using the proposed control scheme

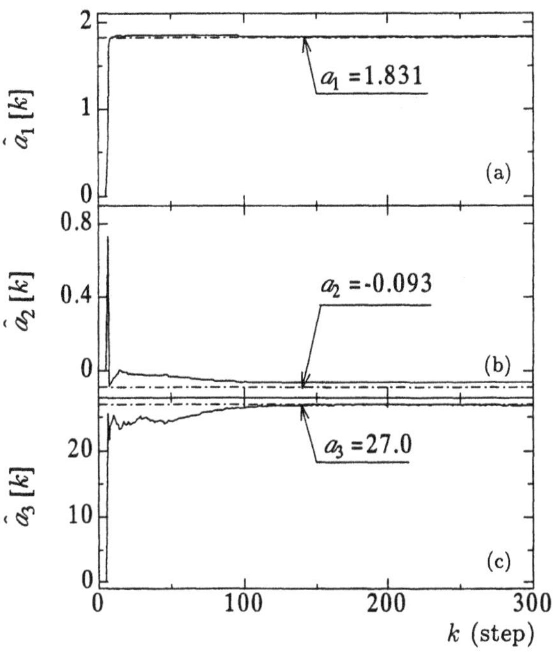

Fig. 7. Trajectories of estimation parameters

input was given by PRB sequence. The amplitude of $\theta[k]$ decreases as time passes. At 290(step), the control has normally completed, because $|\theta_{pol}|$ is smaller than 0.02(rad). After stopping the swing control, $\alpha_W[k]$ has decreased to zero. Furthermore, the trajectories of the estimated parameters are shown in Fig.7. In Fig.7, dashed lines show true values calculated by using the parameters shown in Table 1. The system parameters are adequately estimated.

Next, the experimental result in the case where disturbances are added, is shown in Fig.8. The initial conditions are equal to them in the case of Fig.6. The disturbances are put at 150(step) and 210(step). From Fig.8, the control can be carried out without diverging of the control outputs, even though disturbances are put. From this result, it is verified that the appropriate input is generated by q_θ shown in Eq.(27), and the good control performance can be obtained. Furthermore, the trajectories of the estimated parameters are shown in Fig.9. Fig.9 shows that the parameter estimation is effectively worked.

From these control results, it is clear that the newly proposed self-tuning regulator considering disturbances works well for the VLP system.

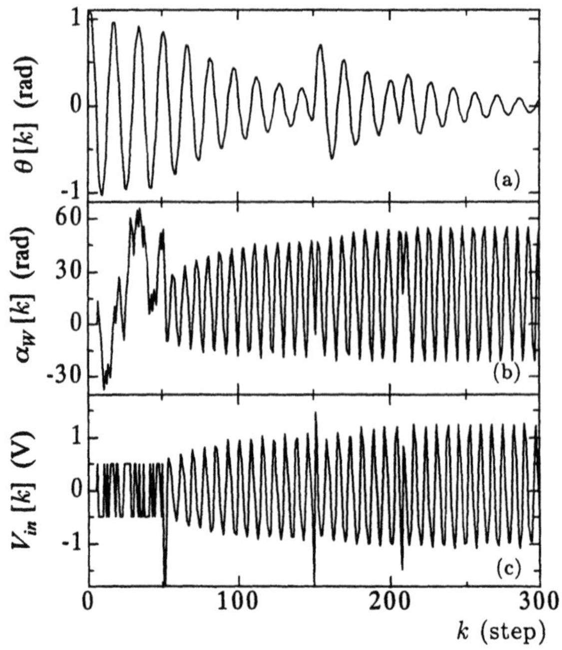

Fig. 8. Control results by using the proposed control scheme in the case where disturbances were put

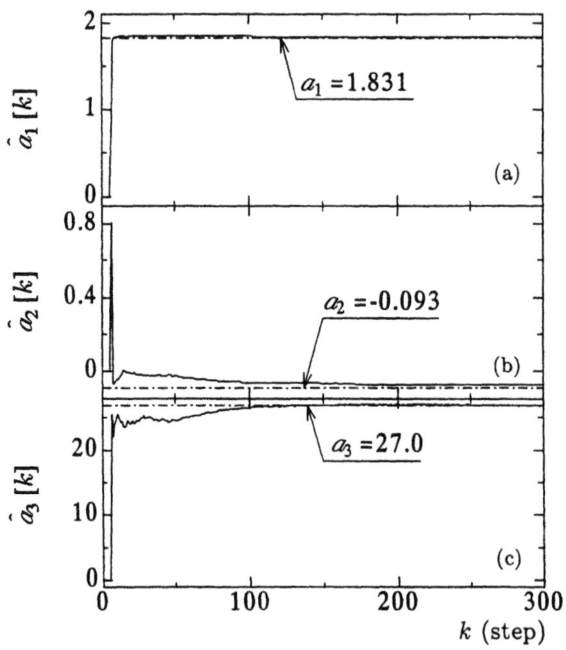

Fig. 9. Trajectories of estimation parameters in the case where disturbances were put

6. CONCLUSIONS

In this paper, the swing control of the VLP using the self-tuning regulator considering disturbances has been discussed, which is designed by minimizing the quadratic cost function. The main conclusions obtained in this paper are summarized as follows:

(1) According to the proposed scheme, the controller can be easily designed without a priori information with respect to system parameters and phase conditions.
(2) Discretizing the continuous-time equation of motion makes easy to deal with Coriolis force.
(3) By introducing of the variable weighting factor q_θ in quadratic cost function J, the control performance is further improved.

The newly proposed self-tuning regulator was experimentally evaluated on the swing control of VLP system.

7. REFERENCES

Åström,K.J. and Wittenmark,B. (1989). *Adaptive Control*, Addison - Wesly.

Clarke,D.W. and Gawthrop,P.J. (1975). Self-Tuning Controller. *Proc.IEE*, 122D-9, pp. 929-934.

Shah,S.L. and Cluett,W.R. (1991). Recursive Least Squares Based Estimation Schemes for Self-Tuning Control. *Canadian Journal of Chemical Engineering*, 69, pp. 89-96.

Wellstead,P.E. and Zarrop,M.B. (1991). *Self-Tuning Systems*, John Wiley Sons Ltd..

Yoshida,K., Kawabe,H. and Nishimura,Y. (2000a). Vibration Control for a Pendulum by Moving the Center of Gravity (in Japanese). *Trans. of ISICE*, Vol. 13, No. 10, pp. 470-477.

Yoshida,K., Uchino,T. and Nishimura,Y. (2000b). Vibration Control for a Pendulum by Sinusoidally Moving the Center of Gravity(in Japanese). *Proc. of SICE*, pp. 38-39.

PARAMETER IDENTIFICATION FOR A SCALAR LINEAR SYSTEM WITH FRACTIONAL BROWNIAN MOTION

T. E. Duncan *,[1] **B. Pasik-Duncan** *

* *Department of Mathematics*
University of Kansas
Lawrence, KS 66045, USA
duncan@math.ukans.edu
bozenna@math.ukans.edu

Abstract: In this paper, a family of estimators for an unknown parameter in the drift term of a scalar linear stochastic differential equation is given. The linear stochastic differential equation has a Brownian motion replaced by a fractional Brownian motion with the Hurst parameter in the interval $(1/2, 1)$. Such a fractional Brownian motion is not a semimartingale, so a different stochastic calculus and some different methods for the analysis of the estimators are required. The estimators are obtained by the formal use of the least squares estimators for the case of Brownian motion. It is shown that this family of estimators is strongly consistent. *Copyright © 2001 IFAC*

Keywords: identification, fractional Brownian motion, linear stochastic systems, estimation.

1. INTRODUCTION

Fractional Brownian motion is a family of Gaussian processes that are indexed by the Hurst parameter H in the interval $(0, 1)$. These processes were introduced by Kolmogorov [10]. The first application of these processes was made by Hurst [7], [8] who used them to model the long term storage capacity of reservoirs along the Nile River. Mandelbrot [12] used these processes to model some economic time series and, most recently, these processes have been used to model telecommunication traffic (e.g., [11]). Two important properties of these Gaussian processes for modeling are self similarity and, for $H \in (1/2, 1)$, a long range dependence. The self similarity means that if $a > 0$ then $(B^H(at), t \geq 0)$ and $(a^H B^H(t), t \geq 0)$ have the same probability law where $(B^H(t), t \geq 0)$ is a (standard) fractional Brownian motion. The long range dependence means that if $r(n) = \mathbb{E}[B^H(1)(B^H(n+1) - B^H(n))]$ then $\sum_{n=1}^{\infty} r(n) = \infty$.

Now a (standard) fractional Brownian motion is defined. For each $H \in (0, 1)$, a real-valued Gaussian process $(B^H(t), t \geq 0)$ is defined such that $\mathbb{E}[B^H(t)] = 0$ and

$$\mathbb{E}[B^H(t)B^H(s)] = \frac{1}{2}[t^{2H} + s^{2H} - |t - s|^{2H}]$$

for all $s, t \in \mathbb{R}_+$. If $H = 1/2$ then the fractional Brownian motion is a standard Brownian motion (Wiener process). These processes have a version with continuous sample paths. In this paper H is restricted to the interval $(1/2, 1)$ so the fractional Brownian motion has a long range dependence. The pth variation of such a process is nonzero and finite for $p = 1/H$, that is, if $(P_n, n \in \mathbb{N})$ is sequence of partitions of $[0, 1]$ that are refinements of the previous ones and become dense in [0,1] then

$$\lim_{n\to\infty} \sum \left| B^H(t_i^{(n)}) - B^H(t_{i-1}^{(n)}) \right|^p = c(p) \qquad \text{a.s.}$$

[1] Research supported in part by NSF Grant DMS 9971790.

where $P_n = \{t_0^{(n)}, \dots, t_n^{(n)}\}$ and $c(p) = \mathbb{E}|B^H(1)|^p$ (e.g., [13]). For $H > 1/2$, $(B^H(t), t \geq 0)$ is not a semimartingale and is not Markov. These facts require that a different stochastic calculus be used.

In this paper, a scalar stochastic system is described by a linear stochastic differential equation with a fractional Brownian motion with $H \in (1/2, 1)$ replacing a Brownian motion. An unknown parameter appears in the drift term of the stochastic differential equation. A family of estimators is given that arise from a least squares method for the parameter identification problem with a Brownian motion (e.g., [4], [6]). The family of estimators is shown to be strongly consistent. The methods of verification of strong consistency are different from the martingale methods for a Brownian motion because these fractional Brownian motions are not (semi)martingales. It is necessary to use a different stochastic calculus for these fractional Brownian motions with $H \in (1/2, 1)$. In this paper, the calculus described in [5] is used. Some alternative approaches are given in [1], [2]. After submission of this work, the authors became aware of [9] where the strong consistency of a family of maximum likelihood estimates for the estimation problem here is verified. The verification and the estimators in [9] are different from the verification and the estimators that are given here. There seems to be no other work on this identification problem.

2. MAIN RESULT

Let $\Omega = C_0(\mathbb{R}_+, \mathbb{R})$ be the Fréchet space of real-valued continuous functions on $\mathbb{R}_+$ with the initial value zero and the topology of local uniform convergence. There is a probability measure, P^H, on $(\Omega, \mathcal{F})$ where $\mathcal{F}$ is the Borel σ-algebra such that on the probability space $(\Omega, \mathcal{F}, P^H)$, the coordinate process is a fractional Brownian motion, that is,

$$B^H(t, \omega) = \omega(t)$$

for each $t \in \mathbb{R}_+$ and (almost) all $\omega \in \Omega$.

Fix $H \in (1/2, 1)$ and let $\phi : \mathbb{R} \to \mathbb{R}_+$ be given by

$$\phi(t) = H(2H - 1)|t|^{2H-2}.$$

It follows directly that

$$\mathbb{E}[B^H(t)B^H(s)] = \int_0^t \int_0^s \phi(u - v)\,du\,dv.$$

Let $(X(t), t \geq 0)$ be the real-valued Gaussian process that is the solution of

$$\begin{aligned} dX(t) &= \alpha_0 X(t)dt + dB^H(t) \\ X(0) &= X_0 \end{aligned} \qquad (1)$$

where $(B^H(t), t \geq 0)$ is a real-valued fractional Brownian motion with Hurst parameter $H \in (1/2, 1)$, X_0 is a zero mean Gaussian random variable that is independent of $(B^H(t), t \geq 0)$, and $\alpha_0 < 0$ is an unknown parameter. It is desired to determine a family of estimators, $(\hat{\alpha}(t), t \geq 0)$, of α_0 given the observations of the system $(X(t), t \geq 0)$ that is strongly consistent. It is elementary to verify that the solution of (1) is

$$X(t) = e^{\alpha_0 t} X_0 + \int_0^t e^{\alpha_0 (t-s)} dB^H(s) \qquad (2)$$

for $t \geq 0$.

Based on a least squares or maximum likelihood approach for a linear system with Brownian motion [6], an estimator $\hat{\alpha}(t)$ is defined as

$$\hat{\alpha}(t) = \frac{\int_0^t X(s)dX(s)}{\int_0^t X^2(s)ds} = \alpha_0 + \frac{\int_0^t X(s)dB^H(s)}{\int_0^t X^2(s)ds} \qquad (3)$$

for $t > 0$ and $\hat{\alpha}(0) = 0$. To verify strong consistency of $(\hat{\alpha}(t), t \geq 0)$, it is sufficient to show that

$$\limsup_{t \to \infty} \frac{\left| \int_0^t X(s)dB^H(s) \right|}{\int_0^t X^2(s)ds} = 0 \quad \text{a.s.} \qquad (4)$$

If $H = 1/2$, that is, the fractional Brownian motion is a Brownian motion, then the verification of (4) is elementary by a time change of the stochastic integral in the numerator by its increasing process in the denominator and the Strong Law of Large Numbers for Brownian motion. In this case of Brownian motion, the Law of the Iterated Logarithm for Brownian motion provides a rate of convergence. However, for the fractional Brownian motion with $H > 1/2$, such a direct procedure is not available.

Let $(Y(t), t \geq 0)$ be the process given by

$$Y(t) = \int_0^t X(s)dB^H(s) \qquad (5)$$

where $(X(s), s \geq 0)$ is the solution of (1). It is useful to bound the growth of $|Y(t)|$ as $t \to \infty$. Initially, it is convenient to bound an absolute moment of $Y(t)$.

Lemma 1. For $p \geq 1$

$$\mathbb{E}|Y(t)|^p \leq M t^{pH} \qquad (6)$$

where M is a constant that only depends on p and $Y(t)$ is given by (5).

Proof. Since the solution of (1) is (2), it easily follows that the ϕ-derivative [5] of $X(t)$, $D_s^\phi X(t)$, is

$$D_s^\phi X(t) = \int_0^t e^{\alpha_0 (t-u)} \phi(s - u)du.$$

Applying the rule for the ϕ-derivative of a stochastic integral (Theorem 4.2 [5]), it follows that

$$\begin{aligned} D_s^\phi Y(t) &= \int_0^t (D_s^\phi X(u))dB^H(u) \\ &\quad + \int_0^t X(u)\phi(s-u)du \\ &= \int_0^t \int_0^u e^{\alpha_0(u-v)}\phi(s-v)dvdB^H(u) \\ &\quad + \int_0^t X(u)\phi(s-u)du. \end{aligned} \tag{7}$$

Initially, an upper bound is given for the second moment of the Gaussian random variable, $D_s^\phi Y(s)$.

$$\begin{aligned} &|D_s^\phi Y(s)|^2 \\ &\quad \leq 2\left(\int_0^s \int_0^u e^{\alpha_0(u-v)}\phi(s-v)dvdB^H(u)\right)^2 \\ &\quad + 2\left(\int_0^s X(u)\phi(s-u)du\right)^2 \end{aligned}$$

$$\begin{aligned} &\mathbb{E}\left[\left(\int_0^s \int_0^u e^{\alpha_0(u-v)}\phi(s-v)dvdB^H(u)\right)^2\right] \\ &= \int_0^s \int_0^s \int_0^u e^{\alpha_0(u-\sigma)}\phi(s-\sigma)d\sigma \\ &\qquad \times \int_0^v e^{\alpha_0(v-\tau)}\phi(s-\tau)d\tau \\ &\qquad \times \phi(u-v)dudv \\ &= \int_0^s \int_0^s \int_0^u e^{\alpha_0\sigma}\phi(s-u+\sigma)d\sigma \\ &\qquad \times \int_0^v e^{\alpha_0\tau}\phi(s-v+\tau)d\tau \\ &\qquad \times \phi(u-v)dudv \\ &\leq \int_0^s \int_0^s \int_0^u e^{\alpha_0\sigma}\phi(s-u)d\sigma \\ &\qquad \times \int_0^v e^{\alpha_0\tau}\phi(s-v)d\tau \\ &\qquad \times \phi(u-v)dudv \\ &\leq K s^{6H-4}. \end{aligned}$$

Since $(X(t),\, t \geq 0)$ has a limiting distribution [3], $\mathbb{E}X^2(t)$ is uniformly bounded and

$$\begin{aligned} &\mathbb{E}\left(\int_0^s X(u)\phi(s-u)du\right)^2 \\ &\quad \leq M\left(\int_0^s \phi(s-u)du\right)^2 \\ &\quad \leq \tilde{M}s^{4H-2}. \end{aligned}$$

Thus,

$$\mathbb{E}|D_s^\phi Y(s)|^2 \leq K_1 s^{4H-2}. \tag{8}$$

By Theorem 5.2 [5], there is the inequality

$$\begin{aligned} &\mathbb{E}|Y(t)|^p \\ &\quad \leq \left[\mathbb{E}|Y(t)|^{2p}\right]^{1/2} \\ &\quad \leq \left[(2p)^{2p}\left(\int_0^t \left(\mathbb{E}|X(s)D_s^\phi Y(s)|^p\right)^{1/p} ds\right)^p\right]^{1/2} \end{aligned} \tag{9}$$

and by the Schwarz inequality

$$\begin{aligned} &\mathbb{E}\left|X(s)D_s^\phi Y(s)\right|^p \\ &\quad \leq \left[\mathbb{E}|X(s)|^{2p}\right]^{1/2}\left[\mathbb{E}\left|D_s^\phi Y(s)\right|^{2p}\right]^{1/2}. \end{aligned} \tag{10}$$

Since $D_s^\phi Y(s)$ is a zero mean Gaussian random variable, it follows by (8) that there is an M_2 that only depends on p such that

$$\left[\mathbb{E}\left|D_s^\phi Y(s)\right|^{2p}\right]^{1/2} \leq M_2 s^{2pH-p}. \tag{11}$$

Since $(X(t),\, t \geq 0)$ has a limiting Gaussian distribution, it follows that there is an M_1 that only depends on p such that

$$\left[\mathbb{E}|X(s)|^{2p}\right]^{1/2} \leq M_1.$$

It follows from (9), (10) and (11) that

$$\mathbb{E}|Y(t)|^p \leq M t^{pH}. \qquad \square$$

The following lemma provides a bound for the growth of $|Y(t)|$.

Lemma 2. For each $\beta > H$ and each $\varepsilon > 0$ the following equality is satisfied

$$\limsup_{t\to\infty} \frac{|Y(t)|}{t^{\beta+\varepsilon}} = 0 \qquad \text{a.s.} \tag{12}$$

where $Y(t)$ is given by (5).

Proof. Let $\beta > H$ and $\varepsilon > 0$ be fixed. Fix $n \in \mathbb{N}$ and consider the sequence of random variables $(|Y(k/2^n)|,\, k \in \mathbb{N})$. For $k \in \mathbb{N}$ let

$$\Lambda_k = \left\{ \frac{\left|Y\left(\frac{k}{2^n}\right)\right|}{\left(\frac{k}{2^n}\right)^\beta} \geq 1 \right\}.$$

By Markov's inequality for $p > 1$ and (5), there is the inequality

$$\begin{aligned} P(\Lambda_k) &\leq \frac{\mathbb{E}\left|Y\left(\frac{k}{2^n}\right)\right|^p}{\left(\frac{k}{2^n}\right)^{p\beta}} \\ &\leq M\left(\frac{k}{2^n}\right)^{pH-p\beta} = \tilde{M}k^{p(H-\beta)}. \end{aligned}$$

Since $H - \beta < 0$, choose $p > 1$ so that

$$\sum_{k=1}^{\infty} \frac{1}{k^{p(\beta-H)}} < \infty.$$

By the Borel-Cantelli Lemma

$$P(\Lambda_k \text{ infinitely often}) = 0.$$

Thus

$$\limsup_{k\to\infty} \frac{\left|Y\left(\frac{k}{2^n}\right)\right|}{\left(\frac{k}{2^n}\right)^{\beta+\varepsilon}} = 0 \qquad \text{a.s.}$$

where $\varepsilon > 0$. There is a null set Γ, that is $P(\Gamma) = 0$, such that if $\omega \in \Gamma^c$ then

$$\limsup_{k\to\infty} \frac{\left|Y\left(\frac{k}{2^n},\omega\right)\right|}{\left(\frac{k}{2^n}\right)^{\beta+\varepsilon}} = 0$$

for all $n \in \mathbb{N}$.

Since $(Y(t)/t^{\beta+\varepsilon}, t \geq 0)$ has continuous sample paths, it follows that

$$\limsup_{t\to\infty} \frac{|Y(t)|}{t^{\beta+\varepsilon}} = 0 \qquad \text{a.s.} \qquad \square$$

Lemma 3. Let $(X(t), t \geq 0)$ be the solution of (1). The following inequality is satisfied

$$\liminf_{t\to\infty} \frac{1}{t}\int_0^t X^2(s)ds > 0 \qquad \text{a.s.} \qquad (13)$$

Proof. Apply a small generalization of the Itô formula (Theorem 4.3, [5]) to the process $(X^2(t), t \geq 0)$ to obtain for $t \geq 0$

$$\begin{aligned} X^2(t) - X^2(0) &= 2\alpha_0 \int_0^t X^2(s)ds + 2\int_0^t X(s)dB^H(s) \\ &+ \int_0^t \int_0^s e^{\alpha_0(s-u)}\phi(s-u)duds. \qquad (14) \end{aligned}$$

Since $(X(t), t \geq 0)$ has a limiting distribution [3], it follows that

$$\limsup_{t\to\infty} \frac{X^2(t)}{t} = 0 \qquad \text{a.s.}$$

By Lemma 2 it follows that

$$\limsup_{t\to\infty} \frac{1}{t}\left|\int_0^t X(s)dB^H(s)\right| = 0 \qquad \text{a.s.} \qquad (15)$$

and by direct computation

$$\begin{aligned} &\lim_{t\to\infty} \frac{1}{t}\int_0^t \int_0^s e^{\alpha_0(s-u)}\phi(s-u)duds \\ &= \int_0^\infty e^{\alpha_0 v}\phi(v)dv. \qquad (16) \end{aligned}$$

It follows from (14–16) that (13) is satisfied. $\square$

The previous two lemmas provide the essential results to verify the strong consistency of $(\hat{\alpha}(t), t \geq 0)$.

Theorem 1. Let $(\hat{\alpha}(t), t \geq 0)$ be the family of estimators of α_0 given by (3). This family of estimators is strongly consistent, that is,

$$\lim_{t\to\infty} \hat{\alpha}(t) = \alpha_0 \qquad \text{a.s.} \qquad (17)$$

Proof. Recall that it suffices to show that

$$\limsup_{t\to\infty} \frac{\left|\int_0^t X(s)dB^H(s)\right|}{\int_0^t X^2(s)ds} = 0 \qquad \text{a.s.}$$

Since

$$\frac{\int_0^t X(s)dB^H(s)}{\int_0^t X^2(s)ds} = \frac{\frac{1}{t}\int_0^t X(s)dB^H(s)}{\frac{1}{t}\int_0^t X^2(s)ds} \qquad (18)$$

apply Lemmas 2 and 3 to the numerator and the denominator of the right hand side of (18) to obtain (17). $\square$

Remark To consider a stochastic adaptive control problem, it is necessary to have some results for the ergodic control of a controlled linear stochastic system with a fractional Brownian motion.

3. REFERENCES

[1] E. Alos, O. Mazet, and D. Nualart, Stochastic calculus with respect to fractional Brownian motion with Hurst parameter lesser than 1/2, *Stochastic Processes Appl.* **86** (2000) 121–139.

[2] L. Decreusefond and A. S. Üstünel, Stochastic analysis of the fractional Brownian motion, *Potential Analysis* **10** (1999) 177–214.

[3] T. E. Duncan, Some applications of fractional Brownian motion to linear systems. In *System Theory: Modeling, Analyses, and Control* (T. E. Djaferis and J. C. Schick, eds.), Kluwer 2000, 97–105.

[4] T. E. Duncan, L. Guo, and B. Pasik-Duncan, Adaptive continuous time linear quadratic Gaussian control, *IEEE Trans. Autom. Control* **44** (1999) 1653–1662.

[5] T. E. Duncan, Y. Z. Hu, and B. Pasik-Duncan, Stochastic calculus for fractional Brownian motion. I: Theory, *SIAM J. Control Optim.* **38** (2000) 582–612.

[6] T. E. Duncan and B. Pasik-Duncan, Adaptive control of continuous-time linear stochastic systems, *Math. Control Signals Systems* **3** (1990) 45–60.

[7] H. E. Hurst, Long-term storage capacity in reservoirs, *Trans. Amer. Soc. Civil Eng.* **116** (1951) 400–410.

[8] H. E. Hurst, Methods of using long-term storage in reservoirs, *Proc. Inst. Civil Engineers, Part I, Chapter 5* (1956) 519–590.
[9] M. L. Kleptsyna and A. Le Breton, Statistical analysis of the fractional Ornstein–Uhlenbeck type process. Preprint.
[10] A. N. Kolmogorov, Wienersche Spiralen und einige andere interessante Kurven im Hilbertschen Raum. *C. R. (Doklady) Acad. URSS (N.S.)* **26** (1940) 115–118.
[11] W. E. Leland, M. S. Taqqu, W. Willinger, and D. V. Wilson, On the self-similar nature of ethernet traffic, *IEEE/ACM Trans. Networking* **2** (1994) 1–15.
[12] B. B. Mandelbrot, The variation of certain speculative prices, *Journal of Business* **36** (1963) 394–419. Reprinted in P. H. Cootner, ed., *The Random Character of Stock Market Prices*, MIT Press, Cambridge, Mass., 1964, 297–337.
[13] L. C. G. Rogers, Arbitrage with fractional Brownian motion, *Math. Finance* **7** (1997) 95–105.

AN ITERATIVE SUBSPACE IDENTIFICATION WITH OBSERVATION OUTLIERS BASED ON WEIGHTED ORTHOGONAL DECOMPOSITION

Hideyuki TANAKA * Tohru KATAYAMA *

* *Department of Applied Mathematics and Physics,
Graduate School of Informatics, Kyoto University,
Yoshida-Hommachi, Sakyo, Kyoto, 606-8501, Japan.
htanaka@amp.i.kyoto-u.ac.jp, katayama@amp.i.kyoto-u.ac.jp*

Abstract: A subspace identification method subject to observation with outliers is considered. Based on a weighted least-squares method, a noise attenuation procedure for subspace identification is developed. Then, a rejection method of outliers is derived by changing a weighting matrix iteratively. A numerical example demonstrates the effectiveness of the proposed method.

Keywords: Identification algorithms, Subspace methods, Least-squares estimation, Robust estimation

1. INTRODUCTION

Subspace identification method has extensively been studied since early 1990s (Moonen *et al.*, 1989; Verhaegen and Dewilde, 1992). The operations involved in subspace identification are performed numerically sound way based on the singular value decomposition and the QR decomposition (LQ decomposition). One of the great advantages is that the difficult model selection problem inherent in the classical parameter optimization approaches to MIMO identification (Ljung, 1987) is completely avoided. It is however pointed out that subspace methods are sometimes sensitive to noise signals. In order to circumvent this difficulty, several methods have been proposed (Overschee and Moor, 1994; Verhaegen, 1994; Okada and Sugie, 1996); they are mainly based on an instrumental variable approach. Studies on the stochastic realization with exogenous inputs have been also reported (Picci and Katayama, 1996; Katayama and Picci, 1999).

In order to reduce the effect due to noise, the least-squares estimation has been often used under the assumption that noise obeys normal distribution with zero mean. In real world, however, this assumption is not always appropriate. In fact, there are some cases where large errors are contained in data sequences with a low probability. The least-squares estimation based on finite length of data sets is rather sensitive to the this type of non-Gaussian disturbances; what we call outliers. In statistics, robust estimation methods have been proposed to overcome this difficulty (Nakagawa and Koyanagi, 1982). System identification subject to missing data has also been studied in ARX model setting (Isaksson, 1991; Issakson, 1993; Pintelon and Schoukens, 1998).

To the best of our knowledge, subspace identification with observation outliers has not been studied in the literature. We therefore consider a subspace identification method subject to large errors with a low probability. We propose an iterative identification procedure, based on a weighted least-squares estimation method. A numerical example is also given to evaluate the effectiveness of the proposed method.

2. PRELIMINARY

Let the vector space $\mathcal{X}$ be generated by row vectors $x \in \mathbf{R}^{1\times\nu}$, and $\mathcal{Y}$ be generated by $y \in \mathbf{R}^{1\times\iota}$. Define a linear map P from $\mathcal{X}$ to $\mathcal{Y}$ as $P : \mathcal{X} \to \mathcal{Y}$, $x \mapsto xP$. Let us denote the image of P as Image(P). We define the space spanned by vectors $x_s(s = 1, 2, \cdots, t)$ as span$\{x_s \,|\, s = 1, \cdots, t\} := \{\sum_{s=1}^{t} \alpha_s x_s \,|\, \alpha_s \in \mathbf{R}\}$.

Given a non-negative definite matrix $\Theta \in \mathbf{R}^{\nu\times\nu}$, define an inner product of x and $y \in \mathbf{R}^{1\times\nu}$ as

$$(x, y)_\Theta = x\Theta y^T. \quad (1)$$

Suppose that x and y are elements belonging to the vector space with inner product $(\cdot,\cdot)_\Theta$. We say that x and y are orthogonal under inner product $(\cdot,\cdot)_\Theta$ or Θ-orthogonal when $(x,y)_\Theta = 0$ holds. We describe such a relation as $x \perp_\Theta y$.

Given elements $\{x_i, y \in \mathbf{R}^{1\times\nu} \,|\, i = 1, 2, \cdots, \kappa\}$ belonging to the space with inner product $(\cdot,\cdot)_\Theta$, we define $\hat{y}$ to be Θ-orthogonal projection onto span$\{x_i \,|\, i = 1, 2, \cdots, \kappa\}$ if

$$y = \hat{y} + \tilde{y}, \quad (2)$$

where $\hat{y} \in$ span$\{x_1, x_2, \cdots, x_\kappa\}$ and $\tilde{y}$ satisfies Θ-orthogonality condition $\tilde{y} \perp_\Theta$ span$\{x_1, \cdots, x_\kappa\}$ or equivalently $(\tilde{y}, x_i)_\Theta = 0$ for $i = 0, \cdots, \kappa$.

Define a product of two matrices $X \in \mathbf{R}^{p\times\nu}$ and $Y \in \mathbf{R}^{q\times\nu}$ as

$$\langle X, Y\rangle_\Theta := X\Theta Y^T. \quad (3)$$

We say X and Y are Θ-orthogonal if $\langle X, Y\rangle_\Theta = 0$.

3. PROBLEM SETTING AND REVIEW

In this section, we give a formulation of our identification problem. We also review a former study of subspace method identification (Picci and Katayama, 1996). A part of the procedure proposed therein is also used in this paper.

3.1 *Problem setting*

Let data sequences $u_t \in \mathbf{R}^p$ and $y_t \in \mathbf{R}^q$ be given. Suppose that y_t is the output of an unknown stable linear time-invariant discrete-time system G, corrupted by a measurement noise v_t:

$$y_t = Gu_t + v_t. \quad (4)$$

Let unknown system G be described by

$$\begin{bmatrix} x_{t+1} \\ \xi_t \end{bmatrix} = \begin{bmatrix} A & B \\ C & D \end{bmatrix} \begin{bmatrix} x_t \\ u_t \end{bmatrix}, \quad x_t \in \mathbf{R}^n, \quad (5)$$

where u_t and ξ_t satisfy $\xi_t = Gu_t$. It is assumed that the representation of system G in (5) is minimal. The identification setup is shown in Figure 1.

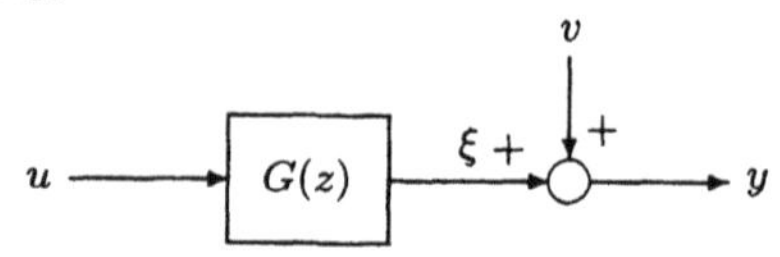

Fig. 1. System identification set-up

Assume that the noise sequence v_t obeys

$$v_t \sim \alpha\, \mathrm{N}(0, \zeta^2) + (1-\alpha)\, \mathrm{N}(0, \eta^2), \quad (6)$$

where $\mathrm{N}(0, \sigma^2)$ describes the normal distribution with zero mean and variance σ^2, and α, η and ζ are unknown parameters subject to $0 < \alpha < 0.5$, $\eta \ll \zeta$. In this setting, we can consider the case where v_t contains large errors with a low probability. Consider the following problem:

Problem *Assume that the data sequences u_t and y_t are given ($t = 1, 2, \cdots, M$). Estimate the realization of $G(z)$ within the freedom of equivalent transformation.*

3.2 *Stochastic realization and subspace method identification*

We review one of the identification method based on stochastic realization (Picci and Katayama, 1996).

Let $\{u(t), t = 0, \pm1, \cdots\}$ and $\{y(t), t = 0, \pm1, \cdots\}$ be infinite time data sets with zero mean and finite variance. Assume that there is no feedback from $u(t)$ to $y(t)$, then there is a decomposition of $y(t)$ such that

$$y(t) = \hat{y}(t) + \tilde{y}(t), \quad \hat{y}(t) \perp \tilde{y}(t), \quad (7)$$

where $\hat{y}(t) = \mathrm{E}\{y(t) \,|\, \mathcal{U}\}$ and $\tilde{y}(t) = y(t) - \hat{y}(t)$, and where $\hat{y}(t)$ and $\tilde{y}(t)$ are deterministic and stochastic parts of $y(t)$, respectively. See (Picci and Katayama, 1996) for more detail.

A realization of stochastic part $\tilde{y}(t)$ is given by

$$\begin{bmatrix} \tilde{x}(t+1) \\ \tilde{y}(t) \end{bmatrix} = \begin{bmatrix} \tilde{A} & \tilde{B} \\ \tilde{C} & I \end{bmatrix} \begin{bmatrix} \tilde{x}(t) \\ e(t) \end{bmatrix}, \quad (8)$$

where $e(t)$ is the innovation, or the one step prediction of $\tilde{y}(t)$. Also, a realization of deterministic part $\hat{y}(t)$ is described by

$$\begin{bmatrix} \hat{x}(t+1) \\ \hat{y}(t) \end{bmatrix} = \begin{bmatrix} \hat{A} & \hat{B} \\ \hat{C} & \hat{D} \end{bmatrix} \begin{bmatrix} \hat{x}(t) \\ u(t) \end{bmatrix}. \quad (9)$$

The identification method based on finite data sets is described as follows (Picci and Katayama, 1996):

Step 1. Estimate the deterministic part $\hat{y}(t)$ and the stochastic part $\tilde{y}(t)$ from the LQ decomposition.

Step 2. Compute a realization G based on $\hat{y}$, by using a subspace method such as MOESP (Verhaegen and Dewilde, 1992) or 4SID (Overschee and Moor, 1994).

This method is appropriate for the noise with $\alpha = 0$ in equation (6), because the orthogonal projection is the LS (least-squares) estimate for the output corrupted by a normally distributed noise. The noise in our setting however is not normally distributed.

In the following, by a slight modification of Step 1, we propose a new identification method for the observation with outliers.

4. WEIGHTED LEAST-SQUARES ESTIMATION

In this section, we give a WLS estimation method, based on Θ-orthogonal projection and LQ decomposition.

4.1 *Description of a sub-problem*

Suppose that ν is sufficiently large. Define the following matrices:

$$U_t := \begin{bmatrix} u_t \; u_{t+1} \; \cdots \; u_{t+\nu-1} \end{bmatrix} \in \mathbf{R}^{q\times\nu}, \qquad (10)$$

$$Y_t := \begin{bmatrix} y_t \; y_{t+1} \; \cdots \; y_{t+\nu-1} \end{bmatrix} \in \mathbf{R}^{p\times\nu}, \qquad (11)$$

$$U := \begin{bmatrix} U_1^T \; \cdots \; U_\kappa^T \end{bmatrix}^T \in \mathbf{R}^{q\kappa\times\nu}, \qquad (12)$$

where $\kappa + \nu - 1 \leq M$. The Sub-problem is then given by a WLS problem as follows:

WLS problem *Given $0 \leq \Theta \in \mathbf{R}^{\nu\times\nu}$ and U in (12) and Y_κ in (11), suppose that $\langle U, U\rangle_\Theta > 0$ holds. Find $\hat{Y}_*$ that satisfies*

$$\hat{Y}_* = \arg\min_{\hat{Y}_\kappa} \mathrm{Tr}\,\langle Y_\kappa - \hat{Y}_\kappa, Y_\kappa - \hat{Y}_\kappa\rangle_\Theta \qquad (13)$$

based on U and Y_κ, where

$$\hat{Y}_\kappa \in \mathrm{span}\,\{U(i,:)\,|\,i = 1,\cdots,q\kappa\}, \qquad (14)$$

and where $U(i,:)$ means the i-th row of U.

4.2 *Optimization by projection*

The WLS problem can be solved by projection in the space with inner product. We first give a definition of the matrix version of Θ-orthogonal projection.

Suppose that matrices $Y \in \mathbf{R}^{\upsilon\times\nu}$ and $U \in \mathbf{R}^{\kappa\times\nu}$ are given, and that $\mathcal{U}$ is the space spanned by row vectors of matrix U. We define $\hat{Y}$ to be Θ-orthogonal projection onto $\mathcal{U}$ from Y if there exists a $P \in \mathbf{R}^{\upsilon\times\kappa}$ such that

$$Y = \hat{Y} + \tilde{Y}, \quad \hat{Y} = PU, \quad \langle\tilde{Y}, U\rangle_\Theta = 0. \qquad (15)$$

We also denote $\hat{Y} = \mathrm{E}_\Theta[Y\,|\,\mathcal{U}]$. The following propositions are known.

Proposition 1 *(Hassibi* et al.*, 1996) There exists a Θ-orthogonal projection from Y onto $\mathcal{U}$ if and only if* $\mathrm{Image}\,(\langle U, Y\rangle_\Theta) \subseteq \mathrm{Image}\,(\langle U, U\rangle_\Theta)$ *holds. Also, Θ-orthogonal projection is given by*

$$\hat{Y} = P_* U, \qquad (16)$$

if it exists, where P_* is an arbitrary solution which satisfies $\langle Y, U\rangle_\Theta = P_*\langle U, U\rangle_\Theta$.

Proposition 2 *If $\Theta \geq 0$, it follows that (Kreindler and Jameson, 1972):*

$$\mathrm{Image}\,(\langle U, Y\rangle_\Theta) \subseteq \mathrm{Image}\,(\langle U, U\rangle_\Theta). \qquad (17)$$

When $\Theta \geq 0$, using Proposition 1 and 2, there always exists a minimizer.

Theorem 1 *Suppose that appropriate size of matrices $\Theta \geq 0$, U and Y are given. Define a quadratic form as*

$$\Phi(P) := \mathrm{Tr}\,\langle Y - PU, Y - PU\rangle_\Theta. \qquad (18)$$

Then, there always exists a P that minimize $\Phi(P)$. If $\langle U, U\rangle_\Theta$ is non-singular, there is a unique minimizer P_* satisfying

$$P_* = \arg\min_P \Phi(P), \qquad (19)$$

where $\mathrm{E}_\Theta\,[Y\,|\,\mathcal{U}] = P_* U$.
Let $\langle U, U\rangle_\Theta$ be non-singular. Suppose $\begin{bmatrix} U^T \; Y^T \end{bmatrix}^T$ is decomposed as

$$\begin{bmatrix} U \\ Y \end{bmatrix} = \begin{bmatrix} L_{11} & 0 \\ L_{21} & L_{22} \end{bmatrix} \begin{bmatrix} Q_1^T \\ Q_2^T \end{bmatrix}, \qquad (20)$$

$$\left\langle \begin{bmatrix} Q_1^T \\ Q_2^T \end{bmatrix}, \begin{bmatrix} Q_1^T \\ Q_2^T \end{bmatrix} \right\rangle_\Theta = \begin{bmatrix} I & 0 \\ 0 & J \end{bmatrix}, \qquad (21)$$

where J = block-diag$(I, 0)$ and L_{11} is square and the size of it is the same as $\langle U, U\rangle_\Theta$. Then, L_{11} is non-singular, and it follows that

$$\mathrm{E}_\Theta[Y\,|\,\mathcal{U}] = L_{21}Q_1^T = L_{21}L_{11}^{-1}U. \qquad (22)$$

Proof There exists a P_* satisfying $U\Theta U^T P_* = U\Theta Y^T$ by Proposition 2. It then follows that

$$\Phi(P) = \text{Tr}\,\langle (P-P_*)U, (P-P_*)U\rangle_\Theta + \text{Tr}\,\langle Y - P_*U, Y - P_*U\rangle_\Theta .$$

If $\langle U,U\rangle_\Theta > 0$, P_* is a unique minimizer and L_{11} is non-singular because of $\langle U,U\rangle_\Theta = L_{11}L_{11}^T > 0$. Equation (22) is obtained by noting that the rows of Q_1^T form a Θ-orthonormal basis for the row space of U. □

4.3 *Algorithm for weighted LQ decomposition*

We present an algorithm for "WLQ" (weighted LQ) decomposition in (20) and (21) when Θ is singular and $\langle U,U\rangle_\Theta > 0$ holds.

Step 1. Calculate an eigenvalue decomposition of $\Theta \in \mathbf{R}^{\nu\times\nu}$ as

$$\Theta = V\begin{bmatrix}\Theta_1 & 0\\ 0 & 0_{\nu_2\times\nu_2}\end{bmatrix}V^T, \qquad (23)$$

where $\Theta_1 = \Theta_1^{\frac{1}{2}}\Theta_1^{\frac{T}{2}} > 0$, $V^TV = I$, $\Theta_1 \in \mathbf{R}^{\nu_1\times\nu_1}$and $\nu = \nu_1 + \nu_2$.

Step 2. Define matrices Λ_u and Λ_y as

$$\begin{bmatrix}\Lambda_u\\ \Lambda_y\end{bmatrix} := \begin{bmatrix}U\\ Y\end{bmatrix}V\begin{bmatrix}\Theta_1^{\frac{1}{2}}\\ 0\end{bmatrix}. \qquad (24)$$

Step 3. Calculate LQ decomposition of $\begin{bmatrix}\Lambda_u\\ \Lambda_y\end{bmatrix}$:

$$\begin{bmatrix}\Lambda_u\\ \Lambda_y\end{bmatrix} = \begin{bmatrix}L_{uu} & 0\\ L_{yu} & L_{yy}\end{bmatrix}\begin{bmatrix}\Omega_u^T\\ \Omega_y^T\end{bmatrix}. \qquad (25)$$

Step 4. Define matrices L_{uz} and L_{yz} as

$$\begin{bmatrix}L_{uz}\\ L_{yz}\end{bmatrix} := \begin{bmatrix}U\\ Y\end{bmatrix}V\begin{bmatrix}0_{\nu_1\times\nu_2}\\ I_{\nu_2}\end{bmatrix}. \qquad (26)$$

Step 5. Finally, define matrices Q_u, Q_y, Q_z as

$$\begin{bmatrix}Q_u^T\\ Q_y^T\\ Q_z^T\end{bmatrix} := \begin{bmatrix}\Omega_u^T\Theta_1^{-\frac{1}{2}} & L_{uu}^{-1}L_{uz}\\ \Omega_y^T\Theta_1^{-\frac{1}{2}} & 0\\ 0 & I_{\nu_2}\end{bmatrix}V^T. \qquad (27)$$

Theorem 2 *It follows that*

$$\begin{bmatrix}U\\ Y\end{bmatrix} = \begin{bmatrix}L_{uu} & 0 & 0\\ L_{yu} & L_{yy} & L_{yz} - L_{yu}L_{uu}^{-1}L_{uz}\end{bmatrix}\begin{bmatrix}Q_u^T\\ Q_y^T\\ Q_z^T\end{bmatrix},$$

$$\left\langle \begin{bmatrix}Q_u^T\\ Q_y^T\\ Q_z^T\end{bmatrix}, \begin{bmatrix}Q_u^T\\ Q_y^T\\ Q_z^T\end{bmatrix}\right\rangle_\Theta = \begin{bmatrix}I & 0 & 0\\ 0 & I & 0\\ 0 & 0 & 0_{\nu_2\times\nu_2}\end{bmatrix},$$

under definition of **Steps 1-5.**

Proof See Appendix **A.** □

5. IDENTIFICATION WITH OUTLIERS

In order to deal with the problem in section 3.1, we need to remove outliers by solving WLS (weighted least-squares) problems iteratively, since we never know when outliers occur.

5.1 *Identification based on WLS*

Suppose that $\kappa = 2\ell$. Define U and Y_κ as (10)-(12). The procedure for identification based on WLS is summarized as follows:

Step 1. Calculate "WLQ decomposition" as in (20) and (21):

$$\begin{bmatrix}U\\ Y_\kappa\end{bmatrix} = \begin{bmatrix}L_{11} & 0\\ L_{21} & L_{22}\end{bmatrix}\begin{bmatrix}Q_1^T\\ Q_2^T\end{bmatrix},$$

$$\left\langle \begin{bmatrix}Q_1^T\\ Q_2^T\end{bmatrix}, \begin{bmatrix}Q_1^T\\ Q_2^T\end{bmatrix}\right\rangle_\Theta = \begin{bmatrix}I & 0\\ 0 & J\end{bmatrix},$$

where $J = \text{block-diag}(I, 0)$.

Estimate deterministic part $\hat{y}_\kappa, \cdots, \hat{y}_{\kappa+\nu-1}$ by Θ-orthogonal projection:

$$\hat{Y}_\kappa = \text{E}_\Theta[Y \,|\, \mathcal{U}] = L_{21}Q_1^T.$$

Step 2. Estimate A, B, C and D, based on U and $\hat{Y}_\kappa$. For example, define $Y_f \in \mathbf{R}^{p\ell\times(\nu-\kappa+1)}$ as

$$Y_f := \begin{bmatrix}\hat{y}_{\kappa+\ell} & \hat{y}_{\kappa+\ell+1} & \cdots & \hat{y}_{\nu+\ell}\\ \vdots & \vdots & & \vdots\\ \hat{y}_{2\kappa-1} & \hat{y}_{2\kappa} & \cdots & \hat{y}_{\nu+\kappa-1}\end{bmatrix}.$$

Define also U_f as in the same way (Okada and Sugie, 1996). MOESP type algorithm is then available (Verhaegen and Dewilde, 1992). See also Appendix **B**.

Note that Step 1 of the procedure is the modification of the one introduced in section 3.2.

5.2 *Iterative algorithm to attenuate outliers*

It is not known when outliers take place. We must therefore iteratively solve WLS problems by changing a weighting matrix Θ at each iteration. Let Θ be a diagonal matrix:

$$\Theta = \text{diag}(\theta_1, \theta_2, \cdots, \theta_\nu), \quad \theta_i \in \mathbf{R}. \qquad (28)$$

The following IWLS (iterative weighted least-squares) algorithm can be used for modifying Θ to attenuate outliers (Nakagawa and Koyanagi, 1982).

Step 1. Define $\Theta = I$. Estimate $\hat{y}_t$.
Step 2. Compute residual $z_t = y_t - \hat{y}_t$.
Step 3. Obtain a median of the residual

$$s = \text{median}\,\{|z_t|\}.$$

Step 4. Define $\Theta = \text{diag}\,(\theta_i)$, where

$$\theta_i = \begin{cases}\left(\frac{cs}{z_i}\right)\sin\left(\frac{z_i}{cs}\right) & \text{for } |z_i| < cs\pi,\\ 0 & \text{for } |z_i| \geq cs\pi.\end{cases} \qquad (29)$$

Note that c is a constant, for which $c \approx 2.1$ is recommended in (Nakagawa and Koyanagi, 1982).

Step 5. Compute a new estimate $\hat{y}_t$ based on Θ.
Step 6. Repeat 2-5, until θ_i converges.

6. NUMERICAL EXAMPLE

This section describes a numerical example. The simulated data is generated by

$$y_t = G(z)(u_t + r_t) + w_t, \tag{30}$$

where u_t is known exogenous input and y_t is measured output corrupted by noises r_t and w_t. Assume $u_t \sim \mathrm{N}(0,1)$, $r_t \sim \mathrm{N}(0, 0.01^2)$, $w_t \sim 0.99\mathrm{N}(0, 0.05^2) + 0.01\mathrm{N}(0,1)$. Note that outliers are contained in only measured output. Define $v_t = Gr_t + w_t$, then the setting in (30) becomes the same one as in (4).

Let $G(z)$ be a fifth-order system described by

$$\begin{aligned} G(z) &= \frac{G_n(z)}{G_d(z)}, \\ G_n(z) &= 0.0275z^{-4} + 0.0551z^{-5} \\ G_d(z) &= 1 - 2.3443z^{-1} + 3.081z^{-2} - 2.5274z^{-3} \\ &\quad +1.2415z^{-4} - 0.3686z^{-5}. \end{aligned}$$

A sample of realizations of noise w_t and output y_t is shown in Figure 2.

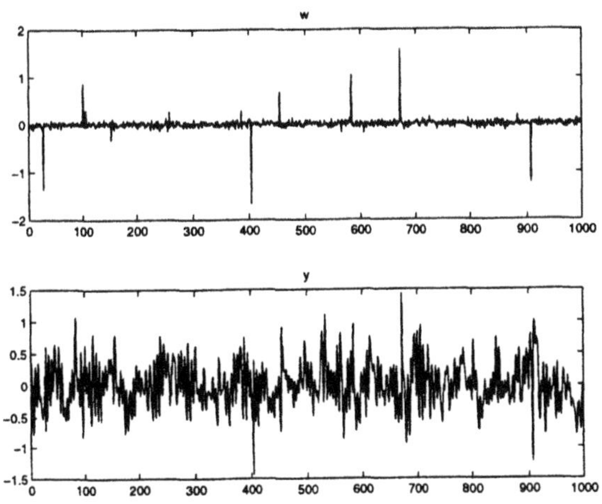

Fig. 2. Noise w_t and output y_t

The size of U in (12) is chosen as $\nu = 1000$ and $\kappa = 30$. The order of realized system is fixed as $n = 5$.

In our simulations, we compare two different identification algorithms:

LS Compute only Step 1.
IWLS Compute Step 1 and repeat Step 2-5, six times.

Figures 3 and 4 describe bode plots of the system estimated by LS and WLS, respectively. Note that we have performed 30 simulations for different noise realizations.

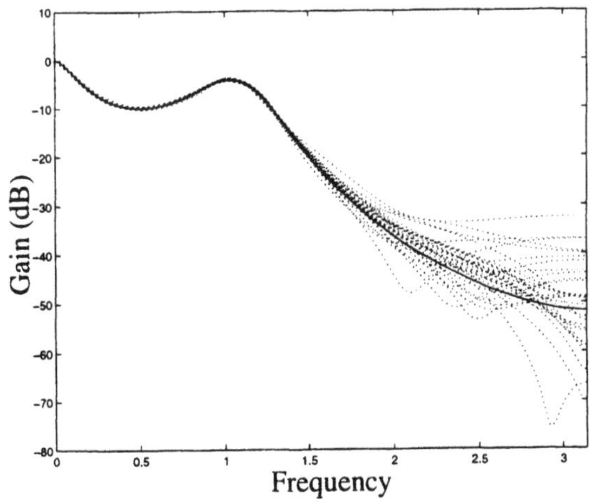

Fig. 3. LS method

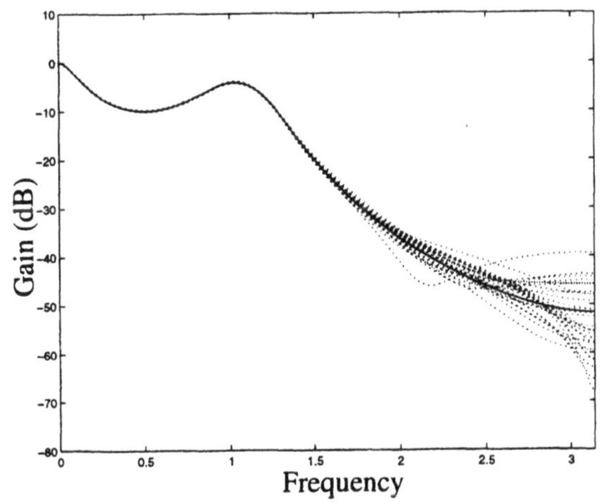

Fig. 4. IWLS method (proposed)

These figures show the effectiveness of the proposed method in the high frequency range where the estimates of gain characteristic are naturally affected by abrupt large noise.

7. CONCLUSION

In this paper, we have considered the problem of the identification with outliers in the framework of subspace method. We have first defined inner product based on a non-negative definite matrix Θ, and a weighted least-squares problem is described in terms of the inner product. We have next developed an algorithm to compute weighted LQ decomposition. Finally, we have proposed an iterative algorithm to attenuate outliers based on weighted least-squares. A numerical example has demonstrated effectiveness of the proposed method.

8. REFERENCES

Hassibi, B., A.H. Sayed and T. Kailath (1996). Linear estimation in Krein spaces–Part I:

Theory. *IEEE Transactions on Automatic Control* **41**(1), 18–33.

Isaksson, A. (1991). System identification subject to missing data. *Proc. of American Control Conference* pp. 693–698.

Issakson, A. (1993). Identification of ARX-models subject to missing data. *IEEE Transactions on Automatic Control* **38**(5), 813–819.

Katayama, T. and G. Picci (1999). Realization of stochastic systems with exogenous inputs and subspace identification methods. *Automatica* **35**(10), 1635–1652.

Kreindler, E. and A. Jameson (1972). Conditions for nonnegativeness of partitioned matrices. *IEEE Transactions on Automatic Control* **17**(1), 147–148.

Ljung, L. (1987). *System Identification - Theory for the User.* Prentice-Hall.

Moonen, M., B. De Moor and J. Vandewalle (1989). On- and off-line identification of linear state-space models. *Int. J. Control* **49**(1), 219–232.

Nakagawa, T. and Y. Koyanagi (1982). *Data Analysis by the Least-Squares Method.* The University of Tokyo Press (in Japanese).

Okada, M. and T. Sugie (1996). Subspace system identification considering both noise attenuation and use of prior knowledge. *Proc. of IEEE Conference on Decision and Control* pp. 3662–3667.

Overschee, P. Van and B. De Moor (1994). N4SID - Subspace algorithms for the identification of combined deterministic – stochastic systems. *Automatica* **30**(1), 75–93.

Picci, G. and T. Katayama (1996). Stochastic realization with exogenous inputs and "subspace-methods" identification. *Signal Processing* **2**(52), 145–160.

Pintelon, R. and J. Schoukens (1998). Frequency domain system identification with missing data. *Proc. of Conference on Decision and Control* pp. 701–705.

Verhaegen, M. (1994). Identification of the deterministic part of MIMO state space models given in innovations form from input-output data. *Automatica* **30**(1), 61–74.

Verhaegen, M. and P. Dewilde (1992). Subspace model identification - Part 1. The output-error state-space model identification class of algorithms. *Int. J. Control* **56**(5), 1187–1210.

Appendix A. PROOF OF THEOREM 2

Equations (24), (25) and (26) give

$$\begin{bmatrix} U \\ Y \end{bmatrix} V \begin{bmatrix} \Theta_1^{\frac{1}{2}} & 0 \\ 0 & I \end{bmatrix}$$
$$= \begin{bmatrix} L_{uu} & 0 & 0 \\ L_{yu} & L_{yy} & L_{yz} - L_{yu}L_{uu}^{-1}L_{uz} \end{bmatrix} \begin{bmatrix} \Omega_u^T & L_{uu}^{-1}L_{uz} \\ \Omega_y^T & 0 \\ 0 & I \end{bmatrix}.$$

Results are obtained by this equation, immediately. □

Appendix B. MOESP SCHEME [2]

Step 1. Compute LQ decomposition of $\begin{bmatrix} U_f \\ Y_f \end{bmatrix}$:

$$\begin{bmatrix} U_f \\ Y_f \end{bmatrix} = \begin{bmatrix} L_{11} & 0 \\ L_{21} & L_{22} \end{bmatrix} \begin{bmatrix} Q_1^T \\ Q_2^T \end{bmatrix}. \qquad \text{(B.1)}$$

$$U_f \in \mathbf{R}^{q\ell \times (\cdot)}, \quad Y_f \in \mathbf{R}^{p\ell \times (\cdot)}$$

Step 2. Compute singular value decomposition of L_{22}:

$$L_{22} = \begin{bmatrix} U_1^T \\ U_2^T \end{bmatrix}^T \begin{bmatrix} \Sigma_1 & 0 \\ 0 & \Sigma_2 \end{bmatrix} \begin{bmatrix} V_1^T \\ V_2^T \end{bmatrix}$$
$$\approx U_1 \Sigma_1 V_1^T. \qquad \text{(B.2)}$$

Step 3. Let n be the size of Σ_1. Define $\mathcal{O}_\ell$, $\mathcal{C}$ and A as

$$\mathcal{O}_\ell := U_1 \Sigma_1^{\frac{1}{2}}, \quad C = \mathcal{O}_1, \quad A = \underline{\mathcal{O}}_\ell^+ \overline{\mathcal{O}}_\ell, \text{(B.3)}$$

where $\mathcal{O}_1 := \mathcal{O}_\ell(1:p, 1:n)$, $\underline{\mathcal{O}}_\ell := \mathcal{O}_\ell(1:p(\ell-1), 1:n)$, $\overline{\mathcal{O}}_\ell := \mathcal{O}_\ell(p+1:p\ell, 1:n)$, $\underline{\mathcal{O}}_\ell^+$ means pseudo-inverse of $\underline{\mathcal{O}}_\ell$.

Step 4. Estimate B and D by solving $U_2^T L_{21} = U_2^T \Gamma_\ell L_{11}$, where Γ_ℓ is defined as

$$\Gamma_\ell = \begin{bmatrix} D & 0 & \cdots & 0 \\ CB & D & \cdots & 0 \\ \vdots & \vdots & \ddots & 0 \\ CA^{\ell-2}B & CA^{\ell-3}B & \cdots & D \end{bmatrix}. \qquad \text{(B.4)}$$

IDENTIFICATION OF LINEAR PARAMETER VARYING MODELS USING KALMAN FILTERING

Marco Lovera, Fabio Previdi

Dipartimento di Elettronica e Informazione, Politecnico di Milano
32, Piazza Leonardo da Vinci, 20133 Milano, Italy
email: {lovera, previdi }@elet.polimi.it

Abstract: The problem of estimating affine linear parametrically varying models is considered. An algorithm based on a Kalman filtering approach is proposed and its performance evaluated in a simulation study.

1. INTRODUCTION

LPV models are linear ones, in which a vector of scheduling variables enters the system matrices in an affine or linear fractional way. The advantage of such a representation for general nonlinear models is easily recognized in the connection with modern robust control theory (Zhou *et al.*, 1996; Apkarian *et al.*, 1995; Apkarian and Gahinet, 1995) and with gain-scheduling nonlinear control techniques (Apkarian and Adams, 1998; Hunt and Johansen, 1997; Previdi, 1999; Kaminer *et al.*, 1995; Shamma and Athans, 1990).

LPV models can also be useful in modeling the dynamics of nonlinear plants. In fact, even though identification using nonlinear models is advisable when little or no a priori information about the system to be studied is available, in many practical situations it would be possible and fruitful to take into account such priors in the structure of the model.

A typical example of this kind of situation is given by a nonlinear plant which is led to operate in a number of different equilibria. In such a case it would be useful to separate the role of conventional input variables from the role of scheduling variables (i.e., variables defining the operating point of the plant), by letting them enter the model in distinct ways (Skeppstedt *et al.*, 1992; Johansen and Foss, 1995).

Surprisingly enough, the literature on LPV identification remains fairly limited up to date, and different lines of work can be easily identified within the existing contributions:

- General prediction error minimisation (PEM) algorithms for LPV systems. In (Lee and Poolla, 1996; Lee and Poolla, 1999) a general algorithm for the PEM identification of LPV models in LFT form has been proposed. The algorithm is based on an iterative optimisation procedure. Similarly, in (Previdi and Lovera, 1999) a different structure based on the LFT interconnection of an ARX and a neural network. The model structure, in this case, can be exploited in order to work out an efficient parameter estimation scheme.
- PEM algorithms for affine LPV models in ARX form (Bamieh and Giarre', 1999; Bamieh and Giarre', 2000). The approach proposed in these papers is based on the least squares method and therefore lends itself to simple recursive implementations. It is worth pointing out that along this line interesting theoretical results concerning persistency of excitation conditions have been derived.
- Identification algorithms for state space models (Nemani *et al.*, 1995; Lovera *et al.*, 1996; Verdult and Verhaegen, 2000). The main interest in the state space approach to the problem lies in the ease by which the multivariable problem can be dealt with in this framework. On the other hand, the technical complexity of the state space representation

for LPV models appears as a considerable drawback.

This paper describes an algorithm for the estimation of affine Linear Parametrically Varying (LPV) models via a Kalman filtering approach. Specifically, in this work the estimation of LPV models is carried out using Kalman filtering by assuming that the dynamics of the parameter vector is described by a random walk state equation fed by a suitable input noise.

The paper is organized as follows: in the Section 2 the LPV model class is shortly outlined. Section 3 is devoted to the presentation of the estimation algorithm for the model class and Section 4 proposes an approach to the convergence study of the algorithm. In Section 5 simulation results are presented. Finally, in Section 6 discussion of the results and future work will be presented.

2. LPV MODELS

In the following, discrete-time SISO (*Single Input Single Output*) systems are considered. The simbol y_k indicates the value of the variable y at the time step $k \in Z^+$. So, $u_k \in \Re$ and $y_k \in \Re$, for all $k = 0, 1, \ldots$. Finally, if A is a generic matrix, A' is its transpose.

Consider the following model structure:

$$y_k = a_{1k} y_{k-1} + \cdots + a_{n_a k} y_{k-n_a} + \\ + b_{0k} u_k + \cdots + b_{n_b k} u_{k-n_b} + v_k \quad (1)$$

where n_a is the maximum output lag and n_b is the maximum input lag in the model. Notice that the parameters a_{ik}, $i = 1, \ldots, n_a$ and b_{jk}, $j = 0, \ldots, n_b$ are also time-varying. In this work, an affine structure for each of them is assumed, i.e.,

$$a_{ik} = a_{i0} + a_{i1}\delta_{1k} + \ldots + a_{ir}\delta_{rk}$$

and similarly for the b_{ik}s, where

$$\delta_k = [\delta_{1k}, \ldots, \delta_{rk}]' \in \Re^r$$

is the vector of time-varying parameters affecting the system's dynamics. A delay of n_k time steps can be introduced by forcing to zero the coefficients b_{jk} for $j = 0, \ldots, n_k - 1$. The term v_k is supposed to be a zero-mean i.i.d. noise process with covariance matrix $V_k = E[v_k v_k']$.

Eq. (1) can be given the following representation in form of a linear regression with time varying parameters

$$y_k = \theta_k' \psi_k + v_k \quad (2)$$

where $\theta_k \in \Re^{n_a+n_b+1}$, for all k, is a time-varying model parameter vector, i.e.

$$\theta_k = [a_{1k}, \ldots, a_{n_a k}, b_{0k}, \ldots, b_{n_b k}]' \quad (3)$$

and $\psi_k \in \Re^{n_a+n_b+1}$, for all k, is the model regressor, i.e. the collection of the past measurements of input and output used to predict the current output value

$$\psi_k = [y_{k-1}, \ldots, y_{k-n_a}, u_k, \ldots, u_{k-n_b}]' \quad (4)$$

By taking into account the affine structure of the time-varying parameters, the model can be written as an ordinary linear regression, with time-invariant coefficients.

For, let:

$$\Theta = \begin{bmatrix} a_{10} & \ldots & a_{1r} \\ \vdots & \vdots & \vdots \\ a_{n_a 0} & \ldots & a_{n_a r} \\ b_{00} & \ldots & b_{0r} \\ \vdots & \vdots & \vdots \\ b_{n_b 0} & \ldots & b_{n_b r} \end{bmatrix} \quad (5)$$

then

$$\theta_k = \Theta \bar{\delta}_k \quad (6)$$

where

$$\bar{\delta}_k = [1, \delta_k']' \quad (7)$$

and therefore the regression model can be written as:

$$y_k = \theta_k' \psi_k = \bar{\delta}_k' \Theta' \psi_k = \phi_k \mathrm{vec}(\Theta') \quad (8)$$

where, denoting by $\otimes$ the Kronecker product:

$$\phi_k = (\psi_k' \otimes \bar{\delta}_k) \quad (9)$$

and the last equality comes from the identity

$$\mathrm{vec}(ABC) = (C' \otimes A)\mathrm{vec}(B) \quad (10)$$

3. MODEL IDENTIFICATION ALGORITHM

Consider a state–space description of the system (2), using the parameter vector

$$\bar{\theta} = \mathrm{vec}(\Theta') \quad (11)$$

as a state vector the dynamics of which is described by the following state equation

$$\bar{\theta}_{k+1} = \bar{\theta}_k + w_k \tag{12}$$

where w_k is a zero mean i.i.d. process with covariance matrix $W_k = E[w_k w_k']$ and it is uncorrelated with v_k.

The observed output for this system is trivially given by Eq. (2).

Now, the Kalman filter equations can be used for the estimation of the state vector in Eq. (12), i.e., of the parameter vector, in analogy with the approach of (Chowdhury, 2000) for the estimation of nonlinear model parameters.

Specifically, the parameter vector estimate at time k is updated according to the following rule

$$\hat{\bar{\theta}}_{k|k} = \hat{\bar{\theta}}_{k|k-1} + K_k(y_k - \hat{y}_{k|k-1}) \tag{13}$$

where the Kalman gain is given by

$$K_k = Q_{k|k-1}\phi_k'(V_k + \phi_k Q_{k|k-1}\phi_k')^{-1} \tag{14}$$

and the covariance $Q_k = E[(\bar{\theta}_k - \hat{\bar{\theta}}_{k|k})(\bar{\theta}_k - \hat{\bar{\theta}}_{k|k})']$ of the parameter estimate is updated as usual, i.e.

$$Q_{k|k} = (I - K_k\phi_k)\, Q_{k|k-1} \tag{15}$$

4. CONVERGENCE ISSUES

General results on the convergence of the Kalman filter under stochastic excitation conditions are available in the literature. As a matter of fact, the following Theorem holds (see (Guo, 1990)):

Theorem

Suppose that the following conditions hold:

(1) v_k and w_k satisfy

$$\sigma_r = \sup_k E\left[||v_k||^r + ||w_k||^r\right] < \infty \tag{16}$$

for some $r > 4$.

$$\mu_4 = \limsup_{n\to\infty} \frac{1}{n}\sum_{k=0}^{n-1}\left(||v_k||^4 + ||w_k||^4\right) < \infty \tag{17}$$

(2) Let $\{\phi_k, F_k\}$ be an adapted sequence, i.e., ϕ_k is a measurable set in the σ-algebra F_k for any k and F_k is a sequence of nondecreasing σ-algebras. Then, there exist a constant $\delta > 0$ and an integer $h > 0$ such that

$$E\left[\sum_{k=mh}^{mh+h-1} \frac{\phi_k\phi_k'}{1+||\phi_k||^2}|F_{mh-1}\right] \geq \delta I \;\forall m \geq 0 \tag{18}$$

Under these conditions, for $\hat{\bar{\theta}}$ defined in Eq. (11) the following results follow:

(1) $\limsup_{n\to\infty} E\left[||\hat{\bar{\theta}}_n - \bar{\theta}_n||^2\right] \leq A(\sigma_r)^{(2/r)}$

(2) $\limsup_{n\to\infty} \frac{1}{n}\sum_{i=0}^{n-1} ||\hat{\bar{\theta}}_i - \bar{\theta}_i|| \leq B(\mu_4)^{(1/4)}$

with A and B finite deterministic constants.

It is worth stressing that explicit expressions for A and B are also given in (Guo, 1990). So, by using conditions (16) and (17) it is possible to explicitly evaluate the asymptotic behaviour of the estimates by means of this result.

The behaviour of the proposed identification algorithm, however, clearly depends on both the excitation conditions on the input and on the variation of the parameter vector, which are jointly expressed by assumption (18) in the above Theorem. It would be desirable to work out separate excitation requirements for u and δ.

5. SIMULATION RESULTS

The proposed approach has been tested in a number of simulation experiments in order to assess at least experimentally the performance and rate of convergence which can be achieved as a function of the signal-to-noise ratio and of the type of parametric excitation, together with the sensitivity with respect to structure selection problems.

The system which has been considered for the simulation study is described by the input/output equation:

$$y_k = (0.2 + 0.3\delta_{k-1})y_{k-1} + 2u_k + v_k \tag{19}$$

where $u_k = WGN(0,1)$ and $v_k = WGN(0,V)$ uncorrelated from u.

5.1 *Effect of signal-to-noise ratio*

The first issue of interest is the accuracy of the parameter estimates as a function of the signal to noise ratio, assuming that $\delta_k = WGN(0,1)$. In order to assess this, the identification procedure has been carried out repeatedly by increasing the variance of the measurement noise v. The results are reported in Figures 1-3.

5.2 *Effect of parametric excitation*

Perhaps of greater interest is the analysis of the convergence properties of the proposed algorithm for different spectral characteristics of the parametric excitation δ. In order to assess the role of δ in determining the behaviour of the algorithm, a multiharmonic structure for this signal has been selected:

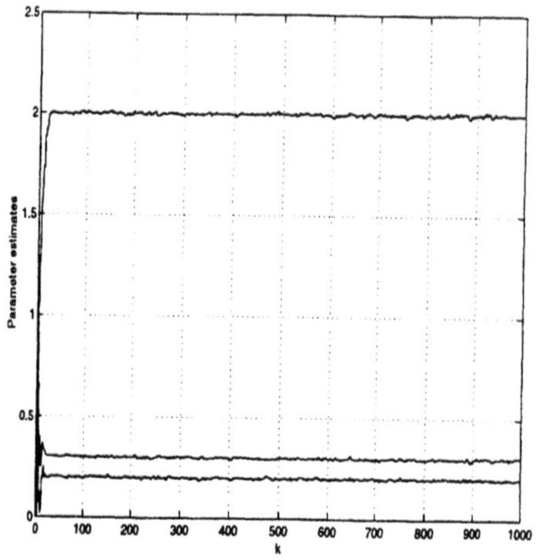

Fig. 1. Convergence of parameter estimates, $V = 0.01^2$

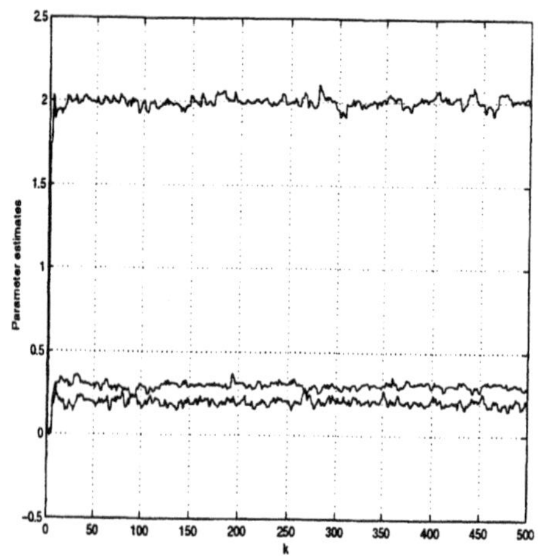

Fig. 2. Convergence of parameter estimates, $V = 0.05^2$

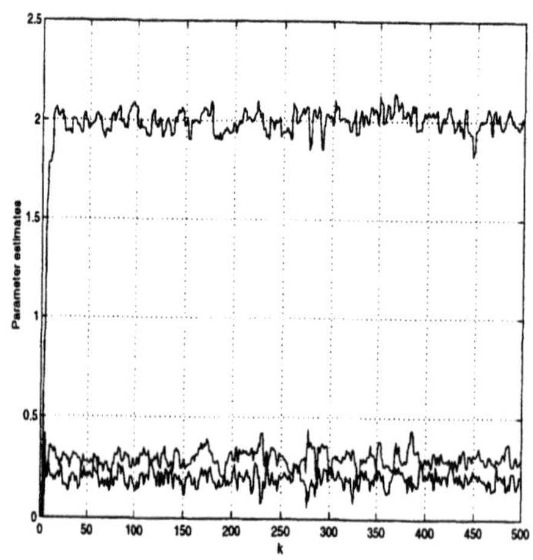

Fig. 3. Convergence of parameter estimates, $V = 0.1^2$

$$\delta_k = \sum_{n=1}^{N_a} A_n cos(\omega_n k + \varphi_n) \qquad (20)$$

with $\omega_n = \omega_0 n$ and φ_n constructed according to the low-peak design algorithm for multiharmonic signals proposed in (Schroeder, 1970). By considering increasing values of N_a one can construct deterministic sequences with a flat spectrum of increasing bandwidth, therefore approximating with improving accuracy the spectral characteristics of a white noise sequence.

As can be seen from Figures 4-6, and according to what one expects from obvious qualitative arguments, the convergence of the identification algorithms improves considerably with the bandwidth of the considered parametric excitation, so one expects that a characterisation of the persistency of excitation similar to the one proposed in (Bamieh and Giarre', 2000) holds also for the Kalman based approach. Such a characterisation, however, remains a subject for future work.

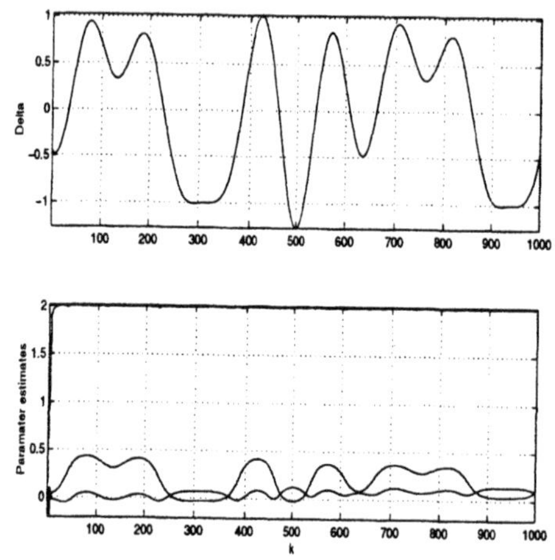

Fig. 4. Convergence of parameter estimates, $N_a = 5$

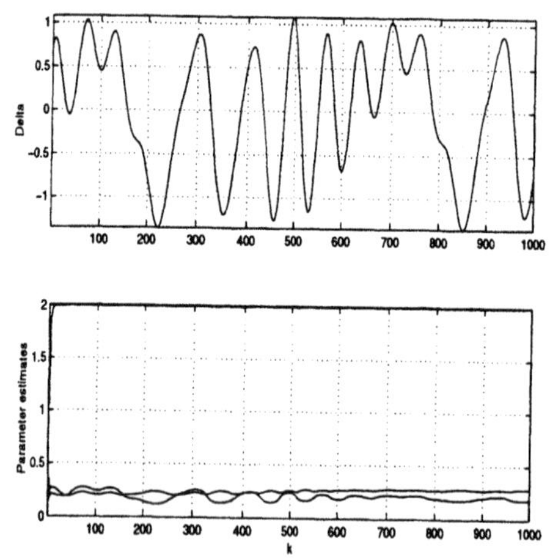

Fig. 5. Convergence of parameter estimates, $N_a = 10$

6. CONCLUDING REMARKS

An algorithm based on Kalman filtering has been proposed, for the estimation of affine linear parametrically varying models. The convergence of the proposed algorithm is guaranteed under suitable

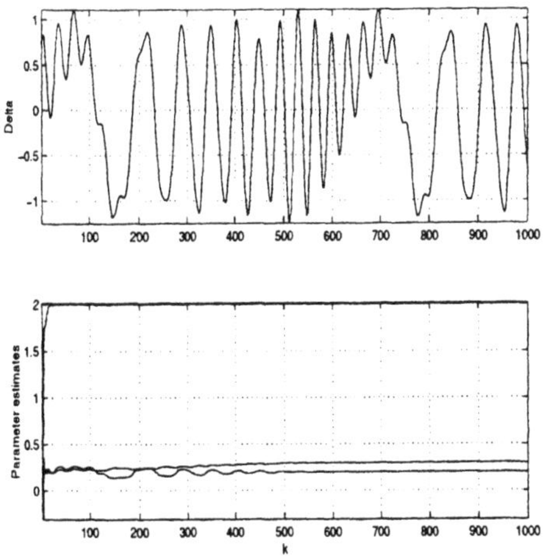

Fig. 6. Convergence of parameter estimates, $N_a = 20$

assumptions on the input and parameter excitation conditions. Simulation results show that accurate estimates can be obtained by this approach.

7. REFERENCES

Apkarian, P. and P. Gahinet (1995). A Convex Characterization of Gain-Scheduling $\mathcal{H}_\infty$ Controllers. *IEEE Trans. Automatic Control* **40**, 853–864.

Apkarian, P. and R. J. Adams (1998). Advanced Gain-Scheduling Techniques for Uncertain Systems. *IEEE Trans. Control System Technology* **6**, 21–32.

Apkarian, P., P. Gahinet and G. Becker (1995). Self-scheduled $\mathcal{H}_\infty$ Control of Linear Parameter-varying Systems: a Design Example. *Automatica* **31**, 1251–1261.

Bamieh, B. and L. Giarre' (1999). Identification of linear parameter varying models. In: *Proc. of the Conference on Decision and Control.* Phoenix, AZ, USA.

Bamieh, B. and L. Giarre' (2000). Identification for a general class of LPV models. In: *Proc. of the IFAC System Identification Symposium, Santa Barbara, USA.*

Chowdhury, F. N. (2000). Input-Output Modeling of Nonlinear Systems with Time-Varying Linear Models. *IEEE Trans. on Automatic Control* **45**, 1355–1358.

Guo, L. (1990). Estimating time-varying parameters by the kalman filter based algorithm: stability and convergence. *IEEE Transactions on Automatic Control* **35**(2), 141–147.

Hunt, K. J. and T. A. Johansen (1997). Design and Analysis of Gain-Scheduled Control using Local Controller Networks. *Int. Journal of Control* **66**, 619–651.

Johansen, T. A. and B. A. Foss (1995). Identification of Non-Linear System Structure and Parameters using Regime Decomposition. *Automatica* **31**, 321–326.

Kaminer, I., A. M. Rascoal, P. P. Karghonekar and E. E. Coleman (1995). A Velocity Algorithm for the Implementation of Gain-Scheduled Controllers. *Automatica* **31**, 1185–1191.

Lee, L. and K. Poolla (1999). Identification of linear parameter-varying systemsusing nonlinear programming. *ASME Journal of Dynamic Systems, Measurement and Control* **121**, 71–78.

Lee, L. H. and K. Poolla (1996). Identification of Linear Parameter-Varying Systems via LFT. In: *Proc. of the 35th Conference on Decision and Control.* Japan. pp. 1545–1550.

Lovera, M., M. Verhaegen and C. T. Chou (1996). State space identification of MIMO linear parameter varying models. In: *Proc. of the International Conference on Mathematical Theory of Networks and Systems.* Padova, Italy.

Nemani, M., R. Ravikanth and B. Bamieh (1995). Identification of linear parameter varying systems. In: *Proc. of the 1995 Conference on Decision and Control.*

Previdi, F. (1999). *Identification and Control using Local Linear Models.* PhD Thesis. Politecnico di Milano, Italy.

Previdi, F. and M. Lovera (1999). Identification of a class of linear models with nonlinearly varying parameters. In: *Proc. of the European Control Conference.* Karlsruhe, Germany.

Schroeder, M.R. (1970). Synthesis of low-peak-factor signals and binary sequences with low autocorrelation. *IEEE Transactions on Information Theory* **16**(1), 85–89.

Shamma, J. S. and M. Athans (1990). Analysis of Gain-Scheduled Control for Nonlinear Plants. *IEEE Trans. Automatic Control* **35**, 898–907.

Skeppstedt, A., L. Ljung and M. Millnert (1992). Construction of composite models from observed data. *Int. J. of Control* **55**, 141–152.

Verdult, V. and M. Verhaegen (2000). Identification of multivariable linear parameter varying systems based on subspace techniques. In: *Proc. of the IEEE Conference on Decision and Control, Sydney, Australia.*

Zhou, K., J. U. Doyle and K. Glover (1996). *Robust and optimal control.* Prentice-Hall, New Jersey.

A NEW ADAPTIVE IDENTIFICATION METHOD OF CRITICAL POINT USING FREQUENCY ESTIMATOR

Masami Saeki * and Hiromichi Takeuchi **

* *Faculty of Engineering, Hiroshima University, Higashi-Hiroshima, 739-8527, Japan*
E-mail: saeki@hiroshima-u.ac.jp
** *Graduate School of Engineering, Hiroshima University, Higashi-Hiroshima, 739-8527, Japan*

Abstract: In this paper, a new adaptive identification method of a frequency response $G(j\omega)$ that satisfies a phase condition $\angle G(j\omega) = \theta, \theta \in [-\pi, -\pi/2)$ is proposed. It corresponds to the critical point of a feedback system composed of the plant $G(s)$, a saturation nonlinearity, and a first-order lag element $k/(1 + sT)$ where k is tuned so that a sinusoidal oscillation can continue and T is tuned so that the phase condition can be met at the oscillation frequency. The frequency response is given by $G(j\omega) = -(1 + j\omega T)/k$ from the convergent values. In the tuning algorithm, a globally convergent frequency estimator is used for estimating the amplitude and the frequency of the oscillation. *Copyright ©2001 IFAC*

Keywords: Critical points, Non-parametric identification, Adaptive algorithm, PID control, Frequency estimation, Saturation

1. INTRODUCTION

PID controllers are widely used in industry, but many are poorly tuned in practice (see Åström and Hägglund, 1984). Since the Ziegler-Nichols tuning rule only uses the critical gain and the critical frequency, it is simple and robust against disturbances (e.g. Ziegler and Nichols, 1943). Satisfactory tuning of PID gains requires accurate identification of process dynamics. A simple method of identifying the critical gain is to gradually increase the loop gain until an oscillation occurs. The estimate obtained by this method tends to be larger than the true value, because it takes much time for the oscillation becomes apparent. And also, this method is not suitable for online tuning, because the response of the closed-loop system will become unbounded during the experiment.

Åström *et al.* (1984) proposed the relay feedback auto-tuning method using the critical gain and the critical frequency which are estimated by the describing function method. This method guarantees that the response is bounded for the stable plants, and it gives a good estimate for a certain class of plants. The following requirements have been pointed out to improve the identification method and several studies have been reported, but they have not yet been completely solved.

1) More accurate estimate is desired, because the estimate by relay feedback is essentially an approximation.
2) Multiple points on the frequency response are necessary, because crude controller settings are obtained from the single identified point. See, for example, Åström *et al.* (1984) and Schei (1994).
3) Extension to multi-input multi-output case is desired, because a large industrial plant has many decentralized PID controllers. See, for example, Zgorzelski, Unbehauen, and Niederlinski (1990) and Palmor, Halevi and Krasney (1995).

One way of coping with the requirement 1 is not to use the describing function method. Recently, a few studies follow this line. Atherton and Majhi (1998) derived an exact relation between the oscillation parameters and the model parameters for parametric identification. This method use the oscillation of the relay feedback system.

The following two methods use the oscillation of a different feedback system shown in Fig. 1 where the relay function is replaced with the set of a saturation function ϕ and a gain k. Friman

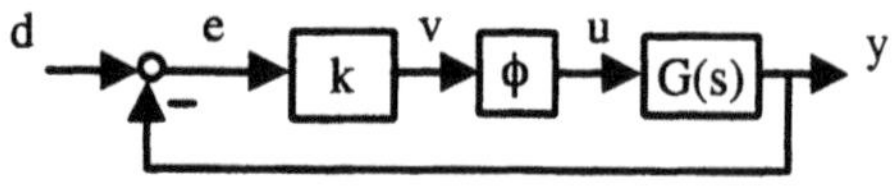

Fig. 1. Nonlinear feedback system with a saturation function ϕ

and Waller (1995) proposed an iterative procedure of selecting the gain where the gain is fixed in each experiment. Saeki (1998) proposed a simple adaptive tuning algorithm: $dk(t)/dt = -\alpha(u(t) - v(t))^2$ where $v(t)$ and $u(t)$ are the input and the output of the saturation function. This method can be easily applied to multi-input multi-output plants, and the selection of α requires a few experiments but it is easy. Thus, this method satisfies the requirements 1 and 3.

The third method proposed by Crowe and Johnson (1999) uses another feedback system where the relay function is replaced with a phase-locked loop element in order to generate a sinusoidal signal with a specified phase.

In this paper, a new adaptive identification method that satisfies the requirements 1 and 2 is given. This method is based on the oscillation of a feedback system composed of a first order element $k/(1+sT)$, a saturation function and the plant, and a globally convergent frequency estimator is used for the estimation of the amplitude and the frequency of the oscillation and the estimates are used for tuning k and T. A tuning method of k will be given in Section 2, and that of T Section 3.

2. IDENTIFICATION OF CRITICAL POINT

Let us consider a feedback system described by

$$y = G(s)u \quad (1)$$

$$u = ke \quad (2)$$

$$e = d - y \quad (3)$$

where $G(s)$ is single-input single-output and $k \in R$. Assume that this system is stable for $0 < k < K_a$ and unstable for $K_a < k$. Then, the critical gain K_a satisfies $1 + K_a G(j\omega) = 0$ at the critical frequency $\omega = \omega_a$, and the critical point is given by $G(j\omega_a) = -1/K_a$. Our problem in this section is to give a method of identifying K_a and ω_a.

Let us insert a saturation function ϕ into the feedback system as shown in Fig. 1 where

$$\phi(\sigma) = \begin{cases} \sigma, & (|\sigma| < U_0) \\ U_0, & (\sigma \geq U_0) \\ -U_0, & (\sigma \leq -U_0) \end{cases} \quad (4)$$

As k approaches infinity, the limit of $\phi(k\sigma)$ is a relay function. Therefore, this system behaves like a relay feedback system for a sufficiently large k. Namely, when k is sufficiently large, the unstable mode increases rapidly and soon a self oscillation occurs. Here note that V (the amplitude of $v(t)$ in the steady state) is much larger than the saturation bound U_0 because k is sufficiently large.

Then, decrease k, and as k comes closer to K_a, V approaches U_0. In other words, the system approaches the unsaturated state. If this oscillation with $V = U_0$ can be maintained by tuning k, the gain and the frequency give the critical gain and the critical frequency, respectively.

In order to realize this procedure, introduce a new tuning rule: starting from a sufficiently large $k(0)$, tune $k(t)$ by

$$\frac{dk}{dt} = -\alpha(k(t)\tilde{E}(t) - U_0) \quad (5)$$

where $\tilde{E}(t)$ is the estimate of the amplitude of $e(t)$ obtained by the globally convergent frequency estimator (see Appendix, Hsu, Ortega, and Damm (1999)). The configuration is shown in Fig. 2.

Remark 1: *If E and ω are constant for $e(t) = E\sin\omega t$, the amplitude of $e(t)$ is obviously E. But, in our case, they are time-varying as shown in Fig. 3, and the amplitude $E(t)$ cannot be defined. Hence, the rule (5) is described by using $\tilde{E}(t)$ which can be defined as the output of the frequency estimator.*

Now, let us examine the behavior of $\tilde{E}(t)$. First, examine the transient behavior. For a sufficiently large k, $\tilde{E}(t)$ increases rapidly and will saturate soon. Fig. 3 shows a typical response of $e(t)$, which is taken from Example 1 of Section 4, where the dotted line shows $e(t)$ and the solid line $\tilde{E}(t)$. The globally convergent frequency estimator works very well, and $\tilde{E}(t)$ increases rapidly and converges to a constant $\tilde{E}_{max}$, while $k(t)$ decreases from $k(0)$ rapidly as shown in Fig. 4. Note that $e(t)$ and $\tilde{E}(t)$ are not sensitive to the change of $k(t)$ and that it can be expected that $\tilde{E}_{max}$ is bounded as $E_1 \geq \tilde{E}(t) \geq E_2 > 0$. This is because $u(t)$ is a simply periodic function with the fixed amplitude U_0 and $e(t)$ is almost sinusoidal by the low pass property of $G(s)$.

Theorem 1 If $E_1 \geq \tilde{E}(t) \geq E_2$ for positive constants E_1 and E_2, then $k(t)$ lies in the interval.

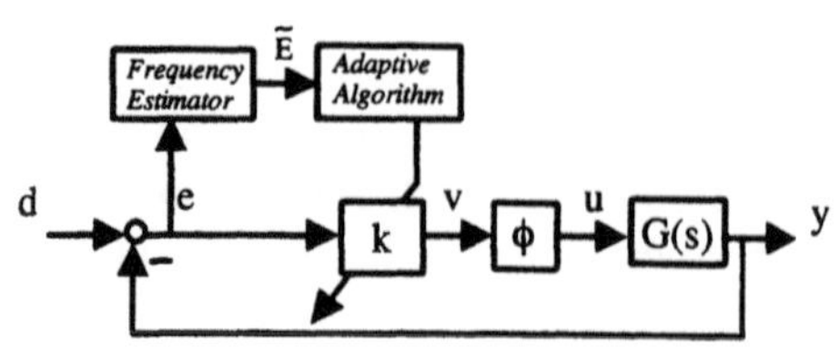

Fig. 2. Configuration for the identification of critical point

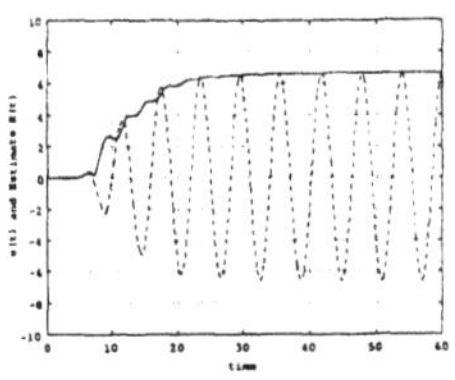

Fig. 3. $e(t)$ and $\tilde{E}(t)$

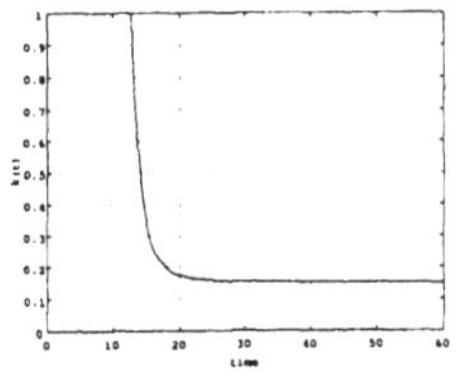

Fig. 4. $k(t)$

$$k_2(t) \geq k(t) \geq k_1(t) \tag{6}$$

where for $i = 1, 2$

$$k_i(t) = k(0)exp(-E_i\alpha t) + \frac{U_0}{E_i}(1 - exp(-E_i\alpha t)) \tag{7}$$

Proof) The solution of $dk/dt = -\alpha(k\tilde{E} - U_0)$ is given by

$$k(t) = k(0)exp(\int_0^t -\alpha\tilde{E}(t)dt)$$
$$+\alpha U_0 \int_0^t exp(\int_t^\tau -\alpha\tilde{E}(t)dt)d\tau \tag{8}$$

Then,

$$k(t) \leq k(0)exp(\int_0^t -\alpha E_2 dt)$$
$$+\alpha U_0 \int_0^t exp(\int_t^\tau -\alpha E_2 dt)d\tau \tag{9}$$
$$= k(0)exp(-E_2\alpha t)$$
$$+\frac{U_0}{E_2}(1 - exp(-E_2\alpha t)) \tag{10}$$

The lower bound can be derived in the same way. This completes the proof.

The bounds $k_1(t)$ and $k_2(t)$ decrease exponentially from k(0) and converge to U_0/E_1 and U_0/E_2, respectively, as shown in Fig. 5. Therefore, $k(t)$ is greater than $U_0/E_2 > 0$ if $k(0) > U_0/E_2$. This guarantees that $k(t)$ does not become negative under the assumption. This is important, because once $k(t)$ becomes negative, the identification procedure fails.

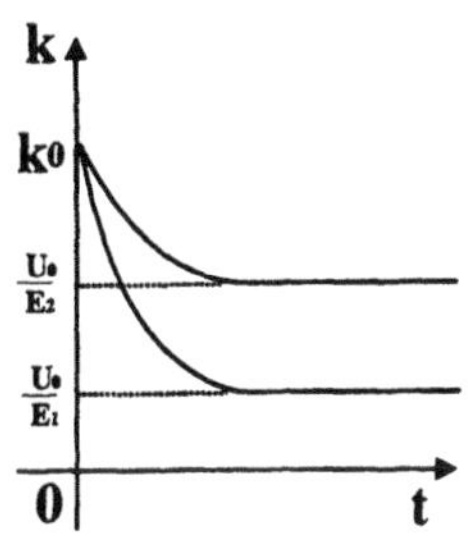

Fig. 5. Bound of $k(t)$

Remark 2: *Let $\tilde{V}(t)$ denote the estimate of the amplitude of $v(t)$ by the frequency estimator, and consider a slightly different rule $dk/dt = -\alpha(\tilde{V}(t) - U_0)$. The computer simulation shows that $k(t)$ becomes negative depending on the estimator parameters by this rule, whereas it does not by Eq.(5), and therefore, the selection of the estimator parameters is much easier for (5). This implies that the similar property given by Theorem 1 is not satisfied for the slightly different rule. This difference may be caused by the fact that the estimation of $V(t)$ is more difficult than that of $E(t)$, because $V(t)$ changes rapidly according to the change of $k(t)$ whereas $e(t)$ is insensitive to the change of $k(t)$.*

As shown in Fig. 4, $k(t)$ converges to a steady state exponentially. Let us examine the stability condition of the steady state. For this purpose, first, make a model on condition that k is constant. Since the dominant pole of $1 + kG(s) = 0$ is $s = j\omega_a$ at $k = K_a$, it can be approximated as

$$s = a(k - K_a) + j\omega_a \tag{11}$$

around $k = K_a$. The amplitude of the corresponding mode is described by $E(t) = exp(a(k - K_a)t)E(0)$, which gives a model

$$\frac{dE}{dt} = a(k - K_a)E, a > 0 \tag{12}$$

Here, a is positive, because E increases for $k > K_a$ and vice versa.

Now, consider the next system composed of the model and the tuning rule (5).

$$\frac{dk}{dt} = -\alpha(kE - U_0) \tag{13}$$
$$\frac{dE}{dt} = a(k - K_a)E \tag{14}$$

where E denotes the amplitude of $e(t)$. The equilibrium is given by $k = K_a, E = E_0(:= U_0/K_a)$. Define new variables x_1, x_2 as

$$x_1 = k - K_a \tag{15}$$
$$x_2 = E - E_0 \tag{16}$$

and linearize the equation around the equilibrium point, then

$$\frac{d}{dt}\begin{pmatrix} x_1 \\ x_2 \end{pmatrix} = \begin{pmatrix} -E_0\alpha & -\alpha K_a \\ aE_0 & 0 \end{pmatrix}\begin{pmatrix} x_1 \\ x_2 \end{pmatrix} \tag{17}$$

Since the characteristic equation is

$$s^2 + E_0\alpha s + aK_aE_0\alpha = 0 \tag{18}$$

, this system is stable if and only if $\alpha > 0$.

Remark 3: *Unfortunately, computer simulation shows that $k(t)$ oscillates with extremely small amplitude near the equilibrium, and it does not converge to a constant. Probably, this may be caused by the dynamics of the frequency estimator, which is not cosidered in the above stability analysis. Further study is required.*

3. EXTENSION TO CONSIDER PHASE CONSTRAINT

The phase of $G(j\omega)$ is $-\pi$ at the critical point. Other points, especially whose phases are between $-\pi/2$ and $-\pi$, give us more information for controller design. In this section, the previous result will be extended to the identification of another $G(j\omega)$ that satisfies the phase constraint $\angle G(j\omega) = -\pi + \theta_b$ where θ_b is specified by the user.

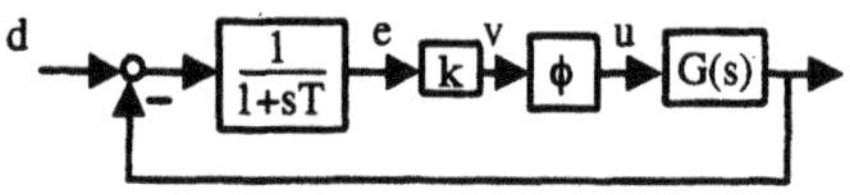

Fig. 6. Identifier with specified phase

Let us consider the system shown in Fig. 6 where the element $1/(1+sT), T \geq 0$ is newly inserted in the loop. Suppose that $k = K_a$ and $\omega = \omega_a$ are obtained for this system by applying the tuning algorithm (5). Then,

$$1 + \frac{K_a}{1 + j\omega_a T} G(j\omega_a) = 0 \qquad (19)$$

and therefore,

$$G(j\omega_a) = -\frac{1 + j\omega_a T}{K_a} \qquad (20)$$

Thus, another point with the phase

$$\angle G(j\omega_a) = -\pi + tan^{-1}(\omega_a T) \qquad (21)$$

can be obtained, but unfortunately, ω_a does not almost always satisfy

$$\theta_b = tan^{-1}(\omega_a T) \qquad (22)$$

for a given $\theta_b \in (0, \pi/2)$.

In order to obtain such $T = T_b$ that satisfies

$$\omega_b T = tan\theta_b \qquad (23)$$

at another critical frequency ω_b, introduce the next error function

$$V = \frac{1}{2}(\omega T - tan\theta_b)^2 \qquad (24)$$

and derive a tuning rule of T that satisfies $dV/dt = -2\beta V$. Then, it can be easily obtained as

$$\frac{dT}{dt} = \left(-\frac{1}{\omega}\frac{d\omega}{dt} - \beta\right) T + \frac{\beta}{\omega} tan\theta_b \qquad (25)$$

The initial value $T(0)$ is chosen as a small positive number in the later examples in order to search for T from a small number.

The rules (5) and (25) are used simultaneously for tuning k and T in Fig. 7 where ω and $d\omega/dt$ can be estimated by the frequency estimator.

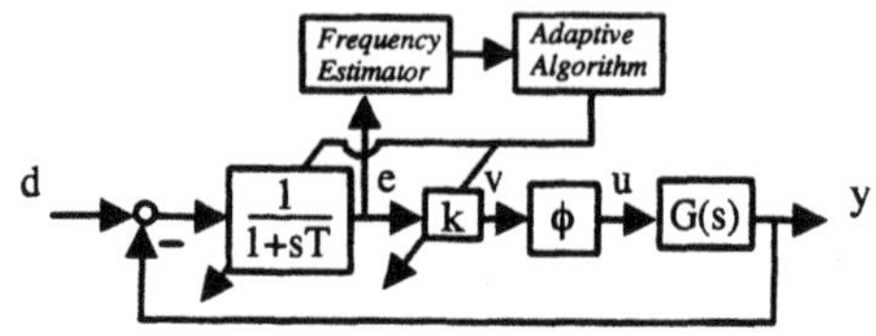

Fig. 7. Identifier with specified phase

4. NUMERICAL EXAMPLES

Let us apply our algorithm to

$$G(s) = \frac{s+1}{s^2 + 0.2s + 1} e^{-2s} \qquad (26)$$

in order to identify the critical point in Example 1 and the point that satisfies $\theta_b = 40°$ in Example 2.

Example 1: The parameters of the adaptive algorithm and the frequency estimator are $\alpha = 0.2, U_0 = 1$ and $\epsilon = 0.6, \zeta = 0.6, \mu = 0.2, N = 1, \alpha_2 = 2$, respectively. The system of Fig. 2 is simulated by SIMULINK, and $k(t)$ and $\omega(t)$ enter the intervals (0.15069, 0.15075) and (1.03216, 1.03238), respectively, for the initial conditions $k(0) = 10, z_1(0) = 1, z_2(0) = 1, z_3(0) = 10$. The responses of $k(t)$, $\omega(t)$, $e(t)$ and $\tilde{E}(t)$, and $v(t)$ are shown in Figs. 8, 9, 10, and 11, respectively.

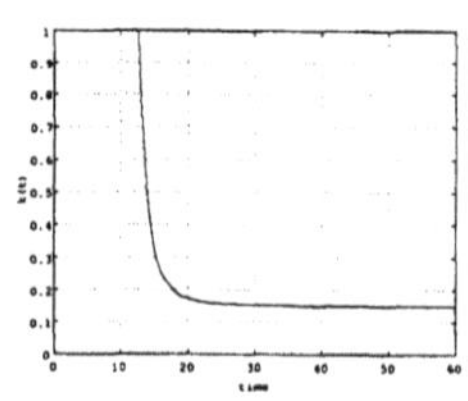

Fig. 8. $k(t)$

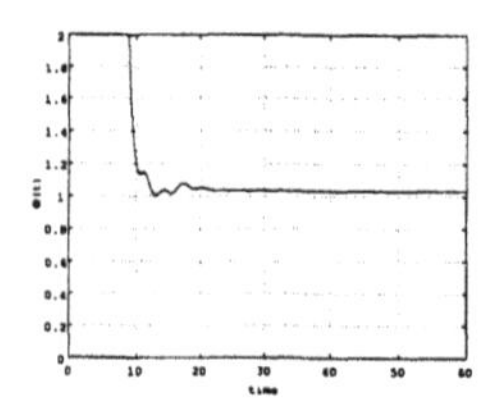

Fig. 9. $\omega(t)$

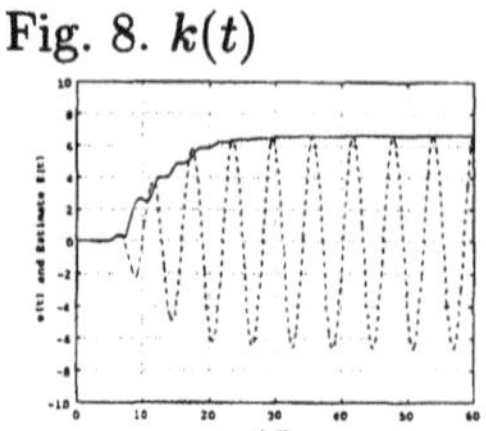

Fig. 10. $e(t)$ and $\tilde{E}(t)$

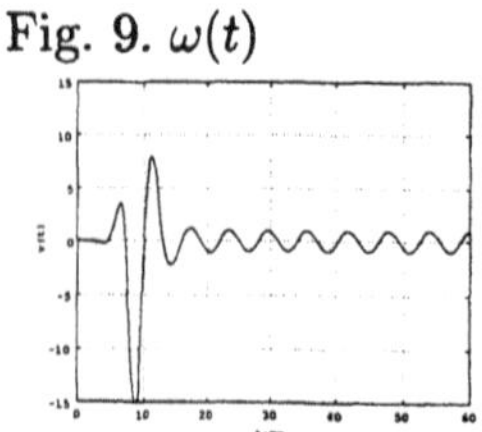

Fig. 11. $v(t)$

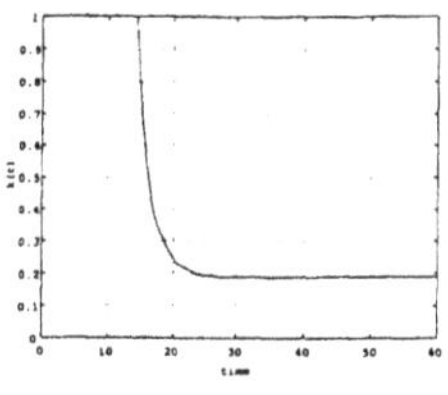

Fig. 12. $k(t)$

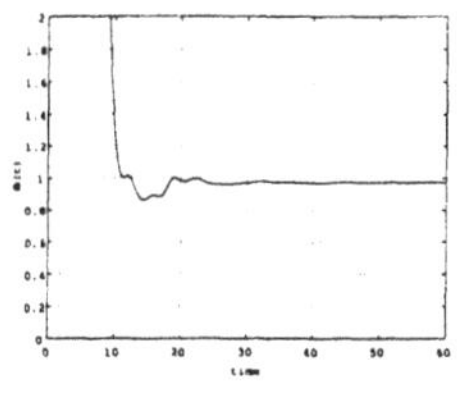

Fig. 13. $\omega(t)$

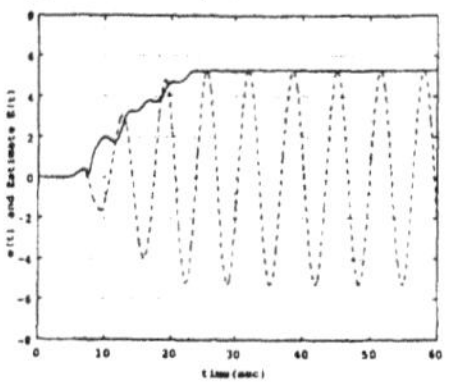

Fig. 14. $e(t)$ and $\tilde{E}(t)$

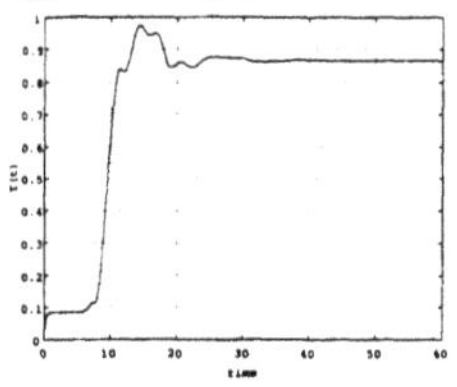

Fig. 15. $T(t)$

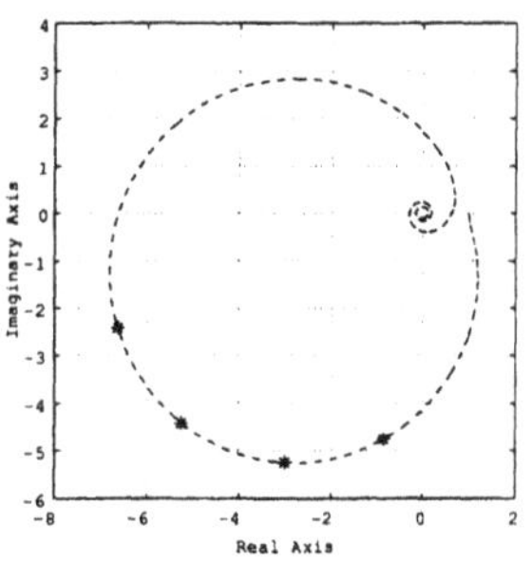

Fig. 16. Nyquist diagram of $G(j\omega)$ and identified points for $\theta_b = 20°, 40°, 60°, 80°$

Example 2: The parameters of the adaptive algorithm and the frequency estimator are $\alpha = 0.3, u_0 = 1, \beta = 5.0$ and $\epsilon = 0.8, \zeta = 0.8, \mu = 0.2, N = 1, \alpha_2 = 2$, respectively. The system of Fig. 7 is simulated by SIMULINK, and $k(t)$, $\omega(t)$ and $T(t)$ enter the intervals $(0.19001, 0.19002)$, $(0.96958, 0.97015)$, and $(0.86492, 0.86542)$, respectively, for the initial conditions $k(0) = 10, z_1(0) = 1, z_2(0) = 1, z_3(0) = 10, T(0) = 0.001$. The responses of $k(t)$, $\omega(t)$, $e(t)$ and $\tilde{E}(t)$, and $T(t)$ are shown in Figs. 12, 13, 14, and 15, respectively.

From the steady state values $T = T_b, k = K_b, \omega = \omega_b$, the frequency response can be estimated as

$$G(j\omega_b) = -\frac{1 + j\omega_b T_b}{K_b} \tag{27}$$

The Nyquist diagram of $G(j\omega)$ and the identified points for $\theta_b = 20°, 40°, 60°, 80°$ are shown by '$*$' in Fig. 16.

5. CONCLUSION

The proposed adaptive identification method gives the frequency response of the plant $G(j\omega)$ that satisfies a phase condition. This method uses the oscillation of the feedback loop composed of the plant, $k/(1+sT)$, and a saturation function. The gain k is tuned to stabilize the oscillation and at the same time, T is tuned so that the phase condition can be met. The amplitude and the frequency of the oscillation are estimated by the globally convergent frequency estimator and are used for the tuning. The estimated values are given as the values of the variables, which is convenient in the realization of online tuning algorithm for PID controller.

The method of Saeki (1998) is still superior to the method of this paper in simplicity of the tuning rule and applicability to multi-input multi-output case. These two methods are complementary, and our future work is to combine these to make a more efficient and accurate identification method for multi-input multi-output plants.

6. REFERENCES

Åström, K.J. and T. Hägglund (1984). Automatic tuning of simple regulators with specifications on phase and amplitude margins. *Automatica*, **20**, 645-651.

Atherton, D.P. and Majhi, S (1998). Plant parameter identification under relay control *Proc. of IEEE CDC*, 1272-1277.

Crowe, J. and M.A. Johnson (1999). A new nonparametric identification procedure for online controller tuning *Proc. of ACC*, 3337-3341.

Friman, M. and Waller, K.V. (1995). Closed-loop identification by use of single-valued nonlinearities. *Industrial & Engineering Chemistry Research*, **34**, 3052-3058.

Hsu, L., R. Ortega, and G. Damm (1999). A globally convergent frequency estimator. *IEEE Trans. Automat. Contr.*, **44**, 698-713.

Palmor,Z.J., Y. Halevi and N. Krasney (1995). Automatic tuning of decentralized PID controllers for TITO processes. *Automatica*, **31**, 1001-1010.

Saeki, M. (1998). A new adaptive identification method of critical loop gain for multi-input multi-output plants. *Proc. of IEEE CDC*, 3984-3989.

Schei, T.S. (1994). Automatic tuning of PID controllers based on transfer function estimation. *Automatica*, **30**, 1983-1989.

Zgorzelski,P., H. Unbehauen, and A. Niederlinski (1990). A new simple decentralized adaptive multivariable regulator and its application to multivariable plants. *Proc. of IFAC World Congress*, 226-231.

Ziegler, J.G. and N.B. Nichols (1943). Optimum settings for automatic controllers. *Trans. ASME*, **65**, 433-444.

APPENDIX

In this appendix, the frequency estimator proposed by Hsu *et al.* (1999) is briefly explained. Let the input to the frequency estimator be a

ONE APPROACH TO ADAPTIVE CONTROL OF NONLINEAR PROCESSES

Petr Dostál, Vladimír Bobál, Martin Blaha

Department of Control Theory, Institute of Information Technologies,
Tomas Bata University
Nám. T. G. Masaryka 275, 762 72 Zlín, Czech Republic
Fax: +420 67 7211521; E-mail: dostalp@ft.utb.cz

Abstract: The paper deals with the adaptive control of nonlinear processes. The nonlinear model of the plant is approximated by an external linear model with recursively estimated parameters. The 2DOF control system configuration is considered. The presented controller design procedure provides the internal properness and stability of the control system. The asymptotic tracking of a step reference is ensured. In addition, the resulting control law penalizes control input changes. The resulting controller is obtained via solutions of the polynomial Diophantine equations with the right side given by spectral factorizations. The adaptive control is tested on the model of a continuous stirred tank reactor. *Copyright © 2001 IFAC*

Keywords: Nonlinear systems, continuous time filters, parameter estimation, polynomial methods, LQ control method, adaptive control.

1. INTRODUCTION

The most part of technological processes belong to a class of nonlinear systems where both steady-state and dynamic behaviour are nonlinear. This fact often causes conventional control methods to be unsatisfactory. Moreover, provided that strong nonlinearities take the place in the process properties, the state and subsequently the output variables may exhibit high sensitivity to control input changes. Therefore, the control system should be able to accept both the process nonlinearity as well as the sensitivity. This requirement can be met by using adaptive control with parallel restriction of control input changes.

The paper presents one approach to continuous-time adaptive control of a nonlinear process. The nonlinear model of the process is approximated by an external linear model (ELM) with recursively estimated parameters. The procedure of the continuous-time ELM parameter estimation is described in detail e. g. in the work of Wahlberg (1990). Since the controlled process input and output derivations cannot be directly measured, the differential filters and filtered variables are established that substitute primary variables. The filtered variables inclusive of their derivations then are measured at discrete time intervals and used in the recursive identification procedure.

The 2DOF control system configuration is considered where a controller contains together with the feedback also the feedforward part, see e. g. (Grimble, 1994). The reference signal is considered to be from a class of step or exponential functions with a finite limit at infinity.

For the controller design, the polynomial approach is used (Kučera, 1993). The general requirements on the control system properties are formulated as the internal properness and stability of the control system, the asymptotic tracking of a reference and a disturbance attenuation. Using the polynomial

approach, the resulting controller is obtained by a solution of polynomial Diophantine equations with a stable polynomial on the right sides.
Further, the LQ control technique is applied that enables to restrict the control input changes (Dostál, and Bobál, 1999). The right sides of the polynomial equations then are given as results of spectral factorizations and they are at all events stable. The procedure yields strictly proper controllers.
The control of a nonlinear SISO process based on the above procedure is tested on the model of a continuous stirred tank reactor described by four nonlinear differential equations (Luyben, 1989; Ingham *et al.*, 1994).

2. EXTERNAL LINEAR MODEL

The external linear model (ELM) is described in the time domain by differential equation

$$a(\sigma)y(t) = b(\sigma)u(t) \qquad (1)$$

where σ is the derivative operator and a, b are polynomials in σ. Generally, nonzero initial conditions for both output and input variables are considered. Using the Laplace transform, the ELM is represented in the complex domain by

$$Y(s) = \frac{b(s)}{a(s)}U(s) + \frac{o_1(s)}{a(s)} \qquad (2)$$

where s is the complex argument, o_1 is the transform of initial conditions and Y and U denote the transforms of y and u. Both a and b now are coprime polynomials in s. Assuming the condition deg $b \leq$ deg a, transfer function $G(s) = b(s)/a(s)$ is proper. The inequality deg $o_1 <$ deg a results from properties of the Laplace transform.

3. ELM PARAMETER ESTIMATION

The method of ELM parameter estimation has been analyzed e.g. by Wahlberg (1990). Since the derivatives of both input and output variables cannot be directly measured, filtered variables u_f and y_f are established as the outputs of filters

$$c(\sigma)u_f(t) = u(t)\,,\; c(\sigma)y_f(t) = y(t) \qquad (3)$$

where $c(\sigma)$ is a stable polynomial in σ which fulfills the condition $\deg c(\sigma) \geq \deg a(\sigma)$. Note that the time constants of the filters must be smaller than the time constants of the model. Since the latter are unknown at the beginning of the estimation procedure, it is necessary to make the filter time constants, selected a priori, sufficiently small.

It can easily be proved that the transfer behaviour between the filtered and between nonfiltered variables are equivalent. This fact enables to employ the filtered variables for the ELM parameter estimation. If these are computed via filters (3) in discrete time intervals $t_k = k\,T_s$, $k = 0,1,2,\ldots$, where T_s is the sampling period, then the parameters of the ELM can be recursively estimated from the equation

$$y_f^{(n)}(t_k) = -a_0 y_f(t_k) - a_1 y_f^{(1)}(t_k) - \ldots - a_{n-1} y_f^{(n-1)}(t_k) +$$
$$+ b_0 u_f(t_k) + b_1 u_f^{(1)}(t_k) + \ldots + b_m u_f^{(m)}(t_k) + \varepsilon(t_k) \quad (4)$$

where ε expresses a difference between the initial conditions of filtered and nonfiltered variables.

4. CONTROL SYSTEM DESCRIPTION

The 2DOF control system is shown in Fig. 1. Here, w is the reference, G is the controlled plant, Q and R are the controller feedback and feedforward parts and F is the compensator. The compensator F is the component of both feedback and feedforward parts, respectively. However, regarding some following mathematical manipulations, the compensator may be formally separated from the controller.

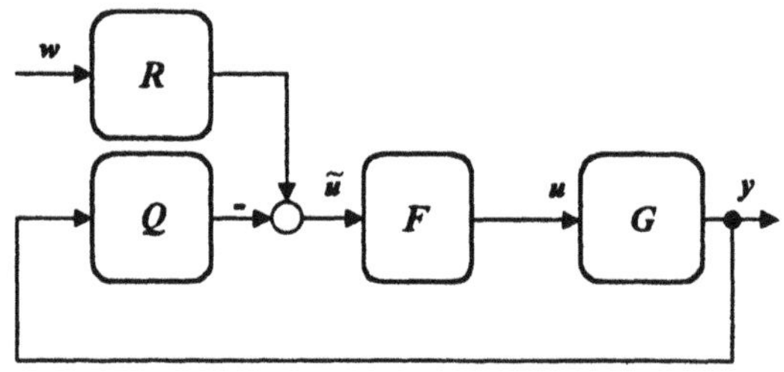

Fig. 1. The 2DOF control system configuration.

For nonzero initial conditions, the signals in the control system can be described in the complex domain as

$$f(s)U(s) = \tilde{U}(s) + o_2(s) \qquad (5)$$
$$p(s)\tilde{U}(s) = r(s)W(s) - q(s)Y(s) + o_3(s) \qquad (6)$$

where q, p and r, p are coprime polynomials in s and o_2, o_3 are the transforms of initial conditions.

The transfer functions of both controller parts and the compensator are

$$Q(s) = \frac{q(s)}{p(s)}\,,\; R(s) = \frac{r(s)}{p(s)}\,,\; F(s) = \frac{1}{f(s)} \qquad (7)$$

where $p(0) \neq 0$.

The reference w is considered to be from a class of step or exponential functions with a finite limit at $t \to \infty$ and with a transform expressed in the form

$$W(s) = \frac{h_w(s)}{\tilde{f}_w(s)} = \frac{h_w(s)}{s f_w(s)} \qquad (8)$$

where h_w, f_w are coprime polynomials in s so that deg $h_w \leq$ deg f_w and f_w is a stable polynomial with $f_w(0) \neq 0$. For nonzero initial conditions, the signals $\tilde{u}$ and e can be obtained in the complex domain (to simplify arguments s in polynomials are omitted) as

$$\tilde{U}(s) = \frac{1}{d}\left\{a f r \frac{h_w}{s f_w} - q\,(f o_1 + b o_2) + a f o_3\right\} \quad (9)$$

$$E(s) = \frac{1}{d}\left\{(d - br)\frac{h_w}{s f_w} - p\,(f o_1 + b o_2) - b o_3\right\} \quad (10)$$

where $d = a f p + b q$.

The general conditions required to govern the control system properties are

- internal properness and stability of the control system
- asymptotic tracking of a reference.

The above requirements are fulfilled for the controller feedback and feedforward parts given by solutions of the couple of polynomial equations

$$a f p + b q = d \qquad (11)$$

$$z f + b r = d \qquad (12)$$

with a stable polynomial d on the right sides where z is an unknown additional polynomial.

The control system satisfies the condition of internal properness only when the transfer functions of all its components are proper. The degrees of polynomials of the controller transfer functions (inclusive of the compensator) must then fulfill inequalities

$$\deg q \leq \deg p + \deg f\,,\ \deg r \leq \deg p + \deg f\,. \quad (13)$$

For a reference with the transform (8), the good control properties can be expected for $f(s)$ in (9) divisible by s. Then, the compensator is an integrator where $f(s) = s$ and in Fig. 1 the equality $\tilde{u} = \dot{u}$ holds. The degrees of polynomials q, p and r in (11) and (12) can be derived to be

$$\deg q = \deg a,\ \deg p \geq \deg a - 1,\ \deg r = 0\,. \quad (14)$$

Moreover, the equality $r_0 = q_0$ can be easily proved.

Considering (7), the complete transfer functions of both parts of a controller inclusive the integrator take forms

$$\tilde{Q}(s) = Q(s)\,F(s) = \frac{q(s)}{s\,p(s)} \qquad (15)$$

$$\tilde{R}(s) = R(s)\,F(s) = \frac{r(s)}{s\,p(s)}\,. \qquad (16)$$

The controller parameters then follow from solutions of polynomial equations (11) and (12) and depend upon coefficients of polynomial d. The next problem here means to find a stable polynomial d that enables to obtain the acceptable stabilizing controller.

5. APPLICATION OF LQ CONTROL THEORY

It may be remark that an objective of the procedure here is not to derive an optimal control. However, it is useful to find the technique that makes possible to obtain always stable polynomial on the right sides of polynomial equations (11) and (12). The LQ control technique exploiting polynomial spectral factorizations provides this possibility.

The goal of optimal deterministic LQ tracking is to design controllers that enable the control system to satisfy the above basic requirements and, in addition, the control law minimizes the cost function in the complex domain

$$J = \frac{1}{2\pi j}\int_{-j\infty}^{j\infty}\left\{E^*(s)\,\mu\,E(s) + \tilde{U}^*(s)\,\varphi\,\tilde{U}(s)\right\}ds \quad (17)$$

where $\mu \geq 0$ and $\varphi > 0$ are weighting coefficients. It has been proved by Kučera and Šebek (1984) that the solution of this problem results in controllers where the parameters depend upon the initial conditions of the signals in the control system (in this case o_1, o_2 and o_3). Clearly, these controllers are ill-posed and unacceptable for the control purposes.

An acceptable controller may only be obtained for the above determined class of references with a transform in the form (8). The design procedure can now be realized by taking the following steps:

Calculate stable polynomials g and n as the results of spectral factorizations

$$(af)^*\varphi\,a f + b^*\mu b = g^* g \qquad (18)$$

$$n^* n = a^* a \qquad (19)$$

where $f(s) = s$.

Polynomials q, r and p of the controller transfer functions are then given by solutions of polynomial equations

$$a f p + b q = g n \qquad (20)$$

$$z f + b r = g n\,. \qquad (21)$$

The transfer functions of both controller parts (inclusive the integrator) are strictly proper. The degree right sides of (20) and (21) is given by $\deg[g(s)n(s)] = 2\deg a(s) + 1$. Taking into account (14) and the relation $\deg[s\,p(s)] = \deg[g(s)n(s)] - \deg a(s) = \deg a(s) + 1$, the strict properness of both (15) and (16) is evident.

6. ADAPTIVE CONTROL DESCRIPTION

The adaptive control system is shown in Fig. 2. According to the scheme, the adaptive control can be realized in the following steps:

- The ELM (1) of a controlled nonlinear process is chosen using a previous analysis of the process dynamic behaviour.
- Both control input and controlled output enter into filters (3).
- The filtered variables inclusive of their derivatives are measured at filter outputs, sampled and subsequently used for the ELM parameter estimation according to (4).
- The coefficients of stable polynomials g and n are computed by using formulas which can be derived from spectral factorizations (18) and (19).
- The controller parameters are computed from a set of equations developed from solutions of (20) and (21) and periodically readjusted.

7. SIMULATION EXAMPLE

Continuous stirred tank reactors (CSTRs) are the central component of many plants in the chemical industry. From the system engineering point of view, they belong to a class of nonlinear systems where both steady-state and dynamic behaviour are nonlinear. From this reason, the model of a CSTR has been used to verification of the above described procedure.

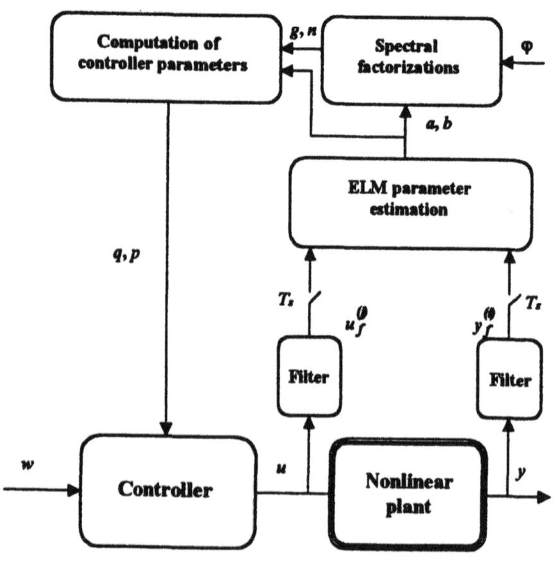

Fig. 2. Adaptive control scheme.

7.1 Model of the plant

Consider a CSTR with the first order consecutive exothermic reactions according to scheme $A \xrightarrow{k_1} B \xrightarrow{k_2} C$ and with a perfectly mixed cooling jacket. Using the usual simplifications, the model of the CSTR is described by four nonlinear differential equations

$$\frac{dc_A}{dt} = -\left(\frac{Q_r}{V_r} + k_1\right)c_A + \frac{Q_r}{V_r}c_{Af}$$

$$\frac{dc_B}{dt} = -\left(\frac{Q_r}{V_r} + k_2\right)c_B + k_1 c_A + \frac{Q_r}{V_r}c_{Bf}$$

$$\frac{dT_r}{dt} = \frac{h_r}{(\rho c_p)_r} + \frac{Q_r}{V_r}(T_{rf} - T_r) + \frac{A_h U}{V_r(\rho c_p)_r}(T_c - T_r)$$

$$\frac{dT_c}{dt} = \frac{Q_c}{V_c}(T_{cf} - T_c) + \frac{A_h U}{V_c(\rho c_p)_c}(T_r - T_c) \quad (22)$$

with initial conditions $c_A(0) = c_A^s$, $c_B(0) = c_B^s$, $T_r(0) = T_r^s$ and $T_c(0) = T_c^s$. Here, t is the time, c are concentrations, T are temperatures, V are volumes, ρ are densities, c_p are specific heat capacities, Q are volumetric flow rates, A_h is the heat exchange surface area and U is the heat transfer coefficient. The subscripts are $(\cdot)_r$ for the reactant mixture, $(\cdot)_c$ for the coolant, $(\cdot)_f$ for feed (inlet) values and the superscript $(\cdot)^s$ for steady-state values.

The reaction rates and the reaction heat are expressed as

$$k_j = k_{0j}\exp\left(\frac{-E_j}{RT_r}\right),\ j = 1,2 \quad (23)$$

$$h_r = h_1 k_1 c_A + h_2 k_2 c_B \quad (24)$$

where k_0 are pre-exponential factors, E are activation energies and h are reaction entalpies. The values of all parameters, feed values and steady-state values are given in Table 1.

Table 1. Parameters, inlet values and initial conditions

$V_r = 1.2\ \mathrm{m^3}$ $V_c = 0.64\ \mathrm{m^3}$ $\rho_r = 985\ \mathrm{kg\ m^{-3}}$ $\rho_c = 998\ \mathrm{kg\ m^{-3}}$ $A = 5.5\ \mathrm{m^2}$	$Q_r = 0.08\ \mathrm{m^3 min^{-1}}$ $Q_c^s = 0.03\ \mathrm{m^3 min^{-1}}$ $c_{pr} = 4.05\ \mathrm{kJ\ kg^{-1}K^{-1}}$ $c_{pc} = 4.18\ \mathrm{kJ\ kg^{-1}K^{-1}}$ $U = 43.5\ \mathrm{kJ\ m^{-2}min^{-1}K^{-1}}$
$k_{10} = 5.616 \cdot 10^{16}\ \mathrm{min^{-1}}$ $k_{20} = 1.128 \cdot 10^{18}\ \mathrm{min^{-1}}$ $h_1 = 4.8 \cdot 10^4\ \mathrm{kJ\ kmol^{-1}}$	$E_1/R = 13477\ \mathrm{K}$ $E_2/R = 15290\ \mathrm{K}$ $h_2 = 2.2 \cdot 10^4\ \mathrm{kJ\ kmol^{-1}}$
$c_{Af} = 2.85\ \mathrm{kmol\ m^{-3}}$ $T_f = 323\ \mathrm{K}$	$c_{Bf} = 0\ \mathrm{kmol\ m^{-3}}$ $T_{cf} = 293\ \mathrm{K}$
$c_A^s = 0.1649\ \mathrm{kmol\ m^{-3}}$ $T_r^s = 350.19\ \mathrm{K}$	$c_b^s = 0.9435\ \mathrm{kmol\ m^{-3}}$ $T_c^s = 330.55\ \mathrm{K}$

For the control purposes, the controlled output and the control input are defined as

$$y(t) = T_r(t) - T_r^s,\ u(t) = 10\,\frac{Q_c(t) - Q_c^s}{Q_c^s}. \qquad (25)$$

These expressions enable to obtain variables of approximately the same magnitude.

7.2 Simulation results

For the adaptive control simulation, the following references were tracked: $w(t) = 4\cdot(1-\exp(-0.1\cdot t))$ K for $0 \le t < 150$ min, $w(t) = -2$ K for $150 \le t < 270$ min and $w(t) = 4$ K for $t \ge 270$ min. The boundary of the control input were used within $-6.5 \le u(t) \le 3.5$. Using some results of the CSTR previous dynamic analysis, the 2nd order ELM expressed by transfer function

$$G(s) = \frac{b(s)}{a(s)} = \frac{b_0}{s^2 + a_1 s + a_0}$$

has been chosen. The unit weighting coefficient μ and various φ coefficients were used.
The control simulation was realized in the following steps:

The filtered variables and their derivatives were obtained from filters

$$\ddot{y}_f + c_1\dot{y}_f + c_0 y_f = y,\ \ddot{u}_f + c_1\dot{u}_f + c_0 u_f = u$$

with filter parameters $c_1 = 0.5$, $c_0 = 0.0625$.

The ELM parameters were recursively estimated from equation

$$\ddot{y}_f(t_k) = b_0 u_f(t_k) - a_0 y_f(t_k) - a_1\dot{y}_f(t_k) + \varepsilon(t_k)$$

in discrete time intervals $t_k = k\,T_s$ with the sampling period T_s = 0.5 min. Here, the identification procedure with exponential forgetting was used.

The polynomials g and n in spectral factorizations (18) and (19) have the form

$$g(s) = g_3 s^3 + g_2 s^2 + g_1 s + g_0$$
$$n(s) = s^2 + n_1 s + n_0$$

and their coefficients were computed as

$$g_0 = \sqrt{\mu b_0^2},\ g_3 = \sqrt{\varphi},\ g_1 = \sqrt{2 g_2 g_0 + \varphi a_0^2}$$
$$g_2 = \sqrt{2 g_3 g_1 + \varphi(a_1^2 - 2a_0)}$$

$$n_0 = \sqrt{a_0^2},\ n_1 = \sqrt{2n_0 + a_1^2 - 2a_0}.$$

The polynomial $d = gn$ on the right sides of polynomial equations (20) and (21) has the form

$$d(s) = d_5 s^5 + d_4 s^4 + d_3 s^3 + d_2 s^2 + d_1 s + d_0$$

with coefficients

$$d_5 = g_3,\ d_4 = g_3 n_1 + g_2,\ d_3 = g_3 n_0 + g_2 n_1 + g_1$$
$$d_2 = g_2 n_0 + g_1 n_1 + g_0,\ d_1 = g_1 n_0 + g_0 n_1,\ d_0 = g_0 n_0.$$

The degrees of polynomials q, r and p were determined in accordance with (14) as

$$\deg q = 2,\ \deg r = 0,\ \deg p = 2$$

and the transfer functions of the resulting controller then take the form

$$\tilde{Q}(s) = Q(s)F(s) = \frac{q_2 s^2 + q_1 s + q_0}{s(p_2 s^2 + p_1 s + p_0)}$$

$$\tilde{R}(s) = R(s)F(s) = \frac{r_0}{s(p_2 s^2 + p_1 s + p_0)}.$$

The controller parameters were computed as

$$p_2 = d_5,\ p_1 = d_4 - a_1 p_2,\ p_0 = d_3 - a_1 p_1 - a_0 p_2$$
$$q_2 = \frac{1}{b_0}(d_2 - a_1 p_0 - a_0 p_1),\ q_1 = \frac{1}{b_0}(d_1 - a_0 p_0)$$
$$q_0 = \frac{1}{b_0} d_0,\ r_0 = q_0.$$

The control law is in the time domain given as

$$p_2\dddot{u} + p_1\ddot{u} + p_0\dot{u} = r_0 w - q_2\ddot{y} - q_1\dot{y} - q_0 y.$$

At the start of the identification procedure (adaptation phase), a P-controller was used.

The controlled output, control input and control input derivative responses to the reference are shown in Figs. 3-5. These document an influence of the coefficient φ (for μ = 1) upon the responses. An increasing value of φ strongly reduces the control input changes. This fact can be important in case of the control of a realistic plant.

In addition, the presence of an integrative part in the controller transfer function enables a step disturbance attenuation. The controlled output responses to a setpoint reference and a step disturbance is shown in Fig. 6. Here, a step change of the component A inlet concentration $\Delta c_{Af} = -0.05$ was injected into the system at time 150 min.

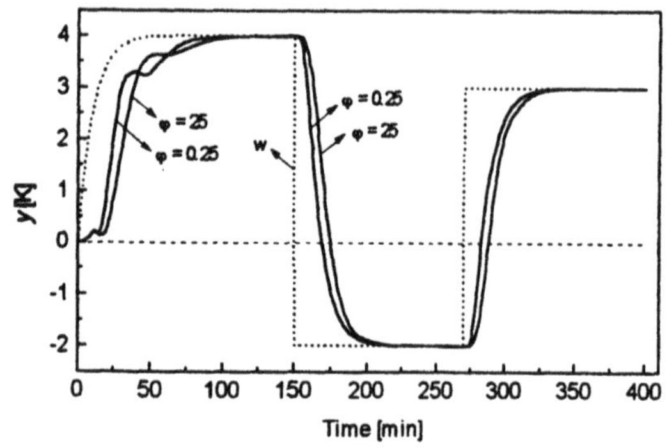

Fig. 3. Controlled output responses.

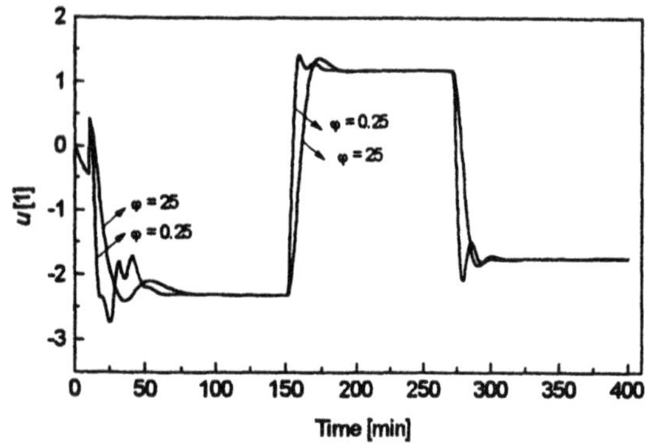

Fig. 4. Control input responses.

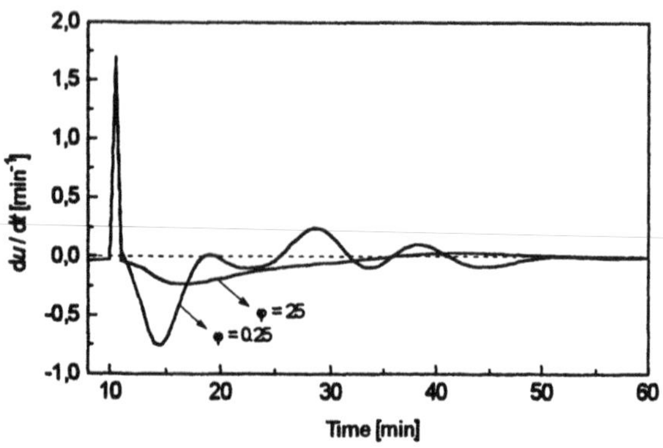

Fig. 5. Control input derivative responses.

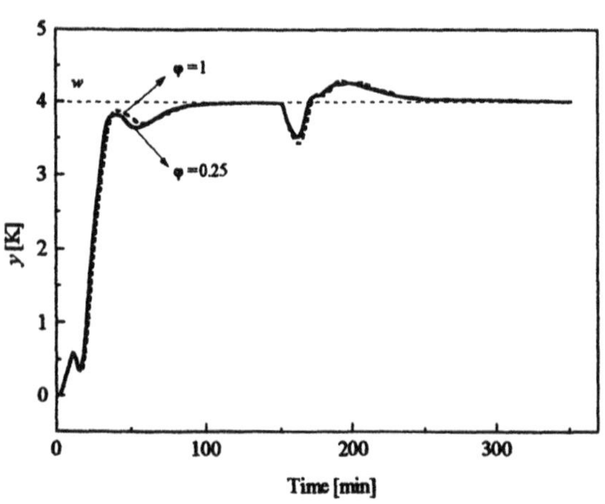

Fig. 6. Controlled output responses to step reference and step disturbance.

8. CONCLUSIONS

In this paper, one possible efficient approach to nonlinear process control design was proposed. The presented strategy enables to create an effective control algorithm. This algorithm is based on an alternative linear model which corresponds to the process primary nonlinear model. The parameters of the resulting continuous-time controller are periodically readjusted according to the estimated model parameters. The idea of an adaptive control is based on the optimal LQ control theory. The procedures in a polynomial ring are used to derive the resulting controller structure. The controller parameters can be tuned by a single selectable parameter φ.

The resulting control system has been tested by computer simulation on the nonlinear model of a CSTR. The results demonstrate the applicability of the presented control strategy. It can be deduced that the described adaptive strategy is also suitable for other technological processes.

ACKNOWLEDGEMENTS

This work was supported in part by the Grant Agency of the Czech Republic under grants No. 102/00/0526 and No. 102/99/1292 and by the Ministry of Education of the Czech Republic under grant MSM 2811 00001.

REFERENCES

Dostál, P. and V. Bobál (1999). The suboptimal tracking problem in linear systems. In: *7th IEEE Mediterannean Conference on Control and Automation,* Haifa, Israel, 667-673.

Grimble, M.J. (1994). *Robust industrial control. Optimal design approach for polynomial systems.* Prentice Hall, London.

Ingham, J., I. J. Dunn, E. Heinzle and J. E. Přenosil (1994). *Chemical Engineering Dynamics. Modelling with PC Simulation.* VCH Verlagsgesellschaft, Weinheim.

Kučera, V and M. Šebek (1984). A polynomial solution to regulation and tracking. Part I. Deterministic problem. *Kybernetika,* 20, 177-188.

Kučera, V. (1993). Diophantine equations in control – A survey. *Automatica,* 29, 1361-1375.

Luyben, W.L. (1989). *Process modelling, simulation and control for chemical engineers.* McGraw-Hill, New York.

Wahlberg, B. (1990). On the identification of continuous-time dynamical systems. In: *Report LiTH-ISY-I-0905,* Linköping.

DESIGN OF CONTROLLER FOR NONLINEAR SYSTEMS BY THE GAIN SCHEDULING TECHNIQUE

Petru DOBRA

Technical University of Cluj, Department of Automation,
St. C. Daicoviciu, No. 15, 3400 Cluj-Napoca, ROMANIA,
Tel: +40-64195777, Fax: +40-64192055,
E-mail: Petru.Dobra@aut.utcluj.ro

Abstract: A gain scheduling controller is a linear regulator whose parameters are changed as a function of the varying operating conditions. By treating nonlinearities as uncertainties, the methods can be used in gain scheduling design. In the discussion, emphasis is put on how to take into consideration different characteristics of the time-varying properties of the system to be controlled. Also robustness and its relation with gain scheduling are treated. In order to handle systems with time-invariant uncertainties, both linear systems and constant parameters, a set of scalings and multipliers are introduced. These are matched to the properties of the uncertainties. Also, multipliers for treating uncertainties that are slowly varying, such that the rate of change is bounded, are introduced. Using these multipliers the applicability of the analysis and synthesis results are greatly extended. *Copyright ©2001 IFAC*

Keywords: Gain scheduling; Linear parameter-varying systems; Nonlinear control system; Linear fractional transformation; Inverted pendulum.

1. INTRODUCTION

This paper considers the analysis of systems with uncertainties and the design of controllers to such systems. Uncertainties are treated in a relatively broad sense covering gain-bounded elements that are not known a priori but could be available to the controller in real time.

The uncertainties are in the most general case norm-bounded operators with a given block-diagonal structure. The structure includes parameters, linear time-invariant and time-varying systems as well as nonlinearities. In some applications the controller may have access to the uncertainty, e.g. a parameter that depends on some known condition.

Gain scheduling is here defined as a linear parameter varying (LPV) feedback regulator whose parameters are changed as a function of operating conditions. It is employed in many applications when one linear controller is not sufficient for achieving the required performance. A normal way to design, e.g. inverted pendulum system, is to select a number of operating conditions for which a linear controller is designed and analyzed. To verify that the closed-loop system behaves as expected and required between the operating conditions, extensive simulations and analysis are usually needed. As an alternative or complement it can be used robustness analysis for verifying performance.

It is review linearization based of gain scheduling, focusing on recently developed analytical aspects. Linear parameter varying (LPV) and linear fractional transformation (LFT) approaches are treated in this paper. It is reviewed how linear matrix inequality (LMI) based construction of H^{∞} optimal controllers can be adapted to compute gain scheduled controllers for LPV systems that guarantee stability and performance (Gahinet, 1996).

Robustness and gain scheduling are closely related. Stability and performance of gain scheduling controllers can be analyzed using robustness criteria if the parameters used for gain scheduling are treated as uncertain parameters available to the controller in real time.

It can be distinguished between three different aspects of robustness that are relevant for such a design:

(a) Robustness to model uncertainties;
(b) Robustness to interpolation errors between operating conditions;
(c) Robustness to dynamic effects when changing between operating conditions.

The methods presented in the thesis can be used both to analyze these aspects and to facilitate the design.

In this paper there will be treated various ways of how to employ these methods. Also, it will be developed a discussion on how to convert a nonlinear model into a linear parameter varying model, in which the parameters are functions of the states.

2. GAIN SCHEDULING

Gain scheduling is a nonlinear feedback of a special type. There is probably no common and strict definition for *gain scheduling.* Åström and Wittenmark (1989) defined *gain scheduling* as "a nonlinear feedback of special type; it has a linear regulator whose parameters are changed as a function of operating conditions in a preprogramming way".

In this paper a slightly broader scope is taken:

Gain scheduling: a linear parameter varying (LPV) feedback regulator whose parameters are changed as a function of operating conditions.

Operating conditions are interpreted to include also dependency on the states of the plant or regulator, i.e. nonlinearities.

The design of a gain scheduled controller for a nonlinear plant can be described as following procedure, see (Rugh and Shamma, 2000).

The first step is to compute a linear parameter varying model for the plant. The most common approach is based on: local linearization (Jacobian linearization), global linearization with fixed equilibrium and global linearization with moving reference. This yields a parametrized family of linearized plants and forms the basis for what it call *linearization scheduling.* Another approach is *quasi-LPV scheduling*, in which the plant dynamics are rewritten to disguise nonlinearities as time-varying parameters that are used as scheduling variables.

The second step is to use linear design methods to design linear controllers for the linear parameter-varying plant model that arises in either linearization or quasi-LPV scheduling. This design process may result directly in a family of linear controllers corresponding to the linear parameter dependent plant or there may be an interpolation process to arrive at a family of linear controllers from a set of controller designs at isolated values of the scheduling variables. The design is such that for each fixed value of the parameter, the linear closed loop system exhibits desirable performance. More recent design approaches attempt to assure desirable performance for time-varying parameter trajectories in a specified class.

The third step, the gain scheduling involves implementing the family of linear controllers such that the controller coefficients (gains) are varied (scheduled) according to the current value of the scheduling variables.

The fourth step consists in performance assessment. In the best case, where analytical performance guarantees are part of the design process, this may be relatively simple. More typically, the local stability and performance properties of the gain scheduled controller might be subject to analytical investigation, while the nonlocal performance evaluation is based on simulation studies.

3. LINEARIZING NONLINEAR MODELS

If the system is truly linear or at least close to linear, it can be in most case obtain good performance using a linear controller. The real need for gain scheduling is for nonlinear systems:

$$\frac{dx}{dt} = \mathbf{f}(x; u);$$
$$y = h(x; u),$$

where u is the plant's input and y its output. Rugh discusses various methods for coming from a nonlinear model to a linearized one, see (Rugh, 1991; Rugh and Shamma, 2000).

I will here discuss some methods for converting a nonlinear model into an LPV model or into an uncertainty model by an LFT. The uncertainties will become functions of the states and possibly also of the reference inputs. The difficulty lies in finding useful models that have bounded uncertainties. There is always a compromise to make when selecting a model for analysis and design. A model is useful if it represents the true system faithfully and if it allows constructive computations to be performed on it. This means that I need models, here represented as uncertainty models, that are not too conservative and that do not deviate essentially from the true plant.

It is important to distinguish between the different types of linearizations. Three types can be distinguished:

(a) a *locally* linearized model;
(b) a *globally* linearized model with *fixed equilibrium*;
(c) a *globally* linearized model with *moving reference* point.

3.1 *Local Linearization*

The local linearization is a model that fits accurately to the nonlinear model close to its equilibrium or reference point, which is assumed to be at $x = 0$. The model is obtained by differentiating the nonlinear model at the equilibrium point. Thus

$$\mathbf{f}(x) \approx Ax$$

with $A = \frac{\partial \mathbf{f}}{\partial x}\big|_{x=0}$. If $\mathbf{f}$ is sufficiently smooth (it must at least be differentiable at $x = 0$) it is can expect it to be accurate close to the equilibrium. As the states leave the equilibrium there is generally no guarantee on how well the linear model will fit the true system.

3.2 *Global Linearization with Fixed Equilibrium*

I will now extend the linearization to include the nonlinear model using differential inclusion. If the plant only operates around a given fixed equilibrium or reference point, here set to $x = 0$, some simplification can be made. In the scalar case ($f : \mathbb{R} \to \mathbb{R}$) a unique model exists:

$$f(x) = \frac{f(x)}{x} x = \delta_f(x)\, x$$

where $f(x) = \frac{f(x)}{x}$ assuming that $f(0) = 0$ and $f(x)$ is differentiable at $x = 0$. If $\frac{f(x)}{x}$ is limited from below and above, it can be find a norm-bounded representation.

3.3 *Global Linearization with Moving Reference*

In many applications the reference point r is not given in advance but it is an input to the controller. As for the fixed-equilibrium case it can be included the system into a differential inclusion. In the scalar case results

$$f(x) = \frac{f(x) - f(r)}{x - r}(x - r) + f(r) = \\ = \delta_f(x, r)(x - r) + f(r).$$

Provided that the reference input is an external signal not depending on x it will not affect stability, since it only enters as a bias term. In this case the uncertainty element

$$\delta_f(x, r) = \frac{f(x) - f(r)}{x - r}$$

is a function of both x and r. It will be bounded if f is differentiable in the interval $[x,\ r]$; specifically $\delta_f(x, r) = f_x(\xi)$ for some $\xi \in [x,\ r]$.

4. ROBUSTNESS ASPECTS OF GAIN SCHEDULING

A gain scheduling controller should be robust with respect to

(a) model uncertainties;
(b) interpolation errors between operating conditions;
(c) dynamic effects when changing between operating conditions.

4.1 *Robustness*

The first aspect, which also applies to time-invariant systems, is how well the controller can handle variations in the plant while maintaining robust stability and performance specifications. This is the "standard" robustness problem. If the system was not time-varying I could design one controller for each specific operating condition. This controller should probably work well also close to the operating condition it was designed for.

4.2 *Interpolation*

The second aspect refers to the need of modifying the controller in order to fit the actual operating condition. This can be implemented by either switching between a set of controllers or by interpolating between them. Interpolation, which is most often linear, can be represented as a linear fractional transformation.

The interpolation between two controllers requires a particular structure. For instance, if two H^∞ controllers are designed at two operating conditions it is not guaranteed that interpolation can be performed between them with expected performance. The two H^∞ controllers may not even be of the same order. One way to overcome this problem is to specify the structure of the controller in a way that reduces the freedom and facilitates interpolation.

To be able to analyze the robustness and performance properties between the operating points I need not only to know the interpolation scheme of the controller but also a global model of the plant. Representing both the controller and the plant as LFT models robust stability and performance can be verified by techniques.

4.3 *Dynamic Effects*

The third aspect concerns the dynamic effects when moving in the operational envelope. In many applications the rate at which the operating conditions are changed cannot be neglected. The controller structure as a linear parameter varying (LPV) system becomes relevant. Using for instance the uncertainty augmentation technique, time-varying parameters can be included in the standard μ framework.

5. INVERTED PENDULUM SYSTEM

Consider the equation of motion for an idealized pendulum about its stable equilibrium:

$$\frac{d^2y}{dt^2} = -\sin(y) + u, \qquad (1)$$

where y is the angular deflection and u is the torque input. Assume that the inverted pendulum has to follow a given reference $r(t)$, which is unknown a priori. Rewrite (1) as

$$\begin{aligned}\frac{d^2y}{dt^2} &= -\sin(y) + u = \\ &= -\frac{\sin(y) - \sin(r)}{y - r}(y - r) + u - \sin(r) = \\ &= \delta(y - r) + \tilde{u},\end{aligned}$$

where $\delta = -\frac{\sin(y)-\sin(r)}{y-r}$ and $\tilde{u} = u - \sin(r)$. That $|\delta| \leq 1$ for all y and r, since $\left|\frac{d\sin(y)}{dy}\right| \leq 1$, $\forall y$. If it is assumed that r enters as an exogenous input, i.e. it does not dependent on y, it will not affect the closed-loop stability.

The examples show that it is sometimes possible to convert a nonlinear problem into a linear model with parametric uncertainties. The uncertainties are functions of the states and reference input, and can be treated either as unknown parameters or as available to the controller.

5.1 *Gain Scheduling Synthesis*

This example shows an application of LFT gain scheduling synthesis, for a very simple system

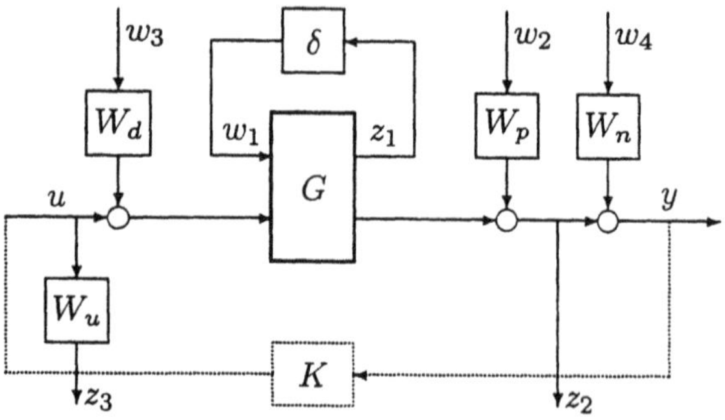

Fig. 1. Augmented inverted pendulum system

with a parameter dependency, see (Packard and Wu, 1993). The system is defined by

$$\begin{aligned}\frac{dx}{dt} &= \begin{pmatrix} 0 & 1 \\ -3.25 - 2.75\delta & 0 \end{pmatrix} x + \begin{pmatrix} 0 \\ 1 \end{pmatrix} u \\ y &= \begin{pmatrix} 1 & 0 \end{pmatrix} x\end{aligned}$$

where $\delta \in [-1,\ 1]$. The model describes a pendulum on a vertically accelerating platform. The parameter contains both the gravitational constant and the positive acceleration of the platform. The control objectives are to stabilize the system, and to reject significant disturbances at the input of the plant. The objectives are depicted in figure 1. The values of the weighting functions from (Packard and Wu, 1993) are $W_p = 0.5$, $W_n(s) = 10\frac{s+1}{s+200}$, $W_d = 1.33$ and $W_u = 0.32$. The augmented system is of third order. The aim of the design is to reduce the $\mathcal{L}_2$-induced norm from w to z without any constraint on the rate of change of δ.

5.2 *Gain Scheduling Controller*

I will study some different controllers for this system based on LFT design techniques. This means that the controller is linear and parametrized with respect to what as a linear fractional transformation. It is assume that δ is known in real time and that there are no bounds on how fast it may vary; it is only bounded to the interval $[-1,\ 1]$. During the design the $\mathcal{L}_2$-induced norm from w to z is minimized. The upper bound of the gain is denoted by γ. For a given constant δ this is equivalent to minimizing the H^∞ norm.

Two LFT gain scheduling designs were made: one assuming δ to be parametric (real) and the second with dynamic (complex) δ. Both controllers are parametrized by δ and give equivalent performance. Below given the controller, assuming dynamic δ, see (Apkarian, 1997; Apkarian and Gahinet, 1995; Scherer *et al.*, 1997; Boyd *et al.*, 1994).

$$K_{GS} = \begin{pmatrix} A_{GS} & B_{GS} \\ C_{GS} & D_{GS} \end{pmatrix} =$$

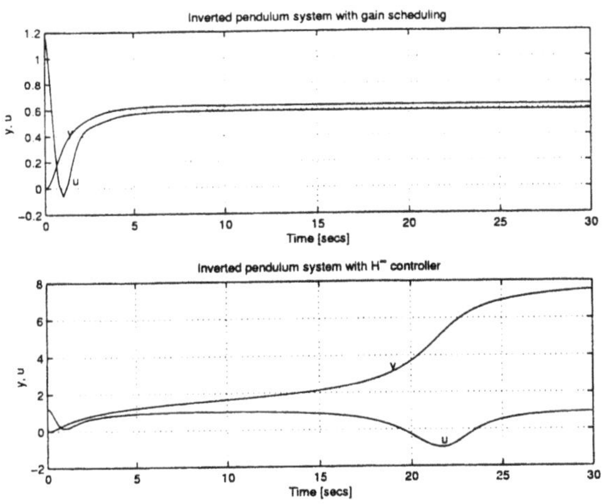

Fig. 2. The step response of inverted pendulum system with *gain scheduling* and H^∞ controller

$$= \left(\begin{array}{cc|cc} -7.83 & -.32 & -12.21 & .02 \\ -8,85 & -6.51 & -2.39 & -.06 \\ \hline 4.40 & 4.62 & -.35 & -.87 \\ 1.90 & .67 & -.13 & -.02 \end{array}\right)$$

5.3 *Robust Controller*

A robust controller without gain scheduling was also designed, see (Boyd and Barratt, 1991; Gahinet, 1996; Scherer *et al.*, 1997; Zhou *et al.*, 1996):

$$K_{H^\infty} = \begin{pmatrix} A_{H^\infty} & B_{H^\infty} \\ C_{H^\infty} & D_{H^\infty} \end{pmatrix} =$$

$$= \left(\begin{array}{cc|c} -5.14 & 1.33 & -16.69 \\ -13.56 & -10.76 & -3.82 \\ \hline 7.50 & 6.99 & -.83 \end{array}\right)$$

which yields $\gamma = 2.67$. This controller is also of second order but does not require any knowledge about δ.

Figure 2 shows the responses for the reference input $r = 1.2 * 1\,(t)$ ($1\,(t)$-unit step). The closed-loop system, with H^∞ controller, becomes unstable already for $|\delta| > 1$.

6. CONCLUSIONS

In the example it was compared some design of a simple control problem with an uncertainty parameter. Using LFT gain scheduling, a significant improvement in performance was obtained in comparison with the robust controller, which has no access to the uncertainty parameter. Even if the LFT controller is designed for a varying uncertainty, it performances almost as well as the optimal static H^∞ controller designed for a fixed parameter value.

When designing controllers based on H^∞ or methods, it is very important to explicitly and correctly state and include all requirements of the closed-loop system. If the actual plant is not within the range used for the design, performance bounds or even stability cannot be guaranteed. These methods can therefore be described as non-forgiving and requires good experience and knowledge both of the design methods and the underlying physics for and system aspects.

The contribution in this paper is a new synthesis algorithm for finding controllers when parametric (real) uncertainties are present. This extends previous results on μ synthesis involving dynamic (complex) uncertainties.

7. REFERENCES

Apkarian, P. (1997). On the discretization of LMI-synthesized linear parameter-varying controllers. *Automatica* **33**(4), 655–661.

Apkarian, P. and P. Gahinet (1995). A convex characterization of gain-scheduled $\mathcal{H}_\infty$ controllers. *IEEE Trans. on Control Systems Technology* **6**(1), 21–32.

Åström, K. J. and B. Wittenmark (1989). *Adaptive Control.* Addison Weseley.

Boyd, S. and C. Barratt (1991). *Linear Controller Design: Limits of Performance.* Prentice-Hall.

Boyd, S., L. El Ghaoui, E. Feron and V. Balakrishnan (1994). *Linear Matrix Inequalities in System and Control Theory.* Vol. 15 of *Studies in Applied Mathematics.* SIAM. Philadelphia, PA.

Gahinet, P. (1996). Explicit controller formulas for LMI-based H_∞ synthesis. *Automatica* **32**(7), 1007–1014.

Packard, A. and F. Wu (1993). Control of linear fractional transformations. In: *IEEE Proceedings of the 32nd Conference on Decision and Control.* Vol. 1. pp. 1036–1041.

Rugh, W. J. (1991). Analytical framework for gain-scheduling. *IEEE Trans. on Control Systems Magazine* **11**(1), 79–84.

Rugh, W. J. and J. S. Shamma (2000). Research on gain scheduling. *Automatica* **36**(10), 1401–1425.

Scherer, C., P. Gahinet and M. Chilali (1997). Multiobjective output-feedback control via LMI optimization. *IEEE Trans. on Automatic Control* **42**(7), 896–911.

Zhou, K., J. Doyle and K. Glover (1996). *Robust and Optimal Control.* Prentice-Hall.

ADAPTIVE NONLINEAR H_∞ CONTROL FOR PROCESSES WITH BOUNDED VARIATIONS OF PARAMETERS - GENERAL FORMS AND GENERAL RELATIVE DEGREE CASE -

Yoshihiko Miyasato*

**The Institute of Statistical Mathematics*
4-6-7 Minami-Azabu, Minato-ku, Tokyo 106-8569, JAPAN

Abstract : A new class of adaptive nonlinear H_∞ control systems for processes with bounded variations of time-varying parameters, is proposed in this manuscript. Those control schemes are derived as solutions of particular nonlinear H_∞ control problems, where unknown system parameters are regarded as exogenous disturbances to the processes, and thus, in the resulting control systems, the $\mathcal{L}^2$ gains from system parameters to generalized outputs are made less than the prescribed $\gamma(>0)$. General forms of the proposed control schemes and two explicit descriptions of those are given in this manuscript.

Keywords: adaptive control; nonlinear systems; H_∞ control; time-varying systems; optimal control

1. INTRODUCTION

In the study of adaptive control, the main topics have been an asymptotic stability of adaptive control systems. So much attention has not been paid on the control performances such as transient performance and other performances (Åström, *et al.*, 1989; Ioannou, *et al.*, 1996). On the contrary, the recent researches on nonlinear H_∞ control and inverse optimality, could derive several adaptive or nonlinear control systems which are optimal to certain meaningful cost functionals (Krstić, *et al.*, 1998; Miyasato, 1999).

Additionally, in the past study of adaptive control, there has been another problem that the stability analysis of adaptive control systems have been focused on time-invariant processes mainly. Although several methods have been examined for time-varying processes in the study of robust adaptive control schemes (Ioannou, *et al.*, 1996), only sufficiently small variations of time-varying parameters are accepted in those approaches.

The purpose of the present paper is to provide a new class of adaptive nonlinear H_∞ control systems for processes with bounded variations of parameters, where the control performances are discussed explicitly, and the stability analysis for time-varying systems are carried out successfully. Those control schemes are derived as solutions of particular nonlinear H_∞ control problems, where unknown system parameters are regarded as exogenous disturbances to the processes, and thus, in the resulting control systems, the $\mathcal{L}^2$ gains from system parameters to generalized outputs are made less than $\gamma(>0)$ (the prescribed positive constants). It is shown that the proposed method can be applied to any time-varying systems, and the resulting control systems are bounded for arbitrarily large bounded variations of time-varying parameters. Also, the control schemes are shown to be sub-optimal to some H_∞ cost functionals, when the high- frequency gains are time-invariant. The present manuscript extends the former versions (Miyasato, 2000a; 2000b; 2000c) to general relative degree cases, and provides general forms of the control schemes with two explicit descriptions.

2. PROBLEM STATEMENT

A single-input single-output nonlinear system is given by

$$\frac{d}{dt}e(t) = \mathcal{L}(e(t)) + \mathcal{L}(f(e,t)) + b_0 u_{fn^*-1}(t) + \Phi^T \omega_1(t), \tag{1}$$

$$u_{fi}(t) = \frac{1}{(s+\lambda)^i}u(t), \quad (\lambda > 0), \tag{2}$$

where $e(t)$ is a control variable to be regulated, $u(t)$ is a control input, and $f(e,t)$ is an unknown nonlinear term. $\omega_1(t)$ is a vector composed of measurable signals, b_0 and Φ are unknown system parameters which can be time-varying, and $\mathcal{L}(\cdot)$ is an unstructured element defined by

$$\mathcal{L}(v(t)) = G_0(s)v(t), \quad (G_0 \in \mathcal{RH}^\infty). \tag{3}$$

$\lambda\,(> 0)$ is a design parameter which is known, and n^* is a relative degree of the controlled process. The following assumptions are introduced.

Assumption 1 : AS-1-1) *The relative degree n^* is known.* **AS-1-2)** *The sign of b_0 remains unchanged ($b_0 > 0$ or $b_0 < 0$), and is known a priori. It is assumed that $b_0 > 0$ without loss of generality.* **AS-1-3)** *$f(e,t)$ is evaluated by*

$$f(e,t)^2 \le f_0 \cdot \phi(e) \cdot e^2, \quad \phi(0) = 0, \tag{4}$$

where f_0 is an unknown positive constant, and $\phi(e)(> 0)$ is a known function of e which is $n^* - 1$ times differentiable with respect to its argument. **AS-1-4)** *The magnitude of $\omega_1(t)$ is evaluated by*

$$\|\omega_1(t)\| \le M_1 \sup_{t\ge\tau}|e(\tau)| + M_2, \quad (M_1, M_2 > 0). \tag{5}$$

AS-1-5) *The upper bound on the norm of parameter vector Φ_0, which will be introduced later, is known. Also, the upper bound of $\bar{b}_0$ and lower bound of $\underline{b}_0$ on the high-frequency gain b_0 are known such that*

$$0 < \delta \le \underline{b}_0 \le b_0 \le \bar{b}_0 \le M < \infty, \tag{6}$$

and the upper bound of $\bar{p}$ and lower bound of $\underline{p}$ on the parameter p are known, too.

$$p \equiv \frac{1}{b_0}, \quad \bar{p} \equiv \frac{1}{\underline{b}_0}, \quad \underline{p} \equiv \frac{1}{\bar{b}_0}. \tag{7}$$

The control problem of this paper is to determine a control input $u(t)$ adaptively such that the overall system is stabilized for arbitrary but bounded time-varying system parameters, and additionally, the control variable $e(t)$ converges to zero asymptotically in the ideal case (stability condition), while the resulting control system becomes optimal or sub-optimal to some meaningful cost functionals (optimality condition).

3. ADAPTIVE NONLINEAR CONTROL SCHEME

The design is based on the backstepping (Krstić, *et al.*, 1996) composed of step 1 $\sim$ step n^*, and in each steps, the control signals $v_i(t)$ are determined by applying nonlinear H_∞ control scheme.

Step 1) Define $z_1(t)$, $z_2(t)$ by

$$z_1(t) \equiv e(t), \tag{8}$$

$$z_2(t) \equiv u_{fn^*-1}(t) - \alpha_1(t). \tag{9}$$

The virtual control input $\alpha_1(t)$ is determined so as to stabilize $z_1(t)$.

$$\alpha_1(t) = -\hat{p}(t)\hat{\Phi}(t)^T\omega_1(t) + v_1(t) \equiv -\hat{p}(t)v_0(t) + v_1(t), \tag{10}$$

where $v_1(t)$ is to be determined later based on nonlinear H_∞ control strategy. In this manuscript, the projection type adaptive laws, where tuning parameters $\hat{\theta}$ are constrained to belong to certain closed regions $\mathcal{S}$, are defined by

$$\dot{\hat{\theta}} = \Pr(\Gamma\phi\epsilon) \equiv \begin{cases} \Gamma\phi\epsilon & \text{Case 1} \\ \Gamma\phi\epsilon - \Gamma\frac{\nabla g \nabla g^T}{\nabla g^T \Gamma \nabla g}\Gamma\phi\epsilon & \text{Case 2,} \end{cases} \tag{11}$$

$$(\Gamma = \Gamma^T > 0), \tag{12}$$

where Case 1 : $\hat{\theta} \in \mathcal{S}^o$, or $\hat{\theta} \in \partial\mathcal{S}$ & $(\Gamma\phi\epsilon)^T\nabla g \le 0$, Case 2 : otherwise, $\mathcal{S} = \{\hat{\theta} : g(\hat{\theta}) \le 0\}$, $\mathcal{S}^o$ = interior of $\mathcal{S}$, $\partial\mathcal{S}$ = boundary of $\mathcal{S}$. By utilizing those descriptions, $\hat{\theta}_1(t)$, $\hat{\theta}_2(t)$, $\hat{p}(t)$ are tuned by

$$\begin{aligned} \dot{\hat{\theta}}_1(t) &= \Pr\{g_{11}z_1(t)^2\}, \quad (g_{11} > 0), \\ \dot{\hat{\theta}}_2(t) &= \Pr\{g_{12}\phi(z_1(t))z_1(t)^2\}, \quad (g_{12} > 0), \\ \dot{\hat{p}}(t) &= \Pr\{g_{13}v_0(t)z_1(t)\}, \quad (g_{13} > 0), \end{aligned} \tag{13}$$

where each constraints are given such that

$$g_{\theta_1}(\hat{\theta}_1) = \hat{\theta}_1^2 - M^2, \quad g_{\theta_2}(\hat{\theta}_2) = \hat{\theta}_2^2 - M^2, \quad g_p(\hat{p}) = \left(\hat{p} - \frac{\delta + M}{2}\right)^2 - \left(\frac{M-\delta}{2}\right)^2. \tag{14}$$

M and δ are properly selected positive constants (**AS-1-5**). Hereafter, the input signal $v_1(t)$ is to be obtained by applying nonlinear H_∞ control strategy. For this purpose, define $V_1(t)$ by

$$\begin{aligned} V_1(t) &= \frac{1}{2}z_1(t)^2 + \frac{1}{2}\sum_{i=1}^{2}\{\hat{\theta}_i(t) - \theta_i^*\}^2/g_{1i} \\ &+ \frac{\underline{b}_0}{2}\{\hat{p}(t) - \bar{p}\}^2/g_{13} \\ &+ \frac{1}{2}\{\hat{\Phi}(t) - \Phi^*\}^T G_{14}^{-1}\{\hat{\Phi}(t) - \Phi^*\} \\ &+ \frac{1}{2}\{\hat{b}_0(t) - \underline{b}_0\}^2/g_{15}, \end{aligned} \tag{15}$$

where $G_{14} = G_{14}{}^T > 0$, $g_{15} > 0$, Φ^* is a nominal value (time-invariant) of the parameter Φ (time-varying), and θ_i^* are also nominal values (time-invariant) of the parameters θ_i (time-varying) determined later. Take the time derivative of $V_1(t)$.

$$\dot{V}_1(t) = z_1(t)\{\mathcal{L}(z_1(t)) + \mathcal{L}(f(z_1,t)) + \Phi^T\omega_1(t)\}$$

$$-\Phi_0^{*T}\omega_0(t)z_1(t) + (\underline{b}_0 - b_0)\hat{p}(t)v_0(t)z_1(t)$$
$$+\hat{\theta}_1(t)z_1(t)^2 + \hat{\theta}_2(t)\phi(z_1(t))z_1(t)^2$$
$$+\{\hat{b}_0(t) + (b_0 - \underline{b}_0)\}v_1(t)z_1(t) + b_0 z_1(t)z_2(t)$$
$$+\{\hat{\Phi}(t) - \Phi^*\}^T G_{14}^{-1}\{\dot{\hat{\Phi}}(t) - \tau_{\phi 1}(t)\}$$
$$+\{\hat{b}_0(t) - \underline{b}_0\}\{\dot{\hat{b}}_0(t) - \tau_{b1}(t)\}/g_{15}, \tag{16}$$
$$\omega_0(t) = [z_1(t), \phi(z_1(t))z_1(t), \omega_1(t)^T]^T, \tag{17}$$
$$\Phi_0^* = [\theta_1^*, \theta_2^*, \Phi^{*T}]^T, \tag{18}$$
$$\tau_{\phi 1}(t) = G_{14}\omega_1(t)z_1(t), \tag{19}$$
$$\tau_{b1}(t) = g_{15}v_1(t)z_1(t). \tag{20}$$

Since $\mathcal{L}(\cdot)$ is defined by (3), there exist bounded θ_{11}, θ_{12} (positive) satisfying the next inequality.

$$\int_0^t z_1(\tau)\{\mathcal{L}(z_1(\tau)) + \mathcal{L}(f(z_1,\tau))\}d\tau$$
$$\leq \int_0^t \{\theta_{11}(\tau)z_1(\tau)^2 + \theta_{12}(\tau)\phi(z_1(\tau))z_1(\tau)^2\}d\tau. \tag{21}$$

Then, the following relation is obtained by integrating $\dot{V}_1(t)$.

$$V_1(t) - V_1(0)$$
$$\leq \int_0^t [\hat{\theta}_1(\tau)z_1(\tau) + \hat{\theta}_2(\tau)\phi(z_1(\tau))z_1(\tau)]z_1(\tau)d\tau$$
$$+\int_0^t \bar{\Theta}_1^T\bar{\omega}_1(\tau)z_1(\tau)d\tau$$
$$+\int_0^t \{\hat{b}_0(\tau) + \tilde{b}_0(\tau)\}v_1(\tau)z_1(\tau)d\tau$$
$$+\int_0^t b_0 z_1(\tau)z_2(\tau)d\tau$$
$$+\int_0^t \{\hat{\Phi}(\tau) - \Phi^*\}^T G_{14}^{-1}\{\dot{\hat{\Phi}}(\tau) - \tau_{\phi 1}(\tau)\}d\tau$$
$$+\int_0^t \{\hat{b}_0(\tau) - \underline{b}_0\}\{\dot{\hat{b}}_0(\tau) - \tau_{b1}(\tau)\}d\tau/g_{15}, \tag{22}$$
$$\Phi_{01} \equiv [\theta_{11}, \theta_{12}, \Phi^T]^T, \tag{23}$$
$$\bar{\Theta}_1 \equiv [(\Phi_{01} - \Phi_0^*)^T, \underline{b}_0 - b_0]^T, \tag{24}$$
$$\tilde{b}_0 \equiv b_0 - \underline{b}_0\ (\geq 0), \tag{25}$$
$$\bar{\omega}_1(t) \equiv [\omega_0(t)^T, \hat{p}(t)v_0(t)\}]^T. \tag{26}$$

From that evaluation of $V_1(t)$, the virtual process is introduced

$$\dot{z}_1 = \hat{\theta}_1 z_1 + \hat{\theta}_2\phi(z_1)z_1 + \bar{\omega}_1^T\bar{\Theta}_1 + (\hat{b}_0 + \tilde{b}_0)v_1$$
$$\equiv f_1(z_1) + g_{11}d_1 + g_{12}v_1, \tag{27}$$
$$f_1(z_1) = \hat{\theta}_1 z_1 + \hat{\theta}_2\phi(z_1)z_1,$$
$$g_{11} = \bar{\omega}_1^T,\ d_1 = \bar{\Theta}_1,\ g_{12} = \hat{b}_0 + \tilde{b}_0, \tag{28}$$

and stabilize it via v_1 by utilizing nonlinear H_∞ control strategy, where unknown parameters $\bar{\Theta}_1$ are regarded as exogenous disturbances to the process. For this purpose, consider the Hamilton-Jacobi-Isaacs (HJI) equation

$$\frac{\partial\tilde{V}_1}{\partial z_1}f_1 + \frac{1}{4}\left(\frac{\|g_{11}\|^2}{\gamma_1^{*2}} - \frac{g_{12}^2}{r_1}\right)\left(\frac{\partial\tilde{V}_1}{\partial z_1}\right)^2 + h_1 z_1^2 \leq 0, \tag{29}$$

where the solution $\tilde{V}_1$ is given by the next equation

$$\tilde{V}_1(t) = \frac{1}{2}z_1(t)^2. \tag{30}$$

h_1 and r_1 are positive functions to be determined from the inequality (29) based on inverse optimality for the given solution $\tilde{V}_1$ and the positive constant γ_1^*. The substitution of (30) into (29) yields

$$\hat{\theta}_1 z_1^2 + \hat{\theta}_2\phi(z_1)z_1^2 + \left\{\frac{\|\bar{\omega}_1\|^2}{\gamma_1^{*2}} - \frac{(\hat{b}_0 + \tilde{b}_0)^2}{r_1}\right\}\frac{z_1^2}{4}$$
$$+h_1 z_1^2 \leq 0. \tag{31}$$

Since the unknown $\tilde{b}_0(\geq 0)$ is included in (31), h_1 and r_1 satisfying (32) are to be obtained.

$$\hat{\theta}_1 z_1^2 + \hat{\theta}_2\phi(z_1)z_1^2 + \left(\frac{\|\bar{\omega}_1\|^2}{\gamma_1^{*2}} - \frac{\hat{b}_0^2}{r_1}\right)\frac{z_1^2}{4} + h_1 z_1^2 \leq 0. \tag{32}$$

Then, the control signal v_1 is derived as a solution of the H_∞ control problem.

$$v_1^* = -\frac{1}{2r_1}g_{12}\frac{\partial\tilde{V}_1}{\partial z_1} = -\frac{1}{2r_1}(\hat{b}_0 + \tilde{b}_0)z_1. \tag{33}$$

Since the unknown $\tilde{b}_0$ is included in (33), the actual input signal is replaced by

$$v_1^* = -\frac{\hat{b}_0}{2r_1}z_1. \tag{34}$$

Hence, the following relation is obtained for the original process (1), (2) and $V_1(t)$ (15).

$$V_1(t) - V_1(0) \leq \int_0^t r_1\left\{\frac{\hat{b}_0(\tau)z_1(\tau)}{2r_1} + v_1(\tau)\right\}$$
$$-\int_0^t \{h_1 z_1(\tau)^2 + r_1 v_1(\tau)^2\}d\tau$$
$$-\int_0^t \gamma_1^{*2}\left\|\bar{\Theta}_1(\tau) - \frac{1}{2\gamma_1^{*2}}\bar{\omega}_1(\tau)z_1(\tau)\right\|^2 d\tau$$
$$+\gamma_1^{*2}\int_0^t \|\bar{\Theta}_1(\tau)\|^2 d\tau$$
$$+\int_0^t \tilde{b}_0(\tau)v_1(\tau)z_1(\tau)d\tau + \int_0^t b_0 z_1(\tau)z_2(\tau)d\tau$$
$$+\int_0^t \{\hat{\Phi}(\tau) - \Phi^*\}^T G_{14}^{-1}\{\dot{\hat{\Phi}}(\tau) - \tau_{\phi 1}(\tau)\}d\tau$$
$$+\int_0^t \{\hat{b}_0(\tau) - \underline{b}_0\}\{\dot{\hat{b}}_0(\tau) - \tau_{b1}(\tau)\}d\tau/g_{15}. \tag{35}$$

Step i) $(2 \leq i \leq n^*)$ Take the time derivative of $z_i(t)$.

$$z_i(t) \equiv u_{fn^*-i+1}(t) - \alpha_{i-1}(t) \tag{36}$$
$$\dot{z}_i(t) = -\lambda u_{fn^*-i+1}(t) + u_{fn^*-i}(t) - \beta_{i-1}(t)$$
$$-\gamma_{i-1}(t)\{\mathcal{L}(z_1(t)) + \mathcal{L}(f(z_1,t))$$
$$+b_0 u_{fn^*-1}(t) + \Phi^T\omega_1(t)\}$$
$$-\gamma_{Ki-1}(t)\Pr\{G_1\bar{v}_1(t)z_1(t)\}$$
$$-\gamma_{\phi i-1}(t)\dot{\hat{\Phi}}(t) - \gamma_{bi-1}(t)\dot{\hat{b}}_0(t), \tag{37}$$

$(i \geq 3)$

$$\beta_{i-1}(t) = \frac{\partial\alpha_{i-1}}{\partial\omega_{i-1}}\dot{\omega}_{i-1}(t)$$
$$+\sum_{j=2}^{i-1}\frac{\partial\alpha_{i-1}}{\partial z_j}\{-\lambda u_{fn^*-j+1}(t)$$
$$+u_{fn^*-j}(t) - \beta_{j-1}(t)\},$$

$$\gamma_{i-1}(t) = \frac{\partial \alpha_{i-1}}{\partial z_1} - \sum_{j=2}^{i-1} \frac{\partial \alpha_{i-1}}{\partial z_j} \gamma_{j-1}(t), \tag{38}$$

$$\gamma_{Ki-1}(t) = \frac{\partial \alpha_{i-1}}{\partial \hat{K}_1} - \sum_{j=2}^{i-1} \frac{\partial \alpha_{i-1}}{\partial z_j} \gamma_{Kj-1}(t), \tag{39}$$

$$\gamma_{\phi i-1}(t) = \frac{\partial \alpha_{i-1}}{\partial \hat{\Phi}} - \sum_{j=2}^{i-1} \frac{\partial \alpha_{i-1}}{\partial z_j} \gamma_{\phi j-1}(t), \tag{40}$$

$$\gamma_{bi-1}(t) = \frac{\partial \alpha_{i-1}}{\partial \hat{b}_0} - \sum_{j=2}^{i-1} \frac{\partial \alpha_{i-1}}{\partial z_j} \gamma_{bj-1}(t), \tag{41}$$

$\omega_{i-1}(t) \equiv$ [vector composed of the elements
$\{u_{fn^*-1}(t) \sim u_{fn^*-i+2}, \omega_{i-2}(t), \dot{\omega}_{i-2}(t)\}$], (42)

$$\Big(\beta_1(t) = \frac{\partial \alpha_1}{\partial \omega_1}\dot{\omega}_1(t),\ \gamma_1(t) = \frac{\partial \alpha_1}{\partial z_1},\ \gamma_{K1}(t) = \frac{\partial \alpha_1}{\partial \hat{K}_1},\ \gamma_{\phi 1}(t) = \frac{\partial \alpha_1}{\partial \hat{\Phi}},\ \gamma_{b1}(t) = \frac{\partial \alpha_1}{\partial \hat{b}_0}\Big), \tag{43}$$

$$\hat{K}_1(t) = [\hat{\theta}_1(t), \hat{\theta}_2(t), \hat{p}(t)]^T, \tag{44}$$

$$\tilde{v}_1(t) = [z_1(t), \phi(z_1(t))z_1(t), v_0(t)]^T, \tag{45}$$

$$G_1 = \mathrm{diag}(g_{11}, g_{12}, g_{13}). \tag{46}$$

For $z_i(t)$, introduce $z_{i+1}(t)$ and determine the virtual control $\alpha_i(t)$ so as to stabilize $z_i(t)$.

$$z_{i+1}(t) \equiv u_{fn^*-i}(t) - \alpha_i(t), \quad (3 \le i \le n^* - 1), \tag{47}$$

$$\alpha_i(t) = \lambda u_{fn^*-i+1}(t) + \beta_{i-1}(t) - c_{i1}(t)z_{i-1}(t) + \gamma_{i-1}(t)\dot{\hat{\Phi}}(t)^T\omega_1(t) + \dot{\hat{b}}_0(t)\gamma_{i-1}(t)u_{fn^*-1}(t) + v_i(t) + \tilde{\alpha}_i(t), \tag{48}$$

$$c_{i1}(t) = \begin{cases} \hat{b}_0(t) & (i=2) \\ 1 & (i \ge 3), \end{cases} \tag{49}$$

where $\tilde{\alpha}_i(t)$ is an auxiliary signal to be determined later, and the input signal $v_i(t)$ is to be obtained by applying nonlinear H_∞ control scheme. For this purpose, introduce the virtual process

$$\dot{z}_i = k_{i1}\gamma_{i-1}^2 z_i + k_{i2}\|\gamma_{Ki-1}\|^2\|\tilde{v}_1\|^2 z_i + \tilde{\omega}_i^T\tilde{\Theta}_2 + v_i \equiv f_i(z_1, \cdots, z_i) + g_{i1}d_i + g_{i2}v_i, \tag{50}$$

$$f_i(z_1, \cdots, z_i) = k_{i1}\gamma_{i-1}^2 z_i + k_{i2}\|\gamma_{Ki-1}\|^2\|\tilde{v}_1\|^2 z_i,$$
$$g_{i1} = \tilde{\omega}_i^T,\ d_i = \tilde{\Theta}_2,\ g_{i2} = 1, \tag{51}$$

$$\tilde{\Theta}_2 = [b_0 - \underline{b}_0, (\Phi - \Phi^*)^T]^T, \tag{52}$$

$$\tilde{\omega}_i(t) = [-\gamma_{i-1}(t)u_{fn^*-1}(t), -\gamma_{i-1}(t)\omega_1(t)^T]^T, \quad (i \ge 3), \tag{53}$$

$$\tilde{\omega}_2(t) = [z_1(t) - \gamma_1(t)u_{fn^*-1}(t), -\gamma_1(t)\omega_1(t)^T]^T, \tag{54}$$

and stabilize it via v_i by applying nonlinear H_∞ control strategy, where unknown parameters $\tilde{\Theta}_2$ are regarded as exogenous disturbances to the process. For this purpose, consider the HJI equation

$$\frac{\partial \tilde{V}_i}{\partial z_i}f_i + \frac{1}{4}\left(\frac{\|g_{i1}\|^2}{\gamma_i^{*2}} - \frac{g_{i2}^2}{r_i}\right)\left(\frac{\partial \tilde{V}_i}{\partial z_i}\right)^2 + h_i z_i^2 \le 0, \tag{55}$$

where the solution $\tilde{V}_i$ is given by the next equation

$$\tilde{V}_i(t) = \frac{1}{2}z_i(t)^2. \tag{56}$$

h_i and r_i are positive functions to be determined from the inequality (55) for the given solution $\tilde{V}_i$ and the positive constant γ_i^*. The substitution of (56) into (55) yields

$$k_{i1}\gamma_{i-1}^2 z_i^2 + k_{i2}\|\gamma_{Ki-1}\|^2\|\tilde{v}_1\|^2 z_i^2 + \left\{\frac{\|\tilde{\omega}_i\|^2}{\gamma_i^{*2}} - \frac{1}{r_i}\right\}\frac{z_i^2}{4} + h_i z_i^2 \le 0. \tag{57}$$

The control inputs $v_i(t)$ are obtained as solutions for the nonlinear H_∞ control problem.

$$v_i^* = -\frac{1}{2r_i}g_{i2}\frac{\partial \tilde{V}_i}{\partial z_i} = -\frac{1}{2r_i}z_i. \tag{58}$$

Then, the following relation is derived for the original process (1), (2) and $V_1(t) \sim V_i(t)$ ((15), (61)).

$$\begin{aligned}
\sum_{j=1}^{i}\{V_j(t) - V_j(0)\} \le& \int_0^t r_1\left\{\frac{\hat{b}_0(\tau)z_1(\tau)}{2r_1} + v_1(\tau)\right\}d\tau \\
&- \int_0^t \{h_1 z_1(\tau)^2 + r_1 v_1(\tau)^2\}d\tau \\
&- \int_0^t \gamma_1^{*2}\left\|\tilde{\Theta}_{1i}(\tau) - \frac{1}{2\gamma_1^{*2}}\tilde{\omega}_1(\tau)z_1(\tau)\right\|^2 d\tau \\
&+ \gamma_1^{*2}\int_0^t \|\tilde{\Theta}_{1i}(\tau)\|^2 d\tau + \int_0^t \tilde{b}_0(\tau)v_1(\tau)z_1(\tau)d\tau \\
&+ \sum_{j=2}^{i}\Bigg[\int_0^t r_j\left\{\frac{z_j(\tau)}{2r_j} + v_j(\tau)\right\}^2 d\tau \\
&- \int_0^t \{h_j z_j(\tau)^2 + r_j v_j(\tau)^2\}d\tau \\
&- \int_0^t \gamma_j^{*2}\left\|\tilde{\Theta}_2(\tau) - \frac{1}{2\gamma_j^{*2}}\tilde{\omega}_j(\tau)z_j(\tau)\right\|^2 d\tau \\
&+ \gamma_j^{*2}\int_0^t \|\tilde{\Theta}_2(\tau)\|^2 d\tau\Bigg] \\
&+ \int_0^t z_i(\tau)z_{i+1}(\tau)d\tau \\
&+ \int_0^t \{\hat{\Phi}(\tau) - \Phi^*\}^T G_{14}^{-1}\{\dot{\hat{\Phi}}(\tau) - \tau_{\phi i}(\tau)\}d\tau \\
&+ \int_0^t \{\hat{b}_0(\tau) - \underline{b}_0\}\{\dot{\hat{b}}_0(\tau) - \tau_{bi}(\tau)\}d\tau / g_{15} \\
&- \sum_{j=2}^{i}\int_0^t \gamma_{\phi j-1}(\tau)\dot{\hat{\Phi}}(\tau)z_j(\tau)d\tau \\
&- \sum_{j=2}^{i}\int_0^t \gamma_{bj-1}(\tau)\dot{\hat{b}}_0(\tau)z_j(\tau)d\tau \\
&+ \sum_{j=2}^{i}\int_0^t \tilde{\alpha}_j(\tau)z_j(\tau)d\tau,
\end{aligned} \tag{59}$$

$$\tilde{\Theta}_{1i} = [(\Phi_{0i} - \Phi_0^*)^T, \underline{b}_0 - b_0]^T, \tag{60}$$

$$V_i(t) = \frac{1}{2}z_i(t)^2, \tag{61}$$

$$\tau_{\phi i}(t) = \tau_{\phi i-1}(t) - G_{14}\gamma_{i-1}(t)\omega_1(t)z_i(t), \tag{62}$$

$$\tau_{bi}(t) = \tau_{bi-1}(t) - g_{15}\gamma_{i-1}(t)u_{fn^*-1}(t)z_i(t), \tag{63}$$

$$\Phi_{0i} = \left[\theta_{11} + \sum_{j=2}^{i}\left(\frac{\theta_{21}}{k_{j1}} + \frac{\|G_1\|^2}{4k_{j2}}\right),\ \theta_{12} + \sum_{j=2}^{i}\frac{\theta_{22}}{k_{j1}},\ \Phi^T\right]^T, \tag{64}$$

where θ_{21}, θ_{22} are positive functions, and k_{i1}, k_{i2} are arbitrary positive constants satisfying

$$-\int_0^t \gamma_{i-1}(\tau)\{\mathcal{L}(z_1(\tau)+\mathcal{L}(f(z_1,\tau))\}z_i(\tau)d\tau$$
$$\leq k_{i1}\int_0^t \gamma_{i-1}(\tau)^2 z_i(\tau)^2 d\tau$$
$$+\frac{1}{k_{i1}}\int_0^t \{\theta_{21}(\tau)z_1(\tau)^2$$
$$+\theta_{22}(\tau)^2\phi(z_1(\tau))z_1(\tau)^2\}d\tau, \quad (65)$$

$$-\int_0^t \gamma_{Ki-1}(\tau)\Pr\{G_1\bar{v}_1(\tau)z_1(\tau)\}z_i(\tau)d\tau$$
$$\leq \int_0^t \|\gamma_{Ki-1}(\tau)\|\|G_1\|\|\bar{v}_1(\tau)\|\|z_1(\tau)\||z_i(\tau)|d\tau$$
$$\leq k_{i2}\int_0^t \|\gamma_{Ki-1}(\tau)\|^2\|\bar{v}_1(\tau)\|^2 z_i(\tau)^2 d\tau$$
$$+\frac{1}{4k_{i2}}\|G_1\|^2\int_0^t z_1(\tau)^2 d\tau. \quad (66)$$

In the last **Step** n^*), the actual control input is

$$u(t)=\alpha_{n^*}(t), \quad (67)$$

and v_{n^*} is derived in the same way as (58).

Step n* + 1) The tuning laws of $\hat{\Phi}(t)$, $\hat{b}_0(t)$ and $\tilde{\alpha}_i(t)$ are determined such that

$$\dot{\hat{\Phi}}(t)=\Pr\{\tau_{\phi n^*}(t)\},\quad \dot{\hat{b}}_0(t)=\Pr\{\tau_{bn^*}(t)\}, \quad (68)$$
$$\tilde{\alpha}_i(t)=-k_{i3}\|\gamma_{\phi i-1}(t)\|^2\|\omega_1(t)\|^2 z_i(t)$$
$$-(n^*-1)k_{i4}\|\gamma_{\phi i-1}(t)\|^2\|\omega_1(t)\|^2 z_i(t)$$
$$-\sum_{j=2}^{n^*}\frac{\|G_{14}\|^2}{4k_{j4}}\gamma_{i-1}(t)^2 z_i(t)$$
$$-k_{i5}\gamma_{bi-1}(t)^2 v_1(t)^2 z_i(t)-k_{i6}\gamma_{bi-1}(t)^2 z_2(t)^2 z_i(t)$$
$$-(n^*-1)k_{i7}\gamma_{bi-1}(t)^2 u_{fn^*-1}(t)^2 z_i(t)$$
$$-\sum_{j=2}^{n^*}\frac{g_{15}^2}{4k_{j7}}\gamma_{i-1}(t)^2 z_i(t), \qquad (2\leq i\leq r+1), \quad (69)$$

where the constraints are defined by

$$g_\phi(\hat{\Phi})=\|\hat{\Phi}\|^2-M^2,\quad g_b(\hat{b}_0)=\|\hat{b}_0\|^2-M^2. \quad (70)$$

$M(>0)$ is determined similarly to (14). Then, the following evaluation of $V_i(t)$ ((15), (61)) is derived.

$$\sum_{i=1}^{n^*}\{V_i(t)-V_i(0)\}\leq\int_0^t r_1\left\{\frac{\hat{b}_0(\tau)z_1(\tau)}{2r_1}+v_1(\tau)\right\}d\tau$$
$$-\int_0^t\{h_1z_1(\tau)^2+r_1v_1(\tau)^2\}d\tau$$
$$-\int_0^t\gamma_1^{*2}\left\|\tilde{\Theta}_{1n^*}(\tau)-\frac{1}{2\gamma_1^{*2}}\tilde{\omega}_1(\tau)z_1(\tau)\right\|^2 d\tau$$
$$+\gamma_1^{*2}\int_0^t\|\tilde{\Theta}_{1n^*}(\tau)\|^2d\tau+\int_0^t\tilde{b}_0(\tau)v_1(\tau)z_1(\tau)d\tau$$
$$+\sum_{i=2}^{n^*}\left[\int_0^t r_i\left\{\frac{z_i(\tau)}{2r_i}+v_i(\tau)\right\}^2 d\tau\right.$$
$$-\int_0^t\{h_iz_i(\tau)^2+r_iv_i(\tau)^2\}d\tau$$
$$-\int_0^t\gamma_i^{*2}\left\|\tilde{\Theta}_2(\tau)-\frac{1}{2\gamma_i^{*2}}\tilde{\omega}_i(\tau)z_i(\tau)\right\|^2 d\tau$$
$$\left.+\gamma_i^{*2}\int_0^t\|\tilde{\Theta}_2(\tau)\|^2d\tau\right], \quad (71)$$

$$\tilde{\Theta}_{1n^*}=[(\Phi_{0n^*}-\Phi_0^*)^T,\ \underline{b}_0-b_0]^T, \quad (72)$$
$$\Phi_{0n^*}=\left[\theta_1,\ \theta_2,\ \Phi^T\right]^T, \quad (73)$$

$$\theta_1=\theta_{11}+\sum_{j=2}^{n^*}\left(\frac{\theta_{21}}{k_{j1}}+\frac{\|G_1\|^2}{4k_{j2}}+\frac{\|G_{14}\|^2}{4k_{j3}}\right.$$
$$\left.+\frac{g_{15}^2}{4k_{j5}}+\frac{g_{15}^2}{4k_{j6}}\right),$$
$$\theta_2=\theta_{12}+\sum_{j=2}^{n^*}\frac{\theta_{22}}{k_{j1}}. \quad (74)$$

θ_1^*, θ_2^* in (15), are nominal values (time-invariant) of θ_1, θ_2 in (74). Then, the following main theorems are derived from (71).

Theorem 1 : *The adaptive control system described above (where $v_1^*(t)\sim v_{n^*}^*(t)$ ((34), (58)) are included) is uniformly bounded for arbitrary bounded variation of system parameters b_0, p, Φ, θ_{11}, θ_{12}, θ_{21}, θ_{22}.*

Theorem 2 : *For that adaptive control systems, assume that b_0 is time-invariant. Then, $v_1^*\sim v_{n^*}^*$ are sub-optimal control inputs which minimize the upper bound on the following cost functional .*

$$J(t)\equiv\sup_{\tilde{\Theta}_{1n^*},\tilde{\Theta}_2\in\mathcal{L}_2}\left[\sum_{i=1}^{n^*}\int_0^t\{h_iz_i(\tau)^2+r_iv_i(\tau)^2\}d\tau\right.$$
$$+\sum_{i=1}^{n^*}V_i(t)-\gamma_1^{*2}\int_0^t\|\tilde{\Theta}_{1n^*}(\tau)\|^2d\tau$$
$$\left.-\sum_{i=2}^{n^*}\gamma_i^{*2}\int_0^t\|\tilde{\Theta}_2(\tau)\|^2d\tau\right]. \quad (75)$$

Also the next inequality is obtained.

$$\sum_{i=1}^{n^*}\int_0^t\{h_iz_i(\tau)^2+r_iv_i^*(\tau)^2\}d\tau+\sum_{i=1}^{n^*}V_i(t)$$
$$\leq\sum_{i=1}^{n^*}V_i(0)+\gamma_1^{*2}\int_0^t\|\tilde{\Theta}_{1n^*}(\tau)\|^2d\tau$$
$$+\sum_{i=2}^{n^*}\gamma_i^{*2}\int_0^t\|\tilde{\Theta}_2(\tau)\|^2d\tau. \quad (76)$$

Especially, if $\tilde{\Theta}_{1n^*}$, $\tilde{\Theta}_2\in\mathcal{L}_2$, then it holds that $z_1(t)\sim z_{n^*}(t)\to 0$.

Theorem 3 : *For that adaptive system, assume that b_0 is not time-invariant. Then, the following inequality is derived.*

$$\sum_{i=1}^{n^*}\int_0^t\{h_iz_i(\tau)^2+r_iv_i^*(\tau)^2\}d\tau+\sum_{i=1}^{n^*}V_i(t)$$

$$+\int_0^t \frac{\tilde{b}_0(\tau)\hat{b}_0(\tau)}{2r_1} z_1(\tau)^2 d\tau$$
$$\le \sum_{i=1}^{n^*} V_i(0) + \gamma_1^{*2} \int_0^t \|\tilde{\Theta}_{1n^*}(\tau)\|^2 d\tau$$
$$+\sum_{i=2}^{n^*} \gamma_i^{*2} \int_0^t \|\tilde{\Theta}_2(\tau)\|^2 d\tau, \tag{77}$$

where $\frac{\tilde{b}_0(t)\hat{b}_0(t)}{2r_1} z_1(t)^2 \ge 0$. *Especially, if* $\tilde{\Theta}_{1n^*}$, $\tilde{\Theta}_2 \in \mathcal{L}_2$, *then, it holds that* $z_1(t) \sim z_{n^*}(t) \to 0$.

Theorem 4 : *Assume that* $(\Phi_{0n^*} - \Phi_0^*)$, $(b_0 - b_0^*)$, $(p-p^*) \in \mathcal{L}_2$ *and* $b_0 \to b_0^*$, $p \to p^*$ *for certain constant* Φ_0^*, b_0^*, p^* $(b_0^* p^* = 1)$. *Then, it holds that* $z_1(t) \sim z_{n^*}(t) \to 0$ *(as* $t \to \infty$*).*

Up to now, the general forms of the control schemes were provided. Next, h_i and r_i are solved, and the explicit descriptions of the control schemes are given by assuming specified forms.

Solution I : From (32), (57), r_1, $r_i (i \ge 2)$ can be chosen such that

$$r_1 = r_{01} \cdot [k_{18} + k_{19}\|\bar{\omega}_1\|^2 + k_{110}\{\hat{\theta}_1 + \hat{\theta}_2\phi(z_1)\}]^{-1}$$
$$r_i = r_{0i} \cdot [k_{i8} + k_{i9}\|\bar{\omega}_i\|^2$$
$$+k_{i10}\{k_{i1}\gamma_{i-1}^2 + k_{i2}\|\gamma_{Ki-1}\|^2\|\bar{v}_1\|^2\}]^{-1}, \tag{78}$$

where $k_{18} \sim k_{110}$, r_{01}, $k_{i8} \sim k_{i10}$, $r_{0i}(>0)$ are design parameters. Then, h_1, $h_i\,(\ge 2)$ are

$$h_1 \le \{\hat{b}_0^2 k_{18}\gamma_1^{*2} + (\hat{b}_0^2 k_{19}\gamma_1^{*2} - r_{01})\|\bar{\omega}_1\|^2$$
$$+(\hat{b}_0^2 k_{110}\gamma_1^{*2} - 4r_{01}\gamma_1^{*2})(\hat{\theta}_1 + \hat{\theta}_2\phi(z_1))\} \cdot$$
$$\cdot(4r_{01}\gamma_1^{*2})^{-1},$$
$$h_i \le [k_{i8}\gamma_i^{*2} + (k_{i9}\gamma_i^{*2} - r_{0i})\|\bar{\omega}_i\|^2$$
$$+(k_{i10}\gamma_i^{*2} - 4r_{0i}\gamma_i^{*2}) \cdot$$
$$\cdot\{k_{i1}\gamma_{i-1}^2 + k_{i2}\|\gamma_{Ki-1}\|^2\|\bar{v}_1\|^2\}] \cdot (4r_{0i}\gamma_i^{*2})^{-1}, \tag{79}$$

In order that h_1, $h_i\,(\ge 2)$ are positive definite, k_{i9} and k_{i10} should be chosen such that

$$k_{19} \ge \frac{r_{01}}{\delta^2\gamma_1^{*2}}, \quad k_{110} \ge \frac{4r_{01}}{\delta^2}, \quad (0 < \delta \le b_0, \hat{b}_0),$$
$$k_{i9} \ge \frac{r_{0i}}{\gamma_i^{*2}}, \quad k_{i10} \ge 4r_{0i}, \quad (i \ge 2). \tag{80}$$

The explicit descriptions of v_1, v_i are given by

$$v_1 = -\hat{b}_0\left[k_{18} + k_{19}\|\bar{\omega}_1\|^2 + k_{110}\{\hat{\theta}_1 + \hat{\theta}_2\phi(z_1)\}\right] z_1 \cdot$$
$$\cdot(2r_{01})^{-1},$$
$$v_i = -[k_{i8} + k_{i9}\|\bar{\omega}_i\|^2$$
$$+k_{i10}\{k_{i1}\gamma_{i-1}^2 + k_{i2}\|\gamma_{Ki-1}\|^2\|\bar{v}_1\|^2\}] z_i \cdot (2r_{0i})^{-1},$$
$$(i \ge 2). \tag{81}$$

Solution II : r_i and h_i are obtained by setting

$$h_i = a_i r_i, \tag{82}$$

where $a_i\,(0 < a_i < \infty)$ prescribes the ratio between r_i and h_i. Then, for equality condition of (32), (57), r_i are given as the positive solutions of

$$a_1 r_1^2 + G_1 r_1 - \frac{\hat{b}_0^2}{4} = 0,$$
$$G_1 = \hat{\theta}_1 + \hat{\theta}_2\phi(z_1) + \frac{\|\bar{\omega}_1\|^2}{4\gamma_1^{*2}},$$
$$a_i r_i^2 + G_i r_i - \frac{1}{4} = 0,$$
$$G_i = k_{i1}\gamma_{i-1}^2 + k_{i2}\|\gamma_{Ki-1}\|^2\|\bar{v}_1\|^2 + \frac{\|\bar{\omega}_i\|^2}{4\gamma_i^{*2}}. \tag{83}$$

Hence, r_i and h_i are

$$r_1 = \frac{\hat{b}_0^2}{2\left\{\sqrt{G_1^2 + a_1\hat{b}_0^2} + G_1\right\}}, \quad h_1 = a_1 r_1 \tag{84}$$
$$r_i = \frac{1}{2\left\{\sqrt{G_i^2 + a_i} + G_i\right\}}, \quad h_i = a_i r_i, \quad (i \ge 2), \tag{85}$$

and the explicit descriptions of v_1, v_i are given by

$$v_1 = -\frac{\hat{b}_0}{2r_1} z_1 = -\frac{1}{\hat{b}_0}\left\{\sqrt{G_1^2 + a_1\hat{b}_0} + G_1\right\} z_1,$$
$$v_i = -\frac{1}{2r_i} z_i = -\left\{\sqrt{G_i^2 + a_i} + G_i\right\} z_i, \quad (i \ge 2). \tag{86}$$

Theorem 5 : *For* **Solution I** *and* **Solution II**, *the residual regions of* z_i *can be made arbitrarily small by sufficiently large* k_{i8}, k_{i9}, k_{i10}, a_i, *and sufficiently small* r_{0i}, γ_i^*.

REFERENCES

Åström, K.J. and B. Wittenmark (1989). *Adaptive Control.* Addison-Wesley.

Ioannou, P.A. and J. Sun (1996). *Robust Adaptive Control.* Prentice-Hall, New Jersey.

Krstić, M., I. Kanellakopoulos and P.V. Kokotović (1995). *Nonlinear and Adaptive Control Design.* John Wiley & Suns, New York.

Krstić, M. and H. Deng (1998). *Stabilization of Nonlinear Uncertain Systems.* Springer.

Miyasato, Y. (1999). Redesign of adaptive control systems based on the notion of optimality. *Proceedings of 38th IEEE Conference on Decision and Control,* 3315-3320, Phoenix.

Miyasato, Y. (2000a). Adaptive nonlinear H_∞ control for processes with bounded variations of parameters. *Proceedings of 2000 American Control Conference,* 297-391, Chicago.

Miyasato, Y. (2000b). General forms of adaptive nonlinear H_∞ control for processes with bounded variations of parameters. *Preprints of IFAC SYSID 2000,* Santa Barbara.

Miyasato, Y. (2000c). Adaptive nonlinear H_∞ control for processes with bounded variations of parameters – general relative degree case – . *Proceedings of the 39th IEEE Conference on Decision and Control,* 1453-1458, Sydney.

ADAPTIVE ITERATIVE REFINEMENT OF AN OPTIMAL *2DF* NONLINEAR CONTROLLER

L. Keviczky and **Cs. Bányász**

Computer and Automation Research Institute, Hungarian Academy of Sciences
H-1111 Budapest, Kende u 13-17, HUNGARY
Phone: +361-466-5435; Fax: +361-466-7503
e-mail: keviczky@sztaki.hu; banyasz@sztaki.hu

Abstract: The paper extends the generic two-degree of freedom controller scheme introduced for linear plants to factorable nonlinear processes. The simple cases of the Hammerstein and Wiener type block-oriented nonlinear models are analyzed and an iterative controller refinement scheme is presented. Simulation examples are also given to present the efficiency of the new methodology. *Copyright© 2001 IFAC*

Keywords: control of nonlinear discrete-time systems, nonlinear two-degree of freedom controller, Hammerstein and Wiener models, iterative controller refinement

1. INTRODUCTION

Because of the infinite variety of nonlinear dynamics it can not be expected that a unique optimal control solution can be found for all of these complex systems. However, the history and development of nonlinear control systems show that one can expect relatively simple methods, which can be close or even reach the effectiveness of linear methodology. A very wide class of approaches are based on the usual Jacobean linearization of a nonlinear system (*NS*). A large class of nonlinear systems can be made to have linear input-output behavior through a choice of nonlinear state feedback control low (Sastry, 1999). Other approaches assume special topology, when the structure of the *NS* makes the linearization possible. Such *NS* classes are e.g. the bilinear and the block-oriented factorable (cascade) systems (Haber and Keviczky, 1999).

Inspite of it is a simple task to compute the resulting nonlinear transfer operator of two serially connected nonlinear blocks, shown in Fig. 1, most of our tests failed when students were asked, because of the almost automatic thinking with linear dynamic blocks.

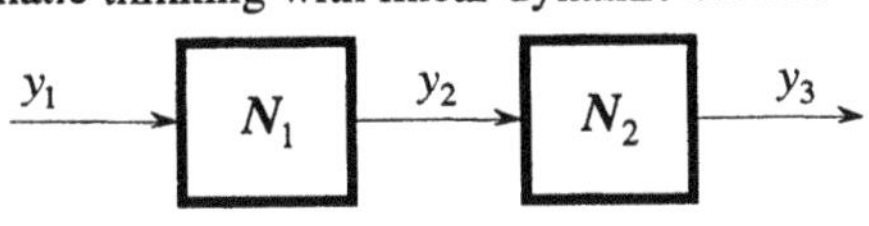

Fig. 1 Serially connected cascade nonlinear blocks

The resulting nonlinear transfer operator is given by

$$y_3 = N_2 N_1 y_1 = N_2 y_2 \qquad (1)$$

and not by $y_3 = N_1 N_2 y_1$, which is the usual procedure if a block oriented scheme is available. So note that the order of the nonlinear operators in the formulas is opposite to the order of the blocks shown in the figures. (This is a usual source of mistakes calculating transfer characteristics of open- and closed-loop nonlinear schemes.)

The most well known factorable models are the simple Hammerstein and Wiener models. The simple Wiener model N_W (shown in Fig. 2) is a cascade structure of a linear dynamic $Y^{W,dyn}$ and a nonlinear static $N^{W,stat}$ terms connected in series, i.e.

$$N_W = N^{W,stat}\, Y^{W,dyn} \text{ simply } N_W = N^W\, Y^W \qquad (2)$$

The simple Hammerstein model N_H (shown in Fig. 3) is a cascade structure of a nonlinear static $N^{H,stat}$ and a linear dynamic $Y^{H,dyn}$ terms connected in series, i.e.

$$N_H = Y^{H,dyn}\, N^{H,stat} \text{ or simply } N_H = Y^H\, N^H \qquad (3)$$

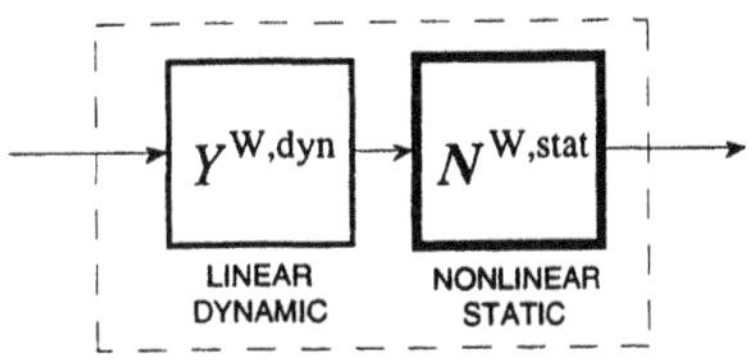

Fig. 2 The simple Wiener model

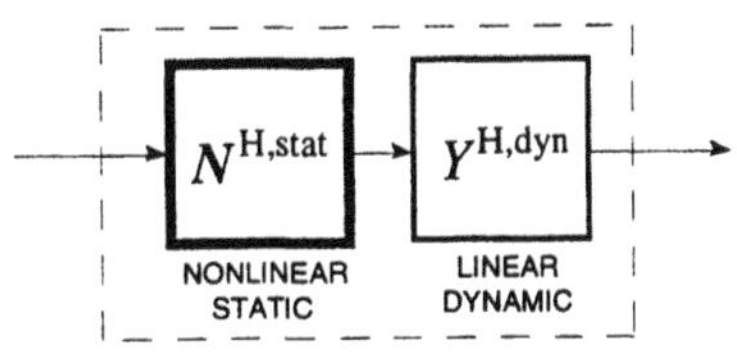

Fig. 3 The simple Hammerstein model

2. THE GENERIC *2DF* LINEAR CONTROL SCHEME

The *Youla- (Y* or *Q) parametrization* is a classical method to characterize all realizable stabilizing regulators (*ARS*) by

$$R = Q/(1 - QS) \qquad (4)$$

for open-loop stable plant $S \in \mathcal{S}$, where $\mathcal{S}$ is the closed set of all stable proper real-rational systems, having all poles within the closed unit disc. The "parameter"

$$Q = R/(1+RS) \quad ; \quad Q \in \mathcal{S} \tag{5}$$

ranges over all proper ($Q(\omega = \infty)$ is finite), stable transfer functions (Maciejowski, 1989).

Using the *K-B-parametrization* (Keviczky and Bányász, 1999b, 2000), the authors introduced a generic two-degree of freedom (*G2DF*) control system providing that it virtually opens the closed-loop

$$y = Q_r S y_r - (1 - QS) w = y_t + y_d \tag{6}$$

where the tracking properties $y_t = Q_r S y_r$ can independently be designed from the regulating behavior $y_d = (1 - QS) w$ by Q_r.

The optimal *ARS* regulator of the *G2DF* scheme (Keviczky and Bányász, 1999b, 2000) is given by

$$R_{\text{opt}} = \frac{P_w K_w}{1 - P_w K_w S} = \frac{Q_o}{1 - Q_o S} = \frac{P_w G_w S_+^{-1}}{1 - P_w G_w S_- z^{-d}} \tag{7}$$

where

$$Q_o = Q_w = P_w K_w = P_w G_w S_+^{-1} \tag{8}$$

is the associated *Y-parameter* (Maciejowski, 1989) furthermore

$$\begin{aligned} Q_r &= P_r K_r = P_r G_r S_+^{-1} \quad ; \quad K_w = G_w S_+^{-1} \\ K_r &= G_r S_+^{-1} \end{aligned} \tag{9}$$

assuming that the process is factorable as

$$S = S_+ \bar{S}_- = S_+ S_- z^{-d} \tag{10}$$

where S_+ means the inverse stable (*IS*) and S_- the inverse unstable (*IU*) factors, respectively. z^{-d} corresponds to the discrete time delay, which is the integer multiple of the sampling time.

It was shown (Keviczky and Bányász, 1999a) that the optimization of the *G2DF* scheme can be performed in $\mathcal{H}_2$ and $\mathcal{H}_\infty$ norm spaces by the proper selection of the serial K_r and embedded K_w filters (compensators). These optimizations will be reduced to the optimal computation of the G_r and G_w filters. If G_r and G_w are optimally selected, then R_{opt} denotes the optimal *ARS* regulator in (7). It is interesting to see how the transfer characteristics of this system looks like:

$$\begin{aligned} y &= P_r K_r S y_r - (1 - P_w K_w S) w = \\ &= P_r G_r S_- z^{-d} y_r - (1 - P_w G_w S_- z^{-d}) w = y_t + y_d \end{aligned} \tag{11}$$

Here P_r and P_w are stable and proper transfer functions, what are partly capable to place desired poles in the servo and the regulatory transfer functions, furthermore they are usually referred as reference signal and output disturbance predictors. They can even be called as reference models, so reasonably $P_r(\omega = 0) = 1$ and $P_w(\omega = 0) = 1$ are selected.

3. *K-B* PARAMETRIZATION OF NONLINEAR *2DF* CONTROL SYSTEMS

It is an interesting question how the above parametrization can be extended for nonlinear systems. The formal generalization of the *Y-parametrization* for nonlinear systems is relatively simple, introducing the nonlinear operator

$$\boldsymbol{Q} = \boldsymbol{R}(1 - \boldsymbol{NR})^{-1} = (1 - \boldsymbol{RN})^{-1}\boldsymbol{R} \tag{12}$$

and using the nonlinear regulator

$$\boldsymbol{R} = (1 - \boldsymbol{QN})^{-1}\boldsymbol{Q} = \boldsymbol{Q}(1 - \boldsymbol{NQ})^{-1} \tag{13}$$

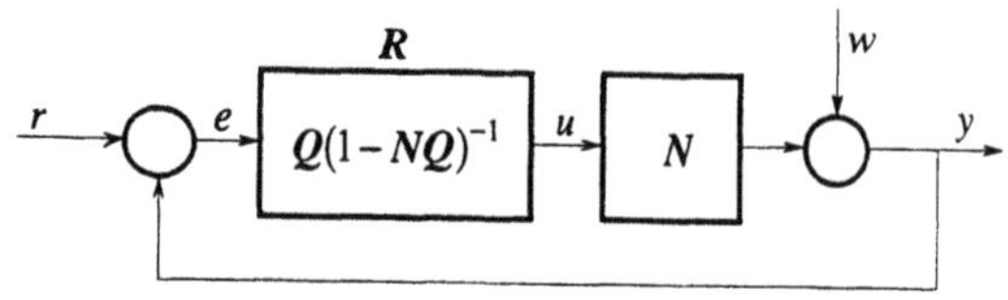

Fig. 4. Generalization of the *Y-parametrization*

Here N is the nonlinear dynamic process (see Fig. 4.), furthermore the *Y-parameter* $\boldsymbol{Q}$ and regulator $\boldsymbol{R}$ are also nonlinear dynamic operators. It is not very difficult to prove that the left and right side formulation of $\boldsymbol{Q}$ and $\boldsymbol{R}$ are equivalent. (Bold means nonlinear transfer operators, whose product is generally not interchangeable.) The next two relationships play important role in the closed-loop characteristics of the systems:

$$(1 + \boldsymbol{NR})^{-1} = \left[1 + \boldsymbol{NQ}(1 - \boldsymbol{NQ})^{-1}\right]^{-1} = 1 - \boldsymbol{NQ} \tag{14}$$

$$(1 + \boldsymbol{RN})^{-1} = \left[1 + (1 - \boldsymbol{QN})^{-1}\boldsymbol{QN}\right]^{-1} = 1 - \boldsymbol{QN} \tag{15}$$

Here the following identities (valid for matrices and nonlinear operators) were used:

$$\begin{aligned} &\left[1 + (1 - \boldsymbol{A})^{-1}\boldsymbol{A}\right]^{-1} = 1 - \boldsymbol{A} \\ &\left[1 + \boldsymbol{B}(1 - \boldsymbol{B})^{-1}\right]^{-1} = 1 - \boldsymbol{B} \end{aligned} \tag{16}$$

After some straightforward (do not miss the interchangeability in nonlinear operator products) manipulations

$$y = \boldsymbol{N} r - (1 - \boldsymbol{NQ}) w \tag{17}$$

is obtained, i.e., the same principles applied in the linear case can be followed, resulting in the Fig. 5.

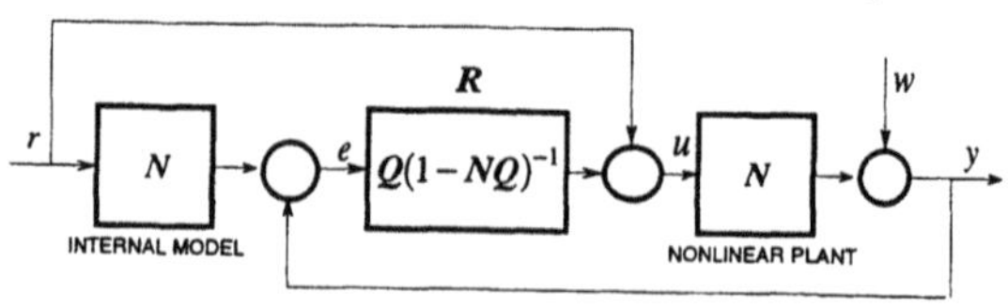

Fig. 5 The general form of the *K-B-parametrized* nonlinear *2DF* control system

4. GENERALIZATION OF THE LINEAR *2DF* CONTROL FOR NONLINEAR SYSTEMS

This generalization of the linear 2DF control for

nonlinear dynamic plants is formally not straightforward, but not so difficult, as Fig. 6 shows. $\boldsymbol{Q}_{\mathrm{w}}$ and $\boldsymbol{K}_{\mathrm{r}}$ are the nonlinear generalized versions of the linear *Y-parameter* Q_{w} and serial filter K_{r}, respectively. Note that the inverses mean functional inverses and they are not simple fractions. It is interesting to see how the transfer characteristics of this nonlinear control system look like:

$$\begin{aligned} y &= \boldsymbol{N}\boldsymbol{Q}_{\mathrm{r}}\, y_{\mathrm{r}} - \left(1 - \boldsymbol{N}\boldsymbol{Q}_{\mathrm{w}}\right) w = \\ &= \boldsymbol{N}\boldsymbol{K}_{\mathrm{r}} P_{\mathrm{r}}\, y_{\mathrm{r}} - \left(1 - \boldsymbol{N}\boldsymbol{K}_{\mathrm{w}} P_{\mathrm{w}}\right) w = y_{\mathrm{t}} + y_{\mathrm{d}} \end{aligned} \tag{18}$$

So the scheme introduced in Fig. 6 can be referred as a *nonlinear G2DF* (*NG2DF*) control system.

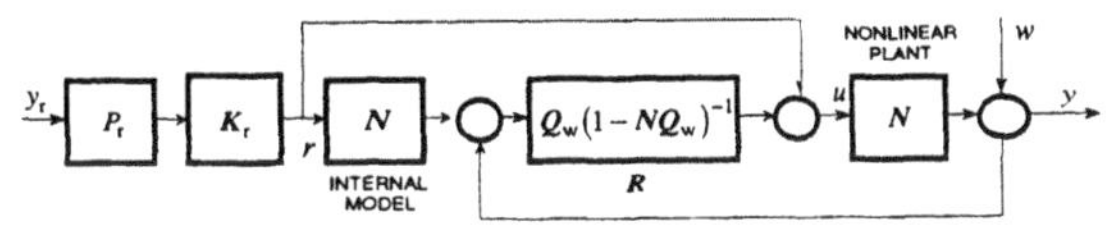

Fig. 6 Generalization of the *generic 2DF* control system for nonlinear processes

If the nonlinear process $\boldsymbol{N}$ is factorable to an invertible $\boldsymbol{N}_{+}$ (*IS*) and to a noninvertable $\boldsymbol{N}_{-}$ (*IU*) operator

$$\boldsymbol{N} = \boldsymbol{N}_{-}\boldsymbol{N}_{+} \neq \boldsymbol{N}_{+}\boldsymbol{N}_{-} \tag{19}$$

then using the proper selection of

$$\boldsymbol{Q} = \boldsymbol{Q}_{\mathrm{w}} = \boldsymbol{K}_{\mathrm{w}} P_{\mathrm{w}} = \boldsymbol{N}_{+}^{-1}\boldsymbol{G}_{\mathrm{w}} P_{\mathrm{w}} \tag{20}$$

and

$$\begin{aligned} &\boldsymbol{Q}_{\mathrm{r}} = \boldsymbol{K}_{\mathrm{r}} P_{\mathrm{r}} = \boldsymbol{N}_{+}^{-1}\boldsymbol{G}_{\mathrm{r}} P_{\mathrm{r}} \\ &\boldsymbol{K}_{\mathrm{w}} = \boldsymbol{N}_{+}^{-1}\boldsymbol{G}_{\mathrm{w}} \quad ; \quad \boldsymbol{K}_{\mathrm{r}} = \boldsymbol{N}_{+}^{-1}\boldsymbol{G}_{\mathrm{r}} \end{aligned} \tag{21}$$

the following nonlinear regulator is obtained

$$\begin{aligned} \boldsymbol{R} &= \boldsymbol{Q}_{\mathrm{w}}\left(1 - \boldsymbol{N}\boldsymbol{Q}_{\mathrm{w}}\right)^{-1} = \boldsymbol{K}_{\mathrm{w}} P_{\mathrm{w}}\left(1 - \boldsymbol{N}\boldsymbol{K}_{\mathrm{w}} P_{\mathrm{w}}\right)^{-1} = \\ &= \boldsymbol{N}_{+}^{-1}\boldsymbol{G}_{\mathrm{w}} P_{\mathrm{w}}\left(1 - \boldsymbol{N}_{-}\boldsymbol{G}_{\mathrm{w}} P_{\mathrm{w}}\right)^{-1} \end{aligned} \tag{22}$$

Using the previous formulas the transfer characteristics of the closed-loop is now obtained as:

$$y = \boldsymbol{N}_{-}\boldsymbol{G}_{\mathrm{r}} P_{\mathrm{r}}\, y_{\mathrm{r}} - \left(1 - \boldsymbol{N}_{-}\boldsymbol{G}_{\mathrm{w}} P_{\mathrm{w}}\right) w = y_{\mathrm{t}} + y_{\mathrm{d}} \tag{23}$$

Here, similarly to the linear case, $\boldsymbol{K}_{\mathrm{r}}$ and $\boldsymbol{K}_{\mathrm{w}}$ contain the $\boldsymbol{N}_{+}^{-1}$ inverse (in functional sense) of $\boldsymbol{N}_{+}$, i.e., the invertible part of $\boldsymbol{N}$, and apply $\boldsymbol{G}_{\mathrm{r}}$ and $\boldsymbol{G}_{\mathrm{w}}$ to attenuate the influence of the invariant nonlinear process factor $\boldsymbol{N}_{-}$.

5. FACTORABLE BLOCK-ORIENTED MODELS OF NONLINEAR PLANTS

Wiener model

Assume that the transfer characteristics of the Wiener model is given by

$$\begin{aligned} N_{\mathrm{W}} &= N^{\mathrm{W}} Y_{-}^{\mathrm{W}} Y_{+}^{\mathrm{W}} = N_{+}^{\mathrm{W}} Y_{-}^{\mathrm{W}} = N^{\mathrm{W}} Y_{+}^{\mathrm{W}} \overline{Y}_{-}^{\mathrm{W}} z^{-d} \\ &= N_{+} N_{-} \end{aligned} \tag{24}$$

$$N_{+} = N_{+}^{\mathrm{W}} = N^{\mathrm{W}} Y_{+}^{\mathrm{W}} \quad ; \quad N_{-} = Y_{-}^{\mathrm{W}} = \overline{Y}_{-}^{\mathrm{W}} z^{-d}$$

which obviously does not correspond to the required nonlinear dynamic process structure (19), with one exception, when $\overline{Y}_{-}^{\mathrm{W}} = 1$, i.e., when $Y_{+}^{\mathrm{W}} = Y^{\mathrm{W}}$ is *IS*. Here Y_{+}^{W} means the *IS* and Y_{-}^{W} the *IU* factors, respectively. z^{-d} corresponds to the discrete time delay, similarly to (10). Assume that the inverse static characteristic $\left(N^{\mathrm{W}}\right)^{-1}$ is realizable on the operational range, i.e., $\left(N^{\mathrm{W}}\right)^{-1} \neq \infty$. This special Wiener model is

$$\begin{aligned} N_{\mathrm{W}} &= \left(N^{\mathrm{W}} Y^{\mathrm{W}}\right) z^{-d} = N_{+}^{\mathrm{W}} N_{-} = z^{-d}\left(N^{\mathrm{W}} Y^{\mathrm{W}}\right) \\ &= N_{-} N_{+} \end{aligned} \tag{25}$$

$$N_{+} = N^{\mathrm{W}} Y^{\mathrm{W}} \quad ; \quad N_{-} = z^{-d} = N_{-}$$

where the non realizable factor $Y_{-}^{\mathrm{W}} = z^{-d}$ is arbitrarily interchangeable in the operator order. Following the general design procedure (20)-(22) the nonlinear regulator is

$$\begin{aligned} \boldsymbol{R}_{\mathrm{W}} &= \boldsymbol{Q}_{\mathrm{w}}^{\mathrm{W}}\left(1 - N_{\mathrm{W}}\boldsymbol{Q}_{\mathrm{w}}^{\mathrm{W}}\right)^{-1} = \\ &= \left(N^{\mathrm{W}} Y^{\mathrm{W}}\right)^{-1} G_{\mathrm{w}}^{\mathrm{W}} P_{\mathrm{w}}\left(1 - N_{-} G_{\mathrm{w}}^{\mathrm{W}} P_{\mathrm{w}}\right)^{-1} = \\ &= \frac{1}{Y^{\mathrm{W}}}\left(N^{\mathrm{W}}\right)^{-1} \frac{P_{\mathrm{w}}}{1 - P_{\mathrm{w}} z^{-d}} \end{aligned} \tag{26}$$

and serial nonlinear compensator is

$$\begin{aligned} \boldsymbol{K}_{\mathrm{r}}^{\mathrm{W}} &= \left(N_{+}^{\mathrm{W}}\right)^{-1} G_{\mathrm{r}}^{\mathrm{W}} = \frac{1}{Y^{\mathrm{W}}}\left(N^{\mathrm{W}}\right)^{-1} \\ \boldsymbol{K}_{\mathrm{w}}^{\mathrm{W}} &= \left(N_{+}^{\mathrm{W}}\right)^{-1} G_{\mathrm{w}}^{\mathrm{W}} = \frac{1}{Y^{\mathrm{W}}}\left(N^{\mathrm{W}}\right)^{-1} \end{aligned} \tag{27}$$

Obviously $\left(N^{\mathrm{W}}\right)^{-1}$ cancels the static nonlinearity in N_{W}, now $G_{\mathrm{w}}^{\mathrm{W}}$ and $G_{\mathrm{r}}^{\mathrm{W}}$ are linear, attenuating the influence of the *IU* $\overline{Y}_{-}^{\mathrm{W}}$. Because $\overline{Y}_{-}^{\mathrm{W}} = 1$, i.e., is *IS*, the very simple serial compensators $G_{\mathrm{w}}^{\mathrm{W}} = 1$ and $G_{\mathrm{r}}^{\mathrm{W}} = 1$ were used here according to the $\mathcal{H}_2$ and $\mathcal{H}_\infty$ optimal selection principle of $G_{\mathrm{w}}^{\mathrm{H}}$ and $G_{\mathrm{r}}^{\mathrm{H}}$, see in (Keviczky and Bányász, 1999a). The reachable best transfer characteristics are equal now to the best reachable linear ones:

$$\begin{aligned} y &= z^{-d}\overline{Y}_{-}^{\mathrm{W}} G_{\mathrm{r}}^{\mathrm{W}} P_{\mathrm{r}}\, y_{\mathrm{r}} - \left(1 - z^{-d}\overline{Y}_{-}^{\mathrm{W}} G_{\mathrm{w}}^{\mathrm{W}} P_{\mathrm{w}}\right) w = \\ &= P_{\mathrm{r}} z^{-d} y_{\mathrm{r}} - \left(1 - P_{\mathrm{w}} z^{-d}\right) w = y_{\mathrm{t}} + y_{\mathrm{d}} \end{aligned} \tag{28}$$

Observe that no nonlinearity remained in (28) and different Hammerstein model substructures were used in the obtained formulas. (If $\overline{Y}_{-}^{\mathrm{W}} \neq 1$, i.e., is *IU*, then other controller design method should be applied.)

Hammerstein model

Assume that the transfer characteristics of the Hammerstein model is given by

$$\begin{aligned} N_{\mathrm{H}} &= Y_{-}^{\mathrm{H}} Y_{+}^{\mathrm{H}} N^{\mathrm{H}} = Y_{-}^{\mathrm{H}} N_{+}^{\mathrm{H}} = z^{-d}\overline{Y}_{-}^{\mathrm{H}} N_{+}^{\mathrm{H}} \\ &= N_{-} N_{+} \end{aligned} \tag{29}$$

$$N_{+} = N_{+}^{\mathrm{H}} = Y_{+}^{\mathrm{H}} N^{\mathrm{H}} \quad ; \quad N_{-} = Y_{-}^{\mathrm{H}} = z^{-d}\overline{Y}_{-}^{\mathrm{H}}$$

It can be well seen, that the Hammerstein model (29) corresponds to the required nonlinear dynamic process structure (19). Assume furthermore that the inverse static characteristic $\left(N^{\mathrm{H}}\right)^{-1}$ is realizable on the operational range, i.e., $\left(N^{\mathrm{H}}\right)^{-1} \neq \infty$. Following the general design procedure (20)-(22) the nonlinear regulator is

$$\begin{aligned} \boldsymbol{R}_{\mathrm{H}} &= Q_{\mathrm{w}}^{\mathrm{H}}\left(1 - N_{\mathrm{H}} Q_{\mathrm{w}}^{\mathrm{H}}\right)^{-1} = \\ &= \left(Y_{+}^{\mathrm{H}} N^{\mathrm{H}}\right)^{-1} G_{\mathrm{w}}^{\mathrm{H}} P_{\mathrm{w}}\left(1 - N_{-} G_{\mathrm{w}}^{\mathrm{H}} P_{\mathrm{w}}\right)^{-1} = \\ &= \left(N^{\mathrm{H}}\right)^{-1} \frac{P_{\mathrm{w}} G_{\mathrm{w}}^{\mathrm{H}}}{\left(1 - P_{\mathrm{w}} G_{\mathrm{w}}^{\mathrm{H}} \overline{Y}_{-}^{\mathrm{H}} z^{-d}\right) Y_{+}^{\mathrm{H}}} \end{aligned} \tag{30}$$

and serial nonlinear compensator is

$$\begin{aligned} \boldsymbol{K}_{\mathrm{r}}^{\mathrm{H}} &= \left(N_{+}^{\mathrm{H}}\right)^{-1} G_{\mathrm{r}}^{\mathrm{H}} = \left(N^{\mathrm{H}}\right)^{-1} \frac{G_{\mathrm{r}}^{\mathrm{H}}}{Y_{+}^{\mathrm{H}}} \\ \boldsymbol{K}_{\mathrm{w}}^{\mathrm{H}} &= \left(N_{+}^{\mathrm{H}}\right)^{-1} G_{\mathrm{w}}^{\mathrm{H}} = \left(N^{\mathrm{H}}\right)^{-1} \frac{G_{\mathrm{w}}^{\mathrm{H}}}{Y_{+}^{\mathrm{H}}} \end{aligned} \tag{31}$$

The reachable best transfer characteristics are:

$$\begin{aligned} y &= z^{-d} \overline{Y}_{-}^{\mathrm{H}} G_{\mathrm{r}}^{\mathrm{H}} P_{\mathrm{r}}\, y_{\mathrm{r}} - \left(1 - z^{-d} \overline{Y}_{-}^{\mathrm{H}} G_{\mathrm{w}}^{\mathrm{H}} P_{\mathrm{w}}\right) w = \\ &= P_{\mathrm{r}} G_{\mathrm{r}}^{\mathrm{H}}\, \overline{Y}_{-}^{\mathrm{H}} z^{-d}\, y_{\mathrm{r}} - \left(1 - P_{\mathrm{w}} G_{\mathrm{w}}^{\mathrm{H}} \overline{Y}_{-}^{\mathrm{H}} z^{-d}\right) w = y_{\mathrm{t}} + y_{\mathrm{d}} \end{aligned} \tag{32}$$

It can be easily seen that $\left(N^{\mathrm{H}}\right)^{-1}$ cancels the static nonlinearity in N_{H}, furthermore the linear $G_{\mathrm{w}}^{\mathrm{W}}$ and $G_{\mathrm{r}}^{\mathrm{W}}$ are now attenuating the influence of the *IU* $\overline{Y}_{-}^{\mathrm{W}}$, which means that the same optimality principles can be followed what were obtained for the $\mathcal{H}_2$ and $\mathcal{H}_\infty$ optimal selection of $G_{\mathrm{w}}^{\mathrm{W}}$ and $G_{\mathrm{r}}^{\mathrm{W}}$, see in (Keviczky and Bányász, 1999a). (Observe that no nonlinearity remained in (32) and different Wiener model substructures were used in the obtained formulas.)

It may be interesting to see how the design formulas change in case of an *IS* linear dynamic part of the Hammerstein model, when $Y_{+}^{\mathrm{H}} = Y^{\mathrm{H}}$ and $\overline{Y}_{-}^{\mathrm{H}} = 1$:

$$\boldsymbol{R}_{\mathrm{H}} = \left(N^{\mathrm{H}}\right)^{-1} \frac{P_{\mathrm{w}}}{\left(1 - P_{\mathrm{w}} z^{-d}\right) Y^{\mathrm{H}}} \tag{33}$$

$$\boldsymbol{K}_{\mathrm{r}}^{\mathrm{H}} = \left(N^{\mathrm{H}}\right)^{-1} \frac{1}{Y^{\mathrm{H}}} \quad ; \quad \boldsymbol{K}_{\mathrm{w}}^{\mathrm{H}} = \left(N^{\mathrm{H}}\right)^{-1} \frac{1}{Y^{\mathrm{H}}} \tag{34}$$

Similarly to the *IS* Wiener model, the overall transfer functions also correspond to the best ideal linear forms:

$$y = P_{\mathrm{r}} z^{-d}\, y_{\mathrm{r}} - \left(1 - P_{\mathrm{w}} z^{-d}\right) w = y_{\mathrm{t}} + y_{\mathrm{d}} \tag{35}$$

The regulators and compensators are summarized in Table 1 for the different process models and process types.

Table 1: Nonlinear regulators and related compensators for different factorable models

Process model	Process type	Model	Regulator	Compensator
Wiener	*IS*	$z^{-d} N^{\mathrm{W}} Y^{\mathrm{W}}$	$\frac{1}{Y^{\mathrm{W}}}\left(N^{\mathrm{W}}\right)^{-1} \frac{P_{\mathrm{w}}}{1 - P_{\mathrm{w}} z^{-d}}$	$\frac{1}{Y^{\mathrm{W}}}\left(N^{\mathrm{W}}\right)^{-1}$
Hammerstein	*IU*	$z^{-d} \overline{Y}_{-}^{\mathrm{H}}\, Y_{+}^{\mathrm{H}} N^{\mathrm{H}}$	$\left(N^{\mathrm{H}}\right)^{-1} \frac{P_{\mathrm{w}} G_{\mathrm{w}}^{\mathrm{H}}}{\left(1 - P_{\mathrm{w}} G_{\mathrm{w}}^{\mathrm{H}} \overline{Y}_{-}^{\mathrm{H}} z^{-d}\right) Y_{+}^{\mathrm{H}}}$	$\left(N^{\mathrm{H}}\right)^{-1} \frac{G_{\mathrm{r}}^{\mathrm{H}}}{Y_{+}^{\mathrm{H}}}$
Hammerstein	*IS*	$z^{-d}\, Y^{\mathrm{H}} N^{\mathrm{H}}$	$\left(N^{\mathrm{H}}\right)^{-1} \frac{P_{\mathrm{w}}}{\left(1 - P_{\mathrm{w}} z^{-d}\right) Y^{\mathrm{H}}}$	$\left(N^{\mathrm{H}}\right)^{-1} \frac{1}{Y^{\mathrm{H}}}$

6. ITERATIVE NONLINEAR CONTROLLER REFINEMENT

Fig. 7, using the *NG2DF structure* (see Fig. 5), shows, an obvious way how to perform the identification (ID) step in a combined ID and control scheme (Keviczky, 1995), i.e., we should use a regular ID algorithm between the auxiliary variable r and the measured controlled variable y. Note that r and y must be obtained from the closed-loop operated by the *generic* optimal controller structure with the optimal regulator $\hat{\boldsymbol{R}}$ and compensator $\hat{\boldsymbol{K}}_{\mathrm{r}}$ (which are now also model based: ^ means the estimation).

Because r depends on the model $\hat{N}$, only an iterative control refinement procedure can be performed. It's simplest - so-called relaxation type - iteration can be built in the following way for an off-line case using n samples (i-th iteration is shown):

1. Calculate the auxiliary variable r based on the

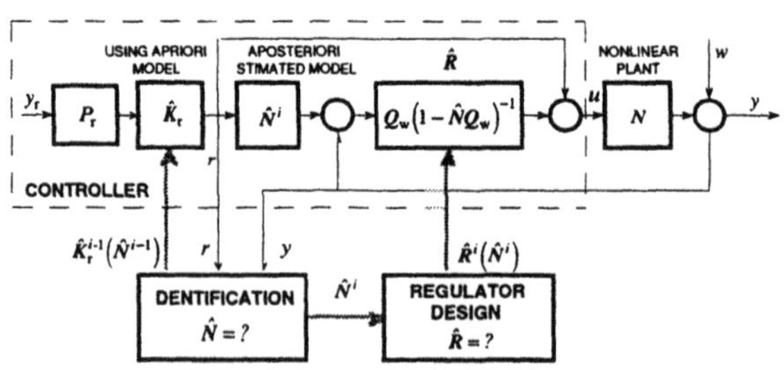

Fig. 7 Model based version of the *NG2DF* scheme for iterative controller refinement

available model $\hat{N}^{i-1}$, using the reference model P_{r} and $\hat{\boldsymbol{K}}_{\mathrm{r}}\left(\hat{N}^{i-1}\right)$, furthermore the applied reference signal series $\boldsymbol{y}_{\mathrm{r}}^{i} = \left\{y_{\mathrm{r}}^{i}(k); k = 1, \ldots, n\right\}$:

$$\begin{aligned} r(k) &= \hat{\boldsymbol{K}}_{\mathrm{r}} P_{\mathrm{r}}\, y_{\mathrm{r}}(k) = \hat{\boldsymbol{N}}_{+}^{-1} \hat{\boldsymbol{G}}_{\mathrm{r}} P_{\mathrm{r}}\, y_{\mathrm{r}}(k) = \\ &= \left[\hat{\boldsymbol{N}}_{+}^{i-1}\right]^{-1} \hat{\boldsymbol{G}}_{\mathrm{r}}\left(\hat{\boldsymbol{N}}_{-}^{i-1}\right) y_{\mathrm{r}}(k) \quad ; \quad k = 1, \ldots, n \end{aligned} \tag{36}$$

Here i denotes the index of the iteration and note that the series y_r^i does not necessarily change by iteration.

If the linear dynamic part of the factorable nonlinear model is IU, then we have to solve the Diophantine equation or the Nevanlinna-Pick problem (Keviczky and Bányász, 1999b) to obtain the filter $\hat{G}_r(\hat{N}_-^{i-1})$ in $\hat{K}_r(\hat{N}^{i-1})$, otherwise $\hat{G}_r = \hat{G}_r = 1$.

2. Identify a model $\hat{N}$ using $r_i = \{r_i(k); k = 1, \dots, n\}$ and $y_{i-1} = \{y_{i-1}(k); k = 1, \dots, n\}$ based on the modeling step

$$\hat{N}^i = \arg \min_{\hat{N} \in \mathcal{N}} J_{ic}\left(\mathcal{M}, \Delta, r_i, \hat{R}_{i-1}, P_w\right) \tag{37}$$

where now J_{ic} is a closed-loop identification criterion with model class $\mathcal{N}$.

3. Calculate the output of the optimal nonlinear regulator as

$$r_i(k) = \left\{\left[\hat{N}_+^{i-1}\right]^{-1} + \hat{R}^i \hat{N}_-^{i-1}\right\} \hat{G}_r\left(\hat{N}_-^{i-1}\right) P_r y_r(k) - \\ - \hat{R}^i y_i(k) \quad ; \quad k = 1, \dots, n \tag{38}$$

and apply to the process in closed-loop.

If the linear dynamic part of the factorable nonlinear model is IU, then we have to solve the Diophantine equation or the Nevanlinna-Pick problem (Keviczky and Bányász, 1999b) to obtain the filter $\hat{G}_w(\hat{N}_-^{i-1})$ in $\hat{K}_w(\hat{N}^{i-1})$, otherwise $\hat{G}_w = \hat{G}_w = 1$.

4. (There is a possibility to optimize the applied reference signal series in this step by a proper input design procedure (Keviczky and Bányász, 1999a))

5. Once $\hat{N}^i$ and $\hat{R}_i$ are found the closed-loop bandwidth can be increased by repeating the procedure. The iterative process is continued from step 1, while a stop condition is not fulfilled (until the ultimate control objective is achieved or it is terminated by reaching some vital constraints).

7. EXAMPLE

Let the nonlinear dynamic plant is a Wiener type process with

$$N_W = N^W Y_+^W Y_-^W = \left(-0.125u^2 + u\right) \frac{0.2z^{-1}}{1 - 0.8z^{-1}} z^{-5} \tag{39}$$

with

$$N^W = -0.125u^2 + u \quad ; \quad Y_+^W = \frac{0.2z^{-1}}{1 - 0.8z^{-1}} \\ Y_-^W = \overline{Y}_-^W z^{-5} = z^{-5} \quad ; \quad \overline{Y}_-^W = 1 \tag{40}$$

Apply the unity gain tracking and disturbance rejection reference models

$$P_r = \frac{0.7z^{-1}}{1 - 0.3z^{-1}} \quad \text{and} \quad P_w = \frac{0.8z^{-1}}{1 - 0.2z^{-1}} \tag{41}$$

Using the design procedure (26)-(27) the following regulator and serial compensators are obtained ($G_w^W = 1$ and $G_r^W = 1$, because the linear part is IS (Keviczky and Bányász, 1999b)):

$$R_W = \left(5 - 4z^{-1}\right)\left(N^W\right)^{-1} \frac{0.8}{1 - 0.2z^{-1} - 0.8z^{-5}} \\ \left(N^W\right)^{-1} = 4\left(1 - \sqrt{1 - 0.5u}\right) \tag{42}$$

$$K_r^W P_r = \left(5 - 4z^{-1}\right)\left(N^W\right)^{-1} \frac{0.7z^{-1}}{1 - 0.3z^{-1}} \tag{43}$$

A unity amplitude square wave reference input signal $y_r(k)$ with periodic time 80 samples was applied. Figure 8 shows the output response $y(k)$ of the closed-loop for the square wave excitation $y_r(k)$ with the linear optimal regulator, when the plant itself is nonlinear. The performance degradation caused by the nonlinearity can be well seen.

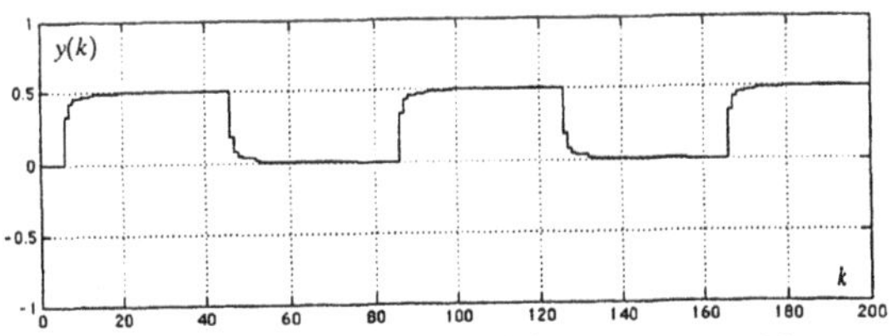

Fig. 8 The output response $y(k)$ of the closed-loop for the square wave excitation with the linear optimal regulator, when the plant is nonlinear

Start the iterative control refinement with the initial model

$$N_W = N^W Y_+^W Y_-^W = \left(-0.15u^2 + 0.8u\right) \frac{0.4z^{-1}}{1 - 0.6z^{-1}} z^{-5} \tag{44}$$

and apply an off-line parameter estimation procedure using $n = 200$ samples.

The static characteristics of the true nonlinear plant N^W and inverse regulator $\left(N^W\right)^{-1}$ and their initial apriori estimates N_o^W and $\left(N_o^W\right)^{-1}$ are shown in Fig. 9.

Figure 10 shows the control and identification loss functions (variances) by the iteration. It can be well seen that the iteration is quite fast reaching the optimal values after 4 steps.

$$N_o^W = -0.15u^2 + 0.8u \\ \left(N_o^W\right)^{-1} = 2.667 - \sqrt{7.111 - 6.667u} \tag{45}$$

In the simulation runs an additive white noise was used as output disturbance with a standard deviation $\lambda = 0.01$. We used a simple off-line LS method for parameter estimation only to demonstrate the operation of the iterative algorithm. (Our experience showed that this method works very well in this scheme while the noise level is low. However, in a noisy case the proper parameter estimation method (*ELS*, *ML*, etc.) should be applied corresponding to the given process and noise model structure.) The outputs of the tracking reference model (continuous) and the controlled process (dashed) are shown in Fig. 11 before and after the iterative control refinement.

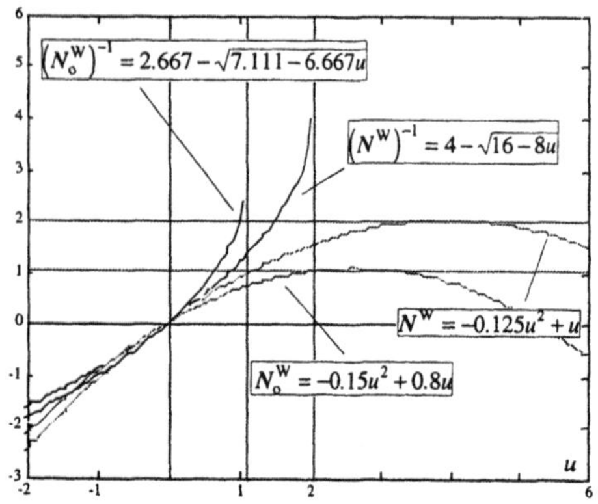

Fig. 9 The nonlinear static characteristics of the plant and the regulator

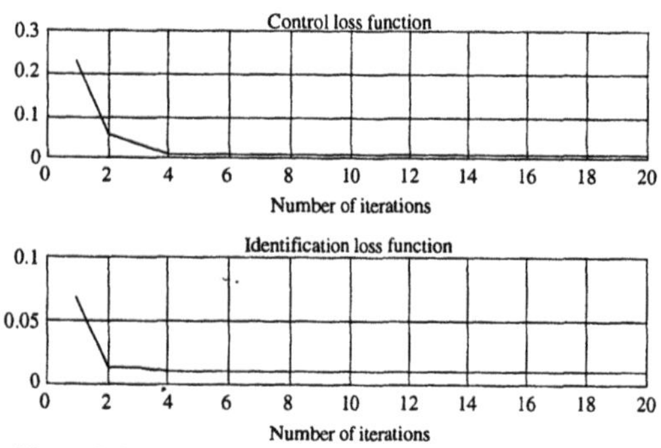

Fig. 10 Loss functions in an iterative controller refinement procedure

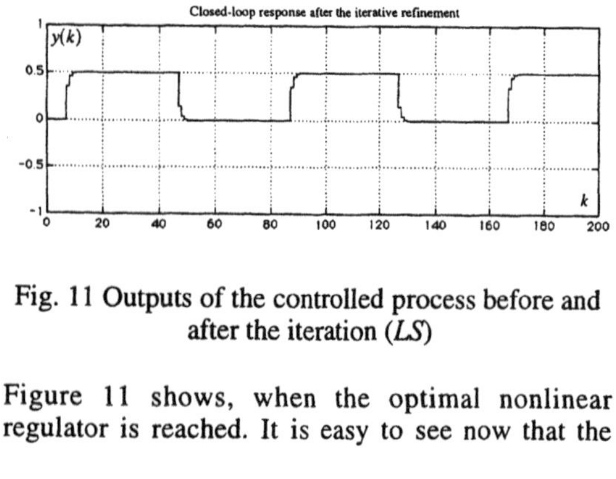

Fig. 11 Outputs of the controlled process before and after the iteration (*LS*)

Figure 11 shows, when the optimal nonlinear regulator is reached. It is easy to see now that the optimal linearized characters from the transient response correspond to the original linear case.

(Please note that in case of the Wiener model the applied LS method, in one step of the off-line iterative nonlinear controller tuning method, itself is an iterative algorithm, because of the strongly nonlinear character of the process model. No explicit one-step solution is known for this identification task (Haber and Keviczky, 1999).)

8. CONCLUSIONS

The paper extends the generic two-degree of freedom (*G2DF*) controller scheme introduced by the authors for linear dynamic processes for factorable nonlinear plants. The introduced new scheme (*NG2DF*) is further analyzed for Hammerstein and Wiener type block-oriented nonlinear models. The design formulas for the regulator and the serial feedforward compensators are also given. It is shown that this generalization is quite straightforward for both *IS* and *IU* Hammerstein models, however the generalization can be performed only for *IS* Wiener models. An adaptive iterative regulator refinement scheme is also presented to identify the optimal nonlinear controller. A simple simulation example is shown to present the efficiency of the new scheme and the developed design formulas.

REFERENCES

Haber, R. and L. Keviczky (1999). *Nonlinear System Identification — Input-Output Modeling Approach.* Kluwer Academic Publishers.

Keviczky, L. (1995). Combined identification and control: another way. (Invited plenary paper.) *5th IFAC Symp. on Adaptive Control and Signal Processing, ACASP'95*, 13-30, Budapest, H.

Keviczky, L. and Cs. Bányász (1999a). Dialectics of identification and control, *Report ESF COSY Course*, pp. 107, Valencia, ES.

Keviczky, L. and Cs. Bányász (1999b). Optimality of two-degree of freedom controllers in $\mathcal{H}_2$- and $\mathcal{H}_\infty$-norm space, their robustness and minimal sensitivity. *14th IFAC World Congress*, **F**, 331-336, Beijing, PRC.

Keviczky, L. and Cs. Bányász (2000). Closed-loop parametrization schemes: Identification of the K-B and d-Y parameters. *3rd IFAC Symposium on Robust Control Design, ROCOND'2000*, CD printing, Prague, CZ.

Maciejowski, J.M. (1989). *Multivariable Feedback Design*, Addison Wesley.

Sastry, S. (1999). *Nonlinear Systems - Analysis, Stability and Control.* Springer.

This work was supported by the Hungarian NSF (OTKA).

Copyright © IFAC Adaptation and Learning in Control and
Signal Processing, Cernobbio-Como, Italy, 2001

REMARKS ON THE SAMPLE COMPLEXITY FOR LINEAR CONTROL SYSTEMS IDENTIFICATION

Pirkko Kuusela * Daniel Ocone ** Eduardo D. Sontag **

* *Networking Laboratory, Helsinki University of Technology, P.O.Box 3000, FIN-02015 HUT, Finland*
** *Dept. of Mathematics, Rutgers, The State University of New Jersey, 110 Frelinghuysen Rd, NJ 08854-8019, USA*

Abstract: The identification of continuous-time control systems is considered as a learning problem. As the general problem is too rich to be learnable, input signals are assumed to have only a finite number k of frequency components. When the system being identified has a dimension at most n, the required sample size scales polynomially with n and logarithmically with k. *Copyright*© 2001 IFAC

Keywords: linear systems, identification, learning, sample sizes, combinatorial mathematics

1. INTRODUCTION

Systems identification, a central problem in control theory, can be seen as a learning problem, in which one aims at learning an unknown functions based on observations. Lately, techniques from Computational Learning Theory ("CLT", see (Vidyasagar, 1997) and papers in (Bartlett and Vidyasagar, 1998)) have been applied to provide new insight into this problem. Discrete-time linear systems on finite-window data were studied in (DasGupta and Sontag, 1996*a*; DasGupta and Sontag, 1996*b*). For other CLT related results see also (Ljung, 1996),(Johansen and Weyer, 1998), (Fiechter, 1997),(Koiran and Sontag, 1998), and (Sontag, 1998).

For continuous-time control systems the identification problem is complicated by the fact that, even for finite-length inputs, learnability is impossible when formulated in the CLT framework; this was proved in (Zador and Pearlmutter, 1996), and, alternatively, can be seen by applying the discrete-time results (through sampling) from (DasGupta and Sontag, 1996*a*; DasGupta and Sontag, 1996*b*).

Thus, in this paper, we suppose that all inputs to be used, in the learning as well as validation stages, belong to the linear span of a fixed number k of sinusoidal basic functions. This band-limiting assumption allows us to obtain a precise result: the sample size needed for identification scales polynomially on an upper bound on the dimension of the systems being identified, and logarithmically with k. This provides a tight analogy to the earlier discrete-time results, in which k appeared as the length of the discrete-time window employed.

This paper contains only a few main sample size results, with emphasis on learning using sign-observations. Far more results together with examples and detailed proofs can be found in (Kuusela *et al.*, 2000).

2. DEFINITIONS

We review our learning setting and, as main results, we state bounds for the number of samples needed in order to identify systems.

Assume that an input space X and a collection $\mathcal{C}$ of mappings $X \to \{0,1\}$ is given. Let W be the set of all sequences $w = (u_1, \phi(u_1)), \ldots, (u_s, \phi(u_s))$ over all $s \geq 1$, $(u_1, \ldots, u_s) \in X^s$ and let $\phi \in \mathcal{C}$.

An *identifier* is a mapping $\psi : W \to \mathcal{C}$ and the value of ψ on a sequence w above is denoted by ψ_w. The "error" of ψ is the probability that ψ will misclassify a future sample. More formally, the *error of ψ with respect to a probability measure P on X, a $\phi \in \mathcal{C}$, and a sequence $(u_1, \ldots, u_s) \in X^s$*, is $\mathrm{Err}(P, \phi, u_1, \ldots, u_s) := P\{u \in X \;;\; \psi_w(u) \neq \phi(u)\}$. The class $\mathcal{C}$ is said to be *learnable* if there is some identifier ψ with the following property: For each *accuracy parameter* $\epsilon > 0$ and *confidence parameter* $\delta > 0$ there is some s so that, for every probability P and every $\phi \in \mathcal{C}$, $P^s\{(u_1, \ldots, u_s) \in X^s \;;\; \mathrm{Err}(P, \phi, u_1, \ldots, u_s) > \epsilon\} < \delta$, where P^s is the s-fold product of P.

In the learnable case, the *sample complexity* $s(\epsilon, \delta)$ provides the smallest s achieving this bound for any positive ϵ and δ. It can be proved that learnability of a class $\mathcal{C}$ is equivalent to the finiteness of the *Vapnik-Chervonenkis dimension* $\mathrm{VC}(\mathcal{C})$ (Vapnik and Chervonenkis, 1971), which is a combinatorial quantity describing the richness of $\mathcal{C}$. Moreover, for consistent learning algorithms (i.e. ones that classify the observed samples correctly), $s(\epsilon, \delta)$ is bounded by

$$\max\left\{\frac{8\mathrm{VC}(\mathcal{C})}{\epsilon}\log_2\left(\frac{8e}{\epsilon}\right), \frac{4}{\epsilon}\log_2\left(\frac{2}{\delta}\right)\right\}.$$

In addition, a similar lower bound for $s(\epsilon, \delta)$ exists.

Definition 1. (VC-dimension) *We say that a set $S = \{x_1, \ldots, x_n\} \subseteq X$ is shattered by $\mathcal{C}$ if, for every subset $B \subseteq S$, there exists a set $A \in \mathcal{C}$ with $S \cap A = B$. The Vapnik-Chervonenkis dimension of $\mathcal{C}$, denoted VC($\mathcal{C}$), equals the largest integer n such that there exists a set of cardinality n that is shattered by $\mathcal{C}$, or infinity otherwise.*

Fat-shattering dimension is related to the sample complexity for learning real-valued functions:

Definition 2. (Fat-shattering dimension) *Let $\mathcal{F}$ be a set of real-valued functions. A set of points X is γ-shattered by $\mathcal{F}$ if there are real numbers r_x indexed by $x \in X$ such that for all binary vectors b_x indexed by X, there is a function $f_b \in \mathcal{F}$ satisfying*

$$f_b(x) \geq r_x + \gamma, \quad \text{if } b_x = 1, \text{ and}$$
$$f_b(x) \leq r_x - \gamma, \quad \text{otherwise.}$$

The fat-shattering dimension $\mathrm{fat}_\gamma(\mathcal{F})$ is a function from positive real numbers to integers which maps a value γ to the size of the largest fat-shattered set, if it is finite, or infinity otherwise.

2.1 *Linear Systems*

In the context of learning we discuss continuous-time linear control systems:

$$\dot{\mathbf{x}} = \mathbf{A}\mathbf{x} + \mathbf{B}\mathbf{u}, \qquad \mathbf{x}(0) = \mathbf{x}^0, \qquad \mathbf{y} = \mathbf{C}\mathbf{x}, \quad (1)$$

where $\mathbf{A}$, $\mathbf{B}$, and $\mathbf{C}$ are $n \times n$, $n \times m$, and $p \times n$ real matrices, and the time interval is [0,1]. We study sign-observations:

$$\mathrm{sign}\, \mathbf{y}(1) = (\mathrm{sign}\, y_1(1), \ldots, \mathrm{sign}\, y_p(1))^T,$$

where sign $z = 0$, if $z \leq 0$, sign $z = 1$, if $z > 0$ and ^{T} stands for the transpose. For scalar observations this is a classification problem and hence related to the VC-dimension. When $p > 1$, a generalization of the VC-dimension or a loss function is needed.

As was said earlier, the general identification problem for continuous-time control systems is not learnable, as the associated VC-dimension is unbounded. Therefore we restrict the class of admissible controls to $\mathbf{u} = (u_1, \ldots, u_m)$ such that

$$\mathbf{u} = \mathbf{G}\omega,$$

where $\mathbf{G}$ is a $m \times k$ matrix that parameterizes the control. The set of basis input functions $\Omega = \{\omega_1, \ldots, \omega_k\}$ is fixed. The bounds for the VC-dimension or other complexity dimensions will depend on the properties of the set Ω.

For scalar inputs (i.e., $m = 1$) the VC-dimension associated to the mapping from inputs $\mathbf{G}$ to scalar sign-observations is bounded by k, which in fact can be very large in applications. This bound is tight (see Section 6). However, by considering band-limited controls a better bound can be achieved and we consider the following set of basis input functions

$$\Omega = \Big\{\omega_1, \ldots, \omega_k \;;\; \omega_1, \ldots, \omega_k \text{ lin. independent,}$$
$$\omega_j = t^{\ell_j} e^{\alpha_j t} \sin(\beta_j t) \text{ or } \omega_j = t^{\ell_j} e^{\alpha_j t} \cos(\beta_j t)$$
$$\text{with } \ell_j \in \mathbb{N}, \alpha_j, \beta_j \in \mathbb{R}, j = 1, \ldots, k\Big\}, \quad (2)$$

and let

$$\ell_{\max} = \max\{\ell_1, \ldots, \ell_k\}. \quad (3)$$

Definition 3. *Order the set of basis input functions Ω and denote $\boldsymbol{\omega} = (\omega_1, \ldots, \omega_k)^T$. Let*

$$X_\Omega = \{\mathbf{G}\boldsymbol{\omega} : [0, 1] \to \mathbb{R}^m \;;\; \mathbf{G} \in \mathbb{R}^{mk}\},$$

and for each linear system $\Sigma = (\mathbf{A}, \mathbf{B}, \mathbf{C}, \mathbf{x}^0)$ of dimension n define the mapping $\Phi_\Sigma : X_\Omega \to \mathbb{R}^p$ by $\Phi_\Sigma(\mathbf{G}\boldsymbol{\omega}) = \mathbf{y}(1)$, where $\mathbf{y}(1)$ is the solution of Σ with control $\mathbf{u} = \mathbf{G}\boldsymbol{\omega}$. Similarly we define the mapping for sign-observations,

$$S_\Sigma : X_\Omega \to \{0, 1\}^p \quad \mathbf{G}\boldsymbol{\omega} \mapsto \mathrm{sign}(\Phi_\Sigma(\mathbf{G}\boldsymbol{\omega})).$$

The class of above mappings is the sign system concept class, $\mathcal{C}_{m,p} = \{S_\Sigma \;;\; \Sigma \text{ linear system of dimension } n\}$.

3. MAIN RESULTS

In this paper we state results only with scalar observations, see (Kuusela *et al.*, 2000) for case $p > 1$. We formulate two theorems in terms of sample complexities as main results. Those statements are immediate corollaries of learning complexity bounds, and hence no separate proofs are given. Also, we sketch in this paper only the VC-dimension proofs. In the following theorems SC stands for the sample complexity.

Theorem 1. (SC for concept learning) *For sign systems concept class $\mathcal{C}_{m,1}$ with scalar observations, the sample complexity for consistent identifiers can be bounded by*

$$\max\left\{\frac{8\,VC(\mathcal{C}_{m,1})}{\epsilon}\log_2\left(\frac{8e}{\epsilon}\right), \frac{4}{\epsilon}\log_2\left(\frac{2}{\delta}\right)\right\},$$

where

$$VC(\mathcal{C}_{m,1}) \leq 2(2mn^2+4n+1)\times$$
$$\log_2\left[8e\left(8mn^2k(n+\ell_{\max})+1\right)\left(2nk+2(1+2k)^n\right)\right]$$

and $\ell_{\max}$ is given by (3).

In terms of n (the dimension of the state space) and k (the band-width) the upper bound for the VC-dimension is of the form $O(n^3\log_2(nk))$. We also state a corresponding VC-dimension lower bound. These bounds provide a $O(\log(k))$ estimate for the number of samples needed in learning, thus improving considerably the k bound obtained by elementary analysis.

The VC-dimension lower bound is for systems with scalar controls and it is in terms of n and k:

Theorem 2. (VC-dimension lower bound)

$$VC(\mathcal{C}_{1,1}) \geq \max\left\{m'\left\lfloor\log_2\left\lfloor\frac{k}{m'}\right\rfloor\right\rfloor, m'\right\}$$

where $m' = \min\{n,k\}$.

In above, $\lfloor x\rfloor$ stands for the integer part of x.

3.1 *Additional results*

Agnostic learning (see (Bartlett and Long, 1998)) formulates the learning paradigm for real-valued functions. We state here a corresponding sample complexity result; formal definitions and proofs can be found in (Kuusela *et al.*, 2000).

For the system (1) we can write $\mathbf{C}e^{\mathbf{A}t}\mathbf{B} = [\gamma_1,\ldots,\gamma_m]$, where each γ_i is a linear combination of n functions $\xi_1,\ldots,\xi_n$. Each ξ_i is of the form $t^\ell e^{at}\sin(bt)$ or $t^\ell e^{at}\cos(bt)$ with $\ell \in \{0,\ldots,n-1\}$ and $a+ib$ an eigenvalue of $\mathbf{A}$. Assume that $\mathbf{A}$ has a fixed Jordan block structure and let a_k+ib_k be an eigenvalue of $\mathbf{A}$. We take $\alpha_{11},\ldots,\alpha_{nm},a_1,b_1,\ldots,a_r,b_r$ to be the system parameters, where $\gamma_i(t) = \sum_{j=1}^n \alpha_{ij}\xi_j(t)$ for $i=1,\ldots,m$ and $a_1,b_1,\ldots,a_r,b_r$ are n eigenvalue parameters.

Let $U \subset \mathbb{R}^{mk}$ be a bounded set. Define a mapping $F: \lambda\times U \to \mathbb{R}$ such that $F(\lambda, G) = y(\tau)$, where $\tau \geq 1$ and $y(\tau)$ is a solution of (1) with system parameters $\boldsymbol{\lambda}$, and initial condition $\mathbf{x}(0) = \mathbf{0}$.

Definition 4. *Assume that the system* (1) *with $\mathbf{x}(0) = \mathbf{0}$ can be parameterized by $\boldsymbol{\lambda} \in \mathbb{R}^{n(m+1)}$ as above and $\|\boldsymbol{\lambda}\|_\infty = \max_{1\leq i\leq n(m+1)}\lambda_i < 1$. Let $F(\boldsymbol{\lambda},\mathbf{u}) = \mathbf{y}(\tau)$ be the solution of* (1) *with parameters $\boldsymbol{\lambda}$ and control $\mathbf{u} = \mathbf{G}\boldsymbol{\omega} \in U = \{\mathbf{u} = (u_1,\ldots,u_m)\ ;\ \int_0^\tau u_i(t)dt \leq M, i = 1,\ldots,m\}$. Denote $B_\infty^k(c) := \{\mathbf{x}\in\mathbb{R}^k\ ;\ \|\mathbf{x}\|_\infty < c\}$ and define*

$$\mathcal{F}_B = \{F(\boldsymbol{\lambda},\cdot)\colon U\to\mathbb{R}\ ;\ \boldsymbol{\lambda}\in B_\infty^{n(m+1)}(1)\}.$$

Theorem 3. (SC for agnostic learning) *Let $\kappa > 0$, then the class $\mathcal{F}_B$, given in Definition 4, is (properly) agnostically learnable from*

$$O\left(\frac{1}{\epsilon^2}\left(\text{fat}_{(1/4-\kappa)\epsilon}(\mathcal{F}_B)\log^2\frac{1}{\epsilon}+\log\frac{1}{\delta}\right)\right)$$

samples, where $\text{fat}_{(1/4-\kappa)\epsilon}(\mathcal{F}_B)$ is bounded by

$$\min\begin{cases}(m+1)n\log_2\left\lfloor\dfrac{n^2m\tau^n e^\tau kM}{(1/4-\kappa)\epsilon}\right\rfloor,\\ 2(m+4)n\log_2\Big(8e\big(nmk4(n+\ell_{\max})+1\big)\times\\ \qquad\qquad\big(2nk+2(2k+1)^n\big)\Big),\end{cases}$$

together with $\ell_{\max}$ given by (3), *and M a constant satisfying*

$$\int_0^\tau |u_i(\tau-t)|dt \leq kM$$

for all $i = 1,\ldots,m$.

4. TECHNIQUES FOR PROVING VC-DIMENSION RESULTS

Our main results are based on a fact that the basis input functions satisfy a certain rationality condition. In this section we first formulate this rationality condition and then we summarize existing results that are used in proving upper and lower bounds for the complexity dimensions.

We illustrate below only the idea of the rationality condition. Technical details are in (Kuusela *et al.*, 2000).

Definition 5. (Rationality condition (RAT)) *Entries of $e^{\mathbf{A}t}$ are functions of the form $t^s e^{at}\cos(bt)$ and $t^s e^{at}\sin(bt)$, where $a+ib$ is an eigenvalue of $\mathbf{A}$. Solving* (1) *involves convolutions of $e^{\mathbf{A}t}$ and the basis input functions ω_j, and we require those to be rational functions of four variables – a, b, $e^a\cos b$, and $e^a\sin b$. Moreover, those rational functions are defined in pieces, so that the denominator does not vanish in the domain of the appropriate piece. Given values of the variables, the correct piece of the rational expression can be obtained by evaluating a boolean formula consisting of polynomial expressions.*

We denote by $d_{\max}$ the maximum degree of any polynomial appearing in the rationality condition.

Review of VC-Dimension Techniques Assume that a function $F : \Lambda \times X \to \{0,1\}$ is given. This induces two function classes

$$\mathcal{F} := \{F(\lambda,\cdot) : X \to \{0,1\}\ ;\ \lambda \in \Lambda\}$$

and

$$\mathcal{F}^* := \{F(\cdot,x) : \Lambda \to \{0,1\}\ ;\ x \in X\}.$$

The complexity dimension $\mathrm{VC}(\mathcal{F}^*)$ is called the *dual VC-dimension* of $\mathcal{F}$, which is related to $\mathrm{VC}(\mathcal{F})$ by $\mathrm{VC}(\mathcal{F}) \geq \lfloor \log_2 \mathrm{VC}(\mathcal{F}^*) \rfloor$.

A sharper estimate was obtained in (DasGupta and Sontag, 1996*a*) for those dichotomies that are defined on "rectangular" $L = L_1 \times \cdots \times L_n$ subsets of Λ. The bound is based on the notion of axis shattering:

Theorem 4. (Axis shattering bound (DasGupta and Sontag, 1996*a*)) *If $L = L_1 \times \cdots \times L_n \subset \Lambda$ can be axis shattered and each L_i has cardinality $r_i > 0$, then $VC(\mathcal{F}) \geq \lfloor \log_2(r_1) \rfloor + \cdots + \lfloor \log_2(r_n) \rfloor$.*

Axis shattering is used in proving VC-dimension lower bounds whereas the upper bounds utilize the following well known result:

Theorem 5. (Goldberg-Jerrum bound (Goldberg and Jerrum, 1995)) *Consider a function $F : \Lambda \times X \to \{0,1\}$ and its associated concept class $\mathcal{F} := \{F(\lambda,\cdot) : X \to \{0,1\}\ ;\ \lambda \in \Lambda\}$, where $\Lambda = \mathbb{R}^\ell$ and $X = \mathbb{R}^k$. Let F be defined in terms of a Boolean formula involving at most s polynomial equalities and inequalities in $\ell + k$ variables, each polynomial being of degree at most d in λ for all $\mathbf{x} \in \mathbb{R}^k$. Then, $VC(\mathcal{F}) \leq 2\ell \log_2(8eds)$.*

5. PROOFS OF THE VC-DIMENSION BOUNDS

5.1 *An Upper Bound for the VC-Dimension with Scalar Observations*

We begin by stating a few technical statements without proofs. After that, as a lemma, we develop an upper bound for the VC-dimension induced by the control system (1) with its initial state fixed to be zero. Theorem 1 with an arbitrary initial condition is then a simple modification of the argument.

Lemma 6. *Each $\omega_j \in \Omega$ given by (2) satisfies the rationality condition. Further, the maximum degree of polynomials in (RAT) is at most $4(n + \ell_{\max})$, where $\ell_{\max}$ is given by (3).*

The next proposition indicates how control systems are parameterized and later the concept or function classes associated to control systems are obtained by varying the parameter vector.

Proposition 7. *Assume that each ω_i, $i = 1,\ldots,k$ satisfies the rationality condition (RAT) and let $\Lambda = \mathbb{R}^{2pn^2m} \times \mathbb{R}^{4n} \times \mathbb{R}^p$. Then there exists a mapping $H : \Lambda \times \mathbb{R}^{mk} \to \mathbb{R}^p$ (depending on $\omega = (\omega_1,\ldots,\omega_k)^T$) such that for each $\Sigma = (\mathbf{A},\mathbf{B},\mathbf{C},\mathbf{x}^0)$ there exists a $\boldsymbol{\lambda} \in \Lambda$ satisfying $\Phi_\Sigma(\mathbf{G}\omega) = H(\boldsymbol{\lambda},\mathbf{G})$ for all $\mathbf{G} \in \mathbb{R}^{mk}$.*

Next we study the VC-dimension of the sign system concept class, $\mathcal{C}_{m,1}$, where each control parameterized by $\mathbf{G}$ gives rise to $\operatorname{sign} y(1)$.

Lemma 8. *Sign system concept class $\mathcal{C}_{m,1}$ with initial condition $\mathbf{x}(0) = \mathbf{0}$ satisfies*

$$VC(\mathcal{C}_{m,1}) \leq 2\left(2mn^2 + 4n\right) \times$$
$$\log_2\left[8e\left(8mn^2k(n+\ell_{\max})+1\right)\left(2nk+2(1+2k)^n\right)\right],$$

where $\ell_{\max}$ is given by (3).

Proof. By Proposition 7 $y(1) = H(\boldsymbol{\lambda},\mathbf{G})$, where $\boldsymbol{\lambda} \in \mathbb{R}^{2mn^2} \times \mathbb{R}^{4n}$ are considered as parameters. In fact, $y(1) = \frac{P}{Q}$, where P and Q denote piecewise polynomial functions. As in the statement of Goldberg-Jerrum bounds we have a function $F : \Lambda \times \mathbb{R}^{mk} \to \{0,1\}$ defined by $F(\boldsymbol{\lambda},\mathbf{G}) = \operatorname{sign} H(\boldsymbol{\lambda},\mathbf{G})$. The concept class associated to the system identification problem is $\mathcal{F} := \{F(\boldsymbol{\lambda},\cdot) : \mathbb{R}^{mk} \to \{0,1\}\ ;\ \boldsymbol{\lambda} \in \Lambda\}$, where $\Lambda = \mathbb{R}^{2mn^2+4n}$. Before applying the Goldberg-Jerrum bound we need to determine the possible degrees of P and Q with respect to the parameters.

Using the parameterization we can write

$$y(1) = \sum_{i=1}^{m}\sum_{r=1}^{n}\sum_{\ell=1}^{2n} \lambda_{ir\ell} \sum_{j=1}^{k} g_{ij} \frac{\hat{P}_{\ell j}}{\hat{Q}_{\ell j}},$$

where

$$\max_{\substack{i\leq \ell \leq n \\ 1\leq j\leq k}} \left\{\deg(\hat{P}_{\ell j}), \deg(\hat{Q}_{\ell j})\right\} \leq d_{\max}$$

due to (RAT). (In above, arguments of $\hat{P}_{\ell j}$ and $\hat{Q}_{\ell j}$ contain additional $4n$ parameters.) Then

$$\frac{\widetilde{P}_{i\ell}}{\widetilde{Q}_{i\ell}} = \sum\nolimits_{j=1}^{k} g_{ij} \frac{\hat{P}_{\ell j}}{\hat{Q}_{\ell j}}$$

so $\deg(\widetilde{Q}_{i\ell}) \leq kd_{\max}$ and $\deg(\widetilde{P}_{i\ell}) \leq kd_{\max}$. Note here that we are calculating the degree with respect to the system parameters, and the inputs g_{ij} do not contribute. By continuing in a similar fashion we can conclude that $\deg(Q) \leq m2n^2kd_{\max}$ and $\deg(P) \leq m2n^2kd_{\max} + 1$.

By carrying out the summations we get $y(1) = P/Q$, where Q consists of powers of polynomials f_{ij1}, f_{ij2} with

$$f_{ij1} = (x_{i1} + \alpha_j)^2 + (x_{i2} + \beta_j)^2,$$
$$f_{ij2} = (x_{i1} + \alpha_j)^2 + (x_{i2} - \beta_j)^2,$$

where α_j, β_j are fixed parameters of the basis input function ω_j, and $i = 1,\ldots,n$, $j = 1,\ldots,k$.

Our final step before applying the Goldberg-Jerrum bound is finding out the number of polynomial inequalities s needed in the Boolean formula evaluating the sign of the final state output.

This is done by studying the number of different P/Q expressions without poles.

Note that varying x_{i1} and x_{i2} we can make at most one of the $2k$ polynomials $f_{ij1}, f_{ij2}, j = 1, \ldots, k$ to be zero. The γ zeros among $f_{ij1}, f_{ij2}, i = 1, \ldots, n$ and $j = 1, \ldots, k$ can be obtained in $(2k)^\gamma \binom{n}{\gamma}$ ways and the number of possible sign assignments is obtained by summing over γ yielding

$$\sum_{\gamma=0}^{n} (2k)^\gamma \binom{n}{\gamma} = (1+2k)^n.$$

Thus the number of P/Q expressions without poles is $(1+2k)^n$, which gives rise to $2(1+2k)^n$ polynomials.

Note that in order to write $\operatorname{sign} y(1)$ as a Boolean formula evaluating polynomial inequalities and equalities it also has to include the $2nk$ polynomials $f_{ij1}, f_{ij2}, i = 1, \ldots, n, j = 1, \ldots, k$. Values of these polynomials determine which P/Q expression is the valid one to determine $\operatorname{sign} y(1)$. The Boolean formula for $\operatorname{sign} y(1)$ can be given as a truth table involving polynomial inequalities of $2nk$ f_{ij1}, f_{ij2} expressions and $2(1+2k)^n$ different P and Q expressions.

Using Lemma 6 for bound on $d_{\max}$, we apply the Goldberg-Jerrum bound with $s = 2nk+2(2k+1)^n$, $d = m2n^2k4(n+\ell_{\max})+1$ and $\ell = 2mn^2+4n$. □

Proof of Theorem 1. By using the previous notation $y = \mathbf{C}e^{\mathbf{A}}\mathbf{x}^0 + \mathbf{C}\int_0^1 e^{\mathbf{A}(1-t)}\mathbf{B}\mathbf{u}(t)dt$. Let $\tilde{x} = \mathbf{C}e^{\mathbf{A}}\mathbf{x}^0$. Then $y = \tilde{x} + P/Q = (\tilde{x}Q + P)/Q = \tilde{P}/Q$. This has $2mn^2 + 4n + 1$ parameters and $\deg(\tilde{P}) \leq m2n^2kd_{\max} + 1$. □

5.2 *Lower Bounds for the VC-Dimension*

The lower bounds for the VC-dimension are developed for a single-input single-output system with initial state zero. The control is

$$u = \sum_{j=1}^{k} g_j\omega_j : [0,1] \to \mathbb{R}.$$

We derive lower bounds by fixing the structure of $\mathbf{A}$, $\mathbf{B}$, and $\mathbf{C}$, and using the dual VC-dimension and axis shattering. A lower bound for a restricted class of control systems provides a lower bound for more general classes. Lemmata given in this section together prove Theorem 2. Note that to prove the lower bounds actually only linear independence and continuity of the input basis functions is needed.

The next proposition is a standard fact utilized in the following proofs.

Proposition 9. *Let $\omega_j : [0,1] \to \mathbb{R}$, $j = 1, \ldots, k$ be continuous and linearly independent. Then the functions*

$$h_j(\lambda) = \int_0^1 e^{\lambda t}\omega_j(t)dt, \quad j = 1, \ldots, k$$

are linearly independent.

Lemma 10. (Lower bound 1) *Sign system concept class $\mathcal{C}_{1,1}$ (scalar inputs and scalar outputs) satisfies*

$$VC(\mathcal{C}_{1,1}) \geq m' \left\lfloor \log_2 \left\lfloor \frac{k}{m'} \right\rfloor \right\rfloor,$$

where $m' = \min\{n, k\}$.

Proof. Let $\omega_j(t)$, $j = 1, \ldots, k$ be continuous and linearly independent. Let $\mathbf{A}$ have n distinct real eigenvalues $-\lambda_1, \ldots, -\lambda_n$, and take $\mathbf{B}$ and $\mathbf{C}$ so that

$$\mathbf{C}e^{\mathbf{A}(1-t)}\mathbf{B} = \sum_{i=1}^{m'} e^{\lambda_i t},$$

where $m' = \min\{n, k\}$. Then the final output of the system is

$$y(1) = \int_0^1 \mathbf{C}e^{\mathbf{A}(1-t)}\mathbf{B} \sum_{j=1}^{k} g_j\omega_j(t)dt = \sum_{i=1}^{m'} \sum_{j=1}^{k} g_j \int_0^1 e^{\lambda_i t}\omega_j(t)dt.$$

Define $h_j(\lambda) = \int_0^1 e^{\lambda t}\omega_j(t)dt$. By Proposition 9 we can find $\lambda_1, \ldots, \lambda_k$ such that the matrix

$$\begin{bmatrix} h_1(\lambda_1) & \cdots & h_k(\lambda_1) \\ \vdots & & \vdots \\ h_1(\lambda_k) & \cdots & h_k(\lambda_k) \end{bmatrix}$$

has rank k.

The control system with sign observations gives the mapping $F : \mathbb{R}^{m'} \times \mathbb{R}^k \to \{0,1\}$ by

$$(\lambda_1, \ldots, \lambda_{m'}, g_1, \ldots, g_k) \mapsto \operatorname{sign}\left[\sum_{i=1}^{m'}\sum_{j=1}^{k} g_j h_j(\lambda_i)\right].$$

The proof is completed by showing that the mapping from parameters $\lambda_1, \ldots, \lambda_{m'}$ to $\{0,1\}$ can be axis shattered (see (Kuusela *et al.*, 2000)). Then it follows that

$$\tilde{\mathcal{F}} = \{F(\lambda_1, \ldots, \lambda_{m'}, \cdot) : \mathbb{R}^k \to \{0,1\} \; ; \; \boldsymbol{\lambda} \in \mathbb{R}^{m'}\}$$

satisfies

$$\mathrm{VC}(\tilde{\mathcal{F}}) \geq m' \left\lfloor \log_2 \left\lfloor \frac{k}{m'} \right\rfloor \right\rfloor.$$

Observing that $\mathrm{VC}(\mathcal{C}_{1,1}) \geq \mathrm{VC}(\tilde{\mathcal{F}})$ gives the result. □

Lemma 11. (Lower bound 2) *If $k \leq n$ then*

$$VC(\mathcal{C}_{1,1}) \geq k.$$

Proof. Assume that $k \leq n$ and let $\mathbf{A}$ have n real eigenvalues $\lambda_1, \ldots, \lambda_n$. Next we take $\mathbf{B}$ and $\mathbf{C}$ so that $\mathbf{C}e^{\mathbf{A}(1-t)}\mathbf{B} = \sum_{i=1}^{n} e^{\lambda_i t}\beta_i$, where $(\beta_1, \ldots, \beta_n, \lambda_1, \ldots, \lambda_n)$ are considered as system parameters.

We study the mapping

$$(\beta_1, \ldots, \beta_n, \lambda_1, \ldots, \lambda_n, g_1, \ldots, g_k) \mapsto$$

$$\text{sign}\left[\sum_{i=1}^{n}\sum_{j=1}^{k} g_j h_j(\lambda_i)\beta_i\right] =$$

$$\text{sign}\left[\sum_{j=1}^{k} g_j \underbrace{\sum_{i=1}^{n} h_j(\lambda_i)\beta_i}_{\gamma_j}\right] = \text{sign}\left[\sum_{j=1}^{k} g_j\gamma_j\right].$$

Given $(\gamma_1, \ldots, \gamma_k)$ and using Proposition 9 we can find $\lambda_1, \ldots, \lambda_n, \beta_1, \ldots, \beta_n$ such that $\sum_{i=1}^{n} h_j(\lambda_i)\beta_i = \gamma_j$, $j = 1, \ldots, k$. But $(\gamma_1, \ldots, \gamma_k)$ can be viewed as a normal vector for a hyperplane through the origin in $\mathbb{R}^k$ and the concept class associated to the mapping $(g_1, \ldots, g_k) \mapsto \text{sign}\left[\sum_{j=1}^{k} g_j\gamma_j\right]$ as $(\gamma_1, \ldots, \gamma_k)$ varies has VC-dimension k. Hence $\text{VC}(\mathcal{C}_{1,1}) \geq k$. □

Lemma 12. (Lower bound 3) *If $n \leq k$ then*

$$VC(\mathcal{C}_{1,1}) \geq n.$$

Proof. The argument is similar to the previous one. □

6. A CLASS OF SYSTEMS WITH VC-DIMENSION *K*

For the control system (1) with scalar control $u(t) = \sum_{i=1}^{k} g_i\omega_i(t)$ and unrestricted $\omega_1, \ldots, \omega_k$, the standard half-space argument gives an upper bound k. This bound is tight.

Consider a control system

$$\dot{x}_1 = x_2$$
$$\dot{x}_2 = -\lambda^2 x_1 + u \qquad (4)$$
$$y = -x_1.$$

For time interval [0,1] and initial condition $(x_1, x_2) = (0, 0)$, the output is given by $y(1) = \int_0^1 \sin(\lambda t)u(1-t)dt$.

Lemma 13. *Controls $\{\omega_1, \ldots, \omega_k\}$ such that $\omega_i(1-t) = 1_{[2^{-i}, 2^{-i}+2^{\alpha}]}$, where $\alpha = -2(k+1)$, are shattered by the control system (4) with sign-observations.*

Proof. See (Kuusela *et al.*, 2000). □

Acknowledgments This research has been financed in part by a grant from US Air Force (grant F47620-97-1-0159).

7. REFERENCES

Bartlett, P. and P. Long (1998). Prediction, learning, uniform congergence and scale-sensitive dimensions. *Journal of Computer and System Sciences* **56**(2), 174–190.

Bartlett, P. and Vidyasagar, M., Eds.) (1998). *Learning theory special issue of Systems and Control Letters.* Vol. 34.

DasGupta, B. and E. D. Sontag (1996*a*). Sample complexity for learning recurrent perceptron mappings. *IEEE Trans. Inform. Theory* **42**(5), 1479–1487.

DasGupta, B. and E. D. Sontag (1996*b*). Sample complexity for learning recurrent perceptron mappings. In: *Advances in Neural Information Prosessing Systems (NIPS'95).* MIT Press, Cambridge, MA. pp. 204–210.

Fiechter, C.-N. (1997). PAC adaptive control of linear systems. In: *Proceedings of Tenth Annual Conference on Computational Learning Theory.* ACM Press. pp. 72–80.

Goldberg, P. and M. Jerrum (1995). Bounding the Vapnik-Chervonenkis dimension of concept classes parametrized by real numbers. *Machine Learning* **18**, 131–148.

Johansen, T. and E. Weyer (1998). On convergence proofs in system identification – a general principle using ideas from learning theory. *Systems and Contol Letters* **34**, 85–92.

Koiran, P. and E. D. Sontag (1998). Vapnik-Chervonenkis dimension of recurrent neural networks. *Discrete Applied Mathematics* **86**(1), 63–80.

Kuusela, P., D. Ocone and E.D Sontag (2000). Learning complexity dimensions for a continuous-time control system. Extended version, 33 pages, available from arXiv.org e-Print archive, `http://xxx.lanl.gov/abs/math.OC/0012163`.

Ljung, L. (1996). PAC-learning and symptotic system identification theory. In: *Proceedings of 35th IEEE Conf. on Decision and Control, Kobe, Japan.* IEEE. pp. 2303–2307.

Sontag, E.D. (1998). A learning result for continuous-time recurrent neural networks. *Systems and Control Letters* **34**, 151–158.

Vapnik, V. and A. Chervonenkis (1971). On the uniform convergence of relative frequencies of events to their probabilities. *Theory Probab. Appl.* **16**, 264–280.

Vidyasagar, M. (1997). *A theory of learning and generalization; with applications to neural networks and control systems.* Springer-Verlag.

Zador, A. and B. Pearlmutter (1996). VC dimension of an integrate-and-fire neuron model. *Neural Computation* **8**(3), 611–624.

THE EM ALGORITHM FOR MULTIVARIABLE DYNAMIC SYSTEM ESTIMATION

Brett Ninness *,[1] **Stuart Gibson** *

* *Dept. of Elec. & Comp. Eng, Uni. Newcastle, Australia.*
*email:*brett@ee.newcastle.edu.au, *FAX: +61 49 21 69 93*

Abstract: This paper presents novel algorithms for the estimation of dynamic systems. These new methods offer several advantages. They are parameterisation free, numerically robust, convergent to statistically optimal estimates, and applicable in a simple fashion to a wide range of multivariable, non-linear and time varying problems. The key tool underlying the new techniques presented here is the 'Expectation-Maximisation' (EM) algorithm.

Keywords: Parameter Estimation, Maximum Likelihood Estimators.

1. INTRODUCTION

The science of System Identification is directed towards finding models for dynamic behaviour on the basis of experimental observations.

Since this problem is so fundamental to manifold areas of engineering, it has been vigorously studied in that context for the last several decades. A seminal development was the Åström and Eykhoff survey paper of 1971 (Åström and Eykhoff 1971) which contained the famous quotation: '*New methods are suggested en masse, and, on the surface, the field appears to look more like a bag of tricks than a unified subject*'.

Since that time, the combined efforts of many researchers have resulted in cohesive approaches to the topic, and there are now substantial texts (Ljung 1999, T.Söderström and P.Stoica 1989, Caines 1988, Hannan and Deistler 1988, Goodwin and Payne 1977) together with sophisticated software packages (Ljung 2000*a*) available. All of this progress has been firmly founded on stochastic principles together with mathematical-statistical roots that are almost a century old (Fisher 1912).

In particular, the so-called Maximum Likelihood principle and its relations, such as prediction error techniques, play a key role. There are several reasons for this. Firstly, there is a very large and sophisticated body of theory supporting these methods (Ljung 1999, Caines 1988, Hannan and Deistler 1988, T.Söderström and P.Stoica 1989). This allows important practical issues such as error analysis and performance tradeoffs to be addressed. Secondly, via this latter theory, it is understood that Maximum Likelihood methods are provably statistically optimal in that they (at least asymptotically) achieve the Cramér–Rao Lower Bound (Ljung 1999, Caines 1988, Hannan and Deistler 1988, T.Söderström and P.Stoica 1989). That is, in some sense they provide the most accurate estimates. Finally, Maximum Likelihood type methods provide a general framework which is applicable to a very wide range of estimation problems (Ljung 1999, Caines 1988, Hannan and Deistler 1988, T.Söderström and P.Stoica 1989).

Balancing this, it should be recognised, that despite these features recommending the Maximum Likelihood approach, it is not a panacea. As a result there are also significant bodies of work directed towards alternative approaches, including non-parametric (Ljung 1999), bounded-error (Norton 1987, Milanese and Vicino 1991), and state space subspace-based estimation methodologies (van Overschee and Moor 1996, Larimore 1990).

Furthermore, despite the theoretical advantages of Maximum Likelihood methods, their practical deployment is not always straightforward. This is largely due to the non-convex optimisation problems that are often implied. Typically, these are solved via a gradient-based search strategy such as a Newton type method or one of its derivatives (Ljung 1999, T.Söderström and P.Stoica 1989, Dennis and Schnabel 1983). The success of such approaches depends on the curvature of the Maximum Likelihood cost being optimised, and this is dependent on the chosen system parameterisation. Selecting this can be difficult, particularly in the multivariable case where the cost contours resulting from natural canonical state-space parameterisations imply poor numerical conditioning during gradient-based search (Deistler 2000, McKelvey 1998).

Indeed, the possibility of avoiding these parameterisation-based difficulties is one of the key reasons for the recent intense interest in the new State Space Subspace based System Identification (4SID) methods (van Overschee and Moor 1996, Larimore 1990).

This paper, motivated by all these issues, presents new methods for gradient-search free computation of Maximum Likelihood dynamic system estimates. These new techniques can employ state space model structures (like 4SID methods), but they do not require explicit parameterisation of the system matrices. Furthermore, the numerical procedures involved here can be implemented very efficiently and reliably via well known methods such as QR decomposition. Finally, while these new algorithms are introduced here for the estimation of linear and time-invariant systems, as will be seen in the companion paper (), they are very simply extended to more complicated scenarios of non-linear, time varying and missing-data estimation problems.

[1] This work was supported by the Australian Research Council and the Centre for Integrated Dynamics and Control (CIDAC).

The central technique employed in this paper is that of the so-called Expectation-Maximisation (EM) algorithm which, for certain classes of Maximum Likelihood estimation problems, has proven to be a robust alternative to gradient-based search for the estimate (Dempster *et al.* 1977).

Despite the successful application of the these EM methods in many other fields such as image processing (Starck *et al.* 1998), speech recognition (Rabiner 1989), and various problems of applied statistics such as epidemiology (Barndorff–Nielsen *et al.* 1999), their potential utility with regard to dynamic system identification problems, particularly those with relevance to control applications, seems to have been largely unappreciated.

2. DYNAMIC SYSTEM ESTIMATION

The estimation problems considered in this paper are ones in which an observed discrete time data record of N samples $Y_N \triangleq \{y_1, y_2, \cdots, y_N\}$ is postulated to depend causally on another data record $U_N \triangleq \{u_1, u_2, \cdots, u_N\}$, and also upon external influences that will be modelled here as realisations of random variables.

A very general way of formulating this scenario (for finite dimensional systems) is via a state space description such as

$$x_t = f(x_t, u_t, w_t) \tag{1}$$

$$y_t = g(x_t, u_t, e_t). \tag{2}$$

Here, $y_t \in \mathbf{R}^p$ and $u_t \in \mathbf{R}^m$ are the vector valued data records just mentioned, while $x_t \in \mathbf{R}^n$ is the system state sequence, and $\{w_t\}$, $\{e_t\}$ are independent and identically distributed vector stochastic process such that $\mathsf{E}\{w_t\} = 0$, $\mathsf{E}\{e_t\} = 0$, $\mathsf{E}\{w_t w_t^T\} \triangleq Q \geq 0$ and $\mathsf{E}\{e_t e_t^T\} \triangleq R \geq 0$ where $\mathsf{E}\{\cdot\}$ denotes expectation over the probability space that w_t and e_t are defined on. Together, the unknown functions $f(\cdot,\cdot,\cdot)$ and $g(\cdot,\cdot,\cdot)$ together with Q and R constitute the model that is to be determined on the basis of the data records Y_N and U_N.

A Maximum Likelihood solution to this estimation problem requires the specification of the probability density functions $p_e(\cdot), p_w(\cdot)$ for the random variables e_t and w_t. Based on this the joint probability

$$p(Y_N, U_N \mid f, g) \tag{3}$$

which is dependent on f and g is calculated, and known as a 'likelihood function'. The Maximum Likelihood estimates of f and g are then defined as those which maximise (3). That is, they are such that they maximise the probability that the observed data is consistent with the estimated model.

Typically, this process is formulated slightly differently by proposing specific forms for f and g that depend on some vector of parameters $\theta \triangleq [\theta_1, \theta_2, \cdots, \theta_\ell]$. In this case, the Maximum Likelihood estimate based on the N observations is defined as

$$\widehat{\theta}_N \triangleq \arg\max_{\theta} \; p(Y_N, U_N \mid \theta). \tag{4}$$

This method of system estimation enjoys a wide acceptance and popularity, in large part due to the following well-known and desirable properties of the scheme. They have been established in a range of works, such as (Hannan and Deistler 1988, Lehmann 1983, Caines 1988, Caines 1988, Ljung 1999) and and they apply under fairly mild regularity assumptions on f, g, w_t, e_t (for simplicity, in what follows, it has been assumed that a true parameter vector $\theta_\circ$ exists that can exactly describe the observed data via the model (1), (2)).

Strong Consistency As $N \to \infty$

$$\widehat{\theta}_N \xrightarrow{a.s.} \theta_\circ = \arg\max_{\theta} \lim_{N\to\infty} \mathsf{E}\{p(Y_N, U_N \mid \theta)\}$$

Asymptotic Normality

$$\sqrt{N} P_N^{-\frac{1}{2}} (\widehat{\theta}_N - \theta_\circ) \xrightarrow{\mathcal{D}} \mathcal{N}(0, I), \quad \text{as } N \to \infty$$

where

$$N \cdot P_N^{-1} = -\frac{\mathrm{d}^2}{\mathrm{d}\theta \mathrm{d}\theta^T}\bigg|_{\theta=\theta_\circ} \mathsf{E}\{\log p(Y_N, U_N N \mid \theta)\} = \mathcal{I}_N,$$

and $\mathcal{I}_N$ is the Fisher Information Matrix associated with $p(Y_N, U_N \mid \theta)$ and evaluated at $\theta = \theta_\circ$.

Asymptotic efficiency The above formulation for P_N, due to its indicated relation to the Fischer Information matrix $\mathcal{I}_N$, allows the Maximum Likelihood estimate to asymptotically achieve the Cramér-Rao lower bound

$$\mathrm{Cov}\{\widehat{\theta}_N\} \geq \mathcal{I}_N^{-1}$$

and hence it is asymptotically (statistically) efficient.

Balancing these attractive features that recommend a Maximum Likelihood approach, there is the significant disadvantage that the equation (4) defining the Maximum Likelihood estimate $\widehat{\theta}_N$ is, in general, a non-convex optimisation problem. As a result, calculation of $\widehat{\theta}_N$ requires some sort of numerical search technique.

Since $p(Y_N, U_N|\theta)$ is typically smooth, any gradient-based search technique such as Steepest-Descent or Newton iteration (Dennis and Schnabel 1983, Nocedal and Wright 1999) may be employed for this purpose, and indeed this is the usual approach (Ljung 1999, Ljung 2000*a*). In this case, an approximation θ_k for $\widehat{\theta}_N$ is repeatedly updated to a new approximation θ_{k+1} according to

$$\theta_{k+1} = \theta_k - \mu_k J_k \left[\frac{\mathrm{d}}{\mathrm{d}\theta} \log p(Y_N, U_N \mid \theta)\bigg|_{\theta=\theta_k} \right] \tag{5}$$

where μ_k is a scalar 'step-length' and J_k is a matrix that may be chosen in various ways (Dennis and Schnabel 1983), but is often related to the Hessian of the cost function, and hence also related to its curvature relative to its parameterisation.

Importantly though, the search strategy (5), by way of requiring a gradient (with respect to a parameterisation θ), in fact also *forces* the use of a parameterisation of the state-space model structure (1), (2). This can lead to important difficulties.

For example, in the case where f and g describe a linear, time invariant (LTI), and multivariable system, it is well known that no surjective mapping exists (hence allowing the description of all possible input-output responses) that is also bijective and therefore ensures that the estimate $\widehat{\theta}_N$ is uniquely defined (Deistler 2000, McKelvey 1998).

Furthermore, in this same LTI case, it is also well known that any simple parameterisation based on canonical forms leads to problems in which Hessian-based choices for J_k become ill-conditioned and lead to slow convergence of the search (5) ().

These difficulties, combined with the fact that subspace-based system identification methods do not require parametrisation of the system matrices (van Overschee and Moor 1996, Larimore 1990), are one of the key features leading to the recent intense interest in them. However, the price paid there is that it is not yet clear what cost function is being optimised by subspace-based estimates. As a result, the theory supporting such approaches is still developing (Deistler *et al.* 1995, Bauer *et al.* 1999).

The contribution of this paper is to show how the theoretical advantages of a Maximum Likelihood approach may be combined with the parameterisation free advantages of a subspace-based method by employing the so-called Expectation Maximisation (EM) algorithm.

3. THE EXPECTATION MAXIMISATION (EM) ALGORITHM

The Expectation Maximisation (EM) algorithm is a technique that, in certain circumstances, can be used to compute Maximum Likelihood estimates without resort to gradient-based search. The method arose in the mathematical statistics community (Dempster *et al.* 1977, Titterington 1984) but has found wide engineering application in areas such as signal processing, pattern recognition and speech recognition (Rabiner 1989, Starck *et al.* 1998).

The key feature of the technique is to exploit the concavity of the log function (together with the fact that the area under a probability density function is one) so as to guarantee iterations of non-decreasing likelihood whilst avoiding the need to calculate derivatives of the likelihood.

To explain these ideas, note that an essential feature of the EM algorithm is the postulate of an unobserved 'complete data set' Z that contains what is actually observed Y, plus other observations X which one might wish were available, but in fact are not, and are termed the 'incomplete' data. That is $Z = (Y, X)$ so that by Bayes' rule

$$p(Z \mid Y) = \frac{p(Z,Y)}{p(Y)} = \frac{p(Z)}{p(Y)}.$$

which implies that

$$\log p(Y \mid \theta) = \log p(Z \mid \theta) - \log p(Z \mid Y,\theta) \quad (6)$$

where we note that since $\log x$ is monotonic in x, then finding θ which maximises $\log p(Y \mid \theta)$ is equivalent to finding θ maximising $p(Y \mid \theta)$.

As a consequence of (6), by taking expectations with respect to probabilities defined by an approximation of the parameters θ', and conditional on the observed data $Y = Y_N$, then leads to ($L(\theta) \triangleq \log p(Y_N \mid \theta)$)

$$\begin{aligned} L(\theta) &= \mathbf{E}\{\log p(Y \mid \theta) \mid Y = Y_N, \theta = \theta'\} \\ &= \mathbf{E}\{\log p(Z \mid \theta) \mid Y = Y_N, \theta = \theta'\} - \\ &\quad \mathbf{E}\{\log p(Z \mid Y,\theta) \mid Y = Y_N, \theta = \theta'\} \\ &= \mathcal{Q}(\theta,\theta') - \mathcal{V}(\theta,\theta') \end{aligned}$$

where the following definitions have clearly been made

$$\mathcal{Q}(\theta,\theta') \triangleq \mathbf{E}\{\log p(Z \mid \theta) \mid Y = Y_N, \theta = \theta'\}, \quad (7)$$
$$\mathcal{V}(\theta,\theta') \triangleq \mathbf{E}\{\log p(Z \mid Y,\theta) \mid Y = Y_N, \theta = \theta'\}. \quad (8)$$

In this case, the difference in log-likelihood corresponding to two different parameter vectors θ and θ' may be written as

$$\begin{aligned} L(\theta) - L(\theta') = &[\mathcal{Q}(\theta,\theta') - \mathcal{Q}(\theta',\theta')] + \\ &[\mathcal{V}(\theta',\theta') - \mathcal{V}(\theta,\theta')]. \end{aligned} \quad (9)$$

The key point now is the following inequality for $\mathcal{V}(\theta,\theta')$ that guarantees non-negativity of the second term in (9).

Lemma 3.1.

$$\mathcal{V}(\theta',\theta') \geq \mathcal{V}(\theta,\theta')$$

with equality if, and only if, $p(Z|Y,\theta) = p(Z|Y,\theta')$ for all Z.

Therefore, as a consequence of this, the decomposition (9) shows that if a value for θ is found that increases $\mathcal{Q}(\theta,\theta')$, then this must also increase the log-likelihood $L(\theta)$. This suggests the following algorithm for iteratively updating an approximation θ_k of the Maximum Likelihood estimate $\widehat{\theta}_N$, to a better one θ_{k+1}.

(1) **E Step** (Compute Expectation)

$$\mathcal{Q}(\theta,\theta_k) \triangleq \mathbf{E}\{\log p(Z \mid \theta) \mid Y = Y_N, \theta = \theta_k\} \quad (10)$$

(2) **M Step** (Maximise)

$$\text{Compute: } \theta_{k+1} = \arg\max_{\theta} \mathcal{Q}(\theta,\theta_k) \quad (11)$$

This is the the Expectation-Maximisation (EM) algorithm, in which the iterative procedure (11) replaces the gradient based one (5) as a method for finding Maximum Likelihood estimates. The principle underlying it is shown in figure 1. There it is illustrated that the function $\mathcal{Q}(\theta,\theta_k)$ acts as an approximant to the likelihood $L(\theta)$ which is exact at $\theta = \theta_k$, and also (locally) follows the contours of $L(\theta)$ in that $L(\theta)$ increases in directions that $\mathcal{Q}(\theta,\theta_k)$ increases.

Clearly, it is only sensible to employ this approach in cases where maximising $\mathcal{Q}(\theta,\theta')$ is straightforward, and certainly easier than maximising $L(\theta)$ directly. In turn, this will depend on what is chosen as the incomplete data set X. As such, an EM algorithm approach is not always appropriate, but as will now be demonstrated, it is very suitable for a wide range of dynamic system estimation problems of engineering relevance.

4. APPLICATION TO LINEAR TIME INVARIANT SYSTEMS

This section illustrates the application of the preceding methods by deriving a new algorithm for the estimation of linear, time invariant and multivariable systems that may be represented via the following specialisation of (1), (2).

$$x_{t+1} = Ax_t + Bu_t + w_t, \quad (12)$$
$$y_t = Cx_t + Du_t + e_t. \quad (13)$$

The estimation of the system model $f(x_t, u_t, w_t) = Ax_t + Bu_t + w_t$ and $g(x_t, u_t, e_t) = Cx_t + Du_t + e_t$

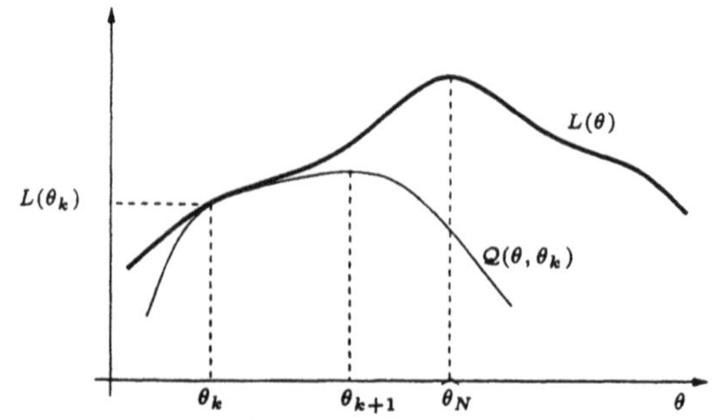

Fig. 1. *Illustration of the principle underlying the EM-algorithm. The function $\mathcal{Q}(\theta, \theta_k)$ acts as a local approximant of the likelihood $L(\theta)$.*

then amounts to estimation of the constant matrices $A \in \mathbf{R}^{n\times n}, B \in \mathbf{R}^{n\times m}, C \in \mathbf{R}^{p\times n}, D \in \mathbf{R}^{p\times m}, Q \in \mathbf{R}^{n\times n}$ and $R \in \mathbf{R}^{p\times p}$. That is, the collected vector of quantities to be estimated is $\theta = \mathrm{Vec}\{A, B, C, D, Q, R\}$ where the $\mathrm{Vec}\{\cdot\}$ operator creates a vector from a matrix by stacking its columns on top of one another.

For the purpose of estimating θ, notice that if x_t where observed, then (12) and (13) would be linear regressions in $[A, B]$ and $[C, D]$ (respectively) which could then be very directly estimated via least-squares. This suggests, that in the interests of defining the incomplete data X so that $\mathcal{Q}(\theta, \theta')$ is simple to maximise, then X should be taken as the unobserved state sequence $X_N \triangleq \{x_0, x_1, \ldots, x_N\}$ (hence the use of the symbol X). That is, the complete data could be taken as

$$Z \triangleq (X_N, Y_N).$$

According to (10), the expection step of the EM-algorithm then requires the calculation of

$$\mathcal{Q}(\theta, \theta') = \mathbf{E}\{\log p_\theta(X_N, Y_N) \mid Y = Y_N, \theta = \theta'\}$$

which in turn, as in all Maximum Likelihood estimation scenarios, requires the specification of the probability density functions governing the random disturbances w_t and e_t. Here we will assume these are Gaussian as follows

$$w_t \sim \mathcal{N}(0, Q), \qquad e_t \sim \mathcal{N}(0, R) \tag{14}$$

which then allows the computation of $\mathcal{Q}(\theta, \theta')$ via the following Lemma.

Lemma 4.1. For the model structure (12), (13) and the Gaussian assumptions (14) the function $-2\mathcal{Q}(\theta, \theta')$ defined in (7) may be computed as

$$\begin{aligned}
&\log|P_0| + N\log|Q| + N\log|R| + \\
&\mathrm{Tr}\left\{P_0^{-1}\left[(\widehat{x}_{0|N}-\mu)(\widehat{x}_{0|N}-\mu)^T + P_0\right]\right\} + \\
&\mathrm{Tr}\left\{Q^{-1}\left[\Phi - \Psi[A, B]^T - [A, B]\Psi^T + [A, B]\Gamma[A, B]^T\right]\right\} + \\
&\mathrm{Tr}\left\{R^{-1}\left[\Omega - \Lambda[C, D]^T - [C, D]\Lambda^T + [C, D]\Pi[C, D]^T\right]\right\}
\end{aligned} \tag{15}$$

with the following definitions applying

$$\widehat{x}_{t|N} \triangleq \mathbf{E}\{x_t \mid Y_N, \theta'\}, \quad z_t \triangleq \begin{bmatrix} x_t \\ u_t \end{bmatrix}, \tag{16}$$

$$\Lambda \triangleq \sum_{t=1}^{N} y_t \mathbf{E}\{z_t^T \mid Y_N, \theta'\}, \quad \Omega \triangleq \sum_{t=1}^{N} y_t y_t^T$$

$$\Pi \triangleq \sum_{t=1}^{N} \mathbf{E}\{z_t z_t^T \mid Y_N, \theta'\}, \Phi \triangleq \sum_{t=1}^{N} \mathbf{E}\{x_t x_t^T \mid Y_N, \theta'\}$$

$$\Psi \triangleq \sum_{t=1}^{N} \mathbf{E}\{x_t z_{t-1}^T \mid Y_N, \theta'\}, \Gamma \triangleq \sum_{t=1}^{N} \mathbf{E}\{z_{t-1} z_{t-1}^T \mid Y_N, \theta'\}$$

and where it has been assumed that the initial distribution on x_0 is

$$x_0 \sim \mathcal{N}(\mu, P_0). \tag{17}$$

This takes care of the Expectation step (10). The particular choice $X = X_N = \{x_0, \cdots, x_N\}$ of the incomplete data that is made here then allows the Maximisation step (11) to be achieved via the expressions of the following Lemma.

Lemma 4.2. The function $\mathcal{Q}(\theta, \theta')$ defined in (15) of Lemma 4.1 is maximised ($-2\log \mathcal{Q}(\theta, \theta')$ is minimised) by the choices

$$[A, B] = \Psi\Gamma^{-1}, \qquad [C, D] = \Lambda\Phi^{-1}, \tag{18}$$

$$Q = N^{-1}(\Phi - \Psi\Gamma^{-1}\Psi^T), R = N^{-1}(\Omega - \Lambda\Pi^{-1}\Lambda^T) \tag{19}$$

$$\mu = \mathbf{E}\{x_0 \mid Y_N, \theta'\}, P_0 = \mathbf{E}\{x_{t-1} x_{t-1}^T \mid Y_N, \theta'\} \tag{20}$$

There are several points to note here. Firstly, the computations (18) for updating of the estimates of the system matrices A, B, C, D may, via their relationship to least-squares solutions, be computed in very efficient and numerically robust fashions (Golub and Loan 1989).

Secondly, note that the estimate (19) for Q is positive semi-definite by construction since it is a Schur complement of

$$\sum_{t=1}^{N} \mathbf{E}_{\theta'}\left\{\begin{bmatrix} z_t \\ z_{t-1} \end{bmatrix}\begin{bmatrix} z_t^T & z_{t-1}^T \end{bmatrix}\right\} \geq 0.$$

A similar argument indicates that the estimate update (19) is also guaranteed to always yield an $R \geq 0$.

Finally, Lemma 4.2 indicates the the implementation of the Maximisation step of the EM-algorithm for the lineary time-invariant scenario requires the computation of the quantities $\widehat{x}_{t|N} = \mathbf{E}\{x_t \mid Y_N, \theta'\}$ and

$$\mathbf{E}\{x_t x_t^T \mid Y_N, \theta'\}, \mathbf{E}\{x_t x_{t-1}^T \mid Y_N, \theta'\} \tag{21}$$

which are essential to the definition of Λ, Π, Φ, Ψ and Γ.

4.1 *Computation of Conditional Expectations*

In the case considered in this paper where the distributions on the random components e_t and w_t are Gaussian, then recursive expressions exist for the computation of the quantities in (21), as specified in the following Lemma.

Lemma 4.3. For the system (12), (13) and with the definition $\widehat{x}_{t|N} \triangleq \mathbf{E}\{x_t \mid Y_N, \theta'\}$ together with

$$P_{t|s} \triangleq \mathbf{E}\{(\widehat{x}_{t|s} - x_t)(\widehat{x}_{t|s} - x_t)^T | Y_s\}, S_t \triangleq P_{t|t} A^T P_{t+1|t}^{-1}$$

then the first two quantities in equation (21) may be computed via the (reverse time) recursions

$$\widehat{x}_{t|N} = \widehat{x}_{t|t} + S_t\left[\widehat{x}_{t+1|N} - Bu_t - A\widehat{x}_{t|t}\right] \tag{22}$$

EM-based algorithm for Maximum Likelihood estimation

(1) Initialise estimates at $\theta_k = [A, B, C, D, Q, R]$. For example, a subspace-based estimation method could be employed.
(2) Using the system specification $\theta_k = [A, B, C, D, Q, R]$, run the Run Kalman-Filter recursions (25)-(29) followed by the Kalman Smoother (type) recursions (22), (23), (32) (30) in order to compute the quantities defined in Lemma 4.2.
(3) Maximise $\mathcal{Q}(\theta, \theta_k)$ over θ via the choices (18) and (19) in order to provide an improved estimate θ_{k+1}.
(4) Return to step 2 and repeat until termination.

$$P_{t|N} = P_{t|t} + S_t \left[P_{t+1|N} - P_{t+1|t}\right] S_t^T, \quad (23)$$

$$\mathbf{E}\left\{x_t x_t^T \mid Y_N, \theta'\right\} = P_{t|N} + \widehat{x}_{t|N}\widehat{x}_{t|N}^T \quad (24)$$

where the quantities $\widehat{x}_{t|t}, P_{t|t}, P_{t|t-1}$ involved in these expressions are pre-computed from the (forward in time) Kalman Filter recursions

$$P_{t|t-1} = AP_{t-1|t-1}A^T + Q \quad (25)$$

$$K_t = P_{t|t-1}C^T \left(CP_{t|t-1}C^T + R\right)^{-1} \quad (26)$$

$$P_{t|t} = P_{t|t-1} - K_t CP_{t|t-1} \quad (27)$$

$$\widehat{x}_{t|t-1} = A\widehat{x}_{t-1|t-1} + Bu_{t-1} \quad (28)$$

$$\widehat{x}_{t|t} = \widehat{x}_{t|t-1} + K_t \left(y_t - Du_t - C\widehat{x}_{t|t-1}\right) \quad (29)$$

which are initialised at

$$\widehat{x}_{0|0} = \mu, \qquad P_{0|0} = P_0$$

and where the system matrices A, B, C, D, Q, R used in (25)- (23) are those corresponding to θ'. The final quantity $\mathbf{E}\{x_t x_{t-1}^T \mid Y_N, \theta'\}$ in (21) may be computed via the (reverse time) recursions

$$M_{t|N} = P_{t|t}S_{t-1}^T + S_t(M_{t+1|N} - AP_{t|t})S_{t-1}^T \quad (30)$$

$$\mathbf{E}\left\{x_t x_{t-1}^T \mid Y_N, \theta'\right\} = M_{t|N} + \widehat{x}_{t|N}\widehat{x}_{t-1|N}^T \quad (31)$$

where

$$M_{t|N} \triangleq \mathbf{E}\left\{(\widehat{x}_{t|s} - x_t)(\widehat{x}_{t-1|s} - x_{t-1})^T | Y_N, \theta'\right\}$$

and (30) is initialised at

$$M_{N|N} = (I - K_N C)AP_{N-1|N-1}. \quad (32)$$

4.2 *Estimation Algorithm*

The previous developments may now be summarised in the following estimation algorithm definition.

At the risk of over-emphasis, the key point of the above algorithm for finding Maximum Likelihood estimates is that, in contrast to the more common gradient based approach, *no* parameterisation of the state-space model structure (12), (13) is required.

Notice too, that from a computational point of view, the above algorithm is comparable to a gradient based approach in that the Recursive Kalman Smoothing operations take the place of the recursive filtering operations necessary for gradient computation.

Finally, on the issue of judging convergence, and hence terminating the above iterative search, an immediately obvious strategy is to monitor the likelihood function $p(y_1, \cdots, y_N|\theta_k)$, and when its rate of increase drops below a threshold, convergence can be declared. This is the method used in the simulation examples of the following section.

5. SIMULATION EXAMPLE

This section provides two brief simulation examples in order to illustrate the utility of the EM-algorithm approach to Maximum–Likelihood estimation proposed in this paper.

In both cases, the observed data is generated according to a system

$$y_t = G(q)u_t + e_t$$

with $G(q)$ given by

$$\begin{bmatrix} \dfrac{0.0355q + 0.02465}{(q-0.3679)(q-0.9084)} & \dfrac{0.2364q + 0.1038}{(q-0.1353)(q-0.6065)} \\ \dfrac{0.07601q + 0.05447}{(q-0.4966)(q-0.7408)} & \dfrac{0.1087q + 0.07286}{(q-0.4493)(q-0.6703)} \end{bmatrix}$$

and u_t is an i.i.d. zero mean and unit variance Gaussian process while e_t is also i.i.d. zero mean and Gaussian, but has variance $\mathbf{E}\{e_t^2\} = \sigma^2 = 0.01$.

For this scenario, $N = 200$ data samples were collected and Maximum-Likelihood estimates were computed via the EM-algorithm described in this paper and initialised with the starting estimate

$$G(q) = \begin{bmatrix} \dfrac{0.1}{(q-0.5)^2} & \dfrac{0.1}{(q-0.7)^2} \\ \dfrac{0.1}{(q-0.6)^2} & \dfrac{0.1}{(q-0.4)^2} \end{bmatrix}.$$

The results of this estimation experiment are shown in figure 2. On the left, the relationship between initial and EM-derived ML estimates is shown together with the true response. On the right, the evolution of the log mean-square cost

$$\frac{1}{N}\sum_{t=1}^{N}\left(y_t - C\widehat{x}_{t|t-1}\right)^2$$

is shown as the EM-algorithm iteration progresses. Clearly, the algorithm converges to estimates close to the true system.

In relation to this simulation, the previous section has raised the possibility of initialising the EM iterations with a subspace-based method, and the results of this strategy for the experimental conditions just outlined are shown in figure 3. There, Overschee and DeMoor's N4SID variant (van Overschee and Moor 1996) of the general class of subspace-based methods is used to provide the initial estimate shown as the dash-dot line. The EM-algorithm of this paper is then used to refine this to be closer to the Maximum Likelihood estimate, with concomitant cost function evolution shown in the right hand diagram of figure 3 and final estimate shown as the dashed line on the left in 3, together (again) with the true system shown as a solid line.

Clearly, the final estimate is significantly improved from the initial subspace-based one, and the key point is that this is achieved in a very simple manner by the parameterisation free method proposed here, while it would be very difficult to implement using a more standard gradient based method that imposed a parameterisation.

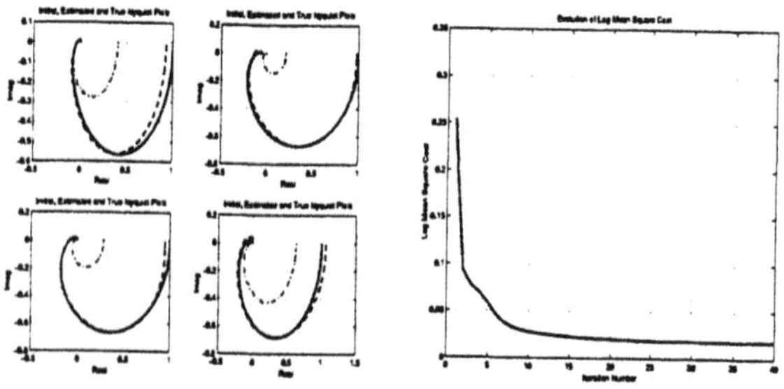

(a) Nyquist Plots (b) LS cost evolution

Fig. 2. *Results when using the EM-algorithm method for computing Maximum Likelihood estimates. The left hand figure shows initial estimates (via Nyquist plots) as dash-dot line, true systems as solid lines, and EM-derived ML estimates as dashed lines. The right hand figure shows the evolution of the means square cost as the EM-algorithm is iterated.*

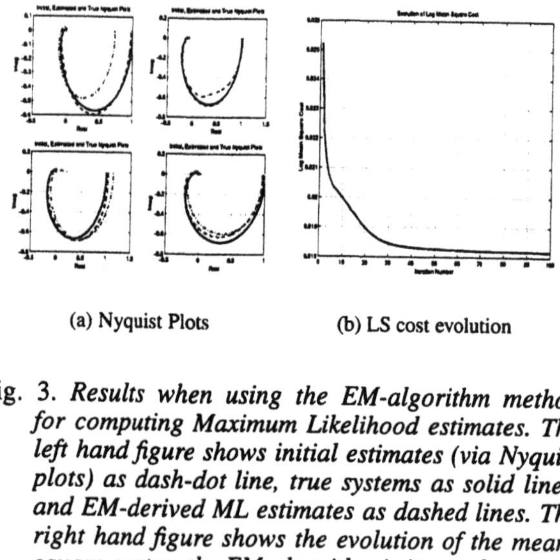

(a) Nyquist Plots (b) LS cost evolution

Fig. 3. *Results when using the EM-algorithm method for computing Maximum Likelihood estimates. The left hand figure shows initial estimates (via Nyquist plots) as dash-dot line, true systems as solid lines, and EM-derived ML estimates as dashed lines. The right hand figure shows the evolution of the means square cost as the EM-algorithm is iterated.*

6. CONCLUSIONS

The contribution of this paper was to suggest a novel, EM-algorithm based approach to Maximum Likelihood estimation of dynamic systems. The key features recommending the approach are that it avoids the need for a particular parameterisation of a state-space model structure, and it is simple to implement.

Although this method is novel in the context considered here, the EM-algorithm itself is quite old, being very well known in (for example) the speech-recognition community as the Baum–Welch method for Hidden Markov Model estimation (Rabiner 1989).

This paper represents only a very preliminary study of this whole topic, and there is much more that needs to be studied in terms of (again, only for example) convergence analysis and extension to more sophisticated model structures

7. REFERENCES

Åström, K. and Peter Eykhoff (1971). 'System identification - a survey'. *Automatica* **7**, 123–162.

Barndorff–Nielsen, O. E., David R. Cox and Claudia Klüppelberg (1999). *Complex Stochastic Systems*. Chapman and Hall.

Bauer, D., M. Deistler and W. Scherrer (1999). 'Consistency and asymptotic Normality of some subspace algorithms for systems without observed inputs'. *Automatica* **35**, 1243–1254. L

Caines, P. (1988). *Linear Stochastic Systems*. John Wiley and Sons. New York.

Deistler, M. (2000). *Model Identification and Adaptive Control*. Springer-Verlag. chapter System Identification - General Aspects and Structure.

Deistler, M., K. Peternell and W. Scherrer (1995). 'Consistency and relative efficiency of subspace methods'. *Automatica* **31**(12), 1865–1875.

Dempster, A., N.M. Laird and D.B. Rubin (1977). 'Maximum likelihood from incomplete data via the EM algorithm'. *Journal of the Royal Statistical Society, Series B* **39**, 1–38.

Dennis, J. and Robert B. Schnabel (1983). *Numerical Methods for Unconstrained Optimization and Nonlinear Equations*. Prentice Hall.

Fisher, R. (1912). 'On an absolute criterion for fitting frequency curves'. *Mess. Math.* pp. 41–155.

Golub, G. and Charles Van Loan (1989). *Matrix Computations*. Johns Hopkins University Press.

Goodwin, G. and R.L. Payne (1977). *Dynamic System Identification*. Academic Press.

Hannan, E. and Manfred Deistler (1988). *The Statistical Theory of Linear Systems*. John Wiley.

Larimore, W. (1990). Canonical variate analysis in identification, filtering and adaptive control. In 'Proceedings of the 29th IEEE Conference on Decision and Control, Hawaii'. pp. 596–604.

Lehmann, E. (1983). *Theory of Point Estimation*. John Wiley & Sons.

Ljung, L. (1999). *System Identification: Theory for the User, (2nd edition)*. Prentice-Hall, Inc.. New Jersey.

Ljung, L. (2000*a*). *MATLAB System Identification Toolbox Users Guide, Version 5*. The Mathworks.

McKelvey, T. (1998). 'Discussion: 'on the use of min. param. in multivariable ARMAX identification' by R.P. Guidorzi'. *European J. Control* **4**, 93–98.

Milanese, M. and A. Vicino (1991). 'Optimal estimation theory for dynamic systems with set membership uncertainty:An overview'. *Automatica* **27**(6).

Nocedal, J. and Stephen Wright (1999). *Numerical Optimization*. Springer-Verlag, New York.

Norton, J. (1987). 'Identification & application of bounded param. models'. *Automatica* **23**, 497–507.

Norton, J. P. (1985). *An Introduction to Identification*. Academic Press.

Rabiner, L. (1989). 'A tutorial on hidden Markov models and selected applications in speech recognition'. *Proceedings of the IEEE* **77**(2), 257–285.

Starck, J.-L., F. Murtagh and A. Bijaoui (1998). *Image processing and data analysis*. Cambridge University Press. Cambridge. The multiscale approach.

Titterington, D. (1984). 'Recursive parameter estimation using incomplete data'. *Journal of the Royal Statistical Society, Series B* **46**(2), 257–267.

T.Söderström and P.Stoica (1989). *System Identification*. Prentice Hall. New York.

van Overschee, P. and Bart De Moor (1996). *Subspace Identification for Linear Systems*. Kluwer Academic Publishers.

NEUROFUZZY STATE SPACE MODELLING AND CONTROL USING KALMAN FILTERING STATE FEEDBACK WITH COLOURED NOISE

C. J. Harris * X. Hong *

* *ISIS Research Group, Dept of Electronics and Computer Science, University of Southampton, Southampton, SO17 1BJ, UK*

Abstract: A new controller scheme is introduced for unknown nonlinear dynamical processes that are modelled by an operating point neurofuzzy system. A neurofuzzy state space model of controllable form is initially constructed, based on a neurofuzzy design and model construction algorithm for nonlinear dynamical process, followed by a controller design based on closed loop pole assignment that ensures the system stability under some mild assumptions. This intelligent controller approach also includes a generalised Kalman filtering algorithm with coloured noise, based on the neurofuzzy state space model, to obtain an optimal state vector estimation for use in the state feedback stabilisation. The application of the algorithm to typical output tracking problem is used to demonstrate the applications of this new approach to intelligent control. *Copyright © 2001 IFAC*

Keywords: neurofuzzy network, system identification, nonlinear modelling, Kalman filtering, state feedback.

1. NEUROFUZZY STATE SPACE MODELLING

Consider a general discrete stochastic nonlinear time series represented by

$$y(t) = f(y(t-1), \cdots, y(t-n_y), u(t-1), \cdots, u(t-n_u), \Theta) + \xi(t) \qquad (1)$$

where $(n_u, n_y)(n_y + n_u = n), n_y \geq n_u$ are positive interger representing the system known lags, $f(\bullet)$ is *a priori* unknown. The observation noise $\xi(t)$ is uncorrelated with zero mean and variance σ^2. Θ is an unknown parameter vector associated with the appropriate but yet to be determined model structure. Consider using a fuzzy system with functional rule basis to model the above system. The operating point of a fuzzy system is an n_y-dimensional measurement vector O_t, given by

$$O_t = [x_1, x_2, \cdots, x_{n_y}]^T$$

$$= [y(t-1), \cdots, y(t-n_y)]^T \in \Re^{n_y} \qquad (2)$$

Here an input affine model structure is used to represent a large class within (1) in order to simplify modelling and control task, but maintain sufficient nonlinear approximation capability, given by

$$\hat{y}(t) = a_1(O_t)y(t-1) + \cdots + a_{n_y}(O_t)y(t-n_y) + c_1 u(t-1) + \cdots + c_{n_u} u(t-n_u) \qquad (3)$$

where $c_j, j = 1, \cdots, n_u$ are constant weights, $a_j(O_t), j = 1, \cdots, n_y$ are modelled by neurofuzzy networks as

$$a_j(O_t) = \sum_{n_i=1}^{M_0} \omega_{jn_i}^{(a)} N^{(n_i)}(O_t), \qquad j = 1, \cdots, n_y \qquad (4)$$

where $\omega_{jn_i}^{(a)}$'s are weights of the neural networks, and $N^{(n_i)}(O_t)$'s are known basis functions with a basis set size of M_0. If Eq.(4) is based on B-spline polynomial, the resulting system is known as B-spline neurofuzzy modelling structure of a generalised Takagi-Sugeno fuzzy input-output model, based on widely used assumptions(Brown and Harris, 1994),(Harris *et al.*, 1996), (Takaga and Sugeno, 1985), (Feng and Harris, 1998) of (i) that fuzzy operations are algebraic sum and product ; (ii) defuzzification is by centre of area and (iii) the fuzzy membership functions $N^{(n_i)}(O_t)$'s are B-splines with a partition of unity. (B-spline theory can be found in (Schumaker, 1981)). Denote the number of rules of x_i as L_i, $k_i \in \{1, \cdots, L_i\}$. With input vector O_t, using algebraic (product/sum) operator for fuzzy reasoning, the truth value of the antecedent part of the (n_i)th fuzzy rule, denoted as $N^{(n_i)}(O_t)$, is given by the multivariate B-spline basis functions formed by the tensor product of n univariate basis functions $B_i^{k_i}(x_i)$, denoted by $N^{(n_i)}(O_t)$:

$$N^{(n_i)}(O_t) = \prod_{i=1}^{n_y} B_i^{k_i}(x_i). \tag{5}$$

the total number of $N^{(n_i)}(O_t)$ is $M_0 = L_1 L_2 \cdots L_{n_y}$, that is $n_i = 1, 2, \cdots, M_0$. Alternatively, to reduce the curse of dimensionality, a construction algorithm using an extended additive decomposition approach based on the analysis of variance (ANOVA) expansion (Bossley, 1997) can be applied.

Based on Eq.(3), Eq.(1) can be written as

$$y(t) = \frac{C(z^{-1})}{A_t(z^{-1})} u(t) + w(t) \tag{6}$$

where

$$A_t(z^{-1}) = 1 - \sum_{j=1}^{n_y} a_j(O_t) z^{-j} \tag{7}$$

$$C(z^{-1}) = \sum_{j=1}^{n_u} c_j z^{-j} \tag{8}$$

$$w(t) = \frac{\xi(t)}{A_t(z^{-1})} \tag{9}$$

Denote

$$x(t) = \frac{1}{A_t(z^{-1})} u(t)$$

as a state variable and define a state vector

$$\mathbf{x}(t) = [x(t-1), x(t-2), \cdots, x(t-n_y)]^T$$

System of Eq.(6) can be represented by a state space model in *controller canonical form* given by

$$\mathbf{x}(t+1) = \mathbf{A}(t)\mathbf{x}(t) + \mathbf{B}u(t) \tag{10}$$

$$y(t) = \mathbf{C}\mathbf{x}(t) + w(t) \tag{11}$$

where

$$\mathbf{A}(t) = \begin{bmatrix} a_1(O_t) & a_2(O_t) & \cdots & a_{n_y}(O_t) \\ 1 & 0 & \cdots & 0 \\ 0 & 1 & \cdots & 0 \\ \cdots & \cdots & \cdots & \cdots \\ 0 & \cdots & 0 & 0 \end{bmatrix} \in \Re^{n_y \times n_y}$$

$$\mathbf{B} = [1, 0, \cdots, 0]^T \in \Re^{n_y}$$

$$\mathbf{C} = \begin{bmatrix} c_1, c_2, \cdots, c_{n_u}, \underbrace{0, \cdots, 0}_{(n_y - n_u)} \end{bmatrix} \in \Re^{n_y} \tag{12}$$

Clearly, the construction of the state space model of Eq.(10) and (11) is readily determined via the identification of the input-output model given by (3). Note that Eq.(3) has a linear in the weight model structure, written as

$$y(t) = \sum_{k=1}^{M} p_k(t)\theta_k + \xi(t) \tag{13}$$

where $p_k(t)$'s are regressors given by

$$p_k(t) = N^{(n_i)}(O_t) x_i, \qquad k = 1, 2, \cdots, n_y M_0$$

or

$$p_k(t) = u(t-j), \qquad k = n_y M_0 + 1 \cdots, M. \tag{14}$$

where $n_i = 1, \cdots, M_0$, $j = 1, \cdots, n_u$, $i = 1, \cdots, n_y$, $M = n_y M_0 + n_u$.

For the general linear in the parameter model for N observations $\mathbf{y} = [y(1), \cdots, y(N)]^T$

$$\mathbf{y} = \mathbf{P}\Theta + \Xi \tag{15}$$

where $\Theta = [\theta_1, \cdots, \theta_M]^T$ is an unknown parameter vector, $\Xi = [\xi(1), \cdots, \xi(N)]^T$ is the residual vector, and $\mathbf{P} \in \Re^{N \times M}$ is the regression matrix fromed on the basis function of the complete neurofuzzy model. For example if $n_y + n_u = 6$, and $L_i = 5$) (number of basis functions used in each input) $M = n \sum_{i=1}^{n} L_i = 93750$, or $M = 17430$ for an ANOVA expansion of degree 3 with no redundant inputs. Generally for large M the complete model contains many redundant terms due to the underlying lattice structure used in a neurofuzzy model, and $\mathbf{P}^T\mathbf{P}$ is generally singular. Here NeuDec generates a reduced rule base with $n_\theta \ll M$ (see Theorem 1, for proof see (Hong and Harris, 2000)), then utilies OLS to find a final model with $n_f \ll n_\theta$ regressors.

Theorem 1 Given some finite number $n_\theta < M$, sort the first n_θ largest regressors into magnitude order from the initial M regressor candidates, and rank them in descending order of magnitude, such that $||\mathbf{p}_{o_1}|| \geq ||\mathbf{p}_{o_2}|| \geq ||\mathbf{p}_{o_{n_\theta}}|| \geq ||\mathbf{p}_j||, j \notin \{o_1, o_2, \cdots, o_{n_\theta}\}$, for an index vector $\mathbf{o} = [o_1, o_2, \cdots, o_{n_\theta}]$. The resultant subset $\mathbf{P}_k$, minimises the lower bound of an A-optimality design criterion(Atkinson and Donev, 1992) defined as the sum of the variance of a parameter estimate vector.

In practice, a n_θ is preselected as a a few hundred so as to form a moderate candidate regressor initial base size for further refinement via OLS. It is assumed that n_θ is sufficiently large so that the estimation set has sufficient modelling capability to approximate the underlying data generating process. The most relevant (or significant) n_f ($\ll n_\theta$) regressors can be determined by the forward OLS scheme (Chen *et al.*, 1989). The OLS starts with an orthogonal decomposition of $\mathbf{P} = \mathbf{WT}$, where the regression matrix has been decomposed into an upper triangular matrix $\mathbf{T}$ and matrix $\mathbf{W}$ with orthogonal columns that satisfy $\mathbf{W}^T\mathbf{W} = diag\{\kappa_1, \cdots, \kappa_{n_\theta}\}$, and which defines an auxillary vector $\Gamma = [\gamma_1, \cdots, \gamma_\theta]^T = \mathbf{T}\Theta$. It can be shown (Chen *et al.*, 1989) that by defining

$$[ERR]_k = \frac{\gamma_k^2 \mathbf{w}_k^T \mathbf{w}_k}{\mathbf{y}^T\mathbf{y}}, \qquad k = 1, \cdots, M \quad (16)$$

The most relevant n_θ regressors can be forward selected according to the value of the error reduction ratio $[ERR]_k$ in a forward regression manner.

Alternatively the A-optimality design criterion can be used to select the n_f regressors, so as to minimise the variance of the parameter estimate vector $\Gamma = [\gamma_1, \cdots, \gamma_{n_f}]^T$ ensuring model adequacy. Minimising this cost function does not also ensure good model approximation, hence V_θ needs an amended or composite cost functional, for some small positive α such that

$$V_f = \frac{1}{N}(\mathbf{y}^T\mathbf{y} - \sum_{k=1}^{n_f} \gamma_k^2 \kappa_k) + \alpha \sum_{k=1}^{n_f} \frac{1}{\kappa_k} \quad (17)$$

Eq.(17) can be incorporated into the forward OLS algorithm to select the most relevant kth regressor at the kth forward regression stage, expressed as

$$V^{(k)} = V^{(k-1)} - \frac{1}{N}\gamma_k^2 \kappa_k + \frac{\alpha}{\kappa_k} \quad (18)$$

At the kth regression stage, a candidate regressor is selected if it produces the smallest $V^{(k)}$ and a further reduction in $V^{(k-1)}$. The procedure terminates if $V^{(k)} \geq V^{(k-1)}$ at the derived model size n_f. Consequently the NeuDec algorithm automatically detects and determins a parsimonious model from data.

2. POLE ASSIGNMENT CONTROLLER VIA STATE FEEDBACK USING KALMAN FILTERING WITH COLOURED NOISE

Suppose that at time t, a system input, observable system output data set $\{u(t-1), \cdots, y(t), y(t-1), \cdots\}$ are available for system identification. Here the control objective is to design a system input $u(t)$ so that the actual system output tracks a desired output trajectory to a required accuracy, that is, to force the system output $y(t+1)$ to follow a specified trajectory $y_d(t+1)$, without prior knowledge of the underlying process that generated the data set.

2.1 *System stabilisation by closed loop pole assignment*

Initially assume that $\mathbf{x}(t)$ is available, by designing a feedback control law

$$u(t) = -\mathbf{K}(t)\mathbf{x}(t) + v^*(t) \quad (19)$$

where

$$\mathbf{K}(t) = [k_1(t), k_2(t), \cdots, k_{n_y}(t)] \in \Re^{n_y} \quad (20)$$

is state variable feedback gain for state space model (10) of the operating point dependent model (3). $v^*(t)$ is system closed loop input signal. Substituting Eq.(19) to Eq.(10) gives

$$\mathbf{x}(t+1) = \tilde{\mathbf{A}}\mathbf{x}(t) + \mathbf{B}v^*(t) \quad (21)$$

where

$$\tilde{\mathbf{A}} = \begin{bmatrix} a_1(O_t) - k_1(t) & a_2(O_t) - k_2(t) & \cdots & a_{n_y}(O_t) - k_{n_y}(t) \\ 1 & 0 & \cdots & 0 \\ 0 & 1 & 0 & 0 \\ \cdots & \cdots & \cdots & \cdots \\ 0 & \cdots & 0 & 0 \end{bmatrix}$$

Now the closed-loop characteristic polynomial is given by

$$\begin{aligned} &\det(z\mathbf{I} - \tilde{\mathbf{A}}) \\ &= z^{n_y} - (a_1(O_t) - k_1(t))z^{n_y-1} - \cdots \\ &\quad -(a_{n_y}(O_t) - k_{n_y}(t)) \\ &\triangleq z^{n_y} + \tilde{a}_1 z^{n_y-1} + \cdots + \tilde{a}_{n_y} \end{aligned} \quad (22)$$

It follows that the closed-loop poles can be arbitrarily assigned by appropriate choice of $k_1(t)$, $\cdots$, $k_{n_y}(t)$. By setting

$$k_j(t) = a_j(O_t) + \tilde{a}_j \quad (23)$$

Then

$$\tilde{\mathbf{A}} = \begin{bmatrix} -\tilde{a}_1 & -\tilde{a}_2 & \cdots & -\tilde{a}_{n_y-1} & -\tilde{a}_{n_y} \\ 1 & 0 & \cdots & 0 & 0 \\ 0 & 1 & \cdots & 0 & 0 \\ \cdots & \cdots & \cdots & \cdots & \cdots \\ 0 & \cdots & 0 & 1 & 0 \end{bmatrix}$$

The closed loop system state space model is given by

$$\mathbf{x}(t+1) = \tilde{\mathbf{A}}\mathbf{x}(t) + \mathbf{B}v^*(t) \quad (24)$$
$$y(t) = \mathbf{C}\mathbf{x}(t) + w(t) \quad (25)$$

Lemma 1 Consider the system (24)-(25). Provided that (i) all of the eigenvalues of $\tilde{\mathbf{A}}$, $\lambda_i(\tilde{\mathbf{A}}) < 1$, $i = 1, \cdots, n_y$. (ii) All of the basis functions in (4) are uniformly bounded such that the $\mathbf{C}$, $A_t(z^{-1})$ are bounded and $\frac{1}{A_t(z^{-1})}$ is nonsigular. Then there exist constants m_1 and m_2, which are independent of N, such that

$$\sum_{t=1}^{N} \|y(t)\|^2 \leq m_1 + m_2 \sum_{t=1}^{N} \|\xi(t)\|^2 + m_3 \sum_{t=0}^{N-1} \|v^*(t)\|^2 \quad (26)$$

where $0 \leq m_1, m_2, m_3 < \infty$, $N > 0$. That is, the system is finite gain stable.

Proof: see (Harris and Hong, 2000)

The control law of Eq.(19) is based on the knowledge of the state vector $\mathbf{x}(t)$ and a model reference signal $v^*(t)$ that represents the desired output trajectory $y_d(\bullet)$. In the following, the optimal estimation of the state vector $\mathbf{x}(t)$ is based on Kalman filtering with coloured noise, followed by the derivation of $v^*(t)$ for the completeness of the controller solution.

2.2 *State estimation using Kalman filtering with coloured noise*

Assume at time index t, $v^*(t)$ (see the derivation in Section 2.3) is available. Upon the arrival of a new system observation $y(t)$, consider state estimation based on the state space model of (24),(25). Note that because the noise term $w(t)$ is given as a colour noise term with an identified structure as given by Eq.(9), the standard Kalman filtering technique is not directly applicable. However, based on an equivalent model representation derived in the following, Kalman filtering can be applied.

Lemma 2 The state space model of Eq.(24)(25) is equivalent to the following state space model

$$\mathbf{x}(t - n_y + 1) = \tilde{\mathbf{A}}\mathbf{x}(t - n_y) + \mathbf{B}v^*(t - n_y) \quad (27)$$

$$z(t) = \tilde{\mathbf{C}}(t)\mathbf{x}(t - n_y) + \tilde{D}(t) + \xi(t) \quad (28)$$

where

$$z(t) = y(t) - \sum_{j=1}^{n_y} a_j(O_t) y(t-j) \quad (29)$$

$$\tilde{\mathbf{C}}(t) = \mathbf{C}\tilde{\mathbf{A}}^{n_y} - \sum_{j=1}^{n_y} a_j(O_t)\mathbf{C}\tilde{\mathbf{A}}^{n_y - j} \quad (30)$$

$$\tilde{D}(t) = -\sum_{j=0}^{n_y-1} a_j(O_t)\mathbf{C} \sum_{\tau=j}^{n_y-1} \tilde{\mathbf{A}}^{\tau-j}\mathbf{B}v^*(t-\tau-1) \quad (31)$$

in which $a_0(O_t) \triangleq -1$.

Proof. See (Harris and Hong, 2000).

For ease of exposition, denote $\tilde{\mathbf{x}}(t) = \mathbf{x}(t - n_y)$. The state space model Eq.(27) and Eq.(28) can be written as

$$\tilde{\mathbf{x}}(t+1) = \tilde{\mathbf{A}}\tilde{\mathbf{x}}(t) + \mathbf{B}v^*(t - n_y) \quad (32)$$
$$z(t) = \tilde{\mathbf{C}}(t)\tilde{\mathbf{x}}(t) + \tilde{D}(t) + \xi(t) \quad (33)$$

where $\tilde{D}(t)$ and $\mathbf{B}v^*(t - n_y)$ are deterministic terms, $\xi(t)$ is uncorrelated model residual with variance σ^2, such as the following recursive Kalman filter algorithm can be applied for optimal state estimation:

Prediction:

$$\hat{\tilde{\mathbf{x}}}(t|t-1) = \tilde{\mathbf{A}}\hat{\tilde{\mathbf{x}}}(t-1|t-1) + \mathbf{B}v^*(t - n_y - 1)$$
$$\mathbf{P}(t|t-1) = \tilde{\mathbf{A}}\mathbf{P}(t-1|t-1)\tilde{\mathbf{A}}^T \quad (34)$$

State estimation (correction):

$$\mathbf{G}(t) = \mathbf{P}(t|t-1)\tilde{\mathbf{C}}^T(t)[\tilde{\mathbf{C}}(t) \mathbf{P}(t|t-1)\tilde{\mathbf{C}}^T(t) + \sigma^2]^{-1}$$
$$\hat{\tilde{\mathbf{x}}}(t|t) = \hat{\tilde{\mathbf{x}}}(t|t-1) + \mathbf{G}(t) \bullet ([z(t) - \tilde{\mathbf{C}}(t)\tilde{\mathbf{x}}(t|t-1) - \tilde{D}(t)]$$
$$\mathbf{P}(t|t) = [I_{n_y} - \mathbf{G}(t)\tilde{\mathbf{C}}(t)]\mathbf{P}(t|t-1) \quad (35)$$

where $I_{n_y} \in \Re^{n_y \times n_y}$ is unit matrix, $\hat{\tilde{\mathbf{x}}}(t|t) = E(\tilde{\mathbf{x}}(t)|t) = E(\mathbf{x}(t - n_y)|t) = \hat{\mathbf{x}}(t - n_y|t)$ is the estimate of the state vector $\mathbf{x}(t - n_y)$ given information up to t. $\mathbf{P}(t|t)$ is the state estimate covariace matrix, and $\mathbf{G}(t)$ is the Kalman gain matrix.

From Eq.(27), gives

$$\mathbf{x}(t) = \tilde{\mathbf{A}}^{n_y}\mathbf{x}(t - n_y) + \sum_{\tau=1}^{n_y} \tilde{\mathbf{A}}^{\tau}\mathbf{B}v^*(t-\tau) \quad (36)$$

The term $\sum_{\tau=1}^{n_y} \tilde{\mathbf{A}}^{\tau}\mathbf{B}v^*(t-\tau)$ is a deterministic term known at time t, hence the optimal estimate of the state vector $\mathbf{x}(t)$ given information up to t can be calculated as

$$\hat{\mathbf{x}}(t|t) = \tilde{\mathbf{A}}^{n_y}\hat{\mathbf{x}}(t|t) + \sum_{\tau=1}^{n_y} \tilde{\mathbf{A}}^{\tau}\mathbf{B}v^*(t-\tau) \quad (37)$$

which can be used in deriving the state feedback control law, based on Eq. (19), as

$$u(t) = -\mathbf{K}(t)\hat{\mathbf{x}}(t|t) + v^*(t) \quad (38)$$

Given a known desired output trajectory sequence $y_d(t)$, the derivation of $v^*(t)$ is based on direct minimisation of the mean square tracking error, as derived in the sequel.

2.3 *Determining closed loop optimal input $v^*(t)$ via output trajectory tracking*

Denote

$$\tilde{A}(z^{-1}) = 1 + \sum_{j=1}^{n_y} \tilde{a}_j z^{-j} \quad (39)$$

From Eq.(24)(25), the closed-loop system is then given by

$$y(t) = \frac{C(z^{-1})}{\tilde{A}(z^{-1})}v^*(t) + w(t) \quad (40)$$

based on Eq.(9) and (40), at time t, the actual system output can be written as

$$\begin{aligned}
&y(t+1)\\
&= -\sum_{j=1}^{n_y} \tilde{a}_j y(t-j+1) + \sum_{j=1}^{n_u} c_j v^*(t-j+1)\\
&+\tilde{A}(z^{-1})w(t+1)\\
&= -\sum_{j=1}^{n_y} \tilde{a}_j y(t-j+1) + \sum_{j=1}^{n_u} c_j v^*(t-j+1)\\
&+\left[\tilde{A}(z^{-1}) - A_{t+1}(z^{-1})\right] w(t+1)\\
&+A_{t+1}(z^{-1})w(t+1)\\
&= -\sum_{j=1}^{n_y} \tilde{a}_j y(t-j+1) +\\
&\sum_{j=1}^{n_u} c_j v^*(t-j+1) + K(z^{-1})w(t+1) + \xi(t+1)\\
&\triangleq \hat{y}(t+1|t) + \xi(t+1)
\end{aligned} \quad (41)$$

where

$$K(z^{-1}) \triangleq \sum_{j=1}^{n_y} [\tilde{a}_j + a_j(O_{t+1})] z^{-j+1} \quad (42)$$

$$= \sum_{j=1}^{n_y} k_j(t+1) z^{-j+1} \quad (43)$$

In order to generate a closed loop optimal input $v^*(t)$ such that the system output $y(t+1)$ following a desired reference signal $y_d(t+1)$, the generalised minimum variance control that minimises $J = E\left\{[y(t+1) - y_d(t+1)]^2 + \beta v^*(t)^2\right\}$ $= E\left\{[\hat{y}(t+1|t) - y_d(t+1)]^2 + \beta v^*(t)^2\right\} + \sigma^2$ can be used, where β is a predetermined weight to control $v^*(t)$. By setting

$$\frac{\partial J}{\partial v^*(t)} = 0 \quad (44)$$

Hence

$$[\hat{y}(t+1|t) - y_d(t+1)]c_1 + \beta v^*(t) = 0 \quad (45)$$

and the solution of Eq.(45) is given as

$$\begin{aligned}
v^*(t) = \frac{c_1}{c_1^2 + \beta} \Bigg\{ & y_d(t+1) + \sum_{j=1}^{n_y} \tilde{a}_j y(t-j+1)\\
& -\sum_{j=2}^{n_u} c_j v^*(t-j+1) -\\
& \sum_{j=1}^{n_y} k_j(t+1) w(t-j+1) \Bigg\}
\end{aligned} \quad (46)$$

Note that Eq.(46) requires the recursive calculation of $w(t)$ which is given by

$$w(t) = \sum_{j=1}^{n_y} a_j(O_t) w(t-j) + [y(t) - \hat{y}(t|t-1)] \quad (47)$$

where $\hat{y}(t|t-1)$, the one-step ahead model prediction, is used recurrently to calculate past model error $\xi(t)$.

3. ILLUSTRATIVE EXAMPLE

The dynamical system used (Obradovic, 1996) in the illustrative example is

$$\begin{aligned}
y(t) =\,& 0.01y(t-1) + 0.4\sin(3y(t-1))\\
&+0.5u(t-1)
\end{aligned} \quad (48)$$

500 input-output pairs were generated for an input of $u(t) = 1.25\sin(\pi t/25)$. The data set was used to identify a neurofuzzy operating point model of (3) based on the NeuDec algorithm. The system lag was set as $n_y = 2$, $n_u = 1$; the input vector was $O_t = [y(t-1), y(t-2)]^T$. The additive decompostion based on ANOVA was used to generate an initial model basis set, by setting the input components interaction degree up to 2. The knot vectors for the input domain were defined as

$$\{-15, -10, -5, -0.5, 0, 0.5, 5, 10, 15\}$$

for $y(t-1)$, $y(t-2)$, and piecewise quadratic B-spline($m = 2$) were used for all the input components, resulting 6 one-dimensional basis functions for each of the 2 input components. The initial model has 169 basis functions.

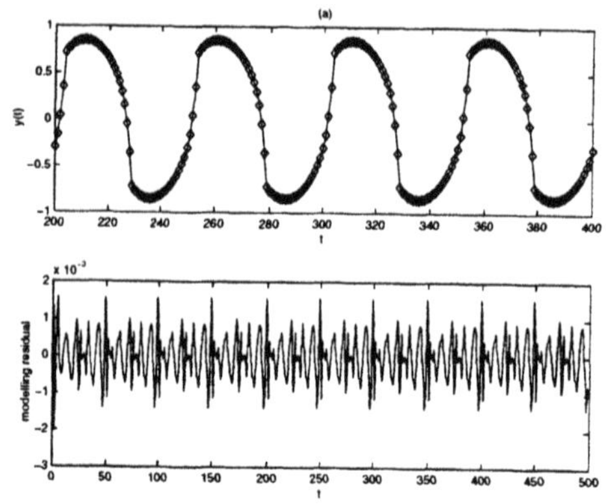

Fig. 1. Identification results in example 1; (a) 'Diamond'- model predictions and 'Solid Line'-actual system output; (b) modelling residuals;

The forward OLS combining the experimental design criterion for model structure detection as introduced in Section 3 was then applied to the full model with 169 basis functions, in which $\alpha = 1e - 12$ is applied. The NeuDec algorithm automatically detected a model size of $n_f = 20$ basis functions with a MSE of $4e - 7$. The model approximation performance is demonstrated in Fig.1. For the same example, the results presented in (Obradovic, 1996) used a 203 centers gaussian RBF model.

From $t = 500$ to $t = 1000$, a reference signal is defined as the rectangular signal

$$y_d(t) = \begin{cases} 20.65 & \text{if } 0 \le \text{rem}(t/200) < 100 \\ -0.65 & \text{if } 100 \le \text{rem}(t/200) < 200 \end{cases}$$

where rem(•) denotes remainder. The system output is required to track this signal. The state vector was obtained based on the Kalman filtering with coloured noise as in Section 4.2, by assigning two poles closed to zero, $z_{1,2} = \pm 1e^{-5}j$. The derivation of the closed loop input reference from Eq.(46), in which $\beta = 0.3$. Based on the identified neurofuzzy system with 20 basis functions, the system input is calculated from Eq.(38) recursively at each step, and by using Eq.(48) to generate the actual system output from $t = 500 \sim 1000$. The system tracking performance is shown in Fig.2. It is seen that the system is stable and the tracking error converges.

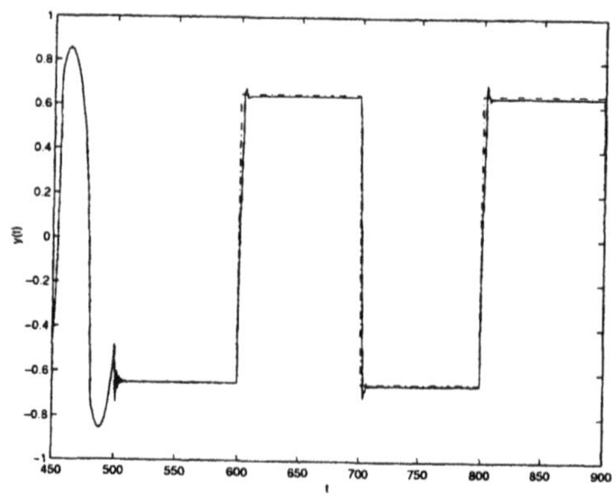

Fig. 2. System tracking performance in example 1; (a) 'Dashed Line'- model reference signal and 'Solid Line'-actual system output.

4. REFERENCES

Atkinson, A. C. and A. N. Donev (1992). *AOptimum Experimental Designs*. Clarendon Press. Oxford.

Bossley, K. M. (1997). Neurofuzzy modelling approaches in system identification. Ph.D Thesis, Dept of Electronics and Computer Science, University of Southampton. Southampton, UK.

Brown, M. and C. J. Harris (1994). *Neurofuzzy Adaptive Modelling and Control.* Prentice Hall. Hemel Hempstead.

Chen, S., S. A. Billings and W. Luo (1989). Orthogonal least squares methods and their application to non-linear system identification. *Int. J. of Control* **50**, 1873–1896.

Feng, M. and C. J. Harris (1998). Adaptive neurofuzzy control for a class of state dependent nonlinear processes. *Int. J. of Systems Science* **29**, 759–771.

Harris, C. J. and X. Hong (2000). Neurofuzzy control using kalman filtering state feedback with coloured noise for unknown nonlinear processes. *ISIS Technical Report,ISIS Group, Dept of Electronics and Computer Science, University of Southampton* p. Submitted for Publication.

Harris, C. J., M. Brown, K. M. Bossley, D. J. Mills and M. Feng (1996). Advances in neurofuzzy algorithms for real time modelling and control. *J. Engineering Applications. AI.* **9**, 1–16.

Hong, X. and C. J. Harris (2000). Neurofuzzy design and model construction of nonlinear dynamical processes from data. *ISIS Technical Report,ISIS Group, Dept of Electronics and Computer Science, University of Southampton* p. Submitted for Publication.

Obradovic, D. (1996). On-line training of recurrent neural networks with continuous topology adaptation. *IEEE Trans. on Neural Networks* **7**, 222–228.

Schumaker, L. L. (1981). *Spline Functions: Basic Theory.* Wiley. New York.

Takaga, T. and M. Sugeno (1985). Fuzzy identification of fuzzy systems and its application to modelling and control. *IEEE Trans on Systems, Man and Cybernestics* **15**, 116–132.

PENALIZED IDENTIFICATION FOR SELF-TUNING CONTROL: AN OVERVIEW [1]

Sergio Bittanti * M.C. Campi ** Maria Prandini **

* *DEI - Politecnico di Milano*
Piazza Leonardo da Vinci 32, 20133 Milano, Italy
bittanti@elet.polimi.it
** *DEA - Università degli Studi di Brescia*
Via Branze 38, 25123 Brescia, Italy
{campi,prandini}@ing.unibs.it

Abstract: In this paper we discuss the use of penalized identification in adaptive control. In particular, we consider self-tuning control incorporating the least squares identification method, and discuss various modifications all based on the introduction of suitable penalization terms in the least squares cost. The objective is to enforce desired properties on the parameter estimates, while preserving the valuable identification properties of the least squares method. We point out that the penalized approach proves to be effective in many situations, ranging from the exploitation of *a-priori* knowledge on the system parameter to the achievement of self-optimality.

Keywords: least squares estimation, constraint satisfaction, adaptive control.

1. INTRODUCTION

In the certainty equivalence approach to adaptive control, the controller applied to the unknown system is continuously tuned to the best representative of the system in a given parameterized model class (Goodwin and Sin, 1984). The choice of what is a representative model is based on an *identification criterion* which depends on the input-output data collected on-line from the closed-loop operating system.
In general, standard identification procedures such as the least squares (LS) algorithm return an estimated model that reproduces the observed closed-loop behavior of the true system. This property is important, since it entails that, by imposing a certain behavior to the estimated model, one can impose the same behavior to the unknown system. On the other hand, there is a main drawback in the use of the raw form of the LS algorithm: it does not incorporate any possible *a-priori* desire that the parameter estimate lies in a certain region of the parameter space. There are many possible motivations for desiring that the estimates belong to certain feasible sets. Some instances are described below.

- It is not rare the case when the system parameters have a physical interpretation and, consequently, must satisfy certain constraints, or, more generally, some *a-priori* knowledge on the system parameter can be available and one would like to embed it in the parameter estimates.
- In other cases, imposing constraints on the values assumed by the parameter estimates is even a requirement for the adaptive control system to work properly. For example, the possible occurrence of unstable pole-zero cancellations in the estimated model leads to a paralysis in the certainty equivalence control law selection.

[1] Research supported by the Ministero dell'Università e della Ricerca Scientifica e Tecnologica under the projects on "New techniques for the identification and adaptive control of industrial systems" and "Synthesis of adaptive and robust controllers".

- The ideal goal of adaptive control is that the adaptively tuned control system attains an optimal performance, i.e., the performance that would be achieved if the system were known from the beginning. As it is better discussed in Section 4, even the *self-optimality* objective can be reformulated as the requirement that the parameter estimates lie in certain regions of the parameter space.

In this paper, we try to give an unifying perspective on the use of penalized identification methods in self-tuning control. In penalized identification, a 'penalization term' is introduced in the standard LS cost so as to bias the criterion in favor of the parameterizations which belong to the desired region. Such a term is appropriately designed so as to preserve the fundamental identification properties of the standard LS algorithm.
Solving a constrained optimization problem where the LS cost is minimized over the desired parameter region, though apparently equivalent to the penalized approach, is not generally effective in self-tuning control, due to the fact that:

- The constrained optimization approach does not ensure that the closed-loop identification property of the original unconstrained identification method is preserved.
- It is generally true that minimizing a penalized cost over the entire parameter space may be easier than minimizing a non-penalized cost over a given feasible region, and in practice in mathematical programming it is often the case that the constraints are suitably embedded into the cost function (Luenberger, 1984).
- It might be the case that the constrained minimization approach is not even conceivable. For example, one might desire that the parameter estimate satisfies a condition that depends on some characteristic of the true system, and hence the constraint parameter set cannot be *a-priori* specified. (This aspect, which may be somehow hermetic at this stage, is given an explicit example in Section 4.2.)

The rest of the paper is organized as follows. In Section 2, we describe our setting and the properties of the LS identification algorithm which are relevant for adaptive control applications. In Section 3, we focus on the case where the constraint region is not known *a-priori*, and precisely describe the penalized approach. We then show that the penalized LS estimate preserves the fundamental properties of the LS algorithm, while satisfying the desired additional constraints. In Section 4 we consider the controllability and self-optimality issues and show how they can be tackled in practice by resorting to the penalized identification approach. Concluding remarks are given in Section 5.

2. SYSTEM DESCRIPTION AND LS IDENTIFICATION

We consider single input-single output systems whose dynamics is governed by:

$$\mathcal{A}(\vartheta^\circ; q^{-1})\, y_t = \mathcal{B}(\vartheta^\circ; q^{-1})\, u_{t-1} + w_t, \qquad (1)$$

with $\mathcal{A}(\vartheta^\circ; q^{-1}) = 1 - \sum_{i=1}^{n} a_i^\circ q^{-i}$ and $\mathcal{B}(\vartheta^\circ; q^{-1}) = \sum_{i=1}^{m} b_i^\circ q^{-i+1}$ polynomials in the unit-delay operator q^{-1} that depend on the *unknown* system parameter vector $\vartheta^\circ = [a_1^\circ\ a_2^\circ \dots a_n^\circ\ b_1^\circ\ b_2^\circ \dots b_m^\circ]^T$. We suppose that the autoregressive part of the system is not trivial, i.e., $n > 0$.

The disturbance $\{w_t\}$ is described as a martingale difference sequence with respect to a filtration $\{\mathcal{F}_t\}$. We make the commonly adopted assumption in adaptive control that

Assumption 1. $\{w_t, \mathcal{F}_t\}$ satisfies almost surely (a.s.)

i) $\sup_t E[|w_{t+1}|^\beta / \mathcal{F}_t] < \infty, \quad \text{for some } \beta > 2,$

ii) $\lim_{N\to\infty} \frac{1}{N} \sum_{t=1}^{N} w_t^2 = \sigma^2 > 0.$

Setting $\varphi_t := [y_t \dots y_{t-(n-1)}\ u_t \dots u_{t-(m-1)}]^T$, system (1) can be given the regression-like form

$$y_t = \varphi_{t-1}^T \vartheta^\circ + w_t. \qquad (2)$$

Based on this representation of system (1), the LS identification cost for estimating ϑ° can be expressed as follows (Ljung, 1999)

$$V_t(\vartheta) = \sum_{s=1}^{t} (y_s - \varphi_{s-1}^T \vartheta)^2. \qquad (3)$$

The properties of the LS estimate $\hat{\vartheta}_t^{LS}$ minimizing (3) have been largely studied. We report below one of the most useful result for adaptive control application (given a square matrix P, $\lambda_{\max}(P)$ denotes its maximum eigenvalue).

Theorem 2. (Lai and Wei, 1982) Suppose that u_t is $\mathcal{F}_t$-measurable. Then,

$$(\vartheta^\circ - \hat{\vartheta}_t^{LS})^T \sum_{s=1}^{t} \varphi_{s-1}\varphi_{s-1}^T (\vartheta^\circ - \hat{\vartheta}_t^{LS})$$
$$= O\big(\log \lambda_{\max}(\sum_{s=1}^{t} \varphi_{s-1}\varphi_{s-1}^T)\big) \ a.s. \qquad (4)$$

Note that this result holds not only in open-loop but also in closed-loop. Indeed, the only assumption is that the input signal is measurable with respect to the past σ-algebra $\mathcal{F}_t$, and this allows for any causal control law.

As better clarified by Proposition 3, one of the distinctive features of result (4) is that it indeed provides a valuable bound on the estimation error $e_t := \varphi_t^T[\vartheta^\circ - \hat{\vartheta}_t]$, *without requiring any particular excitation condition.*

Proposition 3. (Prandini and Campi, 2001) Suppose that the sequence of estimates $\{\hat{\vartheta}_t\}$ satisfies

$$(\vartheta^\circ - \hat{\vartheta}_t)^T \sum_{s=1}^{t} \varphi_{s-1}\varphi_{s-1}^T(\vartheta^\circ - \hat{\vartheta}_t) = o(\sum_{s=1}^{t} \|\varphi_{s-1}\|^2), \quad a.s. \tag{5}$$

Then, $e_t = \varphi_t^T[\vartheta^\circ - \hat{\vartheta}_t]$ satisfies

$$\sum_{t=0,\ t\notin\mathcal{B}_N}^{N} e_t^2 = o(\sum_{t=0}^{N} \|\varphi_t\|^2 + N), \quad a.s., \tag{6}$$

where $\mathcal{B}_N$ is a set of instant points which depends on N, with bounded cardinality: $|\mathcal{B}_N| \leq C_{\mathcal{B}}, \forall N$.

Roughly speaking equation (6) says that the estimation error e_t is negligible compared to the signals involved in the loop. Equation (6) plays a fundamental role in establishing stability and performance results in self-tuning control (Guo, 1996; Prandini and Campi, 1997). The reason for this is easily understood if one notes that (i) the true system can be seen as a variational system with respect to the estimated model, with e_t entering it as a perturbation term:

$$y_{t+1} = \varphi_t^T\hat{\vartheta}_t + e_t + n_{t+1},$$

and (ii) the control law is tuned to the estimated model. From now on, we refer to the key property (5) as the "closed-loop identification property" to emphasize that it holds even in closed-loop.

3. PENALIZED APPROACH

As explained in the introduction, it is often the case that one needs to enforce some constraints in the estimate so as to satisfy specific requirements. Unfortunately, the standard LS criterion is not suitable in this respect, since it does not incorporate any possible *a-priori* desire that the parameter estimate lies in a certain region of the parameter space. The challenge is then to introduce an identification method capable of driving the estimate into some feasible set while preserving the closed-loop identification property (5).
With this twofold objective in mind, we consider penalized identification costs of the form:

$$D_t(\vartheta) = V_t(\vartheta) + P_t(\vartheta), \tag{7}$$

where $P_t(\vartheta)$ is a *penalization term* added to the standard LS cost so as to favor those parameterization satisfying the desired property. Typically,

$$P_t(\vartheta) = \alpha_t\, p(\vartheta). \tag{8}$$

The time invariant term $p(\vartheta)$ in (8) may take a variety of forms depending on the problem at hand, but in any case it is chosen so as to satisfy:

- $p(\vartheta) \geq 0$, for all $\vartheta \in \mathbb{R}^{n+m}$;
- $p(\vartheta)$ is 'small' for all ϑ belonging to the desired region in the parameter space.

The time-varying coefficients $\{\alpha_t\}_{t\geq 0}$ in (8) are positive real numbers adaptively selected on the basis of the data collected from the controlled system so as to balance the contributions of the LS cost and the penalization term in (7).

In this paper, we focus on the case when the desired parameter set can be expressed as

$$\Theta := \{\vartheta \in \mathbb{R}^{n+m} : f(\vartheta) \leq f(\vartheta^\circ)\}, \tag{9}$$

where $f(\vartheta)$ is a positive function of ϑ and $f(\vartheta^\circ)$ is bounded. Indeed, a constraint of the form (9) on the parameter set is most effective in many instances of interest, as it will be better explained in Section 4 by means of a few examples.

Obviously, Θ defined in (9) is not *a-priori* known, since it depends on ϑ°. A possible way of addressing this problem is to replace Θ with the set obtained by substituting ϑ° in (9) with the LS estimate evaluated on a large number of — say— N data, in the hope to get a value for $f(\hat{\vartheta}_N^{LS})$ close to $f(\vartheta^\circ)$. The flaw of this procedure is that in adaptive control there is no guarantee of parameter consistency, and hence the estimate of Θ may be poor.
The alternative we propose here is to resort to the penalized approach with cost (7)-(8). We shall in fact show that by choosing $p(\vartheta) = f(\vartheta)$ and

$$\alpha_t := \log^{1+\delta} \lambda_{max}(\sum_{s=1}^{t} \varphi_{s-1}\varphi_{s-1}^T), \tag{10}$$

$\delta > 0$ being an arbitrary constant, the minimizer of $D_t(\vartheta)$ is enforced to get arbitrarily close to the desired set Θ, without explicitly estimating it.

We are now in a position to supply the fundamental theorem characterizing the remarkable properties of the penalized LS approach. The proof given herein unifies and generalizes the proofs in (Prandini *et al.*, 1998) and (Prandini and Campi, 1998) where specific $f(\vartheta)$ were considered. Denote by $\hat{\vartheta}_t$ the minimizer of $D_t(\vartheta)$, i.e.,

$$\hat{\vartheta}_t := \arg \min_{\vartheta\in\mathbb{R}^{n+m}} D_t(\vartheta) \tag{11}$$

(if $D(\vartheta)$ has multiple minima, pick any of them).

Theorem 4. Suppose that u_t is $\mathcal{F}_t$-measurable. Then, the penalized LS estimate $\hat{\vartheta}_t$ satisfies:

i) $(\vartheta^\circ - \hat{\vartheta}_t)^T \sum_{s=1}^{t} \varphi_{s-1}\varphi_{s-1}^T (\vartheta^\circ - \hat{\vartheta}_t)$

$$= O(\log^{1+\delta} \lambda_{max}(\sum_{s=1}^{t} \varphi_{s-1}\varphi_{s-1}^T), \quad a.s.,$$

ii) $\limsup_{t\to\infty} f(\hat{\vartheta}_t) \le f(\vartheta^\circ), \quad a.s.$

Proof. *i)* The LS estimate $\hat{\vartheta}_t^{LS}$ is such that $\sum_{s=1}^{t} \varphi_{s-1} y_s = \sum_{s=1}^{t} \varphi_{s-1}\varphi_{s-1}^T \hat{\vartheta}_t^{LS}$. By using this expression, it is easy to show that

$$D_t(\vartheta) - V_t(\hat{\vartheta}_t^{LS}) = \alpha_t p(\vartheta)$$
$$+ (\vartheta - \hat{\vartheta}_t^{LS})^T \sum_{s=1}^{t} \varphi_{s-1}\varphi_{s-1}^T (\vartheta - \hat{\vartheta}_t^{LS}). \quad (12)$$

By the definition of $\hat{\vartheta}_t$ we have $D_t(\hat{\vartheta}_t) - V_t(\hat{\vartheta}_t^{LS}) \le D_t(\vartheta) - V_t(\hat{\vartheta}_t^{LS})$, $\vartheta \in \mathbb{R}^{n+m}$. By choosing $\vartheta = \vartheta^\circ$ and using expression (12), we then get

$$(\hat{\vartheta}_t - \hat{\vartheta}_t^{LS})^T \sum_{s=1}^{t} \varphi_{s-1}\varphi_{s-1}^T (\hat{\vartheta}_t - \hat{\vartheta}_t^{LS}) + \alpha_t p(\hat{\vartheta}_t)$$
$$\le (\vartheta^\circ - \hat{\vartheta}_t^{LS})^T \sum_{s=1}^{t} \varphi_{s-1}\varphi_{s-1}^T (\vartheta^\circ - \hat{\vartheta}_t^{LS}) + \alpha_t p(\vartheta^\circ)$$
$$= O(\alpha_t), \quad a.s., \quad (13)$$

where the last equality follows from Theorem 2, the definition (10) of α_t, and the fact that $p(\vartheta^\circ)$ is bounded. Since $\alpha_t p(\hat{\vartheta}_t) \ge 0$, we then have $(\hat{\vartheta}_t - \hat{\vartheta}_t^{LS})^T \sum_{s=1}^{t} \varphi_{s-1}\varphi_{s-1}^T (\hat{\vartheta}_t - \hat{\vartheta}_t^{LS}) = O(\alpha_t)$, a.s. Hence, from definition (10) of α_t and Theorem 2, we get

$$(\hat{\vartheta}_t - \vartheta^\circ)^T \sum_{s=1}^{t} \varphi_{s-1}\varphi_{s-1}^T (\hat{\vartheta}_t - \vartheta^\circ)$$
$$\le 2\Big[(\hat{\vartheta}_t - \hat{\vartheta}_t^{LS})^T \sum_{s=1}^{t} \varphi_{s-1}\varphi_{s-1}^T (\hat{\vartheta}_t - \hat{\vartheta}_t^{LS}) +$$
$$(\hat{\vartheta}_t^{LS} - \vartheta^\circ)^T \sum_{s=1}^{t} \varphi_{s-1}\varphi_{s-1}^T (\hat{\vartheta}_t^{LS} - \vartheta^\circ)\Big] = O(\alpha_t),$$

a.s., thus concluding the proof of *i)*.

ii) A simple elaboration of (13) gives

$$p(\hat{\vartheta}_t) \le \frac{(\vartheta^\circ - \hat{\vartheta}_t^{LS})^T \sum_{s=1}^{t} \varphi_{s-1}\varphi_{s-1}^T (\vartheta^\circ - \hat{\vartheta}_t^{LS})}{\alpha_t}$$
$$+ p(\vartheta^\circ) = \frac{O(\log \lambda_{max}(\sum_{s=1}^{t} \varphi_{s-1}\varphi_{s-1}^T))}{\log^{1+\delta} \lambda_{max}(\sum_{s=1}^{t} \varphi_{s-1}\varphi_{s-1}^T)} + p(\vartheta^\circ),$$

a.s., where we have used definition (10) of α_t and Theorem 2. To conclude, it suffices to show that

$$\lim_{t\to\infty} \log \lambda_{max}(\sum_{s=1}^{t} \varphi_{s-1}\varphi_{s-1}^T) = \infty, \text{ a.s.,} \quad (14)$$

and recall that $p(\vartheta) = f(\vartheta)$.
Observe that from (2) it follows that $w_s^2 \le 2\max\{\|\vartheta^\circ\|^2, 1\}[y_s^2 + \|\varphi_{s-1}\|^2]$. Taking into account that the autoregressive part of model (2) is not trivial ($n > 0$), this in turn implies that $w_s^2 \le 2\max\{\|\vartheta^\circ\|^2, 1\}[\|\varphi_s\|^2 + \|\varphi_{s-1}\|^2]$, from which it is easily shown that $\sum_{s=1}^{t-1} w_s^2 \le h \sum_{s=1}^{t} \|\varphi_{s-1}\|^2$, where h is a suitable constant. From Assumption 1, equation (14) finally follows. □

This result shows that the penalized approach has notable properties. First of all, *i)* entails that the fundamental closed-loop identification property (5) remains valid. Moreover, if $f(\vartheta)$ is a continuous function of ϑ, *ii)* shows that the penalization term $P_t(\vartheta)$ is effective in pushing the estimate arbitrarily close to the desired region Θ.

4. ADAPTIVE CONTROL APPLICATIONS

4.1 *The controllability issue*

A system is controllable if its transfer function does not present pole-zero cancellations. Unfortunately, standard parameter estimation techniques do not guarantee the estimated model to be controllable, thus leading to a paralysis in the certainty equivalence control law selection.
For a given parameterization ϑ, a standard measure of the controllability of the corresponding model $\mathcal{A}(\vartheta; q^{-1})\, y_t = \mathcal{B}(\vartheta; q^{-1})\, u_{t-1}$ is expressed by the determinant of the Sylvester matrix associated with $q^s\mathcal{A}(\vartheta; q^{-1})$ and $q^{s-1}\mathcal{B}(\vartheta; q^{-1})$, $s = \max\{n, m\}$, which is known as the *Sylvester resultant* $Sylv(\vartheta)$ (Kailath, 1980). $|Sylv(\vartheta)|$ is in fact zero if and only if the model $\mathcal{A}(\vartheta; q^{-1})y_t = \mathcal{B}(\vartheta; q^{-1})u_{t-1}$ is not controllable and $|Sylv(\vartheta)| \ge c > 0$ means that the model has a certain degree of controllability.

The estimated model controllability can be secured by an *a-posteriori modification* of the estimate consists in either adding a suitable term to the parameter estimate (Larminat, 1984; Lozano and Zhao, 1994; Campi, 1996) or projecting it onto an *a-priori* known closed convex region of all controllable models (Goodwin and Sin, 1984; Wen and Hill, 1992). Alternatively, one can modify the performance index so as to guarantee that the corresponding estimated model is asymptotically controllable. This is in fact the approach followed herein. Early contributions in this direction are (Kreisselmeier, 1986; Ossman and Kamen, 1987; Middleton *et al.*, 1988), where deterministic systems eventually affected by bounded noise are treated.

In (7), consider the penalization term

$$P_t(\vartheta) = \alpha_t \frac{1}{|Sylv(\vartheta)|},$$

where α_t is given by (10). If $|Sylv(\vartheta^\circ)| \ge \bar{c}$, then by Theorem 4, we get that the estimate $\hat{\vartheta}_t$ satisfies the closed-loop identification property (5) and has a controllability degree at least equal to that of the true system, i.e., $\liminf_{t\to\infty} |Sylv(\hat{\vartheta}_t)| \geq \bar{c}$. Based on these results, it is possible to conceive self-tuning control procedures ensuring closed-loop stability even in the non-minimum phase case (Prandini and Campi, 1997; Prandini *et al.*, 1998). The described penalized method is suitable for adaptive control applications where no *a-priori* knowledge on the system parameter is available besides the fact that it corresponds to a controllable model. On the other hand, the designed identification criterion has, in general, multiple local minima, hence its minimization requires a global optimization algorithm. We next show that, under a certain structured knowledge on the uncertainty region for ϑ°, one can adopt an alternative penalized criterion which is easy to minimize.

Suppose that a nominal value $\bar{\vartheta}$ for ϑ° is available. For example, $\bar{\vartheta}$ has been obtained either by a physical knowledge of the system or by some coarse off-line identification procedure. If $\bar{\vartheta}$ corresponds to a controllable model, we can design the control law that stabilizes such a model based on the chosen control methodology and apply it to system (1). However, this generally results in a closed-loop system with a performance inferior to the one that would be achieved if ϑ° were known, and it might even be the case that the closed-loop system is not stable. It is then natural to try to refine the parameter estimate during the normal on-line operating condition of the control system, so as to get a controller better tuned to the actual plant characteristics.

Assume that there exists an $(n+m)$-sphere centered in $\bar{\vartheta}$, say $\mathcal{S}(\bar{\vartheta})$, such that (i) all the parameterizations in the sphere correspond to controllable models and (ii) ϑ° is an interior point of $\mathcal{S}(\bar{\vartheta})$. In order to get a better estimate of ϑ° than $\bar{\vartheta}$, without loosing the controllability property, we then suggest to adopt the penalized criterion (7) with penalization term

$$P_t(\vartheta) = \alpha_t \|\vartheta - \bar{\vartheta}\|^2. \tag{15}$$

From Theorem 4 it follows that, by adaptively choosing α_t according to (10), we get an estimate $\hat{\vartheta}_t$ that satisfies a.s. both the closed-loop identification property (5) and the condition

$$\limsup_{t\to\infty} \|\hat{\vartheta}_t - \bar{\vartheta}\|^2 \leq \|\vartheta^\circ - \bar{\vartheta}\|^2. \tag{16}$$

Since ϑ° is an interior point of $\mathcal{S}(\bar{\vartheta})$, from (16) it easily follows that there exists a time instant $\bar{t}$ such that $\hat{\vartheta}_t \in \mathcal{S}(\bar{\vartheta})$, $t \geq \bar{t}$, and hence it corresponds to a controllable model. Also, given that $Sylv(\vartheta)$ is a continuous function of ϑ for the values of ϑ corresponding to controllable models, the controllability degree of the estimated model is greater or equal to

$$c := \min_{\{\vartheta:\ \|\vartheta-\bar{\vartheta}\|^2 \leq \|\vartheta^\circ-\bar{\vartheta}\|^2\}} |Sylv(\vartheta)| \ (>0), \text{ a.s.}$$

Moreover, $\hat{\vartheta}_t$ can be computed by the following recursive procedure (Prandini and Campi, 1998):

1. Compute P_t according to the following steps: set $T_0 = P_{t-1}$; for $i = 1$ to $n+m$, set $\phi_i = [0 \ldots \underset{\overset{\downarrow}{i}}{1} \ldots 0]^T$ and compute
$$T_i = T_{i-1} - \frac{(\alpha_t - \alpha_{t-1})T_{i-1}\phi_i\phi_i^T T_{i-1}}{1 + (\alpha_t - \alpha_{t-1})\phi_i^T T_{i-1}\phi_i};$$
then, $P_t = T_{n+m} - \dfrac{T_{n+m}\varphi_{t-1}\varphi_{t-1}^T T_{n+m}}{1 + \varphi_{t-1}^T T_{n+m}\varphi_{t-1}}$.
2. Compute $\hat{\vartheta}_t$ according to the equation:
$$\hat{\vartheta}_t = \hat{\vartheta}_{t-1} + P_t\varphi_{t-1}(y_t - \varphi_{t-1}^T\hat{\vartheta}_{t-1}) + P_t(\alpha_t - \alpha_{t-1})(\bar{\vartheta} - \hat{\vartheta}_{t-1}).$$

Actually, the above recursive algorithm minimizes the identification cost

$$\sum_{s=1}^{t}(y_s - \varphi_{s-1}^T\vartheta)^2 + [\alpha_t + \bar{\alpha}]\|\vartheta - \bar{\vartheta}\|^2, \tag{18}$$

provided that the algorithm is initialized with: $\hat{\vartheta}_0 = \bar{\vartheta}$ and $P_0 = [\alpha_0 + \bar{\alpha}]^{-1}I$, $\bar{\alpha} > 0$. However, the term $\bar{\alpha}\|\vartheta - \bar{\vartheta}\|^2$ is a regularization term, which does not hamper the validity of Theorem 4 for the minimizer of (18).

4.2 *The self-optimality issue*

It is well-known that classical self-tuning control schemes can suffer from an identifiability problem, which can degrade the control system performance. The basic reason for this phenomenon is briefly explained hereafter.

Let us introduce the notation $\mathcal{C}(\vartheta_1, \vartheta_2)$ for the closed-loop system where the system with parameterization ϑ_1 is controlled by the control law which is optimal for the model with parameterization ϑ_2. Suppose that there is a parameter value, say $\bar{\vartheta}$, different from the true value ϑ°, such that $\mathcal{C}(\vartheta^\circ, \bar{\vartheta})$ has a behavior identical to $\mathcal{C}(\bar{\vartheta}, \bar{\vartheta})$. Then, if the parameter estimate is equal to $\bar{\vartheta}$ at some time t, subsequent data will tend to reinforce this erroneous estimate, since applying to the true system a control which is optimal for the estimated model results in observations which concur with those that would have obtained if the estimated system were the true system. As a result, a situation may develop where the identifier locks onto a wrong parameter value, regardless of how long information is collected. Consequently, the certainty equivalence control law gets stuck in the control law designed for $\bar{\vartheta}$. This situation can actually happen when using standard identification methods such as the LS algorithm (Kumar, 1983*b*; van Schuppen, 1994; Campi, 1997).

On the other hand, the fact that $C(\vartheta^\circ, \tilde{\vartheta})$ behaves the same as $C(\tilde{\vartheta}, \tilde{\vartheta})$ implies that the achieved control performance will be the same as the performance of the closed-loop estimated system. However, $C(\tilde{\vartheta}, \tilde{\vartheta})$ is the optimal configuration for the *estimated system*, while $C(\vartheta^\circ, \tilde{\vartheta})$ is not an optimal configuration for the *true system*. Thus the achieved control performance is not the optimal one, because of the identifiability problem.

Consider the case when the control performance of a closed-loop system is measured by a control cost such that the lower the value of the cost, the more satisfactory is the control performance. Then the discussion above can be rephrased by saying that *there is a natural tendency of the LS algorithm to return estimates with larger optimal control cost than the optimal cost associated with the true system*. The idea is then to introduce a penalization term in the LS estimator based on the control cost so as to counteract the above tendency, while preserving the LS identification properties ("cost-biased approach"). The cost-biased approach has been originally introduced in (Kumar and Becker, 1982) and successfully applied in a number of different settings (Kumar, 1983*a*; Kumar, 1983*b*; Campi and Kumar, 1998). Here, we extend it to systems in the input-output form (1).

Denote by $J^*(\vartheta)$ the optimal value for the control cost when the parameter of the system to be controlled is ϑ, i.e., the control cost incurred by $C(\vartheta, \vartheta)$. Suppose that $J^*(\vartheta)$ is a positive continuous function of ϑ and $J^*(\vartheta^\circ)$ is bounded. Then, by setting $P_t(\vartheta) = \alpha_t J^*(\vartheta)$ in the identification criterion (7), from Theorem 4 we have that the parameter estimate (11) gets arbitrarily close to the parameter region $\Theta = \{\vartheta : J^*(\vartheta) \leq J^*(\vartheta^\circ)\}$, and the LS closed-loop identification property (5) is preserved. This is used in (Prandini and Campi, 2001) to prove the self-optimality of an infinite horizon LQG adaptive control scheme for system (1). However, again, the identification cost has no recursive algorithm associated with it.

5. CONCLUSION

We have provided an overview of the penalized approach in self-tuning control. A bottom line of our discussion is that adopting penalized techniques —so extensively used in mathematical programming to deal with constrained optimization problems— turns out to be attractive in the area of adaptive control. The penalized approach is in fact shown to be effective for solving important issues in self-tuning control. Still many open problems remain. Among them, the lack of recursive algorithms for the on-line implementation of some penalized identification procedures.

6. REFERENCES

Campi, M.C. (1996). The problem of pole-zero cancellation in transfer function identification and application to adaptive stabilization. *Automatica* **32**, 849–857.

Campi, M.C. (1997). Achieving optimality in adaptive control: the Bet On the Best approach. In: *Proc. 36th Conf. on Decision and Control*. San Diego, CA, USA.

Campi, M.C. and P.R. Kumar (1998). Adaptive linear quadratic Gaussian control: the cost-biased approach revisited. *SIAM J. Control and Optim.* **36**(6), 1890–1907.

Goodwin, G.C. and K.S. Sin (1984). *Adaptive filtering prediction and control*. Englewood Cliffs, NJ: Prentice-Hall.

Guo, L. (1996). Self-convergence of weighted least-squares with applications to stochastic adaptive control. *IEEE Trans. on Aut. Control* **41**, 79–89.

Kailath, T. (1980). *Linear Systems Theory*. Englewood Cliffs, NJ: Prentice-Hall.

Kreisselmeier, G. (1986). A robust indirect adaptive control approach. *Int. J. Contr.* **43**, 161–175.

Kumar, P.R. (1983*a*). Simultaneous identification and adaptive control of unknown systems over finite parameter sets. *IEEE Trans. on Aut. Control* **AC-28**, 68–76.

Kumar, P.R. (1983*b*). Optimal adaptive control of linear-quadratic-Gaussian systems. *SIAM J. Control and Optim.* **21**, 163–178.

Kumar, P.R. and A. Becker (1982). A new family of optimal adaptive controllers for Markov chains. *IEEE Trans. on Aut. Control* **AC-27**, 137–146.

Lai, T.L. and C.Z. Wei (1982). Least squares estimates in stochastic regression models with applications to identification and control of dynamic systems. *Ann. Statist.* **10**, 154–166.

Larminat, P. de (1984). On the stabilizability condition in indirect adaptive control. *Automatica* **20**, 793–795.

Ljung, L. (1999). *System Identification: Theory for the User*. Prentice Hall, Englewood Cliffs.

Lozano, R. and X.H. Zhao (1994). Adaptive pole placement without excitation probing signals. *IEEE Trans. on Aut. Control* **AC-39**, 47–58.

Luenberger, D.G. (1984). *Linear and Nonlinear Programming*. 2nd ed.. Addison-Wesley.

Middleton, R.H., G.C. Goodwin, D.J. Hill and D.Q. Mayne (1988). Design issues in adaptive control. *IEEE Trans. on Aut. Control* **AC-33**, 50–58.

Ossman, K.A. and E.D. Kamen (1987). Adaptive regulation of MIMO linear discrete-time systems without requiring a persistent excitation. *IEEE Trans. on Aut. Control* **AC-32**, 397–404.

Prandini, M. and M.C. Campi (1997). Adaptive pole placement by means of a simple, singularity free, identification algorithm. In: *Proc. 36th Conf. on Decision and Control*. San Diego, CA, USA.

Prandini, M. and M.C. Campi (1998). A new recursive identification algorithm for singularity free adaptive control. *Syst. & Contr. Lett.* **34**(4), 177–183.

Prandini, M. and M.C. Campi (2001). Adaptive LQG control of input-output systems - A cost-biased approach. *SIAM J. Control and Optim.* **39**(5), 1499–1519.

Prandini, M., S. Bittanti and M.C. Campi (1998). A penalized identification criterion for securing controllability in adaptive control. *J. of Math. Syst., Estim. and Control* **8**, 491–494. Full paper access code: 29460.

van Schuppen, J.H. (1994). Tuning of Gaussian stochastic control systems. *IEEE Trans. on Aut. Control* **AC-39**, 2178–2190.

Wen, C. and D.J. Hill (1992). Global boundedness of discrete-time adaptive control using estimator projection. *Automatica* **28**, 1143–1157.

OVERCOMING THE DETECTABILITY OBSTACLE IN CERTAINTY EQUIVALENCE ADAPTIVE CONTROL*

Romeo Ortega°† Elena Panteley‡

‡IPME, Acad. of Sc. of Russia
61, Bolshoi Ave,
St. Petersburg, Russia
elena@ieee.org

°Lab. des Signaux et Systèmes
CNRS-SUPELEC
Gif-sur-Yvette 91192
FRANCE
rortega[moya]@lss.supelec.fr

Abstract: In this paper we are interested in global adaptive stabilization of nonlinear systems where we know a linearly parametrized stabilizer and a Lyapunov function for the ideal closed-loop system such that the standard estimator is implementable, but this Lyapunov function is not strict, i.e., its derivative is only negative semi-definite. Our motivation to consider this situation stems from the fact that, for many physical systems, the natural Lyapunov function candidate is the total energy, which is always not strict. To complete the stability proof in this case it is necessary to add a—rather restrictive—detectability assumption. Our main contribution is to show that it is possible to overcome this obstacle by adding to the parameter estimator an identification error coming from an indirect identifier. We thus replace the detectability assumption by a new condition that essentially requires that the negative-definite terms appearing in the Lyapunov function derivative dominate, the uncertain terms that cannot be matched by the controller. No matching, or persistency of excitation assumptions are needed.

1 Introduction and problem formulation

In this paper we address the problem of global adaptive stabilization of nonlinear systems of the general form

$$\dot{x} = f(x) + g(x)u,$$

where $x \in \mathbf{R}^n$, $u \in \mathbf{R}^m$ and full state is available for measurement. The must straightforward procedure to solve this problem goes through the following steps. First, we assume that there exists a linearly parametrized control

$$u_*(x,\theta) = u_0(x) + \Psi(x)\theta$$

where $\theta \in \mathbf{R}^q$ is a vector of unknown parameters, such that

$$\dot{x} = f_*(x,\theta) \stackrel{\triangle}{=} f + g(u_0 + \Psi\theta)$$

has a globally asymptotically stable equilibrium at $x = 0$. Second, to ensure that a suitable parameter estimator is realizable, we assume that we know a Lyapunov function $V(x,\theta)$ for the ideal system $\dot{x} = f_*$ such that $g^\top \frac{\partial V}{\partial x}$ is independent of the parameters.[1] Then, invoking standard techniques, it is reasonable to propose a certainty-equivalent adaptive control of the form

$$\begin{aligned} u &= u_0 + \Psi\hat{\theta} \\ \dot{\hat{\theta}} &= -\Gamma\Psi^T g^T \tfrac{\partial V}{\partial x} \end{aligned}$$

where $\Gamma = \Gamma^T > 0$ is the adaptation gain. Denoting the parameter error as $\tilde{\theta} \stackrel{\triangle}{=} \hat{\theta} - \theta$, we can define the error dynamics as

$$\begin{aligned} \dot{x} &= f_* + g\Psi\tilde{\theta} \\ \dot{\tilde{\theta}} &= -\Gamma\Psi^T g^T \frac{\partial V}{\partial x} \end{aligned} \qquad (1)$$

Let us postulate now the classical separate Lyapunov function candidate $V_1(x,\tilde{\theta}) = V + \frac{1}{2}\tilde{\theta}^T\Gamma^{-1}\tilde{\theta}$, whose derivative is given by

$$\dot{V}_1 = \frac{\partial V^T}{\partial x} f_* \leq 0.$$

*This work is partially supported by the CNRS through the "GdR Automatique".

†Corresponding author.

[1] Notice that, in general, V will depend on θ. Of course, f, and possibly g, will also depend on θ, and are therefore only partially known. A precise characterization of this dependence will be given in Section 3.

If the Lyapunov function V is strict, i.e., $\frac{\partial V^T}{\partial x} f_*$ is a negative definite function of x, then La Salle's invariance principle allows us to conclude global regulation to zero of the system state. Our main interest here is to study the case when V is non-strict. This situation appears in physical systems, where the natural Lyapunov function candidate is the total energy, which is always not strict. To complete the stability proof in this situation it is necessary to add a—rather restrictive—detectability assumption, namely that, along the trajectories of the overall system (1), the following implication is true

$$\frac{\partial V^T}{\partial x} f_*[x(t)] \equiv 0 \Rightarrow \lim_{t\to\infty} x(t) = 0 \qquad (2)$$

We should underscore that the detectability condition (2) should hold for the system (1), which is defined in the extended space $(x,\tilde{\theta})$. Even though the implication may hold for the trajectories of the ideal system $\dot{x} = f_*$, it may fail for the overall adaptive system. This happens, for instance, in regulation of mechanical systems with PD plus adaptive gravity compensation, see e.g., [3, Section 4.2, pp. 77]. The solution of this problem, first reported in [9], required the ingenious modification of the systems energy function to include normalized cross-terms. In this paper we take an alternative route, and instead of looking for a new Lyapunov function—a well-known Gordian knot of nonlinear control theory—we propose to modify the parameter estimator.

Our main contribution is the proof that it is possible to overcome the detectability obstacle by adding to the parameter estimator of the direct adaptive controller above an identification error coming from an indirect identifier.[2] The inclusion of this error adds to the derivative of the Lyapunov function W a negative quadratic term in the parameter error that, under some suitable assumptions, allows us to recover global convergence of x without the detectability assumption. The latter is replaced then by a new condition that essentially requires that the negative-definite terms appearing in the Lyapunov function derivative dominate, the uncertain terms that cannot be matched by the controller. No matching, or persistency of excitation assumptions are needed.

Caveat: Even though relaxing the need of a strict Lyapunov function is of practical interest, the remaining assumptions made above (i.e., linear parametrization of the control and estimator realizability) are still very restrictive. However, it is our belief that to come to terms with the theoretical limitations of adaptive control, it is important to show that the

[2] Combining direct and indirect schemes has been explored already in the adaptive control literature with the objective of improving transient performance, see e.g. [1, 2].

detectability assumption (2) is not necessary. (This should not be confused with the necessity shown in the fundamental paper [4], where a different notion of detectability is used.)

2 Background: Global attractivity of perturbed systems

In this section we give two technical lemmas which are instrumental for our convergence proofs. These lemmas may be derived from the much more general results reported in [7]. Througouht this paper, we will assume all functions are sufficiently smooth to guarantee the existence of the solutions of the differential equations considered.

Motivated by the discussion of the previous section we consider an scenario where the zero equilibrium of the system

$$\Sigma: \; \dot{x} = F(x), \tag{3}$$

is globally stable—though not asymptotically—and we want to find conditions under which some of the coordinates actually converge to zero. Namely, introducing the following partition of the state $x = [x_1^T, \; x_2^T]^T$, we want to study the global attractivity of the set

$$\mathcal{A} \triangleq \{x_1 = 0\} \tag{4}$$

In our adaptive control context x_1 will be the actual plant states and x_2 the parameter errors which, as is well-known, don't need to converge to zero to achieve the regulation objective.

Similarly to the problem of identification of linear systems (see, e.g., [5]), it turns out that it is convenient to study this problem adopting an "output injection" perspective. More precisely, an identifier for the linear regression model

$$\dot{\theta} = 0, \quad y = \phi^T \theta \tag{5}$$

with θ the unknown parameter vector, y the plant output and ϕ the regressor, is given by

$$\dot{\hat{\theta}} = \phi(\phi^T \hat{\theta} - y).$$

The fundamental "Uniform Complete Observability under Output Injection Lemma", [5, Lemma 2.5.2, p. 73] establishes the asymptotic stability of the corresponding error equation

$$\dot{\tilde{\theta}} = \phi\phi^T \tilde{\theta} \tag{6}$$

where $\tilde{\theta} \triangleq \hat{\theta} - \theta$, by proving that it is equivalent to the uniform complete observability of the system (5), which results from an output–injection of (6). In the same spirit, we will view our adaptive closed-loop system as a perturbation to a nominal globally convergent system (3) of the form

$$\Sigma_\Delta: \; \dot{x} = F(x) + \Delta(x) \tag{7}$$

In Lemma 2.2 we will characterize perturbations which preserve global convergence. Roughly speaking, admissible perturbations must be bounded by a "detectable output", which in its turn must satisfy an integral inequality. To establish this result we need the following preliminary lemma.

Lemma 2.1 *Consider system (3) and the following partition of the state $x = [x_1^T, \; x_2^T]^T$. Assume $x = 0$ is a globally stable equilibrium point, and that for each $r, \delta > 0$, there exists a locally bounded function $\rho_{r\delta} : \mathbb{R}_{\geq 0} \to \mathbb{R}_{\geq 0}$, such that the following inequality holds*[3]

$$\int_0^\infty [k(|x_1(\tau, x_\circ)|) - \delta] \, d\tau \leq \rho_{r\delta}(|x_\circ|). \tag{8}$$

for all initial conditions $x_\circ \in B(r) \triangleq \{x \in R^n : |x| \leq r\}$ and some class $\mathcal{K}$ function[4] *k. Then, the set $\mathcal{A}$ (4) is globally attractive, that is, for all trajectories of (3) we have*

$$\lim_{t \to \infty} x_1(t) = 0$$

Conversely, assume the system (3) is globally stable and the set $\mathcal{A}$ (4) globally attractive. Then, for each $r, \delta > 0$, there exists a locally bounded function $\rho'_{r\delta} : \mathbb{R}_{\geq 0} \to \mathbb{R}_{\geq 0}$, such that the following inequality holds

$$\int_0^\infty [|x_1(\tau, x_\circ)| - \delta] \, d\tau \leq \rho'_{r\delta}(|x_\circ|). \tag{9}$$

for all initial conditions $x_\circ \in B(r)$.

Proof: To prove global convergence of x_1 to 0 we must show that $\forall \; r, \epsilon > 0$ and for all $x_\circ \in B(r)$ there exists $T(r, \epsilon) > 0$ such that

$$|x_1(t, x_0)| \leq \epsilon, \; \forall t \geq T$$

The proof will be done by contradiction using a slight variation of the well-known Barbalat lemma (see also [6] for a similar argument). Towards this end, assume the opposite; that is, that there exist constants $r, \epsilon > 0$, a subsequence $\{t_i\}_{i=1}^\infty$, satisfying $t_i \to \infty$ as $i \to \infty$, and initial conditions $x_\circ \in B_r$ such that

$$|x_1(t_i, x_\circ)| > \epsilon \tag{10}$$

Now, as x_1 is solution of an ordinary differential equation, it is a continuous function. Furthermore, x_1 is bounded because the system (3) is globally stable. In addition, since x_1 is bounded and $F(x)$ is locally Lipschitz then $\dot{x}_1$ is also bounded. We can conclude then that x_1, being a bounded function with bounded derivative, is uniformly continuous. This means that, given any constant $c > 0$ we can find another constant $\sigma > 0$, such that the bound $|x_1(t + \tau) - x_1(t)| < c$ holds $\forall \; \tau \in [0, \; \sigma]$, $\forall \, t \geq 0$.

We now fix the constant c as $c \triangleq k^{-1}(\frac{1}{4}k(\frac{\epsilon}{2}))$, with k the class $\mathcal{K}$ function of the lemma and ϵ the constant defined in (10). Uniform continuity of x_1 thus ensures the existence of $\sigma > 0$ such that

$$|x_1(t + \tau) - x_1(t)| < k^{-1}(\frac{1}{4}k(\frac{\epsilon}{2})), \; \forall t \geq 0, \; \forall \, \tau \in [0, \; \sigma] \tag{11}$$

Consider the time window $t \in [t_i, \; t_i + \sigma]$. We have the following series of inequalities:

$$\begin{aligned} k(|x_1(t)|) &= k(|x_1(t_i) + x_1(t) - x_1(t_i)|) \\ &\geq k(|x_1(t_i)| - |x_1(t) - x_1(t_i)|) \\ &\geq k(\tfrac{1}{2}|x_1(t_i)|) - k(|x_1(t) - x_1(t_i)|) \\ &\geq k(\tfrac{\epsilon}{2}) - k \circ k^{-1}(\tfrac{1}{4}k(\tfrac{\epsilon}{2})) = \tfrac{3}{4}k(\tfrac{\epsilon}{2}) \end{aligned}$$

where the last inequality follows from (10) and (11).

Define a new subsequence $\{t_i'\}_{i=1}^\infty \subset \{t_i\}_{i=1}^\infty$, which is also diverging, i.e., $t_i' \to \infty$ as $i \to \infty$, but with the additional property that $t_{i+1}' - t_i' \geq \sigma$. (Notice that, as $\{t_i\}_{i=1}^\infty$ is diverging, it is always possible to define such a subsequence.) Fix the constant δ of (8) as $\delta \triangleq \frac{1}{4}k(\frac{\epsilon}{2})$. Using the last inequality above we obtain the bound

$$\begin{aligned} \Sigma_{i=1}^\infty \int_{t_i'}^{t_i'+\sigma} [k(|x(\tau)|) - \delta] d\tau &= \\ & \Sigma_{i=1}^\infty \int_{t_i'}^{t_i'+\sigma} [k(|x(\tau)|) - \tfrac{1}{4}k(\tfrac{\epsilon}{2})] d\tau \\ &\geq \Sigma_{i=1}^\infty \int_{t_i'}^{t_i'+\sigma} [\tfrac{3}{4}k(\tfrac{\epsilon}{2}) - \tfrac{1}{4}k(\tfrac{\epsilon}{2})] d\tau \\ &= \Sigma_{i=1}^\infty \tfrac{1}{2}k(\tfrac{\epsilon}{2})\sigma = \infty \end{aligned}$$

[3] Where we denote by $x(t, x_\circ)$ (and sometimes simply by x if there is no ambiguity from the context) the solution at time t of (3) with $x(0) = x_\circ$.

[4] A function $k : \mathbb{R}_{\geq 0} \to \mathbb{R}_{\geq 0}$ is said to belong to class-$\mathcal{K}$ if it is continuous, zero at zero and strictly increasing. It is said to belong to class-$\mathcal{K}_\infty$ if, furthermore, it is unbounded. For ease of reference we will use the symbols k, γ and β to denote class-$\mathcal{K}$, class-$\mathcal{K}_\infty$ and nondecreasing positive functions $\mathbb{R}_{\geq 0} \to \mathbb{R}_{\geq 0}$, respectively.

On the other hand, from the fact that $t'_{i+1} - t'_i \geq \sigma$, we can upperbound the summation above as

$$\int_0^\infty [k(|x(\tau)|) - \delta]d\tau \geq \Sigma_{i=1}^{\infty} \int_{t'_i}^{t'_i+\sigma} [k(|x(\tau)|) - \delta]d\tau$$

We have shown above that the right hand side is infinity. On the other hand, the function $\rho_{r\delta}$ in (8) is bounded for all bounded r, which establishes our contradiction. This completes the first part of the lemma.

Conversely, if the set $\mathcal{A}$ (4) is globally attractive, then, for each $r, \delta > 0$, exits $T(r, \delta)$ such that

$$|x_1(t, x_0)| \leq \delta \tag{12}$$

for each $t \geq T$ and $x_\circ \in B(r)$. Fix $r, \delta > 0$, then

$$\begin{aligned}
\int_0^\infty [|x_1(\tau, x_\circ)| - \delta]\, d\tau &= \\
\int_0^{T(r,\delta)} [|x_1(\tau, x_\circ)| - \delta]\, d\tau &+ \int_{T(r,\delta)}^{\infty} [|x_1(\tau, x_\circ)| - \delta]\, d\tau \\
\leq \int_0^{T(r,\delta)} |x_1(\tau, x_\circ)| d\tau & \qquad (13)\\
\leq \int_0^{T(r,\delta)} k(|x_\circ|)d\tau = k(|x_\circ|)T(r,\delta) &\triangleq \rho'_{r\delta}(|x_\circ|) \qquad (14)
\end{aligned}$$

where, to get the first inequality we use the bound (12), and for the second one we have used the assumption of global stability of the system (3), being k a class $\mathcal{K}$–function. This completes the proof of the lemma.

We now characterize a class of perturbations to the system (3) of the form (7) which preserves global attractivity of the set $\mathcal{A}$.

Lemma 2.2 *Consider the perturbed system Σ_Δ described by (7) where F, Δ are locally Lipschitz in x, and introduce the following partition of the state $x = [x_1^T,\ x_2^T]^T$. Assume the nominal system Σ (3) has a globally stable equilibrium point at $x = 0$ and the set $\mathcal{A}$ is globally attractive. Then, the set $\mathcal{A}$ will remain globally attractive for Σ_Δ for all functions Δ that satisfy the following conditions:*

(i) Global stability of the trivial equilibrium is preserved.

(ii) There exists nondecreasing functions $\beta_1, \beta_2, \beta_3$, a function $k \in \mathcal{K}$ and a function $\gamma_1 \in \mathcal{K}_\infty$ such that Δ is norm bounded by

$$|\Delta(x)| \leq \beta_2(|x|)\gamma_1(|y|) \tag{15}$$

where $y = h(x)$ is an output function which verifies $h(0) = 0$,

$$|h(x)| \leq \beta_3(|x|) \tag{16}$$

and, along the trajectories of Σ_Δ, y satisfies the integral bound

$$\int_0^\infty k(|y(s)|)ds \leq \beta_1(|x_\circ|),\ \forall\, x_\circ \in \mathbb{R}^n \tag{17}$$

Proof: Since the trivial equilibrium of the system (3) is globally stable and the set $\mathcal{A}$ is globally attractive, it follows that there exists a continuously differentiable function $V(x)$ with the properties

$$V(x) \leq \gamma_3(|x|) \tag{18}$$

$$\dot{V}_{(3)} \leq -\gamma_4(|x_1|) \tag{19}$$

$$\left|\tfrac{\partial V}{\partial x}\right| \leq \gamma_5(|x|) \tag{20}$$

where the functions $\gamma_i \in \mathcal{K}_\infty$, $i = 3, \cdots, 5$.

Taking the derivative of V along the trajectories of (7) we obtain

$$\dot{V}\,|_{(7)} \leq -\gamma_4(|x_1|) + \gamma_5(|x|)\beta_2(|x|)\gamma_1(|y|)$$

where we have used (19), (20) and (15). Now, since the system Σ_Δ is globally stable, there exist $\gamma_6 \in \mathcal{K}_\infty$ such that $|x(t)| \leq \gamma_6(|x_\circ|)$. Therefore, there also exists a nondecreasing function β_4 such that, for all $x_\circ \in B(r)$, we have $\gamma_5(|x(t)|)\beta_2(|x(t)|) \leq \beta_4(r)$ for all $t \geq 0$. Replacing this bound above we get that

$$\dot{V}\,|_{(7)}\ (x(t)) \leq -\gamma_4(|x_1(t)|) + \beta_4(r)\gamma_1(|y(t)|) \tag{21}$$

for all $x_\circ \in B(r)$

We claim that for each r, $\delta > 0$ and $\mu > 0$, there exists $\sigma > 0$ such that

$$\gamma_1(|y|) \leq \frac{\delta}{\beta_4(r)} + \sigma k(|y|), \qquad \forall\, |y| \leq \mu\,.$$

Indeed, noticing that $\gamma_1(s) - \frac{\delta}{\beta_4} < 0$ for $s \in (0, \gamma_1^{-1}(\frac{\delta}{\beta_4}))$, we can take

$$\sigma \triangleq \sup_{s \in (0,\mu]} \left\{ \frac{\gamma_1(s) - \frac{\delta}{\beta_4}}{k(s)} \right\} = \max_{s \in [\gamma_1^{-1}(\frac{\delta}{\beta_4}),\mu]} \left\{ \frac{\gamma_1(s) - \frac{\delta}{\beta_4}}{k(s)} \right\}.$$

Fix $r > 0$ and let $\mu \triangleq \beta_3 \circ \beta_4(r)$. Using all these definitions in (21) we obtain that for all $|y| \leq \mu$ and for any $\delta > 0$ there exists $\sigma = \sigma(r, \delta, \mu)$ such that

$$\dot{V}\,|_{(7)}\ (x(t)) \leq -[\gamma_4(|x_1(t)|) - \delta] + \sigma\beta_4(r)k(|y(t)|)\,.$$

Integrating on both sides of the inequality above from zero to ∞ and using (18), we obtain that for any $\delta > 0$

$$\begin{aligned}
\int_0^\infty [\gamma_4(|x_1(\tau)|) - \delta]\, d\tau &\leq \gamma_3(|x_\circ|) + \sigma\beta_4(r) \int_0^\infty k(|y(\tau)|)d\tau \\
&\leq \gamma_3(|x_\circ|) + \sigma\beta_4(r)\beta_1(|x_\circ|) \\
&\triangleq \rho_{r\delta}(|x_\circ|)
\end{aligned}$$

where the first inequality results from the upperbound of (18), while the last inequality is obtained invoking the assumption (17). The proof is completed invoking Lemma 2.1.

3 Main result

As mentioned in Section 1 we are interested in adaptive control problems where the assumptions of linearity in the parameters (LP) and estimator realizability (ER) hold, but the detectability condition (DC) (2) is not satisfied. Specifically, we assume that the Lyapunov function that we know for the ideal closed–loop system is not strict, i.e., its derivative is only negative semi–definite. In this section we show that adding to the parameter estimator of the standard direct adaptive controller an identification error coming from an indirect identifier it is possible to overcome the detectability obstacle in certain nonlinear adaptive control problems. We will require that the negative–definite terms appearing in the Lyapunov function derivative dominate, the uncertain terms that cannot be matched by the controller—we call this the new detectability condition (NDC). We give in the proposition two alternative conditions—corresponding to two particular definitions of the disturbance term Δ—but as will become clear from the proof, it is easy to work out alternative factorizations that might be more suitable for some specific examples.

We consider linearly parametrized, nonlinear systems of the form

$$\dot{x} = f_1(x) + \Phi(x)\theta + g(x)u \tag{22}$$

where $x \in \mathbf{R}^n$, $u \in \mathbf{R}^m$, and the vector of unknown parameters $\theta \in \mathbf{R}^q$. The identification error of the indirect identifier is designed using standard filtering techniques and, for the sake of clarity, is described in a separate lemma below.

Lemma 3.1 *Consider the system (22) and the filtered signals*

$$\begin{aligned}
\dot{z} &= -\lambda(x + z) - (f_1 + gu) \\
\dot{\Phi}_f &= -\lambda\Phi_f + \Phi
\end{aligned} \tag{23}$$

where $\lambda > 0$ is a design parameter. Then, the identification error

$$e_I = \Phi_f \hat{\theta} - (x+z) \quad (24)$$

where $\hat{\theta}$ is a parameter estimate, exponentially converges to $\Phi_f \tilde{\theta}$, where $\tilde{\theta} \triangleq \hat{\theta} - \theta$, at a speed determined by the filters time constant λ. More precisely, we have

$$\begin{array}{rcl} e_I &=& \Phi_f \tilde{\theta} + \epsilon \\ \dot{\epsilon} &=& -\lambda \epsilon \end{array} \quad (25)$$

Proof: Proceeding from the first equation in (25) we have

$$\begin{array}{rcll} \epsilon &=& e_I - \Phi_f \tilde{\theta} & (26) \\ &=& \Phi_f \hat{\theta} - (x+z) - \Phi_f \hat{\theta} + \Phi_f \theta & (27) \\ &=& -(x+z) + \Phi_f \theta & (28) \end{array}$$

where we have used (24) to obtain the second identity. The derivative of ϵ, along the trajectories of (23), is

$$\begin{array}{rcl} \dot{\epsilon} &=& -\dot{x} - \dot{z} + \dot{\Phi}_f \theta \\ &=& \lambda(x+z) - \Phi\theta + (\Phi - \lambda\Phi_f)\theta \\ &=& -\lambda\epsilon \end{array}$$

and the proof is completed.

We are now ready to present the main result of this paper.

Proposition 3.1 *Consider the system (22). Assume*

(LP) There exists a linearly parametrized control

$$u_*(x,\theta) = u_0(x) + \Psi(x)\theta$$

such that

$$\dot{x} = f_*(x,\theta) \triangleq f_1 + (\Phi + g\Psi)\theta + g u_0 \quad (29)$$

has a globally asymptotically stable equilibrium at $x=0$.

(ER) We know a Lyapunov function $V(x,\theta)$ such that $g^T \frac{\partial V}{\partial x}$ is independent of the parameters.

(NDC) The Lyapunov function V is nonstrict, but there exists an output function $y = h(x)$, with $h(0) = 0$, a function $k_1 \in \mathcal{K}$, and nondecreasing functions β_1, β_5 such that

(i) $\dot{V}|_{(29)} \leq -k_1(|y|)$

(ii) $|\frac{\partial V}{\partial x}| \leq \beta_1(|x|)$

(iii) $|h(x)| \leq \beta_5(|x|)$

Furthermore, there exist nondecreasing functions $\beta_2, \beta_3, \beta_4$, and $\gamma_1, \gamma_2, \gamma_3 \in \mathcal{K}_\infty$ such that the regressor Ψ and the function Φ satisfy the bounds

(iv) $\|\Phi - g\Psi\| \leq \beta_2(|x|)\gamma_1(|y|)$.

(v) $|\frac{\partial \Phi_i}{\partial x}(f_* + g\Psi\tilde{\theta})| \leq \beta_3\left(\left|\left[\begin{smallmatrix} x \\ \tilde{\theta} \end{smallmatrix}\right]\right|\right)\gamma_2(|y|)$, $i = 1,\ldots,q$, *where* $\Phi = [\Phi_1, \cdots \Phi_q]$, $\Phi_i \in \mathbb{R}^{n\times q}$; *or*

(v)' $|\frac{\partial \eta_i}{\partial x}(f_* + g\Psi\tilde{\theta})| \leq \beta_4\left(\left|\left[\begin{smallmatrix} x \\ \tilde{\theta} \end{smallmatrix}\right]\right|\right)\gamma_3(|y|)$, $i = 1,\ldots,q$, *where* $g\Psi = [\eta_1, \cdots, \eta_q]$, $\eta_i \in \mathbb{R}^{n\times q}$.

Under these conditions, the adaptive control law

$$\begin{array}{rcl} u &=& u_0 + \Psi\hat{\theta} \\ \dot{\hat{\theta}} &=& -\Gamma(\Psi^T g^T \frac{\partial V}{\partial x} + \Phi_f^T e_I) \end{array} \quad (30)$$

with e_I, Φ_f given by (23), (24), and $\Gamma = \Gamma^T > 0$ is the adaptation gain, achieves global regulation to zero of the state of the system (22). That is, the closed-loop has a globally stable equilibrium at $(x, \tilde{\theta}) = 0$, and the set $\{x = 0\}$ is globally attractive.

Proof: The proof is established invoking Lemma 2.2 and is divided into four parts. First, using standard arguments we prove that the closed-loop system, which we identify with the perturbed dynamics Σ_Δ, is globally stable. Second, we introduce a change of coordinates that allows us to express Σ_Δ into the perturbed form (7), for some suitably defined F and Δ. Third, we verify that, under the assumptions ER and NDC (i)-(v) of the proposition, the unperturbed system is globally convergent and the perturbation Δ satisfies the conditions of Lemma 2.2. While proving global convergence of the unperturbed system under assumption NDC (v) is relatively straightforward, the establishment of this property with NDC $(v)'$ requires a new mathematical machinery that we borrow from [8]. This is done is the fourth step of the proof.

(1) *(Global stability)* From (29) and invoking Lemma 3.1 we can write the error dynamics of the closed-loop system (22), (23), (24), (30) as

$$\begin{array}{rcl} \dot{x} &=& f_* + g\Psi\tilde{\theta} \\ \dot{\Phi}_f &=& -\lambda\Phi_f + \Phi \\ \dot{\tilde{\theta}} &=& -\Gamma[\Psi^T g^T \frac{\partial V}{\partial x} + \Phi_f^T(\Phi_f\tilde{\theta} + \epsilon)] \\ \dot{\epsilon} &=& -\lambda\epsilon \end{array} \quad (31)$$

Let

$$V_1(x,\tilde{\theta},\epsilon) = V(x) + \frac{1}{2}\tilde{\theta}^T\Gamma^{-1}\tilde{\theta} + \frac{1}{2\lambda}\epsilon^2$$

Using NDC (i) we can bound its derivative as

$$\begin{array}{l} \dot{V}_1|_{(31)} \leq \\ -k_1(|y|) + \frac{\partial V^T}{\partial x} g\Psi\tilde{\theta} - \tilde{\theta}^T\left[\Psi^T g^T \frac{\partial V}{\partial x} + \Phi_f^T(\Phi_f\tilde{\theta} + \epsilon)\right] - \epsilon^2 \\ \leq -k_1(|y|) - \frac{1}{2}|\Phi_f\tilde{\theta}|^2 - \frac{1}{2}\epsilon^2 \end{array} \quad (32)$$

where we have completed a square to get the last inequality. From the inequality above, and the fact that the Φ_f is generated from a linear exponentially stable filter with $\Phi \in \mathcal{C}^0$, we conclude that the system (31) is globally stable.

(2) *(System decomposition)* To express the error dynamics (31) in the perturbed form (7) we define first a change of coordinates

$$\chi = \lambda\Phi_f - \Phi \quad (33)$$

which yields

$$\Sigma_\Delta : \left\{ \begin{array}{rcl} \dot{x} &=& f_* + g\Psi\tilde{\theta} \\ \dot{\chi} &=& -\lambda\chi - \dot{\Phi} \\ \dot{\tilde{\theta}} &=& -\Gamma\left[\Psi^T g^T \frac{\partial V}{\partial x} + \frac{1}{\lambda}(\chi + \Phi)^T(\frac{1}{\lambda}(\chi+\Phi)\tilde{\theta} + \epsilon)\right] \\ \dot{\epsilon} &=& -\lambda\epsilon \end{array} \right.$$

If we define the "disturbance" term

$$\Delta \triangleq - \left[\begin{array}{c} (\Phi - g\Psi)\tilde{\theta} \\ \dot{\Phi} \\ (g\Psi - \Phi)^T \frac{\partial V}{\partial x} \\ 0 \end{array} \right] \quad (34)$$

we obtain the system

$$\Sigma : \left\{ \begin{array}{rcl} \dot{x} &=& f_* + \Phi\tilde{\theta} \\ \dot{\chi} &=& -\lambda\chi \\ \dot{\tilde{\theta}} &=& -\Gamma\left[\Phi^T \frac{\partial V}{\partial x} + \frac{1}{\lambda}(\chi + \Phi)^T(\frac{1}{\lambda}(\chi+\Phi)\tilde{\theta} + \epsilon)\right] \\ \dot{\epsilon} &=& -\lambda\epsilon \end{array} \right.$$

(3) *(Conditions of Lemma 2.2)* First, we are going to check the assumptions (i) and (ii) of Lemma 2.2. Notice that the change of coordinates (33) is globally defined, thus global stability of Σ_Δ follows immediately from step **(1)** of the proof above. In addition, it is possible to establish a bound of the form (15) for the output injection Δ as follows

$$\begin{array}{rcl} |\Delta| &\leq& \|\Phi - g\Psi\|(|\tilde{\theta}| + |\frac{\partial V}{\partial x}|) + \|\dot{\Phi}\| \\ &\leq& \beta_4\left(\left|\left[\begin{smallmatrix} x \\ \tilde{\theta} \end{smallmatrix}\right]\right|\right)\gamma_4(|y|) \end{array}$$

where $\gamma_4 \in \mathcal{K}_\infty$ and we have used NDC $(ii), (iv), (v)$ to get the last bound. The integral bound (17) is also easily obtained integrating (32) to get

$$\int_0^t k_1(|y(\tau)|)d\tau \quad \leq \quad V_1(x(0), \tilde{\theta}(0), \epsilon(0))$$

Finally, assumption NDC (iii) directly gives (16).

In the following, we are going to prove that Σ is globally stable with the set $\{x = 0\}$ globally attractive. To this end, consider the Lyapunov function candidate

$$V_2(x, \tilde{\theta}, \epsilon, \chi) = V_1 + \frac{1}{2} tr\{\chi^T \chi\}$$

where $tr\{\cdot\}$ denotes the trace of a matrix. Using NDC (i) the derivative may be bounded as

$$\dot{V}_2\mid_\Sigma \leq$$

$$k_1(|y|) - \tilde{\theta}^T \frac{1}{\lambda}(\chi + \Phi)^T \left[\frac{1}{\lambda}(\chi + \Phi)\tilde{\theta} + \epsilon\right] - \epsilon^2 - tr\{\chi^T \dot{\chi}\} \quad (35)$$

$$\leq k_1(|y|) - \frac{1}{2}|\frac{1}{\lambda}(\chi + \Phi)\tilde{\theta}|^2 - \frac{1}{2}\epsilon^2 - \lambda tr\{\chi^T \chi\} \quad (36)$$

where we have, again, completed a square and used the properties of the trace to obtain the last bound. From the latter inequality we immediately conclude global stability of Σ. Further, from La Salle's invariance principle we also have that $y, \chi, (\chi + \Phi)\tilde{\theta}, \epsilon \to 0$. Since $\chi \to 0$ and $(\chi + \Phi)\tilde{\theta} \to 0$ we have that $\Phi\tilde{\theta} \to 0$ as well. From the first equation of the Σ dynamics, and given the fact that $\dot{x} = f_*$ has zero as a GAS equilibrium, we have that $x \to 0$. Therefore, we conclude that the set $\{x = 0\}$ is globally attractive for the system Σ. This completes the proof of the proposition with NDC (v).

(4) *(Proof with NDC $(v)'$)* To establish the proof with the second bound NDC $(v)'$, we define a new disturbance term

$$\Delta' \triangleq - \begin{bmatrix} 0 \\ g\Psi - \Phi \\ 0 \\ 0 \end{bmatrix} \quad (37)$$

From step (1) above, we know that Σ_Δ is globally stable and from step (3) we have that the integral bound (16) is satisfied. In addition, given the inequality NDC (i), we obtain directly a bound of the form (15) for the output injection Δ' (37). Therefore, it is only necessary to prove that under the assumptions of the proposition, the associated unperturbed system—which we call now Σ'—is globally stable with the set $\{x = 0\}$ globally attractive. The dynamics of this system are given by

$$\begin{aligned} \dot{x} &= f_* + g\Psi\tilde{\theta} \\ \dot{\Phi}_f &= -\lambda\Phi_f + g\Psi \\ \dot{\tilde{\theta}} &= -\Gamma\left[\Psi^T g^T \frac{\partial V}{\partial x} + \Phi_f^T(\Phi_f\tilde{\theta} + \epsilon)\right] \\ \dot{\epsilon} &= -\lambda\epsilon \end{aligned}$$

which after the change of coordinates

$$\chi = \lambda\Phi_f - g\Psi \quad (38)$$

yields the system

$$\Sigma' : \left\{ \begin{aligned} \dot{x} &= f_* + g\Psi\tilde{\theta} \\ \dot{\chi} &= -\lambda\chi - \widehat{g\dot{\Psi}} \\ \dot{\tilde{\theta}} &= -\Gamma\left[\Psi^T g^T \frac{\partial V}{\partial x} + \frac{1}{\lambda}(\chi + g\Psi)^T(\frac{1}{\lambda}(\chi + g\Psi)\tilde{\theta} + \epsilon)\right] \\ \dot{\epsilon} &= -\lambda\epsilon \end{aligned} \right. \quad (39)$$

To prove global stability of Σ', we consider that all initial conditions belong to the ball, $z_0 \in B_r$, being z the state of the adaptive closed-loop system, that is, $z = [x^T, \ \tilde{\theta}^T, \ \epsilon, \ \chi_v^T]^T$ where we have defined the vector $\chi_v \triangleq [\chi_1^T, \ldots, \chi_q^T]^T$ with χ_i, $i = 1, \ldots, q$, the columns of the matrix χ.

First, let us define the function

$$W \triangleq |\chi_v|^2 \quad (40)$$

Its derivative along the trajectories of (39) may be bounded as

$$\begin{aligned} \dot{W} &\leq -2\lambda W + 2|\chi_v|\|\widehat{g\dot{\Psi}}\| \\ &\leq -\lambda W + \frac{1}{\lambda}\|\widehat{g\dot{\Psi}}\|^2 \\ &\leq -\lambda W + \frac{1}{\lambda}\beta_4^2\left(\left|\begin{bmatrix} x \\ \tilde{\theta} \end{bmatrix}\right|\right)\gamma_3^2(|y|) \\ &\leq -\lambda W + \bar{c}\gamma_3^2(|y|) \end{aligned} \quad (41)$$

where, to get the second inequality, we have completed a square, used the assumption NDC $(v)'$ for the third one, and for the last inequality defined the constant

$$\bar{c} \triangleq \frac{1}{\lambda} \max_{z \in B_r} \beta_4^2\left(\left|\begin{bmatrix} x \\ \tilde{\theta} \end{bmatrix}\right|\right)$$

We now postulate the following Lyapunov function candidate

$$V_2(\omega) = 2V_1 + W_1 \quad (42)$$

where

$$W_1 \triangleq \frac{1}{\tilde{c}}\alpha(W), \quad (43)$$

$\tilde{c}$ is a constant defined as

$$\tilde{c} \triangleq \max_{z \in B_r} \bar{c}\ \gamma_3^2(|h(x)|)$$

and

$$\begin{aligned} \alpha(s) &\triangleq \int_0^s \varrho(t)dt \\ \varrho(t) &\triangleq k_1 \circ (\gamma_3^2)^{-1}\left(\frac{\lambda t}{2\bar{c}}\right) \end{aligned} \quad (44)$$

In the following, we are going to prove the boundedness of the derivative of W_1 which, taken along the trajectories of (39), yields

$$\begin{aligned} \dot{W}_1 &= \frac{1}{\tilde{c}}\varrho(W)\dot{W} \\ &\leq \frac{1}{\tilde{c}}\varrho(W)\left(-\lambda W + \bar{c}\gamma_3^2(|y|)\right) \end{aligned} \quad (45)$$

where to get the last inequality we have used (41). Next, we study the two possible cases:

(a) If $\bar{c}\ \gamma_3^2(|y|) \leq \frac{1}{2}\lambda W$ then using this inequality in (45) we get

$$\dot{W}_1 \leq -\frac{\lambda}{2\tilde{c}}\varrho(W)W$$

(b) Otherwise, that is, if $\bar{c}\ \gamma_3^2(|y|) > \frac{1}{2}\lambda W$, this implies that

$$W \leq \frac{2}{\lambda}\ \bar{c}\ \gamma_3^2(|y|) \quad (46)$$

Using (46) and the definition (44), it is possible to find the following bound of ϱ

$$\begin{aligned} \varrho(W) &\leq \varrho(\frac{2}{\lambda}\ \bar{c}\ \gamma_3^2(|y|)) \\ &= k_1 \circ (\gamma_3^2)^{-1}\left(\frac{2\ \bar{c}\ \gamma_3^2(|y|)\ \lambda}{2\ \bar{c}\ \lambda}\right) = k_1(|y|) \end{aligned} \quad (47)$$

Now, we can obtain a bound of $\dot{W}_1$ as follows

$$\begin{aligned} \dot{W}_1 &\leq -\frac{\lambda}{\tilde{c}}\varrho(W)W + \frac{1}{\tilde{c}}\varrho(W)\bar{c}\gamma_3^2(|y|) \\ &\leq -\frac{\lambda}{\tilde{c}}\varrho(W)W + \varrho(W) \quad (48) \\ &\leq -\frac{\lambda}{\tilde{c}}\varrho(W)W + k_1(|y|) \quad (49) \end{aligned}$$

where, we have used the properties $\frac{\bar{c}}{\tilde{c}}\gamma_3^2(|y|) \leq 1$ and (47) in order to get the inequalities (48) and (49) respectively.

From these two cases, we get the following bound for $\dot{W}_1$

$$\dot{W}_1 \leq -\frac{\lambda}{2\bar{c}}\varrho(W)W + k_1(|y|) \tag{50}$$

Finally, we are going to obtain a bound for the derivative of the Lyapunov function (42). For ease of reference, we write again the expression of V_1,

$$V_1(x,\tilde{\theta},\epsilon) = V(x) + \frac{1}{2}\tilde{\theta}^T\Gamma^{-1}\tilde{\theta} + \frac{1}{2\lambda}\epsilon^2$$

Using NDC (i) we have that

$$\begin{aligned}\dot{V}_1|_{(39)} &\leq -k_1(|y|) - \tilde{\theta}^T\frac{1}{\lambda}(\chi + g\Psi)^T\left[\frac{1}{\lambda}(\chi + g\Psi)\tilde{\theta} + \epsilon\right] - \epsilon^2 \\ &\leq -k_1(|y|) - \frac{1}{2}|\frac{1}{\lambda}(\chi + g\Psi)\tilde{\theta}|^2 - \frac{1}{2}\epsilon^2 \qquad (51)\end{aligned}$$

where we have completed a square to get the last inequality. Therefore, replacing (51) and (50) in the derivative of (42), we obtain finally

$$\dot{V}_2(\omega) \leq -k_1(|y|) - |\frac{1}{\lambda}(\chi + g\Psi)\tilde{\theta}|^2 - \epsilon^2 - \frac{\lambda}{2\bar{c}}\varrho(|\chi_v|^2)|\chi_v|^2$$

From the latter inequality we immediately conclude global stability of Σ'. Again, from La Salle's invariance principle we also have that $y, \chi, (\chi + g\Psi)\tilde{\theta}, \epsilon \to 0$. Since $\chi \to 0$ and $(\chi + g\Psi)\tilde{\theta} \to 0$ we have that $g\Psi\tilde{\theta} \to 0$ as well. From the first equation of the Σ' dynamics we then have that $x \to 0$. Therefore, we conclude that the set $\{x = 0\}$ is globally attractive for the system Σ'. This completes the proof for the second bound.

References

[1] M. Duarte and K. Narendra. Combined direct and indirect approach to adaptive control. *Tech. Rep. No. 8711, Center for System Science, Yale University, September.*

[2] K. Narendra. Adaptive Systems: Past, Present and Future. *Plenary session Procs. ACC'2000.*

[3] R. Ortega, A. Loria, H. Sira-Ramirez, and P.J. Nicklasson. Passivity-based control of euler-lagrange systems: mechanical, electrical and electromechanical applications. *Springer-Verlag, Berlin,* 1998.

[4] J.P. Hespanha and A.S. Morse. Certainty equivalence implies detectability. *Systems & Control Letters,* 36:1–13, 1999.

[5] S. Sastry and M. Bodson. Adaptive control. *Prentice Hall, Englewood Cliffs, N.J.,* 1989.

[6] G. Tao. A simple alternative to the Barbalat Lemma. *IEEE Transactions on Automatic Control,* 42(5):698, 1997.

[7] A. Teel, E. Panteley and A. Loria. Integral characterizations of uniform asymptotic and exponential stability with applications. *Submitted to Maths. Control Signals and Systems,* 2000.

[8] E. Sontag and A. Teel. Changing supply functions in input/state stable systems. *IEEE Transactions on Automatic Control,* 40:1476–1478, 1995.

[9] P. Tomei. Adaptive PD control for robot manipulators. *IEEE Transactions on Robotics Automat.,* 7(4):565–570, 1991.

ON THE DESIGN OF DIRECT ADAPTIVE CONTROLLERS

Felipe M Pait [*,1]

* *pait@lac.usp.br*
Universidade de São Paulo
Laboratório de Automação e Controle – PTC
Av. Prof. Gualberto trav 3–158
São Paulo SP 05508–900 BRAZIL

Abstract: Direct adaptive control without the familiar reference models is considered. A framework for design using quadratic cost functions is presented, and corresponding error equations are derived using ideas from linear-quadratic optimal control. Convexity of a suitably defined cost function is studied, indicating that parameter estimation can be accomplished using standard optimization techniques. Conditions for existence of a global minimum are presented, and shown to depend crucially on the time variation of the actual feedback control used in the loop. A good starting point is to take a feeback gain random in intervals, and feed data from these intervals into a standard minimization algorithm. *Copyright ©2001 IFAC*

Keywords: Adaptive Control; Direct Control; Linear-Quadratic Optimal Control.

1. DIRECT ADAPTIVE CONTROL

The first step in the design of a *direct* adaptive control system is to decide on the underlying control design methodology. The adjustable control parameters, the shape of the error equation, the ever-present adaptive observer, all follow from this initial choice. This contrasts with indirect adaptive control systems. Indirect controllers typically comprise a parameterized observer which generates an identification error; a certainty-equivalence feedback regulator; and a tuner or adaptive law. These components can be designed in a modular fashion, more or less independently, provided each possesses some properties which are indeed satisfied by typical control and estimation algorithms.

The overwhelming majority of the direct adaptive control literature uses reference models as the design paradigm. This is because the control error between a plant's output and that of a suitably defined reference model can be expressed in a convenient form in which the control parameters appear linearly — provided, of course, that a number of restrictive hypothesis are satisfied. Another class of direct controllers are non-identifier based universal controllers (see (Mareels and Polderman, 1996) for a current presentation of direct, indirect, and non-identifier based adaptive control).

However, reference models are just one possibility in indirect adaptive control, and they are used sparingly outside adaptive control if at all. In this paper we explore the feasibility of another design technique: linear-quadratic optimal control. Design using a quadratic objective is perhaps the most transparent and best understood paradigm that can be applied to detectable, stabilizable linear dynamical systems in general. (Interestingly, universal adaptive controllers often employ quadratic cost functions.) In the sequel we show that a parameterized controller can be constructed using tools from linear-quadratic optimal control.

[1] Research supported by FAPESP – State of São Paulo Research Council, under grant 97/04668-1, and by CNPq – Brazilian Research Council, grant 300932/97-9.

In §2 the direct adaptive control design problem we wish to solve is posed. An error equation is derived in §3, and shown to depend nonlinearly on the feedback control parameter estimates, an obstacle that may have delayed the development of direct adaptive controllers using the linear-quadratic design paradigm. Convexity conditions for existence of a global minimum of a suitably defined cost function of the estimates, using data obtained during a given interval, are stated in §4. These conditions are shown to depend on the time variation of the actual control used to close the feedback loop. A random walk algorithm that fulfills the convexity requirements is presented in §5. The results presented here are useful to construct a complete adaptive control algorithm, which is the subject of continuing investigation.

2. FRAMEWORK

We wish to control a plant with measured output $y \in \mathbb{R}^{n_y}$ and control input $u \in \mathbb{R}^{n_u}$ using an n_x-dimensional identifier-based direct adaptive controller of the form

$$\dot{x} = A_I x + B_I u + D_I y \tag{1}$$

$$u = Fx. \tag{2}$$

The "adaptive observer" or identifier given by (1) can be constructed as spelled out in (Morse, 1992; Morse and Pait, 1994).[2] The plant to be controlled must satisfy the

Hypothesis: There exist an $n_y \times n_x$ matrix C_* and an $n_y \times n_y$ singular, strictly lower triangular matrix G_* such that the pair $(A_I + D_I(I - G_*)^{-1}C_*, B_I)$ is stabilizable and moreover

$$\tilde{y} = C_* x - (I - G_*)y \tag{3}$$

is small in some well-defined sense (for instance, $\tilde{y} \in \mathcal{L}^2$ on some time interval; or $\tilde{y}$ has bounded power; or yet $\tilde{y} = \Delta(s)u(s)$ with the stable transfer function Δ subject to some operator norm bound). ♣♣

The point of this hypothesis is that if the construction in (Morse and Pait, 1994) is followed there must exist an $n_D \times n_x$ matrix $E_I(C_*, G_*)$ such that $\hat{x} = E_I x$ is a nonminimal asymptotic observer for the n_D-dimensional stabilizable, detectable "nominal plant" or design model

$$\dot{x}_D = (A_D + D_I(I - G_*)^{-1}C_D)x_D + B_D u \tag{4}$$

$$y = (I - G_*)^{-1}C_D x_D. \tag{5}$$

Thus if $u = \bar{F}x_D$ is a state feedback that stabilizes (4), the separation principle tells us to use as dynamic output feedback

$$u = \bar{F}\hat{x} = \bar{F}E_I x.$$

That is to say, in order to stabilize the process (4), (5) we might use controller (1), (2) with $F = \bar{F}E_I$. Of course the hypothesis is somewhat restrictive; it implies that the real plant's input-output behavior is close to that of the nominal model.

We may substitute the value of y given by (3) into (1):

$$\dot{x} = (A_I + D_I(I - G^*)^{-1}C^*)x + B_I u - D_I(I - G^*)^{-1}\tilde{y}.$$

If we can choose $F(t)$ in a way that keeps the trajectories of the system above bounded, then u and y must both remain bounded; detectability and stabilizability of the plant then guarantee that *all* signals in the overall system remain bounded. Naturally we are interested in the case where neither C_* nor G_* are known! From now on we shall write $A = (A_I + D_I(I - G_*)^{-1}C_*)$, $B = B_I$, and $w = -D_I(I - G_*)^{-1}\tilde{y}$, and concern ourselves with the problem of adaptively stabilizing a process

$$\dot{x} = Ax + Bu + w \tag{6}$$

using state feedback control (2), where A, B, and w are unknown, A and B are fixed, and w is "small." The signal $\dot{x}$ is available because it is constructed as spelled out in (1), and will be used to tune parameter estimates.

3. ERROR EQUATION

Choose constant, symmetric, nonsingular matrices Q and R and define

$$z = x^\top Q x + u^\top R u. \tag{7}$$

We emphasize that this quantity takes positive values only. Stabilizability of (A, B) in (6) ensures the existence of a positive-definite solution to the algebraic Riccati equation

$$A^\top P + PA - PBR^{-1}B^\top P + Q = 0$$

for any choice of positive-definite matrices Q and R. Thus

$$\begin{aligned}\frac{d}{dt}x^\top P x &= x^\top(A^\top P + PA)x + u^\top B^\top P x \\ &\quad + x^\top P B u + w^\top P x + x^\top P w \\ &= -x^\top Q x - u^\top R u + x^\top(F + R^{-1}B^\top P)^\top \\ &\quad R(F + R^{-1}B^\top P)x + 2x^\top P w.\end{aligned}$$

[2] If the plant is m-dimensional, SISO ($n_y = n_u = 1$), one could pick the matrix triple (A_I, B_I, D_I) as follows: choose a controllable pair $(\bar{A}_{m\times m}, \bar{b}_{m\times 1})$ with $\bar{A}$ stable. Then define $A_I = \begin{bmatrix}\bar{A} & 0\\ 0 & \bar{A}\end{bmatrix}$, $B_I = \begin{bmatrix}0\\ \bar{b}\end{bmatrix}$, and $D_I = \begin{bmatrix}\bar{b}\\ 0\end{bmatrix}$. In this case $n_x = 2m$ and $G_* = 0$.

The following identity results:

$$z = -\dot{x}^\top P x - x^\top P \dot{x} + |R(F - F_*)x|^2 + 2x^\top P w, \quad (8)$$

where F_* denotes the (unknown) stabilizing feedback gain matrix $-R^{-1}B^\top P$.

The expression above is affine in P; thus it might appear reasonable to tune an estimate $\hat{P}$ using a standard instantaneous gradient or recursive least-squares algorithm. On the other hand, if we abstract for a moment the terms in P and w, the term in F looks like the error in a traditional parameter estimation problem. It would be amenable to treatment by a least-squares algorithm, except that in the present case the error is known in magnitude only. The need arises for tuners with capabilities comparable to those of traditional least-squares or gradient-type, which can function with information on the magnitude of the error only.

Expanding on ideas presented in (Pait, 1999), we now choose as the tuning error

$$e_T = -\dot{x}^\top \hat{P} x - x^\top \hat{P} \dot{x} + |R(\hat{F} - F)x|^2 - z. \quad (9)$$

While the expression above establishes e_T as a computable quantity (remember that by construction $\dot{x}$ is available), using (8) we obtain an identity that is useful for analysis purposes:

$$e_T = -\dot{x}^\top(\hat{P} - P)x - x^\top(\hat{P} - P)\dot{x} + |R(\hat{F} - F)x|^2 - |R(F - F_*)x|^2 - 2x^\top P w. \quad (10)$$

Motivation for the definition of e_T above stems out of the observation that $|(\hat{F} - F)x|^2 - |(F - F_*)x|^2 = x^\top(\hat{F} + F_* - 2F)^\top(\hat{F} - F_*)x$ is proportional to $\hat{F} - F_*$, the parameter estimate error (Pait, 1999). From now on for simplicity we take $R = I$ (to treat the general case where R is an arbitrary positive-definite matrix $R_1^\top R_1$, simply substitute $R_1 F$, $R_1 \hat{F}$, and $R_1 F_*$ for F, $\hat{F}$, and F_* respectively in what follows). Notice that e_T is a function of time-varying data as well as of $\hat{P}$ and $\hat{F}$.

If we can adjust the direct adaptive control parameter F and the state cost matrix estimate $\hat{P}$ in a manner that keeps e_T small in some sense without allowing $\hat{P}$ to blow up, then we have achieved a form of direct stabilization *without a reference model*. How one might deal with this problem is the subject the following sections.

4. CONVEXITY

In this section we approach the problem of searching for estimates $\hat{P}$ and $\hat{F}$ by studying conditions for convexity of the cost function

$$J(\hat{P}, \hat{F}) = \frac{1}{2}\operatorname{tr} \hat{P}^2 + \frac{1}{4}\left(\operatorname{tr}(\hat{F} - F_i)(\hat{F} - F_i)^\top\right)^2 + \frac{1}{2}\int_{t_i}^{t_i+T} d\tau\, e_T^2(\hat{P}, \hat{F}, \tau) \quad (11)$$

in the variables $\hat{P}$ and $\hat{F}$ on the interval $[t_i, t_i + T)$. F_i can be interpreted as an a priori guess for $\hat{F}$ based on the data available up to time t_i only. The following proposition is proved in the full-length version of this paper.

Proposition 1. A critical point $(\hat{P}, \hat{F})$ of $J(\hat{P}, \hat{F})$ will be a minimum if

$$\begin{aligned} h = &\operatorname{tr} \pi^\top \pi + 2\left(\operatorname{tr} \phi(\hat{F} - F_i)^\top\right)^2 \\ &+ \left(\operatorname{tr}(\hat{F} - F_i)(\hat{F} - F_i)^\top\right)\left(\operatorname{tr} \phi\phi^\top\right) \\ &+ \int_{t_i}^{t_i+T} d\tau \left(\dot{x}^\top \pi x + x^\top \pi \dot{x} - 2x^\top(\hat{F} - F)^\top \phi x\right)^2 \\ &+ 2\int_{t_i}^{t_i+T} d\tau\, e_T x^\top \phi^\top \phi x > 0 \quad (12) \end{aligned}$$

for all $\pi \in \mathbb{R}^{n_x \times n_x}$, $\phi \in \mathbb{R}^{n_u \times n_x}$.

The first integral in (12) is nonnegative. As e_T is affine in $\hat{P}$, one cannot assure that the second integral in the expression for h remains positive unless $\hat{P}$ is bounded, a fact that needs to be taken into account when formulating a tuning rule. For the time being assume that this is the case and concentrate on the conditions F has to satisfy to ensure convexity. In the remainder of this section the integration variable τ and the limits of integration t_i, $t_i + T$ are suppressed to lighten notation. Let

$$\bar{x} := x|\phi x|;$$

$$\bar{z} := (\dot{x}^\top \hat{P} x + x^\top \hat{P} \dot{x} + z)x^\top \phi^\top \phi x; \text{ and}$$

$$\bar{h} := \left(\operatorname{tr} \phi(\hat{F} - F_i)^\top\right)^2 + \frac{1}{2}\left(\operatorname{tr}(\hat{F} - F_i)(\hat{F} - F_i)^\top\right)\left(\operatorname{tr} \phi\phi^\top\right) + \int e_T x^\top \phi^\top \phi x.$$

We are looking for a time function $F(t)$ that ensures that $\bar{h} > 0$, and thus that every critical point of $J(\cdot, \hat{F})$ is a minimum, implying that the cost function has a unique global minimum. To obtain such conditions, use (9) to write

$$\begin{aligned} \bar{h} = &\operatorname{tr}(\hat{F} - F_i)\left(\phi^\top \phi + I_n \frac{\operatorname{tr} \phi^\top \phi}{2n}\right)(\hat{F} - F_i)^\top \\ &+ \int \left(\bar{x}^\top(\hat{F} - F)^\top(\hat{F} - F)\bar{x} - \bar{z}\right) \\ = &\operatorname{tr}\left(\hat{F}\left(M + \int \bar{x}\bar{x}^\top\right)\hat{F}^\top - \hat{F}\left(MF_i^\top + \int \bar{x}\bar{x}^\top F^\top\right)\right. \\ &\left. - \left(F_i M + \int F\bar{x}\bar{x}^\top\right)\hat{F}^\top + F_i M F_i^\top + \int F\bar{x}\bar{x}^\top F^\top\right) \\ &- \int \bar{z}, \end{aligned}$$

where $M = \phi^\top\phi + I_n \mathrm{tr}\,\phi^\top\phi/2n$; then

$$\bar{h} = \mathrm{tr}\left(\hat{F} - \left(F_i M + \int F\bar{x}\bar{x}^\top\right)\right.$$
$$\left.\left(M + \int \bar{x}\bar{x}^\top\right)^{-1}\right)\left(M + \int \bar{x}\bar{x}^\top\right)$$
$$\left(\hat{F}^\top - \left(M + \int \bar{x}\bar{x}^\top\right)^{-1}\left(MF_i^\top + \int \bar{x}\bar{x}^\top F^\top\right)\right)$$
$$+ F_i M F_i^\top$$
$$+\mathrm{tr}\int F\bar{x}\bar{x}^\top F^\top - \mathrm{tr}\left(F_i M + \int F\bar{x}\bar{x}^\top\right)\left(M + \int \bar{x}\bar{x}^\top\right)$$
$$\left(MF_i^\top + \int \bar{x}\bar{x}^\top F^\top\right) - \int \bar{z} \quad (13)$$

The first parcel in expression (13) is nonnegative, thus a feedback F may ensure uniqueness of the minimum of J by making $h_2 > \int \bar{z}$, where

$$h_2 = \mathrm{tr}\left(F_i M F_i^\top + \int F\bar{x}\bar{x}^\top F^\top\right)$$
$$- \mathrm{tr}\left(F_i M + \int F\bar{x}\bar{x}^\top\right)\left(M + \int \bar{x}\bar{x}^\top\right)^{-1}$$
$$\left(MF_i^\top + \int \bar{x}\bar{x}^\top F^\top\right).$$

(14)

Remark 1. Intuition on what is required of F may be developed if we consider the single input case, where F is a row vector, and take $F_i = 0$ for simplicity. Then $h_2 > \int \bar{z}$ if and only if the matrix

$$\begin{bmatrix} \int F\bar{x}\bar{x}^\top F^\top - \bar{z} & \int F\bar{x}\bar{x}^\top \\ \int \bar{x}\bar{x}^\top F^\top & M + \int \bar{x}\bar{x}^\top \end{bmatrix}$$

is positive definite, which in turn is equivalent to

$$M + \int \bar{x}\bar{x}^\top - \left(\int \bar{x}\bar{x}^\top F^\top\right)\left(\int F\bar{x}\bar{x}^\top F^\top - \bar{z}\right)^{-1}$$
$$\left(\int F\bar{x}\bar{x}^\top\right) > 0,$$

or, since $F\bar{x}\bar{x}^\top F^\top$ is a scalar,

$$\left(\mathrm{tr}\,M + \int \bar{x}^\top\bar{x}\right)\left(\int F\bar{x}\bar{x}^\top F^\top - \bar{z}\right) > \left|\int F\bar{x}\bar{x}^\top\right|^2.$$

Using the Cauchy-Schwartz inequality it can be shown that the above expression always holds for null $\bar{z}$.

What is required in general, roughly speaking, is to find an F such that $F\bar{x}$ is largely orthogonal to $\bar{x}$ (in the sense that their inner product is small); and moreover that the energy of $F\bar{x}$ is larger than the energy of $\bar{z}$. One could search for algorithms that pick F at each instant to guarantee this property, however the tricky part is it must be made to hold for all matrices ϕ appearing in the definition $\bar{x} = x|\phi x|$. One thing that certainly would *not* do is to keep F constant in the interval. In §5 we argue that a simple, effective strategy is to choose $F(t)$ as a random walk.

5. A STOCHASTIC ALGORITHM

In keeping with the idea of spreading the energy of F orthogonally to that of the state $\bar{x}$, it is tempting to move F in each of successive intervals $[t_i, t_i + T)$ in a random walk. Although a continuous-time Brownian motion may not be realized perfectly, a simplified analysis shown in the sequel indicates that this idea is on the right track.

We shall choose $F(t)$ to be a continuous random variable such that $F(t_i) = F_i$, the increments of $F(t)$ over disjoint intervals are independent random variables, and

$$E\left[(F(t+s) - F(s))^\top (F(t+s) - F(s))\right] = \sigma_i^2 t I_{n_u};$$

that is to say, in the interval $[t_i, t_i + T)$ the set of row vectors of F realize an n-dimensional Brownian motion with expected value F_i and variance σ_i^2. Further knowledge about the probability distribution of the random variables is not used in the argument. The proposition below shows that the expected value of h_2 is proportional to F's variance. It holds under the assumption that $x(\cdot)$ is independent of $F(\cdot)$ in the interval $t_i, t_i + T)$, a useful approximation that is akin to what is called "independence theory" in adaptive filtering (see for instance (Haykin, 1997)).

Proposition 2. Let F be a random variable independent of x that realizes a Brownian motion as described above, and suppose that $x(\cdot)$ is independent of $F(\cdot)$ in the interval $t_i, t_i + T)$; then a lower bound for $E[h_2]$ is given by

$$\sigma_i^2\,\mathrm{tr}\left(M + \int_{t_i}^{t_i+T} d\tau\,\bar{x}\bar{x}^\top(\tau)\right)^{-1}$$
$$\left(\int_{t_i}^{t_i+T} d\tau\,\bar{x}\bar{x}^\top(\tau)\int_{t_i}^{T} d\tau_1\left(\int_{t_i}^{\tau_1} d\tau_2\,\bar{x}\bar{x}^\top(\tau_2)\right)\right).$$

A proof of the proposition, under the simplifying assumptions, appears in the full version of this paper. It can be seen readily that in average h_2 is proportional to σ_i^2, and roughly speaking increases with the square of T and linearly with the average value of $|\bar{x}|^2$. The design problem that still needs to be treated carefully is how to choose σ_i^2 to keep $h_2 > \int \bar{z}$ during the interval $[t_i, t_i + T)$.

6. CONCLUDING REMARKS

We posed a direct adaptive control design method based on linear-quadratic optimal control ideas. The resulting error equations resemble somewhat those that appear in traditional parameter estimation problems, and would then be amenable to treatment by a least-squares algorithm, except that in the present case the error is known in magnitude only. This precludes the use of standard tuning methods like least-squares or instantaneous gradient. The need arises for tuners with capabilities comparable to the traditional least-squares or gradient-type, which can function with information on the magnitude of the error only. With the help of such a tuner one can close a direct adaptive feedback loop without making the usual restrictive assumptions of minimum phase, known relative degree, etc.

We approach the tuning problem by finding conditions under which a suitably defined error function has a unique minimum, and show that these conditions depend crucially on the time variation of the actual feedback control used in the loop. A rough and ready stochastic analysis indicates that a good starting point is to take F as a Brownian motion in each interval, and feed data from these intervals into a standard minimization algorithm. The full control algorithm also includes such parameter adjustment laws for $\hat{P}$ and $\hat{F}$, which are the subject of continuing investigation.

Acknowledgements The author is grateful to Yoshiharu Kohayakawa, Wagner Truppel, Oswaldo Luiz do Valle Costa, and Luiz C. Trintinalia for helpful explanations. This paper is dedicated to the dear memory of "tio" Reinaldo Rizzo, who told the author to sit down and do the math.

7. REFERENCES

Haykin, Simon (1997). *Adaptive Filter Theory.* 3rd ed.. Prentice Hall. Englewood Cliffs.

Mareels, Iven and Jan Willem Polderman (1996). *Adaptive Systems: An Introduction.* Birkhäuser. Boston.

Morse, A Stephen (1992). Towards a unified theory of parameter adaptive control – Part 2: Certainty equivalence and implicit tuning. IEEE *Trans. Automatic Control* **37**(1), 15–29.

Morse, A Stephen and Felipe M Pait (1994). MIMO design models and internal regulators for cyclically-switched parameter-adaptive control systems. IEEE *Trans. Automatic Control* **39**(9), 1809–1818.

Pait, Felipe M (1999). Tuning via measurements of the squared error. In: *7th* IEEE *Mediterranean Conference on Control and Automation.* Haifa, Israel. pp. 369–373.

ADAPTIVE CONTROL OF TECHNOLOGICAL PROCESSES BASED ON DUAL YOULA-KUCERA PARAMETRIZATION

František Gazdoš, Petr Dostál

Department of Control Theory, Institute of Information Technologies,
Tomas Bata University in Zlin, nam. TGM 275, 762 72 Zlin, Czech Republic
e-mail: gazdos@ft.utb.cz

Abstract: The paper deals with adaptive control of multivariable technological processes. An iterative scheme of the closed-loop identification and control design based on the dual Youla-Kucera parametrization is employed. Modifying the original algorithm, it is not necessary to do a model reduction. In addition, the number of identified coefficients decrease significantly. The method is illustrated and proved on a simulation study. *Copyright © 2001 IFAC*

Keywords: adaptive control, closed-loop identification, parametrization, multivariable systems, iterative methods, control system design, parameter estimation.

1. INTRODUCTION

Increasing complexity of technological processes has necessitated development of sophisticated control systems, such as robust and adaptive control. The former deals with controller design based on a nominal model of the process with predetermined uncertainty (Morari and Zafiriou, 1989). This was the motivation for investigation of identification techniques estimating upper bounds of the deviations between a controlled system and its nominal model (Helmicki, *et al.*, 1991; Gevers, 1993). Knowledge of these deviations enables to ideally design a parametrized controller for the controlled process.

The latter approach for control of systems with uncertainties represents adaptive control (Astrom and Wittenmark, 1989). In this case, the control systems changes (adapts) according to the actual state of the controlled process. For this method, however, it is necessary to use an efficient identification algorithm (quick, stable and precise) estimating (usually recursively) all parameters of the process. Since it is usually not possible to interrupt running technological processes, closed-loop identification is essential. Works of Gevers (1993), and also Van den Hof and Schrama (1995) presenting a collection of recently developed control design methods based on iteration of plant model identification and controller design seem to be suitable for these purposes. Iteration, with its gradual character, helps to overcome initial instability and enables both, fluent control process and identification.

The presented paper employs an iterative scheme of the closed-loop identification and control design based on the dual Youla-Kucera parametrization (e.g. Anderson, 1998). First, the classical approach involving a model reduction is shown and then, its modification solving the reduction problem is introduced.

Technological processes are usually multi-input/multi-output, then, it is necessary to identify many coefficients and parameters for the whole system. For the introduced approach, it is sufficient to identify only one, the Youla-Kucera (Y-K) parameter, which can be perceived as a deviation between a nominal model of the system and the real process. It is shown that although it is a matrix parameter, it can have very simple structure, so that the number of identified coefficients decrease significantly. The approach is illustrated on a simulation study concerned with the control of a simple nonlinear system with two inputs and two outputs.

2. THEORETICAL BACKGROUND

Consider a feedback control configuration as depicted in Fig. 1,

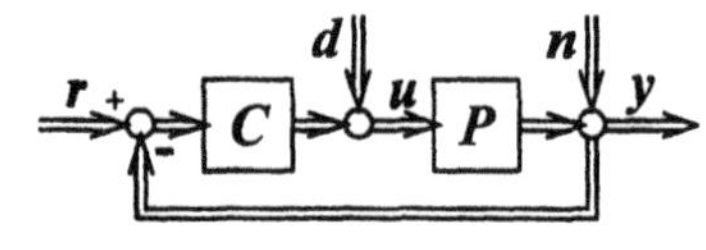

Fig. 1. Classical feedback control set-up

where $y \in \Re^n$, $u \in \Re^m$, $r \in \Re^n$ are vectors of controlled output, control input and the reference; $n \in \Re^n$ represents a vector of additive output noise and $d \in \Re^n$, a vector of disturbance signals. Next, let P stands for a transfer function matrix of a plant and C for a matrix of a controller

$$P = D_L^{-1} N_L = N_R D_R^{-1} \tag{1}$$

$$C = Y_L^{-1} X_L = X_R Y_R^{-1} \tag{2}$$

with D_L, N_L; Y_L, X_L left coprime and N_R, D_R; X_R, Y_R right coprime matrices from φ, where φ denotes the set of stable proper rational transfer functions (matrices). Then, it is well-known that the set of all stabilizing controllers for P is given by this following relationship

$$\begin{aligned} C_S &= (X_R + D_R S)(Y_R - N_R S)^{-1} \\ &= (Y_L - SN_L)^{-1}(X_L + SD_L) \end{aligned} \tag{3}$$

for an arbitrary parameter S from φ, and a stabilizing controller is found using so-called (double) Bezout identity (Vidyasagar, 1985)

$$\begin{bmatrix} Y_L & X_L \\ -N_L & D_L \end{bmatrix} \cdot \begin{bmatrix} D_R & -X_R \\ N_R & Y_R \end{bmatrix} = \begin{bmatrix} I & 0 \\ 0 & I \end{bmatrix} \tag{4}$$

as a particular solution of one of these matrix Diophantine equations

$$\begin{aligned} D_L Y_R + N_L X_R &= I \\ Y_L D_R + X_L N_R &= I \end{aligned} \tag{5}$$

This approach, called Youla-Kucera parametrization (e.g. Anderson, 1998), enables to find all stabilizing controllers if such one is known. This method greatly facilitates design of controllers.

Dual problem to finding all stabilizing controllers if only one is known is finding all plants stabilized by the one controller. Again, consider the control set-up from Fig. 1 with fractional description of a nominal plant and a controller according to (1), (2) and suppose that Bezout equation (4) holds. Then the set of all plant models stabilized by the one controller C is given by

$$\begin{aligned} P_M &= (N_R + Y_R S)(D_R - X_R S)^{-1} \\ &= (D_L - SX_L)^{-1}(N_L + SY_L) \end{aligned} \tag{6}$$

for an arbitrary $S \in \varphi$. This approach, so-called dual Youla-Kucera parametrization, is near to closed-loop identification in the presence of noise, and consequently, to adaptive control.

The basics of closed-loop identification with the help of dual Y-K parametrization were independently given by several authors (Hansen, *et al.*, 1989; Hansen, 1989; Tay, *et al.*, 1989; Schrama, 1991). The key-idea is to directly identify Y-K parameter instead of the classical identification of model coefficients. The most important fact is that this is a standard open-loop identification problem. In addition, estimated models of the process are guaranteed to be stabilized by the current controller.

Substitution of the plant in Fig. 1 by the plant model (6), for example in the right matrix fraction, gives an alternative set-up of the closed loop as shown in Fig. 2.

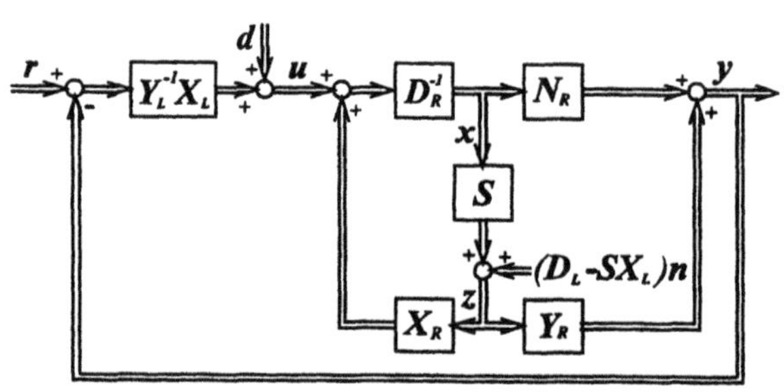

Fig. 2. Alternative set-up of a closed loop

It can easily be proved that the substitution with input u, n and output y implies

$$y = P_M \cdot u + n \tag{7}$$

The task is to identify a true nonlinear plant in the closed-loop from noise-contaminated measurements of u and y. Now, as stated above, the key-identification idea is to identify the parameter S rather than the coefficients of P_M. To see that it is a standard open-loop identification problem observe from Fig. 2 following relationships concerned with signals x and z (used for identification of the parameter S)

$$x = Y_L \cdot d + X_L \cdot r \\ = Y_L \cdot u + X_L \cdot y \qquad (8)$$

$$z = D_L \cdot y - N_L \cdot u \qquad (9)$$

$$z = S \cdot x + (D_L - S \cdot X_L) \cdot n \qquad (10)$$

The last equation represents the most significant relationship for the identification. In this equation, signals x and z are measurable on the closed loop via (8) - (9), and if n is independent of r and d, then x and n are independent processes. Also $S \in \varphi$, i.e. it is stable. Hence, the identification of S is a standard open-loop identification problem.

3. CLASSICAL APPROACH TO ADAPTIVE CONTROL

First, let the time axis be divided into intervals, so that during the *k-th* interval, the control input into the plant is driven by a controller C^k. Next, let there be a nominal model P^k of the plant P stabilized by the controller. Employing $P^k = P$, which leads to $S = 0$ (and shows that the coprime factors of P^k approx. equal to those of the real system), this algorithm of adaptive control based on iteration of following steps can be prepared:

1) For the model of the process, $\dot{P}^k$, a stabilizing controller is computed using Bezout identity (4)

$$Y_L^k D_R^k + X_L^k N_R^k = I \qquad (11)$$

2) The auxiliary signals x and z are constructed

$$x = Y_L^k \cdot u + X_L^k \cdot y \\ z = D_L^k \cdot y - N_L^k \cdot u \qquad (12)$$

3) These signals are used for the identification of S (in a least squares algorithm) with an output error

$$\varepsilon = z - \hat{S} \cdot x \qquad (13)$$

where x is the input and z the output signal.

4) A new transfer function of the identified model is computed

$$P_M^{k+1} = (N_R^k + Y_R^k \hat{S})(D_R^k - X_R^k \hat{S})^{-1} \qquad (14)$$

5) $$P^k = P_M^{k+1} \qquad (15)$$

6) The algorithm continues by step 1 with the new transfer function of the model, P^k.

The fact that parameters of the estimated model are close to the real process is easily recognizable from a record of S: If coefficients of this (matrix) parameter converge to the zero value, then, the identified model is a good approximation of the real system.

However, the computation of a new transfer function using rel. (14) can give rise to a large degree for P_M, such that use of a reduction method is necessary. In order to avoid the reduction, a modification of the presented algorithm follows.

4. MODIFIED APGORITHM OF ADAPTIVE CONTROL

The modification is based on works of Schrama (1991), Van den Hof, *et al.* (1995), and the method is directed towards the identification of coprime factors of P_M employing the auxiliary signal x (measurable on the closed-loop and uncorrelated with the noise signal). Considering $n = 0$, it is possible (from Fig. 2) to derive these formulae

$$y = (N_R + Y_R S) = N_{MR} \cdot x \\ u = D_{MR} \cdot x \qquad (16)$$

where N_{MR}, D_{MR} describes a right matrix fraction of P_M

$$P_M = N_{MR} \cdot D_{MR}^{-1} \qquad (17)$$

Then, the combination of (9) and (10) with (16) yields the following equation

$$S = D_L N_{MR} - N_L D_{MR} \qquad (18)$$

used for the modification of the presented algorithm. All steps are identical to the original version except for step 4. Here, a new transfer function of the identified model is computed by (18), thus

$$\hat{S} = D_L^k \cdot N_{MR}^{k+1} - N_L^k \cdot D_{MR}^{k+1} \qquad (19)$$

If orders of D_L and D_{MR}, and also N_L and N_{MR} are the same, then it is not necessary to use a reduction method as in the original algorithm. This approach, concerned with single-input/single-output systems, was introduced by Kozka, *et al.* (2000). In that paper, the method was utilized for close-loop identification of a laboratory chemical reactor.

Next section illustrates the modified algorithm on a simulation study of controlling a simple nonlinear multivariable system. It is shown that the whole system can be identified using only very few parameters and results are satisfactory.

5. SIMULATION STUDY

Consider the system of two spherical tanks in series as shown in Fig. 3. The task is to control levels of liquid h_1, h_2 by means of the inlets q_{1V}, q_{2V}.

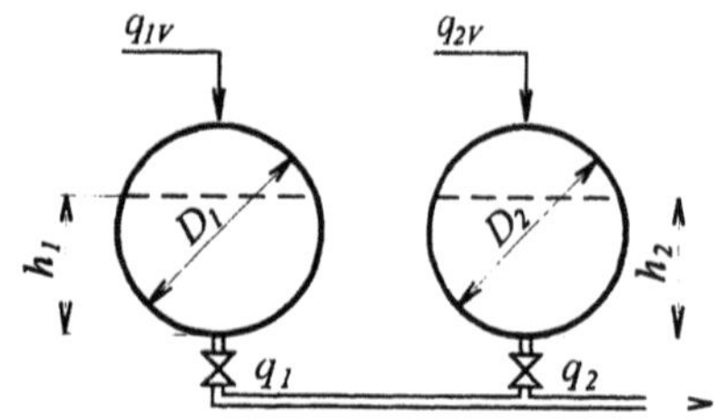

Fig. 3. Controlled system

For this interconnected system with two inputs and two outputs, the following (simplified) mathematical model was derived

$$\pi h_1 (D_1 - h_1)\frac{dh_1}{dt} + q_1 = q_{1V} \qquad (20)$$
$$\pi h_2 (D_2 - h_2)\frac{dh_2}{dt} - q_1 + q_2 = q_{2V}$$

The outflows q_1, q_2 were described using these relationships

$$q_1 = k_1\sqrt{|h_1 - h_2|}\ , \ q_2 = k_2\sqrt{h_2}\ , \qquad (21)$$
$$\text{if } (h_1 - h_2) \prec 0\text{, then } q_1 = -q_1$$

and constants k_1, k_2 were derived from the steady states. Using a linearization technique and conversion of the nonlinear model (20)-(21) into the stable proper rational transfer functions yields this continuous-time linear model of the system (in the right matrix fraction)

$$\boldsymbol{P} = \boldsymbol{N}_R \boldsymbol{D}_R^{-1} \qquad (22)$$

$$\boldsymbol{N}_R = \begin{bmatrix} \dfrac{n_{01}}{s+\alpha} & 0 \\ 0 & \dfrac{n_{04}}{s+\alpha} \end{bmatrix} \quad \boldsymbol{D}_R = \begin{bmatrix} \dfrac{s+d_{01}}{s+\alpha} & \dfrac{d_{02}}{s+\alpha} \\ \dfrac{d_{03}}{s+\alpha} & \dfrac{s+d_{04}}{s+\alpha} \end{bmatrix} \qquad (23)$$

where $\alpha > 0$ is a real constant. An estimate of gains and time-constants of the system from an input-output simulation record served as an initial approximation of the coefficients in (23). Then, the control algorithm consisted of the steps described in detail in sections 3 and 4. First, a stabilizing controller was computed using (11) as

$$\boldsymbol{C} = \boldsymbol{Y}_L^{-1} \boldsymbol{X}_L \qquad (24)$$

$$\boldsymbol{Y}_L = \begin{bmatrix} \dfrac{s}{s+\alpha} & 0 \\ 0 & \dfrac{s}{s+\alpha} \end{bmatrix} \quad \boldsymbol{X}_L = \begin{bmatrix} \dfrac{x_{11}s + x_{01}}{s+\alpha} & \dfrac{x_{12}s}{s+\alpha} \\ \dfrac{x_{13}s}{s+\alpha} & \dfrac{x_{14}s + x_{04}}{s+\alpha} \end{bmatrix} \qquad (25)$$

Next, the auxiliary signals $\boldsymbol{z}$ and $\boldsymbol{x}$ were constructed via (12)

$$\begin{bmatrix} x_1(t) \\ x_2(t) \end{bmatrix} = \begin{bmatrix} \dfrac{s}{s+\alpha} & 0 \\ 0 & \dfrac{s}{s+\alpha} \end{bmatrix} \cdot \begin{bmatrix} u_1(t) \\ u_2(t) \end{bmatrix} + \begin{bmatrix} \dfrac{x_{11}s + x_{01}}{s+\alpha} & \dfrac{x_{12}s}{s+\alpha} \\ \dfrac{x_{13}s}{s+\alpha} & \dfrac{x_{14}s + x_{04}}{s+\alpha} \end{bmatrix} \cdot \begin{bmatrix} y_1(t) \\ y_2(t) \end{bmatrix} \qquad (26)$$

$$\begin{bmatrix} z_1(t) \\ z_2(t) \end{bmatrix} = \begin{bmatrix} \dfrac{s+d_{01}}{s+\alpha} & \dfrac{d_{02}}{s+\alpha} \\ \dfrac{d_{03}}{s+\alpha} & \dfrac{s+d_{04}}{s+\alpha} \end{bmatrix} \cdot \begin{bmatrix} y_1(t) \\ y_2(t) \end{bmatrix} - \begin{bmatrix} \dfrac{n_{01}}{s+\alpha} & 0 \\ 0 & \dfrac{n_{04}}{s+\alpha} \end{bmatrix} \begin{bmatrix} u_1(t) \\ u_2(t) \end{bmatrix} \qquad (27)$$

These signals, representing filtered quantities $\boldsymbol{u}$ and $\boldsymbol{y}$, were then used for identification of the parameter $\boldsymbol{S}$ (step 3), and finally, a new plant model was obtained using equation (19), where its structure was chosen the same as at previous iteration step. For this model, again, a new controller was designed in step 1, etc.

Structure of the identified parameter $\boldsymbol{S}$ was chosen the simplest, thus

$$\boldsymbol{S} = \begin{bmatrix} s_{01} & 0 \\ 0 & s_{04} \end{bmatrix} \qquad (28)$$

This choice enables the identification of parameter $\boldsymbol{S}$, and consequently identification of the model, using only 2 constants: s_{01} and s_{04}, i.e. for the whole multivariable system, it is sufficient to identify only these two coefficients. A simulation record of the control is provided in Fig. 4.

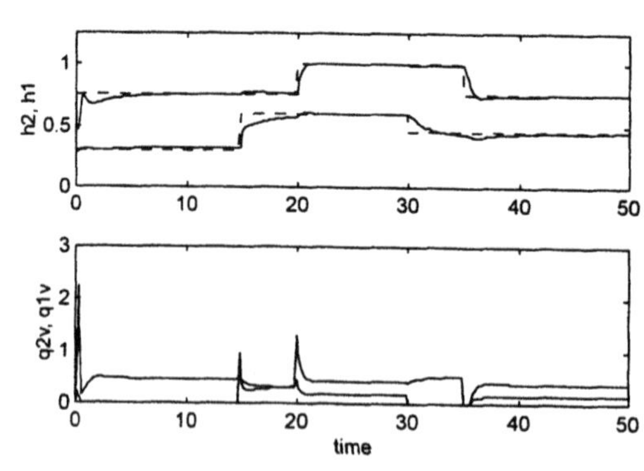

Fig. 4. Simulation control response

As displays Fig. 4, the control response is fluent, without initial instability, where the dynamics can appropriately be affected by the positive constant α, which was chosen as: $\alpha = 0.65$.

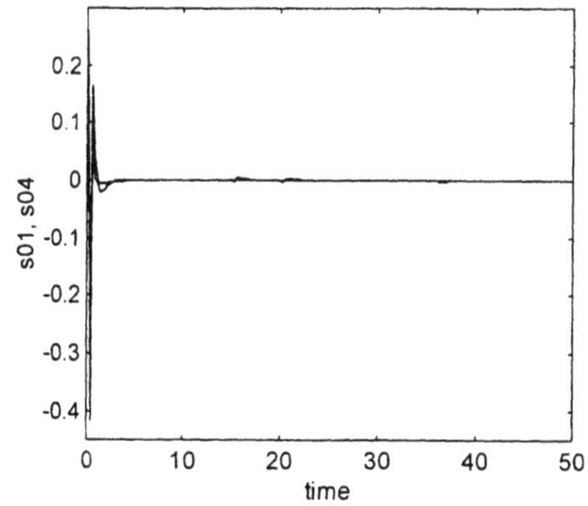

Fig. 5. Identification of Y-K parameter S

Apparently, in Fig. 5, the identified parameter S (its two coefficients) converges to zero, i.e. after a few iteration steps, estimated coefficients of the model (23) are good approximation of the real process. The adaptation of these coefficients is depicted in Fig. 6.

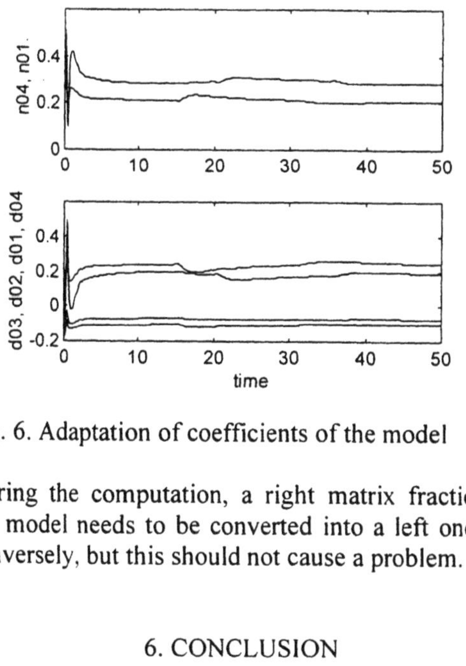

Fig. 6. Adaptation of coefficients of the model

During the computation, a right matrix fraction of the model needs to be converted into a left one and conversely, but this should not cause a problem.

6. CONCLUSION

The paper deals with adaptive control of multivariable technological processes. An iterative scheme of identification and controller design based on the dual Y-K parametrization is utilized. The iteration, with its gradual character, helps to overcome initial instability and provides good convergence for both, identification and controller redesign. Using presented modification of the original algorithm, it is not necessary to do a model reduction. Moreover, it is shown that a whole multi-input/multi-output system can be identified using only very few coefficients (representing the deviation between a nominal model and the real process). Although the identification needs a high signal-to-noise ratio, simulation results class this method as a robust one.

ACKNOWLEDGEMENTS

This work was supported in part by the Grant Agency of the Czech Republic under grants No. 102/00/0526 and No. 102/99/1292, and by the Ministry of Education of the Czech Republic under grant MSM 2811 00001.

REFERENCES

Anderson, B.D.O. (1998). From Youla-Kucera to identification, adaptive and nonlinear control. *Automatica*, **34**, 1485-1506.

Astrom, K.J. and B. Wittenmark (1989). *Adaptive Control.* Addison Wesley, Reading, MA.

Gevers, M. (1993). Towards a joint design of identification and control? In: *Essays on Control: Perspectives in the Theory and its Applications* (H.L. Trentelman and J.C. Willems, Ed.), pp. 111-151. Birkhauser, Boston.

Hansen, F.R. (1989). A fractional representation approach to closed-loop system identification and experiment design. PhD Thesis. Stanford University, CA.

Hansen, F.R., G.F. Franklin and R.L. Kosut (1989). Closed-loop identification via the fractional representation: experiment design. In: *Proc. Am. Control Conf,* pp. 1422-1427. Pittsburgh, PA.

Helmicki, A.J., C.A. Jacobson and C.N. Nett (1991). Control oriented system identification: a worst-case/deterministic approach in H_∞. In: *IEEE Trans. Automat. Contr.*, pp. 1163-1176.

Kozka, S., J. Mikles, M. Fikar, F. Jelenciak and J. Dzivak (2000). Closed-loop identification of a laboratory chemical reactor. In: *CD-ROM Proceedings of 3rd IFAC Symp. on Robust Control Design.* Prague, Czech Republic.

Morari, M and E. Zafiriou (1989). *Robust Process Control.* Prentice Hall, Englewood Cliffs, NY.

Schrama, R.J.P. (1991). An open-loop solution to the approximate closed-loop identification problem. In: *Proc. 9th IFAC/IFORS symp. on Identification andSystems Parameter Estimation*, pp. 1602-1607. Budapest, Hungary.

Tay, T.T., J.B. Moore and R. Horowitz (1989). Indirect adaptive techniques for fixed controller performance enhancement. *Int. J. Control*, **50**, 1941-1989.

Van den Hof, P.M.J. and R.J.P. Schrama (1995). Identification and Control – Closed-loop Issues. *Automatica*, **31**, 1751-1770.

Van den Hof, P.M.J., R.J.P. Schrama, R.A de Callafon and O.H. Bosgra (1995). Identification of normalized coprime plant factors from closed-loop experimental data. *European Journal of Control*, **1**, 62-74.

Vidyasagar, M. (1985). *Control Systems Synthesis: A factorization Approach.* MIT Press. Cambridge, MA.

TRANSIENT PERFORMANCE IMPROVEMENT IN DISCRETE-TIME MODEL REFERENCE ADAPTIVE CONTROLLERS BY FILTERING ESTIMATED PARAMETERS

Naoki Mizuno* and Yasuyuki Fukui **

**Nagoya Institute of Technology, Dept. of Mechanical Engineering, Gokiso-cho, Showa-ku, Nagoya, 466-8555 Japan*
***Nippon Sharyo, LTD, 2-20 Honohara, Toyokawa, 422-0061, Japan*

Abstract: In this paper we propose a extended model reference adaptive controllers in the discrete-time, single-input single-output linear time-invariant plant for the purpose of improving the transient performance. In the proposed method, the estimated parameters in the adaptive system are filtered before using in the control scheme. The fixed filter averages the estimated parameters and reduces the transient input. Furthermore a design method for the time-varying filter is also given, and stability analysis of the proposed adaptive system is examined based on the Lyapnov theorem. Finally a numerical example is illustrated in order to show the effectiveness of the proposed method. *Copyright © 2001 IFAC*

Key words: adaptive control, discrete-time control, transient performance, filtering

1. INTRODUCTION

Thus far, many different approaches to adaptive control system have been proposed and various properties of such systems are being clarified. In these attempts, the model reference adaptive control systems (MRACS) for both continuous and discrete time systems have made remarkable progress in theory and practice.

However, from the practical point of view, one problem in the adaptive methods is that the transient performance of the control system is not directly taken into consideration. Many simulation results based on these methods show that the control input to the plant is usually very large and the control system seems to be unstable during the transient phase of the adaptive control.

These phenomena are mainly due to the following two reasons. One is the difference between the estimated parameters and its true values. If the estimated parameters converge to its true values, the transient property of the adaptive system becomes to the ideal one.

In order to reduce the parameter error, the fast estimated algorithms have been developed for both continuous and discrete time cases.

On the other hands, transient unstable behavior sometimes can be observed when the estimated parameters converge to the true values in vary short instant and it can be concluded that the update of the estimated parameters also affects the transient property of the adaptive system.

From this point of view, much interest has been given to the dynamic certainty equivalent (DyCE) adaptive controller proposedby Morse (Morse, 1992). Since the derivation of the parameter update is taken into account in the design of the control input, the DyCE adaptive controllers are considered to be ones of efficient schemes to improve the transient performance.

Indeed performance analysis of DyCE schemes has been studied (Ortega, 1993; Datta and Ioannou, 1995) and the modifications of the controller structureare also proposed (Ohmori and Shrivastava, 1995; Sun, 1993; Datta and Ioannou, 1994; Masuda et al., 1994). Unfortunately the design methods of DyCE adaptive controller and the modified control schemes are limited to the continuous-time.

In case of discrete-time adaptive algorithms, we can not use the derivation (future value) of the parameter update, so, we propose the filtering approach of the estimated parameters. Using the filtering technique, the stability proof of the proposed control system is very simple and the controller is easy to implement. Extensive simulation results show the effectiveness of the proposed methods.

2. PROBLEM STATEMENTS

Consider the following single-input, single-output, linear-time-invariant system

$$y(k) = \frac{z^{-d}B(z^{-1})}{A(z^{-1})}u(k) \qquad (1)$$

$$A(z^{-1}) = 1 + a_1 z^{-1} + \cdots + a_n z^{-n} \qquad (2)$$

$$B(z^{-1}) = b_0 + b_1 z^{-1} + \cdots + b_m z^{-m} \qquad (3)$$

where $y(k)$ and $u(k)$ denote the plant output and input respectively, and $A(z^{-1})$ and $B(z^{-1})$ are coprime polynomials in the unit delay operator z^{-1} of degrees n and m.
Let the reference model is described as

$$y_m(k) = \frac{z^{-dm}B_M(z^{-1})}{A_M(z^{-1})}u_m(k) \qquad (4)$$

$$A_M(z^{-1}) = 1 + a_{m1} z^{-1} + \cdots + a_{mn'} z^{-n'} \qquad (5)$$

$$B_M(z^{-1}) = b_{m0} + b_{m1} z^{-1} + \cdots + b_{mm'} z^{-m'} \qquad (6)$$

where $u_m(k)$ and $y_m(k)$ are the reference input and output respectively, and $u_m(k)$ is a bounded piecewise continuous function of time.

The following assumptions, which are typical in model reference adaptive control system, are made for the plant and the reference model.

(A.1) The dead time d is known.

(A.2) The upper bounds on the order n and m are known.

(A.3) The dead time d_m of reference model is greater than or equal to that of the plant.

(A.4) The plant has a stable inverse.

The control objective is to determine a discrete-time controller so that all the signals in the closed-loop system remain bounded and the tracking error tends to zero asymptotically.

3. NON-MINIMAL REPRESENTATION OF THE PLANT

In the design of model reference adaptive control system, to estimate the parameters of the controller directly, Eq. (1) is usually rewritten to the non-minimal representation, which consists of the controller parameters.
Let introduce the following stable monic polynomial.

$$D(z^{-1}) = 1 + d_1 z^{-1} + \cdots + d_n z^{-n} \qquad (7)$$

In this case, there exist unique polynomials $R(z^{-1})$ and $S(z^{-1})$ of order $d-1$ and $n-1$ respectively, such that:

$$D(z^{-1}) = A(z^{-1})R(z^{-1}) + z^{-d}S(z^{-1}) \qquad (8)$$

By multiplying $y(k)$ to the both sides of the Eq. (8), then by using Eq. (1), we obtain the following non-minimal representation of the plant.

$$\begin{aligned} D(z^{-1})y(k) &= A(z^{-1})R(z^{-1})y(k) + z^{-d}S(z^{-1})y(k) \\ &= B(z^{-1})R(z^{-1})u(k-d) + S(z^{-1})y(k-d) \\ &= \theta^T \xi(k-d) \qquad (9) \end{aligned}$$

$$\theta^T = [b_0 \; b_0 r_1 + b_1 \cdots b_m r_{d-1} \; h_0 \; h_1 \cdots h_{n-1}] \qquad (10)$$

$$\xi(k) = [u(k) \; u(k-1) \cdots u(k-m-d+1) \; y(k) \cdots y(k-n+1)] \qquad (11)$$

4. STRUCTURE OF CONVENTIONAL MODEL REFERENCE ADAPTIVE CONTROLLERS

In this section, we give the brief description of the conventional Model Reference Adaptive Controllers. In the case of adaptive control, we introduce the following identification model corresponding the non-minimal representation of the plant.

$$D(z^{-1})\hat{y}(k) = \hat{\theta}^T(k)\xi(k-d) \qquad (12)$$

where, $\hat{y}(k)$ and $\hat{\theta}(k)$ denote the estimated plant output and parameter vector respectively. Replacing the estimated plant output $\hat{y}(k)$ to the reference output $y_m(k)$, the following control law can be obtained.

$$u(k) = \frac{D(z^{-1})y_m(k+d) - \hat{\bar{\theta}}^T(k+d)\bar{\xi}(k)}{\hat{b}_0(k+d)} \quad (13)$$

$$\hat{\theta}^T(k) = [\hat{b}_0(k) \quad \hat{\bar{\theta}}(k)] \quad (14)$$

$$\xi^T(k) = [u(k) \quad \bar{\xi}^T(k)] \quad (15)$$

The tracking error $e(t)$ can be calculated in the following way when the control law (13) is utilized.

$$e(k) = D(z^{-1})(y(k) - y_m(k)) = \tilde{\theta}^T(k)\xi(k-d) \quad (16)$$

Now, we introduce the auxiliary signal $\rho(k)$ as

$$\rho(k) = \left(\tilde{\theta}(k+d) - \tilde{\theta}(k)\right)^T \xi(k-d) \quad (17)$$

and we define the augmented error $\eta(k)$ as:

$$\eta(k) = e(k) + \rho(k) \quad (18)$$

Then the augmented error equation becomes

$$\eta(k) = \tilde{\theta}^T(k+d)\xi(k-d) \quad (19)$$

Based on the above error equation, the following standard recursive parameter estimation algorithm can be adopted to estimate the unknown parameters in Eq. (4.2) and the adaptive control system can be constructed using the estimated parameters.

$$\hat{\theta}_0(k+d) = \hat{\theta}_0(k+d-1)$$
$$-\Gamma(k+d-1)\xi(k-d)\,\eta(k) \quad (20)$$

$$\Gamma(k+d) = \frac{1}{\lambda_1(k)}\Big[\Gamma(k+d-1)$$
$$-\frac{\lambda_2(k)\Gamma(k+d-1)\xi(k-d)\xi^T(k-d)\Gamma(k-d-1)}{\lambda_1(k) + \lambda_2(k)\xi^T(k-d)\Gamma(k-d-1)\xi(k-d)}\Big] \quad (21)$$

$\Gamma(d) = \Gamma^T(d) > 0$, $0 < \lambda_1(k) \le 1$, $0 \le \lambda_2(k) < 2$

Suppose that assumptions (A.1) - (A.4) can be met, then, it is proofed that for any bounded $y_m(i)$, all signals in the above model reference adaptive control system are uniformly bounded, and that the tracking error converges to zero asymptotically.

However, from the structure of control law (Eq. (13)), if the transient value of the estimated parameter $\hat{b}_0(k+d)$ approaches to zero, the amplitude of control input increase and the control system may become unstable. The similar unstable behaviors often observed during the transient phase of many types adaptive control systems.

These phenomena may occur when the estimated parameters converge to the true values in vary short instant, depending on the trend or combination of the parameter values. Thus, the control of the trend of estimated parameters is very important.

5. DESIGN OF EXTENDED MODEL REFERENCE ADAPTIVE CONTROLLER WITH ESTIMATED PARAMETER FILTERING

In conventional model reference adaptive control system mentioned above, if $\lambda_1(k)$ and $\lambda_2(k)$ in the adaptive law can be selected so that $\hat{b}_0(k)$ does not approaches to zero for all instant. However, it is difficult to control the trend of the estimated parameters directly by selecting the design parameters in the adaptation algorithm.

To control the trend of the estimated parameters, we introduce the modified non-minimal representation of the plant as follows.

$$D(z^{-1})y(k) = g(\theta/g)^T\xi(k-d) = g\theta_0^T\xi(k-d) \quad (22)$$

where $\theta_0 = \theta/g$ and $g \ge 1$ ($g = 1$: conventional MRACS) denotes the arbitrary positive constant assigned by the designer. In the case where the parameters of the plant are known, the representation Eq. (22) is identical to Eq. (9). However, corresponding to the new representation, we can modify the identification model as follows:

$$D(z^{-1})\hat{y}(k) = \left(\alpha_0(k)\hat{\theta}_0(k) + \alpha_1(k)\hat{\theta}_0(k-1) + \cdots\right.$$
$$\left. + \alpha_{g-1}(k)\hat{\theta}_0(k-g+1)\right)^T\xi(k-d) \quad (23)$$

$$\alpha_0(k) + \alpha_1(k) + \cdots + \alpha_{g-1}(k) = g \quad (24)$$

Now, we define the estimated error $\varepsilon(k)$ as:

$$\varepsilon(k) = D(z^{-1})(\hat{y}(k) - y(k)) \quad (25)$$

Then the estimated error equation becomes

$$\varepsilon(k) = \alpha_0(k)\left(\hat{\theta}_0(k) - \theta_0\right)^T\xi(k-d)$$
$$\cdots + \alpha_{g-1}(k)\left(\hat{\theta}_0(k-g+1) - \theta_0\right)^T\xi(k-d)$$
$$= \left(\alpha_0(k)\tilde{\theta}_0(k) + \alpha_1(k)\tilde{\theta}_0(k-1) + \cdots\right.$$
$$\left. + \alpha_{g-1}(k)\tilde{\theta}_0(k-g+1)\right)^T\xi(k-d) \quad (26)$$

where

$$\widetilde{\theta}_0(k) = \hat{\theta}_0(k) - \theta_0 \tag{27}$$

To obtain the control law, we assume that

$$D(z^{-1})\hat{y}(k) = D(z^{-1})y_m(k) \tag{28}$$

Then the following relation holds.

$$\begin{aligned}\varepsilon(k) &= D(z^{-1})(\hat{y}(k) - y(k)) \\ &= D(z^{-1})(y_m(k) - y(k)) = e(k)\end{aligned} \tag{29}$$

This means that the control input which brings the error between the estimated output $\hat{y}(k)$ and the reference output $y_m(k)$ to approach to zero, makes the tracking error $e(k)$ tend to zero. The Equation (28) can be rewritten as follows.

$$\begin{aligned}D(z^{-1})y_m(k+d) &= D(z^{-1})\hat{y}(k+d) \\ &= \left(\alpha_0(k+d)\hat{\theta}_0(k+d) + \alpha_1(k+d)\hat{\theta}_0(k+d-1) + \cdots \right. \\ &\quad \left. + \alpha_{g-1}(k+d)\hat{\theta}_0(k+d-g+1)\right)^T \xi(k) \\ &= \left(\alpha_0(k+d)\hat{q}_0(k+d) + \alpha_1(k+d)\hat{q}_0(k+d-1) + \cdots \right. \\ &\quad \left. + \alpha_{g-1}(k+d)\hat{q}_0(k+d-g+1)\right)u(k) \\ &\quad + \left(\alpha_0(k+d)\hat{\bar{\theta}}_0(k+d) + \alpha_1(k+d)\hat{\bar{\theta}}_0(k+d-1) + \cdots \right. \\ &\quad \left. + \alpha_{g-1}(k+d)\hat{\bar{\theta}}_0(k+d-g+1)\right)^T \bar{\xi}(k)\end{aligned} \tag{30}$$

$$\hat{\theta}_0^T(k) = [\hat{q}_0(k) \quad \hat{\bar{\theta}}_0^T(k)] \tag{31}$$

$$\xi^T(k) = [u(k) \quad \bar{\xi}^T(k)] \tag{32}$$

From the above relation, we can obtain the following control law.

$$u(k) = \frac{D(z^{-1})y_m(k+d) - \left(\alpha_0(k+d)\hat{\bar{\theta}}_0(k+d) + .. \right.}{\alpha_0(k+d)\hat{b}_0(k+d) + ..} \quad \frac{\left. .. + \alpha_{g-1}(k+d)\hat{\bar{\theta}}_0(k+d-g+1)\right)}{.. + \alpha_{g-1}(k+d)\hat{b}_0(k+d-g+1)} \tag{33}$$

In the above control law, the estimated parameters at past g steps of time instant are weighted by $\alpha_0(k+d)$-$\alpha_{g-1}(k+d)$, and by selecting $g, \alpha_i(k+d)$ we can control the transient property of the adaptive control system.

In this case, we modify the auxiliary signal $\rho(k)$ as

$$\begin{aligned}\rho(k) &= g\widetilde{\theta}_0^T(k+d)\xi(k-d) - \left(\alpha_0(k)\widetilde{\theta}_0(k) + .. \right. \\ &\quad \left. + \alpha_{g-1}(k)\widetilde{\theta}_0(k-g+1)\right)^T \xi(k-d)\end{aligned} \tag{34}$$

then the augmented error equation becomes

$$\eta(k) = g\widetilde{\theta}_0^T(k+d)\xi(k-d) \tag{35}$$

For this error system, we can adopt the following adaptation algorithm.

$$\begin{aligned}\hat{\theta}_0(k+d) &= \hat{\theta}_0(k+d-1) \\ &\quad - g\gamma(k)\Gamma(k+d-1)\xi(k-d)\,\eta(k)\end{aligned} \tag{36}$$

$$\Gamma(k+d) = \frac{1}{\lambda_1(k)}\left[\Gamma(k+d-1) - \frac{g^2\lambda_2(k)\Gamma(k+d-1)\xi(k-d)\xi^T(k-d)\Gamma(k+d-1)}{\lambda_1(k) + g^2\lambda_2(k)\xi^T(k-d)\Gamma(k+d-1)\xi(k-d)}\right] \tag{37}$$

$$\gamma(k) = \frac{\lambda_2(k)}{\lambda_1(k) + g^2\lambda_2(k)\xi^T(k-d)\Gamma(k+d-1)\xi(k-d)} \tag{38}$$

$$\eta(k) \quad g\hat{\theta}_0^T(k+d-1)\xi(k-d) - D(z^{-1})y(k) \tag{39}$$

$$\Gamma(d) = \Gamma^T(d) > 0,\ 0 < \lambda_1(k) \le 1,\ 0 < \lambda_2(k) < \infty$$

6. STABILITY ANALYSIS OF PROPOSED CONTROL SYSTEM

The stability of the proposed control system can be examined by similar way to the conventional model reference adaptive control system. For simplicity, only the case where the recursive least squares algorithm is used to update the estimated parameters. In this case, we consider the following quadratic function.

$$V(k) = \widetilde{\theta}_0^T(k)\Gamma^{-1}(k)\widetilde{\theta}_0(k) \tag{40}$$

Calculating the difference of $V(k)$ along the trajectory of (35), we obtain

$$\begin{aligned}\Delta V(k) &= V(k+d) - V(k+d-1) \\ &= \widetilde{\theta}_0^T(k+d)\Gamma^{-1}(k+d)\widetilde{\theta}_0(k+d) \\ &\quad - \widetilde{\theta}_0^T(k+d-1)\Gamma^{-1}(k+d-1)\widetilde{\theta}_0(k+d-1) \\ &= -g\eta(k)\,\xi^T(k-d)[2\widetilde{\theta}_0(k+d) + \\ &\quad g\Gamma(k+d-1)\xi(k-d)\,\eta(k)] \\ &\quad + [g\widetilde{\theta}_0^T(k+d)\xi(k-d)]^2 \\ &= -\eta^2(k) - g^2\eta^2(k) \\ &\quad \xi^T(k-d)\Gamma(k+d-1)\xi(k-d) \le 0\end{aligned} \tag{41}$$

This means that $\eta(k) \to 0$ as $k \to \infty$, and if the estimated parameters coverage to the constant, then $\rho(k) \to 0$, finally $e(k) \to 0$ under the assumption that the plant has a stable inverse.

7. NUMERICAL EXAMPLE

To investigate the effectiveness of the proposed methods, extensive computer simulations are performed. Only the typical examples are shown as follows. First, for the following plant with sudden change of parameters, the proposed method is applied.

$$y(k) = \frac{z^{-3}(1.5000 - 0.7970z^{-1})}{1.0000 - 1.5282z^{-1} + 0.8891z^{-2}} u(k) \quad (\text{Stable}: 0 \le k < 100)$$

$$y(k) = \frac{z^{-3}(1.5000 - 0.7970z^{-1})}{1.0000 - 1.5282z^{-1} + 1.0991z^{-2}} u(k) \quad (\text{Unstable}: 100 \le k \le 300)$$

The reference model and its input are set at :

$$y_m(k) = \frac{z^{-3}}{1.0000 - 0.6065z^{-1}} u_m(k)$$

$$u_m(k) = sin\left(\frac{\pi}{20}\right)k + sin\left(\frac{\pi}{10}\right)k$$

Moreover, the initial values of the estimated parameters and the gain matrix are set as follows.

$$\hat{\theta}_0^T(3) = [100 \quad 0 \quad 0 \quad 0 \quad 0 \quad 0]$$

$$\hat{\theta}_0(0)\ \hat{\theta}_0(2) = 0 \, , \ \Gamma(3) = 10^{10} I$$

The closed loop poles are assigned by the stable polynomial: $D(z^{-1}) = 1 - 0.8z^{-1} - 0.09z^{-2}$.
In this case, the parameters of the estimated parameters filter, $\alpha_i(k) \equiv 1$ and only the time horizon of the moving average "g" are changed. Fig. 1 and Fig. 2 show the control input in the case where g is equal to 1(conventional) -3 and 4-6 respectively. Fig. 3 and Fig. 4 show the tracking error for each case.

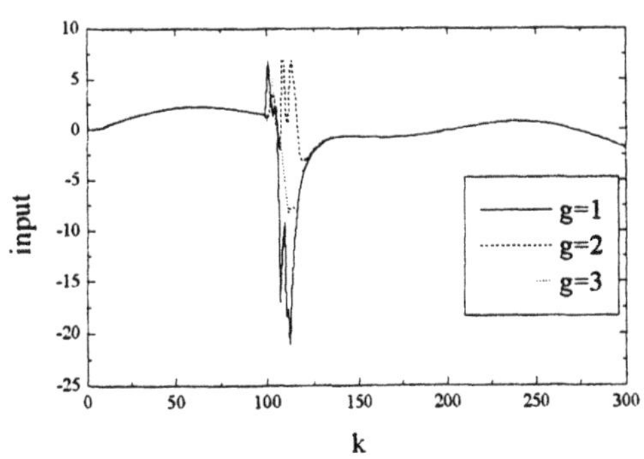

Fig. 1 Control input of proposed controller (g=1-3)

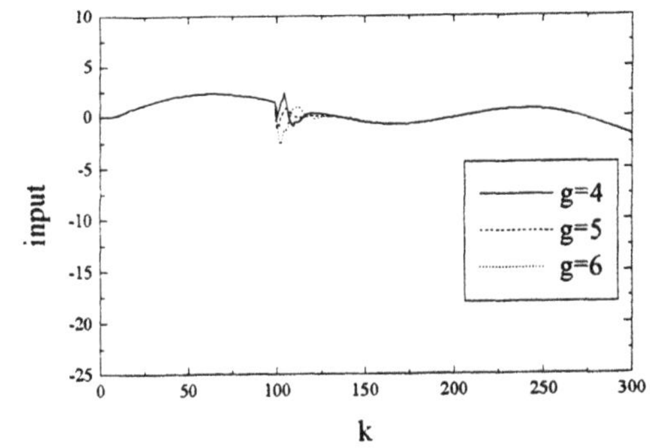

Fig.2 Control input of proposed controller (g= 4 – 6)

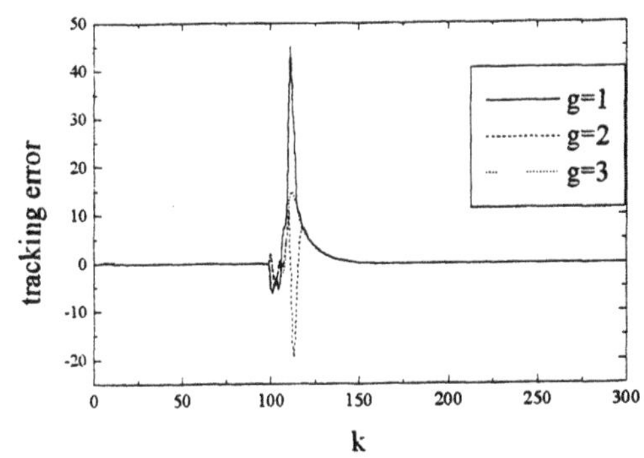

Fig.3 Tracking error in proposed control system (g=1-3)

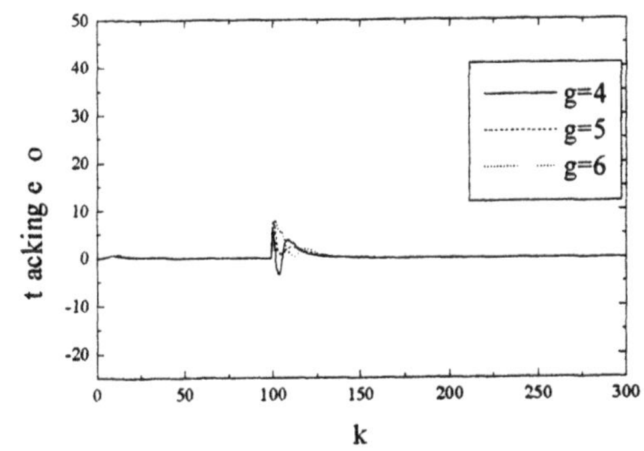

Fig. 4 Tracking error in proposed control system (g=4-6)

From these results, although the coefficients of the estimated parameter filter are fixed, it can be seen that by designing the estimated parameter filter properly, we improve the transient response of the adaptive system. Moreover, the proposed method is effective to the case when the measurement noise contains the observed signals. In this case, the long time horizon of the moving average "g" gives more small variation of the steady state tracking error(see Fig.5).

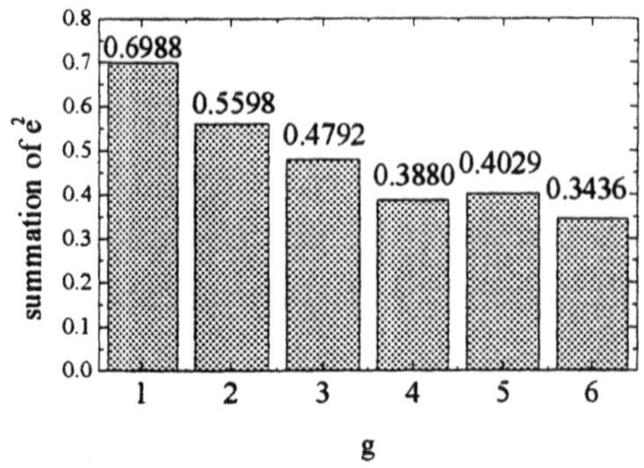

Fig. 5 Sum of the squared error in tracking control under measurement noise.

In the case, where the information about the parameter change or the disturbance can be obtained, the time varying estimated filter is efficient. Next example shows the effectiveness of the time varying filter to reduce the affect of the disturbance for the adaptive system. In this case, the plant is

$$y(k)=\frac{z^{-3}(1.5000-0.7970z^{-1})}{1.0000+1.5020z^{-1}+0.7821z^{-2}}u(k)$$

and the time horizon is fixed at: $g=6$.
The output disturbance $v(k)$ is added to the plant output $y(k)$ at the known instant as follows.

$$v(k)=\begin{cases}0 & (0\le k\le 100)\\ 0.05 & (100\le k\le 105)\\ 0 & (105\le k\le 200)\end{cases}$$

Regarding this information, the time varying parameters of the filter are set at:

$$\alpha_0(k)\text{-}\alpha_5(k)=1 \quad (0\le k\le 96)$$
$$\alpha_0(k)\text{-}\alpha_4(k)=0,\ \alpha_5(k)=6 \quad (97\le k\le 101)$$
$$\alpha_0(k)=6,\ \alpha_1(k)\text{-}\alpha_5(k)=0 \quad (102\le k\le 200)$$

Table 1 shows the comparison in tracking error between the fixed filter case and the time varying filter case.

Table 1 Comparison in tracking error

Filter coeff. (g=5)	$\alpha_i(k)\equiv 1$	variable $\alpha_i(k)$
$\sum_{i=80}^{200} e^2(i)$	3.1589	1.6667

From this result, the time varying filter has the possibility to improve the transient performance of the adaptive system.

8. CONCLUDING REMARKS

In this research, we have proposed the extended discrete-time model reference adaptive controller in order to improve the transient performance. In the proposed method, the estimated parameters in the adaptive system are filtered before using in the control scheme. From the numerical examples, it can be concluded that the fixed estimated parameters filter with proper coefficients can reduces the transient input and the tracking error under many conditions. Furthermore the time-varying filter can be effective for more general case including the case where the information about the changing of the plant characteristic and the disturbances. Finding the optimum structure of the estimated parameter filter for more real situation is one of the most important problems left for future study.

REFERENCES

Morse, A. S.(1992). A Comparative Study of Normalized and Unnormalized Tuning Errors in Parameter-Adaptive Control, *Int. J. Adaptive Control* and Signal Processing, VO1.6, 309-318.

Ortega, R.(1993). On Morse's New Adaptive Controller: Parameter Convergence and Transient Perfor-mance', *IEEE* Trans. *Automat.* Contr., VO1.38, no.8, 1191–1202.

Datta, A. Datta and P. A. Ioannou (1995). Directly Com-putable *Lz* and *Lm* Performance Bounds for Morse's Dynamic Certainty Equivalence Adaptive Control", *Proc. 1995* American *Contr. Conf.*,589-591.

Ohmori, H. and Y. Shrivastava(1995). Robust Adaptive Control System with Fixed Compensator Attaining Robust Positive Realness", *Proc. 1995* American *Contr. Conf.*, 592-596.

Sun, J.(1993). A Modified Model Reference Adaptive Control Scheme for Improved Transient Performance", *IEEE Trans. on Automat.* Contr., vol. AC-38, no. 8, 1255-1259.

Datta, A. and P. A. Ioannou (1994). Performance Anal-ysis and Improvement in Model Reference Adaptive Control, *IEEE Trans.* on *Automat. Contr.*, vol. AC-39, no. 12, 2370-2387.

Masuda, S., A. Inoue and K. Sugimoto (1994). A Construction of Multivariable MRACS with Fixed Compensator Using Coprime Factorization Approach, *Proc. 33rd IEEE Conf. Dec.* Contr., vol. 2, 1688-1693.

ADAPTIVE POLE ASSIGNMENT CONTROL UNDER CONSTRAINTS

Andrzej KRÓLIKOWSKI, Roman STAWSKI

Poznań University of Technology, Department of Control, Robotics and Computer Science, 60-965 Poznań, ul.Piotrowo 3A, POLAND, akrol@ar-kari.put.poznan.pl

Abstract: The problem addressed in this paper is the input-constrained indirect adaptive tracking control of a discrete-time system perturbed by a bounded disturbance. In the pole assignment framework the desired closed-loop polynomial can be considered as a design polynomial. The analysis is focussed on the interplay between the location of desired closed-loop poles, tracking accuracy and imposed constraints.

Keywords: Adaptive control. Pole assignment. Input constraints.

1 Introduction

In practice, the control signal is subjected to prespecified constraints due to technical restrictions. Neglecting these constraints in the controller design may lead to performance deterioration or even instability of the control system. This problem is even more crucial in adaptive control systems. In this case, some stability results for linear systems subject to input amplitude constraint are given in [1] where indirect pole assignment adaptive case is considered. The direct adaptive control with one pole at $z = 1$ is considered in [2], where the boundedness of all closed-loop signals is proved. In [3] an indirect adaptive control system with input rate saturation constraint is analyzed and the bibo stability is shown. The tracking control under amplitude and rate constraints is presented in [4] where the bibo stability is also demonstrated. Usually, the control aim is to design a controller such that the system output y_t follows a reference signal y_t^*, and that the specified closed-loop poles are achieved. The problem addressed in this paper is the indirect adaptive pole assignment tracking control under input constraints. The influence of desired closed-loop poles and constraints on the tracking accuracy is analyzed via simulations.

2 Problem formulation

The single-input single-output plant is described by a linear discrete-time equation of the form

$$y_t = \frac{B(z^{-1})}{A(z^{-1})} u_t^c + w_t \tag{1}$$

where y_t and u_t^c are plant output and constrained (applied) input, respectively, and polynomials $A(z^{-1}) = 1 + a_1 z^{-1} + ... + a_{na} z^{-na}$, $B(z^{-1}) = b_1 z^{-1} + b_2 z^{-2} + ... + b_{nb} z^{-nb}$ are relatively prime and $A(z^{-1})$ is considered to be stable and minimumphase. It is assumed that the disturbance w_t is bounded

$$| w_t | \leq w \tag{2}$$

and consequently, denoting $v_t = A(z^{-1}) w_t$

$$| v_t | \leq v \tag{3}$$

The amplitude and rate constraints of applied control input are realized by the saturation function

$$u_t^c = sat(u_t; \alpha) \tag{4}$$

$$\Delta u_t^c = sat(\Delta u_t; \beta) \tag{5}$$

where u_t is an unconstrained (computed) control and $\Delta = 1 - z^{-1}$. The task of the controller to be designed is to force the plant output y_t to follow a given bounded reference signal y_t^* as close as possible under the given closed-loop characteristics specified by the polynomial $A^*(q^{-1})$, and taking the amplitude or/and rate constraints of control input (4), (5) into account. The parameters of the plant put in the vector $\theta = (a_1, \cdots, a_{na}, b_1, \cdots, b_{nb})^T$ are assumed to be unknown, thus a use of adaptive controller is suggested.

3 Parameter estimation

The following least-squares algorithm with a dead zone is used to estimate θ

$$\hat{\theta}_t = \hat{\theta}_{t-1} + \frac{d_t P_{t-2} \varphi_{t-1} \epsilon_t}{1 + \varphi_{t-1}^T P_{t-2} \varphi_{t-1}} \tag{6}$$

$$P_{t-1} = P_{t-2} - \frac{d_t P_{t-2}\varphi_t\varphi_t^T P_{t-2}}{1+\varphi_{t-1}^T P_{t-2}\varphi_{t-1}} \tag{7}$$

where
$\varphi_{t-1} = (-y_{t-1},\cdots,-y_{t-na},u_{t-1}^c,\cdots,u_{t-nb}^c)^T$,
$\epsilon_t = y_t - \hat{\theta}_{t-1}^T\varphi_{t-1}$, and

$$d_t = \begin{cases} 0 & \text{if } |\epsilon_t| \le \frac{v}{\sqrt{1-\psi}} \\ \psi f[\frac{v}{\sqrt{1-\psi}},\epsilon_t]/\epsilon_t & \text{otherwise} \end{cases} \tag{8}$$

with $0 < \psi < 1$. The dead-zone is defined as follows

$$f(g,\epsilon) = \begin{cases} \epsilon - g & \text{if } \epsilon > g \\ 0 & \text{if } |\epsilon| < g \\ \epsilon + g & \text{if } \epsilon < -g \end{cases} \tag{9}$$

Usually, a bound w is given, however in the estimation algorithm (5), (6) the bound v is needed. If this bound is overestimated (notice that polynomial $A(z^{-1})$ is unknown) then this could "frozen" the algorithm. Therefore, it would be reasonable to update the bound v, e.g. $\hat{v} \ge| \hat{A}(z^{-1},t)w |$ at some estimation time t.

4 Adaptive control under amplitude constraint

The amplitude-constrained indirect adaptive pole assignment controller with integral action is given by (4), where

$$u_t = u_{t-1} + \Delta(z^{-1})\hat{L}'(z^{-1},t)u_t + \hat{P}(z^{-1},t)e_t \tag{10}$$

with $e_t = y_t^* - y_t$, and the controller polynomials $\hat{P}(z^{-1},t)$, $\hat{L}'(z^{-1},t)$ are solutions to the pole assignment identity

$$\Delta(z^{-1})\hat{A}(z^{-1},t)\hat{L}(z^{-1},t) + \hat{B}(z^{-1},t)\hat{P}(z^{-1},t) = \\ = A^*(z^{-1}) \tag{11}$$

where $\hat{L}(z^{-1},t) = 1 - \hat{L}'(z^{-1},t)$ and $A^*(z^{-1})$ is a given n^*th order stable polynomial which determines the desired closed-loop poles. This equation has a unique solution if $A(q^{-1}), B(q^{-1})$ are coprime and the degrees of $\hat{L}(z^{-1},t), \hat{P}(z^{-1},t)$ are selected to satisfy $nl = nb, np = na, n^* \le na + nb + 1$ where the polynomial $\hat{L}(z^{-1},t)$ is monic.

5 Adaptive control under rate constraint

The rate-constrained controller is now given by (5), where [3]

$$\Delta u_t = \hat{L}'(z^{-1},t)\Delta u_t + \hat{P}(z^{-1},t)e_t \tag{12}$$

which is an incremental form of (10) for $nl = nb + 1, np = na + 1$ and $n^* \le na + nb + 2$.

6 Adaptive control under rate and amplitude constraints

In [4], a pole assignment controller is proposed where both rate and amplitude constraints are taken into account. Its equation in adaptive version can be written as (12) where $nl = nb - 1, np = na$ and $n^* \le na + nb$, and

$$u_t = \Delta u_t^c + u_{t-1}^c \tag{13}$$

Then the applied control signal u_t^c is given by (4) with Δu_t^c given by (5).

7 Stability

As it was shown in [1], [3], the pole assignment adaptive control under amplitude or rate constraints assures the boundedness of the system output. A similar result is given in [4] for disturbance free plant (1) when both constraints are imposed on the control input. The stability results given in [1], [3] are valid even if a bounded input-output uncertainty takes place in the plant description (1).

8 Pole assignment analysis

Despite of the closed-loop stability, the system performance determined by the closed-loop poles cannot be achieved because of the potential modelling uncertainty, disturbances and control saturation. However, as pointed out in [1], in the case of $w_t = 0$, and if the adaptive control law (4),(10) satisfies

$$u_{t-i} = \hat{P}(z^{-1},t-i)(y_{t-i}^* - y_{t-i}) + \hat{L}'(z^{-1},t-i)u_{t-i} \tag{14}$$

or similarly for the adaptive control law (12)

$$\Delta u_{t-i} = \hat{P}(z^{-1},t-i)(y_{t-i}^* - y_{t-i}) + \hat{L}'(z^{-1},t-i)\Delta u_{t-i} \tag{15}$$

for $i = 1,2,\cdots,na$, then [1]

$$\lim_{t\to\infty}\{\hat{B}(z^{-1},t)\hat{P}(z^{-1},t)y_t^* - A^*(z^{-1})y_t\} = 0 \tag{16}$$

This happens if for some period of time, the control signal remains unsaturated, i.e. the output y_t and the reference y_t^* signals are in the "quasi steady-state". From (1),(10) it can be found that the closed-loop response is

$$y_t = \frac{\hat{P}(z^{-1},t)\hat{B}(z^{-1},t)}{A^*(z^{-1})}y_t^* + \frac{\Delta(z^{-1})\hat{L}(z^{-1},t)}{A^*(z^{-1})}\hat{v}_t \tag{17}$$

where $\hat{v}_t$ is the equation error of the model

$$\hat{A}(z^{-1},t)y_t = \hat{B}(z^{-1},t)u_t^c + \hat{v}_t \tag{18}$$

By choosing the polynomial $A^*(z^{-1})$ one can try to keep the control signal within saturation limits, especially at the step change of reference signal. In general, the unconstrained control signal should improve the criterion J_N like

$$J_N = \frac{1}{N}\sum_{t=1}^{N} e_t^2 \qquad (19)$$

On the other hand, tuning the polynomial $A^*(z^{-1})$ can in turn affect the control performance given by the criterion J_N, because from (1) and (10) one obtains

$$e_t = \frac{\Delta(z^{-1})\hat{L}(z^{-1},t)}{A^*(z^{-1})A(z^{-1}) - A^c(z^{-1})\hat{A}(z^{-1},t)} \times$$
$$\times\ [\hat{A}(z^{-1},t)v_t - A(z^{-1})\hat{v}_t] \qquad (20)$$

where $A^c(z^{-1}$ is the actual closed-loop characteristic polynomial defined as in (11) where polynomials $\hat{A}(z^{-1},t), \hat{B}(z^{-1},t)$ are replaced by polynomials $A(z^{-1}), B(z^{-1})$. Thus, for a fixed model $\hat{A}(z^{-1},t), \hat{B}(z^{-1},t)$ (notice that $\hat{L}(z^{-1},t)$ is a function of $\hat{A}(z^{-1},t), \hat{B}(z^{-1},t)$ as well) e_t can be seen as a function of parameters θ, their estimates $\hat{\theta}_t$ and parameters a^* of design polynomial $A^*(z^{-1})$, i.e. $e_t(\theta, \hat{\theta}_t, a^*)$. This interplay is analyzed via simulations in the next Section. It is noteworthy that in the context of model reference control when the desired response is defined as

$$y_t^d = \frac{\hat{P}(z^{-1},t)\hat{B}(z^{-1},t)}{A^*(z^{-1})} y_t^* \qquad (21)$$

then from (17), the control performance error e_t^d follows

$$e_t^d = y_t^d - y_t = \frac{\Delta(z^{-1})\hat{L}(z^{-1},t)}{A^*(z^{-1})}\hat{v}_t \qquad (22)$$

Now the error e_t^d is a function only of $\hat{\theta}_t$ and a^*, i.e. $e_t^d(\hat{\theta}_t, a^*)$.

9 Simulation example

Consider the second order plant (1) with $B(z^{-1}) = z^{-1} - 0.9z^{-2}, A(z^{-1}) = (1-0.7z^{-1})(1-0.8z^{-1})$. The reference signal y_t^* is a square sequence switching from -1 to $+1$, and $A^*(z^{-1}) = (1-a^*z^{-1})^{n^*}$. The initial estimates are taken close to zero, and $v = 0.002$. Plots of output signals for amplitude constrained control with $\alpha = 0.8$ are shown in Figs.1,2,3 where $a^* = 0.55, 0.7, 0.85$, respectively while the corresponding control signals are shown in Figs.4,5,6. For $a^* = 0.85$ the control signal remains unsaturated. The corresponding plot of sample criterion $\bar{J}_N$ obtained for $N = 1000$ and averaged over 50 trials is shown in Fig.7 where the minimal value of $\bar{J}_N$ is $\bar{J}_N^{min} = 0.1444$ at $a^* = 0.7405$. It can be observed that for $a^* = 0.85$ despite of desaturated control signal, the sample criterion attains the bigger value. On the other hand, for $a^* = 0.55$, the value of the sample criterion is lower but the control signal is hard saturated. The best trade-off takes place for $a^* = 0.7405$. Plots of output and rate constrained input signals given in Figs.8,9,10,11,12,13 are obtained for rate constrained control with $\beta = 0.2$. The corresponding plot of sample criterion is presented in Fig.14 where $\bar{J}_N^{min} = 0.1226$ at $a^* = 0.6229$. Plots of output and input signals for both amplitude and rate constrained control are shown in Figs.15,16,17,18,19,23, and the corresponding plot of sample criterion in Fig.24 where $\alpha = 0.8, \beta = 0.1$. Again, the minimal value of sample criterion is $\bar{J}_N^{min} = 0.3701$ at $a^* = 0.8044$.

10 Conclusions

The input/output behavior of adaptive input constrained pole assignment control is considered for a linear stable discrete-time system. The desired poles location and constraints have an essential influence on the control performance. Generally, for all considered constrained control algorithms and for given constraints, when the desired poles approach the unit circle then the saturation of control signal can be avoided, however the value of sample criterion $\bar{J}_N$ increases. Successively, for given constraints there is a minimum of sample criterion with respect to a^*, so in all cases an interplay between a^*, α, β and $\bar{J}_N$ can be observed.

References

[1] Gang Feng, Zhang C. and Palaniswami M.: Stability of Input Amplitude Constrained Adaptive Pole Placement Control Systems. *Automatica*, Vol.30, No.6, 1994, pp.1065-1070.

[2] Chaoui F.Z., Giri F., Dion J.M., M'Saad M. and Dugard L.: Direct Adaptive Control Subject to Input Amplitude Constraints. *Proc.3rd Europ.Contr.Conf.*, Rome, Sept. 1995.

[3] Gang Feng, Palaniswami M. and Zhu Y.: Stability of Rate Constrained Robust Pole Placement Adaptive Control Systems. *Systems & Control Letters*, 18, 1992, pp.99-107.

[4] Chaoui F.Z., Giri F. and M'Saad M.: Magnitude and Rate Constrained Linear System Control. *14th World Congress of IFAC*, Beijing, P.R.China, 1999, pp.379-384.

Fig.1. Output response for a*=0.55

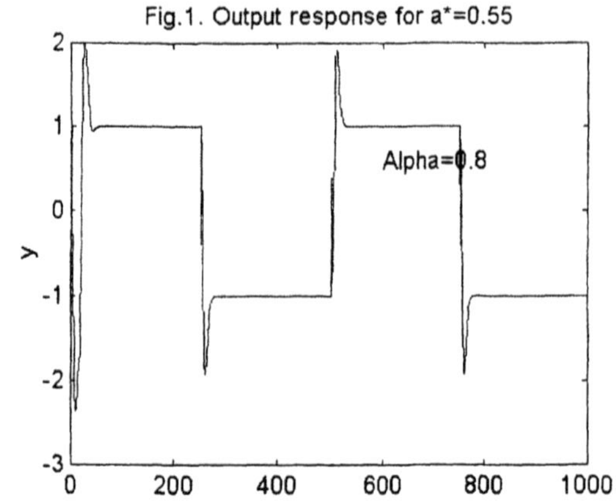

Fig.2. Output response for a*=0.7

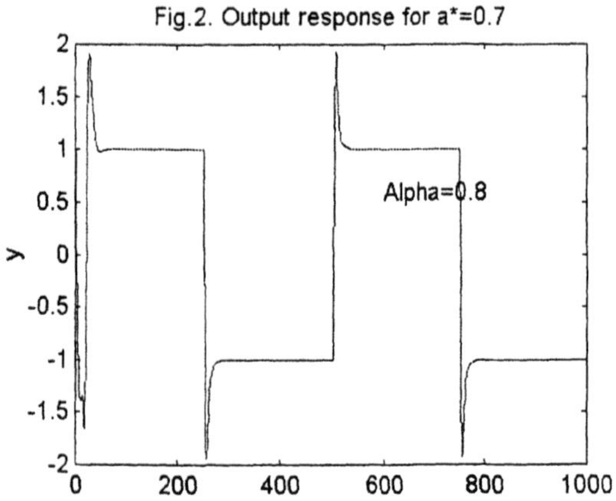

Fig.3. Output response for a*=0.85

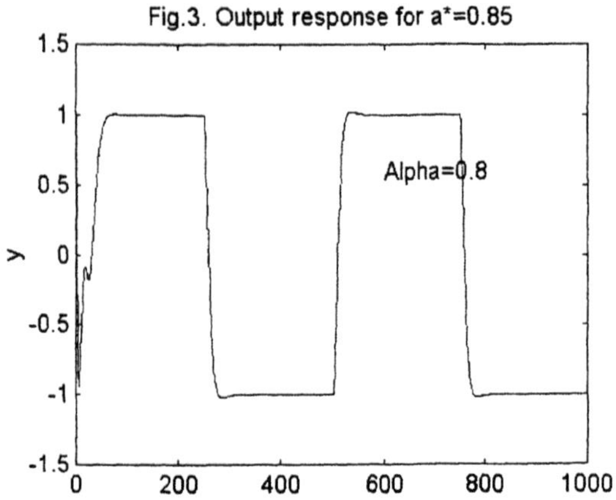

Fig.4. Control signal for a*=0.55

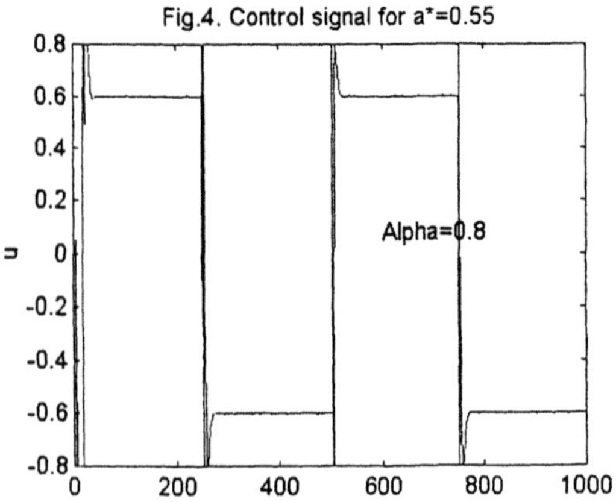

Fig.5. Control signal for a*=0.7

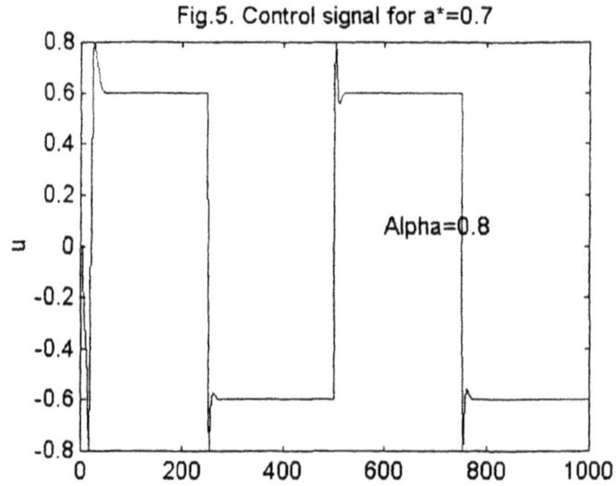

Fig.6. Control signal for a*=0.85

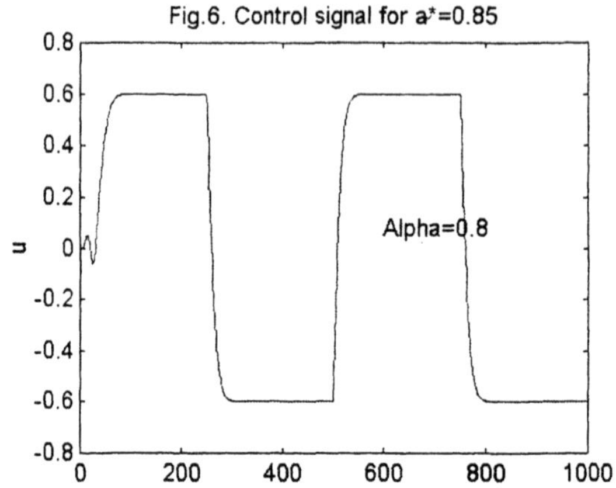

Fig.7. Plot of sample criterion

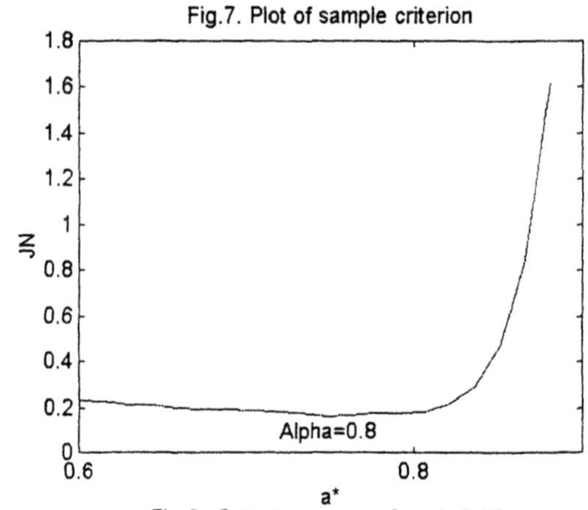

Fig.8. Output response for a*=0.55

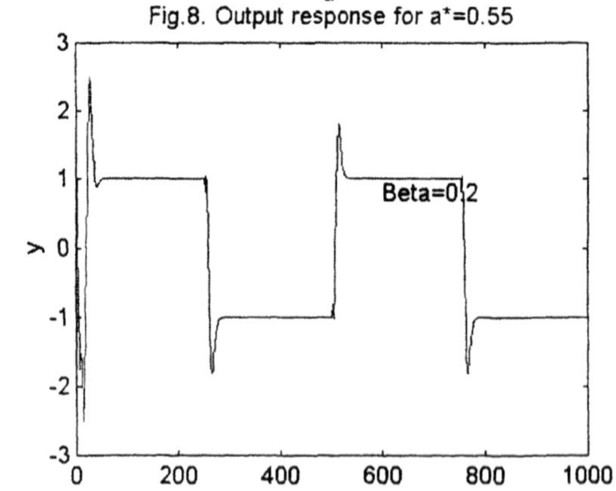

Fig.9. Output response for a*=0.7

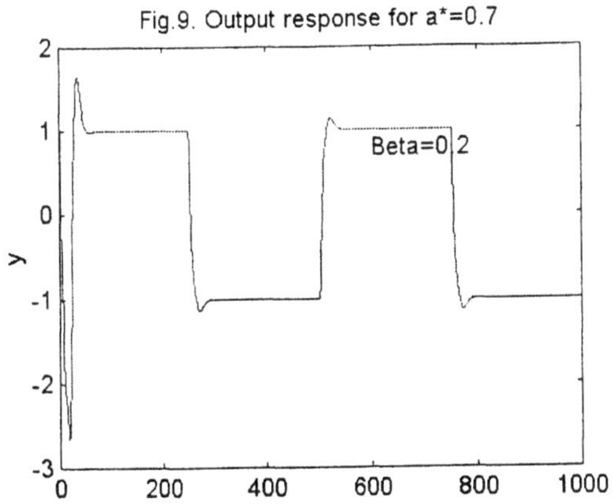

Fig.10. Output response for a*=0.85

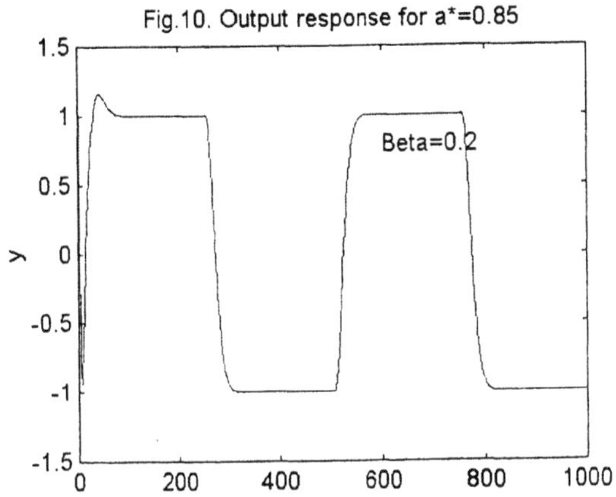

Fig.11. Rate-constrained control signal (Delta(u)) for a*=0.55

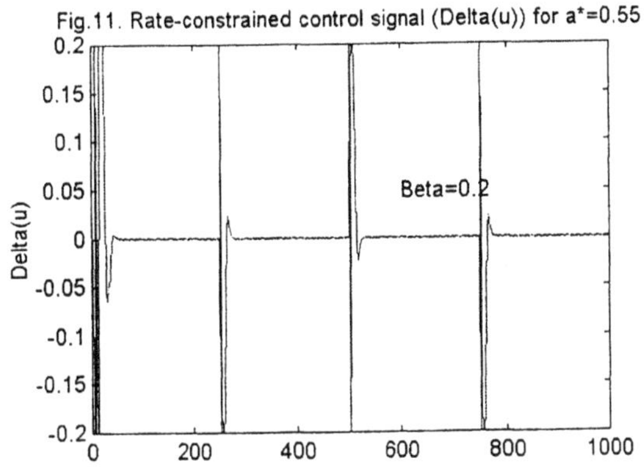

Fig.12. Rate-constrained control signal (Delta(u)) for a*=0.7

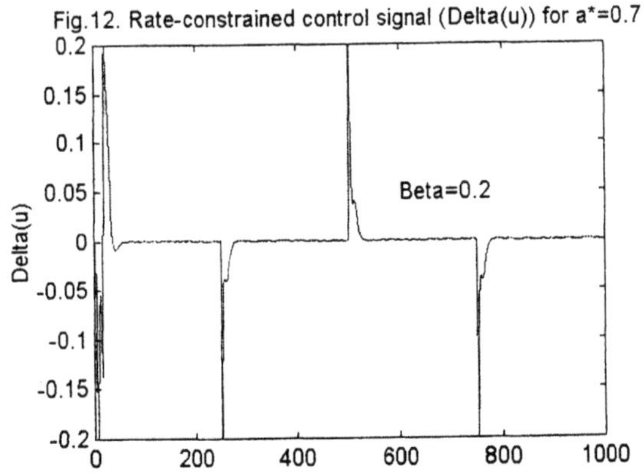

Fig.13. Rate-constrained control signal (Delta(u)) for a*=0.85

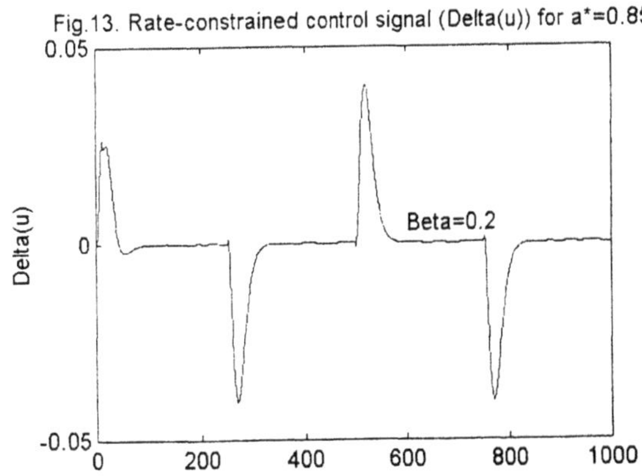

Fig.14. Plot of sample criterion

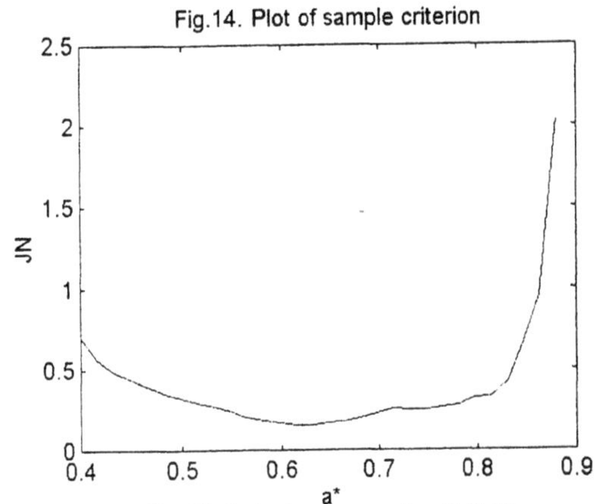

Fig.15. Output response for a*=0.55

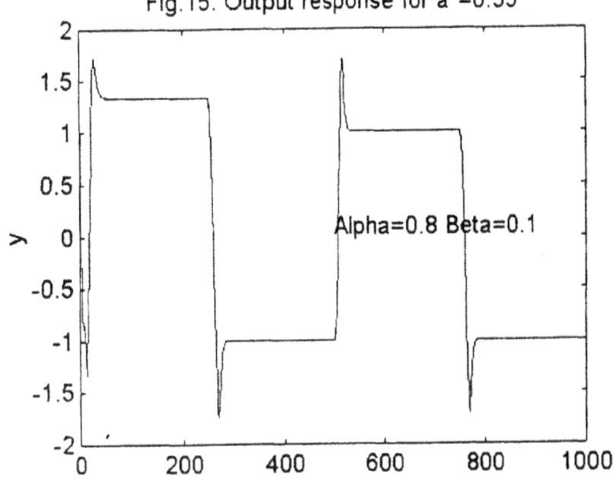

Fig.16. Output response for a*=0.7

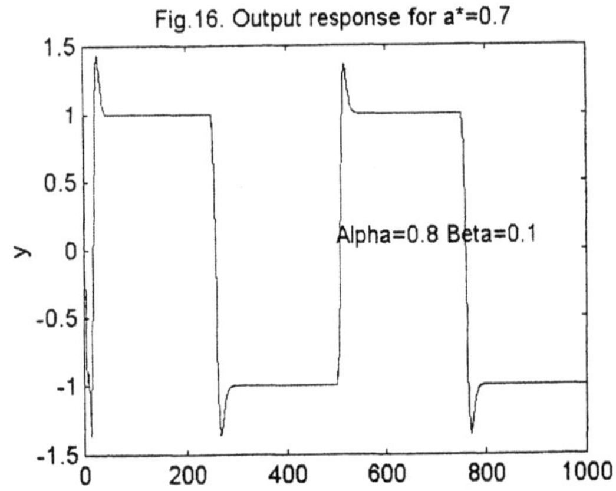

Fig.17. Output response for a*=0.85

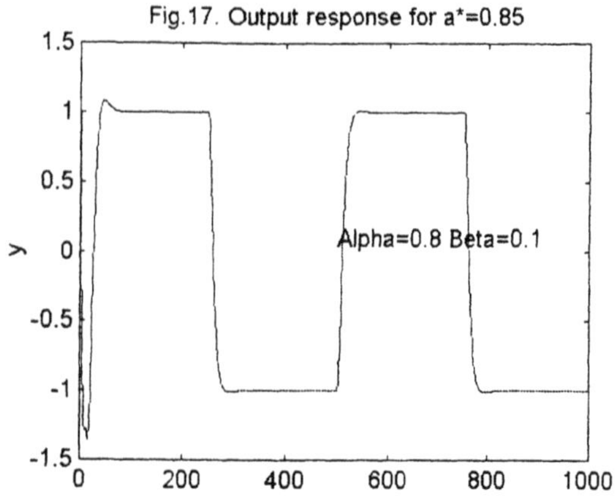

Fig.18. Control signal for a*=0.55

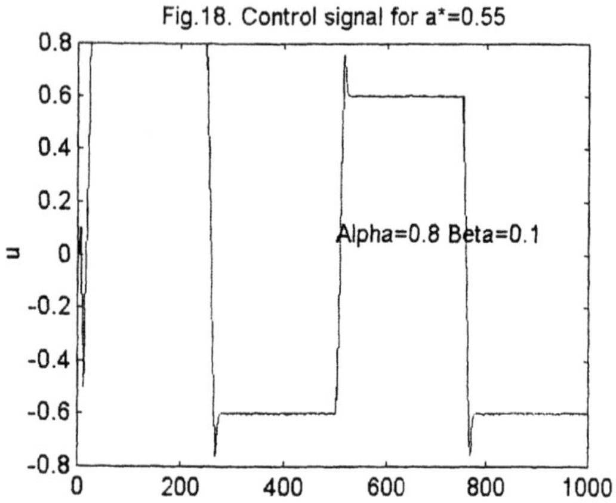

Fig.19. Control signal for a*=0.7

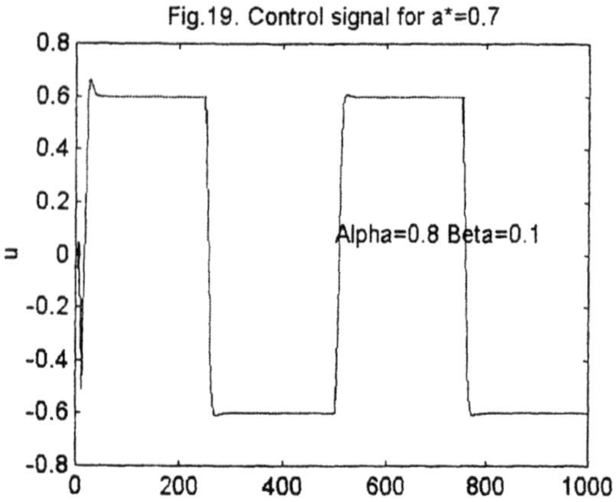

Fig.20. Control signal for a*=0.85

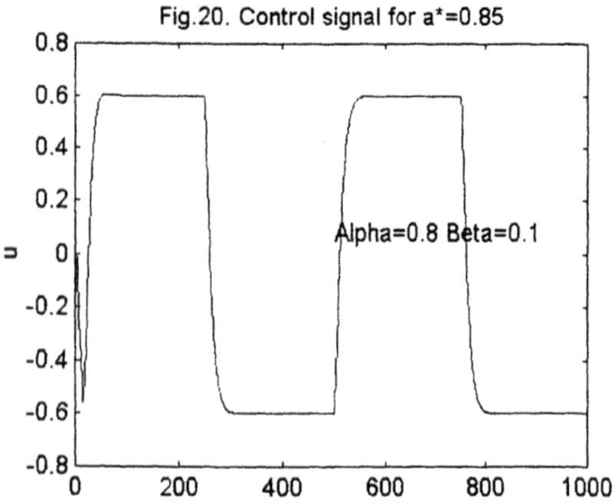

Fig.21. Rate-constrained control signal (Delta(u)) for a*=0.55

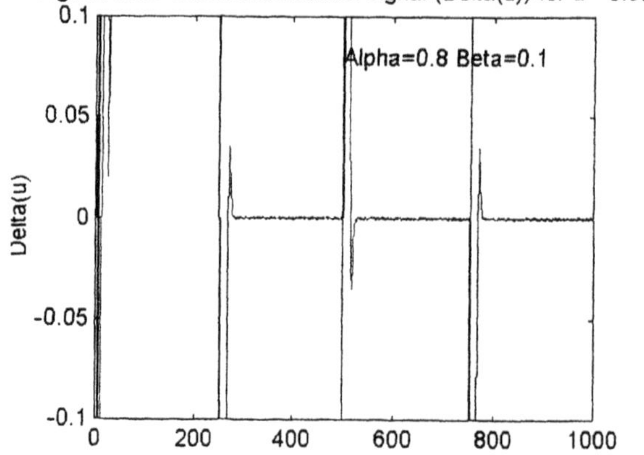

Fig.22. Rate-constrained control signal (Delta(u)) for a*=0.7

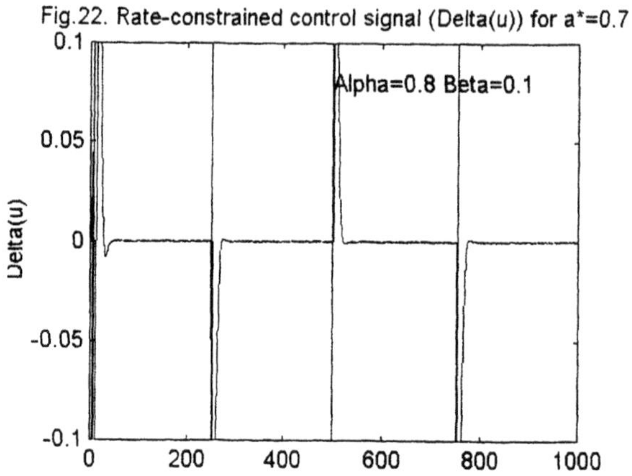

Fig.23. Rate-constrained control signal (Delta(u)) for a*=0.85

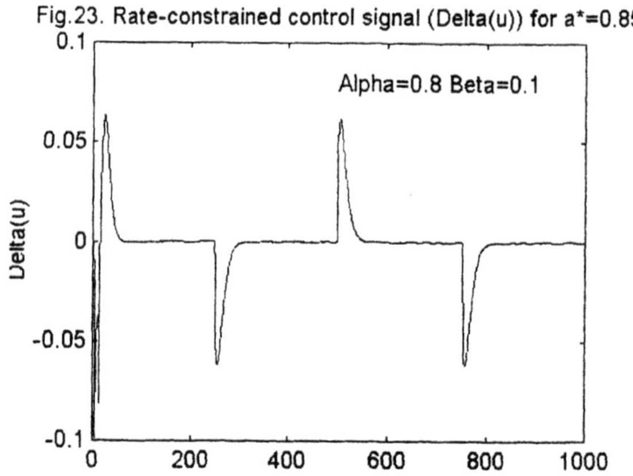

Fig.24. Plot of sample criterion

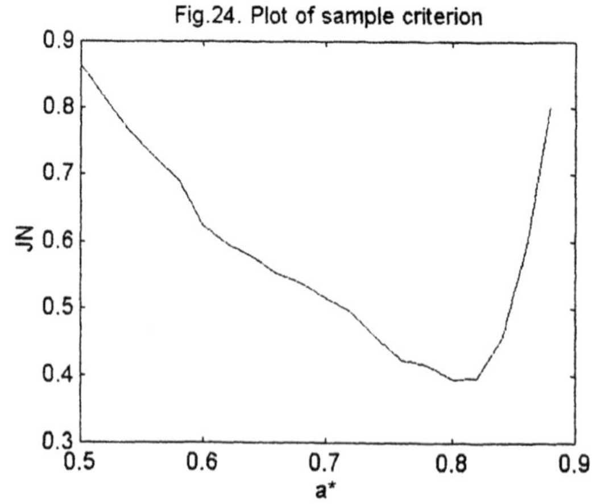

DESIGN AND IMPLEMENTATION OF A CYCLIC ONE-SPOT TUNING PID CONTROLLER

Toru Yamamoto* Kenzo Fujii**

* *Dept. of Technology & Information Education, Graduate School of Education, Hiroshima University, 1-1-1 Kagamiyama, Higashi-Hiroshima, Hiroshima 739-8524, Japan. Tel: +81-824-24-7160, Fax: +81-824-22-7171, E-mail:yama@hiroshima-u.ac.jp*
** *Tokuyama Refinery, Idemitsu Kosan Co. Ltd., Miyamae-Cho, Tokuyama, Yamaguchi, 745-8613, Japan. Tel: +81-834-21-6748, Fax: +81-834-21-0850, E-mail:kenzo.fujii@si.idemitsu.co.jp*

Abstract: In petroleum refineries and petrochemical processes, it is important to reduce the production costs in order to increase competitiveness. The author have proposed the one-spot tuning PID control scheme, and applied it to real atmospheric distillation tower. The effectiveness of the proposed control scheme has been verified in the steady operation mode. However, the satisfactory performance could not be obtained in the transition operation mode in which crude oil and operation mode were switched. This paper presents a new multiloop one-spot tuning PID control scheme. According to the newly proposed control scheme, the satisfactory control performance can be obtained in both the steady and the transition operation modes.

Keywords: PID control, Self-tuning control, Multiloop control, Process control, Decoupling compensator

1. INTRODUCTION

Recently, many control techniques have been actively employed in process industries by reason of cost reduction or saving energy. In these techniques, there are various advanced multivariable control methods and the method which directly manages a lot of constraints using the mathematical programming. However, the rapid progress of the control techniques has brought about packaging of control systems and making them black boxes, and it becomes to be difficult for operators to understand their principle and mechanisms. In implementing control schemes to real processes, it is necessary to make the best use of operators' current experiences. That is to say, advanced control methods such as Dynamic Matrix Control(DMC)(Culter and Ramaker, 1979) schemes are discontinuous to the conventional PID controllers which are widely used in real process industries, and make difficult for operators to intuitively understand the control systems. Especially, model predictive control schemes(Richalet, *et al.*, 1978; Cutler and Ramaker, 1979) have been often employed for distillation towers. However, even if the good control performance can be obtained in each distillation tower, whole plant which is constructed by some distillation towers is not always globally stabilized.

On the other hand, almost process control systems are designed using a distributed control system, so-called DCS. However, the sophisticated control method like DMC is installed in an exclusive computer system, because of its program load. Consequently, DCS does not make full use of its func-

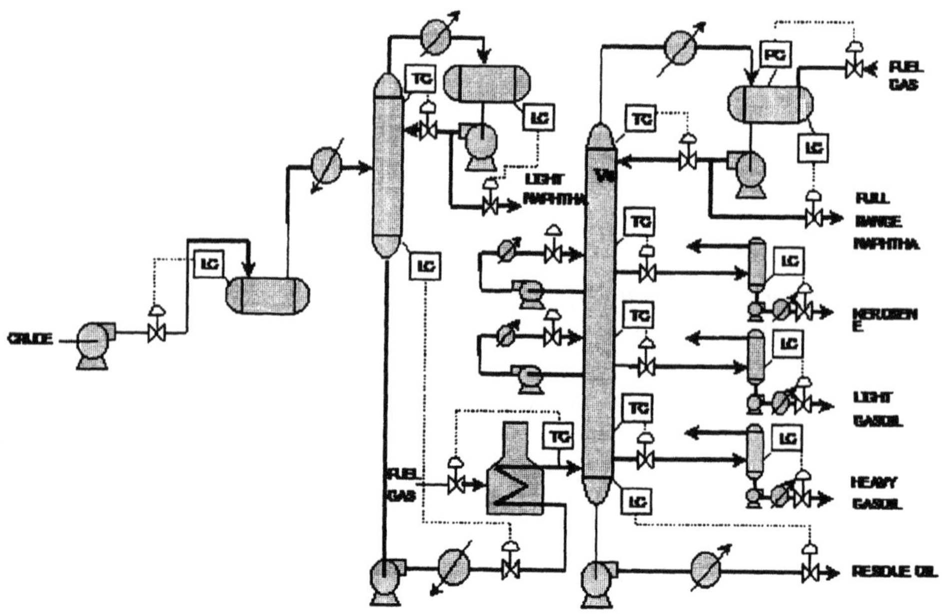

Fig. 1. Scematic figure of atmospheric distillation process.

tions. The authors have proposed one-spot tuning PID control scheme(Fujii and Yamamoto, 2000), which make it possible to stabilize whole plant by effectively utilizing DCS functions. Moreover, not only it is effective in asset management but drastically there is an advantage that a development period can be also shortened. However, the satisfactory performance could not be obtained in transition state in which operation mode was switched. Therefore, it was a problem to be solved requested from operators

In this paper, the new multiloop one-spot tuning PID control scheme is proposed, which the previous proposed one-spot tuning PID control scheme is extended for multivariable systems. The newly proposed control scheme makes it possible to regulate the control performance in both steady and transition operation modes. This paper is organized as follows. First, the atmospheric distillation process to be considered in this work is briefly explained. Next, a design scheme of pre-compensators to decouple interactions of the multivariable system, is discussed. The multiloop one-spot tuning PID controller is designed for subsystems which are approximately decoupled by the pre-compensator. PID parameters are adjusted based on the partial model matching(PMM)(Kitamori, 1979; 1981) scheme. Finally, the control behavior by the newly proposed control scheme is examined by employing for the real atmospheric distillation process.

2. ATMOSPHERIC DISTILATION PROCESS

An atmospheric distillation process is the first stage process for refining petroleum, and plays important role to separate crude oil into various distillation components. It is necessary to switch the operation modes in the distillation tower corresponding to the kind of the crude oil and the market conditions in order to stabilize the distillation tower. Fig.1 shows the schematic figure of the atmospheric distillation process. The light components are firstly separated in the first tower, and other components are supplied to the second tower. By heating them with the inlet furnace, they are separated in each boiling point, that is, naphtha, kerosene, light gas oil, heavy gas oil and residue oil are distilled in the second tower. Note that the level control in the flush drum which is equipped at the exit of the second tower and the boiling point control in drawing naphtha, kerosene, *et. al.* are interfered each other. Furthermore, there are also strong interactions among every component. Therefore, it is well known that it is difficult to control the atmospheric distillation tower.

3. MULTILOOP ONE-SPOT TUNING PID CONTROLLER

3.1 *Outline of controller*

One-spot tuning PID control(OST-PID) scheme for SISO system has been proposed(Fujii and Yamamoto,2000), and evaluated on a real atmospheric distillation process. Especially, high control performances were verified in the steady state. However, in the transition state in which oil properties and running conditions are changed, the proposed one-spot tuning PID control scheme does not effectively work due to process interactions.

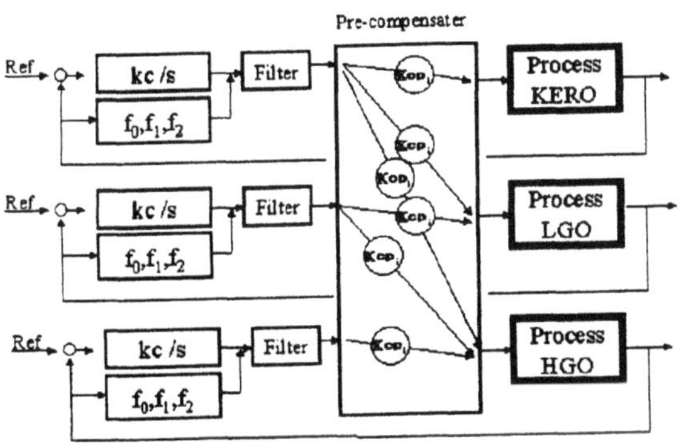

Fig. 2. Block diagram of the cut-point control system.

In this paper, a design scheme of multiloop one-spot tuning PID(MOST-PID) controllers is proposed, which stably works in both steady and transition states. Here, process interactions are approximately decoupled by using the static pre-compensator as shown in Fig.2 based on Nyquist array techniques(Rosenbrock, 1970). The newly proposed MOST-PID control scheme has the following features:

1) The previously proposed control scheme(OST-PID) and the newly proposed control scheme (MOST-PID) can be smoothly switched in steady state and transition state, respectively.
2) It is relatively easy for operators to understand the control system, because control schemes are very simple.

3.2 *Decoupling process interactions*

3.2.1. *System description* The controlled object shown in Fig.1 are described as the following multivariable system with three inputs and three outputs:

$$\mathbf{A}(z^{-1})\mathbf{y}(t) = D\mathbf{B}(z^{-1})\mathbf{u}(t-1), \qquad (1)$$

where

$$\left.\begin{aligned}\mathbf{u}(t) &= [u_1(t), u_2(t), u_3(t)]^T \\ \mathbf{y}(t) &= [y_1(t), y_2(t), y_3(t)]^T.\end{aligned}\right\} \qquad (2)$$

y_1, y_2 and y_3 denote cut-points of kerosene, light gas oil and heavy gas oil, respectively. Furthermore, u_1, u_2 and u_3 denote drawing rates, light gas oil and heavy gas oil corresponding to y_1, y_2 and y_3.

$\mathbf{A}(z^{-1})$ and $\mathbf{B}(z^{-1})$ included in (1) are defined as

$$\mathbf{A}(z^{-1}) = \mathrm{diag}\{A_1(z^{-1}), A_2(z^{-1}), A_3(z^{-1})\} \quad (3)$$

$$\boldsymbol{B}(z^{-1}) = \begin{bmatrix} B_{1,1}(z^{-1}) & 0 & 0 \\ B_{2,1}(z^{-1}) & B_{2,2}(z^{-1}) & 0 \\ B_{3,1}(z^{-1}) & B_{3,2}(z^{-1}) & B_{3,3}(z^{-1}) \end{bmatrix} \quad (4)$$

and $A_i(z^{-1})$ and $B_{i,j}(z^{-1})$ are given by the following equations:

$$A_i(z^{-1}) = 1 + a_{i,1}z^{-1} \qquad (5)$$

$$B_{i,j}(z^{-1}) = b_{i,j,0} + b_{i,j,1}z^{-1} + \cdots + b_{i,j,m}z^{-m} \quad (6)$$

Furthermore, D included in (1) is the time-delay matrix of the form:

$$D = \mathrm{diag}\{z^{-k_{m_1}}, z^{-k_{m_2}}, z^{-k_{m_3}}\}. \qquad (7)$$

k_{m_i} denotes the minimum value of estimated time-delays for the ith row. If the true time-delays, k_i, are exactly known in advance, it is better to set

$$k_{m_i} = \min_j\{k_{i,j}\},$$

where $k_{i,j}$ denotes the time-delay between the i-th output and the j-th input signals. On the other hand, where the information about time-delays is not available, then k_{m_i} is set to 0.

For the system (1), suppose the following assumptions:

[Assumptions]

[A.1] The polynomial matrix $\mathbf{A}(z^{-1})$ is stable.

[A.2] The degree of $\mathbf{B}(z^{-1})$, m is known, and the following relationship is satisfied:

$$k_{m_i} \le k_i \le m \quad (i = 1,2,3).$$

[A.3] $\mathbf{B}(1)$ is non-singular.

[A.4] Reference input $w_i(t)$ consists of piecewise constant signals.

3.2.2. *Pre-compensator* In designing multiloop controllers for multivariable systems, it is important to first remove or compensate for interactions. The simplest pre-compensator that can be designed is a static parameter, H is with the form(Rosenbrock, 1970):

$$H := \mathbf{B}^{-1}(1)\mathbf{A}(1)G. \qquad (8)$$

$\mathbf{A}(1)$ and $\mathbf{B}(1)$ denote the static gains corresponding to $\mathbf{A}(z^{-1})$ and $\mathbf{B}(z^{-1})$, respectively. Furthermore, G is given by

$$G = \mathrm{diag}\left\{\frac{B_{1,1}(1)}{A_1(1)}, \frac{B_{2,2}(1)}{A_2(1)}, \frac{B_{3,3}(1)}{A_3(1)}\right\}. \qquad (9)$$

Then, the pre-compensator (8) is rewritten by the matrix of the form:

$$H = \begin{bmatrix} 1 & 0 & 0 \\ h_{2,1} & 1 & 0 \\ h_{3,1} & h_{3,2} & 1 \end{bmatrix}. \quad (10)$$

Such a pre-compensator essentially looks after the low frequency interaction. The pre-compensator is utilized in only the transition state, and this matrix is replaced with the identity matrix in the steady state. That is to say, the proposed MOST-PID and the previously proposed OST-PID are smoothly switched at each state. In order to realize this switching smoothly, the pre-compensator H is designed so that the diagonal elements of eqn.(10) are given by unity.

The augmented system constructed by eqns.(1) and (8) can then be described as

$$\mathbf{A}(z^{-1})\mathbf{y}(t) = D\mathbf{B}(z^{-1})H\mathbf{v}(t-1), \quad (11)$$

where $\mathbf{v}(t)$ denotes the input signal vector to the augmented or pre-compensated system. The idea behind static decoupling is that diagonal dominance is a suitably weaker requirement in comparison with complete or strong decoupling, via dynamic pre-compensators, in the design of controllers for multivariate systems. In general, careful design of a static decoupler can achieve diagonal dominance. Many industrial processes are controlled in a multi-loop manner without any pre-compensation whatsoever.

By regarding the augmented system (11) as the approximately or almost decoupled system, the following model can be obtained for each diagonal element:

$$A_i(z^{-1})y_i(t) = z^{-k_{mi}}\bar{B}_i(z^{-1})v_i(t-1) \quad (12)$$

where

$$\bar{B}_i(z^{-1}) := \sum_{k=1}^{3} B_{i,k}(z^{-1})h_{k,i} = \bar{b}_{i,0} + \bar{b}_{i,1}z^{-1} + \cdots + \bar{b}_{i,m}z^{-m} \quad (13)$$

and $A_i(z^{-1})$ is given by eqn.(3).

3.3 *One-spot tuning PID controller*

3.3.1. *Controller design* Each loop in the augmented system is described as the discrete model (12). In order to consider a design scheme of PID controllers, (12) is rewritten by the following continuous-time model, that is, the first-order system with time-delay:

$$G(s) = \frac{K}{1+Ts}e^{-Ls}, \quad (14)$$

where T, L and k denote the time constant, the time-delay and the system gain, respectively.

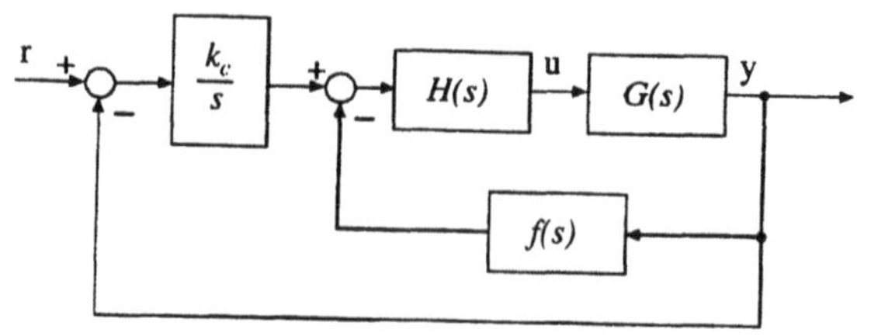

Fig. 3. Block diagram of the I-PD control system.

First, the following first-order filter is equipped with each subsystem:

$$H(s) = \frac{1}{1+\tau s}. \quad (15)$$

The easiness for controlling the system is usually evaluated on the ratio between the time constant and the time-delay, *i.e.*, L/T. Especially, in the case of employing PID control scheme, it is better that the ratio is within 0.2 thru 0.5. Therefore, in the case where the ratio is over 0.5, the pre-filter given by eqn.(15) is inserted, and τ is practically set based on operators' experiences and results from the event analyzer mentioned later. Thus, the pre-filter is utilized in order to cope with the time-delay. On the other hand, more stable control system can be designed, because the excessive large control input is put down by the pre-filter. The stability of process control systems must be firstly considered in designing the control systems. Therefore, it is important to insert the pre-filter (15) in the control system from such reasons.

Next, the control system as shown in Fig.3 is constructed for the augmented system which consists of the controlled object and the pre-filter. $F(s)$ in Fig.3 is a local feedback compensator given by

$$f(s) = f_0 + f_1 s + f_2 s^2 + f_3 s^3 + \cdots. \quad (16)$$

If only f_0 and f_1 are adopted, *i.e.*, $f_2 = f_3 = \cdots = 0$, the PID control system can be designed. The control gain k_c and the local feedback compensator $f(s)$ are designed based on the partial model matching scheme(Kitamori, 1979; 1981). The detailed design scheme is discussed below.

First, consider the reference model of the form:

$$G_m(s) = \frac{e^{-Ls}}{\alpha(s)} = \frac{e^{-Ls}}{1+\sigma s+\alpha_2\sigma^2 s^2+\alpha_3\sigma^3 s^3+\alpha_4\sigma^4 s^4+\cdots}, \quad (17)$$

where σ denotes the parameter corresponding to the rise time of the reference model output. And also, $\{\alpha_2, \alpha_3, \alpha_4, \cdots\}$ are set to $\{3/8, 1/16, 1/256, \cdots\}$ so that the overshoot does not yield(Kitamori, 1979; 1981).

Next, let the transfer function of the closed-loop system shown in Fig.3 be equal to that of the reference model (17), and then the following relation can be obtained:

$$\frac{k_c K e^{-Ls}}{s(1+Ts)(1+\tau s)+e^{-Ls}\{Kf(s)s+k_c K\}} = \frac{e^{-Ls}}{\alpha(s)} \quad (18)$$

Solving eqn.(18) with respect to $f(s)$ yields the following equation:

$$f(s) = e^{Ls}\left\{\frac{k_c}{s}\alpha(s) - \frac{1}{K}(1+Ts)(1+\tau s)\right\} - \frac{k_c}{s}. \quad (19)$$

Furthermore, by expanding e^{Ls} in eqn.(19) based on Maclaurin's expansion, and calculating the coefficient f_i of each term with respect to s, the following relations are obtained.

$$\left.\begin{aligned} f_0 &= k_c(\sigma+L) - 1/K \\ f_1 &= k_c(\alpha_2\sigma^2 + L\sigma + L^2/2) \\ &\quad - (T+L+\tau)/K \\ f_2 &= k_c(\alpha_3\sigma^3 + L\alpha_2\sigma^2 + L^2\sigma/2 + L^3/6) \\ &\quad - \{T\tau + L(T+\tau) + L^2/2\}/K \\ f_3 &= k_c(\alpha_4\sigma^4 + L\alpha_3\sigma^3 + L^2\alpha_2\sigma^2/2 \\ &\quad + L^3\sigma/6 + L^4/24) - \{LT\tau \\ &\quad + L^2(T+\tau)/2 + L^3/6\}/K \\ &\vdots \end{aligned}\right\} \quad (20)$$

Since PID controller is designed in this work, the following equation with respect to σ has to be solved based on the partial model matching scheme, which is derived by eliminating k_c from the relations $f_2 = 0$ and $f_3 = 0$:

$$g\alpha_4\sigma^4 + (gL\alpha_3 - h\alpha_3)\sigma^3 + (gL^2\alpha_2 - hL\alpha_2)\sigma^2 + (gL^3/6 - hL^2/2)\sigma + (gL^4/24 - hL^3/6) = 0, \quad (21)$$

where

$$\left.\begin{aligned} g &:= T\tau + L(T+\tau) + L^2/2 \\ h &:= LT\tau + L^2(T+\tau)/2 + L^3/6. \end{aligned}\right\} \quad (22)$$

Among the solutions of eqn.(21), the positive least value is adopted as σ. Substituting σ into the following equation;

$$k_c = \frac{g}{\alpha_3\sigma^3 + L\alpha_2\sigma^2 + L^2\sigma/2 + L^3/6}, \quad (23)$$

the suitable k_c can be obtained. Moreover, by substituting σ and k_c into f_0 and f_1 in eqn.(20), all the control parameters can be calculated.

3.3.2. *One-spot tuning scheme* The MOST-PID control system can be constructed in the case where systems parameters T, K and L are known. However, these parameters can not be exactly obtained in advance. Then, systems parameters included in eqn.(12) which is corresponding to eqn.(14), are estimated by using the least squares algorithm. T and K are calculated based on these estimates by using the following equations:

$$\hat{T}(t) = -\frac{Ts}{\log(-\hat{a}_1(t))} \quad (24)$$

$$\hat{K}(t) = \frac{\hat{b}_0(t) + \hat{b}_1(t) + \cdots + \hat{b}_m(t)}{1 + \hat{a}_1(t)} \quad (25)$$

where $\hat{a}_1(t)$ and $\hat{b}_i(t)$ are the estimated parameters corresponding to a_1 and b_i. Furthermore, the time-delay L is calculated by the following equation(Miller, *et al.*, 1995):

$$\hat{L}(t) = \{k_m + \hat{k}_i(t)\} \cdot Ts, \quad (26)$$

where $\hat{k}_i(t)$ is defined as

$$\hat{k}_i(t) := \sum_{i=0}^{m} i \cdot \hat{b}_i(t) / \sum_{i=0}^{m} \hat{b}_i(t) \quad (27)$$

Based on the estimates, $\hat{T}(t)$, $\hat{K}(t)$ and $\hat{L}(t)$, PID parameters are calculated by using the PMM scheme mentioned above.

3.3.3. *Event analyzer* The event analyzer is tool that operation condition is analyzed based on event in formations(operator intervention and alarm) which are outputted from the DCS. The detailed items in the event analyzer are as follows:

(1) The frequencies of operation intervention and alarm in every units
(2) Time series analysis of operator intervention when disturbances are inputted
(3) The analysis of the relations between operator intervention and alarm.

Analyzing in these items makes some problems clear in the process system, the control system and the operation skills, and gives suitable control strategies in order to improve the control performance. Concretely, the influences by interactions among every loops and every units, and the effects of control parameters for control performances are evaluated by the event analyzer. Based on these results, control parameters are adjusted adequately. Therefore, the control parameter tuning and the event analyzing are alternately carried out, and then whole plant can be gradually stabilized.

4. CONTROL RESULTS

The effectiveness of the newly proposed MOST-PID control scheme is evaluated on the real atmospheric distillation tower. First, Fig.4 shows the control result in the light gas oil 90% boiling point control by the OST-PID control scheme , where the drawing rate of the kerosene is manually changed by the operator. It is clear that this change strongly influences the performance in the light gas oil boiling point control. Even though the operator carefully manipulated the drawing rate of the kerosene, the relatively large fluctuation occurred in the control performance of the light gas oil.

On the other hand, the newly proposed MOST-PID control scheme was employed, and then the control result in the light gas oil 90% boiling point control as shown in Fig.5 could be obtained, where the desired boiling point of the kerosene was automatically changed. It is clear that the newly proposed control scheme works well, even though the desired boiling point of the kerosene is changed. By employing the MOST-PID control scheme, the influences by the interactions can be effectively reduced, and it enables us to realize the automatic control and stabilization of the control system in the transition mode.

5. CONCLUSION

The multiloop one-spot tuning PID control scheme has been proposed, and the effectiveness in the transition control mode has been verified by employing it for the real atmospheric distillation process. The proposed MOST-PID control scheme and the previously proposed OST-PID control scheme can be smoothly switched in the transition and the steady states. That is, the MOST-PID control scheme is employed in the transition mode, and the pre-compensator is removed from the MOST-PID control scheme in the steady mode. Though the MOST-PID control scheme can be consecutively utilized in the steady state, the satisfactory control performance can not be obtained.

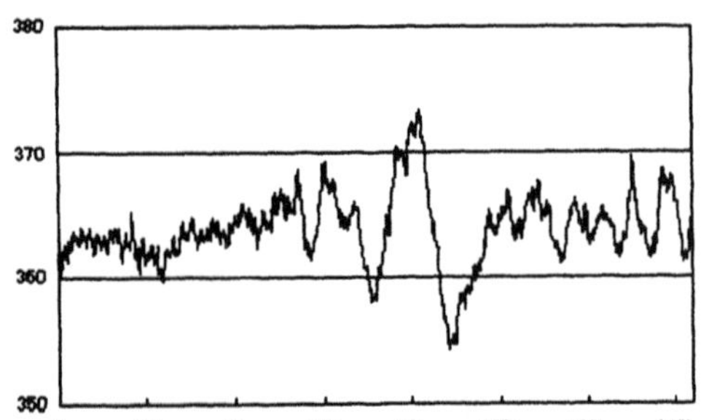

Fig. 4. Control result of LOG90% cut-point without using the proposed control scheme.

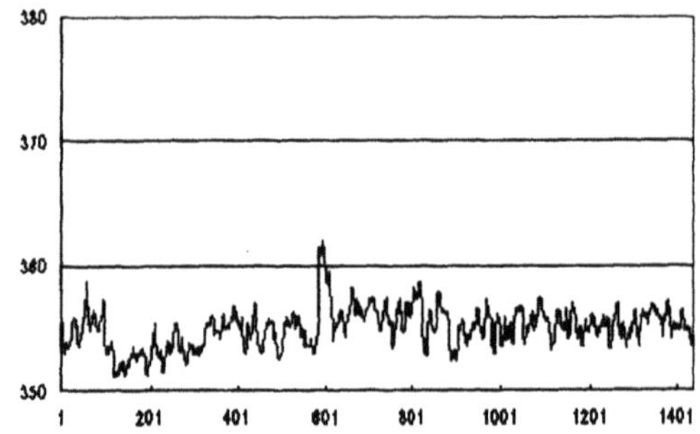

Fig. 5. Control result of LOG90% cut-point using the proposed control scheme.

Therefore, the switching of these control schemes plays an important role in real systems.

6. REFERENCES

Cutler, C.R. and B.L. Ramaker (1979). Dynamic Matrix Control - A Computer Control Algorithm. *Proc. Joint Automatic Control Conference.*

Fujii, K. and T. Yamamoto (2000). Design of a Cyclic One-Spot Tuning PID Control and Its Application for an Atmospheric Distillation Process. *Proc. of IEEE Int.* 285–290, Anchorage.

Kitamori, T. (1979). A Method of Control System Design Based upon Partial Knowledge About Controlled Processes(in Japanese). *Trans. SICE.* **15-4**, 549–555.

Kitamori, T. (1981). A Design Method for Nonlinear Control System Design Based upon Partial Knowledge About Controlled Objects. *Preprints of 8th IFAC World Congress.* **4**, 25–30, Kyoto.

Miller, R.M., S.L. Shah and R.K. Wood (1995). Adaptive Predictive Control Employing On-Line Time Delay Estimation. *Proc. of ISA.* 857–864, New Orleans.

Richalet, J., A. Rault, J.L. Testud and J. Papon (1978). Model Predictive Heuristic Control, Application to Industrial Process. *Automatica.* **14-5**, 413–428.

Rosenbrock, H.H. (1970). State-Space and Multivariable Theory. *Wiley-Intersience.*

AUTHOR INDEX

www.ingramcontent.com/pod-product-compliance
Ingram Content Group UK Ltd.
Pitfield, Milton Keynes, MK11 3LW, UK
UKHW052301280926
15471UKWH00006B/1045

9 780080 436838